T0249902

TIME SERIES WITH MIXED SPECTRA

Ta-Hsin Li

CRC Press
Taylor & Francis Group
Boca Raton London New York

CRC Press is an imprint of the
Taylor & Francis Group, an **informa** business
A CHAPMAN & HALL BOOK

CRC Press
Taylor & Francis Group
6000 Broken Sound Parkway NW, Suite 300
Boca Raton, FL 33487-2742

First issued in paperback 2019

ISBN-13: 978-1-58488-176-6 (hbk)
ISBN-13: 978-1-138-37495-9 (pbk)

Library of Congress Cataloging-in-Publication Data

Li, Ta-Hsin.
Time series with mixed spectra / Ta-Hsin Li.
pages cm
Includes bibliographical references and index.
ISBN 978-1-58488-176-6 (hardback)
1. Time-series analysis. 2. Spectrum analysis. I. Title.

QA280.L49 2013
519.5'5--dc23 2013014918

Visit the Taylor & Francis Web site at
http://www.taylorandfrancis.com

and the CRC Press Web site at
http://www.crcpress.com

To My Family

Contents

Preface

This book focuses on the methods and theory for statistical analysis of time series with mixed spectra. A time series is said to have a mixed spectrum if it comprises a finite number of sinusoids with different frequencies plus random noise. The research on such time series has a long history, and it remains active to this day, especially in the signal processing community where the interest is driven in part by the everlasting desire for fast algorithms to reduce the computational cost. Despite its importance, the subject often receives limited coverage in standard textbooks for understandable reasons. The objective of this book is to provide a more comprehensive and in-depth treatment of the subject. Needless to say, it is impossible to cover every aspect of the subject, not only because of the huge body of literature which keeps growing, but also due to the limited ability and capacity of the author. The topics in this book are selected to reflect what the author thinks are most interesting and relevant.

The intended audience of the book includes graduate students, researchers, engineers, and other professionals who work in the fields of time series analysis and signal processing. For the most part, the book only requires basic knowledge of probability, statistics, and time series analysis. Some theoretical results, especially their proofs, require more advanced knowledge of asymptotic analysis. For this reason, the proofs are deferred to the last section of each chapter in order not to interrupt the flow of intuitive interpretations which are more easily accessible to most readers. To serve the interests of a broader audience, the book deals with both real- and complex-valued time series. Except for the treatment of the proofs, the main style of the book is influenced by the highly successful textbook of Brockwell and David (1991) entitled *Time Series: Theory and Methods.* Other excellent textbooks that influenced the treatment of this book include *Spectral Analysis* by Priestley (1984), *Introduction to Fourier Analysis of Time Series* by Bloomfield (2000), *Spectral Analysis* by Stoica and Moses (1997), and *Modern Spectral Analysis* by Kay (1987).

This book is not merely a survey of the existing literature — it also includes original material which is either unavailable or cannot be easily found in the literature. For example, many results for complex-valued time series are derived in this book by extending the existing results developed in the literature for the real case; some theoretical results, such as the asymptotic theory for closely spaced frequencies and the proof of the asymptotic normality of the nonlinear least-absolute-deviations frequency estimator, are not available in the literature. The book also contains the author's most recent work on the quantile regression method for spectral analysis.

The infancy of this book was a thesis proposal written in 1990 at the University of Maryland in College Park. The idea of writing a book based on the material came about in 1997 when I was on the faculty of the statistics department at Texas A&M University in College Station. It became a prolonged undertaking as priorities changed with my career moves from Texas A&M to the University of California at Santa Barbara and then to the IBM T. J. Watson Research Center at Yorktown Heights, New York. I am therefore very grateful to Bob Stern and his successor David Grubbs, editors of CRC Press, for their infinite patience.

Over the years with the project, I received the unwavering support from Dr. Benjamin Kedem and Dr. Emanuel Parzen, to whom I am deeply indebted. I am also grateful for the help and encouragement from Dr. H. Joseph Newton, Dr. Gerald R. North, Dr. Jerry D. Gibson, Dr. Keh-Shin Lii, Dr. David V. Hinkley, Dr. Emmanuel Yashchin, Dr. David A. Harville, and Dr. Yasuo Amemiya. My especial thanks go to Dr. Kai-Sheng Song for his collaboration on several papers that closed some important gaps in the literature and enriched the content of this book. I would also like to express my gratitude to Dr. Hee-Seok Oh for many helpful comments and suggestions on an earlier draft of the book. Last but not least, I would like to thank Dr. Steve M. Kay for reviewing the proposal and the first draft of the book, and an anonymous reviewer for reviewing the last draft of the book with many thoughtful and constructive suggestions that have been incorporated into the final product.

Ta-Hsin Li
Yorktown Heights, New York

Chapter 1

Introduction

A time series is a sequence of data points obtained over successive uniform time intervals. The word "time" can also be interpreted loosely to mean a time-like variable that provides the ordering and spacing for the data points. In this book, we are mainly interested in a particular type of time series, known as time series with mixed spectra, which can be expressed as a sum of sinusoids plus random noise. Time series of this type are abundant in a variety of science and engineering applications, including astronomy, biology, econometrics, geophysics, meteorology, rotating machinery, radar, sonar, and telecommunications. A main objective for spectral analysis of such time series is to detect and estimate the hidden periodicities represented by the sinusoidal functions.

1.1 Periodicity and Sinusoidal Functions

Periodicity is one of the most important and useful natural phenomena, and widely observable. The earth revolves periodically around its own axis and the sun, giving us different days and seasons. The displacement of a vibrating string or a swinging pendulum exhibits periodic patterns over time.

By definition, a periodic function repeats its values over intervals of a fixed length called the *period*. Sinusoidal, or trigonometric, functions, i.e., sines and cosines, are perfect examples of periodic functions. For any fixed constant $f > 0$, the sinusoidal function $\sin(2\pi f t)$, defined on the real line $\mathbb{R} := (-\infty, \infty)$, has a period $T := 1/f$, because

$$\sin(2\pi f(t+T)) = \sin(2\pi f t) \qquad \forall\, t \in \mathbb{R}.$$

The parameter f, measured in cycles per unit time, is called the *frequency* of the sinusoidal function $\sin(2\pi f t)$. The parameter $\omega := 2\pi f$, measured in radians per unit time, is called the *angular frequency* of the sinusoid.

Not only the sinusoidal functions are periodic, they can also be superimposed to represent any periodic function. In fact, according to the theory of Fourier series, any piecewise continuous function $x(t)$ with period T can be expressed as a sum of sinusoids with frequencies $f_k := k/T$ $(k = 1, 2, \ldots)$, i.e.,

$$x(t) = A_0 + \sum_{k=1}^{\infty} \{A_k \cos(2\pi f_k t) + B_k \sin(2\pi f_k t)\}, \tag{1.1.1}$$

where

$$\begin{aligned} A_0 &:= \frac{1}{T}\int_0^T x(t)\,dt, \\ A_k &:= \frac{2}{T}\int_0^T x(t)\cos(2\pi f_k t)\,dt, \\ B_k &:= \frac{2}{T}\int_0^T x(t)\sin(2\pi f_k t)\,dt. \end{aligned}$$

The convergence of this infinite series takes place at every continuous point of $x(t)$ for $t \in \mathbb{R}$. Observe that the sinusoidal functions in (1.1.1) are orthogonal to each other and to the constant function 1 (i.e., the cosine function with frequency zero) in the sense that

$$\begin{aligned} &\int_0^T \cos(2\pi f_k t)\cos(2\pi f_{k'} t)\,dt = 0 \qquad \forall\, k \neq k', \\ &\int_0^T \sin(2\pi f_k t)\sin(2\pi f_{k'} t)\,dt = 0 \qquad \forall\, k \neq k', \\ &\int_0^T \cos(2\pi f_k t)\sin(2\pi f_{k'} t)\,dt = 0 \qquad \forall\, k, k', \end{aligned}$$

and

$$\int_0^T \cos(2\pi f_k t)\,dt = \int_0^T \sin(2\pi f_k t)\,dt = 0 \qquad \forall\, k.$$

Therefore, the sinusoidal functions in (1.1.1), together with the constant function, form an orthogonal basis for T-periodic functions.

Although the sinusoidal representation (1.1.1) is an infinite sum in general, it can be approximated by a finite sum when $x(t)$ is sufficiently smooth so that the coefficients decay rapidly as k grows. Given such an approximation,

$$\tilde{x}(t) := A_0 + \sum_{k=1}^{K} \{A_k \cos(2\pi f_k t) + B_k \sin(2\pi f_k t)\}, \tag{1.1.2}$$

the total squared error can be expressed as

$$\int_0^T |x(t) - \tilde{x}(t)|^2\,dt = \sum_{k=K+1}^{\infty} \tfrac{1}{2} T(A_k^2 + B_k^2). \tag{1.1.3}$$

An important reason why the sinusoids are the preferred basis for representing periodic functions is that an approximation of the form (1.1.2) is time-invariant:

for any constant τ, the function $\tilde{x}(t+\tau)$ remains a finite sum of sinusoids with the same frequencies, and the error of $\tilde{x}(t+\tau)$ for approximating $x(t+\tau)$ is the same as the error of $\tilde{x}(t)$ for approximating $x(t)$. In fact, it is easy to verify that

$$\tilde{x}(t+\tau) = A_0 + \sum_{k=1}^{K} \{A'_k \cos(2\pi f_k t) + B'_k \sin(2\pi f_k t)\},$$

where

$$A'_k := A_k \cos(2\pi f_k \tau) + B_k \sin(2\pi f_k \tau),$$
$$B'_k := -A_k \sin(2\pi f_k \tau) + B_k \cos(2\pi f_k \tau).$$

A similar expression can be obtained for $x(t+\tau)$. Because $A'^2_k + B'^2_k = A^2_k + B^2_k$ for all k, the identity (1.1.3) remains true for the error $\int_0^T |x(t+\tau) - \tilde{x}(t+\tau)|^2\, dt$.

1.2 Sampling and Aliasing

When continuous-time functions are observed only at discrete time instants, the problem of aliasing arises. This may lead to difficulties in interpreting the sinusoidal components of the resulting time series.

Consider the periodic function $x(t)$ in (1.1.1), for example. If samples are taken with sampling interval Δ at equally spaced time instants Δt for $t \in \mathbb{Z} := \{0, \pm 1, \pm 2, \ldots\}$, where $f_s := 1/\Delta$ is known as the *sampling rate* (measured in samples per unit time), then the resulting time series can be expressed as

$$x_t := x(\Delta t) = A_0 + \sum_{k=1}^{\infty} \{A_k \cos(2\pi f_k \Delta t) + B_k \sin(2\pi f_k \Delta t)\} \quad (t \in \mathbb{Z}). \tag{1.2.1}$$

For any $f_k \in (f_s/2, f_s]$, define $f'_k := f_s - f_k \in [0, f_s/2)$. The 2π-periodicity of the sinusoidal functions implies that for all $t \in \mathbb{Z}$,

$$\cos(2\pi f'_k \Delta t) = \cos(2\pi f_k \Delta t), \quad \sin(2\pi f'_k \Delta t) = -\sin(2\pi f_k \Delta t). \tag{1.2.2}$$

Therefore, the frequency f_k becomes indistinguishable from the frequency f'_k in the discrete-time representation (1.2.1). For this reason, the frequency f'_k is called an *alias* of f_k. Similarly, for any $f_k > f_s$, there exists an integer u such that $\tilde{f}_k := f_k - u f_s \in [0, f_s)$. Let $f'_k := f_s - \tilde{f}_k$ if $\tilde{f}_k \in (f_s/2, f_s)$, and let $f'_k := \tilde{f}_k$ otherwise. In the first case, (1.2.2) is true for all $t \in \mathbb{Z}$. In the second case, we have

$$\cos(2\pi f'_k \Delta t) = \cos(2\pi f_k \Delta t), \quad \sin(2\pi f'_k \Delta t) = \sin(2\pi f_k \Delta t).$$

In both cases, the frequency f_k' is an alias of f_k. As we can see, the aliasing effect in the time series makes it impossible to correctly identify the original frequencies which are greater than $f_s/2$.

If the function $x(t)$ does not contain sinusoidal components whose frequencies are greater than a known constant $f_c > 0$, then an alias-free time series can be obtained by setting the sampling rate f_s higher than $2f_c$ so that $f_c < f_s/2$. The lower bound $2f_c$ is known as the Nyquist rate for alias-free sampling, named after the American physicist and electrical engineer Harry Nyquist (1889–1976).

In many applications, the periodic functions of interest are smooth enough to be well approximated by a finite sum of the form (1.1.2) with sufficiently large K such that the approximation error is comparable to the noise in the measurements. In this case, the aliasing problem can be ignored in the time series obtained with a sampling rate higher than the effective Nyquist rate $2f_K = 2K/T$. In other applications, such as radar and telecommunications, where the continuous-time signals are available, the aliasing problem can be mitigated by filtering the continuous-time signals, before sampling at rate f_s, with an analog device to remove the frequency content higher than $f_c := f_s/2$, which is known as the Nyquist frequency. Of course, the sampling rate f_s needs to be sufficiently high in order to minimize the distortion introduced by anti-alias filtering.

Instead of taking the instantaneous values, which leads to (1.2.1), one can also take the average values of a continuous-time function over regular intervals of length Δ. For the periodic function $x(t)$ in (1.1.1), this sampling technique leads to a time series that can be expressed as

$$x_t := \frac{1}{\Delta}\int_{\Delta(t-1/2)}^{\Delta(t+1/2)} x(s)\,ds = A_0 + \sum_{k=1}^{\infty}\{\tilde{A}_k \cos(2\pi f_k \Delta t) + \tilde{B}_k \sin(2\pi f_k \Delta t)\},$$

where $\tilde{A}_k := \mathrm{sinc}(2\pi f_k\Delta/2)A_k$, $\tilde{B}_k := \mathrm{sinc}(2\pi f_k\Delta/2)B_k$, and $\mathrm{sinc}(t) := \sin(t)/t$. As we can see, the time series retains the original form of $x(t)$ in (1.1.1) as a sum of sinusoids. This is another advantage of the sinusoidal representation of periodic functions. Because the time series takes the same form as that in (1.2.1), the aliasing problem and the alias-free sampling condition remain the same.

1.3 Time Series with Mixed Spectra

The sinusoidal representation (1.1.1) can be generalized to include functions that can be expressed as a sum of periodic functions with different periods. If each periodic component is a finite sum of sinusoids, then, with a sufficiently high sampling rate, the resulting time series can be expressed as

$$y_t = A_0 + \sum_{k=1}^{q} \{A_k \cos(2\pi f_k t) + B_k \sin(2\pi f_k t)\} + \epsilon_t \qquad (t \in \mathbb{Z}),$$

where $\{\epsilon_t\}$ is a random process representing the noise in the observations and $\{f_k\}$ is a set of normalized frequencies in $(0, 1/2)$ which may not be harmonically related as they are in (1.2.1). A time series of this form has a mixed-type spectrum, because the frequency content of the sinusoids concentrates on discrete values in the interval $(0, 1/2)$ with infinite density, whereas the frequency content of the noise spreads over the interval with finite density.

Figure 1.1 shows an example of the sinusoid-plus-noise model for a real-world time series. The time series, shown in the top panel, is known as the light curve. It comprises 1,639 brightness measurements of a variable star over a period of about 34 days. It is part of a large data set produced by the Kepler Mission* of the National Aeronautics and Space Administration (NASA). In the Kepler Mission, a spacecraft carrying a simple aperture photometer (SAP) was launched in March 2009 to monitor the brightness of stars in the Milky Way galaxy. The photometer comprises an array of charge-coupled devices (CCDs), which convert light into electrical signals, and measures the average flux of electrons per second over 30-minute intervals (known as long cadence). The raw flux data are corrected for systematic and other errors by a procedure called presearch data conditioning (PDC) [398]. The measurements shown in the top panel of Figure 1.1 are the corrected flux for a variable star with Kepler ID 8073767 (quarter 1).†

As we can see, the time series exhibits a strong sinusoid-like periodic pattern which repeats approximately 19.5 times. A rough estimate for the periodicity is therefore $1,639/19.5 \approx 84.1$ samples, or $84.1/2 = 42.05$ hours. However, a single sinusoid would not be able to capture the large variations of the peak values. The middle panel of Figure 1.1 depicts a model which employs eight sinusoids plus a constant. The frequencies in this model are estimated by a periodogram maximization technique to be discussed later in the book (Chapter 6). The revised estimate for the dominating periodicity is 84.7 samples or $84.7/2 = 42.35$ hours. The coefficients of the resulting sinusoids are estimated by least-squares regression together with the constant term (Chapter 5). As we can see, the model is able to capture the main oscillatory patterns of the time series, leaving just small random-looking variation in the residuals (bottom panel). A further analysis of the residual time series will be given in Chapter 7.

* http://kepler.nasa.gov/.

† Available at http://archive.stsci.edu/kepler/ and http://exoplanetarchive.ipac.caltech.edu/.

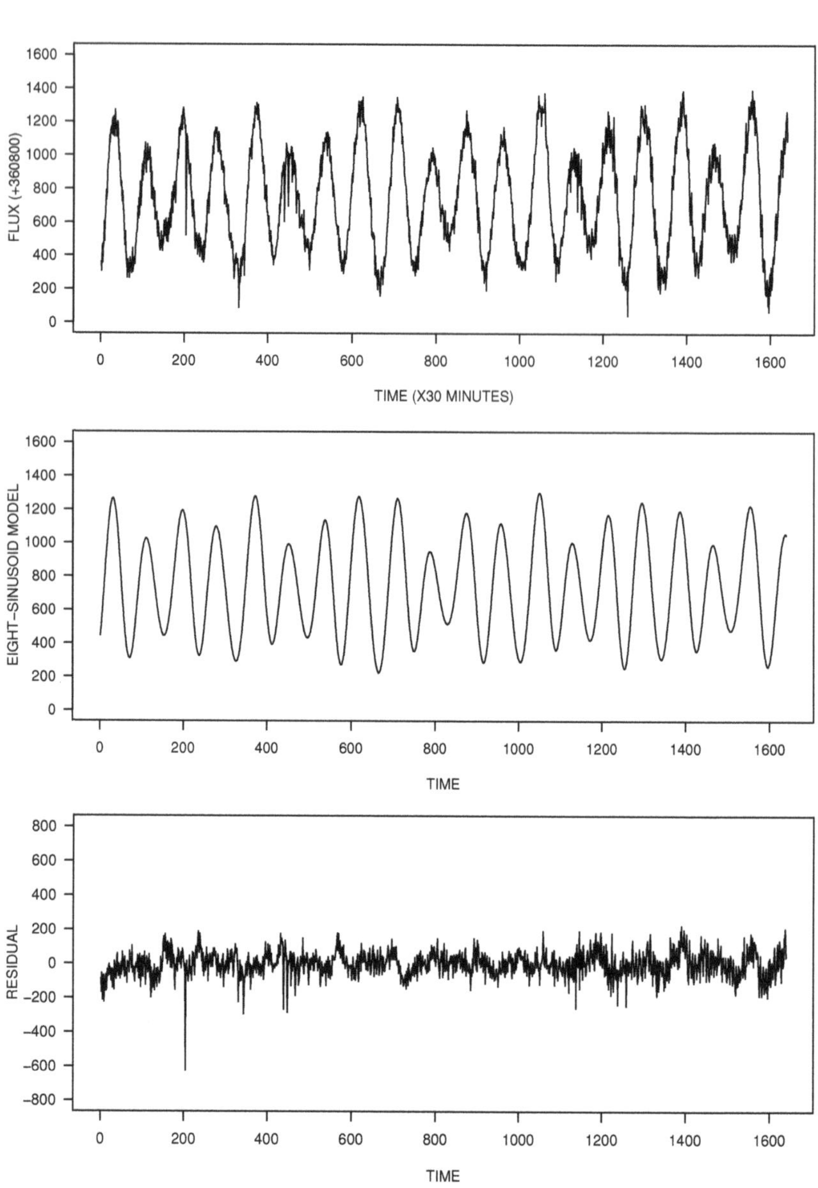

Figure 1.1. Brightness variation of a variable star obtained from NASA's Kepler Mission. Top to bottom: light curve of the variable star observed over 30-minute intervals, an eight-sinusoid model for the time series, and the residuals.

1.4 Complex Time Series with Mixed Spectra

A fundamental and useful property of sinusoidal functions is their invariance to linear time-invariant (LTI) systems. An LTI system is a convolution operator which transforms the input signal $x(t)$ into $y(t)$ given by

$$y(t) := \int_{-\infty}^{\infty} h(\tau)x(t-\tau)\,d\tau,$$

where $h(t)$ is called the impulse response of the system. If $x(t)$ is a sinusoidal function of the form $x(t) = A\sin(2\pi f t) + B\cos(2\pi f t)$, then

$$\begin{aligned} y(t) &= \int_{-\infty}^{\infty} h(\tau)\{A\cos(2\pi f(t-\tau)) + B\sin(2\pi f(t-\tau))\}\,d\tau \\ &= A'(f)\cos(2\pi f t) + B'(f)\sin(2\pi f t), \end{aligned}$$

where

$$\begin{aligned} A'(f) &:= \int_{-\infty}^{\infty} h(\tau)\{A\cos(2\pi f\tau) - B\sin(2\pi f\tau)\}\,d\tau, \\ B'(f) &:= \int_{-\infty}^{\infty} h(\tau)\{A\sin(2\pi f\tau) + B\cos(2\pi f\tau)\}\,d\tau. \end{aligned}$$

In other words, the output $y(t)$ remains to be a sinusoidal function with the same frequency f and the LTI system only changes the coefficients of the sinusoid from A and B to $A'(f)$ and $B'(f)$. Note that the linearity of an LTI system also implies that if the input $x(t)$ is a sum of sinusoids, the output $y(t)$ remains to be a sum of sinusoids with unaltered frequencies.

By taking advantage of the invariance to LTI systems, sinusoidal waves, in electrical or acoustical forms, are widely used for transmission of signals in applications such as radar, sonar, biomedical imaging, and telecommunications. Complex sinusoidal functions, which, by Euler's formula, take the form

$$\exp(i\omega t) = \cos(\omega t) + i\sin(\omega t)$$

with $i := \sqrt{-1}$, appear naturally in these applications. Complex sinusoidal functions are also known as complex exponentials.

In radar [416], for example, sinusoidal signals of the form

$$A\cos(2\pi f_0 t) + B\sin(2\pi f_0 t)$$

are transmitted as electromagnetic waves which propagate at the speed of light c. When hitting a target, the wave is reflected with the Doppler effect that shifts its frequency. The echo, or return, received by the radar takes the form

$$r(t) := A'\cos(2\pi(f_0 - f)t) + B'\sin(2\pi(f_0 - f)t), \tag{1.4.1}$$

where $f := 2vf_0/c$ is the Doppler shift caused by the motion of the target. The parameter v denotes the radial velocity of the target relative to the radar, which is positive when the target moves toward the radar and negative when it moves away from the radar. Although the Doppler shift is unknown, an upper bound, denoted by $b > 0$, can be determined *a priori* for its absolute value. The carrier frequency f_0 is much greater than b, so the frequency in $r(t)$ resides in a narrow band between $f_0 - b$ and $f_0 + b$. If $r(t)$ were sampled at its Nyquist rate $2(f_0 + b)$, which is very large, it would produce an enormous amount of unnecessary data. A more efficient way is to bring the frequency in $r(t)$ down to the neighborhood of zero and sample the resulting baseband function at a much lower rate. This process is known as frequency demodulation.

More precisely, consider the function

$$r_I(t) := \cos(2\pi f_0 t)\, r(t).$$

It follows from (1.4.1) that

$$\begin{aligned} r_I(t) &= \tfrac{1}{2}\{A' \cos(2\pi f t) - B' \sin(2\pi f t)\} \\ &\quad + \tfrac{1}{2}\{A' \cos(2\pi(2f_0 - f)t) + B' \sin(2\pi(2f_0 - f)t)\}. \end{aligned}$$

Observe that the frequency of the second term in $r_I(t)$, which equals $2f_0 - f$, is far above the frequency of the first term, which equals f (assume $f > 0$ for simplicity of discussion). Therefore, the second term can be removed by an analog lowpass filter with impulse response $h(t)$ and cutoff frequency

$$f_c \in (b, 2f_0 - b).$$

This produces the so-called in-phase signal

$$x_I(t) := \tfrac{1}{2}\{A_I' \cos(2\pi f t) + B_I' \sin(2\pi f t)\},$$

where

$$\begin{aligned} A_I' &:= \int_{-\infty}^{\infty} h(\tau)\{A' \cos(2\pi f \tau) + B' \sin(2\pi f \tau)\}\, d\tau, \\ B_I' &:= \int_{-\infty}^{\infty} h(\tau)\{A' \sin(2\pi f \tau) - B' \cos(2\pi f \tau)\}\, d\tau. \end{aligned}$$

Because $x_I(t)$ does not have frequency content higher than f_c, an alias-free time series can be obtained by sampling $x_I(t)$ at a rate $f_s \geq 2f_c$. With instantaneous sampling, the time series takes the form

$$x_I(\Delta t) = \tfrac{1}{2}\{A_I' \cos(2\pi f \Delta t) + B_I' \sin(2\pi f \Delta t)\} \qquad (t \in \mathbb{Z}),$$

where $\Delta := 1/f_s$. Similarly, applying the lowpass filter to the function

$$r_Q(t) := \sin(2\pi f_0 t)\, r(t)$$

followed by instantaneous sampling yields the so-called quadrature signal

$$x_Q(\Delta t) := \tfrac{1}{2}\{A'_Q \cos(2\pi f \Delta t) + B'_Q \sin(2\pi f \Delta t)\} \qquad (t \in \mathbb{Z}),$$

where

$$A'_Q := \int_{-\infty}^{\infty} h(\tau)\{B' \cos(2\pi f \tau) - A' \sin(2\pi f \tau)\}\, d\tau,$$
$$B'_Q := \int_{-\infty}^{\infty} h(\tau)\{B' \sin(2\pi f \tau) + A' \cos(2\pi f \tau)\}\, d\tau.$$

Combining the in-phase and quadrature signals as the real and imaginary parts of a complex-valued signal leads to the complex sinusoid

$$x_t := x_I(\Delta t) + i x_Q(\Delta t) = \beta \exp(i\omega t) \qquad (t \in \mathbb{Z}), \tag{1.4.2}$$

where $\omega := 2\pi f \Delta \in (0, \pi)$ and $\beta := \frac{1}{2}(A' + iB') \int h(\tau) \exp(-i\omega\tau)\, d\tau$.

The same argument can be used to show that (1.4.2) remains valid when f is negative because $\cos(2\pi f t) = \cos(2\pi |f|)$ and $\sin(2\pi f t) = -\sin(2\pi |f|)$. Therefore, for complex sinusoids, the frequency can be negative. However, due to the 2π-periodicity, a negative frequency ω in the interval $(-\pi, 0)$ is an alias of the positive frequency $\omega' := 2\pi - \omega$ in the interval $(\pi, 2\pi)$. Therefore, the frequencies of complex sinusoids can also be restricted to the interval $(0, 2\pi)$.

The model (1.4.2) can be extended to include multiple complex sinusoids and the noise. The general model takes the form

$$y_t = \sum_{k=1}^{p} \beta_k \exp(i\omega_k t) + \epsilon_t \qquad (t \in \mathbb{Z}).$$

In radar applications, this model represents superimposed returns from multiple targets (or scatterers), each moving at a different speed relative to the radar.

Let us consider an example with real-world data. Figure 1.2 shows a 100-sample segment of a radar signal together with the complex sinusoid extracted from it and the residuals. The radar signal is part of a large data set collected in November 1993 by a team of researchers at McMaster University using a high resolution radar overlooking the Atlantic Ocean from a clifftop near Dartmouth, Nova Scotia, Canada.[‡] The target is a spherical block of styrofoam, one meter in diameter and wrapped with wire mesh. The radar signal[§] represents the demodulated returns in the 2,685-meter range bin, sampled at the rate $f_s = 1{,}000$ Hertz (hence $\Delta := 1/f_s = 1$ millisecond).

[‡]http://soma.ece.mcmaster.ca/ipix.

[§]Available at http://soma.ece.mcmaster.ca/ipix/dartmouth/datasets.html as Clutter + Target Data File #283 (range bin 10, vertical polarization).

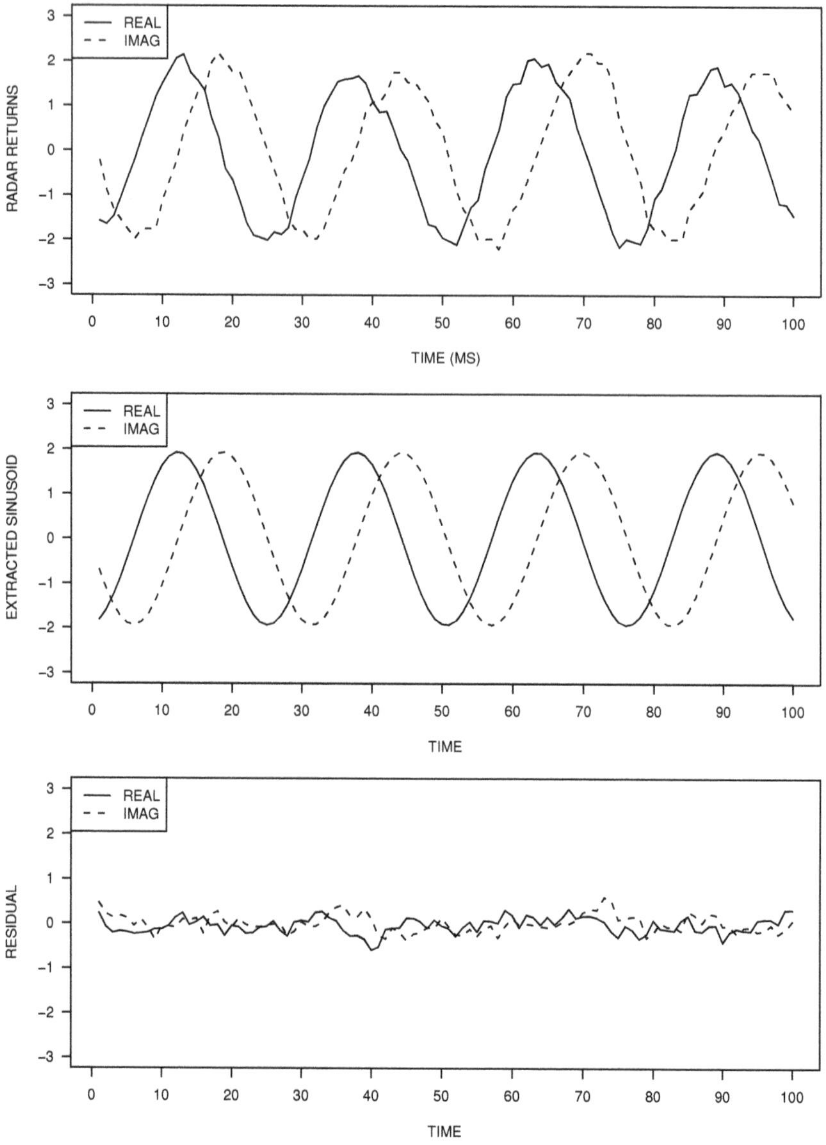

Figure 1.2. Radar returns from a target. Top to bottom: real (solid line) and imaginary (dashed line) parts of the radar data, the extracted complex sinusoid, and the residuals. Time is in milliseconds.

Processing the entire signal requires more sophisticated techniques due to its time-varying nature (see Chapter 6 for details). Figure 1.2 simply shows that for a sufficiently short segment (100 milliseconds in this case) the signal can be well represented by a single complex sinusoid plus random noise.

The frequency of the complex sinusoid is estimated by the periodogram maximization technique to be discussed in Chapter 6, and the complex amplitude is obtained by least-squares regression to be discussed in Chapter 5. The estimated frequency is equal to 0.03912442. It is an estimate for the normalized frequency $f_n := \Delta f = f/f_s$. To get an estimate for the velocity of the target, it suffices to observe that $f = 2vf_0/c$, where $f_0 = 9.39 \times 10^9$ Hertz is the carrier frequency of the radar and $c = 3 \times 10^8$ meters per second is the speed of light. Therefore, the relative velocity of the target is given by $v = f_n \times f_s c/(2f_0)$, or approximately 0.62 meters per second toward the radar.

Chapter 2

Basic Concepts

This chapter discusses two types of parameterization for real and complex sinusoids together with some basic assumptions. It also reviews some basic concepts of random processes, linear prediction theory, and asymptotic statistical theory. Most of these results can be found easily in the literature. Therefore, they are stated without proof but with reference to standard textbooks.

2.1 Parameterization of Sinusoids

As shown in Chapter 1, there are two types of models for sinusoidal functions in practice: the real sinusoid model (RSM) and the complex sinusoid model (CSM). Both models can be further parameterized in two different forms: the Cartesian (or rectangular) form and the polar form.

Consider the real case first. The Cartesian RSM can be expressed as

$$x_t = \sum_{k=1}^{q} \{A_k \cos(\omega_k t) + B_k \sin(\omega_k t)\} \qquad (t \in \mathbb{Z}), \tag{2.1.1}$$

where the $\omega_k \in (0, \pi)$ are the frequency parameters, the $A_k \in \mathbb{R}$ and $B_k \in \mathbb{R}$ are the amplitude parameters (or coefficients) satisfying $A_k^2 + B_k^2 > 0$ $(k = 1, \ldots, q)$. The polar RSM takes the form

$$x_t = \sum_{k=1}^{q} C_k \cos(\omega_k t + \phi_k) \qquad (t \in \mathbb{Z}), \tag{2.1.2}$$

where the $C_k > 0$ are the amplitude parameters and the $\phi_k \in (-\pi, \pi]$ are the phase parameters. Note that in both (2.1.1) and (2.1.2) we exclude a possible constant term A_0 for convenience as it can be easily estimated and removed in practice.

While the C_k in (2.1.2) can be easily interpreted as representing the strength of the sinusoids, the phase parameters ϕ_k represent the advance (if positive) or delay (if negative) of the sinusoidal waves relative to their zero-phase counterparts. This is illustrated in Figure 2.1. In this example, the sinusoidal functions

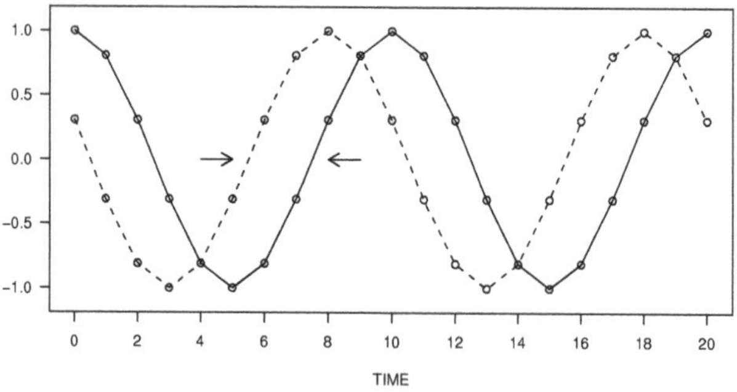

Figure 2.1. Two sinusoids with common frequency $\omega_0 = 2\pi \times 0.1$ and different phases. Solid line, $\phi_1 = 0$; dashed line, $\phi_2 = 2\pi \times 0.2$. The second sinusoid leads the first sinusoid by $(\phi_2 - \phi_1)/\omega_0 = 2$ samples.

are $x_1(t) := \cos(\omega_0 t + \phi_1)$ and $x_2(t) := \cos(\omega_0 t + \phi_2)$, where $\omega_0 = 2\pi \times 0.1$, $\phi_1 = 0$, and $\phi_2 = 2\pi \times 0.2$. As we can see, the first sinusoid (solid line) is just a shifted copy of the second sinusoid (dashed line). At $t = 0$, the first sinusoid takes the value $\cos(\phi_1) = 1$. The second sinusoid takes the same value at $t = (\phi_1 - \phi_2)/\omega_0 = -2$. Hence the second sinusoid leads the first sinusoid by 2 time units.

It is straightforward to verify that the Cartesian and polar forms of the RSM are equivalent under the following parameter transformation:

$$\begin{cases} A_k = C_k \cos(\phi_k), & B_k = -C_k \sin(\phi_k), \\ C_k = \sqrt{A_k^2 + B_k^2}, & \phi_k = \arctan(-B_k, A_k), \end{cases} \tag{2.1.3}$$

where

$$\arctan(B, A) := \begin{cases} \arctan(B/A) & \text{if } A > 0, \\ \arctan(B/A) + \pi & \text{if } A < 0, B \geq 0, \\ \arctan(B/A) - \pi & \text{if } A < 0, B < 0, \\ \pi/2 & \text{if } A = 0, B > 0, \\ -\pi/2 & \text{if } A = 0, B < 0, \\ 0 & \text{if } A = B = 0. \end{cases}$$

This relationship is illustrated in Figure 2.2(a). The equivalence explains why a linear combination of $\cos(\omega t)$ and $\sin(\omega t)$ is considered as a single real sinusoid with frequency ω rather than two sinusoids. For this reason, we say that ω_k, A_k, and B_k in (2.1.1) are the parameters of the kth sinusoid.

Without loss of generality, we always assume that the frequencies ω_k are arranged in an ascending order such that

$$0 < \omega_1 < \cdots < \omega_q < \pi.$$

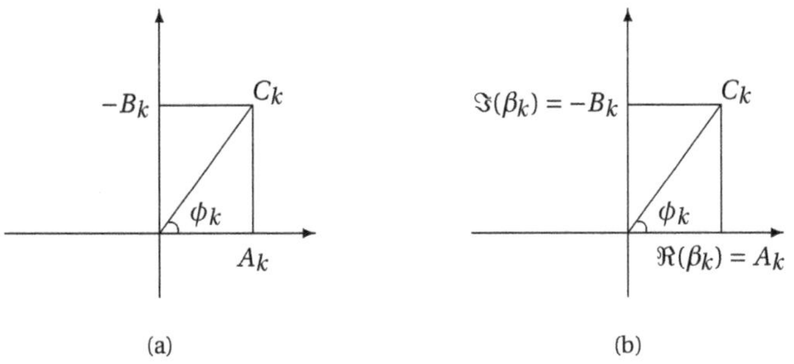

Figure 2.2. Illustration of Cartesian and polar parameters for real and complex sinusoids. (a) Real case. (b) Complex case.

We also exclude the possibility for the frequency parameter to take value zero because it leads to a constant term which is typically removed (by subtracting the sample mean) prior to spectral analysis. Moreover, instead of the angular frequencies ω_k (measured in radians per unit time), it is often more meaningful to consider the normalized frequencies

$$f_k := \omega_k/(2\pi), \tag{2.1.4}$$

which are measured in cycles per unit time. For example, if the time t is measured in seconds, then the normalized frequency f_k is measured in cycles per second, or Hertz; if t is in years, then f_k is in cycles per year. So physical interpretation becomes easier using the normalized frequencies. The angular frequencies are more convenient for mathematical manipulation.

In the complex case, the Cartesian CSM takes the form

$$x_t = \sum_{k=1}^{p} \beta_k \exp(i\omega_k t) \qquad (t \in \mathbb{Z}), \tag{2.1.5}$$

where $\omega_k \in (-\pi, \pi] \setminus \{0\}$ is the frequency of the kth sinusoid and $\beta_k \in \mathbb{C}$ is its complex amplitude. Let β_k be further parameterized by real-valued parameters $A_k \in \mathbb{R}$ and $B_k \in \mathbb{R}$ such that

$$\begin{cases} \beta_k := A_k - iB_k, \\ A_k := \Re(\beta_k), \quad B_k := -\Im(\beta_k). \end{cases} \tag{2.1.6}$$

Then, the polar CSM can be expressed as

$$x_t = \sum_{k=1}^{p} C_k \exp\{i(\omega_k t + \phi_k)\} \qquad (t \in \mathbb{Z}). \tag{2.1.7}$$

where $C_k > 0$ and $\phi_k \in (-\pi, \pi]$ are related to A_k and B_k through the transformation (2.1.3). Note that B_k is defined as $-\Im(\beta_k)$ rather than $\Im(\beta_k)$ in order for the transformation (2.1.3) to be valid for both RSM and CSM. Figure 2.2(b) illustrates this relationship for the CSM. The polar parameters C_k and ϕ_k can also be expressed in terms of the complex amplitude β_k such that

$$C_k = |\beta_k|, \quad \phi_k = \angle\beta_k.$$

Without loss of generality, the frequencies ω_k are assumed to satisfy

$$-\pi < \omega_1 < \cdots < \omega_p < \pi.$$

The frequencies can also be redefined by the aliasing transformation

$$\omega_k \mapsto \begin{cases} \omega_k + 2\pi & \text{if } \omega_k < 0, \\ \omega_k & \text{otherwise,} \end{cases}$$

so that they take values in the interval $(0, 2\pi)$. In this case, it will be assumed that

$$0 < \omega_1 < \cdots < \omega_p < 2\pi.$$

As with the RSM, it is often more meaningful to consider the normalized frequencies f_k defined by (2.1.4) instead of the angular frequencies $\omega_k = 2\pi f_k$.

A real sinusoid can be expressed as the sum of two conjugate complex sinusoids. Indeed, for any $\omega \in (0, \pi)$, $C > 0$, and $\phi \in (-\pi, \pi]$, let

$$\beta := \tfrac{1}{2} C \exp(i\phi).$$

Then,

$$\begin{aligned} C\cos(\omega t + \phi) &= \beta \exp(i\omega t) + \beta^* \exp(-i\omega t) \\ &= \beta \exp(i\omega t) + \beta^* \exp(i(2\pi - \omega)t). \end{aligned}$$

For this reason, the RSM can be written in the form of (2.1.5) with $p := 2q$ and

$$\begin{cases} \omega_{p-k+1} := -\omega_k \text{ or } 2\pi - \omega_k \\ \beta_{p-k+1} := \beta_k^* := \tfrac{1}{2} C_k \exp(-i\phi_k) \end{cases} \quad (k = 1, \ldots, q), \tag{2.1.8}$$

where C_k and ϕ_k are given by (2.1.2). The RSM can also be written in the form of (2.1.7) with $p := 2q$ and

$$\begin{cases} \omega_{p-k+1} := -\omega_k \text{ or } 2\pi - \omega_k \\ \phi_{p-k+1} := -\phi_k = -\arctan(-B_k, A_k) \\ C_{p-k+1} := C_k := \tfrac{1}{2}\sqrt{A_k^2 + B_k^2} \end{cases} \quad (k = 1, \ldots, q), \tag{2.1.9}$$

where A_k and B_k are given by (2.1.1). With this transformation, results derived for the CSM can be translated into the corresponding results for the RSM.

It is worth pointing out that in the RSM (2.1.1)–(2.1.2) the power of the kth sinusoid is equal to $\frac{1}{2}C_k^2$ asymptotically for large sample sizes, because

$$\lim_{n\to\infty} n^{-1}\sum_{t=1}^{n}|C_k\cos(\omega_k t+\phi_k)|^2=\tfrac{1}{2}C_k^2=\tfrac{1}{2}(A_k^2+B_k^2).$$

In the CSM (2.1.5)–(2.1.7), the power of the kth sinusoid is equal to C_k^2 exactly for all sample sizes, because

$$n^{-1}\sum_{t=1}^{n}|C_k\exp\{i(\omega_k t+\phi_k)\}|^2=C_k^2=|\beta_k|^2=A_k^2+B_k^2.$$

This distinction should be kept in mind when comparing the results derived for the RSM with those derived for the CSM in later chapters.

While both RSM and CSM are discussed in this book, not all methods and analyses are presented for both models. A primary reason is that the original sources in the literature focus on different models. In many cases, one can infer the results for one model from the results for the other. In some cases, however, the inference is not straightforward or even not valid (for example, there is no RSM counterpart for the CSM with $p=1$).

Frequency separation is an important concept in dealing with sinusoids. If two frequencies are not sufficiently separated, the corresponding sinusoids cannot be easily distinguished from each other based on a finite data record. In the complex case, the separation of frequencies can be measured by

$$\Delta:=\min_{k\neq k'}\{|\omega_k-\omega_{k'}|,2\pi-|\omega_k-\omega_{k'}|\},$$

where the second term is necessary because of the aliasing. In the real case, the separation of frequencies is measured by

$$\Delta:=\min_{k\neq k'}\{|\omega_k-\omega_{k'}|,\omega_k,\pi-\omega_k\},$$

where the last two terms are required to take the conjugate frequencies into account (a frequency ω_k near 0 or π is also near its conjugate frequency $-\omega_k$ in the CSM representation).

Once the sample size is given, an inherent limit of frequency resolution is determined. In comparison with this resolution limit, the frequencies can be described as well separated or closely spaced. The amount of frequency separation relative to the resolution limit, and hence the sample size, turns out to be a key factor that determines the technique and accuracy with which the frequencies can be estimated on the basis of a finite data record. It also contributes to the increased complexity in the mathematical analysis. In asymptotic analysis, a common and useful technique is to allow the frequency separation to approach zero at a certain rate as the sample size grows. Such frequency separation conditions are typically expressed in terms of a power function of n^{-1} and in the order of magnitude, for example, $\mathcal{O}(n^{-1})$ or $\mathcal{O}(n^{-1/2})$, rather than the exact amount. The separation conditions play a key role in the asymptotic analysis of the estimation methods discussed in later chapters.

Throughout the book, we are concerned with time series of the form

$$y_t = x_t + \epsilon_t \qquad (t = 1, \ldots, n), \tag{2.1.10}$$

where the ϵ_t represent the random measurement errors or noise. Additional assumptions about $\{\epsilon_t\}$ will be made for specific methods and analytical studies. The simplest assumption is that the ϵ_t are independent and identically distributed, or i.i.d., random variables with mean zero and variance σ^2. As a convention, the noise is always assumed real when referring to the RSM and complex when referring to the CSM. Unless noted otherwise, all parameters in the RSM and CSM are regarded as deterministic but possibly unknown constants that are fixed in all realizations of the random process $\{y_t\}$. However, in some analytical results, the phases ϕ_k of the sinusoids are allowed to be random variables that are independent of the noise $\{\epsilon_t\}$. In some simulations, the frequencies $\{\omega_k\}$ are also generated randomly to mitigate the possible dependence of the results on specific frequency values.

2.2 Spectral Analysis of Stationary Processes

Stationarity is an important property of random processes that makes statistical analysis meaningful. Although not always true in practice, the stationarity assumption is typically applicable to sufficiently short data records taken from slowly-varying nonstationary processes.

Definition 2.1 (Stationary Random Processes). A zero-mean real or complex random process $\{X_t\}$ is said to be (weakly or wide-sense) stationary if the covariance function $c(s,t) := E(X_s X_t^*)$ is finite and time-invariant, i.e., $c(t+u,t)$ does not depend on t. In this case, $r(u) := c(t+u,t) = E(X_{t+u}X_t^*)$ is called the autocovariance function (ACF) of $\{X_t\}$.

The following proposition summarizes some useful properties of the ACF.

Proposition 2.1 (Properties of the Autocovariance Function). *The ACF $r(u)$ of a zero-mean stationary process $\{X_t\}$ is nonnegative definite in the sense that*

$$\sum_{u=1}^{m} \sum_{v=1}^{m} a_u a_v^* r(u-v) \geq 0$$

for any constant sequence $\{a_1, \ldots, a_m\} \subset \mathbb{C}$*. The* ACF *is symmetric in the sense that* $r(-u) = r^*(u)$ *and is bounded by the variance of* X_t*, i.e.,* $|r(u)| \le r(0)$.

In addition to the time-domain characterization, the ACF can also be represented in the frequency domain as a linear combination of infinitely many sinusoids with different frequencies. The subject of spectral analysis of stationary processes is largely built upon the following proposition.

Proposition 2.2 (Spectral Representation of the Autocovariance Function [46, pp. 117–118]). *If* $r(u)$ *is the* ACF *of a stationary process* $\{X_t\}$*, then there exists a unique nondecreasing right-continuous function* $S(\omega)$*, with* $S(-\pi) = 0$*, such that*

$$r(u) = (2\pi)^{-1} \int_{-\pi}^{\pi} \exp(i\omega u)\, dS(\omega).$$

The function $S(\omega)$ *is called the spectral distribution function. If* $r(u)$ *is absolutely summable, i.e.,* $\sum |r(u)| < \infty$*, then there exists a unique uniformly continuous nonnegative function* $f(\omega)$*, called the spectral density function* (SDF)*, such that*

$$r(u) = (2\pi)^{-1} \int_{-\pi}^{\pi} f(\omega) \exp(i\omega u)\, d\omega$$

and

$$f(\omega) = \sum_{u=-\infty}^{\infty} r(u) \exp(-iu\omega),$$

in which case, $S(\omega)$ *can be expressed as* $S(\omega) = \int_{-\pi}^{\omega} f(\lambda)\, d\lambda$ *and therefore is differentiable with* $\dot{S}(\omega) = f(\omega)$*. If* $r(u)$ *is real, then* $f(\omega)$ *and* $S(\omega)$ *are symmetric in the sense that* $f(-\omega) = f(\omega)$ *and* $S(-\omega) = S(\pi) - S(\omega)$*. Both* $f(\omega)$ *and* $S(\omega)$ *can be extended as* 2π*-periodic functions in* $\mathbb{R}$*.*

Remark 2.1 The SDF defined in Proposition 2.2 has the property

$$(2\pi)^{-1} \int_{-\pi}^{\pi} f(\omega)\, d\omega = \int_{-1/2}^{1/2} f(2\pi x)\, dx = r(0).$$

Therefore, it represents the distribution of the total variance $r(0)$ with respect to the normalized frequency $\omega/(2\pi)$ (cycles per unit time). This definition is often used in engineering textbooks such as [369]. In statistical textbooks such as [46], the SDF is often defined as $f(\omega) := 2\pi \sum_{u=-\infty}^{\infty} r(u) \exp(-iu\omega)$. In this case, the SDF satisfies $\int_{-\pi}^{\pi} f(\omega)\, d\omega = r(0)$, so it represents the distribution of the total variance with respect to the angular frequency ω (radians per unit time).

Remark 2.2 The absolute summability of the ACF is a sufficient but not necessary condition for the existence of the spectral density function. The spectral distribution function is also known as the integrated spectrum [298, p. 209].

A typical stationary process can be classified according to the type of its spectrum into one of the following three categories.

Definition 2.2 (Discrete, Continuous, and Mixed Spectra). A stationary process with spectral distribution function $S(\omega)$ is said to have a discrete spectrum if $S(\omega) = \sum a_k^2 \mathscr{I}(\omega_k \le \omega)$ for some finite sequences $\{\omega_k\} \subset (-\pi, \pi)$ and $\{a_k\} \subset \mathbb{R}$, where $\mathscr{I}(\cdot)$ is the indicator function. It is said to have a continuous spectrum if there exists a function $f(\omega) \ge 0$ such that $S(\omega) = \int_{-\pi}^{\omega} f(\lambda)\, d\lambda$. It is said to have a mixed spectrum if $S(\omega) = S_d(\omega) + S_c(\omega)$, where $S_d(\omega)$ is a discrete spectrum and $S_c(\omega)$ is a continuous spectrum.

Remark 2.3 A discrete spectrum of the form $S(\omega) = \sum a_k^2 \mathscr{I}(\omega_k \le \omega)$ is a nondecreasing step function with jumps of magnitude a_k^2 at $\omega = \omega_k$. The spectral distribution function not having a jump at $\omega = 0$ is the necessary and sufficient condition for the process to be ergodic in the mean, i.e., the sample mean converges in mean-square to the expected value of the process as the sample size approaches infinity [298, p. 342].

The following proposition shows that a stationary process itself can be represented as a linear combination of infinitely many sinusoids.

Proposition 2.3 (Spectral Representation of Stationary Processes [46, p. 145]). *If $\{X_t\}$ is a zero-mean stationary process with spectral distribution function $S(\omega)$, then there exists a random process $\{Z(\omega), -\pi \le \omega \le \pi\}$, with $Z(-\pi) = 0$, such that*

$$X_t = \int_{-\pi}^{\pi} \exp(i\omega t)\, dZ(\omega) \tag{2.2.1}$$

almost surely, where $Z(\omega)$ satisfies the following conditions:

(a) $E\{Z(\omega)\} = 0$ for $\omega \in [-\pi, \pi]$,
(b) $E\{|Z(\omega)|^2\} < \infty$ for $\omega \in [-\pi, \pi]$,
(c) $E\{|Z(\omega) - Z(\lambda)|^2\} = (2\pi)^{-1}\{S(\omega) - S(\lambda)\}$ for $-\pi \le \omega < \lambda \le \pi$,
(d) $E\{(Z(\omega) - Z(\lambda))(Z(\omega') - Z(\lambda'))^*\} = 0$ for $-\pi \le \lambda \le \omega < \lambda' \le \omega' \le \pi$.

A process satisfying (a), (b), and (d) is called an orthogonal-increment process. The representation (2.2.1) is unique in the sense that if it also holds with another orthogonal-increment process $Z'(\omega)$, then $P\{Z(\omega) = Z'(\omega)\} = 1$ for all ω.

According to Proposition 2.3, if $\{X_t\}$ is a zero-mean stationary process of discrete spectrum with spectral distribution function

$$S(\omega) = 2\pi \sum_{k=1}^{p} C_k^2 \mathscr{I}(\omega_k \le \omega), \tag{2.2.2}$$

then it can be expressed in the form of (2.1.5) with $\beta_k := Z(\omega_k) - Z(\omega_k^-)$, where $Z(\omega_k^-)$ stands for the limit of $Z(\omega)$ (in mean-square) as ω approaches ω_k from the left. Because $Z(\omega)$ satisfies (a), (c), and (d) in Proposition 2.3, we have

$$E(\beta_k)=0, \quad E(\beta_k\beta_{k'}^*)=C_k^2\delta_{k-k'}, \tag{2.2.3}$$

where $\{\delta_u\}$ $(u\in\mathbb{Z})$ is the Kronecker delta sequence, named after the German mathematician Leopold Kronecker (1823–1891), such that $\delta_0=1$ and $\delta_u=0$ for all $u\neq 0$. Therefore, a zero-mean stationary process of discrete spectrum is nothing but a sum of complex sinusoids of the form (2.1.5) with uncorrelated zero-mean random coefficients β_k. Using the notation of Dirac delta $\delta(\omega)$, named after the British physicist Paul Dirac (1902–1984), the SDF that corresponds to $S(\omega)$ in (2.2.2) can be formally expressed as

$$f(\omega)=2\pi\sum_{k=1}^{p}C_k^2\delta(\omega-\omega_k), \tag{2.2.4}$$

which is an impulsive function consisting of discrete impulses located at the ω_k. An SDF of this form is also known as a line spectrum [369]. By Proposition 2.2, the corresponding ACF can be expressed as

$$r(u)=\sum_{k=1}^{p}C_k^2\exp(i\omega_k u), \tag{2.2.5}$$

which is a weighted sum of complex sinusoids with frequencies ω_k.

On the other hand, if $\{X_t\}$ takes the form (2.1.5) with $\beta_k=C_k\exp(i\phi_k)$ and $\omega_k\in(-\pi,\pi)$, and if the ϕ_k are i.i.d. random variables with uniform distribution in $(-\pi,\pi]$ and the C_k are real constants, then the condition (2.2.3) is satisfied. In this case, $\{X_t\}$ can be expressed as (2.2.1) with

$$Z(\omega):=\sum_{k=1}^{p}\beta_k\mathscr{I}(\omega_k\le\omega),$$

which implies that $\{X_t\}$ is a zero-mean stationary process of discrete spectrum whose spectral distribution and density functions and whose ACF are given by (2.2.2), (2.2.4), and (2.2.5), respectively.

More generally, consider the random process $\{y_t\}$ which is given by (2.1.10) with $\{x_t\}$ taking the form (2.1.5). If the ϕ_k are i.i.d. random variables with uniform distribution in $(-\pi,\pi]$ and the noise $\{\epsilon_t\}$ is a zero-mean stationary process of continuous spectrum with spectral distribution function $S_\epsilon(\omega)$ and is independent of the ϕ_k, then $\{y_t\}$ is a zero-mean stationary process of mixed spectrum with spectral distribution function

$$S_y(\omega)=S_x(\omega)+S_\epsilon(\omega),$$

where $S_x(\omega)$ takes the form (2.2.2). The corresponding SDF can be expressed as

$$f_y(\omega)=f_x(\omega)+f_\epsilon(\omega),$$

where $f_x(\omega)$ takes the form (2.2.4) and $f_\epsilon(\omega) := \dot{S}_\epsilon(\omega)$ is the SDF of $\{\epsilon_t\}$.

If the ϕ_k are constants, then $\{y_t\}$ is no longer a stationary process in the sense of Definition 2.1. But, we can still interpret $\{y_t\}$ as a stationary process of mixed spectrum by considering the sample autocovariance function of $\{y_1,\ldots,y_n\}$ as $n \to \infty$. As will be explained in Chapter 4, the sample ACF of $\{y_1,\ldots,y_n\}$, defined as $\hat{r}_y(u) := n^{-1}\sum_{t=1}^{n-|u|} y_{t+|u|}y_t^*$ for $|u| < n$, converges (in a suitable sense) to a finite limit which can be expressed as the sum of the ACF given by (2.2.5) and the ACF of $\{\epsilon_t\}$. In other words, this limit of the sample ACF coincides with the ordinary ACF in the previous case where the phases are random. Therefore, it has a mixed spectrum of the same form as a stationary process.

In practice, the noise process $\{\epsilon_t\}$ in (2.1.10) can be assumed to have a continuous spectrum, which is in contrast to the signal $\{x_t\}$ in (2.1.1)–(2.1.5) that has a discrete spectrum. This is the key feature that distinguishes the signal from the noise. A special case is where $\{\epsilon_t\}$ can be modeled as a white noise process, i.e., the ACF of $\{\epsilon_t\}$ takes the form $r_\epsilon(u) = \sigma^2\delta_u$. This will be denoted as $\{\epsilon_t\} \sim \mathrm{WN}(0,\sigma^2)$. Linear processes constitute a more general model for $\{\epsilon_t\}$, which is useful especially in asymptotic analysis.

Definition 2.3 (Linear Processes). A random process $\{X_t\}$ is said to be a linear process if there exists a sequence of constants $\{\psi_j\}$ and a white noise process $\{\zeta_t\} \sim \mathrm{WN}(0,\sigma^2)$ such that

$$X_t = \sum_{j=-\infty}^{\infty} \psi_j \zeta_{t-j} \tag{2.2.6}$$

converges in mean square. A linear process of the form (2.2.6) can be regarded as the output of a linear time-invariant (LTI) filter with transfer function $\psi(z) := \sum \psi_j z^{-j}$ ($z \in \mathbb{C}$) and input $\{\zeta_t\}$. If z and z^{-1} are interpreted as the forward-shift and backward-shift operators such that $z\zeta_t = \zeta_{t+1}$ and $z^{-1}\zeta_t = \zeta_{t-1}$, then (2.2.6) can also be expressed as $X_t = \psi(z)\zeta_t$.

Useful properties of linear processes are given in the following proposition.

Proposition 2.4 (Properties of Linear Processes [46, pp. 122, 154–155]). *Let $\{\psi_j\}$ be a sequence of constants. If $\sum|\psi_j|^2 < \infty$, then the infinite series in (2.2.6) converges in mean square. If $\sum|\psi_j| < \infty$, then the infinite series in (2.2.6) converges almost surely and the filter $\psi(z) := \sum \psi_j z^{-j}$ is said to be bounded-in-bounded-out (BIBO) stable. In both cases, the SDF of $\{X_t\}$ takes the form*

$$f(\omega) = \sigma^2|\Psi(\omega)|^2,$$

where $\Psi(\omega) := \psi(\exp(i\omega)) = \sum \psi_j \exp(-ij\omega)$, and the ACF of $\{X_t\}$ is given by

$$r(u) = \sigma^2 \sum_{j=-\infty}^{\infty} \psi_{j+u}\psi_j^*.$$

Moreover, if the filter $\psi(z)$ is BIBO stable, then $f(\omega)$ is a uniformly continuous function and $\{r(u)\}$ is an absolutely summable sequence.

The following proposition shows that linear processes encompass a large class of stationary processes with continuous spectra.

Proposition 2.5 (Wold Decomposition [46, p. 187]). *Any purely nondeterministic zero-mean stationary process $\{X_t\}$ can be expressed as $X_t = \sum_{j=0}^{\infty} \psi_j \zeta_{t-j}$, where $\psi_0 := 1$, $\sum |\psi_j|^2 < \infty$, and $\{\zeta_t\} \sim \mathrm{WN}(0, \sigma^2)$ for some constant $\sigma^2 > 0$. Moreover, for any given t, the random variable ζ_t is a member of the closed linear space that comprises all linear combinations of $X_t, X_{t-1}, \ldots$ and their mean-square limits.*

Proposition 2.5 is part of a general theorem [46, Theorem 5.7.2, pp. 187–188] which asserts that any stationary process can be decomposed into a purely nondeterministic component as a linear process, and a deterministic component, such as a sum of sinusoids with i.i.d. random amplitudes, that can be predicted perfectly by a linear combination of its past values.

The process $\{\zeta_t\}$ in Proposition 2.5 is merely guaranteed to be white noise, i.e., its ACF takes the form $\sigma^2 \delta_u$. In the linear process model (2.2.6), stronger assumptions, such as $\{\zeta_t\}$ being a sequence of i.i.d. random variables, are often needed for some asymptotic analyses. Moreover, because Proposition 2.5 only guarantees the square-summability of $\{\psi_j\}$, the corresponding spectrum may not be smooth (or may not even be well-defined at some frequencies). The assumption of BIBO stability ensures that the corresponding SDF is a continuous function (which should not be confused with the concept of continuous spectra). Some analyses require the assumption of strong BIBO stability in the sense that $\sum |j|^r |\psi_j| < \infty$ for some constant $r > 1$, in which case the SDF can be thought of as being smoother than a continuous function. An important example of linear processes with BIBO-stable filters is the autoregressive (AR) process that satisfies the difference equation

$$X_t + \sum_{j=1}^{m} \varphi_j X_{t-j} = \zeta_t, \quad \{\zeta_t\} \sim \mathrm{WN}(0, \sigma^2), \tag{2.2.7}$$

where $\varphi(z) := 1 + \sum_{j=1}^{m} \varphi_j z^{-j}$ has all its roots inside the unit circle of the complex plane $z \in \mathbb{C}$. In this case, $\psi(z) = 1/\varphi(z)$.

2.3 Gaussian Processes and White Noise

The concept of Gaussian or normal distribution is well known for real-valued random variables and vectors. For a real random vector $\mathbf{X}$, the Gaussian distribution is specified completely by the mean $E(\mathbf{X})$ and the covariance matrix $\mathrm{Cov}(\mathbf{X}) := E\{(\mathbf{X}-\boldsymbol{\mu})(\mathbf{X}-\boldsymbol{\mu})^T\}$. In the complex case, the covariance between random variables X and Y is defined as

$$\mathrm{Cov}(X,Y) := E\{(X-E(X))(Y-E(Y))^*\}.$$

Therefore, for a complex random vector $\mathbf{X}$, not only do we need to consider the mean $\boldsymbol{\mu} := E(\mathbf{X})$ and the covariance matrix

$$\boldsymbol{\Sigma} := \mathrm{Cov}(\mathbf{X},\mathbf{X}) := E\{(\mathbf{X}-\boldsymbol{\mu})(\mathbf{X}-\boldsymbol{\mu})^H\},$$

but also the *complementary* covariance matrix, which is defined as

$$\tilde{\boldsymbol{\Sigma}} := \mathrm{Cov}(\mathbf{X},\mathbf{X}^*) := E\{(\mathbf{X}-\boldsymbol{\mu})(\mathbf{X}-\boldsymbol{\mu})^T\}.$$

Given these quantities, the mean and the covariance matrix of the real-valued random vector $\mathbf{X}_r := [\Re(\mathbf{X})^T, \Im(\mathbf{X})^T]^T$ are completely specified by

$$\boldsymbol{\mu}_r := E(\mathbf{X}_r) = [\Re(\boldsymbol{\mu})^T, \Im(\boldsymbol{\mu})^T]^T \tag{2.3.1}$$

and

$$\boldsymbol{\Sigma}_r := \mathrm{Cov}(\mathbf{X}_r) = \frac{1}{2}\begin{bmatrix} \Re(\boldsymbol{\Sigma}+\tilde{\boldsymbol{\Sigma}}) & -\Im(\boldsymbol{\Sigma}-\tilde{\boldsymbol{\Sigma}}) \\ \Im(\boldsymbol{\Sigma}+\tilde{\boldsymbol{\Sigma}}) & \Re(\boldsymbol{\Sigma}-\tilde{\boldsymbol{\Sigma}}) \end{bmatrix}. \tag{2.3.2}$$

On the other hand, if the mean $\boldsymbol{\mu}_r$ and the covariance matrix $\boldsymbol{\Sigma}_r$ of a real-valued random vector $\mathbf{X}_r := [\mathbf{X}_1^T, \mathbf{X}_2^T]^T$ are given in the form of

$$\boldsymbol{\mu}_r = \begin{bmatrix} \boldsymbol{\mu}_1 \\ \boldsymbol{\mu}_2 \end{bmatrix}, \quad \boldsymbol{\Sigma}_r = \begin{bmatrix} \boldsymbol{\Sigma}_{11} & \boldsymbol{\Sigma}_{12} \\ \boldsymbol{\Sigma}_{12}^T & \boldsymbol{\Sigma}_{22} \end{bmatrix}, \tag{2.3.3}$$

then $\boldsymbol{\mu}$, $\boldsymbol{\Sigma}$, and $\tilde{\boldsymbol{\Sigma}}$ of the complex random vector $\mathbf{X} := \mathbf{X}_1 + i\mathbf{X}_2$ are determined by

$$\begin{cases} \boldsymbol{\mu} = \boldsymbol{\mu}_1 + i\boldsymbol{\mu}_2, \\ \boldsymbol{\Sigma} = \boldsymbol{\Sigma}_{11} + \boldsymbol{\Sigma}_{22} + i(\boldsymbol{\Sigma}_{12}^T - \boldsymbol{\Sigma}_{12}), \\ \tilde{\boldsymbol{\Sigma}} = \boldsymbol{\Sigma}_{11} - \boldsymbol{\Sigma}_{22} + i(\boldsymbol{\Sigma}_{12}^T + \boldsymbol{\Sigma}_{12}). \end{cases} \tag{2.3.4}$$

Note that the matrix $\Im(\boldsymbol{\Sigma})$ is antisymmetric, i.e., $\Im(\boldsymbol{\Sigma})^T = -\Im(\boldsymbol{\Sigma})$. This implies in particular that the diagonal elements of $\Im(\boldsymbol{\Sigma})$ are all equal to zero.

With these properties in mind, let us define complex Gaussian distributions and processes as follows.

Definition 2.4 (Complex Gaussian Distributions and Processes). A random vector $\mathbf{X}$ is said to have a general complex Gaussian distribution with mean $\boldsymbol{\mu}$, covariance matrix $\boldsymbol{\Sigma}$, and complementary covariance matrix $\tilde{\boldsymbol{\Sigma}}$, denoted by $\mathbf{X} \sim \mathrm{N}_c(\boldsymbol{\mu}, \boldsymbol{\Sigma}, \tilde{\boldsymbol{\Sigma}})$, if $\mathbf{X}_r := [\Re(\mathbf{X})^T, \Im(\mathbf{X})^T]^T$ is Gaussian with mean $\boldsymbol{\mu}_r$ and covariance matrix $\mathbf{\Sigma}_r$, i.e., $\mathbf{X}_r \sim \mathrm{N}(\boldsymbol{\mu}_r, \mathbf{\Sigma}_r)$, satisfying (2.3.1)–(2.3.4). If $\tilde{\mathbf{\Sigma}} = \mathbf{0}$, then $\mathbf{X}$ is said to have a symmetric complex Gaussian distribution with mean $\boldsymbol{\mu}$ and covariance matrix $\mathbf{\Sigma}$, denoted as $\mathbf{X} \sim \mathrm{N}_c(\boldsymbol{\mu}, \mathbf{\Sigma})$. A random process $\{X_t\}$ is said to be general or symmetric complex Gaussian if $\mathbf{X} := [X_{t_1}, \ldots, X_{t_n}]^T$ is general or symmetric complex Gaussian, respectively, for any distinct points $\{t_1, \ldots, t_n\}$ and any n.

A very important property of real Gaussian distributions is their invariance under linear transformation — a linear transform of Gaussian variables remains Gaussian. This property is also possessed by the general and symmetric complex Gaussian variables, respectively, as stated in the following proposition.

Proposition 2.6 (Linear Transform of Complex Gaussian Random Variables). *Let* $\mathbf{A}$ *be a complex matrix and* $\mathbf{b}$ *be a complex vector.*

(a) If $\mathbf{X} \sim \mathrm{N}_c(\boldsymbol{\mu}, \mathbf{\Sigma}, \tilde{\mathbf{\Sigma}})$, *then* $\mathbf{AX} + \mathbf{b} \sim \mathrm{N}_c(\mathbf{A}\boldsymbol{\mu} + \mathbf{b}, \mathbf{A\Sigma A}^H, \mathbf{A}\tilde{\mathbf{\Sigma}}\mathbf{A}^T)$.

(b) If $\mathbf{X} \sim \mathrm{N}_c(\boldsymbol{\mu}, \mathbf{\Sigma})$, *then* $\mathbf{AX} + \mathbf{b} \sim \mathrm{N}_c(\mathbf{A}\boldsymbol{\mu} + \mathbf{b}, \mathbf{A\Sigma A}^H)$.

(c) If $\mathbf{X} \sim \mathrm{N}(\boldsymbol{\mu}, \mathbf{\Sigma})$, *then* $\mathbf{AX} + \mathbf{b} \sim \mathrm{N}_c(\mathbf{A}\boldsymbol{\mu} + \mathbf{b}, \mathbf{A\Sigma A}^H, \mathbf{A\Sigma A}^T)$.

PROOF. The real and imaginary parts of $\mathbf{AX} + \mathbf{b}$ are jointly Gaussian because they are linear transforms of the real and imaginary parts of $\mathbf{X}$ which are jointly Gaussian. Part (a) follows from the fact that $\mathrm{Cov}(\mathbf{AX} + \mathbf{b}, \mathbf{AX} + \mathbf{b}) = \mathbf{A\Sigma A}^H$ and $\mathrm{Cov}(\mathbf{AX} + \mathbf{b}, (\mathbf{AX} + \mathbf{b})^*) = \mathbf{A}\tilde{\mathbf{\Sigma}}\mathbf{A}^T$. Part (b) is a direct result of (a) with $\tilde{\mathbf{\Sigma}} = \mathbf{0}$. Part (c) follows from the fact that the real and imaginary parts of $\mathbf{AX} + \mathbf{b}$ are linear transforms of $\mathbf{X}$ and $\mathrm{Cov}(\mathbf{AX} + \mathbf{b}, (\mathbf{AX} + \mathbf{b})^*) = \mathbf{A\Sigma A}^T$. □

According to (2.3.4), $\tilde{\mathbf{\Sigma}} = \mathbf{0}$ if and only if $\mathbf{\Sigma}_r$ is symmetric in the sense that

$$\mathbf{\Sigma}_{11} = \mathbf{\Sigma}_{22}, \quad \mathbf{\Sigma}_{12} = -\mathbf{\Sigma}_{12}^T. \tag{2.3.5}$$

Hence, a symmetric complex Gaussian random vector is characterized by its real and imaginary parts having a symmetric Gaussian distribution satisfying (2.3.5). The probability density function (PDF) of a symmetric complex Gaussian random vector $\mathbf{X} \sim \mathrm{N}_c(\boldsymbol{\mu}, \mathbf{\Sigma})$ takes the form

$$p(\boldsymbol{x}) = \pi^{-n} |\mathbf{\Sigma}|^{-1} \exp\{-(\boldsymbol{x} - \boldsymbol{\mu})^H \mathbf{\Sigma}^{-1} (\boldsymbol{x} - \boldsymbol{\mu})\} \qquad (\boldsymbol{x} \in \mathbb{C}^n),$$

which should be interpreted as a function of $\Re(\boldsymbol{x})$ and $\Im(\boldsymbol{x})$. Therefore, if $\mathbf{X}$ and $\mathbf{Y}$ are jointly symmetric complex Gaussian with $\mathrm{Cov}(\mathbf{X}, \mathbf{Y}) = \mathbf{0}$, then $\mathbf{X}$ and $\mathbf{Y}$ are mutually independent. A univariate random variable X has a symmetric complex Gaussian distribution with mean zero and variance σ^2, i.e., $X \sim \mathrm{N}_c(0, \sigma^2)$, if and only if $\Re(X)$ and $\Im(X)$ are i.i.d. $\mathrm{N}(0, \frac{1}{2}\sigma^2)$.

By definition, a symmetric complex Gaussian process $\{X_t\}$ has zero complementary covariance function, i.e., $E(X_s X_t) = 0$ for all s and t. This is equivalent to the following symmetry condition for the real and imaginary parts of $\{X_t\}$:

$$\begin{cases} \mathrm{Cov}\{\Re(X_s), \Re(X_t)\} = \mathrm{Cov}\{\Im(X_s), \Im(X_t)\}, \\ \mathrm{Cov}\{\Re(X_s), \Im(X_t)\} = -\mathrm{Cov}\{\Im(X_s), \Re(X_t)\}. \end{cases} \tag{2.3.6}$$

Such processes arise naturally in real-world applications, including radar, sonar, and communications, as a result of modulation. For example, let $\{Y_t\}$ be a zero-mean real Gaussian process and ϕ be a real random variable, independent of $\{Y_t\}$, such that $E\{\exp(i2\phi)\} = 0$. Consider the modulated complex process

$$X_t := Y_t \exp\{i(\omega_c t + \phi)\},$$

where ω_c, a nonzero constant, is known as the carrier frequency. It is easy to show that $E(X_t) = 0$ and

$$\begin{aligned}
E\{\Re(X_s)\Re(X_t)\} &= \tfrac{1}{2}E(Y_s Y_t)\,[\cos(\omega_c(s-t)) + E\{\cos(\omega_c(s+t)+2\phi)\}] \\
&= \tfrac{1}{2}E(Y_s Y_t)\cos(\omega_c(s-t)), \\
E\{\Im(X_s)\Im(X_t)\} &= \tfrac{1}{2}E(Y_s Y_t)\,[\cos(\omega_c(s-t)) - E\{\cos(\omega_c(s+t)+2\phi)\}] \\
&= \tfrac{1}{2}E(Y_s Y_t)\cos(\omega_c(s-t)), \\
E\{\Re(X_s)\Im(X_t)\} &= \tfrac{1}{2}E(Y_s Y_t)\,[\sin(\omega_c(s-t)) + E\{\sin(\omega_c(s+t)+2\phi)\}] \\
&= \tfrac{1}{2}E(Y_s Y_t)\sin(\omega_c(s-t)).
\end{aligned}$$

So the condition (2.3.6) is satisfied. Observe that $\{\Re(X_t)\}$ and $\{\Im(X_t)\}$ are jointly Gaussian, because conditioning on ϕ they are jointly Gaussian with mean zero and with covariance and cross-covariance functions being independent of ϕ. Therefore, by definition, $\{X_t\}$ is a zero-mean symmetric complex Gaussian process with covariance function $c(s,t) = E(Y_s Y_t)\exp\{i\omega_c(s-t)\}$. In addition, if $\{Y_t\}$ is stationary, so is $\{X_t\}$. Note that (2.3.6) is not satisfied if $E\{\exp(i2\phi)\} \neq 0$, in which case, $\{X_t\}$ is only a general complex Gaussian process.

For convenience, we will drop the word "symmetric" when referring to symmetric complex Gaussian variables, distributions, or processes and simply call them *complex Gaussian*. We will retain the word "general" when referring to general complex Gaussian distributions.

White noise is a special type of stationary process that plays an important role in statistical theory. It is a useful model for ambient noise and measurement errors in many applications. It can also be used to simulate stationary processes with a nonwhite, or colored, spectrum through linear filtering.

Definition 2.5 (White Noise). A zero-mean stationary process $\{X_t\}$ with variance σ^2 is said to be white noise, denoted by $\{X_t\} \sim \mathrm{WN}(0,\sigma^2)$, if the ACF of $\{X_t\}$ takes the form $r(u) = \sigma^2\delta_u$, where $\{\delta_u\}$ is the Kronecker delta sequence; or equivalently, if the SDF of $\{X_t\}$ takes the form $f(\omega) = \sigma^2$ for all ω. A sequence $\{X_t\}$ of i.i.d. random variables with mean zero and variance σ^2 is a white noise process, denoted by $\{X_t\} \sim \mathrm{IID}(0,\sigma^2)$. A sequence $\{X_t\}$ is called Gaussian white noise, denoted by $\{X_t\} \sim \mathrm{GWN}(0,\sigma^2)$ in the real case and by $\{X_t\} \sim \mathrm{GWN}_c(0,\sigma^2)$ in the complex case, if the X_t are i.i.d. $\mathrm{N}(0,\sigma^2)$ or $\mathrm{N}_c(0,\sigma^2)$, respectively.

Martingale differences with constant variance constitute another, perhaps less familiar, example of white noise.

Definition 2.6 (Martingale Differences). A random sequence $\{X_t\}$ is said to be a sequence of martingale differences if there exists an increasing sequence of σ-fields (or filtration) $\{\mathfrak{F}_t\}$ such that (a) X_t is measurable with respect to $\mathfrak{F}_t$ for each t and (b) $E(|X_t|^2) < \infty$ and $E(X_t|\mathfrak{F}_{t-1}) = 0$ almost surely for all t. A sequence $\{X_t\}$ of martingale differences with constant variance σ^2 is a white noise process and is denoted by $\{X_t\} \sim \mathrm{MD}(0, \sigma^2)$.

If $\{X_t\}$ is a sequence of martingale differences, then $S_t := \sum_{j=1}^{t} X_j$ is called a *martingale*, characterized by the property [34, p. 458]

$$E(S_t|\mathfrak{F}_{t-1}) = S_{t-1}.$$

The notion of martingale difference stems from the fact that $X_t = S_t - S_{t-1}$.

The σ-field $\mathfrak{F}_t$ in Definition 2.6 can be interpreted as a representation of certain historical information about the random process up to (and including) time t. The requirement that X_t be measurable with respect to $\mathfrak{F}_t$ simply means that X_t is determined completely by the historical information up to time t. If the historical information is known only up to time $t-1$, then the martingale difference X_t remains unpredictable (in the minimum mean-square sense) because the best prediction, $E(X_t|\mathfrak{F}_{t-1})$, is equal to the mean of X_t, which is zero. Given this interpretation, it is not difficult to see that the one-step prediction errors, $X_t := Y_t - E(Y_t|Y_{t-1}, Y_{t-2}, \ldots)$ $(t = 1, 2, \ldots)$, of a random process $\{Y_t\}$ are martingale differences with $\mathfrak{F}_t$ being the σ-field generated by $\{Y_t, Y_{t-1}, \ldots\}$.

Martingale differences in general are uncorrelated but not necessarily independent or identically distributed or even stationary. For example, the variance can change with t. A trivial sequence of martingale differences is a sequence of i.i.d. random variables with mean zero, in which case $\mathfrak{F}_t$ is defined as the σ-field generated by $\{X_t, X_{t-1}, \ldots\}$. Therefore, the concept of martingale differences is a generalization of the concept of zero-mean i.i.d. random variables.

The assertion in Definition 2.6 that a sequence of martingale differences with constant variance is white noise can be justified as follows. By definition, if $\{X_t\}$ is a sequence of martingale differences, then, by the iterated expectation,

$$E(X_t) = E\{E(X_t|\mathfrak{F}_{t-1})\} = 0 \quad \text{for all } t,$$
$$E(X_t X_s^*) = E\{E(X_t|\mathfrak{F}_{t-1}) X_s^*\} = 0 \quad \text{for all } t > s.$$

This means that $\{X_t\}$ is an uncorrelated process with mean zero. If, in addition, $E(|X_t|^2) = \sigma^2$ for all t, then

$$E(X_t X_s^*) = \sigma^2 \delta_{t-s},$$

which, by definition, implies that $\{X_t\} \sim \mathrm{WN}(0, \sigma^2)$. Note that if $E(|X_t|^2|\mathfrak{F}_{t-1}) = \sigma^2$ for all t, then $E(|X_t|^2) = \sigma^2$ for all t.

2.4 Linear Prediction Theory

Linear prediction is a powerful technique for time series analysis. Autoregressive models are directly related to linear prediction. The following example shows that linear prediction is also useful for modeling sinusoids.

Consider a simple case where $y_t = \cos(\omega_0 t)$, with $\omega_0 \in (0, \pi)$ being a constant. Using trigonometric identities, we obtain

$$\begin{aligned}\cos(\omega_0(t-1)) &= \cos(\omega_0)\cos(\omega_0 t) + \sin(\omega_0)\sin(\omega_0 t),\\ \cos(\omega_0(t-2)) &= \{2\cos^2(\omega_0) - 1\}\cos(\omega_0 t) + 2\cos(\omega_0)\sin(\omega_0)\sin(\omega_0 t).\end{aligned}$$

Combining these equations leads to

$$2\cos(\omega_0)\cos(\omega_0(t-1)) - \cos(\omega_0(t-2)) = \cos(\omega_0 t).$$

In other words, we can write

$$y_t = 2\cos(\omega_0)y_{t-1} - y_{t-2}.$$

This expression means that the current value y_t of the sinusoid can be predicted without error by a suitable linear combination of the previous values y_{t-1} and y_{t-2}. Furthermore, the coefficient $c := 2\cos(\omega_0)$ can be used to identify the frequency ω_0 because $\omega_0 = \arccos(c/2)$. This observation has motivated the development of many linear-prediction-based algorithms for frequency estimation which will be discussed later in Chapter 9.

In this section, we provide a brief review of the linear prediction theory to facilitate the later discussions. First, the following proposition describes the properties of the best linear predictor for a zero-mean stationary process.

Proposition 2.7 (Best Linear Prediction [46, p. 64] [177, p. 157]). *Let $\{X_t\}$ be a zero-mean stationary process with ACF $r(u)$. Then, for any $m \ge 1$, the best linear predictor of X_t based on $\{X_{t-1}, \ldots, X_{t-m}\}$, defined as the minimizer of $E\{|X_t - Y_t|^2\}$ with respect to Y_t being any linear function of $\{X_{t-1}, \ldots, X_{t-m}\}$, can be expressed as $\hat{X}_t = -\sum_{j=1}^{m} \varphi_j X_{t-j}$, where $\boldsymbol{\varphi} := [\varphi_1, \ldots, \varphi_m]^T$ satisfies $\mathbf{R}\boldsymbol{\varphi} = -\mathbf{r}$, with*

$$\mathbf{R} := \begin{bmatrix} r(0) & r^*(1) & \cdots & r^*(m-1) \\ r(1) & r(0) & \cdots & r^*(m-2) \\ \vdots & \vdots & \ddots & \vdots \\ r(m-1) & r(m-2) & \cdots & r(0) \end{bmatrix}, \quad \mathbf{r} := \begin{bmatrix} r(1) \\ r(2) \\ \vdots \\ r(m) \end{bmatrix}.$$

The prediction error $X_t - \hat{X}_t$ has the orthogonality property $E\{(X_t - \hat{X}_t)X_{t-j}^*\} = 0$ $(j = 1, \ldots, m)$ and its variance $\sigma^2 := E\{|X_t - \hat{X}_t|^2\}$ satisfies $\sigma^2 = r(0) + \mathbf{r}^H\boldsymbol{\varphi}$.

The next proposition presents a fast algorithm to compute the inverse of a covariance matrix formed by a stationary process. With this algorithm, the coefficients of the best predictor in Proposition 2.7 can be computed efficiently.

Proposition 2.8 (Levinson-Durbin Algorithm [46, p. 169] [177, pp. 171–176]). *Let $\{X_t\}$ be a zero-mean stationary process with ACF $r(u)$. Let $\mathbf{X} := [X_1, \ldots, X_n]^T$ and $\mathbf{\Sigma} := E(\mathbf{X}\mathbf{X}^H) = [r(s-t)]$ $(s, t = 1, \ldots, n)$. Then, the Cholesky decomposition of $\mathbf{\Sigma}^{-1}$ takes the form $\mathbf{\Sigma}^{-1} = \mathbf{U}\mathbf{D}^{-1}\mathbf{U}^H$, where $\mathbf{D} := \mathrm{diag}(\sigma_0^2, \sigma_1^2, \ldots, \sigma_{n-1}^2)$ is a diagonal matrix and $\mathbf{U} := [u_{st}]$ $(s, t = 1, \ldots, n)$ is an upper triangular matrix with $u_{st} := \varphi_{t-1,t-s}^*$ for $1 \le s \le t \le n$ ($\varphi_{t0} := 1$ for all $t = 0, 1, \ldots$). The φ_{tj} and σ_t^2 can be computed recursively as follows: for $t = 1, 2, \ldots$ and with the initial value $\sigma_0^2 = r(0)$,*

$$
\begin{aligned}
\varphi_{tt} &= -\frac{1}{\sigma_{t-1}^2}\left\{ r(t) + \sum_{j=1}^{t-1} \varphi_{t-1,j}\, r(t-j) \right\}, \\
\varphi_{tj} &= \varphi_{t-1,j} + \varphi_{tt}\,\varphi_{t-1,t-j}^* \quad (j = 1, \ldots, t-1), \\
\sigma_t^2 &= \sigma_{t-1}^2 (1 - |\varphi_{tt}|^2).
\end{aligned}
$$

The best linear predictor of X_{t+1} based on $\{X_1, \ldots, X_t\}$ can be expressed as $\tilde{X}_{t+1} := -\sum_{j=1}^{t} \varphi_{tj} X_{t+1-j}$ with the prediction error variance $\sigma_t^2 = E\{|X_{t+1} - \tilde{X}_{t+1}|^2\}$.

Remark 2.4 The best linear prediction of X_t defined in Proposition 2.7 can be expressed as $\hat{X}_t = -\sum_{j=1}^{m} \varphi_{mj} X_{t-j}$ with $\sigma^2 := E\{|X_t - \hat{X}_t|^2\} = \sigma_m^2$.

Remark 2.5 The quantity φ_{mm} is called the mth reflection coefficient. It coincides with the negative of the lag-m partial autocorrelation coefficient defined as the autocorrelation coefficient between the best forward prediction error of X_t based on $\{X_{t-1}, \ldots, X_{t-m+1}\}$ and the best backward prediction error of X_{t-m} based on the same set of predictors. It follows from the third equation in Proposition 2.8 that $\sigma_m^2 > 0$ if and only if $|\varphi_{tt}| < 1$ for all $t = 1, \ldots, m$.

In Proposition 2.8, the φ_{tj} and σ_t^2 are computed from the ACF of $\{X_t\}$. When $\{X_t\}$ is an AR process, they can be computed directly from the AR parameters based on the following algorithm.

Proposition 2.9 (Levinson-Durbin Algorithm for AR Processes [46, p. 242] [177, pp. 172–173]). *Let $\{X_t\}$ be an AR(m) process of the form (2.2.7). Let φ_{tj} and σ_t^2 be defined in Proposition 2.8. Then, for any $t \ge m$, $\varphi_{tj} = \varphi_j$ if $1 \le j \le m$, $\varphi_{tj} = 0$ if $m < j \le t$, and $\sigma_t^2 = \sigma^2$; for $t = m-1, m-2, \ldots$, the φ_{tj} and σ_t^2 can be computed recursively as follows:*

$$
\begin{aligned}
\varphi_{tj} &= \frac{\varphi_{t+1,j} - \varphi_{t+1,t+1}\,\varphi_{t+1,t+1-j}^*}{1 - |\varphi_{t+1,t+1}|^2} \quad (j = 1, \ldots, t), \\
\sigma_t^2 &= \frac{\sigma_{t+1}^2}{1 - |\varphi_{t+1,t+1}|^2}.
\end{aligned}
$$

Moreover, let $\mathbf{\Sigma}$ be the covariance matrix defined in Proposition 2.8 and let $\mathbf{\Sigma}^{-1} := [\eta_{st}]$ $(s,t=1,\ldots,n)$. Then, $\eta_{st}=0$ for all $|s-t|>m$. In other words, $\mathbf{\Sigma}^{-1}$ is a band matrix with bandwidth $2m+1$.

While the Levinson-Durbin algorithm produces the Cholesky decomposition for the inverse of a covariance matrix, the following innovations algorithm provides the Cholesky decomposition for the covariance matrix itself. It also gives an expression for the best predictor in terms of the prediction errors or innovations, hence the name of the algorithm.

Proposition 2.10 (Innovations Algorithm [46, pp. 172, 193, 255] [177, pp. 29–30]). *Let $\{X_t\}$ be a zero-mean stationary process with ACF $r(u)$. Let $\mathbf{X} := [X_1,\ldots,X_n]^T$ and $\mathbf{\Sigma} := E(\mathbf{X}\mathbf{X}^H) = [r(s-t)]$ $(s,t=1,\ldots,n)$. Then, the Cholesky decomposition of $\mathbf{\Sigma}$ takes the form $\mathbf{\Sigma} = \mathbf{L}\mathbf{D}\mathbf{L}^H$, where $\mathbf{D} := \mathrm{diag}(\sigma_0^2,\sigma_1^2,\ldots,\sigma_{n-1}^2)$ is the diagonal matrix defined in Proposition 2.8 and $\mathbf{L}=\mathbf{U}^{-H} := [c_{st}]$ is a lower triangular matrix with $c_{st} := \psi_{s-1,s-t}$ for $1\le t\le s\le n$ ($\psi_{t0} := 1$ for all $t=0,1,\ldots$). The ψ_{tj} and σ_t^2 can be computed recursively as follows: for $t=1,2,\ldots$ and with $\sigma_0^2 = r(0)$,*

$$\psi_{t,t-j} = \frac{1}{\sigma_j^2}\left\{r(t-j) - \sum_{l=0}^{j-1} \psi_{t,t-l}\psi_{j,j-l}^*\sigma_l^2\right\} \quad (j=0,1,\ldots,t-1),$$

$$\sigma_t^2 = r(0) - \sum_{j=0}^{t-1} |\psi_{t,t-j}|^2\sigma_j^2.$$

The best linear predictor of X_{t+1} based on $\{X_1,\ldots,X_t\}$ can be expressed as $\tilde{X}_{t+1} = \sum_{j=1}^{t}\psi_{tj}Z_{t+1-j}$, where $Z_t := X_t - \tilde{X}_t$ ($\tilde{X}_0 := 0$) and $\mathrm{Var}(Z_t) = \sigma_{t-1}^2$. If $\{X_t\}$ is a moving-average (MA) process of order m, i.e., $X_t = \sum_{j=0}^{m}\psi_j\zeta_{t-j}$, where $\psi_0 := 1$ and $\{\zeta_t\} \sim \mathrm{WN}(0,\sigma^2)$, then $\mathbf{\Sigma}$ is a band matrix with bandwidth $2m+1$ such that $\psi_{tj}=0$ for $t\ge j>m$. Moreover, if the filter $\psi(z) := \sum_{j=0}^{m}\psi_j z^{-j}$ is invertible, then, as $t\to\infty$, $E\{|Z_t-\zeta_t|^2\}\to 0$, $\sigma_t^2\to\sigma^2$, and $|\psi_{tj}-\psi_j| = \mathcal{O}(\rho^t)\to 0$ for all $j=1,\ldots,m$ and for some constant $\rho\in(0,1)$.

Remark 2.6 With $c_{st} := \psi_{s-1,s-t}$, the recursions can be written as

$$c_{st} = \frac{1}{\sigma_{t-1}^2}\left\{r(s-t) - \sum_{l=1}^{t-1} c_{sl}c_{tl}^*\sigma_{l-1}^2\right\} \qquad (t=1,\ldots,s-1),$$

$$\sigma_{s-1}^2 = r(0) - \sum_{l=1}^{s-1} |c_{sl}|^2\sigma_{l-1}^2.$$

This is the standard Cholesky decomposition algorithm [177, pp. 29–30].

Remark 2.7 If $\{X_t\}$ is an MA process of order m, then it suffices to calculate $\psi_{t,t-j}$ for $t-m\le j\le t-1$, and hence the complexity of the Cholesky decomposition by the innovations algorithm takes the form $\mathcal{O}(n)$ for large n.

2.5 Asymptotic Statistical Theory

In many estimation problems, it is often difficult to mathematically characterize and quantify the statistical behavior of an estimator for finite sample sizes. A practical solution to the problem is simulation. While simulation does provide valuable insights and is indispensable to practitioners, simulation results are usually inconclusive and cannot be extrapolated under different conditions. Asymptotic analysis serves as an important tool to fill in the gap. Although it does not directly answer questions regarding the finite-sample behavior, asymptotic analysis often yields conclusive and useful results that are valid under very general conditions and for large but finite sample sizes; in many cases it is also the only way to obtain such results.

As a performance criterion, it is desirable that an estimator calculated from a finite data record converges in some sense to the parameter of interest as the sample size approaches infinity. It is also desirable that the unknown distribution of an estimator converges to a known one so that the randomness of the estimator can be easily characterized and quantified. These concerns lead to three widely-used modes of convergence: convergence in probability, almost sure convergence, and convergence in distribution. Associated with the first two modes of convergence are two ways of evaluating the magnitude of a random variable: the boundedness in probability and the almost sure boundedness. These concepts are summarized in the following.

Definition 2.7 (Convergence and Boundedness in Probability). A sequence of random variables $\{X_n\}$ is said to converge to zero in probability (or weakly), denoted by $X_n = \mathcal{O}_P(1)$ or $X_n \xrightarrow{P} 0$, if for any constant $\delta > 0$, $P(|X_n| > \delta) \to 0$ as $n \to \infty$. It is said to converge to a random variable X, denoted by $X_n \xrightarrow{P} X$, if $X_n - X = \mathcal{O}_P(1)$. It is said to be bounded in probability, denoted by $X_n = \mathcal{O}_P(1)$, if for any constant $\delta > 0$ there exists a constant $c > 0$ such that $P(|X_n| > c) < \delta$ for large n. Moreover, for any sequence of positive constants $\{a_n\}$, we write $X_n = \mathcal{O}_P(a_n)$ if $a_n^{-1} X_n = \mathcal{O}_P(1)$, and we write $X_n = \mathcal{O}_P(a_n)$ if $a_n^{-1} X_n = \mathcal{O}_P(1)$. A statement of convergence or boundedness in probability is valid for a sequence of random vectors if it is valid componentwise.

The concept of convergence in probability can be easily understood by interpreting δ in Definition 2.7 as a prescribed *tolerance level* for deviation of X_n from its target value. In so doing, the statement $X_n \xrightarrow{P} 0$ simply means that no matter how small the tolerance level is, the probability of X_n exceeding that level will approach zero as n approaches infinity.

Note that the probability here is calculated on the basis of repeatedly observing X_n from different random experiments (or scenarios) for each given and fixed n. Therefore, the fact that this probability approaches zero does not necessarily imply that for each given experiment the resulting infinite sequence $\{X_n\}$ converges to zero in the ordinary sense. For example, in an infinite sequence $\{X_n\}$ produced by a given experiment, one may find infinitely many instances in which the tolerance level is violated, but the probability that the tolerance level is violated at a fixed instance n when considering all possible outcomes from the repeated experiments can still approach zero as $n \to \infty$.

Similarly, $X_n = \mathcal{O}_P(1)$ does not necessarily imply that the X_n are bounded in the ordinary sense for each given experiment, i.e., it does not necessarily imply that there exists a constant $c > 0$ such that $|X_n| \le c$ for all n and all experiments. The boundedness in probability only means that the probability of X_n exceeding a bound c can be made arbitrarily small for all n if c is sufficiently large.

To ensure the convergence and the boundedness of a random sequence in the ordinary sense for all possible experiments, the concepts of almost sure convergence and almost sure boundedness are needed.

Definition 2.8 (Almost Sure Convergence and Boundedness). A sequence of random variables $\{X_n\}$ is said to converge to zero almost surely, denoted by $X_n = o(1)$ or $X_n \stackrel{a.s.}{\to} 0$, if $P(X_n \to 0) = 1$ as $n \to \infty$, or equivalently, if $P(|X_n| > \delta \text{ i.o.}) = 0$ for any constant $\delta > 0$, where i.o. stands for "infinitely often." The sequence is said to converge to a random variable X, denoted by $X_n \stackrel{a.s.}{\to} X$, if $X_n - X = o(1)$. It is said to be bounded almost surely, denoted by $X_n = \mathcal{O}(1)$, if there is a constant $c > 0$ such that $P(|X_n| > c) = 0$ for all n. For any sequence of positive constants $\{a_n\}$, we write $X_n = o(a_n)$ if $a_n^{-1} X_n = o(1)$, and we write $X_n = \mathcal{O}(a_n)$ if $a_n^{-1} X_n = \mathcal{O}(1)$. A statement of almost sure convergence or boundedness is valid for a sequence of random vectors if it is valid componentwise.

Convergence in distribution is another mode of convergence. It is useful when the probability distribution of an estimator is of interest, as is the case when constructing a confidence interval for the parameter being estimated.

Definition 2.9 (Convergence in Distribution). A sequence of random vectors $\{\mathbf{X}_n\}$ is said to converge in distribution to a random vector $\mathbf{X}$, denoted by $\mathbf{X}_n \stackrel{D}{\to} \mathbf{X}$, if the cumulative distribution function (CDF) of $\mathbf{X}_n$ converges to the CDF of $\mathbf{X}$ at every continuity point of the latter.

Unlike the other modes of convergence, the convergence in distribution does not directly address the convergence of a random sequence itself. Instead, it concerns the convergence of the CDFs. Just like two random variables from completely unrelated experiments can have the same CDF, the random variables in Definition 2.9 may come from unrelated experiments.

The three modes of convergence are related to each other by a hierarchy with the almost sure convergence being the strongest mode and the convergence in distribution the weakest mode.

Proposition 2.11 (Hierarchy of Convergence Modes [34, p. 330]). *If* $X_n \stackrel{a.s.}{\to} X$, *then* $X_n \stackrel{P}{\to} X$. *If* $X_n \stackrel{P}{\to} X$, *then* $X_n \stackrel{D}{\to} X$. *The converses are not true in general. However, if* $X_n \stackrel{D}{\to} c$ *for some constant* c, *then* $X_n \stackrel{P}{\to} c$.

In the asymptotic analysis of an estimator, it often suffices to consider the major terms in an expansion of the estimator that dominate the other terms in magnitude. This requires to combine random variables of different orders of magnitude, for which the following proposition is very useful.

Proposition 2.12 (Arithmetics of Big O and Small O [46, p. 199]). *Let* $\{X_n\}$ *and* $\{Y_n\}$ *be random sequences. Let* $\{a_n\}$ *and* $\{b_n\}$ *be sequences of positive constants. If* $X_n = o_P(a_n)$ *and* $Y_n = o_P(b_n)$, *then* $X_n Y_n = o_P(a_n b_n)$ *and* $X_n + Y_n = o_P(a_n + b_n)$. *The same implication holds if* o_P *is everywhere replaced by* $\mathcal{O}_P$. *If* $X_n = o_P(a_n)$ *and* $Y_n = \mathcal{O}_P(b_n)$, *then* $X_n Y_n = o_P(a_n b_n)$. *All these assertions remain valid if* o_P *and* $\mathcal{O}_P$ *are everywhere replaced by* o *and* $\mathcal{O}$, *respectively.*

The following proposition ensures that the modes of convergence are preserved after certain transformations.

Proposition 2.13 (Convergence after Simple Transformation).

(a) *If* $\mathbf{X}_n \stackrel{P}{\to} \mathbf{X}$, *then* $\mathbf{g}(\mathbf{X}_n) \stackrel{P}{\to} \mathbf{g}(\mathbf{X})$ *for any continuous function* $\mathbf{g}(\cdot)$. *This assertion remains valid if* $\stackrel{P}{\to}$ *is replaced by* $\stackrel{a.s.}{\to}$ *or* $\stackrel{D}{\to}$.

(b) *If* $\mathbf{X}_n \stackrel{D}{\to} \mathbf{X}$ *and* $\mathbf{Y}_n \stackrel{P}{\to} \mathbf{c}$ *for some constant vector* $\mathbf{c}$, *then* $\mathbf{Y}_n^H \mathbf{X}_n \stackrel{D}{\to} \mathbf{c}^H \mathbf{X}$ *and* $\mathbf{X}_n + \mathbf{Y}_n \stackrel{D}{\to} \mathbf{X} + \mathbf{c}$. *This assertion is known as Slutsky's Theorem.*

PROOF. These results are well known in the real case [34, p. 332 and p. 334] [46, pp. 200–201 and 206–207]. In the complex case, part (a) is trivial with regard to the convergence in probability and almost sure convergence. For the convergence in distribution, observe that by definition $\mathbf{X}_n \stackrel{D}{\to} \mathbf{X}$ is the same as $\mathbf{Y}_n := [\Re(\mathbf{X}_n^T), \Im(\mathbf{X}_n^T)]^T \stackrel{D}{\to} \mathbf{Y} := [\Re(\mathbf{X}^T), \Im(\mathbf{X}^T)]^T$. Let $\mathbf{g} = \mathbf{g}_1 + i\mathbf{g}_2$. Because $\mathbf{g}_1$ and $\mathbf{g}_2$ are continuous real functions, it follows that $[\mathbf{g}_1^T(\mathbf{Y}_n), \mathbf{g}_2^T(\mathbf{Y}_n)]^T \stackrel{D}{\to} [\mathbf{g}_1^T(\mathbf{Y}), \mathbf{g}_2^T(\mathbf{Y})]^T$, which is the same as $\mathbf{g}(\mathbf{X}_n) \stackrel{D}{\to} \mathbf{g}(\mathbf{X})$. Part (b) can be proved similarly. □

Taylor expansion is a useful tool in the asymptotic analysis of estimators. The following proposition ensures its validity for random variables.

Proposition 2.14 (Taylor Expansion). *Let* $\{r_n\}$ *be a sequence of positive constants such that* $r_n \to 0$ *as* $n \to \infty$.

(a) *For a sequence of real random variables* $\{X_n\}$, *if* $X_n = a_n + \mathcal{O}_P(r_n)$, *where* $\{a_n\}$ *is a sequence of constants, then for any real function* $g(\cdot)$ *which has* m *continuous derivatives in a neighborhood of* a_n,

$$g(X_n) = \sum_{j=0}^{m} \{g_j(a_n)/j!\}(X_n - a_n)^j + o_P(r_n^m),$$

where $g_j(\cdot)$ *denotes the* j*th derivative of* $g(\cdot)$.

(b) *For a sequence of real random vectors* $\{\mathbf{X}_n\}$, *if* $\mathbf{X}_n = \mathbf{a}_n + \mathcal{O}_P(r_n)$, *where* $\{\mathbf{a}_n\}$ *is a sequence of constant vectors, then for any real function* $g(\cdot)$ *which is continuously differentiable in a neighborhood of* $\mathbf{a}_n$,

$$g(\mathbf{X}_n) = \{\nabla g(\mathbf{a}_n)\}^T (\mathbf{X}_n - \mathbf{a}_n) + o_P(r_n),$$

where ∇ *denotes the gradient operator. Both assertions remain valid if* o_P *and* $\mathcal{O}_P$ *are everywhere replaced by* o *and* $\mathcal{O}$, *respectively.*

PROOF. Consider part (a). From calculus, we have

$$g(x) = \sum_{j=0}^{m} \{g_j(a_n)/j!\}(x-a_n)^j + o(|x-a_n|^m).$$

The assertion follows immediately from the fact that $o(|X_n - a_n|^m) = o_P(r_n^m)$. Part (b) can be proved similarly. □

Asymptotic normality is an important concept in estimation theory.

Definition 2.10 (Asymptotic Normality). A sequence of random vectors $\{\mathbf{X}_n\}$ with mean $\boldsymbol{\mu}_n$ and covariance matrix $\boldsymbol{\Sigma}_n$ is said to be asymptotically Gaussian (complex Gaussian), denoted by $\mathbf{X}_n \stackrel{A}{\sim} \mathrm{N}(\boldsymbol{\mu}_n, \boldsymbol{\Sigma}_n)$ ($\mathbf{X}_n \stackrel{A}{\sim} \mathrm{N}_c(\boldsymbol{\mu}_n, \boldsymbol{\Sigma}_n)$), if $\boldsymbol{\Sigma}_n^{-H/2}(\mathbf{X}_n - \boldsymbol{\mu}_n) \xrightarrow{D} \mathbf{Z}$, where $\mathbf{Z} \sim \mathrm{N}(\mathbf{0}, \mathbf{I})$ ($\mathbf{Z} \sim \mathrm{N}_c(\mathbf{0}, \mathbf{I})$). It is said to be asymptotically general complex Gaussian with mean $\boldsymbol{\mu}_n$, covariance matrix $\boldsymbol{\Sigma}_n$, and complementary covariance matrix $\tilde{\boldsymbol{\Sigma}}_n$, denoted by $\mathbf{X}_n \stackrel{A}{\sim} \mathrm{N}_c(\boldsymbol{\mu}_n, \boldsymbol{\Sigma}_n, \tilde{\boldsymbol{\Sigma}}_n)$, if $[\Re(\mathbf{X}_n)^T, \Im(\mathbf{X}_n)^T]^T \stackrel{A}{\sim} \mathrm{N}(\boldsymbol{\mu}_{nr}, \boldsymbol{\Sigma}_{nr})$, where $(\boldsymbol{\mu}_{nr}, \boldsymbol{\Sigma}_{nr})$ and $(\boldsymbol{\mu}, \boldsymbol{\Sigma}_n, \tilde{\boldsymbol{\Sigma}}_n)$ satisfy (2.3.1) and (2.3.2).

Many estimators enjoy the asymptotic normality. The following proposition ensures that this property is preserved after differentiable transformations. It is a generalization of the result in [46, p. 211].

Proposition 2.15 (Asymptotic Normality after Transformation). *Let* $\{a_n\}$ *and* $\{b_n\}$ *be sequences of positive constants such* $a_n \to 0$ *and* $b_n \to 0$ *as* $n \to \infty$. *Assume that* $[a_n^{-1}(\mathbf{X}_n - \boldsymbol{\mu}_n)^T, b_n^{-1}(\mathbf{Y}_n - \mathbf{v}_n)^T]^T \stackrel{A}{\sim} \mathrm{N}(\mathbf{0}, \boldsymbol{\Sigma}_n)$ *with* $\boldsymbol{\Sigma}_n = \mathcal{O}(1)$ *and* $\boldsymbol{\Sigma}_n \geq \boldsymbol{\Sigma}$ *(i.e.,* $\boldsymbol{\Sigma}_n - \boldsymbol{\Sigma}$ *is nonnegative definite) for all* n, *where* $\boldsymbol{\Sigma}$ *is nonsingular covariance matrix. If* $\mathbf{g}_1(\cdot)$ *and* $\mathbf{g}_2(\cdot)$ *are real functions, continuously differentiable in a neighborhood of* $\boldsymbol{\mu}_n$ *and* $\mathbf{v}_n$ *respectively, with* $\nabla^T \mathbf{g}_1(\boldsymbol{\mu}_n) \neq \mathbf{0}$ *and* $\nabla^T \mathbf{g}_2(\mathbf{v}_n) \neq \mathbf{0}$, *then*

$$[a_n^{-1}(\mathbf{g}_1(\mathbf{X}_n) - \mathbf{g}_1(\boldsymbol{\mu}_n))^T, b_n^{-1}(\mathbf{g}_2(\mathbf{Y}_n) - \mathbf{g}_2(\mathbf{v}_n))^T]^T \stackrel{A}{\sim} \mathrm{N}(\mathbf{0}, \mathbf{J}_n \boldsymbol{\Sigma}_n \mathbf{J}_n^T),$$

where $\mathbf{J}_n := \mathrm{diag}\{\nabla^T \mathbf{g}_1(\boldsymbol{\mu}_n), \nabla^T \mathbf{g}_2(\mathbf{v}_n)\}$ *is the Jacobian matrix.*

PROOF. Because $a_n^{-1}(\mathbf{X}_n - \boldsymbol{\mu}_n) \stackrel{A}{\sim} \mathrm{N}(\mathbf{0}, \boldsymbol{\Sigma}_{n1})$ and $b_n^{-1}(\mathbf{Y}_n - \mathbf{v}_n) \stackrel{A}{\sim} \mathrm{N}(\mathbf{0}, \boldsymbol{\Sigma}_{n2})$ for some $\boldsymbol{\Sigma}_{n1} = \mathcal{O}(1)$ and $\boldsymbol{\Sigma}_{n2} = \mathcal{O}(1)$, we can write

$$\mathbf{X}_n = \boldsymbol{\mu}_n + \mathcal{O}_P(a_n), \quad \mathbf{Y}_n = \mathbf{v}_n + \mathcal{O}_P(b_n).$$

It follows from Proposition 2.14(b) that

$$\begin{aligned}\mathbf{g}_1(\mathbf{X}_n)-\mathbf{g}_1(\boldsymbol{\mu}_n) &= \mathbf{J}_{n1}(\mathbf{X}_n-\boldsymbol{\mu}_n)+\mathcal{O}_P(a_n),\\ \mathbf{g}_2(\mathbf{Y}_n)-\mathbf{g}_2(\boldsymbol{\nu}_n) &= \mathbf{J}_{n2}(\mathbf{Y}_n-\boldsymbol{\nu}_n)+\mathcal{O}_P(b_n),\end{aligned}$$

where $\mathbf{J}_{n1} := \nabla^T\mathbf{g}_1(\boldsymbol{\mu}_n)$ and $\mathbf{J}_{n2} := \nabla^T\mathbf{g}_2(\boldsymbol{\nu}_n)$. Therefore,

$$\begin{bmatrix} a_n^{-1}(\mathbf{g}_1(\mathbf{X}_n)-\mathbf{g}_1(\boldsymbol{\mu}_n)) \\ b_n^{-1}(\mathbf{g}_2(\mathbf{Y}_n)-\mathbf{g}_2(\boldsymbol{\nu}_n)) \end{bmatrix} = \mathbf{J}_n \begin{bmatrix} a_n^{-1}(\mathbf{X}_n-\boldsymbol{\mu}_n) \\ b_n^{-1}(\mathbf{Y}_n-\boldsymbol{\nu}_n) \end{bmatrix} + \mathcal{O}_P(1) \stackrel{A}{\sim} \mathrm{N}(\mathbf{0},\mathbf{J}_n\boldsymbol{\Sigma}_n\mathbf{J}_n^T),$$

where $\mathbf{J}_n := \mathrm{diag}(\mathbf{J}_{n1},\mathbf{J}_{n2})$. The proof is complete. □

The last proposition summarizes some useful techniques for proving different modes of convergence.

Proposition 2.16 (Methods of Proving Convergence).

(a) *If $E(|X_n|^r) = \mathcal{O}(a_n)$ for some constants $a_n > 0$ and $r > 0$, then $X_n = \mathcal{O}_P(a_n^{1/r})$.*

(b) *If $E(|X_n|^r) \to 0$ for some constant $r > 0$, then $X_n \xrightarrow{P} 0$.*

(c) *If $\sum_{n=1}^{\infty} E(|X_n|^r) < \infty$ for some constant $r > 0$, then $X_n \xrightarrow{a.s.} 0$.*

(d) *Let $\{\mathbf{X}_n\}$ be a sequence of r-dimensional real random vectors. Then, $\mathbf{X}_n \xrightarrow{D} \mathbf{X}$ if and only if $\mathbf{a}^T\mathbf{X}_n \xrightarrow{D} \mathbf{a}^T\mathbf{X}$ for all $\mathbf{a} \in \mathbb{R}^r$. This way of proving the asymptotic distribution of random vectors is known as the Cramér-Wold device.*

(e) *Let $\{\mathbf{X}_n\}$ and $\{\mathbf{Y}_{nm}\}$ be sequences of random vectors such that*

$$\lim_{m\to\infty}\limsup_{n\to\infty} P(\|\mathbf{X}_n-\mathbf{Y}_{nm}\| > \delta) = 0$$

for any constant $\delta > 0$. If $\mathbf{Y}_{nm} \xrightarrow{D} \mathbf{Y}_m$ as $n \to \infty$ for each fixed m and $\mathbf{Y}_m \xrightarrow{D} \mathbf{Y}$ as $m \to \infty$, then $\mathbf{X}_n \xrightarrow{D} \mathbf{Y}$ as $n \to \infty$.

PROOF. A simple proof of part (d) using the characteristic function can be found in [34, p. 383] and [46, p. 204]. Part (e) is well known in the real case [34, p. 332] [46, p. 207]. In the complex case, it can be easily proved by considering, as in the proof of Proposition 2.13, the real vectors formed by the real and imaginary parts of $\mathbf{X}_n$, $\mathbf{Y}_{nm}$, $\mathbf{Y}_m$, and $\mathbf{Y}$. Therefore, let us focus on (a)–(c). For any $c > 0$, Markov's inequality [34, p. 80] gives $P(a_n^{-1/r}|X_n| > c) \le E(|X_n|^r)/(a_n c^r) \le b/c^r$, where $b := \sup\{E(|X_n|^r)/a_n\} < \infty$. By taking $c > (b/\varepsilon)^{1/r}$, we obtain $P(a_n^{-1/r}|X_n| > c) < \varepsilon$. Part (a) is thus proved. Part (b) follows from part (a) with $a_n \to 0$. To prove part (c), we use Markov's inequality $P(|X_n| > \delta) \le \delta^{-r}E(|X_n|^r)$ and obtain

$$\sum_{n=1}^{\infty} P(|X_n| > \delta) \le \delta^{-r}\sum_{n=1}^{\infty} E(|X_n|^r) < \infty$$

for any constant $\delta > 0$. This, according to the Borel-Cantelli lemma [34, p. 59], leads to $P(|X_n| > \delta \text{ i.o.}) = 0$. Hence, by definition, $X_n \xrightarrow{a.s.} 0$ as $n \to \infty$. □

Chapter 3

Cramér-Rao Lower Bound

The Cramér-Rao inequality, also known as the information inequality, provides a lower limit, called the Cramér-Rao lower bound (CRLB), for the covariance matrix of unbiased estimators. It is named after the Swedish statistician Harald Cramér (1893–1985) and the Indian American statistician Calyampudi Radhakrishna Rao (1920–). An unbiased estimator that attains the CRLB is called a statistically efficient estimator because it has the smallest variance among all unbiased estimators. The CRLB has been studied for the estimation of sinusoidal parameters under various conditions. It is widely used as a performance benchmark for comparing the accuracy and statistical efficiency of different estimators.

In this chapter, we derive the CRLB under the assumption that the noise has a Gaussian distribution. We also derive some asymptotic expressions of the CRLB for large sample sizes with well-separated or closely spaced frequencies. Finally, we discuss the CRLB under the condition of nonGaussian white noise and its relationship with the CRLB under the Gaussian assumption. We show in particular that the CRLB is maximized by the Gaussian distribution among all noise distributions that have the same or smaller variance. In this sense, the Gaussian distribution can be regarded as the least favorable distribution for the noise in the estimation of sinusoidal parameters. We also show that the Laplace distribution is the least favorable distribution in another family of noise distributions.

3.1 Cramér-Rao Inequality

The Cramér-Rao inequality for the general problem of parameter estimation is stated in the following proposition. A proof can be found in Section 3.5.

Proposition 3.1 (Cramér-Rao or Information Inequality). *Let* $\mathbf{Y}$ *be a real or complex random vector that has a PDF* $p(\mathbf{y}|\boldsymbol{\vartheta})$ *with respect to certain measure* ν, *where* $\boldsymbol{\vartheta}$ *is a real-valued parameter taking on values in* $\Theta \subset \mathbb{R}^r$. *Let* $\boldsymbol{\eta}(\boldsymbol{\vartheta}) \in \mathbb{R}^m$ *be a real-valued differentiable function with Jacobian matrix* $\mathbf{J}(\boldsymbol{\vartheta}) := \nabla^T \boldsymbol{\eta}(\boldsymbol{\vartheta})$, *where* $\nabla := \partial/\partial\boldsymbol{\vartheta}$ *denotes the gradient operator with respect to* $\boldsymbol{\vartheta}$. *Let* $\hat{\boldsymbol{\eta}} := \hat{\boldsymbol{\eta}}(\mathbf{Y})$ *be an* *unbiased estimator of* $\boldsymbol{\eta}(\boldsymbol{\vartheta})$ *on the basis of* $\mathbf{Y}$*. Assume that the following regularity conditions are satisfied.*

(a) $p(\mathbf{y}|\boldsymbol{\vartheta})$ *has a common support for all* $\boldsymbol{\vartheta} \in \Theta$;

(b) $\nabla p(\mathbf{y}|\boldsymbol{\vartheta})$ *exists almost surely in* $\mathbf{y}$ *for any given* $\boldsymbol{\vartheta} \in \Theta$;

(c) $\ell(\boldsymbol{\vartheta}|\mathbf{y}) := \log\{p(\mathbf{y}|\boldsymbol{\vartheta})\}$*, the log likelihood function, has the property*

$$E\{\nabla \ell(\boldsymbol{\vartheta}|\mathbf{Y})\} = \mathbf{0}, \tag{3.1.1}$$

$$E\{\hat{\boldsymbol{\eta}}[\nabla \ell(\boldsymbol{\vartheta}|\mathbf{Y})]^T\} = \mathbf{J}(\boldsymbol{\vartheta}), \tag{3.1.2}$$

$$0 < \mathbf{I}(\boldsymbol{\vartheta}) := E\{[\nabla \ell(\boldsymbol{\vartheta}|\mathbf{Y})][\nabla \ell(\boldsymbol{\vartheta}|\mathbf{Y})]^T\} < \infty. \tag{3.1.3}$$

Then, for any constant vector $\mathbf{a} \in \mathbb{R}^m$,

$$\mathrm{Var}(\mathbf{a}^T \hat{\boldsymbol{\eta}}) = \mathbf{a}^T \mathrm{Cov}(\hat{\boldsymbol{\eta}})\,\mathbf{a} \ge \mathbf{a}^T \mathbf{J}(\boldsymbol{\vartheta})\,\mathbf{I}(\boldsymbol{\vartheta})^{-1} \mathbf{J}(\boldsymbol{\vartheta})^T \mathbf{a}, \tag{3.1.4}$$

where the equality holds for some $\mathbf{a} \neq \mathbf{0}$ *and* $\boldsymbol{\vartheta} \in \Theta$ *if and only if there exists a constant* $c \neq 0$ *such that* $\mathbf{a}^T\{c(\hat{\boldsymbol{\eta}} - \boldsymbol{\eta}(\boldsymbol{\vartheta})) + \mathbf{J}(\boldsymbol{\vartheta})\,\mathbf{I}(\boldsymbol{\vartheta})^{-1} \nabla \ell(\boldsymbol{\vartheta}|\mathbf{Y})\} = 0$ *almost surely.*

Remark 3.1 The following observations are useful in verifying the regularity conditions in Proposition 3.1. First, because

$$E\{\nabla \ell(\boldsymbol{\vartheta}|\mathbf{Y})\} = \int \nabla p(\mathbf{y}|\boldsymbol{\vartheta})\, dv,$$

the condition (3.1.1) is satisfied if the orders of differentiation and integration are interchangeable in the identity $\nabla \int p(\mathbf{y}|\boldsymbol{\vartheta})\, dv = \nabla 1 = \mathbf{0}$. A sufficient condition that allows the interchange of orders is that there exists an integrable function $g(\mathbf{y})$ such that $\|\nabla p(\mathbf{y}|\boldsymbol{\vartheta})\| \le g(\mathbf{y})$ for all $\boldsymbol{\vartheta} \in \Theta$. Similarly, because

$$E\{\hat{\boldsymbol{\eta}}[\nabla \ell(\boldsymbol{\vartheta}|\mathbf{Y})]^T\} = \int \hat{\boldsymbol{\eta}}(\mathbf{y})[\nabla p(\mathbf{y}|\boldsymbol{\vartheta})]^T dv,$$

the condition (3.1.2) is satisfied if the orders of differentiation and integration are interchangeable in the identity $\nabla^T \int \hat{\boldsymbol{\eta}}(\mathbf{y}) p(\mathbf{y}|\boldsymbol{\vartheta}) dv = \nabla^T \boldsymbol{\eta}(\boldsymbol{\vartheta}) = \mathbf{J}(\boldsymbol{\vartheta})$. A sufficient condition for the interchange of orders is that all elements in $\hat{\boldsymbol{\eta}}(\mathbf{y})[\nabla p(\mathbf{y}|\boldsymbol{\vartheta})]^T$ are upper-bounded in absolute value by an integrable function of $\mathbf{y}$ for all $\boldsymbol{\vartheta} \in \Theta$.

Remark 3.2 The regularity conditions in Proposition 3.1 can be replaced, according to [162, p. 73], by the following conditions:

(a) $p(\mathbf{y}|\boldsymbol{\vartheta})$ is a continuous function of $\boldsymbol{\vartheta} \in \Theta$ for almost every $\mathbf{y}$;

(b) for each $\boldsymbol{\vartheta} \in \Theta$, the function $\sqrt{p(\mathbf{y}|\boldsymbol{\vartheta})}$ has a mean-square derivative at $\boldsymbol{\vartheta}$, i.e., there exists a square-integrable function $\boldsymbol{\psi}(\mathbf{y};\boldsymbol{\vartheta})$ such that

$$\int \left|\sqrt{p(\mathbf{y}|\boldsymbol{\vartheta}+\boldsymbol{\delta})} - \sqrt{p(\mathbf{y}|\boldsymbol{\vartheta})} - \boldsymbol{\delta}^T \boldsymbol{\psi}(\mathbf{y};\boldsymbol{\vartheta})\right|^2 dv = \mathcal{O}(\|\boldsymbol{\delta}\|^2), \quad \boldsymbol{\delta} \to 0;$$

(c) $\boldsymbol{\psi}(\boldsymbol{y};\boldsymbol{\vartheta})$ is mean-square continuous in $\boldsymbol{\vartheta}$, i.e.,

$$\int \|\boldsymbol{\psi}(\boldsymbol{y};\boldsymbol{\vartheta}+\boldsymbol{\delta}) - \boldsymbol{\psi}(\boldsymbol{y};\boldsymbol{\vartheta})\|^2\, dv = \mathcal{o}(1), \quad \boldsymbol{\delta} \to 0.$$

Under these conditions, $\mathbf{I}(\boldsymbol{\vartheta}) = 4\int \|\boldsymbol{\psi}(\boldsymbol{y};\boldsymbol{\vartheta})\|^2 dv$. In addition, if $p(\boldsymbol{y}|\boldsymbol{\vartheta})$ is differentiable with respect to $\boldsymbol{\vartheta}$, then $\boldsymbol{\psi}(\boldsymbol{y};\boldsymbol{\vartheta}) = \nabla\sqrt{p(\boldsymbol{y}|\boldsymbol{\vartheta})} = \frac{1}{2}\sqrt{p(\boldsymbol{y}|\boldsymbol{\vartheta})}\,\nabla\ell(\boldsymbol{y}|\boldsymbol{\vartheta})$.

The matrix $\mathbf{J}(\boldsymbol{\vartheta})\mathbf{I}(\boldsymbol{\vartheta})^{-1}\mathbf{J}(\boldsymbol{\vartheta})^T$ in (3.1.4) is called the *Cramér-Rao lower bound* (CRLB) for estimating the parameter $\boldsymbol{\eta}(\boldsymbol{\vartheta})$ on the basis of $\mathbf{Y}$. It will be denoted as CRLB($\boldsymbol{\eta}(\boldsymbol{\vartheta})$). As a convention, (3.1.4) can also be written as

$$\mathrm{Cov}(\hat{\boldsymbol{\eta}}) \ge \mathrm{CRLB}(\boldsymbol{\eta}(\boldsymbol{\vartheta})) := \mathbf{J}(\boldsymbol{\vartheta})\,\mathbf{I}(\boldsymbol{\vartheta})^{-1}\mathbf{J}(\boldsymbol{\vartheta})^T, \tag{3.1.5}$$

which holds for any unbiased estimator $\hat{\boldsymbol{\eta}}$ of $\boldsymbol{\eta}(\boldsymbol{\vartheta})$ based on $\mathbf{Y}$.

In the special case where $\boldsymbol{\eta}(\boldsymbol{\vartheta}) = \boldsymbol{\vartheta}$, the Cramér-Rao inequality reduces to

$$\mathrm{Cov}(\hat{\boldsymbol{\vartheta}}) \ge \mathrm{CRLB}(\boldsymbol{\vartheta}) := \mathbf{I}(\boldsymbol{\vartheta})^{-1}, \tag{3.1.6}$$

where $\hat{\boldsymbol{\vartheta}}$ is any unbiased estimator of $\boldsymbol{\vartheta}$ based on $\mathbf{Y}$. In other words, for any constant vector $\mathbf{a} \in \mathbb{R}^r$, we have

$$\mathrm{Var}(\mathbf{a}^T\hat{\boldsymbol{\vartheta}}) \ge \mathbf{a}^T\mathbf{I}(\boldsymbol{\vartheta})^{-1}\mathbf{a},$$

where the equality holds for some $\mathbf{a} \ne \mathbf{0}$ and $\boldsymbol{\vartheta} \in \Theta$ if and only if $\mathbf{a}^T\{c(\hat{\boldsymbol{\vartheta}} - \boldsymbol{\vartheta}) + \mathbf{I}(\boldsymbol{\vartheta})^{-1}\nabla\ell(\boldsymbol{\vartheta}|\mathbf{Y})\} = 0$ almost surely for some constant $c \ne 0$.

In the more general case where $\boldsymbol{\eta}(\boldsymbol{\vartheta})$ has the same dimension as $\boldsymbol{\vartheta}$ but $\boldsymbol{\eta}(\boldsymbol{\vartheta}) \ne \boldsymbol{\vartheta}$ for some or all $\boldsymbol{\vartheta} \in \Theta$, one can regard $\hat{\boldsymbol{\eta}}$ as a *biased* estimator of $\boldsymbol{\vartheta}$ with the bias given by $\mathbf{b}(\boldsymbol{\vartheta}) := \boldsymbol{\eta}(\boldsymbol{\vartheta}) - \boldsymbol{\vartheta}$. It follows from (3.1.5) that

$$\begin{aligned} E\{(\hat{\boldsymbol{\eta}} - \boldsymbol{\vartheta})(\hat{\boldsymbol{\eta}} - \boldsymbol{\vartheta})^T\} &= \mathrm{Cov}(\hat{\boldsymbol{\eta}}) + \mathbf{b}(\boldsymbol{\vartheta})\mathbf{b}(\boldsymbol{\vartheta})^T \\ &\ge \mathbf{J}(\boldsymbol{\vartheta})\,\mathbf{I}(\boldsymbol{\vartheta})^{-1}\mathbf{J}(\boldsymbol{\vartheta})^T + \mathbf{b}(\boldsymbol{\vartheta})\mathbf{b}(\boldsymbol{\vartheta})^T, \end{aligned}$$

where $\mathbf{J}(\boldsymbol{\vartheta}) := \nabla^T\boldsymbol{\eta}(\boldsymbol{\vartheta}) = \nabla^T\{\mathbf{b}(\boldsymbol{\vartheta}) + \boldsymbol{\vartheta}\} = \nabla^T\mathbf{b}(\boldsymbol{\vartheta}) + \mathbf{I}$. This inequality generalizes (3.1.6) and is valid for any biased estimator of $\boldsymbol{\vartheta}$ under the regularity conditions in Proposition 3.1. It implies that the mean-square error (MSE) of $\mathbf{a}^T\hat{\boldsymbol{\eta}}$ as a biased estimator of $\mathbf{a}^T\boldsymbol{\vartheta}$ satisfies

$$\mathrm{MSE}(\mathbf{a}^T\hat{\boldsymbol{\eta}}) := E\{|\mathbf{a}^T\hat{\boldsymbol{\eta}} - \mathbf{a}^T\boldsymbol{\vartheta}|^2\} \ge \mathbf{a}^T\mathbf{J}(\boldsymbol{\vartheta})\,\mathbf{I}(\boldsymbol{\vartheta})^{-1}\mathbf{J}(\boldsymbol{\vartheta})^T\mathbf{a} + \{\mathbf{a}^T\mathbf{b}(\boldsymbol{\vartheta})\}^2.$$

Note that $\mathbf{a}^T\mathbf{b}(\boldsymbol{\vartheta})$ is nothing but the bias $\mathbf{a}^T\hat{\boldsymbol{\eta}}$ for estimating $\mathbf{a}^T\boldsymbol{\vartheta}$.

The matrix $\mathbf{I}(\boldsymbol{\vartheta})$, defined in (3.1.3), is known as *Fisher's information matrix* (FIM), named after the English statistician Ronald Aylmer Fisher (1890–1962). Owing to its inverse relationship with the CRLB, Fisher's information matrix can be interpreted as a measure of the intrinsic easiness in estimating $\boldsymbol{\vartheta}$, whereas its inverse, the CRLB, measures the intrinsic difficulty.

Proposition 3.1 can be easily generalized to the case where $\boldsymbol{\vartheta}$ is a random vector. Indeed, if the assumptions in Proposition 3.1 are true when $\boldsymbol{\vartheta}$ is treated as a deterministic variable, then the inequality (3.1.5) remains valid for the conditional covariance matrix $\text{Cov}(\hat{\boldsymbol{\eta}}|\boldsymbol{\vartheta})$, i.e.,

$$\text{Cov}(\hat{\boldsymbol{\eta}}|\boldsymbol{\vartheta}) \geq \mathbf{J}(\boldsymbol{\vartheta})\,\mathbf{I}(\boldsymbol{\vartheta})^{-1}\mathbf{J}(\boldsymbol{\vartheta})^T. \tag{3.1.7}$$

Therefore, the unconditional covariance matrix satisfies

$$\text{Cov}(\hat{\boldsymbol{\eta}}) = E\{\text{Cov}(\hat{\boldsymbol{\eta}}|\boldsymbol{\vartheta})\} \geq E\{\mathbf{J}(\boldsymbol{\vartheta})\,\mathbf{I}(\boldsymbol{\vartheta})^{-1}\mathbf{J}(\boldsymbol{\vartheta})^T\}, \tag{3.1.8}$$

where the expected value is taken with respect to $\boldsymbol{\vartheta}$ as a random variable. In this case, we refer to the lower bound in (3.1.7) as the *conditional* CRLB and refer to the lower bound in (3.1.8) as the *unconditional* CRLB, or simply the CRLB.

Equipped with Proposition 3.1, the remainder of this chapter is devoted to the special case of estimating the sinusoidal parameters. For finite sample sizes, the CRLB can be derived easily under the Gaussian assumption as will be discussed in Section 3.2. This result is useful for numerical calculation, but it offers little insight except for some very special cases. An easier way of analyzing the CRLB is to make the assumption of large sample sizes, as will be discussed in Section 3.3. Under this assumption, the finite sample CRLB can be approximated by much simpler expressions from which interesting conclusions can be drawn. Simple expressions can also be obtained under the condition of nonGaussian white noise. These results lead to very interesting findings concerning the performance limit in nonGaussian cases, which will be discussed in Section 3.4.

3.2 CRLB for Sinusoids in Gaussian Noise

Let us begin with the case of finite sample sizes under the assumption that $\{\epsilon_t\}$ is a zero-mean (real or complex) Gaussian process. Let us also assume that the covariance matrix $\mathbf{R}_\epsilon$ of $\boldsymbol{\epsilon} := [\epsilon_1, \ldots, \epsilon_n]^T$ may depend on an unknown auxiliary parameter $\boldsymbol{\eta}$ (for example, the variance of ϵ_t) which is not a function of the sinusoidal parameter $\boldsymbol{\theta}$. The problem is to find the CRLB for estimating the sinusoidal parameter $\boldsymbol{\theta}$ in the presence of the auxiliary parameter $\boldsymbol{\eta}$ from a data record $\mathbf{y} := [y_1, \ldots, y_n]^T$ that satisfies

$$\mathbf{y} = \mathbf{x} + \boldsymbol{\epsilon}, \tag{3.2.1}$$

where $\mathbf{x} := [x_1, \ldots, x_n]^T$ is given by (2.1.1), (2.1.2), (2.1.5), or (2.1.7).

In the real case where $\mathbf{x}$ is given by (2.1.1) or (2.1.2), the following theorem can be established. See Section 3.5 for a proof.

Theorem 3.1 (CRLB for RSM). *Let* $\mathbf{y}$ *be given by (3.2.1) with* $\boldsymbol{\epsilon} \sim \mathrm{N}(\mathbf{0}, \mathbf{R}_\epsilon)$. *If* $\mathbf{x}$ *takes the Cartesian form (2.1.1), then the CRLB for estimating* $\boldsymbol{\theta} := [\boldsymbol{\theta}_1^T, \dots, \boldsymbol{\theta}_q^T]^T$ *with* $\boldsymbol{\theta}_k := [A_k, B_k, \omega_k]^T$ *can be expressed as*

$$\mathrm{CRLB}(\boldsymbol{\theta}) = (\mathbf{X}^T \mathbf{R}_\epsilon^{-1} \mathbf{X})^{-1}, \tag{3.2.2}$$

where $\mathbf{X} := \partial \mathbf{x} / \partial \boldsymbol{\theta}^T = [\mathbf{X}_1, \dots, \mathbf{X}_q]$, $\mathbf{X}_k := \partial \mathbf{x} / \partial \boldsymbol{\theta}_k^T = [\mathbf{x}_{1k}, \mathbf{x}_{2k}, \mathbf{x}_{3k}]$, *and*

$$\begin{cases} \mathbf{x}_{1k} := [\cos(\omega_k), \dots, \cos(\omega_k n)]^T, \\ \mathbf{x}_{2k} := [\sin(\omega_k), \dots, \sin(\omega_k n)]^T, \\ \mathbf{x}_{3k} := -A_k [\sin(\omega_k), \dots, n \sin(\omega_k n)]^T \\ \qquad\quad + B_k [\cos(\omega_k), \dots, n \cos(\omega_k n)]^T. \end{cases} \tag{3.2.3}$$

If $\mathbf{x}$ *takes the polar form (2.1.2), then the CRLB for estimating* $\boldsymbol{\theta} := [\boldsymbol{\theta}_1^T, \dots, \boldsymbol{\theta}_q^T]^T$ *with* $\boldsymbol{\theta}_k := [C_k, \phi_k, \omega_k]^T$ *can be expressed as (3.2.2) with the* $\mathbf{x}_{jk}$ *defined by*

$$\begin{cases} \mathbf{x}_{1k} := [\cos(\omega_k + \phi_k), \dots, \cos(\omega_k n + \phi_k)]^T, \\ \mathbf{x}_{2k} := -C_k [\sin(\omega_k + \phi_k), \dots, \sin(\omega_k n + \phi_k)]^T, \\ \mathbf{x}_{3k} := -C_k [\sin(\omega_k + \phi_k), \dots, n \sin(\omega_k n + \phi_k)]^T. \end{cases} \tag{3.2.4}$$

Similarly, the following theorem can be obtained for the complex case where $\mathbf{x}$ is given by (2.1.5) or (2.1.7). See Section 3.5 for a proof.

Theorem 3.2 (CRLB for CSM). *Let* $\mathbf{y}$ *be given by (3.2.1) with* $\boldsymbol{\epsilon} \sim \mathrm{N}_c(\mathbf{0}, \mathbf{R}_\epsilon)$. *If* $\mathbf{x}$ *takes the Cartesian form (2.1.5), then the CRLB for estimating* $\boldsymbol{\theta} := [\boldsymbol{\theta}_1^T, \dots, \boldsymbol{\theta}_p^T]^T$ *with* $\boldsymbol{\theta}_k := [A_k, B_k, \omega_k]^T$ *can be expressed as*

$$\mathrm{CRLB}(\boldsymbol{\theta}) = \tfrac{1}{2} \{\Re(\mathbf{X}^H \mathbf{R}_\epsilon^{-1} \mathbf{X})\}^{-1}, \tag{3.2.5}$$

where $\mathbf{X} := \partial \mathbf{x} / \partial \boldsymbol{\theta}^T = [\mathbf{X}_1, \dots, \mathbf{X}_p]$, $\mathbf{X}_k := \partial \mathbf{x} / \partial \boldsymbol{\theta}_k^T = [\mathbf{x}_{1k}, \mathbf{x}_{2k}, \mathbf{x}_{3k}]$, *and*

$$\begin{cases} \mathbf{x}_{1k} := [\exp(i\omega_k), \dots, \exp(i\omega_k n)\}]^T, \\ \mathbf{x}_{2k} := -i \mathbf{x}_{1k}, \\ \mathbf{x}_{3k} := i(A_k - iB_k)[\exp(i\omega_k), \dots, n \exp(i\omega_k n)]^T. \end{cases} \tag{3.2.6}$$

If $\mathbf{x}$ *takes the polar form (2.1.7), then the CRLB in (3.2.5) remains valid for estimating* $\boldsymbol{\theta} := [\boldsymbol{\theta}_1^T, \dots, \boldsymbol{\theta}_p^T]^T$ *with* $\boldsymbol{\theta}_k := [C_k, \phi_k, \omega_k]^T$, *provided the* $\mathbf{x}_{jk}$ *are given by*

$$\begin{cases} \mathbf{x}_{1k} := [\exp\{i(\omega_k + \phi_k)\}, \dots, \exp\{i(\omega_k n + \phi_k)\}]^T, \\ \mathbf{x}_{2k} := i C_k \mathbf{x}_{1k}, \\ \mathbf{x}_{3k} := i C_k [\exp\{i(\omega_k + \phi_k)\}, \dots, n \exp\{i(\omega_k n + \phi_k)\}]^T. \end{cases} \tag{3.2.7}$$

Remark 3.3 The Jacobian matrix of the function that transforms the Cartesian parameter $\{A_k, B_k, \omega_k\}$ into the polar parameter $\{C_k, \phi_k, \omega_k\}$ takes the form

$$\mathbf{J}_k := \begin{bmatrix} A_k/C_k & B_k/C_k & 0 \\ B_k/C_k^2 & -A_k/C_k^2 & 0 \\ 0 & 0 & 1 \end{bmatrix}. \tag{3.2.8}$$

Let $\mathbf{\Sigma}_\mathrm{P}$ and $\mathbf{\Sigma}_\mathrm{C}$ denote the CRLB in (3.2.2) or (3.2.5) for the polar and Cartesian parameters, respectively. Then, we can write

$$\mathbf{\Sigma}_\mathrm{P} = \mathbf{J}\,\mathbf{\Sigma}_\mathrm{C}\,\mathbf{J}^T,$$

where $\mathbf{J} := \mathrm{diag}(\mathbf{J}_1, \dots, \mathbf{J}_q)$ for the RSM and $\mathbf{J} := \mathrm{diag}(\mathbf{J}_1, \dots, \mathbf{J}_p)$ for the CSM.

To gain some insights from these results, consider the following examples.

Example 3.1 (Single Complex Sinusoid in Gaussian White Noise: Cartesian). Let $\mathbf{x}$ be given by the Cartesian CSM (2.1.5) with $p = 1$ and let $\boldsymbol{\epsilon} \sim \mathrm{N}_c(\mathbf{0}, \sigma^2\mathbf{I})$. By Theorem 3.2, the CRLB for estimating $\boldsymbol{\theta} := [A_1, B_1, \omega_1]^T$ takes the form

$$\mathrm{CRLB}(\boldsymbol{\theta}) = \tfrac{1}{2}\sigma^2\{\Re(\mathbf{X}^H\mathbf{X})\}^{-1}, \tag{3.2.9}$$

where $\mathbf{X} := [\mathbf{x}_{11}, \mathbf{x}_{21}, \mathbf{x}_{31}]$ is given by (3.2.6). It is easy to verify that

$$\mathbf{x}_{11}^H\mathbf{x}_{11} = \mathbf{x}_{21}^H\mathbf{x}_{21} = n, \quad \mathbf{x}_{11}^H\mathbf{x}_{21} = -i\mathbf{x}_{11}^H\mathbf{x}_{11} = -in, \quad \mathbf{x}_{11}^H\mathbf{x}_{31} = i(A_1 - iB_1)\sum_{t=1}^n t,$$
$$\mathbf{x}_{21}^H\mathbf{x}_{31} = i\mathbf{x}_{11}^H\mathbf{x}_{31} = -(A_1 - iB_1)\sum_{t=1}^n t, \quad \mathbf{x}_{31}^H\mathbf{x}_{31} = (A_1^2 + B_1^2)\sum_{t=1}^n t^2.$$

Therefore,

$$\Re(\mathbf{X}^H\mathbf{X}) = \begin{bmatrix} n & 0 & B_1\sum t \\ 0 & n & -A_1\sum t \\ B_1\sum t & -A_1\sum t & (A_1^2 + B_1^2)\sum t^2 \end{bmatrix}. \tag{3.2.10}$$

Because $\sum t = \frac{1}{2}n(n+1)$, $\sum t^2 = \frac{1}{6}n(n+1)(2n+1)$, and $C_1^2 = A_1^2 + B_1^2$, we obtain

$$\mathrm{CRLB}(\boldsymbol{\theta}) = \frac{1}{\gamma_1}\begin{bmatrix} \frac{1}{2n}A_1^2 + \frac{2n+1}{n(n-1)}B_1^2 & -\frac{3(n+1)}{2n(n-1)}A_1B_1 & -\frac{3}{n(n-1)}B_1 \\ & \frac{2n+1}{n(n-1)}A_1^2 + \frac{1}{2n}B_1^2 & \frac{3}{n(n-1)}A_1 \\ \text{symmetric} & & \frac{6}{n(n^2-1)} \end{bmatrix}, \tag{3.2.11}$$

where $\gamma_1 := C_1^2/\sigma^2$ is the signal-to-noise ratio (SNR) of the complex sinusoid. ◇

Example 3.2 (Single Complex Sinusoid in Gaussian White Noise: Polar). Let $\mathbf{x}$ be given by the polar CSM (2.1.7) with $p = 1$ and let $\boldsymbol{\epsilon} \sim \mathrm{N}_c(\mathbf{0}, \sigma^2\mathbf{I})$. Then, by Theorem 3.2, the CRLB for estimating $\boldsymbol{\theta} := [C_1, \phi_1, \omega_1]^T$ takes the form (3.2.9) with $\mathbf{X} := [\mathbf{x}_{11}, \mathbf{x}_{21}, \mathbf{x}_{31}]$ given by (3.2.7). Observe that

$$\mathbf{x}_{11}^H\mathbf{x}_{11} = n, \quad \mathbf{x}_{11}^H\mathbf{x}_{21} = iC_1\mathbf{x}_{11}^H\mathbf{x}_{11} = iC_1 n, \quad \mathbf{x}_{11}^H\mathbf{x}_{31} = iC_1 \sum_{t=1}^{n} t,$$

$$\mathbf{x}_{21}^H\mathbf{x}_{21} = C_1^2 n, \quad \mathbf{x}_{21}^H\mathbf{x}_{31} = -iC_1\mathbf{x}_{11}^H\mathbf{x}_{31} = C_1^2 \sum_{t=1}^{n} t, \quad \mathbf{x}_{31}^H\mathbf{x}_{31} = C_1^2 \sum_{t=1}^{n} t^2.$$

Therefore,

$$\Re(\mathbf{X}^H\mathbf{X}) = \begin{bmatrix} n & 0 & 0 \\ 0 & C_1^2 n & C_1^2 \sum t \\ 0 & C_1^2 \sum t & C_1^2 \sum t^2 \end{bmatrix}. \tag{3.2.12}$$

Straightforward calculation yields

$$\mathrm{CRLB}(\boldsymbol{\theta}) = \frac{1}{\gamma_1} \begin{bmatrix} \frac{1}{2n} C_1^2 & 0 & 0 \\ 0 & \frac{2n+1}{n(n-1)} & -\frac{3}{n(n-1)} \\ 0 & -\frac{3}{n(n-1)} & \frac{6}{n(n^2-1)} \end{bmatrix}, \tag{3.2.13}$$

where $\gamma_1 := C_1^2/\sigma^2$ is the SNR of the complex sinusoid. ◇

As can be seen from Examples 3.1 and 3.2, the CRLB for the frequency parameter in both models remains the same and depends solely on the SNR and the sample size. The CRLB under the polar CSM (2.1.7) takes a much simpler form than the CRLB under the Cartesian CSM (2.1.5). In particular, the amplitude is decoupled with the frequency and the phase in (3.2.13). Moreover, because (3.2.13) does not depend on the phase of the sinusoid, it remains valid when the phase is a random variable. This can be justified by first conditioning on the phase to obtain $\mathrm{Cov}(\hat{\boldsymbol{\theta}}|\phi_1) \geq \mathrm{CRLB}(\boldsymbol{\theta})$ and then taking the expected value on both sides with respect to ϕ_1 to get $\mathrm{Cov}(\hat{\boldsymbol{\theta}}) = E\{\mathrm{Cov}(\hat{\boldsymbol{\theta}}|\phi_1)\} \geq \mathrm{CRLB}(\boldsymbol{\theta})$. The ability to accommodate random phase is the major advantage of the polar system.

For large n, the CRLB for the amplitude or phase parameters takes the form $\mathcal{O}(n^{-1})$, but the CRLB for the frequency parameter takes the form $\mathcal{O}(n^{-3})$. This indicates that the frequency can be estimated with potentially much higher accuracy than the amplitude and the phase.

The higher rate of accuracy for frequency estimation may seem surprising to some who are accustomed to the usual $\mathcal{O}(n^{-1})$ rate for parameter estimation.

An intuitive explanation of this phenomenon can be obtained by examining the sensitivity of the sinusoidal model to its parameters. It follows from (2.1.7) that

$$|dx_t| = |dC_k|, \quad |dx_t| = C_k|d\phi_k|, \quad |dx_t| = tC_k|d\omega_k|.$$

These expressions reveal that the error in x_t caused by an amplitude or phase offset remains constant over time but the error caused by a frequency offset grows linearly with time t. This means that the sinusoidal signal is more sensitive to the frequency offset than the amplitude or phase offset. As a result, an error in frequency estimation will be amplified much more strongly than the same amount of error in amplitude or phase estimation, sending a message for further improvement on frequency estimation.

The next example illustrates the interaction among multiple sinusoids in the CRLB which is absent in Examples 3.1 and 3.2.

Example 3.3 (Two Complex Sinusoids in Gaussian White Noise). Let $\mathbf{x}$ be given by the polar CSM (2.1.7) with $p = 2$ and let $\boldsymbol{\epsilon} \sim \mathrm{N}_c(\mathbf{0}, \sigma^2\mathbf{I})$. In this case, $\boldsymbol{\theta}_k := [C_k, \phi_k, \omega_k]^T$, $\boldsymbol{\theta} = [\boldsymbol{\theta}_1^T, \boldsymbol{\theta}_2^T]^T$, and $\mathbf{X} = [\mathbf{X}_1, \mathbf{X}_2]$, where $\mathbf{X}_k := [\mathbf{x}_{1k}, \mathbf{x}_{2k}, \mathbf{x}_{3k}]$ is given by (3.2.7). As in Example 3.2, it is easy to show that

$$\Re(\mathbf{X}_1^H \mathbf{X}_1) = \mathbf{M}_{11}, \quad \Re(\mathbf{X}_2^H \mathbf{X}_2) = \mathbf{M}_{22},$$

where $\mathbf{M}_{kk}$ takes the form (3.2.12) with C_k in place of C_1. Similarly,

$$\Re(\mathbf{X}_1^H \mathbf{X}_2) = \mathbf{M}_{12}, \quad \Re(\mathbf{X}_2^H \mathbf{X}_1) = \mathbf{M}_{21},$$

where

$$\mathbf{M}_{kk'} := \begin{bmatrix} \sum c_{kk'}(t) & C_{k'} \sum s_{kk'}(t) & C_{k'} \sum t s_{kk'}(t) \\ -C_k \sum s_{kk'}(t) & C_k C_{k'} \sum c_{kk'}(t) & C_k C_{k'} \sum t c_{kk'}(t) \\ -C_k \sum t s_{kk'}(t) & C_k C_{k'} \sum t c_{kk'}(t) & C_k C_{k'} \sum t^2 c_{kk'}(t) \end{bmatrix}, \tag{3.2.14}$$

with $c_{kk'}(t) := \cos\{d_{kk'}(t)\}$, $s_{kk'}(t) := \sin\{d_{kk'}(t)\}$, and $d_{kk'}(t) := (\omega_k - \omega_{k'})t + \phi_k - \phi_{k'}$. According to Theorem 3.2,

$$\mathrm{CRLB}(\boldsymbol{\theta}) = \frac{1}{2}\sigma^2 \begin{bmatrix} \mathbf{M}_{11} & \mathbf{M}_{12} \\ \mathbf{M}_{21} & \mathbf{M}_{22} \end{bmatrix}^{-1}. \tag{3.2.15}$$

Moreover, using the second matrix inversion formula in Lemma 12.4.1, the CRLB for estimating $\boldsymbol{\theta}_1$ can be expressed as

$$\mathrm{CRLB}(\boldsymbol{\theta}_1) = \tfrac{1}{2}\sigma^2 (\mathbf{M}_{11} - \mathbf{M}_{12}\mathbf{M}_{22}^{-1}\mathbf{M}_{21})^{-1} \ge \tfrac{1}{2}\sigma^2 \mathbf{M}_{11}^{-1}, \tag{3.2.16}$$

where the inequality is due to the fact that $\mathbf{M}_{12}\mathbf{M}_{22}^{-1}\mathbf{M}_{21}$ is nonnegative definite and hence $\mathbf{M}_{11} - \mathbf{M}_{12}\mathbf{M}_{22}^{-1}\mathbf{M}_{21} \le \mathbf{M}_{11}$. Note that the lower bound in (3.2.16) is nothing but the CRLB in Example 3.2 for estimating $\boldsymbol{\theta}_1$ in the absence of the second sinusoid. This means that the presence of the other sinusoid can raise the CRLB for estimating $\boldsymbol{\theta}_1$.

Similarly, it can be shown that (3.2.15) remains valid for estimating the Cartesian parameters $\boldsymbol{\theta}_k := [A_k, B_k, \omega_k]^T$ except that $\mathbf{X}_k$ is given by (3.2.6) and hence

$$\mathbf{M}_{kk'} = \Re \begin{bmatrix} \sum e_{kk'}(t) & -i\sum e_{kk'}(t) & i\beta_{k'}\sum t e_{kk'}(t) \\ i\sum e_{kk'}(t) & \sum e_{kk'}(t) & -\beta_{k'}\sum t e_{kk'}(t) \\ -i\beta_k \sum t e_{kk'}(t) & -\beta_k \sum t e_{kk'}(t) & \beta_k^* \beta_{k'} \sum t^2 e_{kk'}(t) \end{bmatrix}, \tag{3.2.17}$$

where $e_{kk'}(t) := \exp\{i(\omega_{k'} - \omega_k)\}$ and $\beta_k := A_k - iB_k$.

In general, for estimating p complex sinusoids in Gaussian white noise,

$$\text{CRLB}(\boldsymbol{\theta}) = \tfrac{1}{2}\sigma^2 \mathbf{M}^{-1},$$

where $\mathbf{M} := [\mathbf{M}_{kk'}]$ $(k, k' = 1, \dots, p)$, with $\mathbf{M}_{kk'}$ given by (3.2.14) or (3.2.17). ◇

Example 3.3 shows that the CRLB for multiple sinusoids is generally larger than the corresponding CRLB for a single sinusoid. This can be attributed entirely to the interference among multiple sinusoids. Moreover, unlike the CRLB for a single complex sinusoid, the CRLB for multiple complex sinusoids depends on the frequency as well as the phase of the sinusoids. The dependency is through the functions $d_{kk'}(t)$ $(k \neq k')$ which are determined solely by the frequency difference $\omega_k - \omega_{k'}$ and the phase difference $\phi_k - \phi_{k'}$.

Numerical studies in [89] and [317] demonstrate that the CRLB for two complex sinusoids becomes much higher than the corresponding CRLB for a single complex sinusoid as the frequency separation $\Delta := |\omega_1 - \omega_2|$ falls below $2\pi/n$. This is an indication of increased difficulty in estimating closely spaced frequencies. Moreover, because a real sinusoid with frequency $\omega_1 \in (0, \pi)$ can be regarded as two complex conjugate sinusoids with frequencies ω_1 and $\omega_2 := -\omega_1$, the increased difficulty also applies to the estimation of low-frequency real sinusoids. The same is true for real sinusoids with frequencies near π. Further analysis of the CRLB for closely spaced frequencies is provided later in Section 3.3.

Let us consider the computation of the CRLB under colored noise. To be more specific, let us focus on the real case for which the CRLB equals $(\mathbf{X}^T \mathbf{R}_\epsilon^{-1} \mathbf{X})^{-1}$ by Theorem 3.1. Observe that direct calculation of the CRLB requires the inversion of the n-by-n matrix $\mathbf{R}_\epsilon$, which can be burdensome when n is large. The burden can be reduced considerably if we use the Levinson-Durbin algorithm in Proposition 2.8 to compute $\mathbf{X}^T \mathbf{R}_\epsilon^{-1} \mathbf{X}$ without explicitly inverting $\mathbf{R}_\epsilon$.

According to Proposition 2.8, $\mathbf{R}_\epsilon^{-1} = \mathbf{U}\mathbf{D}^{-1}\mathbf{U}^H$, where $\mathbf{D} := \text{diag}(\sigma_0^2, \sigma_1^2, \dots, \sigma_{n-1}^2)$ and $\mathbf{U} := [\varphi_{t-1,t-s}^*]$ $(s, t = 1, \dots n;\ \varphi_{t-1,t-s} := 0$ for $s > t$ and $\varphi_{t-1,0} := 1)$. Observe that $\mathbf{D}$ and $\mathbf{U}$ can be computed recursively from $\mathbf{R}_\epsilon$. Now, let

$$\mathbf{z}_{jk} := [z_{jk}(1), \dots, z_{jk}(n)]^T := \mathbf{U}^H \mathbf{x}_{jk} \qquad (j = 1, 2, 3).$$

Because $\mathbf{U}^H = [\varphi_{s-1,s-t}]$ $(s,t=1,\ldots,n)$, we have

$$z_{jk}(t) = \sum_{u=0}^{t-1} \varphi_{t-1,u}\, x_{jk}(t-u) \qquad (t=1,\ldots,n). \tag{3.2.18}$$

Moreover, let $\mathbf{Z}_k := [\mathbf{z}_{1k}, \mathbf{z}_{2k}, \mathbf{z}_{3k}] = \mathbf{U}^H \mathbf{X}_k$ and $\mathbf{Z} := [\mathbf{Z}_1, \ldots, \mathbf{Z}_q] = \mathbf{U}^H \mathbf{X}$. Then,

$$\mathbf{X}^H \mathbf{R}_\epsilon^{-1} \mathbf{X} = \mathbf{X}^H \mathbf{U} \mathbf{D}^{-1} \mathbf{U}^H \mathbf{X} = \mathbf{Z}^H \mathbf{D}^{-1} \mathbf{Z} = [\mathbf{I}_{kk'}] \quad (k,k'=1,\ldots,q), \tag{3.2.19}$$

where

$$\mathbf{I}_{kk'} := \mathbf{Z}_k^H \mathbf{D}^{-1} \mathbf{Z}_{k'} = [\mathbf{z}_{jk}^H \mathbf{D}^{-1} \mathbf{z}_{j'k'}] \quad (j,j'=1,2,3), \tag{3.2.20}$$

$$\mathbf{z}_{jk}^H \mathbf{D}^{-1} \mathbf{z}_{j'k'} = \sum_{t=1}^{n} z_{jk}^*(t) z_{j'k'}(t)/\sigma_{t-1}^2 \quad (j,j'=1,2,3). \tag{3.2.21}$$

Equations (3.2.18)–(3.2.21), together with the recursion in Proposition 2.8, constitute a fast algorithm for computing $\mathbf{X}^H \mathbf{R}_\epsilon^{-1} \mathbf{X}$. Note that if $\{\epsilon_t\}$ is an AR process, then the matrices $\mathbf{I}_{kk'}$ can be obtained directly from the AR parameters, rather than the covariance matrix $\mathbf{R}_\epsilon$, by using the algorithm in Proposition 2.9.

Finally, let us investigate an important special case where the signal frequencies are known but the amplitude and phase parameters are unknown. In this case, it suffices to use the CRLB given by the following corollary for the amplitude and phase estimation. The result is stated without proof, as it is a direct result of Theorem 3.1 for the RSM and of Theorem 3.2 for the CSM.

Corollary 3.1 (CRLB for Amplitude and Phase Parameters). *Let* $\mathbf{y}$ *be given by (3.2.1). If* $\boldsymbol{\epsilon} \sim \mathrm{N}(\mathbf{0}, \mathbf{R}_\epsilon)$*, then, the CRLB for estimating* $\boldsymbol{\theta} := [A_1, B_1, \ldots, A_q, B_q]^T$ *in the RSM (2.1.1) or* $\boldsymbol{\theta} := [C_1, \phi_1, \ldots, C_q, \phi_q]^T$ *in the RSM (2.1.2) can be expressed as (3.2.2) with* $\mathbf{X}_k := [\mathbf{x}_{1k}, \mathbf{x}_{2k}]$ *given by (3.2.3) or (3.2.4), respectively. If* $\boldsymbol{\epsilon} \sim \mathrm{N}_c(\mathbf{0}, \mathbf{R}_\epsilon)$*, then the CRLB for estimating* $\boldsymbol{\theta} := [A_1, B_1, \ldots, A_p, B_p]^T$ *in the CSM (2.1.5) or* $\boldsymbol{\theta} := [C_1, \phi_1, \ldots, C_p, \phi_p]^T$ *in the CSM (2.1.7) takes the form (3.2.5) with* $\mathbf{X}_k := [\mathbf{x}_{1k}, \mathbf{x}_{2k}]$ *given by (3.2.6) or (3.2.7), respectively.*

Remark 3.4 In the complex case, it is sometimes convenient to arrange the amplitude parameters such that

$$\boldsymbol{\theta} := [A_1, \ldots, A_p, B_1, \ldots, B_p]^T = [\Re(\boldsymbol{\beta})^T, -\Im(\boldsymbol{\beta})^T]^T,$$

where $\beta_k := A_k - iB_k$. For estimating this parameter, the expression in (3.2.5) remains true except that $\mathbf{X} := [\mathbf{F}, -i\mathbf{F}]$, where $\mathbf{F} := [\mathbf{f}(\omega_1), \ldots, \mathbf{f}(\omega_p)]$ and $\mathbf{f}(\omega_k) := [\exp(i\omega_k), \ldots, \exp(in\omega_k)]^T = \mathbf{x}_{1k}$. In the special case of Gaussian white noise, it reduces to $\mathrm{CRLB}(\boldsymbol{\theta}) = \frac{1}{2}\sigma^2 \boldsymbol{\Sigma}$, where

$$\boldsymbol{\Sigma} := \begin{bmatrix} \Re(\mathbf{F}^H\mathbf{F}) & \Im(\mathbf{F}^H\mathbf{F}) \\ -\Im(\mathbf{F}^H\mathbf{F}) & \Re(\mathbf{F}^H\mathbf{F}) \end{bmatrix}^{-1} = \begin{bmatrix} \Re\{(\mathbf{F}^H\mathbf{F})^{-1}\} & \Im\{(\mathbf{F}^H\mathbf{F})^{-1}\} \\ -\Im\{(\mathbf{F}^H\mathbf{F})^{-1}\} & \Re\{(\mathbf{F}^H\mathbf{F})^{-1}\} \end{bmatrix}.$$

The second expression can be verified by simple matrix algebra.

The following example illustrates this result.

Example 3.4 (CRLB for Amplitude Parameters: Single Complex Sinusoid in Gaussian White Noise). Consider the complex case where $\boldsymbol{\epsilon} \sim \mathrm{N}_c(\mathbf{0}, \sigma^2\mathbf{I})$ and $p = 1$. Because $\mathbf{f}^H(\omega_1)\mathbf{f}(\omega_1) = n$, it follows from Remark 3.4 that the CRLB for estimating $\boldsymbol{\theta} := [A_1, B_1]^T$ can be expressed as $\mathrm{CRLB}(\boldsymbol{\theta}) = \frac{1}{2}n^{-1}\sigma^2\mathbf{I}$. ◇

It is interesting to compare this example with Example 3.1 where the signal frequency is unknown and jointly estimated with the amplitude parameters. In Example 3.1, the CRLB for estimating A_1, which is the (1,1) entry of the matrix in (3.2.11), can be different from that for estimating B_1, which is the (2,2) entry. Furthermore, it is easy to verify that the CRLB for estimating A_1 is strictly less than the CRLB for estimating B_1 if $|A_1| > |B_1|$, regardless of the sample size and the SNR. Conversely, the CRLB for estimating B_1 is strictly less than the CRLB for estimating A_1 if $|B_1| > |A_1|$. In other words, the larger of the two parameters has a smaller CRLB. This is in contrast with Example 3.4 where the signal frequency is known. In Example 3.4, the CRLB is the same for both A_1 and B_1 regardless of their magnitude. Moreover, in the case of $A_1 = B_1 \neq 0$ and large n, the CRLB in Example 3.1 for estimating the amplitude parameters is approximately equal to 2.5 times the CRLB in Example 3.4; in the case of $A_1 = 0$ and $B_1 \neq 0$, the CRLB in Example 3.1 for estimating A_1 is approximately 4 times the CRLB in Example 3.4. These comparisons illustrate the increased difficulty in estimating the amplitude parameters when the frequencies are unknown.

3.3 Asymptotic CRLB for Sinusoids in Gaussian Noise

As demonstrated by Example 3.3, the interaction among the sinusoids makes it difficult to grasp the full implication of the CRLB for multiple sinusoids. This is true even in the simple case of a single real sinusoid. However, when the sample size is large, the CRLB can be approximated by much simpler expressions.

Let us begin by continuing the discussion in Examples 3.1 and 3.2.

Example 3.5 (Single Complex Sinusoid in Gaussian White Noise). Consider the case discussed in Example 3.1. let $\mathbf{K}_n := \mathrm{diag}(n^{1/2}, n^{1/2}, n^{3/2})$. Then, for large n, the CRLB in (3.2.11) for estimating $\boldsymbol{\theta} = \boldsymbol{\theta}_1 := [A_1, B_1, \omega_1]^T$ can be expressed as

$$\mathrm{CRLB}(\boldsymbol{\theta}) = \mathbf{K}_n^{-1}\{\boldsymbol{\Gamma}(\boldsymbol{\theta}) + \mathcal{O}(n^{-1})\}\mathbf{K}_n^{-1}, \tag{3.3.1}$$

where

$$\boldsymbol{\Gamma}(\boldsymbol{\theta}) := \tfrac{1}{2}\gamma_1^{-1}\boldsymbol{\Lambda}_{\mathrm{C}}(\boldsymbol{\theta}_1)$$

and

$$\mathbf{\Lambda}_{\mathrm{C}}(\boldsymbol{\theta}_k) := \begin{bmatrix} A_k^2+4B_k^2 & -3A_kB_k & -6B_k \\ -3A_kB_k & 4A_k^2+B_k^2 & 6A_k \\ -6B_k & 6A_k & 12 \end{bmatrix}. \tag{3.3.2}$$

Similarly, for the case discussed in Example 3.2, the CRLB in (3.2.13) for estimating $\boldsymbol{\theta} = \boldsymbol{\theta}_1 := [C_1, \phi_1, \omega_1]^T$ can be expressed as (3.3.1) with

$$\mathbf{\Gamma}(\boldsymbol{\theta}) := \tfrac{1}{2}\gamma_1^{-1}\mathbf{\Lambda}_{\mathrm{P}}(\boldsymbol{\theta}_1),$$

where

$$\mathbf{\Lambda}_{\mathrm{P}}(\boldsymbol{\theta}_k) := \begin{bmatrix} C_k^2 & 0 & 0 \\ 0 & 4 & -6 \\ 0 & -6 & 12 \end{bmatrix}. \tag{3.3.3}$$

In both cases, $\gamma_1 := C_1^2/\sigma^2$ is the SNR of the complex sinusoid. ◇

The matrix $\mathbf{K}_n^{-1}\mathbf{\Gamma}(\boldsymbol{\theta})\mathbf{K}_n^{-1}$ in (3.3.1) is called the asymptotic CRLB or ACRLB. For any unbiased estimator $\hat{\boldsymbol{\theta}}$ of $\boldsymbol{\theta}$, it follows from (3.3.1) that

$$\mathrm{Cov}(\mathbf{K}_n\hat{\boldsymbol{\theta}}) = \mathbf{K}_n\mathrm{Cov}(\hat{\boldsymbol{\theta}})\mathbf{K}_n \geq \mathbf{K}_n\mathrm{CRLB}(\boldsymbol{\theta})\mathbf{K}_n = \mathbf{\Gamma}(\boldsymbol{\theta}) + \mathcal{O}(n^{-1}).$$

Hence the matrix $\mathbf{\Gamma}(\boldsymbol{\theta})$ serves as an asymptotic lower bound for the covariance matrix of the normalized estimator $\mathbf{K}_n\hat{\boldsymbol{\theta}}$. The different orders of magnitude in the CRLB are made very clear through the normalizing factors in $\mathbf{K}_n$: for amplitude and phase estimation, the normalizing factor is equal to $n^{1/2}$; for frequency estimation, it is equal to $n^{3/2}$. When the normalizing factors are self-evident, we may also refer to $\mathbf{\Gamma}(\boldsymbol{\theta})$ as the ACRLB.

Next, consider the more interesting case in Example 3.3. Because two sinusoids are involved in this example, the degree of frequency separation becomes a key factor in determining the ACRLB. The frequency separation is meaningful only if it is measured with respect to the sample size n. To facilitate this analysis, let us assume that the frequencies may depend on n with the possibility that the distance between them approaches zero as $n \to \infty$.

Example 3.6 (Two Complex Sinusoids in Gaussian White Noise). In Example 3.3 for the polar parameters, define $\Delta := \min\{|\omega_1 - \omega_2|, 2\pi - |\omega_1 - \omega_2|\}$ and assume

$$\lim_{n\to\infty} n\Delta = \infty. \tag{3.3.4}$$

By Lemma 12.1.4, the matrix $\mathbf{M}_{12}$, defined by (3.2.14), can be expressed as

$$\mathbf{M}_{12} = \begin{bmatrix} \mathcal{O}(\Delta^{-1}) & \mathcal{O}(\Delta^{-1}) & \mathcal{O}(n\Delta^{-1}) \\ \mathcal{O}(\Delta^{-1}) & \mathcal{O}(\Delta^{-1}) & \mathcal{O}(n\Delta^{-1}) \\ \mathcal{O}(n\Delta^{-1}) & \mathcal{O}(n\Delta^{-1}) & \mathcal{O}(n^2\Delta^{-1}) \end{bmatrix} = \mathbf{K}_n\mathcal{O}(n^{-1}\Delta^{-1})\mathbf{K}_n.$$

Similarly, $\mathbf{M}_{21} = \mathbf{K}_n \mathcal{O}(n^{-1}\Delta^{-1})\mathbf{K}_n$. Moreover, because $\mathbf{M}_{kk}$ takes the form (3.2.12) with C_k in place of C_1, it follows that

$$\mathbf{M}_{kk} = \mathbf{K}_n\{\mathbf{W}_k + \mathcal{O}(n^{-1})\}\mathbf{K}_n,$$

where

$$\mathbf{W}_k := \begin{bmatrix} 1 & 0 & 0 \\ 0 & C_k^2 & \frac{1}{2}C_k^2 \\ 0 & \frac{1}{2}C_k^2 & \frac{1}{3}C_k^2 \end{bmatrix}. \tag{3.3.5}$$

By matrix algebra, we obtain

$$\mathbf{W}_k^{-1} = (1/C_k^2)\mathbf{\Lambda}_{\mathrm{P}}(\boldsymbol{\theta}_k).$$

Let $\mathbf{\Gamma}(\boldsymbol{\theta}_k) := \frac{1}{2}\sigma^2\mathbf{W}_k^{-1} = \frac{1}{2}\gamma_k^{-1}\mathbf{\Lambda}_{\mathrm{P}}(\boldsymbol{\theta}_k)$, where $\gamma_k := C_k^2/\sigma^2$. Then, the CRLB for estimating $\boldsymbol{\theta} := [\boldsymbol{\theta}_1^T, \boldsymbol{\theta}_2^T]^T$ with $\boldsymbol{\theta}_k := [C_k, \phi_k, \omega_k]^T$ can be expressed as

$$\begin{aligned} \mathrm{CRLB}(\boldsymbol{\theta}) &= \begin{bmatrix} \mathbf{K}_n^{-1} & \mathbf{0} \\ \mathbf{0} & \mathbf{K}_n^{-1} \end{bmatrix} \left\{ \begin{bmatrix} \mathbf{\Gamma}(\boldsymbol{\theta}_1) & \mathbf{0} \\ \mathbf{0} & \mathbf{\Gamma}(\boldsymbol{\theta}_2) \end{bmatrix} + \mathcal{O}(n^{-1}\Delta^{-1}) \right\} \begin{bmatrix} \mathbf{K}_n^{-1} & \mathbf{0} \\ \mathbf{0} & \mathbf{K}_n^{-1} \end{bmatrix} \\ &= \mathbf{K}^{-1}\{\mathbf{\Gamma}(\boldsymbol{\theta}) + \mathcal{O}(n^{-1}\Delta^{-1})\}\mathbf{K}^{-1}, \end{aligned} \tag{3.3.6}$$

where $\mathbf{K} := \mathrm{diag}(\mathbf{K}_n, \mathbf{K}_n)$ and $\mathbf{\Gamma}(\boldsymbol{\theta}) = \mathrm{diag}\{\mathbf{\Gamma}(\boldsymbol{\theta}_1), \mathbf{\Gamma}(\boldsymbol{\theta}_2)\}$.

The CRLB for the Cartesian parameters $\boldsymbol{\theta}_k := [A_k, B_k, \omega_k]^T$ also takes the form (3.3.6) except that $\mathbf{\Gamma}(\boldsymbol{\theta}_k) = \frac{1}{2}\gamma_k^{-1}\mathbf{\Lambda}_{\mathrm{C}}(\boldsymbol{\theta}_k)$. Indeed, by Lemma 12.1.5, $\mathbf{M}_{kk'}$ defined by (3.2.17) takes the form $\mathbf{M}_{kk'} = \mathbf{K}_n\mathcal{O}(n^{-1}\Delta^{-1})\mathbf{K}_n$ for $k \neq k'$. Moreover, it follows from (3.2.10) that $\mathbf{M}_{kk} = \mathbf{K}_n\{\mathbf{W}_k + \mathcal{O}(n^{-1})\}\mathbf{K}_n$, where

$$\mathbf{W}_k := \begin{bmatrix} 1 & 0 & \frac{1}{2}B_k \\ 0 & 1 & -\frac{1}{2}A_k \\ \frac{1}{2}B_k & -\frac{1}{2}A_k & \frac{1}{3}(A_k^2 + B_k^2) \end{bmatrix}. \tag{3.3.7}$$

Because

$$\mathbf{W}_k^{-1} = (1/C_k^2)\mathbf{\Lambda}_{\mathrm{C}}(\boldsymbol{\theta}_k),$$

we obtain (3.3.6) with $\mathbf{\Gamma}(\boldsymbol{\theta}_k) := \frac{1}{2}\sigma^2\mathbf{W}_k^{-1} = \frac{1}{2}\gamma_k^{-1}\mathbf{\Lambda}_{\mathrm{C}}(\boldsymbol{\theta}_k)$ and $\gamma_k := C_k^2/\sigma^2$. ◇

This example shows that for large sample sizes the CRLB for two sinusoids is decoupled so that the CRLB obtained for single complex sinusoids in Examples 3.1 and 3.5 can be used separately to approximate the CRLB for each sinusoid. The key prerequisite for the decoupling is the frequency separation condition (3.3.4). It is required to ensure that $\mathbf{K}_n^{-1}\mathbf{M}_{kk'}\mathbf{K}_n^{-1} \to 0$ as $n \to \infty$ for $k \neq k'$. For finite sample sizes, the accuracy of the approximation depends on Δ. In fact,

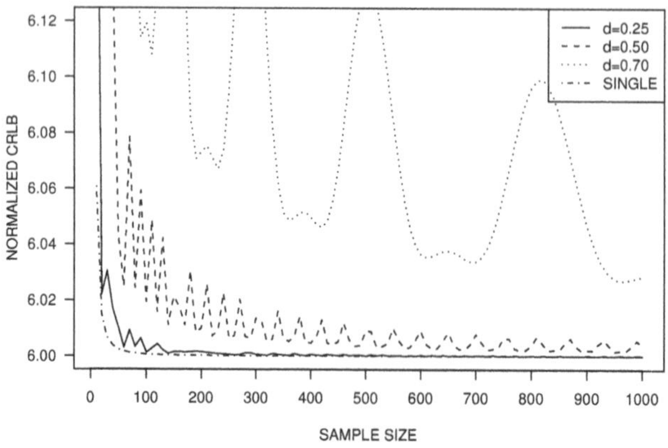

Figure 3.1. Plot of $n^3\text{CRLB}(\omega_1)$ as a function of n in the case of two unit-amplitude complex sinusoids in Gaussian white noise ($\omega_1 = 2\pi \times 0.1$, $\omega_2 = \omega_1 + \Delta$, $\phi_1 = 0$, $\phi_2 = \pi/2$, $\gamma_1 = \gamma_2 = 1$) under different frequency separation conditions of the form $\Delta = 2\pi/n^d$. Solid line, $d = 0.25$; dashed line, $d = 0.5$; dotted line, $d = 0.7$. The dash-dotted line depicts the normalized single-sinusoid CRLB which approaches its asymptotic value 6 as n grows.

if $n\Delta$ does not approach infinity fast enough as n increases, the approximation can be very poor. This point is illustrated by the example shown in Figure 3.1.

In this example, the frequency separation parameter Δ takes the form $\Delta = 2\pi/n^d$ with $d = 0.25, 0.5, 0.7$, so the condition (3.3.4) is always satisfied. But, as we can see from Figure 3.1, the accuracy of the single-sinusoid CRLB as an approximation to the exact CRLB deteriorates rapidly when d gets closer to unity. This result also serves as a confirmation to the earlier comment on Example 3.5 that the CRLB increases rapidly as frequency separation decreases. Note that the oscillations of the CRLB as a function of the sample size are due entirely to the interplay between the sample size and the phases of the sinusoids.

Now, let us investigate the case of real sinusoids through two examples.

Example 3.7 (Single Real Sinusoid in Gaussian White Noise: Cartesian). Let $\mathbf{x}$ be given by (2.1.1) with $q = 1$ and let $\boldsymbol{\epsilon} \sim N(\mathbf{0}, \sigma^2\mathbf{I})$. By Theorem 3.1, the CRLB for estimating $\boldsymbol{\theta} := [A_1, B_1, \omega_1]^T$ takes the form $\text{CRLB}(\boldsymbol{\theta}) = \sigma^2(\mathbf{X}^T\mathbf{X})^{-1}$, where $\mathbf{X} := [\mathbf{x}_{11}, \mathbf{x}_{21}, \mathbf{x}_{31}]$ is given by (3.2.3). Let $c_t := \cos(\omega_1 t)$ and $s_t := \sin(\omega_1 t)$. Then,

$$\mathbf{X}^T\mathbf{X} = \begin{bmatrix} \sum c_t^2 & \sum c_t s_t & \sum t(-A_1 c_t s_t + B_1 c_t^2) \\ & \sum s_t^2 & \sum t(-A_1 s_t^2 + B_1 c_t s_t) \\ \text{symmetric} & & \sum t^2(-A_1 s_t + B_1 c_t)^2 \end{bmatrix}. \tag{3.3.8}$$

This matrix is not as simple as its counterpart (3.2.10) for a single complex sinusoid. Although an explicit expression of the CRLB can be obtained by inverting the matrix, let us derive a simpler expression for large sample sizes.

Toward that end, assume that ω_1 is not too close to 0 or π in the sense that $\Delta := \min\{\omega_1, \pi - \omega_1\}$ satisfies (3.3.4). Then, as $n \to \infty$, Lemma 12.1.5 gives

$$\begin{cases} \sum_{t=1}^{n} t^r c_t s_t = \mathcal{O}(n^r \Delta^{-1}) \quad (r = 0, 1, 2), \\ \sum_{t=1}^{n} c_t^2 = \frac{1}{2}n + \mathcal{O}(\Delta^{-1}), \quad \sum_{t=1}^{n} s_t^2 = \frac{1}{2}n + \mathcal{O}(\Delta^{-1}), \\ \sum_{t=1}^{n} t c_t^2 = \frac{1}{4}n^2 + \mathcal{O}(n\Delta^{-1}), \quad \sum_{t=1}^{n} t s_t^2 = \frac{1}{4}n^2 + \mathcal{O}(n\Delta^{-1}), \\ \sum_{t=1}^{n} t^2 c_t^2 = \frac{1}{6}n^3 + \mathcal{O}(n^2\Delta^{-1}), \quad \sum_{t=1}^{n} t^2 s_t^2 = \frac{1}{6}n^3 + \mathcal{O}(n^2\Delta^{-1}). \end{cases} \tag{3.3.9}$$

Therefore, (3.3.8) can be written as

$$\mathbf{X}^T\mathbf{X} = \mathbf{K}_n\{\tfrac{1}{2}\mathbf{W}_1 + \mathcal{O}(n^{-1}\Delta^{-1})\}\mathbf{K}_n,$$

where $\mathbf{W}_1$ takes the form (3.3.7) with $k = 1$. This result leads to

$$\text{CRLB}(\boldsymbol{\theta}) = \sigma^2(\mathbf{X}^T\mathbf{X})^{-1} = \mathbf{K}_n^{-1}\{\boldsymbol{\Gamma}(\boldsymbol{\theta}) + \mathcal{O}(n^{-1}\Delta^{-1})\}\mathbf{K}_n^{-1}, \tag{3.3.10}$$

where $\boldsymbol{\Gamma}(\boldsymbol{\theta}) := 2\sigma^2\mathbf{W}_1^{-1} = \gamma_1^{-1}\boldsymbol{\Lambda}_{\text{C}}(\boldsymbol{\theta}_1)$, with $\boldsymbol{\Lambda}_{\text{C}}(\boldsymbol{\theta}_1)$ given by (3.3.2) and with $\gamma_1 := \frac{1}{2}(A_1^2 + B_1^2)/\sigma^2$ being the SNR of the real sinusoid. ◇

Example 3.8 (Single Real Sinusoid in Gaussian White Noise: Polar). Let $\mathbf{x}$ be given by (2.1.2) with $q = 1$ and let $\boldsymbol{\epsilon} \sim \text{N}(\mathbf{0}, \sigma^2\mathbf{I})$. By Theorem 3.1, the CRLB for estimating $\boldsymbol{\theta} := [C_1, \phi_1, \omega_1]^T$ is $\text{CRLB}(\boldsymbol{\theta}) = \sigma^2(\mathbf{X}^T\mathbf{X})^{-1}$, where $\mathbf{X} := [\mathbf{x}_{11}, \mathbf{x}_{21}, \mathbf{x}_{31}]$ is given by (3.2.4). Let $c_t := \cos(\omega_1 t + \phi_1)$ and $s_t := \sin(\omega_1 t + \phi_1)$. Then,

$$\mathbf{X}^T\mathbf{X} = \begin{bmatrix} \sum c_t^2 & -C_1 \sum c_t s_t & -C_1 \sum t c_t s_t \\ & C_1^2 \sum s_t^2 & C_1^2 \sum t s_t^2 \\ \text{symmetric} & & C_1^2 \sum t^2 s_t^2 \end{bmatrix}. \tag{3.3.11}$$

Under the assumption that $\Delta := \min\{\omega_1, \pi - \omega_1\}$ satisfies (3.3.4), the results in (3.3.9) remain valid for $c_t = \cos(\omega_1 t + \phi_1)$ and $s_t = \sin(\omega_1 t + \phi_1)$. Substituting these expressions in (3.3.11) yields

$$\mathbf{X}^T\mathbf{X} = \mathbf{K}_n\{\tfrac{1}{2}\mathbf{W}_1 + \mathcal{O}(n^{-1}\Delta^{-1})\}\mathbf{K}_n,$$

where $\mathbf{W}_1$ is defined by (3.3.5) for $k = 1$. Therefore, we obtain (3.3.10) with $\boldsymbol{\Gamma}(\boldsymbol{\theta}) := 2\sigma^2\mathbf{W}_1^{-1} = \gamma_1^{-1}\boldsymbol{\Lambda}_{\text{P}}(\boldsymbol{\theta}_1)$, where $\boldsymbol{\Lambda}_{\text{P}}(\boldsymbol{\theta}_1)$ is defined by (3.3.3) and $\gamma_1 := \frac{1}{2}C_1^2/\sigma^2$ is the SNR of the real sinusoid. ◇

The following theorem summarizes the results for the ACRLB in the general case of multiple real or complex sinusoids. Given Examples 3.5–3.8, it suffices to state the theorem without proof.

Theorem 3.3 (Asymptotic CRLB: Gaussian White Noise). *Let* $\mathbf{y}$ *be given by (3.2.1) with* $\mathbf{x}$ *satisfying (2.1.1), (2.1.2), (2.1.5), or (2.1.7), and with* $\boldsymbol{\epsilon} \sim \mathrm{N}(\mathbf{0}, \sigma^2 \mathbf{I})$ *in the real case and* $\boldsymbol{\epsilon} \sim \mathrm{N}_c(\mathbf{0}, \sigma^2 \mathbf{I})$ *in the complex case. Assume that (3.3.4) is satisfied by*

$$\Delta := \begin{cases} \min_{k \neq k'} \{|\omega_k - \omega_{k'}|, \omega_k, \pi - \omega_k\} & \text{for RSM,} \\ \min_{k \neq k'} \{|\omega_k - \omega_{k'}|, 2\pi - |\omega_k - \omega_{k'}|\} & \text{for CSM.} \end{cases} \tag{3.3.12}$$

Let $\boldsymbol{\theta}_k \in \mathbb{R}^3$ *denote the parameters of the kth sinusoid and define*

$$\boldsymbol{\Gamma}(\boldsymbol{\theta}_k) := \begin{cases} \gamma_k^{-1} \boldsymbol{\Lambda}_\mathrm{C}(\boldsymbol{\theta}_k) & \text{for Cartesian RSM (2.1.1),} \\ \gamma_k^{-1} \boldsymbol{\Lambda}_\mathrm{P}(\boldsymbol{\theta}_k) & \text{for polar RSM (2.1.2),} \\ \frac{1}{2} \gamma_k^{-1} \boldsymbol{\Lambda}_\mathrm{C}(\boldsymbol{\theta}_k) & \text{for Cartesian CSM (2.1.5),} \\ \frac{1}{2} \gamma_k^{-1} \boldsymbol{\Lambda}_\mathrm{P}(\boldsymbol{\theta}_k) & \text{for polar CSM (2.1.7),} \end{cases} \tag{3.3.13}$$

where $\boldsymbol{\Lambda}_\mathrm{C}(\boldsymbol{\theta}_k)$ *and* $\boldsymbol{\Lambda}_\mathrm{P}(\boldsymbol{\theta}_k)$ *are given by (3.3.2) and (3.3.3), respectively, and where* $\gamma_k := \frac{1}{2} C_k^2 / \sigma^2$ *in the real case and* $\gamma_k := C_k^2 / \sigma^2$ *in the complex case. Then, as* $n \to \infty$, *the CRLB for estimating* $\boldsymbol{\theta} := [\boldsymbol{\theta}_1^T, \dots, \boldsymbol{\theta}_r^T]^T$ ($r := q$ *for the RSM and* $r := p$ *for the CSM) can be expressed as*

$$\mathrm{CRLB}(\boldsymbol{\theta}) = \mathbf{K}^{-1} \{ \boldsymbol{\Gamma}(\boldsymbol{\theta}) + \mathcal{O}(n^{-1} \Delta^{-1}) \} \mathbf{K}^{-1},$$

where $\boldsymbol{\Gamma}(\boldsymbol{\theta}) := \mathrm{diag}\{\boldsymbol{\Gamma}(\boldsymbol{\theta}_1), \dots, \boldsymbol{\Gamma}(\boldsymbol{\theta}_r)\}$ *and* $\mathbf{K} := \mathrm{diag}(\mathbf{K}_n, \dots, \mathbf{K}_n)$.

Next, we consider the ACRLB under the condition of Gaussian colored noise with continuous spectrum. In this case, the following theorem can be established as a generalization of Theorem 3.3. See Section 3.5 for a proof.

Theorem 3.4 (Asymptotic CRLB: Gaussian Colored Noise). *Let the conditions of Theorem 3.3 be satisfied except that* $\{\epsilon_t\}$ *is a stationary real or complex Gaussian process with mean zero and SDF* $f_\epsilon(\omega)$, *where* $f_\epsilon(\omega)$ *is a continuous function with* $f_0 := \min_\omega f_\epsilon(\omega) > 0$. *Then, as* $n \to \infty$, *the CRLB for estimating* $\boldsymbol{\theta}$ *defined in Theorem 3.3 can be expressed as*

$$\mathrm{CRLB}(\boldsymbol{\theta}) = \mathbf{K}^{-1} \{ \boldsymbol{\Gamma}(\boldsymbol{\theta}) + o(1) \} \mathbf{K}^{-1}, \tag{3.3.14}$$

where the matrix $\boldsymbol{\Gamma}(\boldsymbol{\theta})$ *is the same as in Theorem 3.3 except that* γ_k *in (3.3.13) is defined with* $f_\epsilon(\omega_k)$ *in place of* σ^2, *i.e.,* $\gamma_k := \frac{1}{2} C_k^2 / f_\epsilon(\omega_k)$ *in the real case and* $\gamma_k := C_k^2 / f_\epsilon(\omega_k)$ *in the complex case.*

Theorem 3.4 reveals that the ACRLB depends on the noise spectrum solely through its values at the signal frequencies, so the noise spectrum elsewhere has zero contribution to the asymptotic performance limit. An intuitive explanation for this interesting result is as follows: Because the sinusoids are extremely well localized in the frequency domain, any reasonable frequency estimator, let alone

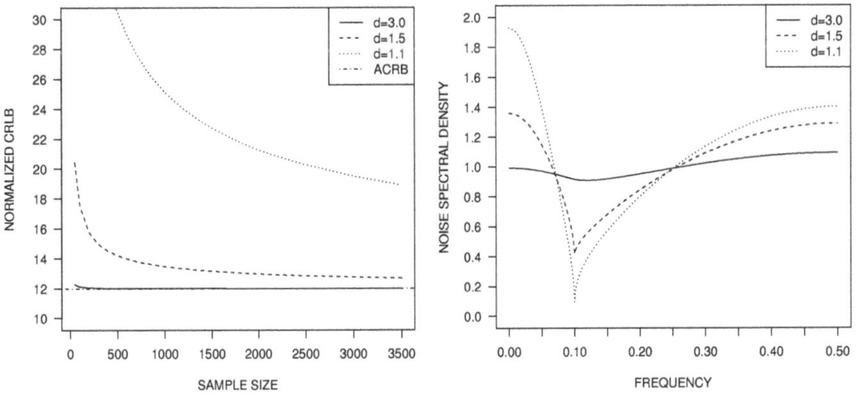

(a) (b)

Figure 3.2. (a) Plot of $n^3\text{CRLB}(\omega_1)$ as a function of n in the case of a single real sinusoid in Gaussian colored noise with SDF $f_\epsilon(\omega) = 1/|\sum_{j=0}^{\infty}\varphi_j \exp(-ij\omega)|^2$, where $\varphi_j := (j+1)^{-d}\cos(j\omega_1)$, $\omega_1 = 2\pi \times 0.1$, and $\gamma_1 = 1$. Solid line, $d = 3$; dashed line, $d = 1.5$; dotted line, $d = 1.1$. The dash-dotted line depicts the asymptotic value of the normalized CRLB. (b) Plot of $f_\epsilon(\omega)$ as a function of the normalized frequency $f := \omega/(2\pi)$.

the optimal ones, must have the capability of suppressing the noise outside a small neighborhood of the signal frequencies. Furthermore, as the sample size grows, the small neighborhood must shrink toward zero in order to suppress the noise most effectively. Therefore, in the limit, only the values of the noise spectrum at the signal frequencies have impact on the estimation accuracy.

It is worth pointing out that although the continuity of the noise spectrum suffices for the validity of Theorem 3.4, the degree of smoothness of the noise spectrum plays an important role in determining the accuracy of the ACRLB as an approximation to the CRLB for finite sample sizes. To demonstrate this point, consider the example shown in Figure 3.2.

In this example, the noise is a linear process of the form

$$\epsilon_t = \sum_{j=-\infty}^{\infty} \psi_j \zeta_{t-j}, \tag{3.3.15}$$

where $\sum|\psi_j| < \infty$ and $\{\zeta_t\} \sim \text{IID}(0, \sigma^2)$. By Proposition 2.4,

$$f_\epsilon(\omega) := \sigma^2 |\Psi(\omega)|^2,$$

where $\Psi(\omega) := \sum \psi_j \exp(-ij\omega)$. Take $\{\zeta_t\} \sim \text{GWN}(0,1)$ and $\Psi(\omega) = 1/\Phi(\omega)$, where $\Phi(\omega) := \sum_{j=0}^{\infty}\varphi_j \exp(-ij\omega)$ with $\varphi_j := (j+1)^{-d}\cos(0.2\pi j)$. In other words, the noise is a special Gaussian AR(∞) process. The smoothness of the noise spectrum is controlled by the parameter d which takes values 1.1, 1.5, and 3. As shown in Figure 3.2(b), the noise spectrum is very smooth with $d = 3$ and becomes less smooth at the signal frequency $\omega_1 = 2\pi \times 0.1$ as d decreases. The exact CRLB shown in Figure 3.2(a) is computed directly from $\{\varphi_j\}$ by using the algorithm in Proposition 2.9 and an AR(6000) truncation of the AR(∞) noise spectrum. As we can see, the ACRLB, which equals 12, is an excellent approximation to the exact CRLB when $d = 3$, even for small sample sizes. But, as d decreases, especially when $d = 1.1$, it becomes poorer even for very large sample sizes. A similar, but less dramatic, effect is also observed (not shown) when the noise is an AR(2) process with a sharp spectral peak located at the signal frequency.

As an important special case, the ACRLB for estimating the amplitude and phase parameters can be easily derived from Theorem 3.3 and Theorem 3.4. The following assertion is stated without proof.

Corollary 3.2 (Asymptotic CRLB for the Amplitude and Phase Parameters). *Let the conditions of Theorem 3.3 or 3.4 be satisfied. For the Cartesian RSM and CSM, let* $\boldsymbol{\theta}_k := [A_k, B_k]^T$. *For the polar RSM and CSM, let* $\boldsymbol{\theta}_k := [C_k, \phi_k]^T$. *Define*

$$\boldsymbol{\Gamma}(\boldsymbol{\theta}_k) := \begin{cases} \boldsymbol{\Gamma}_{\mathrm{C}}(\boldsymbol{\theta}_k) & \text{for Cartesian RSM (2.1.1)}, \\ \boldsymbol{\Gamma}_{\mathrm{P}}(\boldsymbol{\theta}_k) & \text{for polar RSM (2.1.2)}, \\ \frac{1}{4}\boldsymbol{\Gamma}_{\mathrm{C}}(\boldsymbol{\theta}_k) & \text{for Cartesian CSM (2.1.5)}, \\ \frac{1}{4}\boldsymbol{\Gamma}_{\mathrm{P}}(\boldsymbol{\theta}_k) & \text{for polar CSM (2.1.7)}, \end{cases} \tag{3.3.16}$$

where

$$\begin{aligned} \boldsymbol{\Gamma}_{\mathrm{C}}(\boldsymbol{\theta}_k) &:= \mathrm{diag}\{2 f_\epsilon(\omega_k), 2 f_\epsilon(\omega_k)\}, \\ \boldsymbol{\Gamma}_{\mathrm{P}}(\boldsymbol{\theta}_k) &:= \mathrm{diag}\{2 f_\epsilon(\omega_k), 2 f_\epsilon(\omega_k)/C_k^2\}. \end{aligned}$$

Then, as $n \to \infty$, *the CRLB for estimating* $\boldsymbol{\theta} := [\boldsymbol{\theta}_1^T, \ldots, \boldsymbol{\theta}_r^T]^T$ ($r := q$ *for the RSM and* $r := p$ *for the CSM) can be expressed as*

$$\mathrm{CRLB}(\boldsymbol{\theta}) = n^{-1}\{\boldsymbol{\Gamma}(\boldsymbol{\theta}) + \mathcal{O}(1)\}, \tag{3.3.17}$$

where $\boldsymbol{\Gamma}(\boldsymbol{\theta}) := \mathrm{diag}\{\boldsymbol{\Gamma}(\boldsymbol{\theta}_1), \ldots, \boldsymbol{\Gamma}(\boldsymbol{\theta}_r)\}$.

Remark 3.5 By comparing Corollary 3.2 with Theorem 3.3 and Theorem 3.4, we can see that the same remark we made at the end of Section 3.2 regarding the CRLB for estimating the amplitude parameters (A_k, B_k) in the complex case applies to the ACRLB for estimating (A_k, B_k) in the real case.

So far, the signal frequencies are assumed to satisfy (3.3.4). It is under this condition that the sinusoids become decoupled in $\boldsymbol{\Gamma}(\boldsymbol{\theta})$ which takes a block-diagonal form. In the remainder of this section, let us investigate a case where the condition (3.3.4) is not satisfied.

For simplicity, consider the case of $p = 2$ in (2.1.5). Furthermore, let ω_1 be a constant and let ω_2 satisfy

$$\omega_2 = \omega_1 + \Delta, \quad n\Delta \to a \quad \text{as } n \to \infty, \tag{3.3.18}$$

for some constant $a > 0$. This is an example where the signal frequencies are closely spaced relative to the sample size such that

$$\omega_2 - \omega_1 \approx a/n.$$

The following theorem shows that the sinusoids are no longer decoupled in the ACRLB under the condition (3.3.18). A proof is given in Section 3.5.

Theorem 3.5 (Asymptotic CRLB: CSM with Closely Spaced Frequencies). *Let* $\mathbf{y}$ *be given by (3.2.1) with* $\mathbf{x}$ *satisfying the CSM (2.1.5) for* $p = 2$*. Assume that* ω_1 *is a constant and* ω_2 *satisfies (3.3.18). Assume further that* $\{\epsilon_t\}$ *satisfies the conditions of Theorem 3.3 or 3.4. Then, the CRLB for estimating* $\boldsymbol{\theta} := [A_1, B_1, \omega_1, A_2, B_2, \omega_2]^T$ *can be expressed as (3.3.14) except that* $\boldsymbol{\Gamma}(\boldsymbol{\theta})$ *takes the form*

$$\boldsymbol{\Gamma}(\boldsymbol{\theta}) := \tfrac{1}{2} f_\epsilon(\omega_1)\mathbf{W}^{-1},$$

where

$$\mathbf{W} := \begin{bmatrix} \mathbf{W}_1 & \mathbf{W}_{12} \\ \mathbf{W}_{12}^T & \mathbf{W}_2 \end{bmatrix}.$$

In this expression, $\mathbf{W}_1$ *and* $\mathbf{W}_2$ *are defined by (3.3.7) and* $\mathbf{W}_{12}$ *is defined by*

$$\mathbf{W}_{12} := \begin{bmatrix} \Re\{d_0(a)\} & \Im\{d_0(a)\} & \Re\{\beta_2 d_1(a)\} \\ -\Im\{d_0(a)\} & \Re\{d_0(a)\} & -\Im\{\beta_2 \dot{d}_0(a)\} \\ -\Re\{\beta_1^* \dot{d}_0(a)\} & -\Im\{\beta_1^* \dot{d}_0(a)\} & -\Re\{\beta_1^* \beta_2 \ddot{d}_0(a)\} \end{bmatrix}, \tag{3.3.19}$$

where $d_0(a) := \operatorname{sinc}(a/2)\exp(ia/2)$ *and* $\operatorname{sinc}(x) := \sin(x)/x$*. The CRLB for estimating* $\boldsymbol{\theta} := [C_1, \phi_1, \omega_1, C_2, \phi_2, \omega_2]^T$ *can also be expressed as (3.3.14) except that*

$$\boldsymbol{\Gamma}(\boldsymbol{\theta}) := \tfrac{1}{2} f_\epsilon(\omega_1)\mathbf{J}\mathbf{W}^{-1}\mathbf{J}^T,$$

where $\mathbf{J} := \operatorname{diag}(\mathbf{J}_1, \mathbf{J}_2)$ *is the Jacobian matrix with* $\mathbf{J}_k$ *given by (3.2.8).*

Remark 3.6 A similar result can be obtained for the general case of $p > 2$ where $\{\omega_k\}$ consists of well-separated clusters such that the frequencies between clusters satisfy (3.3.4) and the frequencies within clusters satisfy (3.3.18). In the simplest case of a single cluster where all frequencies are close to ω_1 with $\Delta_{kk'} := |\omega_{k'} - \omega_k|$ satisfying $n\Delta_{kk'} \to a_{kk'} > 0$ as $n \to \infty$ for $k \neq k'$, then $\boldsymbol{\Gamma}(\boldsymbol{\theta})$ in (3.3.14) can be expressed as $\frac{1}{2} f_\epsilon(\omega_1)\mathbf{W}^{-1}$ or $\frac{1}{2} f_\epsilon(\omega_1)\mathbf{J}\mathbf{W}^{-1}\mathbf{J}$, where $\mathbf{W} := [\mathbf{W}_{kk'}]$ $(k, k' = 1, \ldots, p)$. For $k < k'$, $\mathbf{W}_{kk'} = \mathbf{W}_{k'k}^T$ takes the same form as $\mathbf{W}_{12}$ in (3.3.19) except that β_1, β_2,

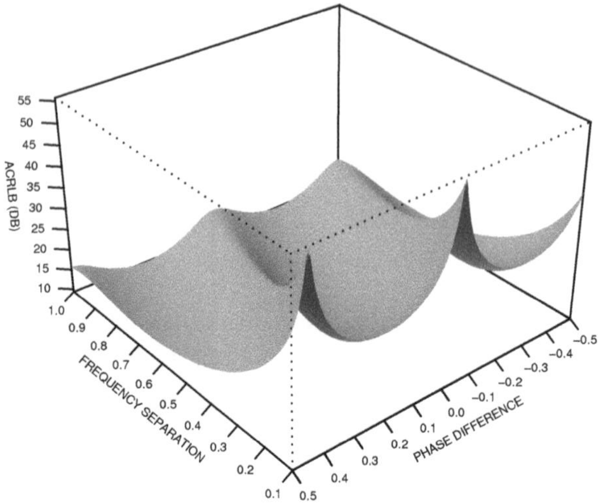

Figure 3.3. Plot of the ACRLB in Theorem 3.5 for estimating ω_1 as a function of the frequency separation parameter $a/(2\pi)$ and the phase difference parameter $\phi/(2\pi)$ in the case of two complex sinusoids in Gaussian white noise with $\omega_1 = 2\pi \times 0.4$, $\omega_2 = \omega_1 + a/n$, $C_1 = C_2 = 1$, $\phi_1 = 0$, $\phi_2 = \phi$, and $\sigma^2 = 1$ $(\gamma_1 = \gamma_2 = 1)$.

and a are replaced by β_k, $\beta_{k'}$, and $a_{kk'}$. In the more general case of multiple clusters, $\mathbf{\Gamma}(\boldsymbol{\theta})$ is a block diagonal matrix where each block corresponds to a cluster and takes the same form as in the case of a single cluster except that the noise spectrum $f_\epsilon(\omega)$ is evaluated at the smallest (or any) frequency in the cluster. In the case of Gaussian white noise, $\mathbf{\Gamma}(\boldsymbol{\theta}) = \frac{1}{2}\sigma^2\mathbf{W}^{-1}$ for the Cartesian system and $\mathbf{\Gamma}(\boldsymbol{\theta}) = \frac{1}{2}\sigma^2\mathbf{J}\mathbf{W}^{-1}\mathbf{J}$ for the polar system.

The phase of the sinusoids plays an important role in the ACRLB when the frequencies are closely spaced. As indicated in Example 3.3, the ACRLB depends on the phase difference $\phi := \phi_2 - \phi_1$ in the case of $p = 2$; by Theorem 3.5, it also depends on the frequency separation parameter a. Figure 3.3 depicts the (3,3) entry of $\mathbf{\Gamma}(\boldsymbol{\theta})$ given by Theorem 3.5 as a function of a and ϕ for estimating ω_1 in the presence of Gaussian white noise.

Figure 3.3 shows that the ACRLB varies dramatically with a and ϕ. In general, the ACRLB decreases as a increases but not monotonically. More importantly, the ACRLB takes much larger values for some combinations of ϕ and a, especially when a is small, indicating the existence of least favorable phases that depend on the frequency separation. The ACRLB shown in Figure 3.3 ranges from 9.4 dB to 55.1 dB, whereas the ACRLB in the absence of the second sinusoid is equal to 7.8 dB. This comparison highlights the increased potential difficulty in estimating closely spaced frequencies.

As a corollary to Theorem 3.5, the following result can be obtained for the estimation of the amplitude and phase parameters when the signal frequencies are known and closely spaced. A proof can be found in Section 3.5.

Corollary 3.3 (Asymptotic CRLB for the Amplitude and Phase Parameters: CSM with Closely Spaced Frequencies). *Under the conditions of Theorem 3.5, let*

$$\mathbf{A} := \begin{bmatrix} \Re\{d_0(a)\} & \Im\{d_0(a)\} \\ -\Im\{d_0(a)\} & \Re\{d_0(a)\} \end{bmatrix}.$$

Then, the CRLB for estimating $\boldsymbol{\theta} := [A_1, B_1, A_2, B_2]^T$ *takes the form (3.3.17) with*

$$\boldsymbol{\Gamma}(\boldsymbol{\theta}) := \tfrac{1}{2} f_\epsilon(\omega_1)\boldsymbol{\Lambda}_0,$$

where

$$\boldsymbol{\Lambda}_0 := \frac{1}{1-|d_0(a)|^2}\begin{bmatrix} \mathbf{I} & -\mathbf{A} \\ -\mathbf{A}^T & \mathbf{I} \end{bmatrix}.$$

The CRLB for estimating $\boldsymbol{\theta} := [C_1, \phi_1, C_2, \phi_2]^T$ *can also be expressed as (3.3.17) with*

$$\boldsymbol{\Gamma}(\boldsymbol{\theta}) := \tfrac{1}{2} f_\epsilon(\omega_1)\mathbf{J}\boldsymbol{\Lambda}_0\mathbf{J}^T,$$

where $\mathbf{J} := \mathrm{diag}(\mathbf{J}_1, \mathbf{J}_2)$ *is the Jacobian matrix with* $\mathbf{J}_k$ *defined by (3.2.8) without the last row and the last column.*

Under the assumption of Gaussian white noise, the following result, more general than Corollary 3.3, can be obtained directly from Remark 3.4.

Corollary 3.4 (Asymptotic CRLB for the Amplitude Parameters: CSM and Gaussian White Noise). *Let* $\mathbf{y}$ *be given by (3.2.1) with* $\mathbf{x}$ *taking the form (2.1.5) and with* $\boldsymbol{\epsilon} \sim \mathrm{N}_c(\mathbf{0}, \sigma^2\mathbf{I})$. *Assume that there exists a nonsingular matrix* $\mathbf{C}$ *such that* $n^{-1}\mathbf{F}^H\mathbf{F} \to \mathbf{C}$ *as* $n \to \infty$. *Then, the CRLB for estimating* $\boldsymbol{\theta} := [\Re(\boldsymbol{\beta})^T, -\Im(\boldsymbol{\beta})^T]^T$ *can be expressed as*

$$\mathrm{CRLB}(\boldsymbol{\theta}) = n^{-1}\{\tfrac{1}{2}\sigma^2\boldsymbol{\Lambda} + \mathcal{O}(1)\},$$

where

$$\boldsymbol{\Lambda} := \begin{bmatrix} \Re(\mathbf{C}) & \Im(\mathbf{C}) \\ -\Im(\mathbf{C}) & \Re(\mathbf{C}) \end{bmatrix}^{-1} = \begin{bmatrix} \Re(\mathbf{C}^{-1}) & \Im(\mathbf{C}^{-1}) \\ -\Im(\mathbf{C}^{-1}) & \Re(\mathbf{C}^{-1}) \end{bmatrix}.$$

Remark 3.7 The assumption that $n^{-1}\mathbf{F}^H\mathbf{F} \to \mathbf{C}$ imposes a constraint on the signal frequencies. It is easy to see that if the signal frequencies take the form $2\pi j/n$ for some integers j, then $\mathbf{F}^H\mathbf{F} = n\mathbf{I}$, so Corollary 3.4 holds with $\mathbf{C} = \mathbf{I}$. It also holds with $\mathbf{C} = \mathbf{I}$ if the frequencies satisfy (3.3.4), because $n^{-1}\mathbf{F}^H\mathbf{F} \to \mathbf{I}$.

3.4 CRLB for Sinusoids in NonGaussian White Noise

Our focus so far has been on the Gaussian noise. Let us now derive the CRLB under the condition of nonGaussian white noise.

First, consider a general signal-plus-noise model in which the signal depends on an unknown vector of parameters and the estimation of the signal parameter is of primary interest. The following proposition provides an expression for Fisher's information matrix with respect to the signal parameter. A proof of this result is given in Section 3.5.

Proposition 3.2 (Fisher's Information Matrix: Real Case). *Consider* $y_t := x_t + \epsilon_t$ $(t = 1, \ldots, n)$, *where* $\{x_t\}$ *is a sequence of real deterministic differentiable functions of a real-valued parameter* $\boldsymbol{\theta}$ *and* $\{\epsilon_t\}$ *is a sequence of i.i.d. real random variables with* PDF $p(x)$. *Assume further that* $p(x)$ *is differentiable everywhere except at a finite number of points such that*

$$0 < \lambda_0 := \int_{p(x)>0} \frac{\{\dot{p}(x)\}^2}{p(x)}\, dx < \infty. \tag{3.4.1}$$

Then, Fisher's information matrix of $\mathbf{y} := [y_1, \ldots, y_n]^T$ *with respect to* $\boldsymbol{\theta}$ *can be expressed as*

$$\mathbf{I}(\boldsymbol{\theta}) = \lambda_0 \sum_{t=1}^{n} (\nabla x_t)(\nabla x_t)^T, \tag{3.4.2}$$

where $\nabla := \partial / \partial \boldsymbol{\theta}$ *denotes the gradient operator with respect to* $\boldsymbol{\theta}$.

If $\{\epsilon_t\}$ is Gaussian white noise with mean zero and variance σ^2, then $p(x) = (\sqrt{2\pi}\,\sigma)^{-1} \exp(-\frac{1}{2} x^2/\sigma^2)$, in which case, we obtain

$$\lambda_0 = \frac{1}{\sigma^4} \int x^2 p(x)\, dx = \frac{1}{\sigma^4} E(\epsilon_t^2) = \frac{1}{\sigma^2}.$$

Thus, under the Gaussian assumption, Fisher's information matrix becomes

$$\mathbf{I}_G(\boldsymbol{\theta}) := \frac{1}{\sigma^2} \sum_{t=1}^{n} (\nabla x_t)(\nabla x_t)^T. \tag{3.4.3}$$

For nonGaussian white noise, we can rewrite (3.4.2) as

$$\mathbf{I}(\boldsymbol{\theta}) = \lambda\, \mathbf{I}_G(\boldsymbol{\theta}), \tag{3.4.4}$$

where

$$\lambda := \sigma^2 \lambda_0 = \sigma^2 \int_{p(x)>0} \frac{\{\dot{p}(x)\}^2}{p(x)}\, dx. \tag{3.4.5}$$

Moreover, if the noise has mean zero and variance σ^2, let $p_0(x)$ denote the PDF of the unit-variance random variable ϵ_t/σ. Then, it is easy to verify that

$$p(x) = (1/\sigma)p_0(x/\sigma), \quad \dot{p}(x) = (1/\sigma^2)p_0(x/\sigma).$$

With this notation, the multiplier λ in (3.4.5) can be rewritten as

$$\lambda = \int_{p_0(x)>0} \frac{\{\dot{p}_0(x)\}^2}{p_0(x)}\, dx. \tag{3.4.6}$$

This expression shows that λ does not depend on the variance of the noise.

According to (3.4.4), different noise distributions are manifested in Fisher's information matrix as different values of the multiplier λ to Fisher's information matrix under the Gaussian assumption. In the Gaussian case, we have $\lambda = 1$. In general, we can interpret λ as a measure of the intrinsic easiness in estimating the unknown parameter under the nonGaussian noise relative to that under the Gaussian noise. For example, consider the Laplace noise with

$$p(x) = (2c)^{-1}\exp(-|x|/c),$$

where $c > 0$ is the scale parameter. In this case, we have

$$\sigma^2 = 2c^2, \quad \lambda = 2.$$

Therefore, under the Laplace assumption, the intrinsic easiness, as measured by Fisher's information, is twice as much as that under the Gaussian assumption.

Three more examples are given in the following to illustrate the range of λ for various noise distributions.

Example 3.9 (Generalized Gaussian White Noise). Let $\{\epsilon_t\}$ be generalized Gaussian white noise with

$$p(x) = \frac{r}{2c\Gamma(1/r)} \exp(-|x/c|^r), \tag{3.4.7}$$

where $r > 0$ is the shape parameter and $c > 0$ is the scale parameter. Note that $r = 2$ corresponds to the Gaussian noise and $r = 1$ corresponds to the Laplace noise. As compared with the Gaussian case, the tails of the generalized Gaussian distribution become lighter as r increases from $r = 2$ and heavier as r decreases from $r = 2$. It can be shown that $\sigma^2 = c^2\Gamma(3/r)/\Gamma(1/r)$ and, for $r > 1/2$,

$$\lambda = \frac{r^2\Gamma(2-1/r)\Gamma(3/r)}{\{\Gamma(1/r)\}^2}. \tag{3.4.8}$$

The top panel of Figure 3.4 depicts λ as a function of r. Observe that λ is minimized at $r = 2$ and that $\lambda \to \infty$ as $r \to 1/2$. ◇

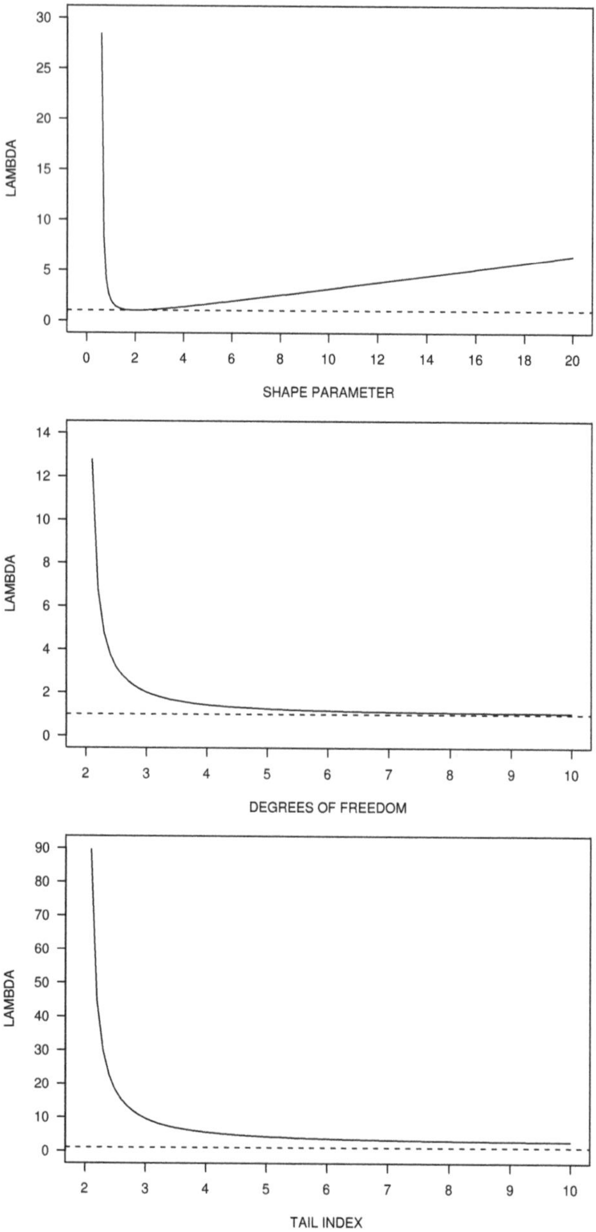

Figure 3.4. Top, plot of λ in (3.4.8) as a function of r for the generalized Gaussian distribution. Middle, plot of λ in (3.4.9) as a function of ν for Student's t-distribution. Bottom, plot of λ in (3.4.10) as a function of α for the Pareto distribution. Dotted horizontal line depicts $\lambda = 1$, the Gaussian case.

Example 3.10 (White Noise with Student's t-Distribution). Assume that the ϵ_t are i.i.d. random variables with PDF

$$p(x) = \frac{\Gamma((\nu+1)/2)}{c\Gamma(\nu/2)\sqrt{\pi\nu}}(1+(x/c)^2/\nu)^{-(\nu+1)/2},$$

which is known as Student's t-distribution with $\nu > 0$ degrees of freedom, where $c > 0$ is the scale parameter. The t-distribution approaches a Gaussian distribution as $\nu \to \infty$, but for finite ν, it has heavier tails than the Gaussian distribution with the same variance. It is easy to verify that for $\nu > 2$, $\sigma^2 = c^2\nu/(\nu-2)$ and

$$\lambda = \frac{\nu(\nu+1)}{(\nu-2)(\nu+3)}. \tag{3.4.9}$$

The middle panel of Figure 3.4 depicts λ as a function of ν. It can be shown that $\lambda > 1$ for all $\nu > 2$, $\lambda \to 1$ as $\nu \to \infty$, and $\lambda \to \infty$ as $\nu \to 2$. ◇

Example 3.11 (Pareto White Noise). Assume that the ϵ_t are i.i.d. with PDF

$$p(x) = \frac{\alpha}{2c}(1+|x|/c)^{-(\alpha+1)},$$

which is known as the Pareto distribution with tail index $\alpha > 0$, where $c > 0$ is the scale parameter. A Pareto distribution has heavier tails than the Gaussian distribution with the same variance. For $\alpha > 2$, $\sigma^2 = 2c^2/((\alpha-2)(\alpha-1)) < \infty$ and

$$\lambda = \frac{2\alpha(\alpha+1)^2}{(\alpha-2)(\alpha-1)(\alpha+2)}. \tag{3.4.10}$$

The bottom panel of Figure 3.4 shows the graph of λ as a function of α. Note that $\lambda > 2$ for all $\alpha > 2$, $\lambda \to 2$ as $\alpha \to \infty$, and $\lambda \to \infty$ as $\alpha \to 2$. ◇

A result similar to Proposition 3.2 can be obtained in the complex case as follows. See Section 3.5 for a proof.

Proposition 3.3 (Fisher's Information Matrix: Complex Case). *Let $y_t := x_t + \epsilon_t$ ($t = 1, \ldots, n$), where $\{x_t\}$ is a sequence of complex-valued deterministic differentiable functions of a real-valued parameter $\boldsymbol{\theta}$ and $\{\epsilon_t\}$ is a sequence of complex-valued random variables whose real and imaginary parts are mutually independent i.i.d. sequences with PDF $p(x)$. Assume further that $p(x)$ is differentiable everywhere except at a finite number of points and satisfies (3.4.1). Then, Fisher's information matrix of $\mathbf{y} := [y_1, \ldots, y_n]^T$ with respect to $\boldsymbol{\theta}$ takes the form*

$$\mathbf{I}(\boldsymbol{\theta}) = \lambda_0 \sum_{t=1}^{n} \Re\{(\nabla x_t)(\nabla x_t)^H\}, \tag{3.4.11}$$

where ∇ denotes the gradient operator with respect to $\boldsymbol{\theta}$.

For the complex Gaussian white noise with mean zero and variance σ^2, i.e., $\{\epsilon_t\} \sim \text{GWN}_c(0,\sigma^2)$, we have $p(x) = (\sqrt{\pi}\sigma)^{-1}\exp(-x^2/\sigma^2)$ and

$$\lambda_0 = \frac{4}{\sigma^4}\int x^2 p(x)\,dx = \frac{4}{\sigma^4}E\{\Re(\epsilon_t)^2\} = \frac{2}{\sigma^2}.$$

Hence the corresponding Fisher's information matrix can be written as

$$\mathbf{I}_G(\boldsymbol{\theta}) := \frac{2}{\sigma^2}\sum_{t=1}^{n}\Re\{(\nabla x_t)(\nabla x_t)^H\}. \tag{3.4.12}$$

For complex nonGaussian white noise, the expression (3.4.4) remains valid with $\mathbf{I}_G(\boldsymbol{\theta})$ given by (3.4.12) if λ is defined as

$$\lambda := \frac{1}{2}\sigma^2\lambda_0 = \frac{1}{2}\sigma^2\int_{p(x)>0}\frac{\{\dot{p}(x)\}^2}{p(x)}\,dx. \tag{3.4.13}$$

If the noise has mean zero and variance σ^2, then λ also takes the form (3.4.6) with $p_0(x)$ denoting the unit-variance PDF such that $p(x) = (\sqrt{2}/\sigma)p_0(\sqrt{2}x/\sigma)$.

Example 3.12 (Complex Generalized Gaussian and Laplace White Noise). The random process $\{\epsilon_t\}$ is said to be complex generalized Gaussian white noise if its real and imaginary parts are mutually independent i.i.d. sequences with PDF $p(x)$ given by (3.4.7) with $c^2 = \frac{1}{2}\sigma^2\Gamma(1/r)/\Gamma(3/r)$ and $r > 0$. In this case, it is easy to verify that λ takes the form (3.4.8) for $r > 1/2$. The special case of $r = 1$ is called the complex Laplace white noise, where $c = \sigma/2$ and $\lambda = 2$. ◇

As can be seen in all these examples, the factor λ in (3.4.4) reaches its minimum value when the noise is Gaussian. This is not a coincidence. Indeed, the following lemma confirms that $\lambda = 1$ is the minimum value of λ over a fairly large family of distributions and is attained if and only if the noise is Gaussian. A proof of this result can be found in Section 3.5.

Proposition 3.4 (Minimization of Fisher's Information). *Assume that the conditions of Proposition 3.2 or 3.3 are satisfied. Assume further that the ϵ_t have mean zero and variance σ^2 and that the support of the PDF $p(x)$ is equal to the entire real line $\mathbb{R}$, i.e., $p(x) > 0$ for all $x \in \mathbb{R}$. Then, Fisher's information matrix of $\mathbf{y}$ with respect to $\boldsymbol{\theta}$ can be expressed as (3.4.4) with $\lambda \geq 1$, where the equality holds if and only if the noise has a Gaussian distribution.*

Remark 3.8 The support of $p(x)$ being the entire real line is a necessary condition for the Gaussian distribution to be the only one such that $\lambda = 1$. A counter example is $p(x) = \exp\{-(x+1)\}$ for $x > -1$ and $p(x) = 0$ otherwise. For this distribution, the support of $p(x)$ is $[-1,\infty)$ and $\lambda = 1$.

For symmetric and unimodal distributions, the factor λ has a tighter lower bound that depends on the noise PDF at zero. This result is stated in the next proposition. A proof can be found in Section 3.5.

Proposition 3.5 (Lower Bound of Fisher's Information for Symmetric Unimodal Noise Distributions). *Let the conditions of Proposition 3.4 be satisfied. Let $p_0(x)$ denote the unit-variance PDF associated with $p(x)$ so that λ takes the form (3.4.6) in both real and complex cases. Assume further that $p_0(x)$ is symmetric about $x = 0$ and monotone decreasing in $|x|$. Then, $\lambda \ge 4\{p_0(0)\}^2$, where the equality holds if and only if the noise has a Laplace distribution.*

Remark 3.9 In the real case, $p_0(0) = \sigma p(0)$, so the inequality in Proposition 3.5 can be written as $\lambda \ge 4\sigma^2\{p(0)\}^2$. In the complex case, $p_0(0) = (\sigma/\sqrt{2})p(0)$ and the inequality becomes $\lambda \ge 2\sigma^2\{p(0)\}^2$.

According to Proposition 3.5, the inequality $\lambda \ge 2$ holds for all symmetric unimodal distributions that satisfy $p_0(0) \ge 1/\sqrt{2}$. Among these distributions, the Laplace distribution, with $p_0(0) = 1/\sqrt{2}$, is the only one that attains the lower bound $\lambda = 2$. The constraint $p_0(0) \ge 1/\sqrt{2}$ simply imposes a minimum on the probability density of the noise at zero.

Equipped with these results, the following theorem can be established for the CRLB under nonGaussian white noise. A proof is given in Section 3.5.

Theorem 3.6 (CRLB: NonGaussian White Noise). *Let $\mathbf{y}$ be given by (3.2.1) with $\mathbf{x}$ satisfying (2.1.1), (2.1.2), (2.1.5), or (2.1.7) and with $\boldsymbol{\epsilon}$ satisfying the conditions of Proposition 3.2 or 3.3. Let $\boldsymbol{\theta}$ denote the vector of parameters in $\mathbf{x}$. Then, the CRLB for estimating $\boldsymbol{\theta}$ on the basis of $\mathbf{y}$ can be expressed as*

$$\mathrm{CRLB}(\boldsymbol{\theta}) = \lambda^{-1}\mathrm{CRLB}_G(\boldsymbol{\theta}), \tag{3.4.14}$$

where λ is given by (3.4.5) or (3.4.13), and $\mathrm{CRLB}_G(\boldsymbol{\theta})$ is the CRLB under the assumption that $\boldsymbol{\epsilon} \sim \mathrm{N}(\mathbf{0}, \sigma^2\mathbf{I})$ or $\boldsymbol{\epsilon} \sim \mathrm{N}_c(\mathbf{0}, \sigma^2\mathbf{I})$. In the real case,

$$\mathrm{CRLB}_G(\boldsymbol{\theta}) := \sigma^2(\mathbf{X}^T\mathbf{X})^{-1}, \tag{3.4.15}$$

where $\boldsymbol{\theta}$ and $\mathbf{X}$ are defined in Theorem 3.1. In the complex case,

$$\mathrm{CRLB}_G(\boldsymbol{\theta}) := \tfrac{1}{2}\sigma^2\{\Re(\mathbf{X}^H\mathbf{X})\}^{-1}, \tag{3.4.16}$$

where $\boldsymbol{\theta}$ and $\mathbf{X}$ are defined in Theorem 3.2.

A similar result for the ACRLB can be obtained under the condition of nonGaussian white noise. Given Theorems 3.3 and 3.6, the following theorem is stated without proof.

Theorem 3.7 (Asymptotic CRLB: NonGaussian White Noise). *Let the conditions of Theorem 3.3 be satisfied except that the noise $\boldsymbol{\epsilon}$ satisfies the assumptions in Theorem 3.6. Then, the asymptotic expressions for the CRLB in Theorem 3.3 remain true except that $\boldsymbol{\Gamma}(\boldsymbol{\theta}_k)$ takes the form*

$$\boldsymbol{\Gamma}(\boldsymbol{\theta}_k) := \lambda^{-1}\boldsymbol{\Gamma}_G(\boldsymbol{\theta}_k), \tag{3.4.17}$$

where λ is given by (3.4.5) or (3.4.13), and $\boldsymbol{\Gamma}_G(\boldsymbol{\theta}_k)$ is given by (3.3.13).

Remark 3.10 With $\boldsymbol{\theta}$ properly redefined, the expressions in (3.4.14) and (3.4.17) remain valid for estimating the amplitude and phase parameters when the signal frequencies are known.

As shown by Theorems 3.6 and 3.7, the only difference the nonGaussian noise makes on the CRLB is the introduction of a multiplier λ^{-1} to the Gaussian CRLB. According to Proposition 3.4, the CRLB is maximized by the Gaussian distribution among all noise distributions that have the same or smaller variance. In this sense, the Gaussian distribution is the *least favorable distribution* for the noise in the estimation of sinusoidal parameters. Similarly, Proposition 3.5 implies that the Laplace distribution is the least favorable distribution for the noise among all finite-variance symmetric unimodal distributions satisfying $p_0(0) \geq 1/\sqrt{2}$.

3.5 Proof of Theorems

This section contains the proof of Propositions 3.1–3.5 and the proof of Theorems 3.1, 3.2, 3.4–3.6, and Corollary 3.3.

Proof of Proposition 3.1. For any constant vectors $\mathbf{a}$ and $\mathbf{b} \neq \mathbf{0}$, define

$$X := \mathbf{a}^T\{\hat{\boldsymbol{\eta}} - \boldsymbol{\eta}(\boldsymbol{\vartheta})\}, \quad Y := \mathbf{b}^T \nabla \ell(\boldsymbol{\vartheta}|\mathbf{Y}).$$

From (3.1.1), (3.1.2), and the unbiasedness of $\hat{\boldsymbol{\eta}}$, we obtain $E(X) = E(Y) = 0$ and

$$E(XY) = \mathbf{a}^T \mathbf{J}(\boldsymbol{\vartheta})\,\mathbf{b}.$$

Applying the Cauchy-Schwarz inequality together with (3.1.3) gives

$$E(X^2) \geq \frac{\{\mathbf{a}^T\mathbf{J}(\boldsymbol{\vartheta})\,\mathbf{b}\}^2}{E(Y^2)} = \frac{\mathbf{c}^T(\mathbf{d}\mathbf{d}^T)\,\mathbf{c}}{\mathbf{c}^T\mathbf{c}},$$

where $\mathbf{c} := \mathbf{I}(\boldsymbol{\vartheta})^{1/2}\mathbf{b}$ and $\mathbf{d} := \mathbf{I}(\boldsymbol{\vartheta})^{-1/2}\mathbf{J}(\boldsymbol{\vartheta})^T\mathbf{a}$. The equality holds if and only if $Y = \kappa_1 X$ for some constant $\kappa_1 \neq 0$ which may depend on $\boldsymbol{\vartheta}$, $\mathbf{a}$, and $\mathbf{b}$. The right-hand side of the foregoing inequality is maximized by taking

$$\mathbf{c} = \kappa_2 \mathbf{d} \tag{3.5.1}$$

for some $\kappa_2 \neq 0$, and the maximum is equal to $\mathbf{d}^T\mathbf{d}$. Therefore,

$$E(X^2) \geq \mathbf{d}^T\mathbf{d} = \mathbf{a}^T\mathbf{J}(\boldsymbol{\vartheta})\,\mathbf{I}(\boldsymbol{\vartheta})^{-1}\mathbf{J}(\boldsymbol{\vartheta})^T\mathbf{a}.$$

The inequality (3.1.4) follows upon noting that $E(X^2) = \mathrm{Var}(\mathbf{a}^T\hat{\boldsymbol{\eta}})$. Moreover, substituting the definition of $\mathbf{c}$ and $\mathbf{d}$ in (3.5.1) gives $\mathbf{b} = \kappa_2\mathbf{I}(\boldsymbol{\vartheta})^{-1}\mathbf{J}(\boldsymbol{\vartheta})^T\mathbf{a}$. Hence, $Y = \kappa_1 X$ if and only if $\mathbf{a}^T\mathbf{J}(\boldsymbol{\vartheta})\,\mathbf{I}(\boldsymbol{\vartheta})^{-1}\nabla\ell(\boldsymbol{\vartheta}|\mathbf{Y}) = (\kappa_1/\kappa_2)\mathbf{a}^T\{\hat{\boldsymbol{\eta}} - \boldsymbol{\eta}(\boldsymbol{\vartheta})\}$. □

Proof of Theorem 3.1. Under the Gaussian assumption, the log likelihood function can be expressed as

$$\ell(\boldsymbol{\vartheta}|\mathbf{y}) = \text{constant} - \tfrac{1}{2}\log|\mathbf{R}_\epsilon| - \tfrac{1}{2}(\mathbf{y}-\mathbf{x})^T\mathbf{R}_\epsilon^{-1}(\mathbf{y}-\mathbf{x}),$$

where $\mathbf{x} := [x_1,\dots,x_n]^T$ and $\boldsymbol{\vartheta} := [\boldsymbol{\theta}^T, \boldsymbol{\eta}^T]^T$, with $\mathbf{x}$ depending solely on $\boldsymbol{\theta}$ and $\mathbf{R}_\epsilon$ depending solely on $\boldsymbol{\eta}$. In this case, the CRLB for $\boldsymbol{\vartheta}$ is decoupled between $\boldsymbol{\theta}$ and $\boldsymbol{\eta}$ in the sense that $\mathbf{I}(\boldsymbol{\vartheta})$ is a block diagonal matrix with two diagonal blocks: one block corresponds to $\boldsymbol{\theta}$, the other corresponds to $\boldsymbol{\eta}$. Therefore, the CRLB for the primary parameter $\boldsymbol{\theta}$ becomes independent of the auxiliary parameter $\boldsymbol{\eta}$.

To prove the decoupling assertion, let θ be any element of $\boldsymbol{\theta}$ and η be any element of $\boldsymbol{\eta}$. Then, with $\boldsymbol{\epsilon} := \mathbf{y} - \mathbf{x}$, we can write

$$\frac{\partial \ell}{\partial \theta} = \frac{\partial \mathbf{x}^T}{\partial \theta}\mathbf{R}_\epsilon^{-1}\boldsymbol{\epsilon}, \quad \frac{\partial \ell}{\partial \eta} = -\frac{1}{2}|\mathbf{R}_\epsilon|^{-1}\frac{\partial|\mathbf{R}_\epsilon|}{\partial \eta} - \frac{1}{2}\boldsymbol{\epsilon}^T\frac{\partial \mathbf{R}_\epsilon^{-1}}{\partial \eta}\boldsymbol{\epsilon}. \tag{3.5.2}$$

Because the ϵ_t are Gaussian with mean zero, all first and third moments of the ϵ_t are zero. This implies that $E\{(\partial\ell/\partial\theta)(\partial\ell/\partial\eta)\} = 0$, which, in turn, leads to

$$\mathbf{I}(\boldsymbol{\vartheta}) = \begin{bmatrix} \mathbf{I}_\theta(\boldsymbol{\vartheta}) & \mathbf{0} \\ \mathbf{0} & \mathbf{I}_\eta(\boldsymbol{\vartheta}) \end{bmatrix},$$

where

$$\mathbf{I}_\theta(\boldsymbol{\vartheta}) := E\{[\nabla_\theta \ell(\boldsymbol{\vartheta}|\mathbf{y})]\,[\nabla_\theta \ell(\boldsymbol{\vartheta}|\mathbf{y})]^T\},$$
$$\mathbf{I}_\eta(\boldsymbol{\vartheta}) := E\{[\nabla_\eta \ell(\boldsymbol{\vartheta}|\mathbf{y})]\,[\nabla_\eta \ell(\boldsymbol{\vartheta}|\mathbf{y})]^T\},$$

with $\nabla_\theta := \partial/\partial\boldsymbol{\theta}$, and $\nabla_\eta := \partial/\partial\boldsymbol{\eta}$. It remains to show that $\mathbf{I}_\theta(\boldsymbol{\vartheta}) = \mathbf{X}^T\mathbf{R}_\epsilon^{-1}\mathbf{X}$.

Consider the RSM (2.1.1). Let $L_{1k} := \partial\ell/\partial A_k$, $L_{2k} := \partial\ell/\partial B_k$, and $L_{3k} := \partial\ell/\partial\omega_k$. Then, it follows from (3.5.2) that $L_{jk} = \mathbf{x}_{jk}^T\mathbf{R}_\epsilon^{-1}\boldsymbol{\epsilon}$ $(j = 1,2,3)$. Therefore, for any $j, j' = 1,2,3$ and $k, k' = 1,\dots,q$, we have

$$E(L_{jk}L_{j'k'}) = \mathbf{x}_{jk}^T\mathbf{R}_\epsilon^{-1}E(\boldsymbol{\epsilon}\boldsymbol{\epsilon}^T)\mathbf{R}_\epsilon^{-1}\mathbf{x}_{j'k'} = \mathbf{x}_{jk}^T\mathbf{R}_\epsilon^{-1}\mathbf{x}_{j'k'}.$$

With $\mathbf{X}_k := [\mathbf{x}_{1k}, \mathbf{x}_{2k}, \mathbf{x}_{3k}]$ given by (3.2.3), we can write

$$\mathbf{I}_{kk'}(\boldsymbol{\vartheta}) := [E(L_{jk}L_{j'k'})] = \mathbf{X}_k^T\mathbf{R}_\epsilon^{-1}\mathbf{X}_{k'} \qquad (k, k' = 1,\dots,q),$$

which leads to $\mathbf{I}_\theta(\boldsymbol{\vartheta}) = [\mathbf{I}_{kk'}(\boldsymbol{\vartheta})] = \mathbf{X}^T\mathbf{R}_\epsilon^{-1}\mathbf{X}$, where $\mathbf{X} := [\mathbf{X}_1,\dots,\mathbf{X}_q]$. The assertion for the RSM (2.1.2) can be proved similarly. □

Proof of Theorem 3.2. Under the complex Gaussian assumption, the log likelihood function takes the form

$$\ell(\boldsymbol{\vartheta}|\mathbf{y}) = \text{constant} - \log|\mathbf{R}_\epsilon| - (\mathbf{y}-\mathbf{x})^H\mathbf{R}_\epsilon^{-1}(\mathbf{y}-\mathbf{x}).$$

Because the decoupling of $\boldsymbol{\theta}$ and $\boldsymbol{\eta}$ remains true in the complex case, it suffices to evaluate $\mathbf{I}_\theta(\boldsymbol{\vartheta})$. For the Cartesian CSM (2.1.5), let $L_{1k} := \partial\ell/\partial A_k$, $L_{2k} := \partial\ell/\partial B_k$, and $L_{3k} := \partial\ell/\partial\omega_k$. Then, it is easy to show that

$$L_{jk} = \mathbf{x}_{jk}^H \mathbf{R}_\epsilon^{-1}\boldsymbol{\epsilon} + \boldsymbol{\epsilon}^H \mathbf{R}_\epsilon^{-1}\mathbf{x}_{jk} = \mathbf{x}_{jk}^H \mathbf{R}_\epsilon^{-1}\boldsymbol{\epsilon} + \mathbf{x}_{jk}^T \mathbf{R}_\epsilon^{-T}\boldsymbol{\epsilon}^* \quad (j = 1,2,3),$$

where the $\mathbf{x}_{jk}$ are defined by (3.2.6). Because $E(\boldsymbol{\epsilon}\boldsymbol{\epsilon}^T) = \mathbf{0}$, we obtain

$$E(L_{jk}L_{j'k'}) = E(L_{jk}L_{j'k'}^H) = 2\Re(\mathbf{x}_{jk}^H \mathbf{R}_\epsilon^{-1}\mathbf{x}_{j'k'}),$$

which gives $\mathbf{I}_\theta(\boldsymbol{\vartheta}) = 2\Re(\mathbf{X}^H \mathbf{R}_\epsilon^{-1}\mathbf{X})$. Similarly, it can be shown that this expression remains valid for the polar CSM (2.1.7) except that $\mathbf{X}$ is defined by (3.2.7). □

Proof of Theorem 3.4. Because the other cases can be proved similarly, let us only consider the assertion for the complex case with $\mathbf{x}$ taking the polar form (2.1.7). According to Theorem 3.2, the CRLB can be expressed as

$$\mathrm{CRLB}(\boldsymbol{\theta}) = \tfrac{1}{2}\{\Re(\mathbf{X}^H \mathbf{R}_\epsilon^{-1}\mathbf{X})\}^{-1}, \tag{3.5.3}$$

where $\mathbf{X} := [\mathbf{X}_1, \ldots, \mathbf{X}_p]$ and $\mathbf{X}_k := [\mathbf{x}_{1k}, \mathbf{x}_{2k}, \mathbf{x}_{3k}]$ are given by (3.2.7). To evaluate $\mathbf{X}^H \mathbf{R}_\epsilon^{-1}\mathbf{X}$, let us consider an autoregressive (AR) approximation for $f_\epsilon(\omega)$.

According to Corollary 4.4.2 in [46, p. 132] (see also Theorem 7.2 in Chapter 7), for any constant $\delta \in (0, f_0/2)$, there exists an AR spectrum of the form

$$f_{\mathrm{AR}}(\omega) := \frac{\sigma^2}{|1 + \sum_{j=1}^m \varphi_j \exp(-ij\omega)|^2}$$

such that

$$|f_\epsilon(\omega) - f_{\mathrm{AR}}(\omega)| < \delta$$

for all ω. This result also implies that $f_{\mathrm{AR}}(\omega) \geq f_0/2 > 0$ for all ω. Define

$$\mathbf{f}(\omega) := [\exp(i\omega), \ldots, \exp(in\omega)]^T.$$

Then, by Proposition 2.2,

$$\mathbf{R}_\epsilon = (2\pi)^{-1} \int_{-\pi}^{\pi} \mathbf{f}(\omega)\mathbf{f}^H(\omega) f_\epsilon(\omega)\, d\omega.$$

A similar expression exists for the covariance matrix associated with the spectrum $f_{\mathrm{AR}}(\omega)$, which we denote by $\mathbf{R}_{\mathrm{AR}}$. Then, for any $\mathbf{a}_1, \mathbf{a}_2 \in \mathbb{C}^n$, we can write

$$\mathbf{a}_1^H(\mathbf{R}_\epsilon - \mathbf{R}_{\mathrm{AR}})\mathbf{a}_2 = (2\pi)^{-1} \int_{-\pi}^{\pi} \mathbf{a}_1^H \mathbf{f}(\omega)\mathbf{f}^H(\omega)\mathbf{a}_2 \{f_\epsilon(\omega) - f_{\mathrm{AR}}(\omega)\}\, d\omega,$$

which, in turn, leads to

$$|\mathbf{a}_1^H(\mathbf{R}_\epsilon - \mathbf{R}_{\mathrm{AR}})\mathbf{a}_2| \leq \delta \|\mathbf{a}_1\|\,\|\mathbf{a}_2\|.$$

Moreover, because $f_\epsilon(\omega)$ and $f_{\mathrm{AR}}(\omega)$ are bounded from below by a positive constant for all ω, it follows from Proposition 4.5.3 in [46, p. 137–138] that the eigenvalues of $\mathbf{R}_\epsilon^{-1}$ and $\mathbf{R}_{\mathrm{AR}}^{-1}$ are all less than a constant, say $\kappa_1 > 0$, which is independent of δ. This result leads to

$$\|\mathbf{R}_\epsilon^{-1}\mathbf{a}\| \le \kappa_1 \|\mathbf{a}\|, \quad \|\mathbf{R}_{\mathrm{AR}}^{-1}\mathbf{a}\| \le \kappa_1 \|\mathbf{a}\|.$$

Let $\mathbf{a} := \mathbf{X}\mathbf{K}^{-1}\mathbf{b}$ for any $\mathbf{b} \in \mathbb{C}^{3r}$, where $r = q$ in the real case and $r = p$ in the complex case. Then,

$$\begin{aligned} |\mathbf{a}^H(\mathbf{R}_\epsilon^{-1} - \mathbf{R}_{\mathrm{AR}}^{-1})\mathbf{a}| &= |\mathbf{a}^H\mathbf{R}_{\mathrm{AR}}^{-1}(\mathbf{R}_{\mathrm{AR}} - \mathbf{R}_\epsilon)\mathbf{R}_\epsilon^{-1}\mathbf{a}| \\ &\le \delta\|\mathbf{R}_{\mathrm{AR}}^{-1}\mathbf{a}\|\,\|\mathbf{R}_\epsilon^{-1}\mathbf{a}\| \\ &\le \delta\kappa_1^2\mathbf{a}^H\mathbf{a} \\ &\le \delta\kappa_1^2\kappa_2, \end{aligned} \tag{3.5.4}$$

where the last inequality is due to the fact that $\mathbf{a} = \mathcal{O}(n^{-1/2})$ so that $\mathbf{a}^H\mathbf{a} \le \kappa_2$ for some constant $\kappa_2 > 0$. We need to calculate $\mathbf{a}^H\mathbf{R}_{\mathrm{AR}}^{-1}\mathbf{a}$.

By Proposition 2.9, $\mathbf{R}_{\mathrm{AR}}^{-1} = \mathbf{C}^H\mathbf{D}^{-1}\mathbf{C}$, where $\mathbf{D}$ is a diagonal matrix and $\mathbf{C}$ is a lower triangular matrix. For $n > m$, we can write $\mathbf{D} = \mathrm{diag}(\sigma_0^2, \sigma_1^2, \ldots, \sigma_{n-1}^2)$, where $\sigma_t^2 = \sigma^2$ for $t \ge m$, and

$$\mathbf{C} := [c_{st}] = \begin{bmatrix} c_{11} & & & & & & & 0 \\ c_{21} & c_{22} & & & & & & \\ \vdots & \vdots & \ddots & & & & & \\ c_{m1} & c_{m2} & \cdots & c_{mm} & & & & \\ \varphi_m & \varphi_{m-1} & \cdots & \varphi_1 & 1 & & & \\ & \varphi_m & \cdots & \varphi_2 & \varphi_1 & 1 & & \\ & & \ddots & & & & \ddots & \\ 0 & & & \varphi_m & \varphi_{m-1} & \cdots & \varphi_1 & 1 \end{bmatrix}$$

where $c_{st} := \varphi_{s-1,s-t}$ if $1 \le s \le m$ and $1 \le t \le s$, $c_{st} := \varphi_{s-t}$ if $m < s \le n$ and $s - m \le t \le s$, and $c_{st} := 0$ otherwise. Let η_{st} $(1 \le s, t \le n)$ denote the entries of $\mathbf{R}_{\mathrm{AR}}^{-1}$. It is easy to verify that $\eta_{st} = 0$ if $|s-t| > m$ so that $\mathbf{R}_{\mathrm{AR}}^{-1}$ is a band matrix with bandwidth $2m+1$, and that for $|s-t| \le m$,

$$\eta_{st} = \sum_{j=a_{st}}^{b_{st}} c_{js}^* c_{jt} / \sigma_{j-1}^2,$$

where $a_{st} := \max(s,t)$ and $b_{st} := \min(n, s+m, t+m)$. Moreover, let

$$h(u) := \begin{cases} \sum_{j=0}^{m-|u|} \varphi_j^* \varphi_{j+|u|} / \sigma_m^2 & \text{if } |u| \le m, \\ 0 & \text{otherwise.} \end{cases}$$

It is easy to verify that

$$\sum_{|u|\le m} h(u)\exp(-iu\omega) = 1/f_{\mathrm{AR}}(\omega). \tag{3.5.5}$$

Consider a fixed s such that $m < s < n-m$. If $0 \le s-t \le m$, then $a_{st} = s > m$, $b_{st} = t+m$, in which case, for any $j \ge a_{st}$, we have $c_{js} = \varphi_{j-s}$, $c_{jt} = \varphi_{j-t}$, and $\sigma_{j-1}^2 = \sigma^2$. Therefore,

$$\eta_{st} = \sum_{j=s}^{t+m} \varphi_{j-s}^* \varphi_{j-t}/\sigma^2 = h(s-t).$$

Similarly, if $0 \le t-s \le m$, then $a_{st} = t > m$ and $b_{st} = s+m$ so that $\eta_{st} = h^*(t-s)$. Furthermore, for any t such that $m < t < n-m$, the Hermitian property of $\mathbf{R}_{\mathrm{AR}}^{-1}$ gives $\eta_{st} = \eta_{ts}^* = h(s-t)$ if $0 \le s-t \le m$ and $\eta_{st} = h^*(t-s)$ if $0 \le t-s \le m$.

Let $\tilde{\mathbf{R}}_{\mathrm{AR}}^{-1} = [\tilde{\eta}_{st}]$ be the Hermitian matrix such that $\tilde{\eta}_{st} = h(s-t)$ if $0 \le s-t \le m$, $\tilde{\eta}_{st} = h^*(t-s)$ if $0 < t-s \le m$, and $\tilde{\eta}_{st} = 0$ if $|s-t| > m$. It is easy to see that $\tilde{\mathbf{R}}_{\mathrm{AR}}^{-1}$ differs from $\mathbf{R}_{\mathrm{AR}}^{-1}$ only on the $m \times m$ blocks in the upper left and bottom right corners. This result, combined with $\mathbf{a} := [a_1, \ldots, a_n]^T = \mathcal{O}(n^{-1/2})$, leads to

$$\mathbf{a}^H(\mathbf{R}_{\mathrm{AR}}^{-1} - \tilde{\mathbf{R}}_{\mathrm{AR}}^{-1})\mathbf{a} = \left\{ \sum_{s,t=1}^{m} + \sum_{s,t=n-m}^{n} \right\} a_s^* a_t (\eta_{st} - \tilde{\eta}_{st}) \to 0 \tag{3.5.6}$$

as $n \to \infty$. Therefore, it suffices to evaluate $\mathbf{a}^H \tilde{\mathbf{R}}_{\mathrm{AR}}^{-1}\mathbf{a}$. In the following, only the polar CSM (2.1.7) is considered. The other cases can be dealt with similarly.

For any k and k', let $d_{kk'}(t) := (\omega_k - \omega_{k'})t + \phi_k - \phi_{k'}$. With the $\mathbf{x}_{jk}$ given by (3.2.7), it is easy to show that

$$\begin{aligned}
\mathbf{x}_{1k}^H \tilde{\mathbf{R}}_{\mathrm{AR}}^{-1} \mathbf{x}_{1k'} &= \sum_{s,t=1}^{n} \tilde{\eta}_{st} \exp\{-i(\omega_k s - \omega_{k'} t + \phi_k - \phi_{k'})\} \\
&= \sum_{\substack{s,t=1 \\ 0 \le s-t \le m}}^{n} h(s-t)\exp\{-i\omega_k(s-t) - i d_{kk'}(t)\} \\
&\quad + \sum_{\substack{s,t=1 \\ 0 < t-s \le m}}^{n} h^*(t-s)\exp\{-i\omega_k(s-t) - i d_{kk'}(t)\} \\
&= \sum_{|u|\le m} h(u)\exp(-iu\omega_k) \sum_{t \in T_n(u)} \exp\{-i d_{kk'}(t)\},
\end{aligned}$$

where $T_n(u) := \{t : \max(1, 1-u) \le t \le \min(n, n-u)\}$. By Lemma 12.1.4, the sum over t takes the form $\mathcal{O}(\Delta^{-1})$ for $k \ne k'$ and equals $(n - |u|)$ for $k = k'$. Combining these results with the assumption (3.3.4) and with (3.5.5) yields

$$\begin{aligned}
n^{-1}\mathbf{x}_{1k}^H \tilde{\mathbf{R}}_{\mathrm{AR}}^{-1} \mathbf{x}_{1k'} &= \sum_{|u|\le m} h(u)\exp(-iu\omega_k)\delta_{k-k'} + \mathcal{O}(1) \\
&= \{1/f_{\mathrm{AR}}(\omega_k)\}\delta_{k-k'} + \mathcal{O}(1).
\end{aligned}$$

Similarly,

$$\begin{aligned}
n^{-1}\mathbf{x}_{1k}^H\tilde{\mathbf{R}}_{\mathrm{AR}}^{-1}\mathbf{x}_{2k'} &= \{i\tfrac{1}{2}C_k/f_{\mathrm{AR}}(\omega_k)\}\delta_{k-k'}+o(1),\\
n^{-2}\mathbf{x}_{1k}^H\tilde{\mathbf{R}}_{\mathrm{AR}}^{-1}\mathbf{x}_{3k'} &= \{i\tfrac{1}{2}C_k/f_{\mathrm{AR}}(\omega_k)\}\delta_{k-k'}+o(1),\\
n^{-1}\mathbf{x}_{2k}^H\tilde{\mathbf{R}}_{\mathrm{AR}}^{-1}\mathbf{x}_{2k'} &= \{C_k^2/f_{\mathrm{AR}}(\omega_k)\}\delta_{k-k'}+o(1),\\
n^{-2}\mathbf{x}_{2k}^H\tilde{\mathbf{R}}_{\mathrm{AR}}^{-1}\mathbf{x}_{3k'} &= \{\tfrac{1}{2}C_k^2/f_{\mathrm{AR}}(\omega_k)\}\delta_{k-k'}+o(1),\\
n^{-3}\mathbf{x}_{3k}^H\tilde{\mathbf{R}}_{\mathrm{AR}}^{-1}\mathbf{x}_{3k'} &= \{\tfrac{1}{3}C_k^2/f_{\mathrm{AR}}(\omega_k)\}\delta_{k-k'}+o(1).
\end{aligned}$$

Therefore, we obtain

$$\mathbf{a}^H\tilde{\mathbf{R}}_{\mathrm{AR}}^{-1}\mathbf{a} = \mathbf{b}^H\{\mathbf{V}+\mathcal{O}(\delta)\}\mathbf{b}+o(1), \tag{3.5.7}$$

where $\mathbf{V} := \mathrm{diag}(\mathbf{V}_1,\dots,\mathbf{V}_p)$ and

$$\mathbf{V}_k := \frac{1}{f_\epsilon(\omega_k)}\begin{bmatrix} 1 & iC_k & iC_k\\ -iC_k & C_k^2 & \frac{1}{2}C_k^2\\ -iC_k & \frac{1}{2}C_k^2 & \frac{1}{3}C_k^2\end{bmatrix}.$$

Combining (3.5.7) with (3.5.4) and (3.5.6) leads to

$$\limsup_{n\to\infty}|\mathbf{b}^H\mathbf{K}^{-1}\mathbf{X}^H\mathbf{R}_\epsilon^{-1}\mathbf{X}\mathbf{K}^{-1}\mathbf{b}-\mathbf{b}^H\mathbf{V}\mathbf{b}| = \mathcal{O}(\delta).$$

Because this expression holds for all $\delta>0$ and all $\mathbf{b}\in\mathbb{C}^{3p}$, we obtain

$$\mathbf{K}^{-1}\mathbf{X}^H\mathbf{R}_\epsilon^{-1}\mathbf{X}\mathbf{K}^{-1} = \mathbf{V}+o(1).$$

Observe that $\Re(\mathbf{V}_k) = \mathbf{W}_k/f_\epsilon(\omega_k)$, where $\mathbf{W}_k$ is defined by (3.3.5). Therefore,

$$\tfrac{1}{2}f_\epsilon(\omega_k)\mathbf{W}_k^{-1} = \mathbf{\Gamma}(\boldsymbol{\theta}_k) := \tfrac{1}{2}\gamma_k^{-1}\mathbf{\Lambda}_{\mathrm{P}}(\boldsymbol{\theta}_k).$$

where $\gamma_k := C_k^2/f_\epsilon(\omega_k)$. Combining these results with (3.5.3) leads to

$$\mathrm{CRLB}(\boldsymbol{\theta}) = \mathbf{K}^{-1}\{\mathbf{\Gamma}(\boldsymbol{\theta})+o(1)\}\mathbf{K}^{-1}$$

with $\mathbf{\Gamma}(\boldsymbol{\theta}) = \mathrm{diag}\{\mathbf{\Gamma}(\boldsymbol{\theta}_1),\dots,\mathbf{\Gamma}(\boldsymbol{\theta}_p)\}$. □

Proof of Theorem 3.5. First, consider the case of Gaussian white noise. By Theorem 3.2, $\mathrm{CRLB}(\boldsymbol{\theta}) = \frac{1}{2}\sigma^2\{\Re(\mathbf{X}^H\mathbf{X})\}^{-1}$, where $\mathbf{X} := [\mathbf{X}_1,\mathbf{X}_2]$ and $\mathbf{X}_k := [\mathbf{x}_{1k},\mathbf{x}_{2k},\mathbf{x}_{3k}]$ are given by (3.2.6). By Example 3.6, we have

$$\Re(\mathbf{X}_k^H\mathbf{X}_k) = \mathbf{K}_n\{\mathbf{W}_k+o(1)\}\mathbf{K}_n,$$

where $\mathbf{W}_k$ is given by (3.3.5). So, it suffices to prove $\mathbf{X}_1^H\mathbf{X}_2 = \mathbf{K}_n\{\mathbf{W}_{12}+o(1)\}\mathbf{K}_n$.

Because $n\Delta \to a$ as $n \to \infty$, where $\Delta := \omega_2 - \omega_1$, it follows from Theorem 3.2 and Lemma 12.1.3(f) that

$$\begin{aligned}
n^{-1}\mathbf{x}_{11}^H\mathbf{x}_{12} &= D_n(\Delta) \to d_0(a), \quad n^{-1}\mathbf{x}_{11}^H\mathbf{x}_{22} = -in^{-1}\mathbf{x}_{11}^H\mathbf{x}_{12} \to -id_0(a),\\
n^{-2}\mathbf{x}_{11}^H\mathbf{x}_{32} &= \beta_2 n^{-1}\dot{D}_n(\Delta) \to \beta_2\dot{d}_0(a),\\
n^{-1}\mathbf{x}_{21}^H\mathbf{x}_{12} &= in^{-1}\mathbf{x}_{11}^H\mathbf{x}_{12} \to id_0(a), \quad n^{-1}\mathbf{x}_{21}^H\mathbf{x}_{22} = n^{-1}\mathbf{x}_{11}^H\mathbf{x}_{12} \to d_0(a),\\
n^{-2}\mathbf{x}_{21}^H\mathbf{x}_{32} &= in^{-2}\mathbf{x}_{11}^H\mathbf{x}_{32} \to i\beta_2\dot{d}_0(a),\\
n^{-2}\mathbf{x}_{31}^H\mathbf{x}_{12} &= -\beta_1^* n^{-1}\dot{D}_n(\Delta) \to -\beta_1^*\dot{d}_0(a), \quad n^{-2}\mathbf{x}_{31}^H\mathbf{x}_{22} = -i\mathbf{x}_{31}^H\mathbf{x}_{12} \to i\beta_1^*\dot{d}_0(a),\\
n^{-3}\mathbf{x}_{31}^H\mathbf{x}_{32} &= -\beta_1^*\beta_2 n^{-2}\ddot{D}_n(\Delta) \to -\beta_1^*\beta_2\ddot{d}_0(a).
\end{aligned}$$

Therefore,

$$\mathbf{K}_n^{-1}\Re(\mathbf{X}_1^H\mathbf{X}_2)\mathbf{K}_n^{-1} \to \Re\begin{bmatrix} d_0(a) & -id_0(a) & \beta_2\dot{d}_0(a)\\ id_0(a) & d_0(a) & i\beta_2\dot{d}_0(a)\\ -\beta_1^*\dot{d}_0(a) & i\beta_1^*\dot{d}_0(a) & -\beta_1^*\beta_2\ddot{d}_0(a)\end{bmatrix} = \mathbf{W}_{12}.$$

The assertion is thus proved for the case of Gaussian white noise.

Now consider the case of Gaussian colored noise. Observe that the continuity of the noise spectrum $f_\epsilon(\omega)$ implies $f_\epsilon(\omega_2) = f_\epsilon(\omega_1) + \mathcal{O}(1)$. Therefore, according to the proof of Theorem 3.4, it suffices to show that

$$\mathbf{K}_n^{-1}\Re(\mathbf{X}_1^H\tilde{\mathbf{R}}_{\mathrm{AR}}^{-1}\mathbf{X}_2)\mathbf{K}_n^{-1} = \{1/f_{\mathrm{AR}}(\omega_1)\}\mathbf{W}_{12} + \mathcal{O}(1),$$

where $f_{\mathrm{AR}}(\omega)$ is an AR spectrum that approximates $f_\epsilon(\omega)$. By Theorem 3.2 and the proof of Theorem 3.4, we have

$$\mathbf{x}_{11}^H\tilde{\mathbf{R}}_{\mathrm{AR}}^{-1}\mathbf{x}_{12} = \sum_{|u|\le m} h(u)\exp(-iu\omega_1)\sum_{t\in T_n(u)}\exp(it\Delta),$$

where $T_n(u) := \{t : \max(1, 1-u) \le t \le \min(n, n-u)\}$. By Lemma 12.1.3(f) gives

$$n^{-1}\sum_{t\in T_n(u)}\exp(it\Delta) = D_n(\Delta) + \mathcal{O}(n^{-1}) \to d_0(a)$$

for any fixed u. Combining this result with (3.5.5) leads to

$$n^{-1}\mathbf{x}_{11}^H\tilde{\mathbf{R}}_{\mathrm{AR}}^{-1}\mathbf{x}_{12} \to \{1/f_{\mathrm{AR}}(\omega_1)\}d_0(a).$$

The remaining elements in $\mathbf{X}_1^H\tilde{\mathbf{R}}_{\mathrm{AR}}^{-1}\mathbf{X}_2$ can be evaluated similarly. □

Proof of Corollary 3.3. For estimating $\boldsymbol{\theta} := [A_1, B_1, A_2, B_2]^T$, it follows from Theorem 3.5 that $\boldsymbol{\Gamma}(\boldsymbol{\theta}) := \frac{1}{2}f_\epsilon(\omega_1)\boldsymbol{\Lambda}_0$, where

$$\boldsymbol{\Lambda}_0 := \begin{bmatrix}\mathbf{I} & \mathbf{A}\\ \mathbf{A}^T & \mathbf{I}\end{bmatrix}^{-1}, \quad \mathbf{A} := \begin{bmatrix}\Re\{d_0(a)\} & \Im\{d_0(a)\}\\ -\Im\{d_0(a)\} & \Re\{d_0(a)\}\end{bmatrix}.$$

Observe that $\mathbf{A}$ is obtained from $\mathbf{W}_{12}$ defined in (3.3.19) by removing the third row and the third column. Applying the same procedure to $\mathbf{W}_k$ $(k=1,2)$ defined in (3.3.7) produces the identity matrix. By using the second matrix inversion formula in Lemma 12.4.1, we obtain

$$\begin{bmatrix} \mathbf{I} & \mathbf{A} \\ \mathbf{A}^T & \mathbf{I} \end{bmatrix}^{-1} = \begin{bmatrix} (\mathbf{I}-\mathbf{A}\mathbf{A}^T)^{-1} & -(\mathbf{I}-\mathbf{A}\mathbf{A}^T)^{-1}\mathbf{A} \\ -\mathbf{A}^T(\mathbf{I}-\mathbf{A}\mathbf{A}^T)^{-1} & (\mathbf{I}-\mathbf{A}^T\mathbf{A})^{-1} \end{bmatrix}.$$

Moreover, because $\mathbf{A}\mathbf{A}^T = \mathbf{A}^T\mathbf{A} = |d_0(a)|^2\mathbf{I}$, it follows that

$$(\mathbf{I}-\mathbf{A}\mathbf{A}^T)^{-1} = (\mathbf{I}-\mathbf{A}^T\mathbf{A})^{-1} = \frac{1}{1-|d_0(a)|^2}\mathbf{I}.$$

Combining these results proves the assertion. □

Proof of Proposition 3.2. By definition, Fisher's information matrix is given by

$$\mathbf{I}(\boldsymbol{\theta}) = E\{[\nabla\ell(\boldsymbol{\theta}|\mathbf{y})][\nabla\ell(\boldsymbol{\theta}|\mathbf{y})]^T\},$$

where $\ell(\boldsymbol{\theta}|\mathbf{y}) := \log\{p(\mathbf{y}|\boldsymbol{\theta})\}$ is the log likelihood function and $p(\cdot|\boldsymbol{\theta})$ is the probability density of $\mathbf{y}$. Because the y_t are independent and the marginal PDF of y_t is $p(y-x_t)$, we can write $p(\mathbf{y}|\boldsymbol{\theta}) = \prod p(y_t - x_t)$ and hence

$$\ell(\boldsymbol{\theta}|\mathbf{y}) = \sum_{t=1}^{n} \log\{p(y_t - x_t)\}.$$

This implies that

$$[\nabla\ell(\boldsymbol{\theta}|\mathbf{y})][\nabla\ell(\boldsymbol{\theta}|\mathbf{y})]^T = \sum_{t=1}^{n}\sum_{s=1}^{n}\left\{\frac{\dot{p}(y_t-x_t)}{p(y_t-x_t)}\right\}\left\{\frac{\dot{p}(y_s-x_s)}{p(y_s-x_s)}\right\}(\nabla x_t)(\nabla x_s)^T \quad (3.5.8)$$

for any $\boldsymbol{\theta}$ and $\mathbf{y}$ such that $y_t - x_t \notin \mathscr{X}$ $(t=1,\dots,n)$, where $\mathscr{X}$ denotes the set of points at which $p(x)$ is not differentiable. For any fixed $\boldsymbol{\theta}$, and hence fixed x_t, all but a finite number of values of y satisfy the condition $y - x_t \notin \mathscr{X}$. Therefore, the identity (3.5.8) is valid almost surely for any given $\boldsymbol{\theta}$. Because $\int \dot{p}(x)\,dx = 0$, taking expectation on both sides of (3.5.8) gives (3.4.2) under the assumption that $\epsilon_1 = y_1 - x_1, \dots, \epsilon_n = y_n - x_n$ are i.i.d. with PDF $p(x)$. □

Proof of Proposition 3.3. Similar to the proof of Proposition 3.2, the assertion follows from the fact that $\ell(\boldsymbol{\theta}|\mathbf{y}) = \sum\log\{p(\Re(y_t - x_t))\} + \sum\log\{p(\Im(y_t - x_t))\}$ and that $\{\Re(\epsilon_t)\}$ and $\{\Im(\epsilon_t)\}$ are independent i.i.d. sequences with PDF $p(x)$. □

Proof of Proposition 3.4. It suffices to consider (3.4.6) in both real and complex cases, with $p_0(x)$ being a unit-variance PDF.

Let $p(y|\theta) := p_0(y-\theta)$ for $(y,\theta) \in \mathbb{R}^2$. Under the assumptions of Proposition 3.4, $p(y|\theta)$ satisfies the regularity conditions (a)–(c) in Proposition 3.1 with Fisher's information $I(\theta) := \lambda$. For any random variable $Y \sim p(y|\theta)$, we have

$$E(Y) = \int y p_0(y-\theta)\, dy = \theta,$$

which means that Y is an unbiased estimator of θ. Therefore, by Proposition 3.1,

$$\text{Var}(Y) \geq I(\theta)^{-1} \tag{3.5.9}$$

with equality if and only if $a(y-\theta) = (d/d\theta)\log\{p_0(y-\theta)\}$ for some constant $a \neq 0$ which may depend on θ. Since $\text{Var}(Y) = 1$ and $I(\theta) = \lambda$, the inequality (3.5.9) can be rewritten as $\lambda \geq 1$. Since $(d/d\theta)\log\{p_0(y-\theta)\} = -\dot{p}_0(y-\theta)/p_0(y-\theta)$, the condition of equality can be rewritten as $ax = -\dot{p}_0(x)/p_0(x)$ or

$$\frac{dz}{dx} = -ax,$$

where $z := \log\{p_0(x)\}$ and $x := y - \theta$. The solution to this differential equation takes the form $z = -\frac{1}{2}ax^2 + bx + c$ for some constants b and c, meaning that $\log\{p_0(x)\}$ is a quadratic function of x. The assertion follows from the fact that the Gaussian distribution with $p_0(x) = (2\pi)^{-1/2}\exp(-\frac{1}{2}x^2)$ is the only distribution of this type that has zero mean and unit variance. □

Proof of Proposition 3.5. Under the symmetry and monotonicity assumptions, we have $\dot{p}_0(x) = -\dot{p}_0(-x) \leq 0$ for $x > 0$. Therefore, an application of the Cauchy-Schwarz inequality gives

$$\begin{aligned}
\int \frac{\{\dot{p}_0(x)\}^2}{p_0(x)}\, dx &= \left\{\int \left|\frac{\dot{p}_0(x)}{p_0(x)}\right|^2 p_0(x)\, dx\right\}\left\{\int p_0(x)\, dx\right\} \\
&\geq \left\{\int \left|\frac{\dot{p}_0(x)}{p_0(x)}\right| p_0(x)\, dx\right\}^2 \\
&= \left\{\int |\dot{p}_0(x)|\, dx\right\}^2 \\
&= \left\{-2\int_0^\infty \dot{p}_0(x)\, dx\right\}^2 \\
&= 4\{p_0(0)\}^2.
\end{aligned}$$

The equality in the second line is true if and only if

$$|\dot{p}_0(x)/p_0(x)| = 1/c \tag{3.5.10}$$

for some constant $c > 0$. Combining this result with (3.4.6) proves the asserted inequality. Note that (3.5.10) can be rewritten as

$$\frac{dz}{dx} = -1/c \quad \text{for } x > 0,$$

where $z := \log\{p_0(x)\}$. The solution to this differential equation is given by $z = -x/c + b$ for some constant b. Therefore, the symmetric PDF that satisfies the equality must take the form $p_0(x) = a\exp(-|x|/c)$ for some constant $a := \exp(b)$. The only distribution of this type that has unit variance is the Laplace distribution with $p_0(x) = (1/\sqrt{2})\exp(-\sqrt{2}|x|)$. □

Proof of Theorem 3.6. Under the RSM (2.1.1), we have $\boldsymbol{\theta} := [\boldsymbol{\theta}_1^T, \ldots, \boldsymbol{\theta}_q^T]^T$ and $\boldsymbol{\theta}_k := [A_k, B_k, \omega_k]^T$. Observe that

$$\sum_{t=1}^{n} (\nabla x_t)(\nabla x_t)^T = \mathbf{X}^T\mathbf{X},$$

where $\mathbf{X} := [\mathbf{X}_1, \ldots, \mathbf{X}_q]$ is given by (3.2.3). This expression remains valid under the polar RSM (2.1.2) except that $\boldsymbol{\theta}_k := [C_k, \phi_k, \omega_k]^T$ and $\mathbf{X}$ is given by (3.2.4). Therefore, according to Proposition 3.2,

$$\mathbf{I}(\boldsymbol{\theta}) = (\lambda/\sigma^2)\mathbf{X}^T\mathbf{X}. \tag{3.5.11}$$

Under the Cartesian CSM (2.1.5), $\boldsymbol{\theta} := [\boldsymbol{\theta}_1^T, \ldots, \boldsymbol{\theta}_p^T]^T$, $\boldsymbol{\theta}_k := [A_k, B_k, \omega_k]^T$, and

$$\sum_{t=1}^{n} (\nabla x_t)(\nabla x_t)^H = \mathbf{X}^H\mathbf{X},$$

where $\mathbf{X} = [\mathbf{X}_1, \ldots, \mathbf{X}_p]$ is given by (3.2.6). The same expression holds for the polar CSM (2.1.7) with $\boldsymbol{\theta}_k := [C_k, \phi_k, \omega_k]^T$ except that $\mathbf{X}$ is given by (3.2.7). Therefore,

$$\mathbf{I}(\boldsymbol{\theta}) = 2(\lambda/\sigma^2)\Re(\mathbf{X}^H\mathbf{X}). \tag{3.5.12}$$

Inverting the matrices in (3.5.11) and (3.5.12) proves the assertion. □

NOTES

As an important performance benchmark, the CRLB has been studied from many angles for different applications. Early references include [162], [316], and [317]. The CRLB under the Gaussian colored noise is evaluated in [114], [121], [368], and [385]. The CRLB based on multiple records (or snapshots) of observations is considered in [328], [372], and [430]. Approximate expressions of the CRLB for closely spaced frequencies are derived in [89], [216], [217], [383], and [384]. The effect of signal correlation and relative phase on the CRLB is investigated in [431]. The CRLB under the condition of finite-variance nonGaussian white noise is discussed in [162], [238], and [381]. The CRLB under the condition of infinite-variance Cauchy white noise is considered in [394]. A more general discussion on least favorable distributions can be found in [85] and [292].

Many numerical and analytical studies show that the CRLB is not a tight bound for the nonlinear problem of frequency estimation at low SNR. Some alternatives, such as the Barankin bound [21] [60] [185], are found useful, especially in explaining the empirical observation of a threshold effect in frequency estimation where the mean-square error of the best available frequency estimators deteriorates rapidly as the SNR drops below a threshold, which the CRLB fails to predict [2] [61] [194] [256] [257] [313] [339]. However, because of computational difficulty, these bounds are not as widely used as the CRLB.

Chapter 4

Autocovariance Function

The autocovariance function is a widely used statistical feature for time series analysis. The statistical behavior of sample autocovariances is well understood when the underlying stationary random process has a continuous spectrum. In this chapter, we examine the statistical behavior of sample autocovariances for time series with mixed spectra. The focus is on the asymptotic properties with regard to the first and second moments and the probability distribution. Attention is paid to both real and complex cases.

4.1 Autocovariances and Autocorrelation Coefficients

Let $\{y_1, \ldots, y_n\}$ be a real or complex time series with zero mean. We make the zero-mean assumption merely to simplify the theoretical analysis. When the mean is nonzero, one can always subtract the sample mean from the original time series and consider the resulting mean-adjusted time series. The asymptotic theory developed in this chapter remains valid for mean-adjusted time series, although the sample sizes at which the asymptotic theory becomes a reasonable approximation may be adversely affected.

The sample autocovariance function of a zero-mean real or complex time series $\{y_1, \ldots, y_n\}$ is defined as

$$\hat{r}_y(u) := \hat{r}_y^*(-u) := \begin{cases} n^{-1} \sum_{t=1}^{n-u} y_{t+u} y_t^* & \text{for } u = 0, 1, \ldots, n-1, \\ 0 & \text{for } u = n, n+1, \ldots. \end{cases} \tag{4.1.1}$$

Normalizing the sample autocovariance function by the sample variance $\hat{r}_y(0)$ produces the sample autocorrelation coefficients

$$\hat{\rho}_y(u) := \hat{r}_y(u) / \hat{r}_y(0). \tag{4.1.2}$$

By definition, the sampling autocovariances and autocorrelation coefficients are symmetric in the sense that

$$\hat{r}_y(-u) = \hat{r}_y^*(u), \quad \hat{\rho}_y(-u) = \hat{\rho}_y^*(u).$$

This is consistent with the symmetry of the autocovariance function and the autocorrelation function of a real or complex random process.

The sample autocovariances and autocorrelation coefficients are the basis of many parameter estimation procedures under the sinusoid-plus-noise model

$$y_t = x_t + \epsilon_t \qquad (t = 1, \ldots, n), \tag{4.1.3}$$

where $\{x_t\}$ is given by (2.1.1), (2.1.2), (2.1.5), or (2.1.7), and $\{\epsilon_t\}$ is a zero-mean random process with continuous spectrum. Therefore, the statistical properties of $\hat{r}_y(u)$ and $\hat{\rho}_y(u)$ under the condition (4.1.3) are of considerable interest. Although finite-sample properties are difficult to obtain, asymptotic results can be derived for large sample sizes under fairly general assumptions.

In the real case and without the sinusoids so that $y_t = \epsilon_t$, the statistical properties of the sample autocovariances and autocorrelation coefficients are well documented in the standard literature such as [46] and [298]. For example, it is well known that in this case the lag-u sample autocovariance $\hat{r}_y(u)$ and the lag-u sample autocorrelation coefficient $\hat{\rho}_y(u)$ converge in a suitable sense to the lag-u autocovariance $r_\epsilon(u)$ and the lag-u autocorrelation coefficient $\rho_\epsilon(u) := r_\epsilon(u)/r_\epsilon(0)$ of the random process $\{\epsilon_t\}$. It is also well known that for any fixed m the random vector comprising $\sqrt{n}\{\hat{r}_y(u) - r_\epsilon(u)\}$ $(u = 0, 1, \ldots, m)$ and the random vector comprising $\sqrt{n}\{\hat{\rho}_y(u) - \rho_\epsilon(u)\}$ $(u = 1, \ldots, m)$ are asymptotically distributed as zero-mean Gaussian random vectors whose covariance matrices are given by the celebrated formulas of Bartlett [25]. In the presence of sinusoids, these classical results are no longer applicable because $\{y_t\}$ does not have a continuous spectrum as required by the classical theory; it is not even a stationary process in the conventional sense when the phases of the sinusoids are constants.

In the remainder of this chapter, the classical results are extended to the case where $\{y_t\}$ contains sinusoidal components. First, it is shown that the sample autocovariances and autocorrelation coefficients converge to well-defined limiting values under fairly general conditions. These limiting values coincide with the ordinary autocovariance and autocorrelation functions of $\{y_t\}$ derived under the assumption that the phases of the sinusoids are i.i.d. random variables with uniform distribution in $(-\pi, \pi]$ so that $\{y_t\}$ becomes a stationary process in the conventional sense. The corresponding spectral density function (SDF), defined conventionally as the Fourier transform of the autocovariance function, is a mixture of discrete and continuous components. Furthermore, it is shown that the sample autocovariances and autocorrelation coefficients are still asymptotically distributed as Gaussian random variables whose variances and covariances can be expressed in closed forms that generalize Bartlett's formulas.

4.2 Consistency and Asymptotic Unbiasedness

Let us first consider the limiting values of $\hat{r}_y(u)$ and $E\{\hat{r}_y(u)\}$ as $n \to \infty$ for fixed u. These values are derived in [223] under the assumption that the frequencies are fixed as $n \to \infty$. To cover the more general case where the signal frequencies are near each other relative to the sample size, let us allow the ω_k to depend on n and define the frequency separation by

$$\Delta := \begin{cases} \min_{k \neq k'}\{|\omega_k - \omega_{k'}|, \omega_k, \pi - \omega_k\} & \text{for RSM,} \\ \min_{k \neq k'}\{|\omega_k - \omega_{k'}|, 2\pi - |\omega_k - \omega_{k'}|\} & \text{for CSM.} \end{cases} \tag{4.2.1}$$

The following theorem summarizes the results. See Section 4.5 for a proof.

Theorem 4.1 (Consistency and Asymptotic Unbiasedness of the Sample Autocovariances). *Let $\{y_t\}$ be given by (4.1.3), where $\{x_t\}$ takes the form (2.1.1), (2.1.2), (2.1.5), or (2.1.7) with $\{\omega_k\}$ satisfying the separation condition*

$$\lim_{n\to\infty} n\Delta = \infty. \tag{4.2.2}$$

Assume that $\{\epsilon_t\}$ is a linear process of the form

$$\epsilon_t = \sum_{j=-\infty}^{\infty} \psi_j \zeta_{t-j}, \tag{4.2.3}$$

where $\sum |\psi_j| < \infty$, $\{\zeta_t\} \sim \mathrm{WN}(0, \sigma^2)$, and $E(|\zeta_t|^4) < \infty$. Assume further that

$$E(\zeta_{t+u}\zeta^*_{t+v}\zeta^*_{t+w}\zeta_t) = r_\zeta(u, v, w),$$

where

$$r_\zeta(u, v, w) := \begin{cases} \kappa\sigma^4 & \textit{if } u = v = w = 0, \\ \xi\sigma^4 & \textit{if } v = w \neq u = 0, \\ \sigma^4 & \textit{if } u = w \neq v = 0 \textit{ or } u = v \neq w = 0, \\ 0 & \textit{otherwise.} \end{cases} \tag{4.2.4}$$

Then, as $n \to \infty$ and for any fixed u,

$$\hat{r}_y(u) - \{r_x(u) + r_\epsilon(u)\} \xrightarrow{a.s.} 0, \tag{4.2.5}$$

where

$$r_x(u) := \begin{cases} \sum_{k=1}^{q} \frac{1}{2} C_k^2 \cos(\omega_k u) & \textit{for RSM,} \\ \sum_{k=1}^{p} C_k^2 \exp(i\omega_k u) & \textit{for CSM,} \end{cases} \tag{4.2.6}$$

$$r_\epsilon(u) := E(\epsilon_{t+u}\epsilon_t^*) = \sigma^2 \sum_{j=-\infty}^{\infty} \psi_j^* \psi_{j+u}. \tag{4.2.7}$$

Moreover, as $n \to \infty$ and for any fixed u,

$$E\{\hat{r}_y(u)\} = r_x(u) + r_\epsilon(u) + \mathcal{O}(n^{-1}\Delta^{-1}). \tag{4.2.8}$$

Remark 4.1 The second expression in (4.2.6) remains valid for the RSM if $p := 2q$, $\omega_{p-k+1} := -\omega_k$, and $C_{p-k+1} := C_k := \frac{1}{2}\sqrt{A_k^2 + B_k^2}$ for $k = 1, \ldots, q$.

Remark 4.2 The frequency separation condition (4.2.2) ensures that the sinusoids are decoupled in the limiting expression of $n^{-1}\sum_{t=1}^n |x_t|^2$ as $n \to \infty$.

Remark 4.3 The assumption about $\{\zeta_t\}$ is satisfied if it is an i.i.d. sequence, $\{\zeta_t\} \sim \text{IID}(0, \sigma^2)$, with $E(|\zeta_t|^4) = \kappa\sigma^4 < \infty$, in which case, $\xi = |E(\zeta_t^2)|^2/\sigma^4 = |\iota|^2$, where $\iota := E(\zeta_t^2)/\sigma^2$. An example is Gaussian white noise where $\{\zeta_t\} \sim \text{GWN}(0, \sigma^2)$ or $\{\zeta_t\} \sim \text{GWN}_c(0, \sigma^2)$. In the real Gaussian case, $\kappa = 3$ and $\xi = 1$; in the complex Gaussian case, $\kappa = 2$ and $\xi = 0$. More general assumptions about $\{\epsilon_t\}$ are discussed, for example, in [94, pp. 489–495].

Remark 4.4 The assumption about $\{\zeta_t\}$ is satisfied if $\{\zeta_t\}$ is a sequence of martingale differences with respect to a filtration $\{\mathfrak{F}_t\}$, i.e., $\{\zeta_t\} \sim \text{MD}(0, \sigma^2)$, such that $E(|\zeta_t|^4) = \kappa\sigma^4 < \infty$, $E(|\zeta_t|^2|\mathfrak{F}_{t-1}) = \sigma^2$, and $E(\zeta_t^2|\mathfrak{F}_{t-1}) = \iota\sigma^2$ for all t. By using the technique of iterated expectation, we obtain $E(\zeta_t^2) = E\{E(\zeta_t^2|\mathfrak{F}_{t-1})\} = \iota\sigma^2$; similarly, for any $v > 0$, we have $E(\zeta_{t+v}^{*2}\zeta_t^2) = E\{E(\zeta_{t+v}^{*2}|\mathfrak{F}_{t+v-1})\zeta_t^2\} = |\iota|^2\sigma^4$, and for any $v < 0$, we have $E(\zeta_{t+v}^{*2}\zeta_t^2) = E\{\zeta_{t+v}^{*2}E(\zeta_t^2|\mathfrak{F}_{t-1})\} = |\iota|^2\sigma^4$. This gives the second equation in (4.2.4) with $\xi := |\iota|^2$. By the same technique, it can be shown that $E(|\zeta_{t+u}|^2|\zeta_t|^2) = \sigma^4$ for all $u \neq 0$, which leads to the third equation in (4.2.4). The last equation in (4.2.4) can be proved similarly. For example, for $u > v > w > 0$, $E(\zeta_{t+u}\zeta_{t+v}^*\zeta_{t+w}^*\zeta_t) = E\{E(\zeta_{t+u}|\mathfrak{F}_{t+u-1})\zeta_{t+v}^*\zeta_{t+w}^*\zeta_t\} = 0$.

In light of Theorem 4.1, it is natural to regard $r_x(u)$ given by (4.2.6) as the ACF of $\{x_t\}$ and define the ACF of $\{y_t\}$ as

$$r_y(u) := r_x(u) + r_\epsilon(u). \tag{4.2.9}$$

Theorem 4.1 ensures that $\hat{r}_y(u)$ is a consistent and asymptotically unbiased estimator of $r_y(u)$ for any u under the frequency separation condition (4.2.2).

It is important to note that because $\{x_t\}$ is deterministic, the function $r_y(u)$ in (4.2.9) is not an ordinary ACF in the sense that $r_y(u) = E(y_{t+u}y_t^*)$. However, it can be transformed into an ordinary ACF if we randomize the phases of the sinusoids. Indeed, if the ϕ_k in (2.1.2) and (2.1.7) are regarded as i.i.d. random variables with uniform distribution in $(-\pi, \pi]$, then $\{x_t\}$ becomes a stationary random process with $E(x_t) = 0$ and $E(x_{t+u}x_t^*) = r_x(u)$. If the ϕ_k are also independent of $\{\epsilon_t\}$, then $\{y_t\}$ is a stationary process with $E(y_t) = 0$ and $E(y_{t+u} y_t^*) = r_x(u) + r_\epsilon(u)$. Theorem 4.1 remains valid under this condition.

Naturally, the SDF of $\{x_t\}$ is defined as the Fourier transform of $r_x(u)$ in the same way as the SDF is defined for the noise process $\{\epsilon_t\}$. In other words, let

$$f_x(\omega) := \sum_{u=-\infty}^{\infty} r_x(u) \exp(-i\omega u).$$

With the help of the following identity in Fourier analysis,

$$\sum_{u=-\infty}^{\infty} \exp(iu\omega) = 2\pi\delta(\omega), \tag{4.2.10}$$

where $\delta(\omega)$ denotes the Dirac delta, we can write

$$f_x(\omega) = \begin{cases} 2\pi \sum_{k=1}^{q} \frac{1}{4} C_k^2 \{\delta(\omega - \omega_k) + \delta(\omega + \omega_k)\} & \text{for RSM,} \\ 2\pi \sum_{k=1}^{p} C_k^2 \delta(\omega - \omega_k) & \text{for CSM.} \end{cases} \tag{4.2.11}$$

Similarly, the SDF of $\{y_t\}$ is defined as the Fourier transform of $r_y(u)$, namely

$$f_y(\omega) := \sum_{u=-\infty}^{\infty} r_y(u) \exp(-iu\omega). \tag{4.2.12}$$

It follows from (4.2.9) that

$$f_y(\omega) = f_x(\omega) + f_\epsilon(\omega), \tag{4.2.13}$$

where $f_\epsilon(\omega) := \sigma^2 |\Psi(\omega)|^2$ is the SDF of the noise $\{\epsilon_t\}$.

Note that $f_\epsilon(\omega)$ is a smooth function but $f_x(\omega)$ is an impulsive function consisting of discrete impulses located at the signal frequencies ω_k. Therefore, as a sum of the two, the function $f_y(\omega)$ contains infinite-amplitude peaks at the signal frequencies ω_k on top of a continuous but not necessarily constant noise floor. The presence of both continuous and discrete (or impulsive) components in (4.2.13) makes $f_y(\omega)$ a mixed spectrum.

As a direct result of Theorem 4.1, the following corollary, stated without proof, can be easily obtained.

Corollary 4.1 (Consistency of the Sample Autocorrelation Coefficients). *Let the conditions of Theorem 4.1 be satisfied. Then, $\hat{\rho}_y(u) - r_y(u)/r_y(0) \stackrel{a.s.}{\to} 0$ as $n \to \infty$ for any fixed u.*

Inspired by this result, let the autocorrelation function of $\{y_t\}$ be defined as

$$\rho_y(u) := r_y(u)/r_y(0). \tag{4.2.14}$$

Then, by Corollary 4.1, $\hat{\rho}_y(u)$ is a consistent estimator of $\rho_y(u)$ under the conditions of Theorem 4.1. An alternative expression for $\rho_y(u)$ is

$$\rho_y(u) = (1-\lambda)\rho_x(u) + \lambda\rho_\epsilon(u),$$

where $\rho_x(u) := r_x(u)/r_x(0)$ can be regarded as the autocorrelation function of $\{x_t\}$, $\rho_\epsilon(u) := r_\epsilon(u)/r_\epsilon(0)$ is the ordinary autocorrelation function of $\{\epsilon_t\}$, and

$$\lambda := r_\epsilon(0)/r_y(0) = r_\epsilon(0)/(r_x(0) + r_\epsilon(0))$$

is the mixing coefficient which takes values in the interval $[0,1]$ and depends solely on the total SNR $r_x(0)/r_\epsilon(0)$.

4.3 Covariances and Asymptotic Normality

In this section, we investigate the second moments and the asymptotic distributions of the sample autocovariances and autocorrelation coefficients.

First, consider the asymptotic expression for the covariances, and in the complex case, the complementary covariances, of $\{\hat{r}_y(u)\}$. The following theorem is the generalization of a classical result [46, p. 226] derived for real time series with continuous spectra to the case of real and complex time series with mixed spectra. It is also an extension of the results in [223] and [235] which assume fixed frequencies. See Section 4.5 for a proof.

Theorem 4.2 (Covariances of the Sample Autocovariance Function). *Let the conditions of Theorem 4.1 be satisfied. In addition, assume that the third moments* $r_\zeta(u,v) := E(\zeta_{t+u}^*\zeta_{t+v}^*\zeta_t)$ *do not depend on t and satisfy* $\sum\sum|r_\zeta(u,v)| < \infty$. *In the complex case, also assume that* $E(\zeta_{t+u}\zeta_t) = \iota\sigma^2\delta_u$ *for all t and u, where* $\{\delta_u\}$ *denotes the Kronecker delta sequence. Then, for any given u and v,*

$$\begin{aligned}\mathrm{Cov}\{\hat{r}_y(u), \hat{r}_y(v)\} &= n^{-1}\gamma(u,v) + \mathcal{O}(n^{-1}),\\ \mathrm{Cov}\{\hat{r}_y(u), \hat{r}_y^*(v)\} &= n^{-1}\gamma(u,-v) + \mathcal{O}(n^{-1}),\end{aligned}$$

where $\gamma(u,v) := \gamma_1(u,v) + \gamma_2(u,v)$. *In the real case,*

$$\gamma_1(u,v) := \sum_{k=1}^{q} 2C_k^2 f_\epsilon(\omega_k)\cos(\omega_k u)\cos(\omega_k v), \tag{4.3.1}$$

$$\begin{aligned}\gamma_2(u,v) := {} & (\kappa-3)r_\epsilon(u)r_\epsilon(v)\\ & + \sum_{\tau=-\infty}^{\infty}\{r_\epsilon(\tau+u-v)r_\epsilon(\tau) + r_\epsilon(\tau+u)r_\epsilon(\tau-v)\}.\end{aligned} \tag{4.3.2}$$

In the complex case,

$$\gamma_1(u,v) := \sum_{k=1}^{p} 2C_k^2 f_\epsilon(\omega_k)\exp\{i\omega_k(u-v)\},$$

$$+ \sum_{\substack{k,k'=1 \\ \omega_k=-\omega_{k'}}}^{p} 2\Re\{\beta_k^*\beta_{k'}^*\iota g_\epsilon(\omega_k)\exp\{i\omega_k(u+v)\}\} \tag{4.3.3}$$

$$\gamma_2(u,v) := (\kappa-2-\xi)r_\epsilon(u)r_\epsilon^*(v)$$

$$+ \sum_{\tau=-\infty}^{\infty}\{r_\epsilon(\tau+u-v)r_\epsilon^*(\tau)+\xi c_\epsilon(\tau+u)c_\epsilon^*(\tau-v)\}, \tag{4.3.4}$$

where $c_\epsilon(u) := \sigma^2\sum_j \psi_j\psi_{j+u}$ *and* $g_\epsilon(\omega) := \sum c_\epsilon(u)\exp(-iu\omega)$. *Moreover,*

$$\mathrm{Var}\{\hat{r}_y(u)\} = \mathcal{O}(n^{-1})$$

uniformly in $|u| < n$.

Remark 4.5 Let $\Psi(\omega) := \sum \psi_j \exp(-ij\omega)$. Then, we have $g_\epsilon(\omega) = \sigma^2\Psi(\omega)\Psi(-\omega)$ and $f_\epsilon(\omega) = \sigma^2|\Psi(\omega)|^2$.

Remark 4.6 The assumption about $\{\zeta_t\}$ is satisfied if it is an i.i.d. sequence, $\{\zeta_t\} \sim \mathrm{IID}(0,\sigma^2)$, with $E(|\zeta_t|^4) = \kappa\sigma^4 < \infty$, in which case, $\xi = |\iota|^2$ and $E(\epsilon_{t+u}\epsilon_t) = \iota c_\epsilon(u)$, where $\iota := E(\zeta_t^2)/\sigma^2$. An important example is Gaussian white noise where $\{\zeta_t\} \sim \mathrm{GWN}(0,\sigma^2)$ or $\{\zeta_t\} \sim \mathrm{GWN}_c(0,\sigma^2)$. In the real Gaussian case, we have $\kappa = 3$, so the first term of $\gamma_2(u,v)$ in (4.3.2) vanishes. This leads to the expression given by [149, Theorem 4.2]. In the complex Gaussian case, we have $\kappa = 2$ and $\xi = \iota = 0$, so the first term of $\gamma_2(u,v)$ in (4.3.2), the second term of $\gamma_1(u,v)$ in (4.3.3), and the third term of $\gamma_2(u,v)$ in (4.3.4) are all equal to zero.

Remark 4.7 The assumption about $\{\zeta_t\}$ is satisfied if $\{\zeta_t\}$ is a sequence of martingale differences with respect to certain filtration $\{\mathfrak{F}_t\}$, i.e., $\{\zeta_t\} \sim \mathrm{MD}(0,\sigma^2)$, such that $E(|\zeta_t|^4) = \kappa\sigma^4 < \infty$, $E(|\zeta_t|^2|\mathfrak{F}_{t-1}) = \sigma^2$, $E(\zeta_t^2|\mathfrak{F}_{t-1}) = \iota\sigma^2$, and $E(|\zeta_t|^2\zeta_t^*) = \alpha\sigma^3$ for all t. To justify this assertion, recall Remark 4.4, where $\{\zeta_t\}$ is shown to satisfy the assumption in Theorem 4.1 with $\xi = |\iota|^2$. In addition, it is easy to show, by iterated expectation, that for any fixed t, $E(\zeta_{t+u}^*\zeta_{t+v}^*\zeta_t) = \alpha\sigma^3$ if $u = v = 0$, and $E(\zeta_{t+u}^*\zeta_{t+v}^*\zeta_t) = 0$ otherwise. This implies that $E(\zeta_{t+u}^*\zeta_{t+v}^*\zeta_t)$ does not depend on t and $\sum\sum|E(\zeta_u^*\zeta_v^*\zeta_0)| = |\alpha|\sigma^3 < \infty$. Similarly, $E(\zeta_{t+u}\zeta_t) = \iota\sigma^2\delta_u$. Therefore, the additional assumption in Theorem 4.2 is satisfied.

The expression of $\gamma_2(u,v)$ in (4.3.2) is known as Bartlett's formula for the sample autocovariances of real time series with continuous spectra [25] [298, p. 326]. The expression of $\gamma_2(u,v)$ in (4.3.4) is a generalization of Bartlett's formula to the complex case. These formulas have alternative spectral representations. Indeed, using the identity (4.2.10), the following spectral representation of $\gamma_2(u,v)$ can be easily obtained: For the real case,

$$\begin{aligned}\gamma_2(u,v) &= (\kappa-3)r_\epsilon(u)r_\epsilon(v)+(2\pi)^{-1}\int_{-\pi}^{\pi}|f_\epsilon(\omega)|^2\{\exp(i\omega(u-v))\\ &\quad +\exp(i\omega(u+v))\}\,d\omega\\ &= (\kappa-3)r_\epsilon(u)r_\epsilon(v)+(2\pi)^{-1}\int_{-\pi}^{\pi}2|f_\epsilon(\omega)|^2\cos(\omega u)\cos(\omega v)\,d\omega,\end{aligned}$$

where the second equality is due to symmetry. For the complex case,

$$\begin{aligned}\gamma_2(u,v) &= (\kappa-2-\xi)r_\epsilon(u)r_\epsilon^*(v)+\ (2\pi)^{-1}\int_{-\pi}^{\pi}\{|f_\epsilon(\omega)|^2\exp(i\omega(u-v))\\ &\quad +\xi|g_\epsilon(\omega)|^2\exp(i\omega(u+v))\}\,d\omega.\end{aligned}$$

Similar spectral representations can be obtained for $\gamma_1(u,v)$: In the real case,

$$\gamma_1(u,v)=(2\pi)^{-1}\int_{-\pi}^{\pi}4f_\epsilon(\omega)f_x(\omega)\cos(\omega u)\cos(\omega v)\,d\omega,$$

where $f_x(\omega)$ is the SDF of $\{x_t\}$ given by the first expression in (4.2.11). In the complex case, we have

$$\begin{aligned}\gamma_1(u,v) &= (2\pi)^{-1}\int_{-\pi}^{\pi}\{2f_\epsilon(\omega)f_x(\omega)\exp(i\omega(u-v))\\ &\quad +2\Re(\iota g_\epsilon(\omega)g_x^*(\omega))\exp(i\omega(u+v))\}\,d\omega,\end{aligned}$$

where $g_x(\omega)$ is defined as

$$g_x(\omega):=2\pi\sum_{\substack{k,k'=1\\ \omega_k=-\omega_{k'}}}^{p}\beta_k\beta_{k'}\delta(\omega-\omega_k) \tag{4.3.5}$$

and $f_x(\omega)$ is given by the second expression in (4.2.11).

In the complex case, the covariances and complementary covariances of the sample autocovariance function can be described equivalently by the covariances of its real and imaginary parts. In fact, with the notation $\hat{r}_y^{(1)}(u):=\Re\{\hat{r}_y(u)\}$ and $\hat{r}_y^{(2)}(u):=\Im\{\hat{r}_y(u)\}$, we can write

$$\mathrm{Cov}\{\hat{r}_y^{(j)}(u),\hat{r}_y^{(l)}(v)\}=n^{-1}\gamma_{jl}(u,v)+\mathcal{O}(n^{-1})\qquad (j,l=1,2), \tag{4.3.6}$$

where the $\gamma_{jl}(u,v)$ are determined by $\gamma(u,v)$ and $\gamma(u,-v)$ such that

$$[\gamma_{jl}(u,v)]_{j,l=1}^2:=\frac{1}{2}\begin{bmatrix}\Re\{\gamma(u,v)+\gamma(u,-v)\} & -\Im\{\gamma(u,v)-\gamma(u,-v)\}\\ \Im\{\gamma(u,v)+\gamma(u,-v)\} & \Re\{\gamma(u,v)-\gamma(u,-v)\}\end{bmatrix}. \tag{4.3.7}$$

Conversely, $\gamma(u,v)$ and $\gamma(u,-v)$ can be expressed in terms of the $\gamma_{jl}(u,v)$:

$$\begin{cases} \gamma(u,v) = \gamma_{11}(u,v) + \gamma_{22}(u,v) + i\{\gamma_{21}(u,v) - \gamma_{12}(u,v)\}, \\ \gamma(u,-v) = \gamma_{11}(u,v) - \gamma_{22}(u,v) + i\{\gamma_{21}(u,v) + \gamma_{12}(u,v)\}. \end{cases} \tag{4.3.8}$$

Therefore, $\{\gamma_{jl}(u,v) : j,l = 1,2\}$ is equivalent to $\{\gamma(u,v), \gamma(u,-v)\}$ for any given u and v. Note that because $\hat{r}_y(0)$ is real, we have $\hat{r}_y^{(2)}(0) = 0$, which implies $\gamma_{21}(0,v) = \gamma_{22}(0,v) = \gamma_{12}(u,0) = \gamma_{22}(u,0) = 0$ and $\gamma_{11}(0,0) = \gamma(0,0)$.

Now let us use the formulas in Theorem 4.2 to investigate the variance of $\hat{r}_y(u)$ in three simple examples.

Example 4.1 (Real Sinusoids in Gaussian White Noise). Consider the RSM and assume that $\{\epsilon_t\} \sim \text{GWN}(0,\sigma^2)$. In this case, $\epsilon_t = \zeta_t$, $\kappa = 3$, $r_\epsilon(u) = \sigma^2\delta_u$, and $f_\epsilon(\omega) = \sigma^2$. It is easy to verify that

$$\gamma(u,v) = \sum_{k=1}^{q} 2C_k^2\sigma^2 \cos(\omega_k u)\cos(\omega_k v) + \sigma^4(\delta_{u-v} + \delta_{u+v}).$$

According to Theorem 4.2, we have

$$\text{Var}\{\hat{r}_y(u)\} \approx n^{-1}\Big\{\sum_{k=1}^{q} 2C_k^2\sigma^2\cos^2(\omega_k u) + \sigma^4(1+\delta_u)\Big\}.$$

In the absence of sinusoids, $\text{Var}\{\hat{r}_y(u)\} \approx n^{-1}\sigma^4(1+\delta_u)$. ◇

Example 4.2 (Real Sinusoids in Gaussian AR(1) Noise). Consider the RSM and assume that $\{\epsilon_t\}$ is a Gaussian AR(1) process given by

$$\epsilon_t + \varphi\epsilon_{t-1} = \zeta_t, \quad \{\zeta_t\} \sim \text{GWN}(0,\sigma^2),$$

where $\varphi \in (-1,1)$. In this case, $f_\epsilon(\omega) = \sigma^2/|\Phi(\omega)|^2$, $\Phi(\omega) := 1 + \varphi\exp(-i\omega)$, and $r_\epsilon(u) = \sigma^2(-\varphi)^{|u|}/(1-|\varphi|^2)$. It is easy to verify that

$$\begin{aligned} \text{Var}\{\hat{r}_y(u)\} \approx n^{-1}\Big\{&\sum_{k=1}^{q} 2C_k^2\sigma^2\cos^2(\omega_k u)/|\Phi(\omega_k)|^2 \\ &+ \sigma^4(1+2\varphi^{2u})(1+\varphi)/(1-\varphi) + 2\sigma^4 u\varphi^{2u}\Big\}. \end{aligned}$$

Under the condition that $u \gg \log n$, this expression can be further simplified as

$$\text{Var}\{\hat{r}_y(u)\} \approx n^{-1}\Big\{\sum_{k=1}^{q} 2C_k^2\sigma^2\cos^2(\omega_k u)/|\Phi(\omega_k)|^2 + \sigma^4(1+\varphi)/(1-\varphi)\Big\}.$$

In the absence of sinusoids, it reduces to $\text{Var}\{\hat{r}_y(u)\} \approx n^{-1}\sigma^4(1+\varphi)/(1-\varphi)$. ◇

Example 4.3 (Complex Sinusoids in Gaussian White Noise). Consider the CSM and assume that $\{\epsilon_t\} \sim \mathrm{GWN}_c(0,\sigma^2)$. Because $\kappa = 2$, $\xi = \iota = 0$, $r_\epsilon(u) = \sigma^2\delta_u$, and $f_\epsilon(\omega) = \sigma^2$, it follows that

$$\gamma(u,v) = \sum_{k=1}^{p} 2C_k^2\sigma^2 \exp\{i\omega_k(u-v)\} + \sigma^4\delta_{u-v}.$$

Therefore, by Theorem 4.2,

$$\mathrm{Var}\{\hat{r}_y(u)\} \approx n^{-1}\Big\{\sum_{k=1}^{p} 2C_k^2\sigma^2 + \sigma^4\Big\}.$$

In the absence of sinusoids, we obtain $\mathrm{Var}\{\hat{r}_y(u)\} \approx n^{-1}\sigma^4$. ◇

The following result, stated without proof, can be easily obtained from Theorem 4.1 and Theorem 4.2.

Corollary 4.2 (Mean-Square Error of the Sample Autocovariance Function). *Let the conditions of Theorem 4.2 be satisfied. Then, for any given u and v,*

$$E\{(\hat{r}_y(u) - r_y(u))(\hat{r}_y(v) - r_y(v))^*\} = n^{-1}\gamma(u,v) + \mathcal{O}(n^{-2}\Delta^{-2}) + o(n^{-1}),$$
$$E\{(\hat{r}_y(u) - r_y(u))(\hat{r}_y(v) - r_y(v))\} = n^{-1}\gamma(u,-v) + \mathcal{O}(n^{-2}\Delta^{-2}) + o(n^{-1}),$$

where $r_y(u)$ is the ACF of $\{y_t\}$ defined by (4.2.9).

According to Corollary 4.2, the mean-square error of $\hat{r}_y(u)$ as an estimator of $r_y(u)$ can be approximated by $n^{-1}\gamma(u,v)$ for large n if Δ satisfies

$$\lim_{n\to\infty} n\Delta^2 = \infty. \tag{4.3.9}$$

This is a stronger condition than (4.2.2) required by Theorems 4.1 and 4.2. It can be attributed entirely to the bias of $\hat{r}_y(u)$, which takes the form $\mathcal{O}(n^{-1}\Delta^{-1})$. If (4.3.9) is violated, the asymptotic expressions in Corollary 4.2 can become inaccurate, even for very large n, as illustrated by the following example.

Example 4.4 (Impact of Frequency Separation on the Mean-Square Error of the Sample Autocovariances). In this example, the normalized mean-square error, $n\mathrm{MSE}\{\hat{r}_y(0)\}$, is computed by a Monte Carlo simulation for time series comprising two real sinusoids in Gaussian white noise with $\omega_1 = 2\pi \times 0.1$, $\omega_2 = \omega_1 + \Delta$, $C_1 = C_2 = 1$, $\phi_1 = \phi_2 = 0$, and $\sigma^2 = 1$ (thus $r_y(0) = 2$). The simulation is carried out for different sample sizes and under different frequency separation conditions of the form $\Delta = 2\pi/n^d$. The results are shown in Table 4.1 and should be compared with the asymptotic value given by Corollary 4.2, which is $\gamma(0,0) = 6$. Note that (4.3.9) is satisfied only if $d < 1/2$.

As can be seen from Table 4.1, the theoretical value $\gamma(0,0) = 6$ is a reasonably good approximation for the normalized mean-square error in the case of $d = 0.4$

Table 4.1. Variance and Mean-Square Error of Sample Variance

		Sample Size n								
	d	200	400	600	800	1000	1200	1400	1600	2000
VAR	0.4	6.02	6.02	5.92	6.05	6.06	6.04	5.93	5.95	5.96
	0.7	6.12	6.10	5.90	6.03	5.97	6.02	5.90	6.04	5.95
	0.8	6.02	6.16	5.90	5.58	6.03	6.15	6.18	5.98	5.92
MSE	0.4	6.02	6.02	5.92	6.05	6.06	6.04	5.94	5.97	5.97
	0.7	6.24	6.12	6.25	6.06	6.03	6.21	6.44	6.30	6.47
	0.8	6.33	6.79	6.34	7.19	6.07	7.14	8.08	6.97	6.46

Results are based on 10,000 Monte Carlo runs.

which satisfies the condition (4.3.9). But, in the case of $d = 0.7$ and $d = 0.8$ where (4.3.9) is violated, the approximation can be very poor, even for very large sample sizes such as $n = 2{,}000$. The poor approximation is largely due to the bias of the sample variance which varies with the frequency separation condition and the sample size. The role of bias can be appreciated by comparing the mean-square error with the variance which is also shown in Table 4.1 (similarly normalized). Note that both $d = 0.7$ and $d = 0.8$ satisfy the weaker condition (4.2.2), so Theorems 4.1 and 4.2 remain valid. Therefore, it is not surprising to see in Table 4.1 that the theoretical value $\gamma(0,0) = 6$ serves as a reasonably good approximation to the normalized variance $n\,\mathrm{Var}\{\hat{r}_y(0)\}$ in all cases. ◇

Equipped with Theorems 4.1 and 4.2, let us now investigate the asymptotic distribution of the sample autocovariances. For the real case, the following result can be obtained. It is the generalization of a classical result for time series with continuous spectra [46, p. 228] and of the result in [235] which is derived under the assumption of fixed frequencies. See Section 4.5 for a proof.

Theorem 4.3 (Asymptotic Normality of the Sample Autocovariances: Real Case). *Let $\{y_t\}$ satisfy (4.1.3), where $\{x_t\}$ is given by (2.1.1) or (2.1.2) and $\{\epsilon_t\}$ is a linear process of the form (4.2.3) with $\sum|\psi_j| < \infty$, $\{\zeta_t\} \sim \mathrm{IID}(0,\sigma^2)$, and $E(\zeta_t^4) = \kappa\sigma^4 < \infty$. Assume that the separation condition (4.3.9) is satisfied. Then, for any fixed integer $m \ge 0$ and as $n \to \infty$, the random variables $\sqrt{n}\{\hat{r}_y(u) - r_y(u)\}$ $(u = 0, 1, \ldots, m)$ are asymptotically jointly Gaussian with mean zero and with covariances $\gamma(u,v)$ defined by (4.3.1) and (4.3.2).*

Remark 4.8 Theorem 4.3 remains valid if $\{\zeta_t\}$ is a sequence of martingale differences satisfying the conditions in Remark 4.7. In the absence of sinusoids, $\gamma(u,v) = \gamma_2(u,v)$ is given by (4.3.2), so Theorem 4.3 reduces to the classical result for time series with continuous spectra [46, p. 228].

The asymptotic normality assertion in Theorem 4.3 can be used to construct confidence intervals for the ACF, as demonstrated by the following example.

Example 4.5 (Confidence Interval for the Variance: Real Case). Consider the RSM (2.1.1) or (2.1.2) and assume that $\{\epsilon_t\} \sim \text{GWN}(0, \sigma^2)$. By Theorem 4.3,

$$\sqrt{n}\{\hat{r}_y(0) - r_y(0)\} \xrightarrow{D} \text{N}(0, \gamma(0,0)),$$

where

$$\gamma(0,0) = \sum_{k=1}^{q} C_k^2 + 2\sigma^4. \tag{4.3.10}$$

Therefore, a symmetric confidence interval for $r_y(0)$ with an asymptotic coverage probability $1-\alpha \in (0,1)$ is given by

$$\hat{r}_y(0) \pm z_{\alpha/2}\sqrt{\hat{\gamma}(0,0)/n},$$

where $z_{\alpha/2}$ denotes the $(1-\alpha/2)$th quantile of the standard Gaussian distribution and $\hat{\gamma}(0,0)$ is a consistent estimator of $\gamma(0,0)$ which can be obtained, for example, by substituting C_k and σ^2 in (4.3.10) with their consistent estimates. ◇

A similar result can be obtained for the complex case. The following theorem is an extension of the result in [223] by allowing the frequencies to depend on n. See Section 4.5 for a proof.

Theorem 4.4 (Asymptotic Normality of the Sample Autocovariances: Complex Case). *Let $\{y_t\}$ satisfy (4.1.3), where $\{x_t\}$ is given by (2.1.5) or (2.1.7) and $\{\epsilon_t\}$ is a linear process of (4.2.3) with $\sum|\psi_j| < \infty$, $\{\zeta_t\} \sim \text{IID}(0, \sigma^2)$, $E(|\zeta_t|^4) = \kappa\sigma^4 < \infty$, and $E(\zeta_t^2) = \iota\sigma^2$. Assume that the condition (4.3.9) is satisfied. Then, for any fixed integer $m \ge 0$ and as $n \to \infty$, the random variables $\sqrt{n}\{\hat{r}_y(u) - r_y(u)\}$ $(u = 0, 1, \ldots, m)$ are asymptotically jointly general complex Gaussian with mean zero and with covariances $\gamma(u,v)$ and complementary covariances $\gamma(u,-v)$ given by (4.3.3) and (4.3.4) where $\xi = |\iota|^2$.*

Remark 4.9 Theorem 4.4 remains valid if $\{\zeta_t\}$ is a sequence of martingale differences satisfying the conditions in Remark 4.7.

The asymptotic distribution of the sample autocorrelation coefficients can be obtained by an application of Proposition 2.15. The results are summarized in the following theorem. See Section 4.5 for a proof.

Theorem 4.5 (Asymptotic Normality of the Sample Autocorrelation Coefficients: Complex Case). *Let the conditions of Theorem 4.4 be satisfied. Then, for fixed integer $m > 0$ and as $n \to \infty$, the random variables $\sqrt{n}\{\hat{\rho}_y(u) - \rho_y(u)\}$ $(u = 1, \ldots, m)$ are asymptotically jointly general complex Gaussian with mean zero, covariances $\sigma(u,v)$, complementary covariances $\sigma(u,-v)$, where for any u and v, $\sigma(u,v) := \sigma_1(u,v) + \sigma_2(u,v)$, with*

$$\sigma_1(u,v) := \lambda^2 \sum_{k=1}^{p} 2\bar{C}_k^{-2}\bar{f}_\epsilon(\omega_k)\{\rho_y(u) - \exp(i\omega_k u)\}\{\rho_y(v) - \exp(i\omega_k v)\}^*$$

$$+\lambda^2 \sum_{\substack{k,k'=1\\ \omega_k=-\omega_{k'}}}^{p} 2\Re\{\bar\beta_k^* \bar\beta_{k'}^* \iota \bar g_\epsilon(\omega_k)\}$$
$$\times \{\rho_y(u) - \exp(i\omega_k u)\}\{\rho_y(v) - \exp(-i\omega_k v)\}^*$$
$$+\lambda^2(\kappa - 2 - \xi)\{\rho_y(u) - \rho_\epsilon(u)\}\{\rho_y(v) - \rho_\epsilon(v)\}^*, \quad (4.3.11)$$

$$\sigma_2(u,v) := \lambda^2 \sum_{\tau=-\infty}^{\infty} \big[\rho_y(u)\rho_y^*(v)\{|\rho_\epsilon(\tau)|^2 + \xi|\bar c_\epsilon(\tau)|^2\}$$
$$-\rho_y(u)\{\rho_\epsilon(\tau - v)\rho_\epsilon^*(\tau) + \xi\,\bar c_\epsilon(\tau)\,\bar c_\epsilon^*(\tau - v)\}$$
$$-\rho_y^*(v)\{\rho_\epsilon(\tau+u)\rho_\epsilon^*(\tau) + \xi\,\bar c_\epsilon(\tau+u)\,\bar c_\epsilon^*(\tau)\}$$
$$+\rho_\epsilon(\tau+u)\rho_\epsilon^*(\tau+v) + \xi\,\bar c_\epsilon(\tau+u)\,\bar c_\epsilon^*(\tau - v)\big]. \quad (4.3.12)$$

In these expressions, $\lambda := r_\epsilon(0)/r_y(0)$, $\bar\beta_k := \beta_k/\sqrt{r_\epsilon(0)}$, $\bar C_k := |\bar\beta_k|^2 = C_k/\sqrt{r_\epsilon(0)}$, $\bar f_\epsilon(\omega) := f_\epsilon(\omega)/r_\epsilon(0)$, $\bar g_\epsilon(\omega) := g_\epsilon(\omega)/r_\epsilon(0)$, *and* $\bar c_\epsilon(u) := c_\epsilon(u)/r_\epsilon(0)$.

Remark 4.10 Define $\hat\rho_y^{(1)}(u) := \Re\{\hat\rho_y(u)\}$, $\hat\rho_y^{(2)}(u) := \Im\{\hat\rho_y(u)\}$, $\rho_y^{(1)}(u) := \Re\{\rho_y(u)\}$, $\rho_y^{(2)}(u) := \Im\{\rho_y(u)\}$. Then, the random variables $\sqrt{n}\{\hat\rho_y^{(j)}(u) - \rho_y^{(j)}(u)\}$ ($u = 1, \ldots, m$; $j = 1, 2$) are asymptotically jointly Gaussian with mean zero and with covariances $\sigma_{jl}(u,v)$ defined by $\sigma(u,v)$ and $\sigma(u,-v)$ in the same way as $\gamma_{jl}(u,v)$ in (4.3.7).

Remark 4.11 The last term of $\sigma_1(u,v)$ makes it depend on the fourth moments of $\{\epsilon_t\}$. This term equals zero when $\{\zeta_t\}$ is Gaussian, in which case $\sigma(u,v)$ depends only on the second-order statistics of $\{\epsilon_t\}$. Using (4.2.14), the last term of $\sigma_1(u,v)$ can be written as $\lambda^2(1-\lambda)^2(\kappa - 2 - \xi)\{\rho_x(u) - \rho_\epsilon(u)\}\{\rho_x(v) - \rho_\epsilon(v)\}^*$.

Note that a spectral representation also exists for $\sigma(u,v)$. In fact, because

$$\rho_\epsilon(u) = (2\pi)^{-1}\int_{-\pi}^{\pi} \bar f_\epsilon(\omega)\exp(i\omega u)\,d\omega,$$
$$\bar c_\epsilon(u) = (2\pi)^{-1}\int_{-\pi}^{\pi} \bar g_\epsilon(\omega)\exp(i\omega u)\,d\omega,$$

it follows from (4.3.11) and (4.3.12) that

$$\sigma_1(u,v) = \lambda^2(2\pi)^{-1}\int_{-\pi}^{\pi} \{2\bar f_\epsilon(\omega)\,\bar f_x(\omega)(\rho_y(u) - \exp(i\omega u))(\rho_y(v) - \exp(i\omega v))^*$$
$$+2\Re(\iota \bar g_\epsilon(\omega)\,\bar g_x^*(\omega))(\rho_y(u) - \exp(i\omega u))(\rho_y(v) - \exp(-i\omega v))^*\}\,d\omega$$
$$+\lambda^2(\kappa - 2 - \xi)\{\rho_y(u) - \rho_\epsilon(u)\}\{\rho_y(v) - \rho_\epsilon(v)\}^*,$$
$$\sigma_2(u,v) = \lambda^2(2\pi)^{-1}\int_{-\pi}^{\pi} \{|\bar f_\epsilon(\omega)|^2(\rho_y(u) - \exp(i\omega u))(\rho_y(v) - \exp(i\omega v))^*$$
$$+\xi\,|\bar g_\epsilon(\omega)|^2(\rho_y(u) - \exp(i\omega u))(\rho_y(v) - \exp(-i\omega v))^*\}\,d\omega,$$

where $\bar f_x(\omega) := f_x(\omega)/r_\epsilon(0)$, $\bar g_x(\omega) := g_x(\omega)/r_\epsilon(0)$, with $g_x(\omega)$ given by (4.3.5).

Theorem 4.5 can be used to construct confidence intervals for the autocorrelation coefficients. This is demonstrated in the following example.

Example 4.6 (Single Complex Sinusoid in Gaussian White Noise). Let $p = 1$ in the CSM and assume that $\{\epsilon_t\} \sim \text{GWN}(0,\sigma^2)$. Then, we have $\kappa = 2$, $\xi = \iota = 0$, $\rho_\epsilon(\tau) = \delta_u$, and $\bar{f}_\epsilon(\omega) = 1$. Therefore, for any $u, v > 0$,

$$\begin{aligned}\sigma_1(u,v) &= 2\lambda^2\bar{C}_1^2\{\rho_y(u) - \exp(i\omega_1 u)\}\{\rho_y(v) - \exp(i\omega_1 v)\}^*,\\ \sigma_2(u,v) &= \lambda^2\{\rho_y(u)\rho_y^*(v) + \delta_{u-v}\},\end{aligned}$$

where $\rho_y(u) = (1-\lambda)\exp(i\omega_1 u) + \lambda\delta_u$, $\lambda = \sigma^2/(C_1^2 + \sigma^2)$, and $\bar{C}_1^2 = C_1^2/\sigma^2 = \gamma_1$ is the SNR. Because $\sigma(u,v) = \sigma_1(u,v) + \sigma_2(u,v)$, it follows that for any $u > 0$,

$$\begin{aligned}\sigma(u,u) &= 2\lambda^2\bar{C}_1^2|\rho_y(u) - \exp(i\omega_1 u)|^2 + \lambda^2\{|\rho_y(u)|^2 + 1\} = \lambda^2(\mu+1),\\ \sigma(u,-u) &= 2\lambda^2\bar{C}_1^2\{\rho_y(u) - \exp(i\omega_1 u)\}^2 + \lambda^2\{\rho_y(u)\}^2 = \lambda^2\mu\exp(i2\omega_1 u),\end{aligned}$$

where $\mu := 2\bar{C}_1^2\lambda^2 + (1-\lambda)^2 = 1 - \lambda^2$. This gives

$$\begin{aligned}\sigma_{11}(u,u) &= \tfrac{1}{2}\Re\{\sigma(u,u) + \sigma(u,-u)\} = \tfrac{1}{2}\lambda^2\{1 + \mu(1 + \cos(2\omega_1 u))\},\\ \sigma_{22}(u,u) &= \tfrac{1}{2}\Re\{\sigma(u,u) - \sigma(u,-u)\} = \tfrac{1}{2}\lambda^2\{1 + \mu(1 - \cos(2\omega_1 u))\},\\ \sigma_{12}(u,u) &= -\tfrac{1}{2}\Im\{\sigma(u,u) - \sigma(u,-u)\} = \tfrac{1}{2}\lambda^2\mu\sin(2\omega_1 u).\end{aligned}$$

Therefore, confidence intervals for $\Re\{\rho_y(u)\}$ and $\Im\{\rho_y(u)\}$ are given by

$$\begin{aligned}&\Re\{\hat{\rho}_y(u)\} \pm z_{\alpha/2}\sqrt{\hat{\sigma}_{11}(u,u)/n},\\ &\Im\{\hat{\rho}_y(u)\} \pm z_{\alpha/2}\sqrt{\hat{\sigma}_{22}(u,u)/n},\end{aligned}$$

where $\hat{\sigma}_{11}(u,u)$ and $\hat{\sigma}_{22}(u,u)$ are obtained by substituting consistent estimators of β_1, σ^2, and ω_1 in the foregoing expression of $\sigma_{11}(u,u)$ and $\sigma_{22}(u,u)$. Note that because $\sigma_{12}(u,u)$ is not identically equal to zero, $\Re\{\hat{\rho}_y(u)\}$ and $\Im\{\hat{\rho}_y(u)\}$ are not asymptotically independent in general. ◇

Because the RSM is a special CSM with constrained parameters, the following result, stated without proof, can be obtained directly from Theorem 4.5. It generalizes a classical result for time series with continuous spectra [46, p. 221].

Corollary 4.3 (Asymptotic Normality of the Sample Autocorrelation Coefficients: Real Case). *Let the conditions of Theorem 4.3 be satisfied. Then, for any given integer $m > 0$ and as $n \to \infty$, the random variables $\sqrt{n}\{\hat{\rho}_y(u) - \rho_y(u)\}$ $(u = 1, \ldots, m)$ are asymptotically jointly Gaussian with mean zero and covariances $\sigma(u,v) := \sigma_1(u,v) + \sigma_2(u,v)$, where*

$$\begin{aligned}\sigma_1(u,v) := \lambda^2\sum_{k=1}^{q} 2\bar{C}_k^2\bar{f}_\epsilon(\omega_k)\{\rho_y(u) - \cos(\omega_k u)\}\{\rho_y(v) - \cos(\omega_k v)\}\\ + \lambda^2(\kappa-3)\{\rho_y(u) - \rho_\epsilon(u)\}\{\rho_y(v) - \rho_\epsilon(v)\},\end{aligned} \tag{4.3.13}$$

$$
\begin{aligned}
\sigma_2(u,v) := \lambda^2 \sum_{\tau=-\infty}^{\infty} \{&2\rho_y(u)\,\rho_y(v)\,\rho_\epsilon^2(\tau) \\
&-2\rho_y(u)\,\rho_\epsilon(\tau-v)\,\rho_\epsilon(\tau) - 2\rho_y(v)\,\rho_\epsilon(\tau+u)\,\rho_\epsilon(\tau) \\
&+\rho_\epsilon(\tau+u)\,\rho_\epsilon(\tau+v) + \rho_\epsilon(\tau+u)\,\rho_\epsilon(\tau-v)\},
\end{aligned}
\tag{4.3.14}
$$

with $\lambda := r_\epsilon(0)/r_y(0)$, $\bar{C}_k := C_k/\sqrt{r_\epsilon(0)}$, *and* $\bar{f}_\epsilon(\omega) := f_\epsilon(\omega)/r_\epsilon(0)$.

The expression of $\sigma_2(u,v)$ in (4.3.14) is known as Bartlett's formula for the sample autocorrelation coefficients [25] [46, p. 221], which is valid for real time series with continuous spectra. The expression of $\sigma_2(u,v)$ in (4.3.12) is a generalization of this formula to the complex case.

Because $\rho_\epsilon(u) = \rho_\epsilon(-u)$, $\sigma_2(u,v)$ in (4.3.14) can be written as

$$
\begin{aligned}
\sigma_2(u,v) &= \lambda^2 \sum_{\tau=1}^{\infty} \{\rho_\epsilon(\tau+u) + \rho_\epsilon(\tau-u) - 2\rho_y(u)\rho_\epsilon(\tau)\} \\
&\qquad \times \{\rho_\epsilon(\tau+v) + \rho_\epsilon(\tau-v) - 2\rho_y(v)\rho_\epsilon(\tau)\}.
\end{aligned}
$$

A spectral representation of $\sigma_1(u,v)$ and $\sigma_2(u,v)$ is given by

$$
\begin{aligned}
\sigma_1(u,v) &= \lambda^2(2\pi)^{-1}\int_{-\pi}^{\pi} 4\bar{f}_\epsilon(\omega)\,\bar{f}_x(\omega)\{\rho_y(u)-\cos(\omega u)\}\{\rho_y(v)-\cos(\omega v)\}\,d\omega \\
&\quad + \lambda^2(\kappa-3)\,\{\rho_y(u)-\rho_\epsilon(u)\}\{\rho_y(v)-\rho_\epsilon(v)\}, \\
\sigma_2(u,v) &= \lambda^2(2\pi)^{-1}\int_{-\pi}^{\pi} 2\bar{f}_\epsilon^2(\omega)\{\rho_y(u)-\cos(\omega u)\}\{\rho_y(v)-\cos(\omega v)\}\,d\omega,
\end{aligned}
$$

where $\bar{f}_x(\omega) := f_x(\omega)/r_\epsilon(0)$.

The following example illustrates Corollary 4.3.

Example 4.7 (Real Sinusoids in Gaussian AR(1) Noise). Consider the case discussed in Example 4.2. By Corollary 4.3, $\hat{\rho}_y(u) \stackrel{A}{\sim} \mathrm{N}(\rho_y(u), n^{-1}\sigma(u,u))$, where $\sigma(u,u) := \sigma_1(u,u) + \sigma_2(u,u)$. Because $\rho_\epsilon(u) = (-\varphi)^{|u|}$ and $\kappa = 3$, it follows that

$$
\begin{aligned}
\sigma_1(u,u) &= \lambda^2 \sum_{k=1}^{q} 2\bar{C}_k^2 \bar{f}_\epsilon(\omega_k)\{\rho_y(u)-\cos(\omega_k u)\}^2, \\
\sigma_2(u,u) &= \lambda^2 \sum_{\tau=1}^{\infty} \{(-\varphi)^{\tau+u} + (-\varphi)^{|\tau-u|} - 2\rho_y(u)(-\varphi)^{\tau}\}^2 \\
&= \lambda^2 \sum_{\tau=1}^{u} \{(-\varphi)^{\tau+u} + (-\varphi)^{u-\tau} - 2\rho_y(u)(-\varphi)^{\tau}\}^2 \\
&\quad + \lambda^2\varphi^2\{1+\varphi^{2u} - 2\rho_y(u)(-\varphi)^u\}^2/(1-\varphi^2).
\end{aligned}
$$

For $u = 1$ in particular,

$$
\sigma_2(1,1) = \lambda^2\{1+\varphi^2+2\rho_y(1)\varphi\}^2/(1-\varphi^2).
$$

In the absence of sinusoids, we have $\rho_y(1) = \rho_\epsilon(1) = -\varphi$ and $\lambda = 1$. This gives $\sigma_2(1,1) = 1-\varphi^2$, which means that $\hat{\rho}_y(1) \stackrel{A}{\sim} \mathrm{N}(-\varphi, n^{-1}(1-\varphi^2))$. ◇

4.4 Autocovariances of Filtered Time Series

Linear filtering is a common technique for processing time-series data. Given a time series of finite length, the boundary effect in filtering must be taken into account in order to ensure that the results obtained in the previous section are applicable to the filtered time series. Intuition tells us that the boundary effect should be negligible if the filter has a finite impulse response (FIR). For filters with infinite impulse response (IIR), a more careful discussion is needed.

Let $\{h_0, h_1, \ldots\}$ be the impulse response of a causal IIR linear filter which is BIBO stable, i.e., $\sum |h_j| < \infty$. By the theory of linear filtering [46, p. 122, Theorem 4.4.1], the filtered process

$$y_t(h) := \sum_{j=0}^{\infty} h_j y_{t-j} = x_t(h) + \epsilon_t(h) \tag{4.4.1}$$

is well defined in the mean-square sense, where $x_t(h) := \sum_{j=0}^{\infty} h_j x_{t-j}$ and $\epsilon_t(h) := \sum_{j=0}^{\infty} h_j \epsilon_{t-j}$ denote the filtered sinusoidal signal and the filtered noise, respectively. It is easy to show that $\{x_t(h)\}$ remains a sum of sinusoids with unchanged frequencies. For example, if $\{x_t\}$ is given by the CSM (2.1.5), then

$$x_t(h) = \sum_{k=1}^{p} \beta_k(h) \exp(i\omega_k t),$$

where $\beta_k(h) := \beta_k H(\omega_k) = C_k(h) \exp(i\phi_k(h))$, $C_k(h) := C_k |H(\omega_k)|$, $\phi_k(h) := \phi_k + \angle H(\omega_k)$, and $H(\omega) := \sum h_j \exp(-ij\omega)$. Moreover, if $\{\epsilon_t\}$ is a linear process of the form (4.2.3), then $\{\epsilon_t(h)\}$ remains a linear process and can be expressed as

$$\epsilon_t(h) = \sum_{j=-\infty}^{\infty} \psi_j(h) \zeta_{t-j},$$

where $\psi_j(h) := \sum_l \psi_l h_{j-l}$. This implies that the ACF and SDF of $\{\epsilon_t(h)\}$ are given by $r_{\epsilon(h)}(u) = \sigma^2 \sum_j \psi_j^*(h) \psi_{j+u}(h)$ and $f_{\epsilon(h)}(\omega) = |H(\omega)|^2 f_\epsilon(\omega)$.

For a given time series $\{y_1, \ldots, y_n\}$ of finite length n, the ideal filter in (4.4.1) can only be approximated by

$$\hat{y}_t(h) := \sum_{j=0}^{t-1} h_j y_{t-j} \qquad (t = 1, \ldots, n).$$

This is equivalent to assuming $y_t = 0$ for all $t \le 0$. As a result, the filtered time series $\{\hat{y}_t(h)\}$ is not a stationary process, even if $\{y_t\}$ is. Therefore, one cannot simply cite the theory of linear filtering to claim that the sample autocovariances and autocorrelation coefficients of $\{\hat{y}_t(h)\}$ have similar properties to those of $\{y_t\}$. To obtain the correct answer, a more careful analysis is called for.

By definition, the sample autocovariances of $\{\hat{y}_1(h),\ldots,\hat{y}_n(h)\}$ are given by

$$\hat{r}_{y(h)}(u) := \hat{r}^*_{y(h)}(-u) := n^{-1}\sum_{t=1}^{n-u} \hat{y}_{t+u}(h)\,\hat{y}^*_t(h) \qquad (u = 0,1,\ldots,n-1)$$

and the corresponding sample autocorrelation coefficients are given by

$$\hat{\rho}_{y(h)}(u) := \hat{r}_{y(h)}(u)/\hat{r}_{y(h)}(0).$$

The following theorem, which extends the result in [235], offers a sufficient condition for $\hat{r}_{y(h)}(u)$ and $\hat{\rho}_{y(h)}(u)$ to behave like their counterparts of the ideally filtered process $\{y_t(h)\}$. A proof of this result can be found in Section 4.5.

Theorem 4.6 (Filtered Time Series). *If the filter is strongly BIBO stable such that $\sum_{j=0}^{\infty} j|h_j| < \infty$, then Theorems 4.1–4.5 and Corollaries 4.1–4.3 remain valid for $\hat{r}_{y(h)}(u)$ and $\hat{\rho}_{y(h)}(u)$, except that $r_y(u)$, C_k, β_k, ϕ_k, $r_\epsilon(u)$, $f_\epsilon(\omega)$, $c_\epsilon(u)$, and $g_\epsilon(\omega)$ should be replaced by their counterparts $r_{y(h)}(u)$, $C_k(h)$, $\beta_k(h)$, $r_{\epsilon(h)}(u)$, $f_{\epsilon(h)}(\omega)$, $c_{\epsilon(h)}(u) := \sigma^2\sum_j \psi_j(h)\psi_{j+u}(h)$, and $g_{\epsilon(h)}(u) := H(\omega)H(-\omega)g_\epsilon(\omega)$.*

Theorem 4.6 shows that the impulse response of the filter must decay sufficiently quickly in order to ensure that the results obtained in the previous section are applicable to the filtered data. Note that the rate of decay required by Theorem 4.6 is higher than that required by the ordinary BIBO stability. For example, if $h_j = \mathcal{O}(j^{-d})$ for large j, then the BIBO stability is satisfied when $d > 1$, but the strong BIBO stability in Theorem 4.6 requires $d > 2$. Moreover, the BIBO stability implies that $H(\omega)$ is a continuous function, but the strong BIBO stability in Theorem 4.6 implies that $H(\omega)$ is also continuously differentiable.

4.5 Proof of Theorems

This section contains the proof of Theorems 4.1–4.6.

Proof of Theorem 4.1. It suffices to prove (4.2.5) and (4.2.8) for $u \geq 0$, because for $u < 0$, we have $\hat{r}_y(u) = \hat{r}^*_y(|u|)$, $r_x(u) = r^*_x(|u|)$, and $r_\epsilon(u) = r^*_\epsilon(|u|)$, so (4.2.5) and (4.2.8) remain valid. For fixed $u \geq 0$, we can write

$$\hat{r}_y(u) = \hat{r}_x(u) + \hat{r}_{\epsilon x}(u) + \hat{r}_{x\epsilon}(u) + \hat{r}_\epsilon(u), \tag{4.5.1}$$

where $\hat{r}_x(u)$ and $\hat{r}_\epsilon(u)$ are the sample autocovariance functions of $\{x_t\}$ and $\{\epsilon_t\}$, respectively, and

$$\hat{r}_{\epsilon x}(u) := n^{-1}\sum_{t=1}^{n-u} \epsilon_{t+u} x^*_t, \quad \hat{r}_{x\epsilon}(u) := n^{-1}\sum_{t=1}^{n-u} x_{t+u}\epsilon^*_t.$$

Therefore, it suffices to evaluate the four terms in (4.5.1).

To evaluate $\hat{r}_\epsilon(u)$ for fixed $u \ge 0$, let $z_t := \epsilon_{t+u}\epsilon_t^* - r_\epsilon(u)$. Then,

$$\begin{aligned}
E(z_{t+s}z_t^*) &= E(\epsilon_{t+s+u}\epsilon_{t+s}^*\epsilon_{t+u}^*\epsilon_t) - |r_\epsilon(u)|^2 \\
&= \sum_{j,j',l,l'} \psi_j\psi_{j'+u}^*\psi_{l+s}^*\psi_{l'+u+s}\, E(\zeta_{t-j}\zeta_{t-j'}^*\zeta_{t-l}^*\zeta_{t-l'}) - |r_\epsilon(u)|^2 \\
&= \sum_{j,j',l,l'} \psi_j\psi_{j'+u}^*\psi_{l+s}^*\psi_{l'+u+s}\, r_\zeta(l'-j, l'-j', l'-l) - |r_\epsilon(u)|^2 \\
&= (\kappa - 2 - \xi)\sigma^4 \sum_j \psi_j\psi_{j+u}^*\psi_{j+s}^*\psi_{j+s+u} \\
&\quad + \sigma^4 \sum_{j,l} \psi_j\psi_{j+s}^*\psi_l^*\psi_{l+s} + \xi\sigma^4 \sum_{j,l} \psi_j\psi_l^*\psi_{l+s-u}^*\psi_{j+s+u} \\
&:= r_z(s).
\end{aligned}$$

This implies that the zero-mean random process $\{z_t\}$ is stationary with ACF $r_z(s)$ which is absolutely summable because $\sum |\psi_j| < \infty$. By Lemma 12.5.3,

$$\hat{r}_\epsilon(u) - r_\epsilon(u) = n^{-1}\sum_{t=1}^{n-u} z_t - r_\epsilon(u)\, u n^{-1} \to 0$$

as $n \to \infty$ in probability and almost surely.

To evaluate $\hat{r}_x(u)$, consider the CSM with $\{x_t\}$ given by (2.1.5). Observe that

$$x_{t+u}x_t^* = \sum_{k,k'=1}^{p} \beta_k\beta_{k'}^* \exp(i\omega_k u)\exp(i d_{kk'}(t)),$$

where $d_{kk'}(t) := (\omega_k - \omega_{k'})t$. By Lemma 12.1.4,

$$n^{-1}\sum_{t=1}^{n-u} \exp(i d_{kk'}(t)) = \begin{cases} \mathcal{O}(n^{-1}\Delta^{-1}) & \text{for } k \ne k', \\ 1 - un^{-1} & \text{for } k = k'. \end{cases}$$

Therefore,

$$\hat{r}_x(u) = n^{-1}\sum_{t=1}^{n-u} x_{t+u}x_t^* = (1 - un^{-1})r_x(u) + \mathcal{O}(n^{-1}\Delta^{-1}), \tag{4.5.2}$$

where $r_x(u)$ is defined by the second expression in (4.2.6) with $C_k := |\beta_k|$. Under the condition (4.2.2), we obtain

$$\hat{r}_x(u) - r_x(u) \to 0.$$

The same result can be obtained for the RSM where $r_x(u)$ is given by the first expression in (4.2.6) with $C_k := \sqrt{A_k^2 + B_k^2}$.

Finally, an application of Lemma 12.5.3 with $X_t := \epsilon_{t+u}$ and $W_{nt} := x_t^*$, together with the fact that $\sum |r_\epsilon(u)| < \infty$, gives

$$\hat{r}_{\epsilon x}(u) \to 0$$

in probability and almost surely. The same result can be obtained for $\hat{r}_{x\epsilon}(u)$. The assertion (4.2.5) is proved. The assertion (4.2.8) follows from (4.5.2) and the fact that $E\{\hat{r}_{\epsilon x}(u)\} = E\{\hat{r}_{x\epsilon}(u)\} = 0$ and $E\{\hat{r}_\epsilon(u)\} = (1-un^{-1})\, r_\epsilon(u)$. □

Proof of Theorem 4.2. First, consider the complex case. For fixed $u \ge 0$ and $v \ge 0$, it follows from (4.5.1) that

$$d_y(u) := \hat{r}_y(u) - E\{\hat{r}_y(u)\} = \hat{r}_{\epsilon x}(u) + \hat{r}_{x\epsilon}(u) + d_\epsilon(u). \tag{4.5.3}$$

where $d_\epsilon(u) := \hat{r}_\epsilon(u) - E\{\hat{r}_\epsilon(u)\}$. Let $z_t(u) := \epsilon_{t+u}\epsilon_t^* - r_\epsilon(u)$ and $\tau := t-s$. Then,

$$\begin{aligned}
&E\{z_t(u) z_s^*(v)\} \\
&= E(\epsilon_{t+u}\epsilon_t^*\epsilon_{s+v}^*\epsilon_s) - r_\epsilon(u)\, r_\epsilon^*(v) \\
&= \sum_{j,j',l,l'} \psi_{j+u}\psi_{j'}^*\psi_{l+v}^*\psi_{l'}\, r_\zeta(l'-j+\tau, l'-j'+\tau, l'-l) - r_\epsilon(u)\, r_\epsilon^*(v) \\
&= (\kappa - 2 - \xi)\,\sigma^4 \sum_j \psi_{j+\tau+u}\psi_{j+\tau}^*\psi_{j+v}^*\psi_j \\
&\quad + r_\epsilon(\tau+u-v)\, r_\epsilon^*(\tau) + \xi\, c_\epsilon(\tau+u)\, c_\epsilon^*(\tau-v) \\
&:= a(\tau,u,v).
\end{aligned}$$

Therefore, we can write

$$\begin{aligned}
n E\{d_\epsilon(u)\, d_\epsilon^*(v)\} &= n^{-1}\sum_{t=1}^{n-u}\sum_{s=1}^{n-v} E\{z_t(u) z_s^*(v)\} \\
&= \sum_{\tau=-n+v+1}^{n-u-1} n^{-1} m(\tau)\, a(\tau,u,v),
\end{aligned} \tag{4.5.4}$$

where

$$m(\tau) := \#\{(t,s):\ t-s=\tau, 1 \le t \le n-u, 1 \le s \le n-v\}.$$

Because

$$\begin{aligned}
m(\tau) &\le \#\{(t,s):\ t-s=\tau, 1\le t,s \le n\} = n - |\tau|, \\
m(\tau) &\ge \#\{(t,s):\ t-s=\tau, 1\le t,s \le n-\max(u,v)\} \\
&= n - \max(u,v) - |\tau|,
\end{aligned}$$

it follows that for any given τ,

$$n^{-1} m(\tau) \le 1 - |\tau| n^{-1} \le 1, \quad n^{-1} m(\tau) \to 1 \quad \text{as } n \to \infty.$$

Furthermore, because $\sum|\psi_j| < \infty$, we have $\sum_\tau |a(\tau,u,v)| < \infty$. Therefore, applying the dominated convergence theorem gives

$$n E\{d_\epsilon(u)\, d_\epsilon^*(v)\} \to \sum_{\tau=-\infty}^{\infty} a(\tau,u,v) = \gamma_2(u,v), \tag{4.5.5}$$

where $\gamma_2(u,v)$ is defined by (4.3.4). Moreover,

$$E\{\hat{r}_{x\epsilon}(u)\,\hat{r}^*_{x\epsilon}(v)\} = n^{-2}\sum_{t=1}^{n-u}\sum_{s=1}^{n-v} r_\epsilon(s-t)\,x_{t+u}x^*_{s+v}.$$

Let $a_{kk'}(u,v) := \beta_k\beta^*_{k'}\exp\{i(\omega_k u - \omega_{k'} v)\}$. Then,

$$\begin{aligned} x_{t+u}x^*_{s+v} &= \sum_{k,k'=1}^{p} a_{kk'}(u,v)\exp\{i(\omega_k t - \omega_{k'} s)\} \\ &= \sum_{k,k'=1}^{p} a_{kk'}(u,v)\exp\{it(\omega_k - \omega_{k'})\}\exp\{-i(s-t)\omega_{k'}\}. \end{aligned}$$

Therefore,

$$\begin{aligned} &n\,E\{\hat{r}_{x\epsilon}(u)\,\hat{r}^*_{x\epsilon}(v)\} \\ &= \sum_{k,k'=1}^{p} a_{kk'}(u,v)\sum_{\tau=-\infty}^{\infty} r_\epsilon(\tau)\exp(-i\tau\omega_{k'})\,D_1(\tau,\omega_k-\omega_{k'}), \end{aligned} \tag{4.5.6}$$

where

$$D_1(\tau,\omega) := \begin{cases} n^{-1}\displaystyle\sum_{t=\max(1,1-\tau)}^{\min(n-u,n-v-\tau)} \exp(it\omega) & \text{if } -n+u+1 \le \tau \le n-v-1, \\ 0 & \text{otherwise.} \end{cases}$$

Note that for any given τ, an application of Lemma 12.1.4 gives $D_1(\tau,\omega_k-\omega_{k'}) = \mathcal{O}(n^{-1}\Delta^{-1}) \to 0$ if $k \neq k'$. Moreover, we have $D_1(\tau,0) \to 1$ as $n\to\infty$, $|D_1(\tau,\omega)| \le 1$, and $\sum|r_\epsilon(\tau)| < \infty$. Therefore, the dominated convergence theorem gives

$$\begin{aligned} n\,E\{\hat{r}_{x\epsilon}(u)\,\hat{r}^*_{x\epsilon}(v)\} &= \sum_{k=1}^{p} a_{kk}(u,v)\sum_{\tau=-\infty}^{\infty} r_\epsilon(\tau)\exp(-i\tau\omega_k) + \mathcal{o}(1) \\ &= \sum_{k=1}^{p} C_k^2 f_\epsilon(\omega_k)\exp\{i\omega_k(u-v)\} + \mathcal{o}(1). \end{aligned} \tag{4.5.7}$$

Similarly, we obtain

$$n\,E\{\hat{r}_{\epsilon x}(u)\,\hat{r}^*_{\epsilon x}(v)\} = \sum_{k=1}^{p} C_k^2 f_\epsilon(\omega_k)\exp\{i\omega_k(u-v)\} + \mathcal{o}(1). \tag{4.5.8}$$

Moreover, it is easy to see that

$$E\{\hat{r}_{\epsilon x}(u)\,\hat{r}^*_{x\epsilon}(v)\} = n^{-2}\sum_{t=1}^{n-u}\sum_{s=1}^{n-v} \iota c_\epsilon(t-s+u)\,x^*_t x^*_{s+v}.$$

Let $b_{kk'}(u,v) := \beta^*_k\beta^*_{k'}\exp\{i(u\omega_k - v\omega_{k'})\}$. Then,

$$x^*_t x^*_{s+v} = \sum_{k,k'=1}^{p} b_{kk'}(u,v)\exp\{-i(\omega_k+\omega_{k'})s\}\exp\{-i\omega_k(t-s+u)\}.$$

Therefore,

$$
\begin{aligned}
&n E\{\hat{r}_{\epsilon x}(u)\, \hat{r}^*_{x\epsilon}(v)\} \\
&= \sum_{k,k'=1}^{p} b_{kk'}(u,v) \sum_{\tau=-\infty}^{\infty} \iota c_\epsilon(\tau) \exp(-i\tau\omega_k)\, D_2(\tau, \omega_k + \omega_{k'}),
\end{aligned} \tag{4.5.9}
$$

where

$$
D_2(\tau,\omega) := \begin{cases} n^{-1} \displaystyle\sum_{t=\max(1,u+1-\tau)}^{\min(n-v,n-\tau)} \exp(-it\omega) & \text{if } -n+v-u+1 \le \tau \le n-1, \\ 0 & \text{otherwise.} \end{cases}
$$

By the dominated convergence theorem and Lemma 12.1.4, we obtain

$$
\begin{aligned}
&n E\{\hat{r}_{\epsilon x}(u)\, \hat{r}^*_{x\epsilon}(v)\} \\
&= \sum_{\substack{k,k'=1 \\ \omega_k = -\omega_{k'}}}^{p} b_{kk'}(u,v) \sum_{\tau=-\infty}^{\infty} \iota c_\epsilon(\tau) \exp(-i\tau\omega_k) + \mathcal{O}(1) \\
&= \sum_{\substack{k,k'=1 \\ \omega_k = -\omega_{k'}}}^{p} \beta^*_k \beta^*_{k'} \iota g_\epsilon(\omega_k) \exp\{i\omega_k(u+v)\} + \mathcal{O}(1).
\end{aligned} \tag{4.5.10}
$$

Similarly, it can be shown that

$$
\begin{aligned}
&n E\{\hat{r}_{x\epsilon}(u)\, \hat{r}^*_{\epsilon x}(v)\} \\
&= \sum_{\substack{k,k'=1 \\ \omega_k = -\omega_{k'}}}^{p} \beta_k \beta_{k'} \iota^*\, g^*_\epsilon(\omega_k) \exp\{i\omega_k(u+v)\} + \mathcal{O}(1).
\end{aligned} \tag{4.5.11}
$$

Note that adding up the leading terms on the right-hand side of (4.5.7)–(4.5.11) gives $\gamma_1(u,v)$ defined by (4.3.3). Furthermore,

$$
E\{\hat{r}_{x\epsilon}(u)\, d^*_\epsilon(v)\} = n^{-2} \sum_{t=1}^{n-u} \sum_{s=1}^{n-v} x_{t+u} E(\epsilon^*_t \epsilon^*_{s+v} \epsilon_s).
$$

It is easy to show that

$$
E(\epsilon^*_t \epsilon^*_{s+v} \epsilon_s) = r_\epsilon(\tau, v) := \sum_{j,j',l} \psi^*_{j+\tau} \psi^*_{j'+v} \psi_l\, r_\zeta(l-j, l-j'),
$$

where $\tau := t - s$. Therefore,

$$
\begin{aligned}
&n E\{\hat{r}_{x\epsilon}(u)\, d^*_\epsilon(v)\} \\
&= \sum_{k=1}^{p} \beta_k \exp\{i(\omega_k u + \phi_k)\} \sum_{\tau=-\infty}^{\infty} D_1(\tau, \omega_k)\, r_\epsilon(\tau, v).
\end{aligned} \tag{4.5.12}
$$

For any τ, $|D_1(\tau,\omega_k)| \le 1$ and, by Lemma 12.1.4, $D_1(\tau,\omega_k) = \mathcal{O}(n^{-1}\Delta^{-1}) \to 0$ as $n \to \infty$. Because $\sum|\psi_j| < \infty$ and $\sum\sum|r_\zeta(j,l)| < \infty$, we have $\sum_\tau |r_\epsilon(\tau,v)| < \infty$. By the dominated convergence theorem, we obtain

$$n\,E\{\hat{r}_{x\epsilon}(u)\,d_\epsilon^*(v)\} \to 0. \tag{4.5.13}$$

Similarly, it can be shown that

$$n\,E\{\hat{r}_{\epsilon x}(u)\,d_\epsilon^*(v)\} \to 0. \tag{4.5.14}$$

Combining this result with (4.5.5)–(4.5.14) and (4.3.9) yields

$$n\,E\{d_y(u)\,d_y^*(v)\} - \gamma(u,v) \to 0,$$

hence the assertion for the covariance between $\hat{r}_y(u)$ and $\hat{r}_y(v)$.

It follows from (4.5.4) that

$$E\{d_\epsilon(u)\,d_\epsilon^*(v)\} = \mathcal{O}(n^{-1})$$

uniformly in $u, v \in \{0,1,\ldots,n-1\}$. The same expression can be obtained from (4.5.6), (4.5.9), and (4.5.12) for $E\{\hat{r}_{x\epsilon}(u)\,\hat{r}_{x\epsilon}^*(v)\}$, $E\{\hat{r}_{\epsilon x}(u)\,\hat{r}_{x\epsilon}^*(v)\}$, $E\{\hat{r}_{x\epsilon}(u)\,d_\epsilon^*(v)\}$, and $E\{\hat{r}_{\epsilon x}(u)\,d_\epsilon^*(v)\}$. Combining these results with (4.5.3) and the fact that $\hat{r}_y(-u) = \hat{r}_y^*(u)$, yields $\mathrm{Var}\{\hat{r}_\epsilon(u)\} = \mathcal{O}(n^{-1})$ uniformly in $|u| < n$.

Now consider the complementary covariance $\mathrm{Cov}\{\hat{r}_y(u), \hat{r}_y^*(v)\}$ for $u \ge 0$ and $v \ge 0$. It is easy to see that

$$E\{z_t(u)z_s(v)\} = E\{z_t(u)z_{s+v}^*(-v)\} = a(\tau - v, u, -v).$$

Therefore, by using an argument similar to that which led to (4.5.5), we obtain

$$n\,E\{d_\epsilon(u)d_\epsilon(v)\} \to \sum_{\tau=-\infty}^{\infty} a(\tau - v, u, -v) = \gamma_2(u, -v).$$

Moreover,

$$E\{\hat{r}_{x\epsilon}(u)\hat{r}_{x\epsilon}(v)\} = n^{-2}\sum_{t=1}^{n-u}\sum_{s=1}^{n-v} \iota^*\, c_\epsilon^*(t-s)x_{t+u}x_{s+v}.$$

Thus, by using an argument similar to that which led to (4.5.11), we can show that $nE\{\hat{r}_{x\epsilon}(u)\hat{r}_{x\epsilon}(v)\}$ has the same asymptotic expression as $nE\{\hat{r}_{x\epsilon}(u)\hat{r}_{\epsilon x}^*(-v)\}$. The rest can be done in the same way. Combining these results gives

$$n\,E\{d_y(u)\,d_y(v)\} - \gamma(u, -v) \to 0,$$

hence the assertion for the complementary covariance.

For $u \ge 0$ and $v < 0$, we have, by definition,

$$\mathrm{Cov}\{\hat{r}_y(u), \hat{r}_y(v)\} = \mathrm{Cov}\{\hat{r}_y(u), \hat{r}_y^*(|v|)\},$$

$$\mathrm{Cov}\{\hat{r}_y(u), \hat{r}_y^*(v)\} = \mathrm{Cov}\{\hat{r}_y(u), \hat{r}_y(|v|)\}.$$

Therefore,

$$n\,\mathrm{Cov}\{\hat{r}_y(u), \hat{r}_y(v)\} - \gamma(u, -|v|) \to 0,$$
$$n\,\mathrm{Cov}\{\hat{r}_y(u), \hat{r}_y^*(v)\} - \gamma(u, |v|) \to 0.$$

Because $\gamma(u,-|v|) = \gamma(u,v)$ and $\gamma(u,|v|) = \gamma(u,-v)$, the assertion in Theorem 4.2 remains valid. We can draw the same conclusion for all other situations. The proof for the complex case is complete.

In the real case, let $p := 2q$, $\omega_{p-k+1} := -\omega_k$, and $\beta_{p-k+1} := \beta_k^* := \frac{1}{2} C_k \exp(-i\phi_k)$ $(k = 1, \ldots, q)$. Then, the RSM (2.1.2) can be written as the CSM (2.1.5). Under this condition, the first term of $\gamma_1(u,v)$ in (4.3.3) becomes

$$\sum_{k=1}^{q} C_k^2 f_\epsilon(\omega) \cos(\omega_k(u-v))$$

upon noting that $f_\epsilon(\omega) = f_\epsilon(-\omega) = f_\epsilon^*(\omega)$. The second term of $\gamma_1(u,v)$ becomes

$$\sum_{k=1}^{q} C_k^2 f_\epsilon(\omega) \cos(\omega_k(u+v))$$

because $g_\epsilon(\omega) = f_\epsilon(\omega)$ and $\iota = 1$. Combining these expressions gives (4.3.1). In addition, because $c_\epsilon(u) = r_\epsilon(u)$ in the real case, (4.3.2) follows from (4.3.4). □

Proof of Theorem 4.3. In the real case, we have

$$\hat{r}_y(u) = \hat{r}_y(-u) = n^{-1} \sum_{t=1}^{n-u} y_{t+u} y_t \qquad (u = 0, 1, \ldots, n-1).$$

For any $|u| < n$, define

$$\tilde{r}_y(u) := n^{-1} \sum_{t=1}^{n} y_{t+u} y_t.$$

Then, we can write $\hat{r}_y(u) = \tilde{r}_y(u) + e_n(u)$, where

$$e_n(u) := \begin{cases} -n^{-1} \sum_{t=1}^{-u} y_{t+u} y_t & \text{if } u < 0, \\ -n^{-1} \sum_{t=n-u+1}^{n} y_{t+u} y_t & \text{if } u > 0. \end{cases}$$

Because $|x_t|$ can be uniformly bounded by a constant $c > 0$, we obtain

$$E(|y_{t+u} y_t|) \le c^2 + 2cE(|\epsilon_t|) + E(|\epsilon_{t+u}\epsilon_t|) \le c^2 + 2c\sqrt{r_\epsilon(0)} + r_\epsilon(0).$$

This implies that $E(|y_{t+u}y_t|) = \mathcal{O}(1)$ uniformly in t and u. Therefore, we obtain $E\{|e_n(u)|\} = \mathcal{O}(n^{-1})$ for any fixed u. By Proposition 2.16(a), we can write $e_n(u) = \mathcal{O}_P(n^{-1})$ for any $u \in \{0, 1, \ldots, m\}$. Therefore, it suffices to show that

$$\sqrt{n}\,[\tilde{r}_y(0) - r_y(0), \ldots, \tilde{r}_y(m) - r_y(m)]^T \stackrel{A}{\sim} \mathrm{N}(\mathbf{0}, \mathbf{\Sigma}),$$

where $\mathbf{\Sigma} := [\gamma(u, v)]$ $(u, v = 0, 1, \ldots, m)$. This, according to Proposition 2.16(d), can be accomplished by proving

$$\sqrt{n} \sum_{u=0}^{m} \lambda_u \{\tilde{r}_y(u) - r_y(u)\} \stackrel{A}{\sim} \mathrm{N}(0, \varsigma^2)$$

for any constants λ_u not all equal to zero, where

$$\varsigma^2 := \sum_{u,v=0}^{m} \lambda_u \lambda_v \gamma(u, v) > 0. \tag{4.5.15}$$

Toward that end, let

$$X_{ut} := x_{t+u}\epsilon_t + \epsilon_{t+u}x_t + \epsilon_{t+u}\epsilon_t - r_\epsilon(u). \tag{4.5.16}$$

Because $n^{-1}\sum_{t=1}^{n} x_{t+u}x_t = r_x(u) + \mathcal{O}(n^{-1}\Delta^{-1})$ by Lemma 12.1.4, it follows that

$$\tilde{r}_y(u) = r_y(u) + n^{-1}\sum_{t=1}^{n} X_{ut} + \mathcal{O}(n^{-1}\Delta^{-1}),$$

which gives

$$\sqrt{n} \sum_{u=0}^{m} \lambda_u \{\tilde{r}_y(u) - r_y(u)\} = n^{-1/2}\sum_{t=1}^{n} X_t + \mathcal{O}(n^{-1/2}\Delta^{-1}),$$

where

$$X_t := \sum_{u=0}^{m} \lambda_u X_{ut}. \tag{4.5.17}$$

Owing to the separation condition (4.2.2), one only needs to show

$$n^{-1/2}\sum_{t=1}^{n} X_t/\varsigma \xrightarrow{D} \mathrm{N}(0, 1). \tag{4.5.18}$$

Let us first prove (4.5.18) under the assumption that $\{\psi_j\}$ is of finite length. This can be done by combining the following three lemmas with Lemma 12.5.4.

Lemma A. *Let the conditions of Theorem 4.3 be satisfied. Let $\{X_t\}$ be defined by (4.5.17). Assume that $\psi_j = 0$ for all $|j| > l$ for some given integer $l > 0$. Then,*

$$\sum_{t=1}^{n} X_t = \sum_{t=1}^{n} M_t + \mathcal{O}_P(1),$$

where $\{M_t\}$ is a sequence of martingale differences with respect to the σ-fields $\mathfrak{F}_t$ generated by $\{\zeta_t, \zeta_{t-1}, \ldots\}$.

Lemma B. *Let the conditions of Lemma A be satisfied. Define*

$$V_n^2 := \sum_{t=1}^{n} E(M_t^2 | \mathfrak{F}_{t-1}), \quad s_n^2 := E(V_n^2).$$

Then, as $n \to \infty$, $n^{-1} V_n^2 / \varsigma^2 \xrightarrow{P} 1$ *and* $n^{-1} s_n^2 / \varsigma^2 \to 1$, *where* ς^2 *is given by (4.5.15).*

Lemma C. *Let the conditions of Lemma A be satisfied and let* s_n^2 *be defined in Lemma B. Then, for any constant* $\delta > 0$,

$$\lim_{n\to\infty} s_n^{-2} \sum_{t=1}^{n} E\{M_t^2 \mathscr{I}(|M_t| \geq \delta s_n)\} = 0.$$

Therefore, by Lemma 12.5.4, $n^{-1/2} \sum_{t=1}^{n} M_t / \varsigma \xrightarrow{D} \mathrm{N}(0,1)$ *as* $n \to \infty$.

Now consider the case where $\{\psi_j\}$ is of infinite length. For each fixed $l > 0$, let

$$\tilde{\epsilon}_t := \sum_{j=-l}^{l} \psi_j \zeta_{t-j},$$

and let $\tilde{X}_t$ and $\tilde{\varsigma}^2$ be defined in the same way as X_t and ς^2, except that $\{\epsilon_t\}$ is everywhere replaced with $\{\tilde{\epsilon}_t\}$. According to Lemma C,

$$n^{-1/2} \sum_{t=1}^{n} \tilde{X}_t / \tilde{\varsigma} \xrightarrow{D} \mathrm{N}(0,1)$$

as $n \to \infty$ for any fixed l. Moreover, it is easy to show that as $l \to \infty$, $\tilde{\varsigma}^2 / \varsigma^2 \to 1$, and uniformly in l as $n \to \infty$,

$$n^{-1} E\Big|\sum_{t=1}^{n} (\tilde{X}_t - X_t)\Big|^2 = \mathscr{O}\Big(\sum_{|j|>l} |\psi_j|\Big), \quad n^{-1} E\Big|\sum_{t=1}^{n} \tilde{X}_t\Big|^2 = \mathscr{O}(1).$$

Therefore,

$$\begin{aligned}
&\lim_{l\to\infty} \limsup_{n\to\infty} n^{-1} E\Big|\sum_{t=1}^{n} (\tilde{X}_t/\tilde{\varsigma} - X_t/\varsigma)\Big|^2 \\
&\leq \lim_{l\to\infty} \limsup_{n\to\infty} 2\varsigma^{-2} \Big\{ n^{-1} E\Big|\sum_{t=1}^{n} (\tilde{X}_t - X_t)\Big|^2 + n^{-1} E\Big|\sum_{t=1}^{n} \tilde{X}_t\Big|^2 (\varsigma/\tilde{\varsigma} - 1)^2 \Big\} \\
&= 0.
\end{aligned}$$

This, combined with Chebyshev's inequality, leads to

$$\lim_{l\to\infty} \limsup_{n\to\infty} P\Big\{ n^{-1/2} \Big|\sum_{t=1}^{n} (\tilde{X}_t/\tilde{\varsigma} - X_t/\varsigma)\Big| > \delta \Big\} = 0$$

for any $\delta > 0$. By Proposition 2.16(e), we obtain (4.5.18). □

It remains to prove Lemmas A, B, and C.

Proof of Lemma A. It follows from (4.5.16) that

$$\sum_{t=1}^{n} X_{ut} = T_1 + T_2 + T_3, \tag{4.5.19}$$

where

$$T_1 := \sum_{t=1}^{n} x_{t+u}\epsilon_t, \quad T_2 := \sum_{t=1}^{n} \epsilon_{t+u}x_t, \quad T_3 := \sum_{t=1}^{n} \{\epsilon_{t+u}\epsilon_t - r_\epsilon(u)\}.$$

The first goal is to show that

$$T_3 = \sum_{t=1}^{n}\sum_{\tau=0}^{\infty} b_\tau(u)\,(\zeta_t\zeta_{t-\tau} - \sigma^2\delta_\tau) + \mathcal{O}_P(1), \tag{4.5.20}$$

where

$$\begin{aligned} b_0(u) &:= r_\epsilon(u)/\sigma^2, \\ b_\tau(u) &:= \{r_\epsilon(u+\tau) + r_\epsilon(u-\tau)\}/\sigma^2 \qquad (\tau = 1,2,\ldots). \end{aligned}$$

Note that $b_\tau(u) = 0$ for $\tau > 2l + u$, so the sum over τ in (4.5.20) is a finite sum. The second goal is to show that

$$T_1 + T_2 = \sum_{t=1}^{n} a_t(u)\zeta_t + \mathcal{O}_P(1), \tag{4.5.21}$$

where

$$a_t(u) := \sum_{j=-l}^{l} \psi_j(x_{t+j+u} + x_{t+j-u}). \tag{4.5.22}$$

These goals are achieved by variable substitution and by adding and subtracting a finite number of terms of the form $\mathcal{O}_P(1)$.

Toward the first goal (4.5.20), let us substitute $\epsilon_t = \sum_{j=-l}^{l}\psi_j\zeta_{t-j}$ and $r_\epsilon(u) = \sigma^2\sum_{j,j'=-l}^{l}\psi_j\psi_{j'}\delta_{j-j'-u}$ in the definition of T_3. This gives

$$\begin{aligned} T_3 &= \sum_{j,j'=-l}^{l} \psi_j\psi_{j'} \sum_{t=1}^{n} (\zeta_{t+u-j}\zeta_{t-j'} - \sigma^2\delta_{j-j'-u}) \\ &= \sum_{j,j'=-l}^{l} \psi_j\psi_{j'} \sum_{t=u-j'+1}^{u-j'+n} (\zeta_t\zeta_{t+j-j'-u} - \sigma^2\delta_{j-j'-u}) \\ &= \sum_{j,j'=-l}^{l} \psi_j\psi_{j'} \sum_{t=1}^{n} (\zeta_t\zeta_{t+j-j'-u} - \sigma^2\delta_{j-j'-u}) + \mathcal{O}_P(1) \end{aligned}$$

$$= \sum_{j,j'=-\infty}^{\infty} \psi_j \psi_{j'} \sum_{t=1}^{n} (\zeta_t \zeta_{t+j-j'-u} - \sigma^2 \delta_{j-j'-u}) + \mathcal{O}_P(1).$$

The third equality is obtained by adding to and subtracting from the second expression a finite number (which is independent of n) of terms of the form

$$\zeta_t \zeta_{t+j-j'-u} - \sigma^2 \delta_{j-j'-u},$$

all of which can be expressed as $\mathcal{O}_P(1)$. The last equality is due to the fact that $\psi_j = 0$ for $|j| > l$. Next, let $\tau := j' - j + u$ in the last expression of T_3, so

$$\begin{aligned} T_3 &= \sum_{\tau=-\infty}^{\infty} \sum_{j=-\infty}^{\infty} \psi_{j+u-\tau} \psi_j \sum_{t=1}^{n} (\zeta_t \zeta_{t-\tau} - \sigma^2 \delta_\tau) + \mathcal{O}_P(1) \\ &= \sigma^{-2} \sum_{\tau=-2l+u}^{2l+u} r_\epsilon(u-\tau) \sum_{t=1}^{n} (\zeta_t \zeta_{t-\tau} - \sigma^2 \delta_\tau) + \mathcal{O}_P(1), \end{aligned} \tag{4.5.23}$$

where the second equality is due to $r_\epsilon(u-\tau) = \sigma^2 \sum_j \psi_{j+u-\tau} \psi_j$ and $r_\epsilon(u-\tau) = 0$ for $|\tau - u| > 2l$. Moreover, the sum over $\tau < 0$ can be rewritten as

$$\begin{aligned} &\sigma^{-2} \sum_{\tau=1}^{2l-u} r_\epsilon(u+\tau) \sum_{t=1}^{n} \zeta_t \zeta_{t+\tau} \\ &= \sigma^{-2} \sum_{\tau=1}^{2l-u} r_\epsilon(u+\tau) \sum_{t=\tau+1}^{\tau+n} \zeta_{t-\tau} \zeta_t \\ &= \sigma^{-2} \sum_{\tau=1}^{2l-u} r_\epsilon(u+\tau) \sum_{t=1}^{n} \zeta_{t-\tau} \zeta_t + \mathcal{O}_P(1) \\ &= \sigma^{-2} \sum_{\tau=1}^{2l+u} r_\epsilon(u+\tau) \sum_{t=1}^{n} \zeta_{t-\tau} \zeta_t + \mathcal{O}_P(1), \end{aligned}$$

where the second equality is obtained by adding and subtracting a finite number of terms $\zeta_{t-\tau} \zeta_t = \mathcal{O}_P(1)$ and the last equality is due to the fact that $r_\epsilon(u+\tau) = 0$ for $\tau > 2l - u$. Substituting this expression in (4.5.23) gives (4.5.20).

To achieve the second goal (4.5.21), let us note that

$$\begin{aligned} T_2 &= \sum_{j=-l}^{l} \psi_j \sum_{t=1}^{n} \zeta_{t+u-j} x_t \\ &= \sum_{j=-l}^{l} \psi_j \sum_{t=u-j+1}^{u-j+n} x_{t+j-u} \zeta_t \\ &= \sum_{j=-l}^{l} \psi_j \sum_{t=1}^{n} x_{t+j-u} \zeta_t + \mathcal{O}_P(1), \end{aligned}$$

where the last equality is obtained by adding and subtracting a finite number of terms $x_{t+j-u} \zeta_t = \mathcal{O}_P(1)$. Similarly,

$$T_1 = \sum_{j=-l}^{l} \psi_j \sum_{t=1}^{n} x_{t+u} \zeta_{t-j}$$

$$= \sum_{j=-l}^{l} \psi_j \sum_{t=j+1}^{j+n} x_{t+j+u}\zeta_t$$

$$= \sum_{j=-l}^{l} \psi_j \sum_{t=1}^{n} x_{t+j+u}\zeta_t + \mathcal{O}_P(1).$$

Adding up these expressions gives (4.5.21).

Now, let us define

$$a_t := \sum_{u=0}^{m} \lambda_u a_t(u), \quad b_\tau := \sum_{u=0}^{m} \lambda_u b_\tau(u),$$

and

$$M_t := a_t\zeta_t + \sum_{\tau=0}^{\infty} b_\tau(\zeta_t\zeta_{t-\tau} - \sigma^2\delta_\tau). \tag{4.5.24}$$

Then, it follows from (4.5.17) and (4.5.19)–(4.5.21) that

$$\sum_{t=1}^{n} X_t = \sum_{u=0}^{m} \lambda_u \sum_{t=1}^{n} X_{ut} = \sum_{t=1}^{n} M_t + \mathcal{O}_P(1).$$

Finally, let $\mathfrak{F}_t$ be the σ-field generated by $\{\zeta_t, \zeta_{t-1}, \ldots\}$. Because M_t is a quadratic function of $\{\zeta_t, \zeta_{t-1}, \ldots\}$ (the a_t and b_τ are constants), it is measurable with respect to $\mathfrak{F}_t$. Furthermore,

$$E(M_t|\mathfrak{F}_{t-1}) = a_t E(\zeta_t) + b_0 E(\zeta_t^2 - \sigma^2) + \sum_{\tau=1}^{\infty} b_\tau E(\zeta_t)\zeta_{t-\tau} = 0.$$

By definition, $\{M_t\}$ is a sequence of martingale differences with respect to $\{\mathfrak{F}_t\}$.

It is important to note that $\{M_t\}$ is not a stationary process because a_t is a function of t. Therefore, the central limit theorems for stationary processes such as Theorem 6.4.2 in [46, p. 213] are not directly applicable. The asymptotic normality of $n^{1/2}\sum_{t=1}^{n} M_t$ is established in Lemmas B and C by the more powerful central limit theorem for martingale differences in Lemma 12.5.4. It is easy to see that $\{M_t\}$ remains to be a sequence of martingale differences if $\{\zeta_t\}$ itself is a sequence of martingale differences instead of a sequence of i.i.d. random variables, provided it satisfies the conditions in Remark 4.7. □

Proof of Lemma B. Straightforward calculation from (4.5.24) gives

$$E(M_t^2|\mathfrak{F}_{t-1}) = \sigma^2 a_t^2 + 2b_0 E(\zeta_t^3) a_t + 2\sigma^2 \sum_{\tau=1}^{\infty} b_\tau a_t \zeta_{t-\tau} + b_0^2 E(\zeta_t^2 - \sigma^2)^2$$
$$+ 2b_0 E(\zeta_t^3) \sum_{\tau=1}^{\infty} b_\tau \zeta_{t-\tau} + \sigma^2 \sum_{\tau,s=1}^{\infty} b_\tau b_s \zeta_{t-\tau}\zeta_{t-s}.$$

Let us first prove $n^{-1}V_n^2/\varsigma^2 \to 1$. By definition,

$$a_t = \sum_{u=0}^{m} \lambda_u a_t(u),$$

where $a_t(u)$ is given by (4.5.22). Because Lemma 12.1.4 gives

$$n^{-1}\sum_{t=1}^{n} x_{t+\tau}x_{t+s} = r_x(\tau - s) + \mathcal{O}(n^{-1}\Delta^{-1}),$$

it follows that

$$\begin{aligned} n^{-1}\sum_{t=1}^{n} a_t^2 &= \sum_{u,v}\lambda_u\lambda_v\sum_{j,j'}\psi_j\psi_{j'}\, n^{-1}\sum_{t=1}^{n}(x_{t+j+u}+x_{t+j-u})(x_{t+j'+v}+x_{t+j'-v}) \\ &= \sum_{u,v}\lambda_u\lambda_v\sum_{j,j'}\psi_j\psi_{j'}\{r_x(j-j'+u-v)+r_x(j-j'+u+v) \\ &\quad + r_x(j-j'-u-v)+r_x(j-j'-u+v)\}+\mathcal{O}(n^{-1}\Delta^{-1}). \end{aligned}$$

By straightforward calculation, we obtain

$$\begin{aligned} \sum_{j,j'}\psi_j\psi_{j'}r_x(j-j'+s) &= \sum_{k=1}^{q}\tfrac{1}{2}C_k^2\Re\Big\{\sum_{j,j'}\psi_j\psi_{j'}\exp(i\omega_k(j-j'+s))\Big\} \\ &= \sigma^{-2}\sum_{k=1}^{q}\tfrac{1}{2}C_k^2 f_\epsilon(\omega_k)\cos(\omega_k s). \end{aligned}$$

Therefore,

$$n^{-1}\sum_{t=1}^{n} a_t^2 = \sigma^{-2}\sum_{u,v}\lambda_u\lambda_v\gamma_1(u,v)+\mathcal{O}(n^{-1}\Delta^{-1}), \tag{4.5.25}$$

where $\gamma_1(u,v)$ is given by (4.3.1) with $\xi = 1$. Moreover, by Lemma 12.1.4 and Proposition 2.16(a), $\sum_{t=1}^{n} x_{t+s} = \mathcal{O}(1)$ and $n^{-1}\sum_{t=1}^{n} x_{t+s}\zeta_{t-\tau} = \mathcal{O}_P(n^{-1/2})$. Hence,

$$n^{-1}\sum_{t=1}^{n} a_t = \sum_{u}\lambda_u\sum_{j}\psi_j\, n^{-1}\sum_{t=1}^{n}(x_{t+j+u}+x_{t+j-u}) = \mathcal{O}(n^{-1}), \tag{4.5.26}$$

and

$$n^{-1}\sum_{t=1}^{n} a_t\zeta_{t-\tau} = \sum_{u}\lambda_u\sum_{j}\psi_j\, n^{-1}\sum_{t=1}^{n}(x_{t+j+u}+x_{t+j-u})\zeta_{t-\tau} = \mathcal{O}_P(n^{-1/2}).$$

Finally, by Proposition 2.16(a), we obtain

$$n^{-1}\sum_{t=1}^{n}\zeta_{t-\tau} = \mathcal{O}_P(n^{-1/2}), \quad n^{-1}\sum_{t=1}^{n}\zeta_{t-\tau}\zeta_{t-s} = \sigma^2\delta_{\tau-s}+\mathcal{O}_P(n^{-1/2}).$$

Combining all these results with $\mathcal{O}(n^{-1}\Delta^{-1}) + \mathcal{O}_P(n^{-1/2}) = \mathcal{O}_P(n^{-1/2})$ gives

$$\begin{aligned} n^{-1} V_n^2 &= n^{-1} \sum_{t=1}^{n} E(M_t^2 | \mathfrak{F}_{t-1}) \\ &= \sum_{u,v=0}^{m} \lambda_u \lambda_v \gamma_1(u,v) + b_0^2 E(\zeta_0^2 - \sigma^2)^2 + \sigma^4 \sum_{\tau=1}^{\infty} b_\tau^2 + \mathcal{O}_P(n^{-1/2}). \end{aligned}$$

Moreover, straightforward calculation shows that

$$b_0^2 E(\zeta_0^2 - \sigma^2)^2 + \sigma^4 \sum_{\tau=1}^{\infty} b_\tau^2 = \sum_{u,v=0}^{m} \lambda_u \lambda_v \gamma_2(u,v), \tag{4.5.27}$$

where $\gamma_2(u,v)$ is given by (4.3.2) with $\xi = 1$. This, combined with $\varsigma^{-2} = \mathcal{O}(1)$, proves $n^{-1} V_n^2 = \varsigma^2 + \mathcal{O}_P(n^{-1/2})$ and hence $n^{-1} V_n^2 / \varsigma^2 - 1 = \mathcal{O}_P(n^{-1/2}) \xrightarrow{P} 0$.

To prove $n^{-1} s_n^2 / \varsigma^2 \to 1$, let us note that

$$\begin{aligned} n^{-1} s_n^2 &= n^{-1} \sum_{t=1}^{n} E\{E(M_t^2 | \mathfrak{F}_{t-1})\} \\ &= \sigma^2 n^{-1} \sum_{t=1}^{n} a_t^2 + 2 b_0 E(\zeta_0^3) n^{-1} \sum_{t=1}^{n} a_t + b_0^2 E(\zeta_0^2 - \sigma^2)^2 + \sigma^4 \sum_{\tau=1}^{\infty} b_\tau^2. \end{aligned}$$

Combining this result with (4.5.25)–(4.5.27) leads to $n^{-1} s_n^2 = \varsigma^2 + \mathcal{O}(n^{-1}\Delta^{-1})$ and hence $n^{-1} s_n^2 / \varsigma^2 - 1 = \mathcal{O}(n^{-1}\Delta^{-1}) \to 0$. Lemma B is thus proved. □

Proof of Lemma C. By Lemma B, Proposition 2.13, and Lemma 12.5.4, the assertion that $n^{-1/2} \sum_{t=1}^{n} M_t / \varsigma \xrightarrow{D} \mathrm{N}(0,1)$ follows from the Lindeberg condition:

$$s_n^{-2} \sum_{t=1}^{n} E\{M_t^2 \mathcal{I}(|M_t| \ge \varepsilon s_n)\} \to 0$$

for any constant $\varepsilon > 0$. Because $n^{-1} s_n^2 / \varsigma^2 \to 1$ and $\varsigma^{-2} = \mathcal{O}(1)$, the Lindeberg condition is implied by $n^{-1} \sum_{t=1}^{n} E\{M_t^2 \mathcal{I}(|M_t| \ge \varepsilon s_n)\} \to 0$. Moreover, because

$$s_n^2 = \sum_{t=1}^{n} E\{E(M_t^2 | \mathfrak{F}_{t-1})\} \ge s_t^2 \qquad (t = 1, \ldots, n),$$

we have $\mathcal{I}(|M_t| \ge \varepsilon s_n) \le \mathcal{I}(|M_t| \ge \varepsilon s_t)$. Therefore, it suffices to show that

$$n^{-1} \sum_{t=1}^{n} E\{M_t^2 \mathcal{I}(|M_t| \ge \varepsilon s_t)\} \to 0.$$

By the Stolz-Cesàro theorem [34, p. 543], this assertion follows immediately if it can be shown that $E\{M_t^2 \mathcal{I}(|M_t| \ge \varepsilon s_t)\} \to 0$ as $t \to \infty$. To prove the latter, let us note that $|a_t| \le c$ for all t and some constant $c > 0$. According to (4.5.24),

$$|M_t| \le U_t := c|\zeta_t| + \left| \sum_{\tau=0}^{\infty} b_\tau (\zeta_t \zeta_{t-\tau} - \sigma^2 \delta_\tau) \right|.$$

Moreover, because $t^{-1/2}s_t/\varsigma \to 1$ and $\varsigma^{-1} = \mathcal{O}(1)$, we have $s_t \geq \varepsilon t^{1/2}$ for sufficiently large t. This implies that

$$\begin{aligned} E\{M_t^2 \mathscr{I}(|M_t| \geq \varepsilon s_t)\} &\leq E\{U_t^2 \mathscr{I}(U_t \geq \varepsilon^2 t^{1/2})\} \\ &= E\{U_0^2 \mathscr{I}(U_0 \geq \varepsilon^2 t^{1/2})\} \to 0, \end{aligned}$$

where the equality is due to the stationarity of $\{U_t\}$ and the zero limit is a result of the finiteness of $E(U_0^2)$. The proof of Lemma C is complete. □

Proof of Theorem 4.4. The proof of this theorem is similar to the proof of Theorem 4.3, but it is more tedious because we have to consider the real and imaginary parts of $\hat{r}_y(u)$. In the following, let superscript 1 denote the real part and superscript 2 denote the imaginary part of all complex variables. In the proof of Theorem 4.3, we have established the fact that $\hat{r}_y(u) = \tilde{r}_y(u) + \mathcal{O}_P(n^{-1})$. Therefore, it suffices to show that

$$\sqrt{n} \sum_{j=1}^{2} \sum_{u=0}^{m} \lambda_{ju} \{\tilde{r}_y^{(j)}(u) - r_y^{(j)}(u)\} \stackrel{A}{\sim} \mathrm{N}(0, \varsigma^2),$$

where

$$\varsigma^2 := \sum_{j,l=1}^{2} \sum_{u,v=0}^{m} \lambda_{ju} \lambda_{lv} \gamma_{jl}(u, v).$$

Note that $\tilde{r}_y^{(1)}(u) = \tilde{r}_y^{(11)}(u) + \tilde{r}_y^{(22)}(u)$ and $\tilde{r}_y^{(2)}(u) = \tilde{r}_y^{(21)}(u) - \tilde{r}_y^{(12)}(u)$, where

$$\tilde{r}_y^{(jl)}(u) := n^{-1} \sum_{t=1}^{n} y_{t+u}^{(j)} y_t^{(l)}.$$

Similar expressions can be obtained for $r_y^{(j)}(u) = r_x^{(j)}(u) + r_\epsilon^{(j)}(u)$ in terms of

$$r_y^{(jl)}(u) := r_x^{(jl)}(u) + r_\epsilon^{(jl)}(u),$$

where $r_\epsilon^{(jl)}(u) := E(\epsilon_{t-u}^{(j)} \epsilon_t^{(l)})$ $(j, l = 1, 2)$,

$$\begin{aligned} r_x^{(11)}(u) &= r_x^{(22)}(u) := \sum_{k=1}^{p} \tfrac{1}{2} \beta_k^2 \cos(\omega_k u), \\ r_x^{(21)}(u) &= -r_x^{(12)}(u) := \sum_{k=1}^{p} \tfrac{1}{2} \beta_k^2 \sin(\omega_k u). \end{aligned}$$

Moreover, because $\epsilon_t^{(1)} = \sum\{\psi_j^{(1)} \zeta_{t-j}^{(1)} - \psi_j^{(2)} \zeta_{t-j}^{(2)}\}$, $\epsilon_t^{(2)} = \sum\{\psi_j^{(1)} \zeta_{t-j}^{(2)} + \psi_j^{(2)} \zeta_{t-j}^{(1)}\}$, and $E\{\zeta_t^{(j)} \zeta_s^{(l)}\} = c_{jl} \delta_{t-s}$, where $c_{jl} := E\{\zeta_t^{(j)} \zeta_t^{(l)}\}$, we have

$$r_\epsilon^{(11)}(u) = \sum_{j=-\infty}^{\infty} \{\psi_{j+u}^{(1)} \psi_j^{(1)} c_{11} - \psi_{j+u}^{(1)} \psi_j^{(2)} c_{12} - \psi_{j+u}^{(2)} \psi_j^{(1)} c_{21} + \psi_{j+u}^{(2)} \psi_j^{(2)} c_{22}\},$$

$$
\begin{aligned}
r_\epsilon^{(22)}(u) &= \sum_{j=-\infty}^{\infty} \{\psi_{j+u}^{(1)}\psi_j^{(1)} c_{22} + \psi_{j+u}^{(1)}\psi_j^{(2)} c_{21} + \psi_{j+u}^{(2)}\psi_j^{(1)} c_{12} + \psi_{j+u}^{(2)}\psi_j^{(2)} c_{11}\},\\
r_\epsilon^{(12)}(u) &= \sum_{j=-\infty}^{\infty} \{\psi_{j+u}^{(1)}\psi_j^{(1)} c_{12} + \psi_{j+u}^{(1)}\psi_j^{(2)} c_{11} - \psi_{j+u}^{(2)}\psi_j^{(1)} c_{22} - \psi_{j+u}^{(2)}\psi_j^{(2)} c_{21}\},\\
r_\epsilon^{(21)}(u) &= \sum_{j=-\infty}^{\infty} \{\psi_{j+u}^{(1)}\psi_j^{(1)} c_{21} - \psi_{j+u}^{(1)}\psi_j^{(2)} c_{22} + \psi_{j+u}^{(2)}\psi_j^{(1)} c_{11} - \psi_{j+u}^{(2)}\psi_j^{(2)} c_{12}\}.
\end{aligned}
$$

As in the proof of Theorem 4.3, it can be shown that

$$
\tilde{r}_y^{(jl)}(u) = r_y^{(jl)}(u) + n^{-1}\sum_{t=1}^{n} X_{ut}^{(jl)} + \mathcal{O}(n^{-1}\Delta^{-1}),
$$

where

$$
X_{ut}^{(jl)} := x_{t+u}^{(j)}\epsilon_t^{(l)} + \epsilon_{t+u}^{(j)}x_t^{(l)} + \epsilon_{t+u}^{(j)}\epsilon_t^{(l)} - r_\epsilon^{(jl)}(u).
$$

Combining these results gives

$$
\begin{aligned}
&\sum_{j=1}^{2}\sum_{u=0}^{m} \lambda_{ju}\{\tilde{r}_y^{(j)}(u) - r_y^{(j)}(u)\}\\
&\quad = \sum_{j,l=1}^{2}\sum_{u=0}^{m} \lambda_u^{(jl)}\{\tilde{r}_y^{(jl)}(u) - r_y^{(jl)}(u)\}\\
&\quad = n^{-1}\sum_{t=1}^{n} X_t + \mathcal{O}(n^{-1}\Delta^{-1}),
\end{aligned}
$$

where $\lambda_u^{(11)} := \lambda_u^{(22)} := \lambda_{1u}$, $\lambda_u^{(21)} := -\lambda_u^{(12)} := \lambda_{2u}$, and

$$
X_t := \sum_{j,l=1}^{2}\sum_{u=0}^{m} \lambda_u^{(jl)} X_{ut}^{(jl)}.
$$

Therefore, it suffices to show that $n^{-1/2}\sum_{t=1}^{n} X_t \overset{A}{\sim} \mathrm{N}(0,\varsigma^2)$.

The remainder of the proof closely follows the proof of Theorem 4.3, because $X_{ut}^{(jl)}$ has a similar structure to X_{ut} in (4.5.16). In particular, the counterpart of $\{M_t\}$ is a sequence of martingale differences with respect to the filtration generated by $\{\zeta_t^{(1)}, \zeta_t^{(2)}, \zeta_{t-1}^{(1)}, \zeta_{t-1}^{(2)}, \ldots\}$. Moreover, in calculating V_n^2 and s_n^2, it is helpful to note that $c_{12} = c_{21}$, $\sigma^2 = c_{11} + c_{22}$, and $|E(\zeta_0^2)|^2 = (c_{11} - c_{22})^2 + 4c_{12}^2$. □

Proof of Theorem 4.5. Because the real and imaginary parts of the sample autocorrelation coefficients are continuous functions of the real and imaginary parts of the sample autocovariances, the asymptotic normality of the former is implied by the asymptotic normality of the latter according to Proposition 2.15. Moreover, because the same functions also transform $r_y(u)$ into $\rho_y(u)$, the asymptotic mean of $\hat{\rho}_y^{(j)}(u)$ is therefore equal to $\rho_y^{(j)}(u)$ by Proposition 2.15. Given these results, it suffices to calculate the asymptotic covariances of the $\hat{\rho}_y^{(j)}(u)$.

Toward that end, observe that

$$\begin{aligned}\partial\rho_y^{(j)}(u)/\partial r_y(0) &= -\{r_y(0)\}^{-1}\rho_y^{(j)}(u),\\ \partial\rho_y^{(j)}(u)/\partial r_y^{(l)}(v) &= \{r_y(0)\}^{-1}\delta_{j-l}\delta_{u-v}.\end{aligned}$$

Therefore, by Proposition 2.15, we obtain

$$\begin{aligned}\sigma_{jl}(u,v) = \{r_y(0)\}^{-2}\{&\rho_y^{(j)}(u)\rho_y^{(l)}(v)\gamma(0,0)\\ &-\rho_y^{(j)}(u)\gamma_{1l}(0,v)-\rho_y^{(l)}(v)\gamma_{j1}(u,0)+\gamma_{jl}(u,v)\}.\end{aligned}$$

Moreover, let

$$\begin{aligned}\sigma(u,v) &:= \sigma_{11}(u,v)+\sigma_{22}(u,v)+i\{\sigma_{21}(u,v)-\sigma_{12}(u,v)\},\\ \alpha(u,v) &:= \sigma_{11}(u,v)-\sigma_{22}(u,v)+i\{\sigma_{21}(u,v)+\sigma_{12}(u,v)\}.\end{aligned}$$

Then, the $\sigma_{jl}(u,v)$ are uniquely determined by $\sigma(u,v)$ and $\alpha(u,v)$ in the same way the $\gamma_{jl}(u,v)$ are determined by $\gamma(u,v)$ and $\gamma(u,-v)$ in (4.3.7). Straightforward calculation gives

$$\begin{aligned}\sigma(u,v) &= \{r_y(0)\}^{-2}\{\rho_y(u)\rho_y^*(v)\gamma(0,0)\\ &\qquad -\rho_y(u)\gamma(0,v)-\rho_y^*(v)\gamma(u,0)+\gamma(u,v)\},\\ \alpha(u,v) &= \{r_y(0)\}^{-2}\{\rho_y(u)\rho_y(v)\gamma(0,0)\\ &\qquad -\rho_y(u)\gamma(0,-v)-\rho_y(v)\gamma(u,0)+\gamma(u,-v)\}.\end{aligned}$$

Combining the first expression with (4.3.3) and (4.3.4) proves $\sigma(u,v)=\sigma_1(u,v)+\sigma_2(u,v)$, where $\sigma_1(u,v)$ and $\sigma_2(u,v)$ are given by (4.3.11) and (4.3.12). Moreover, because $\rho_y(v)=\rho_y^*(-v)$, the second expression leads to $\alpha(u,v)=\sigma(u,-v)$. □

Proof of Theorem 4.6. Because the assertion is true for the sample autocovariance function of $\{y_t(h)\}$, denoted by $\bar{r}_{y(h)}(u)$, it suffices to show that the difference between $\hat{r}_{y(h)}(u)$ and $\bar{r}_{y(h)}(u)$ is negligible under the strong BIBO assumption. More precisely, let

$$\check{r}(u) := \hat{r}_{y(h)}(u)-\bar{r}_{y(h)}(u).$$

It suffices to show that $\check{r}(u)\overset{a.s.}{\to}0$ and $E\{|\check{r}(u)|^2\}=o(n^{-1})$, which implies $\check{r}(u)=o_P(n^{-1/2})\overset{P}{\to}0$, and to show $E\{\check{r}(u)\}=\mathcal{O}(n^{-1})$.

Toward that end, let

$$\tilde{y}_t(h) := y_t(h)-\hat{y}_t(h)=\sum_{j=t}^{\infty}h_j y_{t-j}.$$

Then, we can write

$$\begin{aligned}\check{r}(u) &= n^{-1}\sum_{t=1}^{n-u}\{\hat{y}_{t+u}(h)\,\tilde{y}_t^*(h)+\tilde{y}_{t+u}(h)\,\hat{y}_t^*(h)+\tilde{y}_{t+u}(h)\,\tilde{y}_t^*(h)\}\\ &= K_1+K_2+K_3.\end{aligned}$$

Because $\hat{y}_t(h)=\hat{x}_t(h)+\hat{\epsilon}_t(h)$ and $\tilde{y}_t(h)=\tilde{x}_t(h)+\tilde{\epsilon}_t(h)$, it follows that

$$\begin{aligned}K_1 &= n^{-1}\sum_{t=1}^{n-u}\{\hat{x}_{t+u}(h)\,\tilde{x}_t^*(h)+\hat{x}_{t+u}(h)\,\tilde{\epsilon}_t^*(h)\\ &\quad+\hat{\epsilon}_{t+u}(h)\,\tilde{x}_t^*(h)+\hat{\epsilon}_{t+u}(h)\,\tilde{\epsilon}_t^*(h)\}\\ &= K_{11}+K_{12}+K_{13}+K_{14}.\end{aligned}$$

By definition,

$$\begin{aligned}K_{11} &= n^{-1}\sum_{t=1}^{n-u}\hat{x}_{t+u}(h)\,\tilde{x}_t^*(h)\\ &= n^{-1}\sum_{t=1}^{n-u}\sum_{j=0}^{t+u-1}\sum_{l=t}^{\infty}h_j h_l^*\,x_{t+u-j}x_{t-l}^*.\end{aligned}$$

Because $|x_t|\le a$ for some constant $a>0$ and all t, we have

$$\begin{aligned}|K_{11}|^2 &\le n^{-2}a^2b^2\sum_{t,s=1}^{n}\sum_{j=t}^{\infty}\sum_{l=s}^{\infty}|h_j h_l|\\ &\le n^{-2}a^2b^2\Big\{\sum_{j=1}^{\infty}j|h_j|\Big\}^2,\end{aligned}$$

where $b:=\sum|h_j|$. This, combined with $\sum j|h_j|<\infty$, gives

$$|K_{11}|^2=\mathcal{O}(n^{-2}).$$

Moreover,

$$\begin{aligned}K_{12} &= n^{-1}\sum_{t=1}^{n-u}\hat{x}_{t+u}(h)\,\tilde{\epsilon}_t^*(h)\\ &= n^{-1}\sum_{t=1}^{n-u}\sum_{j=0}^{t+u-1}\sum_{l=t}^{\infty}h_j h_l^*\,x_{t+u-j}\epsilon_{t-l}^*.\end{aligned}$$

Straightforward calculation shows that

$$E(K_{12})=0,\quad E(|K_{12}|^2)\le n^{-2}a^2b^2r_\epsilon(0)=\mathcal{O}(n^{-2}).$$

The same results are valid for K_{13}. Furthermore, $E(|K_{14}|^2)$ does not exceed

$$n^{-2}\sum_{t,s=1}^{n}\sum_{j,j'=0}^{\infty}\sum_{l=t}^{\infty}\sum_{l'=s}^{\infty}|h_j h_{j'} h_l h_{l'}|\,|E(\epsilon_{t+u-j}\epsilon_{t-l}^*\epsilon_{s+u-j'}^*\epsilon_{s-l'})|.$$

As shown in the proof of Theorem 4.2, $|E(\epsilon_{t+u-j}\epsilon^*_{t-l}\epsilon^*_{s+u-j'}\epsilon_{s-l'})|$ can be uniformly bounded by a constant $c > 0$. Therefore, we obtain

$$E(|K_{14}|^2) \le n^{-2}b^2c\Big\{\sum_{j=1}^{\infty} j|h_j|\Big\}^2,$$

which in turn gives

$$E(|K_{14}|^2) = \mathcal{O}(n^{-2}).$$

Similarly, it can be shown that

$$|E(K_{14})| \le n^{-1}b\,r_\epsilon(0)\sum_{j=1}^{\infty} j|h_j| = \mathcal{O}(n^{-1}).$$

Combining these results yields $E(K_1) = \mathcal{O}(n^{-1})$ and $E(|K_1|^2) = \mathcal{O}(n^{-2})$, which, by Proposition 2.16(c), also implies $K_1 \stackrel{a.s.}{\to} 0$. Because the same argument applies to K_2 and K_3, the proof is complete. □

NOTES

The asymptotic expressions for the sample autocovariances and autocorrelations of real time series with continuous spectra are generally attributed to Bartlett [25] (see, for example, [46] and [298]). Besides [223] and [235], additional references for the sample autocovariances of time series with mixed spectra include [87] and [375]. Extensions to multiple time series and two-dimensional random fields can be found in [223] and [191], respectively.

Chapter 5

Linear Regression Analysis

Estimation of the amplitude and phase parameters of the sinusoids when the frequencies are known *a priori* is called linear trigonometric (or harmonic) regression. It is a special linear regression problem that can be solved by standard techniques. Although treatments on basic linear regression analysis are abundant, this chapter has some unique features. For example, we consider both real- and complex-valued data in a unified manner, which is rarely done in other treatments. We also consider the implications of closely spaced frequencies on amplitude and phase estimation.

In this chapter, we first discuss the least-squares (LS) method for estimating the amplitude and phase parameters of the sinusoids. We also investigate the impact of incorrect specification of the frequencies on the amplitude and phase estimation. Then, we discuss LS-based solutions to the problem of identifying the signal frequencies from a set of candidates and determining their number if unknown. Finally, we consider the method of least absolute deviations (LAD) as a robust alternative to the LS method against outliers and heavy-tailed noise.

5.1 Least Squares Estimation

The method of least squares (LS) plays a prominent role in linear regression analysis [261] [338]. It is readily applicable to the problem of amplitude and phase estimation with given signal frequencies.

Consider the complex case where $\mathbf{y} := [y_1, \ldots, y_n]^T$ can be expressed as

$$\mathbf{y} = \sum_{k=1}^{p} \beta_k \mathbf{f}(\omega_k) + \boldsymbol{\epsilon} = \mathbf{F}\boldsymbol{\beta} + \boldsymbol{\epsilon}, \tag{5.1.1}$$

with $\mathbf{F} := [\mathbf{f}(\omega_1), \ldots, \mathbf{f}(\omega_p)]$, $\mathbf{f}(\omega_k) := [\exp(i\omega_k), \ldots, \exp(in\omega_k)]^T$, $\boldsymbol{\beta} := [\beta_1, \ldots, \beta_p]^T$, and $\boldsymbol{\epsilon} := [\epsilon_1, \ldots, \epsilon_n]^T$. The vectors $\mathbf{f}(\omega_k)$ are known as trigonometric regressors and the matrix $\mathbf{F}$ as the design matrix of the regression problem.

Under the assumption that the number of sinusoids p and the signal frequencies ω_k are known *a priori*, the LS estimator of $\boldsymbol{\beta}$ based on $\mathbf{y}$ is defined as the minimizer of the sum of squared residuals

$$\sum_{t=1}^{n} |y_t - \mathbf{x}_t^T \tilde{\boldsymbol{\beta}}|^2 = \|\mathbf{y} - \mathbf{F}\tilde{\boldsymbol{\beta}}\|^2 = (\mathbf{y} - \mathbf{F}\tilde{\boldsymbol{\beta}})^H (\mathbf{y} - \mathbf{F}\tilde{\boldsymbol{\beta}}) \tag{5.1.2}$$

with respect to $\tilde{\boldsymbol{\beta}} \in \mathbb{C}^p$, where $\mathbf{x}_t := [\exp(it\omega_1), \ldots, \exp(it\omega_p)]^T$.

The LS estimator of $\boldsymbol{\beta}$ has a closed-form expression. Indeed, let the gradient of a real-valued function $g(\tilde{\boldsymbol{\beta}})$ of $\tilde{\boldsymbol{\beta}} \in \mathbb{C}^p$ be defined as

$$\nabla g(\tilde{\boldsymbol{\beta}}) := \frac{\partial g(\tilde{\boldsymbol{\beta}})}{\partial \Re(\tilde{\boldsymbol{\beta}})} + i \frac{\partial g(\tilde{\boldsymbol{\beta}})}{\partial \Im(\tilde{\boldsymbol{\beta}})}.$$

Then, with $g(\tilde{\boldsymbol{\beta}}) := \|\mathbf{y} - \mathbf{F}\tilde{\boldsymbol{\beta}}\|^2$, it follows from Lemma 12.3.1 that

$$\nabla g(\tilde{\boldsymbol{\beta}}) = 2\mathbf{F}^H \mathbf{F} \tilde{\boldsymbol{\beta}} - 2\mathbf{F}^H \mathbf{y}.$$

Any minimizer of $g(\tilde{\boldsymbol{\beta}})$ should satisfy $\nabla g(\tilde{\boldsymbol{\beta}}) = \mathbf{0}$, or equivalently,

$$\mathbf{F}^H \mathbf{F} \tilde{\boldsymbol{\beta}} = \mathbf{F}^H \mathbf{y}.$$

Observe that $\mathbf{F}$ has full (column) rank p and hence $\mathbf{F}^H\mathbf{F}$ is nonsingular. As a result, the LS estimator of $\boldsymbol{\beta}$ is unique and can be expressed as

$$\hat{\boldsymbol{\beta}} := [\hat{\beta}_1, \ldots, \hat{\beta}_p]^T := \arg\min_{\tilde{\boldsymbol{\beta}} \in \mathbb{C}^p} \|\mathbf{y} - \mathbf{F}\tilde{\boldsymbol{\beta}}\|^2 = (\mathbf{F}^H\mathbf{F})^{-1}\mathbf{F}^H\mathbf{y}. \tag{5.1.3}$$

The *sum of squared errors* (or residual sum of squares), defined as

$$\text{SSE} := \|\mathbf{y} - \mathbf{F}\hat{\boldsymbol{\beta}}\|^2 = \mathbf{y}^H\{\mathbf{I} - \mathbf{F}(\mathbf{F}^H\mathbf{F})^{-1}\mathbf{F}^H\}\mathbf{y}, \tag{5.1.4}$$

serves as a goodness-of-fit measure for the trigonometric regression model. It is easy to verify that the LS estimator $\hat{\boldsymbol{\beta}}$ in (5.1.3) coincides with the maximum likelihood (ML) estimator of $\boldsymbol{\beta}$ under the assumption that $\{\epsilon_t\}$ is complex Gaussian white noise with $\boldsymbol{\epsilon} \sim \mathrm{N}_c(\mathbf{0}, \sigma^2\mathbf{I})$ for some unknown $\sigma^2 > 0$.

For computational purposes, it is advantageous to reformulate the complex LS problem (5.1.1)–(5.1.3) as an ordinary LS problem with real-valued data and variables. By separating the real and imaginary parts, (5.1.1) can be written as

$$\begin{bmatrix} \Re(\mathbf{y}) \\ \Im(\mathbf{y}) \end{bmatrix} = \begin{bmatrix} \Re(\mathbf{F}) & -\Im(\mathbf{F}) \\ \Im(\mathbf{F}) & \Re(\mathbf{F}) \end{bmatrix} \begin{bmatrix} \Re(\boldsymbol{\beta}) \\ \Im(\boldsymbol{\beta}) \end{bmatrix} + \begin{bmatrix} \Re(\boldsymbol{\epsilon}) \\ \Im(\boldsymbol{\epsilon}) \end{bmatrix}. \tag{5.1.5}$$

This expression facilitates the computation of $\hat{\boldsymbol{\beta}}$ in (5.1.3) using standard regression routines. It also serves as a vehicle for translating a statistical theory developed for real variables into the complex case.

To investigate the statistical properties of the LS estimator $\hat{\boldsymbol{\beta}}$ in (5.1.3), let us substitute $\mathbf{y}$ in (5.1.3) with the expression (5.1.1) and write

$$\hat{\boldsymbol{\beta}} = \boldsymbol{\beta} + (\mathbf{F}^H\mathbf{F})^{-1}\mathbf{F}^H\boldsymbol{\epsilon}. \tag{5.1.6}$$

Let $\mathbf{R}_\epsilon$ denote the covariance matrix of $\boldsymbol{\epsilon}$. Because $E(\boldsymbol{\epsilon}) = \mathbf{0}$ and $\mathrm{Cov}(\boldsymbol{\epsilon}) = \mathbf{R}_\epsilon$, it follows from (5.1.6) that

$$E(\hat{\boldsymbol{\beta}}) = \boldsymbol{\beta}, \quad \mathrm{Cov}(\hat{\boldsymbol{\beta}}) = (\mathbf{F}^H\mathbf{F})^{-1}\mathbf{F}^H\mathbf{R}_\epsilon\mathbf{F}(\mathbf{F}^H\mathbf{F})^{-1}. \tag{5.1.7}$$

Therefore, the LS estimator $\hat{\boldsymbol{\beta}}$ is unbiased for estimating $\boldsymbol{\beta}$, irrespective of the serial dependence or the distribution of the noise, as long as the noise has zero mean. The covariance matrix of $\hat{\boldsymbol{\beta}}$ depends on the second moments of the noise as well as the signal frequencies, but not on the amplitude and phase parameters of the sinusoids. In the special case of white noise with $\mathbf{R}_\epsilon = \sigma^2\mathbf{I}$, we obtain

$$\mathrm{Cov}(\hat{\boldsymbol{\beta}}) = \sigma^2(\mathbf{F}^H\mathbf{F})^{-1}.$$

It further reduces to $\mathrm{Cov}(\hat{\boldsymbol{\beta}}) = n^{-1}\sigma^2\mathbf{I}$ if the vectors $\mathbf{f}(\omega_k)$ are orthogonal to each other, i.e., if the signal frequencies ω_k are integral multiples of $2\pi/n$, known as the Fourier frequencies. In this case, the covariance matrix of $\hat{\boldsymbol{\beta}}$ depends solely on the variance of the noise and the sample size, so the amplitude and phase parameters of all sinusoids are estimated with the same accuracy regardless of their possibly different SNR.

It is easy to show that under the assumption of Gaussian white noise, the LS estimator attains the CRLB and therefore is statistically efficient. For example, consider the Cartesian system with $A_k := \Re(\beta_k)$ and $B_k := -\Im(\beta_k)$. The LS estimator of $\boldsymbol{\theta} := [A_1,\dots,A_p,B_1,\dots,B_p]^T = [\Re(\boldsymbol{\beta})^T, -\Im(\boldsymbol{\beta})^T]^T$ is given by

$$\hat{\boldsymbol{\theta}} := [\Re(\hat{\boldsymbol{\beta}})^T, -\Im(\hat{\boldsymbol{\beta}})^T]^T = [\hat{A}_1,\dots,\hat{A}_p,\hat{B}_1,\dots,\hat{B}_p]^T,$$

where $\hat{A}_k := \Re(\hat{\beta}_k)$ and $\hat{B}_k := -\Im(\hat{\beta}_k)$. It follows from (5.1.7) that

$$E(\hat{\boldsymbol{\theta}}) = \boldsymbol{\theta}, \quad \mathrm{Cov}(\hat{\boldsymbol{\theta}}) = \tfrac{1}{2}\sigma^2\boldsymbol{\Sigma},$$

where

$$\boldsymbol{\Sigma} := \begin{bmatrix} \Re\{(\mathbf{F}^H\mathbf{F})^{-1}\} & \Im\{(\mathbf{F}^H\mathbf{F})^{-1}\} \\ -\Im\{(\mathbf{F}^H\mathbf{F})^{-1}\} & \Re\{(\mathbf{F}^H\mathbf{F})^{-1}\} \end{bmatrix}.$$

By Corollary 3.1 and Remark 3.4, the matrix $\frac{1}{2}\sigma^2\boldsymbol{\Sigma}$ is nothing but the CRLB for estimating $\boldsymbol{\theta}$ under the Gaussian white noise assumption $\boldsymbol{\epsilon} \sim \mathrm{N}_c(\mathbf{0}, \sigma^2\mathbf{I})$.

The distributions of $\hat{\boldsymbol{\beta}}$ and SSE can also be easily derived under the Gaussian white noise assumption. The following theorem summarizes the results whose counterpart in the real case is well known. See Section 5.6 for a proof.

Theorem 5.1 (Distribution of the LS Estimator: Gaussian White Noise). *Let* $\mathbf{y}$ *be given by (5.1.1) with* $\boldsymbol{\epsilon} \sim \mathrm{N}_c(\mathbf{0},\sigma^2\mathbf{I})$. *Let* $\hat{\boldsymbol{\beta}}$ *and* SSE *be defined by (5.1.3) and (5.1.4). Then, the following assertions are true.*

(a) $\hat{\boldsymbol{\beta}} \sim \mathrm{N}_c(\boldsymbol{\beta},\sigma^2(\mathbf{F}^H\mathbf{F})^{-1})$.

(b) $(\hat{\boldsymbol{\beta}}-\boldsymbol{\beta})^H(\mathbf{F}^H\mathbf{F})(\hat{\boldsymbol{\beta}}-\boldsymbol{\beta})/\sigma^2 \sim \frac{1}{2}\chi^2(2p)$.

(c) $\hat{\boldsymbol{\beta}}$ *and* SSE *are independent.*

(d) $\mathrm{SSE}/\sigma^2 \sim \frac{1}{2}\chi^2(2n-2p)$.

Remark 5.1 In the real case where $\{y_t\}$ is given by $y_t = x_t + \epsilon_t$ $(t = 1,\dots,n)$ and $\{x_t\}$ satisfies (2.1.1), we can write (5.1.1) with

$$\begin{cases} \mathbf{F} := [\Re\{\mathbf{f}(\omega_1)\},\Im\{\mathbf{f}(\omega_1)\},\dots,\Re\{\mathbf{f}(\omega_q)\},\Im\{\mathbf{f}(\omega_q)\}], \\ \boldsymbol{\beta} := [A_1,B_1,\dots,A_q,B_q]^T. \end{cases} \tag{5.1.8}$$

The corresponding LS estimator of $\boldsymbol{\beta}$ can be expressed as

$$\hat{\boldsymbol{\beta}} := (\mathbf{F}^T\mathbf{F})^{-1}\mathbf{F}^T\mathbf{y}. \tag{5.1.9}$$

It is an unbiased estimator with covariance matrix $(\mathbf{F}^T\mathbf{F})^{-1}\mathbf{F}^T\mathbf{R}_\epsilon\mathbf{F}(\mathbf{F}^T\mathbf{F})^{-1}$. Under the Gaussian white noise assumption $\boldsymbol{\epsilon} \sim \mathrm{N}(\mathbf{0},\sigma^2\mathbf{I})$, part (a) of Theorem 5.1 remains true for this estimator if the complex Gaussian distribution is replaced by $\mathrm{N}(\mathbf{0},\sigma^2(\mathbf{F}^T\mathbf{F})^{-1})$. Part (b) and part (d) of Theorem 5.1 are also true if the multiplier 1/2 of the chi-square distributions is omitted and the degrees of freedom are replaced by $2q$ and $n-2q$, respectively. Part (c) requires no alteration. These assertions are nothing but special cases of the general theory of linear regression for real-valued data that can be found in standard textbooks such as [261] and [338]. Theorem 5.1 is an extension of this theory to the complex case.

Under the Gaussian white noise assumption $\boldsymbol{\epsilon} \sim \mathrm{N}_c(\mathbf{0},\sigma^2\mathbf{I})$, the ML estimator of σ^2 is given by $n^{-1}\mathrm{SSE}$. According to Theorem 5.1,

$$\hat{\sigma}^2 := n^{-1}\mathrm{SSE} \sim \tfrac{1}{2}n^{-1}\sigma^2\chi^2(2n-2p).$$

Therefore, we obtain

$$E(\hat{\sigma}^2) = \tfrac{1}{2}n^{-1}\sigma^2(2n-2p) = (1-p/n)\sigma^2 < \sigma^2,$$

which means that $\hat{\sigma}^2$ is a biased estimator that underestimates σ^2 on average. An unbiased estimator of σ^2 is given by

$$\tilde{\sigma}^2 := (n-p)^{-1}\mathrm{SSE} = n(n-p)^{-1}\hat{\sigma}^2,$$

where the divisor $n-p$ can be interpreted as the effective sample size left for estimating the noise parameters.

In the general case of nonGaussian colored noise, the statistical properties of the LS estimator can be analyzed asymptotically for large sample sizes. The following theorem asserts that $\hat{\boldsymbol{\beta}}$ in (5.1.3) is a consistent estimator of $\boldsymbol{\beta}$ even if the noise is colored and nonGaussian. See Section 5.6 for a proof.

Theorem 5.2 (Consistency of the LS Estimator). *Let $\hat{\boldsymbol{\beta}}$ be the LS estimator defined by (5.1.3) with* $\mathbf{y}$ *satisfying (5.1.1). Assume that* $n^{-1}\mathbf{F}^H\boldsymbol{\epsilon} \xrightarrow{P} \mathbf{0}$ *and* $n^{-1}\mathbf{F}^H\mathbf{F} \to \mathbf{C}$ *as* $n \to \infty$*, where* $\mathbf{C}$ *is a nonsingular matrix. Then,* $\hat{\boldsymbol{\beta}} \xrightarrow{P} \boldsymbol{\beta}$*. If, in addition,* $n^{-1}\boldsymbol{\epsilon}^H\boldsymbol{\epsilon} \xrightarrow{P} \sigma^2$ *as* $n \to \infty$*, then* $n^{-1}\text{SSE} \xrightarrow{P} \sigma^2$*. The assertions remain valid if convergence in probability is everywhere replaced by almost sure convergence.*

Remark 5.2 If $\{\epsilon_t\}$ is a zero-mean stationary process with absolutely summable ACF $r_\epsilon(u)$, then $n^{-1}\mathbf{F}^H\boldsymbol{\epsilon} \xrightarrow{a.s.} \mathbf{0}$ by Lemma 12.5.3. If $\{\epsilon_t\}$ satisfies the conditions of Theorem 4.1, then $n^{-1}\mathbf{F}^H\boldsymbol{\epsilon} \xrightarrow{a.s.} \mathbf{0}$ and $n^{-1}\boldsymbol{\epsilon}^H\boldsymbol{\epsilon} \xrightarrow{a.s.} \sigma^2 := r_\epsilon(0)$.

Remark 5.3 Because $\beta_k = C_k \exp(i\phi_k)$, one can estimate C_k and ϕ_k by $\hat{C}_k := |\hat{\beta}_k|$ and $\hat{\phi}_k := \angle\hat{\beta}_k$, where $\hat{\boldsymbol{\beta}} := [\hat{\beta}_1, \ldots, \hat{\beta}_p]^T$ is the LS estimator of $\boldsymbol{\beta}$. Under the conditions of Theorem 5.2, $\hat{C}_k$ and $\hat{\phi}_k$ are consistent estimators of C_k and ϕ_k.

Theorem 5.2 requires that $n^{-1}\mathbf{F}^H\mathbf{F}$ approaches a nonsingular matrix as $n \to \infty$. This condition is satisfied if the signal frequencies are well separated such that

$$\lim_{n\to\infty} n\Delta = \infty, \tag{5.1.10}$$

where

$$\Delta := \min_{k \neq k'}\{|\omega_k - \omega_{k'}|, 2\pi - |\omega_k - \omega_{k'}|\}.$$

In this case, Lemma 12.1.4 gives

$$n^{-1}\mathbf{F}^H\mathbf{F} = \mathbf{I} + \mathcal{O}(n^{-1}\Delta^{-1}) \to \mathbf{C} := \mathbf{I}.$$

If the signal frequencies are closely spaced such that $\Delta \to 0$ as $n \to \infty$ but

$$\lim_{n\to\infty} n(\omega_k - \omega_{k'}) = a_{kk'} \neq 0 \quad (k \neq k'), \tag{5.1.11}$$

then $n^{-1}\mathbf{F}^H\mathbf{F} \to \mathbf{C} := [c_{kk'}]$ $(k, k' = 1, \ldots, p)$, where

$$c_{kk} := 1, \quad c_{kk'} := d_0^*(a_{kk'}) \quad (k \neq k'),$$

with $d_0(x) := \text{sinc}(x/2)\exp(ix/2)$. This result can be derived from the fact that $n^{-1}\mathbf{f}^H(\omega_k)\mathbf{f}(\omega_{k'}) = D_n^*(\omega_k - \omega_{k'})$, where $D_n(\omega) := n^{-1}\sum_{t=1}^n \exp(it\omega)$. According to Lemma 12.1.3(f), $D_n(\omega_k - \omega_{k'}) \to d_0(a_{kk'})$, hence the assertion.

The next theorem shows that the LS estimator $\hat{\boldsymbol{\beta}}$ has an asymptotic Gaussian distribution with a diagonal covariance matrix, provided that the noise is a stationary Gaussian process or a linear process of the form

$$\epsilon_t = \sum_{j=-\infty}^{\infty} \psi_j \zeta_{t-j}, \tag{5.1.12}$$

and that the signal frequencies satisfy the separation condition (5.1.10). A proof of this result is given in Section 5.6.

Theorem 5.3 (Asymptotic Normality of the LS Estimator: Case 1). *Let $\hat{\boldsymbol{\beta}}$ be given by (5.1.3) with* $\mathbf{y}$ *given by (5.1.1). Assume that $\{\epsilon_t\}$ is (a) a zero-mean stationary complex Gaussian process with absolutely summable* ACF $r_\epsilon(u)$ *or (b) a complex linear process of the form (5.1.12) with $\sum|\psi_j|<\infty$, $\{\zeta_t\}\sim \mathrm{IID}(0,\sigma^2)$, and $E(\zeta_t^2)=0$. If $\{\omega_k\}$ satisfies (5.1.10) and $\inf\{f_\epsilon(\omega)\}>0$, then $\sqrt{n}\,(\hat{\boldsymbol{\beta}}-\boldsymbol{\beta}) \stackrel{A}{\sim} \mathrm{N}_c(\mathbf{0},\mathbf{S})$ as $n\to\infty$, where $\mathbf{S}:=\mathrm{diag}\{f_\epsilon(\omega_1),\dots,f_\epsilon(\omega_p)\}$.*

Remark 5.4 For the real case discussed in Remark 5.1, the LS estimator $\hat{\boldsymbol{\beta}}$ of $\boldsymbol{\beta}:=[A_1,B_1,\dots,A_q,B_q]^T$ is given by (5.1.9). Assume that $\{\epsilon_t\}$ is (a) a zero-mean stationary Gaussian process with absolutely summable ACF $r_\epsilon(u)$ or (b) a real linear process of the form (5.1.12) with $\sum|\psi_j|<\infty$ and $\{\zeta_t\}\sim\mathrm{IID}(0,\sigma^2)$. If $\{\omega_k\}$ satisfies (5.1.10) with $\Delta:=\min_{k\neq k'}\{|\omega_k-\omega_{k'}|,\omega_k,\pi-\omega_k\}$ and $\inf\{f_\epsilon(\omega)\}>0$, then $\sqrt{n}\,(\hat{\boldsymbol{\beta}}-\boldsymbol{\beta}) \stackrel{A}{\sim} \mathrm{N}(\mathbf{0},2\mathbf{S})$ as $n\to\infty$, where $\mathbf{S}:=\mathrm{diag}\{f_\epsilon(\omega_1),f_\epsilon(\omega_1),\dots,f_\epsilon(\omega_q),f_\epsilon(\omega_q)\}$.

According to Theorem 5.3, the LS estimators $\hat{\beta}_k$ $(k=1,\dots,p)$ are asymptotically independent and for each k,

$$\sqrt{n}\,(\hat{\beta}_k-\beta_k) \stackrel{A}{\sim} \mathrm{N}_c(0,f_\epsilon(\omega_k)).$$

Note that the asymptotic variance of $\hat{\beta}_k$ depends solely on the value of the noise spectrum at the corresponding frequency ω_k.

The LS estimator of the Cartesian parameter $\boldsymbol{\theta}_k:=[A_k,B_k]^T:=[\Re(\beta_k),-\Im(\beta_k)]^T$ is given by $\hat{\boldsymbol{\theta}}_k:=[\hat{A}_k,\hat{B}_k]^T:=[\Re(\hat{\beta}_k),-\Im(\hat{\beta}_k)]^T$. According to Theorem 5.3, the $\hat{\boldsymbol{\theta}}_k$ $(k=1,\dots,p)$ are asymptotically independent and for each k,

$$\sqrt{n}\,(\hat{\boldsymbol{\theta}}_k-\boldsymbol{\theta}_k) \stackrel{A}{\sim} \mathrm{N}(\mathbf{0},\tfrac{1}{2}f_\epsilon(\omega_k)\mathbf{I}).$$

Observe that the asymptotic covariance matrix of $\hat{\boldsymbol{\theta}}_k$ coincides with the ACRLB in Corollary 3.2. Thus the LS estimator is asymptotically efficient under the Gaussian assumption and the conditions of Theorem 5.3, even if the noise is colored. The efficiency assertion remains true in the real case, due to Remark 5.4.

The LS estimator attains the same Gaussian CRLB asymptotically regardless of the actual noise distribution. Because the Gaussian distribution yields the largest CRLB among the class of noise distributions that has the same or smaller variance, as discussed in Proposition 3.4, one should be able to improve the performance in nonGaussian cases with other means. An important example will be discussed in Section 5.5.

Theorem 5.3 can be extended by replacing the frequency separation condition (5.1.10) with a milder condition that can be satisfied by closely spaced signal frequencies with the property (5.1.11). The following theorem is an example of such extensions under the white noise assumption. See Section 5.6 for a proof.

Theorem 5.4 (Asymptotic Normality of the LS Estimator: Case 2). *Let $\hat{\boldsymbol{\beta}}$ be given by (5.1.3) with* $\mathbf{y}$ *satisfying (5.1.1). Assume that $\{\epsilon_t\}\sim\mathrm{IID}(0,\sigma^2)$ and $E(\epsilon_t^2)=0$.*

Assume further that there exists a nonsingular matrix $\mathbf{C}$ *such that* $n^{-1}\mathbf{F}^H\mathbf{F} \to \mathbf{C}$ *as* $n \to \infty$. *Then,* $\sqrt{n}(\hat{\boldsymbol{\beta}} - \boldsymbol{\beta}) \xrightarrow{D} \mathrm{N}_c(\mathbf{0}, \sigma^2\mathbf{C}^{-1})$ *as* $n \to \infty$.

Remark 5.5 If $\{\omega_k\}$ satisfies (5.1.10), then $\mathbf{C} = \mathbf{I}$ and $\sqrt{n}(\hat{\boldsymbol{\beta}} - \boldsymbol{\beta}) \xrightarrow{D} \mathrm{N}_c(\mathbf{0}, \sigma^2\mathbf{I})$, which is a special case of Theorem 5.3.

Remark 5.6 For the real case discussed in Remark 5.1, if $\{\epsilon_t\} \sim \mathrm{IID}(0, \sigma^2)$ and $n^{-1}\mathbf{F}^T\mathbf{F} \to \frac{1}{2}\mathbf{C}$ as $n \to \infty$, then $\sqrt{n}(\hat{\boldsymbol{\beta}} - \boldsymbol{\beta}) \xrightarrow{D} \mathrm{N}(\mathbf{0}, 2\sigma^2\mathbf{C}^{-1})$. Under the separation condition (5.1.10) with $\Delta := \min_{k \neq k'}\{|\omega_k - \omega_{k'}|, \omega_k, \pi - \omega_k\}$, we have $\mathbf{C} = \mathbf{I}$ and hence $\sqrt{n}(\hat{\boldsymbol{\beta}} - \boldsymbol{\beta}) \xrightarrow{D} \mathrm{N}(\mathbf{0}, 2\sigma^2\mathbf{I})$, which is a special case of Remark 5.4.

Under the conditions of Theorem 5.4, the LS estimator $\hat{\boldsymbol{\theta}} := [\Re(\hat{\boldsymbol{\beta}})^T, -\Im(\hat{\boldsymbol{\beta}})^T]^T$ for estimating the parameter $\boldsymbol{\theta} := [\Re(\boldsymbol{\beta})^T, -\Im(\boldsymbol{\beta})^T]^T$ has the property

$$\sqrt{n}(\hat{\boldsymbol{\theta}} - \boldsymbol{\theta}) \xrightarrow{D} \mathrm{N}(\mathbf{0}, \tfrac{1}{2}\sigma^2\boldsymbol{\Lambda}), \tag{5.1.13}$$

where

$$\boldsymbol{\Lambda} := \begin{bmatrix} \Re(\mathbf{C}^{-1}) & \Im(\mathbf{C}^{-1}) \\ -\Im(\mathbf{C}^{-1}) & \Re(\mathbf{C}^{-1}) \end{bmatrix} = \begin{bmatrix} \Re(\mathbf{C}) & \Im(\mathbf{C}) \\ -\Im(\mathbf{C}) & \Re(\mathbf{C}) \end{bmatrix}^{-1}. \tag{5.1.14}$$

As we can see, the asymptotic covariance matrix coincides with the ACRLB in Corollary 3.4. So, the LS estimator attains the Gaussian CRLB asymptotically even if the noise is nonGaussian and the signal frequencies are closely spaced.

To illustrate Theorem 5.4, consider the following example.

Example 5.1 (LS Amplitude Estimation for Two Complex Sinusoids with Closely Spaced Frequencies in White Noise). Let $p = 2$ in (5.1.1) and assume

$$\lim_{n\to\infty} n(\omega_2 - \omega_1) = a > 0.$$

Assume also that $\{\epsilon_t\}$ satisfies the conditions of Theorem 5.4. Then, the asymptotic normality assertion in Theorem 5.4 holds with

$$\mathbf{C}^{-1} = \begin{bmatrix} 1 & d_0(a) \\ d_0^*(a) & 1 \end{bmatrix}^{-1} = \frac{1}{1 - |d_0(a)|^2}\begin{bmatrix} 1 & -d_0(a) \\ -d_0^*(a) & 1 \end{bmatrix}.$$

Note that $\mathbf{C}$ is invertible if $0 < a < 2\pi$, in which case, $|d_0(a)| = \mathrm{sinc}(a/2) < 1$. ◇

It is important to note that the $\sqrt{n}$ rate of convergence, as guaranteed by Theorems 5.3 and 5.4, breaks down in Example 5.1 if $n(\omega_2 - \omega_1) \to 0$ as $n \to \infty$. To see this, we only need to examine the Gaussian case. According to Theorem 5.1, $\sqrt{n}(\hat{\boldsymbol{\beta}} - \boldsymbol{\beta}) \sim \mathrm{N}_c(0, \sigma^2 n(\mathbf{F}^H\mathbf{F})^{-1})$ for any $n \geq p$. It is easy to verify that

$$n(\mathbf{F}^H\mathbf{F})^{-1} = \begin{bmatrix} 1 & D_n(\Delta) \\ D_n^*(\Delta) & 1 \end{bmatrix}^{-1} = \frac{1}{1 - |D_n(\Delta)|^2}\begin{bmatrix} 1 & -D_n(\Delta) \\ -D_n^*(\Delta) & 1 \end{bmatrix},$$

where $D_n(\Delta) := n^{-1}\sum_{t=1}^{n}\exp(it\Delta)$ and $\Delta := \omega_2 - \omega_1$. If $n\Delta \to 0$, then $D_n(\Delta) \to 1$ by Lemma 12.1.3. In this case, we have

$$(1-|D_n(\Delta)|^2)n(\mathbf{F}^H\mathbf{F})^{-1} \to \mathbf{V} := \begin{bmatrix} 1 & -1 \\ -1 & 1 \end{bmatrix}.$$

Let $r_n := \sqrt{n(1-|D_n(\Delta)|^2)}$. Then, it follows from Proposition 2.13(b) that

$$r_n(\hat{\boldsymbol{\beta}} - \boldsymbol{\beta}) \xrightarrow{D} \mathrm{N}_c(\mathbf{0}, \sigma^2\mathbf{V}).$$

Because $D_n(\Delta) \to 1$, we have $r_n/\sqrt{n} \to 0$. In other words, the convergence rate of $\hat{\boldsymbol{\beta}}$ is slower than $\sqrt{n}$. This result is a manifestation of the increased uncertainty in estimating the amplitude parameter when the signal frequencies become too close to each other relative to the sample size.

While the LS estimator given by (5.1.3) does not require any knowledge of the noise, it can be improved when the covariance matrix $\mathbf{R}_\epsilon$ is known up to a constant multiplier by considering the *generalized least-squares* (GLS) estimator

$$\hat{\boldsymbol{\beta}} := (\mathbf{F}^H\mathbf{R}_\epsilon^{-1}\mathbf{F})^{-1}\mathbf{F}^H\mathbf{R}_\epsilon^{-1}\mathbf{y}. \tag{5.1.15}$$

This estimator minimizes $(\mathbf{y}-\mathbf{F}\tilde{\boldsymbol{\beta}})^H\mathbf{R}_\epsilon^{-1}(\mathbf{y}-\mathbf{F}\tilde{\boldsymbol{\beta}})$ with respect to $\tilde{\boldsymbol{\beta}} \in \mathbb{C}^p$. It reduces to the LS estimator in (5.1.3) when $\mathbf{R}_\epsilon = \sigma^2\mathbf{I}$. In general, it can be shown that the LS and GLS estimators are identical if and only if the column space of $\mathbf{F}$ coincides with the column space of $\mathbf{R}_\epsilon^{-1}\mathbf{F}$ or the column space of $\mathbf{R}_\epsilon\mathbf{F}$ [338, p. 63]. The GLS estimator in (5.1.15) does not depend on the unknown multiplier in $\mathbf{R}_\epsilon$ just like the LS estimator in (5.1.3) does not depend on σ^2. The GLS estimator coincides with the ML estimator under the condition $\boldsymbol{\epsilon} \sim \mathrm{N}_c(\mathbf{0}, \mathbf{R}_\epsilon)$.

To investigate the statistical properties of the GLS estimator, it is helpful to combine (5.1.1) with (5.1.15) and write

$$\hat{\boldsymbol{\beta}} = \boldsymbol{\beta} + (\mathbf{F}^H\mathbf{R}_\epsilon^{-1}\mathbf{F})^{-1}\mathbf{F}^H\mathbf{R}_\epsilon^{-1}\boldsymbol{\epsilon}. \tag{5.1.16}$$

From this expression, we obtain

$$E(\hat{\boldsymbol{\beta}}) = \boldsymbol{\beta}, \quad \mathrm{Cov}(\hat{\boldsymbol{\beta}}) = (\mathbf{F}^H\mathbf{R}_\epsilon^{-1}\mathbf{F})^{-1}. \tag{5.1.17}$$

As can be seen, the GLS estimator is also unbiased for estimating $\boldsymbol{\beta}$, just like the the LS estimator. But, the covariance matrix of the GLS estimator is different from the covariance matrix of the LS estimator.

The following theorem asserts that the GLS estimator in (5.1.15) is the best linear unbiased estimator, or BLUE, of $\boldsymbol{\beta}$, in the sense that it has the smallest variance among all unbiased linear estimators of $\boldsymbol{\beta}$, including particularly the LS estimator in (5.1.3). This is the complex-case extension of a classical result for the real case [338]. A proof can be found in Section 5.6.

Theorem 5.5 (Gauss-Markov Theorem). *Let* $\mathbf{y}$ *be given by (5.1.1) with* $\boldsymbol{\epsilon}$ *having mean zero and covariance matrix* $\mathbf{R}_\epsilon$. *Assume that* $\mathbf{R}_\epsilon$ *is known up to a constant multiplier. Let* $\hat{\boldsymbol{\beta}}$ *be the GLS estimator of* $\boldsymbol{\beta}$ *given by (5.1.15). Then, for any constant vector* $\mathbf{a} \in \mathbb{C}^p$, $\mathbf{a}^H\hat{\boldsymbol{\beta}}$ *is the unique estimator with minimum variance in the class of linear unbiased estimators of* $\mathbf{a}^H\boldsymbol{\beta}$.

The GLS estimator $\hat{\boldsymbol{\beta}}$ in (5.1.15) enjoys similar properties as described in Theorem 5.1 under the Gaussian assumption. Indeed, if $\boldsymbol{\epsilon} \sim \mathrm{N}_c(\mathbf{0}, \mathbf{R}_\epsilon)$, then

$$\hat{\boldsymbol{\beta}} \sim \mathrm{N}_c(\boldsymbol{\beta}, (\mathbf{F}^H\mathbf{R}_\epsilon^{-1}\mathbf{F})^{-1}),$$
$$(\hat{\boldsymbol{\beta}} - \boldsymbol{\beta})^H(\mathbf{F}^H\mathbf{R}_\epsilon^{-1}\mathbf{F})(\hat{\boldsymbol{\beta}} - \boldsymbol{\beta}) \sim \tfrac{1}{2}\chi^2(2p).$$

This assertion can be proved by using the *pre-whitening* transformation:

$$\tilde{\mathbf{y}} := \mathbf{R}_\epsilon^{-1/2}\mathbf{y}, \quad \tilde{\mathbf{F}} := \mathbf{R}_\epsilon^{-1/2}\mathbf{F}, \quad \tilde{\boldsymbol{\epsilon}} := \mathbf{R}_\epsilon^{-1/2}\boldsymbol{\epsilon}.$$

Indeed, it is easy to verify that

$$\tilde{\mathbf{y}} = \tilde{\mathbf{F}}\boldsymbol{\beta} + \tilde{\boldsymbol{\epsilon}}, \quad \tilde{\boldsymbol{\epsilon}} \sim \mathrm{N}_c(\mathbf{0}, \mathbf{I}).$$

So, Theorem 5.1 can be applied with $\sigma^2 := 1$. Part (c) and part (d) of Theorem 5.1 remain valid for the GLS estimator if σ^2 is replaced by 1 and SSE replaced by

$$\mathrm{SSE} := (\mathbf{y} - \mathbf{F}\hat{\boldsymbol{\beta}})^H\mathbf{R}_\epsilon^{-1}(\mathbf{y} - \mathbf{F}\hat{\boldsymbol{\beta}}).$$

See [199] for a comprehensive comparison between the LS and GLS estimators under the general linear regression setting and for a discussion on the impact of estimated or misspecified noise covariance matrix on the GLS estimator.

Finally, we remark that when $\mathbf{R}_\epsilon$ contains more unknown parameters, say $\boldsymbol{\eta}$, than just a multiplier, the following iterative procedure can be employed to estimate $\boldsymbol{\beta}$ and $\boldsymbol{\eta}$ alternately: Given an initial value $\hat{\boldsymbol{\eta}}_0$ and for $m = 1, 2, \ldots$,

1. Obtain $\hat{\boldsymbol{\beta}}_m$ using (5.1.15) with $\boldsymbol{\eta}$ in $\mathbf{R}_\epsilon$ replaced by $\hat{\boldsymbol{\eta}}_{m-1}$;
2. Obtain $\hat{\boldsymbol{\eta}}_m$ using an estimator of $\boldsymbol{\eta}$ based on the residual $\hat{\boldsymbol{\epsilon}}_m := \mathbf{y} - \mathbf{F}\hat{\boldsymbol{\beta}}_m$.

As an example, let $\{\epsilon_t\}$ be an AR(1) process satisfying

$$\epsilon_t + \varphi\epsilon_{t-1} = \zeta_t, \quad \{\zeta_t\} \sim \mathrm{IID}(0, \sigma^2)$$

for some $\sigma^2 > 0$ and $\varphi \in \mathbb{C}$ such that $|\varphi| < 1$. In this case, we have

$$\mathbf{R}_\epsilon = r_\epsilon(0)[\rho_\epsilon(t-s)] \quad (t, s = 1, \ldots, n),$$

where $\rho_\epsilon(u) = \rho_\epsilon^*(-u) := (-\varphi)^u$ for $u \ge 0$ and $r_\epsilon(0) = \sigma^2/(1 - |\varphi|^2)$. Because the GLS estimator in (5.1.15) does not depend on $r_\epsilon(0)$, the AR coefficient φ is the only noise parameter that requires estimation, hence $\boldsymbol{\eta} := \varphi \in \mathbb{C}$. The AR coefficient can be estimated in many ways, as will be discussed later in Chapter 7 (Section 7.1.3), and any of them can be applied to $\hat{\boldsymbol{\epsilon}}_m$ in the second step of the iterative procedure.

5.2 Sensitivity to Frequency Offset

Suppose the signal frequencies are not known precisely but are estimated with random error. In this section, let us examine the impact of such error on the LS estimator of the amplitude and phase parameters.

Consider (5.1.1). Assume that some estimated frequencies of the form

$$\hat{\omega}_k = \omega_k + \xi_{nk} \qquad (k = 1, \ldots, p) \tag{5.2.1}$$

are employed to construct the LS estimator of $\boldsymbol{\beta}$ in (5.1.3) instead of the true frequencies ω_k. The ξ_{nk} represent the random error or offset in frequency estimation. The following theorem provides a sufficient condition for consistent estimation of $\boldsymbol{\beta}$ in the presence of frequency offset. See Section 5.6 for a proof.

Theorem 5.6 (Consistency of the LS Estimator with Frequency Offset). *Let* $\mathbf{y}$ *be given by (5.1.1) with* $\{\epsilon_t\}$ *and* $\{\omega_k\}$ *satisfying the conditions of Theorem 5.2. Let* $\tilde{\boldsymbol{\beta}}$ *be the LS estimator of* $\boldsymbol{\beta}$ *based on the frequencies* $\hat{\omega}_k$ *of the form (5.2.1). If* $\xi_{nk} = \mathcal{O}_P(a_n)$ *for a constant sequence* $\{a_n\}$ *such that*

$$\lim_{n\to\infty} n a_n = 0, \tag{5.2.2}$$

then $\tilde{\boldsymbol{\beta}} \xrightarrow{P} \boldsymbol{\beta}$ *as* $n \to \infty$. *If* $\xi_{nk} = \mathcal{O}(a_n)$, *then* $\tilde{\boldsymbol{\beta}} \xrightarrow{a.s.} \boldsymbol{\beta}$. *In either case,*

$$\tilde{\boldsymbol{\beta}} = \hat{\boldsymbol{\beta}} + \mathcal{O}_P(n a_n), \tag{5.2.3}$$

where $\hat{\boldsymbol{\beta}}$ *is the LS estimator of* $\boldsymbol{\beta}$ *based on the true frequencies* ω_k.

Remark 5.7 The assertions in Theorem 5.6 remain true if $\{\omega_k\}$ satisfies the separation condition (5.1.10) and $\boldsymbol{\beta}$ is estimated by $n^{-1}\hat{\mathbf{F}}^H\mathbf{y}$ instead of $(\hat{\mathbf{F}}^H\hat{\mathbf{F}})^{-1}\hat{\mathbf{F}}^H\mathbf{y}$. The former clearly has the computational advantage when the sinusoids with estimated frequencies are not orthogonal.

Theorem 5.6 shows that a sufficient condition for the consistency of $\tilde{\boldsymbol{\beta}}$ is that the accuracy of frequency estimation takes the form $o_P(n^{-1})$. The following example demonstrates that an accuracy of the form $\mathcal{O}_P(n^{-1})$ for frequency estimation is not enough to guarantee the consistency for amplitude estimation.

Example 5.2 (Amplitude Estimation for Single Complex Sinusoid with Frequency Offset). Let $p = 1$. Assume that $\{\epsilon_t\} \sim \text{IID}(0, \sigma^2)$ and $\hat{\omega}_1 = \omega_1 + \xi_n$. Then,

$$\tilde{\beta}_1 = n^{-1} \sum_{t=1}^{n} y_t \exp(-it\hat{\omega}_1) = \beta_1 D_n^*(\xi_n) + e_n,$$

where $D_n(\omega) := n^{-1}\sum_{t=1}^n \exp(it\omega)$ and $e_n := n^{-1}\sum_{t=1}^n \epsilon_t \exp(-it\hat{\omega}_1)$. If ξ_n takes the form $\mathcal{O}_P(a_n)$ and $\{a_n\}$ satisfies (5.2.2), then $n\xi_n \xrightarrow{P} 0$ as $n \to \infty$. In this case,

Lemma 12.1.3(f) ensures $D_n^*(\xi_n) \xrightarrow{P} 1$. By Lemma 12.5.3, we have $e_n \xrightarrow{P} 0$. Combining these results gives $\tilde{\beta}_1 \xrightarrow{P} \beta_1$, as asserted by Theorem 5.6.

Now suppose that the condition (5.2.2) is not satisfied and

$$n\xi_n \xrightarrow{P} a$$

for some constant a such that $0 < |a| < 2\pi$. This means that the frequency is estimated with an accuracy $\mathcal{O}_P(n^{-1})$ but not $o_P(n^{-1})$. In this case, Lemma 12.1.3(f) gives $D_n(\xi_n) \xrightarrow{P} d_0(a) := \text{sinc}(a/2)\exp(ia/2) \neq 1$. Therefore,

$$\tilde{\beta}_1 \xrightarrow{P} \beta_1 d_0^*(a) \neq \beta_1.$$

Note that the ratio $|\tilde{\beta}_1/\beta_1|$ tends to $\text{sinc}(a/2)$ which can take any value between 0 and 1 as $|a|$ varies in the interval $(0, 2\pi)$. The analysis in [314] under the RSM (2.1.1) with $q = 1$ leads to a similar conclusion. Note also that the impact of the frequency offset on the real part of $\tilde{\beta}_1$ may differ from the impact on the imaginary part. In fact, with $\beta_1 = C_1\cos(\phi_1) + iC_1\sin(\phi_1)$, it is easy to see that $\Re(\tilde{\beta}_1) \xrightarrow{P} C_1\text{sinc}(a/2)\cos(\phi_1 - a/2)$ and $\Im(\tilde{\beta}_1) \xrightarrow{P} C_1\text{sinc}(a/2)\sin(\phi_1 - a/2)$. ◇

This analysis shows that the LS estimator of $\boldsymbol{\beta}$ becomes inconsistent when derived from estimated frequencies with poor accuracy. To guarantee the consistency, the accuracy of frequency estimation must be higher than $\mathcal{O}_P(n^{-1})$, and the ordinary consistency $\hat{\omega}_k \xrightarrow{P} \omega_k$, or even the $\sqrt{n}$ consistency $\sqrt{n}(\hat{\omega}_k - \omega_k) \xrightarrow{P} 0$, is not enough. This requirement signifies the need for highly accurate frequency estimation in applications such as detection of hidden periodicities where the estimated amplitude at the estimated frequency plays an important role in determining the presence or absence of a sinusoid.

The inconsistency of amplitude estimation based on poor frequency estimates can be explained intuitively by the fact that the error in the trigonometric regressor $\mathbf{f}(\hat{\omega}_k)$ with a poorly estimated frequency $\hat{\omega}_k$ can be as large as the regressor itself. This error gives rise to a nonvanishing bias in the amplitude estimator.

Figure 5.1 shows an example of the error in the regressor when the frequency offset is equal to π/n^d with $d = 1$ and $d = 1.5$. As can be seen, for $d = 1$, the error in the regressor grows with t and eventually reaches a range of -2 to 2 regardless of the sample size; the range of the true regressor is only -1 to 1. For $d = 1.5$, which satisfies the consistency requirement (5.2.2), the error in the regressor stays small for all t and becomes smaller as the sample size grows.

The requirement on the accuracy of frequency estimation is rather stringent because many frequency estimators, as will be discussed later in the book (e.g., Section 9.1), are only as accurate as $\mathcal{O}_P(n^{-1/2})$. On the other hand, the CRLB suggests that the potential accuracy of frequency estimation can be as high as $\mathcal{O}_P(n^{-3/2})$, which is more than enough to guarantee consistent amplitude and phase estimation. Achieving this optimal accuracy is one of the primary objectives for frequency estimation.

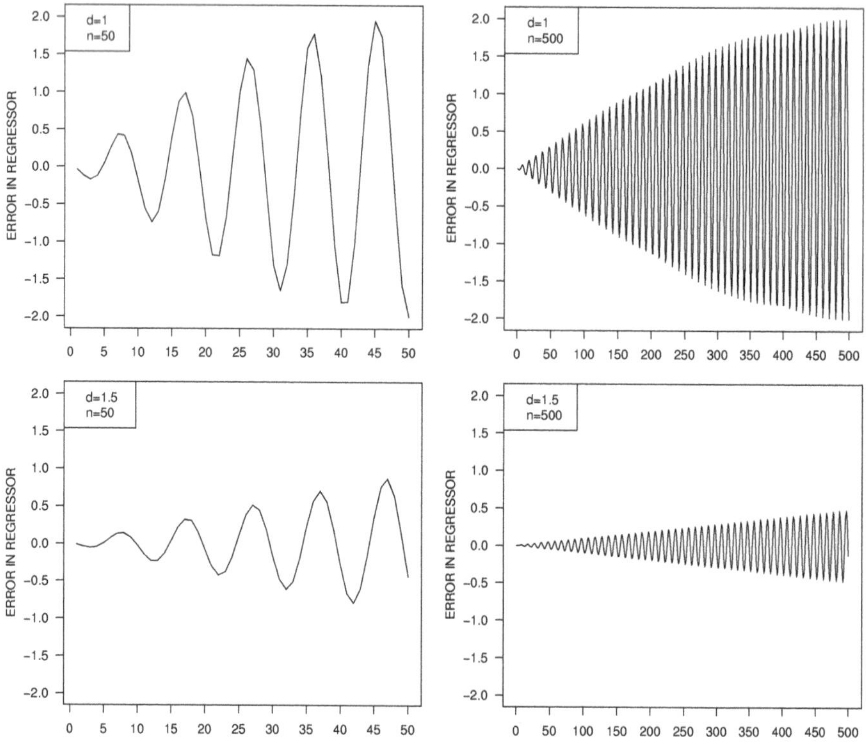

Figure 5.1. Plot of the error in the trigonometric regressor $\cos(\omega t) - \cos(\hat{\omega} t)$ $(t = 1, \ldots, n)$, where the true frequency is $\omega = 2\pi \times 0.1$ and the estimated frequency takes the form $\hat{\omega} = \omega + \pi / n^d$. Top, $d = 1$, $n = 50$ and 500; bottom, $d = 1.5$, $n = 50$ and 500.

5.3 Frequency Identification

In this section, we consider a situation where the signal frequencies in (5.1.1) are members of a finite set $\Omega := \{\lambda_1, \ldots, \lambda_m\}$ $(p < m < n)$, which is completely known *a priori*, and where the objective is to identify which p frequencies in Ω are the actual signal frequencies contained in the data record $\mathbf{y}$. We call this a *frequency identification* problem. This problem can be solved by a hypothesis testing technique in linear regression analysis, as suggested by Hartley [142]. For simplicity, let us focus on the case of Gaussian white noise. For Gaussian colored noise with known (or estimated) covariance matrix, the problem can be transformed into an equivalent one with Gaussian white noise by the pre-whitening procedure discussed at the end of Section 5.1.

Because Ω contains $\{\omega_k\}$, the expression of $\mathbf{y}$ in (5.1.1) can be expanded to include all frequencies in Ω such that

$$\mathbf{y} = \sum_{j=1}^{m} c_j \mathbf{f}(\lambda_j) + \boldsymbol{\epsilon}, \tag{5.3.1}$$

where the c_j are unknown constants and $\boldsymbol{\epsilon} \sim \mathrm{N}_c(\mathbf{0}, \sigma^2\mathbf{I})$. Based on this so-called *complete* model, the frequency identification problem can be handled by deciding whether or not a given p-element subset, or p-subset, of Ω, denoted by Ω', misses any signal frequency. More formally, the hypotheses can be stated as

$$\begin{cases} H_0: c_j = 0 \text{ for all } j \text{ such that } \lambda_j \in \Omega \setminus \Omega', \\ H_1: c_j \neq 0 \text{ for some } j \text{ such that } \lambda_j \in \Omega \setminus \Omega'. \end{cases} \tag{5.3.2}$$

Observe that H_0 is true if and only if $\Omega' = \{\omega_k\}$. Therefore, when a test fails to reject H_0, the p-subset Ω' identifies the signal frequencies.

Without loss of generality, let Ω' comprise the first p frequencies in Ω, i.e., $\Omega' = \{\lambda_1, \ldots, \lambda_p\}$. Given a data record $\mathbf{y}$, let $\{\hat{c}_1, \ldots, \hat{c}_m\}$ and $\hat{\sigma}_m^2$ denote the ML estimates of $\{c_1, \ldots, c_m\}$ and σ^2 under the complete model (5.3.1), and let $\{\tilde{c}_1, \ldots, \tilde{c}_p\}$ and $\hat{\sigma}_p^2$ denote the ML estimates of $\{c_1, \ldots, c_p\}$ and σ^2 under the *reduced* model

$$\mathbf{y} = \sum_{j=1}^{p} c_j \mathbf{f}(\lambda_j) + \boldsymbol{\epsilon}, \tag{5.3.3}$$

which consists only of the frequencies in Ω' under the hypothesis H_0. Moreover, let $p(\mathbf{y} \mid \hat{c}_1, \ldots, \hat{c}_m, \hat{\sigma}_m^2)$ and $p(\mathbf{y} \mid \tilde{c}_1, \ldots, \tilde{c}_p, \hat{\sigma}_p^2)$ denote the maximized likelihood of $\mathbf{y}$ under (5.3.1) and (5.3.3), respectively.

To test the hypotheses, consider the generalized likelihood ratio

$$\mathrm{GLR} := \frac{p(\mathbf{y} \mid \hat{c}_1, \ldots, \hat{c}_m, \hat{\sigma}_m^2)}{p(\mathbf{y} \mid \tilde{c}_1, \ldots, \tilde{c}_p, \hat{\sigma}_p^2)}.$$

It is natural to reject H_0 in favor of H_1 if this statistic is large. Under the Gaussian assumption, $\{\hat{c}_j\}$ and $\{\tilde{c}_j\}$ are the LS estimates and

$$\mathrm{GLR} = (\hat{\sigma}_p^2 / \hat{\sigma}_m^2)^{n/2},$$

where

$$\hat{\sigma}_p^2 = n^{-1}\mathrm{SSE}_p, \quad \hat{\sigma}_m^2 = n^{-1}\mathrm{SSE}_m,$$

with SSE_p and SSE_m being the error sum of squares, defined in the same way as (5.1.4), for the reduced and complete models, respectively. In this case, the GLR test is equivalent to the so-called F-test that rejects H_0 in favor of H_1 if the F-statistic, defined as

$$F := \frac{(\mathrm{SSE}_p - \mathrm{SSE}_m)/(m-p)}{\mathrm{SSE}_m/(n-m)} = \frac{n-m}{m-p}(\mathrm{GLR}^{2/n} - 1), \tag{5.3.4}$$

is greater than a predetermined threshold $\tau > 0$.

The following theorem provides a threshold for the F-test that guarantees a prescribed level of significance $\alpha \in (0,1)$ under the assumption of Gaussian white noise. A proof is given in Section 5.6.

Theorem 5.7 (The F-Test for Frequency Identification). *Let* $\mathbf{y}$ *satisfy (5.1.1) with* $\boldsymbol{\epsilon} \sim \mathrm{N}_c(\mathbf{0}, \sigma^2\mathbf{I})$. *Assume that* $\{\omega_k\}$ *is a subset of* Ω, *where* $\Omega := \{\lambda_1, \ldots, \lambda_m\}$ *is known a priori with* $p < m < n$. *Then, under the hypothesis* H_0 *in (5.3.2) with* Ω' *being a given* p*-subset of* Ω, *the statistic in (5.3.4) has an* F*-distribution with* $2m-2p$ *and* $2n-2m$ *degrees of freedom, denoted by* $F(2m-2p, 2n-2m)$. *Let* $\tau := F_{1-\alpha}(2m-2p, 2n-2m)$ *be the* $1-\alpha$ *quantile of the said* F*-distribution. Then, the hypothesis* H_0 *is rejected at significance level* $\alpha \in (0,1)$ *if* $F > \tau$.

Remark 5.8 A similar F-test can be developed in the real case under the RSM (2.1.1) or (2.1.2) and the Gaussian white noise assumption $\boldsymbol{\epsilon} \sim \mathrm{N}(\mathbf{0}, \sigma^2\mathbf{I})$. In this case, the F-statistic is defined as

$$F := \frac{(\mathrm{SSE}_q - \mathrm{SSE}_m)/(2m-2q)}{\mathrm{SSE}_m/(n-2m)}.$$

Under $H_0 : \Omega' = \{\omega_1, \ldots, \omega_q\}$, it can be shown that the F-statistic is distributed as $F(2m-2q, n-2m)$. Note that there are $2m$ real coefficients for m frequencies in the RSM and that the number of real-valued observations is equal to n. This explains why the denominator degree of freedom is $n-2m$.

By following the proof of Theorem 5.7, it can be shown that if the alternative hypothesis H_1 is true, then the statistic in (5.3.4) has a noncentral F-distribution $F(2m-2q, 2n-2m, \theta)$, where the noncentrality parameter θ takes the form

$$\theta := \sum_{\lambda_j \in \Omega \setminus \Omega'} |c_j|^2 / (\tfrac{1}{2}\sigma^2).$$

Therefore, the power of the F-test with the threshold τ for a given θ can be expressed as [379, Chapter 22]

$$P_D := P(F > \tau), \tag{5.3.5}$$

where $F \sim F(2m-2p, 2n-2m, \theta)$. The power P_D in (5.3.5) represents the detection probability of a false identification, i.e., the probability that the selected subset Ω' is found correctly not to contain all signal frequencies.

Figure 5.2 depicts P_D as a function of θ with different values of m for $p = 2$, $n = 100$, and $\alpha = 0.01$. Because θ can be interpreted as the signal-to-noise ratio in the hypothesis testing problem (5.3.2), it is natural that P_D increases with θ. Moreover, when m is increased, the enlarged set Ω of candidate frequencies raises the likelihood of making false identification and hence reduces P_D.

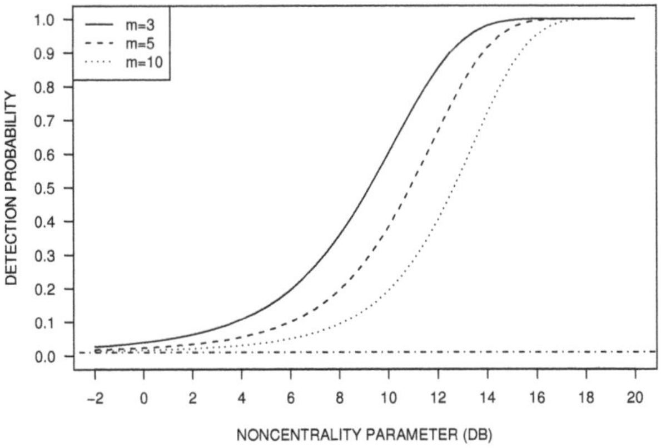

Figure 5.2. Plot of P_D in (5.3.5) as a function of the noncentrality parameter θ (in decibels) for the F-test in the case of two complex sinusoids in Gaussian white noise ($n = 100$) with different choice of m. Solid line, $m = 2$; dashed line, $m = 5$; dotted line, $m = 10$. Dash-dotted line depicts the false alarm probability $\alpha = 0.01$.

With the choice of $\Omega := \{\omega_k\}$ and $\Omega' := \emptyset$, the hypothesis testing problem (5.3.2) reduces to an important special case: detecting the presence of sinusoids with known frequencies. This is discussed in the following example.

Example 5.3 (The F-Test for Detection of Sinusoids with Known Frequencies). In the special case of $\Omega := \{\omega_k\}$ and $\Omega' := \emptyset$, the complete model takes the form (5.1.1), and the hypothesis testing problem (5.3.2) becomes

$$\begin{cases} H_0: \beta_k = 0 \text{ for all } k \text{ (sinusoids are absent)}, \\ H_1: \beta_k \neq 0 \text{ for some } k \text{ (sinusoids are present)}. \end{cases} \tag{5.3.6}$$

The corresponding F-statistic in (5.3.4) can be written as

$$F = \frac{(\text{SSE}_0 - \text{SSE})/p}{\text{SSE}/(n-p)}, \tag{5.3.7}$$

where SSE is given by (5.1.4) and $\text{SSE}_0 := \|\mathbf{y}\|^2$. Under H_0, $F \sim F(2p, 2n-2p)$. Therefore, the threshold $\tau := F_{1-\alpha}(2p, 2n-2p)$ gives a test of significance level α. The power of the test is given by (5.3.5) with $F \sim F(2p, 2n-2p, \theta)$ and

$$\theta := \|\boldsymbol{\beta}\|^2/(\tfrac{1}{2}\sigma^2) = \sum_{k=1}^{p} C_k^2/(\tfrac{1}{2}\sigma^2) = 2\sum_{k=1}^{p} \gamma_k,$$

where $\gamma_k := C_k^2/\sigma^2$ is the SNR of the kth complex sinusoid. ◇

Remark 5.9 A similar F-test can be developed in the real case under the RSM (2.1.1) or (2.1.2) and the assumption $\boldsymbol{\epsilon} \sim \text{N}(\mathbf{0}, \sigma^2\mathbf{I})$. In this case, the F-statistic

Table 5.1. Power of Frequency Identification by Minimizing the F-Statistic or SSE_p Among All Possible p-Subsets of m Candidates ($p = 3, m = 5$)

	SNR per Sinusoid (dB)								
n	−16	−14	−12	−10	−8	−6	−4	−2	0
50	0.199	0.255	0.342	0.469	0.632	0.777	0.887	0.953	0.986
100	0.397	0.559	0.751	0.899	0.976	0.999	1.000	1.000	1.000
200	0.654	0.841	0.955	0.994	1.000	1.000	1.000	1.000	1.000

Results are based on 5,000 Monte Carlo runs.

takes the same form as (5.3.7) except that p is replaced by $2q$ and the threshold is given by $\tau := F_{1-\alpha}(2q, n-2q)$. The detection probability of the test is given by (5.3.5) with $F \sim F(2q, n-2q, \theta)$ and

$$\theta := \sum_{k=1}^{q} (A_k^2 + B_k^2)/\sigma^2 = \sum_{k=1}^{q} C_k^2/\sigma^2 = 2\sum_{k=1}^{q} \gamma_k,$$

where $\gamma_k := \frac{1}{2}C_k^2/\sigma^2$ is the SNR of the kth real sinusoid.

The F-test in (5.3.4) can be used repeatedly for all possible p-subsets of Ω. Because Ω is known to contain the signal frequencies, the p-subset that has the smallest F-statistic provides a solution to the problem of frequency identification. This method is equivalent to minimizing SSE_p over all possible p-subsets of Ω. The best p-subset serves as an estimator of the signal frequencies.

Because the repeated tests are not independent in general, it is difficult to obtain a useful closed-form expression for the power of the best p-subset for frequency identification. For illustration, Table 5.1 provides a simulation result. This table contains the probability that the best p-subset captures all the signal frequencies in a time series that consists of three unit-amplitude random-phase complex sinusoids in Gaussian white noise. The signal frequencies are $\{\omega_k\} = \{2\pi \times (0.20 - 0.5/n, 0.20, 0.22 + 0.5/n)\}$ $(p = 3)$. The candidate frequencies are given by $\Omega = \{\omega_k\} \cup \{2\pi \times (0.18, 0.24 + 0.5/n)\}$ $(m = 5)$. Observe that there are 10 possible subsets of Ω with size 3. Table 5.1 shows that the detection power grows in general as the SNR or the sample size increases.

5.4 Frequency Selection

In the previous section, the focus is on the problem of frequency identification where the number of sinusoids, p, is known *a priori*, and the signal frequencies are identified by finding the best p-subset of Ω, where Ω is known to contain the signal frequencies. In this section, let us discuss a more general problem called *frequency selection* where p also needs to be determined from the data. This problem can be solved by applying the variable selection techniques in linear regression analysis [151] [261, Chapter 9].

To be more specific, let $\mathbf{y}$ be given by (5.1.1) with $\boldsymbol{\epsilon} \sim \mathrm{N}_c(\mathbf{0}, \sigma^2\mathbf{I})$, where p, ω_k, and β_k are unknown parameters. Assume that $\{\omega_k\} \subset \Omega$, where $\Omega := \{\lambda_1, \dots, \lambda_m\}$ is a known set of candidate frequencies with $p < m < n$.

Because the signal frequencies are members of Ω, one can solve the problem by selecting an appropriate subset of Ω such that the corresponding regression model best fits the data in some sense. Toward that end, a straightforward approach is to compare all 2^m subset models (including the complete model and the noise-only model) in terms of their SSE. There are efficient computational procedures for generating these models [338, Section 12.2]. However, unlike the frequency identification problem discussed in Section 5.3, the frequency selection problem cannot be solved by merely minimizing the SSE, because the complete model that contains all frequencies in Ω always produces the smallest SSE regardless of the true value of p. While making the SSE small remains a meaningful criterion, a suitable penalty must be imposed on a large model if it does not provide a significant improvement over a smaller one.

A simple procedure is to compute the SSE of the best $\tilde{p}$-term model for all $\tilde{p} = 0, 1, \dots, m$ and examine the SSE as a monotone decreasing function of $\tilde{p}$. If there is a turning point $\hat{p}$ such that the SSE stops decreasing significantly for $\tilde{p} > \hat{p}$, then $\hat{p}$ serves as an estimate of p.

While turning points can sometimes be extracted by a subjective visual inspection of the SSE curve, a more disciplined and automatic approach is through the minimization of a *generalized information criterion* (GIC) of the form

$$\mathrm{GIC}(\tilde{p}) := 2n \log\{\mathrm{SSE}(\tilde{p})\} + c(n, \tilde{p}), \tag{5.4.1}$$

where $\mathrm{SSE}(\tilde{p})$ denotes the SSE of the best $\tilde{p}$-term model and $c(n, \tilde{p})$ is a suitable function of n and $\tilde{p}$, typically monotone increasing in $\tilde{p}$. The first term in this criterion represents the model's goodness of fit to the data and the second term represents a penalty on the model's complexity. Identification of the turning point in the SSE as an estimator of p is made possible by a suitable tradeoff between the goodness of fit and the model complex. Here is why.

In order for the true value p to be the minimizer of the GIC, it is necessary and sufficient that for all $\tilde{p} \neq p$,

$$\mathrm{GIC}(\tilde{p}) - \mathrm{GIC}(p) = 2n \log\{\mathrm{SSE}(\tilde{p}) / \mathrm{SSE}(p)\} + \{c(n, \tilde{p}) - c(n, p)\} > 0.$$

The first term in this expression can be interpreted as the efficiency loss of the $\tilde{p}$-term model in comparison with the correct p-term model; the second term can be interpreted as the net complexity penalty of a $\tilde{p}$-term model in comparison with a p-term model. Roughly speaking, when $\tilde{p}$ is smaller than p, at least one of the p sinusoids is left out of the the best $\tilde{p}$-subset. In this case, the resulting $\text{SSE}(\tilde{p})$ tends to be much larger than $\text{SSE}(p)$ and hence the efficiency loss $2n\log\{\text{SSE}(\tilde{p})/\text{SSE}(p)\}$ tends to be much larger than 0. Because $c(n,\tilde{p})$ is an increasing function of $\tilde{p}$, the net complexity penalty $c(n,\tilde{p})-c(n,p)$ is negative when $\tilde{p}<p$, in which case the negative penalty $c(n,p)-c(n,\tilde{p})$ represents the net complexity reward. If this reward is not too large as compared to the efficiency loss, then the difference $\text{GIC}(\tilde{p})-\text{GIC}(p)$ will be strictly positive.

On the other hand, when $\tilde{p}$ is greater than p, the efficiency loss will be negative because $\text{SSE}(\tilde{p})$ is a monotone decreasing function of $\tilde{p}$. In this case, the negative efficiency loss $2n\log\{\text{SSE}(p)/\text{SSE}(\tilde{p})\}$ represents the efficiency gain. This efficiency gain tends to be very small because it results from fitting the noise. Therefore, if the net complexity penalty $c(n,\tilde{p})-c(n,p)$ is sufficiently large, then the difference $\text{GIC}(\tilde{p})-\text{GIC}(p)$ can be made strictly positive as well.

In summary, in order for the true value p to be the minimizer of the GIC, the penalty function $c(n,\tilde{p})$ cannot increase too fast with $\tilde{p}$ to overwhelm the efficiency loss when $\tilde{p}<p$; at the same time, it cannot increase too slowly either to be overwhelmed by the efficiency gain when $\tilde{p}>p$.

Note that for the real case the GIC in (5.4.1) should be modified by replacing $2n$ with n and by taking $\tilde{p}:=2\tilde{q}$ for a subset model that contains $\tilde{q}$ real sinusoids of the form (2.1.2). One should also interpret $\text{SSE}(\tilde{p})$ as the SSE of the best subset model that contains $\tilde{q}$ real sinusoids of the form (2.1.2).

The GIC defined by (5.4.1) is a generalization of Akaike's information criterion (AIC) [8] which takes the form

$$\text{AIC}(\tilde{p}) := 2n\log\hat{\sigma}_{\tilde{p}}^2 + 2\tilde{p}, \tag{5.4.2}$$

where $\hat{\sigma}_{\tilde{p}}^2 := n^{-1}\text{SSE}(\tilde{p})$. The guiding principle behind the AIC criterion is to seek a model with PDF $p_1(\cdot)$ for $\mathbf{y}$, whose true PDF is $p_0(\cdot)$, such that the ML estimator of the model on the basis of $\mathbf{y}$, denoted by $\hat{p}_1(\cdot)$, minimizes the mean Kullback-Leibler divergence $E\{D_{\text{KL}}(\hat{p}_1\|p_0)\}$, where

$$D_{\text{KL}}(\hat{p}_1\|p_0) := E_{\mathbf{y}'}\{-2\log[\hat{p}_1(\mathbf{y}')/p_0(\mathbf{y}')]\}, \quad \mathbf{y}' \sim p_0(\cdot).$$

The Kullback-Leibler divergence $D_{\text{KL}}(\hat{p}_1\|p_0)$ measures the deviation of $\hat{p}_1(\cdot)$ with respect to $p_0(\cdot)$ in terms of the extra number bits required to encode samples from $p_0(\cdot)$ based on $\hat{p}_1(\cdot)$ rather than $p_0(\cdot)$ [202]. It is a nonnegative quantity that attains its minimum value zero if and only if $\hat{p}_1(\cdot)=p_0(\cdot)$. Observe that

$$D_{\text{KL}}(\hat{p}_1\|p_0) = H(\hat{p}_1\|p_0) + E_{\mathbf{y}'}\{2\log[p_0(\mathbf{y}')]\},$$

where $H(\hat{p}_1\|p_0) := E_{\mathbf{y}'}\{-2\log[\hat{p}_1(\mathbf{y}')]\}$ is known as the cross entropy of $\hat{p}_1(\cdot)$ with respect to $p_0(\cdot)$. Because the second term in the foregoing expression does not depend on $\hat{p}_1(\cdot)$, minimizing the mean Kullback-Leibler divergence is equivalent to minimizing the mean cross entropy $E\{H(\hat{p}_1\|p_0)\}$. With $p_1(\cdot)$ representing a $\tilde{p}$-term model, the AIC in (5.4.2) is an approximately unbiased estimator of $E\{H(\hat{p}_1\|p_0)\}$ (up to an additive constant) derived by a chi-square approximation technique (see Section 5.6). It is the special case of GIC with $c(n,\tilde{p}) = 2\tilde{p}$.

More generally, the penalty function can take the form

$$c(n,\tilde{p}) = c_n\tilde{p}, \tag{5.4.3}$$

where c_n is a positive constant that may depend on the sample size n. With this choice, the penalty grows linearly with $\tilde{p}$ and the rate of growth is controlled by c_n. Besides the AIC criterion in (7.1.29), which corresponds to $c_n = 2$, Schwarz's Bayesian information criterion (BIC) [337] corresponds to $c_n = \log n$, and so does Rissanen's minimum description length (MDL) criterion [318]. The corrected AIC (or AICC) criterion, which is proposed in [161] and [380], replaces $2\tilde{p}$ in (5.4.2) with a penalty function that takes a slightly more complicated form

$$c(n,\tilde{p}) = 2n(n+\tilde{p})/(n-\tilde{p}-1). \tag{5.4.4}$$

The AICC is an alternative estimator of $E\{H(\hat{p}_1\|p_0)\}$ (up to an additive constant) derived by a refined approximation technique (see Section 5.6).

With $c(n,\tilde{p}) = -2n\log(n-\tilde{p})$, minimizing the GIC in (5.4.1) becomes equivalent to minimizing the residual mean square (RMS) defined as

$$\mathrm{RMS}(\tilde{p}) := \mathrm{SSE}(\tilde{p})/(n-\tilde{p}).$$

In this criterion, a larger model is penalized by a smaller denominator to tradeoff the reduced SSE in the numerator. The model that minimizes $\mathrm{RMS}(\tilde{p})$ also maximizes the so-called adjusted R^2 criterion defined as

$$\bar{R}^2(\tilde{p}) := 1 - \{1 - R^2(\tilde{p})\}(n-1)/(n-\tilde{p}),$$

where $R^2(\tilde{p}) := 1 - \mathrm{SSE}(\tilde{p})/\|\mathbf{y}\|^2$ is the ordinary R^2 statistic, also known as the coefficient of multiple determination, which can be interpreted as the fraction of variability in $\mathbf{y}$ accounted for by the regression model.

In addition to the GIC criterion defined by (5.4.1), one can also minimize the *prediction error criterion* (PEC) of the form

$$\mathrm{PEC}(\tilde{p}) := \mathrm{SSE}(\tilde{p}) + c(n,\tilde{p})\tilde{\sigma}_m^2, \tag{5.4.5}$$

where $\tilde{\sigma}_m^2 := \mathrm{SSE}_m/(n-m)$ is the unbiased estimator of σ^2 under the complete model and $c(n,\tilde{p})$ is again a penalty function. The PEC criterion can be justified in the same way as the GIC criterion as a tradeoff between the goodness of fit and the model complexity. A popular choice for the penalty function is of the

Table 5.2. Distribution of the Estimators of the Number of Sinusoids Obtained by Minimizing All-Subset Based AIC, RMS, BIC, and PEC ($p = 3$ and $m = 5$)

			SNR per Sinusoid (dB)							
	n	$\hat{p}$	−8	−6	−4	−2	0	2	4	6
AIC	50	1	0.002	0.000	0.000	0.000	0.000	0.000	0.000	0.000
		2	0.029	0.006	0.001	0.000	0.000	0.000	0.000	0.000
		3	**0.438**	0.419	0.388	0.367	0.362	0.362	0.362	0.362
		4	0.420	**0.445**	**0.462**	**0.473**	**0.474**	**0.474**	**0.474**	**0.474**
		5	0.111	0.130	0.149	0.160	0.164	0.164	0.164	0.164
RMS	50	1	0.002	0.000	0.000	0.000	0.000	0.000	0.000	0.000
		2	0.033	0.007	0.002	0.000	0.000	0.000	0.000	0.000
		3	**0.471**	**0.455**	0.423	0.402	0.396	0.396	0.396	0.396
		4	0.402	0.431	**0.452**	**0.465**	**0.467**	**0.466**	**0.466**	**0.466**
		5	0.091	0.107	0.123	0.133	0.137	0.138	0.138	0.138
BIC	50	1	0.018	0.004	0.000	0.000	0.000	0.000	0.000	0.000
		2	0.149	0.039	0.005	0.001	0.000	0.000	0.000	0.000
		3	**0.644**	**0.723**	**0.733**	**0.715**	**0.701**	**0.698**	**0.698**	**0.698**
		4	0.177	0.216	0.239	0.257	0.269	0.272	0.272	0.272
		5	0.012	0.018	0.023	0.027	0.030	0.030	0.030	0.030
PEC	50	1	0.160	0.043	0.007	0.000	0.000	0.000	0.000	0.000
		2	**0.470**	0.240	0.055	0.006	0.000	0.000	0.000	0.000
		3	0.357	**0.692**	**0.908**	**0.958**	**0.959**	**0.956**	**0.955**	**0.955**
		4	0.012	0.024	0.029	0.035	0.040	0.043	0.044	0.044
		5	0.001	0.001	0.001	0.001	0.001	0.001	0.001	0.001
AIC	100	1	0.000	0.000	0.000	0.000	0.000	0.000	0.000	0.000
		2	0.001	0.000	0.000	0.000	0.000	0.000	0.000	0.000
		3	0.389	0.389	0.389	0.389	0.389	0.389	0.389	0.389
		4	**0.452**	**0.452**	**0.452**	**0.452**	**0.452**	**0.452**	**0.452**	**0.452**
		5	0.158	0.159	0.159	0.159	0.159	0.159	0.159	0.159
RMS	100	1	0.000	0.000	0.000	0.000	0.000	0.000	0.000	0.000
		2	0.001	0.000	0.000	0.000	0.000	0.000	0.030	0.000
		3	0.405	0.405	0.405	0.405	0.405	0.405	0.405	0.405
		4	**0.449**	**0.450**	**0.450**	**0.450**	**0.450**	**0.450**	**0.450**	**0.450**
		5	0.145	0.145	0.145	0.145	0.145	0.145	0.145	0.145
BIC	100	1	0.002	0.000	0.000	0.000	0.000	0.000	0.000	0.000
		2	0.017	0.001	0.000	0.000	0.000	0.000	0.000	0.000
		3	**0.776**	**0.790**	**0.791**	**0.791**	**0.791**	**0.791**	**0.791**	**0.791**
		4	0.192	0.195	0.195	0.195	0.195	0.195	0.195	0.195
		5	0.013	0.014	0.014	0.014	0.014	0.014	0.014	0.014
PEC	100	1	0.031	0.003	0.000	0.000	0.000	0.000	0.000	0.000
		2	0.138	0.012	0.000	0.000	0.000	0.000	0.000	0.000
		3	**0.813**	**0.963**	**0.978**	**0.978**	**0.978**	**0.978**	**0.978**	**0.978**
		4	0.018	0.022	0.022	0.022	0.022	0.022	0.022	0.022
		5	0.000	0.000	0.000	0.000	0.000	0.000	0.000	0.000

Results are based on 5,000 Monte Carlo runs.
Bold font shows where the largest value of each distribution occurs.

form (5.4.3). The modified Akaike's final prediction error (FPE) criterion [7] [343] corresponds to $c_n = 2$. This is also the choice in Mallows' total squared error (also known as C_p) criterion [251].

To demonstrate the statistical performance of these criteria, the result of a simulation study is shown in Table 5.2. The true model of $\mathbf{y}$ in this example is the same as that for Table 5.1, which consists of three unit-amplitude random-phase complex sinusoids in Gaussian white noise ($p = 3$). The set Ω, which contains $m = 5$ candidate frequencies, is also the same. Table 5.2 shows the simulated probability distribution of $\hat{p}$, the minimizer of AIC($\tilde{p}$), RMS($\tilde{p}$), BIC($\tilde{p}$), and PEC($\tilde{p}$) with $c(n, \tilde{p}) = \tilde{p} \log n$, for different values of SNR and for $n = 50$ and 100.

As can be seen, both the AIC and the RMS have a significant tendency to overestimate the parameter p regardless of the SNR. The BIC is able to reduce the probability of overestimation, thanks to a heavier penalty than the AIC ($c_n \approx 3.9$ for $n = 50$ and $c_n \approx 4.6$ for $n = 100$, as compared to $c_n = 2$). This is achieved at the expense of increased probability of underestimation at low SNR. The PEC works particularly well when the SNR is not too low (≥ -4 dB for $n = 50$), considering the fact that with $n = 50$ the first false frequency, $2\pi \times 0.18$, and the first signal frequency, $2\pi \times 0.19$, are separated by merely π/n. The separation condition is improved when $n = 100$, and so is the performance of the PEC and the BIC. However, the overestimation problem persists for the AIC and the RMS.

Asymptotic analyses of the GIC and PEC criteria are available in the literature under the general linear regression framework [271] [309] [438] [442], where some sufficient conditions are provided for the choice of the penalty function to ensure consistent estimation. The conditions depend crucially on the interplay among the regressors and between the regressors and the noise. Although these analyses do not automatically lead to specific choices of the penalty function, they do provide a unique insight into the variable selection problem.

The following theorem provides the conditions for consistent estimation of p by minimizing the GIC or PEC criterion. A proof is given in Section 5.6.

Theorem 5.8 (Consistency of the GIC and PEC Minimizers). *Let* $\mathbf{y}$ *be given by (5.1.1) with* $\{\omega_k\} \subset \Omega := \{\lambda_1, \ldots, \lambda_m\}$. *Assume that* $\boldsymbol{\epsilon}$ *is a zero-mean random vector such that* $n^{-1}\boldsymbol{\epsilon}^H\boldsymbol{\epsilon} \xrightarrow{P} \sigma^2$ *as* $n \to \infty$ *for some constant* $\sigma^2 > 0$. *Assume that there exists a positive definite matrix* $\mathbf{C}_m$ *such that* $n^{-1}\mathbf{F}_m^H\mathbf{F}_m \to \mathbf{C}_m$ *as* $n \to \infty$, *where* $\mathbf{F}_m := [\mathbf{f}(\lambda_1), \ldots, \mathbf{f}(\lambda_m)]$. *Assume further that* $\mathbf{f}^H(\lambda_j)\boldsymbol{\epsilon} = \mathcal{O}_P(a_n)$ *for all* $\lambda_j \in \Omega$, *where* $\{a_n\}$ *is a constant sequence such that* $n^{-1}a_n \to 0$ *as* $n \to \infty$. *Let* $\hat{p}$ *be the minimizer of* GIC($\tilde{p}$) *or* PEC($\tilde{p}$) *defined by (5.4.1) and (5.4.5), respectively, where* $c(n, \tilde{p})$ *is a strictly increasing function of* $\tilde{p}$ *for each fixed* n *and* $c(n, 0) = 0$. *Then, as* $n \to \infty$, *the following assertions are true. (a) If* $n^{-1}c(n, p) \to 0$, *then* $P(\hat{p} < p) \to 0$. *(b) If* $n\{c(n, p+1) - c(n, p)\}/a_n^2 \to \infty$, *then* $P(\hat{p} > p) \to 0$.

Remark 5.10 Because $\tilde{\sigma}_m^2 = \mathcal{O}_P(1)$, the PEC criterion can also be defined by omitting $\tilde{\sigma}_m^2$ in (5.4.5) and the consistency assertion in Theorem 5.8 remains valid.

Remark 5.11 If $\mathbf{f}^H(\lambda_j)\boldsymbol{\epsilon} = \mathcal{O}_P(a_n)$, then $n^{-1}|\mathbf{f}^H(\lambda_j)\boldsymbol{\epsilon}|^2 = \mathcal{O}_P(n^{-1}a_n^2)$. The function $n^{-1}|\mathbf{f}^H(\omega)\boldsymbol{\epsilon}|^2$ is called the periodogram of $\boldsymbol{\epsilon}$. See Chapter 6 for more details.

Theorem 5.8 shows that in order to avoid underestimation for large sample sizes, the penalty for a $\tilde{p}$-term model with $\tilde{p} \le p$ must not increase faster than $\mathcal{O}(n)$ regardless of the noise properties. On the other hand, to avoid overestimation, the net penalty for a $(p+1)$-term overfitted model relative to a p-term model must grow faster than $n^{-1}a_n^2$, where $\{a_n\}$ depends on the noise. For example, if $\{\epsilon_t\}$ is a zero-mean stationary process with absolutely summable ACF $r_\epsilon(u)$ and SDF $f_\epsilon(\omega) := \sum r_\epsilon(u)\exp(-iu\omega)$. Then,

$$n^{-1}E\{|\mathbf{f}^H(\omega)\boldsymbol{\epsilon}|^2\} = \sum_{|u|<n} (1-|u|/n) r_\epsilon(u)\exp(-iu\omega) = f_\epsilon(\omega) + \mathcal{O}(1).$$

By Proposition 2.16(a), we can write $\mathbf{f}^H(\omega)\boldsymbol{\epsilon} = \mathcal{O}_P(a_n)$ with $a_n = \sqrt{n}$. In this case, the condition for $c(n,p)$ in part (b) of Theorem 5.8 becomes

$$c(n,p+1) - c(n,p) \to \infty.$$

Moreover, if $c(n,p)$ takes the form (5.4.3) for some $c_n > 0$, then the conditions in part (a) and part (b) of Theorem 5.8 become, respectively,

$$n^{-1}c_n \to 0 \quad \text{and} \quad c_n \to \infty.$$

Both conditions are satisfied by $c_n := c\log n$ and $c_n := c\log\log n$ for any constant $c > 0$. The minimizers of the resulting GIC and PEC criteria are consistent estimators of p, i.e., $P(\hat{p} \ne p) \to 0$, or equivalently, $\hat{p} \xrightarrow{P} p$.

If evaluating all possible subset models is computationally prohibitive, which may be the case when m is large, one can simplify the procedure, as suggested in [442], by replacing the error sum of squares of the best $\tilde{p}$-term model for each $\tilde{p} \in \{1,\ldots,m-1\}$ with the $\tilde{p}$-term model that employs the $\tilde{p}$ highest ranked frequencies in Ω. The ranking is based on the magnitude of the absolute standardized coefficients $|\hat{c}_j|/s_j$, where $\hat{c}_j$ denotes the LS estimate of the jth coefficient in the complete model (5.3.1) and s_j^2 denotes the jth diagonal element of the matrix $\hat{\sigma}_m^2(\mathbf{F}_m^H\mathbf{F}_m)^{-1}$ which serves as an estimate for the variance of $\hat{c}_j$. This procedure is not necessarily optimal, but because the signal frequencies tend to be associated with large standardized coefficients, especially for large n, there is a good chance that they appear with the highest ranks.

Alternatively, one can use the *stepwise regression* algorithm [261]. Starting with the noise-only model, the algorithm iteratively adds one frequency at a time to the current model if the best frequency (in the sense of minimum SSE) selected among all remaining frequencies in Ω passes a significance test (e.g., an F-test) for the model with the new frequency versus the model without it. The procedure stops when the remaining frequencies in Ω cannot be added to the current model. The selected frequencies and their total number serve as the estimates of $\{\omega_k\}$ and p. This *forward selection* procedure can be enhanced with the capability of backward elimination. At each iteration, the *backward elimination* procedure drops the worst frequency in the current model if it fails a significance test (e.g., an F-test) for the model with the frequency versus the model without it. The backward elimination procedure is most effective when some of the frequencies in Ω are closely spaced. The performance of the stepwise regression algorithm is controlled by the thresholds for adding and dropping frequencies. A common practice is to use a higher threshold for the backward elimination than for the forward selection.

One can also use a penalized LS regression technique simply known as LASSO (least absolute shrinkage selection operator) [388]. To explain this method, let us reformulate the complete model (5.3.1) in terms of real-valued observations and variables, similar to (5.1.5). In other words, let $\mathbf{y}_r := [\Re(\mathbf{y})^T, \Im(\mathbf{y})^T]^T \in \mathbb{R}^{2n}$, $\mathbf{c}_r := [\Re(c_1), \dots, \Re(c_m), \Im(c_1), \dots, \Im(c_m)]^T \in \mathbb{R}^{2m}$, and

$$\mathbf{X} := [\mathbf{x}_1, \dots, \mathbf{x}_{2m}] := \begin{bmatrix} \Re(\mathbf{F}_m) & -\Im(\mathbf{F}_m) \\ \Im(\mathbf{F}_m) & \Re(\mathbf{F}_m) \end{bmatrix} \in \mathbb{R}^{2n \times 2m},$$

where $\mathbf{F}_m := [\mathbf{f}(\lambda_1), \dots, \mathbf{f}(\lambda_m)]$. To estimate the parameter $\mathbf{c}_r$, the LASSO method minimizes a penalized LS criterion of the form

$$\|\mathbf{y}_r - \mathbf{X}\tilde{\mathbf{c}}_r\|^2 + \rho \sum_{k=1}^{2m} |\tilde{c}_{r,k}| \tag{5.4.6}$$

with respect to $\tilde{\mathbf{c}}_r := [\tilde{c}_{r,1}, \dots, \tilde{c}_{r,2m}]^T \in \mathbb{R}^{2m}$, where $\rho \ge 0$ is a tuning penalty parameter. In other words, the LASSO method imposes an ℓ_1-norm penalty on $\tilde{\mathbf{c}}_r$ while minimizing the LS criterion. When ρ is sufficiently large, the penalty forces some components in the LASSO estimator, denoted as $\hat{\mathbf{c}}_r(\rho)$, to take zero values and thereby selects a reduced model in effect. There exists a large value, say $\rho_{\max}$, that makes all components in the LASSO estimator equal to zero. By gradually reducing the value of ρ from $\rho_{\max}$ to 0, more and more components in $\hat{\mathbf{c}}_r(\rho)$ become nonzero, so that a sequence of reduced models with increasing complexity is created. This is analogous to the forward selection procedure in stepwise regression. By coupling the sequence of reduced models with a criterion such as the GIC or the PEC, a final model can be determined.

As an example, consider the case where the λ_j are Fourier frequencies so that $\mathbf{X}^T\mathbf{X} = \frac{1}{2}n\mathbf{I}$. In this case, the kth component of the LS estimator of $\mathbf{c}_r$, which is $\hat{\mathbf{c}}_r(0)$, can be expressed as $\hat{c}_{r,k}(0) := 2n^{-1}\mathbf{x}_k^T\mathbf{y}_r$ $(k = 1, \dots, 2m)$. For $\rho > 0$, it can be shown [388] that the kth component of $\hat{\mathbf{c}}_r(\rho)$ takes the form

$$\hat{c}_{r,k}(\rho) = \left(1 - \frac{\rho}{n|\hat{c}_{r,k}(0)|}\right)_+ \hat{c}_{r,k}(0). \tag{5.4.7}$$

As we can see, the LASSO estimator sets a component equal to zero if the LS estimate of that component is smaller in absolute value than $n^{-1}\rho$. When ρ takes the value $\rho_{\max} := \max\{n|\hat{c}_{r,k}(0)|\}$, all components are equal to zero. As the value of ρ decreases, the components become nonzero sequentially in a decreasing order by the magnitude of their LS estimates. This generates a sequence of reduced models with increasing complexity but decreasing increment of significance. The corresponding sequence of SSEs can be used to define a model selection criterion such as the GIC in (5.4.1) and the PEC in (5.4.5). Note that the LASSO estimator in (5.4.7) also shrinks the LS estimates that exceed the threshold. A shrinkage estimator of this kind has some intrinsic merits for parameter estimation [388]. Here we only focus on its model selection capability.

For the problem of frequency selection, the LASSO technique just described must be modified to accommodate the requirement that the jth and $(j+m)$th components, which correspond to the same sinusoid with frequency λ_j, should be selected together or set to zero together. This can be accomplished by minimizing the so-called group LASSO criterion [433]

$$\|\mathbf{y}_r - \mathbf{X}\tilde{\mathbf{c}}_r\|^2 + \rho \sum_{j=1}^{m} \sqrt{\tilde{c}_{r,j}^2 + \tilde{c}_{r,j+m}^2}. \tag{5.4.8}$$

Instead of the absolute value of each individual component, the group LASSO method employs the ℓ_2 norm of a group of two components that correspond to the same sinusoid. In terms of the original complex-valued observations and variables, the group LASSO criterion (5.4.8) can be simply expressed as

$$\|\mathbf{y} - \mathbf{F}_m\tilde{\mathbf{c}}\|^2 + \rho \sum_{j=1}^{m} |\tilde{c}_j|,$$

where $\tilde{\mathbf{c}} := [\tilde{c}_1, \ldots, \tilde{c}_m]^T \in \mathbb{C}^m$. Therefore, in effect, the penalty in the group LASSO criterion is imposed on the modulus of the complex coefficients $\tilde{c}_j$.

To demonstrate the statistical performance of the LASSO technique for frequency selection, Table 5.3 contains the result of a simulation study based on the same data set which is used to produce Table 5.2 with the same choice of Ω. The parameter p is estimated by minimizing the same model selection criteria except that the SSEs are obtained from the sequence of reduced models generated by the group LASSO procedure. An R function called `grplasso` in the package of the same name is used to compute the group LASSO estimates with an exponentially decaying set of ρ which takes the form $\rho_{\max}\{(50-u)/49\}^4$ $(u = 1, \ldots, 50)$. An R function called `lambdamax` in the same package provides $\rho_{\max}$.

By comparing Table 5.3 with Table 5.2, we can see that in the first case where $n = 50$ the probability of correct selection, i.e., $P(\hat{p} = p)$, is much lower for the LASSO method than for the all subset method, regardless of the model selection

Table 5.3. Distribution of the Estimators of the Number of Sinusoids Obtained by Minimizing Group LASSO Based AIC, RMS, BIC, and PEC ($p = 3$ and $m = 5$)

			SNR per Sinusoid (dB)							
	n	$\hat{p}$	−8	−6	−4	−2	0	2	4	6
AIC	50	1	0.002	0.000	0.000	0.000	0.000	0.000	0.000	0.000
		2	0.039	0.009	0.002	0.000	0.000	0.000	0.000	0.000
		3	0.340	0.335	0.322	0.319	0.320	0.320	0.322	0.322
		4	**0.370**	**0.385**	**0.392**	**0.399**	**0.401**	**0.401**	**0.399**	**0.399**
		5	0.249	0.271	0.284	0.282	0.279	0.279	0.279	0.279
RMS	50	1	0.003	0.000	0.000	0.000	0.000	0.000	0.000	0.000
		2	0.045	0.010	0.002	0.000	0.000	0.000	0.000	0.000
		3	**0.370**	0.367	0.353	0.349	0.349	0.350	0.352	0.352
		4	0.362	**0.380**	**0.391**	**0.399**	**0.403**	**0.404**	**0.402**	**0.401**
		5	0.220	0.243	0.254	0.252	0.248	0.246	0.246	0.247
BIC	50	1	0.026	0.005	0.000	0.000	0.000	0.000	0.000	0.000
		2	0.169	0.058	0.014	0.003	0.000	0.000	0.000	0.000
		3	**0.525**	**0.585**	**0.591**	**0.573**	**0.568**	**0.572**	**0.575**	**0.577**
		4	0.222	0.271	0.305	0.333	0.345	0.348	0.349	0.348
		5	0.058	0.081	0.090	0.091	0.087	0.080	0.076	0.075
PEC	50	1	0.150	0.041	0.008	0.000	0.000	0.000	0.000	0.000
		2	0.352	0.207	0.072	0.018	0.002	0.000	0.000	0.000
		3	**0.443**	**0.652**	**0.758**	**0.756**	**0.733**	**0.723**	**0.727**	**0.729**
		4	0.050	0.092	0.148	0.204	0.242	0.258	0.260	0.261
		5	0.005	0.008	0.014	0.022	0.023	0.019	0.013	0.010
AIC	100	1	0.000	0.000	0.000	0.000	0.000	0.000	0.000	0.000
		2	0.001	0.000	0.000	0.000	0.000	0.000	0.000	0.000
		3	0.389	0.396	0.400	0.401	0.402	0.402	0.402	0.402
		4	**0.434**	**0.430**	**0.426**	**0.425**	**0.422**	**0.423**	**0.421**	**0.420**
		5	0.176	0.174	0.174	0.174	0.176	0.175	0.177	0.178
RMS	100	1	0.000	0.000	0.000	0.000	0.000	0.000	0.000	0.000
		2	0.001	0.000	0.000	0.000	0.000	0.000	0.000	0.000
		3	0.405	0.411	0.415	0.416	0.417	0.417	**0.418**	**0.418**
		4	**0.430**	**0.426**	**0.422**	**0.421**	**0.417**	**0.418**	0.417	0.415
		5	0.164	0.163	0.163	0.163	0.166	0.165	0.165	0.167
BIC	100	1	0.003	0.000	0.000	0.000	0.000	0.000	0.000	0.000
		2	0.021	0.001	0.000	0.000	0.000	0.000	0.000	0.000
		3	**0.741**	**0.775**	**0.789**	**0.796**	**0.799**	**0.800**	**0.800**	**0.800**
		4	0.210	0.204	0.192	0.184	0.183	0.181	0.180	0.180
		5	0.025	0.020	0.019	0.020	0.018	0.019	0.020	0.020
PEC	100	1	0.035	0.005	0.000	0.000	0.000	0.000	0.000	0.000
		2	0.098	0.012	0.000	0.000	0.000	0.000	0.000	0.000
		3	**0.812**	**0.924**	**0.955**	**0.967**	**0.971**	**0.972**	**0.972**	**0.972**
		4	0.053	0.057	0.043	0.032	0.028	0.027	0.027	0.027
		5	0.002	0.002	0.002	0.001	0.001	0.001	0.001	0.001

Results are based on 5,000 Monte Carlo runs.
Bold font shows where the largest value of each distribution occurs.

criteria. Recall that the first false frequency is very close to the first signal frequency in this case. So the result suggests that the all subset method is better equipped to handle closely spaced frequencies. In the second case where the frequencies are better separated, the results of both methods become quite similar, with the LASSO method yielding slightly higher probabilities of correct selection at sufficiently high SNR for all but the PEC criterion. Of course, the computational burden of the LASSO method can be much lower than that of the all subset method, especially when m is large.

5.5 Least Absolute Deviations Estimation

Although the LS method is statistically efficient only in the Gaussian case, it is widely used in practice because of its computational simplicity and its well-understood numerical and statistical properties. However, there is an important situation where the LS method can fail miserably — that is, when the data are contaminated by outliers or when the noise has a heavy-tailed distribution. To better handle such situations, we need a procedure that is less sensitive, or more robust, to the presence of rare but large errors. There has been a fair amount of research in the statistical literature devoted to the problem of robust estimation in the general context of regression analysis. For comprehensive discussions on the subject, see, for example, [132], [159], and [253].

One way of constructing robust procedures is to modify the LS criterion in (5.1.2) by reducing or curtailing the influence of large residuals. In the LS criterion, large residuals have an enormous influence because they are dramatically magnified by the second power. A simple robust solution is to replace the squared residuals by their absolute values. This modification leads to the so-called *least absolute deviations* (LAD) estimator

$$\begin{aligned}\hat{\boldsymbol{\beta}} &:= \arg\min_{\tilde{\boldsymbol{\beta}}\in\mathbb{C}^p} \sum_{t=1}^{n} \{|\Re(y_t - \mathbf{x}_t^H \tilde{\boldsymbol{\beta}})| + |\Im(y_t - \mathbf{x}_t^H \tilde{\boldsymbol{\beta}})|\} \\ &= \arg\min_{\tilde{\boldsymbol{\beta}}\in\mathbb{C}^p} \|\mathbf{y} - \mathbf{F}\tilde{\boldsymbol{\beta}}\|_1, \end{aligned} \tag{5.5.1}$$

where $\|\cdot\|_1$ denotes the ℓ_1 norm of a complex vector defined as the sum of the absolute values of the real and imaginary parts of all its components. It is easy to see that the LAD estimator in (5.5.1) coincides with the ML estimator when $\{\epsilon_t\}$ is complex Laplace white noise defined in Example 3.12, i.e., when the real and imaginary parts of $\{\epsilon_t\}$ are mutually independent i.i.d. sequences with the Laplace distribution $p(x) = \sigma^{-1}\exp(-2|x|/\sigma)$.

The following theorem establishes the asymptotic normality of the LAD estimator under the white noise assumption. See Section 5.6 for a proof.

Theorem 5.9 (Asymptotic Distribution of the LAD Estimator: White Noise). *Let* $\mathbf{y}$ *be given by (5.1.1) and* $\hat{\boldsymbol{\beta}}$ *be defined by (5.5.1). Assume that there exists a nonsingular matrix* $\mathbf{C}$ *such that* $n^{-1}\mathbf{F}^H\mathbf{F} \to \mathbf{C}$ *as* $n \to \infty$. *Assume also that* $\{\epsilon_t\}$ *is a sequence of independent random variables with independent real and imaginary parts whose CDFs, denoted by* $F_{t1}(x)$ *and* $F_{t2}(x)$, *are differentiable at* $x = 0$ *and satisfy* $F_{tj}(0) = 1/2$, $\dot{F}_{tj}(0) = 1/\kappa > 0$, *and* $F_{tj}(x) - F_{tj}(0) = \dot{F}_{tj}(0)x + \mathcal{O}(x^{d+1})$ *uniformly in* t *for some constant* $d > 0$ *and for* $|x| \ll 1$ $(j = 1, 2)$. *Then, as* $n \to \infty$, $\sqrt{n}(\hat{\boldsymbol{\beta}} - \boldsymbol{\beta}) \xrightarrow{D} \mathrm{N}_c(\mathbf{0}, \eta^2\mathbf{C}^{-1})$, *where* $\eta^2 := \kappa^2/2$.

Remark 5.12 The condition $F_{tj}(0) = 1/2$ simply requires that the real and imaginary parts of the noise have zero median for all t. The condition $F_{tj}(x) - F_{tj}(0) = \dot{F}_{tj}(0)x + \mathcal{O}(x^{d+1})$ is satisfied with $d = 1$ if $F_{tj}(x)$ is twice continuously differentiable with a uniformly bounded second derivative in a neighborhood of $x = 0$. It is also satisfied if the uniformly bounded second derivative exists in $[0, \delta]$ and $[-\delta, 0]$, respectively, for small $\delta > 0$, as is the case with the Laplace distribution.

Remark 5.13 For the real case discussed in Remark 5.1, one can write

$$y_t = \mathbf{x}_t^T\boldsymbol{\beta} + \epsilon_t \qquad (t = 1, \ldots, n),$$

where $\mathbf{x}_t := [\cos(\omega_1 t), \sin(\omega_1 t), \ldots, \cos(\omega_q t), \sin(\omega_q t)]^T$. Therefore, the LAD estimator of $\boldsymbol{\beta} := [A_1, B_1, \ldots, A_q, B_q]^T$ is given by

$$\hat{\boldsymbol{\beta}} := \arg\min_{\tilde{\boldsymbol{\beta}} \in \mathbb{R}^{2q}} \sum_{t=1}^{n} |y_t - \mathbf{x}_t^T\tilde{\boldsymbol{\beta}}| = \arg\min_{\tilde{\boldsymbol{\beta}} \in \mathbb{R}^{2q}} \|\mathbf{y} - \mathbf{F}\tilde{\boldsymbol{\beta}}\|_1.$$

If $\{\epsilon_t\}$ is a sequence of independent real random variables with $\epsilon_t \sim F_t(x)$ such that $F_t(0) = 1/2$, $\dot{F}_t(0) = 1/\kappa > 0$, and $F_t(x) - F_t(0) = \dot{F}_t(0)x + \mathcal{O}(x^{d+1})$ uniformly in t for some constant $d > 0$ and for $|x| \ll 1$, and if $n^{-1}\mathbf{F}^T\mathbf{F} \to \frac{1}{2}\mathbf{C}$ as $n \to \infty$ for some nonsingular matrix $\mathbf{C}$, then $\sqrt{n}(\hat{\boldsymbol{\beta}} - \boldsymbol{\beta}) \xrightarrow{D} \mathrm{N}(\mathbf{0}, 2\eta^2\mathbf{C}^{-1})$, where $\eta^2 := \kappa^2/4$.

Remark 5.14 Let $\hat{\beta}$ denote the sample median of an i.i.d. sequence of real random variables with CDF $F(x)$. If $F(x)$ satisfies the conditions in Remark 5.13, then $\sqrt{n}\,\hat{\beta} \xrightarrow{D} \mathrm{N}(0, \eta^2)$, where $\eta^2 := \kappa^2/4$. This is a classical result for the sample median [218, Theorem 3.4, p. 354].

Because $\dot{F}_{tj}(0)$ is just the probability density of the noise at zero (which is the median of the noise), its reciprocal

$$\kappa := 1/\dot{F}_{tj}(0)$$

is called the *sparsity* of the noise at zero. Theorem 5.9 shows that the accuracy of the LAD estimator is directly related to the sparsity. This parameter does not depend on the tail behavior of the noise at all. Therefore, the LAD estimator is well suited to handle the situation where the noise has a great concentration in the vicinity of zero but with the possibility of large outliers. Moreover, to satisfy the conditions of Theorem 5.9, the noise does not have to be identically distributed over time. Therefore, the LAD estimator is also suitable for handling the situation where the noise exhibits time-varying, or heteroscedastic, volatilities such as large bursty errors over a certain period of time.

Now let us compare Theorem 5.9 with Theorem 5.4. Obviously, for this comparison to be meaningful, the real and imaginary parts of the noise must have both zero median and zero mean. To comply with the requirements of Theorem 5.4, it is also necessary to assume that the noise is an i.i.d. sequence such that $F_{tj}(x) = F(x)$ for $j = 1, 2$ and all t. Under these conditions, an important difference between Theorem 5.9 and Theorem 5.4 is revealed: Theorem 5.4 requires the noise to have finite variance, whereas Theorem 5.9 does not require such an assumption and therefore remains true even if the noise has an infinite variance, or even if the mean of the noise does not exist. A good example is the Cauchy noise. In this case, the LAD estimator, by Theorem 5.9, has an asymptotic normal distribution with finite variance, but the LS estimator, being a linear combination of independent Cauchy random variables, has a Cauchy distribution with infinite variance and Theorem 5.4 does not apply.

When the noise does have a finite variance and meets all the other requirements of Theorem 5.9 and Theorem 5.4, the only difference between the asymptotic distributions of the LAD estimator and the LS estimator is that η^2 replaces σ^2 in the asymptotic covariance matrix of the LAD estimator. As such, the parameter η^2 can be interpreted as a measure of the noise strength for the LAD estimator as the variance σ^2 does for the LS estimator. The LAD estimator is more accurate than the LS estimator if $\eta^2 < \sigma^2$, or equivalently, if

$$\sigma^2/\kappa^2 = \sigma^2\{\dot{F}(0)\}^2 > 1/2. \tag{5.5.2}$$

Let $p_0(x)$ denote the unit-variance PDF such that $\dot{F}(x) = (\sqrt{2}/\sigma)p_0(\sqrt{2}x/\sigma)$. Because $\dot{F}(0) = (\sqrt{2}/\sigma)p_0(0) = 1/\kappa$, we can write

$$\eta^2 = \tfrac{1}{4}\sigma^2/\{p_0(0)\}^2. \tag{5.5.3}$$

Therefore, the condition (5.5.2) can be expressed as

$$p_0(0) > 1/2. \tag{5.5.4}$$

This condition does not depend on the variance of the noise; it simply requires the noise to have sufficiently high probability density at zero.

As an example, consider Student's t-distribution with $\nu > 0$ degrees of freedom and scale parameter $c > 0$ (see Example 3.10). For $\nu > 2$, this distribution has finite variance $\sigma^2/2$ if $c = \sigma\sqrt{(\nu-2)/(2\nu)}$. Because the probability density at zero can be expressed as

$$\dot{F}(0) = \frac{\Gamma((\nu+1)/2)}{c\Gamma(\nu/2)\sqrt{\pi\nu}},$$

the condition (5.5.2) or (5.5.4) becomes

$$\Gamma((\nu+1)/2) - \Gamma(\nu/2)\sqrt{\pi(\nu-2)/4} > 0.$$

Solving this inequality for ν gives $2 < \nu < \nu_G$, where $\nu_G \approx 4.678$. Note that as the value of ν decreases toward 2, the distribution becomes increasingly heavy in the tails and the variance approaches infinity. Hence, the condition (5.5.2) or (5.5.4) favors heavy-tailed noise.

The efficiency of the LAD estimator versus the LS estimator can be measured by the efficiency coefficient defined as

$$\rho := \sigma^2/\eta^2 = 2\sigma^2/\kappa^2 = 2\sigma^2\{\dot{F}(0)\}^2 = 4\{p_0(0)\}^2. \tag{5.5.5}$$

Clearly, if (5.5.4) is satisfied, then $\rho > 1$, meaning that the LAD estimator is more efficient than the LS estimator. On the other hand, it is also possible that $\rho < 1$, in which case the LAD estimator is less efficient than the LS estimator. For example, with the Gaussian noise, we have $p_0(0) = 1/\sqrt{2\pi}$ and hence $\rho = 2/\pi < 1$. This means that under the Gaussian condition the asymptotic variance of the LAD estimator is $\rho^{-1} = \pi/2 \approx 1.57$ times that of the LS estimator. Therefore, the robustness of the LAD estimator against heavy-tailed noise is achieved at the expense of efficiency under Gaussian and other light-tailed noise distributions which do not satisfy (5.5.2) or (5.5.4). This tradeoff is the same as in the choice between the mean and the median of an i.i.d. sample [218, Section 5.3].

The following theorem shows that the efficiency coefficient has a lower bound 1/3 for a large class of noise distributions. A proof is given in Section 5.6.

Theorem 5.10 (Efficiency Lower Bound for the LAD Estimator). *Let the conditions of Theorems 5.4 and 5.9 be satisfied and let $F_{tj}(x) = F(x)$ for all t and j. Assume further that $\dot{F}(x) \le \dot{F}(0) = 1/\kappa$ for all x. Then, the efficiency coefficient ρ of the LAD estimator, defined by (5.5.5), satisfies $\rho \ge 1/3$, where the lower bound is attained by and only by the uniform distribution.*

Remark 5.15 In the real case, let $p_0(x)$ be the unit-variance PDF such that $\dot{F}(x) = (1/\sigma)p_0(x/\sigma)$. Then, the expression for η^2 in (5.5.3) remains true. Owing to Remarks 5.6 and 5.13, the efficiency coefficient of the LAD estimator is given by

$$\rho := \sigma^2/\eta^2 = 4\sigma^2/\kappa^2 = 4\sigma^2\{\dot{F}(0)\}^2 = 4\{p_0(0)\}^2.$$

In this case, $\rho \ge 1/3$ if $\dot{F}(x) \le \dot{F}(0) = 1/\kappa$ for all x, and the lower bound is attained by and only by the uniform distribution. This is a classical result [218, p. 359].

According to Theorem 5.10, the uniform distribution is the least favorable distribution for the LAD estimator in the class of noise distributions whose PDF is maximized at $x = 0$ which is the mean as well as the median.

Now let us compare the asymptotic covariance matrix given by Theorem 5.9 with the CRLB derived under the assumption of Laplace white noise. Toward that end, consider the parameter $\boldsymbol{\theta} := [\Re(\boldsymbol{\beta})^T, -\Im(\boldsymbol{\beta})^T]^T$. According to Corollary 3.4, the ACRLB for estimating $\boldsymbol{\theta}$ under the Gaussian white noise assumption can be expressed as $n^{-1}\boldsymbol{\Gamma}_G(\boldsymbol{\theta})$ with

$$\boldsymbol{\Gamma}_G(\boldsymbol{\theta}) := \tfrac{1}{2}\sigma^2\boldsymbol{\Lambda},$$

where $\boldsymbol{\Lambda}$ is defined by (5.1.14). Therefore, by Theorem 3.7, the ACRLB under the condition of Laplace white noise can be expressed as $n^{-1}\boldsymbol{\Gamma}_L(\boldsymbol{\theta})$, where

$$\boldsymbol{\Gamma}_L(\boldsymbol{\theta}) := \tfrac{1}{2}\boldsymbol{\Gamma}_G(\boldsymbol{\theta}) = \tfrac{1}{4}\sigma^2\boldsymbol{\Lambda}. \tag{5.5.6}$$

On the other hand, if the noise does have a Laplace distribution, then, by Theorem 5.9 and Remark 5.12, the LAD estimator $\hat{\boldsymbol{\theta}} := [\Re(\hat{\boldsymbol{\beta}})^T, -\Im(\hat{\boldsymbol{\beta}})^T]^T$ satisfies

$$\sqrt{n}\,(\hat{\boldsymbol{\theta}} - \boldsymbol{\theta}) \xrightarrow{D} \mathrm{N}(\mathbf{0}, \tfrac{1}{2}\eta^2\boldsymbol{\Lambda}). \tag{5.5.7}$$

Because the Laplace PDF takes the form $\dot{F}(x) = (1/\sigma)\exp(-2|x|/\sigma)$, we obtain $\kappa = 1/\dot{F}(0) = \sigma$, $\eta^2 = \kappa^2/2 = \sigma^2/2$, and hence

$$\tfrac{1}{2}\eta^2\boldsymbol{\Lambda} = \tfrac{1}{4}\kappa^2\boldsymbol{\Lambda} = \tfrac{1}{4}\sigma^2\boldsymbol{\Lambda} = \boldsymbol{\Gamma}_L(\boldsymbol{\theta}).$$

In other words, the asymptotic covariance matrix of the LAD estimator coincides with the Laplace ACRLB. This means that the LAD estimator is asymptotically efficient under the condition of Laplace white noise. Because the LAD estimator is the ML estimator under the Laplace assumption, the asymptotic efficiency assertion should not be too surprising.

An important difference between the LAD estimator and the LS estimator can be observed by comparing (5.5.7) with (5.1.13): unlike the LS estimator which always attains the Gaussian CRLB asymptotically regardless of the actual noise distribution, the LAD estimator can outperform the Laplace CRLB if the noise distribution satisfies the condition $\eta^2 < \sigma^2/2$, or equivalently, if

$$\sigma^2/\kappa^2 = \sigma^2\{\dot{F}(0)\}^2 > 1. \tag{5.5.8}$$

Because $\dot{F}(0) = (\sqrt{2}/\sigma)p_0(0)$, the condition (5.5.8) can also be written as

$$p_0(0) > 1/\sqrt{2}. \tag{5.5.9}$$

This condition is related to the constraint on the symmetric unimodal distributions discussed in Proposition 3.5 with which the Laplace distribution becomes the least favorable one that has the smallest Fisher information. Note that for the Laplace distribution we have $\sigma^2/\kappa^2 = 1$ and $p_0(0) = 1/\sqrt{2}$.

As an example, consider again the t-distribution with ν degrees of freedom for $\nu > 2$. Recall that the condition (5.5.2) or (5.5.4) is satisfied if $2 < \nu < \nu_G \approx 4.678$. To satisfy the condition (5.5.8) or (5.5.9), it is required that

$$\Gamma((\nu+1)/2) - \Gamma(\nu/2)\sqrt{\pi(\nu-2)/2} > 0.$$

Solving this inequality for ν gives $2 < \nu < \nu_L$, where $\nu_L \approx 2.724$. Note that the t-distribution with $\nu \in (2, \nu_L)$ has heavier tails than the t-distribution with $\nu \in (\nu_L, \nu_G)$. This demonstrates that the LAD estimator is able to outperform the Laplace CRLB when the tails of the noise distribution are very heavy.

In summary, the asymptotic variance of the LAD estimator has the following properties, depending on the noise distribution.

1. It is smaller than the Laplace CRLB if $p_0(0) > 1/\sqrt{2}$;
2. It coincides with the Laplace CRLB if the noise has a Laplace distribution, in which case, $p_0(0) = 1/\sqrt{2}$;
3. It is larger than the Laplace CRLB but smaller than the Gaussian CRLB if $1/2 < p_0(0) < 1/\sqrt{2}$;
4. It is larger than the Gaussian CRLB if $p_0(0) < 1/2$, in which case, the efficiency relative to the Gaussian CRLB is given by $\rho := 4\{p_0(0)\}^2$;
5. It is 3 times the Gaussian CRLB in the worst case where the noise has a uniform distribution.

These properties are determined solely by the probability density at zero.

Let us further investigate the LAD estimator by considering its asymptotic distribution under the condition of colored noise. More specifically, let us assume that $\{\epsilon_t\}$ is a random process whose real and imaginary parts $\{\Re(\epsilon_t)\}$ and $\{\Im(\epsilon_t)\}$ are mutually independent with zero median and satisfy

$$P(\Re(\epsilon_t)\Re(\epsilon_s) < 0) = P(\Im(\epsilon_t)\Im(\epsilon_s) < 0) = \gamma(t-s) \tag{5.5.10}$$

for some function $\gamma(\cdot)$, in which case $\{\epsilon_t\}$ is said to be *stationary in zero-crossings* and $\gamma(u)$ is called the lag-u *zero-crossing rate* [224]. It is easy to verify that the condition (5.5.10) can also be stated as

$$P(\Re(\epsilon_t) < 0, \Re(\epsilon_s) < 0) = P(\Im(\epsilon_t) < 0, \Im(\epsilon_s) < 0) = q(t-s)$$

for some function $q(\cdot)$, where $q(u)$ is called the lag-u *orthant probability*, which is related to the lag-u zero-crossing rate by

$$\gamma(u) = 1 - 2q(u).$$

Moreover, it is easy to verify that $1 - 2\gamma(u)$ coincides with the lag-u autocovariance of the zero-mean weakly stationary processes $\{\text{sgn}(\Re(\epsilon_t))\}$ and $\{\text{sgn}(\Im(\epsilon_t))\}$.

Because $\gamma(-u) = \gamma(u)$, the Fourier transform of $\{1 - 2\gamma(u)\}$ can be expressed as

$$h(\omega) := \sum_{u=-\infty}^{\infty} (1 - 2\gamma(u))\cos(u\omega). \tag{5.5.11}$$

This function is called the *zero-crossing spectrum* [224]. It serves as a scale-invariant representation of serial dependence in the frequency domain.

The association of the zero-crossing spectrum with the LAD estimator was discovered in [224] for the real case. The following theorem is an extension of this result to the complex case. It shows that the LAD estimator retains its asymptotic normality established under the white noise condition, but the asymptotic covariance matrix depends on the zero-crossing spectrum of the noise. A proof of this result can be found in Section 5.6.

Theorem 5.11 (Asymptotic Distribution of the LAD Estimator: Colored Noise). *Let* $\mathbf{y}$ *be given by (5.1.1) and* $\hat{\boldsymbol{\beta}}$ *be defined by (5.5.1). Assume that* $\{\omega_k\}$ *satisfies the separation condition (5.1.10) and* $\{\epsilon_t\}$ *is stationary in zero-crossings with the zero-crossing spectrum* $h(\omega)$ *satisfying* $h(\omega_k) > 0$ *for all* k. *Assume further that* $\{\Re(\epsilon_t)\}$ *and* $\{\Im(\epsilon_t)\}$ *satisfy (a) the conditions in Theorem 5.9 regarding the marginal distributions* $F_{tj}(x)$ $(j = 1, 2)$ *and (b) the mixing condition (12.5.5) with* $\sum_{u=1}^{\infty} u^{\delta}\alpha(u) < \infty$ *for some* $\delta > 0$. *Then, as* $n \to \infty$, $\sqrt{n}(\hat{\boldsymbol{\beta}} - \boldsymbol{\beta}) \stackrel{A}{\sim} \mathrm{N}_c(\mathbf{0}, \mathbf{S})$, *where* $\mathbf{S} := \mathrm{diag}\{\ell_\epsilon(\omega_1), \ldots, \ell_\epsilon(\omega_p)\}$, $\ell_\epsilon(\omega) := \eta^2 h(\omega)$, *and* $\eta^2 := \kappa^2/2$.

Remark 5.16 For the real case discussed in Remark 5.13, let $\{\omega_k\}$ satisfy (5.1.10) with $\Delta := \min_{k \neq k'}\{|\omega_k - \omega_{k'}|, \omega_k, \pi - \omega_k\}$ and let $\{\epsilon_t\}$ be stationary in zero-crossings with $\gamma(u) := P(\epsilon_{t+u}\epsilon_t < 0)$ and $h(\omega_k) := \sum(1 - 2\gamma(u))\cos(u\omega_k) > 0$ for all k. If $\{\epsilon_t\}$ also satisfies the mixing condition in Theorem 5.11 and its marginal distributions $F_t(x)$ satisfy the condition in Remark 5.13, then $\sqrt{n}(\hat{\boldsymbol{\beta}} - \boldsymbol{\beta}) \stackrel{A}{\sim} \mathrm{N}(\mathbf{0}, 2\mathbf{S})$ as $n \to \infty$, where $\mathbf{S} := \mathrm{diag}\{\ell_\epsilon(\omega_1), \ell_\epsilon(\omega_1), \ldots, \ell_\epsilon(\omega_q), \ell_\epsilon(\omega_q)\}$, $\ell_\epsilon(\omega) := \eta^2 h(\omega)$, and $\eta^2 := \kappa^2/4$. This result and its generalization can be found in [224].

Remark 5.17 The mixing condition in Theorem 5.11 is a technical one that essentially requires the noise to be weakly dependent. It ensures a central limit theorem for $n^{-1}\sum_{t=1}^{n}\{\mathrm{sgn}(\Re(\epsilon_t)) + i\,\mathrm{sgn}(\Im(\epsilon_t))\}\mathbf{x}_t$. It is also needed to justify a quadratic approximation. These objectives can be achieved under alternative weak-dependence conditions, an example of which can be found in [224].

A comparison of Theorem 5.11 with Theorem 5.3 shows that the function

$$\ell_\epsilon(\omega) := \eta^2 h(\omega) \tag{5.5.12}$$

plays the same role in the asymptotic distribution of the LAD estimator as the SDF does in the asymptotic distribution of the LS estimator. We call $\ell_\epsilon(\omega)$ the *Laplace spectrum* [224]. To compare the Laplace spectrum with the ordinary spectrum $f_\epsilon(\omega)$, recall that if $\{\epsilon_t\}$ is a random process whose real and imaginary parts are mutually independent and weakly stationary with zero mean and common ACF $r_0(u)$, then $r_\epsilon(u) = 2r_0(u)$, which is a real-valued even function of u. In this case, $\rho(u) := r_\epsilon(u)/r_\epsilon(0) = r_0(u)/r_0(0)$ is the autocorrelation function of $\{\epsilon_t\}$, so the ordinary spectrum can be expressed as

$$f_\epsilon(\omega) = \sigma^2 g(\omega), \tag{5.5.13}$$

where $\sigma^2 := r_\epsilon(0)$ and

$$g(\omega) := \sum_{u=-\infty}^{\infty} \rho(u)\cos(u\omega). \tag{5.5.14}$$

The function $g(\omega)$ is known as the autocorrelation spectrum or the normalized power spectrum. It is the traditional scale-invariant representation of serial dependence in the frequency domain. By comparing (5.5.13) with (5.5.12), we can see that the zero-crossing spectrum $h(\omega)$ is the counterpart of the autocorrelation spectrum $g(\omega)$. Moreover, the scaling factor η^2 in (5.5.12) is the counterpart of σ^2 in (5.5.13). With this scaling factor, the Laplace spectrum retains the physical dimension of the ordinary spectrum in magnitude.

As a representation of serial dependence, the zero-crossing spectrum has a close relationship with the autocorrelation spectrum. First, in the special case where $\{\epsilon_t\}$ is an i.i.d. sequence, we have

$$q(u) = \tfrac{1}{4}(1+\delta_u), \quad \gamma(u) = \tfrac{1}{2}(1-\delta_u),$$

and hence $h(\omega) = 1$ for all ω. This result is consistent with the notion of white noise for which $g(\omega) = 1$. Moreover, consider the case where $\boldsymbol{\epsilon}$ has a complex elliptic distribution which is discussed in Lemma 12.2.3. This includes the complex Gaussian distribution as a special case. Under the assumption of zero mean and (weak) stationarity, it can be shown [182, p. 48] that

$$q(u) = \tfrac{1}{4} + (2\pi)^{-1}\arcsin(\rho(u)),$$

where $\rho(u)$ is the lag-u autocorrelation coefficient of $\{\epsilon_t\}$. In this case,

$$\gamma(u) = 1 - 2q(u) = \tfrac{1}{2} - \pi^{-1}\arcsin(\rho(u)),$$

and hence

$$h(\omega) = \sum_{u=-\infty}^{\infty} 2\pi^{-1}\arcsin(\rho(u))\cos(u\omega).$$

In other words, the zero-crossing spectrum $h(\omega)$ of a time series with elliptic distribution is the Fourier transform of the arcsine-transformed autocorrelation function. The autocorrelation spectrum in (5.5.14) is the Fourier transform of the original autocorrelation function.

Now let us turn to the issue of computation. Although the LAD estimator in (5.5.1) can be computed by any optimization routine which is suitable for handling nondifferentiable objective functions, a more efficient approach is to convert the ℓ_1-norm minimization problem into a linear program (LP) and solve it by linear programming techniques.

Toward that end, consider the real-variable representation (5.1.5). Let $\boldsymbol{\theta} := [\Re(\boldsymbol{\beta})^T, -\Im(\boldsymbol{\beta})^T]^T$, $\tilde{\mathbf{x}}_t := [\Re(\mathbf{x}_t)^T, \Im(\mathbf{x}_t)^T]^T$, and $\tilde{\mathbf{x}}_{n+t} := [-\Im(\mathbf{x}_t)^T, \Re(\mathbf{x}_t)^T]^T$ so that $\Re(\mathbf{x}_t^H \boldsymbol{\beta}) = \tilde{\mathbf{x}}_t^T \boldsymbol{\theta}$ and $\Im(\mathbf{x}_t^H \boldsymbol{\beta}) = \tilde{\mathbf{x}}_{n+t}^T \boldsymbol{\theta}$. Similarly, let $\tilde{y}_t := \Re(y_t)$ and $\tilde{y}_{n+t} := \Im(y_t)$. Then, the LAD estimator $\hat{\boldsymbol{\beta}}$ in (5.5.1) can be obtained from

$$\hat{\boldsymbol{\theta}} := [\Re(\hat{\boldsymbol{\beta}})^T, -\Im(\hat{\boldsymbol{\beta}})^T]^T := \arg\min_{\tilde{\boldsymbol{\theta}} \in \mathbb{R}^{2p}} \sum_{t=1}^{2n} |\tilde{y}_t - \tilde{\mathbf{x}}_t^T \tilde{\boldsymbol{\theta}}|. \tag{5.5.15}$$

To convert this problem into a standard LP problem, let $\xi_t^+ \geq 0$ and $\xi_t^- \geq 0$ satisfy $\tilde{y}_t - \tilde{\mathbf{x}}_t^T \tilde{\boldsymbol{\theta}} = \xi_t^+ - \xi_t^-$ $(t = 1, \ldots, 2n)$. Similarly, define $\boldsymbol{\theta}^+ := [\theta_1^+, \ldots, \theta_{2p}^+]^T$ and $\boldsymbol{\theta}^- := [\theta_1^-, \ldots, \theta_{2p}^-]^T$ with $\theta_k^+ \geq 0$ and $\theta_k^- \geq 0$ $(k = 1, \ldots, 2p)$ so that $\tilde{\boldsymbol{\theta}} = \boldsymbol{\theta}^+ - \boldsymbol{\theta}^-$. Then, the LAD problem in (5.5.15) becomes the following LP problem of seeking the optimal values for $\{\xi_t^+\}$, $\{\xi_t^-\}$, $\{\theta_k^+\}$, and $\{\theta_k^-\}$:

$$\begin{aligned}
\text{minimize} \quad & \sum_{t=1}^{2n} (\xi_t^+ + \xi_t^-) \\
\text{subject to} \quad & \tilde{\mathbf{x}}_t^T (\boldsymbol{\theta}^+ - \boldsymbol{\theta}^-) + \xi_t^+ - \xi_t^- = \tilde{y}_t, \\
& \xi_t^+ \geq 0, \xi_t^- \geq 0, \theta_k^+ \geq 0, \theta_k^- \geq 0, \\
& (t = 1, \ldots, 2n; k = 1, \ldots, 2p).
\end{aligned}$$

This problem can be solved by standard LP techniques, such as the simplex-based algorithm of Barrodale and Roberts [23] [24] and the interior-point algorithm of Karmarkar [171]. For comprehensive discussion on these and other algorithms, see, for example, [39], [196], [295], and [421].

In addition to the LP algorithms, one can solve (5.5.15) directly by using an ad hoc technique called the *iteratively re-weighted least squares* (IRWLS), as suggested by Schlossmacher [331]. It is an iterative algorithm of the form

$$\hat{\boldsymbol{\theta}}(m) := \arg\min_{\tilde{\boldsymbol{\theta}} \in \mathbb{R}^{2p}} \sum_{t \in T_n(m)} w_t(m) |\tilde{y}_t - \tilde{\mathbf{x}}_t^T \tilde{\boldsymbol{\theta}}|^2 \qquad (m = 1, 2, \ldots),$$

where $w_t(m) := |\tilde{y}_t - \tilde{\mathbf{x}}_t^T \hat{\boldsymbol{\theta}}(m-1)|^{-1}$ is the weight function and $T_n(m)$ is a subset of $\{1, \ldots, 2n\}$ that excludes the cases with near zero residuals, e.g.,

$$T_n(m) := \{t : |\tilde{y}_t - \tilde{\mathbf{x}}_t^T \hat{\boldsymbol{\theta}}(m-1)| > \delta, 1 \leq t \leq 2n\}$$

for some small constant $\delta > 0$. A convenient choice for the initial value $\hat{\boldsymbol{\theta}}(0)$ is the LS solution that minimizes $\sum_{t=1}^{2n} |\tilde{y}_t - \tilde{\mathbf{x}}_t^T \tilde{\boldsymbol{\theta}}|^2$ with respect to $\tilde{\boldsymbol{\theta}} \in \mathbb{R}^{2p}$.

There is a computationally appealing alternative to the LAD estimator for robust estimation, proposed in [91], [92], and [290]. To motivate this method, observe that when the $\mathbf{f}(\omega_k)$ are orthogonal to each other so that $n^{-1}\mathbf{F}^H\mathbf{F} = \mathbf{I}$, the LS estimator in (5.1.3) becomes the Fourier transform of $\mathbf{y}$, that is, $\hat{\boldsymbol{\beta}} = n^{-1}\mathbf{F}^H\mathbf{y}$. In this case, we can write $\hat{\beta}_k = \hat{A}_k - i\hat{B}_k$ and

$$\begin{cases} \hat{A}_k := n^{-1}\sum_{t=1}^n \Re(y_t \exp(-it\omega_k)), \\ \hat{B}_k := -n^{-1}\sum_{t=1}^n \Im(y_t \exp(-it\omega_k)). \end{cases}$$

Note that $\hat{A}_k$ and $\hat{B}_k$ can be regarded as the sample means of the real-valued time series $\{\Re(y_t \exp(-it\omega_k))\}$ and $\{-\Im(y_t \exp(-it\omega_k))\}$, respectively. Therefore, by replacing the sample mean with the sample median, one obtains a simple robust estimator $\tilde{\beta}_k := \tilde{A}_k - i\tilde{B}_k$, where

$$\begin{cases} \tilde{A}_k := \text{median}\{\Re(y_t \exp(-it\omega_k)) : t = 1, \ldots, n\}, \\ \tilde{B}_k := -\text{median}\{\Im(y_t \exp(-it\omega_k)) : t = 1, \ldots, n\}. \end{cases} \tag{5.5.16}$$

Let this estimator be called the median Fourier transform (MED-FT) estimator. In comparison with the LAD estimator in (5.5.15), an immediate appeal of the MED-FT estimator is its computational simplicity. However, because the interaction among the $\mathbf{f}(\omega_k)$ is ignored, the MED-FT estimator does not perform well when the sinusoids have unequal strengths or closely spaced frequencies.

The following simulation example compares the LAD, MED-FT, and LS estimators in terms of their accuracy under various noise distributions.

Example 5.4 (Amplitude Estimation for Two Complex Sinusoids in Gaussian and Heavy-Tailed White Noise). Let $\mathbf{y}$ consist of two random-phase complex sinusoids plus white noise with $n = 100$. The signal frequencies are: Case 1, $\omega_1 = 2\pi \times 0.12$ and $\omega_2 = 2\pi \times 0.25$; Case 2, $\omega_1 = 2\pi \times 0.12$ and $\omega_2 = \omega_1 + \pi/n = 2\pi \times 0.125$. The amplitude parameters are $C_1 = 1$ and $C_2 = 2$ in both cases.

Table 5.4 contains the mean-square error (MSE) of the LS, LAD, and MED-FT estimators calculated on the basis of 10,000 Monte Carlo runs under different noise distributions. For the first three noise distributions, the variance is always set to unity; for the Cauchy distribution, the scale parameter is set to 1/4. The bursty noise, an example of which is shown in Figure 5.3, is obtained from the Gaussian white noise $\{\zeta_t\} \sim \text{GWN}_c(0,1)$ by applying a nonlinearity $\psi(x)$ to the 30% of the time series in the middle section, i.e., $\epsilon_t := \psi(\Re(\zeta_t)) + i\psi(\Im(\zeta_t))$ for $t = 36, \ldots, 65$ and $\epsilon_t := \zeta_t$ otherwise. The nonlinearity $\psi(x)$ takes the form

$$\psi(x) := \begin{cases} x & \text{if } |x| \le a, \\ b(x-a)+a & \text{if } x > a, \\ b(x+a)-a & \text{if } x < -a, \end{cases}$$

with $a = 1/2$ and $b = 6$. This function amplifies the exceedance by a factor b if the variable exceeds the threshold a in absolute value.

Table 5.4. MSE of Amplitude Estimation ($\times 10^{-3}$)

		LS		LAD		MED-FT	
Noise	Case	β_1	β_2	β_1	β_2	β_1	β_2
Gaussian	1	10	10	16	16	43	22
Laplace	1	10	10	6	6	43	22
Student's $T_{2.1}$	1	5	5	1	1	19	9
Cauchy (1/4)	1	1730000	1730000	3	3	71	25
Bursty	1	47	47	16	16	74	31
Gaussian	2	17	17	26	26	2117	461
Laplace	2	17	17	11	11	2219	494
Student's $T_{2.1}$	2	8	8	2	2	2573	596
Cauchy (1/4)	2	1980000	1980000	6	6	2330	537
Bursty	2	36	36	26	26	1904	437

Results are based on 10,000 Monte Carlo runs.

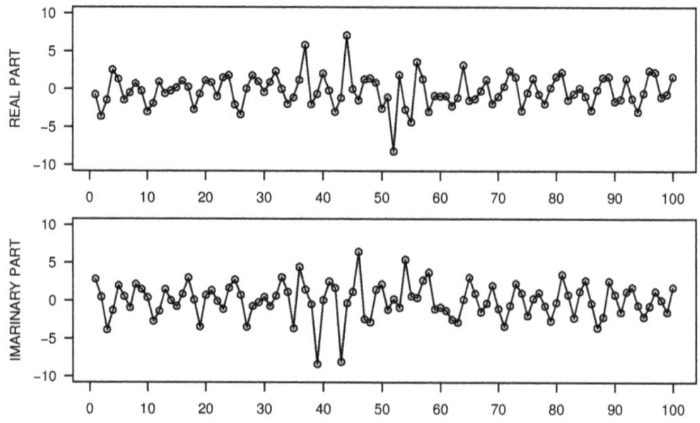

Figure 5.3. Real and imaginary parts of the time series in Example 5.4 with bursty noise due to nonlinear amplification of large errors in the middle section.

As can be seen from Table 5.4, the advantage of the LAD estimator over the LS estimator increases rapidly as the noise becomes more heavy-tailed. Under the condition of Cauchy noise, the LS estimator fails completely, but the LAD estimator remains accurate. In the Gaussian case, the LAD estimator is less accurate than the LS estimator, which signifies the cost one has to pay for the robustness gains. The MED-FT estimator is more robust than the LS estimator, but much less accurate than the LAD estimator, especially in Case 2 where the signal frequencies are closely spaced. Moreover, the accuracy of the MED-FT estimator depends on the strengths of the sinusoids: the weaker signal is estimated less accurately than the stronger one. This is in contrast with the LAD and LS estimators for which both sinusoids are estimated with similar accuracy.

For the first three noise distributions in Table 5.4, the theoretical MSE of the LS estimator is equal to $\sigma^2/n = 0.01$ in Case 1 and equal to $(1-|D_n(\pi/n)|^2)^{-1}\sigma^2/n \approx 0.0168$ in Case 2, where σ^2 is the variance of the noise. The simulation results in Table 5.4 are very close to these theoretical values for the Gaussian and Laplace noise, but not so for the noise with Student's t-distribution with 2.1 degrees of freedom (denoted $T_{2.1}$). The discrepancy in the latter case can be explained by the fact that the sample variance tends to underestimate the theoretical variance when the noise has heavier tails.

Table 5.4 also shows that the LAD estimator achieves the same accuracy under the nonstationary bursty noise as it does under the stationary Gaussian noise. This can be explained by the fact that the nonlinearity applied to the Gaussian noise does not alter the sparsity at zero. As a result, the asymptotic distribution of the LAD estimator remains the same, according to Theorem 5.9. ◇

To close this section, we point out that the LAD estimator is just an example of the so-called M-estimators, "M" for maximum likelihood type [159]. An M-estimator of $\boldsymbol{\beta}$ is defined as the minimizer of an objective function of the form

$$\sum_{t=1}^{n} \{\phi(\Re(y_t - \mathbf{x}_t^H \tilde{\boldsymbol{\beta}})) + \phi(\Im(y_t - \mathbf{x}_t^H \tilde{\boldsymbol{\beta}}))\},$$

where $\phi(x)$ is a nonnegative function which can be designed to reduce the influence of large residuals in different ways. Obviously, the LAD estimator in (5.5.1) corresponds to the choice $\phi(x) = |x|$. Additional choices of $\phi(x)$ can be specified through its derivative, an example being Huber's ψ function of the form

$$\dot{\phi}(x) = \psi(x) := \begin{cases} x & \text{if } |x| < c, \\ c\,\text{sgn}(x) & \text{if } |x| \geq c, \end{cases}$$

where $c > 0$ is a tuning parameter that makes the tradeoff between robustness and efficiency. With this choice, the objective function is nonconvex, so the computation of its minimizer becomes more difficult. Nonetheless, the IRWLS technique, among others, is generally applicable. For comprehensive discussions on robust regression methods in general, see, for example, [132], [159], and [253].

5.6 Proof of Theorems

This section contains the proof of Theorems 5.1–5.11 and the derivation of the AIC and AICC criteria in (5.4.2) and (5.4.4).

Proof of Theorem 5.1. Part (a) follows immediately from Proposition 2.6 upon noting that $\hat{\boldsymbol{\beta}} - \boldsymbol{\beta} = (\mathbf{F}^H\mathbf{F})^{-1}\mathbf{F}^H\boldsymbol{\epsilon}$. To prove part (b), let the eigenvalue decomposition (EVD) of the positive definite matrix $\mathbf{F}^H\mathbf{F}$ be $\mathbf{U}\mathbf{D}\mathbf{U}^H$, where $\mathbf{U}$ is a unitary matrix and $\mathbf{D}$ is a diagonal matrix with positive diagonal elements. Define

$$\mathbf{z} := \mathbf{D}^{1/2}\mathbf{U}^H(\hat{\boldsymbol{\beta}} - \boldsymbol{\beta})/\sigma,$$

Then, we can write

$$(\hat{\boldsymbol{\beta}} - \boldsymbol{\beta})^H(\mathbf{F}^H\mathbf{F})(\hat{\boldsymbol{\beta}} - \boldsymbol{\beta})/\sigma^2 = (\hat{\boldsymbol{\beta}} - \boldsymbol{\beta})^H\mathbf{U}\mathbf{D}\mathbf{U}^H(\hat{\boldsymbol{\beta}} - \boldsymbol{\beta})/\sigma^2 = \|\mathbf{z}\|^2.$$

Because $\mathrm{Cov}(\hat{\boldsymbol{\beta}} - \boldsymbol{\beta}) = \sigma^2(\mathbf{F}^H\mathbf{F})^{-1}$, it follows that

$$E(\mathbf{z}\mathbf{z}^H) = \mathbf{D}^{1/2}\mathbf{U}^H(\mathbf{F}^H\mathbf{F})^{-1}\mathbf{U}\mathbf{D}^{1/2} = \mathbf{D}^{1/2}\mathbf{U}^H(\mathbf{U}\mathbf{D}^{-1}\mathbf{U}^H)\mathbf{U}\mathbf{D}^{1/2} = \mathbf{I}.$$

By combining this result with the fact that $\hat{\boldsymbol{\beta}} - \boldsymbol{\beta} \sim \mathrm{N}_c(\mathbf{0}, \sigma^2(\mathbf{F}^H\mathbf{F})^{-1})$, we obtain $\mathbf{z} \sim \mathrm{N}_c(\mathbf{0}, \mathbf{I})$ according to Proposition 2.6. An application of Lemma 12.2.1 leads to $\|\mathbf{z}\|^2 \sim \frac{1}{2}\chi^2(2p)$. Part (b) is thus proved.

To prove part (c), let $\mathbf{P} := \mathbf{F}(\mathbf{F}^H\mathbf{F})^{-1}\mathbf{F}^H$. Then,

$$\begin{aligned}
\mathrm{Cov}(\hat{\boldsymbol{\beta}}, \mathbf{y} - \mathbf{F}\hat{\boldsymbol{\beta}}) &= \mathrm{Cov}\{(\mathbf{F}^H\mathbf{F})^{-1}\mathbf{F}^H\mathbf{y}, (\mathbf{I} - \mathbf{P})\mathbf{y}\} \\
&= (\mathbf{F}^H\mathbf{F})^{-1}\mathbf{F}^H\mathrm{Cov}(\mathbf{y})(\mathbf{I} - \mathbf{P}^H) \\
&= \sigma^2(\mathbf{F}^H\mathbf{F})^{-1}\mathbf{F}^H(\mathbf{I} - \mathbf{P}^H) = \mathbf{0}.
\end{aligned}$$

This, combined with the fact that $\hat{\boldsymbol{\beta}}$ and $\mathbf{y} - \mathbf{F}\hat{\boldsymbol{\beta}}$ are jointly complex Gaussian, implies that $\hat{\boldsymbol{\beta}}$ is also independent of $\mathbf{y} - \mathbf{F}\hat{\boldsymbol{\beta}}$ and hence of $\hat{\sigma}^2 = n^{-1}\|\mathbf{y} - \mathbf{F}\hat{\boldsymbol{\beta}}\|^2$.

Finally, to prove part (d), observe that

$$\begin{aligned}
Q_1 &:= \boldsymbol{\epsilon}^H\boldsymbol{\epsilon} = (\mathbf{y} - \mathbf{F}\boldsymbol{\beta})^H(\mathbf{y} - \mathbf{F}\boldsymbol{\beta}) \\
&= (\mathbf{y} - \mathbf{F}\hat{\boldsymbol{\beta}} + \mathbf{F}(\hat{\boldsymbol{\beta}} - \boldsymbol{\beta}))^H(\mathbf{y} - \mathbf{F}\hat{\boldsymbol{\beta}} + \mathbf{F}(\hat{\boldsymbol{\beta}} - \boldsymbol{\beta})) \\
&= (\mathbf{y} - \mathbf{F}\hat{\boldsymbol{\beta}})^H(\mathbf{y} - \mathbf{F}\hat{\boldsymbol{\beta}}) + (\hat{\boldsymbol{\beta}} - \boldsymbol{\beta})^H(\mathbf{F}^H\mathbf{F})(\hat{\boldsymbol{\beta}} - \boldsymbol{\beta}) \\
&= \mathrm{SSE} + Q_2,
\end{aligned}$$

where the third line results from the fact that

$$(\mathbf{y} - \mathbf{F}\hat{\boldsymbol{\beta}})^H\mathbf{F}^H(\mathbf{y} - \mathbf{F}\hat{\boldsymbol{\beta}}) = (\mathbf{y} - \mathbf{F}\hat{\boldsymbol{\beta}})^H(\mathbf{F}^H\mathbf{y} - \mathbf{F}^H\mathbf{F}\hat{\boldsymbol{\beta}}) = \mathbf{0}.$$

By Lemma 12.2.1, $2Q_1/\sigma^2 \sim \chi^2(2n)$. As shown in part (b), $2Q_2/\sigma^2 \sim \chi^2(2p)$. Moreover, SSE is independent of Q_2 because Q_2 is a function of $\hat{\boldsymbol{\beta}}$ which is shown in part (c) to be independent of SSE. Therefore, it is easy to show, by using the characteristic function of χ^2 (see [338, p. 20] and Lemma 12.2.1), that $2 \times \mathrm{SSE}/\sigma^2 = 2Q_1/\sigma^2 - 2Q_2/\sigma^2 \sim \chi^2(2n - 2p)$. The proof is complete. □

Proof of Theorem 5.2. The estimator $\hat{\boldsymbol{\beta}}$, which is defined by (5.1.3), satisfies (5.1.6). By assumption, $n^{-1}\mathbf{F}^H\boldsymbol{\epsilon} \xrightarrow{P} \mathbf{0}$ and $n^{-1}\mathbf{F}^H\mathbf{F} \to \mathbf{C}$. Therefore,

$$\hat{\boldsymbol{\beta}} - \boldsymbol{\beta} = (\mathbf{F}^H\mathbf{F})^{-1}\mathbf{F}^H\boldsymbol{\epsilon} = (n^{-1}\mathbf{F}^H\mathbf{F})^{-1}(n^{-1}\mathbf{F}^H\boldsymbol{\epsilon}) \xrightarrow{P} \mathbf{0}.$$

To prove $n^{-1}\text{SSE} \xrightarrow{P} \sigma^2$, observe that

$$n^{-1}\text{SSE} = n^{-1}\boldsymbol{\epsilon}^H\boldsymbol{\epsilon} - 2\Re\{n^{-1}\boldsymbol{\epsilon}^H\mathbf{F}(\hat{\boldsymbol{\beta}} - \boldsymbol{\beta})\} + (\hat{\boldsymbol{\beta}} - \boldsymbol{\beta})^H(n^{-1}\mathbf{F}^H\mathbf{F})(\hat{\boldsymbol{\beta}} - \boldsymbol{\beta}).$$

The first term tends to σ^2 by assumption. The second and third terms approach zero because $n^{-1}\boldsymbol{\epsilon}^H\mathbf{F} \xrightarrow{P} \mathbf{0}$, $n^{-1}\mathbf{F}^H\mathbf{F} \to \mathbf{C}$, and $\hat{\boldsymbol{\beta}} - \boldsymbol{\beta} \xrightarrow{P} \mathbf{0}$. □

Proof of Theorem 5.3. It follows from (5.1.6) that $\sqrt{n}(\hat{\boldsymbol{\beta}} - \boldsymbol{\beta}) = \sqrt{n}(\mathbf{F}^H\mathbf{F})^{-1}\mathbf{F}^H\boldsymbol{\epsilon}$. Under the condition (5.1.10), Lemma 12.1.4 ensures that $n^{-1}\mathbf{F}^H\mathbf{F} = \mathbf{I} + \mathcal{O}(n^{-1}\Delta^{-1})$. Therefore, with $\mathbf{S} := \text{diag}\{f_\epsilon(\omega_1), \dots, f_\epsilon(\omega_p)\}$, it suffices to show that

$$n^{-1/2}\mathbf{S}^{-1/2}\mathbf{F}^H\boldsymbol{\epsilon} \xrightarrow{D} \mathrm{N}_c(\mathbf{0}, \mathbf{I}), \tag{5.6.1}$$

Observe that $n^{-1/2}\mathbf{S}^{-1/2}\mathbf{F}^H\boldsymbol{\epsilon} = [X_1, \dots, X_p]^T$, where

$$X_k := n^{-1/2}\mathbf{f}^H(\omega_k)\boldsymbol{\epsilon}/v_k = n^{-1/2}\sum_{t=1}^{n}\epsilon_t \exp(-it\omega_k)/v_k$$

and $v_k := \sqrt{f_\epsilon(\omega_k)}$. It is easy to verify that

$$\text{Cov}(X_k, X_{k'}) = \sum_{|u|<n} r_\epsilon(u)\exp(-iu\omega_k)D_n^*(u, \omega_k - \omega_{k'})/v_k^2,$$

where $T_n(u) := \{t : \max(1, 1-u) \le t \le \min(n, n-u)\}$ and

$$D_n(u, \omega) := n^{-1}\sum_{t\in T_n(u)} \exp(it\omega).$$

According to Lemma 12.1.4, $D_n(u, \omega_k - \omega_{k'}) \to \delta_{k-k'}$ as $n \to \infty$ for any fixed u. Combining this result with the absolute summability of $\{r_\epsilon(u)\}$ and the dominant convergence theorem leads to

$$\text{Cov}(X_k, X_{k'}) \to \delta_{k-k'}.$$

The assertion (5.6.1) follows immediately in the Gaussian case.

Now consider the case where $\{\epsilon_t\}$ is a linear process. For any given $m > 0$, let $\tilde{\psi}_j := \psi_j \mathscr{I}(|j| \le m)$ and

$$\tilde{\epsilon}_t := \sum_{j=-\infty}^{\infty}\tilde{\psi}_j\zeta_{t-j} = \sum_{j=-m}^{m}\psi_j\zeta_{t-j}.$$

Then, $\{\tilde{\epsilon}_t\}$ is a zero-mean stationary process with ACF $\tilde{r}_\epsilon(u) := E(\tilde{\epsilon}_{t+u}\tilde{\epsilon}_t^*)$ and SDF

$$\tilde{f}_\epsilon(\omega) := \sum_{u=-\infty}^{\infty}\tilde{r}_\epsilon(u)\exp(-iu\omega) = \sigma_\zeta^2|\tilde{\Psi}(\omega)|^2,$$

where $\tilde{\Psi}(\omega) := \sum \tilde{\psi}_j \exp(-ij\omega)$. Let $f_0 := \inf\{f_\epsilon(\omega)\}$. Because $f_\epsilon(\omega) \geq f_0 > 0$ and $\tilde{\Psi}(\omega) - \Psi(\omega) \to 0$ uniformly in ω as $m \to \infty$, there exists a constant $m_0 > 0$ such that $\tilde{f}_\epsilon(\omega) \geq f_0/2$ for all ω if $m \geq m_0$. For any fixed k, let $\tilde{v}_k := \{\tilde{f}_\epsilon(\omega_k)\}^{1/2}$ and

$$\tilde{X}_k := n^{-1/2} \sum_{t=1}^{n} \tilde{\epsilon}_t \exp(-it\omega_k)/\tilde{v}_k = \sum_{t=1-m}^{n+m} c_{nt}\zeta_t,$$

where

$$c_{nt} := n^{-1/2} \sum_{s=1}^{n} \tilde{\psi}_{s-t} \exp(-is\omega_k)/\tilde{v}_k.$$

Because $\sum |\psi_j| < \infty$ and $\tilde{f}_\epsilon(\omega_k) \geq f_0/2$, it follows that

$$\max_{1-m \leq t \leq n+m} |c_{nt}| = \mathcal{O}(n^{-1/2}) \to 0 \quad \text{as } n \to \infty.$$

Moreover, the absolute summability of $\{\tilde{r}_\epsilon(u)\}$, together with the dominant convergence theorem, gives

$$\begin{aligned}
\operatorname{Var}(\tilde{X}_k) &= \sum_{|u|<n} (1 - n^{-1}|u|)\tilde{r}_\epsilon(u) \exp(-iu\omega_k)/\tilde{f}_\epsilon(\omega_k) \\
&= 1 - \sum_{|u| \geq n} \tilde{r}_\epsilon(u) \exp(-iu\omega_k)/\tilde{f}_\epsilon(\omega_k) \\
&\quad - \sum_{|u|<n} n^{-1}|u|\tilde{r}_\epsilon(u) \exp(-iu\omega_k)/\tilde{f}_\epsilon(\omega_k) \\
&\to 1 \quad \text{as } n \to \infty,
\end{aligned}$$

which, in turn, leads to

$$\sum_{t=1-m}^{n+m} |c_{nt}|^2 = \operatorname{Var}(\tilde{X}_k)/\sigma_\zeta^2 \to 1/\sigma_\zeta^2 \quad \text{as } n \to \infty.$$

For any $k' \neq k$, let c'_{nt} be the counterpart of c_{nt} associated with $\omega_{k'}$. Then,

$$\begin{aligned}
\sum_{t=1-m}^{n+m} c^*_{nt} c'_{nt} &= \operatorname{Cov}(\tilde{X}_k, \tilde{X}_{k'})/\sigma_\zeta^2 \\
&= \sum_{|u|<n} D_n(u, \omega_k - \omega_{k'})\tilde{r}_\epsilon(u) \exp(-iu\omega_k)/(\tilde{v}_k \tilde{v}_{k'} \sigma_\zeta^2),
\end{aligned}$$

where

$$D_n(u,\omega) := n^{-1} \sum_{t \in T_n(u)} \exp(-it\omega)$$

and $T_n(u) := \{t : \max(1, 1-u) \leq t \leq \min(n, n-u)\}$. Under the condition (5.1.10), Lemma 12.1.4 ensures that $D_n(u, \omega_k - \omega_{k'}) \to 0$ for any fixed u. Therefore, by the dominant convergence theorem,

$$\sum_{t=1-m}^{n+m} c^*_{nt} c'_{nt} \to 0 \quad \text{as } n \to \infty.$$

Combining these results with Lemma 12.5.7 yields $[\tilde{X}_1,\ldots,\tilde{X}_p]^T \xrightarrow{D} \mathrm{N}_c(\mathbf{0},\mathbf{I})$.

It remains to show that for any constant $\delta > 0$.

$$\lim_{m\to\infty} \limsup_{n\to\infty} P(|X_k - \tilde{X}_k| > \delta) = 0. \tag{5.6.2}$$

Toward that end, observe that

$$\tfrac{1}{2}E|X_k - \tilde{X}_k|^2 f_\epsilon(\omega_k) \le n^{-1}E\Big|\sum_{t=1}^{n} \bar{\epsilon}_t \exp(-it\omega_k)\Big|^2 + (1 - v_k/\tilde{v}_k)^2 n^{-1} E\Big|\sum_{t=1}^{n} \tilde{\epsilon}_t \exp(-it\omega_k)\Big|^2,$$

where $\bar{\epsilon}_t := \epsilon_t - \tilde{\epsilon}_t = \sum \bar{\psi}_j \zeta_{t-j}$ and $\bar{\psi}_j := \psi_j \mathscr{I}(|j| > m)$. Let $\bar{r}_\epsilon(u)$ and $\bar{f}_\epsilon(\omega)$ denote the ACF and SDF of $\{\bar{\epsilon}_t\}$. Then, by the dominant convergence theorem together with the absolute summability of $\{\bar{r}_\epsilon(u)\}$, we obtain

$$\begin{aligned} &n^{-1}E\Big|\sum_{t=1}^{n} \bar{\epsilon}_t \exp(-it\omega_k)\Big|^2 - \bar{f}_\epsilon(\omega_k) \\ &= \sum_{|u|\ge n} \bar{r}_\epsilon(u)\exp(-iu\omega_k) - \sum_{|u|<n} n^{-1}|u|\bar{r}_\epsilon(u)\exp(-iu\omega_k) \\ &\to 0, \quad n\to\infty. \end{aligned}$$

Similarly,

$$n^{-1}E\Big|\sum_{t=1}^{n} \tilde{\epsilon}_t \exp(-it\omega_k)\Big|^2 - \tilde{f}_\epsilon(\omega_k) \to 0, \quad n\to\infty.$$

Moreover, because $\bar{f}_\epsilon(\omega_k) = \sigma_\zeta^2|\bar{\Psi}(\omega_k)|^2$, where $\bar{\Psi}(\omega) := \sum_{|j|>m}\psi_j \exp(-ij\omega)$, it follows that $\bar{f}_\epsilon(\omega) \to 0$ uniformly in ω as $m\to\infty$. Similarly, $\tilde{f}_\epsilon(\omega) \to f_\epsilon(\omega)$ uniformly in ω as $m\to\infty$. Combining these results with $\tilde{v}_k^2 \ge f_0/2 > 0$ yields

$$\lim_{m\to\infty} \limsup_{n\to\infty} n^{-1}E\Big|\sum_{t=1}^{n} \bar{\epsilon}_t \exp(-it\omega_k)\Big|^2 = 0$$

and

$$\lim_{m\to\infty} \limsup_{n\to\infty} (1 - v_k/\tilde{v}_k)^2 n^{-1}E\Big|\sum_{t=1}^{n} \tilde{\epsilon}_t \exp(-it\omega_k)\Big|^2 = 0.$$

An application of Chebyshev's inequality proves (5.6.2). The assertion (5.6.1) follows from (5.6.2) and Proposition 2.16(e). □

Proof of Theorem 5.4. As in the proof of Theorem 5.3, observe that $\sqrt{n}(\hat{\boldsymbol{\beta}} - \boldsymbol{\beta}) = (n^{-1}\mathbf{F}^H\mathbf{F})^{-1}n^{-1/2}\mathbf{F}^H\boldsymbol{\epsilon}$. Let $\mathbf{x}_t := [\exp(-it\omega_1),\ldots,\exp(-it\omega_p)]^T$. Then, $n^{-1/2}\mathbf{F}^H\boldsymbol{\epsilon} = n^{-1/2}\sum \mathbf{x}_t\epsilon_t$. Because $n^{-1}\sum \mathbf{x}_t\mathbf{x}_t^H = n^{-1}\mathbf{F}^H\mathbf{F} \to \mathbf{C}$, and because the elements of

$\mathbf{x}_t$ are uniformly bounded in t, it follows from Lemma 12.5.7 that $n^{-1}\mathbf{F}^H\boldsymbol{\epsilon} \xrightarrow{D} \mathrm{N}_c(\mathbf{0},\sigma^2\mathbf{C})$. Combining this with $(n^{-1}\mathbf{F}^H\mathbf{F})^{-1} \to \mathbf{C}^{-1}$ proves the assertion. □

Proof of Theorem 5.5. The GLS estimator defined by (5.1.15) satisfies (5.1.16) and (5.1.17). This means that $\hat{\theta} := \mathbf{a}^H\hat{\boldsymbol{\beta}}$ is an unbiased estimator of $\theta := \mathbf{a}^H\boldsymbol{\beta}$ with

$$\mathrm{Var}(\hat{\theta}) = \mathbf{a}^H(\mathbf{F}^H\mathbf{R}_\epsilon^{-1}\mathbf{F})^{-1}\mathbf{a}.$$

On the other hand, let $\tilde{\theta} := \mathbf{b}^H\mathbf{y}$ be any linear unbiased estimator of θ. Because $E(\tilde{\theta}) = \mathbf{b}^H E(\mathbf{y}) = \mathbf{b}^H\mathbf{F}\boldsymbol{\beta}$, the unbiasedness of $\tilde{\theta}$ implies that $(\mathbf{b}^H\mathbf{F}-\mathbf{a}^H)\boldsymbol{\beta} = 0$ for all $\boldsymbol{\beta} \in \mathbb{C}^p$, which, in turn, leads to

$$\mathbf{b}^H\mathbf{F} = \mathbf{a}^H. \tag{5.6.3}$$

Moreover, we can write $\tilde{\theta} - \theta = Z_1 + Z_2$, where

$$Z_1 := \tilde{\theta} - \hat{\theta} = \mathbf{b}^H\mathbf{y} - \mathbf{a}^H\hat{\boldsymbol{\beta}}, \quad Z_2 := \hat{\theta} - \theta = \mathbf{a}^H(\hat{\boldsymbol{\beta}} - \boldsymbol{\beta}).$$

Owing to (5.6.3) and the fact that $\mathbf{F}\hat{\boldsymbol{\beta}} = \mathbf{P}\mathbf{y}$, where $\mathbf{P} := \mathbf{F}(\mathbf{F}^H\mathbf{R}_\epsilon^{-1}\mathbf{F})^{-1}\mathbf{F}^H\mathbf{R}_\epsilon^{-1}$, we obtain $Z_1 = \mathbf{b}^H(\mathbf{I}-\mathbf{P})\mathbf{y}$. In addition, we have $E(\boldsymbol{\epsilon}\mathbf{y}^H) = \mathbf{R}_\epsilon$ and $\mathbf{P}\mathbf{R}_\epsilon\mathbf{P}^H = \mathbf{P}\mathbf{R}_\epsilon$. Combining these results with (5.1.16) and (5.6.3) gives

$$\begin{aligned} E(Z_2 Z_1^H) &= \mathbf{a}^H(\mathbf{F}^H\mathbf{R}_\epsilon^{-1}\mathbf{F})^{-1}\mathbf{F}^H\mathbf{R}_\epsilon^{-1}E(\boldsymbol{\epsilon}\mathbf{y}^H)(\mathbf{I}-\mathbf{P}^H)\mathbf{b} \\ &= \mathbf{b}^H\mathbf{P}\mathbf{R}_\epsilon(\mathbf{I}-\mathbf{P}^H)\mathbf{b} = 0. \end{aligned}$$

Therefore, we obtain

$$\mathrm{Var}(\tilde{\theta}) = E(|Z_1|^2) + E(|Z_2|^2) \ge E(|Z_2|^2) = \mathrm{Var}(\hat{\theta}).$$

The equality holds if and only if $Z_1 = 0$, or equivalently, $\tilde{\theta} = \hat{\theta}$, almost surely. □

Proof of Theorem 5.6. Let $\hat{\mathbf{F}} := [\mathbf{f}(\hat{\omega}_1),\ldots,\mathbf{f}(\hat{\omega}_p)]^T$. Then, $\tilde{\boldsymbol{\beta}} = (\hat{\mathbf{F}}^H\hat{\mathbf{F}})^{-1}\hat{\mathbf{F}}^H\mathbf{y}$. Consider the Taylor expansion

$$\exp(it\lambda) = \exp(it\omega) + it(\lambda-\omega)r_t(\lambda,\omega),$$

where $r_t(\lambda,\omega) := \int_0^1 \exp\{it(\omega + s(\lambda-\omega))\}\,ds$. Observe that $|r_t(\lambda,\omega)| \le 1$ for all λ, ω, and t. Therefore, we obtain

$$\begin{aligned} n^{-1}\mathbf{f}^H(\hat{\omega}_{k'})\mathbf{f}(\hat{\omega}_k) &= n^{-1}\sum_{t=1}^{n}\exp(it\hat{\omega}_k)\exp(-it\hat{\omega}_{k'}) \\ &= n^{-1}\sum_{t=1}^{n}\exp\{it(\omega_k-\omega_{k'})\} + \mathcal{O}(n\xi_{nk}) + \mathcal{O}(n\xi_{nk'}) \\ &= n^{-1}\mathbf{f}^H(\omega_{k'})\mathbf{f}(\omega_k) + \mathcal{O}_P(na_n) \end{aligned}$$

and hence

$$n^{-1}\hat{\mathbf{F}}^H\hat{\mathbf{F}} = n^{-1}\mathbf{F}^H\mathbf{F} + \mathcal{O}_P(na_n). \tag{5.6.4}$$

Similarly,

$$\begin{aligned} n^{-1}\mathbf{f}^H(\omega_{k'})\mathbf{f}(\hat{\omega}_k) &= n^{-1}\sum_{t=1}^{n}\exp(it\omega_{k'})\exp(-it\hat{\omega}_k) \\ &= n^{-1}\sum_{t=1}^{n}\exp\{it(\omega_{k'}-\omega_k)\} + \mathcal{O}(n\xi_{nk}) \\ &= n^{-1}\mathbf{f}^H(\omega_{k'})\mathbf{f}(\omega_k) + \mathcal{O}_P(na_n). \end{aligned}$$

Combining this expression with $\mathbf{y} = \mathbf{F}\boldsymbol{\beta} + \boldsymbol{\epsilon}$ leads to

$$n^{-1}\hat{\mathbf{F}}^H\mathbf{y} = n^{-1}\mathbf{F}^H\mathbf{F}\boldsymbol{\beta} + \mathcal{O}_P(na_n) + n^{-1}\hat{\mathbf{F}}^H\boldsymbol{\epsilon}. \tag{5.6.5}$$

According to Lemma 12.5.3, $n^{-1}\hat{\mathbf{F}}^H\boldsymbol{\epsilon} \xrightarrow{P} \mathbf{0}$. Therefore, $n^{-1}\hat{\mathbf{F}}^H\hat{\mathbf{F}} \xrightarrow{P} \mathbf{C}$, $n^{-1}\hat{\mathbf{F}}^H\mathbf{y} \xrightarrow{P} \mathbf{C}\boldsymbol{\beta}$, and hence $\tilde{\boldsymbol{\beta}} \xrightarrow{P} \boldsymbol{\beta}$. The argument remains valid if $\mathcal{O}_P$ is everywhere replaced by $\mathcal{O}$ and if $\xrightarrow{P}$ is everywhere replaced by $\xrightarrow{a.s.}$.

Moreover, because $E\{\sum t|\epsilon_t|\} = \mathcal{O}(n^2)$, we can write $\sum t|\epsilon_t| = \mathcal{O}_P(n^2)$ according to Proposition 2.16(a). Therefore,

$$\begin{aligned} n^{-1}\mathbf{f}^H(\hat{\omega}_k)\boldsymbol{\epsilon} &= n^{-1}\sum_{t=1}^{n}\exp(it\hat{\omega}_k)\epsilon_t \\ &= n^{-1}\sum_{t=1}^{n}\exp(it\omega_k)\epsilon_t + \mathcal{O}_P(n\xi_{nk}) \\ &= n^{-1}\mathbf{f}^H(\omega_k)\boldsymbol{\epsilon} + \mathcal{O}_P(na_n). \end{aligned}$$

Combining this expression with (5.6.5) gives

$$n^{-1}\hat{\mathbf{F}}^H\mathbf{y} = n^{-1}\mathbf{F}^H\mathbf{y} + \mathcal{O}_P(na_n). \tag{5.6.6}$$

Observe that (5.6.4) and (5.6.6) can also be written as

$$n^{-1}\hat{\mathbf{F}}^H\hat{\mathbf{F}} = n^{-1}\mathbf{F}^H\mathbf{F} + \hat{\mathbf{E}}, \quad n^{-1}\hat{\mathbf{F}}^H\mathbf{y} = n^{-1}\mathbf{F}^H\mathbf{y} + \hat{\mathbf{e}},$$

where $\|\hat{\mathbf{E}}\| = \mathcal{O}_P(na_n)$ and $\|\hat{\mathbf{e}}\| = \mathcal{O}_P(na_n)$. It is easy to verify that

$$n^{-1}\mathbf{F}^H\mathbf{F}(\hat{\boldsymbol{\beta}} - \tilde{\boldsymbol{\beta}}) = \hat{\mathbf{E}}\tilde{\boldsymbol{\beta}} - \hat{\mathbf{e}}.$$

Because $n^{-1}\mathbf{F}^H\mathbf{F} \to \mathbf{C}$ and $\tilde{\boldsymbol{\beta}} = \mathcal{O}_P(1)$, we obtain (5.2.3). □

Proof of Theorem 5.7. This assertion is well known in the real case [338, pp. 96–97]. In the complex case, it suffices to reformulate the problem using real variables by separating the real and imaginary parts. Indeed, because

$$\beta\exp(i\omega t) = A\cos(\omega t) + B\sin(\omega t) + i\{A\sin(\omega t) - B\cos(\omega t)\},$$

where $\beta := A - iB$, the CSM of m complex parameters for $\mathbf{y}$ is equivalent to two RSMs of $2m$ common parameters, one for $\Re(\mathbf{y})$ and one for $\Im(\mathbf{y})$. Moreover, by definition, $\boldsymbol{\epsilon} \sim \mathrm{N}_c(\mathbf{0}, \sigma^2\mathbf{I})$ implies $[\Re(\boldsymbol{\epsilon})^T, \Im(\boldsymbol{\epsilon})^T]^T \sim \mathrm{N}(\mathbf{0}, \frac{1}{2}\sigma^2\mathbf{I})$. Therefore, the original linear regression problem of estimating $\{c_j\}$ and σ^2 from $\mathbf{y}$ can be reformulated as a linear regression problem of estimating $\{\Re(c_j), \Im(c_j)\}$ and $\frac{1}{2}\sigma^2$ from $[\Re(\mathbf{y})^T, \Im(\mathbf{y})^T]^T$. In the latter problem, the sample size is $2n$ and the number of real coefficients is twice the number of complex coefficients in the original problem. The hypothesis H_0 can be reformulated as $\Re(c_j) = \Im(c_j) = 0$ for all j such that $\lambda_j \in \Omega \setminus \Omega'$. Therefore, under H_0, the standard theory of F-tests for the real case [338, pp. 96–97] gives

$$F' := \frac{(\mathrm{SSE}'_{2p} - \mathrm{SSE}'_{2m})/(2m-2p)}{\mathrm{SSE}'_{2m}/(2n-2m)} \sim F(2m-2p, 2n-2m),$$

where SSE'_{2p} and SSE'_{2m} are the error sums of squares for the $2p$-parameter reduced model and the $2m$-parameter complete model, respectively. Because the error sum of squares for estimating $\{\Re(c_j), \Im(c_j)\}$ is the same as the error sum of squares for estimating $\{c_j\}$, we obtain

$$\mathrm{SSE}'_{2p} = \mathrm{SSE}_p, \quad \mathrm{SSE}'_{2m} = \mathrm{SSE}_m,$$

which, in turn, implies $F' = F$. Hence the assertion. □

Derivation of (5.4.2) and (5.4.4). The derivation largely follows the lines of [55] which considers the real case rather than the complex case. Let the $\tilde{p}$-term model of $\mathbf{y}$ be denoted by $p_1(\cdot)$, which is $\mathrm{N}_c(\mathbf{F}_{\tilde{p}}\boldsymbol{\beta}_{\tilde{p}}, \sigma_{\tilde{p}}^2\mathbf{I})$, where $\mathbf{F}_{\tilde{p}}$ is a matrix with column vectors $\mathbf{f}(\lambda_j)$ $(\lambda_j \in \Omega_{\tilde{p}})$ and $\Omega_{\tilde{p}}$ is a $\tilde{p}$-subset of $\Omega := \{\lambda_1, \dots, \lambda_m\}$. Under this assumption, the log likelihood function of $\mathbf{y}$ can be expressed as

$$\log\{p(\mathbf{y}|\boldsymbol{\eta})\} = -n\log\pi - n\log\sigma_{\tilde{p}}^2 - \|\mathbf{y} - \mathbf{F}_{\tilde{p}}\boldsymbol{\beta}_{\tilde{p}}\|^2/\sigma_{\tilde{p}}^2,$$

where $\boldsymbol{\eta} := (\Omega_{\tilde{p}}, \boldsymbol{\beta}_{\tilde{p}}, \sigma_{\tilde{p}}^2)$ denotes the set of unknown parameters. The ML estimator of $\boldsymbol{\eta}$ is given by $\hat{\boldsymbol{\eta}} := (\hat{\Omega}_{\tilde{p}}, \hat{\boldsymbol{\beta}}_{\tilde{p}}, \hat{\sigma}_{\tilde{p}}^2)$, where $\hat{\Omega}_{\tilde{p}}$ is the best $\tilde{p}$-subset of Ω, $\hat{\boldsymbol{\beta}}_{\tilde{p}} := (\hat{\mathbf{F}}_{\tilde{p}}^H\hat{\mathbf{F}}_{\tilde{p}})^{-1}\hat{\mathbf{F}}_{\tilde{p}}^H\mathbf{y}$, and $\hat{\sigma}_{\tilde{p}}^2 := n^{-1}\|\mathbf{y} - \hat{\mathbf{F}}_{\tilde{p}}\hat{\boldsymbol{\beta}}_{\tilde{p}}\|^2 = n^{-1}\mathrm{SSE}(\tilde{p})$, with $\hat{\mathbf{F}}_{\tilde{p}}$ being the design matrix corresponding to $\hat{\Omega}_{\tilde{p}}$. Therefore,

$$-2\log\{p(\mathbf{y}|\hat{\boldsymbol{\eta}})\} = 2n\log\pi + 2n\log\hat{\sigma}_{\tilde{p}}^2 + 2n. \tag{5.6.7}$$

For any $\mathbf{y}' \in \mathbb{C}^n$, we can write

$$\begin{aligned} -2\log\{p(\mathbf{y}'|\hat{\boldsymbol{\eta}})\} &= 2n\log\pi + 2n\log\hat{\sigma}_{\tilde{p}}^2 + 2\|\mathbf{y}' - \hat{\mathbf{F}}_{\tilde{p}}\hat{\boldsymbol{\beta}}_{\tilde{p}}\|^2/\hat{\sigma}_{\tilde{p}}^2 \\ &= -2\log\{p(\mathbf{y}|\hat{\boldsymbol{\eta}})\} + 2\|\mathbf{y}' - \hat{\mathbf{F}}_{\tilde{p}}\hat{\boldsymbol{\beta}}_{\tilde{p}}\|^2/\hat{\sigma}_{\tilde{p}}^2 - 2n. \end{aligned} \tag{5.6.8}$$

With fixed $\mathbf{y}$ (and hence fixed $\hat{\boldsymbol{\eta}}$), $\hat{p}_1(\cdot) := p(\cdot|\hat{\boldsymbol{\eta}})$ is the ML estimator for the PDF of the $\tilde{p}$-term model. Moreover, let $\mathbf{y}'$ be a random vector with PDF $p_0(\cdot)$, i.e., $\mathbf{y}' \sim \mathrm{N}_c(\mathbf{F}\boldsymbol{\beta}, \sigma^2\mathbf{I})$. Then, it follows from (5.6.8) that

$$\begin{aligned} H(\hat{p}_1 \| p_0) &= E_{\mathbf{y}'}\{-2\log[p(\mathbf{y}'|\hat{\boldsymbol{\eta}})]\} \\ &= -2\log\{p(\mathbf{y}|\hat{\boldsymbol{\eta}})\} + 2\{n\sigma^2 + \|\mathbf{F}\boldsymbol{\beta} - \hat{\mathbf{F}}_{\tilde{p}}\hat{\boldsymbol{\beta}}_{\tilde{p}}\|^2\}/\hat{\sigma}^2_{\tilde{p}} - 2n. \end{aligned} \tag{5.6.9}$$

From this expression it is clear that $-2\log\{p(\mathbf{y}|\hat{\boldsymbol{\eta}})\}$ is not an unbiased estimator of $E\{H(\hat{p}_1 \| p_0)\}$. To find a correction for the bias, observe that if the true distribution of $\mathbf{y}$ is $\mathrm{N}_c(\mathbf{F}_{\tilde{p}}\boldsymbol{\beta}_{\tilde{p}}, \sigma^2_{\tilde{p}}\mathbf{I})$, i.e., if $\mathbf{F} = \mathbf{F}_{\tilde{p}}$, $\boldsymbol{\beta} = \boldsymbol{\beta}_{\tilde{p}}$, and $\sigma^2 = \sigma^2_{\tilde{p}}$, then, $\hat{\Omega}_{\tilde{p}} = \Omega_{\tilde{p}}$ with high probability for large n or high SNR, in which case, we have $\hat{\mathbf{F}}_{\tilde{p}} = \mathbf{F}_{\tilde{p}}$ so that $\mathbf{F}\boldsymbol{\beta} - \hat{\mathbf{F}}_{\tilde{p}}\hat{\boldsymbol{\beta}}_{\tilde{p}} = -\mathbf{F}_{\tilde{p}}(\hat{\boldsymbol{\beta}}_{\tilde{p}} - \boldsymbol{\beta}_{\tilde{p}})$ and $\hat{\boldsymbol{\beta}}_{\tilde{p}} = (\mathbf{F}^H_{\tilde{p}}\mathbf{F}_{\tilde{p}})^{-1}\mathbf{F}^H_{\tilde{p}}\mathbf{y}$. Under this condition, Theorem 5.1 ensures $\|\mathbf{F}_{\tilde{p}}(\hat{\boldsymbol{\beta}}_{\tilde{p}} - \boldsymbol{\beta}_{\tilde{p}})\|^2 \sim \frac{1}{2}\sigma^2_{\tilde{p}}\chi^2(2\tilde{p})$ and Theorem 5.2 ensures $\hat{\sigma}_{\tilde{p}} \xrightarrow{P} \sigma^2_{\tilde{p}}$ as $n \to \infty$. These observations motivate the employment of $n + \frac{1}{2}\chi^2(2\tilde{p})$ as an approximation to $\{n\sigma^2 + \|\mathbf{F}\boldsymbol{\beta} - \hat{\mathbf{F}}_{\tilde{p}}\hat{\boldsymbol{\beta}}_{\tilde{p}}\|^2\}/\hat{\sigma}^2_{\tilde{p}}$. Combining this approximation with (5.6.9) suggests the use of

$$\mathrm{AIC}(\tilde{p}) := -2\log\{p(\mathbf{y}|\hat{\boldsymbol{\eta}})\} + 2\tilde{p}$$

as an approximately unbiased estimator of $E\{H(\hat{p}_1 \| p_0)\}$. Minimizing this criterion with respect to $\tilde{p}$ is equivalent to minimizing the AIC defined by (5.4.2), because $-2\log\{p(\mathbf{y}|\hat{\boldsymbol{\eta}})\}$ takes the form (5.6.7).

To derive the penalty function given by (5.4.4), it suffices to observe that if the true distribution of $\mathbf{y}$ is $\mathrm{N}_c(\mathbf{F}_{\tilde{p}}\boldsymbol{\beta}_{\tilde{p}}, \sigma^2_{\tilde{p}}\mathbf{I})$, then Theorem 5.1 guarantees $n\hat{\sigma}^2_{\tilde{p}}/\sigma^2_{\tilde{p}} \sim \frac{1}{2}\chi^2(2n-2\tilde{p})$ and $\|\mathbf{F}_{\tilde{p}}(\hat{\boldsymbol{\beta}}_{\tilde{p}} - \boldsymbol{\beta}_{\tilde{p}})\|^2/\hat{\sigma}^2_{\tilde{p}} \sim (n\tilde{p}/(n-\tilde{p}))F(2\tilde{p}, 2n-2\tilde{p})$. This observation motivates the use of $2n^2/\chi^2(2n-2\tilde{p}) + (n\tilde{p}/(n-\tilde{p}))F(2\tilde{p}, 2n-2\tilde{p})$ as an alternative approximation to $\{n\sigma^2 + \|\mathbf{F}\boldsymbol{\beta} - \hat{\mathbf{F}}_{\tilde{p}}\hat{\boldsymbol{\beta}}_{\tilde{p}}\|^2\}/\hat{\sigma}^2_{\tilde{p}}$. Combining this approximation with (5.6.9) and the fact that $E\{1/\chi^2(2n-2\tilde{p})\} = 1/(2n-2\tilde{p}-2)$ and $E\{F(2\tilde{p}, 2n-2\tilde{p})\} = (2n-2\tilde{p})/(2n-2\tilde{p}-2)$ justifies the use of

$$\begin{aligned} \mathrm{AIC}_c(\tilde{p}) &:= -2\log\{p(\mathbf{y}|\hat{\boldsymbol{\eta}})\} + 2\{n^2/(n-\tilde{p}-1) + n\tilde{p}/(n-\tilde{p}-1)\} - 2n \\ &= -2\log\{p(\mathbf{y}|\hat{\boldsymbol{\eta}})\} + 2n(n+\tilde{p})/(n-\tilde{p}-1) - 2n \end{aligned}$$

as an approximately unbiased estimator of $E\{H(\hat{p}_1 \| p_0)\}$. Minimizing this criterion with respect to $\tilde{p}$ is equivalent of minimizing the AIC in (5.4.2) with $2\tilde{p}$ replaced by the penalty function in (5.4.4).

Some caveats in this derivation. First, it conveniently ignores the extra variability in $\hat{\mathbf{F}}_{\tilde{p}}$ which may result in a greater bias than the estimated one without taking it into account. Moreover, it justifies the estimators by assuming that the true distribution $p_0(\cdot)$ is the same as the model $p_1(\cdot)$. Therefore, it may be more precise to describe the resulting $\mathrm{AIC}(\tilde{p})$ and $\mathrm{AIC}_c(\tilde{p})$ as approximately unbiased estimators of $E\{H(\hat{p}_1 \| p_1)\}$ rather than $E\{H(\hat{p}_1 \| p_0)\}$.

Under the true distribution, one can always write $\mathbf{F}\boldsymbol{\beta} = \mathbf{F}_m\boldsymbol{\beta}_m = \mathbf{F}_{\tilde{p}}\boldsymbol{\beta}_{\tilde{p}} + \mathbf{F}'_{\tilde{p}}\boldsymbol{\beta}'_{\tilde{p}}$, where $\mathbf{F}'_{\tilde{p}}$ is the design matrix corresponding to $\Omega \setminus \Omega_{\tilde{p}}$. This implies that

$$\mathbf{F}\boldsymbol{\beta} - \mathbf{F}_{\tilde{p}}\hat{\boldsymbol{\beta}}_{\tilde{p}} = (\mathbf{I} - \mathbf{P}_{\tilde{p}})\mathbf{F}'_{\tilde{p}}\boldsymbol{\beta}'_{\tilde{p}} - \mathbf{P}_{\tilde{p}}\boldsymbol{\epsilon},$$

where $\mathbf{P}_{\tilde{p}} := \mathbf{F}_{\tilde{p}}(\mathbf{F}_{\tilde{p}}^H\mathbf{F}_{\tilde{p}})^{-1}\mathbf{F}_{\tilde{p}}$ is the projection matrix onto the column space of $\mathbf{F}_{\tilde{p}}$ and where $\boldsymbol{\epsilon} := \mathbf{y} - \mathbf{F}\boldsymbol{\beta} \sim \mathrm{N}_c(\mathbf{0}, \sigma^2\mathbf{I})$. Therefore,

$$\|\mathbf{F}\boldsymbol{\beta} - \mathbf{F}_{\tilde{p}}\hat{\boldsymbol{\beta}}_{\tilde{p}}\|^2 = \|(\mathbf{I} - \mathbf{P}_{\tilde{p}})\mathbf{F}'_{\tilde{p}}\boldsymbol{\beta}'_{\tilde{p}}\|^2 + \|\mathbf{P}_{\tilde{p}}\boldsymbol{\epsilon}\|^2.$$

Observe that the second term is distributed as $\frac{1}{2}\sigma^2\chi^2(2\tilde{p})$ but the first term depends on the unknown parameters $\mathbf{F}$ and $\boldsymbol{\beta}$. The latter is ignored (or set to zero) in the derivation of the AIC and the AICC. Moreover, observe that

$$\begin{aligned}\|\mathbf{y} - \mathbf{F}_{\tilde{p}}\hat{\boldsymbol{\beta}}_{\tilde{p}}\|^2 &= \boldsymbol{\epsilon}^H\boldsymbol{\epsilon} + (\hat{\boldsymbol{\beta}}_{\tilde{p}} - \boldsymbol{\beta}_{\tilde{p}})^H(\mathbf{F}_{\tilde{p}}^H\mathbf{F}_{\tilde{p}})(\hat{\boldsymbol{\beta}}_{\tilde{p}} - \boldsymbol{\beta}_{\tilde{p}}) \\ &\quad - 2\Re\{(\hat{\boldsymbol{\beta}}_{\tilde{p}} - \boldsymbol{\beta}_{\tilde{p}})^H\mathbf{F}_{\tilde{p}}^H(\mathbf{F}'_{\tilde{p}}\boldsymbol{\beta}'_{\tilde{p}} + \boldsymbol{\epsilon})\} + 2\Re(\boldsymbol{\epsilon}^H\mathbf{F}'_{\tilde{p}}\boldsymbol{\beta}'_{\tilde{p}}).\end{aligned}$$

If there is a positive definite matrix $\mathbf{C}$ such that $n^{-1}\mathbf{F}_m^H\mathbf{F}_m \to \mathbf{C}$ as $n \to \infty$, then $n^{-1}\mathbf{F}_{\tilde{p}}^H\mathbf{F}_{\tilde{p}} \to \mathbf{C}_{\tilde{p}}$ and $n^{-1}\mathbf{F}_{\tilde{p}}^H\mathbf{F}'_{\tilde{p}} \to \mathbf{C}'_{\tilde{p}}$ for some $\mathbf{C}_{\tilde{p}}$ and $\mathbf{C}'_{\tilde{p}}$. Moreover, because $n^{-1}\mathbf{F}_{\tilde{p}}^H\boldsymbol{\epsilon} \stackrel{P}{\to} \mathbf{0}$, it follows that

$$\hat{\boldsymbol{\beta}}_{\tilde{p}} = \boldsymbol{\beta}_{\tilde{p}} + (\mathbf{F}_{\tilde{p}}^H\mathbf{F}_{\tilde{p}})^{-1}\mathbf{F}_{\tilde{p}}^H\mathbf{F}'_{\tilde{p}}\boldsymbol{\beta}'_{\tilde{p}} + (\mathbf{F}_{\tilde{p}}^H\mathbf{F}_{\tilde{p}})^{-1}\mathbf{F}_{\tilde{p}}^H\boldsymbol{\epsilon} \stackrel{P}{\to} \boldsymbol{\beta}_{\tilde{p}} + \boldsymbol{\delta}_{\tilde{p}},$$

where $\boldsymbol{\delta}_{\tilde{p}} := \mathbf{C}_{\tilde{p}}^{-1}\mathbf{C}'_{\tilde{p}}\boldsymbol{\beta}'_{\tilde{p}}$. In addition, we have $n^{-1}\boldsymbol{\epsilon}^H\boldsymbol{\epsilon} \stackrel{P}{\to} \sigma^2$ and $n^{-1}\boldsymbol{\epsilon}^H\mathbf{F}_{\tilde{p}} \stackrel{P}{\to} \mathbf{0}$. Combining these results leads to

$$\hat{\sigma}_{\tilde{p}}^2 = n^{-1}\|\mathbf{y} - \mathbf{F}_{\tilde{p}}\hat{\boldsymbol{\beta}}_{\tilde{p}}\|^2 \stackrel{P}{\to} \sigma^2 + \boldsymbol{\delta}_{\tilde{p}}^H\mathbf{C}_{\tilde{p}}\boldsymbol{\delta}_{\tilde{p}} - 2\Re(\boldsymbol{\delta}_{\tilde{p}}^H\mathbf{C}'_{\tilde{p}}\boldsymbol{\beta}'_{\tilde{p}}).$$

The last two terms in this expression depend on the unknown parameters $\mathbf{C}_{\tilde{p}}$, $\mathbf{C}'_{\tilde{p}}$, and $\boldsymbol{\beta}'_{\tilde{p}}$. The derivation of the AIC ignores these terms. The derivation of the AICC requires the assumption $\boldsymbol{\beta}'_{\tilde{p}} = \mathbf{0}$. □

Proof of Theorem 5.8. Consider the PEC first. The objective is to show that for any $\tilde{p} \neq p$, $\mathrm{PEC}(\tilde{p}) - \mathrm{PEC}(p) > 0$ with probability tending to unity as $n \to \infty$. Toward that end, observe that

$$\mathrm{PEC}(\tilde{p}) - \mathrm{PEC}(p) = \mathrm{SSE}(\tilde{p}) - \mathrm{SSE}(p) + d(n, \tilde{p})\tilde{\sigma}_m^2, \tag{5.6.10}$$

where $d(n, \tilde{p}) := c(n, \tilde{p}) - c(n, p)$. Hence, we need to evaluate $\mathrm{SSE}(\tilde{p}) - \mathrm{SSE}(p)$. Because $\mathrm{SSE}(p)$ is the error sum of squares for the best p-term model, it follows that $\mathrm{SSE}(p) \leq \mathrm{SSE}$, where SSE is given by (5.1.4). Therefore, we have

$$\mathrm{SSE}(\tilde{p}) - \mathrm{SSE}(p) \geq \mathrm{SSE}(\tilde{p}) - \mathrm{SSE}. \tag{5.6.11}$$

It suffices to consider $\mathrm{SSE}(\tilde{p}) - \mathrm{SSE}$.

Let $\mathbf{P} := \mathbf{F}(\mathbf{F}^H\mathbf{F})^{-1}\mathbf{F}^H$ denote the projection matrix onto the column space of $\mathbf{F}$. Because $\mathbf{x} = \mathbf{F}\boldsymbol{\beta}$, it follows that $\mathbf{P}\mathbf{x} = \mathbf{x}$ or $(\mathbf{I}-\mathbf{P})\mathbf{x} = \mathbf{0}$. Therefore,

$$\mathrm{SSE} = \mathbf{y}^H(\mathbf{I}-\mathbf{P})\mathbf{y} = \boldsymbol{\epsilon}^H(\mathbf{I}-\mathbf{P})\boldsymbol{\epsilon}.$$

Moreover, because $(\mathbf{F}^H\mathbf{F})^{-1} = \mathcal{O}(n^{-1})$ and $\mathbf{F}^H\boldsymbol{\epsilon} = \mathcal{O}_P(a_n)$, we can write $\boldsymbol{\epsilon}^H\mathbf{P}\boldsymbol{\epsilon} = \mathcal{O}_P(n^{-1}a_n^2)$. Combining these expressions yields

$$\mathrm{SSE} = \boldsymbol{\epsilon}^H\boldsymbol{\epsilon} + \mathcal{O}_P(n^{-1}a_n^2). \tag{5.6.12}$$

For any $\tilde{p}$-subset $\Omega' \subset \Omega$, let $\mathbf{F}_{\tilde{p}}$ denote the design matrix corresponding to the trigonometric regressors $\mathbf{f}(\lambda_j)$ for $\lambda_j \in \Omega'$ ($\mathbf{F}_0 := \mathbf{0}$). Let $\mathbf{P}_{\tilde{p}} := \mathbf{F}_{\tilde{p}}(\mathbf{F}_{\tilde{p}}^H\mathbf{F}_{\tilde{p}})^{-1}\mathbf{F}_{\tilde{p}}^H$ ($\mathbf{P}_0 := \mathbf{0}$) be the projection matrix onto the column space of $\mathbf{F}_{\tilde{p}}$. Then, the corresponding error sum of squares can be expressed as

$$\mathrm{SSE}_{\tilde{p}} := \|\mathbf{y} - \mathbf{F}_{\tilde{p}}(\mathbf{F}_{\tilde{p}}^H\mathbf{F}_{\tilde{p}})^{-1}\mathbf{F}_{\tilde{p}}^H\mathbf{y}\|^2 = \mathbf{y}^H(\mathbf{I}-\mathbf{P}_{\tilde{p}})\mathbf{y} = \mathbf{y}^H\mathbf{P}_{\tilde{p}}^{\perp}\mathbf{y},$$

where $\mathbf{P}_{\tilde{p}}^{\perp} := \mathbf{I} - \mathbf{P}_{\tilde{p}}$ is the projection matrix onto the orthogonal complement of the column space of $\mathbf{F}_{\tilde{p}}$. Because $\mathbf{y} = \mathbf{x} + \boldsymbol{\epsilon}$ and $\mathbf{x} = \mathbf{F}\boldsymbol{\beta}$, we obtain

$$\mathrm{SSE}_{\tilde{p}} = \mathbf{x}^H\mathbf{P}_{\tilde{p}}^{\perp}\mathbf{x} + 2\Re(\mathbf{x}^H\mathbf{P}_{\tilde{p}}^{\perp}\boldsymbol{\epsilon}) + \boldsymbol{\epsilon}^H(\mathbf{I}-\mathbf{P}_{\tilde{p}})\boldsymbol{\epsilon}. \tag{5.6.13}$$

Observe that $(\mathbf{F}_{\tilde{p}}^H\mathbf{F}_{\tilde{p}})^{-1} = \mathcal{O}(n^{-1})$ and $\mathbf{F}_{\tilde{p}}^H\boldsymbol{\epsilon} = \mathcal{O}_P(a_n)$ uniformly for all $\tilde{p}$-subsets of Ω. Therefore, we can write $\boldsymbol{\epsilon}^H\mathbf{P}_{\tilde{p}}\boldsymbol{\epsilon} = \mathcal{O}_P(n^{-1}a_n^2)$ and

$$\mathrm{SSE}_{\tilde{p}} = \mathbf{x}^H\mathbf{P}_{\tilde{p}}^{\perp}\mathbf{x} + 2\Re(\mathbf{x}^H\mathbf{P}_{\tilde{p}}^{\perp}\boldsymbol{\epsilon}) + \boldsymbol{\epsilon}^H\boldsymbol{\epsilon} + \mathcal{O}_P(n^{-1}a_n^2) \tag{5.6.14}$$

uniformly for all $\tilde{p}$-subsets of Ω. In the following, let us evaluate (5.6.14) by considering separately the case $\tilde{p} < p$ and the case $\tilde{p} > p$.

For any given $\tilde{p} < p$, the first goal is to show that there exists a strictly positive constant $\kappa_{\tilde{p}} > 0$, which depends solely on $\tilde{p}$, such that

$$\mathbf{x}^H\mathbf{P}_{\tilde{p}}^{\perp}\mathbf{x} \ge n\{\kappa_{\tilde{p}} - \mathcal{O}(1)\}. \tag{5.6.15}$$

To prove this assertion, let $\tilde{\mathbf{F}}$ denote the matrix formed by the common columns in $\mathbf{F}$ and $\mathbf{F}_{\tilde{p}}$, corresponding to the signal frequencies selected by the best $\tilde{p}$-term model; let $\tilde{\boldsymbol{\beta}}$ denote the vector formed by the corresponding elements in $\boldsymbol{\beta}$. If no such columns exist, let $\tilde{\mathbf{F}} := \mathbf{0}$ and $\tilde{\boldsymbol{\beta}} := \mathbf{0}$. Moreover, let $\bar{\mathbf{F}}$ denote the matrix formed by the remaining columns of $\mathbf{F}$, which always exist because $\tilde{p} < p$, and let $\bar{\boldsymbol{\beta}}$ denote the vector formed by the corresponding elements in $\boldsymbol{\beta}$. Then, we can write $\mathbf{x} = \tilde{\mathbf{x}} + \bar{\mathbf{x}}$, where $\tilde{\mathbf{x}} := \tilde{\mathbf{F}}\tilde{\boldsymbol{\beta}}$ and $\bar{\mathbf{x}} := \bar{\mathbf{F}}\bar{\boldsymbol{\beta}} \neq \mathbf{0}$. Because $\mathbf{P}_{\tilde{p}}^{\perp}\tilde{\mathbf{x}} = \mathbf{0}$, we obtain

$$\mathbf{x}^H\mathbf{P}_{\tilde{p}}^{\perp}\mathbf{x} = \bar{\mathbf{x}}^H\mathbf{P}_{\tilde{p}}^{\perp}\bar{\mathbf{x}} = \bar{\boldsymbol{\beta}}^H(\bar{\mathbf{F}}^H\mathbf{P}_{\tilde{p}}^{\perp}\bar{\mathbf{F}})\bar{\boldsymbol{\beta}}. \tag{5.6.16}$$

Observe that the columns of $\mathbf{F}_{\tilde{p}}$ do not overlap with the columns of $\bar{\mathbf{F}}$. Therefore, the matrix $\mathbf{G} := [\mathbf{F}_{\tilde{p}}, \bar{\mathbf{F}}]$ comprises distinct columns of $\mathbf{F}_m$. Under the assumption that $n^{-1}\mathbf{F}_m^H\mathbf{F}_m \to \mathbf{C}_m > 0$, there exists a positive definition matrix $\mathbf{D}$ such that $n^{-1}\mathbf{G}^H\mathbf{G} \to \mathbf{D}$ and hence $n(\mathbf{G}^H\mathbf{G})^{-1} \to \mathbf{D}^{-1}$. By Lemma 12.4.1,

$$(\mathbf{G}^H\mathbf{G})^{-1} = \begin{bmatrix} \mathbf{F}_{\tilde{p}}^H\mathbf{F}_{\tilde{p}} & \mathbf{F}_{\tilde{p}}^H\bar{\mathbf{F}} \\ \bar{\mathbf{F}}^H\mathbf{F}_{\tilde{p}} & \bar{\mathbf{F}}^H\bar{\mathbf{F}} \end{bmatrix}^{-1} = \begin{bmatrix} (\bar{\mathbf{F}}^H\mathbf{P}_{\tilde{p}}^{\perp}\bar{\mathbf{F}})^{-1} & \star \\ \star & \star \end{bmatrix}.$$

Therefore, there exists a positive definite matrix $\bar{\mathbf{D}}$ such that $n^{-1}\bar{\mathbf{F}}^H\mathbf{P}_{\tilde{p}}^{\perp}\bar{\mathbf{F}} \to \bar{\mathbf{D}}$. Combining this result with (5.6.16) and the boundedness of $\bar{\boldsymbol{\beta}}$ leads to

$$\mathbf{x}^H\mathbf{P}_{\tilde{p}}^{\perp}\mathbf{x} = n\{\bar{\boldsymbol{\beta}}^H\bar{\mathbf{D}}\bar{\boldsymbol{\beta}} + \mathcal{o}(1)\}. \tag{5.6.17}$$

Let $\kappa_{\tilde{p}} := \min\{\bar{\boldsymbol{\beta}}^H\bar{\mathbf{D}}\bar{\boldsymbol{\beta}}\}$, where the minimum is taken over all possible $\tilde{p}$-subsets of Ω. Because $\tilde{p} < p$, the constant $\kappa_{\tilde{p}}$ is strictly positive. The assertion (5.6.15) follows immediately.

Moreover, because $(\mathbf{F}_{\tilde{p}}^H\mathbf{F}_{\tilde{p}})^{-1} = \mathcal{O}(n^{-1})$, $\mathbf{F}^H\mathbf{F}_{\tilde{p}} = \mathcal{O}(n)$, $\mathbf{F}_{\tilde{p}}^H\boldsymbol{\epsilon} = \mathcal{O}_P(a_n)$, and $\mathbf{x}^H\boldsymbol{\epsilon} = \mathcal{o}_P(n)$, we obtain

$$\mathbf{x}^H\mathbf{P}_{\tilde{p}}^{\perp}\boldsymbol{\epsilon} = \mathbf{x}^H\boldsymbol{\epsilon} - \boldsymbol{\beta}^H\mathbf{F}^H\mathbf{F}_{\tilde{p}}(\mathbf{F}_{\tilde{p}}^H\mathbf{F}_{\tilde{p}})^{-1}\mathbf{F}_{\tilde{p}}^H\boldsymbol{\epsilon} = \mathcal{o}_P(n) + \mathcal{O}_P(a_n). \tag{5.6.18}$$

Inserting this expression and (5.6.15) into (5.6.14) gives

$$\mathrm{SSE}_{\tilde{p}} \geq n\{\kappa_{\tilde{p}} - \mathcal{o}_P(1) - \mathcal{O}_P(n^{-1}a_n)\} + \boldsymbol{\epsilon}^H\boldsymbol{\epsilon}, \tag{5.6.19}$$

which holds uniformly for all $\tilde{p}$-subsets of Ω such that $\tilde{p} < p$. Because (5.6.19) holds for $\mathrm{SSE}(\tilde{p})$, combining this expression with (5.6.11) and (5.6.12) leads to

$$\mathrm{SSE}(\tilde{p}) - \mathrm{SSE}(p) \geq n\{\kappa_{\tilde{p}} - \mathcal{o}_P(1) - \mathcal{O}_P(n^{-1}a_n)\}. \tag{5.6.20}$$

This expression is valid for all $\tilde{p} < p$.

Now let us turn to (5.6.10). Because $c(n,\tilde{p})$ is monotone increasing in $\tilde{p}$ for each fixed n and $c(n,0) = 0$, we have $0 < -d(n,\tilde{p}) \leq c(n,p)$ for all $\tilde{p} < p$. Moreover, by Remark 5.2, we obtain $\tilde{\sigma}_m^2 \xrightarrow{P} \sigma^2$ as $n \to \infty$. Therefore, $n^{-1}c(n,p)\tilde{\sigma}_m^2 = \mathcal{O}_P(n^{-1}c(n,p))$. Combining these results with (5.6.20) and (5.6.10) yields

$$\mathrm{PEC}(\tilde{p}) - \mathrm{PEC}(p) \geq n\{\kappa - \mathcal{o}_P(1) - \mathcal{O}_P(n^{-1}a_n) - \mathcal{O}_P(n^{-1}c(n,p))\},$$

where $\kappa := \min\{\kappa_{\tilde{p}} : \tilde{p} < p\} > 0$. Because $n^{-1}a_n \to 0$ and $n^{-1}c(n,p) \to 0$ as $n \to \infty$, it follows that $P(\mathrm{PEC}(\tilde{p}) - \mathrm{PEC}(p) > 0) \to 1$ for all $\tilde{p} < p$. Hence, $P(\hat{p} < p) \to 0$.

Next, consider the case $\tilde{p} > p$. Let $\mathbf{F}_{\tilde{p}}$ be formed by the best $\tilde{p}$-subset of Ω. By (5.6.10)–(5.6.14) and the monotonicity of $c(n,\tilde{p})$, we obtain

$$\mathrm{PEC}(\tilde{p}) - \mathrm{PEC}(p) \geq \mathbf{x}^H\mathbf{P}_{\tilde{p}}^{\perp}\mathbf{x} + 2\Re(\mathbf{x}^H\mathbf{P}_{\tilde{p}}^{\perp}\boldsymbol{\epsilon}) - \mathcal{O}_P(n^{-1}a_n^2) + d(n,p+1)\tilde{\sigma}_m^2.$$

If $\mathbf{F}_{\tilde{p}}$ contains all the columns of $\mathbf{F}$, then $\mathbf{P}_{\tilde{p}}^{\perp}\mathbf{x} = \mathbf{0}$. Combining this result with the fact that $\tilde{\sigma}_m^2 = \sigma^2 + \mathcal{o}_P(1)$ leads to

$$\mathrm{PEC}(\tilde{p}) - \mathrm{PEC}(p) \geq d(n,p+1)\sigma^2 - \mathcal{O}_P(n^{-1}a_n^2) - \mathcal{o}_P(d(n,p+1)).$$

If the probability that $\mathbf{F}_{\tilde{p}}$ does not contain all the columns of $\mathbf{F}$ approaches zero, then under the assumption that $nd(n,p+1)/a_n^2 \to \infty$ as $n \to \infty$, we would obtain $P(\text{PEC}(\tilde{p}) - \text{PEC}(p) > 0) \to 1$ for all $\tilde{p} > p$ and hence $P(\hat{p} > p) \to 0$. Observe that $\mathbf{F}_{\tilde{p}}$ not containing all the columns of $\mathbf{F}$ means $\mathbf{x} = \mathbf{x}_{\tilde{p}} + \mathbf{r}$ and $\mathbf{r} \neq \mathbf{0}$. In this case, it follows from (5.6.13), (5.6.17), and (5.6.18) that

$$\text{SSE}_{\tilde{p}} \geq n\{\kappa' - \mathcal{O}_P(1) - \mathcal{O}_P(n^{-1}a_n)\} + \boldsymbol{\epsilon}^H \mathbf{P}_{\tilde{p}}^{\perp} \boldsymbol{\epsilon}, \tag{5.6.21}$$

where $\kappa' := \min\{\bar{\boldsymbol{\beta}}^H \bar{\mathbf{D}} \bar{\boldsymbol{\beta}}\} > 0$, with the minimum taking over all $\tilde{p}$-subset of Ω that does not contain at least one signal frequency. Note that $\boldsymbol{\epsilon}^H \mathbf{P}_{\tilde{p}}^{\perp} \boldsymbol{\epsilon}$ is equal to the error sum of squares when $\mathbf{F}_{\tilde{p}}$ does contain all the columns of $\mathbf{F}$. Therefore, the inequality (5.6.21) implies that the probability of $\mathbf{F}_{\tilde{p}}$ corresponding to the best $\tilde{p}$-subset but not containing all the columns of $\mathbf{F}$ approaches zero as $n \to \infty$. The proof for the PEC criterion is complete.

A similar argument, together with the assumption that $n^{-1}\boldsymbol{\epsilon}^H \boldsymbol{\epsilon} = \sigma^2 + \mathcal{O}_P(1)$, can be used to prove the assertion for the GIC. First, observe that

$$\text{GIC}(\tilde{p}) - \text{GIC}(p) \geq 2n \log(\text{SSE}(\tilde{p})/\text{SSE}) + d(n,\tilde{p}). \tag{5.6.22}$$

According to (5.6.12), we can write

$$\text{SSE} = n\{n^{-1}\boldsymbol{\epsilon}^H \boldsymbol{\epsilon} + \mathcal{O}_P(1)\} = n\{\sigma^2 + \mathcal{O}_P(1)\}. \tag{5.6.23}$$

For $\tilde{p} < p$, it follows from (5.6.19) that

$$\text{SSE}(\tilde{p}) \geq n\{\kappa + n^{-1}\boldsymbol{\epsilon}^H \boldsymbol{\epsilon} - \mathcal{O}_P(1)\} = n\{\kappa + \sigma^2 - \mathcal{O}_P(1)\}.$$

Combining this expression with (5.6.22), (5.6.23), and the fact that $|d(n,\tilde{p})| \leq c(n,p)$ for all $\tilde{p} < p$ leads to

$$\text{GIC}(\tilde{p}) - \text{GIC}(p) \geq 2n\left\{\log \frac{\kappa + \sigma^2 - \mathcal{O}_P(1)}{\sigma^2 + \mathcal{O}_P(1)} - \tfrac{1}{2}n^{-1}c(n,p)\right\}.$$

Because $n^{-1}c(n,p) \to 0$ as $n \to \infty$, we obtain $P(\text{GIC}(\tilde{p}) - \text{GIC}(p) > 0) \to 1$.

For $\tilde{p} > p$, let $\mathbf{F}_{\tilde{p}}$ denote the design matrix corresponding to the best $\tilde{p}$-subset of Ω. If $\mathbf{F}_{\tilde{p}}$ contains all the columns of $\mathbf{F}$, then $\mathbf{P}_{\tilde{p}}^{\perp}\mathbf{x} = \mathbf{0}$, in which case, it follows from (5.6.14) that

$$\text{SSE}(\tilde{p}) = \boldsymbol{\epsilon}^H \boldsymbol{\epsilon} + \mathcal{O}_P(n^{-1}a_n^2) = n\{\sigma^2 + \mathcal{O}_P(1)\}.$$

This, combined with (5.6.22), (5.6.23), and the monotonicity of $d(n,\tilde{p})$, yields

$$\text{GIC}(\tilde{p}) - \text{GIC}(p) \geq 2n\left\{\log \frac{\sigma^2 - \mathcal{O}_P(1)}{\sigma^2 + \mathcal{O}_P(1)} + \tfrac{1}{2}n^{-1}d(n,p+1)\right\}.$$

Because $n^{-1}d(n,p+1) \to \infty$ and because the probability of $\mathbf{F}_{\tilde{p}}$ being the design matrix corresponding to the best $\tilde{p}$-subset but not containing all the columns of $\mathbf{F}$ approaches zero, we obtain $P(\text{GIC}(\tilde{p}) - \text{GIC}(p) > 0) \to 1$. □

Proof of Theorem 5.9. For any $\boldsymbol{\delta} \in \mathbb{C}^p$, define

$$Z_n(\boldsymbol{\delta}) := \sum_{t=1}^{n} \sum_{j=1}^{2} \{|U_{tj} - v_{tj}(\boldsymbol{\delta})| - |U_{tj}|\},$$

where

$$U_{t1} := \Re(\epsilon_t), \quad v_{t1}(\boldsymbol{\delta}) := n^{-1/2}\Re(\mathbf{x}_t^H \boldsymbol{\delta}),$$
$$U_{t2} := \Im(\epsilon_t), \quad v_{t2}(\boldsymbol{\delta}) := n^{-1/2}\Im(\mathbf{x}_t^H \boldsymbol{\delta}).$$

The objective is to show that $Z_n(\boldsymbol{\delta})$ has a quadratic approximation

$$Z_n(\boldsymbol{\delta}) = -\tfrac{1}{2}(\boldsymbol{\delta}^H \boldsymbol{\zeta}_n + \boldsymbol{\zeta}_n^H \boldsymbol{\delta}) + \kappa^{-1}\boldsymbol{\delta}^H \mathbf{C}\boldsymbol{\delta} + R_n(\boldsymbol{\delta}), \tag{5.6.24}$$

where $\boldsymbol{\zeta}_n \overset{A}{\sim} \mathrm{N}_c(\mathbf{0}, 2\mathbf{C})$ and $R_n(\boldsymbol{\delta}) = \mathcal{O}_P(1)$ for any fixed $\boldsymbol{\delta} \in \mathbb{C}^p$. If this can be established, then, owing to the fact that $Z_n(\boldsymbol{\delta})$ is a convex function of $\boldsymbol{\delta}$, the convexity lemma in [293] (see also Lemma 12.5.10) ensures that $R_n(\boldsymbol{\delta}) = \mathcal{O}_P(1)$ uniformly in $\boldsymbol{\delta} \in K$ for any given compact set $K \subset \mathbb{C}^p$. The quadratic function in (5.6.24) has a unique minimum at $\tilde{\boldsymbol{\delta}}_n := \frac{1}{2}\kappa \mathbf{C}^{-1}\boldsymbol{\zeta}_n$. Because $\boldsymbol{\zeta}_n \overset{A}{\sim} \mathrm{N}_c(\mathbf{0}, 2\mathbf{C})$, it follows that

$$\tilde{\boldsymbol{\delta}}_n \xrightarrow{D} \mathrm{N}_c(\mathbf{0}, \eta^2 \mathbf{C}^{-1}),$$

where $\eta^2 := \kappa^2/2$. Let the minimizer of $Z_n(\boldsymbol{\delta})$ be denoted by $\hat{\boldsymbol{\delta}}_n$. Because $Z_n(\boldsymbol{\delta})$ can be reparameterized as a function of $\tilde{\boldsymbol{\beta}} := \boldsymbol{\beta} + \boldsymbol{\delta} n^{-1/2}$ such that

$$Z_n(\boldsymbol{\delta}) = \sum_{t=1}^{n} \{|\Re(y_t - \mathbf{x}_t^H \tilde{\boldsymbol{\beta}})| - |U_{t1}| + |\Im(y_t - \mathbf{x}_t^H \tilde{\boldsymbol{\beta}})| - |U_{t2}|\},$$

it is clear that $\hat{\boldsymbol{\beta}}$ in (5.5.1) minimizes $Z_n(\boldsymbol{\delta})$ with respect to $\tilde{\boldsymbol{\beta}}$, which implies that $\sqrt{n}(\hat{\boldsymbol{\beta}} - \boldsymbol{\beta}) = \hat{\boldsymbol{\delta}}_n$. Therefore, it suffices to show that $\hat{\boldsymbol{\delta}}_n - \tilde{\boldsymbol{\delta}}_n = \mathcal{O}_P(1)$.

Toward that end, we follow the lines of [293]. First, observe that

$$Z_n(\boldsymbol{\delta}) = Z_n(\tilde{\boldsymbol{\delta}}_n) + \kappa^{-1}(\boldsymbol{\delta} - \tilde{\boldsymbol{\delta}}_n)^H \mathbf{C}(\boldsymbol{\delta} - \tilde{\boldsymbol{\delta}}_n) + R_n(\boldsymbol{\delta}) - R_n(\tilde{\boldsymbol{\delta}}_n). \tag{5.6.25}$$

Because $\tilde{\boldsymbol{\delta}}_n = \mathcal{O}_P(1)$, it follows that for any $\varepsilon > 0$ there exists a constant $c > 0$ such that $P(\|\tilde{\boldsymbol{\delta}}_n\| > c) \le \varepsilon$ for large n. Therefore, for any constant $\mu > 0$, we have

$$\begin{aligned} P(\|\hat{\boldsymbol{\delta}}_n - \tilde{\boldsymbol{\delta}}_n\| > \mu) &\le P(\|\tilde{\boldsymbol{\delta}}_n\| > c) + P(\|\hat{\boldsymbol{\delta}}_n - \tilde{\boldsymbol{\delta}}_n\| > \mu, \|\tilde{\boldsymbol{\delta}}_n\| \le c) \\ &\le \varepsilon + P(\|\hat{\boldsymbol{\delta}}_n - \tilde{\boldsymbol{\delta}}_n\| > \mu, \|\tilde{\boldsymbol{\delta}}_n\| \le c). \end{aligned} \tag{5.6.26}$$

Moreover, for any $\boldsymbol{\delta}$ such that $\|\boldsymbol{\delta} - \tilde{\boldsymbol{\delta}}_n\| > \mu$, we can write $\boldsymbol{\delta} = \tilde{\boldsymbol{\delta}}_n + \rho\boldsymbol{\eta}$, where $\rho > \mu$ and $\|\boldsymbol{\eta}\| = 1$. Let $\bar{\boldsymbol{\delta}} := \tilde{\boldsymbol{\delta}}_n + \mu\boldsymbol{\eta}$ and $R_n := \max\{|R_n(\boldsymbol{\delta})| : \|\boldsymbol{\delta}\| \le c, \|\boldsymbol{\delta} - \tilde{\boldsymbol{\delta}}_n\| \le \mu\}$. Then, with $\|\tilde{\boldsymbol{\delta}}_n\| \le c$, the convexity of $Z_n(\boldsymbol{\delta})$, combined with (5.6.25), implies that

$$(\mu/\rho) Z_n(\boldsymbol{\delta}) + (1 - \mu/\rho) Z_n(\tilde{\boldsymbol{\delta}}_n) \ge Z_n(\bar{\boldsymbol{\delta}}) \ge Z_n(\tilde{\boldsymbol{\delta}}_n) + \kappa^{-1}\lambda\mu^2 - 2R_n,$$

where $\lambda > 0$ denotes the smallest eigenvalue of $\mathbf{C}$. By rearranging the foregoing inequality, we obtain

$$\inf\{Z_n(\boldsymbol{\delta}) : \|\boldsymbol{\delta} - \tilde{\boldsymbol{\delta}}_n\| > \mu\} \geq Z_n(\tilde{\boldsymbol{\delta}}_n) + (\rho/\mu)(\kappa^{-1}\lambda\mu^2 - 2R_n).$$

Because $R_n \xrightarrow{P} 0$ as $n \to \infty$, it follows that

$$\lim_{n\to\infty} P(\inf\{Z_n(\boldsymbol{\delta}) : \|\boldsymbol{\delta} - \tilde{\boldsymbol{\delta}}_n\| > \mu\} \leq Z_n(\tilde{\boldsymbol{\delta}}_n), \|\tilde{\boldsymbol{\delta}}_n\| \leq c) = 0,$$

which, in turn, implies that

$$P(\|\hat{\boldsymbol{\delta}}_n - \tilde{\boldsymbol{\delta}}_n\| > \mu, \|\tilde{\boldsymbol{\delta}}_n\| \leq c) \leq \varepsilon$$

for large n. Combining this result with (5.6.26) proves $\hat{\boldsymbol{\delta}}_n - \tilde{\boldsymbol{\delta}}_n \xrightarrow{P} 0$.

To prove (5.6.24), consider Knight's identity [193]

$$|u - v| - |u| = -v\,\mathrm{sgn}(u) + \int_0^v \phi(u,s)\,ds, \tag{5.6.27}$$

where $\phi(u,s) := 2\mathscr{I}(u \leq s) - 2\mathscr{I}(u \leq 0)$. With u and v substituted by U_{tj} and $v_{tj} := v_{tj}(\boldsymbol{\delta})$ $(j = 1,2)$, we can write $Z_n(\boldsymbol{\delta}) = Z_{n1} + Z_{n2}$, where

$$Z_{n1} := -\tfrac{1}{2}(\boldsymbol{\delta}^H\boldsymbol{\zeta}_n + \boldsymbol{\zeta}_n^H\boldsymbol{\delta}), \quad Z_{n2} := \sum_{t=1}^{n}(X_{t1} + X_{t2}),$$

with

$$\boldsymbol{\zeta}_n := n^{-1/2}\sum_{t=1}^{n}\mathbf{x}_t Y_t, \quad Y_t := \mathrm{sgn}(U_{t1}) + i\,\mathrm{sgn}(U_{t2}), \tag{5.6.28}$$

$$X_{tj} := \int_0^{v_{tj}} \phi(U_{tj}, s)\,ds. \tag{5.6.29}$$

Observe that the $\mathrm{sgn}(U_{tj})$ are i.i.d. binary-valued random variables with

$$P\{\mathrm{sgn}(U_{tj}) = 1\} = P\{\mathrm{sgn}(U_{tj}) = -1\} = 1/2,$$
$$E\{\mathrm{sgn}(U_{tj})\} = 0, \quad \mathrm{Var}\{\mathrm{sgn}(U_{tj})\} = 1.$$

This implies that $\{Y_t\} \sim \mathrm{IID}(0,2)$ and $E(Y_t^2) = 0$. Moreover, by assumption, we have $n^{-1}\sum \mathbf{x}_t\mathbf{x}_t^H = n^{-1}\mathbf{F}^H\mathbf{F} \to \mathbf{C}$. Hence, $\boldsymbol{\zeta}_n \xrightarrow{D} \mathrm{N}_c(\mathbf{0}, 2\mathbf{C})$ by Lemma 12.5.7.

To proceed with the proof, observe that

$$\begin{aligned} E(X_{tj}) &= 2\int_0^{v_{tj}} \{F_{tj}(s) - F_{tj}(0)\}\,ds \\ &= 2\int_0^{v_{tj}} \{\dot{F}_{tj}(0)s + \mathscr{O}(s^{d+1})\}\,ds \\ &= \kappa^{-1}v_{tj}^2 + \mathscr{O}(v_{tj}^{d+2}). \end{aligned}$$

Owing to the fact that $\mathbf{x}_t = \mathcal{O}(1)$, we can write $v_n := \max|v_{tj}| = \mathcal{O}(n^{-1/2})$, which, in turn, implies that $\sum \mathcal{O}(v_{tj}^{d+2}) = \mathcal{O}(n^{-d/2})$. Therefore,

$$\begin{aligned} E(Z_{n2}) &= \sum_{t=1}^{n} \{E(X_{t1}) + E(X_{t2})\} \\ &= \kappa^{-1} n^{-1} \sum_{t=1}^{n} \boldsymbol{\delta}^H \mathbf{x}_t \mathbf{x}_t^H \boldsymbol{\delta} + \mathcal{O}(n^{-d/2}) \\ &= \kappa^{-1} \boldsymbol{\delta}^H \mathbf{C} \boldsymbol{\delta} + \mathcal{o}(1). \end{aligned} \tag{5.6.30}$$

Furthermore, observe that if $v \geq 0$, then $\phi(u,s) \in [0,2]$ for all $s \in [0,v]$; if $v < 0$, then $\int_0^v \phi(u,s)\,ds = \int_0^{|v|} \{-\phi(u,-s)\}\,ds$ and $-\phi(u,-s) \in [0,2]$ for all $s \in [0,|v|]$. Therefore, $0 \leq \int_0^v \phi(u,s)\,ds \leq 2|v|$ for any u and v. This implies that

$$\mathrm{Var}(X_{tj}) \leq E(X_{tj}^2) \leq 2|v_{tj}| E(X_{tj}) \leq 2v_n E(X_{tj}).$$

Because U_{t1} and U_{t2} are independent, we obtain

$$\mathrm{Var}(Z_{n2}) = \sum_{t=1}^{n} \{\mathrm{Var}(X_{t1}) + \mathrm{Var}(X_{t2})\} \leq 2v_n E(Z_{n2}) = \mathcal{O}(n^{-1/2}).$$

Combining this result with (5.6.30) leads to $Z_{n2} = \kappa^{-1} \boldsymbol{\delta}^H \mathbf{C} \boldsymbol{\delta} + \mathcal{o}_P(1)$. □

Proof of Theorem 5.10. The proof follows the lines in [218, p. 359]. Because ρ is scale-invariant, it can be assumed without loss of generality that $\kappa = 1$. Then, the problem becomes that of minimizing

$$\rho = 2\sigma^2 = 4 \int x^2 p(x)\,dx$$

subject to $0 \leq p(x) \leq 1$, $p(0) = 1$, and $\int p(x)\,dx = 1$. This problem can be solved by first minimizing

$$\int (x^2 - \lambda) p(x)\,dx \tag{5.6.31}$$

for fixed $\lambda > 0$ without the third constraint and then determining λ so that the third constraint is satisfied. Observe that (5.6.31) is minimized by making $p(x)$ as large as possible for $x^2 < \lambda$ and as small as possible for $x^2 > \lambda$. Therefore, under the remaining constraints, the minimizer is given by

$$p(x) = \mathcal{I}(x^2 < \lambda).$$

Imposing the constraint $\int p(x)\,dx = 1$ leads to $\lambda = 1/4$. This gives a uniform distribution with $\int x^2 p(x)\,dx = 1/12$ and hence $\rho = 1/3$. □

Proof of Theorem 5.11. Similar to the proof of Theorem 5.9, the goal is to establish the quadratic approximation

$$Z_n(\boldsymbol{\delta}) = -\tfrac{1}{2}(\boldsymbol{\delta}^H\boldsymbol{\zeta}_n + \boldsymbol{\zeta}_n^H\boldsymbol{\delta}) + \kappa^{-1}\boldsymbol{\delta}^H\boldsymbol{\delta} + R_n(\boldsymbol{\delta}),$$

where $\boldsymbol{\zeta}_n \stackrel{A}{\sim} \mathrm{N}_c(\mathbf{0}, 2\mathbf{D})$, $\mathbf{D} := \mathrm{diag}\{h(\omega_1),\dots,h(\omega_p)\} = (1/\eta^2)\mathbf{S}$, and $R_n(\boldsymbol{\delta}) = \mathcal{O}_P(1)$ for any fixed $\boldsymbol{\delta} \in \mathbb{C}^p$. Before we prove this assertion, observe that $E\{\mathrm{sgn}(U_{tj})\} = 0$ and

$$E\{\mathrm{sgn}(U_{tj})\,\mathrm{sgn}(U_{sj})\} = 4P(U_{tj} < 0, U_{sj} < 0) - 1 = 1 - 2\gamma(t-s).$$

Therefore, it follows from (12.5.6) that $|1-2\gamma(u)| \le 4\alpha(|u|) = \mathcal{O}(|u|^{-r})$. This implies that $h(\omega) := \sum(1-2\gamma(u))\cos(u\omega)$ is finite for all ω.

Given $\mathbf{a} := [a_1,\dots,a_p]^T \in \mathbb{R}^p$ and $\mathbf{b} := [b_1,\dots,b_p]^T \in \mathbb{R}^p$ such that $\|\mathbf{a}\|^2 + \|\mathbf{b}\|^2 > 0$, let $S_n := \mathbf{a}^T\Re(\boldsymbol{\zeta}_n) + \mathbf{b}^T\Im(\boldsymbol{\zeta}_n)$, where $\boldsymbol{\zeta}_n$ is defined by (5.6.28). The goal is to show that $S_n \stackrel{A}{\sim} \mathrm{N}(0, v^2)$, where $v^2 := \mathbf{a}^T\mathbf{D}\mathbf{a} + \mathbf{b}^T\mathbf{D}\mathbf{b}$. If this can be done, then we obtain $\boldsymbol{\zeta}_n \stackrel{A}{\sim} \mathrm{N}_c(\mathbf{0}, 2\mathbf{D})$ by Proposition 2.16(d). Note that $S_n = S_{n1} + S_{n2}$, where

$$S_{nj} := \sum_{t=1}^{n} c_{jnt}\,\mathrm{sgn}(U_{tj}) \qquad (j = 1, 2)$$

with $c_{1nt} := n^{-1/2}(\mathbf{a}^T\Re(\mathbf{x}_t) + \mathbf{b}^T\Im(\mathbf{x}_t))$ and $c_{2nt} := n^{-1/2}(-\mathbf{a}^T\Im(\mathbf{x}_t) + \mathbf{b}^T\Re(\mathbf{x}_t))$. Note also that $E\{\mathrm{sgn}(U_{tj})\} = 0$ and $E\{\mathrm{sgn}(U_{tj})\,\mathrm{sgn}(U_{sj})\} = 1 - 2\gamma(t-s)$. Therefore, we obtain $E(S_{nj}) = 0$ and

$$\begin{aligned}\mathrm{Var}(S_{nj}) &= \sum_{t,s=1}^{n} c_{jnt}c_{jns}(1-2\gamma(t-s)) \\ &= \sum_{|u|<n} (1-2\gamma(u)) \sum_{t\in T_n(u)} c_{jn,t+u}c_{jnt},\end{aligned}$$

where $T_n(u) := \{t : \max(1, 1-u) \le t \le \min(n, n-u)\}$. Under the frequency separation condition, straightforward calculation, together with Lemma 12.1.5, shows that for fixed u,

$$\begin{aligned}\sum_{t\in T_n(u)} c_{1n,t+u}c_{1nt} &= \tfrac{1}{2}c_1(u) + \mathcal{O}(n^{-1}\Delta^{-1}), \\ \sum_{t\in T_n(u)} c_{2n,t+u}c_{2nt} &= \tfrac{1}{2}c_2(u) + \mathcal{O}(n^{-1}\Delta^{-1}),\end{aligned}$$

where

$$\begin{aligned}c_1(u) &:= \sum_{k=1}^{p}\{(a_k^2 + b_k^2)\cos(\omega_k u) - 2a_kb_k\sin(\omega_k u)\}, \\ c_2(u) &:= \sum_{k=1}^{p}\{(a_k^2 + b_k^2)\cos(\omega_k u) + 2a_kb_k\sin(\omega_k u)\}.\end{aligned}$$

Combining this result with the absolute summability of $\{1-2\gamma(u)\}$ leads to

$$\mathrm{Var}(S_{nj})/s_j^2 \to 1/2 \quad \text{as } n \to \infty,$$

where $s_j^2 := \sum(1-2\gamma(u))c_j(u)$ $(j=1,2)$. By Lemma 12.5.8 and Remark 12.5.5,

$$S_{nj} \stackrel{A}{\sim} \mathrm{N}(0, \tfrac{1}{2}s_j^2) \quad (j=1,2).$$

Note that $s_1^2 + s_2^2 = 2\sum(a_k^2 + b_k^2)h(\omega_k) = 2v^2$. Combining these results with the independence between S_{1n} and S_{2n} yields $S_n \stackrel{A}{\sim} \mathrm{N}(0, \frac{1}{2}(s_1^2+s_2^2)) = \mathrm{N}(0, v^2)$.

As in the proof of Theorem 5.9, we can write $Z_n(\boldsymbol{\delta}) = Z_{n1} + Z_{n2}$ with $Z_{n1} := -\frac{1}{2}(\boldsymbol{\delta}^H\boldsymbol{\zeta}_n + \boldsymbol{\zeta}_n^H\boldsymbol{\delta})$ and $Z_{n2} := \sum X_{t1} + \sum X_{t2} := Z'_{n2} + Z''_{n2}$, where $E(Z_{n2}) = \kappa^{-1}\boldsymbol{\delta}^H\boldsymbol{\delta} + \mathcal{O}(1)$ under the frequency separation condition. By (5.6.29), Z'_{n2} and Z''_{n2} are independent and hence $\mathrm{Var}(Z_{n2}) = \mathrm{Var}(Z'_{n2}) + \mathrm{Var}(Z''_{n2})$. Moreover, because

$$\mathrm{Var}(X_{tj}) \le 2v_n E(X_{tj}),$$

we have

$$|\mathrm{Cov}(X_{tj}, X_{sj})| \le \sqrt{\mathrm{Var}(X_{tj})\mathrm{Var}(X_{sj})} \le 2v_n\sqrt{E(X_{tj})E(X_{sj})}. \tag{5.6.32}$$

Because $0 \le X_{tj} \le 2|v_{tj}| \le 2v_n$, an application of (12.5.6) gives

$$|\mathrm{Cov}(X_{tj}, X_{sj})| \le 16v_n^2\alpha(|t-s|). \tag{5.6.33}$$

For any integer $m \ge 1$, we can write $\mathrm{Var}(Z'_{n2}) = (\sum' + \sum'')\mathrm{Cov}(X_{t1}, X_{t2})$, where $\sum'$ denotes the sum over $\{(t,s) : |t-s| \le m, 1 \le t, s \le n\}$ and $\sum''$ denotes the sum over $\{(t,s) : |t-s| > m, 1 \le t, s \le n\}$. Using (5.6.32), we can bound the first sum by $2v_n\sum'\sqrt{E(X_{t1})E(X_{s1})}$, which, by the Cauchy-Schwartz inequality, is less than or equal to $2v_n\sum' E(X_{t1})$. Note that $\sum' E(X_{t1}) \le (2m+1)\sum_{t=1}^n E(X_{t1}) \le 2nv_n(2m+1)$. Combining these results with $v_n = \mathcal{O}(n^{-1/2})$ gives $\sum'\mathrm{Cov}(X_{t1}, X_{t2}) = \mathcal{O}(mn^{-1/2})$. Moreover, using (5.6.33), we can bound $\sum''\mathrm{Cov}(X_{t1}, X_{t2})$ by

$$\sum_{m<|u|<n} (n-|u|)(16v_n^2\alpha(|u|)) \le 32nv_n^2 \sum_{u=m+1}^{\infty} \alpha(u).$$

The latter can be expressed as $\mathcal{O}(\sum_{u>m}\alpha(u))$ because $nv_n^2 = \mathcal{O}(1)$. Taking m such that $m \to \infty$ and $m = \mathcal{o}(n^{1/2})$ as $n \to \infty$ ensures that $\sum_{u>m}\alpha(u) \to 0$ and $mn^{-1/2} \to 0$, which, in turn, leads to $\mathrm{Var}(Z'_{n2}) = \mathcal{o}(1)$. Similarly, we can show that $\mathrm{Var}(Z''_{n2}) = \mathcal{o}(1)$. Combining these results yields $\mathrm{Var}(Z_{n2}) = \mathcal{o}(1)$ and hence $Z_{n2} = E(Z_{n2}) + \mathcal{o}_P(1) = \kappa^{-1}\boldsymbol{\delta}^H\boldsymbol{\delta} + \mathcal{o}_P(1)$. □

Notes

The LS and LAD methods have a long history [93] [141] [363]. Although it had been used earlier, the LS method was first published in 1805 by the French mathematician Adrien-Marie Legendre (1752–1833) as an appendix entitled "Sur la Méthode des moindres quarrés" in *Nouvelles Méthodes pour la détermination des orbites des comètes* (Paris: Firmin Didot). The German mathematician Carl Friedrich Gauss (1777–1855) explained the LS method in terms of the normal (Gaussian) law of error in his publication *Theoria Motus Corporum Coelestium in Sectionibus Conicis Solem Ambientium* (Hamburg: Perthes and Besser, 1809) where he mentioned that he had made use of the technique since 1795.

The LAD method can be dated back to 1757 by the publication of the Croatian physicist and mathematician Roger Joseph Boscovich (1711–1787) entitled "De litteraria expeditione per pontificiam ditionem, et synopsis amplioris operis, ac habentur plura ejus ex exemplaria etiam sensorum impressa" in *Bononiensi Scientiarum et Artum Instituto Atque Academia Commentarii*, vol. 4, pp. 353–396. The French mathematician Pierre-Simon Laplace (1749–1827) further advanced the LAD method in his publications "Sur quelques points du systèeme du monde" in *Mémoires de l'Acadèmie royale des Sciences de Paris, annèe 1789*, pp. 1–87, and "Application du calcul des probabilités aux operations géodésiques" as the second supplement in 1818 to his monumental work *Théorie Analytique des Probabilités* (Courcier, Paris: 1812). Since then, the LAD method had been overshadowed by the LS method until the computational issues were resolved in our time with the advance of linear programming techniques such as the simplex algorithm [23] [24]. Bassett and Koenker [27] are credited for the development of an asymptotic theory for the LAD linear regression method. Further refinement and extension of the asymptotic theory and the LAD method can be found in [15], [193], [209], [224], [293], [294], and [427].

Chapter 6

Fourier Analysis Approach

The Fourier transform is a powerful tool for signal processing and time series analysis. Solidly grounded in physical science, the basic idea of Fourier analysis is to represent an oscillatory time series as a sum of sinusoidal functions with different frequencies and examine the strength of these components. Since the rediscovery of the fast Fourier transform (FFT) algorithm [148], Fourier analysis has become even more widely used in applications. Fourier analysis is well suited for time series consisting of sinusoids in noise. The tool of choice for analyzing such time series is the periodogram.

In this chapter, we examine some statistical properties of the periodogram for time series with mixed spectra and discuss the application of the periodogram to the estimation of sinusoidal parameters, especially the signal frequencies, and to the detection of hidden periodicities. More specifically, we derive the distribution as well as the mean, variance, and covariance of the periodogram ordinates under the sinusoid-plus-noise model. These results extend the classical theory developed for time series with continuous spectra. We discuss a number of techniques for frequency estimation based on the periodogram. We also discuss several periodogram-based tests for detection of hidden periodicities in white noise. As in the other chapters, we pay attention to both real- and complex-valued time series, although we use the latter to drive most of the exposition.

6.1 Periodogram Analysis

Consider the complex case where $\mathbf{y} := [y_1, \ldots, y_n]^T$ can be expressed as

$$\mathbf{y} = \mathbf{x} + \boldsymbol{\epsilon} = \sum_{k=1}^{p} \beta_k \mathbf{f}(\omega_k) + \boldsymbol{\epsilon} \tag{6.1.1}$$

with $\beta_k := C_k \exp(i\phi_k)$ and $\mathbf{f}(\omega_k) := [\exp(i\omega_k), \ldots, \exp(in\omega_k)]^T$. For convenience, let the signal frequencies ω_k reside in the interval $(0, 2\pi)$ instead of $(-\pi, \pi)$.

Recall that in Chapter 5 (Section 5.3) we discuss a method for identification of the signal frequencies from a finite set Ω of candidates. The method produces the best p-subset of Ω that minimizes the error sum of squares SSE_p defined by (5.1.4). The method is particularly applicable to the situation where Ω comprises n candidate frequencies spread uniformly in the interval $[0,2\pi)$. This choice of Ω reflects a complete lack of knowledge about the signal frequencies, in which case the only viable strategy is to search over the largest possible set without giving preference to any frequency. Note that n is the largest number of candidate frequencies one can take in order for the complete model to have a unique least-squares (LS) solution for amplitude estimation.

Definition 6.1 (Fourier Grid and Fourier Frequencies). The n-point uniform grid in the interval $[0,2\pi)$ is called the Fourier grid for time series of length n. The corresponding grid points of the form $2\pi j/n$, with j being an integer, are called Fourier frequencies. A linear combination of real or complex sinusoids with Fourier frequencies will be referred to as a Fourier model for the time series.

If all the signal frequencies in $\mathbf{y}$ happen to be Fourier frequencies, the p-subset of the Fourier grid that minimizes SSE_p in (5.1.4) represents the best selection of p frequencies among all Fourier frequencies. If the selection is correct, then the selected frequencies become error-free estimates of the signal frequencies. The simulation in Section 5.3 indicates that the error-free estimates can be obtained with a positive probability. However, in the general case where some or all of the signal frequencies are nonFourier, the minimizer of SSE_p no longer produces perfect frequency estimates. Because a nonFourier frequency can appear anywhere between two consecutive Fourier frequencies which are separated by $2\pi/n$, the accuracy of the SSE_p minimizer as a frequency estimator can only be expressed as $\mathcal{O}(n^{-1})$ in general, not good enough to ensure the consistency of the corresponding LS estimator for the amplitude and phase parameters.

The minimizer of SSE_p over all possible p-subsets of the Fourier grid can be found without having to solve the LS problem in (5.1.3) for every p-subset. This is due to the orthogonality of the sinusoids with Fourier frequencies. Indeed, let the Fourier grid in $[0,2\pi)$ be denoted by

$$\Omega_n := \{\lambda_j : j \in \mathbb{Z}_n\}, \quad \lambda_j := 2\pi j/n, \quad j \in \mathbb{Z}_n := \{0,1,\ldots,n-1\}.$$

Here the dependence on n is omitted in the notation λ_j for simplicity. According to Lemma 12.1.2, $\mathbf{F} := [\mathbf{f}(\lambda_0), \mathbf{f}(\lambda_1), \ldots, \mathbf{f}(\lambda_{n-1})]$ is an orthogonal matrix, satisfying

$$\mathbf{F}^H\mathbf{F} = \mathbf{F}\mathbf{F}^H = n\mathbf{I}. \tag{6.1.2}$$

As a result, the minimizer of $\|\mathbf{y} - \mathbf{F}\tilde{\boldsymbol{\beta}}\|^2$ with respect to $\tilde{\boldsymbol{\beta}} \in \mathbb{C}^n$ is given by

$$\mathbf{z} := [z_0, z_1, \ldots, z_{n-1}]^T := n^{-1}\mathbf{F}^H\mathbf{y}, \tag{6.1.3}$$

where

$$z_j = n^{-1}\mathbf{f}^H(\lambda_j)\mathbf{y} = n^{-1}\sum_{t=1}^{n} y_t \exp(-it\lambda_j). \tag{6.1.4}$$

The LS solution $\mathbf{z}$ is nothing but the *discrete Fourier transform* (DFT) of $\mathbf{y}$ and can be calculated efficiently by the FFT algorithms [79] [99] with computational complexity $\mathcal{O}(n\log n)$ instead of $\mathcal{O}(n^2)$.

Owing to (6.1.2), pre-multiplying both sides of (6.1.3) by $\mathbf{F}$ yields

$$\mathbf{y} = \mathbf{F}\mathbf{z} = \sum_{j=0}^{n-1} z_j \mathbf{f}(\lambda_j). \tag{6.1.5}$$

In this expression, the vector $\mathbf{y}$ is decomposed into n orthogonal components of sinusoids with Fourier frequencies. The strength of these components at different frequencies can be measured by their squared norm

$$I_n(\lambda_j) := \|z_j \mathbf{f}(\lambda_j)\|^2 = n|z_j|^2 = n^{-1}\left|\sum_{t=1}^{n} y_t \exp(-it\lambda_j)\right|^2. \tag{6.1.6}$$

The collection $\{I_n(\lambda_j), \lambda_j \in \Omega_n\}$ is called the (discrete) *periodogram* of $\mathbf{y}$. The term "periodogram" was coined by Schuster in 1898 [334], although the method of Fourier analysis was introduced much earlier [111]. Note that the periodogram can be extended periodically with period 2π beyond the domain Ω_n.

The orthogonality of the sinusoidal components in (6.1.5) implies that

$$\|\mathbf{y}\|^2 = \sum_{j=0}^{n-1} I_n(\lambda_j). \tag{6.1.7}$$

This expression is known as *Parseval's identity*, named after the French mathematician Marc-Antoine Parseval (1755–1863). Because $\|\mathbf{y}\|^2$ represents the total variability of the time series $\{y_1, \ldots, y_n\}$ around zero and $I_n(\lambda_j) = \|z_j \mathbf{f}(\lambda_j)\|^2$ represents the variability of its sinusoidal component with frequency λ_j, Parseval's identity (6.1.7) justifies the interpretation of the periodogram as an analysis of variance (ANOVA) decomposition — a statistical method for identifying the contribution of explanatory variables to the total variability.

Given the periodogram, the SSE of a p-subset Fourier model for $\mathbf{y}$ can be easily computed according to the following theorem. See Section 6.6 for a proof.

Theorem 6.1 (SSE of Subset Fourier Models and the Periodogram). *Let Ω_p be a p-subset of the n-point Fourier grid Ω_n ($p < n$). Then, the SSE of the LS regression that fits a data vector $\mathbf{y} \in \mathbb{C}^n$ by a linear combination of complex sinusoids with frequencies in Ω_p can be expressed as*

$$\mathrm{SSE}(\Omega_p) = \sum_{\lambda_j \notin \Omega_p} I_n(\lambda_j) = \|\mathbf{y}\|^2 - \sum_{\lambda_j \in \Omega_p} I_n(\lambda_j), \tag{6.1.8}$$

where $\{I_n(\lambda_j)\}$ is the periodogram of $\mathbf{y}$ defined by (6.1.6).

Remark 6.1 The SSEs in the F-test discussed in Section 5.3 can also be obtained from the periodogram when all candidate frequencies are Fourier frequencies. This result is a special case of the classical theory of linear regression with orthogonal regressors [338] (extended to complex-valued variables).

According to Theorem 6.1, choosing Ω_p to be the p-subset of Ω_n that corresponds to the p largest values in the periodogram gives the smallest SSE in (6.1.8) among all possible p-term Fourier models. Therefore, with $\mathbf{y}$ given by (6.1.1), the Fourier frequencies so obtained serve naturally as estimates of the signal frequencies in $\mathbf{y}$. This procedure, which will be referred to as the *DFT method* for frequency estimation, can be formally expressed as

$$\hat{\Omega}_p := \{\hat{\omega}_1, \ldots, \hat{\omega}_p\} := \arg \max_{\Omega_p \subset \Omega_n} \sum_{\lambda_j \in \Omega_p} I_n(\lambda_j). \tag{6.1.9}$$

Due to its insufficient accuracy to guarantee consistent amplitude estimation (see Section 5.2), the DFT method is often used as a fast way to obtain coarse initial values for subsequent application of more sophisticated methods of frequency estimation [278] [316] [317] [370] [412]. Note that for the purpose of frequency identification and estimation, the raw periodogram in (6.1.6) is found preferable in defining the DFT estimator in (6.1.9) over smoothed periodograms obtained by techniques such as weighted moving average [278] [352].

The DFT method is very effective in identifying and estimating Fourier frequencies but less so for nonFourier frequencies. The following example demonstrates this behavior.

Example 6.1 (Periodogram of Sinusoids in White Noise). Figure 6.1 shows a simulated time series and its periodogram. The time series comprises two unit-amplitude zero-phase complex sinusoids in Gaussian white noise ($\omega_1 = 2\pi \times 6/n = 2\pi \times 0.12$, $\omega_2 = 2\pi \times 10.5/n = 2\pi \times 0.21$, and SNR $= -5$ dB). As we can see, the periodogram takes its largest value exactly at ω_1, so the DFT estimator produces an error-free estimate for this frequency. However, the second largest value of the periodogram does not occur at ω_2 which is not a Fourier frequency. In fact, the second largest value does not even occur at the Fourier frequencies nearest to ω_2. In this case, the DFT estimate for ω_2 is grossly erroneous. ◇

To better understand the performance of the DFT method, let us first focus on the case of Fourier frequencies and consider the following examples.

Example 6.2 (Periodogram of Pure Sinusoid with Fourier Frequency). Let $y_t = \beta_1 \exp(i\omega_1 t)$ $(t = 1, \ldots, n)$ with $\beta_1 := C_1 \exp(i\phi_1)$ and $\omega_1 = 2\pi k/n$ for some integer $k \in \{1, \ldots, n-1\}$. Then, we have

$$z_j = n^{-1} \mathbf{f}^H(\lambda_j)\mathbf{y} = \beta_1 \delta_{j-k}.$$

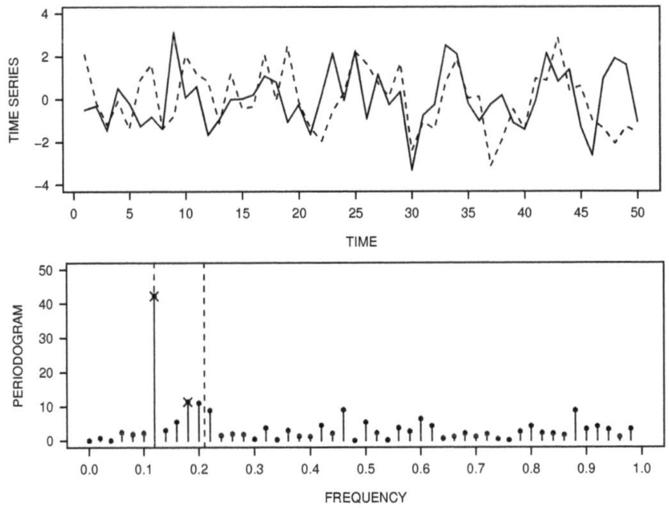

Figure 6.1. Top, a time series consisting of two unit-amplitude zero-phase complex sinusoids in Gaussian white noise: solid line, real part; dashed line, imaginary part. Bottom, periodogram of the time series shown as a function of $\lambda_j/(2\pi)$: dashed vertical lines indicate the signal frequencies; crosses indicate the DFT frequency estimates.

This implies that

$$I_n(\lambda_j) = n|z_j|^2 = nC_1^2\delta_{j-k}.$$

As can be seen, a pure complex sinusoid with Fourier frequency $2\pi k/n = \lambda_k$ manifests itself in the periodogram as a peak at λ_k, whose magnitude is equal to the power of the sinusoid: the squared amplitude multiplied by the sample size. The periodogram ordinates are equal to zero elsewhere, so the signal power is completely concentrated at frequency λ_k. ◇

Example 6.3 (Periodogram of Pure Gaussian White Noise). Let $\mathbf{y} \sim \mathrm{N}_c(\mathbf{0}, \sigma^2\mathbf{I})$. Because $\mathbf{f}^H(\lambda_j)\mathbf{f}(\lambda_j) = n$, it follows from Proposition 2.6 that for all j,

$$z_j = n^{-1}\mathbf{f}^H(\lambda_j)\mathbf{y} \sim \mathrm{N}_c(0, n^{-1}\sigma^2).$$

By Lemma 12.2.1, $I_n(\lambda_j)/\sigma^2 = |z_j|^2/(n^{-1}\sigma^2) \sim \frac{1}{2}\chi^2(2)$, or equivalently,

$$I_n(\lambda_j) \sim \tfrac{1}{2}\sigma^2\chi^2(2).$$

Therefore, the periodogram ordinates of Gaussian white noise are identically distributed random variables. Note that $\frac{1}{2}\sigma^2\chi^2(2)$ is just an exponential distribution with mean σ^2 whose PDF takes the form $(1/\sigma^2)\exp(-x/\sigma^2)\mathscr{I}(x>0)$. ◇

By comparing these examples, we can see that the periodogram of a pure sinusoid with Fourier frequency has a peak of magnitude nC_1^2 at the signal frequency, whereas the periodogram of Gaussian white noise consists of exponentially distributed random variables with mean σ^2. It is this difference in magnitude that makes the identification of signal frequencies possible in the periodogram when the sample size or SNR is sufficiently large.

In general, with $\mathbf{y} = \mathbf{x} + \boldsymbol{\epsilon}$ taking the form (6.1.1), where $\mathbf{x} := \sum_{k=1}^{p} \beta_k \mathbf{f}(\omega_k)$, the DFT of $\mathbf{y}$, which is defined by (6.1.3), can be expressed as

$$\mathbf{z} = \boldsymbol{\mu} + \mathbf{e}, \tag{6.1.10}$$

where

$$\boldsymbol{\mu} := [\mu_0, \mu_1, \ldots, \mu_{n-1}]^T := n^{-1}\mathbf{F}^H\mathbf{x} = n^{-1}\sum_{k=1}^{p} \beta_k \mathbf{F}^H \mathbf{f}(\omega_k), \tag{6.1.11}$$

$$\mathbf{e} := [e_0, e_1, \ldots, e_{n-1}]^T := n^{-1}\mathbf{F}^H\boldsymbol{\epsilon}. \tag{6.1.12}$$

Based on this observation, the following theorem establishes the distribution of the periodogram ordinates under the assumption that the ω_k are Fourier frequencies and $\boldsymbol{\epsilon}$ is Gaussian white noise. A proof is given in Section 6.6.

Theorem 6.2 (Distribution of the Periodogram: Gaussian White Noise). *Let* $\mathbf{y}$ *be given by (6.1.1) with* $\boldsymbol{\epsilon} \sim \mathrm{N}_c(\mathbf{0}, \sigma^2\mathbf{I})$. *Assume that* $\omega_k = \lambda_{j(k)} \in \Omega_n$ *for all* k. *Then, the random variables* $I_n(\lambda_0), I_n(\lambda_1), \ldots, I_n(\lambda_{n-1})$ *are mutually independent and*

$$I_n(\lambda_j) \sim \tfrac{1}{2}\sigma^2\chi^2(2, 2\theta_j), \quad \theta_j := n\sum_{k=1}^{p} \gamma_k \delta_{j-j(k)},$$

where $\chi^2(2, c)$ *denotes the noncentral chi-square distribution with 2 degrees of freedom and noncentrality parameter* c *and where* $\gamma_k := C_k^2/\sigma^2$ *is the SNR of the* k*th complex sinusoid.*

Remark 6.2 In the real case, we can write

$$\mathbf{y} = \sum_{k=1}^{q} (A_k \Re\{\mathbf{f}(\omega_k)\} + B_k \Im\{\mathbf{f}(\omega_k)\}) + \boldsymbol{\epsilon}, \tag{6.1.13}$$

where $\{\omega_k\} \subset (0, \pi)$. Theorem 6.2 remains true if $\boldsymbol{\epsilon} \sim \mathrm{N}(\mathbf{0}, \sigma^2\mathbf{I})$, except that the λ_j are restricted to $\Omega_n \cap [0, \pi]$ because of symmetry, and that

$$I_n(\lambda_j) \sim \begin{cases} \frac{1}{2}\sigma^2\chi^2(2, 2\theta_j) & \text{if } \lambda_j \neq 0 \text{ or } \pi, \\ \sigma^2\chi^2(1) & \text{if } \lambda_j = 0 \text{ or } \pi, \end{cases} \quad \theta_j := n\sum_{k=1}^{q} \tfrac{1}{2}\gamma_k \delta_{j-j(k)},$$

where $\gamma_k := \frac{1}{2}(A_k^2 + B_k^2)/\sigma^2$ is the SNR of the kth real sinusoid. In the absence of sinusoids, we have $I_n(\lambda_j) \sim \frac{1}{2}\sigma^2\chi^2(2)$ for $\lambda_j \neq 0$ or π, and $I_n(\lambda_j) \sim \sigma^2\chi^2(1)$ for $\lambda_j = 0$ or π. This is the classical result for Gaussian white noise [298, p. 398].

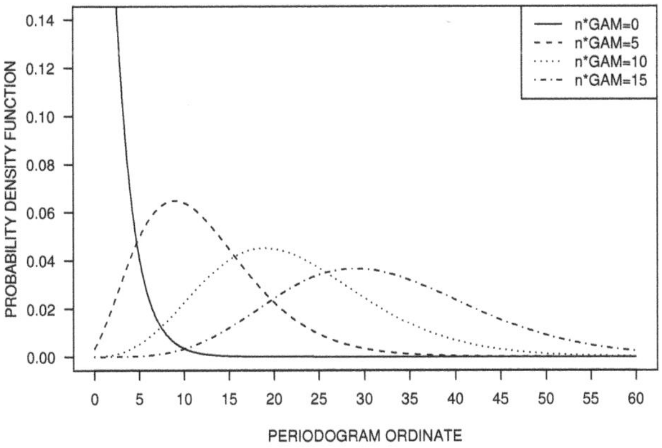

Figure 6.2. Probability density function of the periodogram ordinates at signal and nonsignal frequencies for complex sinusoids with Fourier frequencies in Gaussian white noise ($\sigma^2 = 2$). Solid line, nonsignal frequency; dashed line, signal frequency ($n\gamma = 5$); dotted line, signal frequency ($n\gamma = 10$); dash-dotted line, signal frequency ($n\gamma = 15$), where γ denotes the SNR of a complex sinusoid.

According to Theorem 6.2, the presence of sinusoids is reflected in the periodogram solely by the noncentrality of the chi-square distribution. In this manifestation, it is the product $n\gamma_k$, rather than the SNR γ_k or the sample size n alone, that determines the intrinsic difficulty in distinguishing the periodogram ordinates at signal frequencies from those at nonsignal frequencies. The sample size and the SNR play the same role. For example, a 50% reduction in SNR is equivalent to a 50% decrease in the sample size; by doubling the SNR, one can remedy the performance loss caused by a 50% decrease in the sample size.

Theorem 6.2 implies that the mean of the periodogram can be expressed as

$$E\{I_n(\lambda_j)\} = \sigma^2(1+\theta_j),$$

where $\theta_j = 0$ if $\lambda_j \notin \{\omega_k\}$ and $\theta_j = n\gamma_k$ if $\lambda_j = \omega_k$ for some $k = 1,\ldots,p$. Observe that the mean at the kth signal frequency is equal to $1 + n\gamma_k$ times the mean at a nonsignal frequency. Therefore, on average, the periodogram takes larger values at a signal frequency than it does at nonsignal frequencies.

In addition to the mean, let us now compare the probability distribution of the periodogram at a signal frequency with that at a nonsignal frequency, as illustrated in Figure 6.2. Assuming $\sigma^2 = 2$ without loss of generality, it follows from Theorem 6.2 that the periodogram at a nonsignal frequency is distributed as $\chi^2(2)$, whereas the periodogram at a signal frequency with SNR γ is distributed as $\chi^2(2, 2n\gamma)$. Figure 6.2 depicts the PDF of these distributions for various values of $n\gamma$. As can be seen, while the distribution at a nonsignal frequency is concentrated near zero, the distribution at a signal frequency is concentrated at larger values. As a result, the periodogram at a signal frequency has a greater chance of taking large values than it does at a nonsignal frequency.

Finally, let us compare the periodogram at a signal frequency with the maximum of the periodogram ordinates over all nonsignal frequencies. According to Theorem 6.2, for any $\tau > 0$,

$$P\Big\{I_n(\lambda_j) \le \tfrac{1}{2}\sigma^2\tau \text{ for all } \lambda_j \notin \{\omega_k\}\Big\} = \{1-\exp(-\tfrac{1}{2}\tau)\}^{n-p}. \tag{6.1.14}$$

By taking $\tau := 2\log n - 2\log x$ for any given $x > 0$, the right-hand side of (6.1.14) can be written as $(1-x/n)^{n-p}$ which tends to $\exp(-x)$ as $n \to \infty$. This implies that the maximum of $I_n(\lambda_j)/\sigma^2$ over all nonsignal frequencies is asymptotically equal to $\log n$ plus a random variable that has the standard Gumbel distribution with CDF $\exp(\exp(-x))$. Therefore, we can write

$$\max_{\lambda_j \notin \{\omega_k\}} I_n(\lambda_j) = \mathcal{O}_P(\log n). \tag{6.1.15}$$

On the other hand, at the kth signal frequency, we have

$$I_n(\omega_k) = \mathcal{O}_P(n).$$

This difference in magnitude makes the periodogram stand out at the signal frequencies among all other frequencies.

The statistical performance of the DFT method for frequency identification can be measured by the probability of frequency identification,

$$P_{\text{ID}} := P(\hat{\Omega}_p = \{\omega_k\}).$$

Under the condition of Gaussian white noise, the following theorem provides a formula for P_{ID} as a result of Theorem 6.2. See Section 6.6 for a proof.

Theorem 6.3 (Probability of Frequency Identification: Gaussian White Noise). *Let $\hat{\Omega}_p$ be defined by (6.1.9). If the conditions of Theorem 6.2 are satisfied, then the probability of frequency identification can be expressed as*

$$P_{\text{ID}} = \int_0^\infty \tfrac{1}{2}(n-p)\{1-\exp(-\tfrac{1}{2}x)\}^{n-p-1}\Big\{\prod_{k=1}^{p} \bar{F}(x, 2n\gamma_k)\Big\}\exp(-\tfrac{1}{2}x)\,dx,$$

where $\gamma_k := C_k^2/\sigma^2$ and $\bar{F}(x,\theta) := P\{\chi^2(2,\theta) > x\}$.

Remark 6.3 In the real case where $\mathbf{y}$ takes the form (6.1.13), the DFT method identifies the signal frequencies by locating the q largest values among the periodogram ordinates $I_n(\lambda_j)$ with the restriction $\lambda_j \in (0,\pi)$. Under the assumption that $\boldsymbol{\epsilon} \sim \mathrm{N}(\mathbf{0}, \sigma^2\mathbf{I})$ and $\{\omega_k\} \subset \Omega_n \cap (0,\pi)$, it follows from Remark 6.2 that

$$P_{\text{ID}} = \int_0^\infty \tfrac{1}{2}(m-q)\{1-\exp(-\tfrac{1}{2}x)\}^{m-q-1}\Big\{\prod_{k=1}^{q} \bar{F}(x, n\gamma_k)\Big\}\exp(-\tfrac{1}{2}x)\,dx,$$

where m denotes the cardinality of $\Omega_n \cap (0,\pi)$ and $\gamma_k := \frac{1}{2}(A_k^2 + B_k^2)/\sigma^2$ denotes the SNR of the kth real sinusoid. The special case of this expression for $q = 1$ can be found in [352].

Remark 6.4 By Lemma 12.2.1, the PDF of $\chi^2(2,\theta)$ takes the form

$$p(x) = \tfrac{1}{2}\exp\{-\tfrac{1}{2}(x+\theta)\} \sum_{u=0}^{\infty} \frac{(\theta x/4)^u}{(u!)^2} \mathscr{I}(x>0).$$

Using this expression, we obtain

$$\begin{aligned} \bar{F}(x,\theta) &= \exp\{-\tfrac{1}{2}(x+\theta)\} \sum_{u=0}^{\infty} \sum_{v=0}^{u} \frac{(\theta/2)^u (x/2)^v}{u!\,v!} \\ &= 1 - \exp\{-\tfrac{1}{2}(x+\theta)\} \sum_{v=1}^{\infty} \sum_{u=0}^{v-1} \frac{(\theta/2)^u (x/2)^v}{u!\,v!}. \end{aligned} \tag{6.1.16}$$

The infinite sum in (6.1.16) can be truncated to approximate $\bar{F}(x,\theta)$.

As a corollary to Theorem 6.3, the consistency of the DFT method for frequency identification can be established as follows. See Section 6.6 for a proof.

Corollary 6.1 (Consistency of the DFT Method for Frequency Identification). *Let the conditions of Theorem 6.3 be satisfied. Then, the DFT method in (6.1.9) is consistent in the sense that* $P_{\mathrm{ID}} \to 1$ *if* $n\gamma_k \to \infty$ *for all* k.

Remark 6.5 Corollary 6.1 covers the case where the signal power is weak relative to the sample size such that $\gamma_k \to 0$ as $n \to \infty$ but is compensated for by a sufficiently long data record so that $n\gamma_k \to \infty$.

The following example demonstrates the behavior of P_{ID} as a function of the sample size and the SNR.

Example 6.4 (Single Complex Sinusoid in Gaussian White Noise: Identification). Let $p = 1$ in (6.1.1). Then, using the identity

$$a^{v+1} \int_0^{\infty} x^v \exp(-ax)\,dx = v!$$

and with the notation $\theta := 2n\gamma_1$, one can show that

$$\begin{aligned} P_{\mathrm{ID}} &= (n-1)\exp(-\tfrac{1}{2}\theta) \\ &\quad \times \sum_{u=0}^{\infty} \frac{(\theta/2)^u}{u!} \sum_{v=0}^{u} \frac{1}{v!} \int_0^{\infty} x^v \{1-\exp(-x)\}^{n-2} \exp(-2x)\,dx \\ &= (n-1)\exp(-\tfrac{1}{2}\theta) \sum_{u=0}^{\infty} \frac{(\theta/2)^u}{u!} \sum_{v=0}^{u} \sum_{j=0}^{n-2} \binom{n-2}{j} \frac{(-1)^j}{(j+2)^{v+1}} \end{aligned}$$

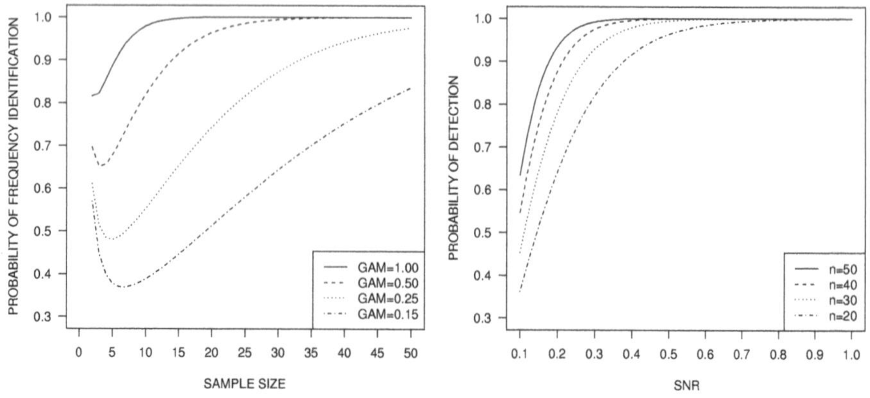

(a) (b)

Figure 6.3. Plot of the probability of frequency identification P_{ID} given by (6.1.17), as a function of (a) the sample size n and (b) the SNR γ_1, for a single complex sinusoid with Fourier frequency in Gaussian white noise. (a) Solid line, $\gamma_1 = 1$; dashed line, $\gamma_1 = 0.5$; dotted line, $\gamma_1 = 0.25$; dash-dotted line, $\gamma_1 = 0.15$. (b) Solid line, $n = 50$; dashed line, $n = 40$; dotted line, $n = 30$; dash-dotted line, $n = 20$.

$$= \sum_{j=0}^{n-2} \binom{n-1}{j+1} (-1)^j \left\{1 - \frac{1}{j+2} \exp\left(-\frac{j+1}{2j+4}\theta\right)\right\}$$

$$= 1 - n^{-1} \sum_{j=2}^{n} \binom{n}{j} (-1)^j \exp\left(-\frac{j-1}{2j}\theta\right). \tag{6.1.17}$$

Figure 6.3 depicts P_{ID} in (6.1.17) as a function of n for various values of γ_1 and as a function of γ_1 for various values of n.

As can be seen from Figure 6.3(a), the probability P_{ID} approaches unity as n grows and the speed of convergence increases with the SNR. When the sample size is very small (≤ 5), the probability P_{ID} suffers from a decline as the sample size increases especially at low SNR, which seems to reflect an increased difficulty for the DFT method to identify the signal frequency as the pool of candidate frequencies expands; the probability P_{ID} recovers from the decline as the sample size increases further and begins to grow toward unity when the noncentrality of the chi-square distributions becomes the dominant factor.

Figure 6.3(b) shows that P_{ID} also approaches unity as the SNR increases. More interestingly, it reveals an important phenomenon called the *threshold effect*: as a function of the SNR, the probability P_{ID} drops rather quickly when the SNR falls below a certain threshold. For $n = 20$, the drop begins at approximately $\gamma_1 = 0.6$; for $n = 50$, it begins at $\gamma_1 = 0.3$. Identification of the turning point is an important step in designing a reliable system. ◇

The probability of frequency identification also plays an important role in the mean-square error (MSE) of the DFT method for frequency estimation. Under the assumption that the ω_k are Fourier frequencies, the total error $\sum(\hat{\omega}_k - \omega_k)^2$ is equal to zero if all the signal frequencies are correctly identified; otherwise, the error is greater than or equal to $(2\pi/n)^2$. Therefore,

$$\mathrm{MSE}(\hat{\Omega}_p) := \sum_{k=1}^{p} E\{(\hat{\omega}_k - \omega_k)^2\} \ge (2\pi/n)^2(1-P_{\mathrm{ID}}).$$

In the special case of $p=1$, the MSE can be computed exactly using the probability P_{ID}, as shown in the following example.

Example 6.5 (Single Complex Sinusoid in Gaussian White Noise: MSE). Consider the situation in Example 6.4. Let $\hat{\omega}_1$ denote the Fourier frequency corresponding to the largest periodogram ordinate. Given that the maximum does not occur at the signal frequency ω_1, the remaining $n-1$ periodogram ordinates are equally likely to be the maximum. Therefore, for $r=1,2$,

$$E(\hat{\omega}_1^r) = \omega_1^r P_{\mathrm{ID}} + (1-P_{\mathrm{ID}}) \sum{}' \lambda_j^r/(n-1),$$

where P_{ID} is given by (6.1.17) and $\sum'$ denotes the sum over $\lambda_j \in \Omega_n \backslash \{\omega_1\}$. Because $\sum' \lambda_j^r = \sum_{j=1}^{n-1} \lambda_j^r - \omega_1^r$ and $\lambda_j = 2\pi j/n$, it follows that

$$\sum{}' \lambda_j = \pi(n-1) - \omega_1, \quad \sum{}' \lambda_j^2 = \tfrac{2}{3}\pi^2(n-1)(2n-1)n^{-1} - \omega_1^2.$$

Therefore, we can write

$$E(\hat{\omega}_1) = \omega_1 + (1-P_{\mathrm{ID}})\Big(\pi - \frac{n}{n-1}\omega_1\Big), \tag{6.1.18}$$

$$\mathrm{MSE}(\hat{\omega}_1) = (1-P_{\mathrm{ID}})\Big\{\frac{n}{n-1}\omega_1^2 + 2\pi\Big(\frac{2n-1}{3n}\pi - \omega_1\Big)\Big\}. \tag{6.1.19}$$

Figure 6.4(a) depicts $\mathrm{MSE}(\hat{\omega}_1)$ and the squared bias $\{E(\hat{\omega}_1) - \omega_1\}^2$ as functions of ω_1 for different values of n. Figure 6.4(b) depicts $\mathrm{MSE}(\hat{\omega}_1)$ together with the corresponding CRLB as functions of γ_1 for various values of n.

As can be seen, the estimator $\hat{\omega}_1$ is biased unless $\omega_1 = \pi(n-1)/n$; the bias approaches zero as $n\gamma_1 \to \infty$ because $P_{\mathrm{ID}} \to 1$. Observe that $\mathrm{MSE}(\hat{\omega}_1)$ is quadratic function of ω_1 with the minimum value $(1/3)(1-P_{\mathrm{ID}})\pi^2(n+1)/n$ attained at $\omega_1 = \pi(n-1)/n$ (in which case $\hat{\omega}_1$ becomes unbiased). As ω_1 departs from this special frequency, $\mathrm{MSE}(\hat{\omega}_1)$ increases monotonically and reaches the maximum value $(2/3)(1-P_{\mathrm{ID}})\pi^2(2n-1)/n$ at $\omega_1 = 0$ and $2\pi(n-1)/n$. The fact that the largest MSE occurs at very low and very high frequencies can be explained by the increased ambiguity among these frequencies due to aliasing (i.e., $\omega = 0$ is the alias of $\omega = 2\pi$, and vice versa).

This example shows that the MSE of the DFT frequency estimator can be less than the CRLB when the SNR is sufficiently high. It should not be too surprising,

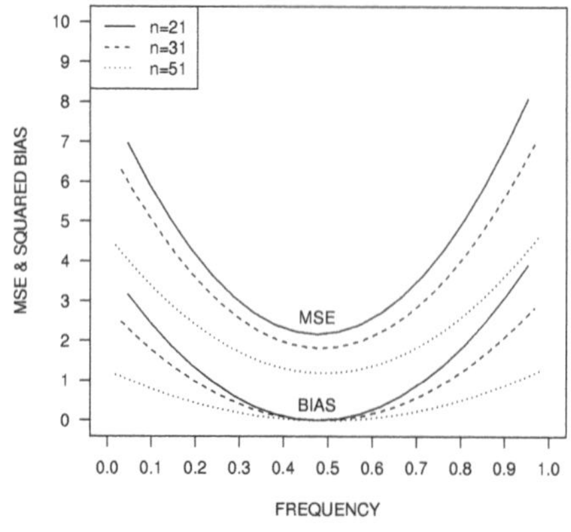

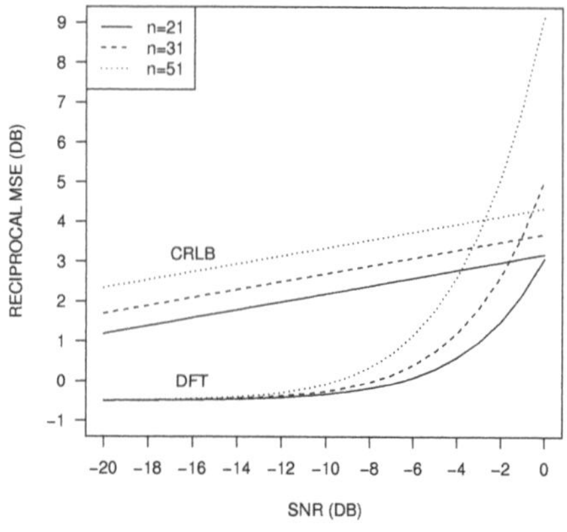

Figure 6.4. Accuracy of the DFT frequency estimator for a single complex sinusoid with Fourier frequency in Gaussian white noise. (a) MSE and squared bias as functions of the signal frequency $\omega_1/(2\pi)$ ($\gamma_1 = 0.1$). (b) Reciprocal MSE and CRLB as functions of γ_1 for $\omega_1 = 2\pi j/n$ with $j = (n-1)/2$. Solid line, $n = 21$; dashed line, $n = 31$; dotted line, $n = 51$.

not only because the CRLB is valid only for unbiased estimators yet the DFT estimator is biased, but also because the CRLB and the DFT method operate in different parameter spaces: the CRLB assumes continuous frequency whereas the DFT method assumes finite and discrete Fourier frequency. The prior knowledge of Fourier frequency restricts the parameter space and makes it possible for the DFT method to achieve higher accuracy than the CRLB suggests.

Note that in light of Remark 6.3, the formula (6.1.17) remains valid in the case of a single real sinusoid, provided that n is replaced by m — the number of Fourier frequencies in the interval $(0,\pi)$ — and θ is defined as $n\gamma_1$. Similarly, the expressions (6.1.18) and (6.1.19) remain valid for the DFT estimator if n is replaced by m. Therefore, at $\omega_1 = \pi(m-1)/m$, the bias equals zero and the mean-square error takes its minimum value. ◇

The distribution of the periodogram becomes more complicated once the assumption of Gaussian white noise is no longer valid. For example, under the condition of Gaussian colored noise, we have the following result as an extension of Theorem 6.2. See Section 6.6 for a proof.

Theorem 6.4 (Distribution of the Periodogram: Gaussian Colored Noise). *Let the conditions of Theorem 6.2 be satisfied except that $\{\epsilon_t\}$ is a zero-mean stationary complex Gaussian process with ACF $r_\epsilon(u)$ and SDF $f_\epsilon(\omega)$ such that $\sum|r_\epsilon(u)| < \infty$ and $\inf\{f_\epsilon(\omega)\} > 0$. Then, for any given $\lambda_j \in \Omega_n$,*

$$I_n(\lambda_j) \sim \tfrac{1}{2}\rho_j^{-1} f_\epsilon(\lambda_j)\,\chi^2(2, 2\rho_j\theta_j), \quad \theta_j := n\sum_{k=1}^{p} \gamma_k \delta_{j-j(k)},$$

where $\gamma_k := C_k^2 / f_\epsilon(\omega_k)$, $\rho_j := n^{-1} f_\epsilon(\lambda_j)/\mathrm{Var}(e_j) = 1 + \mathcal{O}(r_n)$, *and*

$$r_n := \sum_{u=-\infty}^{\infty} \min(1, |u|/n)\, |r_\epsilon(u)| \to 0 \quad \text{as } n \to \infty.$$

The error term $\mathcal{O}(r_n)$ *is uniformly bounded in* $\lambda_j \in \Omega_n$. *Moreover, with* m *fixed as* $n \to \infty$, *the random variables in* $\{I_n(\lambda_j) : \lambda_j \in \Omega_n(m)\}$ *are asymptotically independent for any* m*-subset* $\Omega_n(m) \subset \Omega_n$.

Remark 6.6 Theorem 6.4 remains true in the real case except that the λ_j are restricted to $\Omega_n \cap [0, \pi]$ and that

$$I_n(\lambda_j) \sim \begin{cases} \frac{1}{2}\rho_j^{-1} f_\epsilon(\lambda_j)\, \chi^2(2, 2\rho_j\theta_j) & \text{if } \lambda_j \neq 0 \text{ or } \pi, \\ \rho_j^{-1} f_\epsilon(\lambda_j)\, \chi^2(1) & \text{if } \lambda_j = 0 \text{ or } \pi, \end{cases}$$

$$\theta_j := n \sum_{k=1}^{q} \tfrac{1}{2}\gamma_k \delta_{j-j(k)},$$

where $\gamma_k := \frac{1}{2} C_k^2 / f_\epsilon(\omega_k)$ is the SNR of the kth real sinusoid. In the special case of white noise, we have $f_\epsilon(\lambda_j) = \sigma^2$, $r_n = 0$, and hence $\rho_j = 1$, which leads to the assertion in Theorem 6.2.

According to Theorem 6.4, when r_n is small, the distribution of $I_n(\lambda_j)$ in the absence of sinusoids for any given $\lambda_j \in \Omega_n$ can be approximated by $\frac{1}{2} f_\epsilon(\lambda_j)\chi^2(2)$ whose PDF takes the form $(1/f_\epsilon(\lambda_j)) \exp(-x/f_\epsilon(\lambda_j))\, \mathcal{I}(x > 0)$. Therefore, by ignoring their statistical dependence, the log likelihood of the periodogram ordinates can be approximated by that of the random variables $\frac{1}{2} f_\epsilon(\lambda_j) X_j$ $(j \in \mathbb{Z}_n)$, where the X_j are i.i.d. $\chi^2(2)$. In other words, it can be approximated by

$$L(f_\epsilon) := -\sum_{j=0}^{n-1} \{I_n(\lambda_j)/f_\epsilon(\lambda_j) + \log f_\epsilon(\lambda_j)\}. \tag{6.1.20}$$

This functional of the noise spectrum, known as *Whittle's likelihood* [424], facilitates a pseudo likelihood-based spectral analysis of Gaussian as well as non-Gaussian time series for which the true likelihood is either unavailable or too difficult to work with. Examples of such analysis include parametric and nonparametric inference of the SDF [54] [68] [262] [275] [285] [408] [420] as well as discriminant and cluster analysis of multiple time series [170] [346, Section 7.7]. Note that maximizing Whittle's likelihood in (6.1.20) is equivalent to minimizing the *Kullback-Leibler spectral divergence* [202] [283], which is defined as

$$D_{\mathrm{KL}}(\{I_n(\lambda_j)\}, \{f_\epsilon(\lambda_j)\}) := \sum_{j=0}^{n-1} d_{\mathrm{KL}}(I_n(\lambda_j)/f_\epsilon(\lambda_j)), \tag{6.1.21}$$

where $d_{\mathrm{KL}}(x) := x - \log x - 1$ $(x > 0)$ is a nonnegative function with a unique minimum at $x = 1$. This quantity is widely used in signal processing applications as a measure of discrepancy between two spectral densities [28] [53] [126] [358].

Unlike the case of Gaussian white noise, the periodogram ordinates are no longer independent for finite sample sizes when the noise is colored. The following theorem shows that the periodogram ordinates are asymptotically uncorrelated as $n \to \infty$. It also provides the asymptotic expressions for the mean and variance of the periodogram. A proof of this result can be found in Section 6.6.

Theorem 6.5 (Mean, Variance, and Covariance of the Periodogram: Gaussian Colored Noise). *If the conditions of Theorem 6.4 are satisfied, then*

$$E\{I_n(\lambda_j)\} = n\sum_{k=1}^{p} C_k^2\delta_{j-j(k)} + f_\epsilon(\lambda_j) + \mathcal{O}(r_n), \tag{6.1.22}$$

$$\text{Var}\{I_n(\lambda_j)\} = 2n\sum_{k=1}^{p} C_k^2(f_\epsilon(\omega_k) + \mathcal{O}(r_n))\delta_{j-j(k)} + f_\epsilon^2(\lambda_j) + \mathcal{O}(r_n), \tag{6.1.23}$$

and for $\lambda_l \neq \lambda_j$,

$$\text{Cov}\{I_n(\lambda_j), I_n(\lambda_l)\} = \mathcal{O}(nr_n)\sum_{k,k'=1}^{p} \delta_{j-j(k)}\delta_{l-j(k')} + \mathcal{O}(r_n^2). \tag{6.1.24}$$

All error terms are uniformly bounded in $\lambda_j \in \Omega_n$ *and* $\lambda_l \in \Omega_n$.

Remark 6.7 Theorem 6.5 is applicable to the real case with $\boldsymbol{\epsilon} \in \text{N}(\mathbf{0}, \mathbf{R}_\epsilon)$ if $p := 2q$, $\omega_{p-k+1} := 2\pi - \omega_k$, and $C_{p-k+1} := C_k := \frac{1}{2}\sqrt{A_k^2 + B_k^2}$ $(k = 1, \ldots, q)$, and if λ_j and λ_l are restricted to $\Omega_n \cap [0, \pi]$.

By Theorem 6.5, the correlation coefficient between two periodogram ordinates can be expressed as

$$\text{Corr}\{I_n(\lambda_j), I_n(\lambda_l)\}$$
$$= \begin{cases} \mathcal{O}(r_n) & \text{if both } \lambda_j \text{ and } \lambda_l \text{ are signal frequencies,} \\ \mathcal{O}(n^{-1/2}r_n^2) & \text{if one of them is a signal frequency,} \\ \mathcal{O}(r_n^2) & \text{if none of them is a signal frequency.} \end{cases}$$

Because $r_n \to 0$ as $n \to \infty$, the periodogram ordinates are asymptotically uncorrelated. Observe that the correlation between the periodogram ordinates at the signal frequencies approaches zero at a slower rate than the correlation between other ordinates. This is one of the special properties of time series with mixed spectra as compared to time series with continuous spectra.

The quantity r_n can be regarded as a measure of smoothness for the noise spectrum. The absolute summability of $\{r_\epsilon(u)\}$, as assumed in Theorem 6.5, only ensures that the noise spectrum is a continuous function, in which case, r_n may approach zero very slowly. Now suppose that $r_\epsilon(u)$ satisfies a stronger condition $c := \sum |u|^\delta |r_\epsilon(u)| < \infty$ for some $\delta > 0$. Then,

$$\sum_{|u|\geq n} |r_\epsilon(u)| \leq n^{-\delta} \sum_{|u|\geq n} |u|^\delta |r_\epsilon(u)| \leq cn^{-\delta}.$$

Moreover, if $\delta > 1$, then

$$\sum_{|u|<n} n^{-1}|u|\,|r_\epsilon(u)| \le n^{-1} \sum_{|u|<n} |u|^\delta\,|r_\epsilon(u)| \le cn^{-1},$$

and if $0 < \delta \le 1$, then $|u/n|^{1-\delta} \le 1$ and

$$\begin{aligned}\sum_{|u|<n} n^{-1}|u|\,|r_\epsilon(u)| &= \sum_{|u|<n} |u/n|^\delta\,|u/n|^{1-\delta}\,|r_\epsilon(u)| \\ &\le n^{-\delta} \sum_{|u|<n} |u|^\delta\,|r_\epsilon(u)| \le cn^{-\delta}.\end{aligned}$$

Combining these results gives

$$r_n = \mathcal{O}(n^{-\min(1,\delta)}).$$

In particular, by setting $\delta = 1/2$, we obtain $r_n = \mathcal{O}(n^{-1/2})$, which leads to

$$\begin{aligned}&\mathrm{Corr}\{I_n(\lambda_j), I_n(\lambda_l)\} \\ &\quad = \begin{cases} \mathcal{O}(n^{-1/2}) & \text{if both } \lambda_j \text{ and } \lambda_l \text{ are signal frequencies,} \\ \mathcal{O}(n^{-3/2}) & \text{if one of them is a signal frequency,} \\ \mathcal{O}(n^{-1}) & \text{if none of them is a signal frequency.} \end{cases}\end{aligned}$$

Note that the SDF of the noise with $\delta = 1/2$ is a continuous function that may not be continuously differentiable.

The expressions in Theorem 6.5 for the mean, variance, and covariance of the periodogram can also be generalized to the case where the noise is a linear process. See Section 6.6 for a proof of the following theorem.

Theorem 6.6 (Mean, Variance, and Covariance of the Periodogram: Linear Process Noise). *Let the conditions of Theorem 6.2 be satisfied except that $\{\epsilon_t\}$ is a complex linear process of the form*

$$\epsilon_t = \sum_{j=-\infty}^{\infty} \psi_j \zeta_{t-j}, \tag{6.1.25}$$

where $\sum|\psi_j| < \infty$ and $\{\zeta_t\} \sim \mathrm{WN}(0,\sigma^2)$ with $E(\zeta_t^2) = \iota\sigma^2$. Let $\{\zeta_t\}$ also satisfy the assumptions in Theorems 4.1 and 4.2. Let $\{\delta_u^{(n)}\}$ denote the n-periodic Kronecker delta sequence such that $\delta_u^{(n)} = 1$ if $u = 0$ (mod n) and $\delta_u^{(n)} = 0$ otherwise. Then,

$$E\{I_n(\lambda_j)\} = n\sum_{k=1}^{p} C_k^2 \delta_{j-j(k)} + f_\epsilon(\lambda_j) + \mathcal{O}(r_n), \tag{6.1.26}$$

$$\begin{aligned}\mathrm{Var}\{I_n(\lambda_j)\} &= 2n\sum_{k=1}^{p} \{C_k^2 f_\epsilon(\omega_k) + \Re(\beta_k^{*2}\iota g_\epsilon(\omega_k))\delta_{2j(k)}^{(n)} + \mathcal{O}(r_{n3})\}\delta_{j-j(k)} \\ &\quad + f_\epsilon^2(\lambda_j) + \xi\{|\iota g_\epsilon(\lambda_j)|^2 + \mathcal{O}(r_{n2})\}\delta_{2j}^{(n)} + \mathcal{O}(r_n) + \mathcal{O}(r_{n4}),\end{aligned} \tag{6.1.27}$$

and for $\lambda_l \neq \lambda_j$,

$$\begin{aligned}
&\mathrm{Cov}\{I_n(\lambda_j), I_n(\lambda_l)\} \\
&= 2n \sum_{k,k'=1}^{p} \{\Re(\beta_k^* \beta_{k'}^* \iota g_\epsilon(\omega_k))\delta_{j(k)+j(k')}^{(n)} + \mathcal{O}(r_n) + \mathcal{O}(r_{2,n})\}\delta_{j-j(k)}\,\delta_{l-j(k')} \\
&\quad + \sum_{k=1}^{p} \mathcal{O}(1)(\delta_{j-j(k)} + \delta_{l-j(k)}) + \xi\{|\iota g_\epsilon(\lambda_j)|^2 + \mathcal{O}(r_{n2})\}\delta_{j+l}^{(n)} + \mathcal{O}(r_{n4}),
\end{aligned} \tag{6.1.28}$$

where $r_n := \sum_u \min(1, |u|/n)\,|r_\epsilon(u)|$, $r_{n2} := \sum_u \min(1, |u|/n)\,|\iota c_\epsilon(u)|$, $r_{n3} := \mathcal{O}(r_n) + \mathcal{O}(r_{n2}) + \mathcal{O}(n^{-1})$, $r_{n4} := \mathcal{O}(r_n^2) + \mathcal{O}(r_{n2}^2) + \mathcal{O}(n^{-1})$, *and* $\beta_k := C_k \exp(i\phi_k)$. *All error terms in these expressions are uniformly bounded in* $\lambda_j \in \Omega_n$ *and* $\lambda_l \in \Omega_n$.

Remark 6.8 The assumptions about $\{\zeta_t\}$ are satisfied if $\{\zeta_t\}$ is the i.i.d. sequence discussed in Remark 4.6 or the sequence of martingale differences discussed in Remark 4.7. Theorem 6.6 remains true in the real case as discussed in Remark 6.7 with the additional observation that $\iota := E(\zeta_t^2)/\sigma^2 = 1$, $\xi := |E(\zeta_t^2)|^2/\sigma^4 = |\iota|^2 = 1$, $g_\epsilon(\omega) = f_\epsilon(\omega)$, $r_{n2} = r_n$, and $\phi_{p-k+1} := -\phi_k$ $(k = 1, \ldots, q)$, where $p := 2q$.

Remark 6.9 The term $\mathcal{O}(1)$ in (6.1.28) and the term $\mathcal{O}(n^{-1})$ in r_{n3} are both due to the interaction between the sinusoids and the third moments of the noise process. The term $\mathcal{O}(n^{-1})$ in r_{n4} can be bounded by $n^{-1}|\kappa - 2 - \xi|$, where $\kappa := E(|\zeta_t|^4)/\sigma^4$ and $\xi := |E(\zeta_t^2)|^2/\sigma^4 = |\iota|^2$. All these terms, together with ι and r_{n2}, vanish when $\{\zeta_t\}$ is complex Gaussian white noise, in which case Theorem 6.6 reduces to Theorem 6.5.

Remark 6.10 Under the additional assumption $\sum |j|^{1/2}|\psi_j| < \infty$, it can be shown [46, p. 349] that both $\sum_{|u|<n}(|u|/n)|r_\epsilon(u)|$ and $\sum_{|u|\geq n}|r_\epsilon(u)|$ are less than or equal to $2\sigma^2 n^{-1/2}(\sum |j|^{1/2}|\psi_j|)(\sum |\psi_j|)$. Therefore, we can write $r_n = \mathcal{O}(n^{-1/2})$. Similarly, $r_{n2} = \mathcal{O}(n^{-1/2})$. Combining these expressions leads to $r_{n3} = \mathcal{O}(n^{-1/2})$ and $r_{n4} = \mathcal{O}(n^{-1})$. Therefore, in the real case without sinusoids, we obtain

$$\begin{aligned}
E\{I_n(\lambda_j)\} &= f_\epsilon(\lambda_j) + \mathcal{O}(n^{-1/2}), \\
\mathrm{Var}\{I_n(\lambda_j)\} &= f_\epsilon^2(\lambda_j) + \mathcal{O}(n^{-1/2}), \\
\mathrm{Cov}\{I_n(\lambda_j), I_n(\lambda_l)\} &= \mathcal{O}(n^{-1}) \qquad (\lambda_j \neq \lambda_l),
\end{aligned}$$

where $\lambda_j, \lambda_l \in (0, \pi)$. This is a classical result for the periodogram of linear processes [46, pp. 347–348] [298, pp. 420–426].

A pair of frequencies are said to be conjugate if they add up to zero (mod 2π). Owing to conjugate frequencies, the variance and covariance expressions in Theorem 6.6 are more complicated when $\iota \neq 0$ than they are in Theorem 6.5. Furthermore, when $\iota \neq 0$, the correlation coefficient of the periodogram ordinates at two conjugate frequencies does not necessarily approach zero.

At nonconjugate frequencies, or at any frequencies when $\iota = 0$, the correlation coefficient of the periodogram ordinates can be simply expressed as

$$\mathrm{Corr}\{I_n(\lambda_j), I_n(\lambda_l)\} = \begin{cases} \mathcal{O}(r_{n3}) & \text{if both } \lambda_j \text{ and } \lambda_l \text{ are signal frequencies,} \\ \mathcal{O}(n^{-1/2}) & \text{if only one of them is a signal frequency,} \\ \mathcal{O}(r_{n4}) & \text{if none of them is a signal frequency.} \end{cases}$$

So the periodogram ordinates remain asymptotically uncorrelated in this case. Unlike the Gaussian case, the correlation coefficient at a nonsignal frequency and a signal frequency decays at a slower rate $n^{-1/2}$ instead of $n^{-1/2}r_n^2$. This is due entirely to the possibility of nonvanishing third moment of the noise.

To end this section, let us discuss a test for white noise based on the standardized cumulative periodogram ordinates

$$U_j := \frac{\sum_{l=0}^{j-1} I_n(\lambda_l)}{\sum_{l=0}^{n-1} I_n(\lambda_l)} \qquad (j = 1, \ldots, n-1).$$

According to Theorem 6.2, the periodogram ordinates of Gaussian white noise are distributed as i.i.d. $\frac{1}{2}\sigma^2\chi^2(2)$ random variables. Using the same argument that proves Proposition 10.2.1 in [46, p. 338] for the real case, we can show that the U_j are distributed as the order statistics of an i.i.d. sample of size $n-1$ from the uniform distribution in the interval $(0,1)$, denoted as $\mathrm{U}(0,1)$, regardless of σ^2. Therefore, the function

$$C_n(x) := (n-1)^{-1} \sum_{j=1}^{n-1} \mathcal{I}(U_j \le x) \qquad (0 \le x \le 1)$$

has the same distribution as the empirical distribution function of the i.i.d. sample whose CDF is given by $C(x) := x$ for $x \in (0,1)$. Any significant departure of $C_n(x)$ from this CDF can be regarded as an indication of nonwhite noise.

To measure such departure, we can use the *Kolmogorov-Smirnov statistic*

$$K_n := \sup_{0 \le x \le 1} |C_n(x) - x| = \max_{1 \le j \le n-1} \left\{ U_j - \frac{j-1}{n-1}, \frac{j}{n-1} - U_j \right\}. \qquad (6.1.29)$$

It can be shown [109, p. 394] that under the null hypothesis of Gaussian white noise, the CDF of $\sqrt{n-1}\,K_n$ is approximately (for large n) equal to

$$F_{\mathrm{KS}}(x) := 1 - 2\sum_{u=1}^{\infty} (-1)^{u-1} \exp(-2u^2 x).$$

Therefore, the null hypothesis can be rejected at significance level $\alpha \in (0,1)$ when $\sqrt{n-1}\,K_n > \tau_\alpha := F_{\mathrm{KS}}^{-1}(1-\alpha)$, where $\tau_{0.05} = 1.36$ and $\tau_{0.01} = 1.63$. This is known as the *Kolmogorov-Smirnov test for white noise.*

Due to Parseval's identity (6.1.7), we can write $\sum_{l=0}^{n-1} I_n(\lambda_l) = n\hat{r}_y(0)$. Therefore, the standardized empirical spectral distribution function, defined as

$$\tilde{S}_n(\omega) := \begin{cases} 0 & \text{if } \omega < \lambda_0, \\ U_j & \text{if } \lambda_{j-1} \le \omega < \lambda_j \ (j = 1, \ldots, n-1), \\ 1 & \text{if } \omega \ge \lambda_{n-1}, \end{cases}$$

is an estimator of the standardized spectral distribution function

$$\tilde{S}_\epsilon(\omega) := (2\pi)^{-1} \int_0^\omega \{f_\epsilon(\lambda)/r_\epsilon(0)\}\, d\lambda.$$

Under the white noise assumption, we have $\tilde{S}_\epsilon(\omega) = \omega/(2\pi)$ for $\omega \in [0, 2\pi)$. Therefore, in effect, the Kolmogorov-Smirnov statistic K_n, which is defined by (6.1.29), measures the deviation of $\tilde{S}_n(\lambda_{j-1}) = U_j$ from $\tilde{S}_\epsilon(\lambda_{j-1}) = \lambda_{j-1}/(2\pi) = (j-1)/n$.

In the real case, the Kolmogorov-Smirnov test can be constructed similarly from $U_j := \sum_{l=1}^{j} I_n(\lambda_l) / \sum_{l=1}^{m} I_n(\lambda_l)$ for $j = 1, \ldots, m-1$, provided that n is everywhere replaced by $m := \lfloor (n-1)/2 \rfloor$. In other words, the periodogram ordinates are restricted to the interval $(0, \pi)$ in light of Remark 6.2.

6.2 Detection of Hidden Sinusoids

There are practical situations where the presence or absence of sinusoids in a time series needs to be determined. This problem is known as the *detection of hidden sinusoids* or *periodicities*. Although the problem has a long history, Schuster [333] and Fisher [108] are generally acknowledged as the pioneers for laying the theoretical foundation of statistical analysis in this field. The problem can be formulated as testing statistical hypotheses. In this section, we only consider the simple case of detecting sinusoids in Gaussian white noise, where the data vector $\mathbf{y}$ takes the form (6.1.1) with $\boldsymbol{\epsilon} \sim \mathrm{N}_c(\mathbf{0}, \sigma^2\mathbf{I})$. A similar discussion on the case of colored noise is deferred to Chapter 7. See Priestley's book [298, Section 6.1.4] for additional information. Note that most (if not all) of the tests discussed in this section were originally formulated for the real-valued case, not the complex-valued case as is the central focus here.

First, consider the problem of detecting the presence of a sinusoid at a given frequency $\lambda_j \in \Omega_n$. This problem can be stated as testing the hypotheses

$$\begin{cases} H_0: & \text{sinusoid is absent at frequency } \lambda_j, \\ H_1: & \text{sinusoid is present at frequency } \lambda_j. \end{cases} \tag{6.2.1}$$

According to Theorem 6.2, $I_n(\lambda_j)$ is distributed as $\frac{1}{2}\sigma^2\chi^2(2)$ under H_0 and as $\frac{1}{2}\sigma^2\chi^2(2, 2n\gamma_k)$ under H_1 with $\lambda_j = \omega_k$ for some k. Therefore, when σ^2 is known, it is natural to reject H_0 in favor of H_1 if

$$\tilde{I}_n(\lambda_j) := I_n(\lambda_j)/\sigma^2 > \tfrac{1}{2}\tau, \tag{6.2.2}$$

where τ is set at the $(1-\alpha)$th quantile of $\chi^2(2)$ for some $\alpha \in (0,1)$, i.e.,

$$\tau = -2\log\alpha. \tag{6.2.3}$$

This test is known as *Schuster's test* for hidden sinusoids.

Schuster's test has a significance level α because $\tilde{I}_n(\lambda_j) \sim \frac{1}{2}\chi^2(2)$ under H_0 by Theorem 6.2. Similarly, the power (or probability of detection) of Schuster's test when $\lambda_j = \omega_k$ for some k is equal to

$$\bar{F}(\tau, 2n\gamma_k) := 1 - F(\tau, 2n\gamma_k),$$

where $F(x,\theta)$ denotes the CDF of $\chi^2(2,\theta)$. The power increases with $n\gamma_k$ and approaches unity as $n\gamma_k \to \infty$ because $F(x,\theta) \to 0$ as $\theta \to \infty$ for fixed $x > 0$.

Schuster's test can be applied successively to each frequency in a given m-subset $\Omega_m \subset \Omega_n$ for detection of multiple sinusoids. This procedure is known as *multiple testing*. If Ω_m contains all the signal frequencies, then the frequencies in Ω_m that are tested positive serve as estimates of the signal frequencies. The number of frequencies tested positive, i.e.,

$$\hat{p} := \sum_{\lambda_j \in \Omega_m} \mathscr{I}(\tilde{I}_n(\lambda_j) > \tfrac{1}{2}\tau), \tag{6.2.4}$$

can be used to estimate the parameter p.

To evaluate the statistical performance of $\hat{p}$ in (6.2.4), observe that the random variables $\tilde{I}_n(\lambda_j)$ $(\lambda_j \in \Omega_m)$ are mutually independent with $\tilde{I}_n(\lambda_j) \sim \frac{1}{2}\chi^2(2)$ if $\lambda_j \notin \{\omega_k\}$ and $\tilde{I}_n(\lambda_j) \sim \frac{1}{2}\chi^2(2, 2n\gamma_k)$ if $\lambda_j = \omega_k$. Moreover, if u represents the number of frequencies tested positive and Ω_v denotes the set of true positives among them, which can be any v-subset of $\{\omega_k\}$, then there must be (a) $p - v$ false negatives in $\{\omega_k\} \setminus \Omega_v$, (b) $u - v$ false positives which can be any $(u-v)$-subset of $\Omega_m \setminus \{\omega_k\}$, and (c) $m - p + v - u$ true negatives. Therefore, the probability of underestimation can be expressed as

$$\begin{aligned}
P(\hat{p} < p) &= \sum_{u=0}^{p-1} P(\text{exactly } u \text{ frequencies are tested positive}) \\
&= \sum_{u=0}^{p-1} \sum_{v=0}^{u} \sum_{\Omega_v} \binom{m-p}{u-v} \bar{F}(\tau)^{u-v} F(\tau)^{m-p+v-u} \\
&\quad \times \Big\{ \prod_{\omega_k \in \Omega_v} \bar{F}(\tau, 2n\gamma_k) \Big\} \Big\{ \prod_{\omega_k \notin \Omega_v} F(\tau, 2n\gamma_k) \Big\},
\end{aligned}$$

where $\bar{F}(\tau) := 1 - F(\tau)$. Similarly, the probability of overestimation is given by

$$P(\hat{p} > p) = \sum_{u=p+1}^{m} \sum_{v=0}^{p} \sum_{\Omega_v} \binom{m-p}{u-v} \bar{F}(\tau)^{u-v} F(\tau)^{m-p+v-u}$$

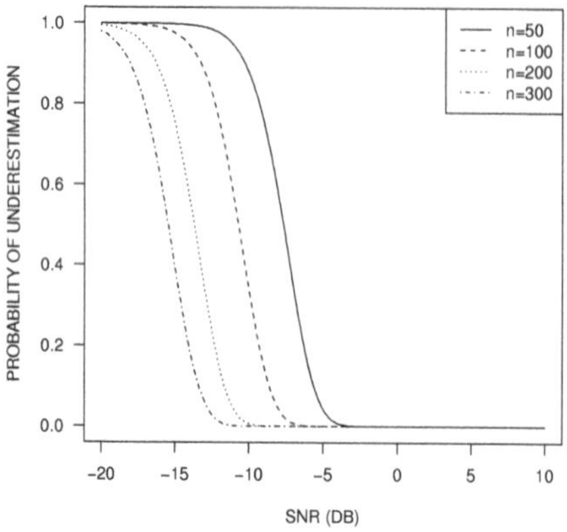

(a)

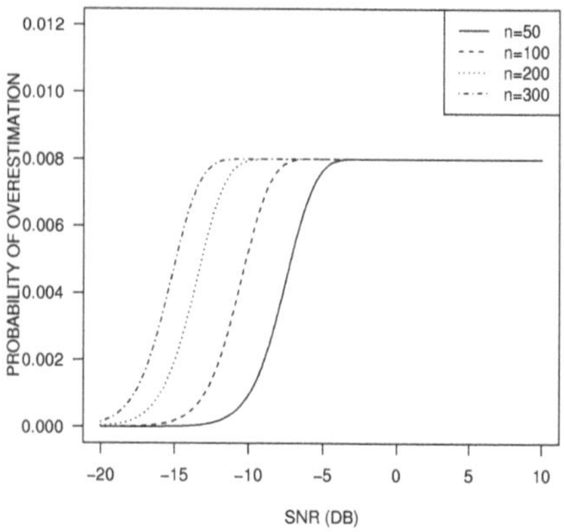

(b)

Figure 6.5. Probability of (a) underestimation of p and (b) overestimation of p by $\hat{p}$ in (6.2.4) for two complex sinusoids with Fourier frequencies and equal amplitudes (hence equal SNR γ) under the independence assumption. Solid line, $n = 50$; dashed line, $n = 100$; dotted line, $n = 200$; dot-dashed line, $n = 300$. In all cases, $m = 10$ and $\alpha = 0.01$.

$$\times \Big\{ \prod_{\omega_k \in \Omega_v} \bar{F}(\tau, 2n\gamma_k) \Big\} \Big\{ \prod_{\omega_k \notin \Omega_v} F(\tau, 2n\gamma_k) \Big\}.$$

As $n \to \infty$ and for fixed m, we obtain $P(\hat{p} < p) \to 0$ and

$$P(\hat{p} > p) \to \sum_{u=p+1}^{m} \binom{m-p}{u-p} \bar{F}(\tau)^{u-p} F(\tau)^{m-u} = 1 - F(\tau)^{m-p} \approx (1 - p/m)\alpha.$$

Observe that $P(\hat{p} > p) \approx \alpha$ for large n when p/m is small.

Example 6.6 (Detection of Two Complex Sinusoids). For illustration, let $p = 2$ and $\gamma_1 = \gamma_2 = \gamma$. In this case,

$$\begin{aligned} P(\hat{p} < p) &= \{F(\tau) + (m-2)\bar{F}(\tau)\} F(\tau)^{m-3} F(\tau, 2n\gamma)^2 \\ &\quad + 2F(\tau)^{m-2} \bar{F}(\tau, 2n\gamma) F(\tau, 2n\gamma), \\ P(\hat{p} > p) &= a F(\tau, 2n\gamma)^2 + 2b \bar{F}(\tau, 2n\gamma) F(\tau, 2n\gamma) + c \bar{F}(\tau, 2n\gamma)^2, \end{aligned}$$

where $a := b - \frac{1}{2}(m-2)(m-3)\bar{F}(\tau)^2 F(\tau)^{m-4}$, $b := c - (m-2)\bar{F}(\tau) F(\tau)^{m-3}$, $c := 1 - F(\tau)^{m-2}$. Figure 6.5 depicts these probabilities as functions of γ for $m = 10$, $\alpha = 0.01$, and for various sample sizes.

As we can see, with the increase of the SNR γ, the probability of underestimation decreases from near unity to zero at roughly the same time when the probability of overestimation increases from near zero to the plateau of 0.008 which is approximately equal to the probability of false alarm $\alpha = 0.01$. The point at which the probabilities change dramatically as the SNR increases depends on the sample size: the larger the sample size, the lower the turning point. It is due to the fact that both probabilities depend on n and γ through the product $n\gamma$. ◇

The multiple testing procedure can be used to test the hypotheses

$$\begin{cases} H_0: & \text{no sinusoids are present in } \Omega_m, \\ H_1: & \text{at least one sinusoid is present in } \Omega_m. \end{cases} \tag{6.2.5}$$

It suffices to reject H_0 in favor of H_1 if at least one frequency in Ω_m is tested positive. The false-alarm probability of this procedure is given by

$$\begin{aligned} P_F &:= P\{\tilde{I}_n(\lambda_j) > \tfrac{1}{2}\tau \text{ for some } \lambda_j \in \Omega_m \mid H_0\} \\ &= 1 - P\{\tilde{I}_n(\lambda_j) \le \tfrac{1}{2}\tau \text{ for all } \lambda_j \in \Omega_m \mid H_0\}. \end{aligned}$$

Under H_0, the $\tilde{I}_n(\lambda_j)$ $(\lambda_j \in \Omega_m)$ are distributed as i.i.d. $\frac{1}{2}\chi^2(2)$ random variables. Therefore, we can write $P_F = 1 - F(\tau)^m$, where $F(x)$ is the CDF of $\chi^2(2)$, i.e.,

$$F(x) := \{1 - \exp(-\tfrac{1}{2}x)\}\mathcal{I}(x > 0). \tag{6.2.6}$$

By taking τ as the solution to the equation $F(\tau) = (1-\alpha)^{1/m}$, i.e.,

$$\tau = -2\log\{1 - (1-\alpha)^{1/m}\}, \tag{6.2.7}$$

we obtain $P_F = \alpha$. For large m, we can write

$$\tau \approx -2\log(\alpha/m).$$

Hence, for multiple testing, the threshold of Schuster's test (6.2.2) can be obtained by simply replacing α in (6.2.3) with α/m. This technique of adjusting the threshold for multiple testing is known as *Bonferroni's procedure*, named after the Italian mathematician Carlo Emillio Bonferroni (1892–1960). Without the adjustment, the original threshold given by (6.2.3) tends to produce excessive false positives when applied to multiple frequencies.

The multiple testing procedure for the hypotheses in (6.2.5) with τ given by (6.2.7) is equivalent to a test that rejects H_0 in favor of H_1 when

$$g := \max_{\lambda_j \in \Omega_m} \{\tilde{I}_n(\lambda_j)\} > \tfrac{1}{2}\tau. \tag{6.2.8}$$

This statistic can be replaced by the $\tilde{p}$th largest value in $\{\tilde{I}_n(\lambda_j) : \lambda_j \in \Omega_m\}$ for a given integer $\tilde{p} \in \{1, \ldots, m\}$ to test the hypotheses

$$\begin{cases} H_0: & \text{no sinusoids are present in } \Omega_m, \\ H_1: & \text{at least } \tilde{p} \text{ sinusoids are present in } \Omega_m. \end{cases} \tag{6.2.9}$$

If $g_{\tilde{p}}$ denotes the resulting statistic, then it follows from by Theorem 6.2 that $2g_{\tilde{p}}$ is distributed under H_0 as the $\tilde{p}$th largest order statistic of m i.i.d. $\chi^2(2)$ random variables, whose CDF takes the form

$$F_{\tilde{p}}(x) := \sum_{u=0}^{\tilde{p}-1} \binom{m}{u} F(x)^{m-u} \bar{F}(x)^u. \tag{6.2.10}$$

The $(1-\alpha)$th quantile of this distribution can be used as the threshold τ so that the test of rejecting H_0 in favor of H_1 when $g_{\tilde{p}} > \frac{1}{2}\tau$ has a significance level α.

The tests which we discussed so far are based on the assumption that the noise variance σ^2 is known *a priori*. When σ^2 is unknown, one should replace it by an estimator $\hat{\sigma}^2$ as suggested in [414]. If $\hat{\sigma}^2$ is a consistent estimator of σ^2, then, as $n \to \infty$ and for fixed m, the i.i.d. $\frac{1}{2}\chi^2(2)$ distribution under H_0 remains valid for the resulting random variables $I_n(\lambda_j)/\hat{\sigma}^2$ $(\lambda_j \in \Omega_m)$. In this case, the resulting tests are guaranteed to have an asymptotic level α without changing the thresholds. A simple choice of $\hat{\sigma}^2$ is the sample variance given by

$$\hat{\sigma}^2 := \hat{r}_y(0). \tag{6.2.11}$$

In the absence of sinusoids, this estimator is consistent for estimating σ^2 under the Gaussian white noise assumption, as ensured by Theorem 4.1.

When the sinusoids are present, the sample variance in (6.2.11) is no longer a consistent estimator of σ^2. In fact, by Theorem 4.1, we have

$$\hat{r}_y(0) \xrightarrow{P} r_y(0) = \sum_{k=1}^{p} C_k^2 + \sigma^2 > \sigma^2,$$

which means that the sample variance tends to overestimate σ^2 in the presence of sinusoids. The overestimation of σ^2 reduces the magnitude of $I_n(\lambda_j)/\hat{\sigma}^2$ and decreases the power of the tests as a result.

To overcome this problem, one can follow Whittle's suggestion [422] and simultaneously estimate σ^2 and the sinusoidal parameters by the maximum likelihood method. Alternatively, one can treat the sinusoids as outliers in the periodogram and employ a more robust estimator of σ^2. The latter approach is taken by Bölviken [40] and Chui [70]. The idea is very simple. Owing to Parseval's identity (6.1.7), the estimator $\hat{\sigma}^2$ in (6.2.11) can also be expressed as

$$\hat{\sigma}^2 = n^{-1} \sum_{j=0}^{n-1} I_n(\lambda_j). \tag{6.2.12}$$

In other words, the sample variance of $\mathbf{y}$ coincides with the sample mean of the periodogram ordinates of $\mathbf{y}$. According to Theorem 6.2, the periodogram ordinates are distributed under H_0 as i.i.d. random variables with mean σ^2, but the presence of sinusoids gives rise to large outliers in the periodogram. A robust estimator of σ^2 can be obtained by simply trimming off a few of the largest values in the periodogram and then computing the average of the remaining values. With $r \in (0,1)$ representing the fraction of remaining values, the *trimmed sample mean* of the periodogram can be expressed as

$$\hat{\sigma}_r^2 := b_r^{-1}[nr]^{-1} \sum_{j=1}^{[nr]} I_{nj}, \tag{6.2.13}$$

where $I_{n1} \le \cdots \le I_{nn}$ denote the ordered periodogram ordinates and

$$b_r := 1 + r^{-1}(1-r)\log(1-r) \tag{6.2.14}$$

is the bias correction factor which makes $\hat{\sigma}_r^2$ a consistent estimator of σ^2 in the absence of sinusoids, just like the untrimmed sample mean in (6.2.12).

The trimmed mean in (6.2.13) is only an example of the so-called *L-estimators* [218, Section 5.5] that take the form

$$\hat{\sigma}_L^2 := b^{-1} \sum_{j=1}^{n} W_{nj} I_{nj}.$$

In this expression, the W_{nj} are constant weights satisfying $W_{nj} \propto K(j/n)$ and $\sum_{j=1}^n W_{nj} = 1$, where $K(x)$ is a nonnegative, bounded, and almost everywhere continuous function of $x \in (0,1]$ such that $\int_0^1 K(x)\,dx = 1$. The trimmed mean in (6.2.13) corresponds to the choice $K(x) = (1/r)\mathscr{I}(0 < x \le r)$.

With regard to the choice of b, we note that the periodogram ordinates are i.i.d. $\frac{1}{2}\sigma^2\chi^2(2)$ random variables in the absence of sinusoids. According to Corollary 5.1 in [218, p. 370] (see also [362]), if the constant b is given by

$$b := \tfrac{1}{2}\int_0^1 K(x) F^{-1}(x)\,dx = \tfrac{1}{2}\int_0^\infty x K(F(x))\,dF(x), \tag{6.2.15}$$

then $\sqrt{n}(\hat{\sigma}_L^2 - \sigma^2) \xrightarrow{D} \mathrm{N}(0, \upsilon^2)$ as $n \to \infty$ for some constant $\upsilon^2 > 0$. This result implies that $\hat{\sigma}_L^2$ is a consistent estimator of σ^2 in the absence of sinusoids. For the trimmed mean in particular, the condition (6.2.15) becomes

$$b = r^{-1}\int_0^{-\log(1-r)} x\exp(-x)\,dx = 1 + r^{-1}(1-r)\log(1-r),$$

which is just the bias correction factor b_r in (6.2.14). An alternative choice for the constant b_r is given by

$$b_r := [nr]^{-1} \sum_{j=1}^{[nr]} \sum_{l=1}^{j} (n-l+1)^{-1}.$$

This choice makes the trimmed mean in (6.2.13) unbiased for all n in the absence of sinusoids [188].

Table 6.1. Power of Schuster's Test with Estimated Noise Variance

	SNR (dB)							
Test	−12	−11	−10	−9	−8	−7	−6	−5
Ordinary	0.29	0.39	0.51	0.64	0.76	0.87	0.94	0.98
Robust	0.32	0.43	0.56	0.69	0.82	0.92	0.97	0.99
Oracle	0.37	0.48	0.61	0.75	0.86	0.94	0.98	1.00

Results are based on 10,000 Monte Carlo runs.

In the presence of sinusoids, it can be shown, along the lines of the proof of Theorem 6.3, that the probability of the periodogram ordinates at all signal frequencies being trimmed in (6.2.13) approaches unity as $n \to \infty$, which means that the trimmed mean $\hat{\sigma}_r^2$ remains consistent for estimating σ^2 with any fixed $r \in (0,1)$. This is where the trimmed mean differs from the untrimmed mean.

The following simulation example demonstrates the power of Schuster's test (6.2.2) based on different estimators of σ^2.

Example 6.7 (Detection of Sinusoids in Gaussian White Noise by Schuster's Test with Estimated Noise Variance). Let $\mathbf{y}$ be given by (6.1.1) with $\boldsymbol{\epsilon} \sim \mathrm{N}_c(0, \sigma^2\mathbf{I})$, $n = 50$, $\sigma^2 = 1$, $p = 2$, $C_1 = C_2 = C$, $\phi_1 = \phi_2 = 0$, $\omega_1 = 2\pi \times 10/n = 2\pi \times 0.2$, and $\omega_2 = 2\pi \times 20/n = 2\pi \times 0.4$. Table 6.1 shows the probability of detection for Schuster's test (6.2.2) at frequency $\lambda_{10} = \omega_1$ with three options for the estimator of σ^2:

1. the ordinary test using $\hat{\sigma}^2$ given by (6.2.11),
2. the robust test using $\hat{\sigma}_r^2$ given by (6.2.13) with $r = 0.9$, and
3. the oracle test using the true value of σ^2.

In all cases, the threshold τ is given by (6.2.3) with $\alpha = 0.01$. The SNR $\gamma := C^2/\sigma^2$ ranges from -12 to -5 dB. As can be seen, the ordinary test experiences a noticeable loss of detection power in comparison with the oracle test. The robust test is able to reduce the power loss by a varying degree, thanks to the trimming of the $(1-r)n = 5$ (or 10%) largest periodogram ordinates. ◇

A similar comparison can be done for testing the hypotheses in (6.2.5) using the g statistic in (6.2.8). Let us consider the most general case where $\Omega_m = \Omega_n$. By taking the sample mean of the periodogram in (6.2.12) as the estimator of σ^2, the resulting test can be expressed as

$$g := \max\{I_n(\lambda_j)/\hat{\sigma}^2\} = n\frac{\max\{I_n(\lambda_j)\}}{\sum_{j=0}^{n-1} I_n(\lambda_j)} > \tfrac{1}{2}\tau. \tag{6.2.16}$$

The exact distribution of g in the absence of sinusoids can be derived under the Gaussian white noise assumption, and the distribution does not depend on the nuisance parameter σ^2. Indeed, by Theorem 6.2, $n^{-1}g$ has the same distribution

Table 6.2. Power of Fisher's Test with Estimated Noise Variance

	SNR (dB)							
Test	−12	−11	−10	−9	−8	−7	−6	−5
Ordinary	0.10	0.15	0.24	0.37	0.54	0.73	0.89	0.97
Robust	0.10	0.16	0.26	0.41	0.59	0.79	0.92	0.99
Oracle	0.15	0.23	0.37	0.54	0.74	0.89	0.97	1.00

Results are based on 10,000 Monte Carlo runs.

as the maximum-to-sum ratio of n i.i.d. $\frac{1}{2}\chi^2(2)$ random variables. Therefore, an application of Fisher's theorem [108] (see also [46, p. 338]) yields

$$F_g(x) := P(n^{-1} g \le x) = 1 - \sum_{u=1}^{n} \binom{n}{u} (-1)^{u-1} (1 - ux)_+^{n-1},$$

where $x_+ := \max\{x, 0\}$. With τ given by the solution to the equation

$$F_g(\tfrac{1}{2} n^{-1} \tau) = 1 - \alpha, \tag{6.2.17}$$

the resulting test (6.2.16) is known as *Fisher's test* for hidden sinusoids. It has an exact significance level α under the assumption of Gaussian white noise.

As with Schuster's test, Fisher's test can also be improved by replacing the untrimmed mean in (6.2.12) with the trimmed mean in (6.2.13). The resulting test can be expressed as

$$\tilde{g} := \max\{I_n(\lambda_j)/\hat{\sigma}_r^2\} = b_r[nr] \frac{\max\{I_n(\lambda_j)\}}{\sum_{j=1}^{[nr]} I_{nj}} > \tfrac{1}{2}\tau. \tag{6.2.18}$$

Under the Gaussian white noise assumption, the exact distribution of $\tilde{g}$ in the absence of sinusoids is derived by Bölviken [41]. The corresponding threshold is evaluated and tabulated in [40] for various sample sizes and for some typical values of α. Obviously, in the special case of $r = 1$ (or zero trimming), the test in (6.2.18) with the exact threshold reduces to Fisher's test in (6.2.16).

The following simulation example compares the power of Fisher's test based on different estimators of σ^2.

Example 6.8 (Detection of Sinusoids in Gaussian White Noise by Fisher's Test with Estimated Noise Variance). From the same data as in Example 6.7, the detection probability of Fisher's test is obtained with the same three options for the estimator of σ^2. The exact threshold for the oracle test that employs the true value of σ^2 is given by (6.2.7) with $m = n$, which equals 17.02 for $\alpha = 0.01$; the exact threshold for the ordinary test (6.2.16) is equal to 15.95; and the exact threshold for the robust test (6.2.18) with $r = 0.9$ is equal to 17.77. Table 6.2 shows the detection probability of these tests for various values of γ. As can be seen, while the power loss is inevitable when σ^2 has to be estimated, the robust test is able to reduce the power loss of the ordinary test. ◇

Fisher's test (6.2.16) can be easily extended to solve the problem in (6.2.9) with $\Omega_m = \Omega_n$ [220] [344]. In fact, it suffices to replace $\max\{I_n(\lambda_j)\} = I_{nn}$ in (6.2.16) by $I_{n,n-\tilde{p}+1}$, the $\tilde{p}$th largest periodogram ordinate. In this case, the threshold τ of an exact α-level test can be determined under the Gaussian white noise assumption by solving the equation [128, p. 93]

$$\frac{n!}{(\tilde{p}-1)!}\sum_{u=\tilde{p}}^{n}\frac{(-1)^{u-\tilde{p}}(1-u\tau)_+^{n-1}}{u(n-u)!(u-\tilde{p})!} = \alpha.$$

Some numerical values of the threshold are tabulated in [344]. The test in (6.2.18) can be generalized in the same way under the restriction $r \le 1 - \tilde{p}/n$ [40] [70]. With the threshold τ obtained by solving $F_{\tilde{p}}(\tau) = 1-\alpha$, where $F_{\tilde{p}}(x)$ is defined by (6.2.10) with $m = n$, the resulting test has an approximate level α under the Gaussian white noise assumption.

Another generalization of Fisher's test is Whittle's sequential test procedure [422]; see also [298, p. 410]. In this procedure, Fisher's test is applied recursively to the remaining periodogram ordinates after removing the largest one that tested positive in the previous recursion. The sample size is reduced by one after each recursion and the threshold is adjusted accordingly. This procedure is able to improve the power of Fisher's test for detecting multiple sinusoids, because the test statistic is properly amplified as a result of the removal of the largest periodogram ordinates that have been tested positive.

For the same purpose of detecting multiple sinusoids, Siegel [347] proposed yet another generalization of Fisher's test that takes the form

$$s_\rho := n^{-1}\sum_{j=0}^{n-1}\mathscr{I}\{I_n(\lambda_j)/\hat{\sigma}^2 > \tfrac{1}{2}\rho\tau_F\} > \tau,$$

where $\hat{\sigma}^2$ is given by (6.2.12), τ_F is the threshold of Fisher's test in (6.2.16), and $\rho \in (0,1]$ is a tuning parameter. This test is equivalent to Fisher's if $\rho = 1$ because in this case $s_\rho > 0$ if and only if $\max\{I_n(\lambda_j)/\hat{\sigma}^2\} > \frac{1}{2}\tau_F$. In general with $\rho < 1$, the test statistic s_ρ is proportional to the total number of periodogram ordinates tested positive by Fisher's test with a reduced threshold that enables an easier detection of multiple sinusoids.

Under the Gaussian white noise assumption, the exact CDF of s_ρ is given by

$$1-\sum_{u=1}^{n}\sum_{v=0}^{u-1}\binom{n}{u}\binom{u-v}{v}\binom{n-1}{v}(-1)^{u+v+1}x^v(1-\tfrac{1}{2}\rho\tau_F u/n - x)_+^{n-v-1}.$$

The $(1-\alpha)$th quantile of this distribution gives the threshold τ for an exact α-level test. Some typical values of τ are given in [347]. While Siegel's test with $\rho < 1$

is less powerful than Fisher's test in (6.2.16) for detecting a single sinusoid, there is evidence that it can outperform Fisher's test for detecting multiple sinusoids if ρ is suitably chosen (somewhere near unity) [347].

Note that for simplicity of presentation the zero frequency has not been excluded in Fisher's test and its generalizations. To exclude this frequency, one can simply use the subset $\{I_n(\lambda_1), \dots, I_n(\lambda_{n-1})\}$ in forming the test statistics as well as in estimating the noise variance. The distribution theory remains valid if the sample size n is everywhere replaced by $n-1$.

Similar tests can be easily obtained for detecting real sinusoids. In this case, it suffices to replace the set of periodogram ordinates involved in the tests for the complex case with the reduced set $\{I_n(\lambda_j) : 0 < \lambda_j < \pi\}$. The sample size in the distribution theory should be adjusted accordingly.

To end this section, let us discuss the consequences of applying the tests developed under the white noise assumption when the noise is actually colored. Consider, for example, Schuster's test (6.2.2) with the threshold τ given by (6.2.3). Suppose that the actual noise has an SDF $f_\epsilon(\omega)$ which is nonconstant. In this case, (6.2.2) can be rewritten as

$$I_n(\lambda_j)/f_\epsilon(\lambda_j) > \tfrac{1}{2}\tau\eta,$$

where $\eta := \sigma^2/f_\epsilon(\lambda_j)$. According to Theorem 6.4, the test would have an asymptotic level α if $\eta = 1$. However, observe that

$$\sigma^2 = (2\pi)^{-1} \int_{-\pi}^{\pi} f_\epsilon(\omega)\, d\omega.$$

In other words, σ^2 is the average value of the SDF. Therefore, when $f_\epsilon(\lambda_j)$ is less than σ^2 so that $\eta > 1$, the actual significance level of the test is less than α. On the other hand, when $f_\epsilon(\lambda_j)$ is greater than σ^2 so that $\eta < 1$, the actual significance level is greater than α. In either case, the actual significance level can be different from the nominal value α, even for large n. Nevertheless, a hidden sinusoid at frequency λ_j will be detected by Schuster's test if n is sufficiently large. Indeed, if a sinusoid is present at λ_j with amplitude C, then, by Theorem 6.4, $I_n(\lambda_j)/\sigma^2$ is approximately distributed as $\frac{1}{2}\chi^2(2, 2n\gamma)\eta^{-1}$, where $\gamma := C^2/\sigma^2$. Because the noncentrality parameter $2n\gamma$ tends to infinity as $n \to \infty$ whereas η is finite, the test statistic $I_n(\lambda_j)/\sigma^2$ has a greater chance of exceeding the threshold when a sinusoid is present at λ_j than when it is absent.

A slightly more sophisticated application of the white-noise-based tests to the case of colored noise is the *grouped periodogram test* [298, pp. 620–621]. In this procedure, one first partitions Ω_n into a number of subsets in which the noise spectrum can be approximated by a constant and then applies Schuster's test, Fisher's test, or any of their variations, to each subset with the threshold adjusted according to the size of the subset. The grouped periodogram test has the appeal of not requiring elaborate estimation of the noise spectrum. However, certain knowledge about the smoothness of the noise spectrum is still required for proper partition of Ω_n: if the subsets are too large, the piecewise-constant assumption may not be valid and the resulting test may have an uncontrolled significance level; if the subsets are too small, the resulting test may not have enough detection power.

6.3 Extension of the Periodogram

Up to now, the signal frequencies are assumed to be Fourier frequencies of the form $2\pi j/n$, where j is an integer and n is the length of $\mathbf{y}$. This assumption, of course, is not always satisfied in practice. For detection and estimation of non-Fourier frequencies, the DFT method, which is based on the Fourier frequencies, may not provide satisfactory results, as demonstrated by Figure 6.1. In this section, let us relax the assumption by allowing some or all signal frequencies to be nonFourier frequencies.

6.3.1 Sinusoids with NonFourier Frequencies

Figure 6.1 shows that in the periodogram the peak value corresponding to a nonFourier signal frequency is lower than the peak value corresponding to a Fourier signal frequency, even though the sinusoids have the same amplitude. To better understand this observation, consider the simple case where

$$\mathbf{y} = \beta_1 \mathbf{f}(\omega_1), \quad \omega_1 \notin \Omega_n.$$

In other words, $\mathbf{y}$ is a noiseless complex sinusoid with nonFourier frequency. To compute the periodogram of $\mathbf{y}$, let

$$D_n(\omega) := n^{-1} \sum_{t=1}^{n} \exp(it\omega), \tag{6.3.1}$$

which is known as the *Dirichlet kernel*, named after the German mathematician Johann Dirichlet (1805–1859). Then, it is easy to see that

$$z_j := n^{-1}\mathbf{f}^H(\lambda_j)\mathbf{y} = \beta_1 n^{-1}\mathbf{f}^H(\lambda_j)\mathbf{f}(\omega_1) = \beta_1 D_n(\omega_1 - \lambda_j).$$

Therefore, with $C_1 := |\beta_1|$, the periodogram of $\mathbf{y}$ can be expressed as

$$I_n(\lambda_j) = n|z_j|^2 = nC_1^2 K_n(\omega_1 - \lambda_j) \qquad (\lambda_j \in \Omega_n), \tag{6.3.2}$$

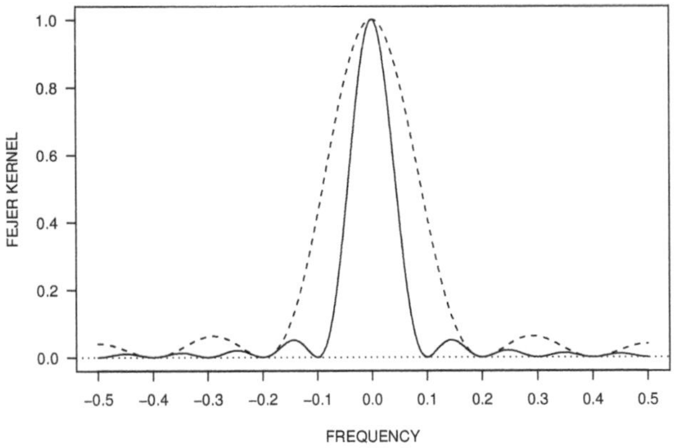

Figure 6.6. Plot of the Fejér kernel $K_n(\omega)$ in (6.3.3) as a function of $f := \omega/(2\pi)$. Solid line, $n = 10$; dashed line, $n = 5$.

where $K_n(\omega) := |D_n(\omega)|^2$ is known as the *Fejér kernel*, named after the Hungarian mathematician Lipót Fejér (1880–1959). By Lemma 12.1.3,

$$K_n(\omega) = \frac{\sin^2(n\omega/2)}{n^2 \sin^2(\omega/2)}. \tag{6.3.3}$$

Hence $K_n(\omega)$ is a nonnegative symmetric 2π-periodic function with $K_n(0) = 1$. Figure 6.6 depicts the Fejér kernel for $n = 5$ and 10.

Because $K_n(\omega) = 0$ if and only if $\omega = 2\pi l/n$ for some integer $l \neq 0$, it follows that the periodogram $I_n(\lambda_j)$ in (6.3.2) is strictly positive for all $\lambda_j \in \Omega_n$. This implies that the power of a sinusoid with nonFourier frequency is no longer concentrated in the periodogram at a single frequency as in Example 6.2 — it is spread to all frequencies. This phenomenon is known as *spectral leakage*.

To investigate the distribution of the periodogram when $\mathbf{y} = \mathbf{x} + \boldsymbol{\epsilon}$ is given by (6.1.1) with the possibility of nonFourier signal frequencies, observe that the DFT of $\mathbf{y}$ defined by (6.1.4) can be written as

$$z_j = \mu_j + e_j \qquad (j = 0, 1, \ldots, n-1), \tag{6.3.4}$$

where

$$\mu_j := n^{-1}\mathbf{f}^H(\lambda_j)\mathbf{x} = \sum_{k=1}^{p} \beta_k D_n(\omega_k - \lambda_j), \tag{6.3.5}$$

$$e_j := n^{-1}\mathbf{f}^H(\lambda_j)\boldsymbol{\epsilon} = n^{-1}\sum_{t=1}^{n} \epsilon_t \exp(-it\lambda_j). \tag{6.3.6}$$

Based on this expression, the following result can be easily obtained as an extension to Theorems 6.2 and 6.4.

Corollary 6.2 (Distribution of the Periodogram: NonFourier Signal Frequency). *Let* $\mathbf{y}$ *be given by (6.1.1) and* μ_j *be defined by (6.3.5).*

(a) *Gaussian White Noise. If* $\{\epsilon_t\} \sim \text{GWN}_c(0,\sigma^2)$, *then the random variables* $I_n(\lambda_0), I_n(\lambda_1), \ldots, I_n(\lambda_{n-1})$ *are independent and* $I_n(\lambda_j) \sim \frac{1}{2}\sigma^2\chi^2(2,2\theta_j)$ *for any* $\lambda_j \in \Omega_n$, *where* $\theta_j := n|\mu_j|^2/\sigma^2$.

(b) *Gaussian Colored Noise. If* $\{\epsilon_t\}$ *is a complex Gaussian process satisfying the conditions of Theorem 6.4, then* $I_n(\lambda_j) \sim \frac{1}{2}\rho_j^{-1} f_\epsilon(\lambda_j)\chi^2(2,2\rho_j\theta_j)$ *for any* $\lambda_j \in \Omega_n$, *where* $\theta_j := n|\mu_j|^2/f_\epsilon(\lambda_j)$ *and* $\rho_j = 1 + \mathcal{O}(r_n)$.

Remark 6.11 If all signal frequencies are Fourier frequencies such that $\omega_k = \lambda_{j(k)}$ for all k, then we have $\mu_j = \sum_{k=1}^{p} \beta_k \delta_{j-j(k)}$, in which case, Corollary 6.2 reduces to Theorems 6.2 and 6.4.

According to Corollary 6.2, the parameter θ_j plays an important role in determining how large the periodogram tends to be at frequency λ_j. To better understand its property, let us compare the magnitude of θ_j when λ_j is near a signal frequency with that when λ_j is far away from all signal frequencies. For simplicity, assume that the signal frequencies satisfy the separation condition

$$\lim_{n\to\infty} n\Delta = \infty, \tag{6.3.7}$$

where

$$\Delta := \min_{k\neq k'}\{|\omega_k - \omega_{k'}|, 2\pi - |\omega_k - \omega_{k'}|\}. \tag{6.3.8}$$

If $\lambda_{j(k)} \in \Omega_n$ denotes the Fourier frequency nearest to ω_k, then $\Delta_k := \omega_k - \lambda_{j(k)}$ satisfies $|\Delta_k| \le \pi/n$. In this case, it follows from (6.3.7) that

$$n \min_{k\neq k'}\{|\omega_{k'} - \lambda_{j(k)}|, 2\pi - |\omega_{k'} - \lambda_{j(k)}|\} \to \infty.$$

By Lemma 12.1.3(d), we can write $D_n(\omega_{k'} - \lambda_{j(k)}) = o(1)$ for $k' \neq k$. Substituting this result in (6.3.5) yields

$$\mu_{j(k)} = \beta_k D_n(\Delta_k) + o(1),$$

which in turn implies that

$$\theta_{j(k)} = n\gamma_k K_n(\Delta_k) + o(n), \tag{6.3.9}$$

where $\gamma_k := C_k^2/\sigma^2$ and $C_k := |\beta_k|$. Owing to the symmetry and monotonicity of $K_n(\omega)$ in the interval $[0, 2\pi/n]$, we have $K_n(\Delta_k) \ge K_n(\pi/n)$. As $n \to \infty$, $K_n(\pi/n) = \{n\sin(\frac{1}{2}\pi/n)\}^{-2} \to 4/\pi^2$. Therefore, there is a constant $c > 0$ such that

$$K_n(\Delta_k) \ge c.$$

On the other hand, if λ_j is far away from all signal frequencies such that

$$n \min_k \{|\lambda_j - \omega_k|, 2\pi - |\lambda_j - \omega_k|\} \to \infty,$$

then we have $\mu_j = \mathcal{O}(1)$ by Lemma 12.1.3(d). This implies that

$$\theta_j = \mathcal{O}(n). \tag{6.3.10}$$

By comparing (6.3.10) with (6.3.9), we can see that θ_j achieves a higher order of magnitude for λ_j near a signal frequency than for λ_j far away from the signal frequencies. This difference ensures that for sufficiently large n the periodogram still has a greater chance of taking larger values near the signal frequencies than elsewhere, even if the signal frequencies are nonFourier.

However, when ω_k is a nonFourier frequency, the spectral leakage reduces the effective SNR in $\theta_{j(k)}$ from its maximum value γ_k, which is attainable only if ω_k is a Fourier frequency, to the value $\gamma_k K_n(\Delta_k)$. Observe that $K_n(\pi/n) \approx 4/\pi^2 \approx 0.4$ for large n. This means that the effective SNR can be reduced by as much as 60% from its peak value. Such reduction degrades the power of the tests discussed in Section 6.2 for detecting hidden sinusoids.

6.3.2 Refined Periodogram

To mitigate the difficulty caused by spectral leakage, one can simply compute the Fourier transform in (6.1.4) on a finer frequency grid and define a new periodogram using the refined Fourier transform in the same way as (6.1.6). This technique is widely used for detection and estimation of hidden periodicities [63] [64] [298, p. 403] [316] [317].

More specifically, let the density of the original n-point Fourier grid Ω_n be increased by a factor of m, for some integer $m > 1$, and let the finer grid be denoted by $\Omega_{mn} := \{\lambda'_0, \lambda'_1, \ldots, \lambda'_{mn-1}\} \supset \Omega_n$, where

$$\lambda'_j := 2\pi j/(mn), \quad j \in \mathbb{Z}_{mn} := \{0, 1, \ldots, mn-1\}.$$

The refined Fourier transform of $\mathbf{y}$ on Ω_{mn} is defined by

$$z'_j := n^{-1}\mathbf{f}^H(\lambda'_j)\mathbf{y} = n^{-1}\sum_{t=1}^{n} y_t \exp(-it\lambda'_j) \qquad (j \in \mathbb{Z}_{mn}). \tag{6.3.11}$$

As an extension of the ordinary DFT in (6.1.4), the refined Fourier transform can be interpreted as the projection of $\mathbf{y}$ on the vectors $\mathbf{f}(\lambda'_0), \mathbf{f}(\lambda'_1), \ldots, \mathbf{f}(\lambda'_{mn-1})$ which constitute a redundant basis of $\mathbb{C}^n$. The refined Fourier transform in (6.3.11) is also known as the mn-point DFT of $\mathbf{y}$. It can be calculated efficiently by the FFT algorithms combined with a technique called *zero-padding*. In fact, if $\mathbf{z}$ in (6.1.4) defines the DFT of the n-dimensional vector $\mathbf{y}$, then $m^{-1}[z'_0, z'_1, \ldots, z'_{mn-1}]^T$ is nothing but the DFT of the mn-dimensional vector $[y_1,\ldots,y_n,0,\ldots,0]^T$ which is formed by padding $(m-1)\times n$ zeros at the end of the time series $\{y_1,\ldots,y_n\}$. Note that zero padding is based on the assumption that $\mathbf{y}$ has mean zero. If not, zero padding must be done after subtracting the sample mean from the data (known as recentering) in order to avoid possible corruption of the entire periodogram.

Similar to the ordinary periodogram in (6.1.6), the *refined periodogram* of $\mathbf{y}$ on Ω_{mn} is defined as

$$I_n(\lambda'_j) := n|z'_j|^2 = n^{-1}\left|\sum_{t=1}^{n} y_t \exp(-it\lambda'_j)\right|^2 \qquad (j \in \mathbb{Z}_{mn}). \tag{6.3.12}$$

Because $\lambda'_{jm} = \lambda_j$ for $j \in \mathbb{Z}_n$, the refined periodogram coincides with the ordinary periodogram on the Fourier grid $\Omega_n \subset \Omega_{mn}$, i.e.,

$$I_n(\lambda'_{jm}) = I_n(\lambda_j) \qquad (j \in \mathbb{Z}_n).$$

This means that the refined periodogram is an interpolation of the ordinary periodogram on the finer frequency grid Ω_{mn}.

As expected, the refined periodogram is able to produce higher peaks near nonFourier signal frequencies than the original periodogram. Indeed, under the assumption of Gaussian white noise, we have

$$I_n(\lambda'_j) \sim \tfrac{1}{2}\sigma^2\chi^2(2, 2\theta'_j),$$

where

$$\theta'_j := n|\mu'_j|^2/\sigma^2, \quad \mu'_j := \sum_{k=1}^{p} \beta_k D_n(\omega_k - \lambda'_j).$$

If $\lambda'_{l(k)}$ denotes the frequency in Ω_{mn} which is nearest to a nonFourier frequency ω_k, then, under the condition (6.3.7), we can write

$$\theta'_{l(k)} = n\gamma_k K_n(\Delta'_k) + \mathcal{O}(n), \tag{6.3.13}$$

where $\Delta'_k := \omega_k - \lambda'_{l(k)}$. In comparison with (6.3.9), the discount factor in (6.3.13) is $K_n(\Delta'_k)$ rather than $K_n(\Delta_k)$. Because $K_n(\omega)$ is symmetric and strictly monotone decreasing in the interval $[0, 2\pi/n]$, and because $|\Delta'_k| < |\Delta_k|$, it follows that

$$K_n(\Delta'_k) > K_n(\Delta_k).$$

Hence, for large n, the effective SNR of the refined periodogram at $\lambda'_{l(k)}$ is higher than that of the ordinary periodogram at $\lambda_{j(k)}$. Moreover, because $|\Delta'_k| \le \pi/(mn)$, we have $K_n(\Delta'_k) \ge K_n(\pi/(mn))$. As $n \to \infty$ and for fixed $m > 1$,

$$K_n(\pi/(mn)) \to (2m/\pi)^2 \sin^2(\tfrac{1}{2}\pi/m).$$

In the case of $m = 2$ for example, the last quantity is equal to $8/\pi^2 \approx 0.8$. Therefore, by simply doubling the points on the frequency grid (i.e., padding n zeros) the worst-case degradation of the SNR is reduced from 60% to 20%. Similarly, by tripling or quadrupling the points, the degradation is further reduced to 10% or 5%, respectively. In practice, a tradeoff between the SNR improvement and the computational cost must be made when choosing a suitable m.

Due to the increased density in Ω_{mn}, more than one frequency will fall into the main lobe of the Fejér kernel produced by a single sinusoid. As a result, the refined periodogram will exhibit clustered large values around the signal frequencies. The clustering effect must be taken into account when using the refined periodogram for frequency identification and estimation. A simple method is to remove all frequencies in a small neighborhood of a peak frequency before searching for the next peak frequency. This leads to the following recursive procedure as a generalization of the DFT method in (6.1.9): For $k = 1, \dots, p$, let

$$\hat{\omega}_k := \arg \max_{\lambda'_j \in \bar{\Omega}_{k-1}} I_n(\lambda'_j), \tag{6.3.14}$$

where $\bar{\Omega}_0 := \Omega_{mn}$ and

$$\bar{\Omega}_k := \bar{\Omega}_{k-1} \setminus \{\lambda'_j \in \bar{\Omega}_{k-1} : |\lambda'_j - \hat{\omega}_k| < \delta\} \qquad (k = 1, \dots, p-1). \tag{6.3.15}$$

The peak frequencies so obtained should be rearranged, if necessary, in an ascending order so that $\hat{\omega}_k$ estimates ω_k for all k. The procedure defined by (6.3.14) and (6.3.15) will be referred to as the *refined* DFT, or RDFT, method.

In the RDFT method, the choice of δ is critical: it should be large enough so that all large values clustered around the kth signal frequency due to spectral leakage are removed from $\bar{\Omega}_k$; it should also be small enough so that the spectral peaks around the remaining signal frequencies are all retained in $\bar{\Omega}_k$. A typical choice of δ takes the form $\delta := cn^{-r}$, where $c > 0$ and $r \in (0, 1)$ are suitable constants. In [63] and [64], for example, $\delta = \pi/\sqrt{n}$ (i.e., $c = \pi$ and $r = 1/2$). With this choice, according to Lemma 12.1.3(d), the magnitude of spectral leakage outside the δ-neighborhood of a signal frequency cannot exceed $(\pi/c)n^{r-1} = \mathcal{O}(1)$ and therefore is asymptotically negligible.

Needless to say, for a successful application of the RDFT method, the signal frequencies must satisfy the separation condition $\Delta > \delta$, where Δ is defined by (6.3.8). This condition is satisfied asymptotically if Δ has the property (6.3.7).

While the RDFT method is expected to improve the accuracy of frequency estimation over the DFT method, mathematical analysis of its statistical performance is difficult because the refined periodogram ordinates are no longer independent of each other, even for Gaussian white noise. The following simulation example serves to demonstrate the statistical performance of the RDFT method for frequency estimation.

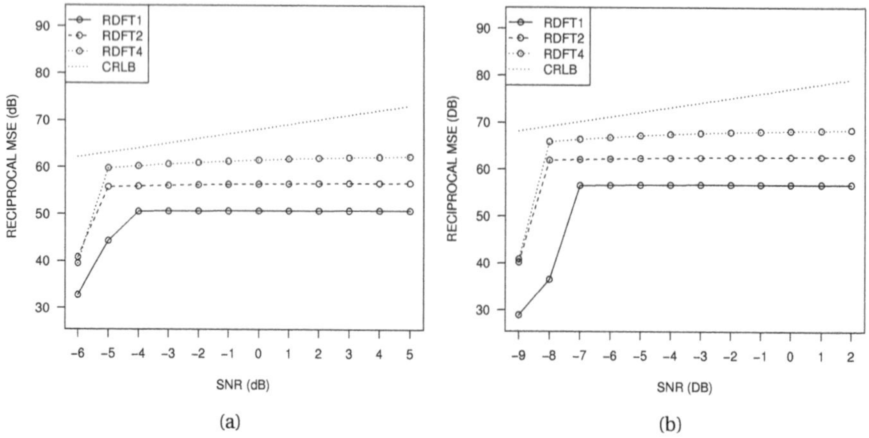

Figure 6.7. Plot of the reciprocal average MSE of the RDFT frequency estimates for the time series of two complex sinusoids in Gaussian white noise discussed in Example 6.9 with different values of SNR and two sample sizes: (a) $n = 100$ and (b) $n = 200$. Solid line with circles, RDFT1; dashed line with circles, RDFT2; dotted line with circles, RDFT4. Dotted line depicts the CRLB. Results are based on 5,000 Monte Carlo runs.

Example 6.9 (Two Complex Sinusoids in Gaussian White Noise: Part 1). Let $\mathbf{y}$ be given by (6.1.1) with $p = 2$, $C_1 = C_2 = 1$, $\{\phi_1, \phi_2\} \sim$ i.i.d. $U(-\pi, \pi)$, and $\boldsymbol{\epsilon} \sim N_c(\mathbf{0}, \sigma^2 \mathbf{I})$. In other words, $\mathbf{y}$ consists of two unit-amplitude random-phase complex sinusoids in Gaussian white noise with mean zero and variance σ^2. To neutralize the favoritism of the RDFT method toward frequencies in Ω_{mn}, the signal frequencies are randomized such that ω_1 and ω_2 are independent random variables with $\omega_1 \sim 2\pi \times U(0.1, 0.11)$ and $\omega_2 \sim 2\pi \times U(0.2, 0.21)$.

Consider the RDFT method defined by (6.3.14) and (6.3.15) with the separation criterion $\delta = \pi/\sqrt{n}$. For $m = 1, 2, 4$, let the resulting frequency estimator be denoted as RDFT1, RDFT2, and RDFT4, respectively. Note that RDFT1 is just the DFT estimator which is based on the ordinary periodogram.

Figure 6.7 depicts the average MSE for estimating the normalized frequencies $f_k := \omega_k/(2\pi)$ $(k = 1, 2)$ under different values of SNR and for two sample sizes. The MSE is calculated on the basis of 5,000 Monte Carlo runs. As a performance benchmark, Figure 6.7 also depicts the average (unconditional) CRLB for estimating the f_k. It is calculated according to (3.2.5) and averaged over the Monte Carlo runs to account for the randomness of the frequencies and phases.

As expected, the RDFT method produces more and more accurate estimates as the zero-padding factor m increases. The accuracy is also improved with the increase of n when m is fixed. Moreover, observe that when the SNR falls below a certain value, the MSE begins to deteriorate rapidly. This threshold effect is closely related to a similar behavior in the probability of frequency identification discussed in Example 6.4. The threshold for RDFT1 is approximately −4 dB when $n = 100$ and −7 dB when $n = 200$. The threshold for RDFT2 and RDFT4 is lowered by 1 dB in both cases, thanks to the refined periodogram. Finally, observe that the MSE curve becomes flat when the SNR increases beyond a certain point. This can be attributed to the finite precision of the DFT and RDFT estimates due to their restriction to the discrete values in Ω_n or Ω_{mn}. ◇

Finally, we remark that if the number of sinusoids p is unknown, the RDFT procedure should be terminated when the next peak value falls below a suitable threshold. In [63] and [64], a threshold of the form $\mathcal{O}(\log n)$ is used together with the choice of $m = 2$. This threshold is motivated by the fact that in the absence of sinusoids the maximum of the refined periodogram can be expressed as $\mathcal{O}(\log n)$ under fairly general conditions (see Remark 12.5.7).

6.3.3 Secondary Analysis

A simple procedure called the *secondary analysis* (SA) [47, pp. 196–199] [298, pp. 413–415] can be used to improve the frequency estimates obtained from the RDFT method defined by (6.3.14) and (6.3.15). With the RDFT method serving as the primary analysis that produces the initial estimates $\hat{\omega}_k$, the secondary analysis can be summarized as follows.

1. Compute the LS estimate $\hat{\boldsymbol{\beta}} := [\hat{\beta}_1, \dots, \hat{\beta}_p]^T$ given by (5.1.3) on the basis of $\{\hat{\omega}_k\}$ and compute the corresponding phase estimates

$$\hat{\phi}_k := \arctan\{\Im(\hat{\beta}_k), \Re(\hat{\beta}_k)\} \qquad (k = 1, \dots, p). \tag{6.3.16}$$

2. Divide $\mathbf{y}$ into $b > 1$ blocks of length $v \gg 1$ and compute

$$\zeta_{kl} := v^{-1} \sum_{t=(l-1)v+1}^{lv} y_t \exp\{-i(\hat{\omega}_k t + \hat{\phi}_k)\} \qquad (l = 1, \dots, b). \tag{6.3.17}$$

3. Compute

$$\phi_{kl} := \arctan\{\Im(\zeta_{kl}), \Re(\zeta_{kl})\} \qquad (l = 1, \dots, b). \tag{6.3.18}$$

4. Define the improved estimate of ω_k by

$$\hat{\hat{\omega}}_k := \hat{\omega}_k + s_k / v, \tag{6.3.19}$$

   where

$$s_k := \frac{\sum_{l=1}^{b} \phi_{kl}(l - (b+1)/2)}{\sum_{l=1}^{b} (l - (b+1)/2)^2}. \tag{6.3.20}$$

To justify the SA procedure, assume that $\{\epsilon_t\}$ is a linear process of the form (6.1.25) with $\sum|\psi_j| < \infty$ and $\{\zeta_t\} \sim \text{IID}(0,\sigma^2)$, and that $\{\omega_k\}$ satisfies the separation condition (6.3.7). It follows from straightforward calculation and an application of Lemma 12.5.9(b) that

$$\zeta_{kl} = \sum_{k'=1}^{p} C_{k'}|D_v(\Delta_{k'})|\exp(i(\Delta_{k'}vl + \phi_{k'} - \hat{\phi}_k + \xi_{k'})) + \mathcal{O}(1),$$

where $\Delta_{k'} := \omega_{k'} - \hat{\omega}_k$ and $\xi_{k'} := \angle D_v(\Delta_{k'})$. Because $n\Delta_{k'} = \mathcal{O}(1)$ for $k' \neq k$, we obtain $|D_v(\Delta_{k'})| = \mathcal{O}(1)$ by Lemma 12.1.3(d). Therefore,

$$\zeta_{kl} = C_k|D_v(\Delta_k)|\exp(i(\Delta_k vl + \phi_k - \hat{\phi}_k + \xi_k)) + \mathcal{O}(1).$$

Assuming $|\Delta_k vl + \phi_k - \hat{\phi}_k + \xi_k| \ll \pi/2$, we can write

$$\phi_{kl} = \angle\zeta_{kl} \approx \Delta_k vl + \phi_k - \hat{\phi}_k + \xi_k \qquad (l = 1, \ldots, b).$$

Observe that the right-hand side of this equation is a linear function of l with slope $\Delta_k v$. Therefore, given the angles ϕ_{kl} $(l = 1, \ldots, b)$, an estimate of the slope, denoted as s_k, can be obtained by performing a simple linear regression of the ϕ_{kl} on l with the intercept. This leads to the expression (6.3.20). With s_k estimating $\Delta_k v$, one can estimate the parameter $\Delta_k = \omega_k - \hat{\omega}_k$ by s_k/v. This leads to the expression (6.3.19) for the improved estimate of ω_k.

Note that the conventional SA procedure, as described in [47, pp. 196–199] and [298, pp. 413–415], begins with step 2, where the ζ_{kl} are defined by (6.3.17) with $\hat{\phi}_k := 0$. The resulting ϕ_{kl}, defined by (6.3.18), are prone to the *phase-wrapping* problem, due to the 2π-periodicity of the phase. To alleviate the problem, an alternative definition of ϕ_{kl} is introduced in [63], i.e.,

$$\phi_{kl} := \arctan\{\Im(\zeta_{kl}\exp(-i\alpha_k)), \Re(\zeta_{kl}\exp(-i\alpha_k))\},$$

where

$$\alpha_k := \arccos(\Re(\bar{\zeta}_k)/|\bar{\zeta}_k|), \quad \bar{\zeta}_k := b^{-1}\sum_{l=1}^{b} \zeta_{kl}/|\zeta_{kl}|.$$

In this method, the ζ_{kl} are rotated by the angle α_k. Because $\bar{\zeta}_k$ is the angular centroid of the ζ_{kl} and α_k is the angle of $\bar{\zeta}_k$, rotating the ζ_{kl} by the angle α_k makes them centered around zero in angles and thereby reduces the chance of phase wrapping in the ϕ_{kl}. Although effective in some cases, we find the method too sensitive to the phases of the sinusoids. As a remedy, we employ the phase estimates $\hat{\phi}_k$ given by (6.3.16) in defining the ζ_{kl}. With the ζ_{kl} given by (6.3.17), further rotation becomes unnecessary, so the ϕ_{kl} can be defined by (6.3.18).

The conventional SA procedure also requires that the block length v be an integer multiple of the estimated period $1/\hat{\omega}_k$ (hence dependent on k) [47, pp.

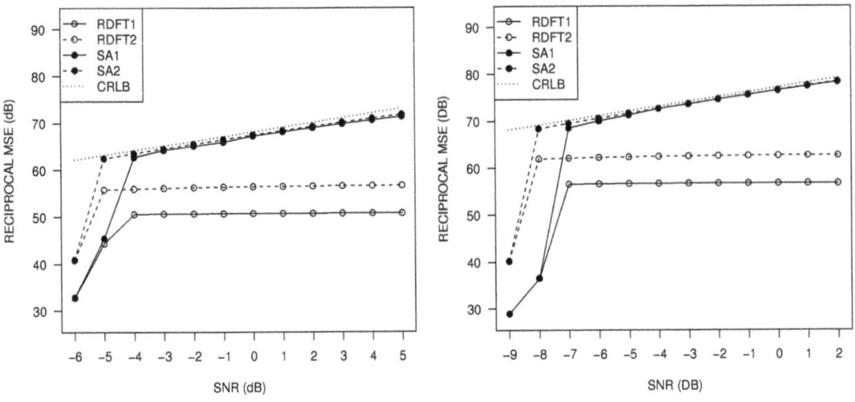

(a) (b)

Figure 6.8. Plot of the reciprocal average MSE of the SA frequency estimates for the time series of two complex sinusoids in Gaussian white noise discussed in Example 6.10 with different values of SNR and two sample sizes: (a) $n = 100$ and (b) $n = 200$. Solid line with circles, RDFT1; dashed line with circles, RDFT2; solid line with dots, SA based on RDFT1; dashed line with dots, SA based on RDFT2. Dotted line depicts the CRLB. Results are based on 5,000 Monte Carlo runs.

196–199] [63] [298, pp. 413–415]. We find this requirement unnecessary in our simulation studies using the SA procedure defined by (6.3.16)–(6.3.20).

The following example demonstrates the effectiveness of the SA method for improving the RDFT frequency estimates.

Example 6.10 (Two Complex Sinusoids in Gaussian White Noise: Part 2). Consider the time series in Example 6.9. Let the signal frequencies be estimated by the SA method defined by (6.3.16)–(6.3.20) on the basis of the RDFT1 and RDFT2 estimates obtained in Example 6.9. Let the resulting estimators be denoted as SA1 and SA2, respectively. In the secondary analysis, let the data record be divided into $b = 5$ blocks of length $v = n/b$ ($v = 20$ for $n = 100$ and $v = 40$ for $n = 200$). Figure 6.8 depicts the average MSE of the SA1 and SA2 estimates and the corresponding RDFT estimates.

As can be seen, the SA method is very effective indeed. When the SNR is sufficiently high (≥ -4 dB for $n = 100$ and ≥ -7 dB for $n = 200$), the SA estimator not only improves the initial RDFT estimator dramatically, but also approaches the CRLB. When the SNR is sufficiently high, the SA1 and SA2 estimators yield almost identical results, so padding zeros becomes unnecessary. The benefit of zero-padding is evident at a lower SNR (-5 dB for $n = 100$ and -8 dB for $n = 200$), where only RDFT2 is able to facilitate a successful secondary analysis that improves the initial estimates. When the SNR is too low, the threshold effect takes place and the SA method no longer improves the RDFT estimates. ◇

6.3.4 Interpolation Estimators

Besides the secondary analysis, the DFT estimator of the signal frequencies can also be improved by simple interpolation of the neighboring periodogram ordinates or DFT coefficients.

For example, let $\hat{\omega}_k = \lambda_j \in \Omega_n$ be the DFT estimator of ω_k. By definition, we have $I_n(\lambda_j) \ge I_n(\lambda_{j\pm 1})$. An improved estimator can be obtained by maximizing the quadratic function that interpolates the periodogram ordinates $I_n(\lambda_{j-1})$, $I_n(\lambda_j)$, and $I_n(\lambda_{j+1})$. The maximizer takes the form

$$\hat{\hat{\omega}}_k := \hat{\omega}_k + 2\pi v_k / n, \tag{6.3.21}$$

where

$$v_k := \frac{\{I_n(\lambda_{j-1}) - I_n(\lambda_{j+1})\}/2}{I_n(\lambda_{j-1}) + I_n(\lambda_{j+1}) - 2I_n(\lambda_j)}. \tag{6.3.22}$$

This method is suggested in [413] (see also [436]) for initializing more sophisticated periodogram maximization algorithms to be discussed in Section 6.4.2.

Another method, proposed in [315] (see also [317]), defines v_k as

$$v_k := \begin{cases} v_+ & \text{if } I_n(\lambda_{j+1}) > I_n(\lambda_{j-1}), \\ v_- & \text{otherwise}, \end{cases} \tag{6.3.23}$$

where

$$v_\pm := \frac{\pm\sqrt{I_n(\lambda_{j\pm 1})}}{\sqrt{I_n(\lambda_{j\pm 1})} + \sqrt{I_n(\lambda_j)}}. \tag{6.3.24}$$

This method is used in [1] to initialize the Newton-Raphson algorithm for periodogram maximization (see Section 6.4.2) and is analyzed in [301] as a standalone estimator. A simple justification for this method is as follows.

Recall that $z_l = \mu_l + e_l$ for all $l \in \mathbb{Z}_n$, where μ_l and e_l are defined by (6.3.5) and (6.3.6). If the signal frequencies satisfy the separation condition $\Delta > 4\pi/n$, then, for any λ_l such that $|\omega_k - \lambda_l| < 4\pi/n$, Lemma 12.1.3(d) ensures that

$$D_n(\omega_{k'} - \lambda_l) = \mathcal{O}(n^{-1}(\Delta - 4\pi/n)^{-1}) \qquad (k' \ne k).$$

Moreover, assuming that $\{\epsilon_t\}$ is a complex linear process which satisfies the conditions of Lemma 12.5.9(b), we can write $e_l = \mathcal{O}(1)$ uniformly for all $l \in \mathbb{Z}_n$. Combining these results gives

$$z_l = \beta_k D_n(\omega_k - \lambda_l) + \mathcal{O}(1). \tag{6.3.25}$$

Now suppose that the DFT estimate $\hat{\omega}_k = \lambda_j$ is sufficiently accurate such that

$$|\omega_k - \lambda_j| < 2\pi/n.$$

In this case, we have $|\omega_k - \lambda_{j\pm1}| < 4\pi/n$, so (6.3.25) is valid for $l \in \{j, j\pm1\}$. Let $\delta := \omega_k - \lambda_j$. Then, we can write

$$\begin{aligned} D_n(\omega_k - \lambda_j) &= \frac{\sin(n\delta/2)}{n\sin(\delta/2)} \exp\{i\tfrac{1}{2}(n+1)\delta\} \\ &= \frac{\sin(n\delta/2)}{n\delta/2} \exp\{i\tfrac{1}{2}(n+1)\delta\} + \mathcal{O}(1). \end{aligned}$$

Similarly, because $\lambda_{j\pm1} = \lambda_j \pm 2\pi/n$ and hence $\omega_k - \lambda_{j\pm1} = \delta \mp 2\pi/n$, we obtain

$$\begin{aligned} D_n(\omega_k - \lambda_{j\pm1}) &= \frac{-\sin(n\delta/2)}{n\sin((\delta\mp 2\pi/n)/2)} \exp\{i\tfrac{1}{2}(n+1)(\delta\mp 2\pi/n)\} \\ &= \frac{-\sin(n\delta/2)}{n(\delta\mp 2\pi/n)/2} \exp\{i\tfrac{1}{2}(n+1)(\delta\mp 2\pi/n)\} + \mathcal{O}(1). \end{aligned}$$

Therefore,

$$\frac{D_n(\omega_k - \lambda_{j\pm1})}{D_n(\omega_k - \lambda_j)} = \frac{\delta}{\delta \mp 2\pi/n} \exp(\mp i\pi/n) + \mathcal{O}(1). \tag{6.3.26}$$

Combining this result with (6.3.25) yields

$$\frac{\sqrt{I_n(\lambda_{j\pm1})}}{\sqrt{I_n(\lambda_j)}} = \frac{|z_{j\pm1}|}{|z_j|} = \frac{|\delta|}{|\delta \mp 2\pi/n|} + \mathcal{O}(1). \tag{6.3.27}$$

Observe that $I_n(\lambda_{j\pm1}) = nC_k^2 K_n(\omega_k - \lambda_{j\pm1}) + \mathcal{O}(n)$. Therefore, when $I_n(\lambda_{j-1}) > I_n(\lambda_{j+1})$, it is most likely (for large n) that $K_n(\omega_k - \lambda_{j-1}) > K_n(\omega_k - \lambda_{j+1})$, in which case $\delta = \omega_k - \lambda_j < 0$. Moreover, because $|\delta| < 2\pi/n$, we have $\delta + 2\pi/n > 0$. Combining these results with (6.3.27) gives

$$\frac{\sqrt{I_n(\lambda_{j-1})}}{\sqrt{I_n(\lambda_j)}} = \frac{-\delta}{\delta + 2\pi/n} + \mathcal{O}(1).$$

Solving this equation for $\delta = \omega_k - \hat{\omega}_k$ while ignoring the error term $\mathcal{O}(1)$ leads to the expression (6.3.21) with v_k equal to v_- defined by (6.3.24). A similar argument justifies the choice of $v_k = v_+$ when $I_n(\lambda_{j-1}) < I_n(\lambda_{j+1})$.

The third method, proposed in [246] (see also [247]), employs not only the periodogram ordinates but also the DFT coefficients. This method simply replaces (6.3.23) and (6.3.24) by

$$v_k := \begin{cases} v_+ & \text{if } \Re(z_{j-1}/z_j) > \Re(z_{j+1}/z_j), \\ v_- & \text{otherwise,} \end{cases} \tag{6.3.28}$$

and

$$v_\pm := \frac{\mp\Re(z_{j\pm1}/z_j)}{1 - \Re(z_{j\pm1}/z_j)}. \tag{6.3.29}$$

Observe that $\Re(z_{j\pm1}/z_j) = \Re(z_{j\pm1}z_j^*)/|z_j|^2$ and $|z_j|^2 = n^{-1}I_n(\lambda_j)$. Therefore, one can also express (6.3.28) and (6.3.29) in terms of $\Re(z_{j\pm1}z_j^*)$ and $I_n(\lambda_j)$.

To justify this method, consider (6.3.26) again. Combining this expression with (6.3.25) and the fact that $\cos(\pi/n) \approx 1$ for large n gives

$$\Re(z_{j\pm1}/z_j) = \frac{\delta}{\delta \mp 2\pi/n} + \mathcal{O}(1). \tag{6.3.30}$$

These equations can be solved for $\delta = \omega_k - \hat{\omega}_k$ by ignoring the error term $\mathcal{O}(1)$. This leads to the expression (6.3.21) with v_k equal to $v_\pm$ defined by (6.3.29). Although both v_+ and v_- are legitimate solutions, the approximation in (6.3.26) tends to be more accurate for $\Re(z_{j+1}/z_j)$ if $\delta > 0$, because in this case λ_{j+1} is closer to ω_k than λ_{j-1}. For this reason, the choice of $v_k = v_+$ is preferred when $\delta > 0$. Similarly, the choice of $v_k = v_-$ is preferred when $\delta < 0$. The criterion in (6.3.23) is a simple test for $\delta > 0$. Indeed, when $\Re(z_{j-1}/z_j) > \Re(z_{j+1}/z_j)$, there is a good chance, according to (6.3.30), that $\delta/(\delta + 2\pi/n) > -\delta/(2\pi/n - \delta)$, which, together with the assumption $|\delta| < 2\pi/n$, implies $\delta > 0$. An alternative criterion, suggested in [302], is that $v_k = v_+$ if both v_+ and v_- are positive. Simulation shows that this criterion is slightly inferior to (6.3.28) at low SNR.

The fourth method, also proposed in [246] (see also [247]), avoids the test for positive δ by replacing (6.3.28) and (6.3.29) with

$$v_k := (4c)^{-1}\{(1+8c^2)^{1/2} - 1\}, \tag{6.3.31}$$

where

$$c := \frac{\Re(z_{j-1}/z_j) - \Re(z_{j+1}/z_j)}{2 + \Re(z_{j-1}/z_j) + \Re(z_{j+1}/z_j)}. \tag{6.3.32}$$

A similar but slightly more complicated alternative is discussed in [302]. Note that (6.3.32) can also be expressed in terms of $\Re(z_{j\pm1}z_j^*)$ and $I_n(\lambda_j)$.

To justify the choice of v_k in (6.3.31), we note that $\Re(z_{j\pm1}/z_j)$ takes the form (6.3.30) and therefore c in (6.3.32) can be expressed as

$$c = -v/(2v^2 - 1) + \mathcal{O}(1)$$

with $v := n\delta/(2\pi)$. Solving this equation for v while ignoring the error term $\mathcal{O}(1)$ leads to (6.3.31). The basic idea in this technique is to use the average of $a_+ := 1 + \Re(z_{j+1}/z_j)$ and $a_- := 1 + \Re(z_{j-1}/z_j)$ to reduce the statistical error in $\Re(z_{j+1}/z_j)$ and $\Re(z_{j-1}/z_j)$. There are two ways to average these quantities: $c_+ := (a_+ + a_-)/2$ and $c_- := (a_+ - a_-)/2$. By (6.3.30), we can write

$$c_+ = (2v^2 - 1)/(v^2 - 1) + \mathcal{O}(1), \quad c_- = v/(v^2 - 1) + \mathcal{O}(1).$$

The common denominator in the leading terms can be eliminated by taking the ratio $c_-/c_+ = v/(2v^2 - 1) + \mathcal{O}(1)$. It is easy to verify that $c_-/c_+ = -c$.

The following example compares the statistical performance of these interpolation estimators.

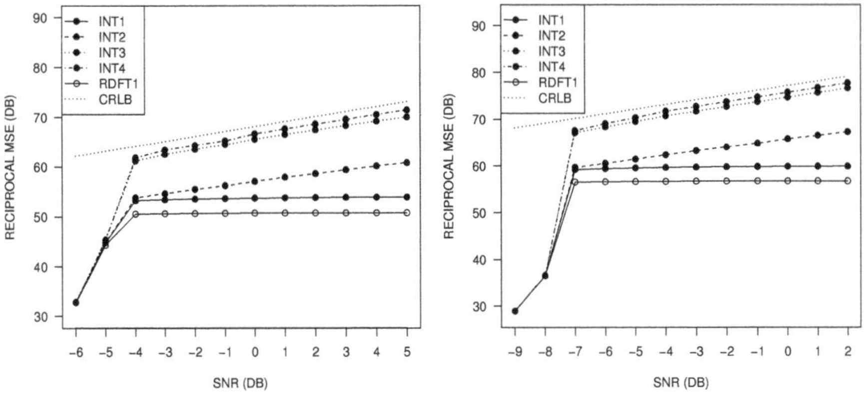

(a) (b)

Figure 6.9. Plot of the reciprocal average MSE of the DFT interpolation frequency estimates for the time series of two complex sinusoids in Gaussian white noise discussed in Example 6.11 with different values of SNR and two sample sizes: (a) $n = 100$. (b) $n = 200$ Solid line with dots, INT1; dashed line with dots, INT2; dotted line with dots, INT3; dash-dotted line with dots, INT4; solid line with circles, RDFT1. Dotted line depicts the CRLB. Results are based on 5,000 Monte Carlo runs.

Example 6.11 (Two Complex Sinusoids in Gaussian White Noise: Part 3). Consider the time series in Example 6.9. Let the frequencies be estimated by the four interpolation estimators of the form (6.3.24) with different choices of v_k:

1. INT1, with v_k defined by (6.3.22);
2. INT2, with v_k defined by (6.3.23) and (6.3.24);
3. INT3, with v_k defined by (6.3.28) and (6.3.29); and
4. INT4, with v_k defined by (6.3.31) and (6.3.32).

Figure 6.9 depicts the average MSE of the interpolation estimates and the DFT (RDFT1) estimates for different values of SNR and two sample sizes.

As we can see, all interpolation methods outperform the DFT method by a large margin. Although INT1 does not improve with the SNR, the other interpolation estimators produce more accurate results as the SNR increases. Both INT3 and INT4 perform significantly better than INT1 and INT2; INT4 is even more accurate than INT3. The SNR threshold of all interpolation methods remains the same as that of the DFT method (-4 dB for $n = 100$ and -7 dB for $n = 200$). While simpler in computation, the best interpolation method, INT4, is still slightly inferior in accuracy to the SA method shown in Figure 6.8. ◇

Note that the quadratic interpolation method can be easily applied to the refined periodogram. In fact, if $\hat{\omega}_k = \lambda'_j = 2\pi j/(mn)$ is the RDFT estimate of ω_k, then the quadratic interpolation estimate can be expressed as $\hat{\hat{\omega}}_k := \hat{\omega}_k + 2\pi v_k/(mn)$, where v_k is defined in the same way as (6.3.22) except that $I_n(\lambda_j)$ and $I_n(\lambda_{j\pm1})$ are replaced by $I_n(\lambda'_j)$ and $I_n(\lambda'_{j\pm1})$. The other interpolation methods are specific for the DFT estimator because they rely on (6.3.26) which is true only for λ_j and $\lambda_{j\pm1}$ in Ω_n. A hybrid method is considered in [4], where the RDFT coefficients at $\lambda_{j\pm1/2} := \lambda_j \pm \pi/n$ are employed to improve the DFT estimate λ_j.

6.4 Continuous Periodogram

The ordinary periodogram in (6.1.6) and the refined periodogram in (6.3.12) can be further generalized to the continuum of frequencies by considering

$$I_n(\omega) := n^{-1}\left|\sum_{t=1}^{n} y_t \exp(-it\omega)\right|^2 \tag{6.4.1}$$

as a 2π-periodic function of $\omega \in \mathbb{R}$. We call this function the *continuous* periodogram. As the notation suggests, the ordinary and refined periodograms can be obtained from the continuous periodogram by evaluating it on the discrete frequency grid Ω_n or Ω_{mn}. The continuous periodogram naturally interpolates the ordinary and refined periodograms beyond Ω_n and Ω_{mn}.

With $\hat{r}_y(u)$ denoting the sample autocovariance function defined by (4.1.1), it is easy to verify, on the basis of (6.4.1), that

$$I_n(\omega) = \sum_{|u|<n} \hat{r}_y(u) \exp(-iu\omega). \tag{6.4.2}$$

In other words, the continuous periodogram is nothing but the Fourier transform of the sample autocovariance function.

6.4.1 Statistical Properties

The continuous periodogram defined by (6.4.1) can be expressed as

$$I_n(\omega) = n|z_n(\omega)|^2, \tag{6.4.3}$$

where $z_n(\omega) := n^{-1}\mathbf{f}^H(\omega)\mathbf{y} = \mu_n(\omega) + e_n(\omega)$, with

$$\mu_n(\omega) := n^{-1}\mathbf{f}^H(\omega)\mathbf{x} = \sum_{k=1}^{p} \beta_k D_n(\omega_k - \omega), \tag{6.4.4}$$

$$e_n(\omega) := n^{-1}\mathbf{f}^H(\omega)\boldsymbol{\epsilon} = n^{-1}\sum_{t=1}^{n} \epsilon_t \exp(-it\omega). \tag{6.4.5}$$

Using these expressions, the following theorem can be established as an extension of the classical theory for real linear processes [46, pp. 347–348] as well as Theorems 6.2 and 6.4–6.6. See Section 6.6 for a proof.

Theorem 6.7 (Statistical Properties of the Continuous Periodogram). *Let* $\mathbf{y}$ *be given by (6.1.1). Let* $\mu_n(\omega)$ *and* $e_n(\omega)$ *be defined by (6.4.4) and (6.4.5).*

(a) *For any* $\omega \in [0, 2\pi)$, *if* $\{\epsilon_t\}$ *satisfies the conditions of Theorem 6.2, then*

$$I_n(\omega) \sim \tfrac{1}{2}\sigma^2 \chi^2(2, 2\theta_n(\omega)), \quad \theta_n(\omega) := n|\mu_n(\omega)|^2/\sigma^2.$$

If $\{\epsilon_t\}$ *satisfies the conditions of Theorem 6.4, then*

$$I_n(\omega) \overset{A}{\sim} \tfrac{1}{2} f_\epsilon(\omega)\, \chi^2(2, 2\theta_n(\omega)), \quad \theta_n(\omega) := n|\mu_n(\omega)|^2/f_\epsilon(\omega).$$

The latter is also true if $\{\epsilon_t\}$ *is a complex linear process of the form (6.1.25) with* $\sum |\psi_j| < \infty$, $\{\zeta_t\} \sim \mathrm{IID}(0, \sigma^2)$, *and* $E(\zeta_t^2) = 0$.

(b) *For any* $\omega, \lambda \in [0, 2\pi)$, *let* $\delta := |\omega - \lambda|$ *if* $|\omega - \lambda| \le \pi$ *and* $\delta := 2\pi - |\omega - \lambda|$ *otherwise. If* $\{\epsilon_t\}$ *satisfies the conditions of Theorem 6.2 or 6.4, then*

$$\begin{aligned} E\{I_n(\omega)\} &= n|\mu_n(\omega)|^2 + f_\epsilon(\omega) + \mathcal{O}(r_n), \\ \mathrm{Cov}\{I_n(\omega), I_n(\lambda)\} &= 2n\Re\{\mu_n^*(\omega)\,\mu_n(\lambda)\,\sigma_n(\omega,\lambda)\} + |\sigma_n(\omega,\lambda)|^2, \end{aligned}$$

where

$$\begin{aligned} \sigma_n(\omega,\lambda) &:= n\,E\{e_n(\omega)\,e_n^*(\lambda)\} \\ &= \begin{cases} \mathcal{O}(r_n) & \text{if } \delta = 2\pi j/n \text{ for some integer } j > 0, \\ \mathcal{O}(n^{-1}\delta^{-1}) + \mathcal{O}(r_n) & \text{if } \delta > \pi/n \text{ and } \delta \ne 2\pi j/n \text{ for all } j, \\ \mathcal{O}(1) & \text{if } 0 < \delta \le \pi/n, \\ f_\epsilon(\omega) + \mathcal{O}(r_n) & \text{if } \delta = 0. \end{cases} \end{aligned}$$

All error terms are uniformly bounded in ω *and* δ.

(c) *For any* $\omega, \lambda \in [0, 2\pi)$, *let* $\tilde{\delta} := \omega + \lambda$ *if* $\omega + \lambda \le \pi$, $\tilde{\delta} := 2\pi - (\omega + \lambda)$ *if* $\pi < \omega + \lambda \le 2\pi$, $\tilde{\delta} := \omega + \lambda - 2\pi$ *if* $2\pi < \omega + \lambda \le 3\pi$, *and* $\tilde{\delta} := 4\pi - (\omega + \lambda)$ *otherwise. If* $\{\epsilon_t\}$ *satisfies the conditions of Theorem 6.6, then*

$$\begin{aligned} E\{I_n(\omega)\} &= n|\mu_n(\omega)|^2 + f_\epsilon(\omega) + \mathcal{O}(r_n), \\ \mathrm{Cov}\{I_n(\omega), I_n(\lambda)\} &= 2n\Re\{\mu_n^*(\omega)\,\mu_n(\lambda)\,\sigma_n(\omega,\lambda) + \mu_n^*(\omega)\,\mu_n^*(\lambda)\,c_n(\omega,\lambda) \\ &\qquad + \mu_n(\omega)\,s_n(\omega,\lambda) + \mu_n(\lambda)\,s_n(\lambda,\omega)\} \\ &\qquad + |\sigma_n(\omega,\lambda)|^2 + \xi|c_n(\omega,\lambda)|^2 + \mathcal{O}(n^{-1}), \end{aligned}$$

where $\sigma_n(\omega,\lambda)$ *is defined in part (b),*

$$\begin{aligned} c_n(\omega,\lambda) &:= n\,E\{e_n(\omega)\,e_n(\lambda)\} \\ &= \begin{cases} \mathcal{O}(r_{n2}) & \text{if } \tilde{\delta} = 2\pi j/n \text{ for some integer } j > 0, \\ \mathcal{O}(n^{-1}\tilde{\delta}^{-1}) + \mathcal{O}(r_{n2}) & \text{if } \tilde{\delta} > \pi/n \text{ and } \tilde{\delta} \ne 2\pi j/n \text{ for all } j, \\ \mathcal{O}(1) & \text{if } 0 < \tilde{\delta} \le \pi/n, \\ \iota g_\epsilon(\omega) + \mathcal{O}(r_{n2}) & \text{if } \tilde{\delta} = 0. \end{cases} \\ s_n(\omega,\lambda) &:= n\,E\{e_n^*(\omega)\,|e_n(\lambda)|^2\} = \mathcal{O}(n^{-1}). \end{aligned}$$

All error terms are uniformly bounded in ω, δ, *and* $\tilde{\delta}$.

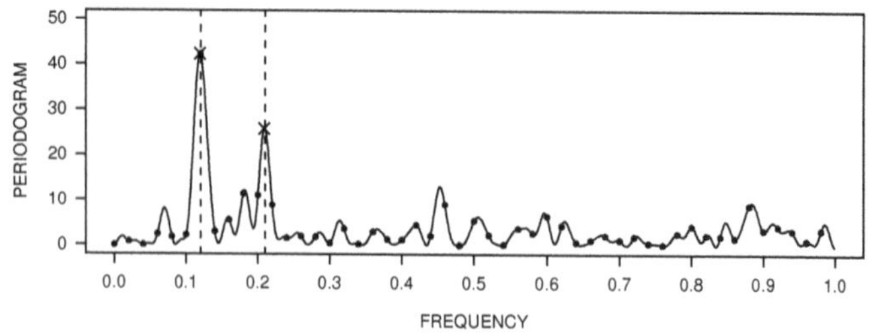

Figure 6.10. Continuous periodogram of the time series shown in Figure 6.1. Dots represent the discrete periodogram ordinates. Dashed vertical lines indicate the signal frequencies. Crosses indicate the two largest local maxima.

(d) *Under the conditions of part (a),* $I_n(\tilde{\omega}_1),\ldots,I_n(\tilde{\omega}_m)$ *are asymptotically independent for any fixed and distinct frequencies* $\tilde{\omega}_1,\ldots,\tilde{\omega}_m \in [0,2\pi)$.

Remark 6.12 The asymptotic independence assertion in part (d) remains valid if $\tilde{\omega}_1,\ldots,\tilde{\omega}_m$ depend on n, provided the vectors $\mathbf{f}(\tilde{\omega}_1),\ldots,\mathbf{f}(\tilde{\omega}_m)$ are asymptotically orthogonal in the sense that $n^{-1}\mathbf{f}^H(\tilde{\omega}_j)\mathbf{f}(\tilde{\omega}_l) \to 0$ as $n \to \infty$ for $j \neq l$.

Under the condition (6.3.7), it follows from (6.4.4) and Lemma 12.1.3(d) that

$$\theta_n(\omega) = \begin{cases} n\gamma_k + \mathcal{O}(n) & \text{if } n|\omega-\omega_k| \to 0 \text{ for some } k, \\ \mathcal{O}(n) & \text{if } n\min_k\{|\omega-\omega_k|, 2\pi-|\omega-\omega_k|\} \to \infty. \end{cases} \tag{6.4.6}$$

This means that the continuous periodogram tends to take large values in the $\mathcal{O}(n^{-1})$-neighborhood of the signal frequencies. As a result, maximizing the continuous periodogram is expected to produce more accurate frequency estimates than maximizing the ordinary or refined periodogram.

Figure 6.10 demonstrates the effectiveness of the continuous periodogram in boosting the spectral peak for a nonFourier signal frequency and thereby improving the accuracy of the local maximum as a frequency estimator. This plot should be compared with the corresponding plot in Figure 6.1 where the ordinary periodogram is depicted for the same data.

6.4.2 Periodogram Maximization

As suggested by Figure 6.10, the signal frequencies in the data record $\mathbf{y}$ of the form (6.1.1) can be estimated by the p largest local maxima of the continuous periodogram in the interval $(0,2\pi)$. This method of frequency estimation will be referred to as the *periodogram maximization* (PM) method.

The PM method can be formulated as a multivariate optimization problem. Indeed, let $\tilde{\boldsymbol{\omega}} := [\tilde{\omega}_1,\ldots,\tilde{\omega}_p]^T \in (0,2\pi)^p$ be the multivariate frequency variable.

Then, the PM estimator $\hat{\boldsymbol{\omega}} := [\hat{\omega}_1, \dots, \hat{\omega}_p]^T$ can be expressed as

$$\hat{\boldsymbol{\omega}} = \arg\max_{\tilde{\boldsymbol{\omega}} \in \Omega_0} \sum_{k=1}^{p} I_n(\tilde{\omega}_k), \tag{6.4.7}$$

where Ω_0 is a suitable neighborhood of $\boldsymbol{\omega} := [\omega_1, \dots, \omega_p]^T$ in $(0, 2\pi)^p$. A sufficient condition for $\hat{\boldsymbol{\omega}} \in \Omega_0$ to be the maximizer comprises

(a) $\dot{I}_n(\hat{\omega}_k) = 0$ for all $k = 1, \dots, p$;

(b) $\ddot{I}_n(\hat{\omega}_k) < 0$ for all $k = 1, \dots, p$; and

(c) $\sum_{k=1}^{p} I_n(\hat{\omega}_k) \ge \sum_{k=1}^{p} I_n(\tilde{\omega}_k)$ for any $\tilde{\boldsymbol{\omega}} \in \Omega_0$ that satisfies (a).

The requirements (a) and (b) together ensure that the $\hat{\omega}_k$ are local maxima of $I_n(\omega)$; the requirement (c) guarantees that the $\hat{\omega}_k$ correspond to the p largest local maxima of $I_n(\omega)$. In order for the sinusoids to appear as distinct local maxima, the signal frequencies should satisfy the separation condition (6.3.7).

A recursive procedure similar to the RDFT method in (6.3.14) and (6.3.15) can be used to produce the PM frequency estimates: For $k = 1, \dots, p$, let

$$\hat{\omega}_k := \arg \max_{\omega \in \bar{\Omega}_{k-1}} I_n(\omega), \tag{6.4.8}$$

where $\bar{\Omega}_0 := (0, 2\pi)$ and

$$\bar{\Omega}_k := \bar{\Omega}_{k-1} \setminus \{\omega \in \bar{\Omega}_{k-1} : |\omega - \hat{\omega}_k| < \delta\} \qquad (k = 1, \dots, p-1) \tag{6.4.9}$$

with $\delta := cn^{-r}$ for some constants $c > 0$ and $r \in (0, 1)$. As with the RDFT estimates, the $\hat{\omega}_k$ should be rearranged, if necessary, in an ascending order.

In general, the local maxima of a continuous periodogram can only be found numerically by iterative procedures, such as the bisection method [3] [435] [436], the secant method [316], and the Newton-Raphson method [1] [412]. The convergence of these procedures is a major concern. There are two issues regarding convergence: the *rate* of convergence and the *region* of convergence, also known as the *basin of attraction* (BOA). Generally speaking, the bisection method has the lowest (linear) rate of convergence, the secant method has a higher (superlinear) rate of convergence, and the Newton-Raphson method has the highest (quadratic) rate of convergence. To achieve such convergence, a typical requirement is that the initial value be "sufficiently close" to the desired solution. This turns out to be a big challenge in periodogram maximization.

As illustrated by Figure 6.10, the continuous periodogram contains many spurious maxima near the desired solution. This is true even in the noiseless case, for which $I_n(\omega) = \sum_{k=1}^{p} nC_k^2 K_n(\omega - \omega_k)$. To avoid being trapped by the spurious maxima, the initial guess for a signal frequency must be well inside the main lobe of the Fejér kernel corresponding to the sinusoid. In other words, it must be within the $2\pi/n$-neighborhood of the signal frequency. The actual BOA of an iterative procedure can be much smaller. This is why the DFT frequency estimates obtained from the ordinary periodogram are not always accurate enough to serve as initial values [314] [412].

For example, consider the Newton-Raphson algorithm (or Newton's algorithm) [365, p. 291]. Starting from an initial value $\hat{\omega}(0)$, the Newton-Raphson algorithm produces a sequence $\{\hat{\omega}(m)\}$ through a fixed point iteration

$$\hat{\omega}(m+1) := \phi(\hat{\omega}(m)) \qquad (m = 0, 1, 2, \ldots). \tag{6.4.10}$$

The fixed point mapping $\omega \mapsto \phi(\omega)$ takes the form

$$\phi(\omega) := \omega - \varrho \dot{I}_n(\omega) / J_n(\omega). \tag{6.4.11}$$

where $\varrho > 0$ is a step-size parameter and $J_n(\omega) := \ddot{I}_n(\omega)$ is a function that serves to normalize the derivative of the periodogram. It follows from (6.4.3) that

$$\begin{aligned} \dot{I}_n(\omega) &= 2n\Re\{z_n(\omega)\dot{z}_n^*(\omega)\}, \\ \ddot{I}_n(\omega) &= 2n\Re\{z_n(\omega)\ddot{z}_n^*(\omega) + |\dot{z}_n(\omega)|^2\}, \end{aligned}$$

where $\dot{z}_n(\omega) := n^{-1}\sum(-it)y_t \exp(-it\omega)$ and $\ddot{z}_n(\omega) := n^{-1}\sum(-t^2)y_t \exp(-it\omega)$.

Because $\dot{I}_n(\hat{\omega}_k) = 0$, the PM estimator $\hat{\omega}_k$ is a fixed point of $\phi(\omega)$, i.e.,

$$\phi(\hat{\omega}_k) = \hat{\omega}_k.$$

The BOA of $\hat{\omega}_k$ is the set of initial values for which the sequence $\{\hat{\omega}(m)\}$ converges to $\hat{\omega}_k$ as $m \to \infty$. Because a local maximum is characterized by having a negative second derivative, the choice of $J_n(\omega) = \ddot{I}_n(\omega)$ as the normalizing function for the gradient $\dot{I}_n(\omega)$ ensures that the iteration converges to a local maximum rather than a local minimum which also satisfies $\dot{I}_n(\omega) = 0$.

While the Newton-Raphson algorithm has a fast quadratic rate of convergence in general [365, p. 295], its BOA for periodogram maximization is usually very small. This means that it requires very good initial values to converge to the desired solutions.

Figure 6.11 depicts the Newton-Raphson fixed point mapping for maximizing the continuous periodogram shown in Figure 6.10 near the signal frequency $\omega_2 = 2\pi \times 0.21$. The mapping has a fixed point $\hat{\omega}_2 = 2\pi \times 0.2093$, which corresponds to the nearest local maximum of the periodogram around ω_2. The BOA of this fixed point is given by the interval $[2\pi \times 0.2037, 2\pi \times 0.2148]$, so the initial values must satisfy $|\hat{\omega}(0) - \omega_2| \le 2\pi \times 0.0048$, which is much more stringent than $2\pi/n = 2\pi \times 0.02$. In this example, the DFT estimate of ω_2 equals $2\pi \times 9/n = 2\pi \times 0.18$ and therefore lies outside the BOA. Consequently, the fixed point iteration using the DFT estimate as initial value will not converge to the correct local maximum. In fact, it will be trapped by the spurious maximum at $2\pi \times 0.1822$.

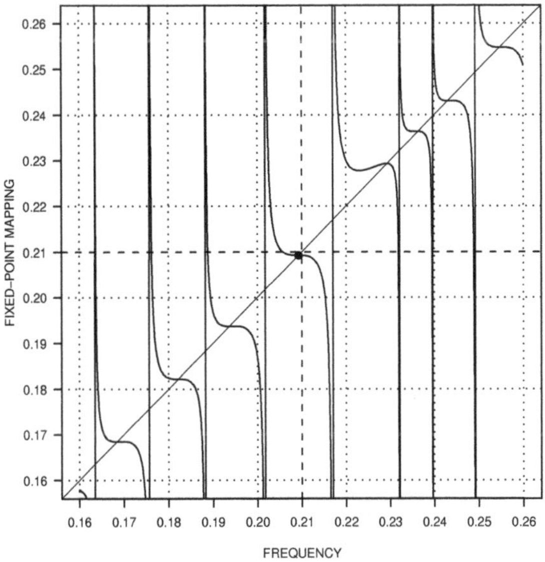

Figure 6.11. The Newton-Raphson fixed point mapping $f \mapsto \phi(2\pi f)/(2\pi)$ ($\varrho = 1$) for maximizing the continuous periodogram shown in Figure 6.10. Thin solid line is the diagonal line. Dashed lines identify the signal frequency $\omega_2 = 2\pi \times 0.21$. Dotted lines depict the grid of Fourier frequencies. The large dot depicts the fixed point corresponding to the peak in the continuous periodogram closest to ω_2.

The Newton-Raphson mapping can be modified by taking

$$J_n(\omega) := -\tfrac{1}{6}(n^2 - 1) I_n(\omega). \tag{6.4.12}$$

This normalizing function is used in [18]. Unlike the second derivative $\ddot{I}_n(\omega)$ in the original Newton-Raphson mapping, the normalizing function in (6.4.12) is always negative, so the resulting fixed point iteration will converge to a local maximum, even in the neighborhood of a local minimum. It is also unaffected by the inflection points of the periodogram where the second derivative equals zero. These properties make the modified mapping more tolerant of poor initial values than the original Newton-Raphson mapping. However, it may take slightly more iterations for the modified Newton-Raphson algorithm to reach the same degree of accuracy as the original one in calculating the fixed point. Note that $J_n(\omega)$ in (6.4.12) coincides with $\ddot{I}_n(\omega)$ when $y_t = \exp(i\omega t)$ ($t = 1, \ldots, n$).

Figure 6.12 depicts the modified Newton-Raphson mapping with $\varrho = 1$ for the continuous periodogram shown in Figure 6.10. As compared with Figure 6.11, the modified mapping clearly exhibits an enlarged BOA around the signal frequency. More precisely, the BOA is given by $[2\pi \times 0.1937, 2\pi \times 0.2292]$, so the required accuracy of initial values becomes $|\hat{\omega}(0) - \omega_2| \le 2\pi \times 0.0163$. This is an

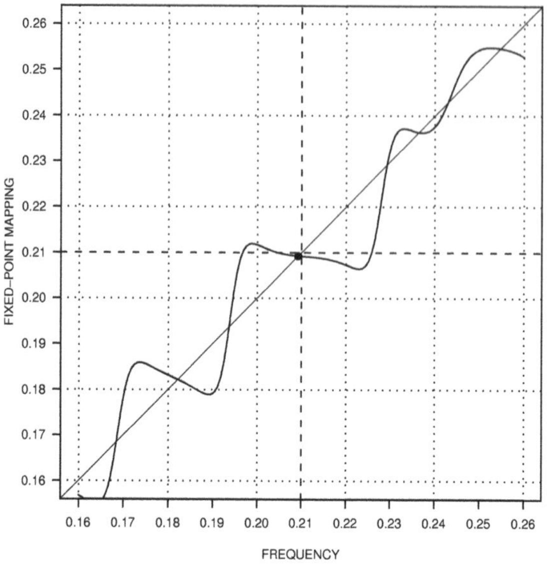

Figure 6.12. The modified Newton-Raphson fixed point mapping defined by (6.4.11) and (6.4.12) ($\varrho = 1$) for maximizing the continuous periodogram shown in Figure 6.10.

improvement over the original mapping, although still not enough to accommodate the poor DFT estimate which equals $2\pi \times 0.18$.

To obtain better initial values, a number of simple techniques have been suggested [1] [316] [317] [412], including the interpolation method discussed in Section 6.3.4 and the RDFT method discussed in Section 6.3.2. The following example demonstrates the statistical performance of the PM method using the RDFT estimates as initial values for the modified Newton-Raphson mapping.

Example 6.12 (Two Complex Sinusoids in Gaussian White Noise: Part 4). Consider the time series in Example 6.9. Let the signal frequencies be estimated by the PM method using the fixed point iteration (6.4.10) with the modified Newton-Raphson mapping defined by (6.4.11) and (6.4.12). Let PM1 and PM2 denote the resulting estimators initialized by RDFT1 and RDFT2, respectively. Figure 6.13 depicts the average MSE for estimating the normalized frequencies $f_k := \omega_k/(2\pi)$ ($k = 1,2$) after 5 iterations with $\varrho = 1$. It also depicts the average MSE of the initial RDFT estimates and the average CRLB for comparison.

This result shows that when the SNR is sufficiently high (≥ -4 dB for $n = 100$ and ≥ -7 dB for $n = 200$), both RDFT1 and RDFT2 are accurate enough to make the modified Newton-Raphson algorithm converge, and the resulting MSE follows the CRLB closely. As the SNR decreases to -5 dB for $n = 100$ and -8 dB for $n = 200$, only RDFT2 provides satisfactory initial values, and RDFT1 becomes in-

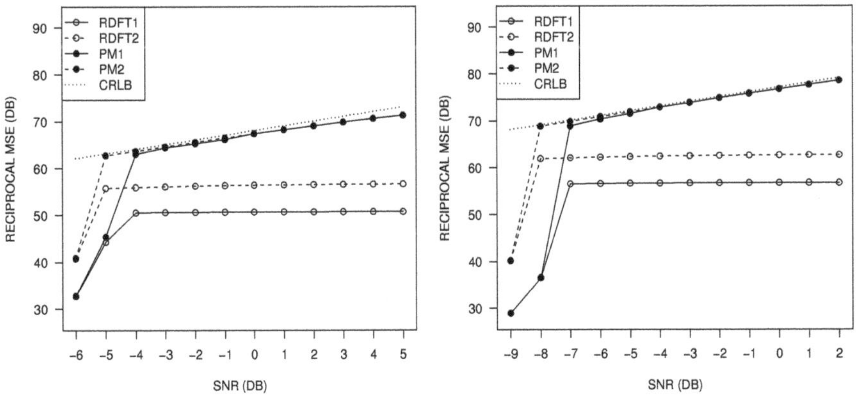

(a) (b)

Figure 6.13. Plot of the reciprocal average MSE of the PM frequency estimates for the time series of two complex sinusoids in Gaussian white noise discussed in Example 6.12 with different values of SNR and two sample sizes: (a) $n = 100$ and (b) $n = 200$. Solid line with circles, RDFT1; dashed line with circles, RDFT2; solid line with dots, PM initialized by RDFT1; dashed line with dots, PM initialized by RDFT2. Dotted line depicts the CRLB. Results are based on 5,000 Monte Carlo runs.

adequate. When the SNR is reduced further by 1 dB for both sample sizes, the threshold effect takes place and all estimators suffer from a rapid deterioration of performance. By comparing Figure 6.13 with Figures 6.8 and 6.9, we can see that the PM method is superior to the interpolation method and the SA method in terms of accuracy, although not dramatically so in the latter case. ◇

Large-sample statistical properties of the PM estimator, including the consistency and asymptotic normality, have been studied for the real case with fixed frequencies, first by Whittle [422] and later by Walker [412] and Hannan [137]. In the following, let us consider a more general situation in which the separation of the signal frequencies may approach zero as the sample size grows.

In the real case where $\mathbf{y} := [y_1, \ldots, y_n]^T$ is given by

$$y_t = \sum_{t=1}^{q} \{A_k \cos(\omega_k t) + B_k \sin(\omega_k t)\} + \epsilon_t \qquad (t = 1, \ldots, n), \tag{6.4.13}$$

it suffices to consider the frequency variable $\tilde{\boldsymbol{\omega}}_r := [\tilde{\omega}_1, \ldots, \tilde{\omega}_q]^T$. The PM estimator of $\boldsymbol{\omega}_r := [\omega_1, \ldots, \omega_q]^T$ can be expressed as

$$\hat{\boldsymbol{\omega}}_r := \arg \max_{\tilde{\boldsymbol{\omega}}_r \in \Omega_0} \sum_{k=1}^{q} I_n(\tilde{\omega}_k), \tag{6.4.14}$$

where Ω_0 is a suitable neighborhood of $\boldsymbol{\omega}_r$ in $(0, \pi)^q$. Moreover, let the amplitude parameters A_k and B_k be estimated simply by the Fourier coefficients

$$\hat{A}_k := 2n^{-1}\sum_{t=1}^{n} y_t \cos(\hat{\omega}_k t), \quad \hat{B}_k := 2n^{-1}\sum_{t=1}^{n} y_t \sin(\hat{\omega}_k t), \tag{6.4.15}$$

which approximate the LS estimator discussed in Chapter 5 (Section 5.1) by ignoring the possible nonorthogonality of the sinusoids. The corresponding polar parameters $C_k := \sqrt{A_k^2 + B_k^2}$ and $\phi_k := \arctan(-B_k, A_k)$ can be estimated by

$$\hat{C}_k := \sqrt{\hat{A}_k^2 + \hat{B}_k^2}, \quad \hat{\phi}_k := \arctan(-\hat{B}_k, \hat{A}_k). \tag{6.4.16}$$

Furthermore, let us assume that the signal frequencies satisfy the separation condition (6.3.7) with

$$\Delta := \min_{k \neq k'}\{\omega_k, \pi - \omega_k, |\omega_k - \omega_{k'}|\}. \tag{6.4.17}$$

It is also required that Ω_0 be a closed neighborhood of $\boldsymbol{\omega}_r$ and satisfy

$$\min_{k \neq k'}\{\tilde{\omega}_k, \pi - \tilde{\omega}_k, |\tilde{\omega}_k - \tilde{\omega}_{k'}|\} \geq c_n, \quad c_n \in (0, \Delta), \quad nc_n \to \infty, \tag{6.4.18}$$

where c_n is a predetermined constant which may depend on n.

Under these conditions, the following theorem can be established for the PM method. See Section 6.6 for a proof which follows the lines of [412].

Theorem 6.8 (Asymptotics of the PM Estimator: Real Case with White Noise). *Let* $\mathbf{y}$ *be given by (6.4.13) and* $\{\hat{\omega}_k, \hat{A}_k, \hat{B}_k, \hat{C}_k, \hat{\phi}_k\}$ *be defined by (6.4.14)–(6.4.16). Define* $\boldsymbol{\theta}_k := [A_k, B_k, \omega_k]^T$ *and* $\hat{\boldsymbol{\theta}}_k := [\hat{A}_k, \hat{B}_k, \hat{\omega}_k]^T$ *for the Cartesian RSM (2.1.1). Define* $\boldsymbol{\theta}_k := [C_k, \phi_k, \omega_k]^T$ *and* $\hat{\boldsymbol{\theta}}_k := [\hat{C}_k, \hat{\phi}_k, \hat{\omega}_k]^T$ *for the polar RSM (2.1.2). Let* $\boldsymbol{\theta} := [\boldsymbol{\theta}_1^T, \ldots, \boldsymbol{\theta}_q^T]^T$ *and* $\hat{\boldsymbol{\theta}} := [\hat{\boldsymbol{\theta}}_1^T, \ldots, \hat{\boldsymbol{\theta}}_q^T]^T$. *Assume that* $\{\omega_k\}$ *satisfies (6.3.7) with* Δ *defined by (6.4.17) and* Ω_0 *satisfies (6.4.18). Assume further that* $\{\epsilon_t\} \sim \mathrm{IID}(0, \sigma^2)$. *Then, as* $n \to \infty$, *the following assertions are true.*

- *(a) For all* k, $n(\hat{\omega}_k - \omega_k) \xrightarrow{P} 0$, $\hat{A}_k \xrightarrow{P} A_k$, $\hat{B}_k \xrightarrow{P} B_k$, $\hat{C}_k \xrightarrow{P} C_k$, *and* $\hat{\phi}_k \xrightarrow{P} \phi_k$.
- *(b) If* $n\Delta^2 \to \infty$, *then* $\mathbf{K}(\hat{\boldsymbol{\theta}} - \boldsymbol{\theta}) \xrightarrow{A} \mathrm{N}(\mathbf{0}, \boldsymbol{\Gamma}(\boldsymbol{\theta}))$, *where* $\boldsymbol{\Gamma}(\boldsymbol{\theta}) := \mathrm{diag}\{\boldsymbol{\Gamma}(\boldsymbol{\theta}_1), \ldots, \boldsymbol{\Gamma}(\boldsymbol{\theta}_q)\}$ *with* $\boldsymbol{\Gamma}(\boldsymbol{\theta}_k)$ *given by (3.3.13),* $\gamma_k := \frac{1}{2}C_k^2/\sigma^2$, $\mathbf{K} := \mathrm{diag}(\mathbf{K}_n, \ldots, \mathbf{K}_n)$, *and* $\mathbf{K}_n := \mathrm{diag}(n^{1/2}, n^{1/2}, n^{3/2})$.

Remark 6.13 Because the distribution $\mathrm{N}(\mathbf{0}, \boldsymbol{\Gamma}(\boldsymbol{\theta}_k))$ does not depend on the phase parameter under the polar RSM (2.1.2), the corresponding assertion in part (b) remains true if the ϕ_k are random variables.

Remark 6.14 In part (b), the stronger separation condition $n\Delta^2 \to \infty$ is needed to eliminate the bias in $\mathbf{K}_n(\hat{\boldsymbol{\theta}}_k - \boldsymbol{\theta}_k)$ which takes the form $\mathcal{O}(n^{-1/2}\Delta^{-1})$. A similar situation arises in Chapter 4 (Section 4.3) when calculating the second-order moments of the sample autocovariances.

Theorem 6.8 shows that the PM estimator of the amplitude or phase parameter is consistent in the ordinary sense, but the PM estimator of the frequency parameter is consistent in a higher order, i.e.,

$$n(\hat{\omega}_k - \omega_k) \xrightarrow{P} 0.$$

According to Theorem 5.6 and Remark 5.7, the higher-order consistency of the frequency estimator is a necessary condition for the consistency of the corresponding amplitude and phase estimator. Moreover, because Theorem 6.8 implies $\mathbf{K}_n(\hat{\boldsymbol{\theta}}_k - \boldsymbol{\theta}_k) \stackrel{A}{\sim} \mathrm{N}(\mathbf{0}, \boldsymbol{\Gamma}(\boldsymbol{\theta}_k))$, we can write $\mathbf{K}_n(\hat{\boldsymbol{\theta}}_k - \boldsymbol{\theta}_k) = \mathcal{O}_P(1)$, or equivalently,

$$\begin{aligned} \hat{A}_k - A_k &= \mathcal{O}_P(n^{-1/2}), \quad \hat{B}_k - B_k = \mathcal{O}_P(n^{-1/2}), \\ \hat{C}_k - C_k &= \mathcal{O}_P(n^{-1/2}), \quad \hat{\phi}_k - \phi_k = \mathcal{O}_P(n^{-1/2}), \end{aligned}$$

and

$$\hat{\omega}_k - \omega_k = \mathcal{O}_P(n^{-3/2}).$$

This means that the amplitude and phase estimator has the ordinary convergence rate $\sqrt{n}$ and the frequency estimator has a higher convergence rate $n\sqrt{n}$. In other words, the amplitude and phase estimator is $\sqrt{n}$-consistent and the frequency estimator is $n\sqrt{n}$-consistent. In the theory of estimation, the convergence rate of an estimator is also known as the error rate of that estimator.

According to Theorem 6.8, the asymptotic covariance matrix of the PM estimator takes the form $\mathbf{K}^{-1}\boldsymbol{\Gamma}(\boldsymbol{\theta})\mathbf{K}^{-1}$. This matrix is nothing but the asymptotic CRLB (ACRLB) in Theorem 3.3 derived under the Gaussian assumption. Therefore, the PM estimator is asymptotically efficient under the condition of Gaussian white noise. An asymptotically efficient estimator enjoys certain optimality properties in terms of the concentration of its asymptotic distribution [404, Chapter 8].

By Theorem 6.8, the PM estimator attains the Gaussian CRLB asymptotically even if the noise is not Gaussian, provided that it is an i.i.d. sequence with zero mean and finite variance. The Gaussian CRLB is the largest lower bound under this condition (Proposition 3.4). Therefore, the PM estimator is not asymptotically efficient in general when the noise is nonGaussian. In the next chapter, we will discuss an alternative estimator which is more efficient under the condition of heavy-tailed noise.

The noise variance σ^2 can be estimated from the residuals by

$$\hat{\sigma}^2 := n^{-1} \sum_{t=1}^{n} \left\{ y_t - \sum_{k=1}^{q} (\hat{A}_k \cos(\hat{\omega}_k t) + \hat{B}_k \sin(\hat{\omega}_k t)) \right\}^2. \tag{6.4.19}$$

The following corollary establishes the consistency and the asymptotic normality of this estimator. See Section 6.6 for a proof.

Corollary 6.3 (Asymptotics of the Estimator of Noise Variance: Real Case). *Let $\hat{\sigma}^2$ be defined by (6.4.19). Then, under the conditions of Theorem 6.8, $\hat{\sigma}^2 \xrightarrow{P} \sigma^2$ as $n \to \infty$. If, in addition, $\kappa := E(\epsilon_t^4)/\sigma^4 < \infty$, then $\sqrt{n}\,(\hat{\sigma}^2 - \sigma^2) \xrightarrow{D} \mathrm{N}(0, (\kappa-1)\sigma^4)$.*

According to Corollary 6.3, the convergence rate of $\hat{\sigma}^2$ is $\sqrt{n}$ if the noise has finite fourth moments. When this assumption is violated, as in the case of some heavy-tailed noise, the convergence rate can be much slower. An example of heavy-tailed noise that satisfies the conditions of Theorem 6.8 but not the condition of finite fourth moments is the noise that has Student's t-distribution with ν degrees of freedom where $2 < \nu < 3$. In this case, the sample variance is a poor estimator, even for very large but finite sample sizes.

So far, the discussion on the PM estimator is restricted to the case of white noise. Let us now consider the case of colored noise. Generally speaking, when the noise is colored, especially when the noise spectrum varies considerably in the frequency domain, it is beneficial to replace the original periodogram $I_n(\omega)$ with the standardized periodogram $\tilde{I}_n(\omega) := I_n(\omega)/\hat{f}_\epsilon(\omega)$, where $\hat{f}_\epsilon(\omega)$ is an estimator of the noise spectrum $f_\epsilon(\omega)$.

Figure 6.14 demonstrates the effect of the standardized periodogram in suppressing the spectral peaks of the noise. In this example, $\mathbf{y}$ takes the form (6.1.1) with $p = 2$, $C_1 = C_2 = 0.4$, $\phi_1 = \phi_2 = 0$, $\omega_1 = 2\pi \times 0.165$, $\omega_2 = 2\pi \times 0.34$, and $n = 100$. The noise $\{\epsilon_t\}$ is a zero-mean unit-variance Gaussian AR(2) process satisfying

$$\epsilon_t + \varphi_1 \epsilon_{t-1} + \varphi_2 \epsilon_{t-2} = \zeta_t, \tag{6.4.20}$$

where $\varphi_1 := -2r_0 \cos\omega_0$, $\varphi_2 := r_0^2$, and $\{\zeta_t\} \sim \mathrm{GWN}(0, \sigma^2)$ with

$$\sigma^2 := \{(1+\varphi_2)^2 - \varphi_1^2\}\,(1-\varphi_2)/(1+\varphi_2).$$

In this example, $r_0 = 0.65$ and $\omega_0 = 2\pi \times 0.24$. The estimated noise spectrum is obtained from an independent realization of the noise process using the Yule-Walker estimator to be discussed in Chapter 7.

As can be seen, the spectral peaks of the sinusoids in the original periodogram are hardly distinguishable from the spectral peaks of the noise, even if the number of sinusoids is known. In fact, the second largest peak in the periodogram is far removed from the two signal frequencies. This illustrates one potential danger of using procedures designed for white noise when colored is the actual case. In the standardized periodogram, the spectral peaks of the noise are suppressed effectively, making the spectral peaks of the sinusoids more pronounced.

The periodogram does not have to be standardized when the sample size is large enough or the signal is strong enough. As will be shown in the next theorem, the PM estimator based on the original periodogram remains asymptotically efficient under the assumption of Gaussian colored noise. This can be explained intuitively by the fact that the noncentrality parameter in the chi-square

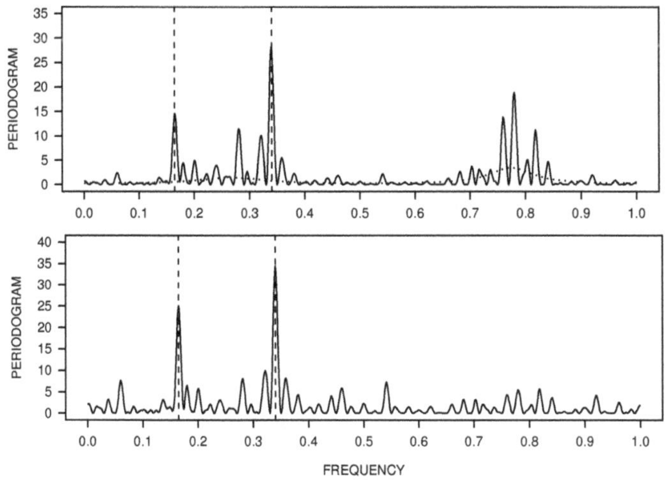

Figure 6.14. Original (top) and standardized (bottom) periodograms of a time series consisting of two equal-amplitude zero-phase complex sinusoids in Gaussian AR(2) noise. Dotted line shows the estimated noise spectrum based on an independent realization of the noise process. Dashed vertical lines indicate the signal frequencies.

distribution of the periodogram takes the form $\mathcal{O}(n)$ in the $\mathcal{O}(n^{-1})$-neighborhood of a signal frequency but takes the form $o(n)$ when ω is far away from the signal frequencies. The difference in magnitude suggests that for large n the largest local maxima in the periodogram would appear in the $\mathcal{O}(n^{-1})$-neighborhood of the signal frequencies. In this neighborhood, the noise spectrum, if sufficiently smooth, can be well approximated by a constant. Therefore, the local maximizer of the standardized periodogram in this neighborhood should be virtually identical to that of the unstandardized periodogram.

With the heuristic argument in mind, the following theorem, which generalizes Theorem 6.8 to the case of colored noise, should come as no surprise. For a rigorous proof, see Section 6.6.

Theorem 6.9 (Asymptotics of the PM Estimator: Real Case with Colored Noise). *Let* **y** *satisfy the conditions of Theorem 6.8 except that* $\{\epsilon_t\}$ *is a zero-mean stationary Gaussian process with ACF* $r_\epsilon(u)$ *such that* $\sum |r_\epsilon(u)| < \infty$, *or a real linear process of the form (6.1.25) with* $\sum |\psi_j| < \infty$ *and* $\{\zeta_t\} \sim \mathrm{IID}(0, \sigma^2)$. *Then, as* $n \to \infty$, *the assertion (a) in Theorem 6.8 remains true. In addition, if* $\inf\{f_\epsilon(\omega)\} > 0$, *then the assertion (b) in Theorem 6.8 remains true except that* $\gamma_k := \frac{1}{2} C_k^2 / f_\epsilon(\omega_k)$.

Theorem 6.9 shows that the asymptotic covariance matrix of the PM estimator can be expressed as $\mathbf{K}^{-1}\boldsymbol{\Gamma}(\boldsymbol{\theta})\mathbf{K}^{-1}$. This matrix is just the ACRLB under the assumption of Gaussian colored noise with SDF $f_\epsilon(\omega)$. Therefore, the PM estimator remains asymptotically efficient in the case of Gaussian colored noise. Moreover, because $\boldsymbol{\Gamma}(\boldsymbol{\theta}_k)$ depends on the noise solely through $f_\epsilon(\omega_k)$, the noise spectrum at nonsignal frequencies does not affect the asymptotic distribution of the PM estimator. By Theorem 6.9, the PM estimator attains the Gaussian CRLB asymptotically even though the noise is nonGaussian.

The asymptotic theory of the PM method in Theorems 6.8 and 6.9 can be extended entirely to the complex case. For complex sinusoids, the PM frequency estimator is given by (6.4.7). The complex amplitudes $\beta_k := C_k \exp(i\phi_k)$ can be estimated by the Fourier coefficients

$$\hat{\beta}_k := \hat{A}_k - i\hat{B}_k := n^{-1} \sum_{t=1}^{n} y_t \exp(-i\hat{\omega}_k t). \tag{6.4.21}$$

The corresponding polar parameters C_k and ϕ_k can be estimated by (6.4.16). Similar to the real case, it is required that Ω_0 in (6.4.7) be a closed neighborhood of $\boldsymbol{\omega}$ and satisfy the separation condition

$$\min_{k\neq k'}\{|\tilde{\omega}_k - \tilde{\omega}_{k'}|, 2\pi - |\tilde{\omega}_k - \tilde{\omega}_{k'}|\} \geq c_n, \quad c_n \in (0,\Delta), \quad nc_n \to \infty. \tag{6.4.22}$$

It is also required that the signal frequencies satisfy (6.3.7). Under these conditions, the following theorem can be established. A proof is given in Section 6.6.

Theorem 6.10 (Asymptotics of the PM Estimator: Complex Case). *Let* $\mathbf{y}$ *be given by (6.1.1) and* $\{\hat{\omega}_k, \hat{A}_k, \hat{B}_k, \hat{C}_k, \hat{\phi}_k\}$ *be defined by (6.4.7), (6.4.21), and (6.4.16). Let* $\boldsymbol{\theta}_k := [A_k, B_k, \omega_k]^T$ *and* $\hat{\boldsymbol{\theta}}_k := [\hat{A}_k, \hat{B}_k, \hat{\omega}_k]^T$ *for the Cartesian CSM (2.1.5). Let* $\boldsymbol{\theta}_k := [C_k, \phi_k, \omega_k]^T$ *and* $\hat{\boldsymbol{\theta}}_k := [\hat{C}_k, \hat{\phi}_k, \hat{\omega}_k]^T$ *for the polar CSM (2.1.7). Let* $\boldsymbol{\theta} := [\boldsymbol{\theta}_1^T, \dots, \boldsymbol{\theta}_p^T]^T$ *and* $\hat{\boldsymbol{\theta}} := [\hat{\boldsymbol{\theta}}_1^T, \dots, \hat{\boldsymbol{\theta}}_p^T]^T$. *Assume that* $\{\omega_k\}$ *satisfies (6.3.7) with* Δ *defined by (6.3.8) and* Ω_0 *satisfies (6.4.22). Assume further that* $\{\epsilon_t\}$ *is a zero-mean stationary complex Gaussian process with ACF* $r_\epsilon(u)$ *such that* $\sum |r_\epsilon(u)| < \infty$, *or a complex linear process of the form (6.1.25) with* $\sum |\psi_j| < \infty$ *and* $\{\zeta_t\} \sim \mathrm{IID}(0, \sigma^2)$. *Then, as* $n \to \infty$, *the following assertions are true.*

(a) For all k, $n(\hat{\omega}_k - \omega_k) \xrightarrow{P} 0$, $\hat{A}_k \xrightarrow{P} A_k$, $\hat{B}_k \xrightarrow{P} B_k$, $\hat{C}_k \xrightarrow{P} C_k$, *and* $\hat{\phi}_k \xrightarrow{P} \phi_k$.

(b) If $n\Delta^2 \to \infty$, $\inf\{f_\epsilon(\omega)\} > 0$, *and* $E(\zeta_t^2) = 0$, *then* $\mathbf{K}(\hat{\boldsymbol{\theta}} - \boldsymbol{\theta}) \stackrel{A}{\sim} \mathrm{N}(\mathbf{0}, \boldsymbol{\Gamma}(\boldsymbol{\theta}))$, *where* $\boldsymbol{\Gamma}(\boldsymbol{\theta}) := \mathrm{diag}\{\boldsymbol{\Gamma}(\boldsymbol{\theta}_1), \dots, \boldsymbol{\Gamma}(\boldsymbol{\theta}_p)\}$ *with* $\boldsymbol{\Gamma}(\boldsymbol{\theta}_k)$ *given by (3.3.13),* $\gamma_k := C_k^2 / f_\epsilon(\omega_k)$, $\mathbf{K} := \mathrm{diag}(\mathbf{K}_n, \dots, \mathbf{K}_n)$, *and* $\mathbf{K}_n := \mathrm{diag}(n^{1/2}, n^{1/2}, n^{3/2})$.

The asymptotic covariance matrix $\mathbf{K}^{-1}\boldsymbol{\Gamma}(\boldsymbol{\theta})\mathbf{K}^{-1}$ coincides with the ACRLB under the assumption of complex Gaussian noise. Therefore, the asymptotic efficiency remarks about the PM estimator in the real case remain valid in the complex case. As in the real case, the PM estimator attains the Gaussian CRLB asymptotically even if the noise is nonGaussian.

The asymptotic normality assertions in Theorems 6.8–6.10 can be used to construct confidence regions for the parameters. For example, let $\hat{\boldsymbol{\Gamma}}_k$ be a consistent

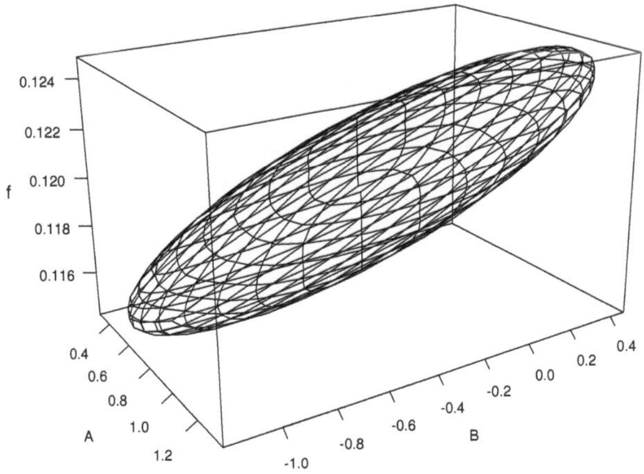

Figure 6.15. An approximately 95% confidence ellipsoid given by (6.4.23) for the parameters of the first complex sinusoid, (A_1, B_1, f_1), in the time series shown in Figure 6.1.

estimator of $\Gamma(\boldsymbol{\theta}_k)$, which can be obtained by substituting $\boldsymbol{\theta}_k$ with the PM estimator $\hat{\boldsymbol{\theta}}_k$ and, in the simple case of white noise, by replacing σ^2 with the sample variance of the residuals

$$\hat{\epsilon}_t := y_t - \sum_{k=1}^{p} \hat{\beta}_k \exp(i\hat{\omega}_k t) \qquad (t = 1, \ldots, n).$$

Then, according to Theorems 6.8–6.10 and Lemma 12.2.1,

$$(\hat{\boldsymbol{\theta}}_k - \boldsymbol{\theta}_k)^T \mathbf{K}_n \hat{\boldsymbol{\Gamma}}_k^{-1} \mathbf{K}_n (\hat{\boldsymbol{\theta}}_k - \boldsymbol{\theta}_k) \stackrel{A}{\sim} \chi^2(3).$$

Therefore, for any $\alpha \in (0,1)$, an approximate $(1-\alpha) \times 100\%$ confidence region (ellipsoid, to be more precise) of $\boldsymbol{\theta}_k$ is given by

$$\Theta_k(\alpha) := \{\tilde{\boldsymbol{\theta}}_k : (\hat{\boldsymbol{\theta}}_k - \tilde{\boldsymbol{\theta}}_k)^T \mathbf{K}_n \hat{\boldsymbol{\Gamma}}_k^{-1} \mathbf{K}_n (\hat{\boldsymbol{\theta}}_k - \tilde{\boldsymbol{\theta}}_k) < c(\alpha)\}, \tag{6.4.23}$$

where $c(\alpha)$ denotes the $(1-\alpha)$th quantile of the $\chi^2(3)$ distribution. The asymptotic efficiency of $\hat{\boldsymbol{\theta}}_k$ implies that $\Theta_k(\alpha)$ is the smallest confidence ellipsoid.

As an example, Figure 6.15 shows a confidence region constructed according to (6.4.23) with $\alpha = 0.05$ for the parameters of the first complex sinusoid in the time series shown in Figure 6.1. The confidence region indeed contains the true parameters, which are $A_1 = 1$, $B_1 = 0$, and $f_1 = 0.12$.

Note that a similar confidence region can be constructed for a subset of $\boldsymbol{\theta}_k$. It suffices to replace the full covariance matrix $\hat{\boldsymbol{\Gamma}}_k$ by the covariance matrix corresponding to the subset of interest and form a quadratic function in the same way

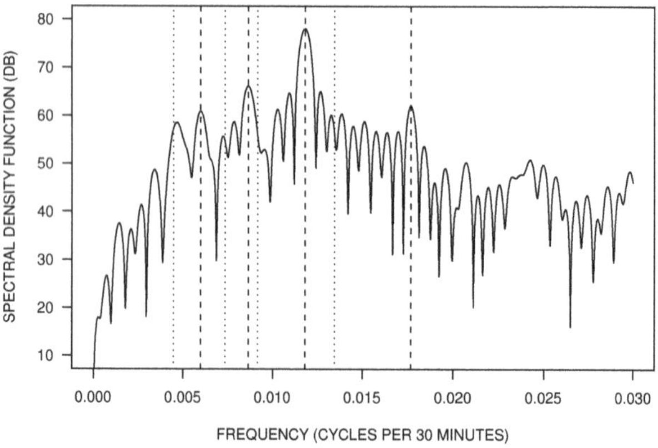

Figure 6.16. The periodogram of the variable star data shown in Figure 1.1 and the frequencies of the eight-sinusoid model. Solid line, periodogram; dashed line, frequencies found in the original data; dotted line, additional frequencies found in the residuals.

Table 6.3. Frequency and Amplitude Estimates for the Variable Star Data

k	f_k	A_k	B_k	k	f_k	A_k	B_k
1	0.004467237	8.4036	−33.8877	5	0.009181022	−15.4682	−37.5248
2	0.005992840	−18.5610	46.0177	6	0.011813066	−214.7132	332.7032
3	0.007354442	−18.1787	4.4174	7	0.013460770	−17.8164	24.0861
4	0.008662235	−20.4151	79.4462	8	0.017682145	−27.9779	−39.4620

The constant term A_0 is estimated as 745.5309.

as (6.4.23). The chi-square distribution theory remains valid except that the degree of freedom should be equal to the size of the subset. The confidence region reduces to an interval if the size of the subset equals 1.

Finally, let us consider an application of the PM method to the variable star data discussed in Chapter 1 (Section 1.3). Recall that the time series, which is shown in Figure 1.1, consists of brightness measurements of a variable star produced by Kepler spacecraft over 30-minute intervals for about 34 days.* Let the time series be modeled as a constant A_0 plus the sum of eight real sinusoids with frequencies $\omega_k := 2\pi f_k$ and coefficients (A_k, B_k) $(k = 1, \ldots, 8)$. Table 6.3 contains the estimated parameters using the PM method which yield the sinusoidal model shown in the middle panel of Figure 1.1.

To obtain the frequency estimates, the RDFT procedure defined by (6.3.14) and (6.3.15) is applied twice with $m = 1$ and $\delta = 5\pi/n$, each followed by 10 iterations of the algorithm (6.4.10) using the modified Newton-Raphson mapping in

*http://kepler.nasa.gov/.

(6.4.12) with $\varrho = 0.5$ and with the DFT estimates as initial values. The first application is to the original time series with $p = 4$, producing the frequency estimates in Table 6.3 for $k = 2,4,6,8$. The second application is to the residuals obtained from least-squares regression using the estimated frequencies from the previous step. This produces the remaining four frequency estimates in Table 6.3. The amplitude estimates in Table 6.3, together with the constant term, are obtained by an application of least-squares regression to the original time series using all eight estimated frequencies.

Figure 6.16 depicts the estimated frequencies with the periodogram of the original time series. Note that the frequencies found in the second step from the residuals (dotted lines) do not necessarily coincide with local maxima in the periodogram of the original data.

As we can see, the strongest component ($k = 6$) has an estimated frequency 0.011813066, corresponding to a periodicity of 84.7 samples, or $84.7/2 = 42.35$ hours. The eight-sinusoid model has an R^2 statistic which equals 94.4%, so it captures the main variability of the data. This can also be seen by comparing the original time series in the top panel of Figure 1.1 with the residuals of the model in the bottom panel. The residuals in this case are not white noise. An autoregressive (AR) model of order 16 seems adequate in describing the serial dependence of the residuals. This will be discussed later in Chapter 7.

6.4.3 Resolution Limit

A major drawback of the periodogram is its inability to resolve frequencies that are too close to each other. The resolution limit is often stated as $2\pi/n$, which is just the half width of the main lobe in the Dirichlet kernel for a time series of length n [254]. The actual resolvability, as it turns out, also depends on the amplitudes and phases of the sinusoids.

Consider the simple example of two complex sinusoids without noise, where

$$y_t = C_1 \exp\{i(\omega_1 t + \phi_1)\} + C_2 \exp\{i(\omega_2 t + \phi_2)\} \qquad (t = 1, \ldots, n).$$

In this example, the periodogram takes the form

$$I_n(\omega) = n|C_1 D_n(\omega_1 - \omega) + C_2 D_n(\omega_2 - \omega) \exp\{i(\phi_2 - \phi_1)\}|^2,$$

where $D_n(\omega) := n^{-1} \sum_{t=1}^{n} \exp(it\omega)$ is the Dirichlet kernel. Figure 6.17 depicts the periodogram under different frequency separation conditions with $C_1 = C_2 = 1$ and $n = 20$ for fixed phase difference $\phi_2 - \phi_1 = 0$.

As we can see, the periodogram is able to show two distinct peaks in cases (a)–(c) where the frequency separation is greater than or equal to the resolution limit $2\pi/n$, but fails to do so in case (d) where the frequency separation falls below the resolution limit. Note that although the sinusoids are resolved in cases

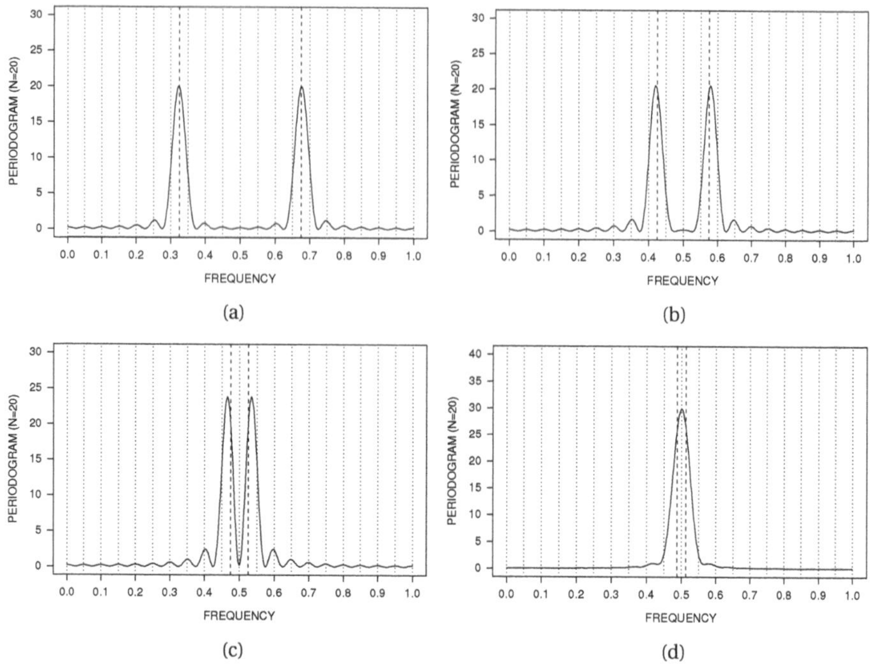

Figure 6.17. Periodogram of two unit-amplitude complex sinusoids with $n = 20$ under different conditions of frequency separation. (a) $\omega_2 - \omega_1 = 2\pi \times 7/n$; (b) $\omega_2 - \omega_1 = 2\pi \times 3/n$; (c) $\omega_2 - \omega_1 = 2\pi \times 1/n$; (d) $\omega_2 - \omega_1 = 2\pi \times 0.5/n$. In all cases, $\phi_2 - \phi_1 = 0$. Vertical dashed lines depict the signal frequencies. Vertical dotted lines represent the Fourier grid.

(a)–(c), the peaks of the periodogram are not necessarily located at the signal frequencies and the discrepancy (or bias) actually increases with the decrease of frequency separation. Therefore, resolvability does not imply unbiasedness.

Figure 6.18 further demonstrates the impact of the phase difference $\phi_2 - \phi_1$ on the resolvability of the sinusoids by the periodogram for fixed frequency separation $\omega_2 - \omega_1 = 2\pi \times 1.2/n$. Although the frequency separation is greater than $2\pi/n$ in all cases, the sinusoids are barely resolved in cases (b) and (d) and totally unresolved in case (c), depending entirely on the phase difference of the sinusoids. Hence $2\pi/n$ should be regarded only as a simplified rule rather than a precise statement about the resolution limit of the periodogram.

It is easier to describe the resolution limit of the periodogram in the asymptotic sense for large sample sizes. According to Theorems 6.8, 6.9, and 6.10, consistent estimation of the sinusoidal parameters requires the frequency separation either to be constant as the sample size grows or to approach zero more slowly than the reciprocal of the sample size. In this sense, $\mathcal{O}(n^{-1})$ can be regarded as the asymptotic resolution limit for consistent parameter estimation.

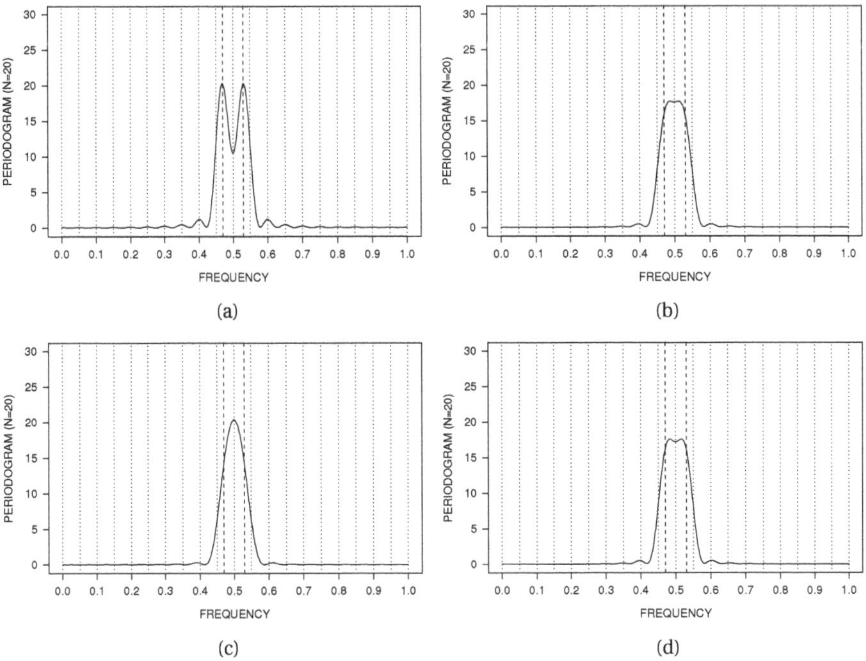

Figure 6.18. Periodogram of two unit-amplitude complex sinusoids with $n = 20$ under different conditions of phase difference. (a) $\phi_2 - \phi_1 = \pi/4$; (b) $\phi_2 - \phi_1 = \pi/2$; (c) $\phi_2 - \phi_1 = 3\pi/4$; (d) $\phi_2 - \phi_1 = \pi$. In all cases, $\omega_2 - \omega_1 = 2\pi \times 1.2/n$. Vertical dashed lines depict the signal frequencies. Vertical dotted lines represent the Fourier grid.

6.5 Time-Frequency Analysis

The techniques of periodogram analysis can be applied to the situation where the sinusoidal parameters change slowly with time. Strictly speaking, a sinusoidal model with constant parameters is no longer valid in this case. However, if the change is gradual relative to the sampling rate, then the parameters can be regarded as approximately constant within a moving window of sufficiently short length. In this case, the periodogram analysis techniques discussed in the previous sections can be applied to the data points within the moving window. This approach is known as *short-time* Fourier analysis.

As the window slides forward in time, one obtains a sequence of periodograms from the short-time Fourier analysis. Plotting these periodograms as a bivariate function of frequency and time produces the so-called *spectrogram*. A spectrogram is a representation of time-varying spectral characteristics of nonstation-

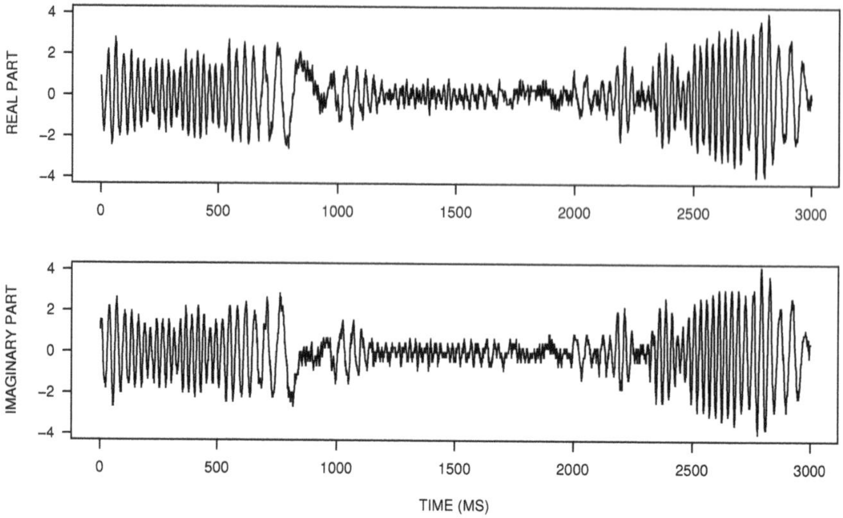

Figure 6.19. Real (in-phase) and imaginary (quadrature) parts of radar returns from a weak target in sea clutters. Time is in milliseconds.

ary time series. It is a powerful tool for *time-frequency analysis.* Note that while spectrograms are widely used in practice, there is a huge body of literature on time-frequency analysis devoted to alternative time-frequency representations. For comprehensive coverage on this subject, see [76] and [129], for example.

To construct a spectrogram in practice, the data points within the window of length m are often weighted by some positive constants $w_1, \ldots, w_m$ which are tapered at both ends to reduce the truncation effect. The weights can be generated by a window function $w(x) \geq 0$ defined in the interval $[0,1]$ such that

$$w_s = w((s-1)/(m-1)) \qquad (s = 1, \ldots, m).$$

A popular choice in applications such as speech signal processing [86] is the Hamming window $w(x) := \{0.54 - 0.46\cos(2\pi x)\}\mathscr{I}(0 \leq x \leq 1)$, named after the American mathematician Richard Wesley Hamming (1915–1998). Obviously, the unweighted case corresponds to the rectangular window $w(x) := \mathscr{I}(0 \leq x \leq 1)$.

Given the weights w_s, and with the center of the sliding window acting as the time index, the spectrogram of $\{y_1, \ldots, y_n\}$ can be formally expressed as

$$I_m(t,\omega) := m^{-1}\left|\sum_{s=1}^{m} w_s y_{t+s-[m/2]} \exp(-i\omega s)\right|^2,$$

where $t = [m/2], [m/2]+1, \ldots, n-[m/2]$ and $m \ll n$.

As an example, consider the time series of radar returns discussed in Chapter 1 (Section 1.4). Recall that the time series contains the noisy echoes reflected by a floating target in the Atlantic Ocean about 2,655 meters from the radar.† The time-varying properties of the time series are due in part to the motion of the target which is not always in the same direction at the same speed [219]. A small segment of this time series is depicted in Figure 1.2 to demonstrate the validity of the sinusoid-plus-noise model. A longer segment, shown in Figure 6.19, reveals its time-varying nonstationary nature.

Figure 6.20 shows the spectrogram of this time series produced by a rectangular window of length $m = 128$. The time-dependent periodograms are calculated by using the $16m$-point DFT with zero padding and subsampled at the rate of $1/64$ (hence a 50% overlap between successive windows). The time-varying peak frequency shown in Figure 6.20 is obtained by applying the RDFT method with $p = 1$ to each (refined) periodogram. It represents the radial velocity, also known as the Doppler frequency, of the moving target due to the Doppler effect.

The spectrogram in Figure 6.20 successfully depicts the moving target as a time-varying spectral peak. The RDFT method provides a good estimate of the peak frequency most of the time when the SNR is sufficiently high. At low SNR, however, the RDFT method becomes much less effective. For example, the RDFT estimates are hugely erroneous between $t = 20$ and $t = 22$ where the spectral peak is noticeably weak.

To overcome this difficulty, one can impose certain smoothness conditions on the frequency trajectory over time. For example, if the RDFT estimate in a given window differs from the estimate in the previous window by $2\pi\delta/m$ for a suitable $\delta > 0$, then the estimate from the previous window is used as the initial value for an iterative algorithm that maximizes the current periodogram to produce an alternative estimate. Figure 6.21 shows the result of this recursive scheme, where the deviation of successive frequency estimates is determined with $\delta = 2$ and the new estimates are produced by the modified Newton-Raphson algorithm in (6.4.11) and (6.4.12) with $\varrho = 1$. As can be seen, the recursive scheme is able track the spectral peak successfully even under low SNR conditions.

In closing, we remark that caution must be exercised when applying the periodogram to sinusoids with time-varying frequencies. Figure 6.22 shows an example where the spectral peak splits as a consequence of time-varying frequency, giving the false impression that multiple sinusoids are present in the data. In this example, the signal is a linear chirp with

$$y_t = \exp\{i\phi(t)\} \qquad (t = 1, \ldots, n), \tag{6.5.1}$$

where $\phi(t) := 2\pi(f_0 t + \frac{1}{2}d(t-2)t/(n-1))$ is called the instantaneous phase and $\omega(t) := \dot{\phi}(t) = 2\pi(f_0 + d(t-1)/(n-1))$ is called the instantaneous frequency, which is a linear function of t. The constant d is called the chirp rate.

†http://soma.ece.mcmaster.ca/ipix.

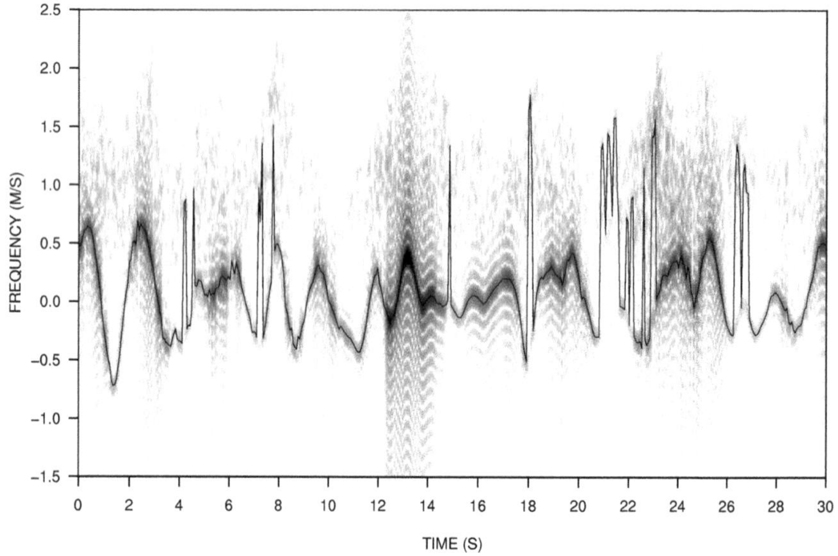

Figure 6.20. Spectrogram of the radar returns (darker shades of grey represent larger values). Solid line depicts the estimated time-varying frequency by the RDFT method. Time is in seconds and Doppler frequency is in meters per second.

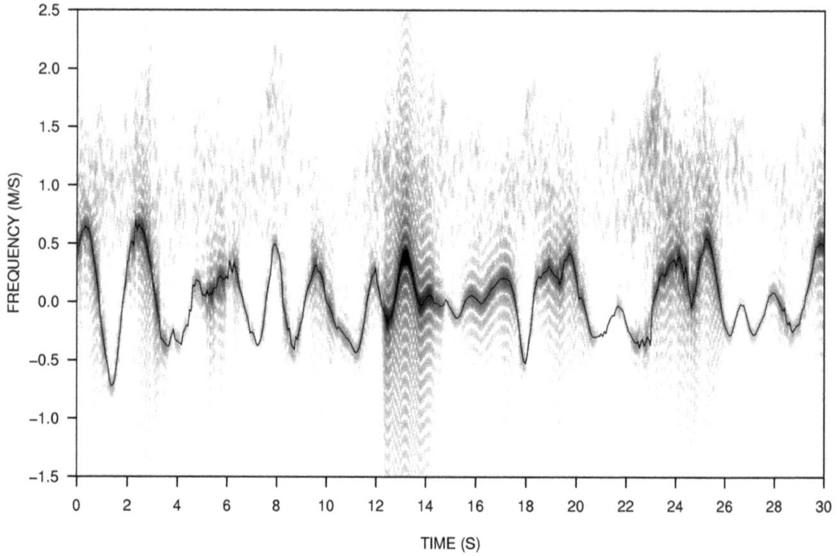

Figure 6.21. Same as Figure 6.20 except that the estimated time-varying frequency is obtained by a recursive scheme that employs the modified Newton-Raphson algorithm defined by (6.4.11) and (6.4.12) with the previous estimate as the initial value when an abrupt change is detected in the current RDFT estimate.

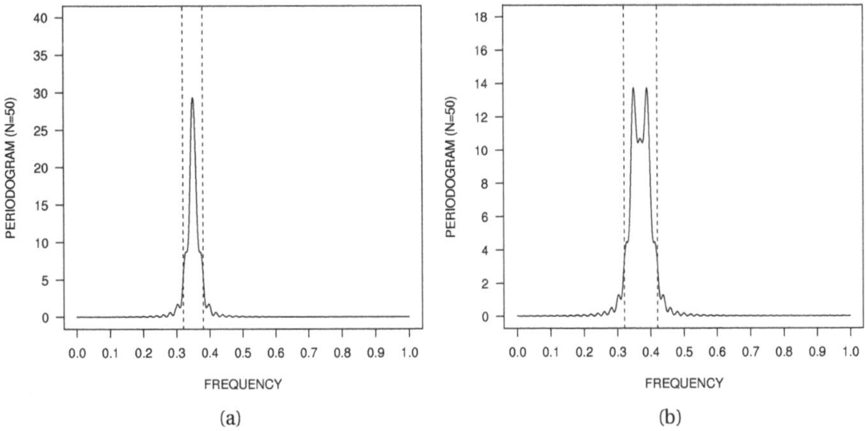

Figure 6.22. Periodogram of a linear chirp signal of the form (6.5.1) with instantaneous frequency $\omega(t) := \dot{\phi}(t) = 2\pi(f_0 + d(t-1)/(n-1))$ ($f_0 = 0.32; n = 50$). (a) $d = 3/n$; (b) $d = 5/n$. Dotted lines depict f_0 and $f_0 + d$.

Figure 6.22 shows that when d is small, the periodogram correctly depicts the signal with a single spectral peak centered between $2\pi f_0$ and $2\pi(f_0+d)$. However, as d increases, the spectral peak begins to split. This represents a situation where the spectral characteristics of the time series change too fast relative to the length of the time series. Therefore, to avoid such misbehavior, choosing a sufficiently short window for the spectrogram is essential. Of course, the downside of a short window is the reduced frequency resolution and detection power. Therefore, in practice, a suitable tradeoff must be made.

6.6 Proof of Theorems

This section contains the proof of Theorems 6.1–6.10 and the proof of Corollaries 6.1 and 6.3.

Proof of Theorem 6.1. Let $\mathbf{F}_p \in \mathbb{C}^{n\times p}$ be the matrix that comprises the column vectors in $\{\mathbf{f}(\lambda_j) : \lambda_j \in \Omega_p\}$. Because $\mathbf{F}_p^H \mathbf{F}_p = n\mathbf{I}$, the corresponding LS solution that minimizes $\|\mathbf{y} - \mathbf{F}_p\tilde{\mathbf{c}}_p\|^2$ with respect to $\tilde{\mathbf{c}}_p \in \mathbb{C}^p$ is given by

$$\hat{\mathbf{c}}_p := (\mathbf{F}_p^H \mathbf{F}_p)^{-1} \mathbf{F}_p \mathbf{y} = n^{-1} \mathbf{F}_p^H \mathbf{y}.$$

On the other hand, it follows from (6.1.3) and the orthogonality of $\mathbf{F}$ that

$$\mathbf{y} = \mathbf{F}\mathbf{z} = \sum_{j=0}^{n-1} z_j \mathbf{f}(\lambda_j). \tag{6.6.1}$$

Therefore,

$$\hat{\mathbf{c}}_p = n^{-1} \sum_{j=0}^{n-1} z_j \mathbf{F}_p^H \mathbf{f}(\lambda_j) = \mathrm{vec}[z_j : \lambda_j \in \Omega_p],$$

which in turn implies that

$$\mathbf{F}_p \hat{\mathbf{c}}_p = \sum_{\lambda_j \in \Omega_p} z_j \mathbf{f}(\lambda_j). \tag{6.6.2}$$

Combining (6.6.2) with (6.6.1) gives

$$\mathrm{SSE}(\Omega_p) := \|\mathbf{y} - \mathbf{F}_p \hat{\mathbf{c}}_p\|^2 = \left\| \sum_{\lambda_j \notin \Omega_p} z_j \, \mathbf{f}(\lambda_j) \right\|^2.$$

This expression leads to (6.1.8) due to the orthogonality of the $\mathbf{f}(\lambda_j)$. □

Proof of Theorem 6.2. Consider the DFT of $\mathbf{y}$ given by (6.1.10)–(6.1.12). Because $\boldsymbol{\epsilon} \sim \mathrm{N}_c(\mathbf{0}, \sigma^2 \mathbf{I})$ and $\mathbf{F}^H \mathbf{F} = n\mathbf{I}$, it follows from Proposition 2.6 that

$$\mathbf{z} = \boldsymbol{\mu} + n^{-1} \mathbf{F}^H \boldsymbol{\epsilon} \sim \mathrm{N}_c(\boldsymbol{\mu}, n^{-1}\sigma^2 \mathbf{I}).$$

In other words, $z_0, z_1, \ldots, z_{n-1}$ are independent complex Gaussian random variables with $z_j \sim \mathrm{N}_c(\mu_j, n^{-1}\sigma^2)$. This implies that $I_n(\lambda_0), I_n(\lambda_1), \ldots, I_n(\lambda_{n-1})$ are independent random variables and by Lemma 12.2.1(b),

$$I_n(\lambda_j)/\sigma^2 = |z_j|^2/(n^{-1}\sigma^2) \sim \tfrac{1}{2}\chi^2(2, 2\theta_j), \tag{6.6.3}$$

where $\theta_j := |\mu_j|^2/(n^{-1}\sigma^2)$. Moreover, because $\omega_k = \lambda_{j(k)}$ for all k and because $\mathbf{f}^H(\lambda_j)\mathbf{f}(\lambda_l) = n\delta_{j-l}$, it follows that

$$\mu_j = n^{-1} \mathbf{f}^H(\lambda_j) \mathbf{x}_t = \sum_{k=1}^{p} \beta_k \delta_{j-j(k)}. \tag{6.6.4}$$

The assertion is thus proved.

In the real case, $\mathbf{y}$ can be expressed as (6.1.1) with $p := 2q$, $\beta_{p-k+1} := \beta_k^* := \frac{1}{2} C_k \exp(-i\phi_k)$, and $\omega_{p-k+1} := 2\pi - \omega_k$ $(k = 1, \ldots, q)$. Therefore, (6.1.10) remains valid. Because $\boldsymbol{\epsilon} \sim \mathrm{N}(\mathbf{0}, \sigma^2 \mathbf{I})$ and $\Re(\mathbf{F})^T \Im(\mathbf{F}) = \mathbf{0}$, it follows that

$$[\Re(\mathbf{z})^T, \Im(\mathbf{z})^T]^T \sim \mathrm{N}([\Re(\boldsymbol{\mu})^T, \Im(\boldsymbol{\mu})^T]^T, n^{-1}\sigma^2 \mathbf{C}),$$

where $\mathbf{C} := n^{-1}\mathrm{diag}\{\Re(\mathbf{F})^T\Re(\mathbf{F}), \Im(\mathbf{F})^T\Im(\mathbf{F})\}$. For any $j \in \{j \in \mathbb{Z}_n : \lambda_j \in \Omega_n \cap [0,\pi]\}$, (6.6.4) can be written as

$$\mu_j = \sum_{k=1}^{q} \beta_k \delta_{j-j(k)}. \tag{6.6.5}$$

Moreover, for any $\lambda_j, \lambda_l \notin \{0,\pi\}$, we have

$$\Re\{\mathbf{f}(\lambda_j)\}^T\Re\{\mathbf{f}(\lambda_l)\} = \Im\{\mathbf{f}(\lambda_j)\}^T\Im\{\mathbf{f}(\lambda_l)\} = \tfrac{1}{2}n\delta_{j-l}.$$

For $\lambda_j = 0$ or π and any λ_l, we have

$$\Re\{\mathbf{f}(\lambda_j)\}^T\Re\{\mathbf{f}(\lambda_l)\} = n\delta_{j-l}, \quad \Im\{\mathbf{f}(\lambda_j)\}^T\Im\{\mathbf{f}(\lambda_l)\} = 0.$$

This result implies that $\mathbf{C}$ is a diagonal matrix, so the z_j and hence the $I_n(\lambda_j)$ are independent random variables. It also implies that for $\lambda_j \notin \{0,\pi\}$, $\Re(z_j)$ and $\Im(z_j)$ are independently distributed as $\mathrm{N}(\Re(\mu_j), \frac{1}{2}n^{-1}\sigma^2)$ and $\mathrm{N}(\Im(\mu_j), \frac{1}{2}n^{-1}\sigma^2)$, respectively, and hence $z_j \sim \mathrm{N}_c(\mu_j, n^{-1}\sigma^2)$. According to Lemma 12.2.1(b) and (6.6.5), we obtain (6.6.3) with $\theta_j := |\mu_j|^2/(n^{-1}\sigma^2) = \frac{1}{2}n\gamma_k\delta_{j-j(k)}$ and $\gamma_k := \frac{1}{2}C_k^2/\sigma^2$. For $\lambda_j = 0$ or π, we have $\mu_j = 0$, $\Re(z_j) \sim \mathrm{N}(0, n^{-1}\sigma^2)$ and $\Im(z_j) = 0$. Therefore, we obtain $I_n(\lambda_j)/\sigma^2 = \Re(z_j)^2/(n^{-1}\sigma^2) \sim \chi^2(1)$. □

Proof of Theorem 6.3. Let $U_n := \max\{(2/\sigma^2)I_n(\lambda_j) : \lambda_j \in \Omega_n \setminus \{\omega_k\}\}$. Because $\{\omega_k\} \subset \Omega_n$, it follows that

$$P_{\mathrm{ID}} = P\{I_n(\omega_k) > \tfrac{1}{2}\sigma^2 U_n, k = 1, \ldots, p\}.$$

According to Theorem 6.2, U_n is independent of the $I_n(\omega_k)$. Therefore, conditional on U_n, the $I_n(\omega_k)$ remain independent random variables with $I_n(\omega_k) \sim \frac{1}{2}\sigma^2\chi^2(2, 2n\gamma_k)$ for all k. This result leads to

$$P_{\mathrm{ID}} = E\left\{\prod_{k=1}^{p} \bar{F}(U_n, 2n\gamma_k)\right\} = \int_0^\infty \left\{\prod_{k=1}^{p} \bar{F}(x, 2n\gamma_k)\right\} dF_n(x),$$

where $F_n(x)$ denotes the CDF of U_n and can be expressed as (6.1.14). □

Proof of Corollary 6.1. For any fixed integer $v > 0$, it is easy to see that

$$\exp(-\tfrac{1}{2}\theta) \sum_{u=0}^{v-1} (\tfrac{1}{2}\theta)^u / u! < 1 \quad \forall \theta > 0$$

and

$$\exp(-\tfrac{1}{2}\theta) \sum_{u=0}^{v-1} (\tfrac{1}{2}\theta)^u / u! \to 0 \quad \text{as } \theta \to \infty.$$

Therefore, it follows from (6.1.16) and the dominated convergence theorem that $\bar{F}(x,\theta) \to 1$ as $\theta \to \infty$. Combining this result with Theorem 6.3 yields

$$P_{\text{ID}} \to \int_0^\infty \tfrac{1}{2}(n-p)\{1-\exp(-\tfrac{1}{2}x)\}^{n-p-1}\exp(-\tfrac{1}{2}x)\,dx = 1$$

as $n\gamma_k \to \infty$ for all k. □

Proof of Theorem 6.4. Consider (6.1.10)–(6.1.12). It is easy to show that

$$\begin{aligned} n\text{Var}(e_j) &= n^{-1}\sum_{t,s=1}^{n} r_\epsilon(t-s)\exp\{-i(t-s)\lambda_j\} \\ &= n^{-1}\sum_{|u|<n}(n-|u|)\,r_\epsilon(u)\exp(-iu\lambda_j) \\ &= f_\epsilon(\lambda_j) - \sum_{|u|<n} n^{-1}|u|\,r_\epsilon(u)\exp(-iu\lambda_j) - \sum_{|u|\ge n} r_\epsilon(u)\exp(-iu\lambda_j) \\ &= f_\epsilon(\lambda_j) + \mathcal{O}(r_n) \end{aligned}$$

uniformly in $\lambda_j \in \Omega_n$, where $r_n := \sum_u \min(1,|u|/n)\,|r_\epsilon(u)| \to 0$ as $n \to \infty$ by the dominated convergence theorem. Therefore, we can write

$$\sigma_j^2 := \text{Var}(e_j) = n^{-1} f_\epsilon(\lambda_j) + \mathcal{O}(n^{-1} r_n) \tag{6.6.6}$$

uniformly in $\lambda_j \in \Omega_n$. Because $\{\epsilon_t\}$ is complex Gaussian, it follows from Proposition 2.6 that the $z_j = \mu_j + e_j$ are jointly complex Gaussian with mean μ_j and variance σ_j^2. By Lemma 12.2.1(b), we obtain

$$|z_j|^2/\sigma_j^2 \sim \tfrac{1}{2}\chi^2(2, 2|\mu_j|^2/\sigma_j^2), \tag{6.6.7}$$

which, in turn, implies that

$$I_n(\lambda_j) = n|z_j|^2 = n\sigma_j^2(|z_j|^2/\sigma_j^2) \sim \tfrac{1}{2}\rho_j^{-1} f_\epsilon(\lambda_j)\chi^2(2, 2\rho_j\theta_j),$$

where $\rho_j := f_\epsilon(\lambda_j)/(n\sigma_j^2)$ and $\theta_j := n|\mu_j|^2/f_\epsilon(\lambda_j)$. It follows from (6.6.6) that $\rho_j = f_\epsilon(\lambda_j)/(f_\epsilon(\lambda_j) + \mathcal{O}(r_n)) = 1 + \mathcal{O}(r_n)$ uniformly in $\lambda_j \in \Omega_n$.

Furthermore, for $j \neq l$, we can write

$$\begin{aligned} \text{Cov}(e_j, e_l) &= n^{-2}\sum_{t,s=1}^{n} r_\epsilon(t-s)\exp\{-i(t\lambda_j - s\lambda_l)\} \\ &= n^{-1}\sum_{|u|<n} D_n(u, \lambda_l - \lambda_j)\,r_\epsilon(u)\exp(-iu\lambda_j), \end{aligned}$$

where

$$D_n(u,\omega) := n^{-1}\sum_{t\in T_n(u)} \exp(it\omega) \tag{6.6.8}$$

and $T_n(u) := \{t : \max(1, 1-u) \le t \le \min(n, n-u)\}$. Because

$$\sum_{t=1}^{n} \exp\{it(\lambda_l - \lambda_j)\} = \mathbf{f}^H(\lambda_j)\mathbf{f}(\lambda_l) = 0,$$

it follows that

$$D_n(u, \lambda_l - \lambda_j) = -n^{-1}\left\{ \sum_{t=\min(n,n-u)+1}^{n} + \sum_{t=1}^{\max(1,1-u)-1} \right\} \exp\{it(\lambda_l - \lambda_j)\},$$

which gives $|D_n(u, \lambda_l - \lambda_j)| \le n^{-1}|u|$ and hence

$$|\text{Cov}(e_j, e_l)| \le n^{-1} \sum_{|u|<n} n^{-1}|u|\,|r_\epsilon(u)| \le n^{-1} r_n.$$

Therefore, we can write

$$\sigma_{jl} := \text{Cov}(e_j, e_l) = \mathcal{O}(n^{-1} r_n) \tag{6.6.9}$$

uniformly in λ_j and λ_l. Because $n\sigma_{jl} = \mathcal{O}(r_n) \to 0$ uniformly in $j, l \in \mathbb{Z}_n$, this result implies that for any m-subset $\mathbb{Z}_n(m)$ of $\mathbb{Z}_n := \{0, 1, \ldots, n-1\}$ with fixed m as $n \to \infty$, the joint CDF of the random variables $\sqrt{n}(\mu_j + e_j)$ ($j \in \mathbb{Z}_n(m)$) can be approximated uniformly in $\mathbb{Z}_n(m)$ by the joint CDF of independent complex Gaussian random variables with mean $\sqrt{n}\mu_j$ and variance $n\sigma_j^2 = f_\epsilon(\lambda_j) + \mathcal{O}(r_n)$. In other words, the random variables $\sqrt{n}(\mu_j + e_j)$ ($j \in \mathbb{Z}_n(m)$) are asymptotically independent. And so are the random variables $I_n(\lambda_j) = |\sqrt{n}(\mu_j + e_j)|^2$ for $j \in \mathbb{Z}_n(m)$. The proof is complete upon noting that $\Omega_n(m) = \{\lambda_j : j \in \mathbb{Z}_n(m)\}$. □

Proof of Theorem 6.5. According to (6.1.10)–(6.1.12), we can write

$$n^{-1} I_n(\lambda_j) = |z_j|^2 = |\mu_j|^2 + \mu_j e_j^* + \mu_j^* e_j + |e_j|^2. \tag{6.6.10}$$

Because $E(e_j) = 0$, it follows from (6.6.10) that

$$n^{-1} E\{I_n(\lambda_j)\} = |\mu_j|^2 + \sigma_j^2, \tag{6.6.11}$$

where $\sigma_j^2 := \text{Var}(e_j)$. Combining this result with (6.6.4) and (6.6.6) yields (6.1.22). Moreover, because $e_j \sim \mathrm{N}_c(0, \sigma_j^2)$ implies $\Re(e_j)$ and $\Im(e_j)$ are independent Gaussian with mean zero and variance $\sigma_j^2/2$, it can be shown [260, p. 82] that

$$E(e_j^2) = 0, \quad E(e_j|e_j|^2) = 0, \quad E\{(|e_j|^2 - \sigma_j^2)^2\} = \sigma_j^4.$$

Combining these results with (6.6.10) yields

$$\begin{aligned} n^{-2}\text{Var}\{I_n(\lambda_j)\} &= E\{|\mu_j e_j^* + \mu_j^* e_j + (|e_j|^2 - \sigma_j^2)|^2\} \\ &= 2|\mu_j|^2\sigma_j^2 + \sigma_j^4. \end{aligned} \tag{6.6.12}$$

This, together with (6.6.4) and (6.6.6), proves (6.1.23).

Moreover, for any $j \neq l$, e_j and e_l are jointly complex Gaussian with mean zero. Therefore, it can be shown [260, p. 82] that

$$E(e_j e_l) = 0, \quad E(e_j^* |e_l|^2) = 0, \quad E(|e_j|^2 |e_l|^2) = \sigma_j^2 \sigma_l^2 + |\sigma_{jl}|^2, \tag{6.6.13}$$

where $\sigma_{jl} := \mathrm{Cov}(e_j, e_l)$. It follows from (6.6.13) and (6.6.10) that

$$n^{-2} \mathrm{Cov}\{I_n(\lambda_j), I_n(\lambda_l)\} = 2\Re(\mu_j^* \mu_l \sigma_{jl}) + |\sigma_{jl}|^2. \tag{6.6.14}$$

Combining this result with (6.6.4) proves (6.1.24). □

Proof of Theorem 6.6. It follows from (6.6.10) and (6.6.11) that

$$\begin{aligned} & n^{-2} \mathrm{Cov}\{I_n(\lambda_j), I_n(\lambda_l)\} \\ &= 2\Re(\mu_j^* \mu_l \sigma_{jl} + \mu_j^* \mu_l^* c_{jl} + \mu_j s_{jl} + \mu_l s_{lj}) + \kappa_{jl}, \end{aligned} \tag{6.6.15}$$

where

$$\sigma_j^2 := E(|e_j|^2), \quad \sigma_{jl} := E(e_j e_l^*), \quad c_{jl} := E(e_j e_l),$$
$$s_{jl} := E(e_j^* |e_l|^2), \quad \kappa_{jl} := E(|e_j|^2 |e_l|^2) - \sigma_j^2 \sigma_l^2.$$

According to (6.6.6) and (6.6.9), we can write

$$\sigma_{jl} = n^{-1}\{f_\epsilon(\lambda_j)\delta_{j-l} + \mathcal{O}(r_n)\}. \tag{6.6.16}$$

Combining this result with (6.6.11) proves (6.1.26). Furthermore,

$$\begin{aligned} c_{jl} &= n^{-2} \sum_{t,s=1}^{n} \iota c_\epsilon(t-s) \exp\{-i(t\lambda_j + s\lambda_l)\} \\ &= n^{-2} \sum_{t,s=1}^{n} \iota c_\epsilon(t-s) \exp\{-i(t\lambda_j - s\lambda_{n-l})\}. \end{aligned}$$

Because c_{jl} takes the same form as σ_{jl} except that $r_\epsilon(u)$ is replaced by $\iota c_\epsilon(u) = E(\epsilon_{t+u}\epsilon_t)$ and λ_l is replaced by λ_{n-l}, the same technique that establishes (6.6.6), (6.6.9), and (6.6.16) can be used to show that

$$c_{jl} = n^{-1}\{\iota g_\epsilon(\lambda_j)\delta_{j+l}^{(n)} + \mathcal{O}(r_{n2})\}. \tag{6.6.17}$$

Moreover, because

$$E(\epsilon_{t+u}^* \epsilon_{t+v}^* \epsilon_t) = \sum_{j,k,l} \psi_{j+u}^* \psi_{k+v}^* \psi_l \, r_\zeta(l-j, l-k) := r_\epsilon(u, v), \tag{6.6.18}$$

it follows that

$$s_{jl} = n^{-3} \sum_{t,s,\tau=1}^{n} r_\epsilon(t-\tau, s-\tau) \exp\{i(t\lambda_j + (s-\tau)\lambda_l)\}$$

$$= n^{-2} \sum_{(u,v)\in T_n} D_n(u,\lambda_j)\, r_\epsilon(u,v) \exp\{i(u\lambda_j + v\lambda_l)\},$$

where $D_n(u,\omega)$ is defined in (6.6.8) and

$$T_n := \{(u,v) : |v| < n, -\min(n, n-v) < u < \min(n, n+v)\}.$$

Because $|D_n(u,\lambda_j)| \le 1 - n^{-1}|u| \le 1$, one can write

$$|s_{jl}| \le n^{-2} \sum_{(u,v)\in T_n} |r_\epsilon(u,v)| \le n^{-2} \sum_{|u|,|v|<n} |r_\epsilon(u,v)|.$$

Because $\sum |\psi_j| < \infty$ and $\sum\sum |r_\zeta(u,v)| < \infty$, it is easy to show from (6.6.18) that $\sum\sum |r_\epsilon(u,v)| < \infty$. Therefore,

$$s_{jl} = \mathcal{O}(n^{-2}). \tag{6.6.19}$$

Finally, it has been shown in the proof of Theorem 4.2 that

$$E(\epsilon_t \epsilon_{t'}^* \epsilon_s^* \epsilon_{s'}) - E(\epsilon_t \epsilon_{t'}^*) E(\epsilon_s^* \epsilon_{s'}) = a(t-t', s-s', t'-s'),$$

where $a(u,v,\tau) := a_1(u,v,\tau) + a_2(u,v,\tau) + a_3(u,v,\tau)$ and

$$\begin{aligned} a_1(u,v,\tau) &:= (\kappa - 2 - \xi)\sigma^4 \sum_j \psi_{j+\tau+u} \psi_{j+\tau}^* \psi_{j+v}^* \psi_j, \\ a_2(u,v,\tau) &:= r_\epsilon(\tau + u - v)\, r_\epsilon^*(\tau), \\ a_3(u,v,\tau) &:= \xi c_\epsilon(\tau + u)\, c_\epsilon^*(\tau - v). \end{aligned}$$

Therefore,

$$\kappa_{jl} = n^{-4} \sum_{t,t',s,s'=1}^{n} a(t-t', s-s', t'-s') \exp\{-i((t-t')\lambda_j - (s-s')\lambda_l)\}.$$

Because $a(u,v,\tau)$ is the sum of $a_1(u,v,\tau)$, $a_2(u,v,\tau)$, and $a_3(u,v,\tau)$, one can decompose κ_{jl} accordingly into three terms, denoted by K_1, K_2, and K_3. It is easy to show that, with $u := t - t'$ and $v := s - s'$,

$$\begin{aligned} |K_1| &\le n^{-4} \sum_{|u|,|v|<n} \sum_{t',s'=1}^{n} |a_1(u,v,t'-s')| \\ &= n^{-3} \sum_{|u|,|v|,|\tau|<n} (1 - |\tau|/n)|a_1(u,v,\tau)| \\ &\le n^{-3} \sum_{|u|,|v|,|\tau|<n} |a_1(u,v,\tau)| \\ &\le n^{-3}|\kappa - 2 - \xi|\sigma^4 \big(\sum |\psi_j|\big)^4. \end{aligned}$$

This implies that

$$K_1 = \mathcal{O}(n^{-3}).$$

Moreover, because $a_2(t-t', s-s', t'-s') = r_\epsilon(t-s)\, r_\epsilon^*(t'-s')$, it follows that

$$K_2 = n^{-4}\left|\sum_{t,s=1}^{n} r_\epsilon(t-s)\exp\{-i(t\lambda_j - s\lambda_l)\}\right|^2 = |\sigma_{jl}|^2.$$

Similarly, because $a_3(t-t', s-s', t'-s') = \xi c_\epsilon(t-s')\, c_\epsilon^*(t'-s)$, one obtains

$$K_3 = n^{-4}\xi\left|\sum_{t,s=1}^{n} c_\epsilon(t-s)\exp\{-i(t\lambda_j + s\lambda_l)\}\right|^2 = \xi|c_{jl}|^2.$$

Combining these results yields

$$\kappa_{jl} = |\sigma_{jl}|^2 + \xi|c_{jl}|^2 + \mathcal{O}(n^{-3}). \tag{6.6.20}$$

With σ_{jl} and c_{jl} given by (6.6.16) and (6.6.17), one can write

$$\begin{aligned}\kappa_{jl} &= n^{-2}\{f_\epsilon^2(\lambda_j) + \mathcal{O}(r_n)\}\delta_{j-l} + n^{-2}\{\xi|\iota g_\epsilon(\lambda_j)|^2 + \mathcal{O}(r_{n2})\}\delta_{j+l}^{(n)} \\ &\quad + \mathcal{O}(n^{-2}r_n^2) + \mathcal{O}(n^{-2}r_{n2}^2) + \mathcal{O}(n^{-3}).\end{aligned}$$

Substituting this expression, together with the expressions in (6.6.16), (6.6.17), and (6.6.19), into (6.6.15) proves (6.1.27) and (6.1.28). □

Proof of Theorem 6.7. Consider (6.4.3)–(6.4.5). Under the conditions of Theorem 6.2, we have $z_n(\omega) \sim \mathrm{N}_c(\mu_n(\omega), n^{-1}\sigma^2)$. An application of Lemma 12.2.1 yields $I_n(\omega) \sim \frac{1}{2}\sigma^2\chi^2(2, 2\theta_n(\omega))$ with $\theta_n(\omega) := n|\mu_n(\omega)|^2/\sigma^2$. Under the conditions of Theorem 6.4, we have $z_n(\omega) \overset{A}{\sim} \mathrm{N}_c(\mu_n(\omega), n^{-1}f_\epsilon(\omega))$. By Lemma 12.2.1, we obtain $I_n(\omega) \overset{A}{\sim} \frac{1}{2}f_\epsilon(\omega)\chi^2(2, 2\theta_n(\omega))$, where $\theta_n(\omega) := n|\mu_n(\omega)|^2/f_\epsilon(\omega)$.

Now, let $\{\epsilon_t\}$ be a complex linear process of the form (6.1.25) with $\sum|\psi_j| < \infty$, $\{\zeta_t\} \sim \mathrm{IID}(0, \sigma^2)$, and $E(\zeta_t^2) = 0$. Then, we can write

$$\begin{aligned} e_n(\omega) &= n^{-1}\sum_{t=1}^{n}\epsilon_t\exp(-it\omega) \\ &= n^{-1}\sum_{j=-\infty}^{\infty}\psi_j\exp(-ij\omega)\sum_{t=1}^{n}\zeta_{t-j}\exp\{-i(t-j)\omega\} \\ &= \sum_{j=-\infty}^{\infty}\psi_j\exp(-ij\omega)\{X_n(\omega) + X_{nj}(\omega)\} \\ &= \Psi(\omega)X_n(\omega) + Y_n(\omega), \end{aligned} \tag{6.6.21}$$

where

$$\begin{aligned} X_n(\omega) &:= n^{-1}\sum_{t=1}^{n}\zeta_t\exp(-it\omega), \\ X_{nj}(\omega) &:= n^{-1}\sum_{t=1-j}^{n-j}\zeta_t\exp(-it\omega) - n^{-1}\sum_{t=1}^{n}\zeta_t\exp(-it\omega), \end{aligned}$$

$$Y_n(\omega) := \sum_{j=-\infty}^{\infty} \psi_j \exp(-ij\omega) X_{nj}(\omega),$$

$$\Psi(\omega) := \sum_{j=-\infty}^{\infty} \psi_j \exp(-ij\omega).$$

Note that $X_{nj}(\omega)$ is a sum of $2\min(|j|,n)$ independent random variables which have mean zero and variance $n^{-2}\sigma^2$. Therefore,

$$E\{|X_{nj}(\omega)|^2\} \le 2\min(|j|,n)n^{-2}\sigma^2.$$

An application of the Cauchy-Schwarz inequality gives

$$\begin{aligned} E\{|Y_n(\omega)|^2\} &\le \sum_{j=-\infty}^{\infty} \sum_{j'=-\infty}^{\infty} |\psi_j \psi_{j'}^* E\{X_{nj}(\omega) X_{nj'}^*(\omega)\}| \\ &\le \left\{ \sum_{j=-\infty}^{\infty} |\psi_j| \sqrt{E\{|X_{nj}(\omega)|^2\}} \right\}^2 \\ &\le 2\sigma^2 \varrho_n^2, \end{aligned}$$

where

$$\varrho_n := n^{-1} \sum_{j=-\infty}^{\infty} \sqrt{\min(|j|,n)}\, |\psi_j|.$$

By Proposition 2.16(a),

$$Y_n(\omega) = \mathcal{O}_P(\varrho_n). \tag{6.6.22}$$

Moreover, for any fixed $m > 0$ and all $n > m$,

$$\begin{aligned} \sqrt{n}\,\varrho_n &= n^{-1/2} \sum_{j=-\infty}^{\infty} \sqrt{\min(|j|,n)}\, |\psi_j| \\ &\le n^{-1/2} \sum_{|j|\le m} |j|^{1/2} |\psi_j| + \sum_{|j|>m} |\psi_j|, \end{aligned}$$

which implies that

$$\limsup_{n\to\infty} \sqrt{n}\,\varrho_n \le \sum_{|j|>m} |\psi_j|.$$

Because $\sum_{|j|>m} |\psi_j| \to 0$ as $m \to \infty$, it follows that $\sqrt{n}\,\varrho_n \to 0$. Combining this result with (6.6.21) and (6.6.22) leads to

$$e_n(\omega) = \Psi(\omega) X_n(\omega) + \mathcal{O}_P(\varrho_n) = \Psi(\omega) X_n(\omega) + o_P(n^{-1/2}). \tag{6.6.23}$$

According to Lemma 12.5.7, we have $\sqrt{n}\, X_n(\omega) \xrightarrow{D} \mathrm{N}_c(0,\sigma^2)$. Therefore, it follows from (6.6.23) and Proposition 2.13(b) that

$$\sqrt{n}\, e_n(\omega) \xrightarrow{D} \mathrm{N}_c(0, f_\epsilon(\omega)),$$

where $f_\epsilon(\omega) = \sigma^2|\Psi(\omega)|^2$. This implies that

$$z_n(\omega) = \mu_n(\omega) + e_n(\omega) \overset{A}{\sim} \mathrm{N}_c(\mu_n(\omega), n^{-1} f_\epsilon(\omega)).$$

Combining this result with Lemma 12.2.1 leads to $I_n(\omega) \overset{A}{\sim} \frac{1}{2} f_\epsilon(\omega)\chi^2(2, 2\theta_n(\omega))$, where $\theta_n(\omega) := n|\mu_n(\omega)|^2 / f_\epsilon(\omega)$.

To prove part (b), observe that the following expressions have been established in the proof of Theorems 6.4 and 6.5:

$$E\{e_n(\omega)\} = 0, \quad n E\{|e_n(\omega)|^2\} = f_\epsilon(\omega) + \mathcal{O}(r_n), \tag{6.6.24}$$

and, for $\omega \neq \lambda$,

$$n E\{e_n(\omega)\, e_n^*(\lambda)\} = \sum_{|u|<n} D_n(u, \lambda - \omega) r_\epsilon(u) \exp(-iu\omega), \tag{6.6.25}$$

where $D_n(u, \omega)$ is defined in (6.6.8). Note that

$$D_n(u, \lambda - \omega) = n^{-1} \mathbf{f}^H(\omega)\mathbf{f}(\lambda) + \mathcal{O}(n^{-1}|u|). \tag{6.6.26}$$

Moreover, according to Lemma 12.1.3,

$$\begin{aligned} n^{-1}\mathbf{f}^H(\omega)\mathbf{f}(\lambda) &= D_n(\lambda - \omega) \\ &= \begin{cases} 0 & \text{if } \delta = 2\pi j/n \text{ for some integer } j > 0, \\ \mathcal{O}(n^{-1}\delta^{-1}) & \text{if } \delta > \pi/n \text{ and } \delta \neq 2\pi j/n \text{ for any } j, \\ \mathcal{O}(1) & \text{if } \delta \leq \pi/n. \end{cases} \end{aligned} \tag{6.6.27}$$

Therefore,

$$\begin{aligned} & n E\{e_n(\omega)\, e_n^*(\lambda)\} \\ &= \begin{cases} \mathcal{O}(r_n) & \text{if } \delta = 2\pi j/n \text{ for some integer } j > 0, \\ \mathcal{O}(n^{-1}\delta^{-1}) + \mathcal{O}(r_n) & \text{if } \delta > \pi/n \text{ and } \delta \neq 2\pi j/n \text{ for any } j, \\ \mathcal{O}(1) & \text{if } \delta \leq \pi/n. \end{cases} \end{aligned} \tag{6.6.28}$$

Substituting this expression and (6.6.24) in the counterparts of (6.6.11)–(6.6.14), with λ_j and λ_l replaced by ω and λ, respectively, proves part (b).

To prove part (c), observe that (6.6.24) and (6.6.28) remain valid under the assumptions of part (c). Moreover, if $\tilde{\delta} = 0$, then $\lambda = -\omega \pmod{2\pi}$ and hence

$$n E\{e_n(\omega)\, e_n(\lambda)\} = \iota g_\epsilon(\omega) + \mathcal{O}(r_{n2}).$$

For $\tilde{\delta} > 0$, we have

$$\begin{aligned} n^{-1}\mathbf{f}^T(\omega)\mathbf{f}(\lambda) &= D_n(\omega + \lambda) \\ &= \begin{cases} 0 & \text{if } \tilde{\delta} = 2\pi j/n \text{ for some integer } j > 0, \\ \mathcal{O}(n^{-1}\tilde{\delta}^{-1}) & \text{if } \tilde{\delta} > \pi/n \text{ and } \tilde{\delta} \neq 2\pi j/n \text{ for any } j, \\ \mathcal{O}(1) & \text{if } \tilde{\delta} \leq \pi/n. \end{cases} \end{aligned} \tag{6.6.29}$$

Therefore,

$$
\begin{aligned}
&n\,E\{e_n(\omega)\,e_n(\lambda)\} \\
&= \begin{cases} \mathcal{O}(r_{n2}) & \text{if } \tilde{\delta} = 2\pi j/n \text{ for some integer } j > 0, \\ \mathcal{O}(n^{-1}\tilde{\delta}^{-1}) + \mathcal{O}(r_{n2}) & \text{if } \tilde{\delta} > \pi/n \text{ and } \tilde{\delta} \neq 2\pi j/n \text{ for any } j, \\ \mathcal{O}(1) & \text{if } \tilde{\delta} \leq \pi/n. \end{cases}
\end{aligned}
$$

Observe that (6.6.19) and (6.6.20) remain valid if λ_j and λ_l are replaced by ω and λ, respectively. Substituting these expressions in the counterparts of (6.6.11) and (6.6.15) proves part (c).

Finally, let us prove part (d). Without loss of generality, consider the case of $m = 2$. For any fixed $\omega \neq \lambda$, it follows from (6.6.25)–(6.6.27) that

$$
\mathrm{Cov}\{\sqrt{n}\,e_n(\omega), \sqrt{n}\,e_n(\lambda)\} = n\,\mathrm{Cov}\{e_n(\omega), e_n(\lambda)\} \to 0. \tag{6.6.30}
$$

Under the conditions of Theorem 6.2 or 6.4, $\sqrt{n}\,e_n(\omega)$ and $\sqrt{n}\,e_n(\lambda)$ are jointly complex Gaussian with mean zero and variances $f_\epsilon(\omega) + o(1)$ and $f_\epsilon(\lambda) + o(1)$, respectively. Therefore, we can write

$$
\sqrt{n}\,[e_n(\omega), e_n(\lambda)]^T \xrightarrow{D} \mathrm{N}_c([0,0]^T, \mathrm{diag}\{f_\epsilon(\omega), f_\epsilon(\lambda)\}). \tag{6.6.31}
$$

This implies that $I_n(\omega)$ and $I_n(\lambda)$ are asymptotically independent. If $\{\epsilon_t\}$ is a complex linear process satisfying the conditions of part (a), then (6.6.23) is valid for both $e_n(\omega)$ and $e_n(\lambda)$. According to Lemma 12.5.7, $\sqrt{n}\,X_n(\omega)$ and $\sqrt{n}\,X_n(\lambda)$ are asymptotically i.i.d. complex Gaussian with mean zero and variance σ^2. This result, combined with (6.6.23) and (6.6.30), leads to (6.6.31). □

Proof of Theorem 6.8. The proof follows the lines of [412] but allows Δ to vary with n. Because $\mathbf{y}$ can be expressed as (6.1.1) with $p := 2q$, $\beta_{p-k+1} := \beta_k^* := \frac{1}{2}C_k \exp(-i\phi_k)$, and $\omega_{p-k+1} := 2\pi - \omega_k$ $(k = 1, \ldots, q)$, we obtain

$$
n^{-1} I_n(\omega) = |\mu_n(\omega)|^2 + \mu_n^*(\omega)\,e_n(\omega) + \mu_n(\omega)\,e_n^*(\omega) + |e_n(\omega)|^2,
$$

where $\mu_n(\omega)$ and $e_n(\omega)$ are defined by (6.4.4) and (6.4.5). It is easy to see that

$$
\mu_n(\omega) = \sum_{k=1}^{q} \{\beta_k D_n(\omega_k - \omega) + \beta_k^* D_n^*(\omega_k + \omega)\}.
$$

With Δ defined by (6.4.17), it follows that for any $\omega \in [0, \pi]$ and any k, $\omega_k + \omega \geq \Delta$ and $2\pi - (\omega_k + \omega) > \pi - \omega_k \geq \Delta$. By Lemma 12.1.3(d), $|D_n(\omega_k + \omega)| = \mathcal{O}(n^{-1}\Delta^{-1})$. Therefore, we can write

$$
\mu_n(\omega) = \sum_{k=1}^{q} \beta_k D_n(\omega_k - \omega) + \mathcal{O}(n^{-1}\Delta^{-1}).
$$

This result, combined with Lemma 12.1.3(b), leads to

$$|\mu_n(\omega)|^2 = \sum_{k,k'=1}^{q} \beta_k \beta_{k'}^* D_n(\omega_k - \omega) D_n^*(\omega_{k'} - \omega) + \mathcal{O}(n^{-1}\Delta^{-1}). \tag{6.6.32}$$

Note that if there is a constant $c > 0$ such that $\omega_k \geq c$ and $\pi - \omega_k \geq c$ for all k, then the error term in (6.6.32) can be replaced by $\mathcal{O}(n^{-1})$.

According to Theorem 5.6 and Remark 5.7, the consistency of the amplitude and phase estimators follows from the higher-order consistency of the frequency estimator in the sense that $n(\hat{\omega}_k - \omega_k) \xrightarrow{P} 0$. Therefore, it suffices to prove the latter. Toward that end, observe that Lemma 12.5.9(a) ensures

$$\max_{\omega} |e_n(\omega)| = \mathcal{O}_P(1). \tag{6.6.33}$$

Combining this result with the fact that $\mu_n(\omega) = \mathcal{O}(1)$ yields

$$n^{-1} I_n(\omega) = |\mu_n(\omega)|^2 + \mathcal{O}_P(1). \tag{6.6.34}$$

Furthermore, because $|\omega_k - \omega| < \pi$ for any $\omega \in [0, \pi]$ and any k, it follows from Lemma 12.1.3(d) that

$$|D_n(\omega_k - \omega)| \leq \pi / (n|\omega_k - \omega|). \tag{6.6.35}$$

For any constant $\delta \in (0, 2\pi)$, let Ω_δ be the subset of Ω_0 in which $|\tilde{\omega}_k - \omega_k| \leq \delta / n$ for all k. Then, for any $\{\tilde{\omega}_k\} \in \Omega_\delta$ and any $k' \neq k$,

$$|\omega_{k'} - \tilde{\omega}_k| \geq |\omega_{k'} - \omega_k| - |\omega_k - \tilde{\omega}_k| \geq \Delta - \delta / n > 0,$$

where the last inequality holds for large n due to the separation condition (6.3.7). This result, combined with (6.6.35), leads to $|D_n(\omega_{k'} - \tilde{\omega}_k)| = \mathcal{O}(n^{-1}(\Delta - \delta/n)^{-1}) = \mathcal{O}(n^{-1}\Delta^{-1})$ for $k' \neq k$. Therefore, it follows from (6.6.32) that

$$|\mu_n(\tilde{\omega}_k)|^2 = C_k^2 K_n(\omega_k - \tilde{\omega}_k) + \mathcal{O}(n^{-1}\Delta^{-1}),$$

where $K_n(\omega) := |D_n(\omega)|^2$ is the Fejér kernel. Combining this expression with (6.6.34) gives

$$\begin{aligned}\max_{\{\tilde{\omega}_k\} \in \Omega_\delta} n^{-1} \sum_{k=1}^{q} I_n(\tilde{\omega}_k) &= \max_{\{\tilde{\omega}_k\} \in \Omega_\delta} \sum_{k=1}^{q} C_k^2 K_n(\omega_k - \tilde{\omega}_k) + \mathcal{O}(n^{-1}\Delta^{-1}) + \mathcal{O}_P(1) \\ &= \sum_{k=1}^{q} C_k^2 + \mathcal{O}(n^{-1}\Delta^{-1}) + \mathcal{O}_P(1), \end{aligned} \tag{6.6.36}$$

which is the first key result of the proof.

Next, consider $\{\tilde{\omega}_k\} \in \Omega_0 \setminus \Omega_\delta$. There are a number of ways in which the points $\tilde{\omega}_1, \ldots, \tilde{\omega}_q$ can reside in the q intervals defined by the midpoints of adjacent signal frequencies together with 0 and π. The most important case is where each interval contains exactly one point, in which case, the kth interval must contain $\tilde{\omega}_k$ due to the order constraint. For a given k, assume that $\tilde{\omega}_k \ge \omega_k$ without loss of generality. Then, for all $k' < k$, $\tilde{\omega}_k - \omega_{k'} \ge \omega_k - \omega_{k'} \ge \Delta$, and for all $k' > k$, $\omega_{k'} - \tilde{\omega}_k \ge \omega_{k+1} - \tilde{\omega}_k \ge (\omega_{k+1} - \omega_k)/2 \ge \Delta/2$. Combining these results with (6.6.35) gives $D_n(\omega_{k'} - \tilde{\omega}_k) = \mathcal{O}(n^{-1}\Delta^{-1})$ for all $k' \ne k$. It follows from (6.6.32) that

$$\sum_{k=1}^{q} |\mu_n(\tilde{\omega}_k)|^2 = \sum_{k=1}^{q} C_k^2 K_n(\omega_k - \tilde{\omega}_k) + \mathcal{O}(n^{-1}\Delta^{-1}). \tag{6.6.37}$$

Observe that $\{\tilde{\omega}_k\} \in \Omega_0 \setminus \Omega_\delta$ implies $|\tilde{\omega}_k - \omega_k| > \delta/n$ for some k, say $k = q$. If δ is sufficiently small, then, by Lemma 12.1.3(e), $K_n(\omega_q - \tilde{\omega}_q) \le K_n(\delta/n) < 1$. This result, combined with (6.6.37) and the fact that $K_n(\omega) \le 1$ for all ω, leads to

$$\sum_{k=1}^{q} |\mu_n(\tilde{\omega}_k)|^2 \le \sum_{k=1}^{q} C_k^2 - \nu_n + \mathcal{O}(n^{-1}\Delta^{-1}), \tag{6.6.38}$$

where $\nu_n := \min\{C_k^2\}(1 - K_n(\delta/n)) > 0$. The same inequality can be established for all other configurations of the points $\{\tilde{\omega}_k\} \in \Omega_0 \setminus \Omega_\delta$. For example, assume that both $\tilde{\omega}_1$ and $\tilde{\omega}_2$ fall into the first interval and for all $k > 2$, $\tilde{\omega}_k$ falls into the kth interval as before. In this case, for all $k \ge 2$, we have $\omega_k - \tilde{\omega}_2 \ge (\omega_2 - \omega_1)/2 \ge \Delta/2$ so that $D_n(\omega_k - \tilde{\omega}_2) = \mathcal{O}(n^{-1}\Delta^{-1})$. Therefore,

$$\sum_{k=1}^{q} |\mu_n(\tilde{\omega}_k)|^2 = C_2^2 K_n(\omega_1 - \tilde{\omega}_2) + \sum_{\substack{k=1 \\ k\ne 2}}^{q} C_k^2 K_n(\omega_k - \tilde{\omega}_k) + \mathcal{O}(n^{-1}\Delta^{-1}).$$

Moreover, the separation constraint $|\tilde{\omega}_1 - \tilde{\omega}_2| \ge c_n$ implies that either $|\omega_1 - \tilde{\omega}_1| \ge c_n/2$ or $|\omega_1 - \tilde{\omega}_2| > c_n/2$. Because $nc_n \to \infty$, we have $c_n/2 > \delta/n$ for large n. Therefore, either $K_n(\omega_1 - \tilde{\omega}_1) \le K_n(\delta/n)$ or $K_n(\omega_1 - \tilde{\omega}_2) \le K_n(\delta/n)$. In both cases, the inequality (6.6.38) remains true. Combining (6.6.38) with (6.6.34) yields

$$\max_{\{\tilde{\omega}_k\}\in\Omega_0\setminus\Omega_\delta} n^{-1} \sum_{k=1}^{q} I_n(\tilde{\omega}_k) \le \sum_{k=1}^{q} C_k^2 - \nu_n + \mathcal{O}(n^{-1}\Delta^{-1}) + \mathcal{O}_P(1), \tag{6.6.39}$$

which is the second key result of the proof.

According to Lemma 12.1.3(f), $K_n(\delta/n) \to \operatorname{sinc}^2(\delta/2)$ and hence

$$\nu_n \to \nu_\delta := \min\{C_k^2\}(1 - \operatorname{sinc}^2(\delta/2)) > 0. \tag{6.6.40}$$

Combining (6.6.40) with (6.6.36) and (6.6.39) leads to the conclusion that the periodogram maximizer $\{\hat{\omega}_k\}$ defined by (6.4.14) must appear in Ω_δ with probability tending to unity as $n \to \infty$, that is, $P(|\hat{\omega}_k - \omega_k| \le \delta/n \text{ for } k = 1, \dots, q) \to 1$. Because it is true for any small $\delta > 0$, this result implies $n(\hat{\omega}_k - \omega_k) \xrightarrow{P} 0$ for all k.

To prove the asymptotic normality under the RSM (2.1.1), let $\tilde{\boldsymbol{\theta}}_k := [\tilde{A}_k, \tilde{B}_k, \tilde{\omega}_k]^T$ and $\tilde{\boldsymbol{\theta}} := [\tilde{\boldsymbol{\theta}}_1^T, \dots, \tilde{\boldsymbol{\theta}}_q^T]^T$. Define

$$U_n(\tilde{\boldsymbol{\theta}}) := \sum_{k=1}^{q} Q_n(\tilde{\boldsymbol{\theta}}_k), \tag{6.6.41}$$

where

$$Q_n(\tilde{\boldsymbol{\theta}}_k) := \tfrac{1}{4}n(\tilde{A}_k^2 + \tilde{B}_k^2) - \sum_{t=1}^{n} y_t\{\tilde{A}_k \cos(\tilde{\omega}_k t) + \tilde{B}_k \sin(\tilde{\omega}_k t)\}. \qquad (6.6.42)$$

For any given $\tilde{\boldsymbol{\omega}} := \{\tilde{\omega}_1, \ldots, \tilde{\omega}_q\}$, let $\hat{A}_k(\tilde{\omega}_k)$ and $\hat{B}_k(\tilde{\omega}_k)$ be defined by (6.4.15) with $\tilde{\omega}_k$ in place of $\hat{\omega}_k$. It is easy to show that $U_n(\tilde{\boldsymbol{\theta}})$ is minimized by $\{\hat{A}_k(\tilde{\omega}_k), \hat{B}_k(\tilde{\omega}_k)\}$ for fixed $\tilde{\boldsymbol{\omega}}$ because it satisfies $\partial U_n(\tilde{\boldsymbol{\theta}})/\partial \tilde{A}_k = 0$ and $\partial U_n(\tilde{\boldsymbol{\theta}})/\partial \tilde{B}_k = 0$ for all k and because the corresponding Hessian matrix of $U_n(\tilde{\boldsymbol{\theta}})$ equals $\frac{1}{2}n\mathbf{I}$, which is positive definite. Substituting $\{\hat{A}_k(\tilde{\omega}_k), \hat{B}_k(\tilde{\omega}_k)\}$ in (6.6.41) gives

$$-\tfrac{1}{4}n \sum_{k=1}^{q} \{\hat{A}_k^2(\tilde{\omega}_k) + \hat{B}_k^2(\tilde{\omega}_k)\} = -\sum_{k=1}^{q} I_n(\tilde{\omega}_k).$$

This result implies that the PM estimator $\hat{\boldsymbol{\theta}} := [\hat{\boldsymbol{\theta}}_1^T, \ldots, \hat{\boldsymbol{\theta}}_q^T]^T$ defined by (6.4.14) and (6.4.15) minimizes $U_n(\tilde{\boldsymbol{\theta}})$ and therefore satisfies

$$\nabla U_n(\hat{\boldsymbol{\theta}}) = \mathbf{0},$$

where ∇ denotes the gradient operator. Let $\boldsymbol{\theta} := [\boldsymbol{\theta}_1^T, \ldots, \boldsymbol{\theta}_q^T]^T$. Then, by the mean value theorem in calculus,

$$\mathbf{H}_n(\hat{\boldsymbol{\theta}} - \boldsymbol{\theta}) = -\nabla U_n(\boldsymbol{\theta}), \quad \mathbf{H}_n := \int_0^1 \nabla^2 U_n(\boldsymbol{\theta} + s(\hat{\boldsymbol{\theta}} - \boldsymbol{\theta}))\, ds, \qquad (6.6.43)$$

where ∇^2 denotes the Hessian operator. Owing to the additivity in (6.6.41),

$$\nabla U_n(\boldsymbol{\theta}) = \begin{bmatrix} \nabla Q_n(\boldsymbol{\theta}_1) \\ \vdots \\ \nabla Q_n(\boldsymbol{\theta}_q) \end{bmatrix}, \quad \nabla^2 U_n(\tilde{\boldsymbol{\theta}}) = \begin{bmatrix} \nabla^2 Q_n(\tilde{\boldsymbol{\theta}}_1) & & \mathbf{0} \\ & \ddots & \\ \mathbf{0} & & \nabla^2 Q_n(\tilde{\boldsymbol{\theta}}_q) \end{bmatrix}.$$

Combining this expression with (6.6.43) yields

$$\hat{\boldsymbol{\theta}}_k - \boldsymbol{\theta}_k = -\mathbf{H}_{kn}^{-1} \nabla Q_n(\boldsymbol{\theta}_k), \quad \mathbf{H}_{kn} := \int_0^1 \nabla^2 Q_n(\boldsymbol{\theta}_k + s(\hat{\boldsymbol{\theta}}_k - \boldsymbol{\theta}_k))\, ds. \qquad (6.6.44)$$

The remaining goal is to show that

$$\mathbf{K}_n^{-1} \nabla Q_n(\boldsymbol{\theta}_k) \xrightarrow{D} \mathrm{N}(\mathbf{0}, \tfrac{1}{2}\sigma^2 \mathbf{W}_k) \qquad (6.6.45)$$

and

$$\mathbf{K}_n^{-1} \mathbf{H}_{kn} \mathbf{K}_n^{-1} \xrightarrow{P} \tfrac{1}{2}\mathbf{W}_k, \qquad (6.6.46)$$

where $\mathbf{W}_k$ is defined by (3.3.7). If this is done, then combining (6.6.44), (6.6.45), and (6.6.46) would lead to

$$\mathbf{K}_n(\hat{\boldsymbol{\theta}}_k - \boldsymbol{\theta}_k) = -\mathbf{K}_n \mathbf{H}_{kn}^{-1} \mathbf{K}_n \mathbf{K}_n^{-1} \nabla Q_n(\boldsymbol{\theta}_k) \xrightarrow{D} \mathrm{N}(\mathbf{0}, \boldsymbol{\Gamma}_{\mathrm{C}}(\boldsymbol{\theta}_k)),$$

where

$$\boldsymbol{\Gamma}_{\mathrm{C}}(\boldsymbol{\theta}_k) := 2\sigma^2 \mathbf{W}_k^{-1} = \gamma_k^{-1} \boldsymbol{\Lambda}_{\mathrm{C}}(\boldsymbol{\theta}_k) \tag{6.6.47}$$

with $\gamma_k := \frac{1}{2}(A_k^2 + B_k^2)/\sigma^2$ and $\boldsymbol{\Lambda}_{\mathrm{C}}(\boldsymbol{\theta}_k)$ given by (3.3.2).

To prove (6.6.45), observe that

$$\nabla Q_n(\boldsymbol{\theta}_k) = \begin{bmatrix} \frac{1}{2}nA_k - \sum y_t \cos(\omega_k t) \\ \frac{1}{2}nB_k - \sum y_t \sin(\omega_k t) \\ \sum t y_t \{A_k \sin(\omega_k t) - B_k \cos(\omega_k t)\} \end{bmatrix}.$$

Let $\mathbf{a} := [a_1, a_2, a_3]^T \neq \mathbf{0}$ be a constant real vector. Then,

$$\begin{aligned} \mathbf{a}^T \mathbf{K}_n^{-1} \nabla Q_n(\boldsymbol{\theta}_k) &= a_1 n^{-1/2} \Big\{ \tfrac{1}{2}nA_k - \sum y_t \cos(\omega_k t) \Big\} \\ &\quad + a_2 n^{-1/2} \Big\{ \tfrac{1}{2}nB_k - \sum y_t \sin(\omega_k t) \Big\} \\ &\quad + a_3 n^{-3/2} \sum t y_t \{A_k \sin(\omega_k t) - B_k \cos(\omega_k t)\}. \end{aligned}$$

With $\{x_t\}$ taking the form (2.1.1) and Δ satisfying (6.3.7), Lemma 12.1.5 gives

$$\begin{aligned} &\sum x_t \cos(\omega_k t) = \tfrac{1}{2}nA_k + \mathcal{O}(\Delta^{-1}), \quad \sum t x_t \cos(\omega_k t) = \tfrac{1}{4}n^2 A_k + \mathcal{O}(n\Delta^{-1}), \\ &\sum x_t \sin(\omega_k t) = \tfrac{1}{2}nB_k + \mathcal{O}(\Delta^{-1}), \quad \sum t x_t \sin(\omega_k t) = \tfrac{1}{4}n^2 B_k + \mathcal{O}(n\Delta^{-1}). \end{aligned}$$

Combining these results with the fact that $y_t = x_t + \epsilon_t$ leads to

$$\mathbf{a}^T \mathbf{K}_n^{-1} \nabla Q_n(\boldsymbol{\theta}_k) = \sum_{t=1}^{n} c_{nt} \epsilon_t + \mathcal{O}(n^{-1/2}\Delta^{-1}), \tag{6.6.48}$$

where

$$\begin{aligned} c_{nt} &:= -n^{-1/2}\{a_1 \cos(\omega_k t) + a_2 \sin(\omega_k t)\} \\ &\quad + a_3 n^{-3/2}\{A_k t \sin(\omega_k t) - B_k t \cos(\omega_k t)\}. \end{aligned}$$

Moreover, it is easy to show that

$$\lim_{n\to\infty} \max_{1\le t\le n} |c_{nt}| = 0, \quad \lim_{n\to\infty} \sum_{t=1}^{n} c_{nt}^2 = \upsilon^2,$$

where υ^2 is defined by

$$\upsilon^2 := \tfrac{1}{2}(a_1^2 + a_2^2 + \tfrac{1}{3}C_k^2 a_3^2 + B_k a_1 a_3 - A_k a_2 a_3) = \tfrac{1}{2}\mathbf{a}^T \mathbf{W}_k \mathbf{a} \tag{6.6.49}$$

and $\mathbf{W}_k$ is defined by (3.3.7). Therefore, by Lemma 12.5.6, we obtain

$$\sum_{t=1}^{n} c_{nt} \epsilon_t \xrightarrow{D} \mathrm{N}(0, \upsilon^2 \sigma^2). \tag{6.6.50}$$

Under the additional assumption $n\Delta^2 \to \infty$, this result, combined with (6.6.48) and Proposition 2.16(d), leads to (6.6.45).

To prove (6.6.46), let $\tilde{\boldsymbol{\theta}}_k := \boldsymbol{\theta}_k + s(\hat{\boldsymbol{\theta}}_k - \boldsymbol{\theta}_k)$ for any given $s \in [0,1]$. Because

$$\nabla^2 Q_n(\tilde{\boldsymbol{\theta}}_k) = \begin{bmatrix} \frac{1}{2}n & 0 & \sum t y_t \sin(\tilde{\omega}_k t) \\ & \frac{1}{2}n & -\sum t y_t \cos(\tilde{\omega}_k t) \\ \text{symmetric} & & \sum t^2 y_t \{\tilde{A}_k \cos(\tilde{\omega}_k t) + \tilde{B}_k \sin(\tilde{\omega}_k t)\} \end{bmatrix},$$

it follows that

$$\begin{aligned} &\mathbf{K}_n^{-1}\{\nabla^2 Q_n(\tilde{\boldsymbol{\theta}}_k)\}\mathbf{K}_n^{-1} \\ &= \begin{bmatrix} \frac{1}{2} & 0 & n^{-2}\sum t y_t \sin(\tilde{\omega}_k t) \\ & \frac{1}{2} & -n^{-2}\sum t y_t \cos(\tilde{\omega}_k t) \\ \text{symmetric} & & n^{-3}\sum t^2 y_t \{\tilde{A}_k \cos(\tilde{\omega}_k t) + \tilde{B}_k \sin(\tilde{\omega}_k t)\} \end{bmatrix}. \end{aligned}$$

Consider the Taylor expansion

$$\sum t x_t \sin(\tilde{\omega}_k t) = \sum t x_t \sin(\omega_k t) + (\tilde{\omega}_k - \omega_k)\sum t^2 x_t \cos(\bar{\omega}_k t),$$

where $\bar{\omega}_k$ denotes an intermediate point between $\tilde{\omega}_k$ and ω_k. With Δ satisfying (6.3.7), it follows from Lemma 12.1.5 that $n^{-2}\sum t x_t \sin(\omega_k t) = \frac{1}{4}B_k + \mathcal{O}(n^{-1}\Delta^{-1})$. In addition, we have $\sum t^2 x_t \cos(\bar{\omega}_k t) = \mathcal{O}(n^3)$. Therefore,

$$\begin{aligned} n^{-2}\sum_{t=1}^{n} t x_t \sin(\tilde{\omega}_k t) &= \tfrac{1}{4}B_k + \mathcal{O}(n(\tilde{\omega}_k - \omega_k)) + \mathcal{O}(n^{-1}\Delta^{-1}) \\ &= \tfrac{1}{4}B_k + o_P(1) + \mathcal{O}(n^{-1}\Delta^{-1}). \end{aligned}$$

The second expression is due to the fact that $n(\hat{\omega}_k - \omega_k) = o_P(1)$, which has been proved earlier, and the fact that $\tilde{\omega}_k = \omega_k + s(\hat{\omega}_k - \omega_k)$ so that $|\tilde{\omega}_k - \omega_k| \le |\hat{\omega}_k - \omega_k|$ for all $s \in [0,1]$. Moreover, by Lemma 12.5.9(a), we have

$$n^{-2}\sum_{t=1}^{n} t\epsilon_t \sin(\tilde{\omega}_k) = \mathcal{O}_P(n^{-1/4})$$

uniformly in $\tilde{\omega}_k$. Combining these results with (6.3.7) proves

$$n^{-2}\sum_{t=1}^{n} t y_t \sin(\tilde{\omega}_k t) \xrightarrow{P} \tfrac{1}{4}B_k,$$

which is the (1,3) entry of $\frac{1}{2}\mathbf{W}_k$. Because the convergence is uniform in $s \in [0,1]$, the (1,3) entry of $\mathbf{K}_n^{-1}\mathbf{H}_{kn}\mathbf{K}_n^{-1}$ in (6.6.46) approaches the same value. A similar argument establishes (6.6.46) for the remaining entries based on the consistency results of $\hat{A}_k$, $\hat{B}_k$, and $\hat{\omega}_k$. Note that the higher-order consistency of the form $n(\hat{\omega}_k - \omega_k) \xrightarrow{P} 0$, rather than the ordinary consistency of the form $\hat{\omega}_k - \omega_k \xrightarrow{P} 0$, is a key requirement for (6.6.46) to hold.

It remains to prove that $\hat{\boldsymbol{\theta}}_k$ and $\hat{\boldsymbol{\theta}}_{k'}$ are asymptotically independent for $k \neq k'$. According to Lemma 12.5.6, it suffices to show that

$$\sum_{t=1}^{n} c_{nt} c'_{nt} \to 0,$$

where c'_{nt} is the counterpart of c_{nt} associated with $\boldsymbol{\theta}_{k'}$. This can be done by applying Lemma 12.1.5 under the condition (6.3.7). For example, there is a term in $\sum c_{nt} c'_{nt}$ that is proportional to $n^{-1} \sum \cos(\omega_k t) \cos(\omega_{k'} t)$. By Lemma 12.1.5, this term can be expressed as $\mathcal{O}(n^{-1}\Delta^{-1})$ which tends to zero under the condition (6.3.7). Another term is proportional to $n^{-3} \sum t^2 \sin(\omega_k t) \sin(\omega_{k'} t)$ which, according to Lemma 12.1.5, can be written as $\mathcal{O}(n^{-1}\Delta^{-1})$ and hence tends to zero under the condition (6.3.7). All other terms can be shown to approach zero by a similar argument.

The asymptotic normality under the RSM (2.1.2) can be proved by an application of Proposition 2.15. Indeed, observe that the function which transforms $[\hat{A}_k, \hat{B}_k, \hat{\omega}_k]^T$ to $[\hat{C}_k, \hat{\phi}_k, \hat{\omega}_k]^T$ is the same function which transforms $[A_k, B_k, \omega_k]^T$ to $[C_k, \phi_k, \omega_k]^T$. The Jacobian matrix of this function takes the form

$$\mathbf{J}_k := \begin{bmatrix} A_k/C_k & B_k/C_k & 0 \\ B_k/C_k^2 & -A_k/C_k^2 & 0 \\ 0 & 0 & 1 \end{bmatrix}. \tag{6.6.51}$$

According to Proposition 2.15, the random vectors $[\hat{C}_k, \hat{\phi}_k, \hat{\omega}_k]^T$ $(k = 1, \ldots, q)$ are asymptotically independent, and for any given k, the asymptotic distribution of $[\hat{C}_k, \hat{\phi}_k, \hat{\omega}_k]^T$ is Gaussian with mean $[C_k, \phi_k, \omega_k]^T$ and covariance matrix

$$\mathbf{K}_n^{-1} \mathbf{J}_k \boldsymbol{\Gamma}_{\mathrm{C}}(\boldsymbol{\theta}_k) \mathbf{J}_k \mathbf{K}_n^{-1} = \frac{1}{\gamma_k} \mathbf{K}_n^{-1} \boldsymbol{\Lambda}_{\mathrm{P}}(\boldsymbol{\theta}_k) \mathbf{K}_n^{-1},$$

where $\boldsymbol{\Gamma}_{\mathrm{C}}(\boldsymbol{\theta}_k)$ is given by (6.6.47) and $\boldsymbol{\Lambda}_{\mathrm{P}}(\boldsymbol{\theta}_k)$ is defined by (3.3.3). This proves that $\mathbf{K}_n(\hat{\boldsymbol{\theta}}_n - \boldsymbol{\theta}_k) \xrightarrow{D} \mathrm{N}(\mathbf{0}, \boldsymbol{\Gamma}_{\mathrm{P}}(\boldsymbol{\theta}_k))$, with $\boldsymbol{\Gamma}_{\mathrm{P}}(\boldsymbol{\theta}_k) := \gamma_k^{-1} \boldsymbol{\Lambda}_{\mathrm{P}}(\boldsymbol{\theta}_k)$. □

Proof of Corollary 6.3. Let $\hat{x}_t := \sum \{\hat{A}_k \cos(\hat{\omega}_k t) + \hat{B}_k \sin(\hat{\omega}_k t)\}$. Then,

$$\hat{\sigma}^2 = n^{-1} \sum_{t=1}^{n} \{(x_t - \hat{x}_t)^2 + 2(x_t - \hat{x}_t)\epsilon_t + \epsilon_t^2\}.$$

Using the Taylor expansion with respect to $\hat{\omega}_k$, we can write

$$\begin{aligned} x_t - \hat{x}_t &= \sum_{k=1}^{q} \{(A_k - \hat{A}_k)\cos(\omega_k t) + (B_k - \hat{B}_k)\sin(\omega_k t) \\ &\quad + (\hat{\omega}_k - \omega_k)(\hat{A}_k t \sin(\tilde{\omega}_k t) - \hat{B}_k t \cos(\tilde{\omega}_k t))\}, \end{aligned}$$

where $\tilde{\omega}_k$ is an intermediate point between $\hat{\omega}_k$ and ω_k. By Theorem 6.8, we have $A_k - \hat{A}_k = \mathcal{O}_P(n^{-1/2})$, $B_k - \hat{B}_k = \mathcal{O}_P(n^{-1/2})$, and $\hat{\omega}_k - \omega_k = \mathcal{O}_P(n^{-3/2})$. Combining these results with Lemma 12.5.9(a) gives

$$n^{-1}\sum_{t=1}^{n}(x_t - \hat{x}_t)\epsilon_t = \mathcal{O}_P(n^{-3/4}). \tag{6.6.52}$$

Moreover, because $x_t - \hat{x}_t = \mathcal{O}_P(n^{-1/2}) + \mathcal{O}_P(n^{-3/2})t$ uniformly in t, we obtain

$$n^{-1}\sum_{t=1}^{n}(x_t - \hat{x}_t)^2 = \mathcal{O}_P(n^{-1}).$$

Combining these results leads to

$$\hat{\sigma}^2 = n^{-1}\sum_{t=1}^{n}\epsilon_t^2 + \mathcal{O}_P(n^{-3/4}). \tag{6.6.53}$$

Citing the weak law of large numbers [404, p. 15] for the i.i.d. sequence $\{\epsilon_t^2\}$ with mean $\sigma^2 = E(\epsilon_t^2) < \infty$ proves the consistency assertion.

Moreover, according to (6.6.53),

$$\sqrt{n}(\hat{\sigma}^2 - \sigma^2) = n^{-1/2}\sum_{t=1}^{n}(\epsilon_t^2 - \sigma^2) + \mathcal{O}_P(n^{-1/4}).$$

Therefore, $\sqrt{n}(\hat{\sigma}^2 - \sigma^2)$ has the same asymptotic distribution as $n^{-1/2}\sum(\epsilon_t^2 - \sigma^2)$. Under the assumption that $\kappa := E(\epsilon_t^4)/\sigma^4 < \infty$, the variance of ϵ_t^2 is finite and equal to $(\kappa - 1)\sigma^4$. Therefore, citing the central limit theorem [404, p. 16] for the i.i.d. sequence $\{\epsilon_t^2\}$ gives $n^{-1/2}\sum(\epsilon_t^2 - \sigma^2) \xrightarrow{D} \mathrm{N}(0, (\kappa - 1)\sigma^4)$. □

Proof of Theorem 6.9. By Lemma 12.5.9, the expression in (6.6.33) remains true when the noise is a linear process or a zero-mean stationary Gaussian process that satisfies the assumption of Theorem 6.9. Therefore, the consistency assertion can be established by the same argument that proves the consistency under the assumption of white noise in Theorem 6.8.

To prove the asymptotic normality, consider the RSM (2.1.1) first. It is easy to see that (6.6.43)–(6.6.48) remain valid in this case. Let $\tilde{c}_{nt} := c_{nt}/\sqrt{f_\epsilon(\omega_k)}$.

$$v_n^2 := \mathrm{Var}\left\{\sum_{t=1}^{n}\tilde{c}_{nt}\epsilon_t\right\} = \sum_{|u|<n} r_\epsilon(u)\sum_{t\in T_n(u)}\tilde{c}_{n,t+u}\tilde{c}_{nt},$$

where $T_n(u) := \{t : \max(1, 1-u) \le t \le \min(n, n-u)\}$. Straightforward calculation, together with (6.3.7), Lemma 12.1.5, and the fact that $f_\epsilon(\omega_k) \ge f_0 := \inf\{f_\epsilon(\omega)\} > 0$, shows that for any fixed u and as $n \to \infty$,

$$\sum_{t\in T_n(u)}\tilde{c}_{n,t+u}\tilde{c}_{nt} - v^2\cos(u\omega_k)/f_\epsilon(\omega_k) \to 0,$$

where v^2 is defined by (6.6.49). This result, combined with the absolute summability of $\{r_\epsilon(u)\}$ and with the dominated convergence theorem, gives

$$v_n^2 - v^2 \sum_{u=-\infty}^{\infty} r_\epsilon(u)\cos(u\omega_k)/f_\epsilon(\omega_k) = v_n^2 - v^2 \to 0. \tag{6.6.54}$$

Therefore, under the Gaussian assumption, we obtain

$$\sum_{t=1}^{n} \tilde{c}_{nt}\epsilon_t \xrightarrow{D} \mathrm{N}(0, v^2). \tag{6.6.55}$$

Combining this result with (6.6.48) yields

$$\mathbf{K}_n^{-1}\nabla Q_n(\boldsymbol{\theta}_k) \overset{A}{\sim} \mathrm{N}(\mathbf{0}, \tfrac{1}{2} f_\epsilon(\omega_k)\mathbf{W}_k).$$

Combining this result with (6.6.46) leads to

$$\mathbf{K}_n(\hat{\boldsymbol{\theta}}_k - \boldsymbol{\theta}_k) = \mathbf{K}_n\mathbf{H}_{nk}^{-1}\mathbf{K}_n\mathbf{K}_n^{-1}\nabla Q_n(\boldsymbol{\theta}_k) \overset{A}{\sim} \mathrm{N}(\mathbf{0}, \boldsymbol{\Gamma}_{\mathrm{C}}(\boldsymbol{\theta}_k)),$$

where $\boldsymbol{\Gamma}_{\mathrm{C}}(\boldsymbol{\theta}_k) := 2 f_\epsilon(\omega_k)\mathbf{W}_k^{-1} = \gamma_k^{-1}\boldsymbol{\Lambda}_{\mathrm{C}}(\boldsymbol{\theta}_k)$ and $\gamma_k := \frac{1}{2}(A_k^2 + B_k^2)/f_\epsilon(\omega_k)$. It remains to show that (6.6.55) remains valid when $\{\epsilon_t\}$ is a linear process.

Toward that end, let ϵ_t be replaced momentarily by $\tilde{\epsilon}_t := \sum \tilde{\psi}_j \zeta_{t-j}$, where

$$\tilde{\psi}_j := \psi_j \mathscr{I}(|j| \le m)$$

for some $m > 0$. Obviously, $\{\tilde{\epsilon}_t\}$ is a zero-mean stationary process with absolutely summable ACF $\tilde{r}_\epsilon(u)$ and SDF

$$\tilde{f}_\epsilon(\omega) := \sum_{u=-\infty}^{\infty} \tilde{r}_\epsilon(u)\cos(u\omega) = \sigma^2|\tilde{\Psi}(\omega)|^2.$$

where $\tilde{\Psi}(\omega) := \sum \tilde{\psi}_j \exp(-it\omega)$. For sufficiently large m, $\tilde{f}_\epsilon(\omega) \ge f_0/2$. Therefore, a counterpart of (6.6.54) can be established, that is, $\tilde{v}_n^2 \to v^2$ as $n \to \infty$, where $\tilde{v}_n^2 := \mathrm{Var}(\sum_{t=1}^{n} \bar{c}_{nt}\tilde{\epsilon}_t)$ and $\bar{c}_{nt} := \{\tilde{f}_\epsilon(\omega_k)\}^{-1/2} c_{nt}$. The goal is to show that

$$\sum_{t=1}^{n} \bar{c}_{nt}\tilde{\epsilon}_t \xrightarrow{D} \mathrm{N}(0, v^2) \quad \text{as } n \to \infty. \tag{6.6.56}$$

If this can be achieved, then the following argument would lead to the desired result (6.6.55). In fact, observe that

$$\begin{aligned} &\tfrac{1}{2}E\left|\sum_{t=1}^{n}(\tilde{c}_{nt}\epsilon_t - \bar{c}_{nt}\tilde{\epsilon}_t)\right|^2 \\ &\le E\left|\sum_{t=1}^{n}\tilde{c}_{nt}(\epsilon_t - \tilde{\epsilon}_t)\right|^2 + E\left|\sum_{t=1}^{n}(\tilde{c}_{nt} - \bar{c}_{nt})\tilde{\epsilon}_t\right|^2. \end{aligned} \tag{6.6.57}$$

By straightforward calculation, the first term in (6.6.57) can be expressed as

$$\sigma^2 \sum_{j=-\infty}^{\infty} \bar{\psi}_j \sum_{|u|<n} \bar{\psi}_{j+u} \sum_{t\in T_n(u)} \tilde{c}_{n,t+u}\tilde{c}_{nt},$$

where $\bar{\psi}_j := \psi_j - \tilde{\psi}_j = \psi_j \mathscr{I}(|j| > m)$. An application of the Cauchy-Schwarz inequality shows that the absolute value of the sum over t is less than or equal to $\sum_{t=1}^n \tilde{c}_{nt}^2 = v^2/f_\epsilon(\omega_k) + \mathscr{O}(1)$ and therefore can be bounded by a positive constant. The second term in (6.6.57) can be expressed as

$$\tilde{v}_n^2 |\tilde{f}_\epsilon(\omega_k)/f_\epsilon(\omega_k) - 1|^2 / \tilde{f}_\epsilon(\omega_k).$$

Combining these results yields

$$\lim_{m\to\infty}\lim_{n\to\infty} E\left|\sum_{t=1}^n (\tilde{c}_{nt}\epsilon_t - \bar{c}_{nt}\tilde{\epsilon}_t)\right|^2 = 0.$$

Using Chebyshev's inequality, we obtain

$$\lim_{m\to\infty}\lim_{n\to\infty} P\left\{\left|\sum_{t=1}^n (\tilde{c}_{nt}\epsilon_t - \bar{c}_{nt}\tilde{\epsilon}_t)\right| > \delta\right\} = 0$$

for any constant $\delta > 0$. This result, together with (6.6.56) and Proposition 2.16(e), proves the assertion (6.6.55).

To prove (6.6.56), observe that

$$\sum_{t=1}^n \bar{c}_{nt}\tilde{\epsilon}_t = \sum_{s=1-m}^{n+m} w_{ns}\zeta_s, \tag{6.6.58}$$

where $w_{ns} := \sum_{t=1}^n \bar{c}_{nt}\tilde{\psi}_{t-s}$. Because $\max\{|\bar{c}_{nt}| : 1 \le t \le n\} \to 0$ as $n \to \infty$ and $\sum |\psi_j| < \infty$, it follows that

$$\max_{1-m\le s\le n+m} |w_{ns}| \to 0 \quad \text{as } n \to \infty.$$

In addition, we have

$$\sum_{s=1-m}^{n+m} w_{ns}^2 = \mathrm{Var}\left\{\sum_{t=1}^n \bar{c}_{nt}\tilde{\epsilon}_t\right\}/\sigma^2 = \tilde{v}_n^2/\sigma^2 \to v^2/\sigma^2 \quad \text{as } n \to \infty.$$

Therefore, an application of Lemma 12.5.6 leads to

$$\sum_{s=1-m}^{n+m} w_{ns}\zeta_s \xrightarrow{D} \mathrm{N}(0, v^2) \quad \text{as } n \to \infty.$$

Combining this result with (6.6.58) yields (6.6.56).

To prove the asymptotic independence between $\hat{\boldsymbol{\theta}}_k$ and $\hat{\boldsymbol{\theta}}_{k'}$ for $k \neq k'$, it suffices to note, according to Lemma 12.5.6, that

$$\sum_{t\in T_n(u)} c_{n,t+u}c'_{nt} \to 0$$

as $n \to \infty$ for any fixed u, where c'_{nt} is the counterpart of c_{nt} associated with $\boldsymbol{\theta}_{k'}$. This result implies that

$$\mathrm{Cov}\left\{\sum_{t=1}^{n} c_{nt}\epsilon_t, \sum_{t=1}^{n} c'_{nt}\epsilon_t\right\} = \sum_{|u|<n} r_\epsilon(u) \sum_{t\in T_n(u)} c_{n,t+u}c'_{nt} \to 0,$$

due to the absolute summability of $\{r_\epsilon(u)\}$ and the dominated convergence theorem. The proof for the RSM (2.1.1) is complete. The asymptotic normality for the RSM (2.1.2) can be proved in the same way as in the case of white noise with the help of the Jacobian matrix $\mathbf{J}_k$ in (6.6.51). □

Proof of Theorem 6.10. The consistency can be established in the same way as the real case, because $\mu_n(\omega)$ can be expressed as

$$\mu_n(\omega) = \sum_{k=1}^{p} \beta_k D_n(\omega_k - \omega)$$

and because the expression in (6.6.33) remains true, by Lemma 12.5.9, under the assumptions of Theorem 6.10. To prove the asymptotic normality under the CSM (2.1.5) with $\boldsymbol{\theta}_k := [A_k, B_k, \omega_k]^T$, consider

$$U_n(\tilde{\boldsymbol{\theta}}) := \sum_{k=1}^{p} Q_n(\tilde{\boldsymbol{\theta}}_k), \quad Q_n(\tilde{\boldsymbol{\theta}}_k) := n|\tilde{\beta}_k|^2 - 2n\Re\{\tilde{\beta}_k^* z_n(\tilde{\omega}_k)\},$$

where $\tilde{\beta}_k := \tilde{A}_k - i\tilde{B}_k$, $z_n(\omega) := n^{-1}\sum y_t \exp(-it\omega)$, $\tilde{\boldsymbol{\theta}} := [\tilde{\boldsymbol{\theta}}_1^T, \dots, \tilde{\boldsymbol{\theta}}_p^T]^T \in \mathbb{R}^{3p}$, and $\tilde{\boldsymbol{\theta}}_k := [\tilde{A}_k, \tilde{B}_k, \tilde{\omega}_k]^T \in \mathbb{R}^3$ $(k = 1, \dots, p)$. For any fixed $\tilde{\boldsymbol{\omega}} = \{\tilde{\omega}_1, \dots, \tilde{\omega}_p\}$, it is easy to show that $U_n(\tilde{\boldsymbol{\theta}})$ is minimized with respect to $\{\tilde{A}_k, \tilde{B}_k\}$ by

$$\hat{A}_k(\tilde{\omega}_k) := n^{-1}\Re\{z_n(\tilde{\omega}_k)\}, \quad \hat{B}_k(\tilde{\omega}_k) := -n^{-1}\Im\{z_n(\tilde{\omega}_k)\}.$$

Substituting these expressions in $U_n(\tilde{\boldsymbol{\theta}})$ yields $-\sum_{k=1}^{p} I_n(\tilde{\omega}_k)$. This means that the PM estimator, $\hat{\boldsymbol{\theta}} := [\hat{\boldsymbol{\theta}}_1^T, \dots, \hat{\boldsymbol{\theta}}_p^T]^T$ with $\hat{\boldsymbol{\theta}}_k := [\hat{A}_k, \hat{B}_k, \hat{\omega}_k]^T$, $\hat{A}_k := \hat{A}_k(\hat{\omega}_k)$, and $\hat{B}_k := \hat{B}_k(\hat{\omega}_k)$, is the minimizer of $U_n(\tilde{\boldsymbol{\theta}})$ and therefore satisfies

$$\nabla U_n(\hat{\boldsymbol{\theta}}) = \mathbf{0}.$$

As in the real case, this result leads to (6.6.43) and (6.6.44) with

$$\nabla Q_n(\boldsymbol{\theta}_k) = \begin{bmatrix} 2nA_k - 2n\Re\{z_n(\omega_k)\} \\ 2nB_k + 2n\Im\{z_n(\omega_k)\} \\ -2n\Re\{\beta_k^* \dot{z}_n(\omega_k)\} \end{bmatrix},$$

and

$$\nabla^2 Q_n(\tilde{\boldsymbol{\theta}}_k) = \begin{bmatrix} 2n & 0 & -2n\Re\{\dot{z}_n(\tilde{\omega}_k)\} \\ & 2n & 2n\Im\{\dot{z}_n(\tilde{\omega}_k)\} \\ \text{symmetric} & & -2n\Re\{\tilde{\beta}_k^* \ddot{z}_n(\tilde{\omega}_k)\} \end{bmatrix}.$$

Also as in the real case, the consistency of the PM estimator gives

$$\mathbf{K}_n^{-1}\mathbf{H}_{nk}\mathbf{K}_n^{-1} \xrightarrow{P} \begin{bmatrix} 2 & 0 & B_k \\ & 2 & -A_k \\ \text{symmetric} & & 2C_k^2/3 \end{bmatrix} = 2\mathbf{W}_k, \tag{6.6.59}$$

where $\mathbf{W}_k$ is defined by (3.3.7) and $C_k^2 := A_k^2 + B_k^2$. Moreover, let $X_n := e_n(\omega_k)$, and $Y_n := \dot{e}_n(\omega_k)$, where $e_n(\omega)$ is defined by (6.4.5). Then, it can be shown, similarly to the real case, that

$$\mathbf{K}_n^{-1}\nabla Q_n(\boldsymbol{\theta}_k) = -2 \begin{bmatrix} n^{1/2}\Re(X_n) \\ -n^{1/2}\Im(X_n) \\ n^{-1/2}\Re(\beta_k^* Y_n) \end{bmatrix} + \mathcal{O}(n^{-1/2}\Delta^{-1}). \tag{6.6.60}$$

The next objective is to show that

$$\mathbf{K}_n^{-1}\nabla Q_n(\boldsymbol{\theta}_k) \stackrel{A}{\sim} \mathrm{N}(\mathbf{0}, 2f_\epsilon(\omega_k)\mathbf{W}_k). \tag{6.6.61}$$

If this is done, then combining (6.6.61) with (6.6.59) would lead to

$$\mathbf{K}_n^{-1}(\hat{\boldsymbol{\theta}}_k - \boldsymbol{\theta}_k) \stackrel{A}{\sim} \mathrm{N}(\mathbf{0}, \boldsymbol{\Gamma}_{\mathrm{C}}(\boldsymbol{\theta}_k)),$$

where $\boldsymbol{\Gamma}_{\mathrm{C}}(\boldsymbol{\theta}_k) := \frac{1}{2}f_\epsilon(\omega_k)\mathbf{W}_k^{-1} = \frac{1}{2}\gamma_k^{-1}\boldsymbol{\Lambda}_{\mathrm{C}}(\boldsymbol{\theta}_k)$, $\gamma_k := C_k^2/f_\epsilon(\omega_k)$, and $\boldsymbol{\Lambda}_{\mathrm{C}}(\boldsymbol{\theta}_k)$ is given by (3.3.2). This proves the assertion for the Cartesian CSM (2.1.5). Using the Jacobian matrix $\mathbf{J}_k$ given by (6.6.51) together with Proposition 2.15, the results for estimating the Cartesian parameters $[A_k, B_k, \omega_k]$ can be transformed into the results for estimating the polar parameters $[C_k, \phi_k, \omega_k]$, in which case $\boldsymbol{\Gamma}_{\mathrm{C}}(\boldsymbol{\theta}_k)$ is replaced by $\boldsymbol{\Gamma}_{\mathrm{P}}(\boldsymbol{\theta}_k) := \frac{1}{2}\gamma_k^{-1}\boldsymbol{\Lambda}_{\mathrm{P}}(\boldsymbol{\theta}_k)$, where $\boldsymbol{\Lambda}_{\mathrm{P}}(\boldsymbol{\theta}_k)$ is given by (3.3.3).

To prove (6.6.61), observe that for any $\mathbf{a} := [a_1, a_2, a_3]^T \neq \mathbf{0}$ and $b := a_1 + i a_2$,

$$\begin{aligned} S_n &:= n^{1/2}\{a_1\Re(X_n) - a_2\Im(X_n)\} + n^{-1/2}a_3\Re(\beta_k^* Y_n) \\ &= \Re(n^{1/2}bX_n + n^{-1/2}a_3\beta_k^* Y_n) \\ &= \Re\Big\{\sum_{t=1}^n c_{nt}\epsilon_t\Big\}, \end{aligned}$$

where $c_{nt} := d_{nt}\exp(-it\omega_k)$ and $d_{nt} := n^{-1/2}b + n^{-3/2}a_3\beta_k^*(-it)$. Therefore,

$$\mathrm{Var}\Big\{\sum_{t=1}^n c_{nt}\epsilon_t\Big\} = \sum_{|u|<n} r_\epsilon(u)\exp(-iu\omega_k) \sum_{t\in T_n(u)} d_{n,t+u}d_{nt}^*.$$

Because

$$\begin{aligned} d_{n,t+u}d_{nt}^* &= n^{-1}|b|^2 + n^{-3}a_3^2C_k^2t^2 - 2n^{-2}a_3\Im(b\beta_k)t \\ &\quad + n^{-3}a_3^2C_k^2ut - n^{-2}a_3b^*\beta_k^*(iu), \end{aligned}$$

it follows that for any fixed $|u| < n$,

$$\sum_{t \in T_n(u)} d_{n,t+u} d_{nt}^* \to v^2 \quad \text{as } n \to \infty,$$

where

$$\begin{aligned} v^2 &:= |b|^2 + \tfrac{1}{3} a_3^2 C_k^2 - a_3 \Im(b\beta_k) \\ &= a_1^2 + a_2^2 + \tfrac{1}{3} C_k^2 a_3^2 + B_k a_1 a_3 - A_k a_2 a_3 = \mathbf{a}^T \mathbf{W}_k \mathbf{a}. \end{aligned}$$

Combining this result with the dominant convergence theorem gives

$$\operatorname{Var}\left\{ \sum_{t=1}^n c_{nt} \epsilon_t \right\} - f_\epsilon(\omega_k) v^2 \to 0.$$

Therefore, under the Gaussian assumption, we obtain

$$\sum_{t=1}^n c_{nt} \epsilon_t \stackrel{A}{\sim} \mathrm{N}_c(0, f_\epsilon(\omega_k) v^2). \tag{6.6.62}$$

This result implies that

$$S_n \stackrel{A}{\sim} \mathrm{N}(0, \tfrac{1}{2} f_\epsilon(\omega_k) v^2).$$

Citing Proposition 2.16(d) proves

$$[n^{1/2}\Re(X_n), -n^{1/2}\Im(X_n), n^{-1/2}\Re(\beta_k^* Y_n)]^T \stackrel{A}{\sim} \mathrm{N}(\mathbf{0}, \tfrac{1}{2} f_\epsilon(\omega_k) \mathbf{W}_k).$$

Combining this result with (6.6.60) leads to (6.6.61). The asymptotic independence of $\hat{\boldsymbol{\theta}}_k$ and $\hat{\boldsymbol{\theta}}_{k'}$ for $k \neq k'$ follows from the fact that

$$\sum_{t=1}^n c_{nt}^* c_{nt}' = \sum_{t=1}^n d_{nt}^* d_{nt}' \exp\{i(\omega_k - \omega_{k'})\} \to 0,$$

where c_{nt}' and d_{nt}' denote the counterpart of c_{nt} and d_{nt} corresponding to $\boldsymbol{\theta}_{k'}$.

Now consider the case of white noise, where $\{\epsilon_t\} \sim \mathrm{IID}(0, \sigma^2)$. Observe that $\max\{|c_{nt}| : 1 \le t \le n\} = \mathcal{O}(n^{-1/2}) \to 0$ and

$$\sum_{t=1}^n |c_{nt}|^2 = \sum_{t=1}^n \{n^{-1}|b|^2 + n^{-3} a_3^2 C_k^2 t^2 - 2n^{-2} a_3 \Im(b\beta_k) t\} \to v^2.$$

Therefore, citing Lemma 12.5.7 gives (6.6.62) with $f_\epsilon(\omega) = \sigma^2$. When $\{\epsilon_t\}$ is a linear process, the argument is similar to that in the proof of Theorem 6.9. First, the infinite sum of the linear process is truncated into a finite sum of $2m+1$ terms. Secondly, a central limit theorem of S_n is established for the truncated noise with fixed m by citing Lemma 12.5.7. Finally, the technique in Proposition 2.16(e) is employed to derive the limiting distribution as the truncation parameter m approaches infinity. The asymptotic independence assertion can also be proved by following the proof of Theorem 6.9. □

Notes

The method of Fourier analysis was introduced in 1822 by the French mathematician and physicist Joseph Fourier (1768–1830) for solving the heat equation [111]. It had been applied to time-series data long before the British physicist Arthur Schuster (1851–1934) coined the term periodogram in 1898 [334]. A year earlier [333], Schuster discussed the probability distribution of the square-root of the periodogram under the Gaussian white noise assumption based on an earlier result of the British physicist John Rayleigh (1842–1919). The distribution, of course, is the Rayleigh distribution. With these papers, Schuster persuasively brought statistical reasoning into Fourier analysis and laid the foundation for statistical spectral analysis of time series. The 1917 textbook [47] of the British meteorologist David Brunt (1886–1965) contains an excellent summary of Schuster's work on the periodogram in addition to a discussion of the periodogram methodology for detection of hidden periodicities.

The first and second moments of the periodogram for real-valued time series with continuous spectra are studied in [12, Sections 8.3.3] under more general conditions than the linear processes assumed in Theorem 6.6.

In his pioneering work [422], Whittle investigates the periodogram maximization method for estimating the amplitude and frequency of a real sinusoid in noise. He notices the unusual property of the asymptotic variance of the frequency estimator as compared to that of the amplitude estimator: the former is of the form $\mathcal{O}(n^{-3})$ whereas the latter is of the usual form $\mathcal{O}(n^{-1})$. He derives the now well-known expression for the asymptotic covariance matrix of the amplitude and frequency estimators based on heuristic arguments. More rigorous analyses are provided later by Walker [412] [413], Hannan [137], and Ivanov [164] [165], where the consistency and the asymptotic normality of these estimators are established. Isokawa [163] considers the case where the observations are unequally spaced random samples.

Chapter 7

Estimation of Noise Spectrum

When the noise is nonwhite (or colored), a proper assessment of the noise spectrum is beneficial to the detection and estimation of the sinusoidal signals, especially for small sample sizes. In some applications, noise-only samples are independently available so that the noise spectrum can be estimated in the absence of sinusoids. In other applications, the noise spectrum has to be estimated from a time series that possibly contains sinusoidal components with unknown parameters. In this chapter, we discuss some techniques of spectral estimation under both scenarios, but with a special emphasis on the latter case. We pay attention to both real- and complex-valued time series.

7.1 Estimation in the Absence of Sinusoids

First, let us consider the situation where a training data set — a time series which shares the same statistical characteristics as the noise that corrupts the sinusoidal signal — is independently available for noise assessment. In this situation, the data record $\{y_1, \ldots, y_n\}$ can be expressed as

$$y_t = \epsilon_t \qquad (t = 1, \ldots, n). \tag{7.1.1}$$

It is a case where standard methods developed for time series with continuous spectra are readily applicable. Because the subject has been covered comprehensively in many excellent textbooks with either statistical or engineering flavor, we only discuss a few exemplary techniques in this section to serve as the basis for the next section which focuses on the estimation of the noise spectrum in the presence of sinusoids and for later chapters on frequency estimation.

For more information about spectral analysis of complex-valued time series with continuous spectra, see [177], [255], [288], and [369]. For extensive mathematical analysis of spectral estimation methods for real-valued time series with continuous spectra, see [12], [45], [46], [128], [135], and [298].

In this section, we briefly discuss three types of spectral estimators — the periodogram smoother (and the related data-taper spectral estimator), the lag-window spectral estimator, and the autoregressive spectral estimator.

7.1.1 Periodogram Smoother

The periodogram smoother is a simple nonparametric spectral estimator for time series of continuous spectra. With $\lambda_j := 2\pi j/n$ ($j = 0, \pm 1, \ldots$) denoting the Fourier frequencies, the periodogram smoother can be expressed as

$$\hat{f}_\epsilon(\lambda_j) := \sum_{|l|\le m} W_{ml}\, I_n(\lambda_{j-l}) = \sum_{|l|\le m} W_{ml}\, I_n(\lambda_j - \lambda_l), \tag{7.1.2}$$

where $\{I_n(\lambda_j)\}$ is the (discrete) periodogram defined by (6.1.6) with periodic extension and $\{W_{ml}\}$ is a sequence of nonnegative symmetric weights satisfying

$$\sum_{|l|\le m} W_{ml} = 1. \tag{7.1.3}$$

In a special case called kernel smoothing, the weights W_{ml} are generated by a nonnegative even function $K(\cdot)$, which is defined on the interval $[-1, 1]$ and equals zero elsewhere, such that $W_{ml} \propto K(l/m)$ for $|l| \le m$.

The periodogram smoother in (7.1.2) is nothing but a weighted moving average of the periodogram ordinates. If the noise spectrum is sufficiently smooth, then the periodogram ordinates in a small frequency interval have approximately the same distribution and therefore averaging them properly gives an estimate of the common mean with reduced statistical variability. Obviously, the size of the moving window cannot be too large; otherwise, peaks and valleys will be smeared, resulting in reduced resolution and increased bias or spectral leakage. On the other hand, the size of the moving window cannot be too small either; otherwise the random fluctuation of the periodogram ordinates will not be effectively tamed. Balancing these contradictory requirements translates into a tradeoff between the bias and the variability of the spectral estimator.

It is well known that the optimal weights which minimize the total MSE across the frequencies depend on certain knowledge about the smoothness of the unknown noise spectrum [298, Section 6.2.4] [423]. In the absence of such knowledge, data-driven techniques, such as cross-validation [29] [160] [275] and bootstrapping [115] [291] [382], are used to guide the weight selection. The weights can also be made dependent on the central frequency and chosen adaptively to fit the local spectral characteristics [107] [264] [354] [407].

As an example, consider the problem of selecting a suitable m in (7.1.2) when the weights are given by $W_{ml} \propto K(l/m)$ for a continuous function $K(\cdot)$ satisfying $K(0) > 0$. Owing to Whittle's likelihood in (6.1.20) and its equivalence to (6.1.21), it is natural to employ the Kullback-Leibler spectral divergence,

$$D_{\mathrm{KL}}(\{I_n(\lambda_j)\},\{\hat{f}_\epsilon(\lambda_j)\}) := \sum_{j=0}^{n-1} d_{\mathrm{KL}}(I_n(\lambda_j)/\hat{f}_\epsilon(\lambda_j)),$$

to measure the goodness of fit for the smoothed periodogram $\{\hat{f}_\epsilon(\lambda_j)\}$ as a non-parametric model for the raw periodogram $\{I_n(\lambda_j)\}$. Clearly, minimization of the Kullback-Leibler spectral divergence favors a small m as it leads to the best fit. Therefore, a suitable penalty needs to be imposed to prevent the smoothed periodogram from overfitting. Inspired by the generalized cross-validation (GCV) criterion for linear smoothers [143, p. 49], a multiplier of the form $(1-W_{m0})^{-2}$ is introduced in [275] as the penalty function, leading to the following criterion:

$$\mathrm{GCV}(m) := \frac{1}{(1-W_{m0})^2} D_{\mathrm{KL}}(\{I_n(\lambda_j)\},\{\hat{f}_\epsilon(\lambda_j)\}). \tag{7.1.4}$$

Because $(1-W_{m0})^{-2}$ approaches infinity with the decrease of m, the multiplier serves as a counterbalance to the Kullback-Leibler spectral divergence which approaches zero with the decrease of m.

To make the periodogram smoother behave well asymptotically as the sample size increases, the weights should satisfy the following conditions:

$$m \to \infty, \quad m/n \to 0, \quad s_m := \sum_{|l|\le m} W_{ml}^2 \to 0, \quad \text{as } n \to \infty. \tag{7.1.5}$$

The first and third conditions require that an increasing number of periodogram ordinates take part in the moving average as the sample size grows, but the second condition restricts the window size so that all frequencies in the window shrink toward the frequency of interest. Under these and some additional conditions, the following theorem shows that the periodogram smoother in (7.1.2) becomes a consistent estimator of the noise spectrum in the absence of sinusoids. A proof of this result can be found in Section 7.4.

Theorem 7.1 (Asymptotic Properties of the Periodogram Smoother in the Absence of Sinusoids). *Let $\{y_t\}$ be given by (7.1.1) and $\hat{f}_\epsilon(\lambda_j)$ be defined by (7.1.2) with $\{W_{ml}\}$ satisfying (7.1.3) and (7.1.5). Assume that $\{\epsilon_t\}$ satisfies the conditions of Theorem 6.5 or 6.6, and that $mr_n^2 \to 0$ and $mr_{n2}^2 \to 0$ as $n \to \infty$, where r_n and r_{n2} are defined in Theorem 6.6. Then, the following assertions are true.*

(a) Asymptotic Uniform Unbiasedness. Let $\Omega_n := \{\lambda_0, \lambda_1, \ldots, \lambda_{n-1}\}$. As $n \to \infty$,

$$\max_{\lambda_j \in \Omega_n} |E\{\hat{f}_\epsilon(\lambda_j)\} - f_\epsilon(\lambda_j)| \to 0.$$

(b) Uniform Consistency in Mean Square. As $n \to \infty$,

$$\max_{\lambda_j \in \Omega_n} E\{|\hat{f}_\epsilon(\lambda_j) - f_\epsilon(\lambda_j)|^2\} \to 0.$$

(c) *Uniform Consistency in Probability. As $n \to \infty$ and for any given $\delta > 0$,*

$$\max_{\lambda_j \in \Omega_n} P(|\hat{f}_\epsilon(\lambda_j) - f_\epsilon(\lambda_j)| > \delta) \to 0.$$

(d) *Uniform Consistency. Let $w_m(\omega) := \sum_{|l| \le m} W_{ml} \exp(i\omega l)$, and assume that $\sum_{\lambda_j \in \Omega_n} |w_m(\lambda_j)| = \mathcal{O}(n^{-1/2})$. Then, as $n \to \infty$,*

$$\max_{\lambda_j \in \Omega_n} |\hat{f}_\epsilon(\lambda_j) - f_\epsilon(\lambda_j)| \xrightarrow{P} 0.$$

(e) *Asymptotic Covariances. As $n \to \infty$,*

$$s_m^{-1} \operatorname{Cov}\{\hat{f}_\epsilon(\lambda_j), \hat{f}_\epsilon(\lambda_l)\} - \{f_\epsilon^2(\lambda_j)\, a_{jl} + \xi |\iota g_\epsilon(\lambda_j)|^2 (b_{jl} + c_{jl})\} \to 0$$

uniformly in $\lambda_j, \lambda_l \in \Omega_n$, where

$$a_{jl} := s_m^{-1} \sum_{|u| \le m} W_{mu} W_{m,l-j+u},$$
$$b_{jl} := s_m^{-1} \sum_{|u| \le m} W_{mu} W_{m,l-j+u} \delta^{(n)}_{2(j-u)},$$
$$c_{jl} := s_m^{-1} \sum_{|u| \le m} \sum_{|v| \le m} W_{mu} W_{mv} \delta^{(n)}_{j-u+l-v},$$

with $\{\delta^{(n)}_u\}$ being the n-periodic delta sequence defined in Theorem 6.6.

Remark 7.1 If $E\{\zeta_t^2\} = 0$, as in the complex Gaussian case, then $\iota = 0$ and

$$\operatorname{Cov}\{\hat{f}_\epsilon(\lambda_j), \hat{f}_\epsilon(\lambda_l)\} = f_\epsilon^2(\lambda_j)\, a_{jl} s_m + \mathcal{O}(s_m).$$

Theorem 7.1 remains true in the real case with $\iota = \xi = 1$ and $g_\epsilon(\omega) = f_\epsilon(\omega)$.

Remark 7.2 If λ_j is sufficiently away from 0, π, and 2π such that $m < j < n/2 - m$ or $n/2 + m < j < n - m$, we have $b_{jj} = c_{jj} = 0$. This, together with $a_{jj} = 1$, gives

$$\operatorname{Var}\{\hat{f}_\epsilon(\lambda_j)\} = f_\epsilon^2(\lambda_j) s_m + \mathcal{O}(s_m).$$

In the real case, this expression holds for all $\lambda_j \in (0, \pi)$; for $\lambda_j = 0$ or π, $f_\epsilon^2(\lambda_j)$ should be replaced with $2f_\epsilon^2(\lambda_j)$. Moreover, in the general complex case, if λ_j and λ_l are sufficiently separated from each other such that $|\lambda_j - \lambda_l| > 4\pi m/n$, then $a_{jl} = b_{jl} = 0$; if λ_j and λ_l are also sufficiently away from being conjugate pairs such that $|\lambda_j + \lambda_l - 2\pi k| > 2\pi m/n$ $(k = 0, 1, 2)$, then $c_{jl} = 0$. Under these conditions, $\operatorname{Cov}\{\hat{f}_\epsilon(\lambda_j), \hat{f}_\epsilon(\lambda_l)\} = \mathcal{O}(s_m) \to 0$ as $n \to \infty$.

Remark 7.3 For any nonFourier frequency $\omega \in (0, 2\pi)$, $f_\epsilon(\omega)$ can be estimated by interpolating the periodogram smoother at neighboring Fourier frequencies. The simplest example is a linear interpolation of $\hat{f}_\epsilon(\lambda_j)$ and $\hat{f}_\epsilon(\lambda_{j+1})$, where λ_j and

λ_{j+1} are the Fourier frequencies satisfying $\lambda_j \le \omega < \lambda_{j+1}$. Under the conditions of Theorem 7.1, the resulting spectral estimator remains asymptotically uniformly unbiased, uniformly mean-square consistent, and uniformly weakly consistent. The reason is that the error of approximation approaches zero uniformly in ω as a result of the uniform continuity of the noise spectrum.

Theorem 7.1 requires that $mr_n^2 \to 0$ and $mr_{n2}^2 \to 0$. This requirement suggests that the weights must be chosen to match the smoothness of the noise spectrum. Generally speaking, a rough spectrum with abruptly changing features, such as sharp peaks and valleys, has a slower rate of decaying for r_n and r_{n2}. In this case, the window size m must grow more slowly with n to ensure the asymptotic properties in Theorem 7.1.

There can be a slight complication near frequency 0 which has been ignored so far. In practice, one often subtracts the sample mean before calculating the periodogram. When this happens, the periodogram ordinate at frequency 0 is equal to zero and therefore contains no information about $f_\epsilon(0)$. To deal with the irregularity, one can redefine $\hat{f}_\epsilon(\lambda_j)$ for $0 \le j \le m$ and $n-m \le j < n$ by excluding $I_n(0)$ in the averaging operation [45, p. 142]. This gives

$$\hat{f}_\epsilon(\lambda_j) = \sum_{\substack{|l|\le m \\ l \ne j \ (\mathrm{mod}\ n)}} \tilde{W}_{jml}\, I_n(\lambda_{j-l}),$$

where the adjusted weights $\tilde{W}_{jl}$ are defined by $\tilde{W}_{jml} := W_{ml}/(1-W_{mj})$ for $0 \le j \le m$ and $\tilde{W}_{jml} := W_{ml}/(1-W_{m,n-j})$ for $n-m \le j < n$.

Besides periodogram smoothing, data tapering is another technique of nonparametric spectral estimation. Recall that the periodogram ordinates are created from the discrete Fourier transform (DFT) of $\mathbf{y}$, defined as

$$z_j := n^{-1} \sum_{t=1}^{n} y_t \exp(-it\lambda_j),$$

such that $I_n(\lambda_j) = n|z_j|^2$. Therefore, the periodogram smoother defined by (7.1.2) can be expressed as

$$\hat{f}_\epsilon(\lambda_j) = n \sum_{|l|\le m} W_{ml} |z_{j-l}|^2. \tag{7.1.6}$$

In other words, the periodogram smoother is nothing but a moving-average filter applied to the squared modulus of the DFT of $\mathbf{y}$. With a proper choice of the impulse response $\{W_{ml}\}$, the choppiness of the DFT can be reduced to achieve the effect of smoothing.

A similar effect can be achieved by applying a filter directly to the DFT instead of the squared modulus. Because convolution in the frequency domain is equivalent to multiplication in the time domain, filtering the DFT can be done by multiplying the time series $\{y_1,\dots,y_n\}$ with a suitable set of weights $\{w_{m1},\dots,w_{mn}\}$, known as the data taper or data window, where m is a certain tuning parameter. With the DFT of the data taper denoted by

$$W_{ml} := n^{-1} \sum_{t=1}^{n} w_{mt} \exp(-it\lambda_l),$$

and with the z_j extended periodically with period n, it is easy to verify that the DFT of the tapered time series $\{w_{m1} y_1, \ldots, w_{mn} y_n\}$ can be expressed as

$$\tilde{z}_j := n^{-1} \sum_{t=1}^{n} w_{mt} y_t \exp(-it\lambda_j) = \sum_{l=0}^{n-1} W_{ml} z_{j-l}.$$

Because smoothing is achieved in this way, the corresponding periodogram

$$\hat{f}_\epsilon(\lambda_j) := n|\tilde{z}_j|^2 = n\left|\sum_{l=0}^{n-1} W_{ml} z_{j-l}\right|^2 \tag{7.1.7}$$

serves directly as a spectral estimator, called the *data-taper spectral estimator*. By comparing (7.1.7) with (7.1.6), one can see the analogy and difference between the data-taper spectral estimator and the periodogram smoother. Discussions on the choice of various data windows can be found in [139] and [298].

An important extension of the data tapering technique is the *multitaper spectral estimator* introduced by Thomson [387]. The idea is to average a set of data-taper spectral estimators constructed from Slepian's discrete prorate spheroidal sequences [348]. With the aim of suppressing the sidelobes of $\{W_{ml}\}$ in (7.1.7), Slepian's discrete prorate spheroidal sequences are given by the eigenvectors of the n-by-n matrix $[\mathrm{sinc}(\Delta(u-v))]$ $(u, v = 1, \ldots, n)$. The constant $\Delta > 0$ is a predetermined bandwidth parameter that controls the frequency resolution of the multitaper estimator; it is typically taken to be a small multiple of $2\pi/n$. In practice, only a small number of principal eigenvectors are useful in the multitaper method, as they have the desired time-domain property of attenuating the endpoints of $\mathbf{y}$ and thereby reducing the truncation effect (i.e., spectral leakage) in the DFT. For a comprehensive treatment on this method, see [288].

7.1.2 Lag-Window Spectral Estimator

An alternative technique of periodogram smoothing, known as the indirect (or Blackman-Tukey) method, is based on the fact that the SDF $f_\epsilon(\omega)$ is a Fourier transform of the ACF $r_\epsilon(u)$ that can be expressed as

$$f_\epsilon(\omega) = \sum_{u=-\infty}^{\infty} r_\epsilon(u) \exp(-iu\omega).$$

By substituting the ACF with the sample autocovariances $\hat{r}_y(u)$ defined in (4.1.1) and by multiplying them with some suitable weights, one obtains the so-called *lag-window spectral estimator* of the form

$$\hat{f}_\epsilon(\lambda_j) := \sum_{|u|\le m} w_{mu}\,\hat{r}_y(u)\exp(-iu\lambda_j). \tag{7.1.8}$$

The weights w_{mu} are usually generated by a window function $w(x)$ such that $w_{mu} := w(u/m)$, and $w(x)$ is an even, piecewise-continuous function satisfying the conditions $w(0) = 1$, $|w(x)| \le 1$ for all x, and $w(x) = 0$ for $|x| > 1$. The window function is typically tapered at both ends so that w_{mu} decreases gradually toward zero as $|u|$ increases. This reduces the boundary effect caused by the abrupt truncation of the infinite series into a finite sum. Owing to the time-frequency duality of Fourier transform, the tapered windowing can also be interpreted as lowpass filtering in the time domain with the effect of reducing the choppiness in the frequency domain.

A connection between the lag-window spectral estimator in (7.1.8) and the continuous periodogram defined by (6.4.1) can be easily established. Indeed, because $I_n(\omega)$ satisfies (6.4.2), it follows that

$$\hat{r}_y(u) = (2\pi)^{-1}\int_{-\pi}^{\pi} I_n(\omega)\exp(i\omega u)\,d\omega \qquad (|u| < n). \tag{7.1.9}$$

Substituting this expression in (7.1.8) gives

$$\hat{f}_\epsilon(\lambda_j) = (2\pi)^{-1}\int_{-\pi}^{\pi} W_m(\omega)\,I_n(\lambda_j - \omega)\,d\omega, \tag{7.1.10}$$

where

$$W_m(\omega) := \sum_{|u|\le m} w_{mu}\exp(-iu\omega). \tag{7.1.11}$$

Clearly, the convolution in (7.1.10) is analogous to the weighted moving average in (7.1.2), both serving as a smoother of the periodogram.

The lag-window spectral estimator was the subject of intensive research in the early days of modern spectral analysis that focused on the effect of different windows on the bias, variance, and spectral resolution of the estimator for time series with continuous spectra. Pioneering works include [26], [37], [82], [127], [279], [280], and [296]. A comprehensive survey of these works and related results can be found, for example, in the books of Anderson [12, Chapter 9] and Priestley [298, Sections 6.2.3 and 6.2.4].

The frequency variable λ_j in (7.1.8) and (7.1.10) can be replaced by $\omega \in [0, 2\pi)$ to obtain the naturally interpolated estimator $\hat{f}_\epsilon(\omega)$. An interesting special case is when ω takes values in $\Omega_{2n} := \{\lambda'_j : j = 0, \pm 1, \pm 2, \ldots\}$, where $\lambda'_j := 2\pi j/(2n)$. In this case, the lag-window spectral estimator can be expressed as a weighted moving average of the refined periodogram defined on Ω_{2n} [288, p. 239]. In fact, let $D_{2n}(\omega) := (2n)^{-1}\sum_{t=1}^{2n}\exp(it\omega)$ denote the Dirichlet kernel and set $w_{mu} := 0$ for $|u| > m$. Then, it follows from (7.1.11) and Lemma 12.1.3 that

$$\begin{aligned}\sum_{l=-n+1}^{n} W_m(\lambda_l')\exp(iv\lambda_l') &= \sum_{u=-n+1}^{n} 2nw_{mu}\exp\{i\pi(u-v)\}D_{2n}^*(\lambda_{u-v}') \\ &= 2nw_{mv} \qquad (v = -n+1,\ldots,n).\end{aligned}$$

Substituting this expression in (7.1.10), together with (6.4.2), yields

$$\hat{f}_\epsilon(\lambda_j') = \sum_{l=-n+1}^{n} W_{ml}' I_n(\lambda_{j-l}'), \quad W_{ml}' := (2n)^{-1}W_m(\lambda_l'). \tag{7.1.12}$$

Note that the convolution form (7.1.12) may not be as desirable computationally as the original form (7.1.8) when m is much smaller than n.

The lag-window spectral estimator in (7.1.8) can be approximated by a periodogram smoother of the form (7.1.2) if the integral in (7.1.10) is replaced by a Riemann sum over the Fourier grid $\Omega_n := \{\lambda_j\}$. In other words,

$$\hat{f}_\epsilon(\lambda_j) \approx \sum_{|l|\le n/2} W_{ml} I_n(\lambda_{j-l}), \quad W_{ml} := n^{-1}W_m(\lambda_l). \tag{7.1.13}$$

In this expression, the weights W_{ml} do not necessarily satisfy (7.1.5) or sum up to unity. But, it is easy to see that

$$\sum_{|l|\le n/2} W_{ml} \approx (2\pi)^{-1}\int_{-\pi}^{\pi} W_m(\omega)\,d\omega = w(0) = 1$$

and

$$\begin{aligned}\sum_{|l|\le n/2} W_{ml}^2 &\approx n^{-1}(2\pi)^{-1}\int_{-\pi}^{\pi} W_m^2(\omega)\,d\omega \\ &= n^{-1}\sum_{|u|\le m} w^2(u/m) \\ &\approx mn^{-1}\int_{-1}^{1} w^2(x)\,dx.\end{aligned}$$

Therefore, it is not too surprising that the assertions in Theorem 7.1 remain valid for the lag-window spectral estimator under the condition that $n\to\infty$, $m\to\infty$, and $m/n\to 0$ [12, Sections 9.3 and 9.4] [298, Section 6.2.4].

Using the expression in (6.4.2), the periodogram smoother in (7.1.2) can be expressed in the form of (7.1.8) with

$$w_{mu} := \sum_{|l|\le m} W_{ml}\exp(iu\lambda_l).$$

Therefore, the periodogram smoother is also a lag-window spectral estimator. The only difference is that the weights w_{mu} do not necessarily take the form $w(u/m)$, so the length of the window is no longer directly controlled by m. The same can be said about the weights W_{ml} in (7.1.13).

7.1.3 Autoregressive Spectral Estimator

In addition to the periodogram-based frequency-domain approach, there are many techniques of spectral estimation that are devised from parametric models in the time domain. A good example is the autoregressive (AR) model

$$\epsilon_t + \sum_{j=1}^{m} \varphi_j \epsilon_{t-j} = \zeta_t, \quad \{\zeta_t\} \sim \text{IID}(0, \sigma_m^2), \tag{7.1.14}$$

whose spectral density takes the form

$$f_{\text{AR}}(\omega) = \frac{\sigma_m^2}{|1 + \sum_{j=1}^{m} \varphi_j \exp(-ij\omega)|^2}. \tag{7.1.15}$$

Owing to its all-pole structure, the AR model is particularly effective in capturing spectral peaks, as recognized by Yule [434] in his pioneering work on hidden periodicities. Generally, the AR spectrum is able to approximate any continuous spectrum as ensured by the following theorem. See Section 7.4 for a proof.

Theorem 7.2 (AR Spectral Approximation). *Let $\{\epsilon_t\}$ be a stationary process with mean zero and SDF $f_\epsilon(\omega)$. If $f_\epsilon(\omega)$ is continuous in $[-\pi, \pi]$, then, for any $\delta > 0$, there exists an AR spectrum $f_{\text{AR}}(\omega)$ such that $|f_\epsilon(\omega) - f_{\text{AR}}(\omega)| < \delta$ for all $\omega \in \mathbb{R}$.*

Besides the spectral approximation property, the AR model can be easily and efficiently estimated from data. There are also simple data-driven techniques for the selection of m. All these properties make the AR spectral estimator generally applicable to non-autoregressive time series [6] [281] [298, pp. 600–602].

If $\{\epsilon_t\}$ is indeed an AR(m) process of the form (7.1.14), then the ACF of $\{\epsilon_t\}$ should satisfy the so-called *Yule-Walker equations*

$$r_\epsilon(u) + \sum_{j=1}^{m} \varphi_j\, r_\epsilon(u-j) = \sigma_m^2 \delta_u \qquad (u = 0, 1, \ldots). \tag{7.1.16}$$

This connection between the ACF of an AR process and the parameters in the difference equation (7.1.14) was made by Walker [415] when analyzing Yule's autoregressive method of periodicity modeling. To prove (7.1.16) for $u \geq 1$, one simply multiplies both sides of (7.1.14) with ϵ_{t-u}^* and takes the expected values while recognizing $E(\zeta_t \epsilon_s^*) = 0$ for all $s < t$ because ϵ_s is a linear function of $\{\zeta_s, \zeta_{s-1}, \ldots\}$. Similarly, the Yule-Walker equation for $u = 0$ can be derived by using the fact that $\epsilon_t = \sum_{j=0}^{\infty} \psi_j \zeta_{t-j}$ ($\psi_0 := 1$), where $\{\psi_j\}$ is determined by the z-transform $\sum_{j=0}^{\infty} \psi_j z^{-j} := 1 / \sum_{j=0}^{m} \varphi_j z^{-j}$ ($\varphi_0 := 1$).

If $\{\epsilon_t\}$ is an arbitrary stationary process with ACF $r_\epsilon(u)$, then the solution to the Yule-Walker equations in (7.1.16) defines the AR(m) model for $\{\epsilon_t\}$. Let

$$\boldsymbol{\varphi} := [\varphi_1, \ldots, \varphi_m]^T.$$

Then, it follows from (7.1.16) that

$$\begin{cases} \mathbf{R}_\epsilon \boldsymbol{\varphi} = -\mathbf{r}_\epsilon, \\ \sigma_m^2 = r_\epsilon(0) + \mathbf{r}_\epsilon^H \boldsymbol{\varphi}, \end{cases} \tag{7.1.17}$$

where

$$\mathbf{R}_\epsilon := \begin{bmatrix} r_\epsilon(0) & r_\epsilon^*(1) & \cdots & r_\epsilon^*(m-1) \\ r_\epsilon(1) & r_\epsilon(0) & \cdots & r_\epsilon^*(m-2) \\ \vdots & \vdots & \ddots & \vdots \\ r_\epsilon(m-1) & r_\epsilon(m-2) & \cdots & r_\epsilon(0) \end{bmatrix}, \quad \mathbf{r}_\epsilon := \begin{bmatrix} r_\epsilon(1) \\ r_\epsilon(2) \\ \vdots \\ r_\epsilon(m) \end{bmatrix}.$$

The equations in (7.1.17) can also be written as

$$\begin{bmatrix} r_\epsilon(0) & \mathbf{r}_\epsilon^H \\ \mathbf{r}_\epsilon & \mathbf{R}_\epsilon \end{bmatrix} \begin{bmatrix} 1 \\ \boldsymbol{\varphi} \end{bmatrix} = \begin{bmatrix} \sigma_m^2 \\ \mathbf{0} \end{bmatrix}.$$

According to Proposition 2.7, the parameters $\boldsymbol{\varphi}$ and σ_m^2 in (7.1.17) are just the coefficient vector of the best linear predictor of ϵ_t based on $\{\epsilon_{t-1},\dots,\epsilon_{t-m}\}$ and the corresponding prediction error variance. Therefore, the Levinson-Durbin algorithm in Proposition 2.8 can be used to compute these parameters efficiently from the first $m+1$ autocovariances $r_\epsilon(0), r_\epsilon(1), \dots, r_\epsilon(m)$. This algorithm circumvents the brute-force inversion of $\mathbf{R}_\epsilon$ in solving (7.1.17) for $\boldsymbol{\varphi}$ and reduces the computational complexity from $\mathcal{O}(m^3)$ to $\mathcal{O}(m^2)$.

The AR model given by (7.1.17) has the following minimum-phase property. See Section 7.4 for a proof.

Theorem 7.3 (Minimum-Phase Property). *Let $\boldsymbol{\varphi}$ and σ_m^2 be defined by (7.1.17). If $\sigma_m^2 > 0$, then the roots of $\varphi_m(z) := 1 + \sum_{j=1}^m \varphi_j z^{-j}$ lie strictly inside the unit circle. A polynomial with this property is called a minimum-phase polynomial.*

Remark 7.4 Let φ_{uu} $(u = 1,\dots,m)$ denote the reflection coefficients given by the Levinson-Durbin algorithm in Proposition 2.8 for the solution of (7.1.17). Then, according to Remark 2.5, the polynomial $\varphi_m(z)$ in Theorem 7.3 is a minimum-phase polynomial if $|\varphi_{uu}| < 1$ for all $u = 1,\dots,m$.

Another interesting property of the AR model given by (7.1.17) is that the corresponding AR spectrum in (7.1.15) maximizes the spectral entropy, defined as

$$H(f) := \int_{-\pi}^{\pi} \log f(\omega)\, d\omega,$$

among all spectral densities whose first $m+1$ autocovariances coincide with $r_\epsilon(u)$ $(u = 0,1,\dots,m)$ [298, pp. 604–605]. The spectral entropy $H(f)$ is a measure of predictability for a stationary process at any given time based on its infinite past values [46, p. 191] [298, pp. 740–741]. Therefore, the AR model defined by (7.1.17) can be regarded as the most unpredictable process under the constraint of the first $m+1$ autocovariances [288, p. 422]. The spectral entropy can also be interpreted as a measure of spectral flatness [177, p. 183].

Moreover, for any $\tilde{\boldsymbol{\varphi}} := [\tilde{\varphi}_1, \ldots, \tilde{\varphi}_m]^T \in \mathbb{C}^m$, define $\tilde{\Phi}(\omega) := 1 + \mathbf{f}_m^H(\omega)\tilde{\boldsymbol{\varphi}}$ and

$$J(\tilde{\boldsymbol{\varphi}}) := (2\pi)^{-1} \int_{-\pi}^{\pi} |\tilde{\Phi}(\omega)|^2 f_\epsilon(\omega)\, d\omega, \tag{7.1.18}$$

where $\mathbf{f}_m(\omega) := [\exp(i\omega), \ldots, \exp(im\omega)]^T$. It is easy to verify that

$$E\Big\{\Big|\epsilon_t + \sum_{j=1}^{m} \tilde{\varphi}_j \epsilon_{t-j}\Big|^2\Big\} = J(\tilde{\boldsymbol{\varphi}}). \tag{7.1.19}$$

Therefore, the parameter $\boldsymbol{\varphi}$ given by (7.1.17), which minimizes the left-hand side of (7.1.19), also minimizes $J(\tilde{\boldsymbol{\varphi}})$, and the minimum value can be expressed as

$$J(\boldsymbol{\varphi}) = (2\pi)^{-1} \int_{-\pi}^{\pi} |\Phi_m(\omega)|^2 f_\epsilon(\omega)\, d\omega = \sigma_m^2,$$

where $\Phi_m(\omega) := 1 + \mathbf{f}_m^H(\omega)\boldsymbol{\varphi}$. Being the minimizer of $J(\tilde{\boldsymbol{\varphi}})$ defined by (7.1.18), the function $|\Phi_m(\omega)|^2$ should take the smallest value possible where $f_\epsilon(\omega)$ is large. This explains why the resulting AR spectrum $f_{\text{AR}}(\omega) = \sigma_m^2 / |\Phi_m(\omega)|^2$ is well suited to capture the spectral peaks in $f_\epsilon(\omega)$. On the other hand, the AR spectrum may not approximate $f_\epsilon(\omega)$ well in places where $f_\epsilon(\omega)$ is small, an extreme example being $f_\epsilon(\omega_0) = 0$ for some $\omega_0 \in (-\pi, \pi)$, in which case the value $|\tilde{\Phi}(\omega_0)|^2$ plays no part in determining the minimum value of $J(\tilde{\boldsymbol{\varphi}})$.

To estimate the parameters in the AR model from a data record $\{y_1, \ldots, y_n\}$ given by (7.1.1), one can simply employ the method of moments, i.e., substitute the ACF $r_\epsilon(u)$ with the sample autocovariances $\hat{r}_y(u)$ defined by (4.1.1) and solve the resulting Yule-Walker equations in (7.1.17). This method is commonly known as the *Yule-Walker method* (also known as the autocorrelation method of linear prediction [249]). To be more specific, let $\hat{\mathbf{R}}_y$ and $\hat{\mathbf{r}}_y$ be defined in the same way as $\mathbf{R}_\epsilon$ and $\mathbf{r}_\epsilon$ except that $r_\epsilon(u)$ is replaced by $\hat{r}_y(u)$. Then, the Yule-Walker estimator of $\boldsymbol{\varphi}$ and σ_m^2 can be expressed as

$$\begin{cases} \hat{\boldsymbol{\varphi}} := -\hat{\mathbf{R}}_y^{-1}\hat{\mathbf{r}}_y, \\ \hat{\sigma}_m^2 := \hat{r}_y(0) + \hat{\mathbf{R}}_y\hat{\boldsymbol{\varphi}}. \end{cases} \tag{7.1.20}$$

An advantage of the Yule-Walker estimator is that it can be computed very efficiently by using the Levinson-Durbin algorithm. Moreover, the Yule-Walker estimator has the minimum-phase property in Theorem 7.3.

In addition to the method of moments that yields the Yule-Walker estimator, alternative estimators can be devised from linear prediction by the method of least squares. One such estimator is given by the forward least-squares linear prediction (also known as the covariance method [249]):

$$\begin{cases} \hat{\boldsymbol{\varphi}} := \arg\min_{\tilde{\boldsymbol{\varphi}} \in \mathbb{C}^m} \|\mathbf{y}_f + \mathbf{Y}_f \tilde{\boldsymbol{\varphi}}\|^2, \\ \hat{\sigma}_m^2 := n^{-1} \|\mathbf{y}_f + \mathbf{Y}_f \hat{\boldsymbol{\varphi}}\|^2, \end{cases} \tag{7.1.21}$$

where

$$\mathbf{Y}_f := \begin{bmatrix} y_m & y_{m-1} & \cdots & y_1 \\ y_{m+1} & y_m & \cdots & y_2 \\ \vdots & \vdots & & \vdots \\ y_{n-1} & y_{n-2} & \cdots & y_{n-m} \end{bmatrix}, \quad \mathbf{y}_f := \begin{bmatrix} y_{m+1} \\ y_{m+2} \\ \vdots \\ y_n \end{bmatrix}.$$

It is easy to verify that $\hat{\boldsymbol{\varphi}}$ and $\hat{\sigma}_m^2$ in (7.1.21) satisfy

$$(\mathbf{Y}_f^H \mathbf{Y}_f)\hat{\boldsymbol{\varphi}} = -\mathbf{Y}_f^H \mathbf{y}_f, \quad \hat{\sigma}_m^2 = n^{-1}(\mathbf{y}_f^H \mathbf{y}_f + \mathbf{y}_f^H \mathbf{Y}_f \hat{\boldsymbol{\varphi}}). \tag{7.1.22}$$

Therefore, the forward least-squares linear prediction method is equivalent to replacing $\mathbf{R}_\epsilon$, $\mathbf{r}_\epsilon$, and $r_\epsilon(0)$ in (7.1.17) by their estimates

$$\hat{\mathbf{R}}_y := n^{-1}\mathbf{Y}_f^H \mathbf{Y}_f, \quad \hat{\mathbf{r}}_y := n^{-1}\mathbf{Y}_f^H \mathbf{y}_f, \quad \hat{r}_y(0) := n^{-1}\mathbf{y}_f^H \mathbf{y}_f.$$

The matrix $\hat{\mathbf{R}}_y$ in this case does not have the Toeplitz property, so the Levinson-Durbin algorithm is no longer applicable. To solve (7.1.22), one can use the Cholesky decomposition technique [122, Section 4.2].

The least-squares estimator defined by (7.1.21) has a maximum likelihood interpretation. Suppose that $\{\zeta_t\}$ in the AR model (7.1.14) is complex Gaussian, i.e., $\{\zeta_t\} \sim \mathrm{GWN}_c(0, \sigma_m^2)$. Then, conditional on $\zeta_1, \ldots, \zeta_m$, the log likelihood function of $\{y_1, \ldots, y_n\}$ is proportional to

$$\text{constant} + (n-m)\log\sigma_m^2 - \|\mathbf{y}_f + \mathbf{Y}_f \boldsymbol{\varphi}\|^2/\sigma_m^2.$$

Maximizing this criterion leads to (7.1.21). The conditional maximum likelihood estimator can be regarded as an approximation to the unconditional maximum likelihood estimator, because the effect of conditioning is negligible when n is large and when the roots of the AR polynomial $\varphi_m(z)$ are far away from the unit circle so that the AR coefficients decay quickly [298, pp. 346–348].

Another estimator, derived by the forward and backward least-squares linear prediction (also known as the modified covariance method), is given by

$$\begin{cases} \hat{\boldsymbol{\varphi}} := \arg\min_{\tilde{\boldsymbol{\varphi}} \in \mathbb{C}^m} \{\|\mathbf{y}_f + \mathbf{Y}_f \tilde{\boldsymbol{\varphi}}\|^2 + \|\mathbf{y}_b + \mathbf{Y}_b \tilde{\boldsymbol{\varphi}}\|^2\}, \\ \hat{\sigma}_m^2 := (2n)^{-1}\{\|\mathbf{y}_f + \mathbf{Y}_f \hat{\boldsymbol{\varphi}}\|^2 + \|\mathbf{y}_b + \mathbf{Y}_b \hat{\boldsymbol{\varphi}}\|^2\}, \end{cases} \tag{7.1.23}$$

where

$$\mathbf{Y}_b := \begin{bmatrix} y_2^* & y_3^* & \cdots & y_{m+1}^* \\ y_3^* & y_4^* & \cdots & y_{m+2}^* \\ \vdots & \vdots & & \vdots \\ y_{n-m+1}^* & y_{n-m+2}^* & \cdots & y_n^* \end{bmatrix}, \quad \mathbf{y}_b := \begin{bmatrix} y_1^* \\ y_2^* \\ \vdots \\ y_{n-m}^* \end{bmatrix}.$$

This estimator coincides with the solution to (7.1.17) when $\mathbf{R}_\epsilon$, $\mathbf{r}_\epsilon$, and $r_\epsilon(0)$ are replaced by $(2n)^{-1}\mathbf{Y}_{fb}^H\mathbf{Y}_{fb}$, $(2n)^{-1}\mathbf{Y}_{fb}^H\mathbf{y}_{fb}$, and $(2n)^{-1}\mathbf{y}_{fb}^H\mathbf{y}_{fb}$, respectively, where

$$\mathbf{Y}_{fb} := \begin{bmatrix} \mathbf{Y}_f \\ \mathbf{Y}_b \end{bmatrix}, \quad \mathbf{y}_{fb} := \begin{bmatrix} \mathbf{y}_f \\ \mathbf{y}_b \end{bmatrix}.$$

The employment of backward prediction can be justified by the fact that the φ_j given by (7.1.17) also minimize the backward prediction error variance

$$E\Big\{\Big|\epsilon_{t-m}^* + \sum_{j=1}^m \tilde{\varphi}_j \epsilon_{t-m+j}^*\Big|^2\Big\}.$$

Due to the incorporation of backward prediction, the estimator in (7.1.23) tends to outperform the Yule-Walker estimator and other linear prediction alternatives for short data records with large spectral peaks [177, p. 227–228].

While the estimators in (7.1.21) and (7.1.23) do not make assumptions about the process $\{y_t\}$ beyond the observed values for $t \in \{1,\ldots,n\}$, the third variation assumes $y_t = 0$ for $t \in \{-m+1,\ldots,0\}$ and solves the so-called complete least-squares linear prediction problem

$$\begin{cases} \hat{\boldsymbol{\varphi}} := \arg\min_{\tilde{\boldsymbol{\varphi}} \in \mathbb{C}^m} \|\mathbf{y} + \mathbf{Y}\tilde{\boldsymbol{\varphi}}\|^2, \\ \hat{\sigma}_m^2 := n^{-1}\|\mathbf{y} + \mathbf{Y}\hat{\boldsymbol{\varphi}}\|^2, \end{cases} \tag{7.1.24}$$

where

$$\mathbf{Y} := \begin{bmatrix} y_0 & y_{-1} & \cdots & y_{-m+1} \\ y_1 & y_0 & \cdots & y_{-m+2} \\ \vdots & \vdots & & \vdots \\ y_{n-1} & y_{n-2} & \cdots & y_{n-m} \end{bmatrix}, \quad \mathbf{y} := \begin{bmatrix} y_1 \\ y_2 \\ \vdots \\ y_n \end{bmatrix}.$$

This estimator is employed in [252] as an approximation to the Gaussian maximum likelihood estimator which is more difficult to compute.

It is interesting to note that the Yule-Walker estimator of $\boldsymbol{\varphi}$ can also be derived from a least-squares problem of the form (7.1.21) if $\mathbf{Y}_f$ and $\mathbf{y}_f$ are replaced by

$$\tilde{\mathbf{Y}} := \begin{bmatrix} y_0 & y_{-1} & \cdots & y_{-m+1} \\ y_1 & y_0 & \cdots & y_{-m+2} \\ \vdots & \vdots & & \vdots \\ y_{n+m-1} & y_{n+m-2} & \cdots & y_n \end{bmatrix}, \quad \tilde{\mathbf{y}} := \begin{bmatrix} y_1 \\ y_2 \\ \vdots \\ y_{n+m} \end{bmatrix},$$

where $y_t := 0$ for $-m+1 \le t \le 0$ and $n+1 \le t \le n+m$. This can be easily verified by the fact that substituting the ACF in $\mathbf{R}_\epsilon$ and $\mathbf{r}_\epsilon$ with the sample autocovariances defined by (4.1.1) gives

$$\hat{\mathbf{R}}_y := n^{-1}\tilde{\mathbf{Y}}^H\tilde{\mathbf{Y}}, \quad \hat{\mathbf{r}}_y := n^{-1}\tilde{\mathbf{Y}}^H\tilde{\mathbf{y}}.$$

While the Yule-Walker estimator has the minimum-phase property, the other least-squares estimators need not. The following example compares these estimators in the case of $m = 1$.

Example 7.1 (Estimation of Complex AR(1) Model: Part 1). In the case of $m = 1$, the Yule-Walker estimator of $\varphi_1 := -r_\epsilon(1)/r_\epsilon(0)$ is given by

$$\hat{\varphi}_{\text{YW}} := -\frac{\sum_{t=1}^{n-1} y_{t+1}y_t^*}{\sum_{t=1}^{n} |y_t|^2},$$

the least-squares estimators in (7.1.21) and (7.1.24) can be expressed as

$$\hat{\varphi}_f := -\frac{\sum_{t=1}^{n-1} y_{t+1}y_t^*}{\sum_{t=1}^{n-1} |y_t|^2},$$

and the least-squares estimator in (7.1.23) takes the form

$$\hat{\varphi}_{fb} := -\frac{\sum_{t=1}^{n-1} y_{t+1}y_t^*}{(\sum_{t=1}^{n-1} |y_t|^2 + \sum_{t=2}^{n} |y_t|^2)/2}.$$

These estimators differ only in their handling of the endpoints y_1 and y_n in the denominator. Moreover, by the Cauchy-Schwarz inequality

$$\left|\sum_{t=1}^{n-1} y_{t+1}y_t^*\right|^2 \le \left(\sum_{t=2}^{n} |y_t|^2\right)\left(\sum_{t=1}^{n-1} |y_t|^2\right),$$

we obtain $|\hat{\varphi}_{\text{YW}}| < 1$ (assuming $y_1 y_n \ne 0$). Therefore, the Yule-Walker estimator gives a stable (minimum-phase) AR model. The same can be said about $\hat{\varphi}_{fb}$ due to the fact that $|ab| \le (a^2 + b^2)/2$. However, it is not necessarily true for $\hat{\varphi}_f$ because y_n in the numerator can be arbitrarily large. ◇

While the linear prediction estimators may have different statistical properties for finite sample sizes, they all share the same asymptotic properties as the sample size grows. The following theorem establishes the consistency of these estimators for estimating the AR parameters and the corresponding AR spectrum under the assumption that $\{\epsilon_t\}$ is a linear process of the form

$$\epsilon_t := \sum_{j=-\infty}^{\infty} \psi_j \zeta_{t-j}. \tag{7.1.25}$$

This result is a direct consequence of Theorem 4.1. See Section 7.4 for a proof.

Theorem 7.4 (Consistency of the Linear Prediction Estimators in the Absence of Sinusoids). *Let $\{y_t\}$ be given by (7.1.1). Assume that $\{\epsilon_t\}$ is a real or complex linear process of the form (7.1.25) with $\sum |\psi_j| < \infty$, $\{\zeta_t\} \sim \mathrm{IID}(0, \sigma^2)$, and $E(|\zeta_t|^4) < \infty$. Then, the estimators defined by (7.1.20), (7.1.21), (7.1.23), and (7.1.24) are consistent for estimating $\boldsymbol{\varphi}$ and σ_m^2 defined by (7.1.17), and the corresponding AR spectra are uniformly consistent for estimating $f_{\mathrm{AR}}(\omega)$ defined by (7.1.15).*

Remark 7.5 When $\{\epsilon_t\}$ satisfies (7.1.14), the AR spectrum $f_{\mathrm{AR}}(\omega)$ in (7.1.15) coincides with $f_\epsilon(\omega)$. Otherwise, $f_{\mathrm{AR}}(\omega)$ is an approximation of $f_\epsilon(\omega)$.

The asymptotic normality of these estimators can also be established for linear processes. In the real case, define

$$\begin{aligned}\sigma(u,v) := &\sum_{\tau=-\infty}^{\infty} \Big\{\sum_{j=0}^{m} \varphi_j r_\epsilon(\tau+u-j)\Big\}\Big\{\sum_{j=0}^{m} \varphi_j r_\epsilon(\tau+v-j)\Big\} \\ &+ \sum_{\tau=-\infty}^{\infty} \Big\{\sum_{j=0}^{m} \varphi_j r_\epsilon(\tau+u-j)\Big\}\Big\{\sum_{j=0}^{m} \varphi_j r_\epsilon(\tau-v+j)\Big\}. \qquad (7.1.26)\end{aligned}$$

where $\varphi_0 := 1$. In the complex case, define

$$\begin{aligned}\sigma(u,v) := &\sum_{\tau=-\infty}^{\infty} \Big\{\sum_{j=0}^{m} \varphi_j r_\epsilon(\tau+u-j)\Big\}\Big\{\sum_{j=0}^{m} \varphi_j r_\epsilon(\tau+v-j)\Big\}^* \\ &+ \sum_{\tau=-\infty}^{\infty} \xi\Big\{\sum_{j=0}^{m} \varphi_j c_\epsilon(\tau+u-j)\Big\}\Big\{\sum_{j=0}^{m} \varphi_j c_\epsilon(\tau-v+j)\Big\}^*, \qquad (7.1.27)\end{aligned}$$

$$\begin{aligned}\tilde{\sigma}(u,v) := &\sum_{\tau=-\infty}^{\infty} \Big\{\sum_{j=0}^{m} \varphi_j r_\epsilon(\tau+u-j)\Big\}\Big\{\sum_{j=0}^{m} \varphi_j r_\epsilon^*(\tau-v+j)\Big\} \\ &+ \sum_{\tau=-\infty}^{\infty} \xi\Big\{\sum_{j=0}^{m} \varphi_j c_\epsilon(\tau+u-j)\Big\}\Big\{\sum_{j=0}^{m} \varphi_j c_\epsilon^*(\tau+v-j)\Big\}, \qquad (7.1.28)\end{aligned}$$

where $\xi := |\iota|^2$, $\iota := E(\zeta_t^2)/\sigma^2$, and $c_\epsilon(u) := \sigma^2 \sum_j \psi_{j+u}\psi_j$. With this notation, the following result can be obtained. See Section 7.4 for a proof.

Theorem 7.5 (Asymptotic Normality of the Linear Prediction Estimators in the Absence of Sinusoids). *Let $\{y_t\}$ be given by (7.1.1) with $\{\epsilon_t\}$ satisfying the conditions of Theorem 7.4. Let $\hat{\boldsymbol{\varphi}}$ and $\hat{\sigma}_m^2$ be defined by (7.1.20), (7.1.21), (7.1.23), or (7.1.24). Then, as $n \to \infty$ and for fixed m,*

$$\sqrt{n}\,(\hat{\boldsymbol{\varphi}} - \boldsymbol{\varphi}) \stackrel{D}{\to} \boldsymbol{\xi}.$$

In the real case, $\boldsymbol{\xi} \sim \mathrm{N}(\mathbf{0}, \mathbf{R}_\epsilon^{-1}\boldsymbol{\Sigma}\mathbf{R}_\epsilon^{-1})$, where $\boldsymbol{\Sigma} := [\sigma(u,v)]$ $(u,v = 1,\dots,m)$ is defined by (7.1.26); in the complex case, $\boldsymbol{\xi} \sim \mathrm{N}_c(\mathbf{0}, \mathbf{R}_\epsilon^{-1}\boldsymbol{\Sigma}\mathbf{R}_\epsilon^{-1}, \mathbf{R}_\epsilon^{-1}\tilde{\boldsymbol{\Sigma}}\mathbf{R}_\epsilon^{-T})$, where $\boldsymbol{\Sigma} := [\sigma(u,v)]$ and $\tilde{\boldsymbol{\Sigma}} := [\tilde{\sigma}(u,v)]$ $(u,v = 1,\dots,m)$ are defined by (7.1.27) and (7.1.28). Moreover, if $\{\epsilon_t\}$ is a real AR(m) process of the form (7.1.14), then $\boldsymbol{\Sigma} = \sigma_m^2\mathbf{R}_\epsilon$; if $\{\epsilon_t\}$ is a complex AR(m) process with $E(\zeta_t^2) = 0$, then $\boldsymbol{\Sigma} = \sigma_m^2\mathbf{R}_\epsilon$ and $\tilde{\boldsymbol{\Sigma}} = \mathbf{0}$.

Remark 7.6 Let $\hat{f}_{\mathrm{AR}}(\omega) := \hat{\sigma}_m^2/|1+\mathbf{f}_m^H(\omega)\hat{\boldsymbol{\varphi}}|^2$ denote the estimated AR spectrum. By Propositions 2.15 and 2.13(b), the asymptotic normality of $\hat{\boldsymbol{\varphi}}$, combined with the consistency of $\hat{\sigma}_m^2$, implies that $\sqrt{n}\{\hat{f}_{\mathrm{AR}}(\omega) - f_{\mathrm{AR}}(\omega)\} \stackrel{D}{\to} \mathrm{N}(0, v_m(\omega))$ as $n \to \infty$ for some $v_m(\omega)$ [213]. See [288, pp. 445–450] for a discussion on the confidence intervals for AR spectra.

To illustrate Theorems 7.4 and 7.5, consider the following example.

Example 7.2 (Estimation of AR(1) Model: Part 2). Let $\{\epsilon_t\}$ satisfy the conditions of Theorem 7.4. According to Example 7.1 and Theorem 4.1,

$$\hat{\varphi}_{\mathrm{YW}} = -\hat{r}_y(1)/\hat{r}_y(0) \stackrel{a.s.}{\to} -r_\epsilon(1)/r_\epsilon(0) = \varphi_1.$$

Similarly,

$$\hat{\varphi}_f = -\hat{r}_y(1)/(\hat{r}_y(0) - |y_n|^2/n) \stackrel{a.s.}{\to} \varphi_1,$$
$$\hat{\varphi}_{fb} = -\hat{r}_y(1)/(\hat{r}_y(0) - \tfrac{1}{2}(|y_1|^2 + |y_n|^2)/n) \stackrel{a.s.}{\to} \varphi_1.$$

This confirms the consistency assertion in Theorem 7.4. Moreover, assume that $\{\epsilon_t\}$ is an AR(1) process satisfying

$$\epsilon_t + \varphi\epsilon_{t-1} = \zeta_t,$$

where $\{\zeta_t\} \sim \mathrm{IID}(0,\sigma^2)$ and $|\varphi| < 1$. Because $\epsilon_t = \sum_{j=0}^\infty \psi_j \zeta_{t-j}$ with $\psi_j := (-\varphi)^j$ for $j \ge 0$, it follows that $r_\epsilon(u) = \sigma^2(-\varphi)^{|u|}/(1-|\varphi|^2)$ and hence $\varphi_1 = \varphi$. If $E(|\zeta_t|^4) < \infty$ and $E(\zeta_t^2) = 0$, then, by Theorem 7.5, the linear prediction estimators of φ are asymptotically distributed as $\mathrm{N}_c(\varphi, (1-|\varphi|^2)/n)$. ◇

Although employed originally by Yule [434] for hidden periodicity analysis, the AR model is advocated by Akaike [6] and Parzen [281] among others as a general-purpose spectral estimator for time series with continuous spectra. This so-called AR approach to spectral estimation is justified by Theorem 7.2. Because the AR coefficients can be estimated easily by the methods discussed before, the remaining question in practice is how to select an appropriate order m for a given time series. There are many proposals in the literature, and a comprehensive survey of these methods is given by [66].

For example, Akaike's information criterion (AIC) [9] takes the form

$$\mathrm{AIC}(m) := 2n\log\hat{\sigma}_m^2 + 2m, \tag{7.1.29}$$

where $\hat{\sigma}_m^2$ denotes the maximum likelihood estimator of the minimum prediction error variance under the Gaussian AR(m) model. Based on this criterion, the best order of the AR model for $\mathbf{y}$ is given by the minimizer of the AIC in (7.1.29) with respect to m. Observe that the first term in (7.1.29) can be made small by choosing a large m, whereas the second term increases with m and serves as a penalize for choosing a large m. Adding these terms together provides a way to balance the goodness of fit and the complexity of the model. Similar to the linear regression analysis discussed in Chapter 5 (Section 5.4), the AIC in (7.1.29) can be viewed as an approximately unbiased estimator (up to an additive constant) of the mean cross entropy $E\{H(\hat{p}_1 \| p_0)\}$, where $p_0(\cdot)$ denotes the true PDF of $\mathbf{y}$ and $\hat{p}_1(\cdot)$ denotes the ML estimator of the PDF $p_1(\cdot)$ that corresponds to the AR(m) model. Minimizing the mean cross entropy is equivalent to minimizing the mean Kullback-Leibler divergence $E\{D_{\mathrm{KL}}(\hat{p}_1 \| p_0)\}$. A derivation of (7.1.29) can be found in Section 7.4. Note that in the real case the multiplier $2n$ in (7.1.29) should be replaced by n, leading to the original formula in [9].

It is well known that the AIC criterion in (7.1.29) has a tendency of overestimating the order of AR processes even for large n [168] [341]. To overcome this difficulty, a heavier penalty term can be used. For example, Schwarz's Bayesian information criterion (BIC) [337] and Rissanen's minimum description length (MDL) criterion [318] employ $m \log n$ instead of $2m$ as the penalty term. The corrected AIC (or AICC) criterion proposed by Hurvich and Tsai [161] takes the form

$$\mathrm{AIC}_c(m) := 2n \log \hat{\sigma}_m^2 + 2n(n+m)/(n-m-1). \tag{7.1.30}$$

This is an alternative estimator (up to an additive constant) of $E\{H(\hat{p}_1 \| p_0)\}$. A derivation of (7.1.30) is given in Section 7.4. In the real case, the penalty term should be $n(n+m)/(n-m-2)$, which is the original formula derived in [161]. Both AIC and AICC are proven asymptotically efficient for modeling AR(∞) processes [46, p. 305] [161] [342].

Another well-known criterion is Parzen's CAT, or criterion for autoregressive transfer function [281] [282], which takes the form

$$\mathrm{CAT}(m) := n^{-1} \sum_{j=1}^{m} 1/\tilde{\sigma}_j^2 - 1/\tilde{\sigma}_m^2. \tag{7.1.31}$$

In this expression, $\tilde{\sigma}_j^2$ ($j = 1, \ldots, m$) denotes an unbiased estimator of the minimum prediction error variance under the AR(j) model. For example, if $\hat{\sigma}_m^2$ is given by (7.1.20), (7.1.21), (7.1.23), or (7.1.24), then take $\tilde{\sigma}_m^2 := n\hat{\sigma}_m^2/(n-m)$. In practice, the CAT criterion often yields similar results as the AIC.

According to Theorem 7.2, an arbitrary SDF can be approximated well by an AR spectrum only if the order of the AR model is allowed to take any value, not just a fixed value. For a data record of length n, the required order may be so large that it becomes comparable to n. To accommodate such cases, the asymptotic analysis must allow m to grow with n. The following theorem, cited from [30] without proof, shows that if m grows at a proper rate then the AR spectral estimator is consistent for estimating the SDF of any real linear process which has an AR(∞) representation.

Theorem 7.6 (Consistency of the AR Spectral Estimator with Growing Order in the Absence of Sinusoids [30]). *Let $\{y_t\}$ be given by (7.1.1). Let $\hat{f}_m(\omega)$ denote the AR spectral estimator of order m obtained by the Yule-Walker method or any of its variations. Assume that $\{\epsilon_t\}$ is a real linear process of the form (7.1.25) with $\psi_j = 0$ for $j \le 0$, $\sum|\psi_j| < \infty$, $\{\zeta_t\} \sim \mathrm{IID}(0,\sigma^2)$, and $E(\zeta_t^4) < \infty$. Assume further that $\varphi(z) := 1/\psi(z) := \sum_{j=0}^{\infty}\varphi_j z^{-j}$ is convergent for $|z| \ge 1$ with $\sum|\varphi_j| < \infty$ so that $\{\epsilon_t\}$ can be expressed as an AR(∞) process of the form $\sum_{j=0}^{\infty}\varphi_j \epsilon_{t-j} = \zeta_t$. If m satisfies*

$$m \to \infty, \quad m^3/n \to 0, \quad m^{1/2}\sum_{j>m}|\varphi_j| \to 0 \quad \textit{as } n \to \infty,$$

then $\hat{f}_m(\omega) \xrightarrow{P} f_\epsilon(\omega)$ for any given $\omega \in \mathbb{R}$.

The next theorem, also cited from [30] without proof, establishes the asymptotic normality of the AR spectral estimator under similar conditions.

Theorem 7.7 (Asymptotic Normality of the AR Spectral Estimator with Growing Order in the Absence of Sinusoids [30]). *Let $\{y_t\}$ be given by (7.1.1) with $\{\epsilon_t\}$ satisfying the conditions of Theorem 7.6. If m satisfies*

$$m \to \infty, \quad m^3/n \to 0, \quad n^{1/2}\sum_{j>m}|\varphi_j| \to 0 \quad \textit{as } n \to \infty,$$

then $(n/m)^{1/2}\{\hat{f}_m(\omega) - f_\epsilon(\omega)\} \xrightarrow{D} \zeta(\omega)$ for any $\omega \in [0,\pi]$, where $\zeta(\omega) \sim \mathrm{N}(0, 2f_\epsilon^2(\omega))$ for $\omega \ne 0$ and π, and $\zeta(\omega) \sim \mathrm{N}(0, 4f_\epsilon^2(\omega))$ for $\omega = 0$ or π. The AR spectral estimates at different frequencies are asymptotically independent.

According to Theorem 7.7, the rate of convergence for the AR spectral estimator is equal to $(n/m)^{-1/2}$ rather than $n^{-1/2}$. The slower rate is a manifestation of the nonparametric nature of the spectral estimation problem. While the condition $m^3/n \to 0$ restricts m from being too large, the conditions $m^{1/2}\sum_{j>m}|\varphi_j| \to 0$ and $n^{1/2}\sum_{j>m}|\varphi_j| \to 0$ require that m be chosen according to the smoothness of the noise spectrum as reflected in the decaying rate of φ_j as $j \to \infty$.

Finally, let us briefly discuss another well-known method for AR spectral estimation called *Burg's estimator* [51] (see also [65]). Recall that the Yule-Walker estimator defined by (7.1.20) can be computed recursively by the Levinson-Durbin algorithm in Proposition 2.8. More specifically, we have $\hat{\boldsymbol{\varphi}}_{\mathrm{YW}} := [\hat{\varphi}_{m1}, \ldots, \hat{\varphi}_{mm}]^T$, where $\hat{\sigma}_0^2 = \hat{r}_y(0)$, $\hat{\varphi}_{11} = -\hat{r}_y(1)/\hat{r}_y(0)$, and for $u = 2, \ldots, m$,

$$\hat{\varphi}_{uu} = -\frac{1}{\sigma_{u-1}^2}\left\{\hat{r}_y(u) + \sum_{j=1}^{u-1}\hat{\varphi}_{u-1,j}\,\hat{r}_y(u-j)\right\}, \tag{7.1.32}$$

$$\hat{\varphi}_{uj} = \hat{\varphi}_{u-1,j} + \hat{\varphi}_{uu}\hat{\varphi}_{u-1,u-j}^* \quad (j = 1, \ldots, u-1), \tag{7.1.33}$$

$$\hat{\sigma}_u^2 = \hat{\sigma}_{u-1}^2(1 - |\hat{\varphi}_{uu}|^2). \tag{7.1.34}$$

Associated with the AR coefficients $\hat{\varphi}_{uj}$ for each fixed u are the forward and backward prediction errors

$$\hat{e}_{f,t}(u) := y_t + \sum_{j=1}^{u} \hat{\varphi}_{uj} y_{t-j}, \quad \hat{e}_{b,t}(u) := y_{t-u} + \sum_{j=1}^{u} \hat{\varphi}_{uj}^* y_{t-u+j}.$$

Owing to (7.1.33), we can write

$$\begin{cases} \hat{e}_{f,t}(u) = \hat{e}_{f,t}(u-1) + \hat{\varphi}_{uu}\, \hat{e}_{b,t-1}(u-1), \\ \hat{e}_{b,t}(u) = \hat{e}_{b,t-1}(u-1) + \hat{\varphi}_{uu}^*\, \hat{e}_{f,t}(u-1). \end{cases} \tag{7.1.35}$$

Instead of (7.1.32) which is based on the sample autocovariances, Burg's method produces the reflection coefficient $\hat{\varphi}_{uu}$ by minimizing the sum of squared forward and backward prediction errors $\sum_{t=u+1}^{n}\{|\hat{e}_{f,t}(u)|^2 + |\hat{e}_{b,t}(u)|^2\}$ with $\{\hat{e}_{f,t}(u)\}$ and $\{\hat{e}_{b,t}(u)\}$ being treated as functions of $\hat{\varphi}_{uu}$ given by (7.1.35). It is easy to verify that the minimizer can be expressed as

$$\hat{\varphi}_{uu} = -\frac{\sum_{t=u+1}^{n} \hat{e}_{f,t}(u-1)\, \hat{e}_{b,t-1}^*(u-1)}{\sum_{t=u+1}^{n}\{|\hat{e}_{f,t}(u-1)|^2 + |\hat{e}_{b,t-1}(u-1)|^2\}/2}. \tag{7.1.36}$$

The entire algorithm comprises (7.1.33)–(7.1.36), where

$$\hat{\sigma}_u^2 := (2n)^{-1} \sum_{t=u+1}^{n} \{|\hat{e}_{f,t}(u)|^2 + |\hat{e}_{b,t}(u)|^2\}.$$

Similar to the reflection coefficient given by the Yule-Walker estimator in (7.1.32), it is not difficult to show (e.g., [369, p. 122]) that the reflection coefficient given by (7.1.36) has the property $|\hat{\varphi}_{uu}| \le 1$ for all $u = 1, \ldots, m$. As compared with the least-squares linear prediction estimators, Burg's estimator has the advantage of computational simplicity. However, it is suboptimal as a linear prediction method because it decouples an m-dimensional optimization problem into m one-dimensional optimization problems [369, p. 123]. Some modifications and extensions of Burg's estimator are discussed in [250].

7.1.4 Numerical Examples

In this section, we demonstrate the spectral estimators using two examples with real-world data. Both are concerned with real-valued time series.

The first example is a time series, shown in Figure 7.1, which comprises 308 yearly sunspot numbers from year 1700 to 2007. The data set is produced by the Royal Observatory of Belgium's Solar Influences Data Analysis Center (SIDC) and known as the international sunspot numbers*.

*R. A. M. Van der Linden and the SIDC team, online catalogue of the sunspot index, http://sidc.oma.be/html/sunspot.html.

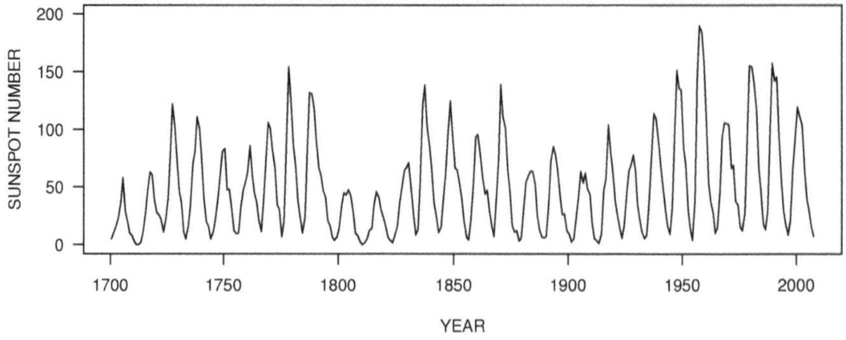

Figure 7.1. Annual sunspot numbers from year 1700 to 2007 produced by the Royal Observatory of Belgium's Solar Influences Data Analysis Center (SIDC).

The sunspot activity has an 11-year cycle. It was first noted by the German astronomer Heinrich Schwabe (1789–1875) in 1843 [336]. Later, the American astronomer George Ellery Hale (1868–1938) and his collaborators discovered a link between sunspots and concentrated solar magnetic activities. The British physicist Arthur Schuster (1851–1934) [334] [335] and the British statistician George Udny Yule (1871–1951) [434] were among the first to use the periodogram and the autoregression techniques, respectively, to investigate the periodicity of the sunspot numbers. Research in this area remains active to this day, especially in the context of climate change [20] [106] [117] [144] [312] [345].

In this example, we simply follow the footsteps of Schuster and Yule. What we have now that they didn't have then are the luxury of modern computing technology, the smoothing and order selection techniques, and an updated time series of the sunspot numbers to year 2007.

Figure 7.2 depicts the periodogram and two spectral estimates for the annual sunspot numbers shown in Figure 7.1 (with the mean subtracted). The Yule-Walker estimate is obtained from an AR(9) model, where the order is selected by the AIC in (7.1.29) (the CAT in (7.1.31) gives the same order). The smoothed periodogram is obtained from the periodogram smoother in (7.1.2), where the weights W_{ml} are proportional to the truncated *Parzen spectral window* [280, p. 226] [298, p. 444], i.e.,

$$W_{ml} \propto \left\{ \frac{\sin(\frac{1}{2}\lfloor n/m \rfloor \lambda_l)}{\sin(\frac{1}{2}\lambda_l)} \right\}^4 \qquad (|l| \le m), \tag{7.1.37}$$

where $\lambda_l = 2\pi l/n$ and $n = 308$. We take $m = 7$ in this example.

As discussed in Section 7.1.1, the bandwidth parameter m can be chosen in many ways. For example, the GCV criterion in (7.1.4) yields $m = 12$, which gives a slightly smoother curve than that shown in Figure 7.2. To facilitate a fair com-

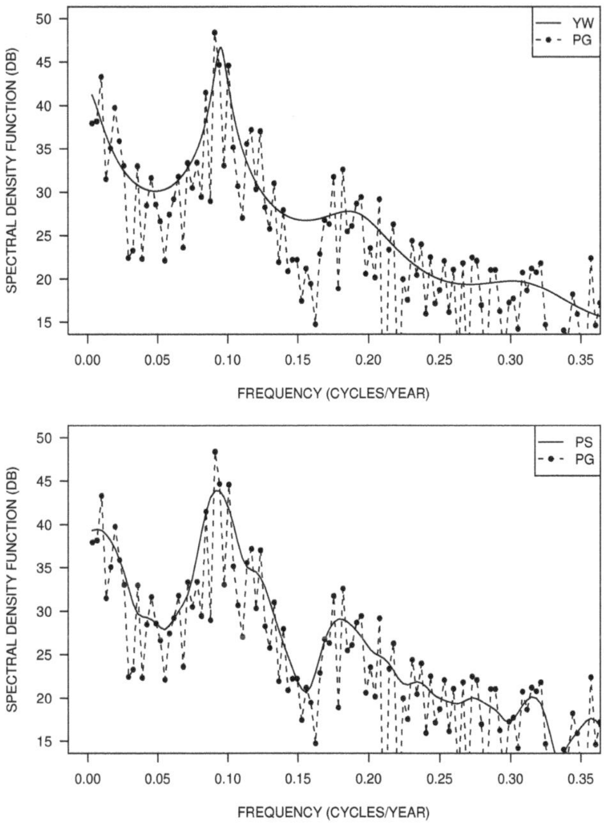

Figure 7.2. The periodogram (dashed line with dots) and the estimated spectrum (solid line) of the sunspot numbers. Top, the Yule-Walker spectral estimator; bottom, the periodogram smoother. Unit of axes: horizontal, in cycles per year (normalized frequency); vertical, in decibels (10 times the common logarithm of the spectrum).

parison with the AR spectral estimator, we choose $m = 7$ so that the smoothed periodogram has approximately the same roughness as the Yule-Walker spectral estimate given by the AIC. The roughness is measured by the ℓ_2 norm of the twice-differenced sequence of log spectral ordinates evaluated at the Fourier frequencies. This quantity is a substitute for the L_2 norm of the second derivative of the log spectrum, a well-known metric for the roughness of functions.

As can be seen from Figure 7.2, the highest peak of the AR spectrum gives an estimated sunspot cycle of 10.62 years (restricted on the Fourier grid). The same estimate is produced by the highest peak of the smoothed periodogram (also restricted on the Fourier grid). The raw periodogram is peaked exactly at the 11-year frequency. This confirms once again the historical finding of an approximately 11-year cycle for the sunspot numbers.

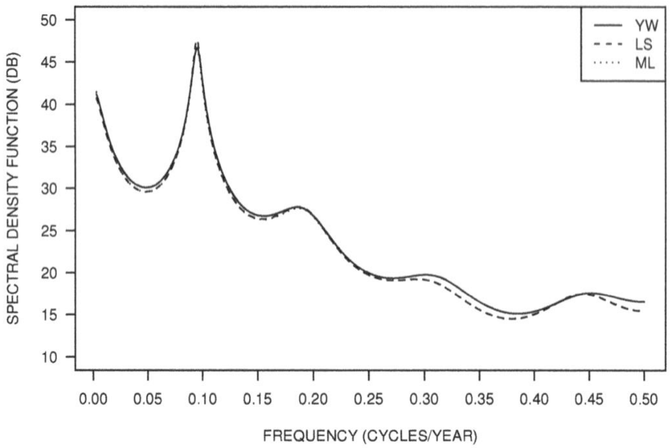

Figure 7.3. Three AR spectral estimates for the sunspot numbers. Solid line, Gaussian maximum likelihood; dashed line, Yule-Walker; dotted line, least-squares.

Figure 7.2 shows that the AR model is very effective in representing sharp spectral peaks. In fact, a good representation only requires the roots of the AR polynomial $\varphi_m(z)$, also known as the poles of the AR filter $1/\varphi_m(z)$, to be placed near the unit circle. The periodogram smoother tends to smear sharp spectral peaks or any other ragged spectral features that are in conflict with the implicit assumption of smoothness. On the other hand, if the spectrum has smooth but deep valleys, as shown in Figure 7.2 for frequencies near 0.15, the periodogram smoother can be more effective than the AR spectral estimator.

Figure 7.3 compares the Yule-Walker AR spectral estimate with the forward least-squares linear prediction estimate given by (7.1.21) and the true Gaussian maximum likelihood estimate obtained by using the R function `ar`. As we can see, the least-squares estimate is almost identical to the Gaussian maximum likelihood estimate, except in a tiny vicinity of the largest peak, where the former is slightly higher. This is not too surprising because the least-squares estimator has long been considered as a substitute for the Gaussian maximum likelihood estimator. The Yule-Walker estimate differs slightly from the others in the valleys and high frequency regions where the spectral values are small. All three methods yield the same 10.50-year sunspot cycle which is calculated numerically without the restriction on the Fourier grid. Finally, we note that Burg's estimate (not shown) is almost identical to the least-squares estimate except that the highest peak is slightly lower.

In the second example, let us consider the residuals of the variable star data discussed in Chapter 1 (Section 1.3) after the removal of the sinusoidal components found in Chapter 6 (Section 6.4.2). The residual time series exhibits con-

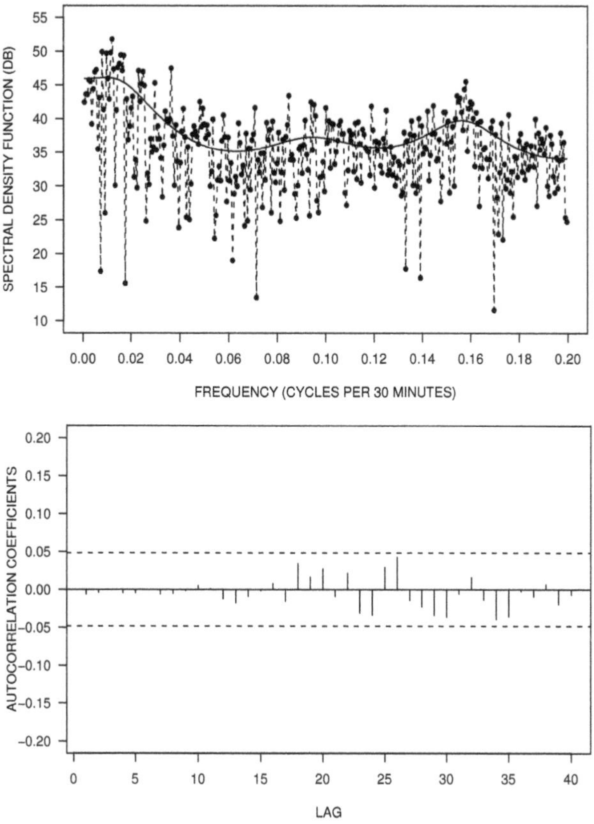

Figure 7.4. Top, the periodogram (dashed line with dots) and the estimated AR spectrum (solid line) of the residuals from the variable star data shown in Figure 1.1 after the removal of eight sinusoidal components. Bottom, the autocorrelation coefficients of the residuals from the AR model with the bounds $\pm 1.96/\sqrt{n}$.

siderable serial dependence, as can be seen from the bottom panel of Figure 1.1. In this exercise, we fit the residuals by an AR model with the parameters estimated by the Yule-Walker estimator. The order of the AR model, which is equal to 16, is determined by the AIC criterion. The top panel of Figure 7.4 depicts the spectrum of the estimated AR model together with the periodogram. As we can see, the AR spectrum has a broad peak in the low-frequency region.

As a simple check for the adequacy of the AR model, the autocorrelation coefficients of the residuals from the AR model are shown in the bottom panel of Figure 7.4 together with the bounds $\pm 1.96/\sqrt{n}$. It is well known that an autocorrelation coefficient from white noise should lie between these bounds with approximately 0.95 probability [46, p. 223]. Because none of the autocorrelation coefficients shown in Figure 7.4 falls outside the bounds, it gives us no grounds on which to reject the hypothesis that the residuals from the AR model no longer have serial dependence. Further evidence in favor of this hypothesis is provided by the Kolmogorov-Smirnov test discussed in Chapter 6 (Section 6.1), which, yields a p-value 0.9999. So the AR model seems adequate.

7.2 Estimation in the Presence of Sinusoids

In this section, let us turn to the situation where the noise spectrum has to be estimated in the presence of sinusoids. To be more specific, consider the complex case for which the data vector $\mathbf{y} := [y_1, \ldots, y_n]^T$ can be expressed as

$$\mathbf{y} = \sum_{k=1}^{p} \beta_k \mathbf{f}(\omega_k) + \boldsymbol{\epsilon}, \tag{7.2.1}$$

where $\beta_k := C_k \exp(i\phi_k)$, $\omega_k \in (0, 2\pi)$, $\mathbf{f}(\omega) := [\exp(i\omega), \ldots, \exp(in\omega)]^T$, and $\{\epsilon_t\}$ is a zero-mean stationary complex random process with SDF $f_\epsilon(\omega)$.

Because the parameters of the sinusoids are unknown, the problem is quite challenging in general. It is particularly difficult when the noise itself contains spectral peaks that are indistinguishable from the spectral peaks generated by the sinusoids based on a finite-length data record. Therefore, successful estimation is possible only if the noise spectrum satisfies certain conditions so that its peaks are much broader than those of the sinusoids for the sample size.

In the special case where the noise is known to be white, the problem of spectral estimation reduces to that of estimating the noise variance, for which some simple and robust methods are available as discussed in Chapter 6 (Section 6.2). If the white noise determination cannot be made *a priori*, then it is still necessary to estimate the entire spectrum as if the noise were colored. The white noise hypothesis can be tested on the basis of the estimated spectrum by using, for example, the Kolmogorov-Smirnov test discussed in Section 6.1.

A joint estimation method was suggested by Whittle [422] and further analyzed by Walker [413]. In this method, the noise spectrum is assumed to take a parametric form, and the noise parameters are estimated jointly with the sinusoidal parameters by maximizing the Gaussian likelihood. The semi-parametric joint estimation method discussed in [197] and [198] is based on the spline regression of the log periodogram combined with stepwise addition and deletion of certain basis functions assisted by a model selection criterion.

While the joint estimation approach will be discussed later in Chapter 8, this section focuses only on a few alternative methods that do not require the estimation of sinusoidal parameters.

7.2.1 Modified Periodogram Smoother

Consider the periodogram smoother in (7.1.2). What happens to this estimator when there is a sinusoidal component in the data record? This question is answered here through a simple example where **y** satisfies (7.2.1) with $p = 1$ and $\omega_1 = \lambda_j$ for some j. By Theorem 6.6, together with (7.1.3), (7.1.5), and the uniform continuity of the noise spectrum, it is easy to show that

$$\begin{aligned} E\{\hat{f}_\epsilon(\lambda_j)\} &= nW_{m0}C_1^2 + \sum_{|l|\le m} W_{ml} f_\epsilon(\lambda_{j-l}) + \mathcal{O}(r_n) \\ &= nW_{m0}C_1^2 + f_\epsilon(\lambda_j) + o(1). \end{aligned} \tag{7.2.2}$$

As we can see, the presence of a sinusoid at frequency λ_j destroys the asymptotic unbiasedness of $\hat{f}_\epsilon(\lambda_j)$ established in Theorem 7.1. The bias, $nW_{m0}C_1^2$, is positive and can be very large indeed. For example, with the uniform weights, $W_{ml} = (2m+1)^{-1}$ $(|l| \le m)$, the bias takes the form $n(2m+1)^{-1}C_1^2$ and approaches infinity as the sample size grows under the condition (7.1.5).

To overcome this problem, Hannan [134] proposed the idea of excluding the periodogram ordinate $I_n(\lambda_j)$ when estimating the noise spectrum at frequency λ_j, whether or not a sinusoid is present at λ_j. Applying this technique to the periodogram smoother in (7.1.2) yields

$$\tilde{f}_\epsilon(\lambda_j) := \frac{\hat{f}_\epsilon(\lambda_j) - W_{m0}\, I_n(\lambda_j)}{1 - W_{m0}}. \tag{7.2.3}$$

A similar technique is used in Section 7.1.1 to deal with the complication at frequency 0 when the sample mean is removed from the data. With $\hat{f}_\epsilon(\lambda_j)$ given by (7.1.2), we can rewrite (7.2.3) as

$$\tilde{f}_\epsilon(\lambda_j) = \sum_{|l|\le m} \tilde{W}_{ml}\, I_n(\lambda_{j-l}), \tag{7.2.4}$$

where

$$\tilde{W}_{ml} := \frac{W_{ml} - W_{m0}\delta_l}{1 - W_{m0}}. \tag{7.2.5}$$

Because $\tilde{W}_{ml} \ge 0$ and $\sum \tilde{W}_{ml} = 1$, the estimator in (7.2.4) remains to be a periodogram smoother, which will be called the *modified* periodogram smoother.

Observe that $s_m \to 0$ implies $W_{m0} \to 0$ and hence

$$\tilde{s}_m := \sum_{|l|\le m} \tilde{W}_{ml}^2 = (s_m - W_{m0}^2)/(1 - W_{m0})^2 \to 0.$$

In other words, the modified weights $\tilde{W}_{ml}$ satisfy the condition (7.1.5). Therefore, the assertions in Theorem 7.1 remain valid for the modified periodogram

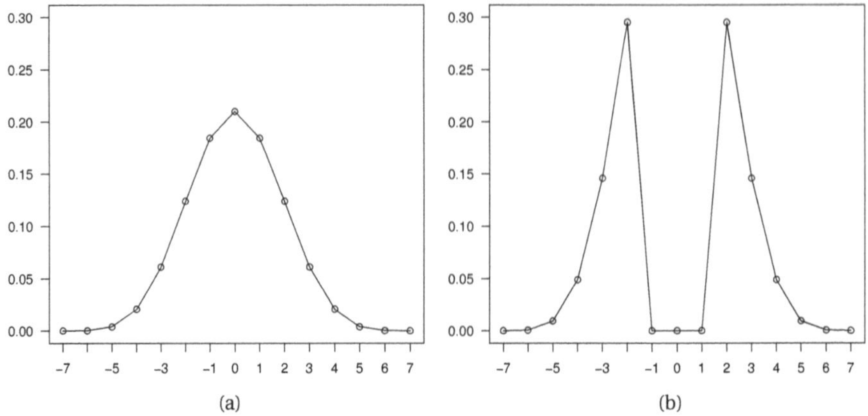

Figure 7.5. (a) Parzen spectral window with $n = 100$ and $m = 7$. (b) Modified Parzen spectral window with $m' = 1$.

smoother in the absence of sinusoids. More importantly, because $\tilde{W}_{m0} = 0$, the presence of a sinusoid at frequency λ_j no longer introduces an asymptotically nonnegligible bias in the modified periodogram smoother at frequency λ_j.

When the sinusoid appears not at λ_j but at a different frequency in the m-neighborhood (more precisely, a neighborhood of radius $2\pi m/n$) of λ_j, then the bias persists. For example, suppose $\omega_1 = \lambda_{j-l}$ for some l such that $0 < |l| \le m$. Then, under the conditions of Theorem 7.1, we can write

$$E\{\tilde{f}_\epsilon(\lambda_j)\} = n\tilde{W}_{ml}C_1^2 + f_\epsilon(\lambda_j) + \mathcal{O}(1). \tag{7.2.6}$$

This expression is the same as (7.2.2) except that the bias becomes $n\tilde{W}_{ml}C_1^2$. Clearly, excluding $I_n(\lambda_j)$ is not enough to achieve asymptotic unbiasedness at λ_j when the sinusoid appears at a frequency other than λ_j.

To alleviate the problem, one may consider excluding several, not just one, periodogram ordinates in the neighborhood of λ_j when estimating $f_\epsilon(\lambda_j)$. This leads to a periodogram smoother of the same form as (7.2.4) but with

$$\tilde{W}_{ml} := \frac{W_{ml} - W_{ml}I(|l| \le m')}{1 - \sum_{|l|\le m'} W_{ml}}, \tag{7.2.7}$$

where m' is a tuning parameter satisfying $0 \le m' < m$ and $m - m' \to \infty$. The estimator in (7.2.3) corresponds to the choice of $m' = 0$.

As an example, Figure 7.5 depicts the weights W_{ml} generated by the Parzen spectral window in (7.1.37) with $n = 100$ and $m = 7$ and the corresponding modified weights $\tilde{W}_{ml}$ defined by (7.2.7) with $m' = 1$.

Because $\tilde{W}_{ml} = 0$ for all $|l| \le m'$, the resulting estimator $\tilde{f}_\epsilon(\lambda_j)$ has zero asymptotic bias even if a sinusoid appears somewhere in the m'-neighborhood of λ_j.

The generalization of (7.2.4) by using the weights in (7.2.7) is essentially a simplified version of the *double window technique* introduced by Priestley [297]; see also [298, pp. 623–624 and 649–652].

The following corollary summarizes the asymptotic properties of the modified periodogram smoother. A proof is given in Section 7.4.

Corollary 7.1 (Asymptotics of the Modified Periodogram Smoother in the Presence of Sinusoids). *Let* $\mathbf{y}$ *satisfy (7.2.1) and let* $\tilde{f}_\epsilon(\lambda_j)$ *be defined by (7.2.4) with* $\{\tilde{W}_{ml}\}$ *given by (7.2.7), where* m' *satisfies* $0 \le m' < m$ *and* $m - m' \to \infty$ *as* $n \to \infty$. *Let* Ω'_n *denote a subset of* Ω_n *such that the* m'*-neighborhood of each frequency in* Ω'_n *contains* $\{\omega_k\}$. *Then, the assertions of Theorem 7.1 remain valid for* $\tilde{f}_\epsilon(\lambda_j)$ *under the constraint that* $\lambda_j \in \Omega'_n$, *i.e.,* Ω_n *is everywhere replaced by* Ω'_n, *and with* W_{ml} *and* s_m *everywhere replaced by* $\tilde{W}_{ml}$ *and* $\tilde{s}_m$.

Corollary 7.1 only guarantees the consistency and asymptotic unbiasedness in the subset Ω'_n rather than the entire Fourier grid Ω_n. Therefore, these results should be regarded as *local consistency* and *local unbiasedness*. Consider a simple example of $p = 2$ with $\omega_1 = \lambda_j$ and $\omega_2 = \lambda_{j+1}$ for some j. If $m' = 1$, then $\Omega'_n = \{\lambda_j, \lambda_{j+1}\}$. According to Corollary 7.1, $\tilde{f}_\epsilon(\lambda_j) - f_\epsilon(\lambda_j) \xrightarrow{P} 0$ and $\tilde{f}_\epsilon(\lambda_{j+1}) - f_\epsilon(\lambda_{j+1}) \xrightarrow{P} 0$ as $n \to \infty$. But, the consistency does not apply to $\tilde{f}_\epsilon(\lambda_{j-1})$ because ω_2 is outside the m'-neighborhood of λ_{j-1} and hence its impact on $\tilde{f}_\epsilon(\lambda_{j-1})$ remains.

The condition that the m'-neighborhood contains all the signal frequencies can be satisfied by making m' sufficiently large. However, the requirement that $m - m' \to \infty$, together with (7.1.5), serves as a constraint that prohibits m' being too large relative to the window size m and the sample size n. Due to this constraint, the assertion in Corollary 7.1 cannot be extended to the entire Ω_n even in the simplest case of a single sinusoid. This should be considered as a limitation of the modified periodogram smoother if the goal is to have a consistent estimator across the entire range of frequencies. It should not be of great concern if one knows *a priori* where the signal frequencies are located and is only interested in the noise spectrum near these frequencies (the ordinary periodogram smoother can be used outside this neighborhood, as suggested in [298, p. 650]).

Finally, we point out that with $\hat{f}_\epsilon(\lambda_j)$ given by (7.1.8), equation (7.2.3) defines a modified lag-window spectral estimator. This is the estimator employed in [64] and [134] for the estimation of noise spectrum in the presence of sinusoids.

7.2.2 M Spectral Estimator

Instead of removing some periodogram ordinates regardless of their magnitude, an alternative approach of spectral estimation treats the sinusoids as outliers in the periodogram and employs robust regression techniques to curtail their influence. This approach is exemplified by the *M spectral estimator* discussed by von Sachs in [326] and [327].

At any given frequency $\lambda_j \in \Omega_n$, the M spectral estimator of $f_\epsilon(\lambda_j)$, denoted by $\tilde{f}_\epsilon(\lambda_j)$, is defined as the solution to the equation

$$\Psi_j(\tilde{f}_\epsilon(\lambda_j)) = 0, \tag{7.2.8}$$

where $\Psi_j(x)$ is a function of $x > 0$ defined as

$$\Psi_j(x) := \sum_{|l| \le m} W_{ml}\, \psi(I_n(\lambda_{j-l})/x),$$

and $\psi(x)$ is a function of $x \ge 0$ that can be chosen to restrict the influence of outliers. It is easy to see that, with the choice of $\psi(x) = x - 1$, equation (7.2.8) produces the periodogram smoother in (7.1.2).

To achieve the robustness such that the influence of outliers is bounded regardless of their magnitude, the ψ function must be bounded. The unboundedness of $\psi(x) = x - 1$ explains the lack of such robustness in the ordinary periodogram smoother. To ensure consistency, the ψ function must also satisfy

$$E\{\psi(X)\} = \int_0^\infty \psi(x) \exp(-x)\, dx = 0, \tag{7.2.9}$$

where $X \sim \frac{1}{2}\chi^2(2)$, which is the asymptotic distribution of $I_n(\lambda_j)/f_\epsilon(\lambda_j)$ in the absence of sinusoids under the conditions of Theorem 6.4.

One way of constructing the required ψ function is through a modification of Huber's ψ functions [159] such that

$$\psi(x) := \begin{cases} \psi_0(x-1) & \text{if } x > 1, \\ \kappa\,\psi_0(x-1) & \text{if } x \le 1, \end{cases} \tag{7.2.10}$$

where $\psi_0(\cdot)$ can be any of Huber's ψ functions and κ is a constant given by

$$\kappa := -\frac{\displaystyle\int_0^\infty \psi_0(x) \exp(-x)\, dx}{\displaystyle\int_{-1}^0 \psi_0(x) \exp(-x)\, dx}. \tag{7.2.11}$$

The role of κ is to handle the asymmetric distribution of the periodogram and ensure that the condition (7.2.9) is satisfied.

The following example shows a simple choice of the ψ function.

Example 7.3 (A Family of Monotone ψ Functions for the M Spectral Estimator). Let $\psi(x)$ be defined by (7.2.10) with

$$\psi_0(x) := \begin{cases} x & \text{if } |x| < c, \\ c\,\mathrm{sgn}(x) & \text{otherwise,} \end{cases} \tag{7.2.12}$$

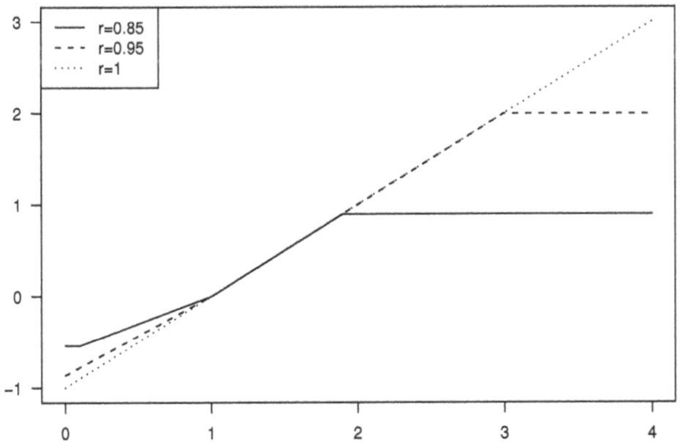

Figure 7.6. Plot of $\psi(x)$ defined by (7.2.10) and (7.2.12), where $c = -\log(1-r) - 1$. Solid line, $r = 0.85$; dashed line, $c = 0.95$; dotted line, $r = 1$ (nonrobust).

where $c > 0$ is a cutoff parameter. With this choice, equation (7.2.11) yields

$$\kappa = \begin{cases} 1 - \exp(-c) & \text{if } c \geq 1, \\ \dfrac{1 - \exp(-c)}{1 + c\exp(1) - \exp(c)} & \text{if } 0 < c < 1. \end{cases}$$

As can be seen, the function $\psi(x)$ imposes a cap c on large values that exceed $1 + c$, so that outliers in the periodogram have only a bounded impact on the resulting spectral estimator. The nonrobust choice of $\psi(x) = x - 1$ corresponds to the limiting case as $c \to \infty$. It is convenient to take $c := -\log(1-r) - 1$ for some constant $r \in (0,1)$, where $1-r$ is approximately equal to the probability that $I_n(\lambda_j)/f_\epsilon(\lambda_j)$ is capped by the mapping $x \mapsto \psi(x)$ in the absence of sinusoids.

Figure 7.6 depicts $\psi(x)$ for different values of r. As we can see, the ψ functions in the uncapped region are close to the nonrobust choice, thus enabling the M spectral estimator to retain a sufficient portion of the statistical efficiency of the periodogram smoother (7.1.2) in the absence of sinusoids. As r decreases, the capped region grows and the uncapped region shrinks, leading to an increased robustness in the presence of sinusoids but decreased efficiency in the absence of sinusoids. Therefore, in practice, one has to balance robustness and efficiency when choosing the value of r. Such a tradeoff is typical in robust estimation problems. Note that the upward bending for $x \leq 1$ results from the mechanism that ensures the condition in (7.2.9). ◇

To obtain the M spectral estimates, the nonlinear equation (7.2.8) must be solved for each frequency of interest. This can be burdensome if the number of these frequencies is large. Nonetheless, one can use standard procedures in numerical analysis to solve (7.2.8) for each given frequency.

For example, the *method of false position* [365, p. 339] produces a sequence $\{u_k\}$ that converges linearly to the solution of (7.2.8) as $k \to \infty$, where the u_k are computed recursively as follows: Given the initial values $0 < u_0 < v_0 < \infty$ such that $\Psi_j(u_0)\Psi_j(v_0) < 0$ and for $k = 0, 1, 2, \ldots$, let

$$u_{k+1} := \frac{v_k \Psi_j(u_k) - u_k \Psi_j(v_k)}{\Psi_j(u_k) - \Psi_j(v_k)}, \tag{7.2.13}$$

and, if $\Psi_j(u_{k+1}) \neq 0$, let

$$v_{k+1} := \begin{cases} v_k & \text{if } \Psi_j(u_{k+1})\Psi_j(u_k) > 0, \\ u_k & \text{if } \Psi_j(u_{k+1})\Psi_j(u_k) < 0. \end{cases}$$

A simple choice for the initial values would be $u_0 := \min\{I_n(\lambda_j)\} - \kappa_1$ and $v_0 := \max\{I_n(\lambda_j)\} + \kappa_2$ for some positive constants κ_1 and κ_2. This, combined with the choice of a monotone nondecreasing ψ function as shown in Figure 7.6, is sufficient to ensure the convergence of (7.2.13) to a unique solution of (7.2.8).

Alternatively, one may replace v_k in (7.2.13) by u_{k-1} to obtain the so-called *secant method* [365, pp. 341]. The secant method has a superlinear rate of convergence if the initial values u_{-1} and u_0 are in a sufficiently small neighborhood of the solution of (7.2.8).

Asymptotic properties of the M spectral estimator in (7.2.8) are studied in [326] and [327] for the real case under the assumption that $\{\epsilon_t\}$ is a linear process (see also [386, Section 6.4.2]). In the following theorem, we consider the complex case in (7.2.1) with Gaussian noise and assume that $\psi(x)$ is a monotone function. The proof of this result can be found in Section 7.4.

Theorem 7.8 (Consistency of the M Spectral Estimator). *Let* $\mathbf{y}$ *be given by (7.2.1) with* $\{\epsilon_t\}$ *being complex Gaussian white noise or complex Gaussian colored noise that satisfies the conditions of Theorem 6.4. Let* $\tilde{f}_\epsilon(\lambda_j)$ *be defined by (7.2.8) with* $\psi(x)$ *being a bounded nondecreasing function which is strictly increasing in the vicinity of* $x = 1$*, Lipschitz continuous (i.e., there is a constant* $c_0 > 0$ *such that* $|\psi(x_1) - \psi(x_2)| \le c_0 |x_1 - x_2|$ *for all* x_1 *and* x_2*), and satisfies (7.2.9). Assume further that* $\{W_{ml}\}$ *satisfies (7.1.3) and (7.1.5). Then, for any given* $\delta > 0$ *and as* $n \to \infty$,

$$\max_{\lambda_j \in \Omega_n} P(|\tilde{f}_\epsilon(\lambda_j) - f_\epsilon(\lambda_j)| > \delta) \to 0$$

regardless of the presence or absence of the sinusoids in (7.2.1).

Remark 7.7 This result can be generalized to the situation where the number of contaminated periodogram ordinates takes the form $\mathcal{O}(s_m^{-1/2})$.

Theorem 7.8 shows that the M spectral estimator has the desired global consistency property irrespective of the sinusoids. This property gives the M spectral estimator a competitive edge over the modified periodogram smoother in (7.2.4) for estimation of the entire noise spectrum. The cost is, of course, the higher computational complexity incurred in solving the nonlinear equation (7.2.8).

7.2.3 Modified Autoregressive Spectral Estimator

Now let us investigate the impact of sinusoids on the AR spectral estimator. Without loss of generality, let us consider the least-squares estimator defined in (7.1.24). In the presence of sinusoids, Theorem 4.1 gives

$$n^{-1}\mathbf{Y}^H\mathbf{Y} \xrightarrow{P} \mathbf{R}_x + \mathbf{R}_\epsilon, \quad n^{-1}\mathbf{Y}^H\mathbf{y} \xrightarrow{P} \mathbf{r}_x + \mathbf{r}_\epsilon.$$

This implies that

$$\hat{\boldsymbol{\varphi}} \xrightarrow{P} \tilde{\boldsymbol{\varphi}} := -(\mathbf{R}_x + \mathbf{R}_\epsilon)^{-1}(\mathbf{r}_x + \mathbf{r}_\epsilon).$$

As we can see, the presence of sinusoids destroys the consistency of the AR estimator, just as it does the periodogram smoother. The discrepancy, called the asymptotic bias, can be expressed as

$$\mathbf{b} := \tilde{\boldsymbol{\varphi}} - \boldsymbol{\varphi} = -(\mathbf{R}_x + \mathbf{R}_\epsilon)^{-1}(\mathbf{R}_x\boldsymbol{\varphi} + \mathbf{r}_x).$$

An evaluation of this quantity is given in the following example.

Example 7.4 (Sinusoid-Induced Asymptotic Bias in AR(1) Model). Consider the simple case where $p = 1$ and $m = 1$ with $\{\epsilon_t\}$ being the complex AR(1) process discussed in Examples 7.1 and 7.2. In this case, $r_x(u) = C_1^2\exp(i\omega_1 u)$, $r_\epsilon(u) = \sigma^2(-\varphi)^u/(1-|\varphi|^2)$, and $\varphi = -r_\epsilon(1)/r_\epsilon(0)$. Let $\hat{\varphi}$ denote any of the estimators discussed in Example 7.1. Then, it follows from Theorem 4.1 that

$$\hat{\varphi} \xrightarrow{a.s.} \tilde{\varphi} := -\frac{r_y(1)}{r_y(0)} = -\frac{r_x(1)+r_\epsilon(1)}{r_x(0)+r_\epsilon(0)} = -\frac{\gamma\exp(i\omega_1)-\varphi}{1+\gamma},$$

where $\gamma := r_x(0)/r_\epsilon(0) = C_1^2(1-|\varphi|^2)/\sigma^2$ is the SNR. Therefore, the asymptotic bias in the estimator $\hat{\varphi}$ is given by

$$b := \tilde{\varphi} - \varphi = -\frac{r_x(0)\varphi + r_x(1)}{r_x(0)+r_\epsilon(0)} = -\frac{\gamma}{1+\gamma}(\varphi + \exp(i\omega_1)).$$

As we can see, the bias depends not only on the SNR γ and the noise parameter φ but also on the frequency ω_1 of the contaminating sinusoid. Moreover, let φ take the form $\varphi = r_0\exp(i\omega_0)$ for some $r_0 \in (0,1)$ and $\omega_0 \in (0,2\pi)$ (the noise spectrum has a peak around ω_0 when r_0 is close to 1). Then, the absolute value of the bias (or absolute bias, in short) can be expressed as

$$|b| = \frac{\gamma}{1+\gamma}\sqrt{1+2r_0\cos(\omega_1-\omega_0)+r_0^2}. \tag{7.2.14}$$

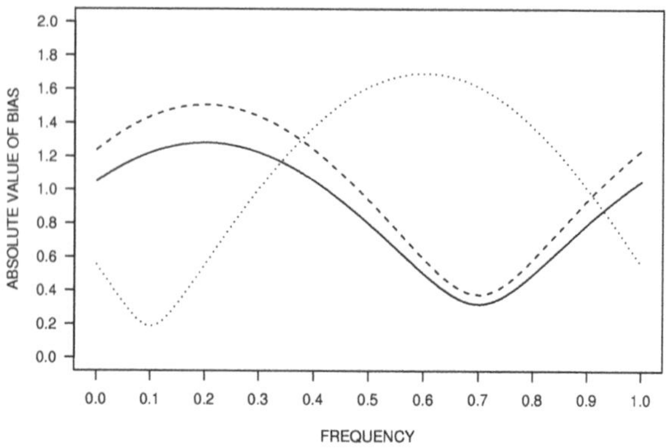

Figure 7.7. Absolute value of the asymptotic bias given by (7.2.14), as a function of $f_1 := \omega_1/(2\pi)$, for estimating $\varphi := r_0 \exp(i\omega_0)$ in the AR(1) model of the noise in the presence of a complex sinusoid with frequency ω_1 and SNR γ. Solid line, $r_0 = 0.6$, $\omega_0 = 2\pi \times 0.2$, $\gamma = 4$; dashed line, $r_0 = 0.6$, $\omega_0 = 2\pi \times 0.2$, $\gamma = 16$; dotted line, $r_0 = 0.8$, $\omega_0 = 2\pi \times 0.6$, $\gamma = 16$.

Clearly, the absolute bias takes its maximum value $(1+r_0)\gamma/(1+\gamma)$ when $\omega_1 = \omega_0$ (i.e., when the signal frequency coincides with the spectral peak of the noise); it takes the minimum value $(1-r_0)\gamma/(1+\gamma)$ when $\omega_1 - \omega_0 = \pi \pmod{2\pi}$. A complete picture of $|b|$ as a function of ω_1 is shown in Figure 7.7 for various values of φ and γ. Note that the absolute bias is always greater than zero but less than $1 + r_0$ regardless of the SNR. Therefore, the sinusoid always has an impact on the AR parameter estimation, but the impact is always bounded. ◇

To overcome the bias problem, a simple method was proposed by Chui in [71]. It is based on the idea of robustifying the sample autocovariances in the Yule-Walker equations. The method only applies to the Yule-Walker spectral estimator in which the AR parameter is estimated from the sample autocovariances by solving the Yule-Walker equations.

It is easy to see that the sample autocovariance function defined by (4.1.1) has a frequency-domain representation as a direct result of (6.4.2). That is,

$$\hat{r}_y(u) = \hat{r}_y^*(-u) = n^{-1} \sum_{j=0}^{n-1} I_n(\lambda_j) \exp(i\lambda_j u) \qquad (u = 0, 1, \ldots, n-1). \quad (7.2.15)$$

Clearly, the sinusoids have unbounded influence on the sample autocovariances through the outliers they produce in the periodogram.

Now suppose that a preliminary AR estimate of the noise spectrum, denoted by $f_{\text{AR}}(\omega, \boldsymbol{\eta})$, is available and suppose that it does not "overfit" the data (i.e., the order of the AR model is not too high as compared to the sample size). Then, there is a good chance that the largest values in the standardized periodogram

$I_n(\lambda_j)/f_{\text{AR}}(\lambda_j, \boldsymbol{\eta})$ are produced solely by the sinusoids. Curtailing the influence of these values is expected to improve the robustness against sinusoids.

This can be achieved by forming a robustified periodogram

$$I_n(\lambda_j, \boldsymbol{\eta}) := \psi(I_n(\lambda_j)/f_{\text{AR}}(\lambda_j, \boldsymbol{\eta}))\, f_{\text{AR}}(\lambda_j, \boldsymbol{\eta}), \tag{7.2.16}$$

where $\psi(x)$ is a bounded nonnegative function. The original periodogram corresponds to $\psi(x) = x$ which is unbounded. Substituting $I_n(\lambda_j)$ with $I_n(\lambda_j, \boldsymbol{\eta})$ in (7.2.15) yields a robustified sample autocovariance function

$$\hat{r}_y(u, \boldsymbol{\eta}) := \hat{r}_y^*(-u, \boldsymbol{\eta}) := n^{-1} \sum_{j=0}^{n-1} I_n(\lambda_j, \boldsymbol{\eta}) \exp(i\lambda_j u), \tag{7.2.17}$$

where $0 \le u < n$. Because large values in the periodogram are curtailed, the Yule-Walker estimator obtained from the $\hat{r}_y(u, \boldsymbol{\eta})$ is expected to be more robust than that obtained from the original sample autocovariances $\hat{r}_y(u)$.

This procedure must be carried out iteratively because the preliminary estimate is usually crude and some better estimates are expected as the iteration proceeds. The iterative algorithm can be stated formally as follows: Given an initial guess $f_{\text{AR}}(\omega, \hat{\boldsymbol{\eta}}_0)$ and for $k = 1, 2, \ldots$,

1. Use $f_{\text{AR}}(\omega, \hat{\boldsymbol{\eta}}_{k-1})$ to obtain $\{I_n(\lambda_j, \hat{\boldsymbol{\eta}}_{k-1})\}$ from (7.2.16);
2. Use $\{I_n(\lambda_j, \hat{\boldsymbol{\eta}}_{k-1})\}$ to obtain $\{\hat{r}_y(u, \hat{\boldsymbol{\eta}}_{k-1})\}$ from (7.2.17);
3. Use $\{\hat{r}_y(u, \hat{\boldsymbol{\eta}}_{k-1})\}$ and the Levinson-Durbin algorithm in Proposition 2.8 to produce a new AR spectrum $f_{\text{AR}}(\omega, \hat{\boldsymbol{\eta}}_k)$.

Symbolically, this *iterative Yule-Walker* (IYW) procedure can be expressed as

$$f_{\text{AR}}(\cdot, \hat{\boldsymbol{\eta}}_k) := \Phi_{\text{IYW}}(f_{\text{AR}}(\cdot, \hat{\boldsymbol{\eta}}_{k-1}), I_n(\cdot), \psi(\cdot)). \tag{7.2.18}$$

Although there is no theoretical analysis that guarantees the convergence of the IYW procedure, the analytical results reported in [71] ensure that the iteration is nondivergent in the sense that the estimation error remains to be of the form $\mathcal{O}_P(n^{-1/2})$ for all k if the initial error takes that form. For further statistical analysis of this procedure, see [386, Section 6.4.2].

The ψ function in (7.2.16) plays the same role as the ψ function in (7.2.8). The counterpart of (7.2.9) is the condition

$$E\{\psi(X)\} = \int_0^\infty \psi(x) \exp(-x)\, dx = 1, \tag{7.2.19}$$

where $X \sim \frac{1}{2}\chi^2(2)$. It is needed to make $\psi(I_n(\lambda_j)/f_\epsilon(\lambda_j)) f_\epsilon(\lambda_j)$ asymptotically unbiased in the absence of sinusoids.

A possible choice of the ψ function in (7.2.16) takes the form

$$\psi(x) := \psi_0(x)/\rho \quad (x \ge 0), \tag{7.2.20}$$

where $\psi_0(x)$ can be any of Huber's ψ functions [159] and

$$\rho := \int_0^\infty \psi_0(x) \exp(-x)\, dx$$

is a normalizing constant that enforces the requirement (7.2.19). If $\psi(x)$ also satisfies the condition

$$0 < \int_0^\infty x \dot{\psi}(x) \exp(-x)\, dx < 2, \tag{7.2.21}$$

then it is shown in [71] that the iteration (7.2.18) is nondivergent.

Example 7.5 (A ψ Function for the Robustified Periodogram). Let $\psi_0(x)$ be given by (7.2.12) with the cutoff value $c := -\log(1-r)$ for some constant $r \in (0,1)$, where $1-r$ is approximately equal to the probability of a periodogram ordinate to be modified by (7.2.16) when $f_{\text{AR}}(\omega, \boldsymbol{\eta})$ is the true noise spectrum. In this case, equation (7.2.20) becomes

$$\psi(x) = \begin{cases} x/\rho & \text{if } 0 \le x < c, \\ c/\rho & \text{if } x \ge c, \end{cases} \tag{7.2.22}$$

where $\rho = 1 - \exp(-c) = r$. Moreover, because

$$\int_0^\infty x \dot{\psi}(x) \exp(-x)\, dx = \frac{1-(c+1)\exp(-c)}{1-\exp(-c)},$$

the condition (7.2.21) is satisfied for all $c > 0$. Note that $r = 1$ (or $c = \infty$) corresponds to the choice of $\psi(x) = x$ which gives the original periodogram.

Figure 7.8 depicts $\psi(x)$ as a function of x for different values of r. As we can see, the ψ function imposes a ceiling on large values of x above the cutoff point, which is critical to achieving robustness; it also raises slightly the values of x below the cutoff point to satisfy (7.2.19). As in Example 7.3, a tradeoff between robustness and efficiency must be made when choosing the value of r. ◇

From the computational point of view, the IYW estimator in (7.2.18) is more appealing than the M spectral estimator in (7.2.8). In fact, the spectral estimates in (7.2.18) at all frequencies can be obtained in closed-form once the parameter $\hat{\boldsymbol{\eta}}_k$ becomes available, whereas the estimates in (7.2.8) require that a nonlinear equation be solved by an iterative algorithm for every frequency.

The initial guess $f_{\text{AR}}(\omega, \hat{\boldsymbol{\eta}}_0)$ can be taken as the AR spectrum estimated by the Yule-Walker method from the original sample autocovariances. Other choices include the smoothed periodogram given by (7.1.2) and, simply, the constant spectrum $\hat{r}_y(0)$. The order of the AR spectrum in the IYW procedure can be determined in the same way as the ordinary Yule-Walker estimator by the AIC in (7.1.29) or the CAT in (7.1.31), for example.

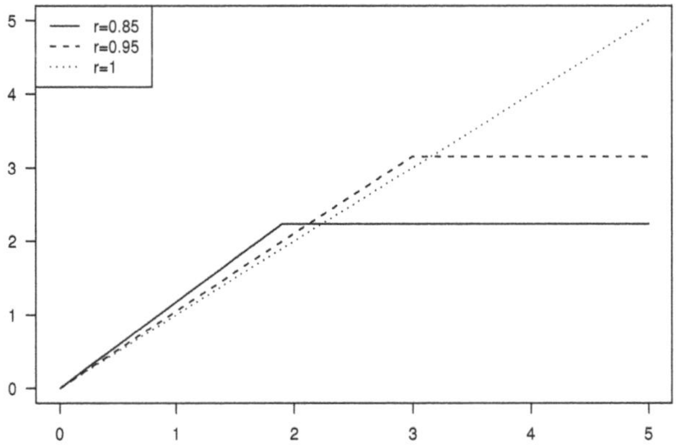

Figure 7.8. Plot of $\psi(x)$ defined by (7.2.20) with $\psi_0(x)$ given by (7.2.12), where $c = -\log(1-r)$. Solid line, $r = 0.85$; dashed line, $r = 0.95$; dotted line, $r = 1$ (nonrobust).

7.2.4 A Comparative Example

The following example is a simulation study in which the spectral estimators discussed previously in this section are compared for estimating an AR(2) noise spectrum in the presence of a sinusoid.

Example 7.6 (Estimation of Noise Spectrum in the Presence of a Sinusoid). Consider a time series $\mathbf{y}$ of length $n = 50$ generated by (7.2.1), where $p = 1$ and $\{\epsilon_t\}$ is a complex Gaussian AR(2) process of the form (6.4.20), with $r_0 = 0.65$, $\omega_0 = 2\pi \times 12/n = 2\pi \times 0.24$, $\sigma^2 = 1$, $\omega_1 := 2\pi \times 20/n = 2\pi \times 0.4$, $C_1 = 0.398$, and $\phi_1 = 0$ (SNR $= -5$ dB). The noise spectrum is estimated by four estimators:

1. PSO, the ordinary periodogram smoother (7.1.2);
2. PSM, the modified periodogram smoother (7.2.4);
3. HUB, the M spectral estimator (7.2.8); and
4. IYW, the iterative Yule-Walker estimator (7.2.18).

The weights for the periodogram smoother and the M spectral estimator are given by the Parzen spectral window in (7.1.37) with $m = 9$. The weights are modified according to (7.2.7) with $m' = 1$ for the modified periodogram smoother. The ψ function for the M spectral estimator is given by (7.2.10) and (7.2.12) with $c = -\log(1-r) - 1$ and $r = 0.85$; the solution of (7.2.8) is produced by 10 iterations of the false position method, initialized by the minimum and maximum of the periodogram. The ψ function for the iterative Yule-Walker estimator is given by (7.2.22) with $c = -\log(1-r)$ and $r = 0.95$; the AR(2) model is employed in all iterations with the AR parameter estimated by the Yule-Walker method; the initial guess is the constant spectrum equal to the sample variance of the observations, and the final estimate is obtained after 10 iterations.

Table 7.1. Performance of Spectral Estimators

	Mean-Square Error				Kullback-Leibler Divergence			
	PSO	PSM	HUB	IYW	PSO	PSM	HUB	IYW
Sinusoid Absent	1.29	2.07	1.37	1.71	0.46	0.74	0.50	0.35
Sinusoid Present	2.46	3.63	1.58	1.76	1.33	1.69	0.66	0.41
Sensitivity Index	1.91	1.75	1.15	1.03	2.89	2.28	1.32	1.17

Results are based on 10,000 Monte Carlo runs.

The performance of these estimators is measured by two criteria of estimation error: the mean-square error (MSE) and the mean Kullback-Leibler spectral divergence (KLD), defined respectively as

$$\mathrm{MSE}(\hat{f}_\epsilon) := 2\pi n^{-1} \sum_{j=0}^{n-1} E\{|\hat{f}_\epsilon(\lambda_j) - f_\epsilon(\lambda_j)|^2\}$$

and

$$\mathrm{KLD}(\hat{f}_\epsilon) := 2\pi n^{-1} \sum_{j=0}^{n-1} E\{d_{\mathrm{KL}}(\hat{f}_\epsilon(\lambda_j)/f_\epsilon(\lambda_j))\},$$

where $d_{\mathrm{KL}}(x) := x - \log x - 1$ for $x > 0$. The robustness is measured by the sensitivity index defined as the estimation error in the presence of sinusoids as a multiple of the estimation error in the absence of sinusoids. Note that the values of r for the M spectral estimator and the iterative Yule-Walker estimator are the empirical minimizers of the KLD calculated on the basis of a preliminary simulation of 1,000 Monte Carlo runs in the presence of the sinusoid.

Table 7.1 shows the MSE and KLD of the estimators based on 10,000 Monte Carlo runs with and without the sinusoid. It also contains the sensitivity index as an indicator of robustness. As we can see from Table 7.1, the accuracy of the periodogram smoother deteriorates significantly when the sinusoid is present, whereas the M spectral estimator and the iterative Yule-Walker estimator suffer only slightly. The modified periodogram smoother has a higher estimation error than the original periodogram smoother in the absence of the sinusoid and offers only slight improvement on the robustness against the sinusoid. ◇

Figure 7.9 provides some further insight into these estimators. As shown in the top panel of this figure, the sinusoid inflates the ordinary periodogram smoother considerably in a neighborhood of the signal frequency $\omega_1 = 2\pi \times 0.4$. The inflation is almost eliminated by the modified periodogram smoother in the small m'-neighborhood of ω_1, but is exacerbated significantly at frequencies just outside that neighborhood. The M spectral estimator is able to reduce (but not eliminate) the inflation in a larger neighborhood of the signal frequency without amplifying the inflation at other frequencies. The bottom panel of Figure 7.9 shows

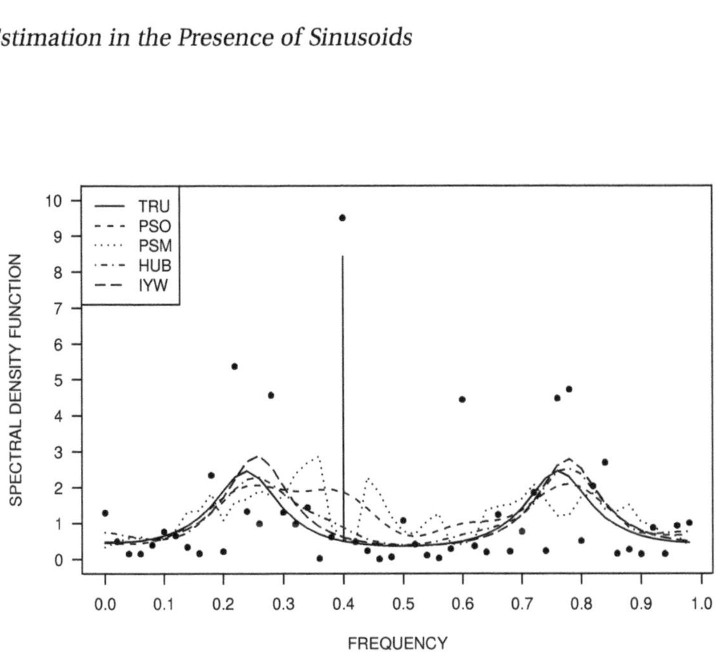

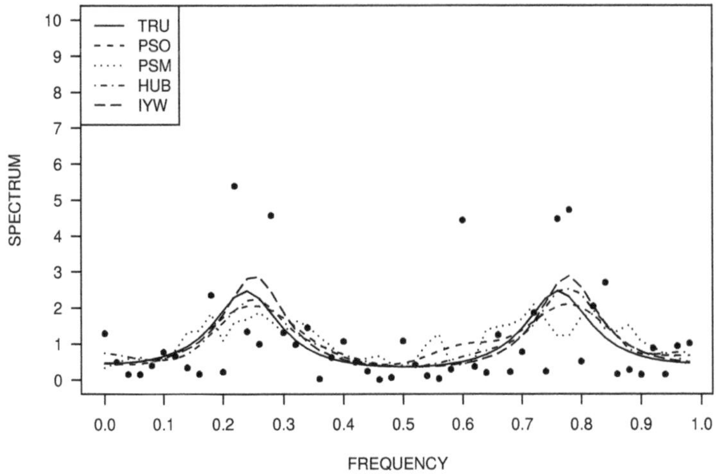

Figure 7.9. Estimation of the complex AR(2) spectrum in Example 7.6 from 50 observations in the presence of a sinusoid (top) and in the absence of the sinusoid (bottom). Solid line, the true spectrum; dashed line, the periodogram smoother (7.1.2); dotted line, the modified periodogram smoother (7.2.3); dot-dashed line, the M spectral estimator (7.2.8); long dashed line, the iterative Yule-Walker estimator (7.2.18); dots, the raw periodogram. All spectra are shown as functions of $f := \omega/(2\pi)$. Vertical line in the top panel shows the location of the contaminating sinusoid with the height equal to $nC_1^2 + f_\epsilon(\omega_1)$, the asymptotic mean of the periodogram ordinate at ω_1.

that in the absence of the sinusoid the modified periodogram smoother gives a much rougher estimate than the original periodogram smoother, which explains the higher estimation error shown in Table 7.1. The M spectral estimator, on the other hand, yields a smooth estimate similar to the estimate produced by the periodogram smoother. The iterative Yule-Walker method performs similarly to the M spectral estimator, with or without the sinusoid.

Note that due to occasional overestimation of spectral peaks (as can be seen from Figure 7.9), the iterative Yule-Walker estimator is slightly outperformed by the M spectral estimator in Table 7.1 in terms of the MSE which measures the absolute estimation error. However, the converse becomes true when the error is measured in relative terms by the KLD, which is related to Whittle's likelihood in (6.1.20), or by the MSE of the ratio $\hat{f}_\epsilon(\lambda_j)/f_\epsilon(\lambda_j)$ (not shown). Note also that the noise spectrum in this example happens to be symmetric about $\omega = \pi$ because the AR model has real-valued coefficients. In general, the spectra of complex-valued AR models are not symmetric.

7.3 Detection of Hidden Sinusoids in Colored Noise

Equipped with the estimators of the noise spectrum, let us return to the problem of detecting hidden sinusoids discussed in Chapter 6 (Section 6.2). The focus now is on the detection of sinusoids in colored noise by using specific estimators for the noise spectrum. With $\hat{f}_\epsilon(\omega)$ denoting an estimate of the noise spectrum, the statistical theory of the periodogram (e.g., Theorem 6.4) justifies the consideration of the standardized periodogram

$$\tilde{I}_n(\omega) := I_n(\omega)/\hat{f}_\epsilon(\omega). \tag{7.3.1}$$

As shown in Figure 6.14, the standardized periodogram is able to attenuate spectral peaks of the noise and thereby enhance the detectability of the sinusoids.

As discussed in Section 6.2, the problem of detecting a sinusoid at a known frequency λ_j can be formulated as testing the hypotheses

$$\begin{cases} H_0: & \text{sinusoid is absent at frequency } \lambda_j, \\ H_1: & \text{sinusoid is present at frequency } \lambda_j. \end{cases}$$

Given the standardized periodogram in (7.3.1), Schuster's test (6.2.2) becomes

$$\tilde{I}_n(\lambda_j) > \tfrac{1}{2}\tau, \tag{7.3.2}$$

where τ is given by (6.2.3). Similarly, if the signal frequency is unknown, then, with $\Omega_n := \{\lambda_j : j = 0, 1, \ldots, n-1\}$ denoting the Fourier grid, the problem can be formulated as testing the hypotheses

$$\begin{cases} H_0: & \text{no sinusoid is present,} \\ H_1: & \text{at least one sinusoid with frequency in } \Omega_n \text{ is present.} \end{cases}$$

In this case, Fisher's test (6.2.16) becomes

$$g := n\frac{\max\{\tilde{I}_n(\lambda_j)\}}{\sum_{j=0}^{n-1}\tilde{I}_n(\lambda_j)} > \tfrac{1}{2}\tau, \tag{7.3.3}$$

where τ is given by (6.2.17).

When the noise spectrum can be estimated from a set of training data that contains only the noise samples, any estimator in Section 7.1 can be used to standardize the periodogram for these tests. In this case, the power calculations in Section 6.2 remain valid. When such training data are not available, the noise spectrum has to be estimated from the same time series on which the tests are to be performed, and the time series may contain hidden sinusoids. In this case, the sinusoid-induced inflation in the ordinary spectral estimators, such as the periodogram smoother in (7.1.2), can significantly reduce the power of the tests. To alleviate the difficulty, the noise spectrum could be estimated after the removal of some suspicious frequencies by regression; it could also be estimated by the robust estimators discussed in Section 7.2.

An example of robust estimators, as suggested in [134] (see also [64] and [298]), is the modified periodogram smoother $\tilde{f}_\epsilon(\lambda_j)$ defined in (7.2.3). This estimator seems particularly viable for Schuster's test (7.3.2) due to its local consistency in the presence of sinusoids. Indeed, according to Corollary 7.1, $\tilde{f}_\epsilon(\lambda_j) = f_\epsilon(\lambda_j) + \mathcal{O}_P(1)$ for large n, whether or not a sinusoid is present at λ_j, provided the frequencies of other sinusoids (if any) are separated from λ_j by more than $2\pi m/n$. Under this condition, $I_n(\lambda_j)/\tilde{f}_\epsilon(\lambda_j)$ has the same asymptotic distribution as $I_n(\lambda_j)/f_\epsilon(\lambda_j)$, so it seems reasonable to expect an increase in detection power by using $\tilde{f}_\epsilon(\lambda_j)$ in (7.2.3) instead of $\hat{f}_\epsilon(\lambda_j)$ in (7.1.2).

Unfortunately, this asymptotic argument is misleading. In fact, using $\tilde{f}_\epsilon(\lambda_j)$ is mathematically equivalent to using $\hat{f}_\epsilon(\lambda_j)$ in Schuster's test. To justify this assertion, observe that

$$I_n(\lambda_j)/\tilde{f}_\epsilon(\lambda_j) = G(I_n(\lambda_j)/\hat{f}_\epsilon(\lambda_j)),$$

where $G(x) := (1-W_{m0})(1-W_{m0}x)^{-1}x$. Because $G(x)$ is monotone increasing,

$$I_n(\lambda_j)/\tilde{f}_\epsilon(\lambda_j) > \tfrac{1}{2}\tilde{\tau} \iff I_n(\lambda_j)/\hat{f}_\epsilon(\lambda_j) > \tfrac{1}{2}\tau,$$

where $\tilde{\tau}$ is uniquely determined by τ such that

$$\tilde{\tau} = 2G(\tfrac{1}{2}\tau) = \frac{1-W_{m0}}{1-\frac{1}{2}\tau W_{m0}}\tau. \tag{7.3.4}$$

In other words, replacing $\hat{f}_\epsilon(\lambda_j)$ with $\tilde{f}_\epsilon(\lambda_j)$ does not change the rejection region of the test if the threshold is adjusted to maintain the same level of significance (i.e., the probability of false alarm). With τ given by (6.2.3), it is easy to show from (7.3.4) that $\tilde{\tau} > \tau$ for all $\alpha < \exp(-1) \approx 0.37$. Therefore, in typical situations where α is very small, the threshold must be raised properly when $\hat{f}_\epsilon(\lambda_j)$ is replaced by $\tilde{f}_\epsilon(\lambda_j)$ in order to retain the level of significance. If the threshold is kept unchanged, the test using $\tilde{f}_n(\lambda_j)$ will appear more powerful because it only needs to cross a lower threshold; but this increase in power is achieved at the expense of an increase in the probability of false alarm. The need for raising the threshold is completely lost in the asymptotic argument because $\tilde{\tau} \to \tau$ as $n \to \infty$.

Note that with the threshold given by (6.2.3) Schuster's test (7.3.2) only has an approximate α level. Under the assumption of Gaussian white noise and with $\hat{f}_\epsilon(\lambda_j)$ given by (7.1.2), it is possible to derive a closed-form expression for the threshold that gives an exact level α. This is shown in the following example.

Example 7.7 (Exact Schuster's Test with Estimated Noise Spectrum). Under the assumption of complex Gaussian white noise and in the absence of sinusoids, it follows from Theorem 6.2 that the periodogram ordinates are distributed as i.i.d. $\frac{1}{2}\sigma^2\chi^2(2)$. Hence, with $\hat{f}_\epsilon(\lambda_j)$ given by (7.1.2) and with $\{X_j\}$ denoting a sequence of i.i.d. $\chi^2(2)$ random variables, the exact threshold τ corresponding to a significance level α for Schuster's test (7.3.2) using $\hat{f}(\lambda_j)$ satisfies

$$\begin{aligned}
\alpha &= P\{I_n(\lambda_j)/\hat{f}_\epsilon(\lambda_j) > \tfrac{1}{2}\tau \mid \text{no sinusoids}\} \\
&= P\Big\{X_j > \tfrac{1}{2}\tau(1-\tfrac{1}{2}\tau W_{m0})^{-1} \sum_{0<|l|\le m} W_{ml} X_{j-l}\Big\} \\
&= E\Big\{\bar{F}\Big(\tfrac{1}{2}\tau(1-\tfrac{1}{2}\tau W_{m0})^{-1} \sum_{0<|l|\le m} W_{ml} X_{j-l}\Big)\Big\} \\
&= \prod_{0<|l|\le m} E\{\exp(-\tfrac{1}{4}\tau(1-\tfrac{1}{2}\tau W_{m0})^{-1} W_{ml} X_{j-l})\} \\
&= \prod_{0<|l|\le m} \frac{1}{1+\tfrac{1}{2}\tau W_{ml}/(1-\tfrac{1}{2}\tau W_{m0})}.
\end{aligned} \tag{7.3.5}$$

Note that the second equality requires $W_{m0} < 2/\tau$ which can be satisfied for large n because $W_{m0} \to 0$ as $n \to \infty$. In the special case of a rectangular window where $W_{ml} = (2m+1)^{-1}$ for $|l| \le m$, equation (7.3.5) reduces to $\alpha = (1-\frac{1}{2}\tau W_{m0})^{2m}$, which in turn yields $\tau = 2(2m+1)(1-\alpha^{1/2m})$. ◇

Besides the modified periodogram smoother in (7.2.3), Section 7.2 discusses several alternative estimators for the noise spectrum, including the modified periodogram smoother given by (7.2.4) and (7.2.7) with $m' \ge 1$, for which the equivalence assertion no longer holds. All these spectral estimators can be employed to standardize the periodogram for the tests (7.3.2) and (7.3.3). The following simulation example demonstrates the performance of these estimators in detecting a single sinusoid in Gaussian colored noise.

Table 7.2. Thresholds of Schuster's Test (7.3.2): Gaussian AR(2) Noise

α	Theoretical Threshold (6.2.3)	Empirical Threshold with True and Estimated Noise Spectrum PSO	PSM	HUB	IYW	CNT	TRU
0.01	9.21	6.21	9.56	6.79	9.04	4.89	9.43
0.05	5.99	4.58	5.72	4.28	5.67	3.22	6.31
0.10	4.61	3.74	4.19	3.47	4.52	2.45	4.87

Table 7.3. Thresholds of Fisher's Test (7.3.3): Gaussian AR(2) Noise

α	Theoretical Threshold (6.2.17)	Empirical Threshold with True and Estimated Noise Spectrum PSO	PSM	HUB	IYW	CNT	TRU
0.01	15.95	11.30	26.92	18.07	18.22	25.59	15.86
0.05	13.13	9.84	20.51	14.09	14.02	20.51	13.13
0.10	11.87	9.21	17.77	12.47	12.38	18.18	11.88

Results are based on 10,000 Monte Carlo runs.

Example 7.8 (Detection of Hidden Sinusoids in Gaussian AR(2) Noise). Let **y** be the same as in Example 7.6. The objective now is to detect the sinusoid by Schuster's test (7.3.2) and Fisher's test (7.3.3) using the true noise spectrum and five different estimators:

1. PSO, the ordinary periodogram smoother (7.1.2);
2. PSM, the modified periodogram smoother (7.2.4);
3. HUB, the M spectral estimator (7.2.8);
4. IYW, the iterative Yule-Walker estimator (7.2.18);
5. CNT, the constant spectral estimator (6.2.11); and
6. TRU, the true noise spectrum.

The weights and other tuning parameters of the spectral estimators remain the same as in Example 7.6. The threshold of each test is estimated empirically by the $(1-\alpha)$th sample quantile of the test statistic calculated from 10,000 Monte Carlo runs under H_0 (no sinusoids). Tables 7.2 and 7.3 contain the simulated thresholds for several values of α. They also contain the theoretical thresholds given by (6.2.3) for Schuster's test and by (6.2.17) for Fisher's test.

As we can see, the theoretical thresholds are very close to the empirical thresholds of the oracle tests that employ the true noise spectrum. However, the empirical thresholds of the tests with estimated noise spectrum differ from the theoretical thresholds to various degrees. For example, Table 7.2 shows that the thresholds of Schuster's test (7.3.2) with the estimated noise spectrum tend to be lower than the theoretical thresholds. Table 7.3 shows that the opposite tends to be true for Fisher's test. Therefore, if the theoretical thresholds are used with

Table 7.4. Power of Schuster's Test (7.3.2) for Detecting a Sinusoid with Known Frequency in Gaussian AR(2) Noise ($\alpha = 0.01$)

Spectral	SNR (dB)								
Estimator	−20	−18	−16	−14	−12	−10	−8	−6	−4
PSO	0.03	0.05	0.08	0.15	0.26	0.46	0.72	0.92	0.99
PSM	0.03	0.05	0.08	0.14	0.25	0.44	0.69	0.90	0.99
HUB	0.03	0.05	0.08	0.14	0.26	0.45	0.71	0.91	0.99
IYW	0.03	0.06	0.09	0.16	0.28	0.48	0.75	0.94	0.99
CNT	0.04	0.06	0.10	0.18	0.33	0.56	0.81	0.97	1.00
TRU	0.04	0.07	0.12	0.21	0.37	0.62	0.86	0.98	1.00

Table 7.5. Power of Fisher's Test (7.3.3) for Detecting a Sinusoid with Unknown Frequency in Gaussian AR(2) Noise ($\alpha = 0.01$)

Spectral	SNR (dB)								
Estimator	−9	−8	−7	−6	−5	−4	−3	−2	−1
PSO	0.04	0.06	0.10	0.17	0.28	0.44	0.62	0.79	0.92
PSM	0.01	0.01	0.03	0.10	0.23	0.42	0.66	0.85	0.95
HUB	0.03	0.06	0.13	0.24	0.41	0.63	0.82	0.94	0.99
IYW	0.05	0.11	0.24	0.43	0.66	0.85	0.95	0.99	1.00
CNT	0.01	0.01	0.01	0.01	0.02	0.07	0.20	0.44	0.73
TRU	0.12	0.25	0.46	0.70	0.88	0.98	1.00	1.00	1.00

Results are based on 10,000 Monte Carlo runs.

an estimated noise spectrum, the actual false-alarm probability could be lower than the nominal value for Schuster's test and higher for Fisher's test.

Table 7.4 contains the detection probability of Schuster's test (7.3.2) at $\alpha = 0.01$ employing the true noise spectrum and different spectral estimators. The empirical thresholds shown in Table 7.2 are used for the detection. As we can see, using the modified periodogram smoother instead of the ordinary periodogram smoother does not improve the detection power (actually the detection power is reduced slightly for some SNR). It seems that the benefit of local consistency is overshadowed by the increased statistical variability due to under-smoothing. The M spectral estimator behaves similarly. Only the iterative Yule-Walker estimator is able to improve the detection power across the range of the SNR considered. Interestingly, the constant spectral estimator derived under the assumption of white noise produces the best result across the SNR range, although the other alternatives correctly assume colored noise. This is possible only because Schuster's test is carried out locally at a single frequency, so an accurate estimation of the entire noise spectrum is not required.

The necessity for accurate estimation of the entire noise spectrum is evident in Table 7.5 where the detection probability is shown for Fisher's test (7.3.3) using the same spectral estimators. In this case, the constant spectral estimator is con-

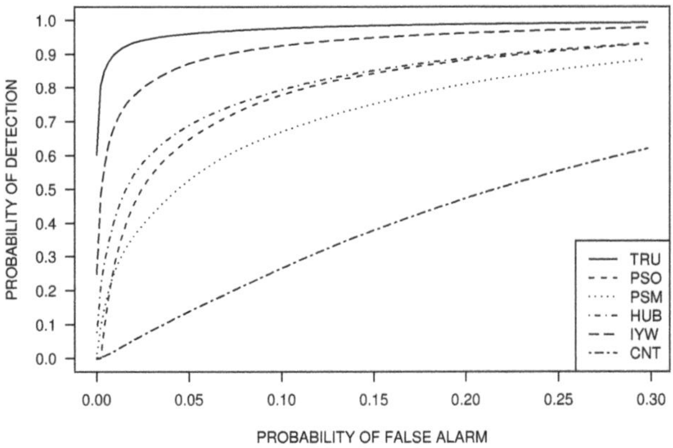

Figure 7.10. ROC curve of Fisher's test (7.3.3) with true and estimated noise spectrum for detecting a hidden sinusoid in complex Gaussian AR(2) noise (SNR = −5 dB). Solid line, true noise spectrum; dashed line, periodogram smoother; dotted line, modified periodogram smoother; dash-dotted line, M spectral estimator; long-dashed line, iterative Yule-Walker estimator; long dash-dotted line, constant spectral estimator.

siderably inferior to the other estimators. The ordinary periodogram smoother is outperformed by the M spectral estimator and the iterative Yule-Walker estimator. The latter works particularly well, owing to its use of the correct parametric form of the noise spectrum. The modified periodogram smoother performs worse than the ordinary periodogram smoother when the SNR is low.

These estimators are further compared in Figure 7.10, where the receiver operating characteristic (ROC) curve is depicted for the case of $\gamma = -5$ dB. As can be seen, the iterative Yule-Walker estimator outperforms the others by a large margin across the range of α considered, thanks to its use of the correct parametric model for the noise spectrum. The M spectral estimator performs much better than the ordinary periodogram smoother for small α. The modified periodogram smoother improves the ordinary periodogram smoother when α is very small but becomes much inferior as α increases. The constant spectral estimator performs very poorly across the range of α. ◇

To end this section, let us again consider the variable star data described in Chapter 1 (Section 1.3). Recall that the time series is well represented by a linear model which consists of eight real sinusoids with the estimated frequencies and amplitudes shown in Table 6.3. In Section 7.1.3, an AR model of order 16 is found adequate for the residuals. The estimated spectrum is shown in Figure 7.4.

Now let us use the spectrum of the residuals to standardize the periodogram of the original time series and perform Schuster's test (7.3.2) on the estimated

Table 7.6. Significance Test for the Frequencies in the Variable Star Data

k	f_k	p-value*	p-value†	k	f_k	p-value*	p-value†
1	0.004467237	0.0000	0.0006	5	0.009181022	0.0003	0.1529
2	0.005992840	0.0000	0.0000	6	0.011813066	0.0000	0.0000
3	0.007354442	0.0014	0.1885	7	0.013460770	0.0002	0.2345
4	0.008662235	0.0000	0.0000	8	0.017682145	0.0000	0.0003

* Noise spectrum is estimated from the residuals (see Figure 7.4).
† Noise spectrum is estimated from the original time series by IYW.

(possibly nonFourier) frequencies to see how significant they really are when the noise factor is taken into account. Table 7.6 contains the p-values of the test calculated on the basis of the theory that the standardized periodogram at a given frequency is distributed as $\frac{1}{2}\chi^2(2)$ under H_0 (for large sample sizes). As we can see, all the p-values (shown in columns with "$*$") are very small indeed.

For comparison, Table 7.6 also contains the p-values of the test (shown in columns with "†") when the noise spectrum is estimated not from the residuals but from the original time series using an AR(16) model produced by the robust IYW method in (7.2.18). The AR spectrum obtained from the residuals is used to initialize the IYW algorithm, where the ψ function is given by (7.2.22) with $c = -\log(1-r)$ and $r = 0.95$. The final spectral estimate is produced after 3 iterations. As we can see from Table 7.6, the interference from the sinusoids still inflates the estimated noise spectrum to varying degrees and thereby reduces the significance of some frequencies, especially the 3rd, 5th, and 7th frequencies. The 6th frequency, which corresponds to the most dominant sinusoidal component, remains significant in spite of the inflated noise spectrum.

7.4 Proof of Theorems

This section contains the proof of Theorems 7.1–7.5, 7.8, and Corollary 7.1, and the derivation of (7.1.29) and (7.1.30).

Proof of Theorem 7.1. By (7.1.2), (7.1.5), and Theorem 6.6,

$$\begin{aligned} E\{\hat{f}_\epsilon(\lambda_j)\} - f_\epsilon(\lambda_j) &= \sum_{|l|\le m} W_{ml}\,[E\{I_n(\lambda_{j-l})\} - f_\epsilon(\lambda_j)] \\ &= \sum_{|l|\le m} W_{ml}\,\{f_\epsilon(\lambda_{j-l}) - f_\epsilon(\lambda_j)\} + \mathcal{O}(r_n). \end{aligned} \tag{7.4.1}$$

For $|l| \le m$, $|\lambda_{j-l} - \lambda_j| \le 2\pi m/n \to 0$. This, combined with the uniform continuity of $f_\epsilon(\omega)$, proves that the first term in (7.4.1) approaches zero uniformly in λ_j as

$n \to \infty$. The second term also tends to zero uniformly in λ_j because $r_n \to 0$. Combining these results proves part (a).

To prove part (e), observe that for any given $\lambda_j, \lambda_l \in \Omega_n$,

$$
\begin{aligned}
&\mathrm{Cov}\{\hat{f}_\epsilon(\lambda_j), \hat{f}_\epsilon(\lambda_l)\} \\
&= \Big\{ \sum_{(u,v)\in\mathscr{D}_1} + \sum_{(u,v)\in\mathscr{D}_2} + \sum_{(u,v)\in\mathscr{D}_3} \Big\} W_{mu} W_{mv} \mathrm{Cov}\{I_n(\lambda_{j-u}), I_n(\lambda_{l-v})\} \\
&:= T_1 + T_2 + T_3,
\end{aligned}
$$

where $\mathscr{D}_1 := \{(u,v) : |u|, |v| \le m, j-u = l-v\}$, $\mathscr{D}_2 := \{(u,v) : |u|, |v| \le m, j-u+l-v = 0 \pmod{n}\}$, and $\mathscr{D}_3 := \{(u,v) : |u|, |v| \le m\} \setminus (\mathscr{D}_1 \cup \mathscr{D}_2)$. By Theorem 6.6,

$$
\begin{aligned}
&\mathrm{Cov}\{I_n(\lambda_{j-u}), I_n(\lambda_{l-v})\} \\
&= \begin{cases}
f_\epsilon^2(\lambda_{j-u}) + \mathcal{O}(r_n) + \mathcal{O}(r_{n4}) & \\
\quad + \xi\{|\iota g_\epsilon(\lambda_{j-u})|^2 + \mathcal{O}(r_{n2})\}\delta_{2(j-u)}^{(n)} & \text{if } (u,v) \in \mathscr{D}_1, \\
\xi\{|\iota g_\epsilon(\lambda_{j-u})|^2 + \mathcal{O}(r_{n2})\} + \mathcal{O}(r_{n4}) & \text{if } (u,v) \in \mathscr{D}_2, \\
\mathcal{O}(r_{n4}) & \text{if } (u,v) \in \mathscr{D}_3.
\end{cases}
\end{aligned}
$$

By the Cauchy-Schwarz inequality, $1 = (\sum W_{ml})^2 \le (2m+1)s_m$, so $s_m^{-1} \le 2m+1$. Thus, under the assumptions of Theorem 7.1, $s_m^{-1} r_n \to 0$, $s_m^{-1} r_{n2} \to 0$, $s_m^{-1} n^{-1} \to 0$, and hence $s_m^{-1} r_{n4} \to 0$. This result, combined with the uniformity of the error terms, implies that there exists a constant $c > 0$ such that

$$
s_m^{-1} |T_3| \le c\, s_m^{-1} r_{n4} \to 0 \tag{7.4.2}
$$

uniformly in λ_j and λ_l. Similarly, it can be shown that

$$
s_m^{-1}(T_1 - T_1') \to 0, \quad s_m^{-1}(T_2 - T_2') \to 0, \tag{7.4.3}
$$

where

$$
\begin{aligned}
T_1' &:= \sum_{(u,v)\in\mathscr{D}_1} W_{mu} W_{mv} \{f_\epsilon^2(\lambda_{j-u}) + \xi |\iota g_\epsilon(\lambda_{j-u})|^2 \delta_{2(j-u)}^{(n)}\}, \\
T_2' &:= \sum_{(u,v)\in\mathscr{D}_2} W_{mu} W_{mv} \xi |\iota g_\epsilon(\lambda_{j-u})|^2.
\end{aligned}
$$

In addition, because

$$
\begin{aligned}
a_{jl} &= s_m^{-1} \sum_{(u,v)\in\mathscr{D}_1} W_{mu} W_{mv}, \\
b_{jl} &= s_m^{-1} \sum_{(u,v)\in\mathscr{D}_1} W_{mu} W_{mv} \delta_{2(j-u)}^{(n)}, \\
c_{jl} &= s_m^{-1} \sum_{(u,v)\in\mathscr{D}_2} W_{mu} W_{mv},
\end{aligned}
$$

the Cauchy-Schwarz inequality gives $|a_{jl}| \le 1$, $|b_{jl}| \le 1$, and $|c_{jl}| \le 1$. It follows from (7.1.5) and the uniform continuity of $f_\epsilon(\omega)$ and $g_\epsilon(\omega)$ that

$$s_m^{-1} T_1' - \{f_\epsilon^2(\lambda_j)\, a_{jl} + \xi |\iota g_\epsilon(\lambda_j)|^2\, b_{jl}\} \to 0, \tag{7.4.4}$$
$$s_m^{-1} T_2' - \xi |\iota g_\epsilon(\lambda_j)|^2 c_{jl} \to 0. \tag{7.4.5}$$

Combining the expressions in (7.4.2)–(7.4.5), which hold uniformly in λ_j and λ_l, with the fact that $W_{mu} = 0$ for all $|u| > m$ proves part (e).

Because the a_{jl}, b_{jl}, and c_{jl} are bounded, it follows that

$$E\{\hat{f}_\epsilon(\lambda_j)\} - f_\epsilon(\lambda_j) \to 0, \quad \mathrm{Var}\{\hat{f}_\epsilon(\lambda_j)\} = \mathcal{O}(s_m) \to 0,$$

uniformly in $\lambda_j \in \Omega_n$, which proves part (b). Part (c) can be proved by combining this result with an application of Chebyshev's inequality.

Finally, it follows from (6.4.2) that

$$\begin{aligned}
\hat{f}_\epsilon(\lambda_j) - E\{\hat{f}_\epsilon(\lambda_j)\} &= \sum_{|l|\le m} W_{ml} \sum_{|u|<n} \{\hat{r}_y(u) - E\{\hat{r}_y(u)\} \exp(-iu\lambda_{j-l}) \\
&= \sum_{|u|<n} \{\hat{r}_y(u) - E\{\hat{r}_y(u)\} w_m(\lambda_u) \exp(-iu\lambda_j),
\end{aligned}$$

where $w_m(\omega) := \sum_{|l|\le m} W_{ml} \exp(il\omega)$. Therefore, we obtain

$$\max_{\lambda_j \in \Omega_n} |\hat{f}_\epsilon(\lambda_j) - E\{\hat{f}_\epsilon(\lambda_j)\}| \le \sum_{|u|<n} |\hat{r}_y(u) - E\{\hat{r}_y(u)\}|\, |w_m(\lambda_u)|.$$

According to Theorem 4.2, there is a constant $c > 0$ such that

$$E\{|\hat{r}_y(u) - E\{\hat{r}_y(u)\}|\} \le \sqrt{\mathrm{Var}\{\hat{r}_y(u)\}} \le c n^{-1/2}.$$

Therefore, it follows that

$$E\Big\{\max_{\lambda_j \in \Omega_n} |\hat{f}_\epsilon(\lambda_j) - E\{\hat{f}_\epsilon(\lambda_j)\}|\Big\} \le 2cn^{1/2} \sum_{|u|<n} |w_m(\lambda_u)|.$$

Combining this result with part (a) and Proposition 2.16(b) proves part (d). □

Proof of Theorem 7.2. For the real case, a proof can be found in [46, p. 132]. For the complex case, the proof follows the same argument except that it relies on a more general result given by Lemma 12.1.8.

To prove the assertion in the complex case, define $g(\omega) := \max\{f_\epsilon(\omega), \delta/2\}$, $\kappa := \max_\omega f_\epsilon(\omega) < \infty$, and $\nu := \min\{(2\kappa)^{-2}\delta, (2\kappa)^{-1}\} > 0$. Because $g(\omega) \le \kappa$, we obtain

$$0 < \kappa^{-1} \le 1/g(\omega) \le 2\delta^{-1} < \infty.$$

According to Lemma 12.1.7 and Lemma 12.1.8, there exist a constant $1/\sigma_m^2 > 0$ and a minimum-phase polynomial $\varphi_m(z) := 1 + \sum_{j=1}^m \varphi_j z^{-j}$ such that

$$|1/f_{\mathrm{AR}}(\omega) - 1/g(\omega)| < \nu,$$

where $f_{\text{AR}}(\omega) := \sigma_m^2 / |\Phi_m(\omega)|^2$ and $\Phi_m(\omega) := \varphi_m(\exp(i\omega))$. Observe that

$$1/f_{\text{AR}}(\omega) > 1/g(\omega) - \nu \geq 1/\kappa - \nu \geq (2\kappa)^{-1}.$$

Therefore, we obtain $f_{\text{AR}}(\omega) < 2\kappa$ and

$$|f_{\text{AR}}(\omega) - g(\omega)| < \nu g(\omega) f_{\text{AR}}(\omega) < 2\nu\kappa^2 = \delta/2.$$

On the other hand, $0 \leq g(\omega) - f_\epsilon(\omega) \leq \delta/2$. Combining these inequalities yields

$$|f_{\text{AR}}(\omega) - f_\epsilon(\omega)| \leq |f_{\text{AR}}(\omega) - g(\omega)| + g(\omega) - f_\epsilon(\omega) < \delta,$$

which is the assertion. □

Proof of Theorem 7.3. The assertion can be proved by following [402]. First, let $\boldsymbol{\epsilon}_{t-1} := [\epsilon_{t-1}, \ldots, \epsilon_{t-m}]^T$ and $e_t := \epsilon_t + \boldsymbol{\varphi}^T \boldsymbol{\epsilon}_{t-1}$, where $\boldsymbol{\varphi} := [\varphi_1, \ldots, \varphi_m]^T$ satisfies (7.1.17). Then, it is easy to verify that

$$E(e_t \boldsymbol{\epsilon}_{t-1}^*) = \mathbf{r}_\epsilon + \mathbf{R}_\epsilon \boldsymbol{\varphi} = \mathbf{0},$$

which is known as the orthogonality property of the best linear prediction.

Let z_0 be any root of $\varphi_m(z) := 1 + \sum_{j=1}^m \varphi_j z^{-j}$. Then, we can write

$$\varphi_m(z) = (1 - z_0 z^{-1})\theta(z), \tag{7.4.6}$$

where $\theta(z) := 1 + \sum_{j=1}^{m-1} \theta_j z^{-j}$ for some $\{\theta_j\}$. Define

$$\xi_t := \theta(z)\epsilon_t = \epsilon_t + \sum_{j=1}^{m-1} \theta_j \epsilon_{t-j}.$$

Then, by the orthogonality property, we have $E(e_t \xi_{t-1}^*) = 0$. On the other hand, it follows from (7.4.6) that

$$e_t = \varphi_m(z)\epsilon_t = (1 - z_0 z^{-1})\xi_t = \xi_t - z_0 \xi_{t-1}. \tag{7.4.7}$$

Therefore, we can write

$$0 = E(e_t \xi_{t-1}^*) = E\{(\xi_t - z_0 \xi_{t-1})\xi_{t-1}^*\} = r_\xi(1) - z_0 r_\xi(0), \tag{7.4.8}$$

where $r_\xi(u)$ denotes the ACF of $\{\xi_t\}$. Combining (7.4.7) and (7.4.8) leads to

$$\begin{aligned} \sigma_m^2 &= E(|e_t|^2) = E(|\xi_t - z_0 \xi_{t-1}|^2) \\ &= r_\xi(0) - z_0^* r_\xi(1) - z_0 r_\xi^*(1) + |z_0|^2 r_\xi(0) \\ &= r_\xi(0)(1 - |z_0|^2). \end{aligned}$$

Because $\sigma_m^2 > 0$ by assumption, we obtain $|z_0| < 1$, and hence the assertion. Note that if $\sigma_m^2 = 0$ and $\sigma_{m-1}^2 > 0$ then we have $|z_0| = 1$ because $r_\xi(0) \ge \sigma_{m-1}^2 > 0$, in which case, all the roots of $\varphi_m(z)$ are on the unit circle. □

Proof of Theorem 7.4. Consider the Yule-Walker estimator given by (7.1.20). According to Theorem 4.1, $\hat{r}_y(u) \stackrel{a.s.}{\to} r_\epsilon(u)$ as $n \to \infty$ for any fixed u. Therefore,

$$\hat{\mathbf{R}}_y \stackrel{a.s.}{\to} \mathbf{R}_\epsilon, \quad \hat{\mathbf{r}}_y \stackrel{a.s.}{\to} \mathbf{r}_\epsilon. \tag{7.4.9}$$

This implies that

$$\hat{\boldsymbol{\varphi}} \stackrel{a.s.}{\to} -\mathbf{R}_\epsilon^{-1}\mathbf{r}_\epsilon = \boldsymbol{\varphi}, \quad \hat{\sigma}_m^2 \stackrel{a.s.}{\to} r_\epsilon(0) + \mathbf{R}_\epsilon\boldsymbol{\varphi} = \sigma_m^2.$$

The same result can be obtained for the estimator in (7.1.22) because

$$n^{-1}\mathbf{Y}_f^H\mathbf{Y}_f = \hat{\mathbf{R}}_y + \mathcal{O}(n^{-1}), \quad n^{-1}\mathbf{Y}_f^H\mathbf{y}_f = \hat{\mathbf{r}}_y + \mathcal{O}(n^{-1}),$$

and $n^{-1}\mathbf{y}_f^H\mathbf{y}_f = \hat{r}_y(0)$. A similar argument leads to the consistency of the estimators defined by (7.1.23) and (7.1.24).

Let $\Phi_m(\omega) := 1 + \mathbf{f}_m^H(\omega)\boldsymbol{\varphi}$ and $\hat{\Phi}_m(\omega) := 1 + \mathbf{f}_m^H(\omega)\hat{\boldsymbol{\varphi}}$. Then,

$$\hat{f}_{\mathrm{AR}}(\omega) - f_{\mathrm{AR}}(\omega) = \frac{|\Phi_m(\omega)|^2\hat{\sigma}_m^2 - |\hat{\Phi}_m(\omega)|^2\sigma_m^2}{|\Phi_m(\omega)|^2|\hat{\Phi}_m(\omega)|^2}.$$

Because $|\hat{\Phi}_m(\omega) - \Phi_m(\omega)| \le c\|\hat{\boldsymbol{\varphi}} - \boldsymbol{\varphi}\|$ for some constant $c > 0$ and for all ω, it follows that $\hat{\Phi}_m(\omega) \stackrel{a.s.}{\to} \Phi_m(\omega)$ uniformly in ω. Combining these results with the fact that $|\Phi_m(\omega)|$ is uniformly bounded away from zero proves $\hat{f}_{\mathrm{AR}}(\omega) \stackrel{a.s.}{\to} f_{\mathrm{AR}}(\omega)$ as $n \to \infty$ uniformly in ω. □

Proof of Theorem 7.5. Because the other estimators can be handled in a similar way, it suffices to consider the Yule-Walker estimator given by (7.1.20). In the complex case, let $\hat{\mathbf{r}} := [\hat{r}_y^*(m-1), \ldots, \hat{r}_y^*(1), \hat{r}_y(0), \hat{r}_y(1), \ldots, \hat{r}_y(m)]^T \in \mathbb{C}^{2m}$. Then, it is easy to verify that

$$\hat{\mathbf{R}}_y\boldsymbol{\varphi} + \hat{\mathbf{r}}_y = \boldsymbol{\Phi}\hat{\mathbf{r}},$$

where

$$\boldsymbol{\Phi} := \begin{bmatrix} \varphi_m & \varphi_{m-1} & \cdots & \varphi_0 & 0 & \cdots & 0 \\ 0 & \varphi_m & \cdots & \varphi_1 & \varphi_0 & & 0 \\ & & \ddots & & & \ddots & \\ 0 & \cdots & & 0 & \varphi_m & \cdots \; \varphi_1 & \varphi_0 \end{bmatrix} \quad (\varphi_0 := 1).$$

Let $\mathbf{r}$ be defined in the same way as $\hat{\mathbf{r}}$ using $r_\epsilon(u)$ instead of $\hat{r}_y(u)$. Then,

$$\mathbf{0} = \mathbf{R}_\epsilon\boldsymbol{\varphi} + \mathbf{r}_\epsilon = \boldsymbol{\Phi}\mathbf{r}.$$

Combining these expressions yields

$$\hat{\boldsymbol{\varphi}} - \boldsymbol{\varphi} = -\hat{\mathbf{R}}_y^{-1}(\hat{\mathbf{R}}_y\boldsymbol{\varphi} + \hat{\mathbf{r}}_y) = -\hat{\mathbf{R}}_y^{-1}\boldsymbol{\Phi}(\hat{\mathbf{r}} - \mathbf{r}).$$

According to Theorem 4.4,

$$\sqrt{n}\,(\hat{\mathbf{r}} - \mathbf{r}) \xrightarrow{D} \mathrm{N}(\mathbf{0}, \boldsymbol{\Gamma}, \tilde{\boldsymbol{\Gamma}}) \tag{7.4.10}$$

where

$$\boldsymbol{\Gamma} := \lim_{n\to\infty} n\,\mathrm{Cov}(\hat{\mathbf{r}}, \hat{\mathbf{r}}), \quad \tilde{\boldsymbol{\Gamma}} := \lim_{n\to\infty} n\,\mathrm{Cov}(\hat{\mathbf{r}}, \hat{\mathbf{r}}^*).$$

According to Theorem 4.1, $\hat{\mathbf{R}}_y \xrightarrow{P} \mathbf{R}_\epsilon$. Hence, by Propositions 2.6(c) and 2.13(b),

$$\sqrt{n}\,(\hat{\boldsymbol{\varphi}} - \boldsymbol{\varphi}) \xrightarrow{D} \mathrm{N}_c(\mathbf{0}, \mathbf{R}_\epsilon^{-1}\boldsymbol{\Phi}\boldsymbol{\Gamma}\boldsymbol{\Phi}^H\mathbf{R}_\epsilon^{-1}, \mathbf{R}_\epsilon^{-1}\boldsymbol{\Phi}\tilde{\boldsymbol{\Gamma}}\boldsymbol{\Phi}^T\mathbf{R}_\epsilon^{-T}). \tag{7.4.11}$$

Furthermore, observe that

$$\begin{aligned}
\tilde{\boldsymbol{\Sigma}} &:= \boldsymbol{\Phi}\boldsymbol{\Gamma}\boldsymbol{\Phi}^T \\
&= \lim_{n\to\infty} n\,\mathrm{Cov}(\boldsymbol{\Phi}\hat{\mathbf{r}}, \boldsymbol{\Phi}^*\hat{\mathbf{r}}^*) \\
&= \lim_{n\to\infty} n\,\mathrm{Cov}(\hat{\mathbf{R}}_y\boldsymbol{\varphi} + \hat{\mathbf{r}}_y, \hat{\mathbf{R}}_y^*\boldsymbol{\varphi}^* + \hat{\mathbf{r}}_y^*).
\end{aligned}$$

For $u = 1, \ldots, m$, the uth entry of $\hat{\mathbf{R}}_y\boldsymbol{\varphi} + \hat{\mathbf{r}}_y$ is $X_u := \sum_{j=0}^m \varphi_j \hat{r}_y(u-j)$. As $n \to \infty$, Theorem 4.2 ensures that

$$\begin{aligned}
n\,\mathrm{Cov}(X_u, X_v^*) &= \sum_{j=0}^m\sum_{l=0}^m \varphi_j\varphi_l\, n\,\mathrm{Cov}\{\hat{r}_y(u-j), \hat{r}_y^*(v-l)\} \\
&\to \sum_{j=0}^m\sum_{l=0}^m \varphi_j\varphi_l\gamma_2(u-j, -v+l) \\
&= (\kappa - 2 - \xi)\Big\{\sum_{j=0}^m \varphi_j r_\epsilon(u-j)\Big\}\Big\{\sum_{j=0}^m \varphi_j r_\epsilon(v-j)\Big\} + \tilde{\sigma}(u,v),
\end{aligned}$$

where $\tilde{\sigma}(u,v)$ is defined by (7.1.28). Because $\sum_{j=0}^m \varphi_j r_\epsilon(u-j)$ is the uth entry of $\mathbf{R}_\epsilon\boldsymbol{\varphi} + \mathbf{r}_\epsilon$ and hence equal to zero, it follows that $\tilde{\boldsymbol{\Sigma}} = [\tilde{\sigma}(u,v)]$. Similarly, it can be shown that $\boldsymbol{\Sigma} := \boldsymbol{\Phi}\boldsymbol{\Gamma}\boldsymbol{\Phi}^H = [\sigma(u,v)]$, where $\sigma(u,v)$ is defined by (7.1.27). Combining these expressions with (7.4.11) proves the assertion.

In the real case, let $\hat{\mathbf{r}}_0 := [\hat{r}_y(0), \hat{r}_y(1), \ldots, \hat{r}_y(m)]^T \in \mathbb{R}^{m+1}$ and let $\mathbf{r}_0$ be defined in the same way using $r_\epsilon(u)$. With this notation, we can write

$$\hat{\boldsymbol{\varphi}} - \boldsymbol{\varphi} = -\hat{\mathbf{R}}_y^{-1}\boldsymbol{\Phi}\mathbf{D}(\hat{\mathbf{r}}_0 - \mathbf{r}_0).$$

where $\mathbf{D} \in \mathbb{R}^{2m\times(m+1)}$ is a constant matrix such that $\hat{\mathbf{r}} - \mathbf{r} = \mathbf{D}(\hat{\mathbf{r}}_0 - \mathbf{r}_0)$. By Theorem 4.3 and Proposition 2.13(b), we obtain

$$\sqrt{n}\,(\hat{\boldsymbol{\varphi}} - \boldsymbol{\varphi}) \xrightarrow{D} \mathrm{N}(\mathbf{0}, \mathbf{R}_\epsilon^{-1}\boldsymbol{\Phi}\mathbf{D}\boldsymbol{\Gamma}_0\mathbf{D}^T\boldsymbol{\Phi}^T\mathbf{R}_\epsilon^{-1}), \tag{7.4.12}$$

where

$$\mathbf{\Gamma}_0 := \lim_{n\to\infty} n\,\mathrm{Cov}(\hat{\mathbf{r}}_0).$$

As in the complex case, we can write

$$\mathbf{\Sigma} := \mathbf{\Phi}\mathbf{D}\mathbf{\Gamma}_0\mathbf{D}^T\mathbf{\Phi}^T = \lim_{n\to\infty} n\,\mathrm{Cov}(\hat{\mathbf{R}}_y\boldsymbol{\varphi}+\hat{\mathbf{r}}_y).$$

The uth entry of $\hat{\mathbf{R}}_y\boldsymbol{\varphi}+\hat{\mathbf{r}}_y$ is equal to $X_u := \sum_{j=0}^m \varphi_j \hat{r}_y(u-j)$. By Theorem 4.2,

$$\begin{aligned} n\,\mathrm{Cov}(X_u, X_v) &= \sum_{j=0}^m \sum_{l=0}^m \varphi_j\varphi_l n\,\mathrm{Cov}\{\hat{r}_y(u-j), \hat{r}_y(v-l)\} \\ &\to (\kappa-3)\Big\{\sum_{j=0}^m \varphi_j r_\epsilon(u-j)\Big\}\Big\{\sum_{j=0}^m \varphi_j r_\epsilon(v-j)\Big\} + \sigma(u,v) \\ &= \sigma(u,v), \end{aligned}$$

where $\sigma(u,v)$ is given by (7.1.26). Therefore, $\mathbf{\Sigma} = [\sigma(u,v)]$. Combining this result with (7.4.12) proves the assertion.

When $\{\epsilon_t\}$ is a complex AR(m) process of the form (7.1.14), it can be expressed as a linear process (7.1.25) with $\psi_j = 0$ for $j < 0$ and $\psi_0 = 1$. Because $\sigma^2 = \sigma_m^2$, $E(\zeta_t\zeta_s^*) = \sigma^2\delta_{t-s}$, it follows that for any u,

$$E(\zeta_t\epsilon_{t-u}^*) = \sigma^2\psi_{-u}^*.$$

Therefore, by multiplying both sides of (7.1.14) with ϵ_{t-u}^* and then taking the expected values, we obtain

$$\sum_{j=0}^m \varphi_j r_\epsilon(u-j) = \sigma^2\psi_{-u}^*. \tag{7.4.13}$$

Moreover, by assumption, $\xi = |E(\zeta_t^2)|^2/\sigma^4 = 0$. Substituting these expressions in (7.1.27) and (7.1.28) yields

$$\begin{aligned} \sigma(u,v) &= \sigma^4 \sum_{\tau=-\infty}^{\infty} \psi_{-\tau-u}^*\psi_{-\tau-v} \\ &= \sigma^4 \sum_{j=0}^{\infty} \psi_{j+u-v}\psi_j^* = \sigma^2 r_\epsilon(u-v) \end{aligned}$$

and

$$\begin{aligned} \tilde{\sigma}(u,v) &= \sigma^4 \sum_{\tau=-\infty}^{\infty} \psi_{-\tau-u}^*\psi_{\tau-v}^* \\ &= \sigma^4 \sum_{j=0}^{\infty} \psi_{-j-u-v}^*\psi_j^* = 0. \end{aligned}$$

Therefore, $\boldsymbol{\Sigma} = \sigma^2 \mathbf{R}_\epsilon$ and $\tilde{\boldsymbol{\Sigma}} = \mathbf{0}$. This proves $\sqrt{n}(\hat{\boldsymbol{\varphi}} - \boldsymbol{\varphi}) \xrightarrow{D} \mathrm{N}_c(\mathbf{0}, \sigma^2 \mathbf{R}_\epsilon^{-1})$.

Similarly, when $\{\epsilon_t\}$ is a real AR(m) process of the form (7.1.14), it follows from (7.1.26) and (7.4.13) that

$$\begin{aligned}
\sigma(u,v) &= \sigma^4 \sum_{\tau=-\infty}^{\infty} (\psi_{-\tau-u}\psi_{-\tau-v} + \psi_{-\tau-u}\psi_{\tau-v}) \\
&= \sigma^4 \sum_{j=0}^{\infty} (\psi_{j+u-v}\psi_j + \psi_{-j-u-v}\psi_j) \\
&= \sigma^2 r_\epsilon(u-v).
\end{aligned}$$

Therefore, $\boldsymbol{\Sigma} = \sigma^2 \mathbf{R}_\epsilon$. This proves $\sqrt{n}(\hat{\boldsymbol{\varphi}} - \boldsymbol{\varphi}) \xrightarrow{D} \mathrm{N}(\mathbf{0}, \sigma^2 \mathbf{R}_\epsilon^{-1})$. □

Derivation of (7.1.29) and (7.1.30). Assuming $y_t = 0$ for $t \le 0$, the log likelihood function under a complex Gaussian AR(m) model of the form (7.1.14) with $\{\zeta_t\} \sim \mathrm{GWN}_c(0, \sigma_m^2)$ is given by

$$\log\{p(\mathbf{y}|\boldsymbol{\eta})\} = -n\log\pi - n\log\sigma_m^2 - \|\mathbf{y} + \mathbf{Y}\boldsymbol{\varphi}\|^2/\sigma_m^2,$$

where $\boldsymbol{\eta} := (\boldsymbol{\varphi}, \sigma_m^2)$. Let $\hat{\boldsymbol{\eta}} := (\hat{\boldsymbol{\varphi}}, \hat{\sigma}_m^2)$ denote the maximizer of this function. It is easy to see that $\hat{\boldsymbol{\varphi}}$ and $\hat{\sigma}_m^2$ are given by (7.1.24) and hence,

$$\log\{p(\mathbf{y}|\hat{\boldsymbol{\eta}})\} = -n\log\pi - n\log\hat{\sigma}_m^2 - n. \tag{7.4.14}$$

For any $\mathbf{y}' \in \mathbb{C}^n$, let $\mathbf{Y}'$ be defined from $\mathbf{y}'$ in the same way as $\mathbf{Y}$ in (7.1.24). Then,

$$-2\log\{p(\mathbf{y}'|\hat{\boldsymbol{\eta}})\} = -2\log\{p(\mathbf{y}|\hat{\boldsymbol{\eta}})\} + 2\|\mathbf{y}' + \mathbf{Y}'\hat{\boldsymbol{\varphi}}\|^2/\hat{\sigma}_m^2 - 2n. \tag{7.4.15}$$

For fixed $\mathbf{y}$, let $\hat{p}_1(\cdot) := p(\cdot|\hat{\boldsymbol{\eta}})$ and $\mathbf{y}' \sim p_0(\cdot)$. Then, it follows from (7.4.15) that

$$\begin{aligned}
H(\hat{p}_1 \| p_0) &:= E_{\mathbf{y}'}\{-2\log[p(\mathbf{y}'|\hat{\boldsymbol{\eta}})]\} \\
&= -2\log\{p(\mathbf{y}|\hat{\boldsymbol{\eta}})\} + 2E_{\mathbf{y}'}\{\|\mathbf{y}' + \mathbf{Y}'\hat{\boldsymbol{\varphi}}\|^2\}/\hat{\sigma}_m^2 - 2n.
\end{aligned}$$

Moreover, it is easy to verify that

$$\begin{aligned}
E_{\mathbf{y}'}\{\|\mathbf{y}' + \mathbf{Y}'\hat{\boldsymbol{\varphi}}\|^2\} &= n\{r_\epsilon(0) + \mathbf{r}_\epsilon^H\hat{\boldsymbol{\varphi}} + \hat{\boldsymbol{\varphi}}^H\mathbf{r}_\epsilon + \hat{\boldsymbol{\varphi}}^H\mathbf{R}_\epsilon\hat{\boldsymbol{\varphi}}\} \\
&= n\sigma_m^2 + n(\hat{\boldsymbol{\varphi}} - \boldsymbol{\varphi})^H\mathbf{R}_\epsilon(\hat{\boldsymbol{\varphi}} - \boldsymbol{\varphi}),
\end{aligned}$$

where the second expression is due to (7.1.17). Therefore, the cross entropy $H(\hat{p}_1 \| p_0)$ can be expressed as

$$-2\log\{p(\mathbf{y}|\hat{\boldsymbol{\eta}})\} + 2n\{\sigma_m^2/\hat{\sigma}_m^2 + (\hat{\boldsymbol{\varphi}} - \boldsymbol{\varphi})^H\mathbf{R}_\epsilon(\hat{\boldsymbol{\varphi}} - \boldsymbol{\varphi})/\hat{\sigma}_m^2\} - 2n. \tag{7.4.16}$$

Let the true model of $\mathbf{y}$ be a complex linear process that satisfies the conditions of Theorem 7.5 with $E(\zeta_t^2) = 0$. Then, by Theorem 7.5, $\sqrt{n}(\hat{\boldsymbol{\varphi}} - \boldsymbol{\varphi}) \xrightarrow{D} \mathrm{N}_c(\mathbf{0}, \sigma_m^2\mathbf{R}_\epsilon^{-1})$. According to Lemma 12.2.1(a), we have

$$n(\hat{\boldsymbol{\varphi}} - \boldsymbol{\varphi})^H\mathbf{R}_\epsilon(\hat{\boldsymbol{\varphi}} - \boldsymbol{\varphi}) \stackrel{A}{\sim} \tfrac{1}{2}\sigma_m^2\chi^2(2m). \tag{7.4.17}$$

Combining these results with (7.4.16) leads to

$$H(\hat{p}_1 \| p_0) \stackrel{A}{\sim} -2\log\{p(\mathbf{y}|\hat{\boldsymbol{\eta}})\} + 2\{n + \tfrac{1}{2}\chi^2(2m)\}\sigma_m^2/\hat{\sigma}_m^2 - 2n.$$

Moreover, because $\hat{\sigma}_m^2 \stackrel{a.s.}{\to} \sigma_m^2$ by Theorem 7.4, the right-hand side of the foregoing expression can be further approximated by $-2\log\{p(\mathbf{y}|\hat{\boldsymbol{\eta}})\} + \chi^2(2m)$. Therefore, an approximately unbiased estimator of $E\{H(\hat{p}_1 \| p_0)\}$ is given by

$$\text{AIC}(m) := -2\log\{p(\mathbf{y}|\hat{\boldsymbol{\eta}})\} + 2m. \tag{7.4.18}$$

This is the AIC criterion introduced by Akaike [9]. Owing to (7.4.14), the AIC defined by (7.4.18) is equivalent to the AIC defined by (7.1.29).

To derive (7.1.30), we follow the basic idea of [161] which only considers the real case. First, we approximate $n\hat{\sigma}_m^2/\sigma_m^2$ by $\frac{1}{2}\chi^2(2n-2m)$ (see Theorem 5.1(d) for inspiration). This, together with (7.4.17), suggests that $(\hat{\boldsymbol{\varphi}}-\boldsymbol{\varphi})^H \mathbf{R}_\epsilon(\hat{\boldsymbol{\varphi}}-\boldsymbol{\varphi})/\hat{\sigma}_m^2$ can be approximated by $m(n-m)^{-1}F(2m, 2n-2m)$, which is the ratio of two independent chi-square random variables with $2m$ and $2n-2m$ degrees of freedom. With these approximations, the distribution of $H(\hat{p}_1 \| p_0)$ becomes

$$-\log\{p(\mathbf{y}|\hat{\boldsymbol{\eta}}) + 2n\{2n/\chi^2(2n-2m) + (m/(n-m))F(2m, 2n-2m)\} - 2n.$$

Observe that $E\{1/\chi^2(2n-2m)\} = 1/(2n-2m-2)$ and $E\{F(2m, 2n-2m)\} = (2n-2m)/(2n-2m-2)$. Therefore, an approximately unbiased estimator of the mean cross entropy $E\{H(\hat{p}_1 \| p_0)\}$ is given by

$$\text{AIC}_c(m) := -2\log\{p(\mathbf{y}|\hat{\boldsymbol{\eta}})\} + 2n(n+m)/(n-m-1) - 2n. \tag{7.4.19}$$

Owing to (7.4.14), minimizing this criterion with respect to m is equivalent to minimizing the AICC defined by (7.1.30). □

Proof of Corollary 7.1. For any $\lambda_j \in \Omega_n'$, the m'-neighborhood of λ_j contains all the signal frequencies by definition. This, combined with the fact that $\tilde{W}_{ml} = 0$ for all $|l| \le m' < m$, annihilates the contributions from the sinusoids in the mean and variance of $I_n(\lambda_{j-u})$ for $|u| \le m$. If both λ_j and λ_l belong to Ω_n', then the same annihilation takes place in the covariance between $I_n(\lambda_{j-u})$ and $I_n(\lambda_{l-v})$ for $|u| \le m$ and $|v| \le m$. Therefore, the assertions can be proved in the same way as Theorem 7.1. □

Proof of Theorem 7.8. Let $f_j := f_\epsilon(\lambda_j)$, $f_L := \min\{f_\epsilon(\omega)\}$, and $f_U := \max\{f_\epsilon(\omega)\}$. The first objective of the proof is to show that for any given $\delta \in (0, f_L)$,

$$E\{\Psi_j(f_j \pm \delta)\} = M_j(f_j \pm \delta) + \mathcal{O}(1) \tag{7.4.20}$$

uniformly in $j \in \mathbb{Z}_n := \{0, 1, \ldots, n-1\}$, where

$$M_j(x) := E\{\psi(X f_\epsilon(\lambda_j)/x)\}, \quad X \sim \tfrac{1}{2}\chi^2(2).$$

Toward that end, let us first consider the case where $\{\epsilon_t\}$ is Gaussian white noise, so that $f_\epsilon(\omega) = \sigma^2$ for all ω. Note that for any $x > 0$,

$$E\{\Psi_j(x)\} = \sum_{|l|\le m} W_{ml} E\{\psi(I_n(\lambda_{j-l})/x)\}. \tag{7.4.21}$$

Let $\psi(\cdot)$ be bounded by a constant $c > 0$. Then, we have

$$|E\{\psi(I_n(\lambda_{j-l})/x)\}| \le c$$

for all λ_{j-l} and all $x > 0$. Moreover, if $\lambda_{j-l} \notin \{\omega_k\}$, then, by Theorem 6.2, we have $I_n(\lambda_{j-l})/f_\epsilon(\lambda_{j-l}) \sim \frac{1}{2}\chi^2(2)$, and hence,

$$E\{\psi(I_n(\lambda_{j-l})/x)\} = E\{\psi(X f_\epsilon(\lambda_{j-l})/x)\}.$$

Combining this result with the fact that $f_\epsilon(\lambda_{j-l}) = f_\epsilon(\lambda_j)$ gives

$$E\{\psi(I_n(\lambda_{j-l})/x)\} = M_j(x) \quad \forall\, \lambda_{j-l} \notin \{\omega_k\}. \tag{7.4.22}$$

As a result, (7.4.21) can be written as

$$E\{\Psi_j(f_j \pm \delta)\} = M_j(f_j \pm \delta) + R_j,$$

where $R_j := \sum' W_{ml}[E\{\psi(I_n(\lambda_{j-l})/(f_j \pm \delta))\} - M_j(f_j \pm \delta)]$ and $\sum'$ denotes the sum over $|l| \le m$ such that $\lambda_{j-l} \in \{\omega_k\}$. Clearly, $|R_j| \le 2c\sum' W_{ml}$. Furthermore, because the number of terms in this sum is no more than p, we obtain $|R_j| \le 2cp\,W_{m*}$, where W_{m*} denotes the largest weight. Combining this result with the fact that $W_{m*}^2 \le s_m \to 0$ as $n \to \infty$ yields

$$|R_j| \le 2cp\sqrt{s_m} = \mathcal{o}(1) \tag{7.4.23}$$

uniformly in $j \in \mathbb{Z}_n$. The expression (7.4.20) follows.

Now let $\{\epsilon_t\}$ be Gaussian colored noise with $r_n := \sum_u \min(1, |u|/n)|r_\epsilon(u)| \to 0$ as $n \to \infty$. In this case, $f_\epsilon(\omega)$ is a uniformly continuous function so that under the condition (7.1.5), we can write

$$f_\epsilon(\lambda_{j-l}) - f_\epsilon(\lambda_j) = \mathcal{o}(1) \tag{7.4.24}$$

uniformly in λ_j and λ_{j-l} such that $|l| \le m$. If $\lambda_{j-l} \notin \{\omega_k\}$, then, by Theorem 6.4,

$$I_n(\lambda_{j-l})/(f_\epsilon(\lambda_{j-l}) + \mathcal{O}(r_n)) \sim \tfrac{1}{2}\chi^2(2),$$

in which case,

$$E\{\psi(I_n(\lambda_{j-l})/x)\} = E\{\psi(X(f_\epsilon(\lambda_{j-l}) + \mathcal{O}(r_n))/x)\}.$$

The Lipschitz continuity of $\psi(\cdot)$, combined with (7.4.24), implies that

$$E\{\psi(X(f_\epsilon(\lambda_{j-l}) + \mathcal{O}(r_n))/(f_j \pm \delta))\} = M_j(f_j \pm \delta) + \mathcal{o}(1)$$

uniformly in λ_j and λ_{j-l} such that $|l| \le m$. Therefore, we obtain

$$E\{\Psi_j(f_j \pm \delta)\} = M_j(f_j \pm \delta) + R_j + \mathcal{O}(1),$$

uniformly in j. Combining this result with (7.4.23) yields (7.4.20).

The second objective of the proof is to show that

$$\mathrm{Var}\{\Psi_j(f_j \pm \delta)\} = \mathcal{O}(1) \tag{7.4.25}$$

uniformly in j. Toward that end, we first note that

$$\mathrm{Var}\{\Psi_j(x)\} = \sum_{|l|,|l'|\le m} W_{ml} W_{ml'} \,\mathrm{Cov}\{\psi(I_n(\lambda_{j-l})/x), \psi(I_n(\lambda_{j-l'})/x)\}.$$

If $\{\epsilon_t\}$ is Gaussian white noise, then, by Theorem 6.1, $I_n(\lambda_{j-l})$ and $I_n(\lambda_{j-l'})$ are independent for $l \ne l'$. This implies that

$$\mathrm{Var}\{\Psi_j(x)\} = \sum_{|l|\le m} W_{ml}^2 \,\mathrm{Var}\{\psi(I_n(\lambda_{j-l})/x)\}.$$

Moreover, we have $\mathrm{Var}\{\psi(I_n(\lambda_{j-l})/x)\} \le 2c^2$ for all λ_{j-l} and all $x > 0$. This result, combined with (7.1.5), gives $\mathrm{Var}\{\Psi_j(f_j \pm \delta)\} \le 2c^2 s_m \to 0$ uniformly in j.

In the case of Gaussian colored noise with $\boldsymbol{\epsilon} \sim \mathrm{N}_c(\mathbf{0}, \mathbf{R}_\epsilon)$, let us define $\boldsymbol{\zeta} := [\zeta_0, \zeta_1, \ldots, \zeta_{n-1}]^T := n^{-1/2}\mathbf{F}^H\mathbf{y}$ so that $I_n(\lambda_j) = |\zeta_j|^2$. It follows from the proof of Theorem 6.4 that

$$\boldsymbol{\zeta} \sim \mathrm{N}_c(n^{-1/2}\boldsymbol{\mu}, n^{-1}\mathbf{F}^H\mathbf{R}_\epsilon\mathbf{F}), \quad n^{-1}\mathbf{F}^H\mathbf{R}_\epsilon\mathbf{F} = \mathrm{diag}\{f_\epsilon(\lambda_j)\} + \mathcal{O}(r_n). \tag{7.4.26}$$

Let $F_{j-l,j-l'}(u,v)$, $F_{j-l}(u)$, and $F_{j-l'}(v)$ denote the joint and marginal distribution functions of ζ_{j-l} and $\zeta_{j-l'}$. Then, (7.4.26) implies that for $l \ne l'$,

$$F_{j-l,j-l'}(u,v) - F_{j-l}(u)F_{j-l'}(v) \to 0$$

uniformly in j, l, and l' for any given u and v. Moreover, we can write

$$\begin{aligned}&\mathrm{Cov}\{\psi(I_n(\lambda_{j-l})/(f_j \pm \delta)), \psi(I_n(\lambda_{j-l'})/(f_j \pm \delta))\}\\&= \int h_{j-l}(u)\, h_{j-l'}(v)\, dF_{j-l,j-l'}(u,v),\end{aligned}$$

where $h_j(u) := \psi(|u|^2/(f_j \pm \delta)) - E\{\psi(I_n(\lambda_j)/(f_j \pm \delta))\}$. Because $|h_j(u)| \le 2c$ for all j, the bounded convergence theorem gives

$$\begin{aligned}&\int h_{j-l}(u)\, h_{j-l'}(v)\, dF_{j-l,j-l'}(u,v)\\&\quad - \int h_{j-l}(u)\, h_{j-l'}(v)\, dF_{j-l}(u)\, dF_{j-l'}(v) = \mathcal{O}(1)\end{aligned}$$

uniformly in j, l, and l'. Combining this result with the fact that $\int h_j(u)\,dF_j(u) = 0$ for all j leads to the conclusion that for any $l \neq l'$,

$$\mathrm{Cov}\{\psi(I_n(\lambda_{j-l})/(f_j \pm \delta)), \psi(I_n(\lambda_{j-l'})/(f_j \pm \delta))\} = \mathcal{o}(1)$$

uniformly in λ_{j-l} and $\lambda_{j-l'}$. Therefore,

$$\begin{aligned}\mathrm{Var}\{\Psi_j(f_j \pm \delta)\} &= \sum_{|l| \le m} W_{ml}^2 \mathrm{Var}\{\psi(I_n(\lambda_{j-l})/(f_j \pm \delta))\} + \mathcal{o}(1) \\ &= \mathcal{O}(s_m) + \mathcal{o}(1).\end{aligned}$$

Combining this result with (7.1.5) leads to (7.4.25).

The third objective of the proof is to show that f_j is the unique solution to the equation $M_j(x) = 0$. Toward that end, observe that (7.2.9) implies

$$M_j(f_j) = 0.$$

Furthermore, because $\psi(\cdot)$ is a nondecreasing function, $M_j(x)$ is nonincreasing in $x > 0$. Therefore, for any $\delta > 0$,

$$M_j(f_j + \delta) \le M_j(f_j) = 0$$

and for any $0 < \delta < f_j$,

$$M_j(f_j - \delta) \ge M_j(f_j) = 0.$$

In addition, the monotonicity of the ψ function gives

$$\psi(x f_j/(f_j + \delta)) \le \psi(x f_U/(f_U + \delta)) \le \psi(x)$$

for all $x > 0$, all j, and all $\delta > 0$. Let $\psi(x)$ be strictly increasing in the interval $[1 - c_L, 1 + c_U]$ for some constants $c_L \in (0,1)$ and $c_U \in (0,\infty)$. Then, for any given $0 < \delta < \delta_U := f_U c_L/(1 - c_L)$, we have $1 - c_L < f_U/(f_U + \delta) < 1 + c_U$, so that the last inequality in the foregoing equation is strict at $x = 1$. The continuity of $\psi(x)$ ensures that the strict inequality also holds in a small neighborhood of $x = 1$. This implies that

$$M_j(f_j + \delta) \le M_U(\delta) := E\{\psi(X f_U/(f_U + \delta))\} < E\{\psi(X)\} = 0$$

for any $0 < \delta < \delta_U$ and for all j. Similarly,

$$\psi(x f_j/(f_j - \delta)) \ge \psi(x f_U/(f_U - \delta)) \ge \psi(x),$$

for all $x > 0$, all j, and $0 < \delta < f_L$. If $0 < \delta < \delta_L := \min\{f_L, f_U c_U/(1 + c_U)\}$, then we have $1 - c_L < f_U/(f_U - \delta) < 1 + c_U$, so the last inequality is strict at $x = 1$ as well as in the small neighborhood of $x = 1$. Therefore,

$$M_j(f_j - \delta) \ge M_L(\delta) := E\{\psi(X f_U/(f_U - \delta))\} > E\{\psi(X)\} = 0$$

for any $0 < \delta < \delta_L$ and for all j. Combining these results with the monotonicity of $M_j(x)$ proves that f_j is the unique solution of $M_j(x) = 0$.

The final objective is to show that the solution to the equation $\Psi_j(x) = 0$, which we denote by $\tilde{f}_j$, has the property that $\tilde{f}_j - f_j \xrightarrow{P} 0$ as $n \to \infty$. This can be easily done by following the steps of [404, p. 47]: Because $\Psi_j(x)$ is nonincreasing in x and $\Psi_j(\tilde{f}_j) = 0$, it follows that for any $0 < \delta < f_L$, if $\Psi_j(f_j - \delta) > 0$ and $\Psi_j(f_j + \delta) < 0$ then $f_j - \delta < \tilde{f}_j < f_j + \delta$. Therefore,

$$P(\Psi_j(f_j - \delta) > 0, \Psi_j(f_j + \delta) < 0) \le P(f_j - \delta < \tilde{f}_j < f_j + \delta). \tag{7.4.27}$$

Based on (7.4.20) and (7.4.25), we obtain, by Proposition 2.16(b),

$$P(|\Psi_j(f_j \pm \delta) - M_j(f_j \pm \delta)| \ge \epsilon) \to 0 \tag{7.4.28}$$

uniformly in j for any $\epsilon > 0$. Because $M_j(f_j - \delta) \ge M_L(\delta) > 0$ and $M_j(f_j + \delta) \le M_U(\delta) < 0$ for all j, the left-hand side of (7.4.27) approaches unity uniformly in j as $n \to \infty$. This leads to $P(f_j - \delta < \tilde{f}_j < f_j + \delta) \to 1$ uniformly in j.

Note that the presence of sinusoids only affects R_j through the number of terms included in its definition. Because R_j is bounded by $2cs_m^{1/2}$ times the number of terms included in R_j, we obtain $R_j = \mathcal{O}(1)$ uniformly in $j \in \mathbb{Z}_n$ even if the number of these terms takes the form $\mathcal{O}(s_m^{-1/2})$. Therefore, the theorem can be generalized to the situation where the number of periodogram ordinates that do not have a central chi-square distribution takes the form $\mathcal{O}(s_m^{-1/2})$. □

Notes

Under the assumption that $\{\epsilon_t\}$ is a real linear process with finite fourth moments and strictly absolutely summable ACF satisfying $\sum |u|^\delta |r_\epsilon(u)| < \infty$ for some $\delta > 0$, it is shown in [12, pp. 522–540] that in the absence of sinusoids, the lag-window spectral estimator with $m = \mathcal{O}(n^{1/(2\delta+1)})$ and $1 - w(x) = \mathcal{O}(|x|^r)$ as $x \to 0$ for some $r > \delta$ has an asymptotic normal distribution with mean $f_\epsilon(\omega)$ and with variance $f_\epsilon^2(\omega)c_n$ for $\omega \neq 0$ and π, and variance $2f_\epsilon^2(\omega)c_n$ for $\omega = 0$ or π, where $c_n := (m/n)\int_{-1}^{1} w^2(x)\,dx$. This result can be used to construct asymptotic confidence intervals for the SDF. A chi-square approximation is suggested in [37] for the finite sample distribution of the lag-window spectral estimator, which also gives approximate confidence intervals.

Wavelet techniques are also employed to smooth the periodogram while preserving certain important singularities [78] [118] [263] [411]. Other techniques of periodogram smoothing include the spline method [75] [197] [410], the local polynomial method [104], the penalized Whittle likelihood method [285], and the Bayesian method [67] [423]. An adaptive technique of smoothing parameter selection for the periodogram smoother is considered in [264] with the aim of providing a better estimation for the global spectral peak location.

Further discussions on the maximum entropy method and Burg's algorithm can be found in [50], [208], [250], [400], [401], and [403].

In his 1898 paper [334], Schuster referred to the sunspot number as a variable quantity that has "obvious periodicities." He not only coined the term periodogram but also proclaimed that "the periodogram of sunspots would show a 'band' in the neighborhood of a period of eleven years." A detailed analysis of sunspot numbers was presented later in his 1906 paper [335].

As an alternative to Schuster's periodogram, Yule, in 1927, employed an AR(2) model in his investigation of the periodicity of sunspot numbers [434]. Four years later [415], the British physicist and statistician Gilbert Thomas Walker (1868–1958) derived what is now known as the Yule-Walker equations for the general AR processes and pointed out that an undisturbed AR difference equation can be satisfied not only by a periodic time series but also by the autocovariances of that series. This connection motivated him to recommend the use of a graph of autocovariances, which he called the correlation-periodogram, combined with a Fourier analysis if necessary, to detect hidden periodicities.

The term "Yule-Walker" estimator is a slight misnomer for the solution of the Yule-Walker equations with the sample autocovariances in place of the ACF, because neither Yule nor Walker mentioned this method in their original papers. In fact, Yule in [434] employed the least-squares linear prediction method to estimate the AR parameters. It is not clear which variation of the least-squares estimators was used. Our re-analysis of the sunspot numbers that Yule analyzed (from year 1749 to 1924) cannot exactly reproduce the estimates reported in [434]. The closest result is given by the forward linear prediction estimator with an additional constant on the right-hand side of the AR equation, as suggested by Yule, to account for a possibly non-zero mean of the forcing.

Mann and Wald [252] are credited as the first to rigorously study the consistency and asymptotic normality for fixed-order AR parameters in the real case. Other properties of the AR spectral estimator such as the bias, the confidence bands, and the spectral peaks, are investigated in [52], [101], [200], [243], [269], [270], [340], [364], and [437]. The AR model is also used to help bootstrap the periodogram for analysis of its finite sample properties [201]. Asymptotic properties of the AR model when the order depends on the sample size are investigated in [10] based on the uniform convergence of sample autocovariances.

The AR model belongs to a larger family of ARMA models [42] that can also be employed for spectral estimation. Order selection criteria such as the AIC can be generalized to the ARMA case [66]. The ARMA model can be more parsimonious (i.e., with fewer parameters) than the AR model in approximating a spectrum with both sharp peaks and deep valleys, but a general ARMA model is more difficult to estimate. Nonlinear techniques such as the Gaussian maximum likelihood method are usually required. The likelihood function for a Gaussian ARMA model can be computed efficiently by the innovations algorithm in Proposition 2.10. This facilitates the calculation of the Gaussian maximum likelihood estimator using a general-purpose optimization procedure [46, Section 8.7]. See [259] for a discussion on some computational issues of this estimator.

Chapter 8

Maximum Likelihood Approach

The maximum likelihood (ML) approach is one of the most important statistical methodologies for parameter estimation. It is based on the fundamental assumption that the underlying probability distribution of the observations belongs to a family of distributions indexed by unknown parameters. The ML estimator of the unknown parameters is the maximizer of the likelihood function, corresponding to the probability distribution in the family that gives the observations the highest chance of occurrence. In many cases, the ML estimator enjoys an optimality property called asymptotic efficiency. A manifestation of this property is that the asymptotic variance of the ML estimator attains the corresponding CRLB in addition to having an asymptotic normal distribution.

In this chapter, we investigate the ML estimator of the sinusoidal parameters derived under two types of noise distributions: the Gaussian distribution and the Laplace distribution. The Gaussian assumption leads to a nonlinear least-squares (NLS) procedure that extends the periodogram maximization method discussed in Chapter 6. The Laplace assumption leads to a nonlinear least absolute deviations (NLAD) procedure that offers more robust estimates than the NLS procedure under the conditions of outliers and heavy-tailed noise.

8.1 Maximum Likelihood Estimation

Consider the observations given by

$$y_t = x_t + \epsilon_t \qquad (t = 1, \ldots, n), \tag{8.1.1}$$

where $\{\epsilon_t\}$ is a random process and $\{x_t\}$ takes the form (2.1.1), (2.1.2), (2.1.5), or (2.1.7). In matrix notation, we can express $\mathbf{y} := [y_1, \ldots, y_n]^T$ as

$$\mathbf{y} = \mathbf{x}(\boldsymbol{\theta}) + \boldsymbol{\epsilon}, \tag{8.1.2}$$

where $\boldsymbol{\epsilon} := [\epsilon_1, \ldots, \epsilon_n]^T$ and $\mathbf{x}(\boldsymbol{\theta}) := [x_1, \ldots, x_n]^T$, with $\boldsymbol{\theta}$ denoting the unknown parameters in $\{x_t\}$. Because $\mathbf{x}(\boldsymbol{\theta})$ is deterministic (conditioning on the phase parameter if necessary), the probability distribution of $\mathbf{y}$ is completely determined by the probability distribution of $\boldsymbol{\epsilon}$, given $\boldsymbol{\theta}$. Therefore, the ML approach is applicable when the PDF of $\boldsymbol{\epsilon}$ can be assumed to take a parametric form.

More precisely, suppose that the PDF of $\boldsymbol{\epsilon}$ can be expressed as $p(\cdot|\boldsymbol{\eta})$, where $\boldsymbol{\eta} \in \Xi$ is an unknown parameter and $p(\cdot|\boldsymbol{\eta})$ is a fully specified function in the domain of $\boldsymbol{\epsilon}$ for each fixed $\boldsymbol{\eta} \in \Xi$, with Ξ being a known subset of $\mathbb{R}^r$ for some $r \geq 1$. Under this assumption, the PDF of $\mathbf{y}$ takes the form $p(\mathbf{y}-\mathbf{x}(\boldsymbol{\theta})|\boldsymbol{\eta})$. Therefore, the joint ML estimator of $(\boldsymbol{\theta}, \boldsymbol{\eta})$ is given by

$$(\hat{\boldsymbol{\theta}}, \hat{\boldsymbol{\eta}}) := \arg \max_{\tilde{\boldsymbol{\theta}} \in \Theta, \tilde{\boldsymbol{\eta}} \in \Xi} p(\mathbf{y}-\mathbf{x}(\tilde{\boldsymbol{\theta}})|\tilde{\boldsymbol{\eta}}). \tag{8.1.3}$$

The set Θ depends on the particular form of $\{x_t\}$ and is known in advance: Under the Cartesian CSM (2.1.5), we have $\boldsymbol{\theta} := [\beta_1, \omega_1, \ldots, \beta_p, \omega_p]^T$ and

$$\Theta := \{[\tilde{\beta}_1, \tilde{\omega}_1, \ldots, \tilde{\beta}_p, \tilde{\omega}_p]^T : \tilde{\beta}_1, \ldots, \tilde{\beta}_p \in \mathbb{C}; 0 < \tilde{\omega}_1 < \cdots < \tilde{\omega}_p < 2\pi\}.$$

Under the Cartesian RSM (2.1.1), we have $\boldsymbol{\theta} := [A_1, B_1, \omega_1, \ldots, A_q, B_q, \omega_q]^T$ and

$$\Theta := \{[\tilde{A}_1, \tilde{B}_1, \tilde{\omega}_1, \ldots, \tilde{A}_q, \tilde{B}_q, \tilde{\omega}_q]^T : \tilde{A}_1, \tilde{B}_1, \ldots, \tilde{A}_q, \tilde{B}_q \in \mathbb{R}; 0 < \tilde{\omega}_1 < \cdots < \tilde{\omega}_q < \pi\}.$$

Additional constraints can be imposed to reduce the size of Θ based on prior knowledge about $\boldsymbol{\theta}$. For example, in the complex case, if it is known *a priori* that the signal frequencies satisfy the separation condition

$$\Delta := \min_{k \neq k'} \{|\omega_k - \omega_{k'}|, 2\pi - |\omega_k - \omega_{k'}|\} \geq \delta$$

for some known constant $\delta > 0$, then a constraint of the form

$$\min_{k \neq k'} \{|\tilde{\omega}_k - \tilde{\omega}_{k'}|, 2\pi - |\tilde{\omega}_k - \tilde{\omega}_{k'}|\} \geq c,$$

with $0 < c < \delta$, can be imposed on the frequency variables.

To compute the ML estimator in (8.1.3), it is usually advantageous to deal with the signal parameter $\boldsymbol{\theta}$ and the noise parameter $\boldsymbol{\eta}$ alternately while holding the other fixed. This alternating optimization technique, also known as block coordinate descent/ascent [31, Section 2.7], leads to a generic iterative procedure for the joint ML estimation: Given a suitable initial value $\hat{\boldsymbol{\eta}}_0 \in \Xi$ and for $m = 1, 2, \ldots$,

1. Compute $\hat{\boldsymbol{\theta}}_m := \arg\max_{\tilde{\boldsymbol{\theta}}} p(\mathbf{y}-\mathbf{x}(\tilde{\boldsymbol{\theta}})|\hat{\boldsymbol{\eta}}_{m-1})$;
2. Compute $\hat{\boldsymbol{\eta}}_m := \arg\max_{\tilde{\boldsymbol{\eta}}} p(\mathbf{y}-\mathbf{x}(\hat{\boldsymbol{\theta}}_m)|\tilde{\boldsymbol{\eta}})$.

In this procedure, the first step solves a nonlinear regression problem of estimating $\boldsymbol{\theta}$ under the working assumption that $\mathbf{y} = \mathbf{x}(\boldsymbol{\theta}) + \boldsymbol{\epsilon}_m$ and $\boldsymbol{\epsilon}_m \sim p(\cdot|\hat{\boldsymbol{\eta}}_{m-1})$; the second step yields an ML estimate of $\boldsymbol{\eta}$ based on the observed residual $\hat{\boldsymbol{\epsilon}}_m := \mathbf{y}-\mathbf{x}(\hat{\boldsymbol{\theta}}_m)$ and the working assumption $\hat{\boldsymbol{\epsilon}}_m \sim p(\cdot|\boldsymbol{\eta})$.

In practice, the ML estimator is often derived under a simple noise model, either because the true noise distribution is difficult to determine or because the ML estimator corresponding to the true noise distribution is difficult to compute. In these cases, it becomes necessary to look beyond the optimality of the ML estimator under the assumed model and investigate the statistical properties of the estimator under more general noise conditions.

This kind of analysis is best illustrated by the sample mean of an i.i.d. sample of size n from a population with mean μ and variance σ^2. The sample mean is the ML estimator of μ if the distribution is Gaussian, and it is distributed as $\mathrm{N}(\mu, \sigma^2/n)$ under the Gaussian assumption. Knowing this property of the sample mean is clearly not enough when the true distribution is nonGaussian. Therefore, it is helpful to be assured that, by the central limit theorem, the sample mean behaves similarly under any distribution with finite variance when the sample size n is sufficiently large.

In the remainder of this chapter, we examine several important cases of the ML estimation in terms of statistical properties and computational issues.

8.2 Maximum Likelihood under Gaussian White Noise

The most important example is that of Gaussian white noise with unknown variance. In the real case, let $\boldsymbol{\epsilon} \sim \mathrm{N}(\mathbf{0}, \sigma^2\mathbf{I})$, or equivalently, let $\{\epsilon_t\}$ be an i.i.d. sequence with PDF $p(x) = (2\pi\sigma^2)^{-1/2} \exp(-\frac{1}{2}x^2/\sigma^2)$ $(x \in \mathbb{R})$, where $\sigma^2 > 0$ is an unknown parameter. In the complex case, let $\boldsymbol{\epsilon} \sim \mathrm{N}_c(\mathbf{0}, \sigma^2\mathbf{I})$, or equivalently, let the real and imaginary parts of $\{\epsilon_t\}$ be mutually independent i.i.d. sequences with PDF $p(x) = (\pi\sigma^2)^{-1/2} \exp(-x^2/\sigma^2)$ $(x \in \mathbb{R})$. Under both conditions, the ML estimator of $(\boldsymbol{\theta}, \sigma^2)$ based on $\mathbf{y}$ in (8.1.2) is given by

$$\hat{\boldsymbol{\theta}} := \arg\min_{\tilde{\boldsymbol{\theta}} \in \Theta} \|\mathbf{y} - \mathbf{x}(\tilde{\boldsymbol{\theta}})\|^2, \tag{8.2.1}$$

$$\hat{\sigma}^2 := n^{-1} \|\mathbf{y} - \mathbf{x}(\hat{\boldsymbol{\theta}})\|^2. \tag{8.2.2}$$

As can be seen, the estimation of $\boldsymbol{\theta}$ is conveniently decoupled from the estimation of σ^2, so that knowledge of the noise variance is not required to obtain the ML estimator of the signal parameter. The estimator $\hat{\boldsymbol{\theta}}$ in (8.2.1) is the solution of a nonlinear least-squares problem and the estimator $\hat{\sigma}^2$ in (8.2.2) is simply the minimum SSE divided by the sample size n.

Nothing prevents $\hat{\boldsymbol{\theta}}$ in (8.2.1) from being used to estimate the signal parameter $\boldsymbol{\theta}$ when the noise $\boldsymbol{\epsilon}$ is not Gaussian white noise. In this case, the estimator in (8.2.1) is often referred to as the *nonlinear least-squares* (NLS) estimator or the *Gaussian maximum likelihood* (GML) estimator. Before investigating its statistical properties under the Gaussian and nonGaussian conditions, let us first consider the computation of the NLS estimator.

8.2.1 Multivariate Periodogram

Standard optimization procedures can be used to compute the minimizer in (8.2.1) jointly for the amplitude and frequency parameters. More details on these general procedures can be found [31], [36], and [273]. In the following, let us consider an alternative approach known sometimes as variable projection [123]. By observing that $\mathbf{x}(\tilde{\boldsymbol{\theta}})$ is linear and unconstrained in the amplitude variable, we eliminate the amplitude variable first and reduce the joint minimization problem (8.2.1) into a minimization problem for the frequency variable only.

To be more specific, consider the complex case where $\{x_t\}$ is given by (2.1.5). In this case, we obtain (8.1.2) with $\boldsymbol{\theta} = [\beta_1, \omega_1, \ldots, \beta_p, \omega_p]^T$ and

$$\mathbf{x}(\boldsymbol{\theta}) = \mathbf{F}(\boldsymbol{\omega})\boldsymbol{\beta}, \tag{8.2.3}$$

where

$$\boldsymbol{\beta} := [\beta_1, \ldots, \beta_p]^T \in \mathbb{C}^p, \quad \boldsymbol{\omega} := (\omega_1, \ldots, \omega_p) \in \Omega,$$
$$\mathbf{F}(\boldsymbol{\omega}) := [\mathbf{f}(\omega_1), \ldots, \mathbf{f}(\omega_p)], \quad \mathbf{f}(\omega) := [\exp(i\omega), \ldots, \exp(in\omega)]^T,$$

and

$$\Omega := \{(\tilde{\omega}_1, \ldots, \tilde{\omega}_p) : 0 < \tilde{\omega}_1 < \cdots < \tilde{\omega}_p < 2\pi\}.$$

Therefore, the NLS estimator in (8.2.1) can be expressed as

$$(\hat{\boldsymbol{\beta}}, \hat{\boldsymbol{\omega}}) := \arg \min_{\tilde{\boldsymbol{\beta}} \in \mathbb{C}^p, \tilde{\boldsymbol{\omega}} \in \Omega} \|\mathbf{y} - \mathbf{F}(\tilde{\boldsymbol{\omega}})\tilde{\boldsymbol{\beta}}\|^2. \tag{8.2.4}$$

The NLS estimator of σ^2 in (8.2.2) can be expressed as

$$\hat{\sigma}^2 = n^{-1}\|\mathbf{y} - \mathbf{F}(\hat{\boldsymbol{\omega}})\hat{\boldsymbol{\beta}}\|^2. \tag{8.2.5}$$

Because the ML estimators in general are invariant to parameter transformation, the NLS estimators of (A_k, B_k) and (C_k, ϕ_k) are given by

$$\begin{cases} \hat{A}_k := \Re(\hat{\beta}_k), & \hat{B}_k := -\Im(\hat{\beta}_k), \\ \hat{C}_k := |\hat{\beta}_k|, & \hat{\phi}_k := \arctan\{\Im(\hat{\beta}_k), \Re(\hat{\beta}_k)\}, \end{cases} \tag{8.2.6}$$

where $\hat{\beta}_k$ is the kth component of $\hat{\boldsymbol{\beta}}$ given by (8.2.4).

For any fixed $\tilde{\boldsymbol{\omega}} \in \Omega$, minimizing $\|\mathbf{y} - \mathbf{F}(\tilde{\boldsymbol{\omega}})\tilde{\boldsymbol{\beta}}\|^2$ with respect to $\tilde{\boldsymbol{\beta}} \in \mathbb{C}^p$ is an unconstrained linear least-squares problem that has a simple solution

$$\hat{\boldsymbol{\beta}}(\tilde{\boldsymbol{\omega}}) := \arg \min_{\tilde{\boldsymbol{\beta}} \in \mathbb{C}^p} \|\mathbf{y} - \mathbf{F}(\tilde{\boldsymbol{\omega}})\tilde{\boldsymbol{\beta}}\|^2 = \{\mathbf{F}^H(\tilde{\boldsymbol{\omega}})\mathbf{F}(\tilde{\boldsymbol{\omega}})\}^{-1}\mathbf{F}^H(\tilde{\boldsymbol{\omega}})\mathbf{y}. \tag{8.2.7}$$

The solution to the joint estimation problem in (8.2.4) can be expressed as

$$\hat{\boldsymbol{\omega}} = \arg\min_{\tilde{\boldsymbol{\omega}}\in\Omega} \|\mathbf{y} - \mathbf{F}(\tilde{\boldsymbol{\omega}})\hat{\boldsymbol{\beta}}(\tilde{\boldsymbol{\omega}})\|^2 \tag{8.2.8}$$

and

$$\hat{\boldsymbol{\beta}} = \hat{\boldsymbol{\beta}}(\hat{\boldsymbol{\omega}}) = \{\mathbf{F}^H(\hat{\boldsymbol{\omega}})\mathbf{F}(\hat{\boldsymbol{\omega}})\}^{-1}\mathbf{F}^H(\hat{\boldsymbol{\omega}})\mathbf{y}. \tag{8.2.9}$$

Because $\hat{\boldsymbol{\beta}}$ in (8.2.9) is easy to compute once $\hat{\boldsymbol{\omega}}$ is given, the only challenge that remains is the computation of the frequency estimator $\hat{\boldsymbol{\omega}}$ in (8.2.8).

Obviously, one can employ a general procedure, such as the Gauss-Newton algorithm (GNA) [325], to compute $\hat{\boldsymbol{\omega}}$ in (8.2.8). For any $m = 0,1,2,\ldots$, let $\hat{\boldsymbol{\omega}}_m$ denote an estimate of $\boldsymbol{\omega}$ and let $\mathbf{G}_m$ denote the Jacobian matrix of $\mathbf{F}(\tilde{\boldsymbol{\omega}})\hat{\boldsymbol{\beta}}(\tilde{\boldsymbol{\omega}})$ evaluated at $\hat{\boldsymbol{\omega}}_m$. Then, the GNA iteration can be expressed as

$$\hat{\boldsymbol{\omega}}_{m+1} := \hat{\boldsymbol{\omega}}_m + \{\Re(\mathbf{G}_m^H\mathbf{G}_m)\}^{-1}\Re\{\mathbf{G}_m^H(\mathbf{y} - \hat{\mathbf{x}}_m)\}, \tag{8.2.10}$$

where $\hat{\mathbf{x}}_m := \mathbf{F}(\hat{\boldsymbol{\omega}}_m)\hat{\boldsymbol{\beta}}(\hat{\boldsymbol{\omega}}_m)$ is the estimated sinusoidal signal. In this algorithm, the new frequency estimate $\hat{\boldsymbol{\omega}}_{m+1}$ is obtained by replacing the nonlinear function $\mathbf{F}(\tilde{\boldsymbol{\omega}})\hat{\boldsymbol{\beta}}(\tilde{\boldsymbol{\omega}})$ in (8.2.8) with a linear approximation $\hat{\mathbf{x}}_m + \mathbf{G}_m(\tilde{\boldsymbol{\omega}} - \hat{\boldsymbol{\omega}}_m)$, which is the Taylor expansion at $\hat{\boldsymbol{\omega}}_m$, and then solving the resulting linear least-squares problem with respect to $\tilde{\boldsymbol{\omega}} \in \mathbb{R}^p$ without constraint.

Derivative-based techniques such as the GNA depend heavily on the validity of the local approximation and require not only good initial values but also proper behavior of the objective function near the minimizer. Alternatively, one may employ derivative-free techniques such as the Nelder-Mead algorithm [266]. This algorithm is a direct search technique based on simplex for unconstrained nonlinear optimization. It is implemented in R as the default choice for the function `optim` and in MATLAB® as `fminsearch`. The Nelder-Mead algorithm turns out to be much less sensitive to initial values than the GNA in (8.2.10) for computing the NLS frequency estimates when the signal frequencies are closely spaced, in which case the minimizer tends to appear near the boundary of Ω where the linear approximation in the GNA becomes problematic.

Let us examine the problem (8.2.8) more closely. First, define

$$\mathbf{Q}(\tilde{\boldsymbol{\omega}}) := \mathbf{F}(\tilde{\boldsymbol{\omega}})\{\mathbf{F}^H(\tilde{\boldsymbol{\omega}})\mathbf{F}(\tilde{\boldsymbol{\omega}})\}^{-1}\mathbf{F}^H(\tilde{\boldsymbol{\omega}}). \tag{8.2.11}$$

Then, by simple matrix algebra, we obtain

$$\|\mathbf{y} - \mathbf{F}(\tilde{\boldsymbol{\omega}})\hat{\boldsymbol{\beta}}(\tilde{\boldsymbol{\omega}})\|^2 = \mathbf{y}^H\{\mathbf{I} - \mathbf{Q}(\tilde{\boldsymbol{\omega}})\}\mathbf{y} = \|\mathbf{y}\|^2 - \mathbf{y}^H\mathbf{Q}(\tilde{\boldsymbol{\omega}})\mathbf{y}. \tag{8.2.12}$$

Therefore, the NLS estimator $\hat{\boldsymbol{\omega}}$ in (8.2.8) can be expressed as

$$\hat{\boldsymbol{\omega}} = \arg\max_{\tilde{\boldsymbol{\omega}}\in\Omega} \mathbf{y}^H\mathbf{Q}(\tilde{\boldsymbol{\omega}})\mathbf{y}. \tag{8.2.13}$$

Note that $\mathbf{Q}(\tilde{\boldsymbol{\omega}})$ is the projection matrix onto the column space of $\mathbf{F}(\tilde{\boldsymbol{\omega}})$. It is also known in statistics as the hat matrix that converts the data vector $\mathbf{y}$ into an estimate $\hat{\mathbf{x}} := \mathbf{Q}(\tilde{\boldsymbol{\omega}})\mathbf{y} = \mathbf{F}(\tilde{\boldsymbol{\omega}})\hat{\boldsymbol{\beta}}(\tilde{\boldsymbol{\omega}})$ of the signal vector $\mathbf{x} := \mathbf{x}(\boldsymbol{\theta})$. Therefore, the NLS estimator in (8.2.13) can be interpreted as one that maximizes the correlation between the data vector $\mathbf{y}$ and the estimated signal $\hat{\mathbf{x}}$.

In the special case of $p = 1$ with $\tilde{\boldsymbol{\omega}} = \omega \in (0, 2\pi)$, we have $\mathbf{F}(\omega) = \mathbf{f}(\omega)$ and $\mathbf{Q}(\omega) = n^{-1}\mathbf{f}(\omega)\mathbf{f}^H(\omega)$. Hence,

$$\mathbf{y}^H\mathbf{Q}(\omega)\mathbf{y} = n^{-1}\|\mathbf{f}^H(\omega)\mathbf{y}\|^2 = I_n(\omega), \tag{8.2.14}$$

where

$$I_n(\omega) := n^{-1}\left|\sum_{t=1}^{n} y_t \exp(-it\omega)\right|^2$$

is just the continuous periodogram of $\mathbf{y}$. This means that the NLS estimator given by (8.2.13) with $p = 1$ is nothing but the periodogram maximizer, or the PM estimator, which is discussed in Chapter 6 (Section 6.4.2).

Note that in the case of $p = 1$ the Jacobian matrix $\mathbf{G}_m$ and the estimated signal $\hat{\mathbf{x}}_m$ in (8.2.10) can be expressed as

$$\begin{aligned}
\mathbf{G}_m &= n^{-1}\{\mathbf{f}(\hat{\omega}_m)\mathbf{f}^H(\hat{\omega}_m) + \mathbf{f}(\hat{\omega}_m)\dot{\mathbf{f}}^H(\hat{\omega}_m)\}\mathbf{y}, \\
\hat{\mathbf{x}}_m &= n^{-1}\mathbf{f}(\hat{\omega}_m)\mathbf{f}^H(\hat{\omega}_m)\mathbf{y},
\end{aligned}$$

where $\dot{\mathbf{f}}(\omega) := i[\exp(i\omega), \dots, n\exp(in\omega)]^T$ is the derivative of $\mathbf{f}(\omega)$. Therefore, with the notation $z_n(\omega) := n^{-1}\mathbf{f}^H(\omega)\mathbf{y}$ and $\dot{z}_n(\omega) := n^{-1}\dot{\mathbf{f}}^H(\omega)\mathbf{y}$, the GNA in (8.2.10) can be written as a fixed point iteration $\hat{\omega}_{m+1} = \phi(\hat{\omega}_m)$ for the mapping

$$\phi(\omega) := \omega - \dot{I}_n(\omega)/J_n(\omega),$$

where

$$\begin{aligned}
\dot{I}_n(\omega) &:= 2n\Re\{z_n^*(\omega)\dot{z}_n(\omega)\}, \\
J_n(\omega) &:= -\tfrac{1}{3}(n+1)(2n+1)I_n(\omega) - 2n[(n+1)\Im\{z_n^*(\omega)\dot{z}_n(\omega)\} + |\dot{z}_n(\omega)|^2].
\end{aligned}$$

This is similar to the modified Newton-Raphson algorithm discussed in Chapter 6 (Section 6.4.2) for periodogram maximization, where $J_n(\omega)$ is defined by (6.4.12). Both algorithms have similar requirements on the initial values which are less stringent than the ordinary Newton-Raphson algorithm.

For $p > 1$, consider the case where the $\tilde{\omega}_k$ are Fourier frequencies of the form $2\pi j_k/n$ for some integers j_k. In this case, we obtain $\mathbf{F}^H(\tilde{\boldsymbol{\omega}})\mathbf{F}(\tilde{\boldsymbol{\omega}}) = n\mathbf{I}$ and

$$\mathbf{y}^H\mathbf{Q}(\tilde{\boldsymbol{\omega}})\mathbf{y} = n^{-1}\|\mathbf{F}^H(\tilde{\boldsymbol{\omega}})\mathbf{y}\|^2 = \sum_{k=1}^{p} I_n(\tilde{\omega}_k). \tag{8.2.15}$$

Therefore, if the maximization problem in (8.2.13) is solved with the $\tilde{\omega}_k$ confined on the Fourier grid, then it is equivalent to identifying the location of the p largest ordinates in the ordinary discrete periodogram, which is just the DFT frequency estimator discussed in Chapter 6 (Section 6.1). Thus, the DFT estimator can be regarded as a *confined* NLS estimator on the Fourier grid.

As we can see, the multivariate function

$$P_n(\tilde{\boldsymbol{\omega}}) := \mathbf{y}^H \mathbf{Q}(\tilde{\boldsymbol{\omega}})\mathbf{y} = \|\mathbf{y}\|^2 - \|\mathbf{y} - \mathbf{F}(\tilde{\boldsymbol{\omega}})\hat{\boldsymbol{\beta}}(\tilde{\boldsymbol{\omega}})\|^2 \qquad (\tilde{\boldsymbol{\omega}} \in \Omega) \tag{8.2.16}$$

is a generalization of the ordinary univariate periodogram $I_n(\omega)$ in the sense of (8.2.14) and (8.2.15). For this reason, we call $P_n(\tilde{\boldsymbol{\omega}})$ in (8.2.16) the p-dimensional *multivariate periodogram* of $\mathbf{y}$. The NLS estimator given by (8.2.13) is the global maximizer of the multivariate periodogram. Moreover, it is easy to see that $P_n(\tilde{\boldsymbol{\omega}})$ coincides with the generalized log likelihood ratio statistic for testing the hypothesis $H_0 : \mathbf{y} = \boldsymbol{\epsilon}$ against the hypothesis $H_1 : \mathbf{y} = \mathbf{F}(\tilde{\boldsymbol{\omega}})\boldsymbol{\beta} + \boldsymbol{\epsilon}$ for some unknown $\boldsymbol{\beta} \neq \mathbf{0}$ under the assumption $\boldsymbol{\epsilon} \sim \mathrm{N}_c(\mathbf{0}, \mathbf{I})$. Note that due to the symmetry it suffices to confine the multivariate periodogram to the region $\tilde{\boldsymbol{\omega}} \in \Omega$ instead of $(0, 2\pi)^p$.

For illustration, Figure 8.1 depicts the multivariate periodogram with $p = 2$, or simply the bivariate periodogram, for the time series shown in Figure 6.1 which consists of two unit-amplitude zero-phase complex sinusoids in Gaussian white noise. Note that if the evaluation of the bivariate periodogram were not restricted to Ω, a redundant mirror image would have appeared below the diagonal line. As we can see from Figure 8.1, the highest peak in the bivariate periodogram is located almost exactly at the signal frequencies $(f_1, f_2) = (0.12, 0.21)$. In fact, numerical calculation gives the peak location $(\hat{f}_1, \hat{f}_2) = (0.1198, 0.2093)$. This example suggests that maximizing the multivariate periodogram in general should produce very accurate frequency estimates.

According to (8.2.15), the bivariate periodogram coincides with $I_n(\tilde{\omega}_1) + I_n(\tilde{\omega}_2)$ at the Fourier frequencies and interpolates these values elsewhere. Therefore, the ridges in the bivariate periodogram that are parallel to the frequency axes can be attributed to the spectral peaks in the ordinary periodogram.

By examining the ordinary periodogram shown in Figure 8.1, we can see that the PM estimator, which comprises the location of the $p = 2$ largest local maxima in the ordinary periodogram, should also produce very accurate frequency estimates. Hence, the NLS method and the PM method are both viable choices in this example. The PM estimator has the computational advantage because it only needs to solve univariate, rather than multivariate, maximization problems. This advantage applies to the general case with multiple sinusoids when the signal frequencies are well separated relative to the sample size.

The story is quite different for the next example shown in Figure 8.2, where the frequency separation is less than $2\pi/n$. In this case, the ordinary periodogram is incapable of resolving the signal frequencies. If we insist on finding two, rather

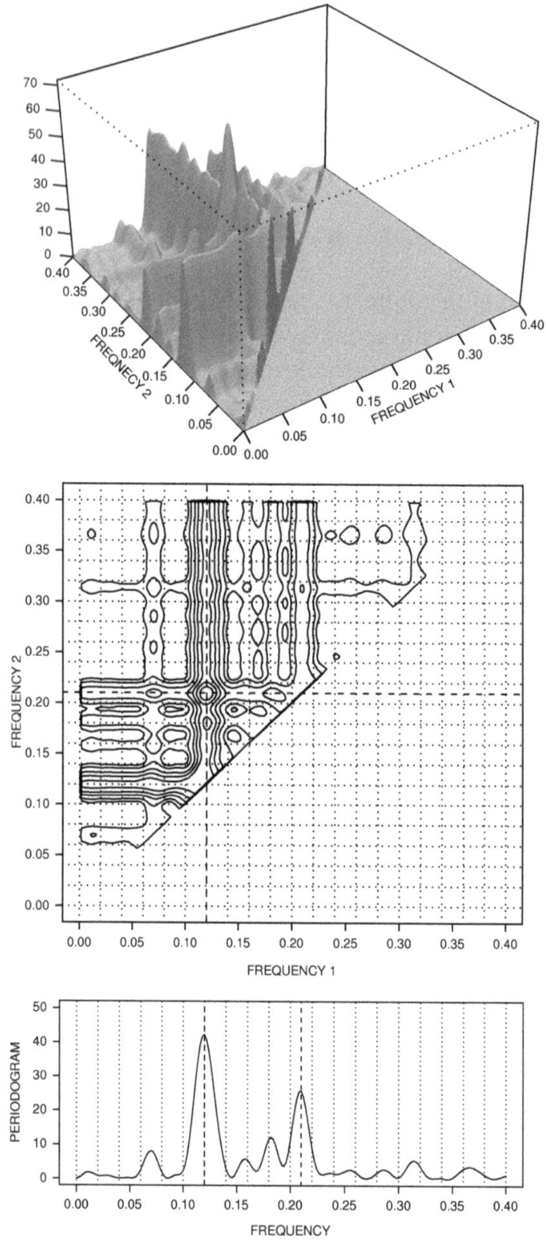

Figure 8.1. Plot of the bivariate periodogram for a time series consisting of two unit-amplitude zero-phase complex sinusoids in Gaussian white noise ($n = 50$, SNR $= -5$ dB per sinusoid). The signal frequencies are $\omega_1 = 2\pi \times 0.12$ and $\omega_2 = 2\pi \times 0.21$. The ordinary periodogram is shown at the bottom. Dashed lines depict the signal frequencies and dotted lines depict the Fourier grid.

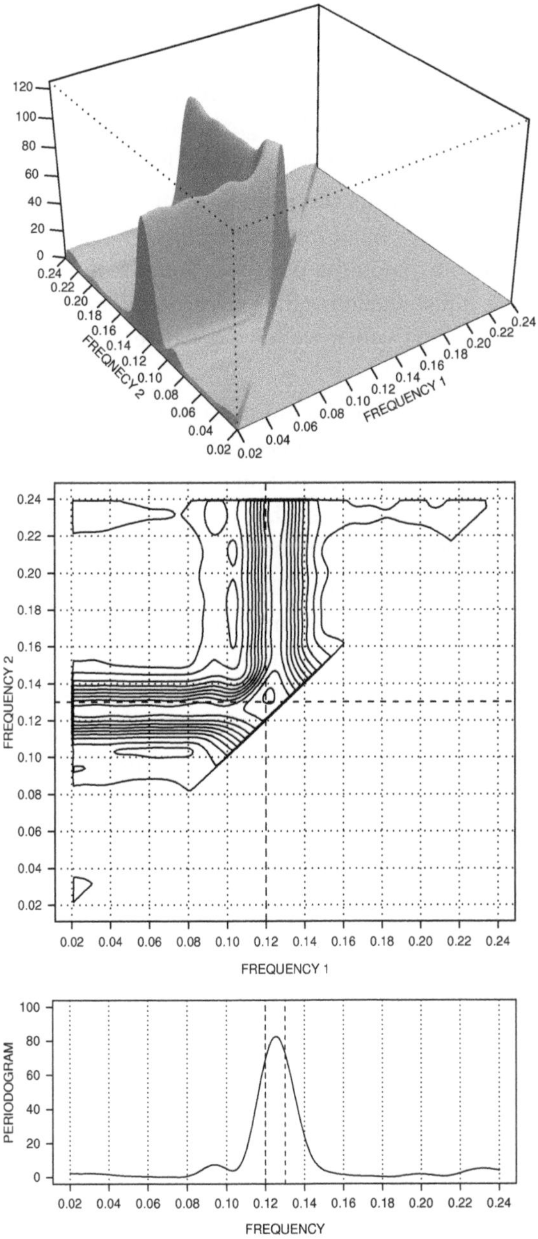

Figure 8.2. Same as Figure 8.1 except that the signal frequencies are closer to each other with $\omega_1 = 2\pi \times 0.12$ and $\omega_2 = 2\pi \times 0.13$.

than one, largest local maxima in the ordinary periodogram, the result will be hugely erroneous for frequency estimation. While a significance test may root out the weaker maximizer, it will only yield a single frequency rather than two frequencies. In cases such as this, the multivariate periodogram shows its advantage over the ordinary periodogram. As can be seen from Figure 8.2, the bivariate periodogram successfully resolves the signal frequencies by producing the highest peak near them but away from the diagonal line.

Numerical calculation yields the peak location $(\hat{f}_1, \hat{f}_2) = (0.1215, 0.1336)$. In comparison with the true signal frequencies $(f_1, f_2) = (0.12, 0.13)$, this result is clearly not as precise as that in the previous example. But it should not come as a surprise, because it just illustrates the increased difficulty in estimating closely spaced frequencies, a fact which we have discussed in Chapter 3 (Section 3.3) when investigating the CRLB.

Figures 8.1 and 8.2 show that the multivariate periodogram has a very rugged terrain with numerous peaks and ridges, just like the ordinary periodogram. In order to climb to the highest peak of a multivariate periodogram, one must be equipped with a good iterative algorithm and a good initial value. In particular, the initial value must be well within the $2\pi/n$-neighborhood of the global maximizer to ensure convergence to the desired solution.

When the signal frequencies comprise clusters that are well separated from each other, the maximizer of the multivariate periodogram can be obtained approximately by searching for the maxima in lower dimensional spaces. To illustrate, suppose $\boldsymbol{\omega} = (\boldsymbol{\omega}_1, \boldsymbol{\omega}_2)$, where $\boldsymbol{\omega}_1$ and $\boldsymbol{\omega}_2$ are well separated from each other but the frequencies in $\boldsymbol{\omega}_1$ or $\boldsymbol{\omega}_2$ may be closely spaced. Let $\tilde{\boldsymbol{\omega}} = (\tilde{\boldsymbol{\omega}}_1, \tilde{\boldsymbol{\omega}}_2)$ and $\mathbf{F}(\tilde{\boldsymbol{\omega}}) = [\mathbf{F}(\tilde{\boldsymbol{\omega}}_1), \mathbf{F}(\tilde{\boldsymbol{\omega}}_2)]$ be partitioned in the same way as $\boldsymbol{\omega}$. Then, with $\tilde{\boldsymbol{\omega}}$ being sufficiently close to $\boldsymbol{\omega}$, the frequency variables $\tilde{\boldsymbol{\omega}}_1$ and $\tilde{\boldsymbol{\omega}}_2$ are also well separated, in which case, we have $\mathbf{F}^H(\tilde{\boldsymbol{\omega}}_1)\mathbf{F}(\tilde{\boldsymbol{\omega}}_2) = \mathcal{O}(n)$ and hence

$$n^{-1}\{\mathbf{F}^H(\tilde{\boldsymbol{\omega}})\mathbf{F}(\tilde{\boldsymbol{\omega}})\}^{-1} = n^{-1}\mathrm{diag}[\{\mathbf{F}^H(\tilde{\boldsymbol{\omega}}_1)\mathbf{F}(\tilde{\boldsymbol{\omega}}_1)\}^{-1}, \{\mathbf{F}^H(\tilde{\boldsymbol{\omega}}_2)\mathbf{F}(\tilde{\boldsymbol{\omega}}_2)\}^{-1}] + \mathcal{O}(1).$$

This implies that the multivariate periodogram is decomposed as

$$P_n(\tilde{\boldsymbol{\omega}}) = \{1 + \mathcal{O}(1)\}\{P_n(\tilde{\boldsymbol{\omega}}_1) + P_n(\tilde{\boldsymbol{\omega}}_2)\}.$$

The decomposition can be justified in the same way as (8.2.25) later. Owing to the decoupling of $P_n(\tilde{\boldsymbol{\omega}})$ with respect to the frequency clusters, maximizing the lower dimensional periodograms $P_n(\tilde{\boldsymbol{\omega}}_1)$ and $P_n(\tilde{\boldsymbol{\omega}}_2)$ near $\boldsymbol{\omega}_1$ and $\boldsymbol{\omega}_2$, respectively, yields an approximate joint maximizer of $P_n(\tilde{\boldsymbol{\omega}})$.

Now let us consider the behavior of the multivariate periodogram in the absence of sinusoids. For any fixed $\tilde{\boldsymbol{\omega}} := (\tilde{\omega}_1, \ldots, \tilde{\omega}_p) \in \Omega$, let $\tilde{\boldsymbol{\lambda}}_n := (\tilde{\lambda}_{j(1)}, \ldots, \tilde{\lambda}_{j(p)})$ denote the point on the p-dimensional Fourier grid Ω_n^p which is nearest to $\tilde{\boldsymbol{\omega}}$.

Then, in the absence of sinusoids, it follows from (8.2.15) and Lemma 12.5.9 that

$$P_n(\tilde{\boldsymbol{\omega}}) = P_n(\tilde{\boldsymbol{\lambda}}_n) + \mathcal{O}_P(1) = \sum_{k=1}^{p} I_n(\tilde{\lambda}_{j(k)}) + \mathcal{O}_P(1).$$

For sufficiently large n, the $\tilde{\lambda}_{j(k)}$ are distinct. Therefore, under the Gaussian assumption, Theorem 6.4 ensures that $I_n(\tilde{\lambda}_{j(1)}),\ldots,I_n(\tilde{\lambda}_{j(p)})$ are asymptotically distributed as $\frac{1}{2}f_\epsilon(\tilde{\lambda}_{j(1)})X_1,\ldots,\frac{1}{2}f_\epsilon(\tilde{\lambda}_{j(p)})X_p$, where $X_1,\ldots,X_p$ are i.i.d. $\chi^2(2)$ random variables. Therefore, upon noting that $f_\epsilon(\tilde{\lambda}_{j(k)}) = f_\epsilon(\tilde{\omega}_k) + \mathcal{O}(1)$, we can write

$$P_n(\tilde{\boldsymbol{\omega}}) \xrightarrow{D} \sum_{k=1}^{p} \tfrac{1}{2} f_\epsilon(\tilde{\omega}_k) X_k. \tag{8.2.17}$$

In other words, the p-dimensional multivariate periodogram is asymptotically distributed as a weighted sum of p independent $\chi^2(2)$ random variables, where the weights are the noise spectrum evaluated at the corresponding frequencies. In the special case of Gaussian white noise, (8.2.17) reduces to

$$P_n(\tilde{\boldsymbol{\omega}}) \xrightarrow{D} \tfrac{1}{2}\sigma^2\chi^2(2p).$$

In other words, for Gaussian white noise, the multivariate periodogram at all frequencies has the same asymptotic distribution, which is $\frac{1}{2}\sigma^2\chi^2(2p)$.

The discussion so far has been focused on the complex case. Let us now briefly comment on the real case. First, observe that under the Cartesian RSM (2.1.1) we have (8.1.2) and (8.2.3) with $\boldsymbol{\theta} = [A_1, B_1, \omega_1, \ldots, A_q, B_q, \omega_q]^T$,

$$\boldsymbol{\beta} := [A_1, B_1, \ldots, A_q, B_q]^T \in \mathbb{R}^{2q}, \quad \boldsymbol{\omega} := (\omega_1, \ldots, \omega_q) \in \Omega,$$
$$\mathbf{F}(\boldsymbol{\omega}) := [\Re\{\mathbf{f}(\omega_1)\}, \Im\{\mathbf{f}(\omega_1)\}, \ldots, \Re\{\mathbf{f}(\omega_q)\}, \Im\{\mathbf{f}(\omega_q)\}],$$

and

$$\Omega := \{(\tilde{\omega}_1, \ldots, \tilde{\omega}_q) : 0 < \tilde{\omega}_1 < \cdots < \tilde{\omega}_q < \pi\}. \tag{8.2.18}$$

Therefore, the NLS estimator of $(\boldsymbol{\beta}, \boldsymbol{\omega})$ can be expressed as

$$(\hat{\boldsymbol{\beta}}, \hat{\boldsymbol{\omega}}) := \arg \min_{\tilde{\boldsymbol{\beta}} \in \mathbb{R}^{2q}, \tilde{\boldsymbol{\omega}} \in \Omega} \|\mathbf{y} - \mathbf{F}(\tilde{\boldsymbol{\omega}})\tilde{\boldsymbol{\beta}}\|^2. \tag{8.2.19}$$

With the notation $\tilde{\boldsymbol{\omega}} := (\tilde{\omega}_1, \ldots, \tilde{\omega}_q)$, the multivariate periodogram is defined as

$$P_n(\tilde{\boldsymbol{\omega}}) := \tfrac{1}{2}\{\|\mathbf{y}\|^2 - \|\mathbf{y} - \mathbf{F}(\tilde{\boldsymbol{\omega}})\hat{\boldsymbol{\beta}}(\tilde{\boldsymbol{\omega}})\|^2\}, \tag{8.2.20}$$

where

$$\hat{\boldsymbol{\beta}}(\tilde{\boldsymbol{\omega}}) := \arg \min_{\tilde{\boldsymbol{\beta}} \in \mathbb{R}^{2q}} \|\mathbf{y} - \mathbf{F}(\tilde{\boldsymbol{\omega}})\tilde{\boldsymbol{\beta}}\|^2 = \{\mathbf{F}^T(\tilde{\boldsymbol{\omega}})\mathbf{F}(\tilde{\boldsymbol{\omega}})\}^{-1}\mathbf{F}^T(\tilde{\boldsymbol{\omega}})\mathbf{y}.$$

It is easy to verify that the multivariate periodogram $P_n(\tilde{\boldsymbol{\omega}})$ in (8.2.20) coincides with the generalized log likelihood ratio statistic for testing $H_0 : \mathbf{y} = \boldsymbol{\epsilon}$ against $H_1 : \mathbf{y} = \mathbf{F}(\tilde{\boldsymbol{\omega}})\boldsymbol{\beta} + \boldsymbol{\epsilon}$ for some unknown $\boldsymbol{\beta} \neq \mathbf{0}$ under the assumption $\boldsymbol{\epsilon} \sim \mathrm{N}(\mathbf{0},\mathbf{I})$. It is also easy to verify that the NLS estimator $\hat{\boldsymbol{\omega}}$ in (8.2.19) maximizes $P_n(\tilde{\boldsymbol{\omega}})$ in (8.2.20) and the NLS estimator $\hat{\boldsymbol{\beta}}$ in (8.2.19) satisfies $\hat{\boldsymbol{\beta}} = \hat{\boldsymbol{\beta}}(\hat{\boldsymbol{\omega}})$.

Moreover, if $\tilde{\boldsymbol{\omega}}$ is restricted on the Fourier grid, then $\mathbf{F}^T(\tilde{\boldsymbol{\omega}})\mathbf{F}(\tilde{\boldsymbol{\omega}}) = \frac{1}{2}n\mathbf{I}$ and

$$P_n(\tilde{\boldsymbol{\omega}}) = n^{-1}\|\mathbf{F}^T(\tilde{\boldsymbol{\omega}})\mathbf{y}\|^2 = \sum_{k=1}^{q} I_n(\tilde{\omega}_k). \tag{8.2.21}$$

Being the counterpart of (8.2.15), this expression justifies the multivariate periodogram defined by (8.2.20) as an extension of the ordinary periodogram in the real case. Note that for $q = 1$ and $\tilde{\boldsymbol{\omega}} = \omega$ the multivariate periodogram in (8.2.20) does not necessarily coincide with $I_n(\omega)$ unless ω is a Fourier frequency. This is different from the complex case with $p = 1$ where (8.2.14) holds for all ω. In the absence of sinusoids, it follows from (8.2.21) and Remark 6.6 that (8.2.17) remains true for the multivariate periodogram defined by (8.2.20) except that p should be replaced by q.

8.2.2 Statistical Properties

Although it is difficult to analyze the statistical properties of the NLS estimator mathematically for finite sample sizes, useful asymptotic results can be obtained under fairly general conditions.

As shown earlier, in the complex case with $p = 1$, the multivariate periodogram coincides with the ordinary periodogram and hence the NLS frequency estimator is the same as the PM frequency estimator discussed in Chapter 6 (Section 6.4.2). In addition, it is easy to see that the NLS amplitude estimator $\hat{\beta}_1$ given by (8.2.9) is equal to $n^{-1}\sum y_t \exp(-it\hat{\omega}_1)$ which coincides with the PM amplitude estimator given by (6.4.21). Therefore, in the complex case with $p = 1$, the consistency and asymptotic normality assertions in Theorem 6.10 are valid for the NLS estimators $[\hat{A}_1, \hat{B}_1, \hat{\omega}_1]^T$ and $[\hat{C}_1, \hat{\phi}_1, \hat{\omega}_1]^T$ given by (8.2.6).

The asymptotic analysis for the complex case with $p > 1$ is facilitated by extending the relationship (8.2.15) between the multivariate periodogram and the ordinary periodogram beyond the Fourier frequencies. Toward that end, assume that the signal frequencies ω_k satisfy the separation condition

$$n\Delta \to \infty, \quad \Delta := \min_{k \neq k'}\{|\omega_k - \omega_{k'}|, 2\pi - |\omega_k - \omega_{k'}|\}. \tag{8.2.22}$$

In this case, the NLS frequency estimator can be obtained as

$$\hat{\boldsymbol{\omega}} = \arg\max_{\tilde{\boldsymbol{\omega}} \in \Omega_0} P_n(\tilde{\boldsymbol{\omega}}), \tag{8.2.23}$$

where $P_n(\tilde{\boldsymbol{\omega}})$ is defined by (8.2.16) and Ω_0 is the closed subset of Ω, satisfying

$$\min_{k\neq k'}\{|\tilde{\omega}_k-\tilde{\omega}_{k'}|, 2\pi-|\tilde{\omega}_k-\tilde{\omega}_{k'}|\}\geq c_n, \quad c_n\in(0,\Delta), \quad nc_n\to\infty, \tag{8.2.24}$$

with c_n being a predetermined constant which may depend on n. By construction, the signal frequency $\boldsymbol{\omega}$ is an interior point of Ω_0. Note that the condition in (8.2.24) is identical to that required by Theorem 6.10 for the PM estimator under the assumption (8.2.22).

With Ω_0 defined by (8.2.24), it can be shown that the p-dimensional multivariate periodogram $P_n(\tilde{\boldsymbol{\omega}})$, for $\tilde{\boldsymbol{\omega}}\in\Omega_0$, is asymptotically equivalent to the sum of p ordinary univariate periodograms $I_n(\tilde{\omega}_1),\ldots,I_n(\tilde{\omega}_p)$ in the sense that

$$P_n(\tilde{\boldsymbol{\omega}})=\{1+R_n(\tilde{\boldsymbol{\omega}})\}\sum_{k=1}^{p} I_n(\tilde{\omega}_k), \tag{8.2.25}$$

where $R_n(\tilde{\boldsymbol{\omega}})=\mathcal{O}(n^{-1}c_n^{-1})\to 0$ uniformly in $\tilde{\boldsymbol{\omega}}\in\Omega_0$. See Section 8.6 for a proof. This expression is an extension of (8.2.15) beyond the Fourier frequencies. It is important to note that (8.2.25) cannot be written as $P_n(\tilde{\boldsymbol{\omega}})=\sum_{k=1}^{p} I_n(\tilde{\omega}_k)+\mathcal{O}_P(1)$ because $I_n(\tilde{\omega}_k)=\mathcal{O}_P(n)$ when $\tilde{\omega}_k$ coincides with a signal frequency. In other words, the approximation error is multiplicative rather than additive.

The asymptotic equivalence given by (8.2.25) forms the basis on which the consistency of the NLS estimator can be deduced from the consistency of the PM estimator. Moreover, because $R_n(\tilde{\boldsymbol{\omega}})$ is uniformly negligible in (8.2.25), it is not surprising that the NLS estimator also has the same asymptotic distribution as the PM estimator. These assertions are stated in the following theorem. A proof can be found in Section 8.6.

Theorem 8.1 (Asymptotics of the NLS Estimator: Complex Case). *Let* $\mathbf{y}$ *satisfy (8.1.2) and* $\mathbf{x}(\boldsymbol{\theta})$ *take the form (8.2.3). Let* $\hat{\boldsymbol{\beta}}:=[\hat{\beta}_1,\ldots,\hat{\beta}_p]^T$ *and* $\hat{\boldsymbol{\omega}}:=(\hat{\omega}_1,\ldots,\hat{\omega}_p)$ *be defined by (8.2.9) and (8.2.23) with* Ω_0 *satisfying (8.2.24). Let* $\{\hat{A}_k,\hat{B}_k,\hat{C}_k,\hat{\phi}_k\}$ *be defined from* $\hat{\boldsymbol{\beta}}$ *according to (8.2.6). Assume that* $\{\omega_k\}$ *satisfies (8.2.22). Assume further that* $\{\epsilon_t\}$ *is a zero-mean stationary complex Gaussian process with ACF* $r_\epsilon(u)$ *such that* $\sum|r_\epsilon(u)|<\infty$, *or that* $\{\epsilon_t\}$ *is a complex linear process of the form* $\epsilon_t=\sum\psi_j\zeta_{t-j}$ *with* $\sum|\psi_j|<\infty$ *and* $\{\zeta_t\}\sim\mathrm{IID}(0,\sigma_\zeta^2)$. *Then, as* $n\to\infty$, *the following assertions are true.*

(a) For all k, $n(\hat{\omega}_k-\omega_k)\xrightarrow{P}0$, $\hat{A}_k\xrightarrow{P}A_k$, $\hat{B}_k\xrightarrow{P}B_k$, $\hat{C}_k\xrightarrow{P}C_k$, *and* $\hat{\phi}_k\xrightarrow{P}\phi_k$.

(b) Let $\boldsymbol{\theta}:=[\boldsymbol{\theta}_1^T,\ldots,\boldsymbol{\theta}_p^T]^T$, *where* $\boldsymbol{\theta}_k:=[A_k,B_k,\omega_k]^T$ *for the Cartesian CSM (2.1.5) and* $\boldsymbol{\theta}_k:=[C_k,\phi_k,\omega_k]^T$ *for the polar CSM (2.1.7). Let* $\hat{\boldsymbol{\theta}}$ *be defined accordingly on the basis of* $\{\hat{A}_k,\hat{B}_k,\hat{C}_k,\hat{\phi}_k\}$. *Assume that there is a constant* $f_0>0$ *such that* $f_\epsilon(\omega)\geq f_0$ *for all* ω. *If* $\{\epsilon_t\}$ *is a complex linear process, assume also that* $E(\zeta_t^2)=0$. *Then,* $\mathbf{K}(\hat{\boldsymbol{\theta}}-\boldsymbol{\theta})\xrightarrow{A}\mathrm{N}(\mathbf{0},\boldsymbol{\Gamma}(\boldsymbol{\theta}))$, *where* $\mathbf{K}:=\mathrm{diag}(\mathbf{K}_n,\ldots,\mathbf{K}_n)$, $\mathbf{K}_n:=\mathrm{diag}(n^{1/2},n^{1/2},n^{3/2})$, *and* $\boldsymbol{\Gamma}(\boldsymbol{\theta}):=\mathrm{diag}\{\boldsymbol{\Gamma}(\boldsymbol{\theta}_1),\ldots,\boldsymbol{\Gamma}(\boldsymbol{\theta}_p)\}$, *with* $\boldsymbol{\Gamma}(\boldsymbol{\theta}_k)$ *given by (3.3.13) and* $\gamma_k:=C_k^2/f_\epsilon(\omega_k)$.

Remark 8.1 Similar assertions can be made in the real case under the assumption that $\{\epsilon_t\}$ is a zero-mean stationary Gaussian process or a real linear process. In this case, $\boldsymbol{\theta} := [\boldsymbol{\theta}_1^T, \ldots, \boldsymbol{\theta}_q^T]^T$, $\hat{\boldsymbol{\beta}} := [\hat{\beta}_1, \ldots, \hat{\beta}_{2q}]^T$, $\hat{\boldsymbol{\omega}} := (\hat{\omega}_1, \ldots, \hat{\omega}_q)$,

$$\hat{A}_k := \hat{\beta}_{2k-1}, \quad \hat{B}_k := \hat{\beta}_{2k},$$
$$\hat{C}_k := \sqrt{\hat{A}_k^2 + \hat{B}_k^2}, \quad \hat{\phi}_k := \arctan(-\hat{B}_k, \hat{A}_k).$$

The frequency separation condition (8.2.22) is replaced by

$$n\Delta \to \infty, \quad \Delta := \min_{k \neq k'}\{\omega_k, \pi - \omega_k, |\omega_k - \omega_{k'}|\}.$$

The frequency estimator in (8.2.23) is redefined with Ω_0 being a closed subset of $\Omega := \{\tilde{\boldsymbol{\omega}} : 0 < \tilde{\omega}_1 < \cdots < \tilde{\omega}_q < \pi\}$ such that

$$\min_{k \neq k'}\{\omega_k, \pi - \omega_k, |\omega_k - \omega_{k'}|\} \geq c_n, \quad c_n \in (0, \Delta), \quad nc_n \to \infty.$$

The matrix $\boldsymbol{\Gamma}(\boldsymbol{\theta}_k)$ is given by (3.3.13) with $\gamma_k := \frac{1}{2}C_k^2 / f_\epsilon(\omega_k)$.

Remark 8.2 The asymptotic properties of the NLS estimator in the real case have been investigated in [136] and [412] under various conditions. A complete proof of consistency and asymptotic normality is given in [357]. The complex case is considered in [310] under the assumption of Gaussian white noise. Theorem 8.1 is an extension of this result. Because the amplitude estimator and the frequency estimator have different rates of convergence, one cannot obtain the results in Theorem 8.1 simply from the standard theory of nonlinear least-squares regression, such as [166] and [426].

It is important to note that although the NLS estimator is derived under the assumption of Gaussian white noise, Theorem 8.1 remains true under more general noise conditions. For example, the noise can be a nonGaussian i.i.d. sequence with zero mean and finite variance, or an autoregressive process.

The matrix $\mathbf{K}^{-1}\boldsymbol{\Gamma}(\boldsymbol{\theta})\mathbf{K}^{-1}$ in Theorem 8.1 is nothing but the asymptotic CRLB under the Gaussian assumption. This means that the NLS estimator is asymptotically efficient under Gaussian noise — white or colored. The asymptotic efficiency is not particularly surprising in the case of Gaussian white noise, because the NLS estimator is the ML estimator under this condition. The asymptotic efficiency under Gaussian colored noise implies that knowing the ACF of the Gaussian noise has little benefit to the estimation accuracy if the sample size is sufficiently large. Of course, this statement is not true for fixed sample sizes, due partly to the Gauss-Markov theorem (Theorem 5.5).

Recall that the PM estimator, discussed in Chapter 6 (Section 6.4.2), is also asymptotically efficient under Gaussian white or colored noise (Theorem 6.10).

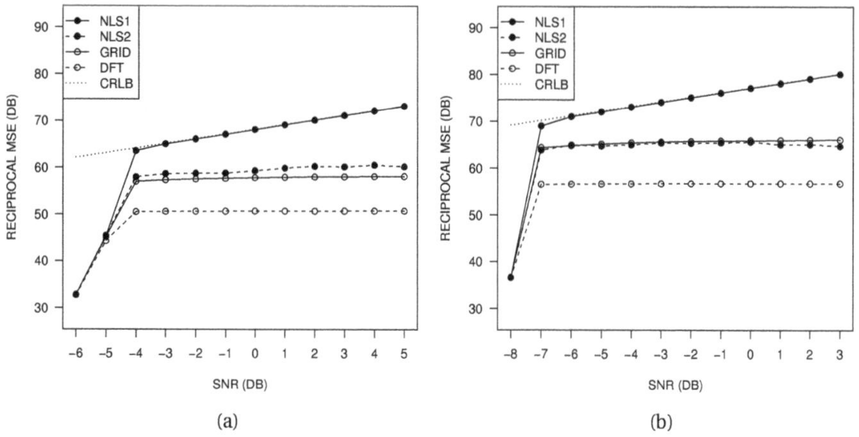

Figure 8.3. Plot of the reciprocal average MSE of the NLS frequency estimates for the time series of two complex sinusoids in Gaussian white noise discussed in Example 8.1 with different values of SNR and two sample sizes: (a) $n = 100$ and (b) $n = 200$. Solid line with dots, NLS initialized by grid search; dashed line with dots, NLS initialized by DFT; solid line with circles, grid search; dashed line with circles, DFT. Dotted line depicts the CRLB. Results are based on 5,000 Monte Carlo runs.

This assertion, however, requires a stronger condition of frequency separation, i.e., $n\Delta^2 \to \infty$, which is not required by the NLS estimator.

According to Theorem 8.1, the NLS estimator always attains the Gaussian CRLB asymptotically regardless of the actual noise distribution. Therefore, the NLS estimator may not be the most efficient estimator when the noise is nonGaussian. This subject will be explored further in Section 8.3.

The following example demonstrates the statistical performance of the NLS estimator for finite sample sizes.

Example 8.1 (Two Complex Sinusoids in Gaussian White Noise). Let $\mathbf{y}$ be the time series in Example 6.9 which consists of two unit-amplitude random-phase complex sinusoids in Gaussian white noise. Consider the NLS frequency estimates which are computed by the Nelder-Mead algorithm [266] using the R function `optim` with default options. The initial values for the algorithm are obtained in two ways: (a) the DFT method defined by (6.3.14) and (6.3.15) with $m = 1$ and $p = 2$ (with $\delta = \pi/\sqrt{n}$, as in Example 6.9), and (b) the exhaustive grid search method conducted within the $\pm 2\pi/n$ neighborhood of the DFT estimates with step size equal to $2\pi n^{-1.2}$ (see [406] for an alternative search strategy).

Figure 8.3 depicts the average MSE of the NLS estimates and their initial values for the same data as used in Example 6.9. It can be seen that the DFT method does not provide sufficiently good initial values for the Nelder-Mead algorithm. As a result, the algorithm tends to be trapped in spurious local maxima, giving rise to a poor MSE in comparison with the CRLB. The initial values from the grid search are good enough to make the Nelder-Mead algorithm converge to the desired solutions. The MSE of the resulting estimates follows the CRLB almost perfectly, so long as the SNR is greater than a threshold determined by the DFT method that initializes the procedure.

A comparison of Figure 8.3 with Figure 6.13 reveals a slight loss of accuracy when the PM estimator replaces the NLS estimator, even though the estimators are asymptotically equivalent. The discrepancy is more pronounced for $n = 100$ or at the higher end of the SNR considered. ◇

Theorem 8.1 shows that the NLS estimator has the same asymptotic distribution as the PM estimator when the signal frequencies satisfy the separation condition (8.2.22). Example 8.1 demonstrates that the NLS estimator and the PM estimator also have similar accuracy when estimating well-separated frequencies for finite sample sizes. Let us now consider the situation where the signal frequencies are closely spaced relative to the sample size. In this case, the NLS estimator, which maximizes the multivariate periodogram, becomes superior to the PM estimator, which maximizes the univariate periodogram.

For simplicity, let $p = 2$, and assume that ω_1 is fixed and ω_2 satisfies

$$\omega_2 = \omega_1 + \Delta, \quad n\Delta \to a > 0 \quad \text{as } n \to \infty.$$

The same situation is considered in Chapter 3 (Section 3.3). Note that the requirement of Theorem 8.1 is not satisfied in this case because $\Delta = \mathcal{O}(n^{-1})$. The NLS frequency estimator is still given by (8.2.23) but with Ω_0 defined as

$$\Omega_0 := \{(\tilde{\omega}_1, \tilde{\omega}_2) : 0 < \tilde{\omega}_1 < \tilde{\omega}_2 < 2\pi, 2\pi - |\tilde{\omega}_2 - \tilde{\omega}_1| \geq c_n, |\tilde{\omega}_2 - \tilde{\omega}_1| \geq \kappa_n\}, \tag{8.2.26}$$

where $c_n > 0$ and $\kappa_n > 0$ satisfy the condition

$$nc_n \to \infty, \quad \kappa_n := b/n, \tag{8.2.27}$$

with $b \in (0, a)$ being a predetermined constant. By construction, the parameter space Ω_0 always contains $\boldsymbol{\omega} = (\omega_1, \omega_2)$ as an interior point.

To gain some intuitive insight into the multivariate periodogram for closely spaced frequencies, consider the noiseless case with $\mathbf{y} = \mathbf{x} := \mathbf{F}(\boldsymbol{\omega})\boldsymbol{\beta}$. Let $P_{nx}(\tilde{\boldsymbol{\omega}})$ denote the multivariate periodogram of $\mathbf{x}$. It is easy to verify that

$$P_{nx}(\tilde{\boldsymbol{\omega}}) = n\boldsymbol{\beta}^H \mathbf{K}_n(\tilde{\boldsymbol{\omega}}, \boldsymbol{\omega})\boldsymbol{\beta}, \tag{8.2.28}$$

where

$$\mathbf{K}_n(\tilde{\boldsymbol{\omega}}, \boldsymbol{\omega}) := n^{-1}\{\mathbf{F}^H(\boldsymbol{\omega})\mathbf{F}(\tilde{\boldsymbol{\omega}})\}\{\mathbf{F}^H(\tilde{\boldsymbol{\omega}})\mathbf{F}(\tilde{\boldsymbol{\omega}})\}^{-1}\{\mathbf{F}^H(\tilde{\boldsymbol{\omega}})\mathbf{F}(\boldsymbol{\omega})\}.$$

Observe that $\mathbf{K}_n(\tilde{\boldsymbol{\omega}},\boldsymbol{\omega})$ is a 2π-periodic function of $\tilde{\boldsymbol{\omega}}$ for fixed $\boldsymbol{\omega}$. We call this function the *multivariate Fejér kernel located at* $\boldsymbol{\omega}$. With the notation $\lambda_{kk'} := \tilde{\omega}_{k'} - \tilde{\omega}_k$, and $\tilde{\lambda}_{kk'} := \omega_{k'} - \tilde{\omega}_k$ $(k,k' = 1,\ldots,p)$, we can write

$$n^{-1}\mathbf{F}^H(\tilde{\boldsymbol{\omega}})\mathbf{F}(\tilde{\boldsymbol{\omega}}) = [D_n(\lambda_{kk'})], \quad n^{-1}\mathbf{F}^H(\tilde{\boldsymbol{\omega}})\mathbf{F}(\boldsymbol{\omega}) = [D_n(\tilde{\lambda}_{kk'})],$$

where $D_n(\lambda) := n^{-1}\sum_{t=1}^{n}\exp(it\lambda)$ is the Dirichlet kernel. This implies that the multivariate Fejér kernel $\mathbf{K}_n(\tilde{\boldsymbol{\omega}},\boldsymbol{\omega})$ depends only on the frequency differences $\lambda_{kk'}$ and $\tilde{\lambda}_{kk'}$. In the special case of $p=1$, we have $\mathbf{K}_n(\tilde{\omega},\omega) = K_n(\omega-\tilde{\omega})$, where $K_n(\lambda) := |D_n(\lambda)|^2$ is just the ordinary Fejér kernel (located at the origin).

Figure 8.4 depicts the contours of $P_{nx}(\tilde{\boldsymbol{\omega}})$ in the case of $p=2$ for a time series consisting of two noiseless complex sinusoids with different sample sizes. The signal frequencies are closely spaced such that $\Delta := \omega_2 - \omega_1 = \pi/n$. It can be seen that the bivariate periodogram has a unique global maximum at $\tilde{\boldsymbol{\omega}} = \boldsymbol{\omega}$ regardless of the sample size. This suggests that maximizing the multivariate periodogram should produce accurate results even for closely spaced frequencies.

In the presence of noise, the NLS estimator can be justified by the following theorem. A proof is given in Section 8.6.

Theorem 8.2 (Asymptotics of the NLS Estimator: CSM with Closely Spaced Frequencies, $p=2$). *Let* $\mathbf{y}$ *be given by (8.1.2) with* $\mathbf{x}(\boldsymbol{\theta})$ *taking the form (8.2.3) with* $p=2$. *Let* ω_1 *be a constant and* $\omega_2 = \omega_1 + \Delta$ *with* $n\Delta \to a$ *as* $n\to\infty$ *for some constant* $a>0$. *Let* $\hat{\boldsymbol{\beta}} := [\hat{\beta}_1,\hat{\beta}_2]^T$ *and* $\hat{\boldsymbol{\omega}} := (\hat{\omega}_1,\hat{\omega}_2)$ *be defined by (8.2.9) and (8.2.23) with* Ω_0 *satisfying (8.2.26) and (8.2.27). Assume that* $\{\epsilon_t\}$ *satisfies the conditions of Theorem 8.1. Then, the assertions (a) and (b) of Theorem 8.1 remain true except that* $\boldsymbol{\Gamma}(\boldsymbol{\theta}) = \frac{1}{2}f_\epsilon(\omega_1)\mathbf{W}^{-1}$ *for estimating* $\boldsymbol{\theta} := [A_1,B_1,\omega_1,A_2,B_2,\omega_2]^T$ *and* $\boldsymbol{\Gamma}(\boldsymbol{\theta}) = \frac{1}{2}f_\epsilon(\omega_1)\mathbf{J}\mathbf{W}^{-1}\mathbf{J}$ *for estimating* $\boldsymbol{\theta} := [C_1,\phi_1,\omega_1,C_2,\phi_2,\omega_2]^T$, *where* $\mathbf{J} := \mathrm{diag}(\mathbf{J}_1,\mathbf{J}_2)$ *is the Jacobian matrix given by (6.6.51) and* $\mathbf{W}$ *is defined as*

$$\mathbf{W} := \begin{bmatrix} \mathbf{W}_1 & \mathbf{W}_{12} \\ \mathbf{W}_{12}^T & \mathbf{W}_2 \end{bmatrix},$$

with $\mathbf{W}_1$ *and* $\mathbf{W}_2$ *given by (3.3.7) and* $\mathbf{W}_{12}$ *given by*

$$\mathbf{W}_{12} := \begin{bmatrix} \Re\{d_0(a)\} & \Im\{d_0(a)\} & \Re\{\beta_2\dot{d}_0(a)\} \\ -\Im\{d_0(a)\} & \Re\{d_0(a)\} & -\Im\{\beta_2\dot{d}_0(a)\} \\ -\Re\{\beta_1^*\dot{d}_0(a)\} & -\Im\{\beta_1^*\dot{d}_0(a)\} & -\Re\{\beta_1^*\beta_2\ddot{d}_0(a)\} \end{bmatrix},$$

where $d_0(a) := \mathrm{sinc}(a/2)\exp(ia/2)$.

Remark 8.3 It is not difficult to extend the asymptotic normality assertion to the general case of $p>2$ where the signal frequencies form well-separated clusters as discussed in Remark 3.6. In this case, the asymptotic covariance matrix $\boldsymbol{\Gamma}(\boldsymbol{\theta})$ takes the same block diagonal form as described in Remark 3.6.

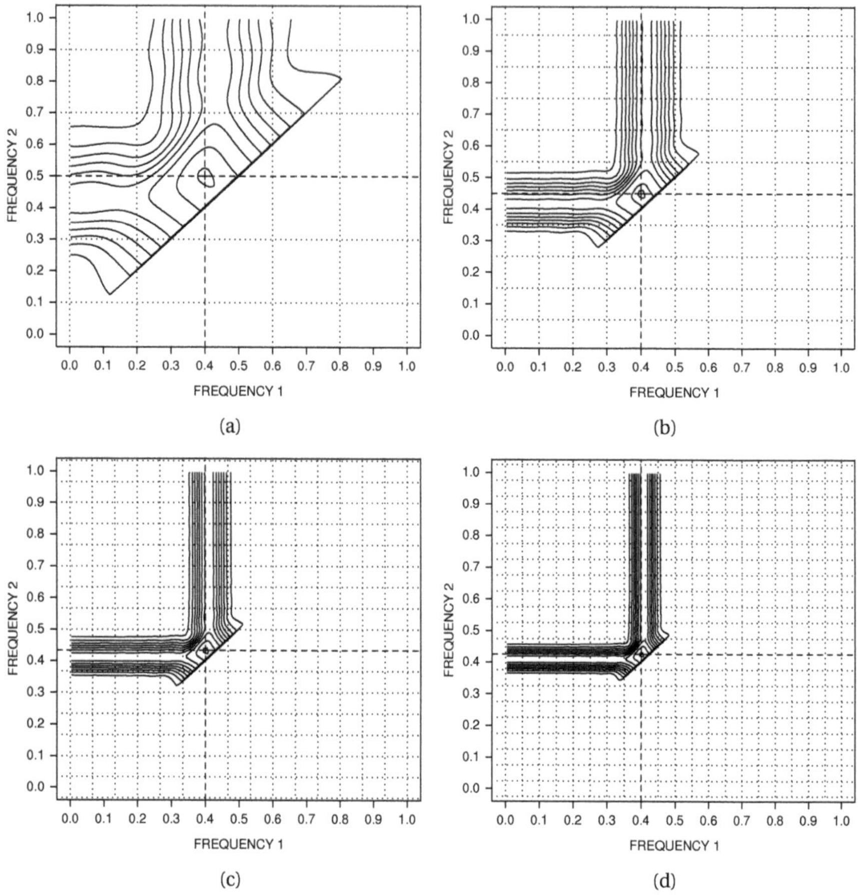

Figure 8.4. Contour plot of the bivariate periodogram in the absence of noise for (a) $n = 5$, (b) $n = 10$, (c) $n = 15$, (d) $n = 20$. The signal parameters are: $\omega_1 = 2\pi \times 0.4$, $\omega_2 = \omega_1 + 2\pi \times 0.5/n$, $\beta_1 = 0.2576776 - 0.7273993i$, $\beta_2 = -0.2256499 - 0.5946214i$. Dashed lines depict the signal frequencies. Dotted lines depict the Fourier grid centered at the signal frequencies. Note that in all cases the signal frequencies cannot be resolved by the ordinary periodogram (not shown).

The main difference between Theorem 8.2 and Theorem 8.1 is the presence of $\mathbf{W}_{12}$ in the former. This matrix depends on the frequency separation parameter a and may take nonzero values. It vanishes as $a \to \infty$ because $d_0(a) \to 0$, $\dot{d}_0(a) \to 0$, and $\ddot{d}_0(a) \to 0$ (see Lemma 12.1.3), in which case $\mathbf{W}$ becomes diag$(\mathbf{W}_1, \mathbf{W}_2)$ and $\boldsymbol{\Gamma}(\boldsymbol{\theta})$ becomes diag$\{\boldsymbol{\Gamma}(\boldsymbol{\theta}_1), \boldsymbol{\Gamma}(\boldsymbol{\theta}_2)\}$, which is the assertion of Theorem 8.1.

The matrix $\mathbf{K}^{-1}\boldsymbol{\Gamma}(\boldsymbol{\theta})\mathbf{K}^{-1}$ in Theorem 8.2 coincides with the Gaussian ACRLB given by Theorem 3.5. Therefore, the NLS estimator remains asymptotically efficient under the Gaussian assumption for estimating closely spaced frequencies that satisfy the condition of Theorem 8.2. It still attains the Gaussian ACRLB regardless of the actual noise distribution.

Theorem 8.2 imposes a restriction on the frequency separation in the form of $n\Delta \to a > 0$ as $n \to \infty$. This is the asymptotic resolution limit for the NLS estimator to maintain the convergence rates that are valid in the case of well-separated frequencies. An open problem is the rate of convergence and the limiting distribution for the NLS estimator when the signal frequencies are more closely spaced such that $n\Delta \to 0$, an example being $\Delta = \mathcal{O}(n^{-r})$ for some constant $r > 1$.

The statistical performance of the NLS estimator for closely spaced frequencies with finite sample sizes is demonstrated by the following example.

Example 8.2 (Estimation of Closely Spaced Frequencies). Consider the time series of two unit-amplitude random-phase complex sinusoids in Gaussian white noise. Let the signal frequencies satisfy $\omega_2 = \omega_1 + \Delta$ with $\Delta = 2\pi \times 0.5/n$ (i.e., $n\Delta = a$ and $a = \pi$), where ω_1 is generated randomly from the uniform distribution $2\pi \times 0.06 + \mathrm{U}(0, 2\pi/n)$. As in Example 8.1, the NLS frequency estimates are computed by the Nelder-Mead algorithm, and the DFT and grid search methods are used for initialization. The only difference is that in the current example we use $p = 1$ rather than $p = 2$ in the DFT method because the periodogram cannot resolve the signal frequencies. In other words, the initial DFT method only gives the largest peak in the periodogram. Moreover, the step size for the grid search is reduced to $2\pi n^{-1.25}$ in response to the decreased separation of the signal frequencies. Figure 8.5 shows the average MSE of the frequency estimates for different values of SNR and two sample sizes based on 5,000 Monte Carlo runs.

It can be seen that the NLS estimator performs well despite the reduced frequency separation for which the PM method is no longer applicable. The MSE of the NLS estimator follows the CRLB closely, provided the SNR is higher than a threshold. As in Example 8.1, the DFT method without grid search is not accurate enough for the Nelder-Mead algorithm. The grid search in the neighborhood of the DFT estimates enables the Nelder-Mead algorithm to produce the desired estimates. The SNR threshold is higher in this example than in Example 8.1, although the amplitudes of the sinusoids remain the same. Therefore, the rising threshold is a consequence of the reduced frequency separation. It signifies the increased difficulty in resolving closely spaced frequencies at low SNR.

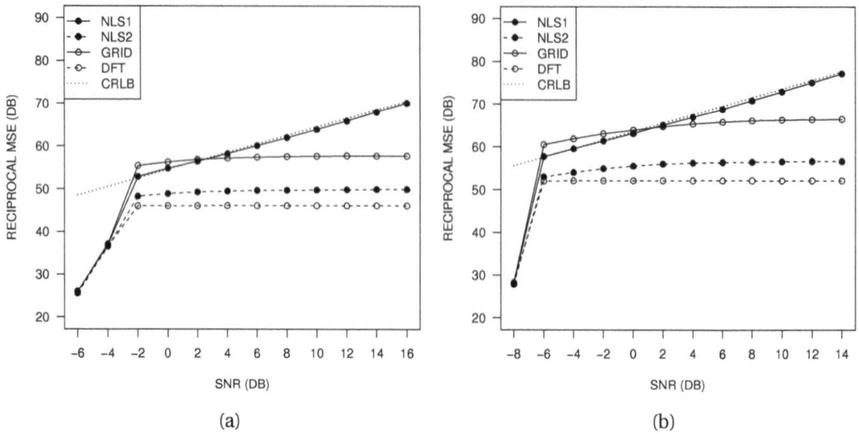

Figure 8.5. Plot of the reciprocal average MSE of the NLS frequency estimates for the time series of two complex sinusoids with closely spaced frequencies in Gaussian white noise, as discussed in Example 8.2, for different values of SNR and two sample sizes: (a) $n = 100$ and (b) $n = 200$. Solid line with dots, NLS initialized by grid search; dashed line with dots, NLS initialized by DFT; solid line with circles, grid search; dashed line with circles, DFT. Dotted line depicts the CRLB. Results are based on 5,000 Monte Carlo runs.

Note that the frequency estimates from grid search are generally biased, so their MSE may be less than the CRLB, which is the lower bound for unbiased estimators. This happens when the reduced variability due to the constraint on the grid outweighs the bias in the MSE which equals the variance plus the squared bias. It is the case in Figure 8.5(a) at SNR = −2 dB, for example. ◇

As demonstrated by Examples 8.1 and 8.2, the NLS method is able to produce very accurate frequency estimates not only for well-separated frequencies but also for closely spaced frequencies that cannot be resolved by the ordinary periodogram. However, in both cases, the optimization procedures, such as the Nelder-Mead algorithm, rely heavily on the availability of good initial values that the simple DFT method cannot provide. The grid search method used in the numerical examples requires additional computation. How to obtain accurate frequency estimates with less stringent initial requirements and lower computational cost remains an important research topic.

Finally, we remark that in the case of white noise the consistency and asymptotic normality assertions in Corollary 6.3 remain valid for the estimator $\hat{\sigma}^2$ in (8.2.5) under the frequency separation condition in Theorem 8.1 or Theorem 8.2. This is due to the fact that the error of the NLS estimator for the amplitude parameter can be expressed as $\mathcal{O}_P(n^{-1/2})$ and the error of the NLS estimator for the frequency parameter can be expressed as $\mathcal{O}_P(n^{-3/2})$.

8.3 Maximum Likelihood under Laplace White Noise

The assumption of Gaussian white noise is often made not because it fits the actual noise distribution but because it leads to a computationally simple and mathematically tractable solution. When the actual noise distribution differs from the Gaussian assumption, especially when it has heavy tails that give rise to rare but large errors in the observed data, the Gaussian solution can fail miserably. This phenomenon is generally known as the lack of robustness [159]. It may have serious consequences in applications if left unattended.

One way to obtain robust solutions, as discussed in Chapter 5 (Section 5.5), is to substitute the least-squares (LS) criterion with the least-absolute-deviations (LAD) criterion so that the influence of large residuals is reduced. This approach is equivalent to the ML estimation under the assumption of Laplace white noise. The LAD technique is employed in Chapter 5 (Section 5.5) to estimate the amplitude parameter when the frequency parameter is known. In this section, we consider the joint estimation of the amplitude and frequency parameters based on the LAD criterion, which is a more difficult nonlinear problem.

In the real case, let $\{\epsilon_t\}$ be an i.i.d. sequence with PDF $p(x) = (2c)^{-1}\exp(-|x|/c)$ ($x \in \mathbb{R}$) for some unknown scale parameter $c > 0$. In the complex case, let the real and imaginary parts of $\{\epsilon_t\}$ be mutually independent i.i.d. sequences with PDF $p(x) = c^{-1}\exp(-2|x|/c)$ ($x \in \mathbb{R}$). Under these assumptions, it is easy to verify that the ML estimator of $\boldsymbol{\theta}$ and c in both cases can be expressed as

$$\hat{\boldsymbol{\theta}} := \arg\min_{\tilde{\boldsymbol{\theta}} \in \Theta} \|\mathbf{y} - \mathbf{x}(\tilde{\boldsymbol{\theta}})\|_1, \tag{8.3.1}$$

$$\hat{c} := n^{-1}\|\mathbf{y} - \mathbf{x}(\hat{\boldsymbol{\theta}})\|_1, \tag{8.3.2}$$

where $\|\cdot\|_1$ denotes the ℓ_1 norm of a real or complex vector defined as the sum of the absolute values of the real and imaginary parts of all components (in the real case the imaginary parts are equal to zero). As can be seen, the estimator $\hat{\boldsymbol{\theta}}$ in (8.3.1) is the solution of a nonlinear least absolute deviations problem and the estimator $\hat{c}$ in (8.3.2) is simply the ℓ_1 norm of the residuals divided by the sample size n. In the real case, we have $c = \sigma/\sqrt{2}$, so the ML estimator of the noise variance σ^2 is given by $\hat{\sigma}^2 := 2\hat{c}^2$. In the complex case, the ML estimator of σ^2 is given simply by $\hat{\sigma}^2 := \hat{c}^2$ because $c = \sigma$.

The ML estimator in (8.3.1) can be used when the actual noise does not have a Laplace distribution. In such cases, it will be called the *nonlinear least absolute deviations* (NLAD) estimator or the *Laplace maximum likelihood* (LML) estimator, just like the ML estimator in (8.2.1) is called the nonlinear least-squares or Gaussian maximum likelihood estimator when applied to nonGaussian cases.

8.3.1 Multivariate Laplace Periodogram

Because $\mathbf{x}(\tilde{\boldsymbol{\theta}}) = \mathbf{F}(\tilde{\boldsymbol{\omega}})\tilde{\boldsymbol{\beta}}$, the NLAD estimator in (8.3.1) can be obtained by first minimizing the ℓ_1-norm criterion with respect to $\tilde{\boldsymbol{\beta}}$ for fixed $\tilde{\boldsymbol{\omega}}$. In the complex case with $\mathbf{F}(\tilde{\boldsymbol{\omega}}) := [\mathbf{f}(\tilde{\omega}_1), \ldots, \mathbf{f}(\tilde{\omega}_p)]$, this method leads to

$$\hat{\boldsymbol{\beta}}(\tilde{\boldsymbol{\omega}}) := \arg\min_{\tilde{\boldsymbol{\beta}} \in \mathbb{C}^p} \|\mathbf{y} - \mathbf{F}(\tilde{\boldsymbol{\omega}})\tilde{\boldsymbol{\beta}}\|_1, \tag{8.3.3}$$

which is a linear LAD problem that can be solved by linear programming (LP) techniques as discussed in Chapter 5 (Section 5.5). With $\hat{\boldsymbol{\beta}}(\tilde{\boldsymbol{\omega}})$ obtained according to (8.3.3), the NLAD frequency estimator $\hat{\boldsymbol{\omega}}$ can be expressed as

$$\hat{\boldsymbol{\omega}} = \arg\min_{\tilde{\boldsymbol{\omega}} \in \Omega} \|\mathbf{y} - \mathbf{F}(\tilde{\boldsymbol{\omega}})\hat{\boldsymbol{\beta}}(\tilde{\boldsymbol{\omega}})\|_1, \tag{8.3.4}$$

and the NLAD amplitude estimator $\hat{\boldsymbol{\beta}}$ is given by

$$\hat{\boldsymbol{\beta}} = \hat{\boldsymbol{\beta}}(\hat{\boldsymbol{\omega}}).$$

This is the counterpart of a similar procedure for the NLS estimation.

The counterpart of the multivariate periodogram in (8.2.16) is given by

$$L_n(\tilde{\boldsymbol{\omega}}) := \|\mathbf{y}\|_1 - \|\mathbf{y} - \mathbf{F}(\tilde{\boldsymbol{\omega}})\hat{\boldsymbol{\beta}}(\tilde{\boldsymbol{\omega}})\|_1. \tag{8.3.5}$$

The NLAD frequency estimator $\hat{\boldsymbol{\omega}}$ in (8.3.4) can be expressed as

$$\hat{\boldsymbol{\omega}} = \arg\max_{\tilde{\boldsymbol{\omega}} \in \Omega} L_n(\tilde{\boldsymbol{\omega}}). \tag{8.3.6}$$

We call $L_n(\tilde{\boldsymbol{\omega}})$ the *multivariate Laplace periodogram*. Analogously, the function $P_n(\tilde{\boldsymbol{\omega}})$ in (8.2.16) may be referred to as the *multivariate Gaussian periodogram*. The multivariate Laplace periodogram coincides with the generalized log likelihood ratio statistic for testing $H_0 : \mathbf{y} = \boldsymbol{\epsilon}$ against $H_1 : \mathbf{y} = \mathbf{F}(\tilde{\boldsymbol{\omega}})\boldsymbol{\beta} + \boldsymbol{\epsilon}$ for some unknown $\boldsymbol{\beta} \neq \mathbf{0}$ under the assumption that $\boldsymbol{\epsilon}$ is complex Laplace white noise with mean 0 and variance 4. According to (8.3.6), the NLAD frequency estimator maximizes the multivariate Laplace periodogram, just like the NLS frequency estimator maximizes the multivariate Gaussian periodogram.

In the special case of $p = 1$, the multivariate Laplace periodogram reduces to the univariate Laplace periodogram

$$L_n(\omega) = \|\mathbf{y}\|_1 - \|\mathbf{y} - \mathbf{f}(\omega)\hat{\beta}(\omega)\|_1, \tag{8.3.7}$$

where

$$\hat{\beta}(\omega) := \arg\min_{\beta \in \mathbb{C}} \|\mathbf{y} - \mathbf{f}(\omega)\beta\|_1.$$

This is an alternative to the Laplace periodogram which is defined by $n|\hat{\beta}(\omega)|^2$ in [228] as the extension of its real-case counterpart discussed in [224]. To distinguish them when necessary, the latter can be referred to as the Laplace periodogram of the first kind and the former as the Laplace periodogram of the second kind. As a function of ω, the Laplace periodogram of the second kind is much smoother than the Laplace periodogram of the first kind, and therefore is more suitable for frequency estimation.

With $\lambda_j := 2\pi j/n$ ($j = 0, 1, \ldots, n-1$) denoting the Fourier frequencies, the LAD solution $\{\hat{\beta}(\lambda_0), \hat{\beta}(\lambda_1), \ldots, \hat{\beta}(\lambda_{n-1})\}$ can be regarded as a robust alternative to the DFT. We call it the *Laplace discrete Fourier transform* (LDFT) of $\mathbf{y}$. Locating the p largest values among the LDFT-based Laplace periodogram ordinates $L_n(\lambda_j)$ constitutes the LDFT method of frequency estimation — a robust alternative to the DFT method discussed in Chapter 6 (Section 6.1). Similarly, by employing the refined Laplace periodogram $L_n(\lambda'_j)$ instead of the refined ordinary periodogram $I_n(\lambda'_j)$, where $\lambda'_j := 2\pi j/(mn)$ ($j = 0, 1, \ldots, mn-1$) for some $m \geq 1$, the refined DFT (RDFT) method of frequency estimation discussed in Section 6.3.2 becomes the refined LDFT (or RLDFT) method. The PM method discussed in Section 6.4.2 can also be modified by replacing $I_n(\omega)$ with $L_n(\omega)$, leading to the Laplace periodogram maximization (LPM) method.

Figures 8.6 and 8.7 present two examples of the multivariate Laplace periodogram with $p = 2$, or the bivariate Laplace periodogram, and the corresponding univariate Laplace periodogram for a time series consisting of two complex sinusoids in white noise under different frequency separation conditions. The noise has Student's t-distribution with 2.1 degrees of freedom, which we call the $T_{2.1}$ white noise. Similar to Figures 8.1 and 8.2 for the multivariate Gaussian periodogram, Figures 8.6 and 8.7 suggest that the global maximizer of the multivariate Laplace periodogram can produce accurate estimates not only for well-separated frequencies but also for closely spaced frequencies that cannot be resolved by the univariate periodogram.

Figure 8.8 demonstrates the robustness advantage of the Laplace periodogram over the ordinary periodogram in the presence of outliers. In this example, the Laplace periodogram behaves similarly to the ordinary periodogram in the absence of outliers, but exhibits stronger resistance to outlier contamination. Observe that the outlier not only alters the location of the two largest spectral peaks of the ordinary periodogram, but also considerably reduces the magnitude of the first peak. The Laplace periodogram is able to keep the location and magnitude of the spectral peaks nearly intact. As shown in [226], the multivariate Laplace periodogram exhibits similar robustness against outliers.

In the real case, we can write $\mathbf{x}(\tilde{\boldsymbol{\theta}}) = \mathbf{F}(\tilde{\boldsymbol{\omega}})\tilde{\boldsymbol{\beta}}$ with $\tilde{\boldsymbol{\omega}} := (\tilde{\omega}_1, \ldots, \tilde{\omega}_q)$ and

$$\mathbf{F}(\tilde{\boldsymbol{\omega}}) := [\Re\{\mathbf{f}(\tilde{\omega}_1)\}, \Im\{\mathbf{f}(\tilde{\omega}_1)\}, \ldots, \Re\{\mathbf{f}(\tilde{\omega}_q)\}, \Im\{\mathbf{f}(\tilde{\omega}_q)\}].$$

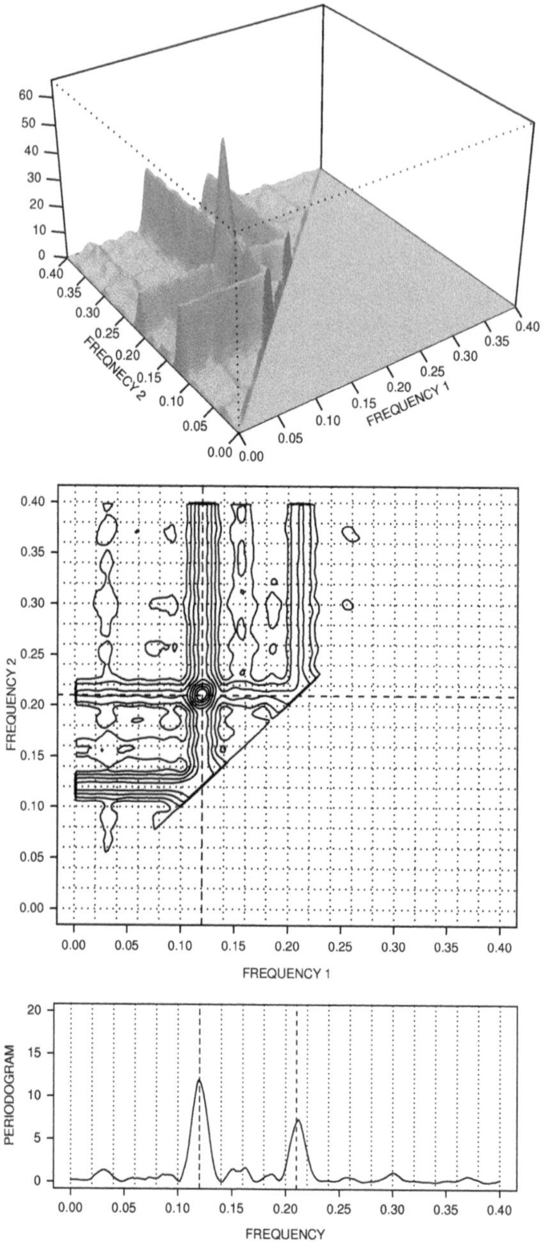

Figure 8.6. Plot of the bivariate Laplace periodogram for a time series consisting of two unit-amplitude zero-phase complex sinusoids in $T_{2.1}$ white noise ($n = 50$, SNR $= -5$ dB per sinusoid). The signal frequencies are $\omega_1 = 2\pi \times 0.12$ and $\omega_2 = 2\pi \times 0.21$. The univariate Laplace periodogram is shown at the bottom. Dashed lines depict the signal frequencies and dotted lines depict the Fourier grid.

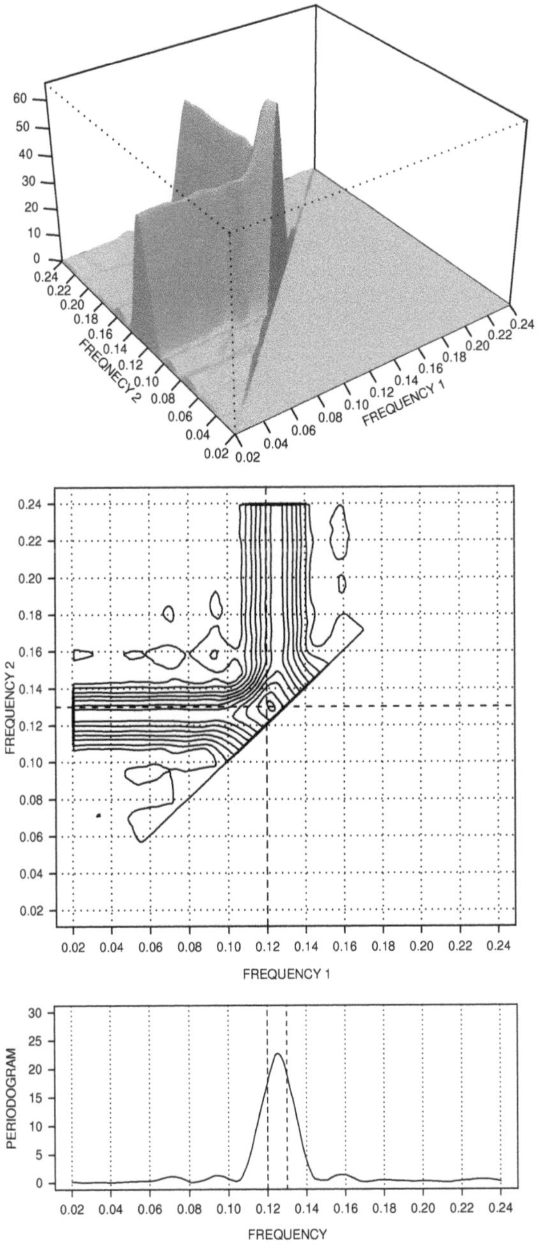

Figure 8.7. Same as Figure 8.6 except that the signal frequencies are closer to each other with $\omega_1 = 2\pi \times 0.12$ and $\omega_2 = 2\pi \times 0.13$.

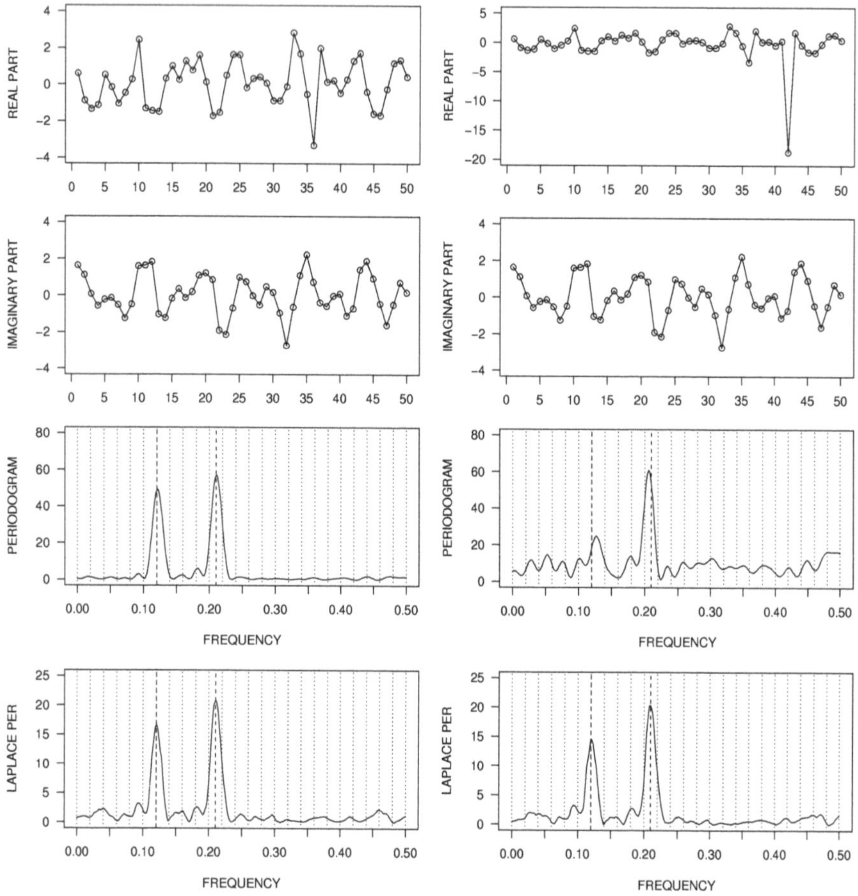

Figure 8.8. Robustness of the Laplace periodogram under the condition of heavy-tailed noise. Top to bottom: real and imaginary parts of a time series which consists of two complex sinusoids in noise, ordinary periodogram, and Laplace periodogram. Left, Gaussian white noise. Right, same as left except the large noise in the real part at $t = 42$. Dashed lines depict the signal frequencies. Dotted lines depict the Fourier grid.

Therefore, (8.3.3) and (8.3.4) remain valid except that the LAD problem in (8.3.3) is solved for the real variable $\tilde{\boldsymbol{\beta}} \in \mathbb{R}^{2q}$. Moreover, the multivariate Laplace periodogram in the real case can be defined in the same way as (8.2.20), i.e.,

$$L_n(\tilde{\boldsymbol{\omega}}) := \tfrac{1}{2}\{\|\mathbf{y}\|_1 - \|\mathbf{y} - \mathbf{F}(\tilde{\boldsymbol{\omega}})\hat{\boldsymbol{\beta}}(\tilde{\boldsymbol{\omega}}))\|_1\}. \tag{8.3.8}$$

This quantity coincides with the generalized log likelihood ratio statistic for testing $H_0 : \mathbf{y} = \boldsymbol{\epsilon}$ against $H_1 : \mathbf{y} = \mathbf{F}(\tilde{\boldsymbol{\omega}})\boldsymbol{\beta} + \boldsymbol{\epsilon}$ for some unknown $\boldsymbol{\beta} \neq \mathbf{0}$ under the assumption that $\boldsymbol{\epsilon}$ is Laplace white noise with mean 0 and variance 8. The NLAD or LML frequency estimator in the real case maximizes $L_n(\tilde{\boldsymbol{\omega}})$ in (8.3.8).

For $q = 1$, we have $\mathbf{F}(\omega) := [\Re\{\mathbf{f}(\omega)\}, \Im\{\mathbf{f}(\omega)\}]$, and the multivariate Laplace periodogram in (8.3.8) reduces to the univariate Laplace periodogram

$$L_n(\omega) = \tfrac{1}{2}\{\|\mathbf{y}\|_1 - \|\mathbf{y} - \mathbf{F}(\omega)\hat{\boldsymbol{\beta}}(\omega)\|_1\},$$

where

$$\hat{\boldsymbol{\beta}}(\omega) := \arg\min_{\tilde{\boldsymbol{\beta}} \in \mathbb{R}^2} \|\mathbf{y} - \mathbf{F}(\omega)\tilde{\boldsymbol{\beta}}\|_1.$$

This is the Laplace periodogram of the second kind introduced in [228]. It serves as an alternative to the Laplace periodogram of the first kind which is defined by $\frac{1}{4}n\|\hat{\boldsymbol{\beta}}(\omega)\|^2$ in [224]. The latter will be further discussed in Chapter 11.

8.3.2 Statistical Properties

It is a very challenging task to investigate the statistical properties of the NLAD estimator, even for asymptotic analysis. The difficulty can be attributed mainly to the nondifferentiability of the ℓ_1-norm criterion compounded with the nonlinearity of the frequency parameter. Nonetheless, the following theorem can be obtained with regard to the asymptotic distribution of the NLAD estimator. A proof of this result is given in Section 8.6.

Theorem 8.3 (Asymptotic Distribution of the NLAD Estimator). *Let* $\mathbf{y}$ *be given by (8.1.2) with* $\mathbf{x}(\boldsymbol{\theta})$ *taking the form (8.2.3). Let* $\boldsymbol{\theta}$ *be defined in the same way as in Theorem 8.1 and let* $\hat{\boldsymbol{\theta}} := \arg\min\{\|\mathbf{y} - \mathbf{x}(\tilde{\boldsymbol{\theta}})\|_1 : \tilde{\boldsymbol{\theta}} \in \Theta_n\}$, *where* Θ_n *is a closed set that contains* $\boldsymbol{\theta}$ *as an interior point. Moreover, let* $\mathbf{X} := \nabla^T \mathbf{x}(\boldsymbol{\theta})$ *denote the Jacobian matrix of* $\mathbf{x}(\tilde{\boldsymbol{\theta}})$ *evaluated at* $\boldsymbol{\theta}$. *Assume that* $\mathbf{K}^{-1}\mathbf{X}^H\mathbf{X}\mathbf{K}^{-1} = \mathbf{C} + \mathcal{O}(1)$ *as* $n \to \infty$, *where* $\mathbf{K}$ *is defined in Theorem 8.1 and* $\mathbf{C}$ *is a constant matrix with nonsingular real part* $\mathbf{W} := \Re(\mathbf{C})$. *Assume that* $\{\epsilon_t\}$ *is a sequence of independent random variables with independent real and imaginary parts whose CDFs, denoted by* $F_{t1}(x)$ *and* $F_{t2}(x)$, *are differentiable at* $x = 0$ *and satisfy* $F_{tj}(0) = 1/2$, $\dot{F}_{tj}(0) = 1/\kappa > 0$, *and* $F_{tj}(x) - F_{tj}(0) = \dot{F}_{tj}(0)x + \mathcal{O}(x^{d+1})$ *uniformly in* t *for some constant* $d > 0$ *and for* $|x| \ll 1$ $(j = 1, 2)$. *Assume further that* $\hat{\boldsymbol{\theta}}$ *is consistent in the sense of Theorem 8.1(a). Then, as* $n \to \infty$, $\mathbf{K}(\hat{\boldsymbol{\theta}} - \boldsymbol{\theta}) \stackrel{A}{\sim} \mathrm{N}(\mathbf{0}, \frac{1}{2}\eta^2\mathbf{W}^{-1})$, *where* $\eta^2 := \kappa^2/2$.

Remark 8.4 In the real case, a similar assertion can be made under the assumption that $\mathbf{K}^{-1}\mathbf{X}^T\mathbf{X}\mathbf{K}^{-1} = \frac{1}{2}\mathbf{W} + \mathcal{O}(1)$ for some nonsingular matrix $\mathbf{W}$ and that the noise $\{\epsilon_t\}$ satisfies the condition in Remark 5.13. In this case, we obtain $\mathbf{K}(\hat{\boldsymbol{\theta}} - \boldsymbol{\theta}) \stackrel{A}{\sim} \mathrm{N}(\mathbf{0}, 2\eta^2\mathbf{W}^{-1})$, where $\eta^2 := \kappa^2/4$. See [238] for more details.

Remark 8.5 The NLAD estimator in the real case has been investigated under various conditions in [145], [186], [238], and [417]. In Theorem 8.3, we consider not only the complex case but also the case where the signal frequencies are closely spaced. By following the basic ideas of [239] and [240], the proof in Section 8.6 makes a critical and nontrivial extension of the existing results which is necessary for handling the different rates of convergence for amplitude and frequency parameters. The consistency of the NLAD estimator is a key assumption in Theorem 8.3 to ensure the asymptotic normality. The consistency is asserted without detailed proof in [186] for the real case with fixed frequencies. In general, a rigorous proof of the consistency remains unknown.

In light of Remark 5.12, Theorem 8.3 is valid under the condition of Laplace white noise with $\dot{F}(x) = \sigma^{-1}\exp(-|x|/\sigma)$. We can compare the asymptotic covariance of the NLAD estimator in Theorem 8.3 with the CRLB derived under the assumption of Laplace white noise. According to Theorem 3.6, the assumption $\mathbf{K}^{-1}\Re(\mathbf{X}^H\mathbf{X})\mathbf{K}^{-1} = \mathbf{W} + \mathcal{O}(1)$ implies that the Laplace CRLB can be expressed as

$$\tfrac{1}{4}\sigma^2\{\Re(\mathbf{X}^H\mathbf{X})\}^{-1} = \tfrac{1}{4}\sigma^2\mathbf{K}^{-1}\{\mathbf{W}^{-1} + \mathcal{O}(1)\}\mathbf{K}^{-1}.$$

Therefore, the asymptotic Laplace CRLB takes the form $\mathbf{K}^{-1}\boldsymbol{\Gamma}_L(\boldsymbol{\theta})\mathbf{K}^{-1}$ with

$$\boldsymbol{\Gamma}_L(\boldsymbol{\theta}) := \tfrac{1}{4}\sigma^2\mathbf{W}^{-1}.$$

On the other hand, we have $\kappa = 1/\dot{F}(0) = \sigma$ for the complex Laplace white noise. Therefore, by Theorem 8.3, the asymptotic covariance matrix of the NLAD estimator can be expressed as $\mathbf{K}^{-1}\mathbf{V}\mathbf{K}^{-1}$, where

$$\mathbf{V} := \tfrac{1}{2}\eta^2\mathbf{W}^{-1} = \tfrac{1}{4}\kappa^2\mathbf{W}^{-1} = \tfrac{1}{4}\sigma^2\mathbf{W}^{-1} = \boldsymbol{\Gamma}_L(\boldsymbol{\theta}).$$

In other words, the NLAD estimator attains the Laplace CRLB asymptotically under the condition of Laplace white noise. This means that the NLAD estimator is asymptotically efficient under the Laplace white noise assumption. The same conclusion is drawn in Chapter 5 (Section 5.5) for the LAD estimator of the amplitude parameter when the frequency is known.

As in Section 5.5, we can also compare the asymptotic covariance matrix of the NLAD estimator given by Theorem 8.3 with the asymptotic covariance matrix of the NLS estimator given by Theorem 8.1 and Theorem 8.2. In order for this comparison to be valid, we need to assume that the real and imaginary parts of $\{\epsilon_t\}$ are mutually independent i.i.d. sequences of zero-mean and zero-median random variables with finite variance $\sigma^2 > 0$ and with CDF $F(x)$ such that $\dot{F}(0) = 1/\kappa > 0$. Under this condition, the NLS estimator attains the asymptotic Gaussian CRLB which can be expressed as $\mathbf{K}^{-1}\mathbf{\Gamma}_G(\boldsymbol{\theta})\mathbf{K}^{-1}$ with

$$\mathbf{\Gamma}_G(\boldsymbol{\theta}) := \tfrac{1}{2}\sigma^2\mathbf{W}^{-1},$$

and the NLAD estimator has an asymptotic covariance matrix $\mathbf{K}^{-1}\mathbf{V}\mathbf{K}^{-1}$ with

$$\mathbf{V} = \tfrac{1}{2}\eta^2\mathbf{W}^{-1} = \rho^{-1}\mathbf{\Gamma}_G(\boldsymbol{\theta}), \tag{8.3.9}$$

where $\rho := \sigma^2/\eta^2$ is the efficiency coefficient discussed in Section 5.5.

Owing to (8.3.9), the efficiency analysis in Section 5.5 for the LAD estimator versus the LS estimator of the amplitude parameter remains valid when comparing the NLAD estimator with the NLS estimator for joint estimation of the amplitude and frequency parameters. In particular, because $\rho = 4\{p_0(0)\}^2$, where $p_0(x)$ is the unit-variance PDF such that $\dot{F}(x) = (\sqrt{2}/\sigma)p_0(\sqrt{2}x/\sigma)$, the NLAD estimator is more efficient than the NLS estimator if $p_0(0) > 1/2$, as is the case when the noise has the Laplace distribution or Student's t-distribution with ν degrees of freedom such that $2 < \nu < 4.678$. Moreover, because $\mathbf{\Gamma}_L(\boldsymbol{\theta}) = \frac{1}{2}\mathbf{\Gamma}_G(\boldsymbol{\theta})$, it follows from (8.3.9) that

$$\mathbf{V} = 2\rho^{-1}\mathbf{\Gamma}_L(\boldsymbol{\theta}) = \tfrac{1}{2}\{p_0(0)\}^{-2}\mathbf{\Gamma}_L(\boldsymbol{\theta}).$$

This implies that the asymptotic variance of the NLAD estimator is less than the Laplace CRLB if the noise satisfies the condition $p_0(0) > 1/\sqrt{2}$ (for the Laplace noise, $p_0(0) = 1/\sqrt{2}$). In contrast, the asymptotic variance of the NLS estimator is always equal to the Gaussian CRLB.

The following two examples demonstrate the statistical performance of the NLAD estimator under the condition that the noise has Student's t-distribution with $\nu = 2.1$ degrees of freedom, i.e., the $T_{2.1}$ white noise.

Example 8.3 (Two Complex Sinusoids with Well-Separated Frequencies in $T_{2.1}$ White Noise). Let $\mathbf{y}$ be a time series consisting of two unit-amplitude random-phase complex sinusoids in $T_{2.1}$ white noise. The signal frequencies are generated randomly such that $\omega_1 \sim 2\pi \times \mathrm{U}(0.1, 0.11)$ and $\omega_2 \sim 2\pi \times \mathrm{U}(0.2, 0.21)$. These are the same frequencies as in Example 8.1. Let the signal frequencies be estimated by the NLAD estimator defined in (8.3.4). To compute the NLAD estimates, we again employ the Nelder-Mead algorithm. The initial values are produced in the same way as in Example 8.1 except that the DFT is replaced by the LDFT and the multivariate Gaussian periodogram is replaced by the multivariate Laplace periodogram. Figure 8.9 shows the average MSE of the NLAD frequency estimates and the average MSE of the initial values for different values of SNR and two sample sizes. The asymptotic variance of the NLAD estimator given by Theorem 8.3

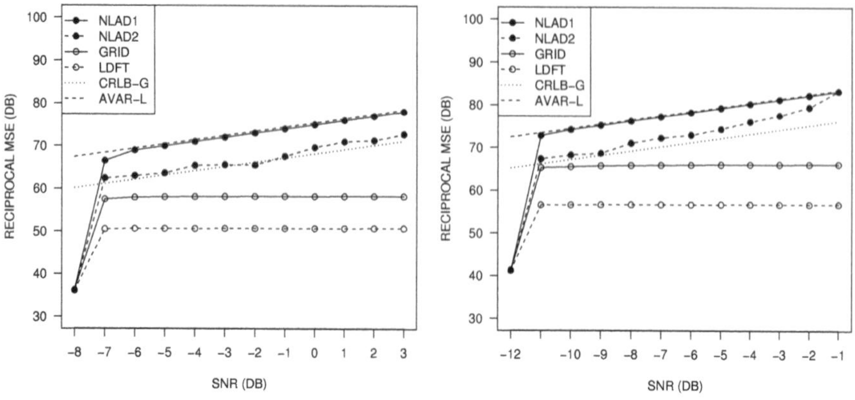

(a) (b)

Figure 8.9. Plot of the reciprocal average MSE of the NLAD frequency estimates for time series of two unit-amplitude random-phase complex sinusoids in $T_{2.1}$ white noise as discussed in Example 8.3. (a) $n = 100$. (b) $n = 200$. Solid line with dots, NLAD initialized by grid search; dashed line with dots, NLAD initialized by LDFT; solid line with circles, grid search; dashed line with circles, LDFT. Dotted line depicts the Gaussian CRLB; dashed line depicts the asymptotic variance of the NLAD estimator under the assumption of $T_{2.1}$ white noise. Results are based on 5,000 Monte Carlo runs.

under the assumption of $T_{2.1}$ white noise is also depicted in Figure 8.9 together with the Gaussian CRLB.

As can be seen, the MSE of the NLAD estimator follows the asymptotic variance closely for both sample sizes when the SNR is higher than a threshold. Observe that the asymptotic variance is much smaller than the Gaussian CRLB. Therefore, the NLAD estimator is much more efficient than the NLS estimator in this case. In fact, for the $T_{2.1}$ noise, we have $\rho = 2\sigma^2\{p(0)\}^2 \approx 5.31$, so the asymptotic variance of the NLAD estimator is approximately 7.25 dB smaller than the corresponding Gaussian CRLB achieved asymptotically by the NLS estimator.

Like the DFT method in the NLS case, the LDFT method is not accurate enough to initialize the Nelder-Mead algorithm in computing the NLAD frequency estimates. With the subsequent grid search around the LDFT estimates, the Nelder-Mead algorithm successfully converges to the desired solutions. ◇

Example 8.4 (Two Complex Sinusoids with Closely Spaced Frequencies in $T_{2.1}$ White Noise). Let $\mathbf{y}$ be a time series consisting of two unit-amplitude random-phase complex sinusoids in $T_{2.1}$ white noise. The signal frequencies are generated randomly such that $\omega_2 = \omega_1 + 2\pi \times 0.5/n$ and $\omega_1 \sim 2\pi \times 0.06 + \mathrm{U}(0, 2\pi/n)$. The same frequencies have been used in Example 8.2. Let the signal frequencies be estimated by the NLAD estimator defined in (8.3.4). The Nelder-Mead algorithm

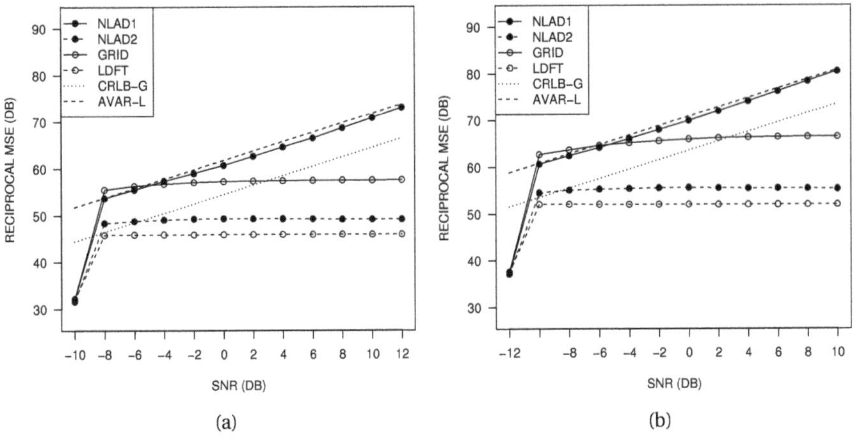

Figure 8.10. Plot of the reciprocal average MSE of the NLAD frequency estimates for time series of two unit-amplitude random-phase complex sinusoids with closely spaced frequencies in $T_{2.1}$ white noise as discussed in Example 8.4. (a) $n = 100$. (b) $n = 200$. Solid line with dots, NLAD initialized by grid search; dashed line with dots, NLAD initialized by LDFT; solid line with circles, grid search; dashed line with circles, LDFT. Dotted line depicts the Gaussian CRLB; dashed line depicts the asymptotic variance of the NLAD estimator under $T_{2.1}$ white noise. Results are based on 5,000 Monte Carlo runs.

is used to compute the NLAD estimates. The initial values are produced in the same way as in Example 8.2 using the LDFT and the multivariate Laplace periodogram. Figure 8.10 depicts the average MSE of the NLAD frequency estimates and the average MSE of the initial values.

Observe that the MSE of the NLAD estimator again follows the asymptotic variance despite the fact that the signal frequencies are closely spaced. As in Example 8.3, the asymptotic variance of the NLAD estimator is approximately 7.25 dB smaller than the corresponding Gaussian CRLB. Also as in Example 8.3, the grid search around the LDFT estimates enables the Nelder-Mead algorithm to converge to the desired solutions. ◇

8.4 The Case of Gaussian Colored Noise

Although the NLS estimator, derived under the Gaussian white noise assumption, remains asymptotically efficient for Gaussian colored noise, it may still be advantageous, given finite sample sizes, to consider the ML estimation under the assumption of Gaussian colored noise. This is motivated in part by the Gauss-

Markov theorem for amplitude estimation (Theorem 5.5).

To be specific, let us consider the complex case with $\boldsymbol{\epsilon} \sim \mathrm{N}_c(\mathbf{0}, \mathbf{R}_\epsilon)$, where $\mathbf{R}_\epsilon$ is a known covariance matrix. Because the PDF of $\boldsymbol{\epsilon}$ takes the form

$$p(\boldsymbol{x}) := \pi^{-n} |\mathbf{R}_\epsilon|^{-1} \exp(-\boldsymbol{x}^H \mathbf{R}_\epsilon^{-1} \boldsymbol{x}) \qquad (\boldsymbol{x} \in \mathbb{C}^n),$$

the ML estimator of $(\boldsymbol{\beta}, \boldsymbol{\omega})$ is given by

$$(\hat{\boldsymbol{\beta}}, \hat{\boldsymbol{\omega}}) = \arg \min_{\tilde{\boldsymbol{\beta}} \in \mathbb{C}^p,\, \tilde{\boldsymbol{\omega}} \in \Omega} (\mathbf{y} - \mathbf{F}(\tilde{\boldsymbol{\omega}})\tilde{\boldsymbol{\beta}})^H \mathbf{R}_\epsilon^{-1} (\mathbf{y} - \mathbf{F}(\tilde{\boldsymbol{\omega}})\tilde{\boldsymbol{\beta}}). \tag{8.4.1}$$

In general, we refer to this estimator as the *nonlinear generalized least-squares* (NGLS) estimator. Apparently, the only difference between the NGLS estimator in (8.4.1) and the NLS estimator in (8.2.4) is the presence of the non-identity covariance matrix $\mathbf{R}_\epsilon$ in the NGLS estimator.

For any fixed $\tilde{\boldsymbol{\omega}} \in \Omega$, let $\hat{\boldsymbol{\beta}}(\tilde{\boldsymbol{\omega}})$ be the generalized least-squares solution to the linear regression problem of estimating $\boldsymbol{\beta}$ for given $\tilde{\boldsymbol{\omega}}$, i.e.,

$$\begin{aligned}
\hat{\boldsymbol{\beta}}(\tilde{\boldsymbol{\omega}}) &:= \arg\min_{\tilde{\boldsymbol{\beta}} \in \mathbb{C}^p} (\mathbf{y} - \mathbf{F}(\tilde{\boldsymbol{\omega}})\tilde{\boldsymbol{\beta}})^H \mathbf{R}_\epsilon^{-1} (\mathbf{y} - \mathbf{F}(\tilde{\boldsymbol{\omega}})\tilde{\boldsymbol{\beta}}) \\
&= \{\mathbf{F}^H(\tilde{\boldsymbol{\omega}}) \mathbf{R}_\epsilon^{-1} \mathbf{F}(\tilde{\boldsymbol{\omega}})\}^{-1} \mathbf{F}^H(\tilde{\boldsymbol{\omega}}) \mathbf{R}_\epsilon^{-1} \mathbf{y}.
\end{aligned}$$

It is easy to verify that

$$\{\mathbf{y} - \mathbf{F}(\tilde{\boldsymbol{\omega}})\hat{\boldsymbol{\beta}}(\tilde{\boldsymbol{\omega}})\}^H \mathbf{R}_\epsilon^{-1} \{\mathbf{y} - \mathbf{F}(\tilde{\boldsymbol{\omega}})\hat{\boldsymbol{\beta}}(\tilde{\boldsymbol{\omega}})\} = \mathbf{y}^H \mathbf{R}_\epsilon^{-1} \mathbf{y} - \mathbf{y}^H \tilde{\mathbf{Q}}(\tilde{\boldsymbol{\omega}}) \mathbf{y},$$

where

$$\tilde{\mathbf{Q}}(\tilde{\boldsymbol{\omega}}) := \mathbf{R}_\epsilon^{-1} \mathbf{F}(\tilde{\boldsymbol{\omega}}) \{\mathbf{F}(\tilde{\boldsymbol{\omega}}) \mathbf{R}_\epsilon^{-1} \mathbf{F}(\tilde{\boldsymbol{\omega}})\}^{-1} \mathbf{F}^H(\tilde{\boldsymbol{\omega}}) \mathbf{R}_\epsilon^{-1}.$$

Therefore, by defining

$$\begin{aligned}
\tilde{P}_n(\tilde{\boldsymbol{\omega}}) &:= \mathbf{y}^H \mathbf{R}_\epsilon^{-1} \mathbf{y} - \{\mathbf{y} - \mathbf{F}(\tilde{\boldsymbol{\omega}})\hat{\boldsymbol{\beta}}(\tilde{\boldsymbol{\omega}})\}^H \mathbf{R}_\epsilon^{-1} \{\mathbf{y} - \mathbf{F}(\tilde{\boldsymbol{\omega}})\hat{\boldsymbol{\beta}}(\tilde{\boldsymbol{\omega}})\} \\
&= \mathbf{y}^H \tilde{\mathbf{Q}}(\tilde{\boldsymbol{\omega}}) \mathbf{y},
\end{aligned} \tag{8.4.2}$$

the NGLS estimator in (8.4.1) can be expressed as

$$\hat{\boldsymbol{\omega}} = \arg\max_{\tilde{\boldsymbol{\omega}} \in \Omega} \tilde{P}_n(\tilde{\boldsymbol{\omega}}), \quad \hat{\boldsymbol{\beta}} = \hat{\boldsymbol{\beta}}(\hat{\boldsymbol{\omega}}). \tag{8.4.3}$$

We call $\tilde{P}_n(\tilde{\boldsymbol{\omega}})$ in (8.4.2) the *standardized multivariate periodogram.* According to (8.4.3), the NGLS frequency estimator is the global maximizer of the standardized multivariate periodogram. In comparison, the NLS estimator is the global maximizer of the ordinary (unstandardized) multivariate periodogram. To compute the standardized multivariate periodogram, one can use a procedure similar to that discussed at the end of Section 3.2, which is based on the Cholesky decomposition of $\mathbf{R}_\epsilon^{-1}$ given by the Levinson-Durbin algorithm in Proposition 2.8.

The standardized multivariate periodogram in (8.4.2) is closely related to the standardized univariate periodogram discussed in Chapter 6 (Section 6.4.2). To see this, let us assume that $\tilde{\boldsymbol{\omega}}$ is away from the diagonal line and the boundary of Ω so that $n^{-1}\mathbf{F}^H(\tilde{\boldsymbol{\omega}})\mathbf{F}(\tilde{\boldsymbol{\omega}}) = \mathbf{I} + \mathcal{O}(1)$ for large n. Then, by following the proof of Theorem 3.4, it is not difficult to show that $\mathbf{R}_\epsilon^{-1}\mathbf{f}(\tilde{\omega}_k) = \{1/f_\epsilon(\tilde{\omega}_k)\}\mathbf{f}(\tilde{\omega}_k) + \mathcal{O}(1)$ for all k and large n. Therefore, with $\mathbf{D} := \mathrm{diag}\{f_\epsilon(\tilde{\omega}_1), \ldots, f_\epsilon(\tilde{\omega}_p)\}$, we can write

$$\mathbf{R}_\epsilon^{-1}\mathbf{F}(\tilde{\boldsymbol{\omega}}) = \mathbf{F}(\tilde{\boldsymbol{\omega}})\mathbf{D}^{-1} + \mathcal{O}(1), \quad n^{-1}\mathbf{F}^H(\tilde{\boldsymbol{\omega}})\mathbf{R}_\epsilon^{-1}\mathbf{F}(\tilde{\boldsymbol{\omega}}) = \mathbf{D}^{-1} + \mathcal{O}(1),$$

and hence

$$\tilde{\mathbf{Q}}(\tilde{\boldsymbol{\omega}}) \approx n^{-1}\mathbf{F}(\tilde{\boldsymbol{\omega}})\mathbf{D}^{-1}\mathbf{F}^H(\tilde{\boldsymbol{\omega}}).$$

Combining this expression with $\mathbf{f}^H(\tilde{\omega}_k)\mathbf{y} = \sum_{t=1}^n y_t \exp(-it\tilde{\omega}_k)$ leads to

$$\tilde{P}_n(\tilde{\boldsymbol{\omega}}) \approx \sum_{k=1}^p I_n(\tilde{\omega}_k)/f_\epsilon(\tilde{\omega}_k). \tag{8.4.4}$$

Observe that $I_n(\omega)/f_\epsilon(\omega)$ is just the standardized univariate periodogram. Therefore, for estimating well-separated signal frequencies, the NGLS frequency estimator given by (8.4.3) is approximately equal to the location of the p largest local maxima in the standardized univariate periodogram.

Note that the estimation procedure described above remains valid if the noise variance $r_\epsilon(0)$ is unknown but the correlation matrix $\tilde{\mathbf{R}}_\epsilon := \mathbf{R}_\epsilon/r_\epsilon(0)$ is known. In this case, it suffices to replace $\mathbf{R}_\epsilon$ by $\tilde{\mathbf{R}}_\epsilon$ because the estimator in (8.4.1) is invariant to the scale parameter in $\mathbf{R}_\epsilon$. The standardized multivariate periodogram in (8.4.2) can also be defined with $\tilde{\mathbf{R}}_\epsilon$ in place of $\mathbf{R}_\epsilon$.

Recall that the standardized univariate periodogram has the capability of suppressing noise spectral peaks and enhancing signal spectral peaks for the benefit of detection and estimation of the sinusoidal signal. This capability is demonstrated by Figure 6.14. The standardized multivariate periodogram has a similar capability, as demonstrated by Figures 8.11 and 8.12. In this example, the noise $\{\epsilon_t\}$ is a unit-variance complex Gaussian AR(2) process satisfying

$$\epsilon_t + \varphi_1\epsilon_{t-1} + \varphi_2\epsilon_{t-2} = \zeta_t, \quad \{\zeta_t\} \sim \mathrm{GWN}_c(0, \sigma^2), \tag{8.4.5}$$

where $\varphi_1 = -2r_0\cos(\omega_0)$, $\varphi_2 = r_0^2$, and $\sigma^2 = \{(1+\varphi_2)^2 - \varphi_1^2\}(1-\varphi_2)/(1+\varphi_1)$, with $r_0 = 0.65$ and $\omega_0 = 2\pi \times 0.08$. The noise has two broad spectral peaks: one around frequency $\omega_0 = 2\pi \times 0.08$ (near the first signal frequency) and the other around its conjugate $2\pi - \omega_0 = 2\pi \times 0.92$. By comparing the standardized bivariate periodogram in Figure 8.11 with the ordinary bivariate periodogram in Figure 8.12, we can see that the former has a more pronounced and better behaved spectral peak in the vicinity of the signal frequencies. A comparison of the univariate

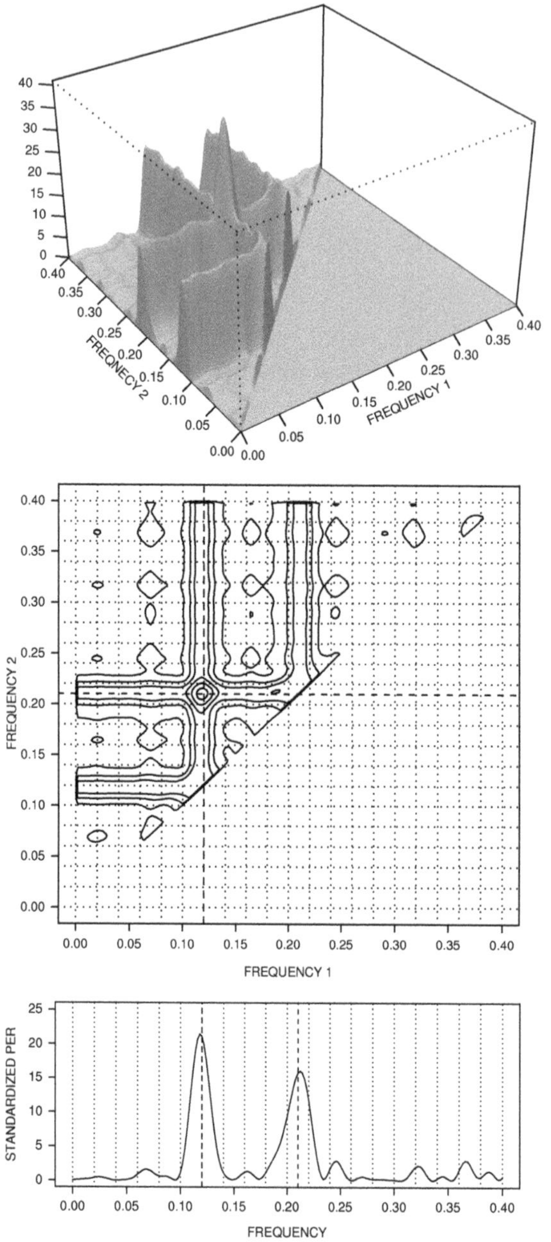

Figure 8.11. Plot of the standardized bivariate periodogram for a time series consisting of two unit-amplitude zero-phase complex sinusoids in Gaussian colored noise ($n = 50$, SNR = −5 dB per sinusoid). The signal frequencies are $\omega_1 = 2\pi \times 0.12$ and $\omega_2 = 2\pi \times 0.21$. The standardized univariate periodogram is shown at the bottom. Dashed lines depict the signal frequencies and dotted lines depict the Fourier grid.

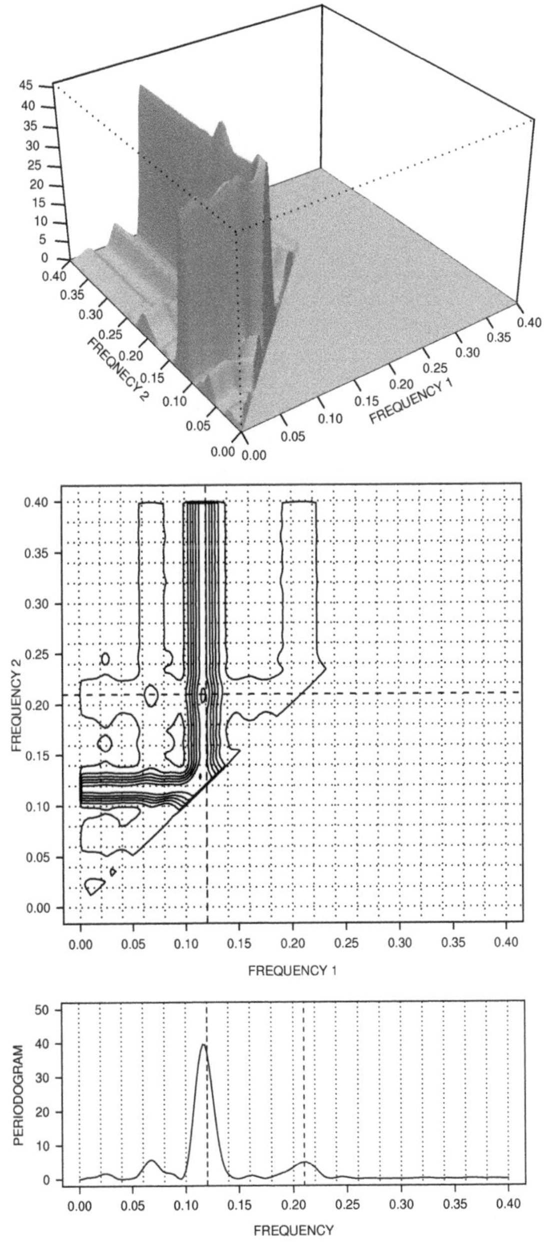

Figure 8.12. Plot of the ordinary (unstandardized) bivariate and univariate periodograms for the same time series as in Figure 8.11.

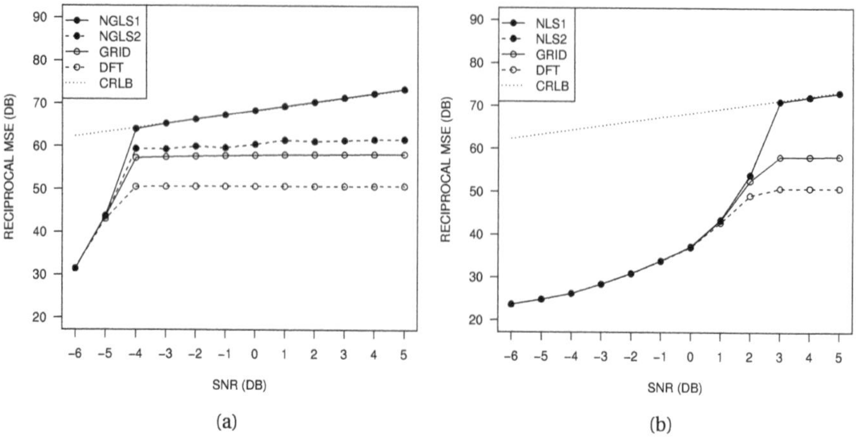

Figure 8.13. Plot of the reciprocal average MSE of (a) the NGLS frequency estimates and (b) the NLS frequency estimates for two complex sinusoids in Gaussian AR(2) noise discussed in Example 8.5 with $n = 100$. Solid line with dots, NGLS and NLS initialized by grid search; dashed line with dots, NGLS and NLS initialized by DFT (the latter is indistinguishable from NLS initialized by grid search); solid line with circles, grid search; dashed line with circles, DFT. Dotted line depicts the Gaussian CRLB.

periodograms suggests that the benefit can be attributed largely to the enhancement of the second sinusoid relative to the first one.

The following example compares the statistical performance of the NGLS estimator with the statistical performance of the NLS estimator.

Example 8.5 (Two Complex Sinusoids in Gaussian AR(2) Noise: Part 1). Let the sinusoidal signal be the same as in Example 8.1 and the noise be a Gaussian AR(2) process given by (8.4.5). To compute the NGLS frequency estimates, we apply the Nelder-Mead algorithm to the standardized bivariate periodogram. The initial values are obtained by the same DFT and grid search methods as discussed in Example 8.1 except that the standardized periodograms are in place of the unstandardized periodograms. For comparison, the NLS frequency estimates are also computed using the method described in Example 8.1. The only exception is that the DFT method is applied to a restricted search interval $[2\pi\times 0.01, 2\pi\times 0.3]$ that contains the signal frequencies. This constraint improves the chance for the DFT method to produce desirable initial values, especially at low SNR.

Figure 8.13 shows the average MSE of the NGLS frequency estimates and the NLS frequency estimates together with the average MSE of the corresponding initial values. Note that 0.2% worst estimates are trimmed in calculating the MSE of the NLS estimator and its initializer. No trimming for the NGLS estimator.

As we can see, the use of noise covariance matrix makes a big difference in the estimates. Figure 8.13(a) shows that the NGLS estimator is able to follow the CRLB closely, just like the NLS estimator in Example 8.1 under the white noise condition, which is shown in Figure 8.3(a). Without using the noise covariance matrix, the NLS estimator performs poorly under the colored noise condition. The huge errors shown in Figure 8.13(b) at low and moderate SNR can be attributed to the poor initial values given by the unstandardized DFT method, despite the favorable restriction on the search interval and the trimming of worst estimates. Figure 8.13(b) also shows that when the SNR is sufficiently high, the NLS estimator is able to produce good estimates that follow the CRLB closely, although still slightly inferior to the NGLS estimates by a closer examination. In this case, the knowledge of the noise covariance matrix becomes less important. ◇

Finally, let us consider the more difficult case where the covariance matrix $\mathbf{R}_\epsilon := \mathbf{R}_\epsilon(\boldsymbol{\eta})$ contains an unknown parameter $\boldsymbol{\eta}$ other than the noise variance that must be estimated together with the signal parameter $\boldsymbol{\theta}$. The difficulty arises especially when the noise has prominent spectral peaks. In this case, the noise model has to be flexible enough to accommodate the spectral peaks; however, the flexibility also gives rise to the potential of overestimating the noise spectrum near the signal frequencies. The consequence of overestimation is the reduced or even diminished contribution of the affected sinusoids in estimating the signal parameter. It is particularly harmful to the determination of initial values based on the standardized univariate periodogram.

Despite the difficulty, one can still achieve improved results in some cases by estimating the noise parameter instead of assuming white noise. A simple method is to use a robust procedure discussed in Chapter 7 (Section 7.2) to produce an estimate $\hat{\boldsymbol{\eta}}$ for the noise parameter and then replace $\mathbf{R}_\epsilon$ with $\mathbf{R}_\epsilon(\hat{\boldsymbol{\eta}})$ in the standardized multivariate periodogram. The standardized univariate periodogram can be constructed similarly with the estimated noise spectrum.

As an example, Figure 8.14 shows the standardized bivariate and univariate periodograms with an estimated noise parameter for the same time series as used in Figures 8.11 and 8.12. Under the assumption that the noise is an AR(2) process, the noise parameter is estimated by the robust iterative Yule-Walker method discussed in Section 7.2 after 5 iterations with $c := -\log(1-r)$ and $r = 0.7$, initialized by a constant spectrum which equals the sample variance.

By comparing Figure 8.14 with Figures 8.11 and 8.12, we can see that the standardized bivariate periodogram with estimated noise parameter behaves similarly to the standardized bivariate periodogram with the true noise parameter in producing a much better defined spectral peak near the signal frequencies than the unstandardized bivariate periodogram. This again can be attributed to the boosted second sinusoid relative to the first one, as can be seen by comparing the univariate periodograms. Note that the standardized bivariate periodogram still has a less prominent spectral peak when the noise parameter is estimated

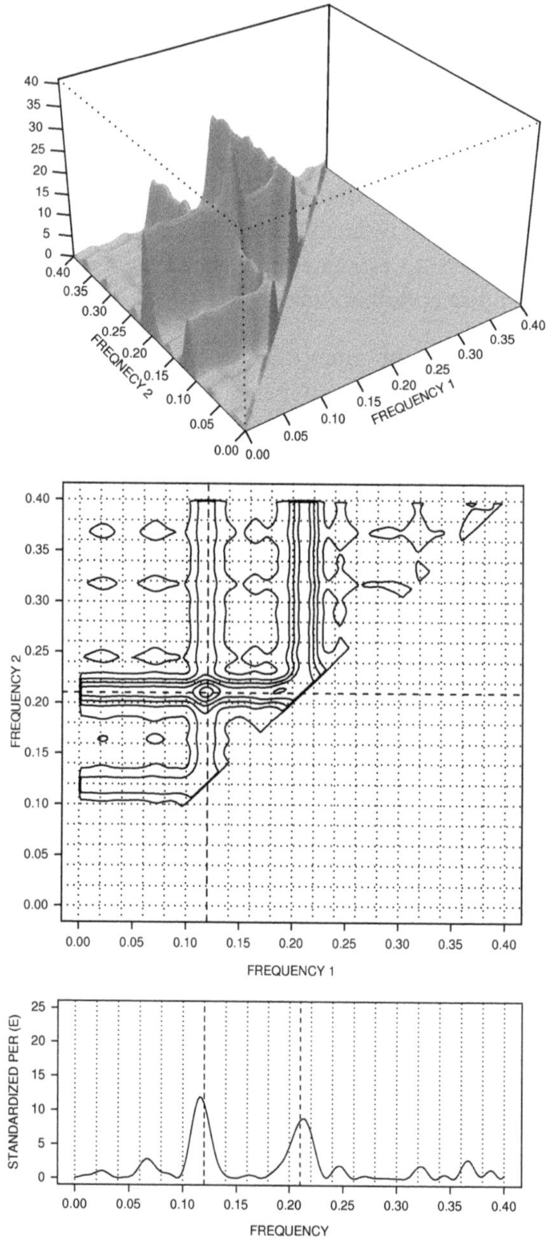

Figure 8.14. Plot of the standardized bivariate and univariate periodograms with estimated noise parameter for the same time series as in Figure 8.11.

than when it is the true value. This is due to the suppression of the sinusoids as a consequence of overestimated noise spectrum near the signal frequencies that cannot be completely avoided even by a robust spectral estimator.

Under the Gaussian assumption, the noise parameter can also be estimated jointly with the sinusoidal parameter by using the iterative procedure described in Section 8.1: Given the initial value $\hat{\boldsymbol{\eta}}_0$ and for $m = 1, 2, \ldots$,

1. Compute $\hat{\boldsymbol{\beta}}_m$ and $\hat{\boldsymbol{\omega}}_m$ according to (8.4.3) with $\mathbf{R}_\epsilon(\hat{\boldsymbol{\eta}}_{m-1})$ in place of $\mathbf{R}_\epsilon$;
2. Compute

$$\hat{\boldsymbol{\eta}}_m := \arg\max_{\tilde{\boldsymbol{\eta}} \in \Xi} \{-\log|\mathbf{R}_\epsilon(\tilde{\boldsymbol{\eta}})| - \hat{\boldsymbol{\epsilon}}_m^H \mathbf{R}_\epsilon^{-1}(\tilde{\boldsymbol{\eta}}) \hat{\boldsymbol{\epsilon}}_m\},$$

   where $\hat{\boldsymbol{\epsilon}}_m := \mathbf{y} - \mathbf{F}(\hat{\boldsymbol{\omega}}_m)\hat{\boldsymbol{\beta}}_m$.

Note that the second step produces the ML estimate of $\boldsymbol{\eta}$ from the residual $\hat{\boldsymbol{\epsilon}}_m$ under the working assumption $\hat{\boldsymbol{\epsilon}}_m \sim \mathrm{N}_c(\mathbf{0}, \mathbf{R}_\epsilon(\boldsymbol{\eta}))$. For some parametric models, there exist efficient procedures to compute $\hat{\boldsymbol{\eta}}_m$. In particular, by modeling the noise as an AR process, one can estimate the AR parameters by the linear prediction method discussed in Chapter 7 (Section 7.1.3) which produces approximate ML estimates with low computational cost. However, given the possibility of contamination from the residual sinusoids in $\hat{\boldsymbol{\epsilon}}_m$ due to poor frequency estimation, a robust method, like the iterative Yule-Walker estimator discussed in Section 7.2, seems more appropriate.

The success of the joint estimation procedure depends crucially on the ability of the first step to produce a reasonably accurate frequency estimate so that the impact of the sinusoids on the residuals used in the second step is sufficiently reduced. No improvement should be expected if the initial frequency estimates are so poor that the sinusoids remain virtually unchanged in the residuals, even if robust estimators are used in the second step.

The following example demonstrates the performance of the NGLS estimator with the noise parameter estimated by the joint estimation procedure.

Example 8.6 (Two Complex Sinusoids in Gaussian AR(2) Noise: Part 2). Continue with Example 8.5. Under the assumption that the noise is a complex AR(2) process, let the noise parameter be estimated from the original data by the iterative Yule-Walker method discussed in Section 7.2 with 5 iterations, initialized by a constant spectrum which equals the sample variance. Let the NGLS estimates be computed as in Example 8.5 except that the true noise covariance matrix is replaced by the estimated noise covariance matrix. For a fair comparison with the result of the NLS estimator shown in Figure 8.13(b), let the restricted search interval described in Example 8.5 be used in this example to produce the initial DFT estimates from the standardized univariate periodogram with estimated noise parameter. Also let the NGLS estimates be trimmed in the same way as

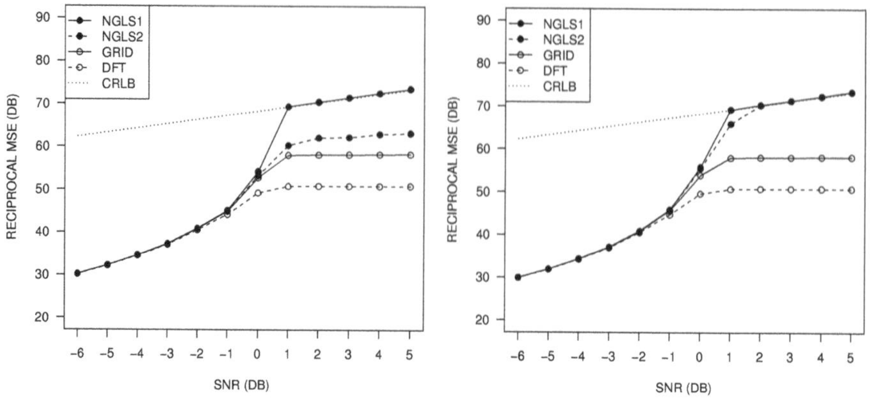

(a) (b)

Figure 8.15. Plot of the reciprocal average MSE of the NGLS frequency estimates with estimated noise parameter for two complex sinusoids in Gaussian AR(2) noise discussed in Example 8.6 with $n = 100$. (a) No iteration. (b) One iteration of the joint estimation procedure. Solid line with dots, NGLS initialized by grid search; dashed line with dots, NGLS initialized by DFT; solid line with circles, grid search; dashed line with circles, DFT. Dotted line depicts the Gaussian CRLB.

the NLS estimates. Figure 8.15(a) shows the MSE of the resulting NGLS frequency estimates and their initial values.

By comparing Figure 8.15(a) with Figure 8.13(b), we can see that the NGLS estimator reduces the SNR threshold of the NLS estimator from 3 dB to 1 dB; it also improves the MSE considerably below the threshold. By comparing Figure 8.15(a) with Figure 8.13(a), we can see that the SNR threshold of the NGLS estimator is much higher when the noise parameter is estimated than when it is the true value. The discrepancy results mainly from poor initial values as a consequence of overestimated noise spectrum particularly near the first signal frequency.

Figure 8.15(b) shows the result of the joint estimation procedure after one iteration: based on the residuals from the previous frequency estimates, an updated estimate of the noise parameter is produced by the iterative Yule-Walker method, and the final NGLS frequency estimates are obtained as before but with the updated noise parameter.

As expected, the iteration does not lower the SNR threshold because it cannot produce better initial values when the original frequency estimates are poor. However, the iteration does improve the numerical conditions of the standardized bivariate periodogram so that the DFT method becomes accurate enough to initialize the NGLS calculation when the SNR is above the 1 dB threshold. ◇

The examples in this section demonstrate that, by incorporating the noise spectrum and covariance matrix, the sinusoids can be more easily identified in the univariate and multivariate periodograms when the noise is colored. Preferably, the noise spectrum and covariance matrix should be estimated from a set of training data in the absence of sinusoids. The most difficult case is when the noise estimation has to be done in the presence of sinusoids with unknown parameters. Certain prior information regarding the location of the signal frequencies, the location of the noise spectral peaks, and the smoothness of the noise spectrum would be helpful in restricting the initial search for the signal frequencies and the choice of a noise model. When the SNR is sufficiently high or the noise spectrum is sufficient flat, the noise modeling becomes unnecessary, because accurate estimates of the signal parameters can be produced by a suitable procedure developed under the white noise assumption.

8.5 Determining the Number of Sinusoids

Determining the number of sinusoids based solely on the observed data adds a new dimension of difficulty, especially when some of the signal frequencies are closely spaced, even at high SNR. Visual inspection of the ordinary periodogram is a common practice in identifying well-separated frequencies. Visual inspection of the bivariate periodogram together with the ordinary (univariate) periodogram can be used to identify frequency clusters consisting of one frequency or two closely spaced frequencies. This is demonstrated by the example in Figure 8.16, where the time series contains three complex sinusoids with frequencies $(\omega_1, \omega_2, \omega_3) = 2\pi \times (0.12, 0.13, 0.23)$ and the sample size is $n = 50$.

Recall that the bivariate periodogram $P_n(\tilde{\omega}_1, \tilde{\omega}_2)$ coincides with $I_n(\tilde{\omega}_1) + I_n(\tilde{\omega}_2)$ on the Fourier grid $\Omega_n \times \Omega_n$. Therefore, a spectral peak in the ordinary periodogram appears in the bivariate periodogram as a vertical ridge and a horizontal ridge which meet at the diagonal line. A pair of well-separated spectral peaks in the ordinary periodogram yields a local maximum in the bivariate periodogram, away from the diagonal line at the intersection of the corresponding vertical and horizontal ridges. More importantly, a pair of closely spaced frequencies produces a local maximum near the diagonal line in the corner formed by the vertical and horizontal ridges that correspond to the unresolved peak in the ordinary periodogram.

In Figure 8.16, the largest spectral peak in the bivariate periodogram appears at $(0.125, 0.229)$, which can be associated with the two largest spectral peaks in the ordinary periodogram. The second largest spectral peak in the bivariate periodogram is located at $(0.121, 0.130)$, which corresponds to the two closely spaced

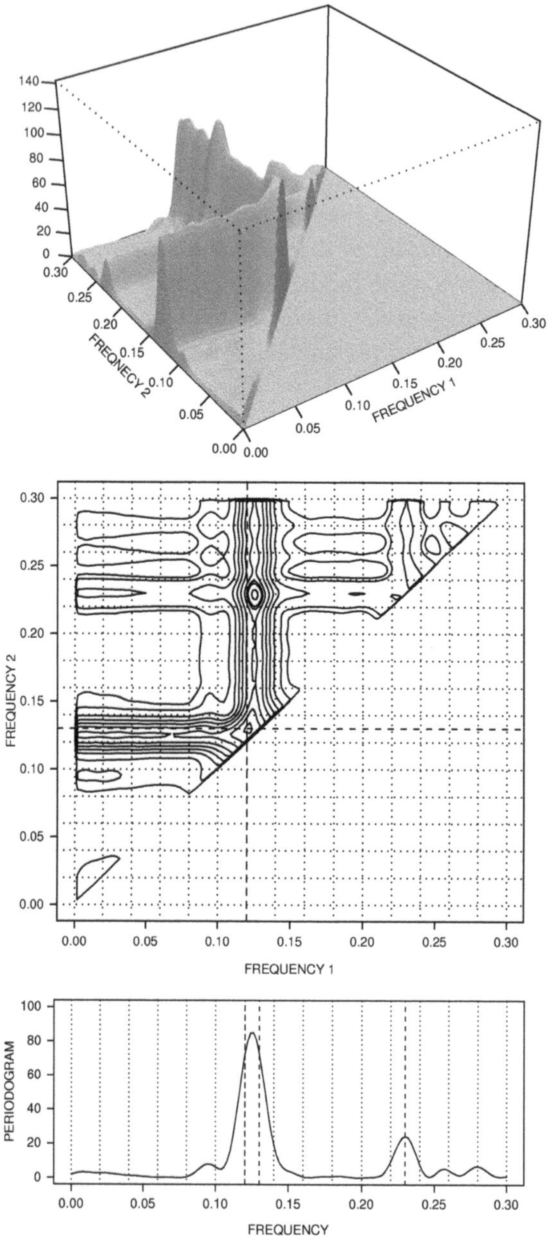

Figure 8.16. Plot of the bivariate periodogram for a time series consisting of three unit-amplitude zero-phase complex sinusoids in Gaussian colored noise ($n = 50$, SNR $= -5$ dB per sinusoid). The signal frequencies are $\omega_1 = 2\pi \times 0.12$, $\omega_2 = 2\pi \times 0.13$, and $\omega_3 = 2\pi \times 0.23$. The univariate periodogram is shown at the bottom. Dashed lines depict the signal frequencies and dotted lines depict the Fourier grid.

signal frequencies that are unresolved in the ordinary periodogram. By comparing Figure 8.16 with Figure 8.1, one can distinguish the case where the time series contains two well-separated frequencies from the case where two frequencies appear in the ordinary periodogram as an unresolved cluster while the third frequency is well separated from them. The clue lies in whether a local maximum occurs in the corner where the main ridges meet with the diagonal line. In Figure 8.1, there is no such local maximum in that corner, whereas in Figure 8.16 a local maximum is observed. The same clue can be used to distinguish a two-frequency cluster as shown in Figure 8.2 from a single frequency (not shown). These properties of the bivariate periodogram lead to the conclusion that the time series in this example contains three sinusoids with estimated frequencies $(\hat{\omega}_1, \hat{\omega}_2, \hat{\omega}_3) = 2\pi \times (0.121, 0.130, 0.229)$.

While visual inspection of the univariate and bivariate periodograms is useful, visualization of higher dimensional periodograms becomes much more difficult. Therefore, in general, one has to rely on some "blind" automatic procedures to help identify clusters of more than two closely spaced frequencies. Let us consider a few examples of such procedures.

Recall that in Chapter 5 (Section 5.4) we discuss the problem of frequency selection under the condition that the signal frequencies belong to a known set of candidate frequencies, where the objective is to determine which and how many candidate frequencies are the signal frequencies contained in a data record. In this case, minimizing the generalized information criterion (GIC) in (5.4.1) or the prediction error criterion (PEC) in (5.4.5) yields consistent estimators for the number of sinusoids when the penalty function is suitably chosen.

This method can be applied directly to the situation where the candidate frequencies are the frequency estimates produced by the NLS method from the observed data on the basis of a complete model with m sinusoids ($m \ge p$). The main challenge is to find a suitable initial value for the NLS estimator, especially when some of the signal frequencies are closely spaced. To guarantee consistent amplitude estimation, the initial value must be chosen such that the resulting m frequency candidates produced by the NLS estimator include p estimates for the signal frequencies with accuracy $\mathcal{O}_P(n^{-1})$.

A multistep procedure is proposed in [432] to deal with the initial value problem. The gist of the method is the following. Start with a set of Fourier frequencies obtained by applying a significance test to the discrete periodogram ordinates based on the n-point DFT of $\mathbf{y}$. Split each initial Fourier frequency estimate into equally spaced multiple frequencies within the corresponding Fourier bin to produce the initial value for multiple frequency estimation by the NLS method (or other alternatives). The number of splits is chosen according to the potential error, determined by the CRLB with estimated SNR, for estimating two closely spaced frequencies near the initial frequency such that a smaller potential error leads to more splits, and no splits when the error is larger than the width of a Fourier bin (i.e., $2\pi/n$). After the NLS estimation, an additional step is performed to combine some of the estimated frequencies deemed too close to each other in comparison with the potential error of estimation. After combining, the NLS method is used again to generate the final candidates for frequency selection using the GIC or PEC criterion. To avoid calculating the residual sum of squares for all subsets, one can sort the candidate frequencies by the corresponding amplitudes and define $\text{SSE}(\tilde{p})$ in (5.4.1) and (5.4.5) as the residual sum of squares corresponding to the $\tilde{p}$ frequencies with the largest amplitudes.

When the signal frequencies are known to be well separated, one can also use the PM method discussed in Chapter 6 (Section 6.4.2) to estimate the signal parameters. The resulting GIC criterion is investigated in [418] and the consistency of its minimizer as an estimator of p is established under the assumption that the penalty function in (5.4.1) takes the form $c(n,\tilde{p}) := c_n\tilde{p}$ with c_n satisfying the conditions of Theorem 5.8. A similar result is given in [300] for the special case where frequency estimation is restricted on the Fourier grid.

For estimating closely spaced signal frequencies, multidimensional joint optimization techniques are unavoidable. But for well-separated frequencies, the following stepwise estimation procedure, proposed in [81], can be used to determine the number of sinusoids and their parameters. It solves a sequence of one-dimensional optimization problems. Starting with $\hat{p} := 0$,

1. Determine the Fourier frequency corresponding to the largest discrete periodogram ordinate based on the n-point DFT of $\mathbf{y}$ and test its significance by Fisher's test discussed in Chapter 6 (Section 6.2) with a suitably small trimming parameter r (such that $n(1-r) \geq p$). If insignificant, stop the procedure; otherwise, increase $\hat{p}$ by 1 and go to the next step.
2. Compute the local maximizer of the continuous periodogram using the Fourier frequency found in step 1 as initial value.
3. Remove the sinusoid with the estimated frequency using the least-squares technique discussed in Chapter 5 (Section 5.1) to produce the residuals and the amplitude estimate.
4. Go to step 1 with the residuals in place of $\mathbf{y}$ and with the trimming parameter r increased by $1/n$.

Because it works with one sinusoid at a time, the stepwise procedure avoids the multidimensional optimization required by joint estimation. This computational advantage is particularly attractive when the time series contains a large number of sinusoids with well-separated frequencies.

The stepwise estimation procedure repeats the steps that can be generally described as sinusoid identification, frequency estimation, and sinusoid removal. These steps can also be realized by supplementing the ordinary periodogram with the bivariate periodogram to identify and estimate frequency clusters that consist of one frequency or two closely spaced frequencies. The univariate and bivariate Laplace periodograms can be used in this procedure to achieve robustness against outliers.

8.6 Proof of Theorems

This section contains the proof of (8.2.25) and Theorems 8.1–8.3.

Proof of (8.2.25). Let $\mathbf{z}(\tilde{\omega}) := n^{-1}\mathbf{F}^H(\tilde{\omega})\mathbf{y}$ be the Fourier transform of $\mathbf{y}$ at frequency $\tilde{\omega}$. Then, it is easy to see that

$$P_n(\tilde{\omega}) = n\mathbf{z}^H(\tilde{\omega})\{n^{-1}\mathbf{F}^H(\tilde{\omega})\mathbf{F}(\tilde{\omega})\}^{-1}\mathbf{z}(\tilde{\omega}). \tag{8.6.1}$$

By Lemma 12.1.5, the condition in (8.2.24) guarantees that

$$n^{-1}\mathbf{F}^H(\tilde{\omega})\mathbf{F}(\tilde{\omega}) = \mathbf{I} + \mathcal{O}(n^{-1}c_n^{-1})$$

uniformly in $\tilde{\omega} \in \Omega_0$. Therefore, we can write

$$\{n^{-1}\mathbf{F}^H(\tilde{\omega})\mathbf{F}(\tilde{\omega})\}^{-1} = \mathbf{I} + \mathbf{D}(\tilde{\omega}), \tag{8.6.2}$$

where $\mathbf{D}(\tilde{\omega}) := \mathbf{I} - \{n^{-1}\mathbf{F}^H(\tilde{\omega})\mathbf{F}(\tilde{\omega})\}^{-1}$ can be expressed as $\mathcal{O}(n^{-1}c_n^{-1})$ uniformly in $\tilde{\omega} \in \Omega_0$. Substituting (8.6.2) in (8.6.1) leads to

$$P_n(\tilde{\omega}) = n\{\|\mathbf{z}(\tilde{\omega})\|^2 + \mathbf{z}^H(\tilde{\omega})\mathbf{D}(\tilde{\omega})\mathbf{z}(\tilde{\omega})\}. \tag{8.6.3}$$

Moreover, let $\varrho_n(\tilde{\omega})$ denote the spectral radius, i.e., the largest eigenvalue in absolute value, of the Hermitian matrix $\mathbf{D}(\tilde{\omega})$. Then, according to the matrix theory [153, pp. 296–297], we have

$$\varrho_n(\tilde{\omega}) \le \|\mathbf{D}(\tilde{\omega})\| = \mathcal{O}(n^{-1}c_n^{-1})$$

and

$$|\mathbf{z}^H(\tilde{\omega})\mathbf{D}(\tilde{\omega})\mathbf{z}(\tilde{\omega})| \le \varrho_n(\tilde{\omega})\|\mathbf{z}(\tilde{\omega})\|^2.$$

Combining these expressions with (8.6.3) yields

$$P_n(\tilde{\omega}) = n\{1 + R_n(\tilde{\omega})\}\|\mathbf{z}(\tilde{\omega})\|^2, \tag{8.6.4}$$

where $R_n(\tilde{\omega}) = \mathcal{O}(n^{-1}c_n^{-1}) \to 0$ uniformly in $\tilde{\omega} \in \Omega_0$. Finally, by straightforward calculation, we obtain

$$\|\mathbf{z}(\tilde{\omega})\|^2 = n^{-1}\sum_{k=1}^{p} I_n(\tilde{\omega}_k).$$

Substituting this expression in (8.6.4) leads to (8.2.25). □

Proof of Theorem 8.1. Consider the frequency estimation first. For any $\delta > 0$, let Ω_δ be the subset of Ω_0 in which $|\tilde{\omega}_k - \omega_k| \le \delta / n$. According to the proof of Theorem 6.8 and especially the counterparts of (6.6.36), (6.6.39), and (6.6.40) for the complex case, we have

$$\max_{\tilde{\omega} \in \Omega_\delta} n^{-1} \sum_{k=1}^{p} I_n(\tilde{\omega}_k) = \sum_{k=1}^{p} C_k^2 + \mathcal{O}_P(1),$$

$$\max_{\tilde{\omega} \in \Omega_0 \setminus \Omega_\delta} n^{-1} \sum_{k=1}^{p} I_n(\tilde{\omega}_k) \le \sum_{k=1}^{p} C_k^2 - \nu_\delta + \mathcal{O}_P(1),$$

where $\nu_\delta > 0$ is a constant. This result, combined with (8.2.25), leads to

$$\max_{\tilde{\omega} \in \Omega_\delta} n^{-1} P_n(\tilde{\boldsymbol{\omega}}) \ge \sum_{k=1}^{p} C_k^2 + \mathcal{O}_P(1) - \mathcal{O}_P(n^{-1} c_n^{-1}),$$

$$\max_{\tilde{\omega} \in \Omega_0 \setminus \Omega_\delta} n^{-1} P_n(\tilde{\boldsymbol{\omega}}) \le \sum_{k=1}^{p} C_k^2 - \nu_\delta + \mathcal{O}_P(1) + \mathcal{O}_P(n^{-1} c_n^{-1}).$$

Therefore, the maximizer of $P_n(\tilde{\boldsymbol{\omega}})$ must appear in Ω_δ with probability tending to unity as $n \to \infty$, i.e., $P(|\hat{\omega}_k - \omega_k| \le \delta / n$ for all $k) \to 1$. This proves the consistency of the frequency estimator $\hat{\boldsymbol{\omega}}$: $n(\hat{\omega}_k - \omega_k) \xrightarrow{P} 0$ for all k.

To show the consistency of $\hat{\boldsymbol{\beta}}$ given by (8.2.9), observe that (8.6.2) leads to

$$\hat{\boldsymbol{\beta}} = n^{-1} \{\mathbf{I} + \mathbf{D}(\hat{\boldsymbol{\omega}})\} \mathbf{F}^H(\hat{\boldsymbol{\omega}}) \mathbf{y}.$$

Because $\hat{\boldsymbol{\omega}} - \boldsymbol{\omega} = \mathcal{O}_P(n^{-1})$, it follows from Theorem 5.6 and Remark 5.7 that

$$n^{-1} \mathbf{F}^H(\hat{\boldsymbol{\omega}}) \mathbf{y} \xrightarrow{P} \boldsymbol{\beta}.$$

Combining this result with $\mathbf{D}(\hat{\boldsymbol{\omega}}) = \mathcal{O}(1)$ yields $\hat{\boldsymbol{\beta}} \xrightarrow{P} \boldsymbol{\beta}$, and hence the consistency of $\hat{A}_k$, $\hat{B}_k$, $\hat{C}_k$, and $\hat{\phi}_k$ due to Proposition 2.13(a).

Now consider the asymptotic normality. For the Cartesian case, we have $\boldsymbol{\theta} := [\boldsymbol{\theta}_1^T, \ldots, \boldsymbol{\theta}_p^T]^T$ and $\boldsymbol{\theta}_k := [A_k, B_k, \omega_k]^T$. Let $\hat{\boldsymbol{\theta}} := [\hat{\boldsymbol{\theta}}_1^T, \ldots, \hat{\boldsymbol{\theta}}_p^T]^T$ be the NLS estimator of $\boldsymbol{\theta}$ given by (8.2.1). Then, we have $\hat{\boldsymbol{\theta}}_k = [\hat{A}_k, \hat{B}_k, \hat{\omega}_k]^T$, where $\hat{A}_k$ and $\hat{B}_k$ are defined by (8.2.6). Because $\hat{\boldsymbol{\theta}}$ minimizes $S_n(\tilde{\boldsymbol{\theta}}) := \|\mathbf{y} - \mathbf{x}(\tilde{\boldsymbol{\theta}})\|^2$, it follows that

$$\nabla S_n(\hat{\boldsymbol{\theta}}) = \mathbf{0},$$

where ∇ denotes the gradient operator with respect to $\boldsymbol{\theta}$. Combining this equation with the mean value theorem in calculus leads to

$$\mathbf{H}_n(\hat{\boldsymbol{\theta}} - \boldsymbol{\theta}) = -\nabla S_n(\boldsymbol{\theta}), \quad \mathbf{H}_n := \int_0^1 \nabla^2 S_n(\boldsymbol{\theta} + s(\hat{\boldsymbol{\theta}} - \boldsymbol{\theta}))\, ds,$$

where ∇^2 denotes the Hessian operator. It suffices to show that

$$\mathbf{K}^{-1}\nabla^2 S_n(\tilde{\boldsymbol{\theta}})\mathbf{K}^{-1} \xrightarrow{P} \mathbf{U} \tag{8.6.5}$$

uniformly in $\tilde{\boldsymbol{\theta}} := \boldsymbol{\theta} + s(\hat{\boldsymbol{\theta}} - \boldsymbol{\theta})$ for $s \in [0,1]$ and

$$\mathbf{K}^{-1}\nabla S_n(\boldsymbol{\theta}) \stackrel{A}{\sim} \mathrm{N}(\mathbf{0}, \boldsymbol{\Sigma}), \tag{8.6.6}$$

where $\mathbf{U}$ and $\boldsymbol{\Sigma}$ are certain positive-definite matrices. If this can be done, then an application of Proposition 2.13(b) would yield $\mathbf{K}(\hat{\boldsymbol{\theta}} - \boldsymbol{\theta}) \stackrel{A}{\sim} \mathrm{N}(\mathbf{0}, \mathbf{U}^{-1}\boldsymbol{\Sigma}\mathbf{U}^{-1})$.

As in the proof of Theorems 6.8 and 6.10, (8.6.5) can be established by applying Lemma 12.1.5 and Lemma 12.5.9(b) to a Taylor expansion, together with the consistency assertion in part (a). In this case, $\mathbf{U} = \mathrm{diag}(\mathbf{U}_1, \ldots, \mathbf{U}_p)$ and

$$\mathbf{U}_k := \begin{bmatrix} 2 & 0 & B_k \\ 0 & 2 & -A_k \\ B_k & -A_k & 2C_k^2/3 \end{bmatrix} = 2\mathbf{W}_k,$$

where $\mathbf{W}_k$ is defined by (3.3.7). Moreover, let ∇_k denote the gradient operator with respect to $\boldsymbol{\theta}_k$ and let $e_n(\omega) := n^{-1}\sum \epsilon_t \exp(-it\omega)$. Then, we can write

$$\boldsymbol{\xi}_{nk} := \mathbf{K}_n^{-1}\nabla_k S_n(\boldsymbol{\theta}) = -2\begin{bmatrix} n^{1/2}\Re(X_{nk}) \\ -n^{1/2}\Im(X_{nk}) \\ n^{-1/2}\Re(\beta_k^* Y_{nk}) \end{bmatrix}, \tag{8.6.7}$$

where $X_{nk} := e_n(\omega_k)$ and $Y_{nk} := \dot{e}_n(\omega_k)$. Because this expression is just the leading term in (6.6.60), it follows that $\mathbf{K}^{-1}\nabla S_n(\boldsymbol{\theta})$ has the same asymptotic distribution as $\mathbf{K}^{-1}\nabla U_n(\boldsymbol{\theta})$ in the proof of Theorem 6.10. Hence (8.6.6) holds with $\boldsymbol{\Sigma} := \mathrm{diag}\{2f_\epsilon(\omega_1)\mathbf{W}_1, \ldots, 2f_\epsilon(\omega_p)\mathbf{W}_p\}$. The assertion follows upon noting that

$$\mathbf{U}^{-1}\boldsymbol{\Sigma}\mathbf{U}^{-1} = \mathrm{diag}\{\boldsymbol{\Gamma}_{\mathrm{C}}(\boldsymbol{\theta}_1), \ldots, \boldsymbol{\Gamma}_{\mathrm{C}}(\boldsymbol{\theta}_p)\},$$

where $\boldsymbol{\Gamma}_{\mathrm{C}}(\boldsymbol{\theta}_k) := \frac{1}{2}f_\epsilon(\omega_k)\mathbf{W}_k^{-1} = \frac{1}{2}\gamma_k^{-1}\boldsymbol{\Lambda}_{\mathrm{C}}(\boldsymbol{\theta}_k)$, $\gamma_k := C_k^2/f_\epsilon(\omega_k)$, and $\boldsymbol{\Lambda}_{\mathrm{C}}(\boldsymbol{\theta}_k)$ is given by (3.3.2). Using the Jacobian matrix $\mathbf{J}_k$ given by (6.6.51), together with Proposition 2.15, we obtain the asymptotic distribution of $\hat{\boldsymbol{\theta}}_k := [\hat{C}_k, \hat{\phi}_k, \hat{\omega}_k]^T$. □

Proof of Theorem 8.2. For the consistency assertion, it suffices to show that $n(\hat{\boldsymbol{\omega}} - \boldsymbol{\omega}) \xrightarrow{P} 0$ because this result leads to $\hat{\boldsymbol{\beta}} \xrightarrow{P} \boldsymbol{\beta}$ by Theorem 5.6.

Given $\mathbf{y} = \mathbf{x} + \boldsymbol{\epsilon}$, it follows that $\mathbf{z}(\tilde{\boldsymbol{\omega}}) := n^{-1}\mathbf{F}^H(\tilde{\boldsymbol{\omega}})\mathbf{y} = \boldsymbol{\mu}(\tilde{\boldsymbol{\omega}}) + \mathbf{e}(\tilde{\boldsymbol{\omega}})$, where $\boldsymbol{\mu}(\tilde{\boldsymbol{\omega}}) := n^{-1}\mathbf{F}^H(\tilde{\boldsymbol{\omega}})\mathbf{x}$ and $\mathbf{e}(\tilde{\boldsymbol{\omega}}) := n^{-1}\mathbf{F}^H(\tilde{\boldsymbol{\omega}})\boldsymbol{\epsilon}$. Let $\mathbf{C}(\tilde{\boldsymbol{\omega}}) := n^{-1}\mathbf{F}^H(\tilde{\boldsymbol{\omega}})\mathbf{F}(\tilde{\boldsymbol{\omega}})$. Then,

$$\begin{aligned} n^{-1}P_n(\tilde{\boldsymbol{\omega}}) &= \boldsymbol{\mu}^H(\tilde{\boldsymbol{\omega}})\mathbf{C}^{-1}(\tilde{\boldsymbol{\omega}})\boldsymbol{\mu}(\tilde{\boldsymbol{\omega}}) + \boldsymbol{\mu}^H(\tilde{\boldsymbol{\omega}})\mathbf{C}^{-1}(\tilde{\boldsymbol{\omega}})\mathbf{e}(\tilde{\boldsymbol{\omega}}) \\ &\quad + \mathbf{e}^H(\tilde{\boldsymbol{\omega}})\mathbf{C}^{-1}(\tilde{\boldsymbol{\omega}})\boldsymbol{\mu}(\tilde{\boldsymbol{\omega}}) + \mathbf{e}^H(\tilde{\boldsymbol{\omega}})\mathbf{C}^{-1}(\tilde{\boldsymbol{\omega}})\mathbf{e}(\tilde{\boldsymbol{\omega}}). \end{aligned}$$

Let $D_n(\lambda) := n^{-1}\sum_{t=1}^n \exp(it\lambda)$ be the Dirichlet kernel. Then, for $p=2$,

$$\mathbf{C}^{-1}(\tilde{\boldsymbol{\omega}}) = \begin{bmatrix} 1 & D_n(\lambda) \\ D_n^*(\lambda) & 1 \end{bmatrix}^{-1} = \frac{1}{1-|D_n(\lambda)|^2}\begin{bmatrix} 1 & -D_n(\lambda) \\ -D_n^*(\lambda) & 1 \end{bmatrix},$$

where $\tilde{\boldsymbol{\omega}} = (\tilde{\omega}_1, \tilde{\omega}_2)$ and $\lambda := \tilde{\omega}_2 - \tilde{\omega}_1$. Clearly, all elements of $\mathbf{C}^{-1}(\tilde{\boldsymbol{\omega}})$ do not exceed $a_n(\lambda) := 1/\{1-|D_n(\lambda)|\}$ in absolute value. Moreover, the eigenvalues of $\mathbf{C}^{-1}(\tilde{\boldsymbol{\omega}})$ are $1/\{1+|D_n(\lambda)|$ and $1/\{1-|D_n(\lambda)|$ and hence for any $\mathbf{a} \in \mathbb{C}^2$ we have

$$\frac{1}{1+|D_n(\lambda)|}\|\mathbf{a}\|^2 \le \mathbf{a}^H \mathbf{C}^{-1}(\tilde{\boldsymbol{\omega}})\mathbf{a} \le \frac{1}{1-|D_n(\lambda)|}\|\mathbf{a}\|^2.$$

By Lemma 12.5.9, $\mathbf{e}(\tilde{\boldsymbol{\omega}}) = \mathcal{O}_P(1)$ uniformly in $\tilde{\boldsymbol{\omega}}$. Combining these results with the fact that $\boldsymbol{\mu}(\tilde{\boldsymbol{\omega}}) = \mathcal{O}(1)$ uniformly in $\tilde{\boldsymbol{\omega}}$ gives

$$n^{-1}P_n(\tilde{\boldsymbol{\omega}}) = n^{-1}P_{nx}(\tilde{\boldsymbol{\omega}}) + \mathcal{O}_P(a_n(\lambda)). \tag{8.6.8}$$

For $\delta \in (0, a]$ and large n such that $\delta/n < \Delta$, we can write

$$\max_{\tilde{\boldsymbol{\omega}} \in \Omega_\delta} n^{-1}P_{nx}(\tilde{\boldsymbol{\omega}}) = n^{-1}P_{nx}(\tilde{\boldsymbol{\omega}}). \tag{8.6.9}$$

The examples shown in Figure 8.4 justify this expression.

To evaluate $P_{nx}(\tilde{\boldsymbol{\omega}})$ in (8.6.9), observe that

$$\mathbf{K}_n(\boldsymbol{\omega}, \boldsymbol{\omega}) = \begin{bmatrix} 1 & D_n(\Delta) \\ D_n^*(\Delta) & 1 \end{bmatrix}.$$

Therefore, it follows from (8.2.28) that

$$n^{-1}P_{nx}(\boldsymbol{\omega}) = |\beta_1|^2 + |\beta_2|^2 + 2\Re\{\beta_1^*\beta_2 D_n(\Delta)\}.$$

Because $n\Delta \to a > 0$ as $n \to \infty$, Lemma 12.1.3(f) guarantees that

$$D_n(\Delta) \to d_0(a) := \operatorname{sinc}(a/2)\exp(ia/2).$$

Observe that $|d_0(a)| < 1$. Therefore,

$$\lim_{n\to\infty} n^{-1}P_{nx}(\boldsymbol{\omega}) = |\beta_1|^2 + |\beta_2|^2 + 2\Re\{\beta_1^*\beta_2 d_0(a)\} > 0. \tag{8.6.10}$$

Moreover, under the condition in (8.2.27), there exists a constant $v_\delta > 0$, which may depend on $\boldsymbol{\beta}$, such that

$$\max_{\tilde{\boldsymbol{\omega}} \in \Omega_0 \setminus \Omega_\delta} n^{-1}P_{nx}(\tilde{\boldsymbol{\omega}}) \le n^{-1}P_{nx}(\boldsymbol{\omega}) - v_\delta. \tag{8.6.11}$$

For illustration, consider the example shown in Figure 8.4. With $c_n = 2\pi \times 0.01$, $\kappa_n = 2\pi \times 0.25/n$, and $\delta = 2\pi \times 0.125$, the maximum of $P_{nx}(\tilde{\boldsymbol{\omega}})/P_{nx}(\boldsymbol{\omega})$ over $\Omega_0 \setminus \Omega_\delta$

equals 0.9975, 0.9970, 0.9969, 0.9968, 0.9963, and 0.9963, respectively, for $n = 5$, 10, 15, 20, 100, and 200, suggesting that the ratio $\{n^{-1}P_{nx}(\tilde{\boldsymbol{\omega}})\}/\{n^{-1}P_{nx}(\boldsymbol{\omega})\}$ is strictly less than unity in $\Omega_0\backslash\Omega_\delta$. Because the denominator converges to a strictly positive constant given by (8.6.10), the existence of a constant $v_\delta > 0$ that satisfies (8.6.11) is confirmed. Finally, because $\tilde{\boldsymbol{\omega}} \in \Omega_0$ implies $\lambda \geq \kappa_n$, we have

$$\max_{\tilde{\boldsymbol{\omega}}\in\Omega_0\backslash\Omega_\delta} a_n(\lambda) = \mathcal{O}(1).$$

Combining this result with (8.6.8)–(8.6.11) proves $n(\hat{\boldsymbol{\omega}} - \boldsymbol{\omega}) \xrightarrow{P} 0$.

Now let us prove the asymptotic normality. As in the proof of Theorem 8.1, it suffices to establish (8.6.5) and (8.6.6) with some $\mathbf{U}$ and $\boldsymbol{\Sigma}$. Let ∇_k denote the gradient operator with respect to $\boldsymbol{\theta}_k$. Then, it is easy to verify that

$$\nabla_1\nabla_2^T S_n(\tilde{\boldsymbol{\theta}}) = 2\sum_{t=1}^{n}[\nabla_1\Re\{x_t(\tilde{\boldsymbol{\theta}})\}\nabla_2^T\Re\{x_t(\tilde{\boldsymbol{\theta}})\} + \nabla_1\Im\{x_t(\tilde{\boldsymbol{\theta}})\}\nabla_2^T\Im\{x_t(\tilde{\boldsymbol{\theta}})\}],$$

where $\tilde{\boldsymbol{\theta}} := \boldsymbol{\theta} + s(\hat{\boldsymbol{\theta}} - \boldsymbol{\theta})$ for $s \in [0,1]$. Because

$$x_t(\boldsymbol{\theta}) = \sum_{k=1}^{2}(A_k - iB_k)\exp(it\omega_k),$$

it follows that

$$\frac{\partial}{\partial A_k}\Re\{x_t(\tilde{\boldsymbol{\theta}})\} = \cos(\tilde{\omega}_k t), \quad \frac{\partial}{\partial A_k}\Im\{x_t(\tilde{\boldsymbol{\theta}})\} = \sin(\tilde{\omega}_k t).$$

Therefore,

$$\frac{\partial^2 S_n(\tilde{\boldsymbol{\theta}})}{\partial A_1 \partial A_2} = 2\sum_{t=1}^{n}\{\cos(\tilde{\omega}_1 t)\cos(\tilde{\omega}_2 t) + \sin(\tilde{\omega}_1 t)\sin(\tilde{\omega}_2 t)\}.$$

Moreover, the Taylor expansion gives

$$\begin{aligned}\cos(\tilde{\omega}_k t) &= \cos(\omega_k t) + \mathcal{O}(\tilde{\omega}_k - \omega_k)t,\\ \sin(\tilde{\omega}_k t) &= \sin(\omega_k t) + \mathcal{O}(\tilde{\omega}_k - \omega_k)t.\end{aligned}$$

Combining these expressions with $\tilde{\omega}_k - \omega_k = s(\hat{\omega}_k - \omega_k) = o_P(n^{-1})$ leads to

$$\begin{aligned}\frac{\partial^2 S_n(\tilde{\boldsymbol{\theta}})}{\partial A_1 \partial A_2} &= 2\sum_{t=1}^{n}\{\cos(\omega_1 t)\cos(\omega_2 t) + \sin(\omega_1 t)\sin(\omega_2 t)\} + o_P(n)\\ &= 2n\Re\{D_n(\lambda)\} + o_P(n)\end{aligned}$$

uniformly in s, where the second equality is due to the trigonometric identity

$$\cos(x)\cos(y) + \sin(x)\sin(y) = \cos(x-y). \tag{8.6.12}$$

Similarly, using (8.6.12) and

$$\sin(x)\cos(y)-\cos(x)\sin(y)=\sin(x-y), \tag{8.6.13}$$

we obtain

$$\begin{aligned}
\frac{\partial^2 S_n(\tilde{\boldsymbol\theta})}{\partial A_1 \partial B_2} &= 2n\Im\{D_n(\lambda)\}+\mathcal{O}_P(n),\\
\frac{\partial^2 S_n(\tilde{\boldsymbol\theta})}{\partial A_1 \partial \omega_2} &= 2n\Re\{\beta_2 \dot{D}_n(\lambda)\}+\mathcal{O}_P(n^2),\\
\frac{\partial^2 S_n(\tilde{\boldsymbol\theta})}{\partial B_1 \partial A_2} &= -2n\Im\{D_n(\lambda)\}+\mathcal{O}_P(n),\\
\frac{\partial^2 S_n(\tilde{\boldsymbol\theta})}{\partial B_1 \partial B_2} &= 2n\Re\{D_n(\lambda)\}+\mathcal{O}_P(n),\\
\frac{\partial^2 S_n(\tilde{\boldsymbol\theta})}{\partial B_1 \partial \omega_2} &= -2n\Im\{\beta_2 \dot{D}_n(\lambda)\}+\mathcal{O}_P(n^2),\\
\frac{\partial^2 S_n(\tilde{\boldsymbol\theta})}{\partial \omega_1 \partial A_2} &= -2n\Re\{\beta_1^* \dot{D}_n(\lambda)\}+\mathcal{O}_P(n^2),\\
\frac{\partial^2 S_n(\tilde{\boldsymbol\theta})}{\partial \omega_1 \partial B_2} &= -2n\Im\{\beta_1^* \dot{D}_n(\lambda)\}+\mathcal{O}_P(n^2),\\
\frac{\partial^2 S_n(\tilde{\boldsymbol\theta})}{\partial \omega_1 \partial \omega_2} &= -2n\Re\{\beta_1^* \beta_2 \ddot{D}_n(\lambda)\}+\mathcal{O}_P(n^3).
\end{aligned}$$

Under the assumption that $n\lambda \to a$, Lemma 12.1.3(f) gives

$$\begin{aligned}
&n^{-1}\frac{\partial^2 S_n(\tilde{\boldsymbol\theta})}{\partial A_1 \partial A_2} \to 2\Re\{d_0(a)\}, \quad n^{-1}\frac{\partial^2 S_n(\tilde{\boldsymbol\theta})}{\partial A_1 \partial B_2} \to 2\Im\{d_0(a)\},\\
&n^{-2}\frac{\partial^2 S_n(\tilde{\boldsymbol\theta})}{\partial A_1 \partial \omega_2} \to 2\Re\{\beta_2 \dot{d}_0(a)\},\\
&n^{-1}\frac{\partial^2 S_n(\tilde{\boldsymbol\theta})}{\partial B_1 \partial A_2} \to -2\Im\{d_0(a)\}, \quad n^{-1}\frac{\partial^2 S_n(\tilde{\boldsymbol\theta})}{\partial B_1 \partial B_2} \to 2\Re\{d_0(a)\},\\
&n^{-2}\frac{\partial^2 S_n(\tilde{\boldsymbol\theta})}{\partial B_1 \partial \omega_2} \to -2\Im\{\beta_2 \dot{d}_0(a)\},\\
&n^{-2}\frac{\partial^2 S_n(\tilde{\boldsymbol\theta})}{\partial \omega_1 \partial A_2} \to -2\Re\{\beta_1^* \dot{d}_0(a)\}, \quad n^{-2}\frac{\partial^2 S_n(\tilde{\boldsymbol\theta})}{\partial \omega_1 \partial B_2} \to -2\Im\{\beta_1^* \dot{d}_0(a)\},\\
&n^{-3}\frac{\partial^2 S_n(\tilde{\boldsymbol\theta})}{\partial \omega_1 \partial \omega_2} \to -2\Re\{\beta_1^* \beta_2 \ddot{d}_0(a)\}.
\end{aligned}$$

Combining this result with the fact that $\mathbf{K}_n^{-1}\nabla_k^2 S_n(\tilde{\boldsymbol\theta})\mathbf{K}_n^{-1} \xrightarrow{P} 2\mathbf{W}_k$, which has been established in the proof of Theorem 8.1, leads to (8.6.5) with

$$\mathbf{U} = 2\mathbf{W}, \tag{8.6.14}$$

where

$$\mathbf{W} := \begin{bmatrix} \mathbf{W}_1 & \mathbf{W}_{12} \\ \mathbf{W}_{12}^T & \mathbf{W}_2 \end{bmatrix}$$

and

$$\mathbf{W}_{12} := [w_{jj'}] := \begin{bmatrix} \Re\{d_0(a)\} & \Im\{d_0(a)\} & \Re\{\beta_2 \dot{d}_0(a)\} \\ -\Im\{d_0(a)\} & \Re\{d_0(a)\} & -\Im\{\beta_2 \dot{d}_0(a)\} \\ -\Re\{\beta_1^* \dot{d}_0(a)\} & -\Im\{\beta_1^* \dot{d}_0(a)\} & -\Re\{\beta_1^* \beta_2 \ddot{d}_0(a)\} \end{bmatrix}.$$

Note that for well-separated frequencies we have $\mathbf{W} = \text{diag}(\mathbf{W}_1, \mathbf{W}_2)$.

Observe that the elements in $\mathbf{K}^{-1}\nabla S_n(\boldsymbol{\theta})$ are linear combinations of $\{\epsilon_1, \ldots, \epsilon_n\}$ with coefficients satisfying the conditions in the proof of Theorem 8.1. Therefore, the normality assertion in (8.6.6) remains true except that $\boldsymbol{\Sigma}$ takes a different form. The next objective is to show that the covariance matrix of $\mathbf{K}^{-1}\nabla S_n(\boldsymbol{\theta})$ can be expressed as $\boldsymbol{\Sigma} + \mathcal{O}(1)$ with

$$\boldsymbol{\Sigma} = 2 f_\epsilon(\omega_1)\mathbf{W}. \tag{8.6.15}$$

First, the covariance matrix of $\mathbf{K}_n^{-1}\nabla_k S_n(\boldsymbol{\theta})$, as shown in the proof of Theorem 8.1, can be expressed as $2f_\epsilon(\omega_k)\mathbf{W}_k + \mathcal{O}(1)$. It can also be expressed as $2f_\epsilon(\omega_1)\mathbf{W}_k + \mathcal{O}(1)$ because $f_\epsilon(\omega_k) = f_\epsilon(\omega_1) + \mathcal{O}(1)$. It remains to show that the covariance between $\mathbf{K}_n^{-1}\nabla_1 S_n(\boldsymbol{\theta})$ and $\mathbf{K}_n^{-1}\nabla_2 S_n(\boldsymbol{\theta})$ takes the form $2f_\epsilon(\omega_1)\mathbf{W}_{12} + \mathcal{O}(1)$. In fact, for any fixed u, combining (8.6.12) with Lemma 12.1.5(f) gives

$$\begin{aligned} D_{n0}(u) &:= n^{-1} \sum_{t \in T_n(u)} \{\cos(\omega_1(t+u))\cos(\omega_2 t) + \sin(\omega_1(t+u))\sin(\omega_2 t)\} \\ &= \Re\{D_n(\lambda)\exp(-iu\omega_1)\} + \mathcal{O}(1), \end{aligned}$$

where $T_n(u) := \{t : \max(1, 1-u) \le t \le \min(n, n-u)\}$. This result, together with the dominant convergence theorem and the fact that $E\{\Re(\epsilon_t)\Re(\epsilon_s)\} = E\{\Im(\epsilon_t)\Im(\epsilon_s)\} = \frac{1}{2}r_\epsilon(t-s)$ and $E\{\Re(\epsilon_t)\Im(\epsilon_s)\} = 0$, leads to

$$\begin{aligned} nE\{\Re(X_{n1})\Re(X_{n2})\} &= \sum_{|u|<n} \tfrac{1}{2} r_\epsilon(u) D_{n0}(u) \\ &= \tfrac{1}{2}\Re\{d_0(a) f_\epsilon(\omega_1)\} + \mathcal{O}(1) \\ &= \tfrac{1}{2} f_\epsilon(\omega_1) w_{11} + \mathcal{O}(1). \end{aligned}$$

Similarly, we obtain

$$\begin{aligned} nE\{\Im(X_{n1})\Im(X_{n2})\} &= \tfrac{1}{2} f_\epsilon(\omega_1) w_{22} + \mathcal{O}(1), \\ nE\{\Re(X_{n1})\Im(X_{n2})\} &= -\tfrac{1}{2} f_\epsilon(\omega_1) w_{12} + \mathcal{O}(1), \\ nE\{\Im(X_{n1})\Re(X_{n2})\} &= -\tfrac{1}{2} f_\epsilon(\omega_1) w_{21} + \mathcal{O}(1). \end{aligned}$$

Moreover, owing to (8.6.12) and (8.6.13), we can write

$$
\begin{aligned}
D_{n1}^{(1)}(u) &:= n^{-2} \sum_{t \in T_n(u)} t\{\sin(\omega_1(t+u))\cos(\omega_2 t) - \cos(\omega_1(t+u))\sin(\omega_2 t)\} \\
&= \Re\{n^{-1}\dot{D}_n(\lambda)\exp(-iu\omega_1)\} + o(1),
\end{aligned}
$$

$$
\begin{aligned}
D_{n1}^{(2)}(u) &:= n^{-2} \sum_{t \in T_n(u)} t\{\cos(\omega_1(t+u))\cos(\omega_2 t) + \sin(\omega_1(t+u))\sin(\omega_2 t)\} \\
&= \Im\{n^{-1}\dot{D}_n(\lambda)\exp(-iu\omega_1)\} + o(1).
\end{aligned}
$$

Combining this result with Lemma 12.1.5(f) gives

$$
\begin{aligned}
E\{\Re(X_{n1})\Re(\beta_2^* Y_{n2})\} &= \sum_{|u|<n} \tfrac{1}{2} r_\epsilon(u)\{A_2 D_{n1}^{(1)}(u) + B_2 D_{n1}^{(2)}(u)\} \\
&= \tfrac{1}{2}\Re\{\beta_2 \dot{d}_0(a) f_\epsilon(\omega_1)\} + o(1) \\
&= \tfrac{1}{2} f_\epsilon(\omega_1) w_{13} + o(1).
\end{aligned}
$$

Similarly, we have

$$
\begin{aligned}
E\{\Re(\beta_1^* Y_{n1})\Re(X_{n2})\} &= \tfrac{1}{2} f_\epsilon(\omega_1) w_{31} + o(1), \\
E\{\Im(X_{n1})\Re(\beta_2^* Y_{n2})\} &= -\tfrac{1}{2} f_\epsilon(\omega_1) w_{23} + o(1), \\
E\{\Re(\beta_1^* Y_{n1})\Im(X_{n2})\} &= -\tfrac{1}{2} f_\epsilon(\omega_1) w_{32} + o(1).
\end{aligned}
$$

Finally, because

$$
\begin{aligned}
D_{n2}^{(1)}(u) &:= n^{-3} \sum_{t \in T_n(u)} (t+u) t\{\cos(\omega_1(t+u))\cos(\omega_2 t) + \sin(\omega_1(t+u))\sin(\omega_2 t)\} \\
&= -\Re\{n^{-2}\ddot{D}_n(\lambda)\exp(-iu\omega_1)\} + o(1),
\end{aligned}
$$

$$
\begin{aligned}
D_{n2}^{(2)}(u) &:= n^{-3} \sum_{t \in T_n(u)} (t+u) t\{\sin(\omega_1(t+u))\cos(\omega_2 t) - \cos(\omega_1(t+u))\sin(\omega_2 t)\} \\
&= \Im\{n^{-2}\ddot{D}_n(\lambda)\exp(-iu\omega_1)\} + o(1),
\end{aligned}
$$

it follows that

$$
\begin{aligned}
n^{-1} E\{\Re(\beta_1^* Y_{n1})\Re(\beta_2^* Y_{n2})\} &= \sum_{|u|<n} \tfrac{1}{2} r_\epsilon(u)\{\Re(\beta_1^* \beta_2)D_{n2}^{(1)}(u) + \Im(\beta_1^* \beta_2)D_{n2}^{(2)}(u)\} \\
&= -\tfrac{1}{2}\Re\{\beta_1^* \beta_2 \ddot{d}_0(a) f_\epsilon(\omega_1)\} + o(1) \\
&= \tfrac{1}{2} f_\epsilon(\omega_1) w_{33} + o(1).
\end{aligned}
$$

Combining these results proves (8.6.15). Combining (8.6.14) with (8.6.15) leads to $\mathbf{U}^{-1}\boldsymbol{\Sigma}\mathbf{U}^{-1} = \tfrac{1}{2} f_\epsilon(\omega_1)\mathbf{W}^{-1}$. The proof is complete. $\square$

Proof of Theorem 8.3. The proof has two parts. In the first part, we assume

$$
\mathbf{K}(\hat{\boldsymbol{\theta}} - \boldsymbol{\theta}) = \mathcal{O}_P(1) \tag{8.6.16}
$$

and derive the asymptotic distribution of $\mathbf{K}(\hat{\boldsymbol{\theta}}-\boldsymbol{\theta})$. In the second part, we prove (8.6.16) under the consistency assumption that can be expressed as $\tilde{\mathbf{K}}(\hat{\boldsymbol{\theta}}-\boldsymbol{\theta}) \xrightarrow{P} 0$, where $\tilde{\mathbf{K}} := n^{-1/2}\mathbf{K} = \mathrm{diag}(1,1,n,\ldots,1,1,n)$.

First, assume that (8.6.16) is satisfied. To derive the asymptotic distribution, we take the same approach in the proof of Theorem 5.9 by considering

$$Z_n(\boldsymbol{\delta}) := \|\mathbf{y}-\mathbf{x}(\boldsymbol{\theta}+\mathbf{K}^{-1}\boldsymbol{\delta})\|_1 - \|\boldsymbol{\epsilon}\|_1$$

for any $\boldsymbol{\delta} \in \mathbb{R}^{3p}$. The objective is to show that

$$Z_n(\boldsymbol{\delta}) = -\boldsymbol{\delta}^T\boldsymbol{\zeta}_n + \kappa^{-1}\boldsymbol{\delta}^T\mathbf{W}\boldsymbol{\delta} + R_n(\boldsymbol{\delta}), \tag{8.6.17}$$

where $\boldsymbol{\zeta}_n \xrightarrow{D} \mathrm{N}(\mathbf{0},\mathbf{W})$ as $n \to \infty$, and $R_n(\boldsymbol{\delta}) = \mathcal{O}_P(1)$ uniformly over $\boldsymbol{\delta}$ in any compact subset of $\mathbb{R}^{3p}$. Because the quadratic function in (8.6.17) has a unique minimizer $\tilde{\boldsymbol{\delta}}_n := \frac{1}{2}\kappa\mathbf{W}^{-1}\boldsymbol{\zeta}_n$, if $\boldsymbol{\zeta}_n \xrightarrow{D} \mathrm{N}(\mathbf{0},\mathbf{W})$, then $\tilde{\boldsymbol{\delta}}_n \xrightarrow{D} \mathrm{N}(\mathbf{0},\frac{1}{2}\eta^2\mathbf{W}^{-1})$, where $\eta^2 := \kappa^2/2$. On the other hand, it is easy to see that $\hat{\boldsymbol{\delta}}_n := \mathrm{arg\,min}\{Z_n(\boldsymbol{\delta}) : \boldsymbol{\delta} \in \Delta_n\} = \mathbf{K}(\hat{\boldsymbol{\theta}}-\boldsymbol{\theta})$, where $\Delta_n := \{\boldsymbol{\delta} : \boldsymbol{\theta}+\mathbf{K}^{-1}\boldsymbol{\delta} \in \Theta_n\}$. It suffices to show that $\hat{\boldsymbol{\delta}}_n - \tilde{\boldsymbol{\delta}}_n \xrightarrow{P} 0$, or $P(\|\hat{\boldsymbol{\delta}}_n - \tilde{\boldsymbol{\delta}}_n\| > \mu) \to 0$ for any constant $\mu > 0$.

Toward that end, observe that substituting $\tilde{\boldsymbol{\delta}}_n = \frac{1}{2}\kappa\mathbf{W}^{-1}\boldsymbol{\zeta}_n$ in (8.6.17) yields

$$Z_n(\boldsymbol{\delta}) = Z_n(\tilde{\boldsymbol{\delta}}_n) + \kappa^{-1}(\boldsymbol{\delta}-\tilde{\boldsymbol{\delta}}_n)^T\mathbf{W}(\boldsymbol{\delta}-\tilde{\boldsymbol{\delta}}_n) + R_n(\boldsymbol{\delta}) - R_n(\tilde{\boldsymbol{\delta}}_n).$$

Because $\tilde{\boldsymbol{\delta}}_n = \mathcal{O}_P(1)$ due to (8.6.17) and $\hat{\boldsymbol{\delta}}_n = \mathcal{O}_P(1)$ due to (8.6.16), it follows that for any $\varepsilon > 0$, there exists a constant $c > 0$ such that

$$P(\|\tilde{\boldsymbol{\delta}}_n\| > c \text{ or } \|\hat{\boldsymbol{\delta}}_n\| > c) < \varepsilon.$$

Define $R_n := \max\{|R_n(\boldsymbol{\delta})| : \|\boldsymbol{\delta}\| \le c\}$. It is easy to see that for any $\boldsymbol{\delta}$ such that $\|\boldsymbol{\delta}-\tilde{\boldsymbol{\delta}}_n\| > \mu$ and $\|\boldsymbol{\delta}\| \le c$, if $\|\tilde{\boldsymbol{\delta}}_n\| \le c$, then

$$Z_n(\boldsymbol{\delta}) \ge Z_n(\tilde{\boldsymbol{\delta}}_n) + \kappa^{-1}a\mu^2 - 2R_n,$$

where $a > 0$ denotes the smallest eigenvalue of $\mathbf{W}$. This result, combined with the assumption $R_n \xrightarrow{P} 0$, implies that

$$P(\|\tilde{\boldsymbol{\delta}}_n\| \le c, \|\hat{\boldsymbol{\delta}}_n\| \le c, \|\hat{\boldsymbol{\delta}}_n - \tilde{\boldsymbol{\delta}}_n\| > \mu) < \varepsilon$$

for sufficiently large n. Therefore, for large n,

$$\begin{aligned} P(\|\hat{\boldsymbol{\delta}}_n - \tilde{\boldsymbol{\delta}}_n\| > \mu) &\le P(\|\tilde{\boldsymbol{\delta}}_n\| > c \text{ or } \|\hat{\boldsymbol{\delta}}_n\| > c) \\ &\quad + P(\|\tilde{\boldsymbol{\delta}}_n\| \le c, \|\hat{\boldsymbol{\delta}}_n\| \le c, \|\hat{\boldsymbol{\delta}}_n - \tilde{\boldsymbol{\delta}}_n\| > \mu) \\ &< 2\varepsilon. \end{aligned}$$

This proves $P(\|\hat{\boldsymbol{\delta}}_n - \tilde{\boldsymbol{\delta}}_n\| > \mu) \to 0$ as $n \to \infty$ for any $\mu > 0$.

To prove (8.6.17), we write $x_t = x_t(\boldsymbol{\theta})$ explicitly as a function of $\boldsymbol{\theta}$ and consider the Taylor expansion $x_t(\boldsymbol{\theta}+\mathbf{K}^{-1}\boldsymbol{\delta}) = x_t + v_t + r_t$, where

$$v_t := (\mathbf{K}^{-1}\boldsymbol{\delta})^T \nabla x_t, \quad r_t := \int_0^1 \tfrac{1}{2}(\mathbf{K}^{-1}\boldsymbol{\delta})^T \{\nabla^2 x_t(\boldsymbol{\theta}+s\mathbf{K}^{-1}\boldsymbol{\delta})\}(\mathbf{K}^{-1}\boldsymbol{\delta})\, ds.$$

We want to show that

$$w_t := v_t + r_t = \mathcal{O}(n^{-1/2}) \tag{8.6.18}$$

uniformly in $\boldsymbol{\delta} \in \mathcal{D}$ and in t, where $\mathcal{D}$ denotes any compact subset of $\mathbb{R}^{3p}$. Toward that end, observe that $\nabla^2 x_t(\tilde{\boldsymbol{\theta}})$ is a block diagonal matrix in which the kth diagonal block is equal to the 3-by-3 Hessian matrix of $x_t(\tilde{\boldsymbol{\theta}})$ with respect to $\boldsymbol{\theta}_k := [C_k, \phi_k, \omega_k]^T$ or $\boldsymbol{\theta}_k := [A_k, B_k, \omega_k]^T$ $(k = 1, \dots, p)$. Let $u_t(\tilde{\boldsymbol{\theta}})$ denote any nonzero element in the kth diagonal block of the matrix $\mathbf{K}^{-1}\{\nabla^2 x_t(\tilde{\boldsymbol{\theta}})\}\mathbf{K}^{-1}$ and consider the Taylor expansion

$$u_t(\boldsymbol{\theta}+s\mathbf{K}^{-1}\boldsymbol{\delta}) = u_t(\boldsymbol{\theta}) + s(\nabla u_t(\bar{\boldsymbol{\theta}}))^T \mathbf{K}^{-1}\boldsymbol{\delta}, \tag{8.6.19}$$

where $\bar{\boldsymbol{\theta}}$ is an intermediate point between $\boldsymbol{\theta}+s\mathbf{K}^{-1}\boldsymbol{\delta}$ and $\boldsymbol{\theta}$. It is not difficult to verify that all elements in $(\nabla u_t(\bar{\boldsymbol{\theta}}))^T\mathbf{K}^{-1}$ can be expressed as $\mathcal{O}(n^{-3/2})$ so that

$$u_t(\boldsymbol{\theta}+s\mathbf{K}^{-1}\boldsymbol{\delta}) = u_t(\boldsymbol{\theta}) + \mathcal{O}(n^{-3/2}) \tag{8.6.20}$$

uniformly in $s \in [0,1]$, $\boldsymbol{\delta} \in \mathcal{D}$, and t. As an example, consider the (3,3) entry in the kth diagonal block of $\mathbf{K}^{-1}\{\nabla^2 x_t(\tilde{\boldsymbol{\theta}})\}\mathbf{K}^{-1}$, which is given by

$$u_t(\tilde{\boldsymbol{\theta}}) = -n^{-3}\tilde{C}_k t^2 \exp\{i(\tilde{\omega}_k t + \tilde{\phi}_k)\}.$$

In this case, the third element in $(\nabla u_t(\bar{\boldsymbol{\theta}}))^T\mathbf{K}^{-1}$ can be expressed as

$$-n^{-3/2} i\bar{C}_k (t/n)^3 \exp\{i(\bar{\omega}_k t + \bar{\phi}_k)\}.$$

It takes the form $\mathcal{O}(n^{-3/2})$ because $\bar{C}_k = C_k + \mathcal{O}(n^{-1/2}) = \mathcal{O}(1)$. Similarly, the first and second elements also take the form $\mathcal{O}(n^{-3/2})$. Given (8.6.20), we can write

$$r_t = \tfrac{1}{2}\boldsymbol{\delta}^T\mathbf{K}^{-1}(\nabla^2 x_t)\mathbf{K}^{-1}\boldsymbol{\delta} + \mathcal{O}(n^{-3/2}) \tag{8.6.21}$$

uniformly in $\boldsymbol{\delta} \in \mathcal{D}$ and t. Moreover, because $\mathbf{K}^{-1}(\nabla^2 x_t)\mathbf{K}^{-1} = \mathcal{O}(n^{-1})$ uniformly in t, it follows that

$$a_t := \tfrac{1}{2}\boldsymbol{\delta}^T\mathbf{K}^{-1}(\nabla^2 x_t)\mathbf{K}^{-1}\boldsymbol{\delta} = \mathcal{O}(n^{-1})$$

and hence $r_t = a_t + \mathcal{O}(n^{-3/2}) = \mathcal{O}(n^{-1})$ uniformly in $\boldsymbol{\delta} \in \mathcal{D}$ and t. Similarly, we have $\mathbf{K}^{-1}\nabla x_t = \mathcal{O}(n^{-1/2})$ and hence $v_t = \mathcal{O}(n^{-1/2})$ uniformly in $\boldsymbol{\delta} \in \mathcal{D}$ and t. Combining these results yields (8.6.18).

Next, define

$$U_{t1} := \Re(\epsilon_t), \quad w_{t1} := \Re(w_t), \quad v_{t1} := \Re(v_t), \quad a_{t1} := \Re(a_t),$$
$$U_{t2} := \Im(\epsilon_t), \quad w_{t2} := \Im(w_t), \quad v_{t2} := \Im(v_t), \quad a_{t2} := \Im(a_t).$$

It is easy to see that

$$Z_n(\boldsymbol{\delta}) = \sum_{t=1}^{n} \sum_{j=1}^{2} \{|U_{tj} - w_{tj}| - |U_{tj}|\}.$$

Define $Y_t := \operatorname{sgn}(U_{t1}) + i \operatorname{sgn}(U_{t2})$. Then, based on Knight's identity [193]

$$|u - v| - |u| = -v \operatorname{sgn}(u) + \int_0^v \phi(u, s)\, ds, \tag{8.6.22}$$

where $\phi(u,s) := 2\{\mathscr{I}(|u| \le s) - \mathscr{I}(|u| \le 0)\}$, we can write

$$Z_n(\boldsymbol{\delta}) = -\boldsymbol{\delta}^T \boldsymbol{\zeta}_n + Z_{n2} + Z_{n3} + Z_{n4}, \tag{8.6.23}$$

where

$$\boldsymbol{\zeta}_n := \Re\Big\{\sum_{t=1}^{n} \mathbf{K}^{-1}(\nabla x_t) Y_t^*\Big\}, \quad Z_{n2} := -\Re\Big\{\sum_{t=1}^{n} r_t Y_t^*\Big\},$$
$$Z_{n3} := \sum_{t=1}^{n} (R_{t1} + R_{t2}), \quad Z_{n4} := \sum_{t=1}^{n} (X_{t1} + X_{t2}),$$
$$R_{tj} := \int_{v_{tj}+a_{tj}}^{w_{tj}} \phi(U_{tj}, s)\, ds, \quad X_{tj} := \int_0^{v_{tj}+a_{tj}} \phi(U_{tj}, s)\, ds.$$

The task is to evaluate the terms in (8.6.23).

In the proof of Theorem 5.9, we have shown that $\{Y_t\}$ is an i.i.d. sequence with mean zero and variance 2 and with independent real and imaginary parts. Moreover, it is easy to verify that $\mathbf{K}^{-1}\nabla x_t = \mathcal{O}(n^{-1/2})$ uniformly in t and

$$\mathbf{C}_n := \sum_{t=1}^{n} \mathbf{K}^{-1}(\nabla x_t)(\nabla x_t)^H \mathbf{K}^{-1} = \mathbf{K}^{-1}\mathbf{X}^H\mathbf{X}\mathbf{K}^{-1}.$$

An application of Lemma 12.5.7, together with the assumption $\mathbf{C}_n \to \mathbf{C}$, gives $\sum \mathbf{K}^{-1}(\nabla x_t) Y_t^* \xrightarrow{D} \mathrm{N}_c(\mathbf{0}, 2\mathbf{C})$ and hence $\boldsymbol{\zeta}_n \xrightarrow{D} \mathrm{N}(\mathbf{0}, \mathbf{W})$, where $\mathbf{W} := \Re(\mathbf{C})$.

Let $\mathbf{U}_n := \mathbf{K}^{-1} \sum (\nabla^2 x_t) Y_t^* \mathbf{K}^{-1}$. Then, it follows from (8.6.21) and the boundedness of $\{Y_t\}$ that

$$Z_{n2} = -\tfrac{1}{2}\boldsymbol{\delta}^T \Re(\mathbf{U}_n)\boldsymbol{\delta} + \mathcal{O}(n^{-1/2}) \tag{8.6.24}$$

uniformly in $\boldsymbol{\delta} \in \mathcal{D}$. Observe that $\mathbf{U}_n$ has the same block-diagonal structure as $\nabla^2 x_t$. It is easy to verify that all nonzero elements in $\mathbf{U}_n$ have mean zero and variance $\mathcal{O}(n^{-1})$. For example, the (3,3) entry in the kth diagonal block of $\mathbf{U}_n$ is equal to $-n^{-3}\sum C_k t^2 \exp\{i(\omega_k t+\phi_k)\} Y_t^*$. This random variable has mean zero because $E(Y_t)=0$; it has variance $2n^{-6}\sum C_k^2 t^4 = \mathcal{O}(n^{-1})$ because the Y_t are independent with $\mathrm{Var}(Y_t)=2$. By Proposition 2.16(a), $\mathbf{U}_n = \mathcal{O}_P(n^{-1/2})$. Combining this expression with (8.6.24) yields $Z_{n2} = \mathcal{O}_P(n^{-1/2})$ uniformly in $\boldsymbol{\delta} \in \mathcal{D}$.

Because $|\phi(u,s)| \le 2$, we have $|R_{tj}| \le 2|r_{tj} - a_{tj}|$ $(j=1,2)$, where $r_{t1} := \Re(r_t)$ and $r_{t2} := \Im(r_t)$. Combining this result with (8.6.21) leads to $|R_{tj}| = \mathcal{O}(n^{-3/2})$ uniformly in $\boldsymbol{\delta} \in \mathcal{D}$ and t. Therefore, $Z_{n3} = \mathcal{O}(n^{-1/2})$ uniformly in $\boldsymbol{\delta} \in \mathcal{D}$.

Finally, because $v_t = \mathcal{O}(n^{-1/2})$, $a_t = \mathcal{O}(n^{-1})$, and $v_t + a_t = \mathcal{O}(n^{-1/2})$, the same argument as in the proof of Theorem 5.9 gives

$$
\begin{aligned}
E(X_{tj}) &= \kappa^{-1}(v_{tj}+a_{tj})^2 + \mathcal{O}((v_{tj}+a_{tj})^{d+2}) \\
&= \kappa^{-1} v_{tj}^2 + \mathcal{O}(n^{-3/2}) + \mathcal{O}(n^{-(d+2)/2}) \\
&= \kappa^{-1} v_{tj}^2 + \mathcal{O}(n^{-\beta-1}),
\end{aligned}
$$

where $\beta := \min(1/2, d/2)$. This, combined with $\sum(v_{t1}^2 + v_{t2}^2) = \boldsymbol{\delta}^T \Re(\mathbf{C}_n)\boldsymbol{\delta}$, gives

$$
E(Z_{n4}) = \kappa^{-1}\boldsymbol{\delta}^T \Re(\mathbf{C}_n)\boldsymbol{\delta} + \mathcal{O}(n^{-\beta}) \tag{8.6.25}
$$

uniformly in $\boldsymbol{\delta} \in \mathcal{D}$. Moreover, for any fixed $\boldsymbol{\delta}$, it follows from the proof of Theorem 5.9, together with $E(Z_{n4}) = \mathcal{O}(1)$, that

$$
\mathrm{Var}(Z_{n4}) \le 2E(Z_{n4}) \max_{1\le t\le n}\{|v_t + a_t|\} = \mathcal{O}(n^{-1/2}). \tag{8.6.26}
$$

By Proposition 2.16(a), we have

$$
Z_{n4} - E(Z_{n4}) = \mathcal{O}_P(n^{-1/4})
$$

for any fixed $\boldsymbol{\delta}$. This, combined with (8.6.25) and $\Re(\mathbf{C}_n) \to \mathbf{W}$, leads to

$$
Z_{n4}(\boldsymbol{\delta}) - \kappa^{-1}\boldsymbol{\delta}^T \mathbf{W}\boldsymbol{\delta} \xrightarrow{P} 0 \tag{8.6.27}
$$

for any fixed $\boldsymbol{\delta}$. In the last expression, we have substituted Z_{n4} with $Z_{n4}(\boldsymbol{\delta})$ to emphasize its dependence on $\boldsymbol{\delta}$. The next step is to strengthen (8.6.27) to uniform convergence in $\boldsymbol{\delta} \in \mathcal{D}$. According to Lemma 12.5.11, it suffices to show that the random functions $Z_{n4}(\boldsymbol{\delta})$ $(n=1,2,\ldots)$ are stochastically equicontinuous in $\mathcal{D}$, i.e., for any constant $\tau > 0$,

$$
\lim_{h\to 0}\limsup_{n\to\infty} P\left\{ \max_{(\boldsymbol{\delta},\boldsymbol{\delta}')\in\mathcal{B}(h)} |Z_{n4}(\boldsymbol{\delta}) - Z_{n4}(\boldsymbol{\delta}')| \ge \tau \right\} = 0, \tag{8.6.28}
$$

where $\mathcal{B}(h) := \{(\boldsymbol{\delta},\boldsymbol{\delta}') \in \mathcal{D}\times\mathcal{D} : \|\boldsymbol{\delta}-\boldsymbol{\delta}'\| \le h\}$. Obviously, a sufficient condition for (8.6.28) to hold is the Lipschitz condition:

$$
|Z_{n4}(\boldsymbol{\delta}) - Z_{n4}(\boldsymbol{\delta}')| \le h\xi_n \tag{8.6.29}
$$

for all $(\boldsymbol{\delta}, \boldsymbol{\delta}') \in \mathcal{B}(h)$, where $\{\xi_n\}$ is a random sequence, independent of $\boldsymbol{\delta}$, $\boldsymbol{\delta}'$, and h, such that $\xi_n = \mathcal{O}_P(1)$.

To verify (8.6.29), observe that

$$\left| \int_a^b \phi(u,s)\, ds \right| \le 2|b-a| \mathcal{I}\{|u| \le \max(|a|,|b|)\}. \tag{8.6.30}$$

This inequality can be obtained by calculating the integral for $u > 0$ and $u \le 0$, respectively, and then assembling the results for $b > a \ge 0$, $b > 0 > a$, and $0 \ge b > a$ (see Lemma 1 in [238]). Now let X'_{tj}, v'_{tj}, and a'_{tj} denote the counterparts of X_{tj}, v_{tj}, and a_{tj} corresponding to $\boldsymbol{\delta}'$ instead of $\boldsymbol{\delta}$. Define $b_{tj} := v_{tj} + a_{tj}$ and $b'_{tj} := v'_{tj} + a'_{tj}$. Then, it follows from (8.6.30) that

$$|X_{tj} - X'_{tj}| = \left| \int_{b'_{tj}}^{b_{tj}} \phi(U_{tj}, s)\, ds \right| \le 2|b_{tj} - b'_{tj}| \mathcal{I}\{|U_{tj}| \le \max(|b_{tj}|, |b'_{tj}|)\}.$$

Because $v_t = \mathcal{O}(n^{-1/2})$ and $a_t = \mathcal{O}(n^{-1})$ uniformly in $\boldsymbol{\delta} \in \mathcal{D}$ and t, we can write $b_{tj} = \mathcal{O}(n^{-1/2})$ and $b'_{tj} = \mathcal{O}(n^{-1/2})$, so there exists a constant $c_1 > 0$ such that

$$\mathcal{I}\{|U_{tj}| \le \max(|b_{tj}|, |b'_{tj}|)\} \le \mathcal{I}(|U_{tj}| < c_n)$$

where $c_n := c_1 n^{-1/2}$. Because $\mathbf{K}^{-1} \nabla x_t = \mathcal{O}(n^{-1/2})$ and $\mathbf{K}^{-1}(\nabla^2 x_t)\mathbf{K}^{-1} = \mathcal{O}(n^{-1})$ uniformly in t, it follows that

$$|v_{tj} - v'_{tj}| \le |(\boldsymbol{\delta} - \boldsymbol{\delta}')^T \mathbf{K}^{-1} \nabla x_t| = \mathcal{O}(hn^{-1/2}),$$
$$|a_{tj} - a'_{tj}| \le |\boldsymbol{\delta}^T \mathbf{K}^{-1} (\nabla^2 x_t) \mathbf{K}^{-1} \boldsymbol{\delta} - \boldsymbol{\delta}'^T \mathbf{K}^{-1} (\nabla^2 x_t) \mathbf{K}^{-1} \boldsymbol{\delta}'| = \mathcal{O}(hn^{-1}),$$

and hence

$$|b_{tj} - b'_{tj}| \le |v_{tj} - v'_{tj}| + |a_{tj} - a'_{tj}| = \mathcal{O}(hn^{-1/2})$$

uniformly in $(\boldsymbol{\delta}, \boldsymbol{\delta}') \in \mathcal{B}(h)$ and t. Combining these results leads to

$$|X_{tj} - X'_{tj}| \le \mathcal{O}(hn^{-1/2}) \mathcal{I}(|U_{tj}| < c_n)$$

uniformly $(\boldsymbol{\delta}, \boldsymbol{\delta}') \in \mathcal{B}(h)$ and t. This implies that there is a constant $c_0 > 0$, which does not depend on $\boldsymbol{\delta}$, $\boldsymbol{\delta}'$, and h, such that

$$|Z_{n4}(\boldsymbol{\delta}) - Z_{n4}(\boldsymbol{\delta}')| \le \sum_{t=1}^{n} \sum_{j=1}^{2} |X_{tj} - X'_{tj}| \le c_0 h \sqrt{n}\, S_n,$$

where

$$S_n := n^{-1} \sum_{t=1}^{n} \sum_{j=1}^{2} \mathcal{I}(|U_{tj}| < c_n).$$

Observe that

$$E\{\mathscr{I}(|U_{tj}| < c_n)\} = F_{tj}(c_n) - F_{tj}(-c_n) = 2\kappa^{-1} c_n + \mathcal{O}(c_n^{d+1}) = \mathcal{O}(c_n)$$

and hence $E(S_n) = \mathcal{O}(c_n) = \mathcal{O}(n^{-1/2})$. By Proposition 2.16(a), $S_n = \mathcal{O}_P(n^{-1/2})$. Therefore, the Lipschitz condition (8.6.29) is satisfied with $\xi_n := c_0\sqrt{n}\, S_n = \mathcal{O}_P(1)$. This proves the stochastic equicontinuity assertion.

Finally, combining the uniform convergence (8.6.27) with the proven fact that $Z_{n2} = \mathcal{O}_P(n^{-1/2})$ and $Z_{n3} = \mathcal{O}(n^{-1/2})$ uniformly in $\boldsymbol{\delta} \in \mathcal{D}$ leads to (8.6.17) with $R_n(\boldsymbol{\delta}) = \mathcal{O}_P(1)$ uniformly in $\boldsymbol{\delta} \in \mathcal{D}$ for any compact subset $\mathcal{D} \subset \mathbb{R}^{3p}$. The first part of the proof is complete.

The goal of the second part is to prove (8.6.16) which establishes the rate of convergence. This can be done by following the steps in [239] to check the conditions of a general theorem in [405] (Theorem 3.4.1 or 3.2.5). A major difference between our problem and the problem considered in [239] is that the estimator in our problem has different rates of convergence for its components, whereas the estimator in [239] is assumed to have the same rate of convergence for all components. So, we cannot simply cite the results in [239].

To apply Theorem 3.4.1 in [405] to our problem, let

$$\mathcal{D}_n(\mu_0) := \{\boldsymbol{\delta} \in \mathbb{R}^{3p} : d_n(\boldsymbol{\delta}, \mathbf{0}) \le \mu_0\},$$

where $\mu_0 > 0$ is a constant and

$$d_n(\boldsymbol{\delta}, \boldsymbol{\delta}') := n^{-1/2} \|\boldsymbol{\delta} - \boldsymbol{\delta}'\|$$

is a distance measure that depends on n. Consider the random function $Z_n(\boldsymbol{\delta})$ for $\boldsymbol{\delta} \in \mathcal{D}_n(\mu_0)$. Observe that (8.6.16) is equivalent to $\hat{\boldsymbol{\delta}} = \mathcal{O}_P(1)$ or $\sqrt{n}\, d_n(\hat{\boldsymbol{\delta}}, \mathbf{0}) = \mathcal{O}_P(1)$, and that $\tilde{\mathbf{K}}(\hat{\boldsymbol{\theta}} - \boldsymbol{\theta}) \xrightarrow{P} \mathbf{0}$ is equivalent to $n^{-1/2}\hat{\boldsymbol{\delta}} \xrightarrow{P} \mathbf{0}$ or $d_n(\hat{\boldsymbol{\delta}}, \mathbf{0}) \xrightarrow{P} 0$. Therefore, according to Theorem 3.4.1 in [405], the rate of convergence in (8.6.16) is implied by the assumption that $\tilde{\mathbf{K}}(\hat{\boldsymbol{\theta}} - \boldsymbol{\theta}) \xrightarrow{P} 0$ if there exist constants $0 < \mu_1 < \mu_0$, $0 < \mu_2 < \mu_0$, $c_1 > 0$, and $c_2 > 0$ such that

$$\liminf_{n\to\infty} \min_{\boldsymbol{\delta}\in\mathcal{D}_n(\mu_1)} [n^{-1} E\{Z_n(\boldsymbol{\delta})\} - c_1 d_n^2(\boldsymbol{\delta}, \mathbf{0})] \ge 0 \tag{8.6.31}$$

and

$$E\left[\max_{\boldsymbol{\delta}\in\mathcal{D}_n(\mu)} n^{-1/2} |Z_n(\boldsymbol{\delta}) - E\{Z_n(\boldsymbol{\delta})\}|\right] \le c_2 \mu \tag{8.6.32}$$

for all $0 < \mu \le \mu_2$. These conditions are verified in the following.

Verification of (8.6.31). Recall that $Z_n(\boldsymbol{\delta})$ can be expressed as (8.6.23). Because $E(\boldsymbol{\zeta}_n) = \mathbf{0}$ and $E(Z_{n2}) = 0$, we have $E\{Z_n(\boldsymbol{\delta})\} = E(Z_{n3}) + E(Z_{n4})$. It has been shown that $|R_{tj}| \le 2|r_{tj} - a_{tj}| \le 2|r_t - a_t|$. Moreover, consider (8.6.19), where

$u_t(\tilde{\boldsymbol{\theta}})$ denotes any nonzero element in $\mathbf{K}^{-1}\{\nabla^2 x_t(\tilde{\boldsymbol{\theta}})\}\mathbf{K}^{-1}$. Although $\boldsymbol{\delta} \in \mathcal{D}_n(\mu_0)$ implies $\boldsymbol{\delta} = \mathcal{O}(\sqrt{n})$, the amplitude parameters in any intermediate point $\bar{\boldsymbol{\theta}}$ between $\boldsymbol{\theta} + s\mathbf{K}^{-1}\boldsymbol{\delta}$ and $\boldsymbol{\theta}$ for some $s \in [0,1]$ can still be expressed as $\mathcal{O}(1)$ and hence $(\nabla u_t(\bar{\boldsymbol{\theta}}))^T \mathbf{K}^{-1} = \mathcal{O}(n^{-3/2})$. Substituting this expression in (8.6.19) gives

$$u_t(\boldsymbol{\theta} + s\mathbf{K}^{-1}\boldsymbol{\delta}) = u_t(\boldsymbol{\theta}) + \mathcal{O}(n^{-1} d_n(\boldsymbol{\delta}, \mathbf{0}))$$

uniformly in $s \in [0,1]$, $\boldsymbol{\delta}$, and t. Therefore, we obtain

$$r_t = a_t + \mathcal{O}(d_n^3(\boldsymbol{\delta}, \mathbf{0}))$$

and hence $|R_{tj}| = \mathcal{O}(d_n^3(\boldsymbol{\delta}, \mathbf{0}))$ uniformly in $\boldsymbol{\delta}$ and t. This implies that

$$E(Z_{n3}) = \mathcal{O}(n d_n^3(\boldsymbol{\delta}, \mathbf{0}))$$

uniformly in $\boldsymbol{\delta} \in \mathcal{D}_n(\mu_0)$. Moreover, because $\mathbf{K}^{-1}(\nabla^2 x_t)\mathbf{K}^{-1} = \mathcal{O}(n^{-1})$, we can write $a_t = \frac{1}{2}\boldsymbol{\delta}\mathbf{K}^{-1}(\nabla^2 x_t)\mathbf{K}^{-1}\boldsymbol{\delta} = \mathcal{O}(d_n^2(\boldsymbol{\delta}, \mathbf{0}))$ uniformly in $\boldsymbol{\delta}$. Similarly, because $\mathbf{K}^{-1}\nabla x_t = \mathcal{O}(n^{-1/2})$, we have $v_t = \mathcal{O}(d_n(\boldsymbol{\delta}, \mathbf{0}))$ uniformly in $\boldsymbol{\delta}$. Therefore, for sufficiently small $\mu_1 \in (0, \mu_0)$,

$$\begin{aligned} E(X_{tj}) &= \kappa^{-1}(v_{tj} + a_{tj})^2 + \mathcal{O}((v_{tj} + a_{tj})^{d+2}) \\ &= \kappa^2 v_{tj}^2 + \mathcal{O}(d_n^{r+2}(\boldsymbol{\delta}, 0)) \end{aligned}$$

uniformly in $\boldsymbol{\delta} \in \mathcal{D}_n(\mu_1)$ and t, where $r := \min(1, d) > 0$. This implies that

$$E(Z_{n4}) = \kappa^{-1}\boldsymbol{\delta}^T \Re(\mathbf{C}_n)\boldsymbol{\delta} + \mathcal{O}(n d_n^{r+2}(\boldsymbol{\delta}, 0)).$$

Because $\Re(\mathbf{C}_n) \to \mathbf{W} > 0$, there exists a constant $c_0 > 0$ such that

$$\kappa^{-1}\boldsymbol{\delta}^T \Re(\mathbf{C}_n)\boldsymbol{\delta} > c_0 \|\boldsymbol{\delta}\|^2 = n c_0 d_n^2(\boldsymbol{\delta}, \mathbf{0}).$$

Combining these results leads to

$$E\{Z_n(\boldsymbol{\delta})\} > n\{c_0 + \mathcal{O}(\mu_1^r)\} d_n^2(\boldsymbol{\delta}, \mathbf{0}).$$

Because $c_0 + \mathcal{O}(\mu_1^r) > 0$ for small $\mu_1 \in (0, \mu_0)$, there exists $c_1 > 0$ such that

$$n^{-1}[E\{Z_n(\boldsymbol{\delta}) - c_1 d_n^2(\boldsymbol{\delta}, \mathbf{0})] > \{c_0 + \mathcal{O}(\mu_1^r) - c_1\} d_n^2(\boldsymbol{\delta}, \mathbf{0}) > 0.$$

Therefore, the condition (8.6.31) is satisfied.

Verification of (8.6.32). This requires repeated application of the maximal inequality in Lemma 12.5.12 which is an extension of Lemma 5 in [239] to accommodate an arbitrary distance measure that may also depend on n. First, let $x_{t1}(\tilde{\boldsymbol{\theta}})$ and $x_{t2}(\tilde{\boldsymbol{\theta}})$ denote the real and imaginary parts of $x_t(\tilde{\boldsymbol{\theta}})$, and let ∇x_{t1} and ∇x_{t2} denote the real and imaginary parts of $\nabla x_t = \nabla x_t(\boldsymbol{\theta})$. Moreover, write

$w_t = w_t(\boldsymbol{\delta}) := x_t(\boldsymbol{\theta} + \mathbf{K}^{-1}\boldsymbol{\delta}) - x_t$ and $w_{tj} = w_{tj}(\boldsymbol{\delta}) := x_{tj}(\boldsymbol{\theta} + \mathbf{K}^{-1}\boldsymbol{\delta}) - x_{tj}$ explicitly as functions of $\boldsymbol{\delta}$, and define

$$V_{tj}(\boldsymbol{\delta}) := \int_0^{w_{tj}(\boldsymbol{\delta})} \phi(U_{tj}, s)\, ds.$$

Then, owing to Knight's identity (8.6.22) and $E\{\operatorname{sgn}(U_{tj})\} = 0$, we can write

$$n^{-1/2}[Z_n(\boldsymbol{\delta}) - E\{Z_n(\boldsymbol{\delta})\}] = T_n(\boldsymbol{\delta}) + U_n(\boldsymbol{\delta}) + V_n(\boldsymbol{\delta}), \tag{8.6.33}$$

where

$$T_n(\boldsymbol{\delta}) := n^{-1/2} \sum_{t=1}^{n} \sum_{j=1}^{2} -\operatorname{sgn}(U_{tj})\, (\nabla x_{tj})^T \mathbf{K}^{-1}\boldsymbol{\delta},$$

$$U_n(\boldsymbol{\delta}) := n^{-1/2} \sum_{t=1}^{n} \sum_{j=1}^{2} -\operatorname{sgn}(U_{tj})\, \{w_{tj}(\boldsymbol{\delta}) - (\nabla x_{tj})^T \mathbf{K}^{-1}\boldsymbol{\delta}\},$$

$$V_n(\boldsymbol{\delta}) := n^{-1/2} \sum_{t=1}^{n} \sum_{j=1}^{2} [V_{tj}(\boldsymbol{\delta})) - E\{V_{tj}(\boldsymbol{\delta})\}].$$

To evaluate these quantities, we need some inequalities for $x_t(\boldsymbol{\theta})$. First, it is easy to verify that $\tilde{\mathbf{K}}^{-1}\nabla x_t(\boldsymbol{\theta}+\mathbf{K}^{-1}\boldsymbol{\delta}) = \mathcal{O}(1)$ uniformly in $\boldsymbol{\delta} \in \mathcal{D}_n(\mu_0)$ and t. This means that there is a constant $c_0 > 0$ such that

$$\|\tilde{\mathbf{K}}^{-1}\nabla x_t(\boldsymbol{\theta} + \mathbf{K}^{-1}\boldsymbol{\delta})\| \le c_0 \tag{8.6.34}$$

for all $\boldsymbol{\delta} \in \mathcal{D}_n(\mu_0)$ and t. Moreover, by the Taylor expansion, we obtain

$$\nabla x_t(\boldsymbol{\theta} + \mathbf{K}^{-1}\boldsymbol{\delta}) - \nabla x_t(\boldsymbol{\theta}) = \int_0^1 \{\nabla^2 x_t(\boldsymbol{\theta} + s\mathbf{K}^{-1}\boldsymbol{\delta})\}\mathbf{K}^{-1}\boldsymbol{\delta}\, ds.$$

It is easy to verify that $\tilde{\mathbf{K}}^{-1}\{\nabla^2 x_t(\boldsymbol{\theta} + s\mathbf{K}^{-1}\boldsymbol{\delta})\}\tilde{\mathbf{K}}^{-1} = \mathcal{O}(1)$ uniformly in $s \in [0,1]$, $\boldsymbol{\delta} \in \mathcal{D}_n(\mu_0)$, and t. Therefore, there is a constant $c_0 > 0$ such that

$$\|\tilde{\mathbf{K}}^{-1}\{\nabla x_t(\boldsymbol{\theta} + \mathbf{K}^{-1}\boldsymbol{\delta}) - \nabla x_t(\boldsymbol{\theta})\}\| \le c_0 d_n(\boldsymbol{\delta}, \mathbf{0}) \tag{8.6.35}$$

for all $\boldsymbol{\delta} \in \mathcal{D}_n(\mu_0)$ and t. Similarly, because

$$x_t(\boldsymbol{\theta} + \mathbf{K}^{-1}\boldsymbol{\delta}) - x_t(\boldsymbol{\theta} + \mathbf{K}^{-1}\boldsymbol{\delta}') = (\nabla x_t(\bar{\boldsymbol{\theta}}))^T \mathbf{K}^{-1}(\boldsymbol{\delta} - \boldsymbol{\delta}'),$$

where $\bar{\boldsymbol{\theta}}$ lies between $\boldsymbol{\theta} + \mathbf{K}^{-1}\boldsymbol{\delta}$ and $\boldsymbol{\theta} + \mathbf{K}^{-1}\boldsymbol{\delta}'$, it follows from (8.6.34) that

$$|x_t(\boldsymbol{\theta} + \mathbf{K}^{-1}\boldsymbol{\delta}) - x_t(\boldsymbol{\theta} + \mathbf{K}^{-1}\boldsymbol{\delta}')| \le c_0 d_n(\boldsymbol{\delta}, \boldsymbol{\delta}') \tag{8.6.36}$$

for all $\boldsymbol{\delta}, \boldsymbol{\delta}' \in \mathcal{D}_n(\mu_0)$ and t.

To evaluate $T_n(\boldsymbol{\delta})$, we observe that $T_n(\boldsymbol{\delta}) = n^{-1/2}\boldsymbol{\xi}_n^T\boldsymbol{\delta}$, where

$$\boldsymbol{\xi}_n := n^{-1/2} \sum_{t=1}^{n} \sum_{j=1}^{2} -\operatorname{sgn}(U_{tj})\tilde{\mathbf{K}}^{-1}\nabla x_{tj}.$$

Because the U_{tj} are independent with $E\{\mathrm{sgn}(U_{tj})\} = 0$ and $E\{|\mathrm{sgn}(U_{tj})|^2\} = 1$, it is straightforward to verify that

$$E(\|\boldsymbol{\xi}_n\|^2) = \sum_{t=1}^{n} \|\mathbf{K}^{-1}\nabla x_t\|^2 = \mathrm{tr}(\mathbf{C}_n).$$

Because $\mathbf{C}_n \to \mathbf{C}$, there exists a constant $b_0 > 0$ such that $E(\|\boldsymbol{\xi}_n\|^2) \le b_0$. Moreover, we have $|T_n(\boldsymbol{\delta})| \le \|\boldsymbol{\xi}_n\|\mu$ for all $\boldsymbol{\delta} \in \mathcal{D}_n(\mu)$ and $0 < \mu < \mu_0$, and

$$E(\|\boldsymbol{\xi}_n\|) \le \{E(\|\boldsymbol{\xi}_n\|^2)\}^{1/2}.$$

Combining these results yields

$$E\Big\{\max_{\boldsymbol{\delta}\in\mathcal{D}_n(\mu)} |T_n(\boldsymbol{\delta})|\Big\} \le E(\|\boldsymbol{\xi}_n\|)\mu \le b_1\mu \tag{8.6.37}$$

for all $0 < \mu < \mu_0$, where $b_1 := \sqrt{b_0}$.

To evaluate $U_n(\boldsymbol{\delta})$, we apply Lemma 12.5.12 to

$$S_{nt}(\boldsymbol{\delta}) := -\mathrm{sgn}(U_{tj})\{w_{tj}(\boldsymbol{\delta}) - (\nabla x_{tj})^T\mathbf{K}^{-1}\boldsymbol{\delta}\}.$$

Observe that

$$\begin{aligned}|S_{nt}(\boldsymbol{\delta}) - S_{nt}(\boldsymbol{\delta}')| &= |x_{tj}(\boldsymbol{\theta} + \mathbf{K}^{-1}\boldsymbol{\delta}) - x_{tj}(\boldsymbol{\theta} + \mathbf{K}^{-1}\boldsymbol{\delta}') - (\nabla x_{tj})^T\mathbf{K}^{-1}(\boldsymbol{\delta} - \boldsymbol{\delta}')| \\ &= |(\nabla x_{tj}(\bar{\boldsymbol{\theta}}) - \nabla x_{tj}(\boldsymbol{\theta}))^T\mathbf{K}^{-1}(\boldsymbol{\delta} - \boldsymbol{\delta}')|,\end{aligned}$$

where $\bar{\boldsymbol{\theta}} := \boldsymbol{\theta} + s\mathbf{K}^{-1}(\boldsymbol{\delta} - \boldsymbol{\delta}')$ for some $s \in [0, 1]$. An application of (8.6.35) gives

$$\|\tilde{\mathbf{K}}^{-1}\{\nabla x_{tj}(\bar{\boldsymbol{\theta}}) - \nabla x_{tj}(\boldsymbol{\theta})\}\| \le c_0 d_n(s(\boldsymbol{\delta} - \boldsymbol{\delta}'), \mathbf{0}) \le c_0 d_n(\boldsymbol{\delta}, \boldsymbol{\delta}').$$

Therefore, we obtain

$$|S_{nt}(\boldsymbol{\delta}) - S_{nt}(\boldsymbol{\delta}')| \le c_0 d_n^2(\boldsymbol{\delta}, \boldsymbol{\delta}').$$

In particular, we have $|S_{nt}(\boldsymbol{\delta})| \le c_0 d_n^2(\boldsymbol{\delta}, \mathbf{0})$ because $S_{nt}(\mathbf{0}) = 0$, and hence

$$\max_{\boldsymbol{\delta}\in\mathcal{D}_n(\mu)} n^{-1}\sum_{t=1}^{n} S_{nt}^2(\boldsymbol{\delta}) \le c_0^2\mu^4.$$

Because $d_n(\boldsymbol{\delta}, \boldsymbol{\delta}') \le 2\mu_0$, we can also write

$$|S_{nt}(\boldsymbol{\delta}) - S_{nt}(\boldsymbol{\delta}')| \le 2c_0\mu_0 d_n(\boldsymbol{\delta}, \boldsymbol{\delta}').$$

An application of Lemma 12.5.12 with $\xi_{nt} := 2c_0\mu_0$ leads to

$$E\Big\{\max_{\boldsymbol{\delta}\in\mathcal{D}_n(\mu)} \Big|n^{-1/2}\sum_{t=1}^{n} S_{nt}(\boldsymbol{\delta})\Big|\Big\} \le \tfrac{1}{2}b_2\mu^{2\alpha}$$

for all $0 < \mu < \mu_0$ and n, where $b_2 > 0$ is a constant that may depend on $\alpha \in (0,1)$. Because this expression is true for $j = 1$ and $j = 2$, it follows that

$$E\left\{\max_{\boldsymbol{\delta}\in\mathcal{D}_n(\mu)} |U_n(\boldsymbol{\delta})|\right\} \le b_2\mu^{2\alpha} \tag{8.6.38}$$

for all $0 < \mu < \mu_0$.

Similarly, to evaluate $V_n(\boldsymbol{\delta})$, we apply Lemma 12.5.12 to

$$S_{nt}(\boldsymbol{\delta}) := V_{tj}(\boldsymbol{\delta}) - E\{V_{tj}(\boldsymbol{\delta})\}.$$

Owing to (8.6.30) and (8.6.36), we can write

$$|V_{tj}(\boldsymbol{\delta}) - V_{tj}(\boldsymbol{\delta}')| \le 2|w_{tj}(\boldsymbol{\delta}) - w_{tj}(\boldsymbol{\delta}')| \le 2c_0 d_n(\boldsymbol{\delta},\boldsymbol{\delta}')$$

for all $\boldsymbol{\delta},\boldsymbol{\delta}' \in \mathcal{D}_n(\mu_0)$. Therefore,

$$\begin{aligned}|S_{nt}(\boldsymbol{\delta}) - S_{nt}(\boldsymbol{\delta}')| &\le |V_{tj}(\boldsymbol{\delta}) - V_{tj}(\boldsymbol{\delta}')| + E\{|V_{tj}(\boldsymbol{\delta}) - V_{tj}(\boldsymbol{\delta}')|\}\\ &\le 4c_0 d_n(\boldsymbol{\delta},\boldsymbol{\delta}').\end{aligned}$$

Moreover, it also follows from (8.6.30) and (8.6.36) that

$$|V_{tj}(\boldsymbol{\delta})| \le 2c_0\mu\mathcal{I}(|U_{tj}| \le c_0\mu)$$

for all $\boldsymbol{\delta} \in \mathcal{D}_n(\mu)$ and $0 < \mu < \mu_0$. Given a sufficiently small constant $\mu_2 \in (0,\mu_0)$, there is a constant $b_0 > 0$ such that

$$E\{\mathcal{I}(|U_{tj}| \le c_0\mu)\} = F_{tj}(c_0\mu) - F_{tj}(-c_0\mu) \le b_0\mu$$

for all $0 < \mu < \mu_2$. Combining these results leads to

$$\begin{aligned}S_{nt}^2(\boldsymbol{\delta}) &\le 2V_{tj}^2(\boldsymbol{\delta}) + 2[E\{|V_{tj}(\boldsymbol{\delta})|\}]^2\\ &\le 8c_0^2\mu^2\mathcal{I}(|U_{tj}| \le c_0\mu) + 8c_0^2 b_0^2\mu^4\end{aligned}$$

for all $\boldsymbol{\delta} \in \mathcal{D}_n(\mu)$ and $0 < \mu < \mu_2$, which in turn gives

$$\begin{aligned}E\left\{\max_{\boldsymbol{\delta}\in\mathcal{D}_n(\mu)} S_{nt}^2(\boldsymbol{\delta})\right\} &\le 8c_0^2\mu^2 E\{\mathcal{I}(|U_{tj}| \le c_0\mu)\} + 8c_0^2 b_0^2\mu^4\\ &\le 8c_0^2 b_0(1 + b_0\mu)\mu^3.\end{aligned}$$

for all $0 < \mu < \mu_2$. Therefore,

$$\begin{aligned}E\left\{\max_{\boldsymbol{\delta}\in\mathcal{D}_n(\mu)} n^{-1}\sum_{t=1}^n S_{nt}^2(\boldsymbol{\delta})\right\} &\le n^{-1}\sum_{t=1}^n E\left\{\max_{\boldsymbol{\delta}\in\mathcal{D}_n(\mu)} S_{nt}^2(\boldsymbol{\delta})\right\}\\ &\le 8c_0^2 b_0(1 + b_0\mu)\mu^3\end{aligned}$$

for all $0 < \mu < \mu_2$. An application of Lemma 12.5.12 with $\xi_{nt} := 4c_0$ gives

$$E\left\{\max_{\boldsymbol{\delta}\in\mathcal{D}_n(\mu)} \left|n^{-1/2}\sum_{t=1}^{n} S_{nt}(\boldsymbol{\delta})\right|\right\} \le \tfrac{1}{2} b_3 \mu^{3\alpha/2}$$

for all $0 < \mu < \mu_2$ and n, where $b_3 > 0$ is a constant. Because this result is valid for $j = 1$ and $j = 2$, we obtain

$$E\left\{\max_{\boldsymbol{\delta}\in\mathcal{D}_n(\mu)} |V_n(\boldsymbol{\delta})|\right\} \le b_3 \mu^{3\alpha/2} \tag{8.6.39}$$

for all $0 < \mu < \mu_2$. To complete the verification of (8.6.32), it suffices to combine (8.6.37), (8.6.38), and (8.6.39) with the decomposition (8.6.33) and take any α between 2/3 and 1. □

Notes

In the literature, there exist some general statistical theories for nonlinear least squares regression [166] [426] and nonlinear least absolute deviations regression [169] [239] [240] [274] [417]. These results are not directly applicable to the estimation of sinusoidal parameters because (8.1.1) cannot be expressed as the nonlinear regression model studied in these references, which takes the form $y_t = g(z_t, \boldsymbol{\theta}) + \epsilon_t$, where $\{z_t\}$ is a known sequence and g is a known function. A ramification of the model mismatch is that the estimators of all parameters in this model have the same rate of convergence, whereas the estimators of the frequency and amplitude parameters in (8.1.1) have different rates of convergence. See [145] and [357] for further comments.

Some GIC-type criteria for estimating the number of sinusoids are discussed in [90], [174], [300], and [418] where the signal parameters are estimated by the NLS method or the PM method.

Chapter 9

Autoregressive Approach

The autoregressive (AR) model plays a key role in modern spectral analysis. As a parametric alternative to the nonparametric periodogram, the AR model is a linear dynamic system in the form of a difference equation with a few unknown parameters. By estimating these parameters and examining the transfer function of the resulting dynamic system, one is able to accomplish the objectives of spectral analysis. In this chapter, we discuss two ways of using the AR model to estimate the signal frequencies in a time series with mixed spectrum. One way is to fit an AR model to the time series and identify the peaks of the resulting AR spectrum like a periodogram. Another way is to reparameterize the sinusoidal signal by an AR model and estimate the corresponding AR parameters instead.

9.1 Linear Prediction Method

Linear prediction, or autoregression, is a general method of spectral estimation, as discussed in Chapter 7 (Section 7.1.3). A theoretical foundation of the method is provided by Theorem 7.2, which asserts that any continuous spectrum can be approximated very well by an autoregressive (AR) spectrum so long as the order of the AR model is free to take any value. Although time series with mixed spectra do not belong to this class in the strict sense, it is still interesting, for both practical and theoretical purposes, to investigate the performance of the linear prediction method when applied to time series with mixed spectra.

Our discussion will focus almost exclusively on the complex case where the sinusoidal signal $\{x_t\}$ takes the form (2.1.5) or (2.1.7). However, owing to the correspondence between the RSM and the CSM, most of the results remain valid in the real case where $\{x_t\}$ is given by (2.1.1) or (2.1.2), so long as the signal parameters are properly defined according to (2.1.8) or (2.1.9).

9.1.1 Linear Prediction Estimators

According to Proposition 2.7, if $\{y_t\}$ is a zero-mean stationary process with ACF $r_y(u)$, then the best linear prediction coefficient $\mathbf{c} := [c_1, \ldots, c_m]^T$, which minimizes the prediction error variance $E\{|y_t + \sum_{j=1}^m \tilde{c}_j y_{t-j}|^2\}$ with respect to $\tilde{\mathbf{c}} := [\tilde{c}_1, \ldots, \tilde{c}_m]^T \in \mathbb{C}^m$, satisfies the Yule-Walker equation

$$\mathbf{R}_y \mathbf{c} = -\mathbf{r}_y, \tag{9.1.1}$$

where

$$\mathbf{R}_y := \begin{bmatrix} r_y(0) & r_y^*(1) & \cdots & r_y^*(m-1) \\ r_y(1) & r_y(0) & \cdots & r_y^*(m-2) \\ \vdots & \vdots & \ddots & \vdots \\ r_y(m-1) & r_y(m-2) & \cdots & r_y(0) \end{bmatrix}, \quad \mathbf{r}_y := \begin{bmatrix} r_y(1) \\ r_y(2) \\ \vdots \\ r_y(m) \end{bmatrix}. \tag{9.1.2}$$

The minimum prediction error variance $\sigma_m^2 := E\{|y_t + \sum_{j=1}^m c_j y_{t-j}|^2\}$ is given by

$$\sigma_m^2 = r_y(0) + \mathbf{r}_y^H \mathbf{c}. \tag{9.1.3}$$

Let $c(z) := 1 + \sum_{j=1}^m c_j z^{-j}$ denote the m-degree AR polynomial associated with $\mathbf{c}$ and let $C(\omega) := c(\exp(i\omega)) = 1 + \sum_{j=1}^m c_j \exp(-ij\omega)$ denote the reciprocal transfer function of the corresponding AR (all-pole) filter $1/c(z)$. Then, the corresponding AR spectrum of order m takes the form

$$f_{\text{AR}}(\omega) := \frac{\sigma_m^2}{|C(\omega)|^2} = \frac{\sigma_m^2}{|c(\exp(i\omega))|^2} = \frac{\sigma_m^2}{|1 + \sum_{j=1}^m c_j \exp(-ij\omega)|^2}. \tag{9.1.4}$$

We can also write $C(\omega) = 1 + \mathbf{f}_m^H(\omega)\mathbf{c}$, where $\mathbf{f}_m(\omega) := [\exp(i\omega), \ldots, \exp(im\omega)]^T$.

As discussed in Chapter 4 (Section 4.2), when $\{y_t\}$ is given by

$$y_t = x_t + \epsilon_t, \tag{9.1.5}$$

where $\{x_t\}$ is a sinusoidal signal of the form (2.1.5) or (2.1.7) and $\{\epsilon_t\}$ is a zero-mean stationary process with SDF $f_\epsilon(\omega)$, the spectrum of $\{y_t\}$ takes the form

$$f_y(\omega) = 2\pi \sum_{k=1}^{p} C_k^2 \delta(\omega - \omega_k) + f_\epsilon(\omega),$$

where $\delta(\omega)$ denotes the Dirac delta. This is a spectrum of mixed type that contains not only a continuous component $f_\epsilon(\omega)$ but also a spike at each signal frequency. By using $f_{\text{AR}}(\omega)$ of the form (9.1.4) to approximate $f_y(\omega)$, one hopes that the AR spectrum will exhibit a large peak near ω_k for each k so that the signal frequencies can be retrieved from an estimated AR spectrum in the same way as from a periodogram. Because the parametric AR model extrapolates the ACF of

$\{y_t\}$ beyond a finite set, one also hopes that the AR spectrum will exhibit a higher resolution than the nonparametric periodogram whose resolution is limited by its employment of only a finite number of autocovariances.

The estimation techniques discussed in Chapter 7 (Section 7.1.3) can be readily used to produce an AR spectrum from a finite data record $\{y_1, \dots, y_n\}$. For example, by replacing the ACF in the Yule-Walker equation (9.1.1) with the sample autocovariances $\hat{r}_y(0), \hat{r}_y(1), \dots, \hat{r}_y(m)$ defined by (4.1.1), we obtain a method-of-moments estimator, better known as the Yule-Walker estimator, of $\mathbf{c}$:

$$\hat{\mathbf{c}}_{\mathrm{YW}} := -\hat{\mathbf{R}}_y^{-1}\hat{\mathbf{r}}_y, \tag{9.1.6}$$

where

$$\hat{\mathbf{R}}_y := \begin{bmatrix} \hat{r}_y(0) & \hat{r}_y^*(1) & \cdots & \hat{r}_y^*(m-1) \\ \hat{r}_y(1) & \hat{r}_y(0) & \cdots & \hat{r}_y^*(m-2) \\ \vdots & \vdots & \ddots & \vdots \\ \hat{r}_y(m-1) & \hat{r}_y(m-2) & \cdots & \hat{r}_y(0) \end{bmatrix}, \quad \hat{\mathbf{r}}_y := \begin{bmatrix} \hat{r}_y(1) \\ \hat{r}_y(2) \\ \vdots \\ \hat{r}_y(m) \end{bmatrix}. \tag{9.1.7}$$

The Yule-Walker estimator of σ_m^2 is given by

$$\hat{\sigma}_{\mathrm{YW}}^2 := \hat{r}_y(0) + \hat{\mathbf{r}}_y^H \hat{\mathbf{c}}_{\mathrm{YW}}.$$

Substituting $\mathbf{c}$ and σ_m^2 in (9.1.4) with their estimates $\hat{\mathbf{c}}_{\mathrm{YW}}$ and $\hat{\sigma}_{\mathrm{YW}}^2$ yields an estimated AR spectrum of order m. The Levinson-Durbin algorithm in Proposition 2.8 provides an efficient way of computing $\hat{\mathbf{c}}_{\mathrm{YW}}$ and $\hat{\sigma}_{\mathrm{YW}}^2$.

Alternatively, one can estimate $\mathbf{c}$ and σ_m^2 by the least-squares (LS) linear prediction method that minimizes the sum of squared prediction errors

$$\sum_{t=m+1}^{n} \Big| y_t + \sum_{j=1}^{m} \tilde{c}_j y_{t-j} \Big|^2 = \|\mathbf{y}_f + \mathbf{Y}_f \tilde{\mathbf{c}}\|^2,$$

where

$$\mathbf{Y}_f := \begin{bmatrix} y_m & y_{m-1} & \cdots & y_1 \\ y_{m+1} & y_m & \cdots & y_2 \\ \vdots & \vdots & & \vdots \\ y_{n-1} & y_{n-2} & \cdots & y_{n-m} \end{bmatrix}, \quad \mathbf{y}_f := \begin{bmatrix} y_{m+1} \\ y_{m+2} \\ \vdots \\ y_n \end{bmatrix}. \tag{9.1.8}$$

To avoid overfitting, the order m must satisfy $m < n - m$, or equivalently,

$$m < n/2.$$

Under this condition, the minimizer can be expressed as

$$\hat{\mathbf{c}}_f := -(\mathbf{Y}_f^H \mathbf{Y}_f)^{-1} \mathbf{Y}_f^H \mathbf{y}_f, \tag{9.1.9}$$

which is called the forward linear prediction (FLP) estimator. The corresponding estimator of σ_m^2 is given by

$$\hat{\sigma}_f^2 := n^{-1}\|\mathbf{y}_f + \mathbf{Y}_f\hat{\mathbf{c}}_f\|^2 = n^{-1}(\mathbf{y}_f^H\mathbf{y}_f + \mathbf{y}_f^H\mathbf{Y}_f\hat{\mathbf{c}}_f).$$

Of course, a disadvantage of the LS method is that it cannot be computed as quickly as the Yule-Walker estimator because the Levinson-Durbin algorithm in Proposition 2.8 is no longer applicable.

Another LS estimator can be obtained by minimizing the sum of squared forward and backward prediction errors

$$\begin{aligned}&\sum_{t=m+1}^{n}\Big|y_t + \sum_{j=1}^{m}\tilde{c}_j y_{t-j}\Big|^2 + \sum_{t=m+1}^{n}\Big|y_{t-m}^* + \sum_{j=1}^{m}\tilde{c}_j y_{t-m+j}^*\Big|^2\\ &= \|\mathbf{y}_f + \mathbf{Y}_f\tilde{\mathbf{c}}\|^2 + \|\mathbf{y}_b + \mathbf{Y}_b\tilde{\mathbf{c}}\|^2 = \|\mathbf{y}_{fb} + \mathbf{Y}_{fb}\tilde{\mathbf{c}}\|^2,\end{aligned}$$

where

$$\mathbf{Y}_b := \begin{bmatrix} y_2^* & y_3^* & \cdots & y_{m+1}^* \\ y_3^* & y_4^* & \cdots & y_{m+2}^* \\ \vdots & \vdots & & \vdots \\ y_{n-m+1}^* & y_{n-m+2}^* & \cdots & y_n^* \end{bmatrix}, \quad \mathbf{y}_b := \begin{bmatrix} y_1^* \\ y_2^* \\ \vdots \\ y_{n-m}^* \end{bmatrix}, \tag{9.1.10}$$

and

$$\mathbf{Y}_{fb} := \begin{bmatrix} \mathbf{Y}_f \\ \mathbf{Y}_b \end{bmatrix}, \quad \mathbf{y}_{fb} := \begin{bmatrix} \mathbf{y}_f \\ \mathbf{y}_b \end{bmatrix}. \tag{9.1.11}$$

In comparison with the forward-only prediction problem, the number of cases in the forward-backward prediction problem is increased from $n-m$ to $2(n-m)$. Therefore, the constraint on m is relaxed to $m < 2(n-m)$, or equivalently,

$$m < 2n/3,$$

which allows m to take larger values than in the FLP method. Under this condition, the minimizer, which is known as the forward-backward linear prediction (FBLP) estimator of $\mathbf{c}$, can be expressed as

$$\hat{\mathbf{c}}_{fb} := -(\mathbf{Y}_{fb}^H\mathbf{Y}_{fb})^{-1}\mathbf{Y}_{fb}^H\mathbf{y}_{fb}. \tag{9.1.12}$$

The corresponding estimator of σ_m^2 is given by

$$\hat{\sigma}_{fb}^2 := (2n)^{-1}\|\mathbf{y}_{fb} + \mathbf{Y}_{fb}\hat{\mathbf{c}}_{fb}\|^2 = (2n)^{-1}(\mathbf{y}_{fb}^H\mathbf{y}_{fb} + \mathbf{y}_{fb}^H\mathbf{Y}_{fb}\hat{\mathbf{c}}_{fb}).$$

The FBLP estimator has been found more effective in modeling spectral peaks than the FLP estimator especially for short data records [177].

With $\hat{\mathbf{c}}$ and $\hat{\sigma}_m^2$ denoting any of the estimators for $\mathbf{c}$ and σ_m^2, the corresponding AR spectral estimator can be expressed as $\hat{f}_{\mathrm{AR}}(\omega) := \hat{\sigma}_m^2/|\hat{C}(\omega)|^2$, where $\hat{C}(\omega) := 1 + \mathbf{f}_m^H(\omega)\hat{\mathbf{c}}$. As an example, Figure 9.1 depicts the estimated AR spectrum by the FBLP method with different values of m for a time series that comprises three complex sinusoids in Gaussian white noise. Observe that the first two signal frequencies are separated by less than $2\pi/n$ and unresolved in the periodogram; the third signal frequency is far away from the first two and responsible for the second spectral peak in the periodogram.

As we can see from Figure 9.1, the AR spectrum with the lowest order $m = p = 3$ is totally inadequate — it does not even capture the two spectral peaks shown in the periodogram. However, when m is increased to 15, two spectral peaks do emerge in the AR spectrum, which coincide approximately with the spectral peaks in the periodogram. In this case, the third signal frequency can be well estimated by the location of the second AR spectral peak, but the two closely spaced signal frequencies remain unresolved. As m is increased further to 27, the closely spaced frequencies are also resolved successfully. This example shows that the linear prediction method is indeed able to produce an AR spectrum whose resolution is higher than the resolution of the periodogram.

Figure 9.1 also reveals a serious drawback of the linear prediction method: the emergence of large spurious spectral peaks when the order is too high. It is troublesome especially in the case of $m = 28$ where the largest spurious peak even overwhelms the spectral peaks associated with the sinusoids, making it difficult to identify and estimate the signal frequencies. Note that the area under the curve of the AR spectrum with $m = 28$ is considerably smaller than the area under the periodogram. This is largely due to the underestimation of σ_m^2 in the AR spectrum, especially when m is close to n.

In Figure 9.1, the linear prediction method works well with $m = 27$ but not with $m = 28$. This illustrates how crucial it is to choose a proper order m for the linear prediction method. Generally speaking, if m is too small, the signal frequencies may appear unresolved; if m is too large (sometimes just slightly), the signal frequencies may become obscured by spurious spectral peaks. Even with a fixed order, the intrinsic statistical variability of the estimated AR spectrum may still cause the signal frequencies to be unresolved in some cases while resolved in others, especially when the SNR is not sufficiently high. Simple order selection criteria such as the AIC discussed in Chapter 7 (Section 7.1.3) are inadequate.

To better understand the AR spectrum in (9.1.4), let us examine the linear prediction coefficient $\mathbf{c}$ defined by (9.1.1) when $\{y_t\}$ is given by (9.1.5). First, because $r_y(u) = r_x(u) + r_\epsilon(u)$, it follows that

$$\mathbf{R}_y = \mathbf{R}_x + \mathbf{R}_\epsilon, \quad \mathbf{r}_y = \mathbf{r}_x + \mathbf{r}_\epsilon, \tag{9.1.13}$$

where $\mathbf{R}_x$, $\mathbf{R}_\epsilon$, $\mathbf{r}_x$, and $\mathbf{r}_\epsilon$ are defined by $r_x(u)$ and $r_\epsilon(u)$ in the same way as $\mathbf{R}_y$

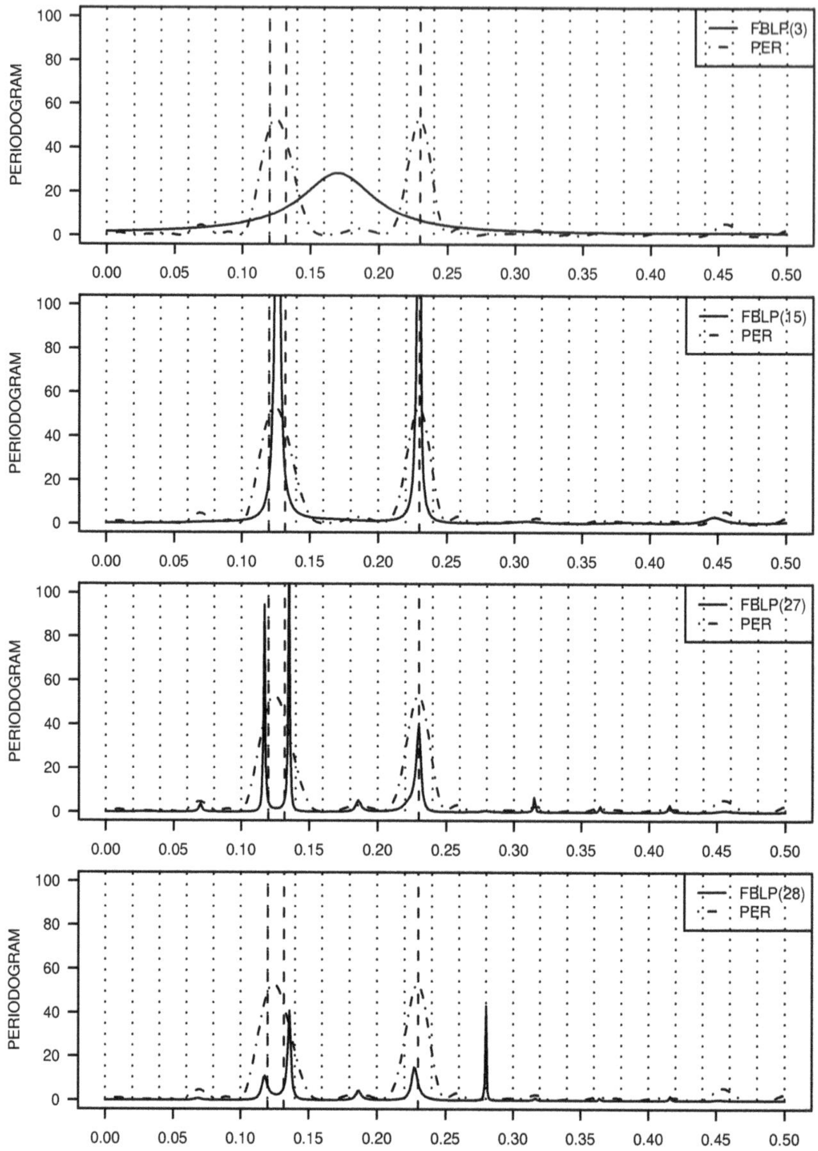

Figure 9.1. Plot of estimated AR spectrum of different orders by the FBLP method for a time series that consists of three unit-amplitude zero-phase complex sinusoids in Gaussian white noise ($n = 50$, $\omega_1 = 2\pi \times 6/n = 2\pi \times 0.12$, $\omega_2 = 2\pi \times 6.6/n = 2\pi \times 0.132$, $\omega_3 = 2\pi \times 11.5/n = 2\pi \times 0.23$, and SNR = 4 dB per sinusoid, i.e., $\sigma^2 = 0.3981072$). Solid line, AR spectrum of order $m = 3, 15, 27, 28$; dashed-dotted line, periodogram. Vertical dashed lines depict the signal frequencies and vertical dotted lines depict the Fourier grid.

and $\mathbf{r}_y$ in (9.1.2). By Theorem 4.1, $r_x(u) = \sum_{k=1}^{p} C_k^2 \exp(i\omega_k u)$. Therefore,

$$\begin{cases} \mathbf{R}_x = \sum_{k=1}^{p} C_k^2 \mathbf{f}_m(\omega_k)\mathbf{f}_m^H(\omega_k) = \mathbf{F}_m \mathbf{G} \mathbf{F}_m^H, \\ \mathbf{r}_x = \sum_{k=1}^{p} C_k^2 \mathbf{f}_m(\omega_k) = \mathbf{F}_m \mathbf{g}, \end{cases} \tag{9.1.14}$$

where

$$\mathbf{G} := \mathrm{diag}(C_1^2, \ldots, C_p^2), \quad \mathbf{g} := [C_1^2, \ldots, C_p^2]^T,$$
$$\mathbf{f}_m(\omega) := [\exp(i\omega), \ldots, \exp(im\omega)]^T, \quad \mathbf{F}_m := [\mathbf{f}_m(\omega_1), \ldots, \mathbf{f}_m(\omega_p)].$$

Substituting (9.1.13) and (9.1.14) in (9.1.1) yields

$$\mathbf{c} = -(\mathbf{R}_x + \mathbf{R}_\epsilon)^{-1}(\mathbf{r}_x + \mathbf{r}_\epsilon) = -(\mathbf{F}_m \mathbf{G} \mathbf{F}_m^H + \mathbf{R}_\epsilon)^{-1}(\mathbf{F}_m \mathbf{g} + \mathbf{r}_\epsilon). \tag{9.1.15}$$

Moreover, by the first matrix inversion formula in Lemma 12.4.1,

$$(\mathbf{F}_m \mathbf{G} \mathbf{F}_m^H + \mathbf{R}_\epsilon)^{-1} = \mathbf{R}_\epsilon^{-1} - \mathbf{R}_\epsilon^{-1}\mathbf{F}_m(\mathbf{G}^{-1} + \mathbf{F}_m^H \mathbf{R}_\epsilon^{-1}\mathbf{F}_m)^{-1}\mathbf{F}_m^H \mathbf{R}_\epsilon^{-1}.$$

Therefore, with $\boldsymbol{\varphi} := -\mathbf{R}_\epsilon^{-1}\mathbf{r}_\epsilon$ denoting the linear prediction coefficient of $\{\epsilon_t\}$, we can rewrite (9.1.15) as

$$\begin{aligned} \mathbf{c} &= \mathbf{R}_\epsilon^{-1}\mathbf{F}_m\{(\mathbf{G}^{-1} + \mathbf{F}_m^H \mathbf{R}_\epsilon^{-1}\mathbf{F}_m)^{-1}\mathbf{F}_m^H \mathbf{R}_\epsilon^{-1}\mathbf{F}_m - \mathbf{I}\}\mathbf{g} \\ &\quad + \{\mathbf{I} - \mathbf{R}_\epsilon^{-1}\mathbf{F}_m(\mathbf{G}^{-1} + \mathbf{F}_m^H \mathbf{R}_\epsilon^{-1}\mathbf{F}_m)^{-1}\mathbf{F}_m^H\}\boldsymbol{\varphi} \\ &= -\mathbf{R}_\epsilon^{-1}\mathbf{F}_m(\mathbf{G}^{-1} + \mathbf{F}_m^H \mathbf{R}_\epsilon^{-1}\mathbf{F}_m)^{-1}\mathbf{G}^{-1}\mathbf{g} \\ &\quad + \{\mathbf{I} - \mathbf{R}_\epsilon^{-1}\mathbf{F}_m(\mathbf{G}^{-1} + \mathbf{F}_m^H \mathbf{R}_\epsilon^{-1}\mathbf{F}_m)^{-1}\mathbf{F}_m^H\}\boldsymbol{\varphi}. \end{aligned} \tag{9.1.16}$$

This expression shows that $\mathbf{c}$ is an affine transformation of $\boldsymbol{\varphi}$.

Consider the special case where $\{\epsilon_t\} \sim \mathrm{WN}(0, \sigma^2)$ and hence $\mathbf{R}_\epsilon = \sigma^2\mathbf{I}$ and $\mathbf{r}_\epsilon = \mathbf{0}$. In this case, $\boldsymbol{\varphi} = \mathbf{0}$. Therefore, upon noting $\mathbf{G}^{-1}\mathbf{g} = \mathbf{1} := [1, \ldots, 1]^T \in \mathbb{R}^p$, the formula in (9.1.16) is simplified as

$$\mathbf{c} = -\mathbf{F}_m(\boldsymbol{\Gamma}^{-1} + \mathbf{F}_m^H \mathbf{F}_m)^{-1}\mathbf{1}, \tag{9.1.17}$$

where $\boldsymbol{\Gamma} := \mathbf{G}/\sigma^2 = \mathrm{diag}(\gamma_1, \ldots, \gamma_p)$ and $\gamma_k := C_k^2/\sigma^2$. This expression facilitates a more detailed analysis of the AR spectrum.

Example 9.1 (AR Spectrum of a Single Complex Sinusoid in White Noise). Let $p = 1$. In this case, $r_x(0) = C_1^2$, $\mathbf{r}_x = C_1^2 \mathbf{f}_m(\omega_1)$, $\mathbf{F}_m = \mathbf{f}_m(\omega_1)$, and $\boldsymbol{\Gamma} = \gamma_1$. Therefore, it follows from (9.1.17) that

$$\mathbf{c} = -\frac{\gamma_1}{m\gamma_1 + 1}\mathbf{f}_m(\omega_1) \qquad (m \geq 1)$$

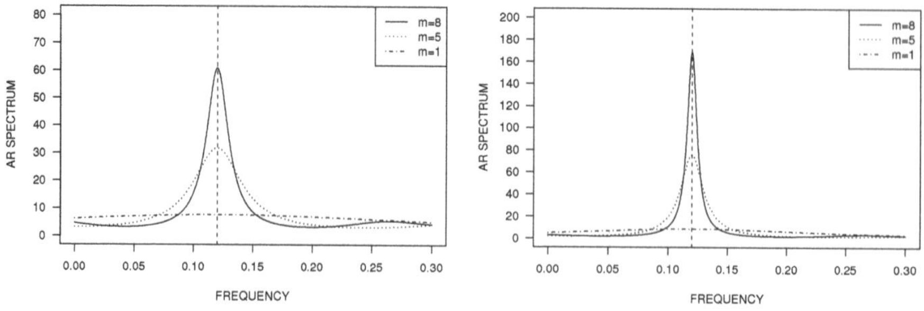

(a) (b)

Figure 9.2. Plot of theoretical AR spectrum with different order for the case of a single complex sinusoid with unit amplitude in white noise ($\omega_1 = 2\pi \times 0.12$). (a) SNR = −5 dB, (b) SNR = −1 dB. Vertical dashed line depicts the signal frequency.

and hence $c(z) = 1 - m^{-1}\sum_{j=1}^{m} d_m \exp(ij\omega_1)z^{-j}$, where

$$d_m := m\gamma_1/(m\gamma_1 + 1).$$

On the unit circle with $z = \exp(i\omega)$, we obtain

$$C(\omega) := c(\exp(i\omega)) = 1 - d_m D_m(\omega_1 - \omega),$$

where $D_m(\omega) := m^{-1}\sum_{t=1}^{m} \exp(it\omega)$ is the Dirichlet kernel. In addition,

$$\sigma_m^2 = r_x(0) + \sigma^2 + \mathbf{r}_x^H \mathbf{c} = \sigma^2(1 + m^{-1}d_m).$$

Therefore, the corresponding AR spectrum takes the form

$$f_{\text{AR}}(\omega) = \frac{\sigma^2(1 + m^{-1}d_m)}{|1 - d_m D_m(\omega_1 - \omega)|^2}. \tag{9.1.18}$$

Observe that for any $\omega \in (0, 2\pi)$ such that $\omega \neq \omega_1$ we have

$$\begin{aligned} |1 - d_m D_m(\omega_1 - \omega)| &\geq 1 - d_m |D_m(\omega_1 - \omega)| \\ &> 1 - d_m = |1 - d_m D_m(0)|. \end{aligned}$$

Therefore, for any $m \geq 1$, the AR spectrum in (9.1.18) has a unique global maximum at the signal frequency ω_1 which equals $\sigma^2(1 + m^{-1}d_m)/(1 - d_m)^2$.

Figure 9.2 depicts the AR spectrum given by (9.1.18) for different values of order m and SNR γ_1. As can be seen, besides the unique maximum that appears exactly at the signal frequency $\omega_1 = 2\pi \times 0.12$, the spectral peak becomes sharper as the order m increases. In fact, as $m \to \infty$, we have $d_m \to 1$ and $D_m(\omega_1 - \omega) \to 0$ for any $\omega \neq \omega_1$. Therefore, $f_{\text{AR}}(\omega) \to \sigma^2$ as $m \to \infty$ for $\omega \neq \omega_1$ and

$$\lim_{m \to \infty} f_{\text{AR}}(\omega_1) = \lim_{m \to \infty} \frac{\sigma^2(1 + m^{-1}d_m)}{(1 - d_m)^2} = \infty.$$

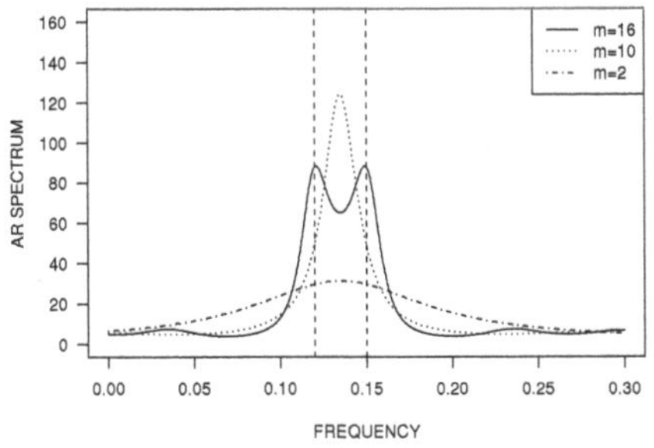

(a)

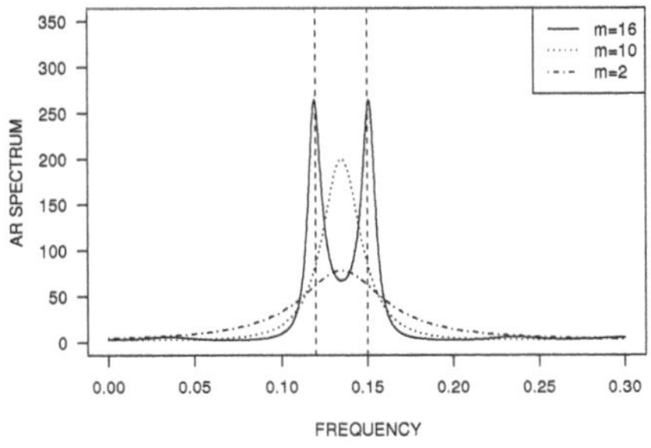

(b)

Figure 9.3. Plot of theoretical AR spectrum with different order for the case of two complex sinusoids with unit amplitude in white noise ($\omega_1 = 2\pi \times 0.12$, $\omega_2 = 2\pi \times 0.15$). (a) SNR = −5 dB per sinusoid, (b) SNR = −1 dB per sinusoid. Vertical dashed lines depict the signal frequencies.

Moreover, as $\sigma \to 0$ and m fixed, we also have $d_m \to 1$. Therefore,

$$\lim_{\sigma\to 0} \frac{1}{\sigma^2} f_{\text{AR}}(\omega_1) = \lim_{\sigma\to 0} \frac{1+m^{-1}d_m}{(1-d_m)^2} = \infty$$

and

$$\lim_{\sigma\to 0} \frac{1}{\sigma^2} f_{\text{AR}}(\omega) = \frac{1+m^{-1}}{|1-D_m(\omega_1-\omega)|^2} < \infty \qquad (\omega \neq \omega_1).$$

This result justifies the observation from Figure 9.2 that the spectral peak at ω_1 also becomes sharper when the SNR increases. ◇

In the general case of multiple sinusoids with $p > 1$, the AR spectrum in (9.1.4) exhibits p spectral peaks only when the order m is sufficiently high; even then, the local maxima may not appear exactly at the signal frequencies. This property is illustrated by Figure 9.3 with $p = 2$. As we can see, the AR spectrum successfully reveals two spectral peaks only when m is increased to 16. In this case, the local maxima appear at $2\pi \times 0.1208$ and $2\pi \times 0.1492$ in Figure 9.3(a) and at $2\pi \times 0.1192$ and $2\pi \times 0.1508$ in Figure 9.3(b), both determined numerically. However, the true signal frequencies are $2\pi \times 0.12$ and $2\pi \times 0.15$.

The discrepancy between the AR spectral peaks and the signal frequencies can be reduced by increasing the order m. Indeed, according to Lemma 12.1.4(b), $\mathbf{F}_m^H \mathbf{F}_m = m\mathbf{I} + \mathcal{O}(1)$ as $m \to \infty$. Substituting this expression in (9.1.17) gives

$$\begin{aligned} \mathbf{c} &= -m^{-1}\mathbf{F}_m(\mathbf{I} + m^{-1}\boldsymbol{\Gamma}^{-1})^{-1}\mathbf{1} + \mathcal{O}(m^{-2}) \\ &= -m^{-1}\mathbf{F}_m(\mathbf{I} + \mathcal{O}(m^{-1}))\mathbf{1} + \mathcal{O}(m^{-2}) \\ &= -m^{-1}\mathbf{F}_m\mathbf{1} + \mathcal{O}(m^{-2}). \end{aligned} \qquad (9.1.19)$$

Because $\mathbf{F}_m\mathbf{1} = \sum_{k=1}^p \mathbf{f}_m(\omega_k)$, it follows that

$$C(\omega) = 1 + \mathbf{f}_m^H(\omega)\mathbf{c} = 1 - \sum_{k=1}^p D_m(\omega_k - \omega) + \mathcal{O}(m^{-1}).$$

Therefore, as $m \to \infty$, $C(\omega) \to 1$ if $\omega \notin \{\omega_k\}$, and $C(\omega_k) \to 0$ for all $k = 1, \ldots, p$. Moreover, it follows from (9.1.14) and (9.1.19) that

$$\begin{aligned}\sigma_m^2 &= r_y(0) + \mathbf{r}_x^H\mathbf{c} = r_y(0) + \mathbf{g}^T\mathbf{F}_m^H\mathbf{c} \\ &= r_y(0) - \mathbf{g}^T\{m^{-1}\mathbf{F}_m^H\mathbf{F}_m\mathbf{1} + \mathcal{O}(m^{-2})\} \\ &= \sigma^2 + \mathbf{g}^T\mathbf{1} - \mathbf{g}^T\{(\mathbf{I} + \mathcal{O}(m^{-1}))\mathbf{1} + \mathcal{O}(m^{-2})\} \\ &= \sigma^2 + \mathcal{O}(m^{-1}).\end{aligned}$$

Hence, $\sigma_m^2 \to \sigma^2$ as $m \to \infty$. Combining these results leads to

$$\lim_{m\to\infty} f_{\text{AR}}(\omega) = \begin{cases} \infty & \text{if } \omega = \omega_k \ (k = 1, \ldots, p), \\ \sigma^2 & \text{otherwise.} \end{cases} \tag{9.1.20}$$

This analysis suggests that for sufficiently large m the AR spectrum will exhibit large peaks near the signal frequencies and the location of these spectral peaks will converge to the location of the signal frequencies as $m \to \infty$.

A similar statement can be made when $\{\epsilon_t\}$ is colored noise with SDF $f_\epsilon(\omega)$. In this case, let $\varsigma_m^2 := r_\epsilon(0) + \mathbf{r}_\epsilon^H\boldsymbol{\varphi}$ denote the minimum prediction error variance of $\{\epsilon_t\}$. Then, the AR spectrum of the noise takes the form

$$f_m(\omega) := \varsigma_m^2/|\Phi_m(\omega)|^2,$$

where $\Phi_m(\omega) := 1 + \mathbf{f}_m^H(\omega)\boldsymbol{\varphi}$. Assume that

$$\lim_{m\to\infty} \max_\omega |f_m(\omega) - f_\epsilon(\omega)| = 0, \tag{9.1.21}$$

in which case, we say that $\{\epsilon_t\}$ has an AR(∞) representation. A similar assumption is required by Theorem 7.6 in order to ensure the consistency of the AR spectral estimator [30]. The assumption is satisfied if $\{\epsilon_t\}$ is a causal linear process of the form $\epsilon_t = \sum_{j=0}^\infty \psi_j \zeta_{t-j}$ ($\psi_0 := 1$) with $\{\zeta_t\} \sim \text{WN}(0, \sigma_\zeta^2)$ and $\sum_{j=0}^\infty |\psi_j| < \infty$, such that $1/\psi(z)$ has an infinite series expansion $1/\psi(z) = \sum_{j=0}^\infty \varphi_j z^{-j}$ ($\varphi_0 := 1$) with the property $\sum_{j=0}^\infty |\varphi_j| < \infty$. In this case,

$$f_\epsilon(\omega) = \frac{\sigma_\zeta^2}{|\sum_{j=0}^\infty \varphi_j \exp(-ij\omega)|^2}.$$

The following theorem summarizes the asymptotic behavior of $f_{\text{AR}}(\omega)$ as $m \to \infty$ under these conditions. It is a generalization of the results obtained under the white noise assumption. See Section 9.6 for a proof.

Theorem 9.1 (Limiting Behavior of the AR Spectrum as Order Grows). *Let* $\{y_t\}$ *be given by (9.1.5), where* $\{x_t\}$ *takes the form (2.1.5) or (2.1.7). Let* $f_{\text{AR}}(\omega)$ *be the AR spectrum of order* m *defined by (9.1.4) with* $\mathbf{c}$ *and* σ_m^2 *given by (9.1.1) and (9.1.3). If* $\{\epsilon_t\}$ *has an AR(∞) representation in the sense of (9.1.21) and* $f_\epsilon(\omega)$ *is bounded away from zero, then, as* $m \to \infty$, *the expression in (9.1.20) remains valid except that* σ^2 *should be replaced by* $f_\epsilon(\omega)$ *for* $\omega \notin \{\omega_k\}$.

Finally, let us investigate the behavior of the linear prediction estimators when the noise is absent. The result of this analysis sheds some light on the performance of these estimators at high SNR. For the FLP estimator, the following theorem can be established. See Section 9.6 for a proof.

Theorem 9.2 (Least-Squares Linear Prediction in the Absence of Noise). *Let* $\mathbf{X}_f$ *and* $\mathbf{x}_f$ *be defined by* $\{x_t\}$ *in the same way as* $\mathbf{Y}_f$ *and* $\mathbf{y}_f$ *in (9.1.8). Then, for any* m *and* n *such that* $p \le m < n - m$, *the* m*-degree polynomial associated with any solution to the problem of minimizing* $\|\mathbf{x}_f + \mathbf{X}_f\tilde{\mathbf{c}}\|^2$ *with respect to* $\tilde{\mathbf{c}} \in \mathbb{C}^m$ *has* p *roots* $z_1, \dots, z_p$ *on the unit circle, where* $z_k := \exp(i\omega_k)$ $(k = 1, \dots, p)$. *For the minimum-norm solution, the remaining* $m - p$ *roots lie strictly inside the unit circle.*

According to Theorem 9.2, if $\hat{c}(z)$ denotes the m-degree polynomial associated with the minimum-norm solution, then we can write

$$\hat{c}(z) = a(z)\hat{b}(z),$$

where

$$a(z) := \prod_{k=1}^{p} (1 - z_k z^{-1})$$

is a p-degree polynomial with roots $z_1, \dots, z_p$ and $\hat{b}(z)$ is a minimum-phase polynomial of degree $m - p$. In this case, the signal frequencies are uniquely determined by the p roots of $\hat{c}(z)$ on the unit circle, or equivalently, by the p smallest local minima of $|\hat{C}(\omega)|^2$ in the interval $(0, 2\pi)$, where $\hat{C}(\omega) := \hat{c}(\exp(i\omega))$. A similar result can be obtained for the FBLP estimator [396].

9.1.2 Statistical Properties

Although finite-sample statistical properties of the linear prediction estimators are difficult to analyze mathematically, some useful large-sample asymptotic properties can be derived under fairly general conditions. In the following, we investigate the limiting value of the linear prediction estimators and their asymptotic distributions as $n \to \infty$ for fixed m.

First, consider the limiting value. With $\{y_t\}$ given by (9.1.5), let $\{\epsilon_t\}$ satisfy the conditions of Theorem 4.1. Then, the sample autocovariance function $\hat{r}_y(u)$, defined by (4.1.1), has the property that for any fixed u as $n \to \infty$,

$$\hat{r}_y(u) \stackrel{a.s.}{\to} r_y(u) := r_x(u) + r_\epsilon(u),$$

where $r_\epsilon(u)$ is the ACF of $\{\epsilon_t\}$ and $r_x(u)$ is the ACF of $\{x_t\}$ defined by (4.2.6). Therefore, for any fixed m, $\hat{\mathbf{R}}_y$ and $\hat{\mathbf{r}}_y$ in (9.1.7) have the property

$$\hat{\mathbf{R}}_y \xrightarrow{a.s.} \mathbf{R}_y, \quad \hat{\mathbf{r}}_y \xrightarrow{a.s.} \mathbf{r}_y.$$

In other words, $\hat{\mathbf{R}}_y$ and $\hat{\mathbf{r}}_y$ are consistent estimators of $\mathbf{R}_y$ and $\mathbf{r}_y$ defined by (9.1.2). The same can be said about the alternative estimators of $\mathbf{R}_y$ and $\mathbf{r}_y$ given by $n^{-1}\mathbf{Y}_f^H\mathbf{Y}_f$, $n^{-1}\mathbf{Y}_f^H\mathbf{y}_f$, $(2n)^{-1}\mathbf{Y}_{fb}^H\mathbf{Y}_{fb}$, and $(2n)^{-1}\mathbf{Y}_{fb}^H\mathbf{y}_{fb}$. Therefore, the linear prediction coefficient $\mathbf{c}$ defined by (9.1.1) can be estimated consistently by the linear prediction estimators in (9.1.6), (9.1.9), and (9.1.12). The corresponding estimators of σ_m^2 are also consistent. Moreover, by Theorem 9.1, the AR spectrum $f_{\text{AR}}(\omega)$ in (9.1.4) has p largest peaks near the signal frequencies when m is sufficiently large. Let the location of these spectral peaks be denoted by

$$\boldsymbol{\omega}_0 := [\omega_{01}, \ldots, \omega_{0p}]^T, \tag{9.1.22}$$

where $0 < \omega_{01} < \cdots < \omega_{0p} < 2\pi$. Let $\hat{\mathbf{c}}$ denote any of the estimators in (9.1.6), (9.1.9), and (9.1.12), and let $\hat{C}(\omega) := 1 + \mathbf{f}_m^H(\omega)\hat{\mathbf{c}}$. Because $\hat{\mathbf{c}} \xrightarrow{P} \mathbf{c}$ as $n \to \infty$, it follows from Lemma 12.5.1 that $|\hat{C}(\omega)|^2$ has a local minimum in the vicinity of ω_{0k} with probability tending to unity for each k. Let these local minima be denoted as

$$\hat{\boldsymbol{\omega}} := [\hat{\omega}_1, \ldots, \hat{\omega}_p]^T, \tag{9.1.23}$$

where $0 < \hat{\omega}_1 < \cdots < \hat{\omega}_p < 2\pi$. Note that a local minimum of $|\hat{C}(\omega)|^2$ is also a local maximum of the corresponding AR spectrum. By Lemma 12.5.1, $\hat{\boldsymbol{\omega}} \xrightarrow{P} \boldsymbol{\omega}_0$ as $n \to \infty$ for sufficiently large but fixed m. This proves the following theorem.

Theorem 9.3 (Consistency of the Linear Prediction Estimators). *Let $\{y_t\}$ be given by (9.1.5), where $\{x_t\}$ takes the form (2.1.5) or (2.1.7). Assume that $\{\epsilon_t\}$ satisfies the conditions of Theorem 4.1. If $\hat{\mathbf{c}}$ denotes any of the estimators of $\mathbf{c}$ in (9.1.6), (9.1.9), or (9.1.12), and $\hat{\sigma}_m^2$ denotes the corresponding estimator of σ_m^2, then, for any fixed m, $\hat{\mathbf{c}} \xrightarrow{a.s.} \mathbf{c}$ and $\hat{\sigma}_m^2 \xrightarrow{a.s.} \sigma_m^2$ as $n \to \infty$. Moreover, let $\boldsymbol{\omega}_0$ and $\hat{\boldsymbol{\omega}}$ be defined by (9.1.22) and (9.1.23) for sufficiently large but fixed m. Then, $\hat{\boldsymbol{\omega}} \xrightarrow{P} \boldsymbol{\omega}_0$ as $n \to \infty$.*

Remark 9.1 Theorem 9.3 remains valid in the real case under the parameter transformation (2.1.8). For frequency estimation in the real case, it suffices to consider $\boldsymbol{\omega}_{0r} := [\omega_{01}, \ldots, \omega_{0q}]^T$ and $\hat{\boldsymbol{\omega}}_r := [\hat{\omega}_1, \ldots, \hat{\omega}_q]^T$ instead of $\boldsymbol{\omega}_0$ and $\hat{\boldsymbol{\omega}}$, where the subscript r is used to distinguish the real case from the complex case.

Theorem 9.3 does not necessarily imply that the signal frequencies can be estimated consistently by the local maxima of the estimated AR spectrum with fixed m because $\boldsymbol{\omega}_0$ differs from $\boldsymbol{\omega}$ in general for $p > 1$. In the case of $p = 1$, the consistency follows from the discussion in Example 9.1.

Similar to Theorem 7.5, the asymptotic normality of the linear prediction estimators of $\mathbf{c}$ can be derived for fixed m as $n \to \infty$. The asymptotic normality of the corresponding frequency estimators follows from Lemma 12.5.1. To present these results, let us define

$$\sigma(u,v) := \sum_{j=0}^{m}\sum_{l=0}^{m} c_j c_l^* \, \gamma(u-j, v-l), \tag{9.1.24}$$

$$\tilde{\sigma}(u,v) := \sum_{j=0}^{m}\sum_{l=0}^{m} c_j c_l \, \gamma(u-j, -v+l), \tag{9.1.25}$$

where $\gamma(\cdot,\cdot)$ is given by (4.3.1) and (4.3.2) in the real case, and by (4.3.3) and (4.3.4) in the complex case. Moreover, let

$$\begin{aligned}\mathbf{d}_k &:= \{2\Re(\ddot{C}(\omega)C^*(\omega)) + 2|\dot{C}(\omega)|^2\}^{-1} \\ &\quad \times \{\dot{C}(\omega_{0k})\mathbf{f}_m(\omega_{0k}) + C(\omega_{0k})\dot{\mathbf{f}}_m(\omega_{0k})\},\end{aligned} \tag{9.1.26}$$

where $C(\omega) := 1 + \mathbf{f}_m^H(\omega)\mathbf{c}$. With this notation, we have the following theorem. See Section 9.6 for a proof.

Theorem 9.4 (Asymptotic Normality of the Linear Prediction Estimators). *Let $\{y_t\}$ be given by (9.1.5), where $\{x_t\}$ takes the form (2.1.1) or (2.1.2) in the real case and the form (2.1.5) or (2.1.7) in the complex case. Let $\hat{\mathbf{c}}$ denote any of the estimators of $\mathbf{c}$ in (9.1.6), (9.1.9), and (9.1.12). Let the conditions of Theorem 4.3 be satisfied in the real case and the conditions of Theorem 4.4 be satisfied in the complex case. Then, for any fixed m as $n \to \infty$,*

$$\sqrt{n}\,(\hat{\mathbf{c}} - \mathbf{c}) \xrightarrow{D} \boldsymbol{\xi},$$

where $\boldsymbol{\xi} \sim \mathrm{N}(\mathbf{0}, \mathbf{R}_y^{-1}\boldsymbol{\Sigma}\mathbf{R}_y^{-1})$ in the real case and $\boldsymbol{\xi} \sim \mathrm{N}_c(\mathbf{0}, \mathbf{R}_y^{-1}\boldsymbol{\Sigma}\mathbf{R}_y^{-1}, \mathbf{R}_y^{-1}\tilde{\boldsymbol{\Sigma}}\mathbf{R}_y^{-T})$ in the complex case, with $\boldsymbol{\Sigma} := [\sigma(u,v)]$ and $\tilde{\boldsymbol{\Sigma}} := [\tilde{\sigma}(u,v)]$ $(u,v = 1,\ldots,m)$ given by (9.1.24) and (9.1.25). Moreover, let $\hat{\boldsymbol{\omega}}$ and $\hat{\boldsymbol{\omega}}_r$ denote the corresponding frequency estimators. Then, in the real case,

$$\sqrt{n}\,(\hat{\boldsymbol{\omega}}_r - \boldsymbol{\omega}_{0r}) \xrightarrow{D} \mathrm{N}(\mathbf{0}, \mathbf{D}_r^T\mathbf{R}_y^{-1}\boldsymbol{\Sigma}\mathbf{R}_y^{-1}\mathbf{D}_r),$$

and in the complex case,

$$\sqrt{n}\,(\hat{\boldsymbol{\omega}} - \boldsymbol{\omega}_0) \xrightarrow{D} \mathrm{N}(\mathbf{0}, 2\Re(\mathbf{D}^H\mathbf{R}_y^{-1}\boldsymbol{\Sigma}\mathbf{R}_y^{-1}\mathbf{D} + \mathbf{D}^H\mathbf{R}_y^{-1}\tilde{\boldsymbol{\Sigma}}\mathbf{R}_y^{-T}\mathbf{D}^*)),$$

where $\mathbf{D}_r := 2\Re[\mathbf{d}_1,\ldots,\mathbf{d}_q]$ and $\mathbf{D} := [\mathbf{d}_1,\ldots,\mathbf{d}_p]$ with $\mathbf{d}_k$ given by (9.1.26).

To illustrate this result, consider a simple example in the complex case.

Example 9.2 (Asymptotic Normality of the Linear Prediction Estimators of Order 1 for a Single Complex Sinusoid in Gaussian White Noise). Let $\{\epsilon_t\} \sim \mathrm{GWN}_c(0,\sigma^2)$ and $p = 1$. Then, for $m = p = 1$, we have

$$\mathbf{c} = c_1 := -\frac{\gamma_1}{\gamma_1 + 1}\exp(i\omega_1) = -\rho_y(1).$$

Consider the Yule-Walker estimator given by (9.1.6). It is easy to see that

$$\hat{\mathbf{c}} = \hat{c}_1 := -\hat{r}_y(1)/\hat{r}_y(0) = -\hat{\rho}_y(1).$$

By Theorem 4.1, $\hat{c}_1 \stackrel{a.s.}{\to} c_1$ as $n \to \infty$. By Theorem 4.5 and Example 4.6,

$$\sqrt{n}\begin{bmatrix} \Re(\hat{c}_1) - \Re(c_1) \\ \Im(\hat{c}_1) - \Im(c_1) \end{bmatrix} \stackrel{A}{\sim} \sqrt{n}\begin{bmatrix} \Re\{\hat{\rho}_y(1)\} - \Re\{\rho_y(1)\} \\ \Im\{\hat{\rho}_y(1)\} - \Im\{\rho_y(1)\} \end{bmatrix} \stackrel{D}{\to} \mathrm{N}(\mathbf{0}, \boldsymbol{\Sigma}_0),$$

$$\boldsymbol{\Sigma}_0 := \tfrac{1}{2}\lambda^2 \begin{bmatrix} 1 + 2\mu\cos^2(\omega_1) & 2\mu\sin(\omega_1)\cos(\omega_1) \\ 2\mu\sin(\omega_1)\cos(\omega_1) & 1 + 2\mu\sin^2(\omega_1) \end{bmatrix},$$

where $\lambda := 1/(\gamma_1 + 1)$ and $\mu := 1 - \lambda^2$. This result implies that the rate of convergence of $\hat{c}_1$ is $\sqrt{n}$ and the asymptotic variance of $\hat{c}_1$ is equal to

$$\tfrac{1}{2}\lambda^2\{(1 + 2\mu\cos^2(\omega_1)) + (1 + 2\mu\sin^2(\omega_1))\}/n = \lambda^4/n.$$

Because $\hat{C}(\omega) = 1 - \hat{\rho}_y(1)\exp(-i\omega)$, a unique minimizer of $|\hat{C}(\omega)|^2$ appears at

$$\hat{\omega}_1 := \angle\hat{\rho}_y(1) = \arctan(\Im\{\hat{\rho}_y(1)\}, \Re(\hat{\rho}_y(1)\}),$$

which defines the frequency estimator. Observe that the Jacobian of $\arctan(y, x)$ can be expressed as $\{1 + (y/x)^2\}^{-1}[-y/x^2, 1/x]^T$. At $x = \Re\{\rho_y(1)\}$ and $y = \Im\{\rho_y(1)\}$, it becomes $(\lambda\gamma_1)^{-1}[-\sin(\omega_1), \cos(\omega_1)]^T$. This result, combined with the asymptotic normality of $\hat{\rho}_y(1)$ and with Proposition 2.15, leads to

$$\sqrt{n}(\hat{\omega}_1 - \omega_1) \stackrel{D}{\to} \mathrm{N}(0, \tfrac{1}{2}\gamma_1^{-2}).$$

The same assertion holds for the other linear prediction estimators.

Of course, the asymptotic distribution of $\hat{\omega}_1$ can be obtained directly from Theorem 9.4. Indeed, with $m = p = 1$, it is easy to verify that

$$\begin{aligned} \mathbf{D} &= \mathbf{d}_1 = \frac{1 + \gamma_1}{2\gamma_1}\, i\exp(i\omega_1), \\ \boldsymbol{\Sigma} &= \sigma(1,1) = \sigma^4\frac{1 + 4\gamma_1 + 2\gamma_1^2}{(1 + \gamma_1)^2}, \\ \tilde{\boldsymbol{\Sigma}} &= \tilde{\sigma}(1,1) = \sigma^4\frac{2\gamma_1 + \gamma_1^2}{(1 + \gamma_1)^2}\exp(i2\omega_1). \end{aligned}$$

Combining these expressions with $\mathbf{R}_y = r_y(0) = \sigma^2(1 + \gamma_1)$ leads to

$$\begin{aligned} &2\Re(\mathbf{D}^H\mathbf{R}_y^{-1}\boldsymbol{\Sigma}\mathbf{R}_y^{-1}\mathbf{D} + \mathbf{D}^H\mathbf{R}_y^{-1}\tilde{\boldsymbol{\Sigma}}\mathbf{R}_y^{-T}\mathbf{D}^*) \\ &= \tfrac{1}{2}\sigma^{-4}\gamma_1^{-2}\Re\{\sigma(1,1) - \tilde{\sigma}(1,1)\exp(-i2\omega_1)\} = \tfrac{1}{2}\gamma_1^{-2}. \end{aligned}$$

This variance is the same as the one obtained previously. ◇

Example 9.2 shows that for estimating the frequency of a single complex sinusoid, the linear prediction method with $m = 1$ produces consistent estimates and its accuracy takes the form $\mathcal{O}_P(n^{-1/2})$. This statement remains true for the linear prediction method with $m > 1$. However, in the case of multiple sinusoids, the linear prediction method with fixed m is no longer consistent for frequency estimation, and hence the accuracy can only be expressed as $\mathcal{O}_P(1)$.

Theorems 9.1–9.4 indicate that the linear prediction estimators do not produce accurate frequency estimates unless the order m is sufficiently large. While a large m is required to reduce the bias, a much larger n is needed to ensure accurate estimation of the $m+1$ parameters in the AR model. A difficult theoretical problem is to determine the proper rate of growth for m as a function of n. This problem is tackled by Mackisack and Poskitt [244] [245] under the condition of a single real sinusoid ($p = 2q = 2$) in white noise. Their study suggests that for consistent estimation of the signal frequency, the order m should satisfy the condition that $m \to \infty$ but $m^2/n \to 0$; in other words, m should not grow faster than $\sqrt{n}$, or equivalently, n should increase faster than m^2. The study also suggests that if m takes the form $\mathcal{O}(n^r)$ for some constant $r \in (1/6, 1/2)$ so that m grows faster than $n^{1/6}$ but more slowly than $\sqrt{n}$, then the resulting frequency estimator can achieve an accuracy of the form $\mathcal{O}_P(n^{-\nu})$, where $\nu := 5/4 - 3(1/2 - r)/2$ ranges from $3/4$ to $5/4$. With $r \in (1/3, 1/2)$, this error rate is superior to the error rate of the DFT estimator, which takes the form $\mathcal{O}_P(n^{-1})$, but inferior to the error rate of the PM and NLS estimators, which takes the form $\mathcal{O}_P(n^{-3/2})$.

The statistical performance of the linear prediction method for multiple frequency estimation with finite sample sizes is demonstrated by the following two examples with different choices of m. In these examples, the frequency estimates are obtained numerically by using a fixed point iteration with the modified Newton-Raphson mapping

$$\omega \mapsto \phi(\omega) := \omega - \varrho g(\omega)/h(\omega), \tag{9.1.27}$$

where $g(\omega) := 2\Re\{\hat{C}(\omega)\dot{\hat{C}}^*(\omega)\}$, $h(\omega) := 2|\dot{\hat{C}}(\omega)|^2$, and $\hat{C}(\omega) := 1 + \mathbf{f}_m^H(\omega)\hat{\mathbf{c}}$, with $\varrho > 0$ being the step-size parameter. Note that $g(\omega)$ is the first derivative of $|\hat{C}(\omega)|^2$ and $h(\omega)$ is an approximation to the second derivative of $|\hat{C}(\omega)|^2$ obtained by dropping a term that involves the second derivative of $\hat{C}(\omega)$. Therefore, the resulting fixed point iteration converges to a local minimum of $|\hat{C}(\omega)|^2$ which is also a local maximum of the AR spectrum $\hat{f}_{\text{AR}}(\omega) := \hat{\sigma}_m^2/|\hat{C}(\omega)|^2$. As compared with the ordinary Newton-Raphson mapping, the fixed point mapping defined by (9.1.27) has a larger basin of attraction. The same advantage is possessed by the modified Newton-Raphson mapping in (6.4.11) for periodogram maximization.

Example 9.3 (Estimation of Well-Separated Frequencies: Part 1). Consider the time series in Example 8.1 which consists of two unit-amplitude complex sinusoids with well-separated frequencies in Gaussian white noise. Let the signal fre-

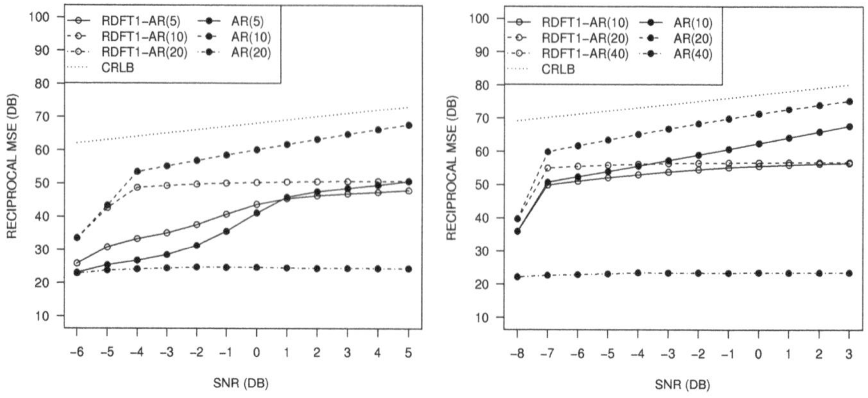

(a) (b)

Figure 9.4. Plot of the reciprocal average MSE of the FBLP frequency estimates for the time series of two complex sinusoids in Gaussian white noise discussed in Example 9.3 with different values of SNR. (a) $n = 100$: solid line, $m = 5$; dashed line, $m = 10$; dash-dotted line, $m = 20$. (b) $n = 200$: solid line, $m = 10$; dashed line, $m = 20$; dash-dotted line, $m = 40$. Lines with dots, final FBLP estimates; lines with circles, initial RDFT1 estimates. Dotted line depicts the CRLB. Results are based on 5,000 Monte Carlo runs.

quencies be estimated by the two largest local maxima of the AR spectrum produced by the FBLP estimator in (9.1.12). The frequency estimates are obtained by the fixed point iteration of the modified Newton-Raphson mapping in (9.1.27) with $\varrho = 0.5$. The initial values for the fixed point iteration are produced by the DFT method discussed in Example 8.1 (or the RDFT1 method discussed in Example 6.9) except that the discrete periodogram is replaced by the estimated AR spectrum evaluated at the Fourier frequencies. Figure 9.4 depicts the average MSE of the FBLP frequency estimates after 5 iterations and of the initial RDFT1 estimates with different choices of m and for different values of SNR.

As can be seen, the estimates are poor when m is either too small or too large (e.g., $m = 5$ or $m = 20$ for $n = 100$) due to the excessive bias or the obstructive spurious spectral peaks, respectively. Good results are produced with $m = 10$ for $n = 100$ and $m = 20$ for $n = 200$, in which case the FBLP estimator outperforms the DFT (RDFT1) estimator, and the amount of improvement increases with the SNR. However, regardless of the choice of m, a large gap remains between the MSE and the CRLB. In comparison with Figures 6.13 and 8.3 for the same data, one can see that the FBLP estimator is much less efficient than the PM and NLS estimators, although the threshold effect occurs at the same SNR. ◇

Example 9.4 (Estimation of Closely Spaced Frequencies: Part 1). Consider the time series in Example 8.2 which consists of two unit-amplitude complex sinusoids with closely spaced frequencies in Gaussian white noise. Let the FBLP

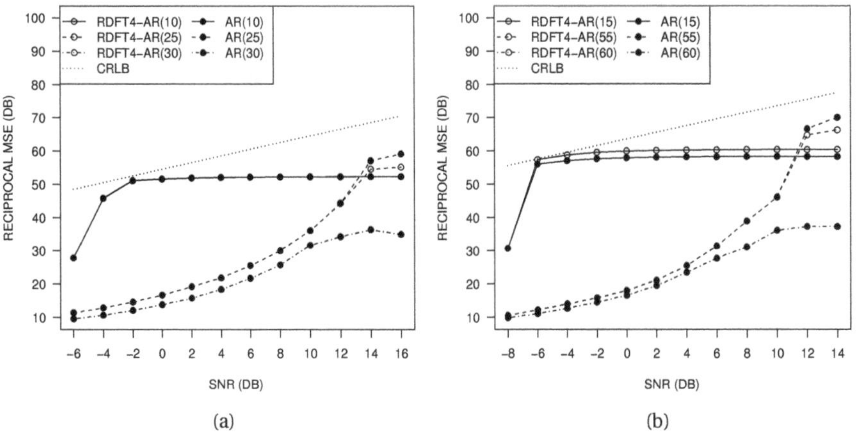

Figure 9.5. Plot of the reciprocal average MSE of the FBLP frequency estimates for the time series of two complex sinusoids with closely spaced frequencies in Gaussian white noise discussed in Example 9.4 with different values of SNR. (a) $n = 100$: solid line, $m = 10$; dashed line, $m = 25$; dash-dotted line, $m = 30$. (b) $n = 200$: solid line, $m = 15$; dashed line, $m = 55$; dash-dotted line, $m = 60$. Lines with dots, final FBLP estimates; lines with circles, initial RDFT4 estimates (in many cases indistinguishable from the final estimates). Dotted line depicts the CRLB. Results are based on 5,000 Monte Carlo runs.

frequency estimates be computed by the modified Newton-Raphson algorithm described in Example 9.3 except that the RDFT1 estimates are replaced by the RDFT4 estimates based on the AR spectrum and with a reduced frequency separation criterion $\delta = \frac{1}{2}\pi/n$. Figure 9.5 depicts the average MSE of the FBLP frequency estimates after 5 iterations and of the initial RDFT4 estimates with different choices of m and for different values of SNR.

Observe that when m is too small ($m = 10$ for $n = 100$ and $m = 15$ for $n = 200$), the MSE does not increase with the SNR after it exceeds a certain threshold (i.e., -2 dB for $n = 100$ and -6 dB for $n = 200$). This is nearly identical to the result of a single-frequency estimator (not shown) that corresponds to the global maximum of the AR spectrum. When m is suitably large ($m = 25$ for $n = 100$ and $m = 55$ for $n = 200$), an improvement begins to occur at the very high end of the SNR range considered; but for the rest of the SNR range, the performance deteriorates considerably. A further increase of m (e.g., $m = 30$ for $n = 100$ and $m = 60$ for $n = 200$) only worsens the situation. A comparison with Figure 8.5 shows that the FBLP estimator with the best choice of m still has a much higher SNR threshold than the NLS estimator and remains less accurate than the NLS estimator when the SNR exceeds the threshold.

The AR spectrum with a small m (e.g., $m = 10$ for $n = 100$) tends to produce only a single peak near the two signal frequencies, but the spectral peak is so broad that the two frequencies obtained from the RDFT procedure remain in the neighborhood. With these frequencies as initial values, the modified Newton-Raphson algorithm converges to a single value corresponding to the global maximum. The AR spectrum produces sharper spectral peaks when m is increased (e.g., $m = 25$ for $n = 100$), making it possible to resolve closely spaced frequencies. But if the signal frequencies are still unresolved, as is often the case when the SNR is not sufficiently high, the sharpened spectral peak near the signal frequencies causes the RDFT procedure to select its second frequency from a spurious peak which is often far away from the signal frequencies. This gives rise to a significant degradation in the average MSE as compared to the single-value estimator. When m is too large (e.g., $m = 30$ for $n = 100$), the chance increases for the global maximum of the AR spectrum to be a spurious peak, leading to the performance deterioration even at relatively high SNR. ◇

9.2 Autoregressive Reparameterization

It was recognized a long time ago, dating back at least to the work of Prony in 1795 [299] and the work of Yule [434] and Walker [415] in modern times, that a sinusoidal signal $\{x_t\}$ of the form (2.1.1) or (2.1.5) satisfies a homogeneous autoregressive (AR) difference equation

$$\sum_{j=0}^{p} a_j x_{t-j} = 0 \qquad (t \in \mathbb{Z}). \tag{9.2.1}$$

In this equation, the AR coefficients a_j are defined by the signal frequencies ω_k through the p-degree polynomial

$$a(z) := \sum_{j=0}^{p} a_j z^{-j} := a_0 \prod_{k=1}^{p} (1 - z_k z^{-1}) \qquad (z \in \mathbb{C}), \tag{9.2.2}$$

where $z_k := \exp(i\omega_k)$ and $a_0 \neq 0$. To prove (9.2.1), it suffices to regard z^{-1} as the backward shift operator such that $z^{-1} x_t = x_{t-1}$ and observe that

$$x_t = \sum_{k=1}^{p} \beta_k z_k^t, \quad \sum_{j=0}^{p} a_j x_{t-j} = a(z) x_t,$$

$$(1 - z_k z^{-1}) z_k^t = z_k^t - z_k z_k^{t-1} = 0.$$

Because the ω_k are uniquely determined by the a_j, and vice versa, the problem of frequency estimation can be reformulated as the problem of estimating the equivalent AR coefficients in (9.2.1).

While a_0 can be any nonzero constant in general, it is useful to consider the case where $a_0 = 1$ (otherwise, consider the rescaled coefficients a_j/a_0). Under this condition, one can express (9.2.2) as a transformation from the frequency parameter $\boldsymbol{\omega} := [\omega_1, \ldots, \omega_p]^T$ to the AR parameter $\mathbf{a} := [a_1, \ldots, a_p]^T$, i.e.,

$$\mathbf{a} = \Phi_{\text{AR}}(\boldsymbol{\omega}). \tag{9.2.3}$$

This transform has a unique inverse $\boldsymbol{\omega} = \Phi_{\text{AR}}^{-1}(\mathbf{a})$ if $\mathbf{a}$ is restricted to the subspace of $\mathbb{C}^p$ such that the corresponding polynomial $a(z)$, defined by (9.2.2), has p distinct roots $z_1, \ldots, z_p$ on the unit circle, satisfying $\{\omega_k\} = \{\angle z_k\} \pmod{2\pi}$ and $0 < \omega_1 < \cdots < \omega_p < 2\pi$. Because the inverse is a continuous function of $\mathbf{a}$ [140], the transform in (9.2.3) establishes a continuous one-to-one correspondence (or homeomorphism) between $\boldsymbol{\omega} \in \Omega := \{\tilde{\boldsymbol{\omega}} : 0 < \tilde{\omega}_1 < \cdots < \tilde{\omega}_p < 2\pi\}$ and $\mathbf{a} \in \mathcal{A} := \Phi_{\text{AR}}(\Omega) \subset \mathbb{C}^p$. Therefore, the AR parameter $\mathbf{a} \in \mathcal{A}$ can be regarded as a reparameterization of the frequency parameter $\boldsymbol{\omega} \in \Omega$.

For example, in the case of $p = 1$, the transform in (9.2.3) reduces to

$$a_1 = -\exp(i\omega_1), \tag{9.2.4}$$

and the inverse transform is given by

$$\omega_1 = \angle z_1 \pmod{2\pi},$$

where $z_1 = -a_1$. For $p = 2$, the transform becomes

$$a_1 = -\{\exp(i\omega_1) + \exp(i\omega_2)\}, \quad a_2 = \exp\{i(\omega_1 + \omega_2)\}, \tag{9.2.5}$$

and its inverse takes the form

$$\{\omega_1, \omega_2\} = \{\angle z_1, \angle z_2\} \pmod{2\pi},$$

where $\{z_1, z_2\} = \frac{1}{2}\{-a_1 \pm (a_1^2 - 4a_2)^{1/2}\}$. The top panel of Figure 9.6 shows the correspondence in the case of $p = 2$ between the frequency parameter $\boldsymbol{\omega} = (\omega_1, \omega_2)$ and the AR parameter a_1 in (9.2.5) (a_2 resides on the unit circle).

A similar AR reparameterization can be obtained in the real case where $\{x_t\}$ is given by (2.1.1) or (2.1.2). It suffices to transform the RSM into a special CSM with $p := 2q$ according to (2.1.8). In this case, the roots $z_1, \ldots, z_p$ of the corresponding polynomial $a(z)$ defined by (9.2.2) appear in q conjugate pairs such that $z_{p-k+1} = z_k^*$ $(k = 1, \ldots, q)$. Therefore, we can rewrite (9.2.2) as

$$a(z) = \sum_{j=0}^{p} a_j z^{-j} = a_0 \prod_{k=1}^{q} (1 - 2\Re(z_k) z^{-1} + z^{-2}), \tag{9.2.6}$$

where $a_0 \neq 0$ and $\Re(z_k) = \cos(\omega_k)$ $(k = 1, \ldots, q)$.

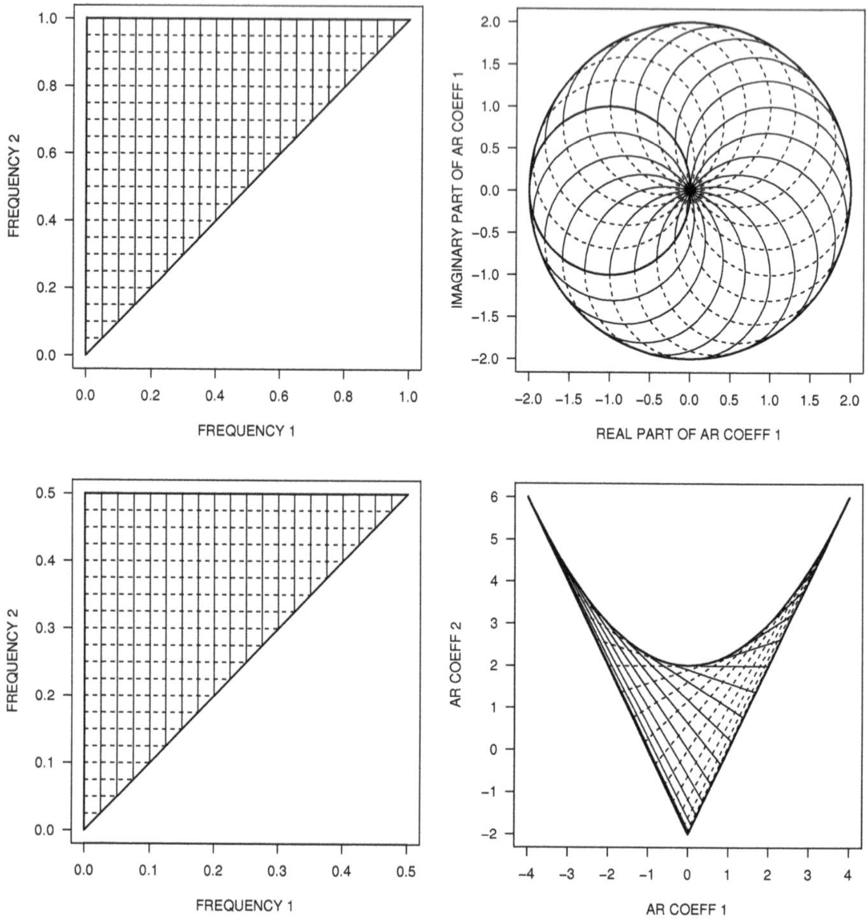

Figure 9.6. Mapping between frequency and AR parameter spaces. Top, complex case with $p = 2$, where the thick big circle on the right corresponds to the diagonal line on the left, and the thick small circle on the right corresponds to the vertical boundary as well as the horizontal boundary of the triangle on the left. Bottom, real case with $q = 2$, where the thick parabola on the right corresponds to the diagonal line on the left, and the left and right halves of the thick "v" on the right correspond, respectively, to the vertical and horizontal boundaries of the triangle on the left. In both cases, the solid lines on the left map to the solid lines on the right, and likewise for the dashed lines.

Under the constraint $a_0 = 1$, the coefficients of $a(z)$ in (9.2.6) are real and symmetric in the sense that

$$a_j = a_{p-j} \qquad (j = 0, 1, \ldots, q-1). \tag{9.2.7}$$

For example, with $q = 1$, we have $a(z) = a_0(1 - 2\Re(z_1)z^{-1} + z^{-2})$ and hence

$$\begin{aligned} a_0 &= a_2 = 1, \\ a_1 &= -2\Re(z_1) = -2\cos(\omega_1). \end{aligned}$$

For $q = 2$, we have $a(z) = a_0(1 - 2\Re(z_1)z^{-1} + z^{-2})(1 - 2\Re(z_2)z^{-1} + z^{-2})$, so

$$\begin{aligned} a_0 &= a_4 = 1, \\ a_1 &= a_3 = -2\{\Re(z_1) + \Re(z_2)\} = -2\{\cos(\omega_1) + \cos(\omega_2)\}, \\ a_2 &= 2\{1 + 2\Re(z_1)\Re(z_2)\} = 2\{1 + 2\cos(\omega_1)\cos(\omega_2)\}. \end{aligned}$$

In general, owing to the symmetry, $\mathbf{a} := [a_1, \ldots, a_p]^T$ is determined completely by

$$\mathbf{a}_r := [a_1, \ldots, a_q]^T$$

such that

$$\mathbf{a} = \mathbf{B}\mathbf{a}_r + \mathbf{b}, \quad \mathbf{B} := \begin{bmatrix} \mathbf{I}_{q-1} & \mathbf{0} \\ \mathbf{0}^T & 1 \\ \tilde{\mathbf{I}}_{q-1} & \mathbf{0} \\ \mathbf{0}^T & 0 \end{bmatrix}, \quad \mathbf{b} := \begin{bmatrix} \mathbf{0} \\ 1 \end{bmatrix}, \tag{9.2.8}$$

where $\mathbf{I}_{q-1}$ denotes the $(q-1)$-dimensional identity matrix and $\tilde{\mathbf{I}}_{q-1}$ denotes the $(q-1)$-dimensional antidiagonal identity matrix obtained by reversing the rows (or columns) of $\mathbf{I}_{q-1}$. Therefore, it suffices to consider the transformation from the frequency parameter $\boldsymbol{\omega}_r := [\omega_1, \ldots, \omega_q]^T$ to the AR parameter $\mathbf{a}_r$, denoted as

$$\mathbf{a}_r = \Phi_{\text{AR}}(\boldsymbol{\omega}_r). \tag{9.2.9}$$

In these expressions, the subscript r is used to distinguish the frequency and AR parameters for the real case from their counterparts for the complex case.

Similar to the complex case, the transform in (9.2.9) defines a one-to-one correspondence between $\boldsymbol{\omega}_r \in \Omega_r := \{\tilde{\boldsymbol{\omega}}_r : 0 < \tilde{\omega}_1 < \cdots < \tilde{\omega}_q < \pi\}$ and $\mathbf{a}_r \in \mathcal{A}_r := \Phi_{\text{AR}}(\Omega_r) \subset \mathbb{R}^q$. The bottom panel of Figure 9.6 depicts the mapping between the frequency parameter (ω_1, ω_2) and the AR parameter (a_1, a_2) in the case of $q = 2$.

When estimating multiple frequencies, the AR coefficients are rarely confined to the minimal parameter space $\mathcal{A}$ (or $\mathcal{A}_r$ in the real case) because the restriction would lead to difficult nonlinear problems. Instead, the estimation is often carried out by embedding the AR parameter in a larger space, such as $\mathbb{C}^p$ (or $\mathbb{R}^q$ in the real case), where simple solutions exist. Given an estimate $\hat{\mathbf{a}}$ in such a larger parameter space, the corresponding polynomial $\hat{a}(z)$ may not have all its roots on the unit circle, but the angles of these roots can still serve as estimates of the signal frequencies. This frequency estimator can be expressed symbolically as

$$\hat{\boldsymbol{\omega}} := \Phi_{\mathrm{AR}}^{-1}(\hat{\mathbf{a}}),$$

provided that the domain of $\Phi_{\mathrm{AR}}^{-1}(\cdot)$ is extended from $\mathcal{A}$ to the embedded parameter space, in which the mapping remains to be continuous [140]. In general, the inverse transform does not have a closed-form expression, so the roots have to be computed numerically with the help of a root-finding algorithm for complex polynomials [365, pp. 316–338]. In the remainder of this chapter, we discuss a number of methods for estimating the AR parameter, directly or indirectly, in such embedded spaces.

9.3 Extended Yule-Walker Method

It has been established in Section 9.1 that an AR spectrum of order p is inadequate for approximating the spectrum of $\{y_t\}$ given by (9.1.5), although it is shown in Section 9.2 that the sinusoidal signal $\{x_t\}$ of the form (2.1.5) or (2.1.7) satisfies the AR equation of order p in (9.2.1). The reason is that $\{y_t\}$ is not an AR process of any order even if $\{\epsilon_t\}$ is white noise. In fact, it is easy to verify, using (9.2.1), that $\{y_t\}$ actually satisfies the equation

$$\sum_{j=0}^{p} a_j y_{t-j} = \sum_{j=0}^{p} a_j \epsilon_{t-j} \qquad (t \in \mathbb{Z}). \tag{9.3.1}$$

Observe that the SDF of the process on the right-hand side of this equation takes the form $|A(\omega)|^2 f_\epsilon(\omega)$ with $A(\omega) := a(\exp(i\omega))$ which cannot be a positive constant for any noise spectrum $f_\epsilon(\omega)$ due to the fact that $A(\omega_k) = 0$ $(k = 1,\ldots,p)$. Therefore, the process on the right-hand side cannot be white noise as required by an AR model of order p regardless of the noise spectrum.

When $\{\epsilon_t\}$ is white noise, the process on the right-hand side of (9.3.1) is a moving-average (MA) process of order p. Hence, approximating $\{y_t\}$ by an AR model of order $m > p$ is equivalent to approximating the process on the right-hand side of (9.3.1) by an AR model of order $m - p$. It is also equivalent to approximating the pth-order all-zero filter $a(z)$, with all its zeros on the unit circle, by an $(m-p)$th-order all-pole filter $1/b(z)$, where $b(z) := \sum_{j=0}^{m-p} b_j z^{-j}$. Clearly, for an accurate approximation, the order m has to be very large indeed.

Throughout this section, we will assume that $\{\epsilon_t\}$ is white noise. Under this assumption, the equation in (9.3.1) can be viewed as defining a special autoregressive moving-average (ARMA) process where the roots of the AR and MA polynomials are on the unit circle. Unlike the general problem of parameter estimation for ARMA processes in which both AR and MA coefficients need to be estimated simultaneously, it suffices to estimate only the AR parameter $\mathbf{a} := [a_1, \ldots, a_p]^T$ because the MA parameter in (9.3.1) is identical to the AR parameter. A simple method for estimating the AR coefficients of an ARMA process is to use the extended (or modified) Yule-Walker equations [53] [58] [116].

9.3.1 Extended Yule-Walker Estimators

For convenience, let us randomize the phases of the sinusoids by assuming that the ϕ_k in (2.1.5) are i.i.d. random variables, uniformly distributed in $(-\pi, \pi]$ and independent of $\{\epsilon_t\}$. Under this assumption, the sinusoidal signal $\{x_t\}$ becomes a zero-mean stationary process with ACF $r_x(u) = \sum_{k=1}^{p} C_k^2 \exp(i\omega_k u)$ and hence $\{y_t\}$ becomes a zero-mean stationary process with ACF

$$r_y(u) := E(y_{t+u} y_t^*) = r_x(u) + r_\epsilon(u),$$

where $r_\epsilon(u)$ is the ACF of $\{\epsilon_t\}$. Because $\{\epsilon_t\}$ is white noise, it follows that for any $u > p$ and $j = 0, 1, \ldots, p$, $E(\epsilon_{t-j} y_{t-u}^*) = E(\epsilon_{t-j}\epsilon_{t-u}^*) = 0$. Therefore, by multiplying both sides of (9.3.1) with y_{t-u}^* and then taking the expected values, we obtain the *extended* Yule-Walker equations

$$\sum_{j=0}^{p} a_j r_y(u-j) = 0 \qquad (u = p+1, p+2, \ldots). \tag{9.3.2}$$

Unlike the ordinary Yule-Walker equations, the extended Yule-Walker equations do not depend on $r_y(0)$ which contains the noise variance.

Based on the extended Yule-Walker equations, a simple estimator of the AR parameter $\mathbf{a} := [a_1, \ldots, a_p]^T$ can be obtained by the method of moments, i.e., by replacing the ACF $r_y(u)$ in (9.3.2) with the sample autocovariances $\hat{r}_y(u)$ defined by (4.1.1) and solving any p of the resulting equations for $\mathbf{a}$. In particular, by taking the first p equations in (9.3.2), one obtains an extended Yule-Walker (EYW) estimator $\hat{\mathbf{a}}$ which satisfies

$$\hat{\mathbf{U}}_p \hat{\mathbf{a}} = -\hat{\mathbf{u}}_p, \tag{9.3.3}$$

where for any $l \ge p$,

$$\hat{\mathbf{U}}_l := \begin{bmatrix} \hat{r}_y(p) & \cdots & \hat{r}_y(1) \\ \vdots & & \vdots \\ \hat{r}_y(p+l-1) & \cdots & \hat{r}_y(l) \end{bmatrix}, \quad \hat{\mathbf{u}}_l := \begin{bmatrix} \hat{r}_y(p+1) \\ \vdots \\ \hat{r}_y(p+l) \end{bmatrix}.$$

This estimator can be computed with complexity $\mathcal{O}(p^2)$ by a fast algorithm, due to Trench [391], which utilizes the Toeplitz property of $\hat{\mathbf{U}}_p$.

Generally, it is helpful to take more than p equations in (9.3.2) to construct an estimator of $\mathbf{a}$. In particular, for any given l such that $p < l < n-p$, an estimator of $\mathbf{a}$ can be obtained from the first l equations by minimizing $\|\hat{\mathbf{U}}_l\,\tilde{\mathbf{a}} + \hat{\mathbf{u}}_l\|^2$ with respect to $\tilde{\mathbf{a}} \in \mathbb{C}^p$. The resulting EYW estimator $\hat{\mathbf{a}}$ satisfies

$$\hat{\mathbf{U}}_l^H \hat{\mathbf{U}}_l\,\hat{\mathbf{a}} = -\hat{\mathbf{U}}_l^H \hat{\mathbf{u}}_l. \tag{9.3.4}$$

Obviously, the EYW estimator defined by (9.3.3) can be regarded as a special case of the EYW estimator in (9.3.4) with $l = p$.

Example 9.5 (EYW Estimator for a Single Complex Sinusoid: Part 1). In the case of $p = 1$, we have $\mathbf{a} = a_1 = -z_1 = -\exp(i\omega_1)$. The EYW estimator of a_1 defined by (9.3.4) with l satisfying $1 \le l < n-1$ can be expressed as

$$\hat{a}_1 = -\frac{\sum_{j=1}^{l} \hat{r}_y^*(j)\hat{r}_y(j+1)}{\sum_{j=1}^{l} |\hat{r}_y(j)|^2}.$$

In the special case of $l = 1$, it becomes

$$\hat{a}_1 = -\hat{r}_y(2)/\hat{r}_y(1).$$

The frequency estimator is given by $\hat{\omega}_1 := \angle \hat{z}_1 \pmod{2\pi}$, where $\hat{z}_1 := -\hat{a}_1$. ◇

A similar EYW estimator can be derived in the real case where $\{x_t\}$ is given by (2.1.1) or (2.1.2). It suffices to impose the symmetry constraint (9.2.8) on $\mathbf{a} \in \mathbb{R}^p$ $(p := 2q)$ and solve the resulting extended Yule-Walker equations for $\mathbf{a}_r := [a_1, \ldots, a_q]^T \in \mathbb{R}^q$ with $\{\hat{r}_y(u)\}$ in place of $\{r_y(u)\}$. In particular, for any l such that $q \le l < n-2q$, the EYW estimator of $\mathbf{a}_r$ that minimizes $\|\hat{\mathbf{U}}_l\,\tilde{\mathbf{a}} + \hat{\mathbf{u}}_l\|^2$ under the constraint (9.2.8) satisfies

$$\mathbf{B}^T \hat{\mathbf{U}}_l^T \hat{\mathbf{U}}_l \mathbf{B}\,\hat{\mathbf{a}}_r = -\mathbf{B}^T \hat{\mathbf{U}}_l^T (\hat{\mathbf{U}}_l \mathbf{b} + \hat{\mathbf{u}}_l). \tag{9.3.5}$$

This is the counterpart of the unconstrained estimator $\hat{\mathbf{a}}$ defined by (9.3.4).

Example 9.6 (EYW Estimator for a Single Real Sinusoid: Part 1). In the case of $q = 1$, the EYW estimator of $a_1 = -2\cos(\omega_1)$ defined by (9.3.5) with $1 \le l < n-2$ can be expressed as

$$\hat{a}_1 = -\frac{\sum_{j=1}^{l} \hat{r}_y(j+1)\,\{\hat{r}_y(j) + \hat{r}_y(j+2)\}}{\sum_{j=1}^{l} |\hat{r}_y(j+1)|^2}.$$

For $l = 1$ in particular, we obtain

$$\hat{a}_1 = -\{\hat{r}_y(1) + \hat{r}_y(3)\}/\hat{r}_y(2).$$

The corresponding frequency estimator is given by $\hat{\omega}_1 := \arccos(-\tfrac{1}{2}\hat{a}_1)$. ◇

9.3.2 Statistical Properties

First, consider the consistency of the EYW estimators. Under the conditions of Theorem 4.1, we have $\hat{r}_y(u) \stackrel{a.s.}{\to} r_y(u)$ as $n \to \infty$. Therefore, for any fixed $l \ge p$,

$$\hat{\mathbf{U}}_l \stackrel{a.s.}{\to} \mathbf{U}_l, \quad \hat{\mathbf{u}}_l \stackrel{a.s.}{\to} \mathbf{u}_l, \tag{9.3.6}$$

where

$$\mathbf{U}_l := \begin{bmatrix} r_y(p) & \cdots & r_y(1) \\ \vdots & & \vdots \\ r_y(p+l-1) & \cdots & r_y(l) \end{bmatrix}, \quad \mathbf{u}_l := \begin{bmatrix} r_y(p+1) \\ \vdots \\ r_y(p+l) \end{bmatrix}.$$

Moreover, under the white noise assumption, $r_y(u) = r_x(u) = \sum_{k=1}^{p} C_k^2 z_k^u$ for all $u \neq 0$. Therefore,

$$\mathbf{U}_l = \sum_{k=1}^{p} C_k^2 z_k^p \mathbf{f}_l(\omega_k) \mathbf{f}_p^H(\omega_k) = \mathbf{G}_l \mathbf{F}_p^H, \tag{9.3.7}$$

where $\mathbf{F}_p := [\mathbf{f}_p(\omega_1), \dots, \mathbf{f}_p(\omega_p)]$ and $\mathbf{G}_l := [C_1^2 z_1^p \mathbf{f}_l(\omega_1), \dots, C_p^2 z_p^p \mathbf{f}_l(\omega_p)]$. Because $\mathbf{F}_p$ is nonsingular and $\mathbf{G}_l$ has full (column) rank p for any $l \ge p$, the matrix $\mathbf{U}_l$ also has full (column) rank p for any $l \ge p$. Thus, the EYW estimators in (9.3.3) and (9.3.4) are uniquely defined almost surely for large n. Moreover, it follows from (9.3.2) that the AR parameter $\mathbf{a}$ is the unique solution to

$$\mathbf{U}_l \mathbf{a} = -\mathbf{u}_l. \tag{9.3.8}$$

Combining these results with (9.3.6) proves the following theorem.

Theorem 9.5 (Consistency of the EYW Estimators). *Let $\{y_t\}$ be given by (9.1.5), where $\{x_t\}$ takes the form (2.1.5) or (2.1.7). Let $\hat{\mathbf{a}}$ be defined by (9.3.4) and let $\hat{\boldsymbol{\omega}} := \Phi_{\text{AR}}^{-1}(\hat{\mathbf{a}})$. Assume that $\{\epsilon_t\} \sim \text{IID}(0, \sigma^2)$ and $E(|\epsilon_t|^4) < \infty$. Then, as $n \to \infty$, $\hat{\mathbf{a}} \stackrel{a.s.}{\to} \mathbf{a}$ and $\hat{\boldsymbol{\omega}} \stackrel{a.s.}{\to} \boldsymbol{\omega}$ for any fixed $l \ge p$.*

Remark 9.2 A similar assertion can be made about $\hat{\mathbf{a}}_r$ in (9.3.5) because (9.3.7) and (9.3.8) remain valid in the real case owing to the correspondence between the RSM and the CSM defined by (2.1.8). However, the matrix $\mathbf{U}_l\mathbf{B}$ does not always have full (column) rank q when $l \ge q$. For example, in the case of $l = q = 1$, we have $\mathbf{U}_l\mathbf{B} = r_x(2) = \frac{1}{2}C_1^2 \cos(2\omega_1)$, which is equal to zero when $\omega_1 = \pi/4$ or $3\pi/4$. Therefore, the EYW estimator in (9.3.5) may not be uniquely defined, even for large n. Nonetheless, when $\mathbf{U}_l\mathbf{B}$ does have full rank q, and if $\{\epsilon_t\}$ satisfies the conditions of Theorem 9.5, then we have $\hat{\mathbf{a}}_r \stackrel{a.s.}{\to} \mathbf{a}_r$ and $\hat{\boldsymbol{\omega}}_r := \Phi_{\text{AR}}^{-1}(\hat{\mathbf{a}}_r) \stackrel{a.s.}{\to} \boldsymbol{\omega}_r$ as $n \to \infty$ for any fixed $l \ge q$.

By comparing Theorem 9.5 with Theorem 9.3, we can see that the EYW method produces consistent frequency estimates for any fixed $l \ge p$, whereas the linear prediction method is generally inconsistent under the same white noise assumption if the order m is fixed as $n \to \infty$. The consistency of the EYW method is due entirely to its use of the correct parametric model (9.3.1) that leads to the extended Yule-Walker equations in (9.3.2).

Example 9.7 (EYW Estimator for a Single Complex Sinusoid: Part 2). In the case of $p = 1$, we have $a_1 = -\exp(i\omega_1)$. Under the conditions of Theorem 9.5, $\hat{r}_y(u) \stackrel{a.s.}{\to} r_x(u) = C_1^2 \exp(i\omega_1 u)$ as $n \to \infty$ for all $u \neq 0$. This implies that for any $l \geq 1$ the EYW estimator in Example 9.5 has the property

$$\hat{a}_1 \stackrel{a.s.}{\to} -\frac{\sum_{j=1}^{l} r_x^*(j) r_x(j+1)}{\sum_{j=1}^{l} |r_x(j)|^2} = -\exp(i\omega_1) = a_1.$$

As a result, we have $\hat{z}_1 := -\hat{a}_1 \stackrel{a.s.}{\to} -a_1 = z_1$ and $\hat{\omega}_1 := \angle \hat{z}_1 \stackrel{a.s.}{\to} \angle z_1 = \omega_1$. ◇

Example 9.8 (EYW Estimator for a Single Real Sinusoid: Part 2). In the real case with $q = 1$, we have $a_1 = -2\cos(\omega_1)$. Under the conditions of Theorem 9.5, $\hat{r}_y(u) \stackrel{a.s.}{\to} r_x(u) = \frac{1}{2} C_1^2 \cos(\omega_1 u)$ as $n \to \infty$ for all $u \neq 0$. Therefore, for any $l \geq 1$, the EYW estimator in Example 9.6 has the property

$$\begin{aligned}
\hat{a}_1 &\stackrel{a.s.}{\to} -\frac{\sum_{j=1}^{l} r_x(j+1)\{r_x(j) + r_x(j+2)\}}{\sum_{j=1}^{l} |r_x(j+1)|^2} \\
&= -\frac{\sum_{j=1}^{l} \cos((j+1)\omega_1)\{\cos(j\omega_1) + \cos((j+2)\omega_1)\}}{\sum_{j=1}^{l} \cos^2((j+1)\omega_1)} \\
&= -\frac{\sum_{j=1}^{l} \cos((j+1)\omega_1)\{2\cos((j+1)\omega_1)\cos(\omega_1)\}}{\sum_{j=1}^{l} \cos^2((j+1)\omega_1)} \\
&= -2\cos(\omega_1),
\end{aligned}$$

which, in turn, implies that $\hat{\omega}_1 := \arccos(-\frac{1}{2}\hat{a}_1) \stackrel{a.s.}{\to} \arccos(-\frac{1}{2}a_1) = \omega_1$. This result requires the condition $\sum_{j=1}^{l} \cos^2((j+1)\omega_1) > 0$, which is always satisfied if $l \geq 2$. For $l = 1$, the condition is satisfied when $\omega_1 \neq \pi/4, 3\pi/4$. ◇

Now let us investigate the asymptotic distribution of the EYW estimators. First, consider $\hat{\mathbf{a}}_r$ defined by (9.3.5). Recall that in the real case $\{x_t\}$ satisfies (9.2.1) with $p := 2q$, $a_0 := 1$, and with $\mathbf{a} := [a_1, \ldots, a_p]^T$ determined by $\mathbf{a}_r$ according to (9.2.8). For any given $l \geq q$, define the l-by-$(p+l)$ matrix

$$\mathbf{A}_l := \begin{bmatrix} a_p & a_{p-1} & \cdots & a_0 & 0 & \cdots & 0 \\ 0 & a_p & \cdots & a_1 & a_0 & & 0 \\ & & \ddots & & & \ddots & \\ 0 & \cdots & & 0 & a_p & \cdots & a_1 \;\; a_0 \end{bmatrix}. \tag{9.3.9}$$

For $k = 1, \ldots, q$, define $z_{p-k+1} := z_k^*$ and

$$\mathbf{d}_k := \cos(\omega_k)\Im(\mathbf{v}_k) - \sin(\omega_k)\Re(\mathbf{v}_k), \tag{9.3.10}$$

where

$$\mathbf{v}_k := \left\{ \prod_{\substack{k'=1 \\ k' \neq k}}^{p} (1 - z_{k'} z_k^{-1}) \right\}^{-1} [1, z_k^{-1}, \ldots, z_k^{-q+1}]^T. \tag{9.3.11}$$

Then, the following theorem can be established as a result of Theorem 4.3 and Lemma 12.5.2. A proof is given in Section 9.6.

Theorem 9.6 (Asymptotic Normality of the EYW Estimators: Real Case). *Let $\{y_t\}$ be given by (9.1.5), where $\{x_t\}$ takes the form (2.1.1) or (2.1.2) and $\{\epsilon_t\}$ satisfies the assumption in Theorem 9.5. Let $\hat{\mathbf{a}}_r$ be defined by (9.3.5). Assume that $\mathbf{U}_l\mathbf{B}$ has full column rank q. Then, as $n \to \infty$,*

$$\sqrt{n}(\hat{\mathbf{a}}_r - \mathbf{a}_r) \xrightarrow{D} \mathrm{N}(\mathbf{0}, \mathbf{V}_l),$$

where

$$\mathbf{V}_l := \sigma^4(\mathbf{B}^T\mathbf{U}_l^T\mathbf{U}_l\mathbf{B})^{-1}\mathbf{B}^T\mathbf{U}_l^T\mathbf{A}_l\mathbf{A}_l^T\mathbf{U}_l\mathbf{B}(\mathbf{B}^T\mathbf{U}_l^T\mathbf{U}_l\mathbf{B})^{-1}$$

and $\mathbf{A}_l$ is defined by (9.3.9). Moreover, let $\hat{\boldsymbol{\omega}}_r := \Phi_{\mathrm{AR}}^{-1}(\hat{\mathbf{a}}_r)$, Then, as $n \to \infty$,

$$\sqrt{n}(\hat{\boldsymbol{\omega}}_r - \boldsymbol{\omega}_r) \xrightarrow{D} \mathrm{N}(\mathbf{0}, \mathbf{D}^T\mathbf{B}\mathbf{V}_l\mathbf{B}^T\mathbf{D}),$$

where $\mathbf{D} := [\mathbf{d}_1, \ldots, \mathbf{d}_q]$ with $\mathbf{d}_k$ given by (9.3.10).

Here is a simple example that illustrates Theorem 9.6.

Example 9.9 (EYW Estimator for a Single Real Sinusoid: Part 3). In the real case with $q = 1$, we have $a_1 = -2\cos(\omega_1)$, $a_2 = a_0 = 1$, and $r_x(u) = \frac{1}{2}C_1^2\cos(\omega_1 u)$. Let $\mathbf{x}_l := [\cos(2\omega_1), \ldots, \cos((l+1)\omega_1)]^T$. Then,

$$\mathbf{U}_l\mathbf{B} = [r_x(2), \ldots, r_x(l+1)]^T = \tfrac{1}{2}C_1^2\mathbf{x}_l.$$

Under the assumption that $\mathbf{x}_l \neq \mathbf{0}$, we obtain

$$\mathbf{V}_l = v_l^2 := \frac{\mathbf{x}_l^T\mathbf{A}_l\mathbf{A}_l^T\mathbf{x}_l}{\gamma_1^2\|\mathbf{x}_l\|^4},$$

where $\gamma_1 := \frac{1}{2}C_1^2/\sigma^2$ is the SNR. According to Theorem 9.6, the EYW estimator $\hat{a}_1$ in Example 9.6 has the property $\sqrt{n}(\hat{a}_1 - a_1) \xrightarrow{D} \mathrm{N}(0, v_l^2)$. Note that $\mathbf{x}_l \neq \mathbf{0}$ if $l \geq 2$. In the case of $l = 1$, $\mathbf{x}_1 = \cos(2\omega_1) \neq 0$ if and only if $\omega_1 \neq \pi/4, 3\pi/4$. Under this condition, together with the fact that $\mathbf{A}_1\mathbf{A}_1^T = 2 + a_1^2 = 2 + 4\cos^2(\omega_1)$, we obtain

$$v_1^2 = \frac{2 + 4\cos^2(\omega_1)}{\gamma_1^2\cos^2(2\omega_1)}.$$

Note that v_1^2 becomes unbounded when ω_1 approaches $\pi/4$ or $3\pi/4$. ◇

A similar result can be obtained in the complex case on the basis of Theorem 4.4 and Lemma 12.5.2. See Section 9.6 for a proof.

Theorem 9.7 (Asymptotic Normality of the EYW Estimators: Complex Case). *Let* $\{y_t\}$ *be given by (9.1.5), where* $\{x_t\}$ *takes the form (2.1.5) or (2.1.7) and* $\{\epsilon_t\}$ *satisfies the assumption in Theorem 9.5 and the additional assumption that* $E(\epsilon_t^2) = 0$. *Let* $\hat{\mathbf{a}}$ *be defined by (9.3.4). Then, as* $n \to \infty$,

$$\sqrt{n}\,(\hat{\mathbf{a}} - \mathbf{a}) \xrightarrow{D} \mathrm{N}_c(\mathbf{0}, \mathbf{V}_l),$$

where

$$\mathbf{V}_l := \sigma^4 (\mathbf{U}_l^H \mathbf{U}_l)^{-1} \mathbf{U}_l^H \mathbf{A}_l \mathbf{A}_l^H \mathbf{U}_l (\mathbf{U}_l^H \mathbf{U}_l)^{-1}.$$

Moreover, let $\hat{\boldsymbol{\omega}} := \Phi_{\mathrm{AR}}^{-1}(\hat{\mathbf{a}})$. *Then, as* $n \to \infty$,

$$\sqrt{n}\,(\hat{\boldsymbol{\omega}} - \boldsymbol{\omega}) \xrightarrow{D} \mathrm{N}(\mathbf{0}, \boldsymbol{\Sigma}_l),$$

where $\boldsymbol{\Sigma}_l := [\sigma_l(k,k')]$ $(k, k' = 1, \ldots, p)$ *and*

$$\begin{aligned}\sigma_l(k,k') := \tfrac{1}{2}\{&\sin(\omega_k)\Re(\mathbf{v}_k^H \mathbf{V}_l \mathbf{v}_{k'}^*)\sin(\omega_{k'}) + \cos(\omega_k)\Re(\mathbf{v}_k^H \mathbf{V}_l \mathbf{v}_{k'}^*)\cos(\omega_{k'}) \\ &+ \sin(\omega_k)\Im(\mathbf{v}_k^H \mathbf{V}_l \mathbf{v}_{k'}^*)\cos(\omega_{k'}) - \cos(\omega_k)\Im(\mathbf{v}_k^H \mathbf{V}_l \mathbf{v}_{k'}^*)\sin(\omega_{k'})\},\end{aligned}$$

with $\mathbf{v}_k$ *defined by (9.3.11).*

According to Theorems 9.6 and 9.7, the error rate of the EYW estimators is merely $\sqrt{n}$ when l is fixed as $n \to \infty$. The following simulation example demonstrates that for finite sample sizes and well-separated signal frequencies the accuracy can be improved by taking a large l which is comparable to n.

Example 9.10 (Estimation of Well-Separated Frequencies: Part 2). Consider the time series in Example 9.3 which consists of two unit-amplitude complex sinusoids with well-separated frequencies in Gaussian white noise. Let the signal frequencies be estimated by the EYW estimator in (9.3.4). Figure 9.7 depicts the average MSE of the EYW frequency estimates with different choices of l and for different values of SNR and two sample sizes.

This result shows that the accuracy of the EYW estimator is very poor when l is too small, but it can be improved considerably by taking a suitably large l. Although a slightly better result can be obtained by fine tuning (e.g., $l = 45$ for $n = 100$ and $l = 90$ for $n = 200$), a simple choice of l is the largest possible value $n - p - 1$ ($l = 97$ for $n = 100$ and $l = 197$ for $n = 200$). With this choice, the EYW estimator outperforms the DFT (RDFT1) estimator shown in Figure 6.7 when the SNR is sufficiently high. It also performs better than the FBLP estimator shown in Figure 9.4 if the latter is obtained with an order m too large or too small, but it performs worse than the FBLP estimator with a suitably selected order (e.g., $m = 10$ for $n = 100$ and $m = 20$ for $n = 200$). ◇

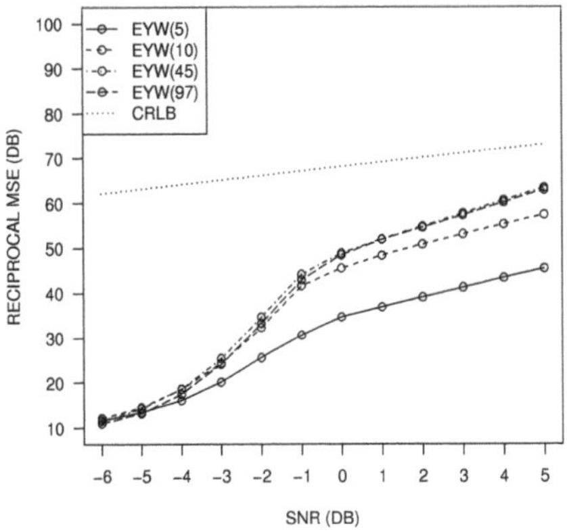

(a)

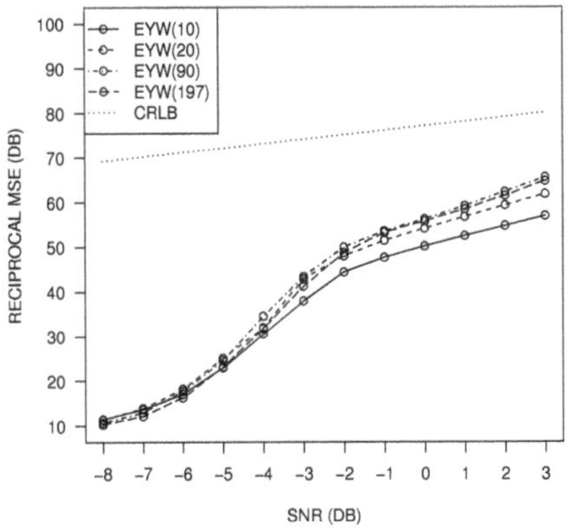

(b)

Figure 9.7. Plot of the reciprocal average MSE of the EYW frequency estimates for the time series of two complex sinusoids in Gaussian white noise discussed in Example 9.10 with different values of SNR. (a) $n = 100$: solid line, $l = 5$; dashed line, $l = 10$; dash-dotted line, $l = 45$; long dashed line, $l = 97$. (b) $n = 200$: solid line, $l = 10$; dashed line, $l = 20$; dash-dotted line, $l = 90$; long dashed line, $l = 197$. Dotted line depicts the CRLB. Results are based on 5,000 Monte Carlo runs.

The next simulation example shows that for estimating closely spaced frequencies, in which case Theorems 9.5–9.7 are no longer valid, the EYW method performs rather poorly, regardless of the choice of l.

Example 9.11 (Estimation of Closely Spaced Frequencies: Part 2). Consider the time series in Example 9.4 which consists of two unit-amplitude complex sinusoids with closely spaced frequencies in Gaussian white noise. Let the signal frequencies be estimated by the EYW estimator in (9.3.4). The simulation results are shown in Figure 9.8.

In this example, the EYW estimator performs poorly. Although the accuracy can be improved by taking a large l, significant improvement is achieved only at very high SNR. A closer examination reveals that one of the two frequencies produced by the EYW estimator stays close to the signal frequencies but the other tends be far away from them (i.e., severely biased) and thereby drives up the average MSE. Were the former regarded as a single-value estimate for both signal frequencies, the resulting MSE would behave similarly to that of the FBLP estimator shown in Figure 9.5 with the choice of $m = 10$ for $n = 100$ and $m = 15$ for $n = 200$, which also yields a single frequency. ◇

These examples show that the EYW method, though consistent and computationally attractive, lacks accuracy for frequency estimation. A considerable improvement can be achieved by employing higher-order AR models, similar to the

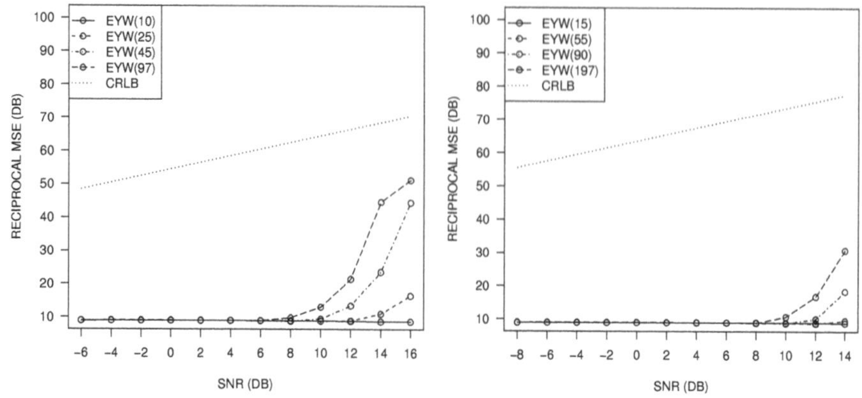

(a) (b)

Figure 9.8. Plot of the reciprocal average MSE of the EYW frequency estimates for the time series of two complex sinusoids with closely spaced frequencies in Gaussian white noise discussed in Example 9.11 with different values of SNR. (a) $n = 100$: solid line, $l = 10$; dashed line, $l = 25$; dash-dotted line, $l = 45$; long dashed line, $l = 97$. (b) $n = 200$: solid line, $l = 15$; dashed line, $l = 55$; dash-dotted line, $l = 90$; long dashed line, $l = 197$. Dotted line depicts the CRLB. Results are based on 5,000 Monte Carlo runs.

linear prediction method, together with a technique to reduce the negative impact of overparameterization. The same technique also helps to improve the linear prediction method. This will be discussed later in Chapter 10.

9.4 Iterative Filtering Method

As discussed in Section 9.3, the noisy signal $\{y_t\}$ satisfies the special ARMA equation (9.3.1) and therefore cannot be adequately handled by an AR model of order p. The extended Yule-Walker method discussed in Section 9.3 explores the ARMA equation by considering the autocovariances of higher lags. In this section, let us consider another method, called iterative filtering, that takes advantage of the ARMA equation from a different perspective.

Observe that the ARMA equation (9.3.1) can be expressed as

$$a(z)y_t = \xi_t, \tag{9.4.1}$$

where $\xi_t := \sum_{j=0}^{p} a_j \epsilon_{t-j} = a(z)\epsilon_t$. An AR model of order p will not fit $\{y_t\}$ very well because $\{\xi_t\}$ is not white noise. In fact, it is an MA process when $\{\epsilon_t\}$ is white noise, as will be assumed throughout this section. Intuitively, one could whiten the right-hand side of (9.4.1) by applying the inverse filter $1/a(z)$ to $\{y_t\}$. Indeed, by regarding $\{y_t\}$ as the output of a linear filter $h(z)$ with $\{\xi_t\}$ as the input, i.e., $y_t = h(z)\xi_t$, one can write

$$\tilde{y}_t := \{1/a(z)\}y_t = \{1/a(z)\}h(z)\xi_t = h(z)\{1/a(z)\}\xi_t = h(z)\epsilon_t.$$

Combining this expression with the fact that $h(z) = 1/a(z)$ leads to

$$a(z)\tilde{y}_t = a(z)h(z)\epsilon_t = \epsilon_t.$$

Observe that the right-hand side is now a white noise process. This intuitive argument suggests that an AR model of order p should fit the filtered process $\{\tilde{y}_t\}$ better than it does the original process $\{y_t\}$. Of course, the above procedure is not practical because the inverse filter $1/a(z)$ depends on the unknown AR parameter. This problem can be solved when a preliminary estimate of the AR parameter is available, in which case an approximate inverse filter can be constructed to produce the filtered data. Given the filtered data, a new, and hopefully better, estimate of the AR parameter can be obtained. This logic leads to the idea of iterative filtering (IFIL) which can be traced to [176] and [360].

In the following, we will discuss the IFIL method separately for the complex and real cases under the assumption of white noise.

9.4.1 Iterative Filtering Estimator: Complex Case

Consider the complex case first. Suppose that $\tilde{\mathbf{a}} := [\tilde{a}_1, \dots, \tilde{a}_p]^T$ is a preliminary estimate of $\mathbf{a} := [a_1, \dots, a_p]^T$ and let $\tilde{a}(z) := \sum_{j=0}^{p} \tilde{a}_j z^{-j}$ ($\tilde{a}_0 := 1$). Applying the inverse filter $1/\tilde{a}(z)$ to $\{y_t\}$ produces the filtered process

$$y_t(\tilde{\mathbf{a}}) := \{1/\tilde{a}(z)\}y_t. \tag{9.4.2}$$

Fitting an AR model of order p to $\{y_t(\tilde{\mathbf{a}})\}$ by the linear prediction techniques discussed in Chapter 7 (Section 7.1.3) yields a new estimate of $\mathbf{a}$, denoted as

$$\tilde{\mathbf{a}}' := \hat{\boldsymbol{\varphi}}(\tilde{\mathbf{a}}). \tag{9.4.3}$$

Repeating the cycle of inverse filtering in (9.4.2) and AR estimation in (9.4.3) constitutes an iterative procedure. Under the right conditions, the sinusoidal signal should be enhanced by the inverse filter and more accurate estimates should be produced as the iteration progresses.

For practical implementation of the IFIL method, two computational issues need to be addressed more carefully. First, because the roots of $\tilde{a}(z)$ tend to appear near the unit circle, the inverse filter in (9.4.2) may not be stable. One way to alleviate the instability problem is to employ suitable AR estimators, such as Burg's estimator discussed in Chapter 7 (Section 7.1.3), for which the roots of

$\tilde{a}(z)$ cannot be outside the unit circle. This is the method adopted in [176]; see also [277] for additional comments. Alternatively, one may replace $\tilde{a}(z)$ by

$$\tilde{a}(\eta^{-1}z) = \sum_{j=0}^{p} \tilde{a}_j \eta^j z^{-j},$$

where $\eta \in (0,1)$ is a shrinkage parameter. If $\tilde{z}_1,\ldots,\tilde{z}_p$ are the roots of $\tilde{a}(z)$, then the roots of $\tilde{a}(\eta^{-1}z)$ are $\eta\tilde{z}_1,\ldots,\eta\tilde{z}_p$, so the poles of the inverse filter $1/\tilde{a}(\eta^{-1}z)$ are shrunk toward the origin along their radius, making the inverse filter more stable. Because shrinking the poles toward the origin widens their bandwidth, one may also refer to η as a bandwidth parameter [234].

The second issue is that with a finite data record $\{y_1,\ldots,y_n\}$ one can only obtain an approximation of the filtered process $\{y_t(\tilde{\mathbf{a}})\}$ through the recursion

$$\hat{y}_t(\tilde{\mathbf{a}}) = -\sum_{j=1}^{p} \tilde{a}_j \eta^j \hat{y}_{t-j}(\tilde{\mathbf{a}}) + y_t \qquad (t=1,\ldots,n), \tag{9.4.4}$$

where the initial values are set to zero, i.e., $\hat{y}_t(\tilde{\mathbf{a}}) := 0$ for $t \le 0$. Therefore, the estimator $\hat{\boldsymbol{\varphi}}(\tilde{\mathbf{a}})$ should be constructed from the filtered data $\{\hat{y}_1(\tilde{\mathbf{a}}),\ldots,\hat{y}_n(\tilde{\mathbf{a}})\}$.

As an example, consider the least-squares forward linear prediction (FLP) estimator of $\mathbf{a}$ based on $\{\hat{y}_1(\tilde{\mathbf{a}}),\ldots,\hat{y}_n(\tilde{\mathbf{a}})\}$. Let

$$\mathbf{Y}_f(\tilde{\mathbf{a}}) := \begin{bmatrix} \hat{y}_p(\tilde{\mathbf{a}}) & \cdots & \hat{y}_1(\tilde{\mathbf{a}}) \\ \vdots & & \vdots \\ \hat{y}_{n-1}(\tilde{\mathbf{a}}) & \cdots & \hat{y}_{n-p}(\tilde{\mathbf{a}}) \end{bmatrix}, \quad \mathbf{y}_f(\tilde{\mathbf{a}}) := \begin{bmatrix} \hat{y}_{p+1}(\tilde{\mathbf{a}}) \\ \vdots \\ \hat{y}_n(\tilde{\mathbf{a}}) \end{bmatrix}. \tag{9.4.5}$$

By minimizing $\|\mathbf{y}_f(\tilde{\mathbf{a}}) + \mathbf{Y}_f(\tilde{\mathbf{a}})\boldsymbol{\varphi}\|^2$ with respect to $\boldsymbol{\varphi} \in \mathbb{C}^p$, one obtains

$$\hat{\boldsymbol{\varphi}}(\tilde{\mathbf{a}}) := -\{\mathbf{Y}_f^H(\tilde{\mathbf{a}})\mathbf{Y}_f(\tilde{\mathbf{a}})\}^{-1}\mathbf{Y}_f^H(\tilde{\mathbf{a}})\mathbf{y}_f(\tilde{\mathbf{a}}). \tag{9.4.6}$$

The corresponding IFIL iteration can be expressed as

$$\hat{\mathbf{a}}_m := \hat{\boldsymbol{\varphi}}(\hat{\mathbf{a}}_{m-1}) \qquad (m=1,2,\ldots). \tag{9.4.7}$$

This procedure is just a fixed point iteration of the mapping

$$\tilde{\mathbf{a}} \mapsto \hat{\boldsymbol{\varphi}}(\tilde{\mathbf{a}}).$$

If the mapping is contractive in a neighborhood of $\mathbf{a}$, then the iteration converges to a unique fixed point, denoted as $\hat{\mathbf{a}}$, which will be referred to as the IFIL estimator of $\mathbf{a}$. The IFIL frequency estimator is given by $\hat{\boldsymbol{\omega}} := \Phi_{\mathrm{AR}}^{-1}(\hat{\mathbf{a}})$. Note that $\hat{\mathbf{a}}_0 := \hat{\boldsymbol{\varphi}}(\mathbf{0})$ is nothing but the FLP estimator of order p, also known as Prony's estimator [179], which can be used to initialize the IFIL iteration.

In general, there is no guarantee that the estimate $\hat{\mathbf{a}}_m$ in (9.4.7) will give an AR polynomial $\hat{a}_m(z)$ with all its roots on the unit circle even if $\hat{\mathbf{a}}_{m-1}$ has such a property. This means that the mapping $\hat{\boldsymbol{\varphi}}(\tilde{\mathbf{a}})$ resides in a bigger space than the minimal parameter space $\mathcal{A}$ discussed in Section 9.2. One may impose the constraint $\hat{\mathbf{a}}_m \in \mathcal{A}$ in each iteration by defining

$$\hat{\mathbf{a}}_m := \mathcal{A}(\hat{\boldsymbol{\varphi}}(\hat{\mathbf{a}}_{m-1})) \qquad (m = 1, 2, \ldots). \tag{9.4.8}$$

In this expression, $\tilde{\mathbf{a}} \mapsto \mathcal{A}(\tilde{\mathbf{a}})$ denotes the projection of $\tilde{\mathbf{a}} \in \mathbb{C}^p$ onto $\mathcal{A}$ which simply resets the amplitude of each root of the corresponding AR polynomial to unity. As a result, the sequence $\{\hat{\mathbf{a}}_m\}$ produced by (9.4.8) remains in $\mathcal{A}$ for any $\hat{\mathbf{a}}_0 \in \mathcal{A}$. Another possibility is to reset the amplitude only for the roots that appear outside the unit circle, called unstable roots, but leave the roots inside the unit circle unchanged. This technique allows the resulting sequence $\{\hat{\mathbf{a}}_m\}$ to travel inside the unit circle while still ensuring the stability of inverse filtering.

Now let us investigate the asymptotic expression of $\hat{\boldsymbol{\varphi}}(\tilde{\mathbf{a}})$ as $n \to \infty$ when η is fixed. Toward that end, let $\mathbf{R}_y(\tilde{\mathbf{a}})$ and $\mathbf{r}_y(\tilde{\mathbf{a}})$ denote the counterparts of $\mathbf{R}_y$ and $\mathbf{r}_y$ defined by (9.1.2) with $m = p$ for the filtered process $\{y_t(\tilde{\mathbf{a}})\}$. According to Theorem 4.6, if $\{\epsilon_t\}$ satisfies the conditions of Theorem 4.1, then, as $n \to \infty$,

$$\hat{\boldsymbol{\varphi}}(\tilde{\mathbf{a}}) \stackrel{a.s.}{\to} \boldsymbol{\varphi}(\tilde{\mathbf{a}}) := -\mathbf{R}_y^{-1}(\tilde{\mathbf{a}})\mathbf{r}_y(\tilde{\mathbf{a}}). \tag{9.4.9}$$

This result can also be strengthened to uniform convergence in a neighborhood of $\mathbf{a}$ along the lines of [233]. Therefore, the IFIL estimator $\hat{\mathbf{a}}$, being a fixed point of $\hat{\boldsymbol{\varphi}}(\tilde{\mathbf{a}})$ in that neighborhood, converges to a fixed point of the deterministic mapping $\boldsymbol{\varphi}(\tilde{\mathbf{a}})$. The latter, however, is not the desired AR parameter $\mathbf{a}$.

To prove this assertion, observe that

$$\mathbf{R}_y(\tilde{\mathbf{a}}) = \mathbf{R}_x(\tilde{\mathbf{a}}) + \mathbf{R}_\epsilon(\tilde{\mathbf{a}}), \quad \mathbf{r}_y(\tilde{\mathbf{a}}) = \mathbf{r}_x(\tilde{\mathbf{a}}) + \mathbf{r}_\epsilon(\tilde{\mathbf{a}}),$$

where $\mathbf{R}_x(\tilde{\mathbf{a}})$, $\mathbf{r}_x(\tilde{\mathbf{a}})$, $\mathbf{R}_\epsilon(\tilde{\mathbf{a}})$, and $\mathbf{r}_\epsilon(\tilde{\mathbf{a}})$ are defined in the same way as $\mathbf{R}_y(\tilde{\mathbf{a}})$ and $\mathbf{r}_y(\tilde{\mathbf{a}})$ but for the filtered sinusoidal signal $\{x_t(\tilde{\mathbf{a}})\}$ and the filtered noise $\{\epsilon_t(\tilde{\mathbf{a}})\}$, respectively. Because $r_x(u) = \sum_{k=1}^p C_k^2 z_k^u$, it follows that for all u,

$$\sum_{j=0}^p a_j r_x(u-j) = \sum_{k=1}^p C_k^2 z_k^u a(z_k) = 0.$$

This equation is still satisfied by the ACF of $\{x_t(\tilde{\mathbf{a}})\}$. Therefore,

$$\mathbf{R}_x(\tilde{\mathbf{a}})\,\mathbf{a} = -\mathbf{r}_x(\tilde{\mathbf{a}}).$$

Moreover, because $\tilde{a}(\eta^{-1}z)\,\epsilon_t(\tilde{\mathbf{a}}) = \epsilon_t$, the filtered noise $\{\epsilon_t(\tilde{\mathbf{a}})\}$ is an AR(p) process with the AR coefficients $\eta^j \tilde{a}_j$ $(j = 1, \ldots, p)$. This implies that the ACF of $\{\epsilon_t(\tilde{\mathbf{a}})\}$ satisfies the Yule-Walker equation

$$\mathbf{R}_\epsilon(\tilde{\mathbf{a}})\mathbf{D}\tilde{\mathbf{a}} = -\mathbf{r}_\epsilon(\tilde{\mathbf{a}}),$$

where $\mathbf{D} := \mathrm{diag}(\eta, \ldots, \eta^p)$. Combining these equations with (9.4.9) yields

$$\begin{aligned}
\boldsymbol{\varphi}(\tilde{\mathbf{a}}) - \mathbf{a} &= -\mathbf{R}_y^{-1}(\tilde{\mathbf{a}})\{\mathbf{r}_y(\tilde{\mathbf{a}}) + \mathbf{R}_y(\tilde{\mathbf{a}})\mathbf{a}\} \\
&= -\mathbf{R}_y^{-1}(\tilde{\mathbf{a}})\{\mathbf{r}_x(\tilde{\mathbf{a}}) + \mathbf{r}_\epsilon(\tilde{\mathbf{a}}) + (\mathbf{R}_x(\tilde{\mathbf{a}}) + \mathbf{R}_\epsilon(\tilde{\mathbf{a}}))\mathbf{a}\} \\
&= \mathbf{R}_y^{-1}(\tilde{\mathbf{a}})\mathbf{R}_\epsilon(\tilde{\mathbf{a}})(\mathbf{D}\tilde{\mathbf{a}} - \mathbf{a}).
\end{aligned}$$

For $\tilde{\mathbf{a}} = \mathbf{a}$ in particular, we obtain

$$\boldsymbol{\varphi}(\mathbf{a}) - \mathbf{a} = -\mathbf{R}_y^{-1}(\mathbf{a})\mathbf{R}_\epsilon(\mathbf{a})(\mathbf{I} - \mathbf{D})\mathbf{a} \neq \mathbf{0}. \tag{9.4.10}$$

This means that the AR parameter $\mathbf{a}$ is not a fixed point of $\boldsymbol{\varphi}(\tilde{\mathbf{a}})$. Therefore, the IFIL estimator $\hat{\mathbf{a}}$ cannot be a consistent estimator of $\mathbf{a}$ for any fixed $\eta \in (0,1)$.

The good news is that the bias $\mathbf{b}(\mathbf{a}) := \boldsymbol{\varphi}(\mathbf{a}) - \mathbf{a}$ approaches zero as $\eta \to 1$. In fact, let the eigenvalue decomposition (EVD) of $\mathbf{R}_\epsilon^{-1}(\mathbf{a})\mathbf{R}_x(\mathbf{a})$ be denoted by $\mathbf{U}\boldsymbol{\Lambda}\mathbf{U}^H$, where $\mathbf{U}$ is a unitary matrix and $\boldsymbol{\Lambda}$ is a diagonal matrix with strictly positive diagonal elements. Because

$$\mathbf{R}_y^{-1}(\mathbf{a})\mathbf{R}_\epsilon(\mathbf{a}) = \{\mathbf{I} + \mathbf{R}_\epsilon^{-1}(\mathbf{a})\mathbf{R}_x(\mathbf{a})\}^{-1} = \mathbf{U}(\mathbf{I} + \boldsymbol{\Lambda})^{-1}\mathbf{U}^H,$$

it follows from (9.4.10) that

$$\|\boldsymbol{\varphi}(\mathbf{a}) - \mathbf{a}\| = \|(\mathbf{I} + \boldsymbol{\Lambda})^{-1}\mathbf{U}^H(\mathbf{I} - \mathbf{D})\mathbf{a}\| < \|\mathbf{U}^H(\mathbf{I} - \mathbf{D})\mathbf{a}\| = \|(\mathbf{I} - \mathbf{D})\mathbf{a}\|.$$

Hence, $\|\boldsymbol{\varphi}(\mathbf{a}) - \mathbf{a}\| \to 0$ as $\eta \to 1$. Moreover, because $\mathbf{R}_y^{-1}(\mathbf{a})\mathbf{R}_\epsilon(\mathbf{a}) \to \mathbf{0}$ as $\sigma^2 \to 0$, the bias also approaches zero with the increase of SNR.

Example 9.12 (Bias of the IFIL Mapping for a Single Complex Sinusoid). Consider the case of $p = 1$ for which $\mathbf{a} = a_1 = -\exp(i\omega_1)$. Because $\mathbf{R}_x(\mathbf{a}) = C_1^2/(1-\eta)^2$ and $\mathbf{R}_\epsilon(\mathbf{a}) = \sigma^2/(1-\eta^2)$, it follows from (9.4.10) that the bias takes the form

$$-\frac{(1-\eta)^2}{1 - \eta + \gamma_1(1+\eta)}\, a_1,$$

where $\gamma_1 := C_1^2/\sigma^2$ is the SNR. Observe that the bias tends to zero as $\eta \to 1$ or $\gamma_1 \to \infty$ and the absolute value of the bias never exceeds $1 - \eta$. ◇

To mitigate the bias problem, especially when η is small or the SNR is low, one can subtract from $\hat{\boldsymbol{\varphi}}(\hat{\mathbf{a}}_{m-1})$ an estimated bias, denoted as $\hat{\mathbf{b}}(\hat{\mathbf{a}}_{m-1})$, so the fixed point iteration (9.4.7) becomes

$$\hat{\mathbf{a}}_m := \hat{\boldsymbol{\varphi}}(\hat{\mathbf{a}}_{m-1}) - \hat{\mathbf{b}}(\hat{\mathbf{a}}_{m-1}) \qquad (m = 1, 2, \ldots). \tag{9.4.11}$$

Motivated by (9.4.10), it is natural to consider

$$\hat{\mathbf{b}}(\tilde{\mathbf{a}}) := -\hat{\mathbf{R}}_y^{-1}(\tilde{\mathbf{a}})\hat{\mathbf{R}}_\epsilon(\tilde{\mathbf{a}})(\mathbf{I} - \mathbf{D})\tilde{\mathbf{a}}, \tag{9.4.12}$$

where $\hat{\mathbf{R}}_y(\tilde{\mathbf{a}}) := n^{-1}\mathbf{Y}_f^H(\tilde{\mathbf{a}})\mathbf{Y}_f(\tilde{\mathbf{a}})$ is obtained from $\{\hat{y}_t(\tilde{\mathbf{a}})\}$ and $\hat{\mathbf{R}}_\epsilon(\tilde{\mathbf{a}})$ is similarly constructed from the linear prediction errors $\{\hat{\epsilon}_t(\tilde{\mathbf{a}})\}$.

The parameter η not only determines the accuracy of the final IFIL estimates but also controls the requirement of initial values and the speed of convergence for the fixed point iteration. To illustrate, consider Figure 9.9. The time series in this example is the same as that used in Figure 8.1, which consists of two unit-amplitude zero-phase complex sinusoids with well-separated frequencies in Gaussian white noise. Figure 9.9 depicts the trajectory of the frequency estimates produced by the fixed point iteration (9.4.7) with different choices of η and for different initial values.

As can be seen from Figures 9.9(a) and (b), the IFIL algorithm with $\eta = 0.995$ not only takes many iterations (up to 50) to converge from all the initial values considered, but also produces spurious fixed points that are far from the signal frequencies. On the other hand, the fixed point which is closest to the signal frequencies turns out to be very accurate for frequency estimation: the average absolute error for estimating $f_k := \omega_k/(2\pi)$ $(k = 1,2)$ is equal to 4.3×10^{-4}. When the value of η is reduced to 0.85, as shown in Figure 9.9(c), all spurious fixed points disappear, and the convergence to the remaining fixed point is much faster. However, the accuracy of the fixed point as a frequency estimator deteriorates considerably: the average error becomes 1.6×10^{-3}.

This particular behavior of the IFIL estimator can be explained by the fact that (9.4.4) defines a bandpass filter centered around the estimated frequencies where the bandwidth is controlled by η (assuming that the frequency estimates are sufficiently separated from each other and η is sufficiently close to unity). A larger η corresponds to a narrower bandwidth which enables more effective suppression of the noise outside the passband, leading to more accurate frequency estimation based on the filtered time series. But the positive feedback takes place only if the filter captures the signal frequencies in its passband to begin with, meaning that a good initial value is needed. On the other hand, a smaller η makes the bandwidth wider so that the filter is able to capture the signal frequencies even with less accurate initial values. However, an increased bandwidth reduces the effectiveness of noise suppression, resulting in a degradation of accuracy for frequency estimation.

To take advantage of this behavior, a variable bandwidth parameter should be used: start with a smaller η to accommodate possibly poor initial values, and increase η gradually toward unity as the iteration progresses. Figure 9.9(d) shows the results obtained with a variable bandwidth parameter. The algorithm begins with $\eta = 0.85$ for 10 iterations, switches to $\eta = 0.99$ for an additional 5 iterations, and finally takes $\eta = 0.995$ for 5 more iterations. We denote this setting of the iteration parameters by $\eta = (0.85, 0.99, 0.995)$ and $m = (10, 5, 5)$. As can be seen from Figure 9.9(d), the algorithm with variable bandwidth parameter is able to

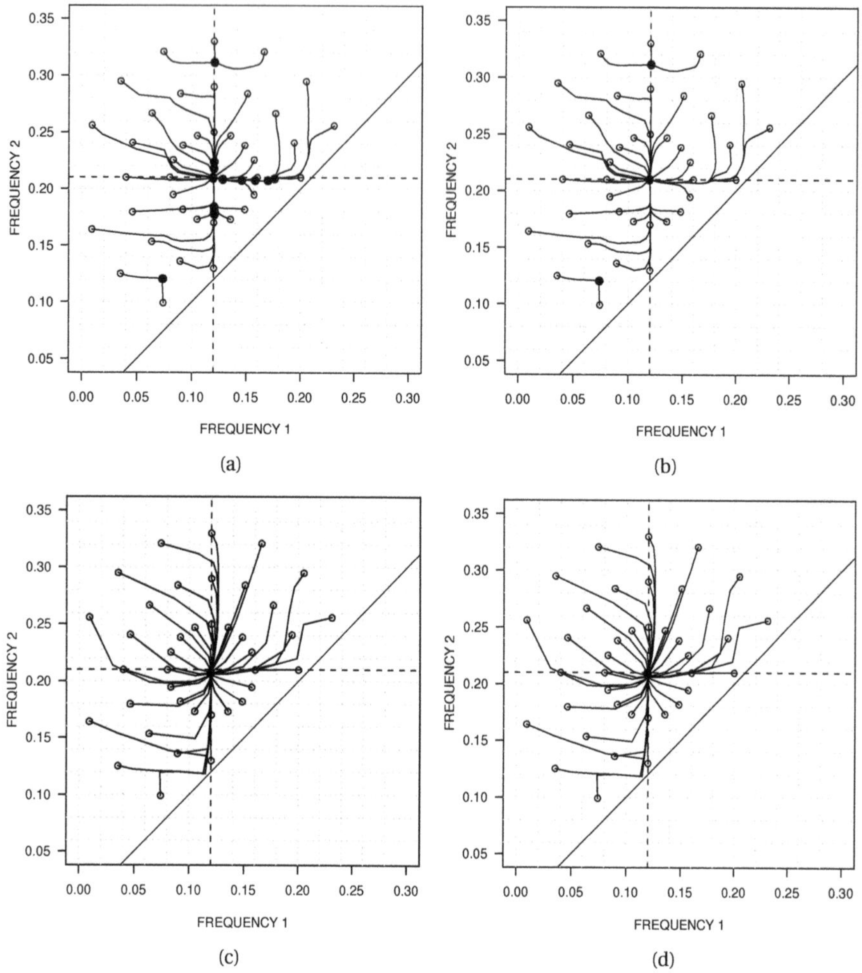

Figure 9.9. Trajectory of the IFIL frequency estimates given by (9.4.7) for a time series of two unit-amplitude zero-phase complex sinusoids in Gaussian white noise ($\omega_1 = 2\pi \times 0.12$, $\omega_2 = 2\pi \times 0.21$, $n = 50$, SNR $= -5$ dB per sinusoid) with different initial values and bandwidth parameters. (a) $\eta = 0.995$, $m = 20$. (b) $\eta = 0.995$, $m = 50$. (c) $\eta = 0.85$, $m = 20$. (d) $\eta = (0.85, 0.99, 0.995)$, $m = (10, 5, 5)$. Circle, initial value; solid dot, final estimate. Dashed lines indicate the signal frequencies and dotted lines define the Fourier grid. Thin solid line depicts the diagonal line. The same time series is used in Figure 8.1.

overcome the problem of spurious fixed points and at the same time produce accurate frequency estimates with fewer iterations: the average error is the same as that achieved with $\eta = 0.995$, but only 20 iterations are sufficient for the algorithm to converge to a single fixed point from all the initial values considered.

Note that in this example one can increase the value of η from 0.85 directly to 0.995 without affecting the result. But, in general, an intermediate value, as employed in Figure 9.9(d), has proven useful to ensure proper convergence (see Theorem 9.10 in Section 9.4.2 for motivation). While choosing suitable values of η for a given time series remains a challenging problem, it can be said, as a general rule of thumb, that the values of η should increase appropriately with n, because the increased accuracy of frequency estimates from a longer time series allows the filter to take a narrower bandwidth for more effective suppression of the noise. Some additional guidelines will be discussed in Section 9.4.2.

The following examples demonstrate the statistical performance of the IFIL estimator with variable bandwidth parameter.

Example 9.13 (Estimation of Well-Separated Frequencies: Part 3). Consider the time series in Example 9.3 which consists of two unit-amplitude complex sinusoids with well-separated frequencies in Gaussian white noise. Let the signal frequencies be estimated by the IFIL estimator in (9.4.7) with the initial values produced by two methods from the unfiltered data: Prony's method (i.e., the FLP estimator of order 2) and the DFT method discussed in Example 8.1 (i.e., the location of the two largest discrete periodogram ordinates). Figure 9.10 depicts the average MSE of the IFIL frequency estimates, together with the average MSE of the initial Prony and DFT estimates, for different values of SNR and two sample sizes. The iteration parameters are: $\eta = (0.85, 0.99, 0.995)$ and $m = (5, 5, 5)$ for $n = 100$; $\eta = (0.90, 0.995, 0.998)$ and $m = (5, 5, 5)$ for $n = 200$.

The IFIL estimator initialized by the DFT method follows the CRLB closely when the SNR exceeds a threshold (-4 dB for $n = 100$ and -7 dB for $n = 200$). This is very similar to the performance of the PM and NLS estimators shown in Figures 6.13 and 8.3. When the SNR is above a higher threshold (-1 dB for $n = 100$ and -3 dB for $n = 200$), Prony's method also produces good initial values for the IFIL iteration and achieves the same final accuracy, although the initial estimates themselves are very poor. The IFIL estimator performs better than the FBLP and EYW estimators shown in Figures 9.4 and 9.7. ◇

Example 9.14 (Estimation of Closely Spaced Frequencies: Part 3). Consider the time series in Example 9.4 which consists of two unit-amplitude complex sinusoids with closely spaced frequencies in Gaussian white noise. Let the signal frequencies be estimated by the IFIL estimator in (9.4.7) with the initial values produced from the unfiltered data by Prony's method as in Example 9.13 and by the DFT method discussed in Example 8.2 (i.e., the location of the largest discrete periodogram ordinate). The iteration parameters are: $\eta = (0.98, 0.99, 0.995)$

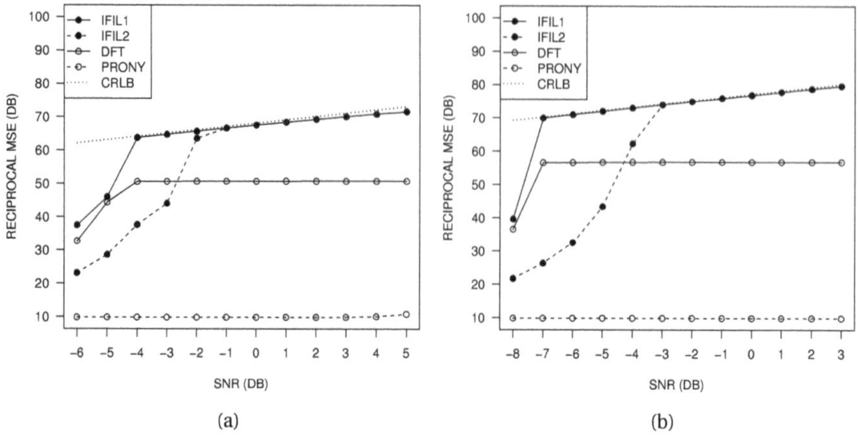

Figure 9.10. Plot of the reciprocal average MSE of the IFIL frequency estimates obtained from (9.4.7) for the time series of two complex sinusoids in Gaussian white noise discussed in Example 9.13 with different values of SNR. (a) $n = 100$. (b) $n = 200$. Solid line with dots, IFIL initialized by DFT; dashed line with dots, IFIL initialized by Prony; solid line with circles, DFT; dashed line with circles, Prony. Dotted line depicts the CRLB. Results are based on 5,000 Monte Carlo runs.

and $m = (5,5,5)$ for $n = 100$; $\eta = (0.99, 0.995, 0.998)$ and $m = (5,5,5)$ for $n = 200$. Figure 9.11 shows the simulation results.

As can be seen, the IFIL estimator is able to improve the DFT estimator when the latter is employed to initialize the iteration; Prony's estimator also produces good initial values when the SNR is sufficiently high. The MSE of the IFIL estimator is close to the CRLB at -2 dB for $n = 100$ and at -6 dB for $n = 200$, but remains unchanged when the SNR increases. This is similar to the behavior of the FBLP estimator shown in Figure 9.5 with order 10 for $n = 100$ and order 15 for $n = 200$. But unlike the FBLP estimator, the IFIL estimator produces two distinct frequencies rather than a single frequency.

A closer examination shows that the MSE of the IFIL estimator plateaus at approximately 4.1×10^{-6} in the case of $n = 100$, whereas that of the FBLP estimator with order 10 stays at approximately 5.9×10^{-6} (a similar observation is made in the case of $n = 200$). So the two-frequency IFIL estimator still outperforms the one-frequency FBLP estimator in terms of the MSE. With a suitably higher order, the FBLP estimator is able to resolve the signal frequencies and reduce the MSE when the SNR is sufficiently high, but at lower SNR it performs much worse than the IFIL estimator due to spurious spectral peaks.

A comparison between Figure 9.11 and Figure 9.8 shows that the IFIL estimator considerably outperforms the EYW estimator which produces two distinct but severely biased frequencies. The bias of the IFIL estimator, which cannot be elim-

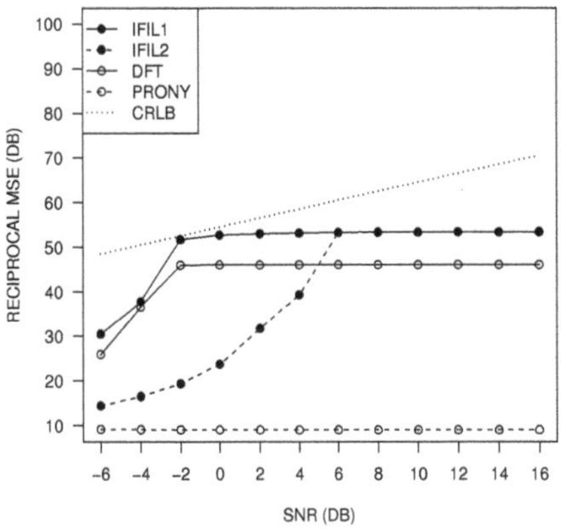

(a)

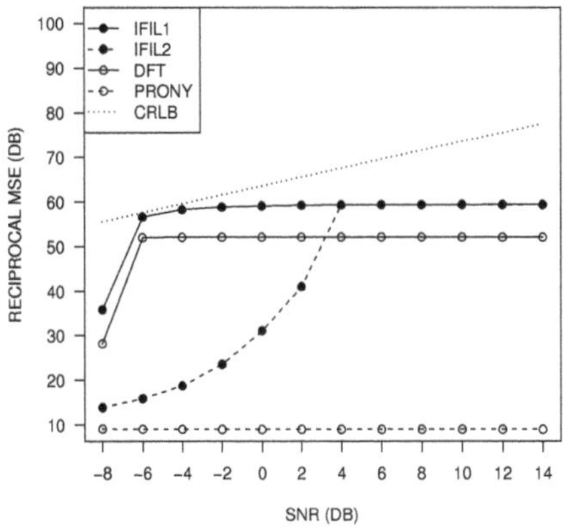

(b)

Figure 9.11. Plot of the reciprocal average MSE of the IFIL frequency estimates for the time series of two complex sinusoids with closely spaced frequencies in Gaussian white noise discussed in Example 9.14 with different values of SNR. (a) $n = 100$. (b) $n = 200$. Solid line with dots, IFIL initialized by DFT; dashed line with dots, IFIL initialized by Prony; solid line with circles, DFT; dashed line with circles, Prony. Dotted line depicts the CRLB. Results are based on 5,000 Monte Carlo runs.

inated by using the modified mapping in (9.4.11), also plays a dominating role in the MSE (see Table 9.3 in Section 9.5 for further details).

Note that the IFIL estimator given by (9.4.8) with the unit-root projection applied in each iteration achieves a slightly smaller MSE (not shown) when initialized by the DFT method. Prony's method in this case is no longer accurate enough to initialize the IFIL iteration, which suggests that the projection worsens the initial requirement. ◇

9.4.2 Iterative Filtering Estimator: Real Case

Now consider the real case where $\{x_t\}$ is given by (2.1.1) or (2.1.2). Because the AR equation (9.2.1) is still satisfied in this case with $p := 2q$, the IFIL procedure discussed in the previous section remains valid for the estimation of $\mathbf{a} := [a_1, \ldots, a_p]^T$. In this section, we consider a more efficient method by observing that in the real case the AR coefficients are real and symmetric, satisfying (9.2.7). Owing to this constraint, it suffices to estimate $\mathbf{a}_r := [a_1, \ldots, a_q]^T$ which resides in the smaller space $\mathbb{R}^q$. Furthermore, in the real case, there is an alternative method of bias correction, originally proposed in [428] and later generalized in [234]. The method removes the bias by judiciously designing the filter rather than subtracting an estimated bias from the linear prediction mapping.

Given $\tilde{\mathbf{a}}_r \in \mathbb{R}^q$ and $\eta \in (0,1)$, consider the filtered time series defined by

$$\hat{y}_t(\tilde{\mathbf{a}}_r) = -\sum_{j=1}^{p} \theta_j(\tilde{\mathbf{a}}_r)\eta^j \hat{y}_{t-j}(\tilde{\mathbf{a}}_r) + y_t \qquad (t = 1,\ldots,n), \tag{9.4.13}$$

where $\hat{y}_t(\tilde{\mathbf{a}}_r) := 0$ for $t \le 0$. The coefficients $\theta_j(\tilde{\mathbf{a}}_r)$ are certain real-valued continuous functions of $\tilde{\mathbf{a}}_r$ (and possibly η) which satisfy the same symmetry constraint as the a_j, i.e.,

$$\theta_j(\tilde{\mathbf{a}}_r) = \theta_{p-j}(\tilde{\mathbf{a}}_r) \qquad (j = 0,1,\ldots,q-1),$$

with $\theta_0(\tilde{\mathbf{a}}_r) := 1$. Similar to the complex case, let $\mathbf{Y}_f(\tilde{\mathbf{a}}_r)$ and $\mathbf{y}_f(\tilde{\mathbf{a}}_r)$ be defined by (9.4.5) on the basis of $\{\hat{y}_1(\tilde{\mathbf{a}}_r),\ldots,\hat{y}_n(\tilde{\mathbf{a}}_r)\}$. Let $\mathbf{B}$ and $\mathbf{b}$ be defined by (9.2.8) so that the symmetry constraint (9.2.7) can be expressed as

$$\mathbf{a} = \mathbf{B}\mathbf{a}_r + \mathbf{b}.$$

Then, the FLP estimator of $\mathbf{a}_r$ that minimizes $\|\mathbf{y}_f(\tilde{\mathbf{a}}_r) + \mathbf{Y}_f(\tilde{\mathbf{a}}_r)(\mathbf{B}\boldsymbol{\varphi}_r + \mathbf{b})\|^2$ with respect to $\boldsymbol{\varphi}_r \in \mathbb{R}^q$ takes the form

$$\hat{\boldsymbol{\varphi}}_r(\tilde{\mathbf{a}}_r) := -\{\mathbf{B}^T\mathbf{Y}_f^T(\tilde{\mathbf{a}}_r)\mathbf{Y}_f(\tilde{\mathbf{a}}_r)\mathbf{B}\}^{-1}\mathbf{B}^T\mathbf{Y}_f^T(\tilde{\mathbf{a}}_r)\{\mathbf{Y}_f(\tilde{\mathbf{a}}_r)\mathbf{b} + \mathbf{y}_f(\tilde{\mathbf{a}}_r)\}. \tag{9.4.14}$$

The IFIL estimator of $\mathbf{a}_r$, denoted as $\hat{\mathbf{a}}_r$, is defined as the limiting value of the sequence $\{\hat{\mathbf{a}}_{r,m}\}$ produced by the fixed point iteration

$$\hat{\mathbf{a}}_{r,m} := \hat{\boldsymbol{\varphi}}_r(\hat{\mathbf{a}}_{r,m-1}) \qquad (m = 1,2,\ldots). \tag{9.4.15}$$

In other words, $\hat{\mathbf{a}}_r$ is a fixed point of the random mapping $\hat{\boldsymbol{\varphi}}_r(\tilde{\mathbf{a}}_r)$.

To eliminate the asymptotic bias of this estimator as $n \to \infty$ for fixed $\eta \in (0,1)$, let $\{y_t(\tilde{\mathbf{a}}_r)\}$ denote the filtered process produced by the filter (9.4.13) without the finite sample constraint. Let $\tilde{\mathbf{a}}_r$ be restricted to a closed subset $\mathcal{A}_0 \subset \mathbb{R}^q$, which contains $\mathbf{a}_r$ as an interior point, such that the poles of the filter lie strictly inside the unit circle. Similar to the complex case, let $\mathbf{R}_y(\tilde{\mathbf{a}}_r)$ and $\mathbf{r}_y(\tilde{\mathbf{a}}_r)$ denote the counterparts of $\mathbf{R}_y$ and $\mathbf{r}_y$ defined by (9.1.2) for $m = p$ with the ACF of $\{y_t(\tilde{\mathbf{a}}_r)\}$ in place of $r_y(u)$. Then, by Theorem 9.3,

$$\hat{\boldsymbol{\varphi}}_r(\tilde{\mathbf{a}}_r) \overset{a.s.}{\to} \boldsymbol{\varphi}_r(\tilde{\mathbf{a}}_r) := -\{\mathbf{B}^T\mathbf{R}_y(\tilde{\mathbf{a}}_r)\mathbf{B}\}^{-1}\mathbf{B}^T\{\mathbf{R}_y(\tilde{\mathbf{a}}_r)\mathbf{b} + \mathbf{r}_y(\tilde{\mathbf{a}}_r)\}.$$

Suppose that the pointwise convergence can be strengthened to uniform convergence and the IFIL estimator $\hat{\mathbf{a}}_r$ converges almost surely to a constant vector $\mathbf{a}_{r0}$ as $n \to \infty$. Then, it follows that

$$\boldsymbol{\varphi}_r(\hat{\mathbf{a}}_r) = \boldsymbol{\varphi}_r(\hat{\mathbf{a}}_r) - \hat{\boldsymbol{\varphi}}_r(\hat{\mathbf{a}}_r) + \hat{\mathbf{a}}_r \overset{a.s.}{\to} \mathbf{a}_{r0}.$$

This result, combined with $\boldsymbol{\varphi}_r(\hat{\mathbf{a}}_r) \overset{a.s.}{\to} \boldsymbol{\varphi}_r(\mathbf{a}_{r0})$, implies that $\mathbf{a}_{r0}$ must be a fixed point of the deterministic mapping $\boldsymbol{\varphi}_r(\tilde{\mathbf{a}}_r)$. Therefore, in order for $\hat{\mathbf{a}}_r$ to be a consistent estimator of $\mathbf{a}_r$, the fixed point $\mathbf{a}_{r0}$ must coincide with $\mathbf{a}_r$.

This objective can be achieved by constructing the coefficients $\theta_j(\tilde{\mathbf{a}}_r)$ based on a general technique discussed in [234]. Toward that end, observe that

$$\boldsymbol{\varphi}_r(\tilde{\mathbf{a}}_r) = \mathbf{a}_r - \{\mathbf{B}^T\mathbf{R}_y(\tilde{\mathbf{a}}_r)\mathbf{B}\}^{-1}\mathbf{B}^T\{\mathbf{R}_y(\tilde{\mathbf{a}}_r)(\mathbf{B}\mathbf{a}_r + \mathbf{b}) + \mathbf{r}_y(\tilde{\mathbf{a}}_r)\}. \tag{9.4.16}$$

Let $\mathbf{R}_x(\tilde{\mathbf{a}}_r)$, $\mathbf{r}_x(\tilde{\mathbf{a}}_r)$, $\mathbf{R}_\epsilon(\tilde{\mathbf{a}}_r)$, and $\mathbf{r}_\epsilon(\tilde{\mathbf{a}}_r)$ be the counterparts of $\mathbf{R}_y(\tilde{\mathbf{a}}_r)$ and $\mathbf{r}_y(\tilde{\mathbf{a}}_r)$ defined for the filtered sinusoidal signal $\{x_t(\tilde{\mathbf{a}}_r)\}$ and the filtered noise $\{\epsilon_t(\tilde{\mathbf{a}}_r)\}$. As in the complex case, we can write

$$\mathbf{R}_x(\tilde{\mathbf{a}}_r)(\mathbf{B}\mathbf{a}_r + \mathbf{b}) = -\mathbf{r}_x(\tilde{\mathbf{a}}_r).$$

By combining this result with the fact that $\mathbf{R}_y(\tilde{\mathbf{a}}_r) = \mathbf{R}_x(\tilde{\mathbf{a}}_r) + \mathbf{R}_\epsilon(\tilde{\mathbf{a}}_r)$ and $\mathbf{r}_y(\tilde{\mathbf{a}}_r) = \mathbf{r}_x(\tilde{\mathbf{a}}_r) + \mathbf{r}_\epsilon(\tilde{\mathbf{a}}_r)$, we can rewrite (9.4.16) as

$$\boldsymbol{\varphi}_r(\tilde{\mathbf{a}}_r) = \mathbf{a}_r - \{\mathbf{B}^T\mathbf{R}_y(\tilde{\mathbf{a}}_r)\mathbf{B}\}^{-1}\mathbf{B}^T\{\mathbf{R}_\epsilon(\tilde{\mathbf{a}}_r)(\mathbf{B}\mathbf{a}_r + \mathbf{b}) + \mathbf{r}_\epsilon(\tilde{\mathbf{a}}_r)\}. \tag{9.4.17}$$

Suppose that the filter (9.4.13) is constructed such that

$$\mathbf{B}^T\mathbf{R}_\epsilon(\tilde{\mathbf{a}}_r)(\mathbf{B}\tilde{\mathbf{a}}_r + \mathbf{b}) = -\mathbf{B}^T\mathbf{r}_\epsilon(\tilde{\mathbf{a}}_r). \tag{9.4.18}$$

Under this condition, (9.4.17) becomes

$$\boldsymbol{\varphi}_r(\tilde{\mathbf{a}}_r) = \mathbf{a}_r + \mathbf{C}(\tilde{\mathbf{a}}_r)\,(\tilde{\mathbf{a}}_r - \mathbf{a}_r), \tag{9.4.19}$$

where

$$\mathbf{C}(\tilde{\mathbf{a}}_r) := \{\mathbf{B}^T\mathbf{R}_y(\tilde{\mathbf{a}}_r)\mathbf{B}\}^{-1}\mathbf{B}^T\mathbf{R}_\epsilon(\tilde{\mathbf{a}}_r)\mathbf{B}.$$

Evaluating (9.4.19) at the fixed point $\mathbf{a}_{r0}$ gives

$$\mathbf{a}_{r0} = \boldsymbol{\varphi}_r(\mathbf{a}_{r0}) = \mathbf{a}_r + \mathbf{C}(\mathbf{a}_{r0})\,(\mathbf{a}_{r0} - \mathbf{a}_r),$$

or equivalently,

$$\{\mathbf{B}^T\mathbf{R}_x(\mathbf{a}_{r0})\mathbf{B}\}(\mathbf{a}_{r0} - \mathbf{a}_r) = \mathbf{0}.$$

Because the matrix $\mathbf{B}^T\mathbf{R}_x(\mathbf{a}_{r0})\mathbf{B}$ is nonsingular, we obtain $\mathbf{a}_{r0} = \mathbf{a}_r$. In summary, if the filter (9.4.13) is parameterized to satisfy (9.4.18), then the mapping $\boldsymbol{\varphi}_r(\tilde{\mathbf{a}}_r)$ has a unique fixed point which equals $\mathbf{a}_r$.

The following theorem establishes the convergence of the fixed point iteration (9.4.15) for the random mapping $\hat{\boldsymbol{\varphi}}_r(\tilde{\mathbf{a}}_r)$ and the consistency of its fixed point as an estimator of $\mathbf{a}_r$ under the condition (9.4.18). See Section 9.6 for a proof.

Theorem 9.8 (Convergence and Consistency of the IFIL Estimator: Real Case). *Let $y_t := x_t + \epsilon_t$ $(t = 1, \dots, n)$, where $\{x_t\}$ is given by (2.1.1) or (2.1.2) and $\{\epsilon_t\} \sim$ $\mathrm{WN}(0, \sigma^2)$ satisfies the conditions of Theorem 4.1. Assume that the filter (9.4.13)* *satisfies the condition (9.4.18). Then, for any fixed* $\eta \in (0,1)$, *there exists a closed neighborhood* $\mathcal{N} := \{\tilde{\mathbf{a}}_r \in \mathcal{A}_0 : \|\tilde{\mathbf{a}}_r - \mathbf{a}_r\| \le \delta\}$ *for some constant* $\delta > 0$ *which is independent of* n *such that almost surely for sufficiently large* n *the mapping* $\hat{\boldsymbol{\varphi}}_r(\tilde{\mathbf{a}}_r)$ *given by (9.4.14) is contractive in* $\mathcal{N}$ *and therefore has a unique fixed point* $\hat{\mathbf{a}}_r$ *to which the sequence* $\{\hat{\mathbf{a}}_{r,m}\}$ *given by (9.4.15) converges for any initial value* $\hat{\mathbf{a}}_{r,0} \in \mathcal{N}$. *Furthermore, as* $n \to \infty$, $\hat{\mathbf{a}}_r \stackrel{a.s.}{\to} \mathbf{a}_r$ *and hence* $\hat{\boldsymbol{\omega}}_r := \Phi_{\mathrm{AR}}^{-1}(\hat{\mathbf{a}}_r) \stackrel{a.s.}{\to} \boldsymbol{\omega}_r$.

Now let us examine the condition (9.4.18). Because $\{\epsilon_t\}$ is white noise, the filtered noise $\{\epsilon_t(\tilde{\mathbf{a}}_r)\}$ is an AR process of order p with the AR coefficients $\theta_j(\tilde{\mathbf{a}}_r)\eta^j$ $(j = 1,\ldots,p)$. The corresponding Yule-Walker equations can be written as

$$\mathbf{R}_\epsilon(\tilde{\mathbf{a}}_r)\mathbf{D}(\mathbf{B}\boldsymbol{\theta}_r + \mathbf{b}) = -\mathbf{r}_\epsilon(\tilde{\mathbf{a}}_r), \tag{9.4.20}$$

where $\boldsymbol{\theta}_r := [\theta_1(\tilde{\mathbf{a}}_r),\ldots,\theta_q(\tilde{\mathbf{a}}_r)]^T$ and $\mathbf{D} := \mathrm{diag}(\eta,\ldots,\eta^p)$. It is easy to verify that

$$\mathbf{B}^T\mathbf{R}_\epsilon(\tilde{\mathbf{a}}_r)\mathbf{D}\mathbf{b} = \eta^p\mathbf{B}^T\mathbf{R}_\epsilon(\tilde{\mathbf{a}}_r)\mathbf{b} = \eta^p\mathbf{B}^T\mathbf{r}_\epsilon(\tilde{\mathbf{a}}_r).$$

Therefore, one can rewrite (9.4.20) as

$$\mathbf{B}^T\mathbf{R}_\epsilon(\tilde{\mathbf{a}}_r)\mathbf{D}\mathbf{B}\boldsymbol{\theta}_r = -(1+\eta^p)\mathbf{B}^T\mathbf{r}_\epsilon(\tilde{\mathbf{a}}_r).$$

For the same reason, one can also rewrite (9.4.18) as

$$\mathbf{B}^T\mathbf{R}_\epsilon(\tilde{\mathbf{a}}_r)\mathbf{B}\tilde{\mathbf{a}}_r = -2\mathbf{B}^T\mathbf{r}_\epsilon(\tilde{\mathbf{a}}_r).$$

Combining these expressions yields

$$\mathbf{B}^T\mathbf{R}_\epsilon(\tilde{\mathbf{a}}_r)\mathbf{D}\mathbf{B}\boldsymbol{\theta}_r = \tfrac{1}{2}(1+\eta^p)\mathbf{B}^T\mathbf{R}_\epsilon(\tilde{\mathbf{a}}_r)\mathbf{B}\tilde{\mathbf{a}}_r. \tag{9.4.21}$$

Furthermore, it is not difficult to verify that

$$\mathbf{B}^T\mathbf{R}_\epsilon(\tilde{\mathbf{a}}_r)\mathbf{D}\mathbf{B} = \tfrac{1}{2}(1+\eta^p)\mathbf{B}^T\mathbf{R}_\epsilon(\tilde{\mathbf{a}}_r)\mathbf{B}\mathbf{H}_\eta^{-1},$$

where $\mathbf{H}_\eta := \mathrm{diag}(h_1,\ldots,h_q)$ and $h_j := (1+\eta^p)/(\eta^j + \eta^{p-j})$ $(j = 1,\ldots,q)$. Inserting this expression in (9.4.21) leads to

$$\boldsymbol{\theta}_r = \mathbf{H}_\eta\tilde{\mathbf{a}}_r \quad \text{or} \quad \theta_j(\tilde{\mathbf{a}}_r) = \frac{1+\eta^p}{\eta^j + \eta^{p-j}}\,\tilde{a}_j \qquad (j = 1,\ldots,q). \tag{9.4.22}$$

Unlike the intuitive choice $\boldsymbol{\theta}_r := \tilde{\mathbf{a}}_r$, as in [96] and [267], the filter design given by (9.4.22) employs a matrix $\mathbf{H}_\eta$ to transform $\tilde{\mathbf{a}}_r$ into $\boldsymbol{\theta}_r$. It is this nontrivial matrix that makes the resulting IFIL estimator consistent for fixed $\eta \in (0,1)$.

The asymptotic distribution of the IFIL estimator $\hat{\mathbf{a}}_r$ can also be derived for fixed $\eta \in (0,1)$ on the basis of Theorem 4.4. The result is given in the following theorem. See Section 9.6 for a proof.

Theorem 9.9 (Asymptotic Normality of the IFIL Estimator: Real Case). *Let the conditions of Theorem 9.8 be satisfied and* $\{\epsilon_t\} \sim \text{IID}(0,\sigma^2)$ *with* $E(\epsilon_t^4) < \infty$. *Then, for any fixed* $\eta \in (0,1)$, $\sqrt{n}(\hat{\mathbf{a}}_r - \mathbf{a}_r) \xrightarrow{D} \text{N}(\mathbf{0},\mathbf{V}_\eta)$ *as* $n \to \infty$. *In this expression,*

$$\mathbf{V}_\eta := \{\mathbf{B}^T\mathbf{R}_x(\mathbf{a}_r)\mathbf{B}\}^{-1}\boldsymbol{\Gamma}_\epsilon(\mathbf{a}_r)\,\{\mathbf{B}^T\mathbf{R}_x(\mathbf{a}_r)\mathbf{B}\}^{-1} \tag{9.4.23}$$

and

$$\boldsymbol{\Gamma}_\epsilon(\mathbf{a}_r) := 4\mathbf{B}^T[\mathbf{0},\mathbf{A}_p]\left\{\sum_{\tau=1}^{\infty}\boldsymbol{\gamma}_\tau\boldsymbol{\gamma}_\tau^T\right\}[\mathbf{0},\mathbf{A}_p]^T\mathbf{B},$$

where $\boldsymbol{\gamma}_\tau := E\{[\epsilon_{\tau-p}(\mathbf{a}_r),\dots,\epsilon_{\tau+p}(\mathbf{a}_r)]^T\epsilon_0(\mathbf{a}_r)\} \in \mathbb{R}^{2p+1}$ *is an autocovariance vector of* $\{\epsilon_t(\mathbf{a}_r)\}$ *and* $\mathbf{A}_p \in \mathbb{R}^{p\times 2p}$ *is defined by (9.3.9) with* $l = p$.

According to Theorem 9.9, the error rate of the IFIL estimator with fixed η is $\sqrt{n}$. The same error rate is achieved by the EYW estimators with fixed l, as ensured by Theorem 9.6. Moreover, just as increasing l helps improve the accuracy of the EYW estimators, increasing η toward unity also helps improve the accuracy of the IFIL estimator. In fact, it can be shown [233] that $\mathbf{V}_\eta = \mathcal{O}((1-\eta)^3)$. However, this result does not imply that the asymptotic variance of the IFIL estimator can be expressed as $\mathcal{O}((1-\eta)^3/n)$ uniformly in $\eta \in (0,1)$. Indeed, when η is very close to unity relative to the sample size n, a different expression for the asymptotic variance should be expected. Derivation of such expressions is very difficult in general. But for the simplest case of $q = 1$, some interesting results are obtained in [235], [237], [304], and [355].

In the special case of $q = 1$, the filter given by (9.4.13) and (9.4.22) becomes

$$\hat{y}_t(\tilde{a}_1) = -\tfrac{1}{2}(1+\eta^2)\tilde{a}_1\hat{y}_{t-1}(\tilde{a}_1) - \eta^2\hat{y}_{t-2}(\tilde{a}_1) + y_t \quad (t = 1,\dots,n), \tag{9.4.24}$$

where $\hat{y}_{-1}(\tilde{a}_1) = \hat{y}_0(\tilde{a}_1) := 0$. This filter has its poles inside the unit circle for all $\tilde{a}_1 \in (-2,2)$. Given $\{\hat{y}_t(\tilde{a}_1)\}$, the FLP mapping in (9.4.14) can be written as

$$\hat{\varphi}(\tilde{a}_1) := -\frac{\sum_{t=3}^n \hat{y}_{t-1}(\tilde{a}_1)\{\hat{y}_t(\tilde{a}_1) + \hat{y}_{t-2}(\tilde{a}_1)\}}{\sum_{t=3}^n\{\hat{y}_{t-1}(\tilde{a}_1)\}^2}. \tag{9.4.25}$$

Hence, the fixed point iteration (9.4.15) takes the form

$$\hat{a}_{1,m} := \hat{\varphi}(\hat{a}_{1,m-1}) \qquad (m = 1,2,\dots). \tag{9.4.26}$$

Observe that $\hat{\rho}(\tilde{a}_1) := -\frac{1}{2}\hat{\varphi}(\tilde{a}_1)$ is just a variation of the ordinary lag-1 sample autocorrelation coefficient of $\{\hat{y}_t(\tilde{a}_1)\}$. For fixed $\eta \in (0,1)$, the following result can be obtained as a corollary to Theorem 9.9. See [235] for a proof.

Corollary 9.1 (Asymptotic Normality of the IFIL Estimator for a Single Real Sinusoid in White Noise: Part 1 [235]). *Let the conditions of Theorem 9.9 be satisfied* *with $q = 1$. Let $\hat{a}_1$ be the fixed point of $\hat{\varphi}(\tilde{a}_1)$ in (9.4.25). Then, for any fixed $\eta \in (0,1)$, $\sqrt{n}(\hat{a}_1 - a_1) \xrightarrow{D} \mathrm{N}(0, v_a^2)$ and $\sqrt{n}(\hat{\omega}_1 - \omega_1) \xrightarrow{D} \mathrm{N}(0, v_\omega^2)$ as $n \to \infty$, where*

$$v_a^2 := (4 - a_1^2)\, v_\omega^2, \quad v_\omega^2 := \frac{1}{\gamma_1^2}\left(\frac{1-\eta^2}{1+\eta^2}\right)^3,$$

with $\gamma_1 := \frac{1}{2}C_1^2/\sigma^2$ being the SNR.

Three observations about the variance v_ω^2 in Corollary 9.1: first, it does not depend on the frequency or the phase of the sinusoid; second, it is inversely related to the SNR; finally, it can be expressed as $\mathcal{O}((1-\eta)^3)$ as $\eta \to 1$. The last observation suggests that the estimation accuracy can be improved by setting η near unity. Of course, when η is very close to unity relative to the sample size n, Corollary 9.1 becomes invalid and more sophisticated analyses are required.

One way of describing the situation where η is very close to unity relative to the sample size is to assume that $\eta \to 1$ as $n \to \infty$. Under this assumption, an analysis is carried out in [355] where the fixed point mapping is defined as

$$\hat{\varphi}(\tilde{a}_1) := -\frac{\sum_{t=1}^n \hat{y}_{t-1}(\tilde{a}_1)\{\hat{y}_t(\tilde{a}_1) + \eta^2 \hat{y}_{t-2}(\tilde{a}_1)\}}{\frac{1}{2}(1+\eta^2)\sum_{t=1}^n \{\hat{y}_{t-1}(\tilde{a}_1)\}^2}. \tag{9.4.27}$$

In this case, $\hat{\rho}(\tilde{a}_1) := -\frac{1}{2}\hat{\varphi}(\tilde{a}_1)$ is another variation of the lag-1 sample autocorrelation coefficient of $\{\hat{y}_t(\tilde{a}_1)\}$. It is also the minimizer of the weighted sum of squared forward and backward prediction errors

$$\sum_{t=1}^n \{\hat{y}_t(\tilde{a}_1) - \rho\, \hat{y}_{t-1}(\tilde{a}_1)\}^2 + \eta^2 \sum_{t=1}^n \{\hat{y}_{t-2}(\tilde{a}_1) - \rho\, \hat{y}_{t-1}(\tilde{a}_1)\}^2$$

with respect to $\rho \in \mathbb{R}$. Although the mapping in (9.4.27) is slightly different from that in (9.4.25), it retains the asymptotic properties described by Theorems 9.8 and 9.9 for fixed η. The following theorem, cited from [355] without proof, reveals two additional rates of convergence for the fixed point of the mapping in (9.4.27), depending on how fast η approaches unity as $n \to \infty$.

Theorem 9.10 (Asymptotic Normality of the IFIL Estimator for a Single Real Sinusoid in White Noise: Part 2 [355]). *Let the conditions of Corollary 9.1 be satisfied.*

(a) If $\eta \in (0,1)$ is chosen such that $n\eta^n = \mathcal{O}(1)$ and $(1-\eta)^{3-2\mu}\log n \to 0$ as $n \to \infty$ for some constant $\mu \in (1, 3/2)$, then the convergence and consistency assertions in Theorem 9.8 remain valid for the mapping in (9.4.27) with $\delta := d(1-\eta)^\mu$ for some constant $d > 0$.

The following assertions are also true under the additional conditions.

(b) If $(1-\eta)^2 n \to \infty$ and $(1-\eta)^5 n \to 0$, then

$$(1-\eta)^{-3/2}\sqrt{n}(\hat{\omega}_1 - \omega_1) \xrightarrow{D} \mathrm{N}(0, \gamma_1^{-2}).$$

Table 9.1. Role of Bandwidth Parameter in IFIL

$1-\eta=\mathcal{O}(n^{-\nu})$	Initial Accuracy	Final Accuracy
$\nu=0$	$\mathcal{O}(1)$	$\mathcal{O}_P(n^{-1/2})$
$\nu\in(1/5,1/2)$	$\mathcal{O}(n^{-\mu\nu})$	$\mathcal{O}_P(n^{-(1+3\nu)/2})$
$\nu\in(1/2,1)$	$\mathcal{O}(n^{-\mu\nu})$	$\mathcal{O}_P(n^{-(2+\nu)/2})$

(c) *If* $(1-\eta)^2 n\to 0$, *then*

$$(1-\eta)^{-1/2} n(\hat{\omega}_1-\omega_1) \xrightarrow{D} \mathrm{N}(0,\gamma_1^{-1}).$$

In these expressions, $\gamma_1 := \frac{1}{2}C_1^2/\sigma^2$ *is the SNR.*

Theorems 9.8–9.10 and Corollary 9.1 highlight the role of η, not only in controlling the accuracy of the IFIL estimator but also in determining the requirement of initial values. For illustration, let

$$1-\eta=\mathcal{O}(n^{-\nu})$$

for some constant $\nu\ge 0$. The case of fixed $\eta\in(0,1)$, as discussed in Corollary 9.1, corresponds to $\nu=0$. For any given $\nu\in(0,1)$, the conditions of Theorem 9.10(a) are satisfied, because

$$n\eta^n = n\{1-\mathcal{O}(n^{-\nu})\}^n \approx n\exp\{-\mathcal{O}(n^{-\nu+1})\} = o(1)$$

and

$$(1-\eta)^{3-2\mu}\log n = \mathcal{O}(n^{-(3-2\mu)\nu})\log n \to 0.$$

In both cases, the radius of the neighborhood $\mathcal{N}$ can be expressed as $\mathcal{O}(n^{-\mu\nu})$ with $\mu\in(1,3/2)$. This expression represents the accuracy required for the initial values of the fixed point iteration. Table 9.1 summarizes the relationship between the required accuracy of initial values and the accuracy of the final IFIL estimator for different choices of ν.

As shown in Table 9.1, the accuracy of the IFIL estimator is inversely related to the accuracy required for the initial values. Indeed, the IFIL estimator with $\nu=0$ achieves the lowest accuracy $\mathcal{O}_P(n^{-1/2})$ but only requires the initial value to be as accurate as $\mathcal{O}(1)$. By taking $\nu\in(1/5,1/2)$, the accuracy of the IFIL estimator is improved to $\mathcal{O}_P(n^{-(1+3\nu)/2})$, but the initial value must be as accurate as $\mathcal{O}(n^{-\mu\nu})$. For example, with $\nu=(1/5)^+$ (i.e., a value just above 1/5), the IFIL estimator reaches an accuracy $o_P(n^{-4/5})$, whereas the required initial accuracy becomes $o(n^{-1/5})$ (taking $\mu=1^+$). If ν is further increased so that $1/2<\nu<1$, then the accuracy of the IFIL estimator becomes $\mathcal{O}_P(n^{-(2+\nu)/2})$ and the required initial accuracy is increased with ν. To achieve the highest final accuracy, take $\nu=1^-$ (i.e., a value just below 1). With this choice, the accuracy of the IFIL estimator is arbitrarily close to $\mathcal{O}_P(n^{-3/2})$, the optimal error rate, but the required accuracy of initial values becomes $\mathcal{O}(n^{-1})$ (taking $\mu = 1/\nu$).

This analysis suggests a three-step procedure which is able to bring an initial guess of accuracy $\mathcal{O}(1)$ to a final estimate whose accuracy is arbitrarily close to the optimal rate $\mathcal{O}_P(n^{-3/2})$.

1. Start with the initial guess and carry out the iteration (9.4.26) with a suitably small $\eta_1 \in (0,1)$ such that $1-\eta_1 = \mathcal{O}(n^{-\nu_1})$ and $\nu_1 = 0$;
2. Use the estimate obtained from Step 1 to initiate the iteration (9.4.26) with a suitably larger $\eta_2 \in (\eta_1, 1)$ such that $1-\eta_2 = \mathcal{O}(n^{-\nu_2})$ and $\nu_2 \in (1/5, 1/2)$;
3. Use the estimate obtained from Step 2 to initiate (9.4.26) with an even larger $\eta_3 \in (\eta_2, 1)$ such that $1-\eta_3 = \mathcal{O}(n^{-\nu_3})$ and $\nu_3 \in (1/2, 1)$.

The estimate from Step 1 has accuracy $\mathcal{O}_P(n^{-1/2})$; it is accurate enough to initiate Step 2 because there exists a constant $\mu_2 \in (1, 3/2)$ such that $\mu_2\nu_2 < 1/2$. The estimate from Step 2 has accuracy $\mathcal{O}_P(n^{-(1+3\nu_2)/2})$; it is good enough to initiate Step 3 because there exists a constant $\mu_3 \in (1, 3/2)$ such that $\mu_3\nu_3 < 1 < (1+3\nu_2)/2$. Therefore, the convergence of the entire procedure is guaranteed by Theorems 9.8–9.10 and Corollary 9.1. By taking $\nu_3 = 1^-$, one can make the accuracy of the estimate from Step 3 arbitrarily close to $\mathcal{O}_P(n^{-3/2})$. Finally, one can use the estimators given by (9.4.25) and (9.4.27) without filtering to produce the initial value in Step 1. They are nothing but variations of Prony's estimator. In this way, the entire procedure becomes self-contained.

For the extreme case of $\eta = 1$, the analysis in [304] suggests that the corresponding IFIL estimator attains the Gaussian CRLB asymptotically, i.e.,

$$n^{-3/2}(\hat{\omega}_1 - \omega_1) \xrightarrow{D} \mathrm{N}(0, 12\gamma_1^{-1}),$$

and the required initial accuracy takes the form $o(n^{-1})$. Therefore, one may also take $\eta_3 = 1$ in Step 3 because the accuracy of the estimate in Step 2 takes the form $o_P(n^{-1})$, which is sufficient to initiate Step 3 with $\eta_3 = 1$. See also [393] for a related algorithm.

Now let us consider the computation required in each step of the procedure. By the mean-value theorem in calculus, we can write

$$\hat{\varphi}(\tilde{a}_1) - \hat{a}_1 = \hat{c}(\tilde{a}_1)(\tilde{a}_1 - \hat{a}_1), \tag{9.4.28}$$

where $\hat{c}(\tilde{a}_1) := \int_0^1 \nabla\hat{\varphi}(\hat{a}_1 + s(\tilde{a}_1 - \hat{a}_1))\,ds$. In the case of $1-\eta = \mathcal{O}(1)$, there exists a constant $c \in (0,1)$ such that

$$\max_{\tilde{a}_1 \in \mathcal{N}} |\hat{c}(\tilde{a}_1)| \le c$$

almost surely for large n. Combining this result with (9.4.26) and (9.4.28) gives

$$|\hat{a}_{1,m} - \hat{a}_1| \le c^m |\hat{a}_{1,0} - \hat{a}_1| \qquad (m = 1, 2, \ldots).$$

Because the final estimate $\hat{a}_1$ has accuracy $\mathcal{O}_P(n^{-1/2})$, there is no significant gain to continue the iteration after $\hat{a}_{1,m}$ falls into a neighborhood of $\hat{a}_1$ with radius $\mathcal{O}(n^{-1/2})$. Therefore, Step 1 can be terminated after $m_1 := \mathcal{O}(\log n)$ iterations. In the case of $1-\eta = \mathcal{O}(n^{-\nu})$ for some $\nu \in (0,1)$, it can be shown [355] that

$$\hat{c}(\tilde{a}_1) = \mathcal{O}(n^{-\nu})$$

uniformly in $\tilde{a}_1 \in \mathcal{N}$. Therefore, the final accuracy $\mathcal{O}_P(n^{-(1+3\nu_2)/2})$ in Step 2 can be reached after $m_2 := \lceil (1+3\nu_2)/(2\nu_2) \rceil \le 4$ iterations, independent of n. Similarly, to achieve the final accuracy $\mathcal{O}_P(n^{-(2+\nu_3)/2})$ in Step 3, it suffices to iterate $m_3 := \lceil (2+\nu_3)/(2\nu_3) \rceil \le 3$ times. These numbers may be rather optimistic and appropriate only for large n. Nevertheless, one can say that the number of iterations required by Steps 2 and 3 takes the form $\mathcal{O}(1)$. In total, the three-step procedure takes $\mathcal{O}(\log n)$ iterations to improve the estimates from an initial accuracy $\mathcal{O}_P(1)$ to a final accuracy which is arbitrarily close to $\mathcal{O}_P(n^{-3/2})$. Because each iteration involves $\mathcal{O}(n)$ multiplications to compute the estimates, the total computational complexity can be expressed as $\mathcal{O}(n \log n)$, which has the same order of magnitude as the DFT method discussed in Chapter 6 (Section 6.1).

The univariate IFIL algorithm (9.4.24)–(9.4.26) can be employed to estimate the frequencies of multiple sinusoids if the frequencies are well separated. In this case, the mapping $\hat{\varphi}(\tilde{a}_1)$ will have an attractive fixed point in the vicinity of each signal frequency and the iteration (9.4.26) will converge to one of the fixed points, depending on the initial value. This is illustrated by Figure 9.12 which depicts the trajectory of frequency estimates produced by the univariate IFIL algorithm for a time series consisting of three real sinusoids in Gaussian white noise. In this example, three increasing values of η are used to obtain the IFIL estimates with the mapping $\hat{\varphi}(\tilde{a}_1)$ given by (9.4.27). Note that the last value of η must be sufficiently close to 1 in order to resolve the first two frequencies.

Due to the presence of multiple sinusoids, the convergence of the univariate IFIL algorithm depends not only on η but also on the frequency separation condition relative to the choice of η. A rigorous investigation of this relationship is carried out in [237]. To present the results, define

$$a_{1k} := -2\cos(\omega_k), \quad \Delta := \min_{k \neq k'}\{|\omega_k - \omega_{k'}|\},$$

and

$$b_k := \frac{C_k^2}{\sum_{k' \neq k} \frac{1}{2} C_{k'}^2 \{\cot(\frac{1}{2}(\omega_k - \omega_{k'})) + \cot(\frac{1}{2}(\omega_k + \omega_{k'}))\}}.$$

Note that b_k can be interpreted as the signal-to-interference ratio for the kth sinusoid. Moreover, let us allow Δ to approach zero as $n \to \infty$. Under this condition, the following theorem, cited from [237] without proof, can be established.

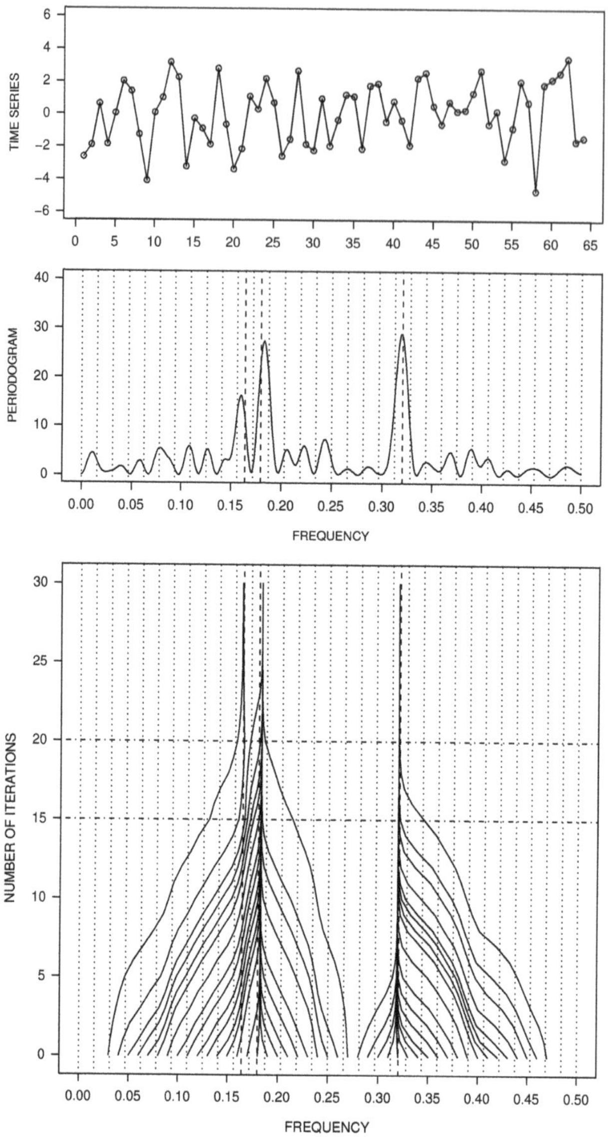

Figure 9.12. Estimation of multiple frequencies by the univariate IFIL algorithm. Top, a time series consisting of three unit-amplitude zero-phase real sinusoids in Gaussian white noise ($n = 64$, $\omega_1 = 2\pi \times 10.5/n = 2\pi \times 0.1640625$, $\omega_2 = 2\pi \times 11.5/n = 2\pi \times 0.1796875$, $\omega_3 = 2\pi \times 20.5/n = 2\pi \times 0.3203125$, SNR = −6 dB per sinusoid). Middle, periodogram. Bottom, trajectory of the IFIL frequency estimates for different initial values, where $\eta = (0.96, 0.98, 0.99)$ and $m = (15, 5, 10)$. Dashed vertical lines depict the signal frequencies. Dotted vertical lines depict the Fourier grid. Dash-dotted horizontal lines indicate the iteration number at which the value of η is increased.

Table 9.2. Role of Bandwidth Parameter in IFIL: Multiple Sinusoids

$1-\eta=\mathcal{O}(n^{-\nu})$	Initial Accuracy	Final Accuracy	Δ^{-1}
$\nu\in(1/5,1/3]$	$\mathcal{O}(n^{-\mu\nu})$	$\mathcal{O}_P(n^{-2\nu})$	$\mathcal{O}(n^{(5\nu-1)/8})$
$\nu\in(1/3,1/2)$	$\mathcal{O}(n^{-\mu\nu})$	$\mathcal{O}_P(n^{-2\nu})$	$o(n^{\nu/4})$
$\nu\in(1/2,2/3)$	$\mathcal{O}(n^{-\mu\nu})$	$\mathcal{O}_P(n^{-2\nu})$	$o(n^{\nu/4})$
$\nu\in(2/3,1)$	$\mathcal{O}(n^{-\mu\nu})$	$\mathcal{O}_P(n^{-1-\nu/2})$	$o(n^{\nu/4})$

Theorem 9.11 (Estimation of Multiple Sinusoids by the Univariate IFIL Algorithm [237]). *Let $y_t:=x_t+\epsilon_t$ $(t=1,\ldots,n)$, where $\{x_t\}$ is given by (2.1.1) or (2.1.2) and $\{\epsilon_t\}\sim\mathrm{IID}(0,\sigma^2)$ with $E(\epsilon_t^4)<\infty$.*

(a) Assume that η satisfies the conditions in Theorem 9.10(a) and the additional condition that $\Delta^{-2}(1-\eta)^{2-\mu}=\mathcal{O}(1)$. Then, for any $k=1,\ldots,q$, there exists a neighborhood $\mathcal{N}_k:=\{\tilde{a}_1\in\mathcal{A}_0:|\tilde{a}_1-a_{1k}|\le\delta\}$ with $\delta:=d_k(1-\eta)^{\mu}$ for some constant $d_k>0$ such that almost surely for sufficiently large n the mapping $\hat{\varphi}(\tilde{a}_1)$ in (9.4.27) has a unique fixed point in $\mathcal{N}_k$, denoted as $\hat{\alpha}_k$, and the sequence $\{\hat{a}_{1,m}\}$ generated by (9.4.26) converges to $\hat{\alpha}_k$ for any $\hat{a}_{1,0}\in\mathcal{N}_k$. Moreover, as $n\to\infty$, $\hat{\alpha}_k\overset{a.s.}{\to}a_{1k}$ and $\hat{\omega}_k:=\arccos(-\frac{1}{2}\hat{\alpha}_k)\overset{a.s.}{\to}\omega_k$.

The following assertions are also true under the additional conditions.

(b) If $(1-\eta)^2n\to\infty$, $(1-\eta)^{5-2r}n=\mathcal{O}(1)$, and $\Delta^{-4}(1-\eta)^r\to0$ for some constant $r\in(0,1]$, then, as $n\to\infty$,

$$(1-\eta)^{-3/2}\sqrt{n}\,(\hat{\omega}_k-\omega_k-(1-\eta)^2\eta^{-1}b_k^{-1})\overset{D}{\to}\mathrm{N}(0,\gamma_k^{-2}).$$

(c) If $(1-\eta)^2n\to0$ and $\Delta^{-4}(1-\eta)\to0$, then

$$(1-\eta)^{-1/2}n\,(\hat{\omega}_k-\omega_k-(1-\eta)^2\eta^{-1}b_k^{-1})\overset{D}{\to}\mathrm{N}(0,\gamma_k^{-1}).$$

In these expressions, $\gamma_k:=\frac{1}{2}C_k^2/\sigma^2$ is the SNR of the kth sinusoid.

For illustration, let $1-\eta=\mathcal{O}(n^{-\nu})$ again. Table 9.2 summarizes the accuracy of the IFIL estimator together with the corresponding requirement of initial values and frequency separation for different choices of ν. Note that the choice $1-\eta=\mathcal{O}(1)$ (or $\nu=0$) is no longer valid because the filter (9.4.24), which is designed under the assumption of a single sinusoid, needs a shrinking bandwidth as n grows in order to suppress the interference among multiple sinusoids.

To derive the results in Table 9.2, let us observe that the conditions in Theorem 9.10(a) are satisfied for any $\nu\in(0,1)$. The additional condition in Theorem 9.11(a) is satisfied if $\Delta^{-1}=\mathcal{O}(n^{(2-\mu)\nu/2})$, which is implied by the assumption $\Delta^{-1}=o(n^{\nu r/4})$ for some $r\in(0,1]$. Theorem 9.11(b) requires $1/(5-2r)\le\nu<1/2$ and $\Delta^{-1}=o(n^{\nu r/4})$, in which case the accuracy of the IFIL estimator takes the form $\mathcal{O}(n^{-2\nu})+\mathcal{O}_P(n^{-(1+3\nu)/2})=\mathcal{O}_P(n^{-2\nu})$. Theorem 9.11(c) requires $1/2<\nu<1$

and $\Delta^{-1} = \mathcal{O}(n^{\nu/4})$, in which case the accuracy of the IFIL estimator can be expressed as $\mathcal{O}(n^{-2\nu}) + \mathcal{O}_P(n^{-1-\nu/2})$, which takes the form $\mathcal{O}_P(n^{-2\nu})$ if $\nu \in (1/2, 2/3)$ and the form $\mathcal{O}_P(n^{-1-\nu/2})$ if $\nu \in (2/3, 1)$.

As we can see in this analysis, when the bandwidth is not small enough, the estimation error is dominated by the bias of the form $\mathcal{O}(n^{-2\nu})$ which occurs as a result of the interference from other sinusoids. The bias becomes negligible when the bandwidth is reduced with the choice of $\nu \in (2/3, 1)$. Because the bandwidth is narrower when η is closer to unity, the required frequency separation becomes less stringent but the required accuracy for initial values becomes more stringent. Note that the frequency separation conditions in Table 9.2 are all satisfied if Δ is bounded away from zero as $n \to \infty$.

The results shown in Table 9.2 suggest that, for estimating the frequencies of multiple sinusoids, the univariate IFIL algorithm should employ a shrinking bandwidth determined by three increasing values of ν, i.e., $\nu_1 \in [1/3, 1/2)$, $\nu_2 \in (1/2, 3/2)$, and $\nu_3 = 1^-$, such that $1-\eta = \mathcal{O}(n^{-\nu})$. This procedure is able to accommodate an initial guess of accuracy $\mathcal{O}(n^{-1/3})$ and produce a final estimate whose accuracy is arbitrarily close to the optimal rate $\mathcal{O}_P(n^{-3/2})$. To achieve such results, the frequency separation must be greater than $\mathcal{O}(n^{-1/12})$. Moreover, it can be shown [237] that the total computational complex of the procedure takes the form $\mathcal{O}(n \log n)$ as in the single sinusoid case.

9.5 Iterative Quasi Gaussian Maximum Likelihood Method

As demonstrated in Sections 9.3 and 9.4, a main benefit of the AR reparameterization is that it transforms the nonlinear problem of frequency estimation into a linear prediction problem that can be solved by relatively simple algorithms. In this section, we discuss two additional algorithms derived from the AR reparameterization on the basis of the Gaussian maximum likelihood principle.

Consider the complex case first. In Chapter 8 (Section 8.2), we show that the Gaussian maximum likelihood (GML) estimator of $\boldsymbol{\omega} := (\omega_1, \ldots, \omega_p)$ is the solution to a nonlinear least-squares (NLS) problem that minimizes the criterion

$$\|\mathbf{y} - \mathbf{F}(\tilde{\boldsymbol{\omega}})\hat{\boldsymbol{\beta}}(\tilde{\boldsymbol{\omega}})\|^2 = \mathbf{y}^H\{\mathbf{I} - \mathbf{Q}(\tilde{\boldsymbol{\omega}})\}\mathbf{y}, \tag{9.5.1}$$

where

$$\mathbf{Q}(\tilde{\boldsymbol{\omega}}) := \mathbf{F}(\tilde{\boldsymbol{\omega}})\{\mathbf{F}^H(\tilde{\boldsymbol{\omega}})\mathbf{F}(\tilde{\boldsymbol{\omega}})\}^T\mathbf{F}^H(\tilde{\boldsymbol{\omega}}),$$
$$\mathbf{F}(\tilde{\boldsymbol{\omega}}) := [\mathbf{f}(\tilde{\omega}_1), \ldots, \mathbf{f}(\tilde{\omega}_p)], \quad \mathbf{f}(\omega) := [\exp(i\omega), \ldots, \exp(in\omega)]^T.$$

This criterion can be reparameterized in terms of the AR parameter

$$\mathbf{a} := [a_1, \ldots, a_p]^T := \Phi_{\text{AR}}(\boldsymbol{\omega}).$$

Toward that end, consider the n-by-$(n-p)$ matrix

$$\mathbf{A} := \begin{bmatrix} a_p^* & 0 & \cdots & 0 \\ a_{p-1}^* & a_p^* & & 0 \\ \vdots & \vdots & & \vdots \\ a_0^* & a_1^* & & 0 \\ 0 & a_0^* & & a_p^* \\ \vdots & & & \vdots \\ 0 & 0 & & a_0^* \end{bmatrix}. \tag{9.5.2}$$

It is easy to verify that

$$\mathbf{A}^H \mathbf{f}(\omega_k) = a(z_k) \begin{bmatrix} z_k^{p+1} \\ \vdots \\ z_k^n \end{bmatrix} = \mathbf{0} \qquad (k = 1, \ldots, p). \tag{9.5.3}$$

Because $\mathbf{A}$ has full (column) rank $n-p$, it follows from (9.5.3) that the column space of $\mathbf{A}$ must coincide with the orthogonal complement of the column space of $\mathbf{F}(\boldsymbol{\omega})$. This assertion remains valid for $\tilde{\mathbf{A}}$ and $\mathbf{F}(\tilde{\boldsymbol{\omega}})$, where $\tilde{\mathbf{A}}$ is defined by (9.5.2) in terms of $\tilde{\mathbf{a}} := \Phi_{\text{AR}}(\tilde{\boldsymbol{\omega}})$ for any $\tilde{\boldsymbol{\omega}} \in \Omega$. The projection matrix onto the column space of $\tilde{\mathbf{A}}$ equals $\tilde{\mathbf{A}}(\tilde{\mathbf{A}}^H\tilde{\mathbf{A}})^{-1}\tilde{\mathbf{A}}^H$ and the projection matrix onto the orthogonal complement of the column space of $\mathbf{F}(\tilde{\boldsymbol{\omega}})$ equals $\mathbf{I} - \mathbf{Q}(\tilde{\boldsymbol{\omega}})$. The coincidence of these spaces and the uniqueness of projection matrix imply that

$$\mathbf{I} - \mathbf{Q}(\tilde{\boldsymbol{\omega}}) = \tilde{\mathbf{A}}(\tilde{\mathbf{A}}^H\tilde{\mathbf{A}})^{-1}\tilde{\mathbf{A}}^H. \tag{9.5.4}$$

Substituting this expression in (9.5.1) leads to

$$\|\mathbf{y} - \mathbf{F}(\tilde{\boldsymbol{\omega}})\hat{\boldsymbol{\beta}}(\tilde{\boldsymbol{\omega}})\|^2 = \mathbf{y}^H\tilde{\mathbf{A}}(\tilde{\mathbf{A}}^H\tilde{\mathbf{A}})^{-1}\tilde{\mathbf{A}}^H\mathbf{y}. \tag{9.5.5}$$

Consequently, the GML estimator of $\mathbf{a}$ is given by

$$\hat{\mathbf{a}} := \underset{\tilde{\mathbf{a}} \in \mathcal{A}}{\arg\min}\, \mathbf{y}^H\tilde{\mathbf{A}}(\tilde{\mathbf{A}}^H\tilde{\mathbf{A}})^{-1}\tilde{\mathbf{A}}^H\mathbf{y}, \tag{9.5.6}$$

where $\mathcal{A} := \Phi_{\text{AR}}(\Omega) \subset \mathbb{C}^p$ is the parameter space. A similar AR reparameterization can be obtained in the real case with $\tilde{\mathbf{a}} := \mathbf{B}\tilde{\mathbf{a}}_r + \mathbf{b}$ and $\tilde{\mathbf{a}}_r \in \mathcal{A}_r := \Phi_{\text{AR}}(\Omega_r) \subset \mathbb{R}^q$.

Note that the GML estimator of $\mathbf{a}$ in (9.5.6) is also the maximizer of

$$S_n(\tilde{\mathbf{a}}) := \|\mathbf{y}\|^2 - \mathbf{y}^H\tilde{\mathbf{A}}(\tilde{\mathbf{A}}^H\tilde{\mathbf{A}})^{-1}\tilde{\mathbf{A}}^H\mathbf{y}. \tag{9.5.7}$$

This function can be regarded as an extension of the multivariate periodogram $P_n(\tilde{\boldsymbol{\omega}}) := \|\mathbf{y}\|^2 - \|\mathbf{y} - \mathbf{F}(\tilde{\boldsymbol{\omega}})\hat{\boldsymbol{\beta}}(\tilde{\boldsymbol{\omega}})\|^2$ for the AR parameter $\tilde{\mathbf{a}}$ from $\mathcal{A}$ to $\mathbb{C}^p$, because

$$S_n(\tilde{\mathbf{a}}) = P_n(\tilde{\boldsymbol{\omega}}) \qquad \forall\, \tilde{\mathbf{a}} \in \mathcal{A} \ \text{ and } \ \tilde{\boldsymbol{\omega}} := \Phi_{\text{AR}}^{-1}(\tilde{\mathbf{a}}).$$

For this reason, we call $S_n(\tilde{\mathbf{a}})$ the multivariate periodogram in the AR domain, or simply the multivariate AR periodogram.

As a function of $\tilde{\mathbf{a}} \in \mathbb{C}^p$, the multivariate AR periodogram $S_n(\tilde{\mathbf{a}})$ is just as rugged as the ordinary multivariate periodogram $P_n(\tilde{\boldsymbol{\omega}})$ for $\tilde{\boldsymbol{\omega}} \in \Omega$. An example is shown in Figure 9.13, where $S_n(\tilde{\mathbf{a}})$ with $p = 1$ is depicted as a function of $\tilde{\mathbf{a}}^* = \tilde{a}_1^* \in \mathbb{C}$ for a time series consisting of a single complex sinusoid in Gaussian white noise. In this case $\mathcal{A}$ is just the unit circle. Clearly, the multivariate AR periodogram exhibits numerous local maxima in addition to a sharp global peak near the true parameter value $\mathbf{a}^* = a_1^* = -\exp(-i\omega_1)$. Standard optimization techniques, such as the Newton-Raphson algorithm, can be used to compute the global maximizer of $S_n(\tilde{\mathbf{a}})$ [359], but the computation remains as challenging as the computation of the NLS estimator in the original frequency domain.

So, where is the benefit? It turns out that the special structure of the AR reparameterization in (9.5.5) can be utilized to devise simple iterative algorithms. Two such algorithms — the iterative generalized least-squares algorithm and the iterative least-eigenvalue algorithm — are discussed in the remainder of this section. Both algorithms approximate the Gaussian maximum likelihood problem in (9.5.6) by a sequence of quadratic minimization problems that have simple solutions in a larger space of the AR parameter. This approximation technique is known in the literature as the iterative quadratic maximum likelihood (IQML) method [44]. To emphasize its statistical underpinning, we call it the iterative quasi Gaussian maximum likelihood method.

9.5.1 Iterative Generalized Least-Squares Algorithm

In the complex case with $\tilde{\mathbf{a}} := [\tilde{a}_1, \ldots, \tilde{a}_p]^T$ and $\tilde{a}_0 := 1$, it is easy to verify that

$$\tilde{\mathbf{A}}^H \mathbf{y} = \mathbf{y}_p + \mathbf{Y}_p \tilde{\mathbf{a}}, \tag{9.5.8}$$

where

$$\mathbf{Y}_p := \begin{bmatrix} y_p & \cdots & y_1 \\ \vdots & & \vdots \\ y_{n-1} & \cdots & y_{n-p} \end{bmatrix}, \quad \mathbf{y}_p := \begin{bmatrix} y_{p+1} \\ \vdots \\ y_n \end{bmatrix}. \tag{9.5.9}$$

Therefore, the criterion in (9.5.5) can be expressed as

$$(\mathbf{y}_p + \mathbf{Y}_p \tilde{\mathbf{a}})^H (\tilde{\mathbf{A}}^H \tilde{\mathbf{A}})^{-1} (\mathbf{y}_p + \mathbf{Y}_p \tilde{\mathbf{a}}). \tag{9.5.10}$$

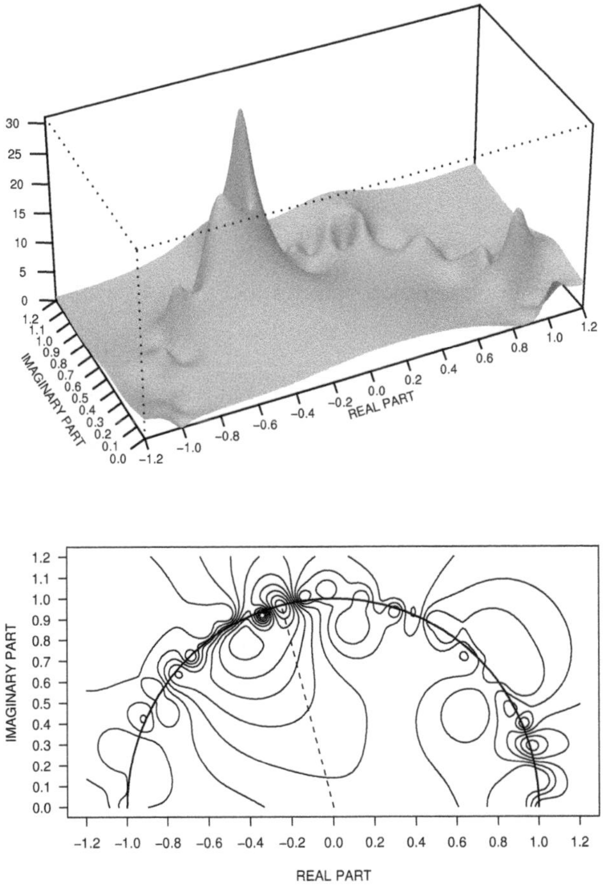

Figure 9.13. Multivariate AR periodogram in (9.5.7) with $p = 1$, as a function of the real and imaginary parts of $\tilde{a}_1^*$, for a time series consisting of a single complex sinusoid in Gaussian white noise ($n = 50$, $\omega_1 = 2\pi \times 0.21$, SNR = −5 dB). Thick solid line depicts the unit circle. Dashed line shows the location on the unit circle that corresponds to the signal frequency. The same time series is used to produce Figures 6.11 and 6.12.

Minimizing this function with respect to $\tilde{\mathbf{a}}$ is analogous to a generalized least-squares (GLS) problem except that the weight matrix $(\tilde{\mathbf{A}}^H\tilde{\mathbf{A}})^{-1}$ depends on the variable $\tilde{\mathbf{a}}$. If the weight matrix is replaced by an estimate, then the resulting GLS problem becomes a linear one which has a simple solution. The GLS solution not only serves as an estimate of $\mathbf{a}$ but also produces a new estimate of the weight matrix. Repeating this procedure leads to an iterative algorithm, which we call the *iterative generalized least-squares* (IGLS) algorithm. The resulting estimates of $\mathbf{a}$ and $\boldsymbol{\omega} = \Phi_{\mathrm{AR}}^{-1}(\mathbf{a})$ are called the IGLS estimates.

More precisely, for any given estimate $\hat{\mathbf{a}}_m := [\hat{a}_{1,m}, \ldots, \hat{a}_{p,m}]^T$ $(m = 0, 1, \ldots)$, let $\tilde{\mathbf{A}}_m$ be defined by (9.5.2) on the basis of $\hat{\mathbf{a}}_m$ and let the matrix $\tilde{\mathbf{A}}^H\tilde{\mathbf{A}}$ in (9.5.10) be replaced by its estimate

$$\mathbf{R}_m := \tilde{\mathbf{A}}_m^H\tilde{\mathbf{A}}_m. \tag{9.5.11}$$

Minimizing the resulting criterion with respect to $\tilde{\mathbf{a}} \in \mathbb{C}^p$ without any constraint yields the GLS solution

$$\hat{\mathbf{a}}_{m+1} := -(\mathbf{Y}_p^H\mathbf{R}_m^{-1}\mathbf{Y}_p)^{-1}\mathbf{Y}_p^H\mathbf{R}_m^{-1}\mathbf{y}_p \qquad (m = 0, 1, \ldots). \tag{9.5.12}$$

The algorithm comprising (9.5.11) and (9.5.12) will be referred to as the *unconstrained* IGLS algorithm, or simply the IGLS algorithm. A procedure of the form (9.5.12) is used in [103] to solve a filter design problem (see also [203]). It is also a variation of the IQML algorithm discussed in [44].

The IGLS estimates given by (9.5.12) are not necessarily in the parameter space $\mathcal{A} := \Phi_{\mathrm{AR}}(\Omega)$, but the constraint can be imposed by projecting the GLS solution onto $\mathcal{A}$ so that the new estimate $\hat{\mathbf{a}}_{m+1}$ takes the form

$$\hat{\mathbf{a}}_{m+1} := \mathcal{A}(-(\mathbf{Y}_p^H\mathbf{R}_m^{-1}\mathbf{Y}_p)^{-1}\mathbf{Y}_p^H\mathbf{R}_m^{-1}\mathbf{y}_p) \qquad (m = 0, 1, \ldots). \tag{9.5.13}$$

This will be referred to as the *constrained* IGLS algorithm. Note that performing the projection in (9.5.13) at every iteration may become burdensome computationally when it requires repeated use of a nontrivial root-finding procedure.

The special structure of $\mathbf{R}_m$ can be exploited to speed up the computation of the GLS estimator in (9.5.12) [73] [154] [203]. Indeed, with $\mathbf{A}$ defined by (9.5.2), $\mathbf{A}^H\mathbf{A}$ is equal to the covariance matrix of $[X_1, \ldots, X_{n-p}]^T$, where $\{X_t\}$ is an MA(p) process of the form $X_t := \sum_{j=0}^{p} a_j \zeta_{t-j}$ with $\{\zeta_t\} \sim \mathrm{IID}(0,1)$. In other words,

$$\mathbf{A}^H\mathbf{A} = [r_{\mathrm{MA}}(t-s)] \quad (t, s = 1, \ldots, n-p),$$

where

$$r_{\mathrm{MA}}(t-s) := E(X_t X_s^*) = \begin{cases} \sum_{j=0}^{p} a_{j+t-s}\, a_j^* & \text{if } t \ge s, \\ r_{\mathrm{MA}}^*(t-s) & \text{if } t < s, \\ 0 & \text{if } |t-s| > p, \end{cases}$$

with $a_j := 0$ for $j \notin \{0,1,\ldots,p\}$. With $\{r_{\mathrm{MA}}(0), r_{\mathrm{MA}}(1),\ldots,r_{\mathrm{MA}}(p)\}$ as input, an application of the innovations algorithm in Proposition 2.10 for $t=1,\ldots,n-p-1$ produces the Cholesky decomposition $\mathbf{A}^H\mathbf{A} = \mathbf{L}\mathbf{D}\mathbf{L}^H$, where $\mathbf{D}$ is a diagonal matrix and $\mathbf{L}$ is a lower triangular band matrix of bandwidth $p+1$. For large n, the complexity of the innovations algorithm takes the form $\mathcal{O}(n)$.

Let $\mathbf{R}_m = \mathbf{L}_m\mathbf{D}_m\mathbf{L}_m^H$ be the Cholesky decomposition computed by the innovations algorithm on the basis of $\hat{\mathbf{a}}_m$, where

$$\begin{aligned}\mathbf{L}_m &:= [\psi_{t-1,t-s}(m)] \quad (t,s=1,\ldots,n-p),\\ \mathbf{D}_m &:= \mathrm{diag}\{v_0(m), v_1(m),\ldots,v_{n-p-1}(m)\}.\end{aligned}$$

Define $\mathbf{U}_m := \mathbf{L}_m^{-1}\mathbf{Y}_p$ and $\mathbf{u}_m := \mathbf{L}_m^{-1}\mathbf{y}_p$. Then, the IGLS estimator in (9.5.12) can be expressed as

$$\hat{\mathbf{a}}_{m+1} = -(\mathbf{U}_m^H\mathbf{D}_m^{-1}\mathbf{U}_m)^{-1}\mathbf{U}_m^H\mathbf{D}_m^{-1}\mathbf{u}_m \qquad (m=0,1,\ldots). \tag{9.5.14}$$

Because $\mathbf{L}_m$ is a lower triangular band matrix, it is possible to compute $\mathbf{U}_m$ and $\mathbf{u}_m$ with complexity $\mathcal{O}(n)$. Indeed, define

$$\mathbf{u}_k(m) := [u_{1k}(m),\ldots,u_{n-p,k}(m)]^T \qquad (k=0,1,\ldots,p),$$

and

$$u_{tk}(m) := y_{t+k} - \sum_{j=1}^{p} \psi_{t-1,j}(m)\, u_{t-j,k}(m) \qquad (t=1,\ldots,n-p), \tag{9.5.15}$$

where $u_{tk}(m) := 0$ for $t \le 0$ and $\psi_{0j}(m) := 0$ for all j. It is easy to verify that

$$\mathbf{u}_m = \mathbf{u}_p(m), \quad \mathbf{U}_m = [\mathbf{u}_{p-1}(m),\ldots,\mathbf{u}_1(m),\mathbf{u}_0(m)]. \tag{9.5.16}$$

Hence the total complexity of computing $\hat{\mathbf{a}}_{m+1}$ in (9.5.14) by means of the innovations algorithm together with (9.5.15) and (9.5.16) can be expressed as $\mathcal{O}(n)$.

Example 9.15 (IGLS Algorithm for a Single Complex Sinusoid). For $p=1$, we have $r_{\mathrm{MA}}(0) = 1+|a_1|^2$, $r_{\mathrm{MA}}(1) = a_1$, and $r_{\mathrm{MA}}(u) = 0$ for $u \ge 2$. The innovations algorithm in Proposition 2.10 gives $v_0 = r_{\mathrm{MA}}(0)$, $\psi_{t1} = r_{\mathrm{MA}}(1)/v_{t-1}$ for $t \ge 1$, $\psi_{tj} = 0$ for $t \ge j \ge 2$, and $v_t = r_{\mathrm{MA}}(0) - |\psi_{t1}|^2 v_{t-1}$ for $t \ge 1$. In other words,

$$\begin{cases}\psi_{t1} = a_1/v_{t-1}\\ v_t = 1+|a_1|^2 - |a_1|^2/v_{t-1}\end{cases} \qquad (t=1,\ldots,n-2), \tag{9.5.17}$$

with $v_0 = 1+|a_1|^2$. Therefore, we obtain $\mathbf{D} = \mathrm{diag}(v_0, v_1,\ldots,v_{n-2})$ and

$$\mathbf{L} = \begin{bmatrix} 1 & & & 0\\ \psi_{11} & 1 & & \\ & \ddots & \ddots & \\ 0 & & \psi_{n-2,1} & 1\end{bmatrix} \in \mathbb{C}^{(n-1)\times(n-1)}.$$

Let $\mathbf{u} := [u_1, \ldots, u_{n-1}]^T$ be the solution to the equation

$$\mathbf{L}\mathbf{u} = \mathbf{d}$$

for some $\mathbf{d} := [d_1, \ldots, d_{n-1}]^T$. The u_t can be obtained recursively:

$$u_t = d_t - \psi_{t-1,1} u_{t-1} \qquad (t = 1, \ldots, n-1) \tag{9.5.18}$$

with $u_0 := \psi_{01} := 0$. As can be seen, the algorithm defined by (9.5.17) and (9.5.18) requires $2(n-2)+(n-1) = 2n-5$ multiplications/divisions and $(n-2)+(n-1) = 2n-3$ additions/subtractions.

Now let $\{\psi_{t1}(m)\}$ and $\{v_t(m)\}$ denote the results from (9.5.17) with a_1 replaced by $\hat{a}_{1,m}$. Let $\{u_{t0}(m)\}$ and $\{u_{t1}(m)\}$ denote the results from (9.5.18) with $\mathbf{d} := [y_1, \ldots, y_{n-1}]^T$ and $\mathbf{d} := [y_2, \ldots, y_n]^T$, respectively, and with $\{\psi_{t1}(m)\}$ in place of $\{\psi_{t1}\}$. In other words, let

$$\begin{cases} u_{t0}(m) = y_t - \psi_{t-1,1}(m)\, u_{t-1,0}(m) \\ u_{t1}(m) = y_{t+1} - \psi_{t-1,1}(m)\, u_{t-1,1}(m) \end{cases} \qquad (t = 1, \ldots, n-1),$$

where $u_{00}(m) := u_{01}(m) := \psi_{01}(m) := 0$. Then, it follows from (9.5.14) that

$$\hat{a}_{1,m+1} = -\frac{\sum_{t=1}^{n-1} u_{t0}(m)\, u_{t1}^*(m)/v_{t-1}(m)}{\sum_{t=1}^{n-1} |u_{t0}(m)|^2/v_{t-1}(m)} \qquad (m = 0, 1, \ldots). \tag{9.5.19}$$

Moreover, projecting the unconstrained IGLS estimate $\hat{a}_{1,m}$ onto the parameter space $\mathcal{A} = \{\tilde{a}_1 : |\tilde{a}_1| = 1\}$ gives the constrained IGLS estimate

$$\hat{a}_{1,m} \mapsto \hat{a}_{1,m}/|\hat{a}_{1,m}|.$$

In both cases, the frequency estimate $\hat{\omega}_{1,m}$ equals the angle of $\hat{z}_{1,m} := -\hat{a}_{1,m}$.

Figure 9.14 depicts the trajectory of the constrained and unconstrained IGLS frequency estimates for a time series comprising a single complex sinusoid in Gaussian white noise. The same time series is used to produce Figure 9.13.

The top panel of Figure 9.14 shows that although converging to a spurious frequency is still possible when the initial value is too far from the signal frequency, the constrained IGLS algorithm has a much larger basin of attraction (BOA) than the periodogram-maximizing algorithms discussed in Chapter 6 (Figures 6.11 and 6.12). The improved convergence property can be attributed largely to the fact that the constrained IGLS algorithm allows the intermediate estimates to go beyond the parameter space $\mathcal{A}$ (which is the unit circle in this case) before projecting them back onto $\mathcal{A}$. The enlarged parameter space provides the possibility for the algorithm to take a more suitable route in reaching the desired destination. The middle panel of Figure 9.14 shows that by ignoring the projection altogether the unconstrained IGLS algorithm exhibits an even better convergence

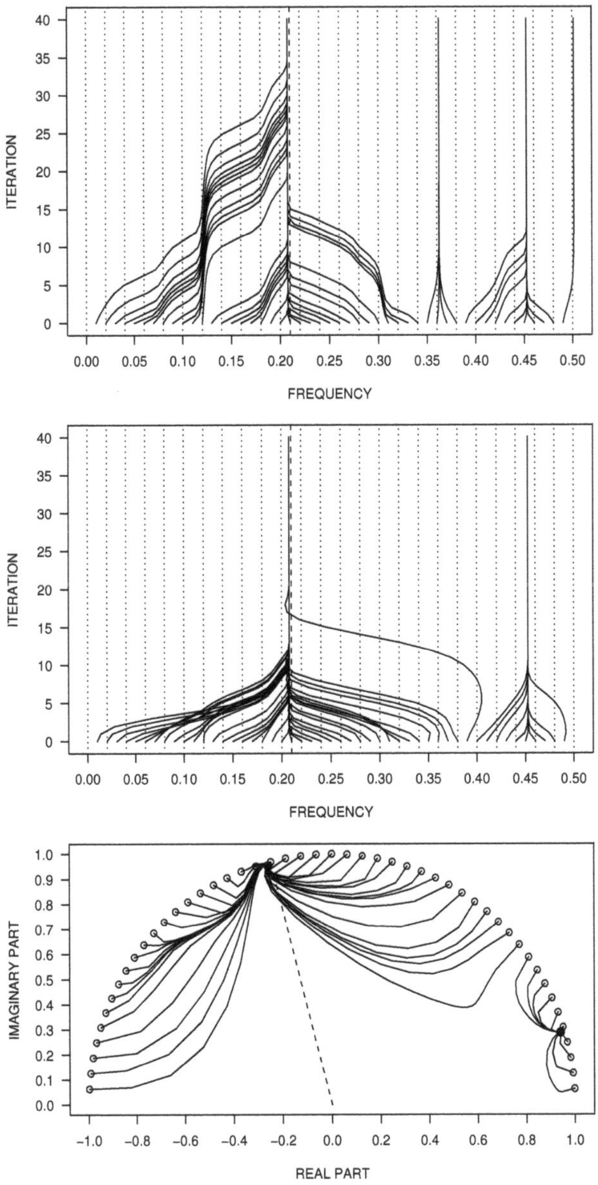

Figure 9.14. Frequency estimation by IGLS with different initial values for the time series used in Figure 9.13. Top, trajectory of frequency estimates by constrained IGLS. Middle, trajectory of frequency estimates by unconstrained IGLS. Bottom, trajectory of unconstrained IGLS estimates for the parameter a_1^* that correspond to the frequency estimates shown in the middle panel: open circle, initial value; solid dot, final estimate. Dashed line shows the signal frequency. Dotted lines depict the Fourier grid.

property than the constrained IGLS algorithm, i.e., faster convergence and fewer spurious fixed points. Hence the unconstrained one is preferred.

The bottom panel of Figure 9.14 depicts the trajectory of the unconstrained IGLS estimates for the AR parameter $a_1^* = -z_1^* = -\exp(-i\omega_1)$. These estimates indeed traverse the interior of the unit circle before reaching a converging point near the unit circle. The angles of the negative complex conjugate of these estimates define the frequency estimates shown in the middle panel. ◇

The IGLS estimator defined by (9.5.19) is closely related to the IFIL estimator defined by (9.4.7). In fact, with the notation $\varrho_t := v_t/(v_t-1)$, the second equation in (9.5.17) can be written as

$$\varrho_t = |a_1|^{-2}\varrho_{t-1} + 1 = |a_1|^{-2t}\varrho_0 + \sum_{j=0}^{t-1}|a_1|^{-2j}.$$

If $|a_1| \le 1$, then, as $t \to \infty$, we obtain $\varrho_t \to \infty$ and hence

$$v_t \to 1, \quad \psi_{t1} \to a_1.$$

This implies that $v_t(m) \to 1$ and $\psi_{t1}(m) \to \hat{a}_{1,m}$ as $t \to \infty$ if $|\hat{a}_{1,m}| \le 1$. Therefore, for large t, we can write

$$u_{t0}(m) \approx \tilde{y}_t, \quad u_{t1}(m) \approx \tilde{y}_{t-1},$$

where $\{\tilde{y}_t\}$ is given recursively by

$$\tilde{y}_t = y_t - \hat{a}_{1,m}\tilde{y}_{t-1} \qquad (t = 1, \ldots, n-1) \tag{9.5.20}$$

with $\tilde{y}_0 := 0$. The last equation is just an AR(1) filter applied to $\{y_t\}$, where the AR coefficient is equal to the estimated parameter $\hat{a}_{1,m}$. Moreover, with $u_{t0}(m)$ and $u_{t1}(m)$ approximated by $\tilde{y}_t$ and $\tilde{y}_{t-1}$, respectively, the IGLS estimator in (9.5.19) can be expressed as

$$\hat{a}_{1,m+1} \approx -\frac{\sum_{t=1}^{n-1}\tilde{y}_t\tilde{y}_{t-1}^*}{\sum_{t=1}^{n-1}|\tilde{y}_t|^2} = -\hat{\rho}_{\tilde{y}}(1), \tag{9.5.21}$$

where $\hat{\rho}_{\tilde{y}}(1) := \sum_{t=1}^{n-1}\tilde{y}_t\tilde{y}_{t-1}^* / \sum_{t=1}^{n-1}|\tilde{y}_t|^2$ is nothing but the lag-1 sample autocorrelation coefficient of $\{\tilde{y}_t\}$. Observe that the IFIL estimator defined by (9.4.4)–(9.4.7) with $p = 1$ and $\eta = 1$ takes the same form as (9.5.21).

A similar relationship exists in the general case with $p > 1$. Indeed, for large t, we have $\psi_{tj} \approx \hat{a}_{j,m}$ and $v_t \approx 1$, hence $u_{tk}(m)$ in (9.5.15) is approximately equal to $\tilde{y}_{t-k}$, where $\{\tilde{y}_t\}$ is the output of an AR(p) filter defined by

$$\tilde{y}_t = y_t - \sum_{j=1}^{p}\hat{a}_{j,m}\tilde{y}_{t-j} \qquad (t = 1, \ldots, n-p)$$

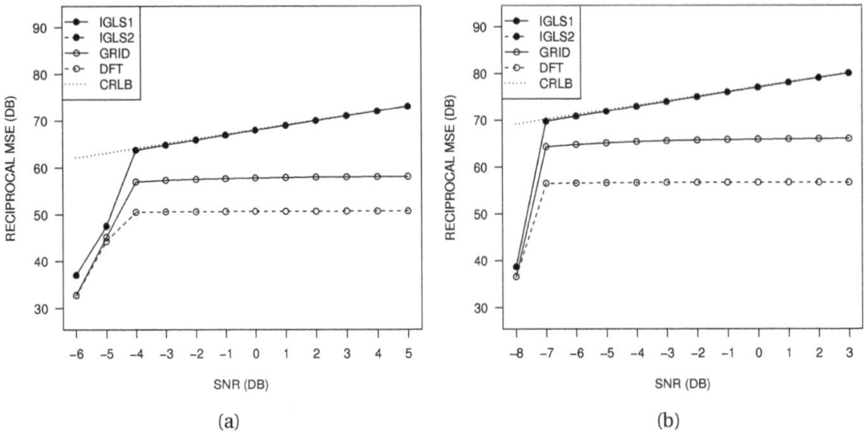

Figure 9.15. Plot of the reciprocal average MSE of the IGLS frequency estimates produced by (9.5.12) for the time series of two complex sinusoids in Gaussian white noise discussed in Example 9.16 with different values of SNR and two sample sizes: (a) $n = 100$ and (b) $n = 200$. Solid line with dots, IGLS initialized by grid search; dashed line with dots (indistinguishable from solid line with dots), IGLS initialized by DFT; solid line with circles, grid search; dashed line with circles, DFT. Dotted line depicts the CRLB. Results are based on 5,000 Monte Carlo runs.

with $\tilde{y}_t := 0$ for $t \le 0$. As a result, (9.5.14) can be written as

$$\hat{\mathbf{a}}_{m+1} \approx -(\tilde{\mathbf{Y}}_p^H \tilde{\mathbf{Y}}_p)^{-1} \tilde{\mathbf{Y}}_p^H \tilde{\mathbf{y}}_p,$$

where $\tilde{\mathbf{Y}}_p$ and $\tilde{\mathbf{y}}_p$ are defined by $\{\tilde{y}_t\}$ in the same way as $\mathbf{Y}_p$ and $\mathbf{y}_p$ are defined by $\{y_t\}$. This expression is nothing but the IFIL procedure defined by (9.4.4)–(9.4.7) with $\eta = 1$. Therefore, the IFIL estimator can be regarded as a simplified approximation to the IGLS estimator in (9.5.12) where the inversion of $\mathbf{R}_m$ is circumvented with the help of inverse filtering.

The statistical performance of the IGLS estimator for estimating well-separated and closely spaced frequencies is demonstrated by the following examples.

Example 9.16 (Estimation of Well-Separated Frequencies: Part 4). Consider the time series in Example 9.3 which consists of two unit-amplitude complex sinusoids with well-separated frequencies in Gaussian white noise. Let the frequency estimates be produced by the IGLS algorithm in (9.5.12), with initial values given by the DFT and grid search methods discussed in Example 8.1. Figure 9.15 depicts the average MSE of the IGLS estimates after 25 iterations and the average MSE of the initial estimates for different values of SNR and two sample sizes.

By comparing Figure 9.15 with Figure 8.3, we can see that the MSE of the IGLS estimator is almost identical to that of the NLS estimator, both following the CRLB

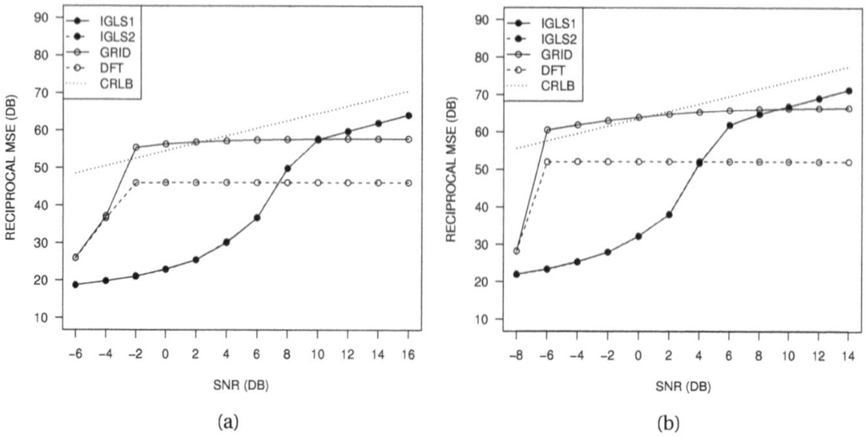

Figure 9.16. Plot of the reciprocal average MSE of the IGLS frequency estimates produced by (9.5.12) for the time series of two complex sinusoids with closely spaced frequencies in Gaussian white noise, as discussed in Example 9.17, with different values of SNR and two sample sizes: (a) $n = 100$ and (b) $n = 200$. Solid line with dots, IGLS initialized by grid search; dashed line with dots (indistinguishable from solid line with dots), IGLS initialized by DFT; solid line with circles, grid search; dashed line with circles, DFT. Dotted line depicts the CRLB. Results are based on 5,000 Monte Carlo runs.

very closely when the SNR is above a threshold which is determined by the initial DFT estimator. In comparison with the Nelder-Mead algorithm that produces the NLS estimates, the IGLS algorithm exhibits a much higher level of tolerance for poor initial values — it is able to accommodate the DFT estimates as initial values, whereas the Nelder-Mead algorithm requires a further grid search.

By comparing Figure 9.15 with Figure 9.10, we can see that the IGLS estimator is slightly more accurate than the IFIL estimator at high SNR for $n = 100$; otherwise they behave similarly, not only in accuracy but also in the ability to accommodate the DFT estimates as initial values. The IGLS estimator performs much better than the FBLP and EYW estimators shown in Figures 9.4 and 9.7. ◇

If the previous example suggests that the IGLS algorithm can be regarded as a better alternative to the Nelder-Mead algorithm for calculating the GML estimates, the next example leads to a different conclusion.

Example 9.17 (Estimation of Closely Spaced Frequencies: Part 4). Consider the time series in Example 9.4 which consists of two unit-amplitude complex sinusoids with closely spaced frequencies in Gaussian white noise. Let the frequency estimates be computed by the IGLS algorithm as in Example 9.16, but with initial values provided by the DFT method with $p = 1$ and by the grid search with a reduced step size as discussed in Example 8.2. Figure 9.16 depicts the average

Table 9.3. Bias, Variance, and MSE for Estimating Closely Spaced Frequencies

	SNR = 4 dB			SNR = 14 dB		
Method	Bias^2	Var	MSE	Bias^2	Var	MSE
NLS	1.0×10^{-9}	1.5×10^{-6}	1.5×10^{-6}	1.4×10^{-11}	1.6×10^{-7}	1.6×10^{-7}
IFIL	2.6×10^{-6}	1.7×10^{-6}	4.4×10^{-6}	2.6×10^{-6}	1.5×10^{-6}	4.1×10^{-6}
IGLS	6.3×10^{-5}	3.7×10^{-3}	3.7×10^{-3}	1.3×10^{-9}	6.4×10^{-7}	6.5×10^{-7}
ILEV	2.7×10^{-6}	2.2×10^{-5}	2.5×10^{-5}	4.3×10^{-9}	5.8×10^{-7}	5.8×10^{-7}

Results are based on 5,000 Monte Carlo runs with $n = 100$.
NLS: Example 8.2; IFIL: Example 9.14; IGLS: Example 9.17; ILEV: Example 9.19.

MSE of the IGLS estimates after 25 iterations and the average MSE of the initial estimates for different values of SNR and two sample sizes.

In comparison with Figure 8.5, we can see that the IGLS estimates are much less accurate than the NLS estimates calculated by the Nelder-Mead algorithm. Even at the higher end of the SNR range considered, the IGLS estimates are still far away from the CRLB which is followed closely by the NLS estimates. In the lower end of the SNR range (e.g., ≤ 6 dB for $n = 100$ and ≤ 2 dB for $n = 200$), the IGLS estimates are even less accurate than the initial DFT estimates with a single frequency. The grid search technique, which helps the Nelder-Mead algorithm to reach the CRLB, does not improve the results of the IGLS algorithm.

A comparison of Figure 9.16 with Figure 9.11 shows that the IGLS estimator outperforms the IFIL estimator at the higher end of the SNR range, and the accuracy of the IGLS estimator increases with the SNR rather than stays constant. At the lower end of the SNR range, the IGLS estimator is inferior to the IFIL estimator. Therefore, inverse filtering instead of matrix inversion reduces the estimation error at low SNR but increases it at high SNR. The IGLS estimator compares favorably to the FBLP and EYW estimators shown in Figures 9.5 and 9.8.

Table 9.3 further compares IGLS with some other estimators in terms of bias, variance, as well as MSE. It shows that the poor accuracy of the IGLS estimator at low SNR is mainly the result of its excessive variance, although its bias is also considerably larger than that of the IFIL and NLS estimators. ◇

This example demonstrates the importance of the interaction between the weight matrix $(\tilde{\mathbf{A}}^H\tilde{\mathbf{A}})^{-1}$ and the other factors in the GML criterion (9.5.10) when the signal frequencies are closely spaced. The IGLS algorithm ignores the interaction and treats them separately in an iterative manner. This technique works well when the signal frequencies are sufficiently separated from each other, as shown in Example 9.16, but not so well when the signal frequencies are closely spaced, as shown in Example 9.17. Therefore, unlike the Nelder-Mead algorithm, the IGLS algorithm should not be regarded as a computational procedure for the GML estimator. In fact, for closely spaced frequencies, the IGLS algorithm produces its own estimates that do not necessarily minimize the GML criterion. For this reason, we regard the IGLS algorithm as a *quasi* GML method.

In closing, let us briefly discuss the real case where the AR parameter satisfies the symmetry constraint (9.2.8). By taking $\tilde{\mathbf{a}} = \mathbf{B}\tilde{\mathbf{a}}_r + \mathbf{b}$ with $\tilde{\mathbf{a}}_r \in \mathbb{R}^q$, we can write

$$\tilde{\mathbf{A}}^T \mathbf{y} = \mathbf{y}_p + \mathbf{Y}_p \tilde{\mathbf{a}} = \mathbf{y}_p + \mathbf{Y}_p \mathbf{b} + \mathbf{Y}_p \mathbf{B}\tilde{\mathbf{a}}_r.$$

Therefore, the GML criterion in (9.5.6) becomes

$$(\mathbf{y}_p + \mathbf{Y}_p \mathbf{b} + \mathbf{Y}_p \mathbf{B}\tilde{\mathbf{a}}_r)^T (\tilde{\mathbf{A}}^T \tilde{\mathbf{A}})^{-1} (\mathbf{y}_p + \mathbf{Y}_p \mathbf{b} + \mathbf{Y}_p \mathbf{B}\tilde{\mathbf{a}}_r).$$

As in the complex case, let $\tilde{\mathbf{A}}^T \tilde{\mathbf{A}}$ be replaced by an estimate $\mathbf{R}_m := \tilde{\mathbf{A}}_m^T \tilde{\mathbf{A}}_m$, where $\tilde{\mathbf{A}}_m$ is defined by (9.5.2) in terms of $\hat{\mathbf{a}}_m := \mathbf{B}\hat{\mathbf{a}}_{r,m} + \mathbf{b}$, with $\hat{\mathbf{a}}_{r,m}$ being an estimate of $\mathbf{a}_r$. Then, by minimizing the resulting criterion with respect to $\tilde{\mathbf{a}}_r \in \mathbb{R}^q$ without constraint, we obtain the GLS solution

$$\hat{\mathbf{a}}_{r,m+1} := -(\mathbf{B}^T \mathbf{Y}_p^T \mathbf{R}_m^{-1} \mathbf{Y}_p \mathbf{B})^{-1} \mathbf{B} \mathbf{Y}_p^T \mathbf{R}_m^{-1} (\mathbf{y}_p + \mathbf{Y}_p \mathbf{b}) \qquad (m = 0, 1, \ldots).$$

This is the (unconstrained) IGLS algorithm for estimating $\mathbf{a}_r$. It is related to the IFIL algorithm defined by (9.4.13)–(9.4.15), similar to the complex case.

9.5.2 Iterative Least-Eigenvalue Algorithm

According to (9.2.2), the roots of $a(z)$ are $z_1, \ldots, z_p$ for any $a_0 \neq 0$. Therefore, the fundamental property (9.5.3) remains valid without the restriction $a_0 = 1$. Given this observation, we can reparameterize the problem of frequency estimation by considering the complete set of the AR coefficients,

$$\mathbf{a}_e := [a_0, a_1, \ldots, a_p]^T,$$

which we call the *extended* AR parameter. The frequency parameter $\boldsymbol{\omega}$ is determined by $\mathbf{a}_e$ according to $\boldsymbol{\omega} = \Phi_{\mathrm{AR}}^{-1}(\mathbf{a})$, where $\mathbf{a} := [a_1/a_0, \ldots, a_p/a_0]^T$.

The relaxation on a_0 in the extended AR parameterization does not alter the identity (9.5.5) because it is invariant to the variable transformation $\tilde{a}_j \mapsto \tilde{a}_0 \tilde{a}_j$ $(j = 0, 1, \ldots, p)$ for any $\tilde{a}_0 \neq 0$. However, an application of the iterative technique behind the IGLS algorithm to the extended AR parameter leads to a different algorithm, which we call the *iterative least-eigenvalue* (ILEV) algorithm.

More precisely, let $\tilde{\mathbf{a}}_e := [\tilde{a}_0, \tilde{a}_1, \ldots, \tilde{a}_p]^T \in \mathbb{C}^{p+1}$ and

$$\mathbf{Y}_{p+1} := [\mathbf{y}_p, \mathbf{Y}_p], \qquad (9.5.22)$$

where $\mathbf{y}_p$ and $\mathbf{Y}_p$ are given by (9.5.9). It is easy to verify that

$$\tilde{\mathbf{A}}^H \mathbf{y} = \mathbf{Y}_{p+1} \tilde{\mathbf{a}}_e.$$

Therefore, the GML criterion in (9.5.6) can be expressed as

$$\tilde{\mathbf{a}}_e^H \mathbf{Y}_{p+1}^H (\tilde{\mathbf{A}}^H \tilde{\mathbf{A}})^{-1} \mathbf{Y}_{p+1} \tilde{\mathbf{a}}_e.$$

Similar to the IGLS algorithm, let the matrix $\tilde{\mathbf{A}}^H \tilde{\mathbf{A}}$ be replaced by its estimate

$$\mathbf{R}_m := \tilde{\mathbf{A}}_m^H \tilde{\mathbf{A}}_m, \qquad (9.5.23)$$

where $\tilde{\mathbf{A}}_m$ is defined by an estimate $\hat{\mathbf{a}}_{e,m}$ of $\mathbf{a}_e$. Minimizing the resulting criterion with respect to $\tilde{\mathbf{a}}_e$ leads to

$$\hat{\mathbf{a}}_{e,m+1} := \arg \min_{\tilde{\mathbf{a}}_e \in \mathcal{A}_e} \tilde{\mathbf{a}}_e^H \mathbf{Y}_{p+1}^H \mathbf{R}_m^{-1} \mathbf{Y}_{p+1} \tilde{\mathbf{a}}_e \qquad (m = 0, 1, \ldots). \qquad (9.5.24)$$

The parameter space $\mathcal{A}_e$ is a suitable subset of $\mathbb{C}^{p+1}$ that excludes the trivial solution $\tilde{\mathbf{a}}_e = \mathbf{0}$. A particularly attractive choice for $\mathcal{A}_e$ is given by

$$\mathcal{A}_e := \{\tilde{\mathbf{a}}_e \in \mathbb{C}^{p+1} : \|\tilde{\mathbf{a}}_e\| = 1\}. \qquad (9.5.25)$$

In this case, the problem in (9.5.24) becomes that of finding a unit-norm eigenvector associated with the smallest eigenvalue of the nonnegative-definite Hermitian matrix $\mathbf{Y}_{p+1}^H \mathbf{R}_m^{-1} \mathbf{Y}_{p+1}$ [276]. The procedure defined by (9.5.23)–(9.5.25) will be referred to as the unconstrained ILEV algorithm, or simply the ILEV algorithm. A constrained ILEV algorithm can be obtained by projecting the roots of the ILEV estimates on the unit circle. But the projection may not always be beneficial.

To compute the ILEV estimates, let $\mathbf{R}_m = \mathbf{L}_m \mathbf{D}_m \mathbf{L}_m^H$ be the Cholesky decomposition produced by the innovations algorithm in Proposition 2.10. Then,

$$\mathbf{Y}_{p+1}^H \mathbf{R}_m^{-1} \mathbf{Y}_{p+1} = \mathbf{U}_m^H \mathbf{D}_m^{-1} \mathbf{U}_m,$$

where $\mathbf{U}_m := \mathbf{L}_m^{-1} \mathbf{Y}_{p+1} = [\mathbf{u}_p(m), \ldots, \mathbf{u}_1(m), \mathbf{u}_0(m)]$ is given by (9.5.15). The ILEV estimate $\hat{\mathbf{a}}_{e,m+1}$ can be obtained as the unit-norm eigenvector associated with the smallest eigenvalue of $\mathbf{U}_m^H \mathbf{D}_m^{-1} \mathbf{U}_m$ by using standard eigenvalue techniques, such as the inverse iteration or the QR method [122, Section 8.2] [365, Section 6.6]. Because it has to solve an eigenvalue problem in each iteration, the ILEV algorithm in general has a higher computational burden than the IGLS algorithm which only needs to solve a generalized least-squares problem. The following example provides the details of the ILEV algorithm for $p = 1$.

Example 9.18 (ILEV Algorithm for a Single Complex Sinusoid). Consider the case of $p = 1$. Similar to Example 9.15, we obtain $r_{\text{MA}}(0) = |a_0|^2 + |a_1|^2$, $r_{\text{MA}}(1) = a_0^* a_1$, and $r_{\text{MA}}(u) = 0$ for $u \geq 2$. Therefore, the innovations algorithm becomes

$$\begin{cases} \psi_{t1} = a_0^* a_1 / v_{t-1} \\ v_t = |a_0|^2 + |a_1|^2 - |a_0 a_1|^2 / v_{t-1} \end{cases} \qquad (t = 1, \ldots, n-2) \qquad (9.5.26)$$

with $v_0 := |a_0|^2 + |a_1|^2$. Moreover, the matrix $\mathbf{U}_m^H \mathbf{D}_m^{-1} \mathbf{U}_m$ can be expressed as

$$\begin{bmatrix} a & c^* \\ c & b \end{bmatrix} := \begin{bmatrix} \sum |u_{t1}(m)|^2 / v_{t-1}(m) & \sum u_{t1}^*(m)\, u_{t0}(m) / v_{t-1}(m) \\ \sum u_{t0}^*(m)\, u_{t1}(m) / v_{t-1}(m) & \sum |u_{t0}(m)|^2 / v_{t-1}(m) \end{bmatrix},$$

where the sum is over $t = 1, \ldots, n-1$. It is easy to verify that the smallest eigenvalue of this matrix is given by

$$\lambda := \tfrac{1}{2}\left\{a + b - \sqrt{(a-b)^2 + 4|c|^2}\right\}$$

and the associated eigenvector takes the form $[d, -dc/(b-\lambda)]^T$ for some $d \neq 0$. Hence, the ILEV estimate $\hat{\mathbf{a}}_{e,m+1} := [\hat{a}_{0,m+1}, \hat{a}_{1,m+1}]^T$ satisfies $\|\hat{\mathbf{a}}_{e,m+1}\| = 1$ and

$$\frac{\hat{a}_{1,m+1}}{\hat{a}_{0,m+1}} = \frac{-c}{b-\lambda} = \frac{-2c/(b-a)}{1 + \sqrt{1 + 4|c/(b-a)|^2}}. \tag{9.5.27}$$

The corresponding frequency estimate $\hat{\omega}_{1,m+1}$ is given by the angle of $\hat{z}_{1,m+1} := -\hat{a}_{1,m+1}/\hat{a}_{0,m+1}$. The unit-circle constraint on the root can be enforced through the transformation $(\hat{a}_{0,m+1}, \hat{a}_{1,m+1}) \mapsto (\hat{a}_{0,m+1}, -\hat{a}_{0,m+1}\hat{z}_{1,m+1}/|\hat{z}_{1,m+1}|)$. ◇

Similar to the IGLS algorithm, the ILEV algorithm can also be viewed through inverse filtering. In fact, it can be shown that $\varrho_t := v_t/(v_t - |a_0|^2) \to \infty$ as $t \to \infty$ if $|a_1/a_0| \le 1$, in which case, $v_t \to |a_0|^2$ and $\psi_{t1} \to a_1/a_0$. Hence, for large t, $u_{t0}(m)$ and $u_{t1}(m)$ can be approximated by $\tilde{y}_t$ and $\tilde{y}_{t-1}$ respectively, with $\{\tilde{y}_t\}$ given by (9.5.20), provided $|\hat{a}_{1,m}/\hat{a}_{0,m}| \le 1$. Under this condition and for large n,

$$c|\hat{a}_{0,m}|^2 \approx \sum_{t=1}^{n-1} \tilde{y}_t \tilde{y}_{t-1}^*,$$

$$(b-a)|\hat{a}_{0,m}|^2 \approx \sum_{t=1}^{n-1} |\tilde{y}_t|^2 - \sum_{t=1}^{n-1} |\tilde{y}_{t-1}|^2 = |\tilde{y}_{n-1}|^2.$$

Hence, $c/(b-a) \approx n\tilde{\rho}_{\tilde{y}}(1)$, where $\tilde{\rho}_{\tilde{y}}(1) := n^{-1}\sum_{t=1}^{n-1} \tilde{y}_t \tilde{y}_{t-1}^* / |\tilde{y}_{n-1}|^2$. Substituting this expression in (9.5.27) gives

$$\frac{\hat{a}_{1,m+1}}{\hat{a}_{0,m+1}} \approx \frac{-2n\tilde{\rho}_{\tilde{y}}(1)}{1 + \sqrt{1 + 4n^2|\tilde{\rho}_{\tilde{y}}(1)|^2}} \approx -\frac{\tilde{\rho}_{\tilde{y}}(1)}{|\tilde{\rho}_{\tilde{y}}(1)|}. \tag{9.5.28}$$

Note that $\tilde{\rho}_{\tilde{y}}(1)$ can be regarded as a variation of the lag-1 sample autocorrelation coefficient of $\{\tilde{y}_t\}$. Because $\{\tilde{y}_t\}$ is an estimate of $\{x_t\}$ whose lag-1 autocorrelation coefficient is equal to $z_1 := \exp(i\omega_1)$, it is natural to use the angle of

$$\hat{z}_{1,m+1} := -\hat{a}_{1,m+1}/\hat{a}_{0,m+1} \approx \tilde{\rho}_{\tilde{y}}(1)/|\tilde{\rho}_{\tilde{y}}(1)|$$

to estimate the frequency ω_1. Unlike the IGLS estimator in (9.5.21), the ILEV estimator in (9.5.28) is associated with the self-normalized autocorrelation coefficient. Therefore, the roots of the ILEV estimates should stay near the unit circle.

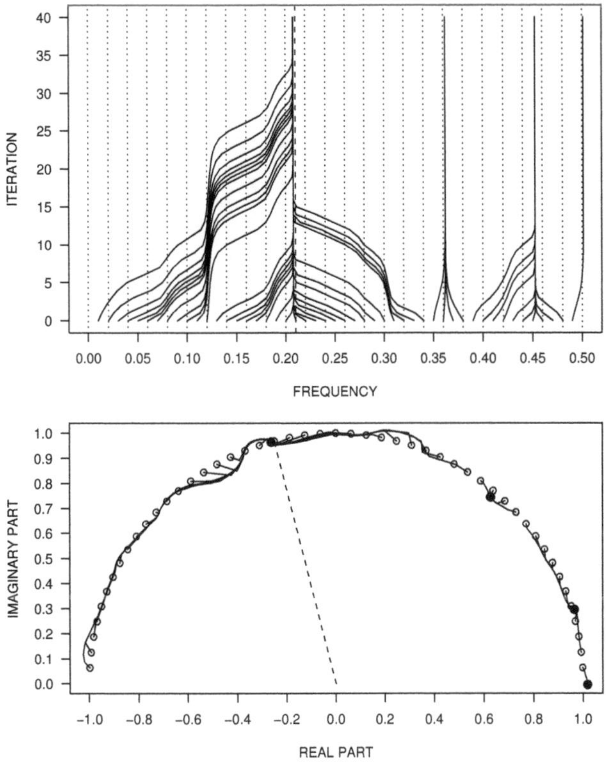

Figure 9.17. Top, trajectory of the unconstrained ILEV frequency estimates from different initial values for the time series which is used in Figure 9.14. Bottom, trajectory of the corresponding estimates for a_1^*/a_0^*: open circle, initial value; solid dot, final estimate. Dashed line shows the signal frequency.

As a result, the ILEV algorithm should behave similarly to the constrained IGLS algorithm. This behavior is confirmed by comparing Figure 9.17 with Figure 9.14. Therefore, it should not be surprising that the ILEV algorithm may take more iterations to converge than the unconstrained IGLS algorithm and may require more precise initial values as well.

In terms of statistical performance, the ILEV algorithm behaves similarly to the constrained IGLS algorithm for estimating the well-separated frequencies in Example 9.16. The following example shows that the ILEV algorithm can achieve a better accuracy for estimating closely spaced frequencies at lower SNR.

Example 9.19 (Estimation of Closely Spaced Frequencies: Part 5). Consider the data in Example 9.17. Let the frequency estimates be obtained by the ILEV algorithm with 25 iterations and with the same initial values as in Example 9.17. For each iteration, the minimum eigenvalue and the associated eigenvector are

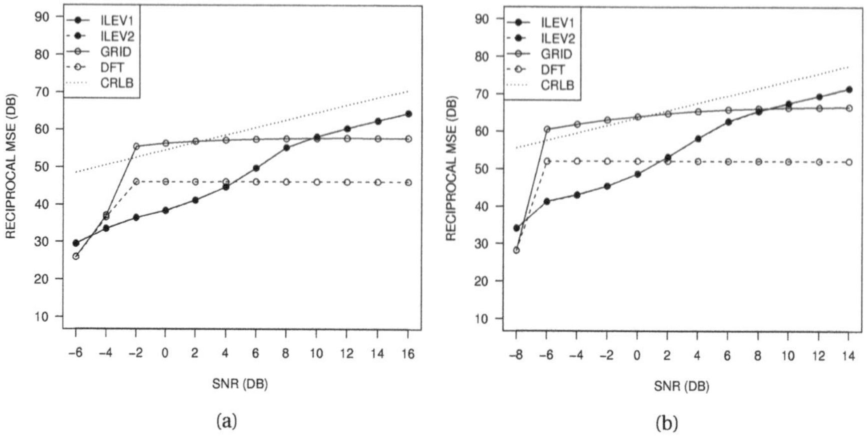

Figure 9.18. Plot of the reciprocal average MSE of the ILEV frequency estimates for the time series of two complex sinusoids with closely spaced frequencies in Gaussian white noise, as discussed in Example 9.19, with different values of SNR and two sample sizes: (a) $n = 100$ and (b) $n = 200$. Solid line with dots, ILEV initialized by grid search; dashed line with dots (indistinguishable from solid line with dots), ILEV initialized by DFT; solid line with circles, grid search; dashed line with circles, DFT. Dotted line depicts the CRLB. Results are based on 5,000 Monte Carlo runs.

computed by the R function `eigen` which in turn calls LAPACK routines. The average MSE of the ILEV estimates is depicted in Figure 9.18.

Comparing this result with Figure 9.16 shows that the main advantage of the ILEV estimator over the IGLS estimator is the significant improvement in accuracy at the lower end of the SNR range considered (≤ 8 dB for $n = 100$ and ≤ 4 for $n = 200$). This is largely due to the reduction of variance, as can be seen from Table 9.3. At the higher end of the SNR range, the MSE of the ILEV estimator is approximately the same as that of the IGLS estimator, or even slightly better (see Table 9.3). As with IGLS, the grid search does not help the ILEV algorithm to reach the CRLB as it does the Nelder-Mead algorithm for the NLS estimator shown in Figure 8.5. This means that the ILEV algorithm should not be regarded as a computational procedure for the GML estimator.

At the lower end of the SNR range considered, the ILEV estimator is still less accurate than the IFIL estimator shown in Figure 9.11; at the higher end of the SNR range, it becomes more accurate than the IFIL estimator. The ILEV estimator outperforms the FBLP and EYW estimators shown in Figures 9.5 and 9.8. ◇

The parameter space $\mathcal{A}_e$ in (9.5.25) can be reduced by observing that for any $\rho > 0$ there exists $\theta \in (0, 2\pi)$ such that $\mathbf{a}_e := [a_0, a_1, \ldots, a_p]^T$ defined by (9.2.2) with

$a_0 := \rho \exp(i\theta)$ satisfies the conjugate symmetry condition

$$a_j = a^*_{p-j} \qquad (j = 0, 1, \ldots, p). \tag{9.5.29}$$

To prove this assertion, observe that (9.2.2) implies

$$a^*(z) := \sum_{j=0}^{p} a^*_j z^{-j} = a^*_0 \prod_{k=1}^{p} (1 - z^*_k z^{-1}).$$

Combining this expression with the fact that $z^*_k = z_k^{-1}$ for all k yields

$$\begin{aligned} z^{-p} a^*(z^{-1}) &= a^*_0 \prod_{k=1}^{p} (z^{-1} - z_k^{-1}) \\ &= \frac{(-1)^p a^*_0}{z_1 \cdots z_p a_0} a_0 \prod_{k=1}^{p} (1 - z_k z^{-1}) \\ &= (-1)^p \exp\{-i(2\theta + (\omega_1 + \cdots + \omega_p)\} a(z). \end{aligned}$$

Taking $\theta := \frac{1}{2}\{p\pi - (\omega_1 + \cdots + \omega_p)\}$ gives

$$z^{-p} a^*(z^{-1}) = a(z).$$

Comparing the coefficients of the polynomial on the right-hand side of the equation with that on the left-hand side leads to (9.5.29).

To reduce the parameter space under the condition (9.5.29), define

$$\boldsymbol{\alpha} := \begin{bmatrix} \Re(\mathbf{a}_q) \\ \Im(\mathbf{a}_q) \end{bmatrix}, \quad \mathbf{P} := \begin{bmatrix} \mathbf{I}_{q+1} & i\mathbf{I}_{q+1} \\ \tilde{\mathbf{I}}_{q+1} & -i\tilde{\mathbf{I}}_{q+1} \end{bmatrix} \quad \text{if } p = 2q+1,$$

$$\boldsymbol{\alpha} := \begin{bmatrix} \Re(\mathbf{a}_{q-1}) \\ a_q \\ \Im(\mathbf{a}_{q-1}) \end{bmatrix}, \quad \mathbf{P} := \begin{bmatrix} \mathbf{I}_q & \mathbf{0} & i\mathbf{I}_q \\ \mathbf{0}^T & 1 & \mathbf{0}^T \\ \tilde{\mathbf{I}}_q & \mathbf{0} & -i\tilde{\mathbf{I}}_q \end{bmatrix} \quad \text{if } p = 2q,$$

where $\mathbf{a}_q := [a_0, a_1, \ldots, a_q]^T$ and $\mathbf{a}_{q-1} := [a_0, a_1, \ldots, a_{q-1}]^T$. Then, we can write

$$\mathbf{a}_e = \mathbf{P}\boldsymbol{\alpha}.$$

The free parameter ρ in a_0 can be made unique by imposing the unit-norm constraint $\|\boldsymbol{\alpha}\| = 1$. With the parameter space $\mathcal{A}_e$ redefined as

$$\mathcal{A}_e := \{\tilde{\mathbf{a}}_e := \mathbf{P}\tilde{\boldsymbol{\alpha}} : \tilde{\boldsymbol{\alpha}} \in \mathbb{R}^{p+1}, \|\tilde{\boldsymbol{\alpha}}\| = 1\},$$

the minimization problem in (9.5.24) can still be solved as a least-eigenvalue problem in terms of the real-valued variable $\tilde{\boldsymbol{\alpha}} \in \mathbb{R}^{p+1}$ [265] [349]. The reduced parameter space does not lead to better results than those shown in Figure 9.18 for the data in Example 9.19, although it is found beneficial elsewhere [349].

9.5.3 Self Initialization

In some applications, it is desirable to have a simplified computational architecture for easier implementation. Initializing the IGLS and ILEV algorithms with the periodogram-based DFT method or the multivariate-periodogram-based grid search, as in Examples 9.16, 9.17, and 9.19, may not fit the bill. In the following, we investigate the feasibility of self-initialization for the IGLS and ILEV algorithms, i.e., producing an initial value from the same algorithm.

Without any prior knowledge about the signal frequencies, the simplest way to obtain the initial values $\hat{\mathbf{a}}_0$ and $\hat{\mathbf{a}}_{e,0}$ for the IGLS and ILEV algorithms is to take $\mathbf{R}_{-1} := \mathbf{I}$ in (9.5.12) and (9.5.24). For IGLS, this self-initialization method leads to the least-squares forward linear prediction (FLP) estimator

$$\hat{\mathbf{a}}_0 := -(\mathbf{Y}_p^H \mathbf{Y}_p)^{-1} \mathbf{Y}_p^H \mathbf{y}_p = \arg\min_{\tilde{\mathbf{a}} \in \mathbb{C}^p} \|\mathbf{y}_p + \mathbf{Y}_p \tilde{\mathbf{a}}\|^2, \tag{9.5.30}$$

which is also known as Prony's estimator. For ILEV, the self-initialization method yields the least-eigenvalue (LEV) estimator

$$\hat{\mathbf{a}}_{e,0} := \arg\min_{\tilde{\mathbf{a}}_e \in \mathcal{A}_e} \tilde{\mathbf{a}}_e^H \mathbf{Y}_{p+1}^H \mathbf{Y}_{p+1} \tilde{\mathbf{a}}_e = \arg\min_{\tilde{\mathbf{a}}_e \in \mathcal{A}_e} \|\mathbf{Y}_{p+1} \tilde{\mathbf{a}}_e\|^2, \tag{9.5.31}$$

which is also known as Pisarenko's estimator [289]. The unified computational architecture is the attraction of the self-initialized IGLS and ILEV algorithms.

Figure 9.19 demonstrates the statistical performance of these algorithms. In this figure, the average MSE of the frequency estimates is depicted for the time series with well-separated frequencies discussed in Example 9.16, and for the time series with closely spaced frequencies discussed in Examples 9.17 and 9.19. The only difference is that the IGLS and ILEV algorithms are now initialized by the FLP and LEV estimates, respectively, rather than by DFT and grid search.

A comparison of the top panel of Figure 9.19 with Figures 9.15(a) and 9.16(a) shows that the IGLS algorithm is able to cope with the initial FLP estimates, which are far less accurate than the DFT estimates, and yield the same MSE as the DFT-initialized alternative, for both well-separated frequencies and closely spaced frequencies. The ILEV algorithm, on the other hand, is less tolerant of the initial LEV estimates. This can be seen by comparing the bottom-right panel of Figure 9.19 with Figures 9.18(a) for the case with closely spaced frequencies, and by comparing the bottom-left panel of Figure 9.19 with Figure 9.15(a) which is nearly identical to the performance of the ILEV algorithm initiated by DFT and grid search for well-separated frequencies.

The poor performance of the self-initialized ILEV algorithm is not surprising, because the ILEV algorithm behaves similarly to the constrained IGLS algorithm and thus demands higher accuracy for the initial values. The example shown in Figure 9.19 suggests that the LEV estimator in (9.5.31) does not always provide the accuracy needed by the ILEV algorithm, even though it is more accurate than the FLP estimator in (9.5.30) that initializes the IGLS algorithm.

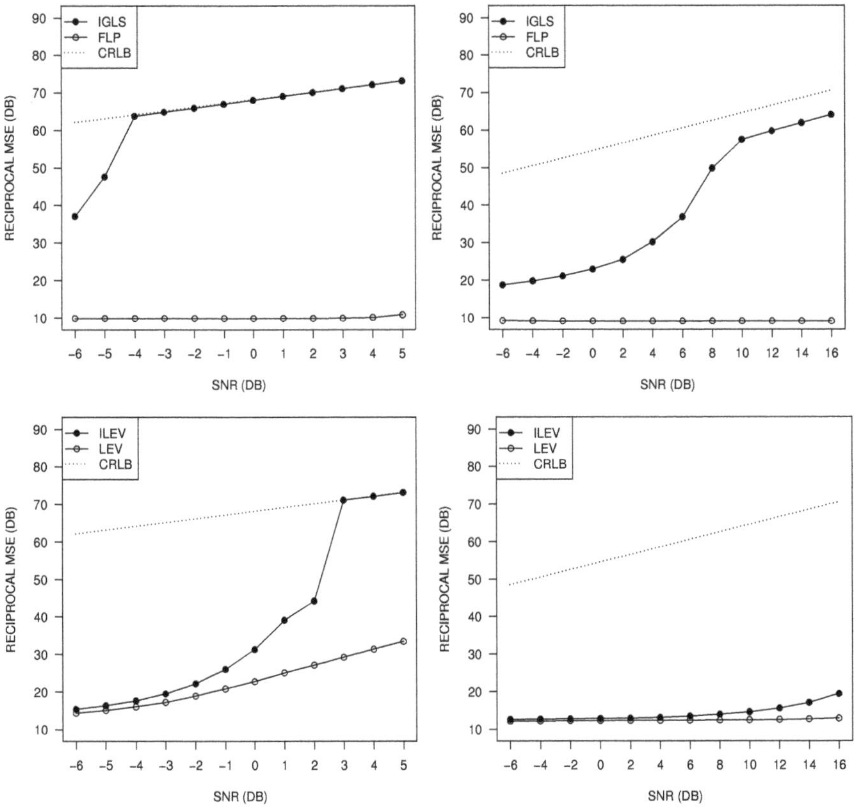

Figure 9.19. Plot of the reciprocal average MSE of the self-initialized IGLS frequency estimates (top) and the self-initialized ILEV frequency estimates (bottom) for the time series with well-separated frequencies discussed in Example 9.16 (left) and for the time series with closely spaced frequencies discussed in Examples 9.17 and 9.19 (right). In all cases, $n = 100$. Solid line with dots, final IGLS and ILEV estimates; solid line with circles, initial FLP and LEV estimates. Dotted line depicts the CRLB.

9.6 Proof of Theorems

This section contains the proof of Theorems 9.1, 9.2, 9.4, and 9.6–9.9.

Proof of Theorem 9.1. According to the proof of Theorem 3.4,

$$m^{-1}\mathbf{F}_m^H \mathbf{R}_\epsilon^{-1} \mathbf{F}_m = \mathbf{S}^{-1} + \mathcal{O}(1), \tag{9.6.1}$$

where $\mathbf{S} := \text{diag}\{f_\epsilon(\omega_1),\ldots,f_\epsilon(\omega_p)\}$. This implies that

$$(\mathbf{G}^{-1} + \mathbf{F}_m^H \mathbf{R}_\epsilon^{-1}\mathbf{F}_m)^{-1}\mathbf{F}_m^H \mathbf{R}_\epsilon^{-1}\mathbf{F}_m = \mathbf{I} + \mathcal{O}(1). \tag{9.6.2}$$

The Hermitian transpose of the kth column of this expression takes the form

$$\mathbf{f}_m^H(\omega_k)\mathbf{R}_\epsilon^{-1}\mathbf{F}_m(\mathbf{G}^{-1} + \mathbf{F}_m^H\mathbf{R}_\epsilon^{-1}\mathbf{F}_m)^{-1} = \mathbf{e}_k^T + \mathcal{O}(1),$$

where $\mathbf{e}_k$ denotes the kth canonical basis vector in $\mathbb{R}^p$. Combining these results with (9.1.16) and the fact that $\mathbf{G}^{-1}\mathbf{g} = \mathbf{1}$ leads to

$$\begin{aligned} C(\omega_k) &= 1 - \{\mathbf{e}_k^T + \mathcal{O}(m^{-1})\}\mathbf{1} + \{\mathbf{f}_m^H(\omega_k) - (\mathbf{e}_k^T + \mathcal{O}(1))\mathbf{F}_m^H\}\boldsymbol{\varphi} \\ &= \mathcal{O}(1) + \mathcal{O}(1)\mathbf{F}_m^H\boldsymbol{\varphi} = \mathcal{O}(1) + \mathcal{O}(1)(\mathbf{F}_m^H\boldsymbol{\varphi} + \mathbf{1}). \end{aligned} \tag{9.6.3}$$

Similarly, for any $\omega \notin \{\omega_k\}$,

$$m^{-1}\mathbf{f}_m^H(\omega)\mathbf{R}_\epsilon^{-1}\mathbf{F}_m = \mathcal{O}(1).$$

Combining this expression with (9.1.16) and (9.6.2) leads to

$$C(\omega) = \Phi(\omega) + \mathcal{O}(1) + \mathcal{O}(1)(\mathbf{F}_m^H\boldsymbol{\varphi} + \mathbf{1}). \tag{9.6.4}$$

Observe that the kth element in $\mathbf{F}_m^H\boldsymbol{\varphi} + \mathbf{1}$ is equal to $\Phi(\omega_k)$.

Because $f_0 := \inf\{f_\epsilon(\omega)\} > 0$, it follows from (9.1.21) that $f_m(\omega) \ge f_0/2$ for sufficiently large m and all ω. Combining this result with $\varsigma_m^2 \le r_\epsilon(0)$ yields

$$|\Phi(\omega)|^2 = \varsigma_m^2 / f_m(\omega) \le 2r_\epsilon(0)/f_0 < \infty$$

for all ω and large m. Therefore,

$$\mathbf{F}_m^H\boldsymbol{\varphi} + \mathbf{1} = \mathcal{O}(1). \tag{9.6.5}$$

Substituting this expression in (9.6.3) and (9.6.4) yields

$$C(\omega) = \begin{cases} \mathcal{O}(1) & \text{if } \omega = \omega_k \ (k = 1,\ldots,p), \\ \Phi(\omega) + \mathcal{O}(1) & \text{if } \omega \notin \{\omega_k\}. \end{cases} \tag{9.6.6}$$

Finally, by (9.1.16)–(9.6.2) and (9.6.5),

$$\begin{aligned}\sigma_m^2 &= r_y(0) + (\mathbf{F}_m\mathbf{g}+\mathbf{r}_\epsilon)^H\mathbf{R}_\epsilon^{-1}\mathbf{F}_m(\mathbf{G}^{-1}+\mathbf{F}_m^H\mathbf{R}_\epsilon^{-1}\mathbf{F}_m)^{-1}\mathbf{1}\\ &\quad + (\mathbf{F}_m\mathbf{g}+\mathbf{r}_\epsilon)^H\{\mathbf{I}-\mathbf{R}_\epsilon^{-1}\mathbf{F}_m(\mathbf{G}^{-1}+\mathbf{F}_m^H\mathbf{R}_\epsilon^{-1}\mathbf{F}_m)^{-1}\mathbf{F}_m^H\}\boldsymbol{\varphi}\\ &= r_\epsilon(0) + \mathbf{r}_\epsilon^H\boldsymbol{\varphi} + \mathcal{O}(1)\mathbf{F}_m^H\boldsymbol{\varphi} - m^{-1}\boldsymbol{\varphi}^H\mathbf{F}_m\{\mathbf{S}+\mathcal{O}(1)\}(\mathbf{F}_m^H\boldsymbol{\varphi}+\mathbf{1})\\ &= \varsigma_m^2 + \mathcal{O}(1).\end{aligned}$$

Combining this result with (9.6.6) proves the assertion. □

Proof of Theorem 9.2. Any solution $\hat{\mathbf{c}} := [\hat{c}_1,\ldots,\hat{c}_m]^T$ to the problem of minimizing $\|\mathbf{x}_f+\mathbf{X}_f\tilde{\mathbf{c}}\|^2$ with respect to $\tilde{\mathbf{c}}\in\mathbb{C}^m$ should satisfy the normal equation

$$(\mathbf{X}_f^H\mathbf{X}_f)\hat{\mathbf{c}} = -\mathbf{X}_f^H\mathbf{x}_f. \tag{9.6.7}$$

Let $\mathbf{Z}_m := \mathrm{diag}(\beta_1 z_1^m,\ldots,\beta_p z_p^m)$. Then, it is easy to verify that

$$\mathbf{X}_f = \sum_{k=1}^p \beta_k z_k^m \mathbf{f}_{n-m}(\omega_k)\mathbf{f}_m^H(\omega_k) = \mathbf{F}_{n-m}\mathbf{Z}_m\mathbf{F}_m^H. \tag{9.6.8}$$

For any m such that $p\le m<n-m$, both $\mathbf{F}_m\in\mathbb{C}^{m\times p}$ and $\mathbf{F}_{n-m}\in\mathbb{C}^{(n-m)\times p}$ have full (column) rank p. Therefore, the rank of $\mathbf{X}_f$ is always equal to p [153, p. 13]. When $m=p$, the normal equation (9.6.7) has a unique solution because $\mathbf{X}_f$ has full column rank. If $p<m<n-m$, then $\mathbf{X}_f$ is rank-deficient, in which case, there are infinitely many solutions to the normal equation (9.6.7).

Because $x_t=\sum_{k=1}^p \beta_k z_k^t$, we obtain

$$\sum_{j=0}^p a_j x_{t-j} = \sum_{k=1}^p \beta_k z_k^t a(z_k) = 0.$$

It is easy to verify that the minimum value of $\|\mathbf{x}_f+\mathbf{X}_f\tilde{\mathbf{c}}\|^2$ equals zero and can be attained at $\tilde{\mathbf{c}}=[a_1,\ldots,a_p,0,\ldots,0]^T$. Therefore, any minimizer $\hat{\mathbf{c}}$ of the criterion should yield a zero value. Combining this result with (9.6.7) gives

$$0 = \|\mathbf{x}_f+\mathbf{X}_f\hat{\mathbf{c}}\|^2 = \mathbf{x}_f^H\mathbf{x}_f + \mathbf{x}_f^H\mathbf{X}_f\hat{\mathbf{c}}. \tag{9.6.9}$$

Equations (9.6.7) and (9.6.9) can be put into a single equation

$$\bar{\mathbf{X}}_f^H\bar{\mathbf{X}}_f\bar{\mathbf{c}} = \mathbf{0},$$

where

$$\bar{\mathbf{X}}_f := [\mathbf{x}_f,\mathbf{X}_f],\quad \bar{\mathbf{c}} := \begin{bmatrix}1\\ \hat{\mathbf{c}}\end{bmatrix}.$$

This result implies that $\bar{\mathbf{c}}^H \bar{\mathbf{X}}_f^H \bar{\mathbf{X}}_f \bar{\mathbf{c}} = 0$ and hence

$$\bar{\mathbf{X}}_f \bar{\mathbf{c}} = \mathbf{0}. \tag{9.6.10}$$

Similar to (9.6.8), it can be shown that

$$\bar{\mathbf{X}}_f = \mathbf{F}_{n-m} \mathbf{Z}_{m+1} \mathbf{F}_{m+1}^H.$$

Combining this expression with (9.6.10) leads to $\mathbf{F}_{m+1}^H \bar{\mathbf{c}} = \mathbf{0}$, or equivalently,

$$\mathbf{f}_{m+1}^H(\omega_k)\,\bar{\mathbf{c}} = z_k^{-1}\hat{c}(z_k) = 0 \qquad (k = 1,\dots,p),$$

where $\hat{c}(z) := \sum_{j=0}^m \hat{c}_j z^{-j}$ ($\hat{c}_0 := 1$). This result implies that $\hat{c}(z)$ has roots $z_1,\dots,z_p$ and therefore can be factored as

$$\hat{c}(z) = a(z)\hat{b}(z)$$

for some $(m-p)$-degree polynomial $\hat{b}(z) := \sum_{j=0}^{m-p} \hat{b}_j z^{-j}$ ($\hat{b}_0 := 1$).

Now, let $\{Y_t\}$ be an MA process defined by $Y_t := a(z)\zeta_t$ and $\{\zeta_t\} \sim \mathrm{WN}(0,1)$. Because $\hat{b}(z)Y_t = \hat{b}(z)a(z)\zeta_t = \hat{c}(z)\zeta_t$, it follows that

$$E\{|\hat{b}(z)Y_t|^2\} = 1 + \|\hat{\mathbf{c}}\|^2.$$

Therefore, minimizing $\|\hat{\mathbf{c}}\|$ over the solutions of (9.6.7) is equivalent to minimizing the prediction error variance $E\{|\hat{b}(z)Y_t|^2\}$ with respect to the $\hat{b}_j$ ($\hat{b}_0 := 1$). The former gives the minimum-norm solution by definition, and the latter yields a minimum-phase polynomial according to Proposition 2.7. □

Proof of Theorem 9.4. The asymptotic normality of $\hat{\mathbf{c}}$ can be proved in the same way as Theorem 7.5 with the exception that by Theorem 4.2,

$$n\,\mathrm{Cov}\{\hat{r}_y(u), \hat{r}_y(v)\} \to \gamma(u,v),$$
$$n\,\mathrm{Cov}\{\hat{r}_y(u), \hat{r}_y^*(v)\} \to \gamma(u,-v),$$

where $\gamma(\cdot,\cdot)$ is given by (4.3.1) and (4.3.2) in the real case, and by (4.3.3) and (4.3.4) in the complex case.

In the complex case, $\sqrt{n}(\hat{\mathbf{c}} - \mathbf{c}) \xrightarrow{D} \mathrm{N}_c(\mathbf{0}, \mathbf{R}_y^{-1}\mathbf{\Sigma}\mathbf{R}_y^{-1}, \mathbf{R}_y^{-1}\tilde{\mathbf{\Sigma}}\mathbf{R}_y^{-T})$ and $\hat{\boldsymbol{\omega}} \xrightarrow{P} \boldsymbol{\omega}_0$. It follows from Lemma 12.5.1 that

$$\sqrt{n}(\hat{\boldsymbol{\omega}} - \boldsymbol{\omega}_0) \xrightarrow{D} \mathrm{N}(\mathbf{0}, 2\Re(\mathbf{D}^H\mathbf{R}_y^{-1}\mathbf{\Sigma}\mathbf{R}_y^{-1}\mathbf{D} + \mathbf{D}^H\mathbf{R}_y^{-1}\tilde{\mathbf{\Sigma}}\mathbf{R}_y^{-T}\mathbf{D}^*)),$$

where $\mathbf{D} := [\mathbf{d}_1,\dots,\mathbf{d}_p]$. In the real case, we have $\sqrt{n}(\hat{\mathbf{c}} - \mathbf{c}) \xrightarrow{D} \mathrm{N}(\mathbf{0}, \mathbf{R}_y^{-1}\mathbf{\Sigma}\mathbf{R}_y^{-1})$ and $\hat{\boldsymbol{\omega}}_r \xrightarrow{P} \boldsymbol{\omega}_{0r}$. Therefore, by Lemma 12.5.1,

$$\sqrt{n}(\hat{\boldsymbol{\omega}}_r - \boldsymbol{\omega}_{0r}) \xrightarrow{D} \mathrm{N}(\mathbf{0}, \mathbf{D}_r^T\mathbf{R}_y^{-1}\mathbf{\Sigma}\mathbf{R}_y^{-1}\mathbf{D}_r),$$

where $\mathbf{D}_r := 2\Re[\mathbf{d}_1,\ldots,\mathbf{d}_q]$. In both cases, $\mathbf{d}_k$ is defined by (9.1.26). □

Proof of Theorem 9.6. It follows from (9.3.5) that

$$\hat{\mathbf{a}}_r - \mathbf{a}_r = -(\mathbf{B}^T\hat{\mathbf{U}}_l^T\hat{\mathbf{U}}_l\mathbf{B})^{-1}\mathbf{B}^T\hat{\mathbf{U}}_l^T\{(\hat{\mathbf{U}}_l\mathbf{b}+\hat{\mathbf{u}}_l)+\hat{\mathbf{U}}_l\mathbf{B}\,\mathbf{a}_r\}.$$

By Theorem 4.1, $\hat{\mathbf{U}}_l \xrightarrow{P} \mathbf{U}_l$ as $n \to \infty$. Therefore, according to Slutsky's theorem in Proposition 2.13(b), it suffices to show that

$$\sqrt{n}\,(\hat{\mathbf{U}}_l\mathbf{b}+\hat{\mathbf{u}}_l+\hat{\mathbf{U}}_l\mathbf{B}\,\mathbf{a}_r) \xrightarrow{D} \mathrm{N}(\mathbf{0},\boldsymbol{\Sigma}_l), \tag{9.6.11}$$

where $\boldsymbol{\Sigma}_l = \sigma^4\mathbf{A}_l\mathbf{A}_l^T$. Toward that end, let $\hat{\mathbf{r}} := [\hat{r}_y(1),\ldots,\hat{r}_y(p+l)]^T$. Then, by Theorem 4.3, we obtain

$$\sqrt{n}\,(\hat{\mathbf{r}}-\mathbf{r}) \xrightarrow{D} \mathrm{N}(\mathbf{0},\boldsymbol{\Gamma}_l), \tag{9.6.12}$$

where $\mathbf{r} := [r_x(1),\ldots,r_x(p+l)]^T$ and $\boldsymbol{\Gamma}_l := [\gamma(u,v)]$ $(u,v=1,\ldots,p+l)$ with

$$\gamma(u,v) := \sum_{k=1}^{q} 2\sigma^2 C_k^2 \cos(\omega_k u)\cos(\omega_k v) + \sigma^4\delta_{u-v}.$$

Define $\mathbf{F}_{p+l} := [\mathbf{f}_{p+l}(\omega_1),\ldots,\mathbf{f}_{p+l}(\omega_q)]$ and $\mathbf{G} := \mathrm{diag}(\frac{1}{2}C_1^2,\ldots,\frac{1}{2}C_q^2)$. Then,

$$\boldsymbol{\Gamma}_l = \sigma^4\mathbf{I} + 4\sigma^2\Re(\mathbf{F}_{p+l})\mathbf{G}\Re(\mathbf{F}_{p+l})^T.$$

It is easy to verify that

$$\hat{\mathbf{U}}_l\mathbf{b}+\hat{\mathbf{u}}_l+\hat{\mathbf{U}}_l\mathbf{B}\,\mathbf{a}_r = \hat{\mathbf{u}}_l+\hat{\mathbf{U}}_l\,\mathbf{a} = \mathbf{A}_l\,\hat{\mathbf{r}}.$$

Combining this expression with (9.6.12) leads to

$$\sqrt{n}\,(\hat{\mathbf{U}}_l\mathbf{b}+\hat{\mathbf{u}}_l+\hat{\mathbf{U}}_l\mathbf{B}\,\mathbf{a}_r-\mathbf{A}_l\,\mathbf{r}) \xrightarrow{D} \mathrm{N}(\mathbf{0},\boldsymbol{\Sigma}_l),$$

where

$$\boldsymbol{\Sigma}_l := \mathbf{A}_l\boldsymbol{\Gamma}_l\mathbf{A}_l^T = \sigma^4\mathbf{A}_l\mathbf{A}_l^T + 4\sigma^2\Re(\mathbf{A}_l\mathbf{F}_{p+l})\mathbf{G}\Re(\mathbf{A}_l\mathbf{F}_{p+l})^T.$$

Moreover, it is easy to verify that

$$\mathbf{A}_l\,\mathbf{f}_{p+l}(\omega_k) = a(z_k)\begin{bmatrix} z_k^{p+1} \\ \vdots \\ z_k^{p+l}\end{bmatrix} = \mathbf{0} \qquad (k=1,\ldots,q).$$

Therefore, we obtain $\mathbf{A}_l\mathbf{F}_{p+l} = \mathbf{0}$ and hence $\boldsymbol{\Sigma}_l = \sigma^4\mathbf{A}_l\mathbf{A}_l^T$. Finally, owing to (9.3.8),

$$\mathbf{A}_l\,\mathbf{r} = \mathbf{U}_l\mathbf{a}+\mathbf{u}_l = \mathbf{0}.$$

Combining these results leads to $\sqrt{n}\,(\hat{\mathbf{a}}_r - \mathbf{a}_r) \xrightarrow{D} \mathrm{N}(\mathbf{0}, \mathbf{V}_l)$.

For the frequency estimator $\hat{\boldsymbol{\omega}}_r$, because

$$\sqrt{n}\,(\hat{\mathbf{a}} - \mathbf{a}) = \sqrt{n}\,\mathbf{B}(\hat{\mathbf{a}}_r - \mathbf{a}_r) \xrightarrow{D} \mathrm{N}(\mathbf{0}, \mathbf{B}\mathbf{V}_l\mathbf{B}^T),$$

it follows from Lemma 12.5.2 that

$$\sqrt{n}\,(\hat{\boldsymbol{\omega}}_r - \boldsymbol{\omega}_r) \xrightarrow{D} \mathrm{N}(\mathbf{0}, \mathbf{D}^T\mathbf{B}\mathbf{V}_l\mathbf{B}^T\mathbf{D}),$$

where $\mathbf{D} := [\Re(\mathbf{v}_1), \Im(\mathbf{v}_1), \dots, \Re(\mathbf{v}_q), \Im(\mathbf{v}_q)]\mathbf{J}^T$ and $\mathbf{J} := \mathrm{diag}(\mathbf{J}_1, \dots, \mathbf{J}_q)$. Observe that $\mathbf{J}_k = [-\sin(\omega_k), \cos(\omega_k)]$. Therefore, $\mathbf{D} = [\mathbf{d}_1, \dots, \mathbf{d}_q]$, where $\mathbf{d}_k := [\Re(\mathbf{v}_k), \Im(\mathbf{v}_k)]\mathbf{J}_k^T = \cos(\omega_k)\Im(\mathbf{v}_k) - \sin(\omega_k)\Re(\mathbf{v}_k)$. □

Proof of Theorem 9.7. Similar to the proof of Theorem 9.6, we can write

$$\begin{aligned}\hat{\mathbf{a}} - \mathbf{a} &= -(\hat{\mathbf{U}}_l^H\hat{\mathbf{U}}_l)^{-1}\hat{\mathbf{U}}_l^H(\hat{\mathbf{u}}_l + \hat{\mathbf{U}}_l\mathbf{a}) \\ &= -(\hat{\mathbf{U}}_l^H\hat{\mathbf{U}}_l)^{-1}\hat{\mathbf{U}}_l^H\mathbf{A}_l\hat{\mathbf{r}},\end{aligned}$$

where $\hat{\mathbf{r}} := [\hat{r}_y(1), \dots, \hat{r}_y(p+l)]^T$. Define $\mathbf{r} := [r_x(1), \dots, r_x(p+l)]^T$. Then, according to Theorem 4.4, we obtain

$$\sqrt{n}\,(\hat{\mathbf{r}} - \mathbf{r}) \xrightarrow{D} \mathrm{N}_c(\mathbf{0}, \boldsymbol{\Gamma}_l, \tilde{\boldsymbol{\Gamma}}_l),$$

where $\boldsymbol{\Gamma}_l := [\gamma(u,v)]$ and $\tilde{\boldsymbol{\Gamma}}_l := [\gamma(u,-v)]$ $(u, v = 1, \dots, p+l)$. Under the white noise assumption with $\kappa := E(|\epsilon_t|^4)/\sigma^2 < \infty$,

$$\gamma(u,v) = \sum_{k=1}^{p} 2C_k^2\sigma^2 z_k^{u-v} + \sigma^4\delta_{u-v} + \sigma^4(\kappa - 2)\delta_u\delta_v \qquad (u, v \in \mathbb{Z}).$$

Let $\mathbf{F}_{p+l} := [\mathbf{f}_{p+l}(\omega_1), \dots, \mathbf{f}_{p+l}(\omega_p)]$ and $\mathbf{G} := \mathrm{diag}(C_1^2, \dots, C_p^2)$. Then, we can write

$$\boldsymbol{\Gamma}_l = \sigma^4\mathbf{I} + 2\sigma^2\mathbf{F}_{p+l}\mathbf{G}\mathbf{F}_{p+l}^H, \quad \tilde{\boldsymbol{\Gamma}}_l = 2\sigma^2\mathbf{F}_{p+l}\mathbf{G}\mathbf{F}_{p+l}^T.$$

It is easy to verify that $\mathbf{A}_l\mathbf{F}_{p+l} = \mathbf{0}$. Therefore,

$$\mathbf{A}_l\boldsymbol{\Gamma}_l\mathbf{A}_l^H = \sigma^4\mathbf{A}_l\mathbf{A}_l^H, \quad \mathbf{A}_l\tilde{\boldsymbol{\Gamma}}_l\mathbf{A}_l^T = \mathbf{0}.$$

Combining these results with $\mathbf{A}_l\mathbf{r} = \mathbf{0}$ yields

$$\sqrt{n}\mathbf{A}_l\hat{\mathbf{r}} = \sqrt{n}\mathbf{A}_l(\hat{\mathbf{r}} - \mathbf{r}) \xrightarrow{D} \mathrm{N}_c(\mathbf{0}, \sigma^4\mathbf{A}_l\mathbf{A}_l^H).$$

The assertion follows from Slutsky's theorem in Proposition 2.13(b) and the fact that $(\hat{\mathbf{U}}_l^H\hat{\mathbf{U}}_l)^{-1}\hat{\mathbf{U}}_l^H \xrightarrow{P} (\mathbf{U}_l^H\mathbf{U}_l)^{-1}\mathbf{U}_l^H$ as $n \to \infty$.

Moreover, because $\sqrt{n}\,(\hat{\mathbf{a}} - \mathbf{a}) \xrightarrow{D} \mathrm{N}_c(\mathbf{0}, \mathbf{V}_l)$, it follows from Lemma 12.5.2 that $\sqrt{n}\,(\hat{\boldsymbol{\omega}} - \boldsymbol{\omega}) \xrightarrow{D} \mathrm{N}(\mathbf{0}, \boldsymbol{\Sigma}_l)$, where $\boldsymbol{\Sigma}_l := [\mathbf{J}_k\boldsymbol{\Lambda}_l(k,k')\mathbf{J}_{k'}^T]$, $\mathbf{J}_k := [-\sin(\omega_k), \cos(\omega_k)]$, and

$$\boldsymbol{\Lambda}_l(k,k') := \frac{1}{2}\begin{bmatrix} \Re(\mathbf{v}_k^H\mathbf{V}_l\mathbf{v}_{k'}^*) & -\Im(\mathbf{v}_k^H\mathbf{V}_l\mathbf{v}_{k'}^*) \\ \Im(\mathbf{v}_k^H\mathbf{V}_l\mathbf{v}_{k'}^*) & \Re(\mathbf{v}_k^H\mathbf{V}_l\mathbf{v}_{k'}^*) \end{bmatrix},$$

with $\mathbf{v}_k$ defined by (9.3.11). Hence the assertion. □

Proof of Theorem 9.8. According to the fixed point theorem in Lemma 12.3.2, it suffices to show that there exist constants $\delta > 0$ and $\kappa \in (0,1)$ such that the following inequalities hold almost surely for sufficiently large n:

(a) $\|\hat{\boldsymbol{\varphi}}_r(\tilde{\mathbf{a}}_r) - \hat{\boldsymbol{\varphi}}_r(\tilde{\mathbf{a}}_r')\| \le \kappa \|\tilde{\mathbf{a}}_r - \tilde{\mathbf{a}}_r'\|$ for all $\tilde{\mathbf{a}}_r, \tilde{\mathbf{a}}_r' \in \mathcal{N}$, and

(b) $\|\hat{\boldsymbol{\varphi}}_r(\mathbf{a}_r) - \mathbf{a}_r\| \le (1-\kappa)\delta < \delta$.

To prove these assertions, let $\nabla^T \hat{\boldsymbol{\varphi}}_r(\tilde{\mathbf{a}}_r)$ and $\nabla^T \boldsymbol{\varphi}_r(\tilde{\mathbf{a}}_r)$ denote the Jacobian matrices of $\hat{\boldsymbol{\varphi}}_r(\tilde{\mathbf{a}}_r)$ and $\boldsymbol{\varphi}_r(\tilde{\mathbf{a}}_r)$, respectively. Then, it follows from (9.4.19) that

$$\nabla^T \boldsymbol{\varphi}_r(\mathbf{a}_r) = \mathbf{C}(\mathbf{a}_r).$$

Observe that $\mathbf{C}(\mathbf{a}_r) = \{\mathbf{I} + \boldsymbol{\Gamma}(\mathbf{a}_r)\}^{-1}$, where $\boldsymbol{\Gamma}(\mathbf{a}_r) := \{\mathbf{B}^T \mathbf{R}_\epsilon(\mathbf{a}_r)\mathbf{B}\}^{-1}\mathbf{B}^T\mathbf{R}_x(\mathbf{a}_r)\mathbf{B}$. It is easy to show that the eigenvalues of $\boldsymbol{\Gamma}(\mathbf{a}_r)$, denoted by λ_j, are strictly positive. Therefore, the eigenvalues of $\mathbf{C}(\mathbf{a}_r)$, which take the form $(1+\lambda_j)^{-1}$, are strictly less than unity. This implies that

$$\|\nabla^T \boldsymbol{\varphi}_r(\mathbf{a}_r)\| := \max_{\|\mathbf{c}\|=1} \|\nabla^T \boldsymbol{\varphi}_r(\mathbf{a}_r)\mathbf{c}\| < 1.$$

Combining this result with the continuity of $\nabla^T \boldsymbol{\varphi}_r(\tilde{\mathbf{a}}_r)$ leads to $\|\nabla^T \boldsymbol{\varphi}_r(\tilde{\mathbf{a}}_r)\| \le c$ for all $\tilde{\mathbf{a}}_r \in \mathcal{N}$, where $\delta > 0$ and $c \in (0,1)$ are some constants. On the other hand, it can be shown [233] that $\nabla^T \hat{\boldsymbol{\varphi}}_r(\tilde{\mathbf{a}}_r) \overset{a.s.}{\to} \nabla^T \boldsymbol{\varphi}_r(\tilde{\mathbf{a}}_r)$ uniformly in $\tilde{\mathbf{a}}_r \in \mathcal{N}$. Hence, there is a constant $\kappa \in (c,1)$ such that for all $\tilde{\mathbf{a}}_r \in \mathcal{N}$,

$$\|\nabla^T \hat{\boldsymbol{\varphi}}_r(\tilde{\mathbf{a}}_r)\| \le \|\nabla^T \hat{\boldsymbol{\varphi}}_r(\tilde{\mathbf{a}}_r) - \nabla^T \boldsymbol{\varphi}(\tilde{\mathbf{a}}_r)\| + \|\nabla^T \boldsymbol{\varphi}_r(\tilde{\mathbf{a}}_r)\| \le \kappa$$

almost surely for large n. Furthermore, by the mean-value theorem,

$$\hat{\boldsymbol{\varphi}}_r(\tilde{\mathbf{a}}_r) - \hat{\boldsymbol{\varphi}}_r(\tilde{\mathbf{a}}_r') = \left\{ \int_0^1 \nabla^T \hat{\boldsymbol{\varphi}}_r(\tilde{\mathbf{a}}_r' + s(\tilde{\mathbf{a}}_r - \tilde{\mathbf{a}}_r'))\, ds \right\} (\tilde{\mathbf{a}}_r - \tilde{\mathbf{a}}_r').$$

Combining these results leads to the assertion (a). The assertion (b) follows from the fact that $\hat{\boldsymbol{\varphi}}(\mathbf{a}_r) \overset{a.s.}{\to} \boldsymbol{\varphi}(\mathbf{a}_r) = \mathbf{a}_r$ as $n \to \infty$.

Finally, according to (a), we can write

$$\begin{aligned} \|\hat{\mathbf{a}}_r - \mathbf{a}_r\| &\le \|\hat{\boldsymbol{\varphi}}_r(\hat{\mathbf{a}}_r) - \hat{\boldsymbol{\varphi}}(\mathbf{a}_r)\| + \|\hat{\boldsymbol{\varphi}}_r(\mathbf{a}_r) - \mathbf{a}_r\| \\ &\le \kappa \|\hat{\mathbf{a}}_r - \mathbf{a}_r\| + \|\hat{\boldsymbol{\varphi}}(\mathbf{a}_r) - \mathbf{a}_r\|. \end{aligned}$$

This implies that

$$\|\hat{\mathbf{a}}_r - \mathbf{a}_r\| \le (1-\kappa)^{-1} \|\hat{\boldsymbol{\varphi}}_r(\mathbf{a}_r) - \mathbf{a}_r\| \overset{a.s.}{\to} 0.$$

Hence the consistency assertion. □.

Proof of Theorem 9.9. Observe that

$$\hat{\mathbf{a}}_r - \mathbf{a}_r = \hat{\boldsymbol{\varphi}}_r(\hat{\mathbf{a}}_r) - \hat{\boldsymbol{\varphi}}_r(\mathbf{a}_r) + \hat{\boldsymbol{\varphi}}_r(\mathbf{a}_r) - \mathbf{a}_r.$$

By the mean-value theorem,

$$\hat{\boldsymbol{\varphi}}_r(\hat{\mathbf{a}}_r) - \hat{\boldsymbol{\varphi}}_r(\mathbf{a}_r) = \hat{\mathbf{C}}(\hat{\mathbf{a}}_r)\,(\hat{\mathbf{a}}_r - \mathbf{a}_r),$$

where

$$\hat{\mathbf{C}}(\hat{\mathbf{a}}_r) := \int_0^1 \nabla^T \hat{\boldsymbol{\varphi}}_r(\mathbf{a}_r + s(\hat{\mathbf{a}}_r - \mathbf{a}_r))\,ds.$$

Therefore, we can write

$$\hat{\mathbf{a}}_r - \mathbf{a}_r = \{\mathbf{I} - \hat{\mathbf{C}}(\hat{\mathbf{a}}_r)\}^{-1}\{\hat{\boldsymbol{\varphi}}_r(\mathbf{a}_r) - \mathbf{a}_r\}.$$

Because $\nabla^T \boldsymbol{\varphi}_r(\hat{\mathbf{a}}_r) \overset{a.s.}{\to} \nabla^T \boldsymbol{\varphi}_r(\mathbf{a}_r) = \mathbf{C}(\mathbf{a}_r)$ and because $\nabla^T \hat{\boldsymbol{\varphi}}_r(\tilde{\mathbf{a}}_r) \overset{a.s.}{\to} \nabla^T \boldsymbol{\varphi}_r(\tilde{\mathbf{a}}_r)$ uniformly in $\tilde{\mathbf{a}}_r \in \mathcal{N}$, it follows that $\hat{\mathbf{C}}(\hat{\mathbf{a}}_r) \overset{a.s.}{\to} \mathbf{C}(\mathbf{a}_r)$. Therefore, by Slutsky's theorem, $\hat{\mathbf{a}}_r - \mathbf{a}_r$ has the same asymptotic distribution as $\{\mathbf{I} - \mathbf{C}(\mathbf{a}_r)\}^{-1}\{\hat{\boldsymbol{\varphi}}_r(\mathbf{a}_r) - \mathbf{a}_r\}$. Note that $\{\mathbf{I} - \mathbf{C}(\mathbf{a}_r)\}^{-1} = \{\mathbf{B}^T\mathbf{R}_x(\mathbf{a}_r)\mathbf{B}\}^{-1}\mathbf{B}^T\mathbf{R}_y(\mathbf{a}_r)\mathbf{B}$. Hence, it suffices to show that

$$\sqrt{n}\{\hat{\boldsymbol{\varphi}}_r(\mathbf{a}_r) - \mathbf{a}_r\} \overset{D}{\to} \mathrm{N}(\mathbf{0}, \mathbf{V}), \tag{9.6.13}$$

with $\mathbf{V} = \{\mathbf{B}^T\mathbf{R}_y(\mathbf{a}_r)\mathbf{B}\}^{-1}\boldsymbol{\Gamma}_\epsilon(\mathbf{a}_r)\,\{\mathbf{B}^T\mathbf{R}_y(\mathbf{a}_r)\mathbf{B}\}^{-1}$. Furthermore, let us define

$$\hat{\mathbf{R}}_y(\mathbf{a}_r) := n^{-1/2}\mathbf{Y}_f^H(\mathbf{a}_r)\mathbf{Y}_f(\mathbf{a}_r), \quad \hat{\mathbf{r}}_y(\mathbf{a}_r) := n^{-1/2}\mathbf{Y}_f^H(\mathbf{a}_r)\mathbf{y}_f(\mathbf{a}_r).$$

Then, it follows from (9.4.14) that

$$\begin{aligned}\hat{\boldsymbol{\varphi}}_r(\mathbf{a}_r) - \mathbf{a}_r &= -\{\mathbf{B}^T\hat{\mathbf{R}}_y(\mathbf{a}_r)\mathbf{B}\}^{-1}\mathbf{B}^T\{\hat{\mathbf{R}}_y(\mathbf{a}_r)(\mathbf{B}\mathbf{a}_r + \mathbf{b}) + \hat{\mathbf{r}}_y(\mathbf{a}_r)\}\\ &= -\{\mathbf{B}^T\hat{\mathbf{R}}_y(\mathbf{a}_r)\mathbf{B}\}^{-1}\mathbf{B}^T\{\hat{\mathbf{R}}_y(\mathbf{a}_r)\mathbf{a} + \hat{\mathbf{r}}_y(\mathbf{a}_r)\},\end{aligned}$$

where $\mathbf{a} = \mathbf{B}\mathbf{a}_r + \mathbf{b}$. Because $\hat{\mathbf{R}}_y(\mathbf{a}_r) \overset{a.s.}{\to} \mathbf{R}_y(\mathbf{a}_r)$, the assertion (9.6.13) can be established by showing that

$$\sqrt{n}\mathbf{B}^T\{\hat{\mathbf{R}}_y(\mathbf{a}_r)\mathbf{a} + \hat{\mathbf{r}}_y(\mathbf{a}_r)\} \overset{D}{\to} \mathrm{N}(\mathbf{0}, \boldsymbol{\Gamma}_\epsilon(\mathbf{a}_r)). \tag{9.6.14}$$

This is the objective for the rest of the proof.

For simplicity of notation, let $r(u)$ denote the ACF of $\{y_t(\mathbf{a}_r)\}$ and define $\mathbf{r} := [r(0), r(1), \ldots, r(p)]^T$. Then, it is easy to verify that

$$\mathbf{R}_y(\mathbf{a}_r)\mathbf{a} + \mathbf{r}_y(\mathbf{a}_r) = \mathbf{h}(\mathbf{r}),$$

where the lth entry in $\mathbf{h}(\mathbf{r})$ is defined as $h_l(\mathbf{r}) := \sum_{j=0}^p a_j r(|l-j|)$ $(l = 1, \ldots, p)$. Similarly, let $\hat{\mathbf{r}}$ denote the counterpart of $\mathbf{r}$ formed by the sample autocovariances of $\{\hat{y}_t(\mathbf{a}_r)\}$. It is easy to show that

$$\hat{\mathbf{R}}_y(\mathbf{a}_r)\mathbf{a} + \hat{\mathbf{r}}_y(\mathbf{a}_r) = \mathbf{h}(\hat{\mathbf{r}}) + \mathcal{O}_P(n^{-1/2}).$$

According to Theorem 4.3, $\sqrt{n}(\hat{\mathbf{r}}-\mathbf{r}) \xrightarrow{D} \mathrm{N}(\mathbf{0},\boldsymbol{\Gamma})$, where the $(u+1, v+1)$ entry of $\boldsymbol{\Gamma}$ for $u, v = 0,1,\dots,p$ is equal to $\gamma(u,v)$ defined by (4.3.1) and (4.3.2) with the amplitudes of the sinusoids in $\{x_t\}$ and the spectrum of the noise $\{\epsilon_t\}$ replaced by the corresponding quantities of $\{x_t(\mathbf{a}_r)\}$ and $\{\epsilon_t(\mathbf{a}_r)\}$. By Proposition 2.15,

$$\begin{aligned}&\sqrt{n}\mathbf{B}^T\{\hat{\mathbf{R}}_y(\mathbf{a}_r)\mathbf{a}+\hat{\mathbf{r}}_y(\mathbf{a}_r)-\mathbf{h}(\mathbf{r})\}\\ &= \sqrt{n}\mathbf{B}^T\{\mathbf{h}(\hat{\mathbf{r}})-\mathbf{h}(\mathbf{r})\}+\mathcal{O}_P(1) \xrightarrow{D} \mathrm{N}(\mathbf{0},\mathbf{B}^T\boldsymbol{\Sigma}_0\mathbf{B}),\end{aligned}$$

where $\boldsymbol{\Sigma}_0 := \{\nabla^T\mathbf{h}(\mathbf{r})\}\boldsymbol{\Gamma}\{\nabla^T\mathbf{h}(\mathbf{r})\}^T$. Furthermore, observe that

$$\begin{aligned}\mathbf{B}^T\mathbf{h}(\mathbf{r}) &= \mathbf{B}^T\{\mathbf{R}_y(\mathbf{a}_r)(\mathbf{B}\mathbf{a}_r+\mathbf{b})+\mathbf{r}_y(\mathbf{a}_r)\}\\ &= \mathbf{B}^T\mathbf{R}_y(\mathbf{a}_r)\mathbf{B}\{\mathbf{a}_r-\boldsymbol{\varphi}(\mathbf{a}_r)\} = \mathbf{0}.\end{aligned}$$

Therefore, (9.6.14) can be established by proving $\mathbf{B}^T\boldsymbol{\Sigma}_0\mathbf{B} = \boldsymbol{\Gamma}_\epsilon(\mathbf{a}_r)$.

Toward that end, observe that $\nabla^T\mathbf{h}(\mathbf{r}) = [\mathbf{g}_0,\mathbf{g}_1,\dots,\mathbf{g}_p]$, where $\mathbf{g}_0 := \mathbf{a}$ and

$$\mathbf{g}_u := [a_{1-u}+a_{1+u},\dots,a_{p-u}+a_{p+u}]^T \qquad (u=1,\dots,p),$$

with $a_j := 0$ for $j<0$ or $j>p$. Therefore,

$$\boldsymbol{\Sigma}_0 = \sum_{u=0}^{p}\sum_{v=0}^{p}\gamma(u,v)\mathbf{g}_u\mathbf{g}_v^T.$$

By breaking the double sum into four pieces, i.e., $u=v=0$, $u=0$ and $v=1,\dots,p$, $u=1,\dots,p$ and $v=0$, and $u,v=1,\dots,p$, one can show, after straightforward but slightly tedious calculation, that the (u,v) entry of $\boldsymbol{\Sigma}_0$ can be expressed as

$$\sigma(u,v) := \sum_{j=0}^{p}\sum_{l=0}^{p} a_j a_l \gamma(u-j, v-l) \qquad (u,v=1,\dots,p).$$

Observe that $\gamma(u,v)$ in (4.3.1) and (4.3.2) consists of three terms: the first term involves the sinusoids and the second term involves the fourth moment of the white noise. The contribution from the first term to $\boldsymbol{\Sigma}_0$ is equal to zero because $a(z_k)=0$ $(k=1,\dots,q)$. The contribution from the second term takes the form

$$\boldsymbol{\Sigma}_2 := (\kappa-3)\{\mathbf{R}_\epsilon(\mathbf{a}_r)\mathbf{a}+\mathbf{r}_\epsilon(\mathbf{a}_r)\}\{\mathbf{R}_\epsilon(\mathbf{a}_r)\mathbf{a}+\mathbf{r}_\epsilon(\mathbf{a}_r)\}^T.$$

Because the ACF of the filtered noise satisfies (9.4.18), it follows that

$$\mathbf{B}^T\boldsymbol{\Sigma}_2\mathbf{B} = \mathbf{0}.$$

The contribution from the remaining term, denoted as $\boldsymbol{\Sigma}_3$, can be expressed as $\boldsymbol{\Sigma}_3 = [\sigma_3(u,v)]$ $(u,v=1,\dots,p)$, where

$$\sigma_3(u,v) := \sum_{\tau=-\infty}^{\infty}\sum_{j=0}^{p}\sum_{l=0}^{p} a_j a_l \gamma_\tau(u-j)\{\gamma_\tau(v-l)+\gamma_\tau(l-v)\}$$

and $\gamma_\tau(s) := E\{\epsilon_{\tau+s}(\mathbf{a}_r)\epsilon_0(\mathbf{a}_r)\}$ $(s = 0, \pm 1, \pm 2, \ldots)$. Define

$$\mathbf{A}_0 := [\mathbf{0}, \mathbf{A}_p] = \begin{bmatrix} 0 & a_p & a_{p-1} & \ldots & a_0 & 0 & \ldots & 0 \\ 0 & 0 & a_p & \ldots & a_1 & a_0 & & 0 \\ & & & \ddots & & & \ddots & \\ 0 & 0 & \cdots & 0 & a_p & \ldots & a_1 & a_0 \end{bmatrix} \in \mathbb{R}^{p\times 2p+1},$$

where $\mathbf{A}_p$ is given by (9.3.9) with $l = p$. Then, it is easy to verify that

$$\boldsymbol{\Sigma}_3 = \sum_{\tau=-\infty}^{\infty} \mathbf{A}_0 \boldsymbol{\gamma}_\tau \{(\mathbf{A}_0 + \mathbf{A}_0 \tilde{\mathbf{I}})\boldsymbol{\gamma}_\tau\}^T = \sum_{\tau=-\infty}^{\infty} \mathbf{A}_0 \boldsymbol{\gamma}_\tau \boldsymbol{\gamma}_\tau^T (\mathbf{A}_0^T + \tilde{\mathbf{I}}\mathbf{A}_0^T).$$

where $\tilde{\mathbf{I}}$ is the antidiagonal matrix obtained by reversing the rows (or columns) of $\mathbf{I}$ and $\boldsymbol{\gamma}_\tau := [\gamma_\tau(-p), \gamma_\tau(-p+1), \ldots, \gamma_\tau(p)]^T$. Furthermore, observe that

$$\tilde{\mathbf{I}}\mathbf{A}_0^T \mathbf{B} = \mathbf{A}_0^T \mathbf{B}.$$

Therefore,

$$\mathbf{B}^T \boldsymbol{\Sigma}_3 \mathbf{B} = 2 \sum_{\tau=-\infty}^{\infty} \mathbf{B}^T \mathbf{A}_0 \boldsymbol{\gamma}_\tau \boldsymbol{\gamma}_\tau^T \mathbf{A}_0^T \mathbf{B}.$$

Finally, for $\tau = 1, 2, \ldots$, we have $\boldsymbol{\gamma}_{-\tau} = \tilde{\mathbf{I}}\boldsymbol{\gamma}_\tau$ and hence

$$\mathbf{B}^T \mathbf{A}_0 \boldsymbol{\gamma}_{-\tau} = \mathbf{B}^T \mathbf{A}_0 \tilde{\mathbf{I}} \boldsymbol{\gamma}_\tau = \mathbf{B}^T \mathbf{A}_0 \boldsymbol{\gamma}_\tau.$$

Moreover,

$$\mathbf{B}^T \mathbf{A}_0 \boldsymbol{\gamma}_0 = \mathbf{B}^T \{\mathbf{R}_\epsilon(\mathbf{a}_r)\mathbf{a} + \mathbf{r}_\epsilon(\mathbf{a}_r)\} = \mathbf{0}.$$

Combining these results leads to

$$\mathbf{B}^T \boldsymbol{\Sigma}_0 \mathbf{B} = \mathbf{B}^T \boldsymbol{\Sigma}_3 \mathbf{B} = 4\mathbf{B}^T \mathbf{A}_0 \left\{ \sum_{\tau=1}^{\infty} \boldsymbol{\gamma}_\tau \boldsymbol{\gamma}_\tau^T \right\} \mathbf{A}_0^T \mathbf{B} = \boldsymbol{\Gamma}_\epsilon(\mathbf{a}_r).$$

Hence the assertion (9.6.14). □

NOTES

Analytical properties of the higher-order AR spectrum are investigated in [208] and [329] under the assumption of multiple sinusoids in white noise. The impact of order selection on the resulting frequency estimates is studied in [130] and [367]. The resolution of the AR spectrum is investigated in [130] and [254]. It is also studied in [439] under the framework of statistical decision theory. The asymptotic normality of the linear prediction estimator $\hat{\mathbf{c}}$ in Theorem 9.4 as well as the asymptotic normality of the corresponding AR spectral estimator at a fixed frequency are derived in [213] for the real case.

A result similar to Theorem 9.6 is obtained in [375] for the EYW estimators in the real case without the symmetric constraint. The case where l increases with the sample size n is also investigated in [375].

A simple filter, called the alpha filter, is employed by He and Kedem [147] to construct an iterative filtering procedure based on the lag-1 autocorrelation coefficient of the filtered time series. It consistently estimates the frequency of a single real sinusoid in white noise. A sufficient condition is provided by Yakowitz [428] to ensure the consistency of the iterative filtering procedure when using other filters. The idea is further generalized in [234] to the case of multiple real sinusoids. The resulting method, called *parametric filtering*, is an extension of the iterative filtering method discussed in Section 9.4 to general linear filters. See also [181], [182], [222], and [232] for more details.

Chapter 10

Covariance Analysis Approach

For a time series that comprises sinusoids in white noise, the autocovariance function (ACF) has a very special structure: it is a weighted sum of sinusoids, whose frequencies are the same as the frequencies of the sinusoidal signal, plus a Kronecker delta sequence. In this chapter, we discuss several methods of frequency estimation that utilize the special structure through the application of a powerful technique of matrix analysis — the eigenvalue decomposition (EVD). The key idea is to construct suitable matrices based on the ACF such that the signal frequencies can be obtained from their eigenvalues and eigenvectors.

10.1 Eigenvalue Decomposition of Covariance Matrix

Consider the time series $\{y_t\}$ which is given by

$$y_t := x_t + \epsilon_t \qquad (t = 1, \ldots, n), \tag{10.1.1}$$

where $\{\epsilon_t\}$ is white noise with mean zero and variance σ^2. In the complex case, $\{x_t\}$ takes the form (2.1.5) or (2.1.7). Therefore, the ACF of $\{y_t\}$, according to Theorem 4.1, can be expressed as

$$r_y(u) = r_x(u) + r_\epsilon(u) = \sum_{k=1}^{p} C_k^2 \exp(i\omega_k u) + \sigma^2 \delta_u. \tag{10.1.2}$$

The same expression is valid in the real case where $\{x_t\}$ is given by (2.1.1) or (2.1.2), provided the parameters are defined by (2.1.8). In this case, the $p := 2q$ frequencies form q conjugate pairs such that $\omega_{p-k+1} = 2\pi - \omega_k$ $(k = 1, \ldots, q)$.

For any given $m \geq 1$, consider the m-dimensional covariance matrix

$$\mathbf{R}_y := \begin{bmatrix} r_y(0) & r_y^*(1) & \cdots & r_y^*(m-1) \\ r_y(1) & r_y(0) & \cdots & r_y^*(m-2) \\ \vdots & \vdots & \ddots & \vdots \\ r_y(m-1) & r_y(m-2) & \cdots & r_y(0) \end{bmatrix}. \tag{10.1.3}$$

By randomizing the phases of the sinusoids so that $\{y_t\}$ becomes as a zero-mean stationary process with $E(y_{t+u}y_t^*) = r_y(u)$, we can write

$$\mathbf{R}_y = E(\mathbf{y}_t \mathbf{y}_t^H),$$

where $\mathbf{y}_t := [y_t, y_{t+1}, \ldots, y_{t+m-1}]^T \in \mathbb{C}^m$. Because $\mathbf{R}_y$ is a positive-definite Hermitian matrix, all its eigenvalues are real and positive [153, p. 104]. Let the eigenvalues of $\mathbf{R}_y$ be denoted as $\lambda_1 \geq \cdots \geq \lambda_m > 0$, which are always arranged in descending order by convention, and let

$$\mathbf{\Lambda} := \mathrm{diag}(\lambda_1, \ldots, \lambda_m).$$

According to the matrix theory [153, p. 171], there exists a unitary matrix

$$\mathbf{V} := [\mathbf{v}_1, \ldots, \mathbf{v}_m]$$

such that

$$\mathbf{R}_y = \mathbf{V}\mathbf{\Lambda}\mathbf{V}^H = \sum_{j=1}^{m} \lambda_j \mathbf{v}_j \mathbf{v}_j^H. \tag{10.1.4}$$

This expression is called an *eigenvalue decomposition* (EVD) of $\mathbf{R}_y$. Because the vectors $\mathbf{v}_1, \ldots, \mathbf{v}_m$ are orthogonal to each other with unit norm, also known as orthonormal, it follows from (10.1.4) that

$$\mathbf{R}_y \mathbf{v}_j = \lambda_j \mathbf{v}_j \qquad (j = 1, \ldots, m). \tag{10.1.5}$$

Hence $\mathbf{v}_j$ is an eigenvector of $\mathbf{R}_y$ associated with the eigenvalue λ_j.

Under the white noise assumption, we have

$$\mathbf{R}_y = \mathbf{R}_x + \sigma^2 \mathbf{I}, \tag{10.1.6}$$

where $\mathbf{R}_x$ denotes the covariance matrix of the form (10.1.3) defined by $\{r_x(u)\}$ instead of $\{r_y(u)\}$. Because $r_x(u) = \sum_{k=1}^{p} C_k^2 \exp(i\omega_k u)$, it is easy to verify that

$$\mathbf{R}_x = \sum_{k=1}^{p} C_k^2 \mathbf{f}_m(\omega_k) \mathbf{f}_m^H(\omega_k) = \mathbf{F}_m \mathbf{G} \mathbf{F}_m^H, \tag{10.1.7}$$

where

$$\begin{cases} \mathbf{f}_m(\omega) := [\exp(i\omega), \ldots, \exp(im\omega)]^T, \\ \mathbf{F}_m := [\mathbf{f}_m(\omega_1), \ldots, \mathbf{f}_m(\omega_p)], \quad \mathbf{G} := \mathrm{diag}(C_1^2, \ldots, C_p^2). \end{cases} \tag{10.1.8}$$

Therefore, for any $m \geq p$, the matrix $\mathbf{R}_x$ has rank p. Owing to this special structure, the following theorem establishes a relationship between the EVD of $\mathbf{R}_y$ and the EVD of $\mathbf{R}_x$. It serves as an important foundation of the covariance analysis approach. A proof of the theorem is given in Section 10.7.

Theorem 10.1 (Eigenvalue Decomposition of the Covariance Matrix). *Given $m > p$, let (10.1.4) be an eigenvalue decomposition of the m-dimensional covariance matrix $\mathbf{R}_y = \mathbf{R}_x + \sigma^2\mathbf{I}$. Then,*

$$\lambda_j = \begin{cases} \mu_j + \sigma^2 & \text{for } j = 1,\ldots,p, \\ \sigma^2 & \text{for } j = p+1,\ldots,m, \end{cases}$$

where $\mu_1 \geq \cdots \geq \mu_p > 0$ are the nonzero eigenvalues of the rank-p matrix $\mathbf{R}_x$. Moreover, the vectors $\mathbf{v}_1,\ldots,\mathbf{v}_p$ are eigenvectors of $\mathbf{R}_x$ associated with $\mu_1,\ldots,\mu_p$, respectively, and the remaining vectors $\mathbf{v}_{p+1},\ldots,\mathbf{v}_m$ are eigenvectors of $\mathbf{R}_x$ associated with the zero eigenvalue.

To better understand the eigenvectors $\mathbf{v}_1,\ldots,\mathbf{v}_m$, observe that the sinusoidal vectors $\mathbf{f}_m(\omega_1),\ldots,\mathbf{f}_m(\omega_p)$, which are linearly independent of each other, decompose the m-dimensional space $\mathbb{C}^m$ into two orthogonal subspaces:

$$\mathbb{C}^m_{\mathrm{S}} := \mathrm{span}\{\mathbf{f}_m(\omega_1),\ldots,\mathbf{f}_m(\omega_p)\},$$
$$\mathbb{C}^m_{\mathrm{N}} := \mathbb{C}^m \ominus \mathbb{C}^m_{\mathrm{S}}.$$

The p-dimensional subspace $\mathbb{C}^m_{\mathrm{S}}$ is called the *signal subspace* and its $(m-p)$-dimensional orthogonal complement $\mathbb{C}^m_{\mathrm{N}}$ is called the *noise subspace.* Theorem 10.1 asserts that $\mathbf{v}_{p+1},\ldots,\mathbf{v}_m$ are eigenvectors of $\mathbf{R}_x$ associated with its zero eigenvalue. This, combined with (10.1.7), leads to

$$\mathbf{R}_x\mathbf{v}_j = \mathbf{F}_m\mathbf{G}\mathbf{F}_m^H\mathbf{v}_j = \mathbf{0} \qquad (j = p+1,\ldots,m).$$

Because $\mathbf{F}_m\mathbf{G}$ has full (column) rank p, it follows that

$$\mathbf{f}_m^H(\omega_k)\,\mathbf{v}_j = 0 \qquad (j = p+1,\ldots,m;\ k = 1,\ldots,p). \tag{10.1.9}$$

In other words, the vectors $\mathbf{v}_{p+1},\ldots,\mathbf{v}_m$ are orthogonal to $\mathbf{f}_m(\omega_1),\ldots,\mathbf{f}_m(\omega_p)$ and therefore reside in the noise subspace $\mathbb{C}^m_{\mathrm{N}}$. Moreover, because $\mathbb{C}^m_{\mathrm{N}}$ has dimension $m-p$, the $m-p$ vectors $\mathbf{v}_{p+1},\ldots,\mathbf{v}_m$, which are linearly independent of each other (and indeed orthonormal), must form a basis of $\mathbb{C}^m_{\mathrm{N}}$, i.e.,

$$\mathrm{span}\{\mathbf{v}_{p+1},\ldots,\mathbf{v}_m\} = \mathbb{C}^m_{\mathrm{N}}.$$

The remaining vectors $\mathbf{v}_1,\ldots,\mathbf{v}_p$, which are orthogonal to $\mathbf{v}_{p+1},\ldots,\mathbf{v}_m$, must form a basis for the orthogonal complement of $\mathbb{C}^m_{\mathrm{N}}$ which is $\mathbb{C}^m_{\mathrm{S}}$. In other words,

$$\mathrm{span}\{\mathbf{v}_1,\ldots,\mathbf{v}_p\} = \mathbb{C}^m_{\mathrm{S}}.$$

Thus the signal and noise subspaces are completely determined by $\mathbf{v}_1,\ldots,\mathbf{v}_p$ and $\mathbf{v}_{p+1},\ldots,\mathbf{v}_m$, respectively, which are the eigenvectors of $\mathbf{R}_y$. This result is summarized in the following corollary to Theorem 10.1.

Corollary 10.1 (Signal and Noise Subspaces). *Let the conditions of Theorem 10.1 be satisfied. Then, the eigenvectors of* $\mathbf{R}_y$ *associated with the p largest eigenvalues span the signal subspace* $\mathbb{C}_{\mathrm{S}}^m := \mathrm{span}\{\mathbf{f}_m(\omega_1),\ldots,\mathbf{f}_m(\omega_p)\}$ *and the remaining eigenvectors of* $\mathbf{R}_y$ *span the noise subspace* $\mathbb{C}_{\mathrm{N}}^m := \mathbb{C}^m \ominus \mathbb{C}_{\mathrm{S}}^m$.

In light of Theorem 10.1 and Corollary 10.1, the EVD of $\mathbf{R}_y$ in (10.1.4) can also be expressed as

$$\mathbf{R}_y = \mathbf{V}_{\mathrm{S}}\boldsymbol{\Lambda}_{\mathrm{S}}\mathbf{V}_{\mathrm{S}}^H + \sigma^2\mathbf{V}_{\mathrm{N}}\mathbf{V}_{\mathrm{N}}^H, \tag{10.1.10}$$

where

$$\mathbf{V}_{\mathrm{S}} := [\mathbf{v}_1,\ldots,\mathbf{v}_p], \quad \mathbf{V}_{\mathrm{N}} := [\mathbf{v}_{p+1},\ldots,\mathbf{v}_m], \tag{10.1.11}$$

and

$$\boldsymbol{\Lambda}_{\mathrm{S}} := \mathrm{diag}(\lambda_1,\ldots,\lambda_p).$$

By construction, the column space of $\mathbf{V}_{\mathrm{S}}$ coincides with the signal subspace $\mathbb{C}_{\mathrm{S}}^m$ and the column space of $\mathbf{V}_{\mathrm{N}}$ coincides with the noise subspace $\mathbb{C}_{\mathrm{N}}^m$. Because $\mathbf{V} = [\mathbf{V}_{\mathrm{S}},\mathbf{V}_{\mathrm{N}}]$ is a unitary matrix, $\mathbf{V}_{\mathrm{S}}$ and $\mathbf{V}_{\mathrm{N}}$ have the following properties:

$$\mathbf{V}_{\mathrm{S}}^H\mathbf{V}_{\mathrm{S}} = \mathbf{I}, \quad \mathbf{V}_{\mathrm{N}}^H\mathbf{V}_{\mathrm{N}} = \mathbf{I}, \quad \mathbf{V}_{\mathrm{S}}^H\mathbf{V}_{\mathrm{N}} = \mathbf{0},$$

and

$$\mathbf{V}_{\mathrm{S}}\mathbf{V}_{\mathrm{S}}^H + \mathbf{V}_{\mathrm{N}}\mathbf{V}_{\mathrm{N}}^H = \mathbf{I}.$$

The last identity can also be written as

$$\mathbf{P}_{\mathrm{S}} + \mathbf{P}_{\mathrm{N}} = \mathbf{I},$$

where $\mathbf{P}_{\mathrm{S}} := \mathbf{V}_{\mathrm{S}}\mathbf{V}_{\mathrm{S}}^H$ is the projection matrix onto the signal subspace $\mathbb{C}_{\mathrm{S}}^m$ and $\mathbf{P}_{\mathrm{N}} := \mathbf{V}_{\mathrm{N}}\mathbf{V}_{\mathrm{N}}^H$ is the projection matrix onto the noise subspace $\mathbb{C}_{\mathrm{N}}^m$.

In the remainder of this chapter, we discuss several frequency estimators that exploit these properties of the covariance matrix.

10.2 Principal Component Analysis Method

According to (10.1.6) and (10.1.7), the matrix $\mathbf{R}_y$ becomes rank-deficient when the noise is absent. In situations like this, one can always enhance an estimate of such a matrix, which is obtained from noisy observations, by a lower rank approximation. This can be accomplished with the help of a general technique called the *principal component analysis* (PCA) [167]. The PCA technique derives the lower rank approximation from an eigenvalue decomposition (EVD) by using the largest eigenvalues only. It minimizes the approximation error subject to the rank constraint. In this section, we employ the PCA technique to improve the linear prediction estimators and the extended Yule-Walker estimators discussed in Chapter 9, resulting in two types of reduced rank estimators.

10.2.1 Reduced Rank Autoregressive Estimators

Recall that the linear prediction estimators discussed in Section 9.1 are solutions to a normal equation of the form

$$\hat{\mathbf{R}}_y\hat{\mathbf{c}} = -\hat{\mathbf{r}}_y, \tag{10.2.1}$$

where $\hat{\mathbf{R}}_y$ and $\hat{\mathbf{r}}_y$ are certain estimators of $\mathbf{R}_y$ and $\mathbf{r}_y$. To enhance the linear prediction estimators, let us replace $\hat{\mathbf{R}}_y$ by a rank-p matrix, which is motivated by the fact that $\mathbf{R}_y$ reduces to $\mathbf{R}_x$ in the absence of noise and therefore has rank p.

For any $m > p$, a rank-p matrix that best approximates $\hat{\mathbf{R}}_y$ can be derived from its EVD. Indeed, let $\hat{\mathbf{R}}_y$ have an EVD of the form

$$\hat{\mathbf{R}}_y = \sum_{j=1}^{m} \hat{\lambda}_j\hat{\mathbf{v}}_j\hat{\mathbf{v}}_j^H, \tag{10.2.2}$$

where $\hat{\lambda}_1 \geq \cdots \geq \hat{\lambda}_m > 0$ are the eigenvalues of $\hat{\mathbf{R}}_y$ and $\hat{\mathbf{v}}_1, \ldots, \hat{\mathbf{v}}_m$ are the associated eigenvectors. In light of Theorem 10.1, it is natural to approximate $\hat{\mathbf{R}}_y$ by

$$\tilde{\mathbf{R}}_y := \sum_{j=1}^{p} \hat{\lambda}_j\hat{\mathbf{v}}_j\hat{\mathbf{v}}_j^H, \tag{10.2.3}$$

which has rank p if $\hat{\lambda}_p > 0$. Because $\hat{\lambda}_1, \ldots, \hat{\lambda}_p$ are the largest eigenvalues of $\hat{\mathbf{R}}_y$, the rank-1 matrices $\hat{\lambda}_1\hat{\mathbf{v}}_1\hat{\mathbf{v}}_1^H, \ldots, \hat{\lambda}_p\hat{\mathbf{v}}_p\hat{\mathbf{v}}_p^H$ are called the *principal components* of $\hat{\mathbf{R}}_y$. By employing the principal components, the matrix $\tilde{\mathbf{R}}_y$ given by (10.2.3) is optimal in the sense that it minimizes the approximation error

$$\|\hat{\mathbf{R}}_y - \tilde{\mathbf{R}}_y\|^2 := \mathrm{tr}\{(\hat{\mathbf{R}}_y - \tilde{\mathbf{R}}_y)^H(\hat{\mathbf{R}}_y - \tilde{\mathbf{R}}_y)\}$$

among all rank-p matrices of dimension m. This assertion is known in the literature as the Schmidt-Mirsky or Eckart-Young theorem [361, p. 208]. Note that approximating $\hat{\mathbf{R}}_y$ by $\tilde{\mathbf{R}}_y$ in (10.2.3) is equivalent to setting the $m-p$ smallest eigenvalues in the EVD of $\hat{\mathbf{R}}_y$ equal to zero and therefore can be interpreted as a form of filtering in the eigenvalue domain.

With $\hat{\mathbf{R}}_y$ in (10.2.1) replaced by the rank-p matrix $\tilde{\mathbf{R}}_y$, the resulting normal equation has multiple solutions. The most interesting solution is the one that has the smallest norm, called the minimum-norm solution. By the matrix theory [153, p. 421], the minimum-norm solution can be expressed as

$$\hat{\mathbf{c}}_{\text{PCA}} := -\tilde{\mathbf{R}}_y^{\dagger}\hat{\mathbf{r}}_y = -\sum_{j=1}^{p} \hat{\lambda}_j^{-1}(\hat{\mathbf{v}}_j^H\hat{\mathbf{r}}_y)\,\hat{\mathbf{v}}_j, \tag{10.2.4}$$

where $\tilde{\mathbf{R}}_y^{\dagger}$ is the pseudo inverse (also known as the Moore-Penrose generalized inverse) of $\tilde{\mathbf{R}}_y$, which takes the form

$$\tilde{\mathbf{R}}_y^{\dagger} := \sum_{j=1}^{p} \hat{\lambda}_j^{-1}\hat{\mathbf{v}}_j\hat{\mathbf{v}}_j^H.$$

We call $\hat{\mathbf{c}}_{\text{PCA}}$ in (10.2.4) a *reduced rank* linear prediction estimator. Note that $\hat{\mathbf{c}}_{\text{PCA}}$ resides in the estimated signal subspace $\hat{\mathbb{C}}_{\text{S}}^m := \text{span}\{\hat{\mathbf{v}}_1,\dots,\hat{\mathbf{v}}_p\}$.

As an example, consider the reduced rank FBLP estimator (also known as the modified FBLP estimator [396]) defined by (10.2.4) with

$$\hat{\mathbf{R}}_y := (2n)^{-1}\mathbf{Y}_{fb}^H\mathbf{Y}_{fb}, \quad \hat{\mathbf{r}}_y := (2n)^{-1}\mathbf{Y}_{fb}^H\mathbf{y}_{fb},$$

where $\mathbf{Y}_{fb}$ and $\mathbf{y}_{fb}$ are given by (9.1.11). Because $\hat{\mathbf{R}}_y$ is an inner product of $\mathbf{Y}_{fb}$, the EVD of $\hat{\mathbf{R}}_y$ in (10.2.3) can be obtained directly from the singular value decomposition (SVD) of $\mathbf{Y}_{fb}$. Indeed, if m satisfies $2(n-m) > m > p$, or equivalently,

$$p < m < 2n/3, \tag{10.2.5}$$

then, with $l := 2(n-m)$, the SVD of $\mathbf{Y}_{fb}$ takes the form

$$\mathbf{Y}_{fb} = \hat{\mathbf{U}}\begin{bmatrix}\hat{\boldsymbol{\Sigma}}\\ \mathbf{0}\end{bmatrix}\hat{\mathbf{V}}^H = \sum_{j=1}^{m} \hat{s}_j\hat{\mathbf{u}}_j\hat{\mathbf{v}}_j^H, \tag{10.2.6}$$

where $\hat{\mathbf{U}} := [\hat{\mathbf{u}}_1,\dots,\hat{\mathbf{u}}_l]$ is an l-by-l unitary matrix, $\hat{\mathbf{V}} := [\hat{\mathbf{v}}_1,\dots,\hat{\mathbf{v}}_m]$ is an m-by-m unitary matrix, and $\hat{\boldsymbol{\Sigma}} := \text{diag}(\hat{s}_1,\dots,\hat{s}_m)$ is an m-by-m diagonal matrix comprising the singular values $\hat{s}_1 \ge \cdots \ge \hat{s}_m \ge 0$ [153, p. 414]. From (10.2.6) we obtain

$$\hat{\mathbf{R}}_y := (2n)^{-1}\mathbf{Y}_{fb}^H\mathbf{Y}_{fb} = \hat{\mathbf{V}}\hat{\boldsymbol{\Lambda}}\hat{\mathbf{V}}^H = \sum_{j=1}^{m} \hat{\lambda}_j\hat{\mathbf{v}}_j\hat{\mathbf{v}}_j^H, \tag{10.2.7}$$

where $\hat{\boldsymbol{\Lambda}} := \text{diag}(\hat{\lambda}_1,\dots,\hat{\lambda}_m)$ and $\hat{\lambda}_j := (2n)^{-1}\hat{s}_j^2$ $(j = 1,\dots,m)$. This expression constitutes an EVD of $\hat{\mathbf{R}}_y$ because the $\hat{\lambda}_j$ are eigenvalues of $\hat{\mathbf{R}}_y$ and the $\hat{\mathbf{v}}_j$ are the associated eigenvectors. Moreover, it follows from (10.2.6) that

$$\hat{\mathbf{r}}_y := (2n)^{-1}\mathbf{Y}_{fb}^H\mathbf{y}_{fb} = (2n)^{-1}\sum_{j=1}^{m} \hat{s}_j(\hat{\mathbf{u}}_j^H\mathbf{y}_{fb})\hat{\mathbf{v}}_j. \tag{10.2.8}$$

Inserting (10.2.7) and (10.2.8) into (10.2.4) yields

$$\hat{\mathbf{c}}_{\mathrm{PCA}} = -\sum_{j=1}^{p} \hat{s}_j^{-1}(\hat{\mathbf{u}}_j^H \mathbf{y}_{fb})\,\hat{\mathbf{v}}_j. \tag{10.2.9}$$

This SVD technique of computing the reduced rank estimator circumvents the matrix multiplication $\mathbf{Y}_{fb}^H \mathbf{Y}_{fb}$ and tends to have better numerical properties.

Figure 10.1 depicts the AR spectrum produced by the reduced rank FBLP estimator with different values of m for the time series used in Figure 9.1. The AR spectrum takes the form $\hat{f}_{\mathrm{AR}}(\omega) := \hat{\sigma}_m^2/|\hat{C}(\omega)|^2$ with $\hat{\sigma}_m^2 := (2n)^{-1}\|\mathbf{y}_{fb} + \mathbf{Y}_{fb}\hat{\mathbf{c}}_{\mathrm{PCA}}\|^2$ and $\hat{C}(\omega) := 1 + \mathbf{f}_m^H(\omega)\hat{\mathbf{c}}_{\mathrm{PCA}}$. As we can see by comparing Figure 10.1 with Figure 9.1, the PCA technique successfully reduces the spurious spectral peaks in the cases of $m = 27$ and $m = 28$ while retaining the ability to resolve the two closely spaced frequencies that cannot be resolved by the periodogram.

Those with a keener eye may have noticed that the reduced rank FBLP estimator makes the spectral peaks sharper in the case of $m = 28$ but less sharp in the case of $m = 27$. A sharper spectral peak does not necessarily lead to a more accurate frequency estimate. The strength of the PCA method lies mainly in its ability to suppress spurious spectral peaks while enhancing the desired ones.

Note that the success of the reduced rank FBLP estimator as illustrated by Figure 10.1 does not imply that the PCA method completely eliminates the problem of spurious spectral peaks or the sensitivity to the choice of m. In fact, when the signal frequencies are not resolved due to low SNR, spurious spectral peaks can still occur if m is not suitably chosen.

The statistical performance of the reduced rank FBLP frequency estimator is demonstrated by the following examples with different choices of m. The impact of PCA can be appreciated by comparing them with Examples 9.3 and 9.4.

Example 10.1 (Estimation of Well-Separated Frequencies: Part 1). Consider the time series in Example 9.3 which consists of two unit-amplitude complex sinusoids with well-separated frequencies in Gaussian white noise. Let the frequency estimates be computed in the same way as in Example 9.3 except that the reduced rank FBLP estimator in (10.2.9) is employed instead of the ordinary FBLP estimator given by (9.1.12). Figure 10.2 depicts the average MSE of the frequency estimates and the initial values for different SNR and two sample sizes.

Recall that the linear prediction estimators need a very large m to reduce the bias but cannot take such values due to the problem of spurious spectral peaks. Figure 9.4 shows that in the case of $n = 100$ the choice of $m = 20$ ($m = 40$ for $n = 200$) already makes the ordinary FBLP estimator behave very poorly. With the PCA method, this choice becomes valid, and the resulting MSE is considerably improved, as can be seen by comparing Figure 9.4 with Figure 10.2. Better still, a much larger value of m can be used in the reduced rank FBLP estimator without running into the problem of spurious spectral peaks. Indeed, Figure 10.2 shows

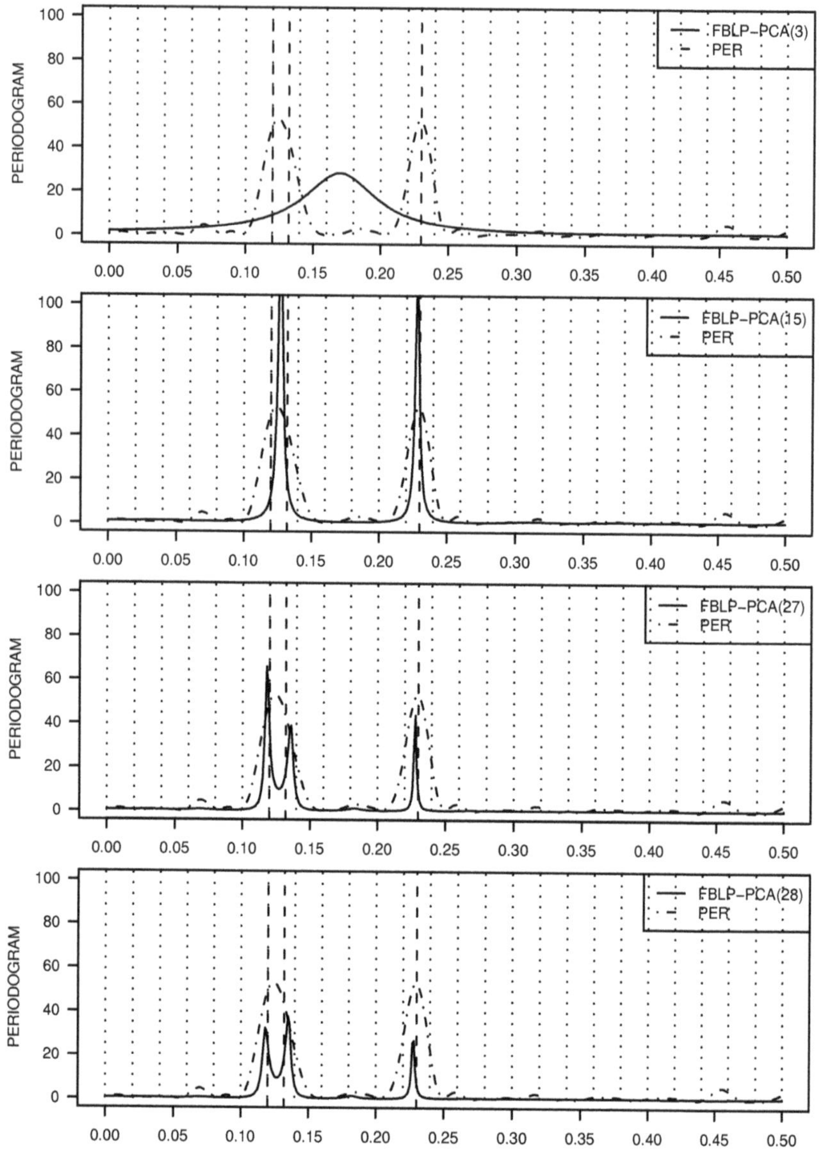

Figure 10.1. Plot of AR spectrum of different orders estimated by the reduced rank FBLP method for a time series that consists of three unit-amplitude zero-phase complex sinusoids in Gaussian white noise ($n = 50$, $\omega_1 = 2\pi \times 6/n = 2\pi \times 0.12$, $\omega_2 = 2\pi \times 6.6/n = 2\pi \times 0.132$, $\omega_3 = 2\pi \times 11.5/n = 2\pi \times 0.23$, and SNR = 4 dB per sinusoid, i.e., $\sigma^2 = 0.3981072$). Solid line, estimated AR spectrum of order $m = 3, 15, 27, 28$; dashed-dotted line, periodogram. Vertical dashed lines depict the signal frequencies and vertical dotted lines depict the Fourier grid. For $m = p = 3$, the reduced rank FBLP estimator is identical to the ordinary FBLP estimator. The same time series is used to produce Figure 9.1.

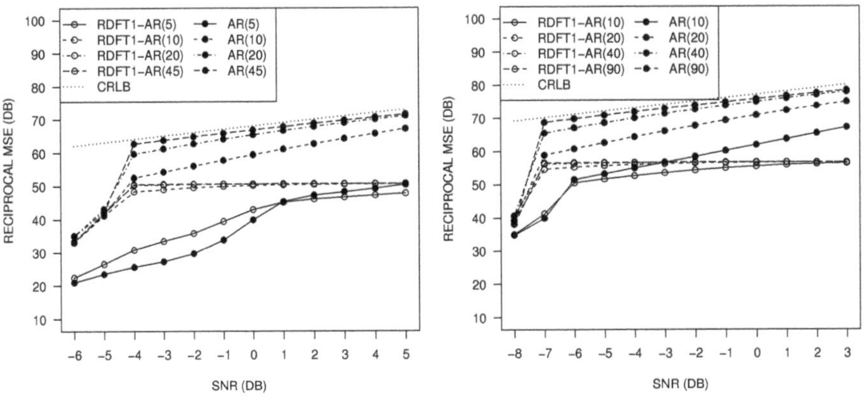

(a) (b)

Figure 10.2. Plot of the reciprocal average MSE of the reduced rank FBLP frequency estimates for the time series of two complex sinusoids in Gaussian white noise discussed in Example 10.1 with different values of SNR. (a) $n = 100$: solid line, $m = 5$; dashed line, $m = 10$; dash-dotted line, $m = 20$; long dashed line, $m = 45$. (b) $n = 200$: solid line, $m = 10$; dashed line, $m = 20$; dash-dotted line, $m = 40$; long dashed line, $m = 90$. Lines with dots, final estimates; lines with circles, initial RDFT1 estimates. Dotted line depicts the CRLB. Results are based on 5,000 Monte Carlo runs.

that it is possible to take $m = 45$ in this example for $n = 100$ ($m = 90$ for $n = 200$) and make the MSE even closer to the CRLB.

However, the PCA method does not eliminate the necessity of selecting a proper order m. In fact, the performance of the reduced rank FBLP estimator still begins to deteriorate when m becomes too large (e.g., $m = 60$ for $n = 100$; not shown). A comparison of Figure 10.2 with Figure 9.4 also shows that the PCA method does not offer any significant benefit when m is not large enough (e.g., $m = 5$ and 10 for $n = 100$; $m = 10$ and 20 for $m = 200$). Moreover, the best results shown in Figure 10.2 remain inferior to the results obtained by the PM, NLS, IFIL, and IGLS estimators shown in Figures 6.13, 8.3, 9.10, and 9.15, respectively. ◇

Example 10.2 (Estimation of Closely Spaced Frequencies: Part 1). Consider the time series in Example 9.4 which consists of two unit-amplitude complex sinusoids with closely spaced frequencies in Gaussian white noise. Let the frequency estimates be computed in the same way as in Example 9.4 but from the reduced rank FBLP estimator given by (10.2.9) instead of the ordinary FBLP estimator given by (9.1.12). Figure 10.3 shows the simulation result.

Comparing this result with Figure 9.5 shows that the PCA method makes $m = 30$ a valid choice for $n = 100$ ($m = 60$ for $n = 200$) and the resulting MSE is greatly improved. The PCA method also allows a larger value $m = 45$, which brings the

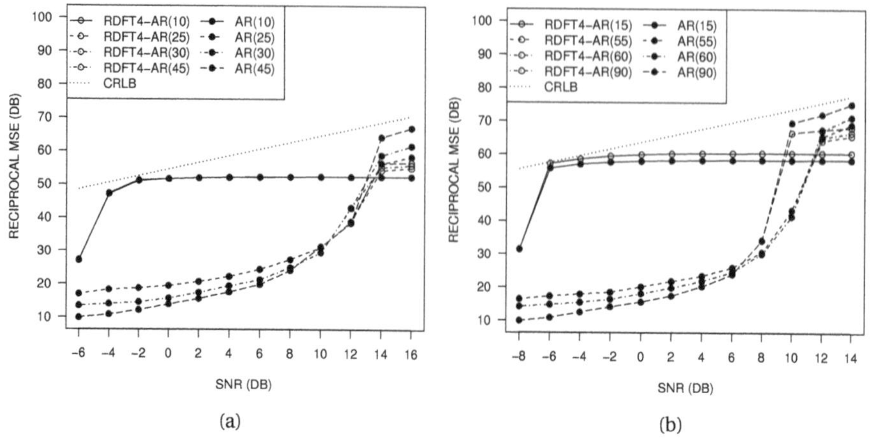

Figure 10.3. Plot of the reciprocal average MSE of the reduced rank FBLP frequency estimates for the time series of two complex sinusoids with closely spaced frequencies in Gaussian white noise discussed in Example 10.2 with different values of SNR. (a) $n = 100$: solid line, $m = 10$; dashed line, $m = 25$; dash-dotted line, $m = 30$; long dashed line, $m = 45$. (b) $n = 200$: solid line, $m = 15$; dashed line, $m = 55$; dash-dotted line, $m = 60$; long dashed line, $m = 90$. Lines with dots, final estimates; lines with circles, initial RDFT4 estimates. Dotted line depicts the CRLB. Results are based on 5,000 Monte Carlo runs.

MSE closer to the CRLB than is possible with the ordinary FBLP estimator. The SNR threshold is reduced with the choice of $m = 90$ in the case of $n = 200$.

These improvements take place only at the very high end of the SNR range considered. The performance remains poor elsewhere, due to spurious spectral peaks. Even at the high end of the SNR range, the best results shown in Figure 10.3 still fall short of the CRLB, so the reduced rank FBLP estimator remains inferior to the NLS estimator whose performance is shown in Figure 8.5. At high SNR, the reduced rank FBLP estimator performs better than the IGLS and ILEV estimators and much better than the IFIL estimator, but at low SNR, it performs worse than these estimators, especially IFIL. This can be seen by comparing Figure 10.3 with Figures 9.11, 9.16, and 9.18. Similar to the previous example, no significant benefit is obtained from the PCA method if m is not large enough (e.g., $m = 10$ and 25 for $n = 100$, $m = 15$ and 55 for $n = 200$). ◇

These examples show that the PCA method is indeed very effective in boosting the performance of the ordinary linear prediction estimators with properly selected order m. The benefit stems mainly from its ability to enhance the signal spectral peaks without suffering too much from the problem of spurious spectral peaks. For estimating closely spaced frequencies, spurious spectral peaks remain troublesome at low SNR in comparison with some alternatives such as the IFIL, IGLS, and ILEV estimators, and especially the NLS estimator.

The reduced rank estimator defined by (10.2.4) can be modified by replacing $\hat{\lambda}_j$ with $\hat{\lambda}_j - \hat{\sigma}^2$ $(j = 1,\ldots,p)$, where $\hat{\sigma}^2$ is an estimator of the noise variance σ^2. Recall that the motivation for using $\tilde{\mathbf{R}}_y$ in (10.2.4) rather than $\hat{\mathbf{R}}_y$ is that $\tilde{\mathbf{R}}_y$, which is defined by (10.2.2), has the same reduced-rank property as the signal covariance matrix $\mathbf{R}_x$. According to Theorem 10.1, the p largest eigenvalues of $\mathbf{R}_y$ can be expressed as $\lambda_j = \mu_j + \sigma^2$ $(j = 1,\ldots,p)$, where the μ_j are the nonzero eigenvalues of $\mathbf{R}_x$. If the $\hat{\lambda}_j$ are regarded as estimates of the λ_j, then a reduced-rank matrix which is more resemblant to $\mathbf{R}_x$ than $\tilde{\mathbf{R}}_y$ is the one with $\hat{\lambda}_j$ in (10.2.2) replaced by $\hat{\lambda}_j - \hat{\sigma}^2$ $(j = 1,\ldots,p)$. Therefore, instead of $\tilde{\mathbf{R}}_y$, consider

$$\tilde{\mathbf{R}}_x := \sum_{j=1}^{p} (\hat{\lambda}_j - \hat{\sigma}^2)\hat{\mathbf{v}}_j\hat{\mathbf{v}}_j^H. \tag{10.2.10}$$

While $\hat{\sigma}^2$ can be any estimator of σ^2, Theorem 10.1 suggests that we take

$$\hat{\sigma}^2 := (m-p)^{-1} \sum_{j=p+1}^{m} \hat{\lambda}_j. \tag{10.2.11}$$

Assuming $\hat{\lambda}_j > \hat{\sigma}^2$ for all $j = 1,\ldots,p$, the matrix $\tilde{\mathbf{R}}_x$ has rank p and its pseudo inverse is given by $\tilde{\mathbf{R}}_x^{\dagger} := \sum_{j=1}^{p} (\hat{\lambda}_j - \hat{\sigma}^2)^{-1}\hat{\mathbf{v}}_j\hat{\mathbf{v}}_j^H$. Therefore, the resulting reduced-rank estimator can be expressed as

$$\tilde{\mathbf{c}}_{\text{PCA}} := -\tilde{\mathbf{R}}_x^{\dagger}\hat{\mathbf{r}}_y = -\sum_{j=1}^{p} (\hat{\lambda}_j - \hat{\sigma}^2)^{-1}(\hat{\mathbf{v}}_j^H\hat{\mathbf{r}}_y)\hat{\mathbf{v}}_j. \tag{10.2.12}$$

We call $\tilde{\mathbf{c}}_{\text{PCA}}$ a *biased-corrected* reduced rank linear prediction estimator.

With $\tilde{\sigma}_m^2 := (2n)^{-1}\|\mathbf{y}_{fb} + \mathbf{Y}_{fb}\tilde{\mathbf{c}}_{\text{PCA}}\|^2$ and $\hat{C}_0(\omega) := 1 + \mathbf{f}_m^H(\omega)\tilde{\mathbf{c}}_{\text{PCA}}$, Figure 10.4 depicts the function $\tilde{\sigma}_m^2/|\hat{C}_0(\omega)|^2$ with different values of m for the same time series as in Figure 10.1. We call this function a *pseudo* AR spectrum because it is no longer a consistent estimator of the AR spectrum discussed in Chapter 9 (Section 9.1), as will be shown later. Nonetheless, the pseudo AR spectrum can be used in the same way as the ordinary AR spectrum to produce the frequency estimates through its local maxima. A comparison of Figure 10.4 with Figure 10.1 shows that the spectral peaks are considerably enhanced by the bias-corrected estimator, especially for $m = 27$ and $m = 28$.

Simulation results based on the same data as in Examples 10.1 and 10.2 indicate that the bias-corrected estimator produces more accurate frequency estimates than the original one for well-separated frequencies when m is not too large. For example, in the case of $n = 100$ and $m = 10$, the MSE of the bias-corrected estimator at 5 dB is equal to 1.54×10^{-7} as compared to 1.88×10^{-7} without bias correction. When m is large (e.g., $m = 45$ for $n = 100$), the difference becomes negligible. Bias correction remains beneficial for estimating closely spaced frequencies. For example, in the case of $n = 100$ and $m = 25$,

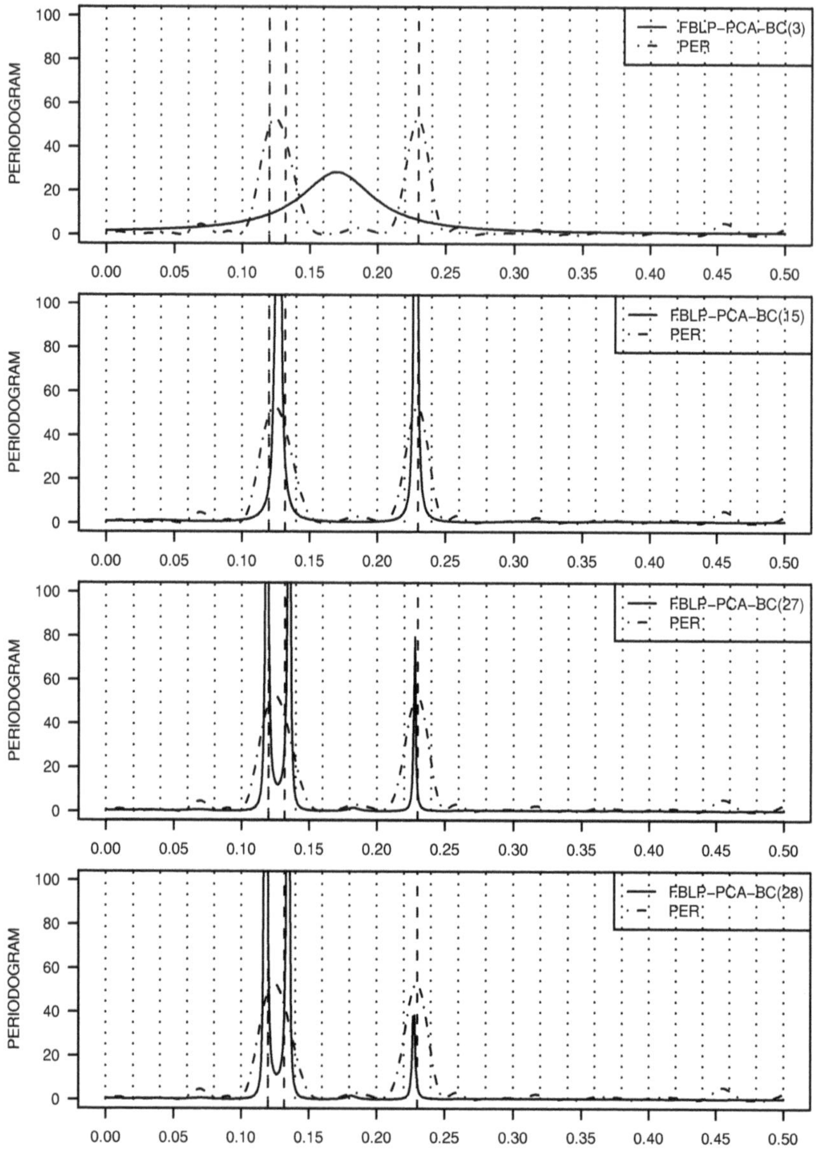

Figure 10.4. Plot of pseudo AR spectrum produced by the bias-corrected reduced rank FBLP estimator of the form (10.2.12) for the same time series as used in Figure 10.1 with the same choices of m.

the MSE of the bias-corrected estimator is 1.14×10^{-6} at 14 dB as compared to 2.35×10^{-6} without bias correction. However, when m is too small and the signal frequencies are not resolved (e.g., $m = 10$ for $n = 100$), the performance can be degraded by bias correction. This is mainly due to the increased risk of spurious spectral peaks as a result of its more aggressive peak enhancement capability.

The reduced rank estimator without bias correction is analogous to ridge regression [261] — a regularization technique commonly used in linear regression analysis to combat the problem of collinearity, as it happens with closely spaced frequencies. Ridge regression has a bias, but it improves the numerical stability of the ordinary regression solution in the presence of collinearity. Therefore, it should not be a surprise that the chance of spurious spectral peaks, which is a manifestation of numerical instability, increases once the bias is corrected in the reduced rank estimator for estimating closely spaced frequencies.

The PCA technique can also be employed to improve the extended Yule-Walker (EYW) estimators discussed in Section 9.3. Toward that end, let us first generalize the EYW method to higher-order AR models.

Recall that the frequency parameter $\boldsymbol{\omega} := [\omega_1, \dots, \omega_p]^T$ is uniquely determined by the AR parameter $\mathbf{a} := [a_1, \dots, a_p]^T$ through the p-degree polynomial

$$a(z) := \sum_{j=0}^{p} a_j z^{-j} := \prod_{k=1}^{p} (1 - z_k z^{-1}) \quad (a_0 := 1),$$

where $z_k := \exp(i\omega_k)$ $(k = 1, \dots, p)$. The EYW estimator of $\mathbf{a}$ is constructed by the method of moments on the basis of the extended Yule-Walker equations

$$\sum_{j=0}^{p} a_j r_y(u - j) = 0 \qquad (u = p+1, p+2, \dots). \tag{10.2.13}$$

For any given $m > p$, consider the m-degree polynomial

$$c(z) := \sum_{j=0}^{m} c_j z^{-j} := a(z) b(z), \tag{10.2.14}$$

where $b(z) := \sum_{j=0}^{m-p} b_j z^{-j}$ $(b_0 := 1)$ is any $(m-p)$-degree polynomial. By setting $a_j := 0$ for $j \notin \{0, 1, \dots, p\}$, we can write

$$c_j = \sum_{j'=0}^{m-p} a_{j-j'} b_{j'} \qquad (j = 0, 1, \dots, m).$$

Combining this expression with (10.2.13) leads to

$$\sum_{j=0}^{m} c_j r_y(u - j) = 0 \qquad (u = m+1, m+2, \dots). \tag{10.2.15}$$

We call these equations the *higher-order* extended Yule-Walker equations.

Let $\mathbf{c} := [c_1, \ldots, c_m]^T$. Then, for any $l \ge m$, the first l equations in (10.2.15) can be expressed in matrix notation as

$$\mathbf{U}_{lm}\,\mathbf{c} = -\mathbf{u}_{lm}, \tag{10.2.16}$$

where

$$\mathbf{U}_{lm} := \begin{bmatrix} r_y(m) & \cdots & r_y(1) \\ \vdots & & \vdots \\ r_y(m+l-1) & \cdots & r_y(l) \end{bmatrix}, \quad \mathbf{u}_{lm} := \begin{bmatrix} r_y(m+1) \\ \vdots \\ r_y(m+l) \end{bmatrix}.$$

To estimate $\mathbf{c}$ by the method of moments, let the ACF in (10.2.16) be replaced by the sample autocovariances $\hat{r}_y(u)$ $(u = 1, \ldots, m+l)$ given by (4.1.1) with

$$p < m \le l \le n - m - 1,$$

and consider the least-squares (LS) solution to the resulting equation by minimizing $\|\hat{\mathbf{u}}_{lm} + \hat{\mathbf{U}}_{lm}\tilde{\mathbf{c}}\|^2$ with respect to $\tilde{\mathbf{c}} \in \mathbb{C}^m$, where $\hat{\mathbf{U}}_{lm}$ and $\hat{\mathbf{u}}_{lm}$ are the counterparts of $\mathbf{U}_{lm}$ and $\mathbf{u}_{lm}$ defined by $\{\hat{r}_y(u)\}$ instead of $\{r_y(u)\}$. Any solution $\hat{\mathbf{c}}$ to the LS problem should satisfy the normal equation

$$\hat{\mathbf{H}}_{lm}\hat{\mathbf{c}} = -\hat{\mathbf{U}}_{lm}^H\hat{\mathbf{u}}_{lm}, \tag{10.2.17}$$

where $\hat{\mathbf{H}}_{lm} := \hat{\mathbf{U}}_{lm}^H\hat{\mathbf{U}}_{lm}$. In this equation, the matrix $\hat{\mathbf{H}}_{lm}$ is a natural estimator of $\mathbf{H}_{lm} := \mathbf{U}_{lm}^H\mathbf{U}_{lm}$. Because $r_y(u) = r_x(u) = \sum_{k=1}^p C_k^2 z_k^u$ for any $u \ne 0$, it follows that

$$\mathbf{U}_{lm} = \mathbf{G}_{lm}\mathbf{F}_m^H, \tag{10.2.18}$$

where $\mathbf{G}_{lm} := [z_1^m\mathbf{f}_l(\omega_1), \ldots, z_p^m\mathbf{f}_l(\omega_p)]\mathbf{G}$. Therefore, the matrix $\mathbf{H}_{lm}$, which can be expressed as $\mathbf{F}_m\mathbf{G}_{lm}^H\mathbf{G}_{lm}\mathbf{F}_m^H$, has rank p for any $l \ge m > p$. Motivated by this observation, a reduced rank EYW estimator is obtained by replacing $\hat{\mathbf{H}}_{lm}$ in (10.2.17) with a rank-p approximation derived from the PCA technique.

Owing to the inner product form, the rank-p approximation of $\hat{\mathbf{H}}_{lm}$ can be computed directly from the SVD of $\hat{\mathbf{U}}_{lm}$. Indeed, if the SVD of $\hat{\mathbf{U}}_{lm}$ is given by

$$\hat{\mathbf{U}}_{lm} = \sum_{j=1}^m \hat{v}_j\hat{\mathbf{p}}_j\hat{\mathbf{q}}_j^H,$$

where $\hat{v}_1 \ge \cdots \ge \hat{v}_m \ge 0$, then $\hat{\mathbf{H}}_{lm}$ has an EVD of the form

$$\hat{\mathbf{H}}_{lm} = \sum_{j=1}^m \hat{v}_j^2\,\hat{\mathbf{q}}_j\hat{\mathbf{q}}_j^H. \tag{10.2.19}$$

In this case, an optimal rank-p approximation of $\hat{\mathbf{H}}_{lm}$ is given by

$$\tilde{\mathbf{H}}_{lm} := \sum_{j=1}^p \hat{v}_j^2\,\hat{\mathbf{q}}_j\hat{\mathbf{q}}_j^H. \tag{10.2.20}$$

With $\hat{\mathbf{H}}_{lm}$ in (10.2.17) replaced by $\tilde{\mathbf{H}}_{lm}$ and with $\tilde{\mathbf{H}}_{lm}^{\dagger} := \sum_{j=1}^{p} \hat{v}_j^{-2} \hat{\mathbf{q}}_j \hat{\mathbf{q}}_j^H$ denoting the pseudo inverse of $\tilde{\mathbf{H}}_{lm}$, the minimum-norm solution to the resulting equation can be expressed as

$$\hat{\mathbf{c}}_{\text{PCA}} := -\tilde{\mathbf{H}}_{lm}^{\dagger} \hat{\mathbf{U}}_{lm}^H \hat{\mathbf{u}}_{lm} = -\sum_{j=1}^{p} \hat{v}_j^{-1} (\hat{\mathbf{p}}_j^H \hat{\mathbf{u}}_{lm}) \hat{\mathbf{q}}_j. \qquad (10.2.21)$$

We call this solution a *reduced rank* EYW estimator (also known as the higher-order Yule-Walker estimator [376]). Observe that

$$\tilde{\mathbf{H}}_{lm}^{\dagger} \hat{\mathbf{U}}_{lm}^H = \sum_{j=1}^{p} \hat{v}_j^{-1} \hat{\mathbf{q}}_j \hat{\mathbf{p}}_j^H = \hat{\mathbf{U}}_{lm}^{\dagger}.$$

Therefore, we can also express $\hat{\mathbf{c}}_{\text{PCA}}$ in (10.2.21) as

$$\hat{\mathbf{c}}_{\text{PCA}} = -\hat{\mathbf{U}}_{lm}^{\dagger} \hat{\mathbf{u}}_{lm}.$$

With $\hat{C}_0(\omega) := 1 + \mathbf{f}_m^H(\omega) \hat{\mathbf{c}}_{\text{PCA}}$, the frequency estimator $\hat{\boldsymbol{\omega}}$ of the reduced rank EYW method comprises the local minima of the function $|\hat{C}_0(\omega)|^2$, or equivalently, the local maxima of the function $1/|\hat{C}_0(\omega)|^2$, near $\boldsymbol{\omega}$.

Figure 10.5 depicts the function $1/|\hat{C}_0(\omega)|^2$, which we call the EYW spectrum, with the choice of $l = m$ for a time series consisting of three complex sinusoids in Gaussian white noise. Observe that only two spectral peaks instead of three appear in the EYW spectrum with $m = 9$; increasing the value of m to 14 gives three peaks, one of which is considerably erroneous. The best result is obtained with $m = 16$, where the two closely spaced frequencies are barely resolved. These frequencies become completely unresolved again when m is increased to 17. This example suggests that the reduced rank EYW method is inferior to the reduced rank linear prediction method in resolving closely spaced frequencies.

The next two examples demonstrate the statistical performance of the reduced rank EYW estimator with $l = m$ for different values of m. The effect of the PCA-enabled higher-order AR models on the EYW method can be appreciated by comparing these examples with Examples 9.10 and 9.11 in Chapter 9.

Example 10.3 (Estimation of Well-Separated Frequencies: Part 2). Consider the time series in Example 10.1. Let the frequency estimates be computed in the same way as in Example 10.1 except that the reduced rank EYW estimator in (10.2.21) with $l = m > p$ is used instead of the reduced rank FBLP estimator. Figure 10.6 depicts the average MSE of the frequency estimates with different choices of m and for different values of SNR. Note that with the choice of $l = m$, the order m cannot be greater than $(n-1)/2$.

In comparison with Figure 9.7, we can see that the reduced rank EYW estimator dramatically improves the ordinary EYW estimator. In particular, with a suitably large m (e.g., $m = 45$ for $n = 100$ and $m = 90$ for $n = 200$), the resulting MSE

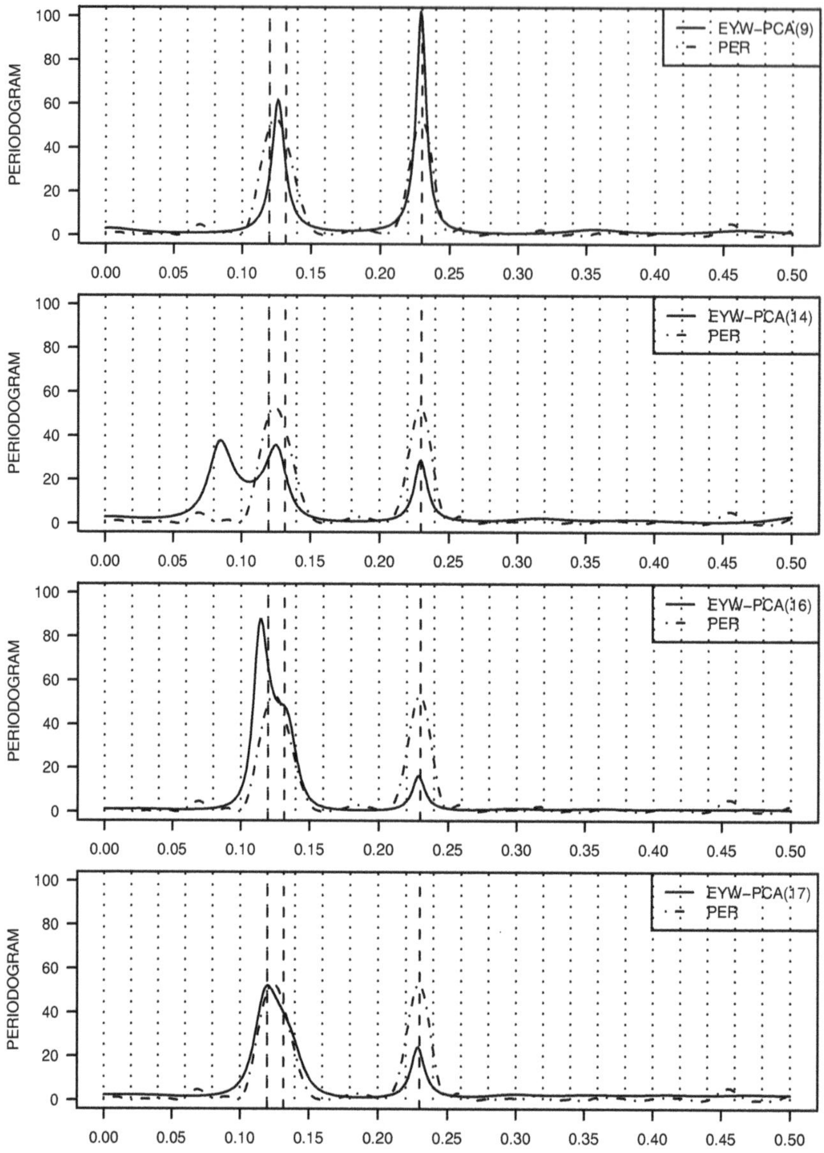

Figure 10.5. Plot of EYW spectrum with different choices of m ($l = m$) for the time series which is used in Figure 10.1. Solid line, EYW spectrum with $m = 9, 14, 16, 17$; dashed-dotted line, periodogram. Vertical dashed lines depict the signal frequencies and vertical dotted lines depict the Fourier grid. The EYW spectrum is rescaled so that the curve covers the same area in the plot as the periodogram.

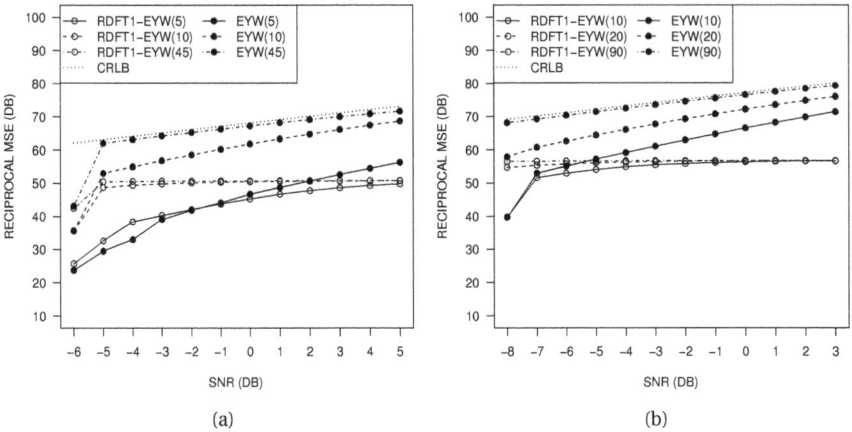

Figure 10.6. Plot of the reciprocal average MSE of the reduced rank EYW frequency estimates for the time series of two complex sinusoids in Gaussian white noise discussed in Example 10.3 with different values of SNR. (a) $n = 100$: solid line, $m = l = 5$; dashed line, $m = l = 10$; dash-dotted line, $m = l = 45$. (b) $n = 200$: solid line, $m = 10$; dashed line, $m = 20$; dash-dotted line, $m = 90$. Lines with dots, final reduced rank EYW estimates; lines with circles, initial RDFT1 estimates. Dotted line depicts the CRLB. Results are based on 5,000 Monte Carlo runs.

approaches the CRLB. This is comparable to the best performance of the reduced rank FBLP estimator shown in Figure 10.2. Moreover, as a pleasant surprise, the SNR threshold of the reduced rank EYW estimator is slightly lower than that of the reduced rank FBLP estimator for both sample sizes, although both are initialized by the DFT method using the corresponding spectrum. Note that unlike the ordinary EYW estimator in Figure 9.7, the choice of $l = m = 97$ for $n = 100$ and $l = m = 197$ for $n = 200$ are no longer valid for the reduced rank EYW estimator because m cannot be greater than $(n-1)/2$. ◇

Example 10.4 (Estimation of Closely Spaced Frequencies: Part 2). Consider the time series in Example 10.2. Let the frequency estimates be computed in the same way as in Example 10.2 except that the reduced rank EYW estimator in (10.2.21) with $l = m$. Figure 10.7 depicts the simulation result.

A comparison of this result with Figure 9.8 shows again a dramatic improvement in accuracy achieved by the reduced rank EYW estimator over the ordinary EYW estimator. However, even with a large m, the accuracy refuses to improve with the SNR beyond a threshold because of its inability to resolve the signal frequencies. This is similar to the behavior of the reduced rank FBLP estimator shown in Figure 10.3 when the order is not large enough. At the lower end of

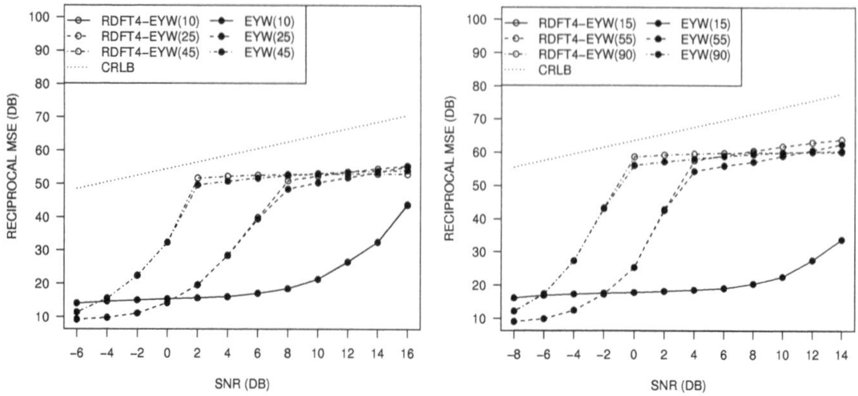

(a) (b)

Figure 10.7. Plot of the reciprocal average MSE of the reduced rank EYW frequency estimates for the time series of two complex sinusoids with closely spaced frequencies in Gaussian white noise discussed in Example 9.4 with different values of SNR. (a) $n = 100$: solid line, $m = l = 10$; dashed line, $m = l = 25$; dash-dotted line, $m = l = 45$. (b) $n = 200$: solid line, $m = 15$; dashed line, $m = 55$; dash-dotted line, $m = 90$. Lines with dots, final reduced rank EYW estimates; lines with circles, initial RDFT1 estimates. Dotted line depicts the CRLB. Results are based on 5,000 Monte Carlo runs.

the SNR range considered, the reduced rank EYW estimator with a suitably large m (e.g., $m = 45$ for $n = 100$ and $m = 90$ for $n = 200$) outperforms the reduced rank FBLP estimator of the same order, due primarily to its greater resistance to spurious spectral peaks. ◇

10.2.2 Statistical Properties

Examples 10.1 and 10.2 suggest that the reduced rank linear prediction estimators are useful only when the order m is large enough to be comparable to the sample size n. Unfortunately, such cases are very difficult for theoretical analysis, so the asymptotic properties of these estimators when m grows with n are still unknown. However, for fixed m as $n \to \infty$, some interesting results can be obtained regarding the consistency and the asymptotic distribution of the reduced rank linear prediction estimators.

First, let us derive the limiting value of $\hat{\mathbf{c}}_{\text{PCA}}$ in (10.2.4) as $n \to \infty$ for fixed m. According to Theorem 9.3, the linear prediction estimators discussed in Section 9.1 all converge to the same higher-order AR parameter

$$\mathbf{c} := -\mathbf{R}_y^{-1}\mathbf{r}_y. \tag{10.2.22}$$

Given the EVD of $\mathbf{R}_y$ in (10.1.4), we can also write

$$\mathbf{c} = -\sum_{j=1}^{m} \lambda_j^{-1}(\mathbf{v}_j^H \mathbf{r}_y)\mathbf{v}_j. \tag{10.2.23}$$

Under the white noise assumption, it follows from (10.1.2) that

$$\mathbf{r}_y = \mathbf{r}_x = \mathbf{F}_m \mathbf{g} \in \mathbb{C}_S^m,$$

where $\mathbf{g} := [C_1^2,\ldots,C_p^2]^T$. Therefore, by Theorem 10.1, we obtain

$$\mathbf{v}_j^H \mathbf{r}_y = 0 \qquad (j = p+1,\ldots,m).$$

Combining this result with (10.2.23) leads to

$$\mathbf{c} = -\sum_{j=1}^{p} \lambda_j^{-1}(\mathbf{v}_j^H \mathbf{r}_y)\mathbf{v}_j = -\bar{\mathbf{R}}_y^{\dagger}\mathbf{r}_y, \tag{10.2.24}$$

where $\bar{\mathbf{R}}_y^{\dagger} := \sum_{j=1}^{p} \lambda_j^{-1}\mathbf{v}_j\mathbf{v}_j^H$ denotes the pseudo inverse of $\bar{\mathbf{R}}_y := \sum_{j=1}^{p} \lambda_j \mathbf{v}_j\mathbf{v}_j^H$. Note that $\bar{\mathbf{R}}_y$ is the EVD-based optimal rank-p approximation of $\mathbf{R}_y$.

According to Lemma 12.4.2, we can write

$$\hat{\lambda}_j = \lambda_j + \mathcal{O}(\|\hat{\mathbf{R}}_y - \mathbf{R}_y\|) \qquad (j = 1,\ldots,m). \tag{10.2.25}$$

Because $\lambda_p = \mu_p + \sigma^2 > \sigma^2 = \lambda_{p+1}$, Lemma 12.4.6 ensures that

$$\tilde{\mathbf{R}}_y^{\dagger} = \bar{\mathbf{R}}_y^{\dagger} + \mathcal{O}(\|\hat{\mathbf{R}}_y - \mathbf{R}_y\|). \tag{10.2.26}$$

Therefore, if $\hat{\mathbf{R}}_y \stackrel{a.s.}{\to} \mathbf{R}_y$ and $\hat{\mathbf{r}}_y \stackrel{a.s.}{\to} \mathbf{r}_y$, then $\hat{\mathbf{c}}_{\text{PCA}} \stackrel{a.s.}{\to} \mathbf{c}$. For sufficiently large m, Theorem 9.1 asserts that the AR spectrum associated with $\mathbf{c}$ has p largest local maxima, denoted by $\boldsymbol{\omega}_0 := [\omega_{01},\ldots,\omega_{0p}]^T$, in the vicinity of $\boldsymbol{\omega} := [\omega_1,\ldots,\omega_p]^T$. For sufficiently large n, Lemma 12.5.1 ensures that the estimated AR spectrum $\hat{f}_{\text{AR}}(\omega) := \hat{\sigma}_m^2/|\hat{C}(\omega)|^2$, with $\hat{C}(\omega) := 1 + \mathbf{f}_m^H(\omega)\hat{\mathbf{c}}_{\text{PCA}}$, has local maxima near $\boldsymbol{\omega}_0$, denoted by $\hat{\boldsymbol{\omega}}$, such that $\hat{\boldsymbol{\omega}} \stackrel{P}{\to} \boldsymbol{\omega}_0$. Combining this discussion with Theorem 4.1, which guarantees $\hat{\mathbf{R}}_y \stackrel{a.s.}{\to} \mathbf{R}_y$ and $\hat{\mathbf{r}}_y \stackrel{a.s.}{\to} \mathbf{r}_y$, proves the following theorem.

Theorem 10.2 (Consistency of the Reduced Rank Linear Prediction Estimators). *Let $\{y_t\}$ be given by (10.1.1) with $\{x_t\}$ taking the form (2.1.5) or (2.1.7). Let $\hat{\mathbf{c}}_{\text{PCA}}$ be defined by (10.2.4) and $\mathbf{c}$ be defined by (10.2.22). Assume that $\{\epsilon_t\} \sim \text{IID}(0,\sigma^2)$ and $E(|\epsilon_t|^4) < \infty$. Then, for any fixed $m > p$, $\hat{\mathbf{c}}_{\text{PCA}} \stackrel{a.s.}{\to} \mathbf{c}$ as $n \to \infty$. Moreover, for sufficiently large $m > p$, let $\boldsymbol{\omega}_0$ comprise the p largest local maxima of the higher-order AR spectrum associated with $\mathbf{c}$ and let $\hat{\boldsymbol{\omega}}$ denote the frequency estimator based on $\hat{\mathbf{c}}_{\text{PCA}}$. Then, $\hat{\boldsymbol{\omega}} \stackrel{P}{\to} \boldsymbol{\omega}_0$ as $n \to \infty$.*

According to Theorem 10.2, the reduced rank linear prediction estimators have the same limiting value as the ordinary linear prediction estimators. Therefore, the resulting frequency estimates remain inconsistent in general when m is fixed as $n \to \infty$. As with the ordinary linear prediction estimators, the asymptotic bias of the reduced rank linear prediction estimators vanishes as $m \to \infty$.

The asymptotic distribution of the reduced rank linear prediction estimators can also be derived for fixed $m > p$ as $n \to \infty$. To present this result, let

$$\mathbf{b} := [b_1, \ldots, b_m]^T = -\mathbf{R}_x^\dagger \mathbf{c}, \quad c_0 := 1, \quad b_0 := 0,$$

and define

$$c_{11}(u,v) := \sum_{j=0}^{m} c_{u-v+j} c_j^*, \quad c_{12}(u,v) := \sum_{j=0}^{m} c_{u-v+j} b_j^*,$$

$$c_{22}(u,v) := \sum_{j=0}^{m} b_{u-v+j} b_j^*,$$

$$\tilde{c}_{11}(u,v) := \sum_{j=0}^{m} c_{u+v-j} c_j, \quad \tilde{c}_{12}(u,v) := \sum_{j=0}^{m} c_{u+v-j} b_j,$$

$$\tilde{c}_{22}(u,v) := \sum_{j=0}^{m} b_{u+v-j} b_j.$$

For $1 \le j \le j' \le 2$, let $\mathbf{C}_{jj'} := [c_{jj'}(u,v)]$ and $\tilde{\mathbf{C}}_{jj'} := [\tilde{c}_{jj'}(u,v)]$ $(u, v = 1, \ldots, m)$. Furthermore, define

$$\boldsymbol{\Sigma}_{11} := \sigma^4 \left\{ \sum_{k=1}^{p} 2\gamma_k |c(z_k)|^2 \mathbf{f}_m(\omega_k) \mathbf{f}_m^H(\omega_k) + \mathbf{C}_{11} + (\kappa - 2)\mathbf{c}\mathbf{c}^H \right\},$$

$$\tilde{\boldsymbol{\Sigma}}_{11} := \sigma^4 \left\{ \sum_{k=1}^{p} 2\gamma_k \{c(z_k)\}^2 \mathbf{f}_m(\omega_k) \mathbf{f}_m^T(\omega_k) + \tilde{\mathbf{C}}_{11} + (\kappa - 2)\mathbf{c}\mathbf{c}^T \right\}$$

$$\boldsymbol{\Sigma}_{12} := \sigma^4 \mathbf{C}_{12}, \quad \boldsymbol{\Sigma}_{22} := \sigma^4 \mathbf{C}_{22}, \quad \tilde{\boldsymbol{\Sigma}}_{12} := \sigma^4 \tilde{\mathbf{C}}_{12}, \quad \tilde{\boldsymbol{\Sigma}}_{22} := \sigma^4 \tilde{\mathbf{C}}_{22},$$

where $\kappa := E(|\epsilon_t|^4)/\sigma^4$. With this notation, the following theorem can be established as a result of Theorem 4.4. See Section 10.7 for a proof.

Theorem 10.3 (Asymptotic Normality of the Reduced Rank Linear Prediction Estimators). *Let the conditions of Theorem 10.2 be satisfied. Assume further that* $E(\epsilon_t^2) = 0$. *Then, as* $n \to \infty$, $\sqrt{n}(\hat{\mathbf{c}}_{\text{PCA}} - \mathbf{c}) \xrightarrow{D} \mathrm{N}_c(\mathbf{0}, \boldsymbol{\Sigma}_0, \tilde{\boldsymbol{\Sigma}}_0)$ *for fixed* $m > p$, *where*

$$\boldsymbol{\Sigma}_0 := [\bar{\mathbf{R}}_y^\dagger, \mathbf{P}_{\mathrm{N}}] \begin{bmatrix} \boldsymbol{\Sigma}_{11} & \boldsymbol{\Sigma}_{12} \\ \boldsymbol{\Sigma}_{12}^H & \boldsymbol{\Sigma}_{22} \end{bmatrix} [\bar{\mathbf{R}}_y^\dagger, \mathbf{P}_{\mathrm{N}}]^H,$$

$$\tilde{\boldsymbol{\Sigma}}_0 := [\bar{\mathbf{R}}_y^\dagger, \mathbf{P}_{\mathrm{N}}] \begin{bmatrix} \tilde{\boldsymbol{\Sigma}}_{11} & \tilde{\boldsymbol{\Sigma}}_{12} \\ \tilde{\boldsymbol{\Sigma}}_{12}^T & \tilde{\boldsymbol{\Sigma}}_{22} \end{bmatrix} [\bar{\mathbf{R}}_y^\dagger, \mathbf{P}_{\mathrm{N}}]^T.$$

Moreover, $\sqrt{n}(\hat{\boldsymbol{\omega}} - \boldsymbol{\omega}_0) \xrightarrow{D} \mathrm{N}(\mathbf{0}, 2\Re(\mathbf{D}^H \boldsymbol{\Sigma}_0 \mathbf{D} + \mathbf{D}^H \tilde{\boldsymbol{\Sigma}}_0 \mathbf{D}^*))$ *for sufficiently large but fixed* $m > p$, *where* $\mathbf{D}$ *is defined in Theorem 9.4.*

Remark 10.1 Under the same conditions, Theorem 9.4 guarantees that the ordinary linear prediction estimator $\hat{\mathbf{c}} := -\hat{\mathbf{R}}_y^{-1}\hat{\mathbf{r}}_y$ has the property

$$\sqrt{n}(\hat{\mathbf{c}} - \mathbf{c}) \xrightarrow{D} \mathrm{N}_c(\mathbf{0}, \mathbf{R}_y^{-1}\boldsymbol{\Sigma}_{11}\mathbf{R}_y^{-1}, \mathbf{R}_y^{-1}\tilde{\boldsymbol{\Sigma}}_{11}\mathbf{R}_y^{-T}).$$

If $\{\epsilon_t\}$ is Gaussian white noise, then $\kappa = 2$, in which case the last terms vanish in the expressions of $\boldsymbol{\Sigma}_{11}$ and $\tilde{\boldsymbol{\Sigma}}_{11}$.

Theorem 10.3 shows that the rate of convergence for the reduce rank linear prediction estimators takes the form $\sqrt{n}$ for fixed m. However, due to the asymptotic bias, the error rate for estimating the signal frequencies with fixed m is only $\mathcal{O}_P(1)$. This confirms the finding that the reduce rank linear prediction estimators produce poor frequency estimates when m is too small. Fortunately, Examples 10.1 and 10.2 show that the accuracy can be considerably improved by taking a suitably large m in comparison with the sample size n.

Next, let us derive the limiting value of the bias-corrected reduced rank estimator $\tilde{\mathbf{c}}_{\mathrm{PCA}}$ in (10.2.12) as $n \to \infty$ for fixed m. Toward that end, observe that

$$\tilde{\mathbf{R}}_x = \tilde{\mathbf{R}}_y - \hat{\sigma}^2\hat{\mathbf{P}}_{\mathrm{S}},$$

where $\hat{\mathbf{P}}_{\mathrm{S}} := \sum_{j=1}^{p}\hat{\mathbf{v}}_j\hat{\mathbf{v}}_j^H$ is the projection matrix onto $\hat{\mathbb{C}}_{\mathrm{S}}^m$. According to Theorem 10.1, Lemma 12.4.6, and Lemma 12.4.7, $\hat{\mathbf{R}}_y \to \mathbf{R}_y$ as $n \to \infty$ implies $\tilde{\mathbf{R}}_y \to \bar{\mathbf{R}}_y$ and $\hat{\mathbf{P}}_{\mathrm{S}} \to \mathbf{P}_{\mathrm{S}}$. In this case, if $\hat{\sigma}^2 \to \sigma^2$, then we obtain

$$\tilde{\mathbf{R}}_x \to \bar{\mathbf{R}}_y - \sigma^2\mathbf{P}_{\mathrm{S}} = \sum_{j=1}^{p}\mu_j\mathbf{v}_j\mathbf{v}_j^H = \mathbf{R}_x.$$

Moreover, because $\hat{\lambda}_j - \hat{\sigma}^2 \to \lambda_j - \sigma^2 = \mu_j > 0$ for $j = 1, \ldots, p$, the matrix $\tilde{\mathbf{R}}_x$ has the same rank p as the matrix $\mathbf{R}_x$ when n is sufficiently large. An application of Lemma 12.4.6 to these matrices yields $\tilde{\mathbf{R}}_x^{\dagger} \to \mathbf{R}_x^{\dagger}$. Therefore, under the additional assumption $\hat{\mathbf{r}}_y \to \mathbf{r}_y$, coupled with the fact that $\mathbf{r}_y = \mathbf{r}_x$, the limiting value of $\tilde{\mathbf{c}}_{\mathrm{PCA}}$ as $n \to \infty$ can be expressed as

$$\mathbf{c}_0 := -\mathbf{R}_x^{\dagger}\mathbf{r}_x, \tag{10.2.27}$$

which, unlike $\mathbf{c}$ in (10.2.22), depends solely on the signal parameters.

To interpret $\mathbf{c}_0$ in (10.2.27), recall that $\{x_t\}$ can be regarded as a zero-mean stationary process with $E(x_{t+u}x_t^*) = r_x(u)$ by randomizing the phases of the sinusoids. Under this condition, let us consider the linear prediction problem of minimizing $E\{|x_t + \sum_{j=1}^{m}\tilde{c}_j x_{t-j}|^2\}$ with respect to $\tilde{\mathbf{c}} := [\tilde{c}_1, \ldots, \tilde{c}_m]^T \in \mathbb{C}^m$. By Proposition 2.7, any solution to the problem should satisfy

$$\mathbf{R}_x\tilde{\mathbf{c}} = -\mathbf{r}_x, \tag{10.2.28}$$

and the parameter $\mathbf{c}_0$ in (10.2.27) is just the minimum-norm solution. An important property of these solutions, especially the minimum-norm solution $\mathbf{c}_0$, is given by the following theorem. See Section 10.7 for a proof.

Theorem 10.4 (Best Linear Predictor in the Absence of Noise). *For any $m \geq p$, the m-degree polynomial associated with any solution to (10.2.28) has p roots $z_1, \ldots, z_p$ on the unit circle, where $z_k := \exp(i\omega_k)$ $(k = 1, \ldots, p)$. For the minimum-norm solution, the remaining $m - p$ roots lie strictly inside the unit circle.*

According to Theorem 10.4, the m-degree polynomial associated with $\mathbf{c}_0$, denoted as $c_0(z)$, can be expressed as

$$c_0(z) = a(z) b_0(z),$$

where $a(z) := \prod_{k=1}^{p} (1 - z_k z^{-1})$ is a p-degree polynomial with roots $z_1, \ldots, z_p$ and $b_0(z)$ is an $(m-p)$-degree minimum-phase polynomial with all its roots strictly inside the unit circle. Let $C_0(\omega) := 1 + \mathbf{f}_m^H(\omega)\mathbf{c}_0 = c_0(\exp(i\omega))$. Then, the p smallest local minima of $|C_0(\omega)|^2$ coincide with the signal frequencies. Let $\hat{C}_0(\omega) := 1 + \mathbf{f}_m^H(\omega)\tilde{\mathbf{c}}_{\text{PCA}}$. If $\tilde{\mathbf{c}}_{\text{PCA}} \stackrel{a.s.}{\to} \mathbf{c}_0$, then, for sufficiently large n, Lemma 12.5.1 guarantees that $|\hat{C}_0(\omega)|^2$ has local minima near $\boldsymbol{\omega}$, denoted as $\hat{\boldsymbol{\omega}}$, such that $\hat{\boldsymbol{\omega}} \stackrel{P}{\to} \boldsymbol{\omega}$. Therefore, an application of Theorem 4.1 leads to the following assertion.

Theorem 10.5 (Consistency of the Bias-Corrected Reduced Rank Linear Prediction Estimators). *Let $\tilde{\mathbf{c}}_{\text{PCA}}$ be defined by (10.2.12) and $\mathbf{c}_0$ be defined by (10.2.27). Assume that $\{\epsilon_t\} \sim \text{IID}(0, \sigma^2)$ and $E(|\epsilon_t|^4) < \infty$. Assume further that $\hat{\sigma}^2 \stackrel{a.s.}{\to} \sigma^2$ as $n \to \infty$. Then, $\tilde{\mathbf{c}}_{\text{PCA}} \stackrel{a.s.}{\to} \mathbf{c}_0$ for fixed $m > p$. Moreover, let $\hat{\boldsymbol{\omega}}$ denote the corresponding frequency estimator. Then, $\hat{\boldsymbol{\omega}} \stackrel{P}{\to} \boldsymbol{\omega}$ for fixed $m > p$.*

Remark 10.2 If $\hat{\mathbf{R}}_y \stackrel{a.s.}{\to} \mathbf{R}_y$ and $\hat{\sigma}^2$ is given by (10.2.11), then $\hat{\sigma}^2 \stackrel{a.s.}{\to} \sigma^2$.

According to Theorem 10.5, $\tilde{\mathbf{c}}_{\text{PCA}}$ is not a consistent estimator of $\mathbf{c}$, which is unlike the estimator $\hat{\mathbf{c}}_{\text{PCA}}$ given by (10.2.4). As a result, the pseudo AR spectrum shown in Figure 10.4, which is derived from $\tilde{\mathbf{c}}_{\text{PCA}}$, is not a consistent estimator of the ordinary AR spectrum $f_{\text{AR}}(\omega) := \sigma_m^2 / |C(\omega)|^2$. However, the local maxima of the pseudo AR spectrum produce consistent frequency estimates.

The asymptotic distribution of $\tilde{\mathbf{c}}_{\text{PCA}}$ in general is more complicated than that of $\hat{\mathbf{c}}_{\text{PCA}}$ when the estimator $\hat{\sigma}^2$ in (10.2.12) is obtained jointly from the same data record $\mathbf{y}$, such as that given by (10.2.11). We only consider a simplified case in which $\hat{\sigma}^2$ is given by the sample variance of an independent realization of $\{\epsilon_t\}$. This assumption not only enables us to derive the asymptotic distribution, but also represents a practical situation where the noise variance can be estimated *a priori* from an independent set of noise-only training data. Let us assume that the sample size of the training data is much larger than n such that

$$\hat{\sigma}^2 = \sigma^2 + \mathcal{O}_P(n^{-1/2}). \tag{10.2.29}$$

For example, if the training sample size is equal to n^{1+r} for some constant $r > 0$, then we can write $\hat{\sigma}^2 = \sigma^2 + \mathcal{O}_P(n^{-(1+r)/2}) = \sigma^2 + o_P(n^{-1/2})$. Moreover, let $\mathbf{C}_{jj'}$ and $\mathbf{C}_{jj'}$ be defined in the same way as in Theorem 10.3 but based on $\mathbf{c}_0$ instead of $\mathbf{c}$. With this in mind, the following theorem can be obtained as the counterpart of Theorem 10.3. A proof is given in Section 10.7.

Theorem 10.6 (Asymptotic Normality of the Bias-Corrected Reduced Rank Linear Prediction Estimators). *Let the conditions of Theorem 10.5 be satisfied. Let $\hat{\sigma}^2$ satisfy the additional assumption (10.2.29). Assume further that $E(\epsilon_t^2) = 0$. Then, as $n \to \infty$, $\sqrt{n}\,(\tilde{\mathbf{c}}_{\text{PCA}} - \mathbf{c}_0) \xrightarrow{D} \mathrm{N}_c(\mathbf{0}, \boldsymbol{\Sigma}_0, \tilde{\boldsymbol{\Sigma}}_0)$ for fixed $m > p$, where $\boldsymbol{\Sigma}_0$ and $\tilde{\boldsymbol{\Sigma}}_0$ take the same forms as in Theorem 10.3 except that $\tilde{\mathbf{R}}_y^\dagger$ is replaced by $\mathbf{R}_x^\dagger$ and that*

$$\boldsymbol{\Sigma}_{11} := \sigma^4\{\mathbf{C}_{11} + (\kappa - 2)\mathbf{c}_0\mathbf{c}_0^H\}, \quad \tilde{\boldsymbol{\Sigma}}_{22} := \sigma^4\{\tilde{\mathbf{C}}_{11} + (\kappa - 2)\mathbf{c}_0\mathbf{c}_0^T\}.$$

Moreover, $\sqrt{n}\,(\hat{\boldsymbol{\omega}} - \boldsymbol{\omega}) \xrightarrow{D} \mathrm{N}(\mathbf{0}, 2\Re(\mathbf{D}^H\mathbf{R}_x^\dagger\boldsymbol{\Sigma}_{11}\mathbf{R}_x^\dagger\mathbf{D} + \mathbf{D}^H\mathbf{R}_x^\dagger\tilde{\boldsymbol{\Sigma}}_{11}\mathbf{R}_x^{*\dagger}\mathbf{D}^*))$ for sufficiently large but fixed $m > p$, where $\mathbf{D} := \frac{1}{2}[\mathbf{f}_m(\omega_1)/\dot{C}_0^*(\omega_1), \dots, \mathbf{f}_m(\omega_p)/\dot{C}_0^*(\omega_p)]$.

To illustrate Theorem 10.6, consider the following example.

Example 10.5 (Asymptotic Normality of the Bias-Corrected Reduced Rank Linear Prediction Estimators for a Single Complex Sinusoid in Gaussian White Noise). Let $p = 1$ and $\{\epsilon_t\} \sim \mathrm{GWN}_c(0, \sigma^2)$ (hence $\kappa = 2$). Then, under the conditions of Theorem 10.6, $\sqrt{n}\,(\hat{\omega}_1 - \omega_1) \xrightarrow{D} \mathrm{N}(0, \sigma_\omega^2)$ as $n \to \infty$ for fixed $m > 1$, where

$$\sigma_\omega^2 := 2\sigma^4\Re(\mathbf{D}^H\mathbf{R}_x^\dagger\mathbf{C}_{11}\mathbf{R}_x^\dagger\mathbf{D} + \mathbf{D}^H\mathbf{R}_x^\dagger\tilde{\mathbf{C}}_{11}\mathbf{R}_x^{*\dagger}\mathbf{D}^*)). \tag{10.2.30}$$

Because $\mathbf{R}_x = C_1^2\mathbf{f}_m(\omega_1)\mathbf{f}_m^H(\omega_1)$, we obtain $\mu_1 = mC_1^2$, $\mathbf{v}_1 = m^{-1/2}\mathbf{f}_m(\omega_1)$, and

$$\mathbf{R}_x^\dagger = \mu_1^{-1}\mathbf{v}_1\mathbf{v}_1^H = (mC_1^2)^{-1}\mathbf{f}_m(\omega_1)\mathbf{f}_m^H(\omega_1).$$

Because $\dot{C}_0(\omega_1) = i(m+1)/2$, we obtain

$$\mathbf{D} = \tfrac{1}{2}\mathbf{f}_m(\omega_1)/\dot{C}_0^*(\omega_1) = i(m+1)^{-1}\mathbf{f}_m(\omega_1).$$

Inserting these expressions in (10.2.30) yields

$$\sigma_\omega^2 = \frac{2}{m^2(m+1)^2\gamma_1^2}\Re\{\mathbf{f}_m^H(\omega_1)\mathbf{C}_{11}\mathbf{f}_m(\omega_1) - \mathbf{f}_m^H(\omega_1)\tilde{\mathbf{C}}_{11}\mathbf{f}_m^*(\omega_1)\}, \tag{10.2.31}$$

where $\gamma_1 := C_1^2/\sigma^2$. Moreover, because $\mathbf{r}_x = C_1^2\mathbf{f}_m(\omega_1)$, we obtain

$$\mathbf{c}_0 = -\mathbf{R}_x^\dagger\mathbf{r}_x = -m^{-1}\mathbf{f}_m(\omega_1).$$

It is not difficult to verify that

$$c_{11}(u,v) = a(u,v)z_1^{u-v}, \quad \tilde{c}_{11}(u,v) = \tilde{a}(u,v)z_1^{u+v},$$

where

$$a(u,v) := \begin{cases} 1+1/m & \text{if } u=v, \\ -|u-v|/m^2 & \text{if } u\neq v, \end{cases}$$

$$\tilde{a}(u,v) := \begin{cases} (2m-u-v+1)/m^2 & \text{if } u+v>m, \\ -(2m-u-v+1)/m^2 & \text{if } u+v\leq m. \end{cases}$$

Straightforward calculation yields

$$\mathbf{f}_m^H(\omega_1)\mathbf{C}_{11}\mathbf{f}_m(\omega_1) = \sum_{u,v=1}^{m} a(u,v) = \frac{2m^2+3m+1}{3m},$$

$$\mathbf{f}_m^H(\omega_1)\tilde{\mathbf{C}}_{11}\mathbf{f}_m^*(\omega_1) = \sum_{u,v=1}^{m} \tilde{a}(u,v) = -\frac{m^2-3m-1}{3m}.$$

Combining these results with (10.2.31) leads to a simple expression

$$\sigma_\omega^2 = \frac{2}{m(m+1)^2\gamma_1^2}.$$

As we can see, σ_ω^2 decreases with the increase of m or γ_1. ◇

In this example, the inverse relationship between the asymptotic variance and the order m is consistent with the finding in Examples 10.1 and 10.2 that the accuracy of the reduced rank linear prediction estimators can be improved by suitably increasing the value of m. However, it does not necessarily mean that the rate of convergence is $\mathcal{O}(m^{3/2}\sqrt{n})$ when m grows with n, because the asymptotic theory is valid only for fixed m as n grows.

Similar to the reduced rank linear prediction estimators, we can analyze the asymptotic properties of the reduced rank EYW estimator under the condition that m and l are fixed as $n \to \infty$.

To investigate the asymptotic behavior of $\hat{\mathbf{c}}_{\text{PCA}}$ defined by (10.2.21), let the SVD of the matrix $\mathbf{U}_{lm}$ be given by

$$\mathbf{U}_{lm} = \sum_{j=1}^{m} \nu_j \mathbf{p}_j \mathbf{q}_j^H.$$

With this notation, an EVD of $\mathbf{H}_{lm} := \mathbf{U}_{lm}^H \mathbf{U}_{lm}$ can be expressed as

$$\mathbf{H}_{lm} = \sum_{j=1}^{m} \nu_j^2 \mathbf{q}_j \mathbf{q}_j^H. \tag{10.2.32}$$

Because $\mathbf{U}_{lm}$ has rank p, we have $\nu_1 \geq \cdots \geq \nu_p > \nu_{p+1} = \cdots = \nu_m = 0$ and hence

$$\mathbf{U}_{lm}\mathbf{q}_j = \nu_j \mathbf{p}_j = \mathbf{0} \qquad (j = p+1, \ldots, m).$$

Combining this result with (10.2.18) and the fact that $\mathbf{G}_{lm}$ has full column rank p leads to $\mathbf{F}_m^H \mathbf{q}_j = \mathbf{0}$ $(j = p+1, \ldots, m)$, or equivalently,

$$\mathbf{f}_m^H(\omega_k)\mathbf{q}_j = 0 \qquad (j = p+1, \ldots, m;\ k = 1, \ldots, p).$$

This result implies that

$$\text{span}\{\mathbf{q}_{p+1}, \ldots, \mathbf{q}_m\} = \mathbb{C}_{\text{N}}^m, \quad \text{span}\{\mathbf{q}_1, \ldots, \mathbf{q}_p\} = \mathbb{C}_{\text{S}}^m.$$

Therefore, the eigenvectors of $\mathbf{H}_{lm}$ play the same role as the eigenvectors of $\mathbf{R}_y$ in describing the signal and noise subspaces. The estimator $\hat{\mathbf{c}}_{\text{PCA}}$ in (10.2.21) is a natural estimator of

$$\mathbf{c}_0 := -\mathbf{H}_{lm}^{\dagger}\mathbf{U}_{lm}^H \mathbf{u}_{lm} = -\sum_{j=1}^{p} v_j^{-1}(\mathbf{p}_j^H \mathbf{u}_{lm})\mathbf{q}_j, \tag{10.2.33}$$

where $\mathbf{H}_{lm}^{\dagger} := \sum_{j=1}^{p} v_j^{-2}\mathbf{q}_j\mathbf{q}_j^H$ is the pseudo inverse of $\mathbf{H}_{lm}$. Observe that

$$\mathbf{H}_{lm}^{\dagger}\mathbf{U}_{lm}^H = \sum_{j=1}^{p} v_j^{-1}\mathbf{q}_j\mathbf{p}_j^H = \mathbf{U}_{lm}^{\dagger}.$$

Therefore, we can also write

$$\mathbf{c}_0 = -\mathbf{U}_{lm}^{\dagger}\mathbf{u}_{lm}. \tag{10.2.34}$$

This parameter is just the minimum-norm solution of (10.2.16) that minimizes $\|\mathbf{u}_{lm} + \mathbf{U}_{lm}\tilde{\mathbf{c}}\|^2$ with respect to $\tilde{\mathbf{c}} \in \mathbb{C}^m$. An important property of the minimum-norm solution is given in the following theorem. See Section 10.7 for a proof.

Theorem 10.7 (Minimum-Norm Solution to Higher-Order EYW Equations). *For any $l \geq m > p$, the m-degree polynomial associated the parameter $\mathbf{c}_0$ in (10.2.33) has p roots $z_1, \ldots, z_p$ on the unit circle, and the remaining $m-p$ roots lie strictly inside the unit circle.*

According to Theorem 10.7, a factorization of the form (10.2.14) remains valid for the m-degree polynomial $c_0(z)$ associated with $\mathbf{c}_0$ in (10.2.33), with $b(z)$ being a minimum-phase polynomial. Therefore, the signal frequencies ω_k are the smallest local minima of $|C_0(\omega)|^2$, where $C_0(\omega) := 1 + \mathbf{f}_m^H(\omega)\mathbf{c}_0 = c_0(\exp(i\omega))$. In addition, because $\hat{r}_y(u) \to r_y(u)$ for any fixed u implies $\hat{\mathbf{U}}_{lm}\hat{\mathbf{u}}_{lm} \to \mathbf{U}_{lm}\mathbf{u}_{lm}$ and $\hat{\mathbf{H}}_{lm} \to \mathbf{H}_{lm}$, an application of Theorem 4.1, Lemma 12.4.2, and Lemma 12.4.6 leads to $\hat{\mathbf{c}}_{\text{PCA}} \to \mathbf{c}_0$. Combining this result with Theorem 4.1 and Lemma 12.5.1 proves the following theorem.

Theorem 10.8 (Consistency of the Reduced Rank EYW Estimators). *Let $\{y_t\}$ be given by (10.1.1) with $\{x_t\}$ taking the form (2.1.5) or (2.1.7). Let $\hat{\mathbf{c}}_{\text{PCA}}$ be defined by (10.2.21) and $\mathbf{c}_0$ be defined by (10.2.33). Let $\hat{\boldsymbol{\omega}}$ denote the corresponding frequency estimator. If $\{\epsilon_t\} \sim \text{IID}(0, \sigma^2)$ and $E(|\epsilon_t|^4) < \infty$, then, $\hat{\mathbf{c}}_{\text{PCA}} \stackrel{a.s.}{\to} \mathbf{c}_0$ and $\hat{\boldsymbol{\omega}} \stackrel{P}{\to} \boldsymbol{\omega}$ as $n \to \infty$ for any fixed $l \geq m > p$.*

The asymptotic distribution of the reduced rank EYW estimators can be derived from Theorem 4.4. Indeed, let $[c_1,\ldots,c_m]^T := \mathbf{c}_0$ ($c_0 := 1$) and

$$[b_1,\ldots,b_l]^T := \mathbf{b} := -\mathbf{U}_{lm}\mathbf{H}_{lm}^{\dagger}\mathbf{c}_0 = -\mathbf{U}_{lm}^{\dagger H}\mathbf{c}_0.$$

Define

$$\mathbf{B}_{lm} := \begin{bmatrix} 0 & \cdots & 0 & b_1 & \cdots & b_l & 0 \\ \vdots & & \iddots & & \iddots & & \vdots \\ 0 & b_1 & \cdots & b_l & 0 & \cdots & 0 \\ b_1 & \cdots & b_l & 0 & 0 & \cdots & 0 \end{bmatrix} \in \mathbb{C}^{m\times(m+l)},$$

and

$$\mathbf{C}_{lm} := \begin{bmatrix} c_m & c_{m-1} & \cdots & c_0 & 0 & \cdots & 0 \\ 0 & c_m & \cdots & c_0 & c_0 & \cdots & 0 \\ & & \ddots & & \ddots & & \\ 0 & \cdots & 0 & c_m & \cdots & c_1 & c_0 \end{bmatrix} \in \mathbb{C}^{l\times(m+l)}.$$

Then, we have the following theorem. See Section 10.7 for a proof.

Theorem 10.9 (Asymptotic Normality of the Reduced Rank EYW Estimators). *Let the conditions of Theorem 10.8 be satisfied. Assume further that $E(\epsilon_t^2) = 0$. Then, for any fixed $l \ge m > p$, $\sqrt{n}(\hat{\mathbf{c}}_{\mathrm{PCA}} - \mathbf{c}_0) \xrightarrow{D} \mathrm{N}_c(\mathbf{0}, \boldsymbol{\Sigma}_0, \tilde{\boldsymbol{\Sigma}}_0)$ as $n \to \infty$, where*

$$\boldsymbol{\Sigma}_0 := [\mathbf{U}_{lm}^{\dagger}, \mathbf{P}_{\mathrm{N}}] \begin{bmatrix} \boldsymbol{\Sigma}_{11} & \mathbf{0} \\ \mathbf{0} & \boldsymbol{\Sigma}_{22} \end{bmatrix} [\mathbf{U}_{lm}^{\dagger}, \mathbf{P}_{\mathrm{N}}]^H,$$

$$\tilde{\boldsymbol{\Sigma}}_0 := [\mathbf{U}_{lm}^{\dagger}, \mathbf{P}_{\mathrm{N}}] \begin{bmatrix} \mathbf{0} & \tilde{\boldsymbol{\Sigma}}_{12} \\ \tilde{\boldsymbol{\Sigma}}_{12}^T & \tilde{\boldsymbol{\Sigma}}_{22} \end{bmatrix} [\mathbf{U}_{lm}^{\dagger}, \mathbf{P}_{\mathrm{N}}]^T,$$

with

$$\boldsymbol{\Sigma}_{11} := \sigma^4 \mathbf{C}_{lm}\mathbf{C}_{lm}^H, \quad \boldsymbol{\Sigma}_{22} := \mathbf{B}_{lm}(\sigma^4\mathbf{I} + 2\sigma^2\mathbf{F}_{m+l}^*\mathbf{G}\mathbf{F}_{m+l}^T)\mathbf{B}_{lm}^H,$$
$$\tilde{\boldsymbol{\Sigma}}_{12} := \sigma^4 \mathbf{C}_{lm}\mathbf{B}_{lm}, \quad \tilde{\boldsymbol{\Sigma}}_{22} := 2\sigma^2\mathbf{B}_{lm}\mathbf{F}_{m+l}^*\mathbf{G}\mathbf{F}_{m+l}^H\mathbf{B}_{lm}^T.$$

Moreover, $\sqrt{n}(\hat{\boldsymbol{\omega}} - \boldsymbol{\omega}) \xrightarrow{D} \mathrm{N}(\mathbf{0}, 2\sigma^4\Re(\mathbf{D}^H\mathbf{U}_{lm}^{\dagger}\mathbf{C}_{lm}\mathbf{C}_{lm}^H\mathbf{U}_{lm}^{\dagger H}\mathbf{D}))$ for fixed $l \ge m > p$ as $n \to \infty$, where $\mathbf{D} := \frac{1}{2}[\mathbf{f}_m(\omega_1)/\dot{C}_0^*(\omega_1), \ldots, \mathbf{f}_m(\omega_p)/\dot{C}_0^*(\omega_p)]$.

The following example serves to illustrate Theorem 10.9.

Example 10.6 (Asymptotic Normality of the Reduced Rank EYW Estimators for a Single Complex Sinusoid). Let $p = 1$. Then, under the conditions of Theorem 10.9, $\sqrt{n}(\hat{\omega}_1 - \omega_1) \xrightarrow{D} \mathrm{N}(0, \sigma_\omega^2)$, where

$$\sigma_\omega^2 := 2\sigma^4\Re(\mathbf{D}^H\mathbf{U}_{lm}^{\dagger}\mathbf{C}_{lm}\mathbf{C}_{lm}^H\mathbf{U}_{lm}^{\dagger H}\mathbf{D}). \tag{10.2.35}$$

Because $\mathbf{G}_{lm} = C_1^2 z_1^m \mathbf{f}_l(\omega_1)$ and $\mathbf{F}_m = \mathbf{f}_m(\omega_1)$, we obtain

$$\mathbf{U}_{lm} = \mathbf{G}_{lm}\mathbf{F}_m = C_1^2 z_1^m \mathbf{f}_l(\omega_1)\mathbf{f}_m^H(\omega_1) = v_1 \mathbf{p}_1 \mathbf{q}_1^H,$$

where $v_1 := \sqrt{lm}\,C_1^2$, $\mathbf{p}_1 := l^{-1/2} z_1^m \mathbf{f}_l(\omega_1)$, and $\mathbf{q}_1 := m^{-1/2}\mathbf{f}_m(\omega_1)$. Therefore,

$$\mathbf{U}_{lm}^\dagger = v_1^{-1}\mathbf{q}_1\mathbf{p}_1^H = (lmC_1^2)^{-1} z_1^{-m} \mathbf{f}_m(\omega_1)\mathbf{f}_l^H(\omega_1).$$

Inserting this expression and $\mathbf{u}_{lm} = C_1^2 z_1^m \mathbf{f}_l(\omega_1)$ in (10.2.34) yields

$$\mathbf{c}_0 = -m^{-1}\mathbf{f}_m(\omega_1).$$

Therefore, as in Example 10.5, we obtain

$$\mathbf{D} = i(m+1)^{-1}\mathbf{f}_m(\omega_1).$$

Substituting these expressions in (10.2.35) leads to

$$\sigma_\omega^2 = \frac{2}{(m+1)^2 l^2 \gamma_1^2} \Re\{\mathbf{f}_l^H(\omega_1)\mathbf{C}_{lm}\mathbf{C}_{lm}^H\mathbf{f}_l(\omega_1)\}, \tag{10.2.36}$$

where $\gamma_1 := C_1^2/\sigma^2$ is the SNR. Moreover, as in Example 10.5, we can write

$$\mathbf{C}_{lm}\mathbf{C}_{lm}^H = [c(u,v)] \quad (u,v = 1,\ldots,l),$$

where

$$c(u,v) := \sum_{j=0}^{m} c_{u-v+j}c_j^* = a(u,v)z_1^{u-v}$$

and

$$a(u,v) := \begin{cases} 1 + 1/m & \text{if } u = v, \\ -|u-v|/m^2 & \text{if } 1 \le |u-v| \le m, \\ 0 & \text{otherwise.} \end{cases}$$

Straightforward calculation gives

$$\mathbf{f}_l^H(\omega_1)\mathbf{C}_{lm}\mathbf{C}_{lm}^H\mathbf{f}_l(\omega_1) = \sum_{u,v=1}^{l} a(u,v) = \frac{(m+1)(2m+1)}{3m}.$$

Inserting this result in (10.2.36) yields a simple expression

$$\sigma_\omega^2 = \frac{2(2m+1)}{3m(m+1)l^2\gamma_1^2}.$$

Observe that σ_ω^2 decreases with the increase of l, m, or γ_1. ◇

This example suggests that the accuracy of the reduced rank EYW estimators can be improved by increasing m or l. But the rate of convergence does not necessarily take the form $\mathcal{O}(l\sqrt{mn})$ when m and l grow with n.

10.3 Subspace Projection Method

Another way of constructing frequency estimators based on the EVD is to utilize the orthogonality between the signal subspace and the noise subspace. Because the signal subspace is spanned by $\mathbf{f}_m(\omega_1),\ldots,\mathbf{f}_m(\omega_p)$, the signal frequencies can be identified as those values of ω that yield minimal projections for the vector $\mathbf{f}_m(\omega)$ onto an estimated noise subspace. We call this method the subspace projection (SSP) method. In this section, we discuss two ways of examining the projections in the SSP method, which lead to the so-called MUSIC estimator and the minimum-norm (MN) estimator.

10.3.1 MUSIC and Minimum-Norm Estimators

For any $m > p$, let $\mathbf{v}_j$ $(j = 1,\ldots,m)$ denote the eigenvectors of $\mathbf{R}_y$ given by the EVD in (10.1.4). Consider the m-by-$(m-p)$ matrix $\mathbf{V}_\mathrm{N}$ defined by (10.1.11). Because $\mathbf{V}_\mathrm{N}$ has full (column) rank $m-p$, the orthogonality property in (10.1.9) implies that the p signal frequencies $\omega_1,\ldots,\omega_p$ are the only values which satisfy

$$\mathbf{V}_\mathrm{N}^H\mathbf{f}_m(\omega) = \mathbf{0},$$

or equivalently,

$$\|\mathbf{V}_\mathrm{N}^H\mathbf{f}_m(\omega)\|^2 = \sum_{j=p+1}^{m} |\mathbf{v}_j^H\mathbf{f}_m(\omega)|^2 = 0. \tag{10.3.1}$$

Let $\hat{\mathbf{R}}_y$ be an estimator of $\mathbf{R}_y$ with an EVD that takes the form

$$\hat{\mathbf{R}}_y = \sum_{j=1}^{m} \hat{\lambda}_j\hat{\mathbf{v}}_j\hat{\mathbf{v}}_j^H, \tag{10.3.2}$$

where $\hat{\lambda}_1 \ge \cdots \ge \hat{\lambda}_m \ge 0$. Define $\hat{\mathbf{V}}_\mathrm{N} := [\hat{\mathbf{v}}_{p+1},\ldots,\hat{\mathbf{v}}_m]$. Then, motivated by (10.3.1), the ω_k can be estimated by the local minima of the objective function

$$\hat{g}_m(\omega) := \|\hat{\mathbf{V}}_\mathrm{N}^H\mathbf{f}_m(\omega)\|^2 = \sum_{j=p+1}^{m} |\hat{\mathbf{v}}_j^H\mathbf{f}_m(\omega)|^2. \tag{10.3.3}$$

This estimator, proposed in [33] and [332], is known in the literature simply as MUSIC (shorthand for multiple signal characterization) [369, p. 155].

In the special case of $m = p+1$, we obtain $\hat{g}_{p+1}(\omega) = |\hat{\mathbf{v}}_{p+1}^H\mathbf{f}_{p+1}(\omega)|^2$. With the notation $[\hat{a}_0,\hat{a}_1,\ldots,\hat{a}_p]^T := \hat{\mathbf{v}}_{p+1}$ and $\hat{A}(\omega) := \sum_{j=0}^{p}\hat{a}_j\exp(-ij\omega)$, we can write

$$\hat{g}_{p+1}(\omega) = |\mathbf{f}_{p+1}^H(\omega)\hat{\mathbf{v}}_{p+1}|^2 = |\hat{A}(\omega)|^2.$$

Because $\hat{\mathbf{v}}_{p+1}$ is an eigenvector associated with the smallest eigenvalue of $\hat{\mathbf{R}}_y$, it is also a minimizer of $\mathbf{v}^H\hat{\mathbf{R}}_y\mathbf{v}$ under the unit-norm constraint $\|\mathbf{v}\| = 1$. Moreover, minimizing $\hat{g}_{p+1}(\omega)$ is equivalent to maximizing $1/|\hat{A}(\omega)|^2$ which is an AR spectrum of order p (assuming $\hat{a}_0\hat{a}_p \neq 0$) associated with the eigenvector $\hat{\mathbf{v}}_{p+1}$. This is known as Pisarenko's harmonic decomposition method [289].

Observe that the vector $\hat{\mathbf{V}}_{\mathrm{N}}^H\mathbf{f}_m(\omega)$ comprises the projection coefficients of the vector $\mathbf{f}_m(\omega)$ onto the estimated noise subspace $\hat{\mathbb{C}}_{\mathrm{N}}^m := \mathrm{span}\{\hat{\mathbf{v}}_{p+1},\dots,\hat{\mathbf{v}}_m\}$ and the function $\hat{g}_m(\omega)$ in (10.3.3) is just the squared norm of the projection coefficient vector. One can also express $\hat{g}_m(\omega)$ in terms of the projection coefficients onto the estimated signal subspace $\hat{\mathbb{C}}_{\mathrm{S}}^m := \mathrm{span}\{\hat{\mathbf{v}}_1,\dots,\hat{\mathbf{v}}_p\}$, i.e., the column space of $\hat{\mathbf{V}}_{\mathrm{S}} := [\hat{\mathbf{v}}_1,\dots,\hat{\mathbf{v}}_p]$. In fact, observe that the projection matrices onto the estimated signal and noise subspaces are given, respectively, by

$$\hat{\mathbf{P}}_{\mathrm{S}} := \hat{\mathbf{V}}_{\mathrm{S}}\hat{\mathbf{V}}_{\mathrm{S}}^H, \quad \hat{\mathbf{P}}_{\mathrm{N}} := \hat{\mathbf{V}}_{\mathrm{N}}\hat{\mathbf{V}}_{\mathrm{N}}^H.$$

Also observe that

$$\hat{g}_m(\omega) = \mathbf{f}_m^H(\omega)\hat{\mathbf{P}}_{\mathrm{N}}\mathbf{f}_m(\omega).$$

Because $\hat{\mathbf{V}} := [\hat{\mathbf{V}}_{\mathrm{S}}, \hat{\mathbf{V}}_{\mathrm{N}}]$ is unitary, we have

$$\mathbf{I} = \hat{\mathbf{V}}\hat{\mathbf{V}}^H = \hat{\mathbf{V}}_{\mathrm{S}}\hat{\mathbf{V}}_{\mathrm{S}}^H + \hat{\mathbf{V}}_{\mathrm{N}}\hat{\mathbf{V}}_{\mathrm{N}}^H = \hat{\mathbf{P}}_{\mathrm{S}} + \hat{\mathbf{P}}_{\mathrm{N}}.$$

Combining these results yields

$$\hat{g}_m(\omega) = m - \mathbf{f}_m^H(\omega)\hat{\mathbf{P}}_{\mathrm{S}}\mathbf{f}_m(\omega) = m - \sum_{j=1}^{p} |\hat{\mathbf{v}}_j^H\mathbf{f}_m(\omega)|^2. \tag{10.3.4}$$

This alternative expression of $\hat{g}_m(\omega)$ has a lower computational burden when the dimension of the signal subspace is less than the dimension of the noise subspace, i.e., when $p < m - p$, or equivalently, $m > 2p$.

Obviously, the MUSIC estimator depends on the choice of $\hat{\mathbf{R}}_y$. A simple example is $\hat{\mathbf{R}}_y := [\hat{r}_y(u-v)]$ $(u, v = 1,\dots,m)$, where the $\hat{r}_y(u)$ are the sample autocovariances defined by (4.1.1). Another example is

$$\hat{\mathbf{R}}_y := (2n)^{-1}\mathbf{Y}_{fb}^H\mathbf{Y}_{fb}, \tag{10.3.5}$$

where $\mathbf{Y}_{fb}$ is defined by (9.1.11) with $2(n-m) > m > p$, or equivalently,

$$p < m < 2n/3.$$

In this case, the eigenvectors required by (10.3.3) or (10.3.4) can be obtained directly from the SVD of $\mathbf{Y}_{fb}$ according to (10.2.6) and (10.2.7).

Figure 10.8 depicts the function $1/\hat{g}_m(\omega)$, called the MUSIC spectrum, based on $\hat{\mathbf{R}}_y$ in (10.3.5) with different values of m for the time series which is used in

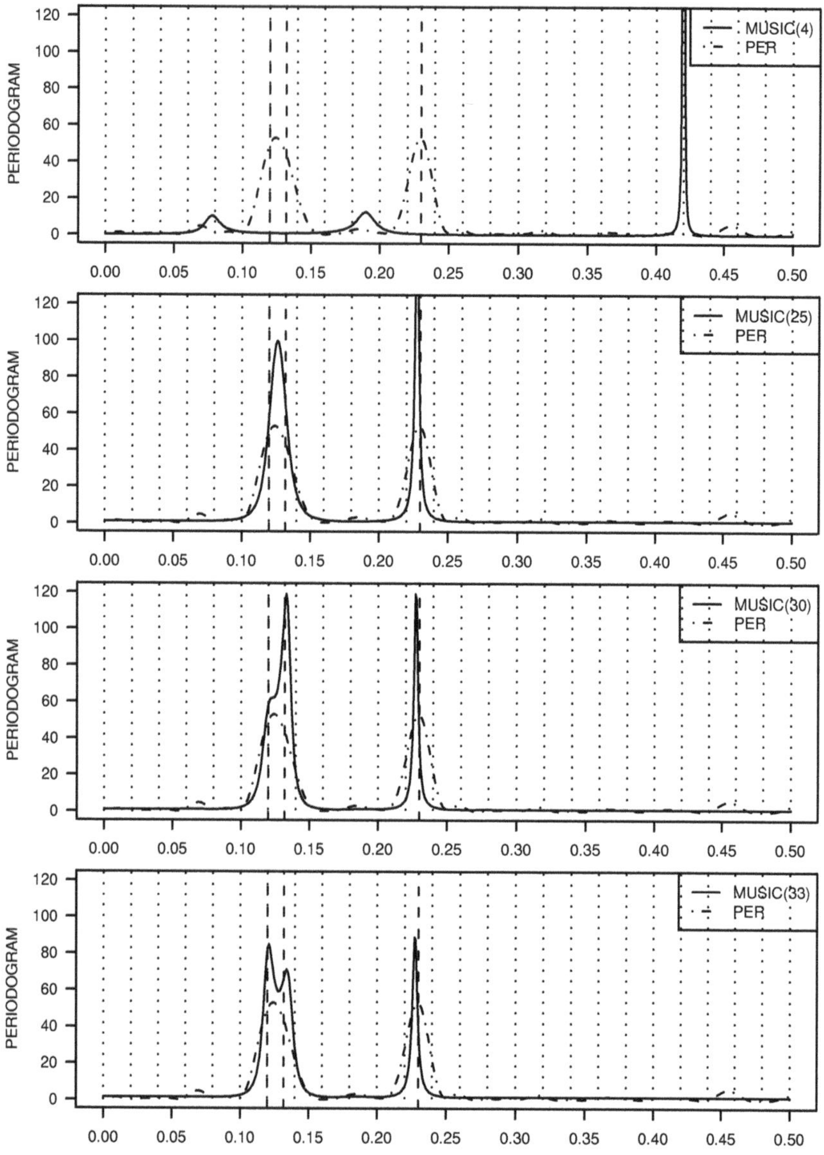

Figure 10.8. Plot of MUSIC spectrum with different choices of m for the time series which is used in Figure 10.1. Solid line, MUSIC spectrum with $m = 4, 25, 30, 33$; dashed-dotted line, periodogram. Vertical dashed lines depict the signal frequencies and vertical dotted lines depict the Fourier grid. The MUSIC spectrum is rescaled so that the curve covers the same area in the plot as the periodogram.

Figures 10.1, 10.4, and 10.5. Although the MUSIC spectrum is not an estimator of the spectral density of $\{y_t\}$, the local maxima of the MUSIC spectrum provide frequency estimates in the same way as the local maxima of the periodogram. As can be seen from Figure 10.8, the choice $m = p+1 = 4$, which gives Pisarenko's estimator, works very poorly. With $m = 25$, the MUSIC spectrum exhibits two large peaks at almost the same location as the peaks of the periodogram, so the third frequency can be identified from the second peak but the two closely spaced frequencies remain unresolved. When m is increased to 30, a third peak begins to emerge, and when m takes its largest possible value $\lfloor 2n/3 \rfloor = 33$, the closely spaced frequencies become fully resolved and all three largest peaks in the MUSIC spectrum occur in the vicinity of the signal frequencies. A comparison with Figure 9.1 shows that m is allowed to take a larger value in the MUSIC estimator than in the FBLP estimator without suffering from spurious spectral peaks. This property is also shared by the reduced rank FBLP estimator shown in Figures 10.1 and 10.4 where the same EVD is employed, albeit in a different way.

To compute the MUSIC frequency estimates which minimize $\hat{g}_m(\omega)$ (or maximize the MUSIC spectrum), a numerical procedure has to be employed, just like maximizing an AR spectrum or a continuous periodogram. In particular, with the notation $\mathbf{u}(\omega) := \hat{\mathbf{V}}_{\mathrm{N}}^H \mathbf{f}_m(\omega)$, one can use a fixed point iteration based the modified Newton-Raphson mapping of the form

$$\phi(\omega) := \omega - \varrho\, \Re\{\mathbf{u}^H(\omega)\, \dot{\mathbf{u}}(\omega)\} / \|\dot{\mathbf{u}}(\omega)\|^2. \qquad (10.3.6)$$

Observe that $2\Re\{\mathbf{u}^H(\omega)\,\dot{\mathbf{u}}(\omega)\}$ is just the derivative of $\hat{g}_m(\omega) = \|\mathbf{u}(\omega)\|^2$ and that $2\|\dot{\mathbf{u}}(\omega)\|^2$ coincides with the second derivative of $\hat{g}_m(\omega)$ when $\hat{g}_m(\omega) = 0$. A similar mapping is used to compute the reduced rank FBLP and EYW estimates in Section 10.2 which minimize different objective functions.

The statistical performance of the MUSIC estimator is demonstrated by the following two simulation examples.

Example 10.7 (Estimation of Well-Separated Frequencies: Part 3). Consider the time series in Example 10.1. Let the MUSIC frequency estimates be computed by the fixed point iteration of the modified Newton-Raphson mapping in (10.3.6) with $\hat{\mathbf{R}}_y$ given by (10.3.5). Similar to Example 10.1, let the initial values be produced by the RDFT1 method using the MUSIC spectrum instead of the AR spectrum. Figure 10.9 depicts the average MSE of the MUSIC frequency estimates after 5 iterations with $\varrho = 1$ and the initial RDFT1 estimates for different choices of m and different values of SNR.

Clearly, the MUSIC estimates are poor when m is too small, but the accuracy can be improved considerably by increasing m. With the choice of $m = 45$ for $n = 100$ and $m = 90$ for $n = 200$, the resulting MSE becomes near the CRLB when the SNR exceeds a threshold which is similar to that of the DFT estimator. This is comparable to the performance of the reduced rank FBLP estimator shown in

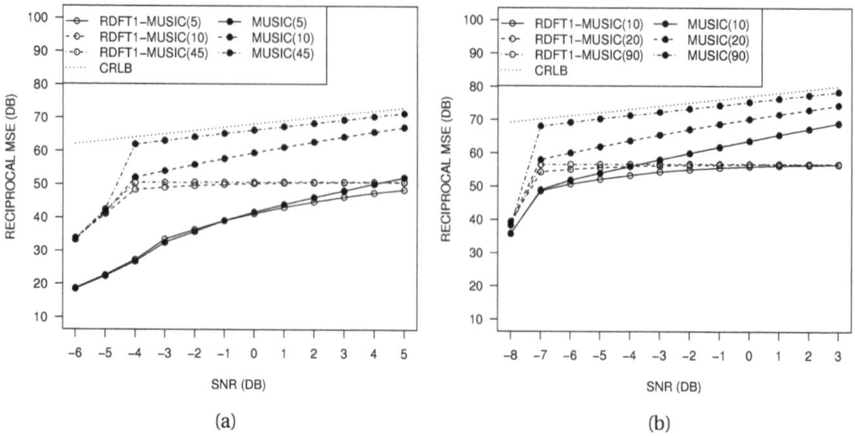

Figure 10.9. Plot of the reciprocal average MSE of the MUSIC frequency estimates for the time series of two complex sinusoids in Gaussian white noise discussed in Example 10.7 with different values of m and SNR. (a) $n = 100$: solid line, $m = 5$; dashed line, $m = 10$; dash-dotted line, $m = 45$. (b) $n = 200$: solid line, $m = 10$; dashed line, $m = 20$; dash-dotted line, $m = 90$. Dotted line depicts the CRLB. Results are based on 5,000 Monte Carlo runs.

Figures 10.2 which employs the same covariance estimator $\hat{\mathbf{R}}_y$. The performance begins to deteriorate when m is too large (e.g., $m = 66$ for $n = 100$; not shown), but the deterioration is rather mild as compared to that of the reduced rank FBLP estimator with the same choice of m. A comparison with Figures 6.13 and 8.3 suggests that the MUSIC estimator with the best choice of m remains inferior to the PM and NLS estimators. ◇

Example 10.8 (Estimation of Closely Spaced Frequencies: Part 3). Consider the time series in Example 10.2. Let the MUSIC frequency estimates be computed by the algorithm described in Example 10.7 except that the initial values are produced by the RDFT4 method as in Example 10.2. Figure 10.10 shows the average MSE of the frequency estimates with several choices of m.

As we can see, the MUSIC estimator with a suitably large m performs similarly to the reduced rank FBLP estimator with the same choice of m which is shown in Figure 10.3. But the MUSIC estimator performs more poorly when m is too small (e.g., $m = 10$ for $n = 100$ and $m = 15$ for $n = 200$). This is due to its more aggressive enhancement of spurious peaks as illustrated by Figure 10.8. Moreover, in comparison with the NLS estimator shown in Figure 8.5, the MUSIC estimator has a much higher SNR threshold and remains less accurate at high SNR. ◇

As an alternative to the MUSIC estimator, the minimum-norm (MN) estimator discussed in [205] also exploits the orthogonality between the signal and noise

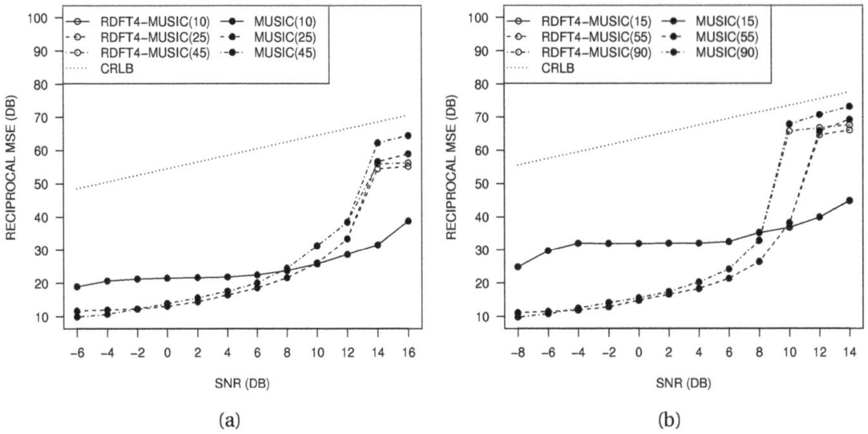

Figure 10.10. Plot of the reciprocal average MSE of the MUSIC frequency estimates for the time series of two complex sinusoids with closely spaced frequencies in Gaussian white noise discussed in Example 10.8 with different values of m and SNR. (a) $n = 100$: solid line, $m = 10$; dashed line, $m = 25$; dash-dotted line, $m = 45$. (b) $n = 200$: solid line, $m = 15$; dashed line, $m = 55$; dash-dotted line, $m = 90$. Dotted line depicts the CRLB. Results are based on 5,000 Monte Carlo runs.

subspaces but with slightly less computational cost. The MN estimator replaces the m-by-$(m-p)$ matrix $\hat{\mathbf{V}}_{\mathrm{N}}$ in (10.3.3) with a single m-by-1 vector

$$\hat{\mathbf{v}}_0 := \begin{bmatrix} 1 \\ \hat{\mathbf{c}} \end{bmatrix}$$

under the condition that it resides in the estimated noise subspace $\hat{\mathbb{C}}_{\mathrm{N}}^m$ and has the smallest norm. With $\hat{\mathbf{g}}^H$ denoting the first row of $\hat{\mathbf{V}}_{\mathrm{N}}$ and with $\tilde{\mathbf{V}}_{\mathrm{N}}$ denoting the remaining $(m-1)$-by-$(m-p)$ submatrix, one can write $\hat{\mathbf{v}}_0 = \hat{\mathbf{V}}_{\mathrm{N}}\mathbf{d}$ for some $\mathbf{d} \in \mathbb{C}^{m-p}$ such that $\hat{\mathbf{g}}^H\mathbf{d} = 1$ and $\hat{\mathbf{c}} = \tilde{\mathbf{V}}_{\mathrm{N}}\mathbf{d}$. Minimizing $\|\hat{\mathbf{v}}_0\|^2 = \|\mathbf{d}\|^2$ with respect to $\mathbf{d}$ subject to $\hat{\mathbf{g}}^H\mathbf{d} = 1$ gives $\mathbf{d} = \hat{\mathbf{g}}/\|\hat{\mathbf{g}}\|^2$. Therefore,

$$\hat{\mathbf{c}} = \tilde{\mathbf{V}}_{\mathrm{N}}\hat{\mathbf{g}}/\|\hat{\mathbf{g}}\|^2.$$

The frequency estimates are given by the local minima of the function

$$\tilde{g}_m(\omega) := |\mathbf{f}_m^H(\omega)\hat{\mathbf{v}}_0|^2. \tag{10.3.7}$$

Because it requires only a single vector $\hat{\mathbf{v}}_0$, this function is slightly easier to evaluate than the objective function of the MUSIC estimator in (10.3.3) or (10.3.4).

As an example, Figure 10.11 depicts the function $1/\tilde{g}_m(\omega)$, which we call the MN spectrum, for the same time series used in Figure 10.8 with the same choice of $\hat{\mathbf{R}}_y$ and m. Comparing these results shows that the MN spectrum has a greater

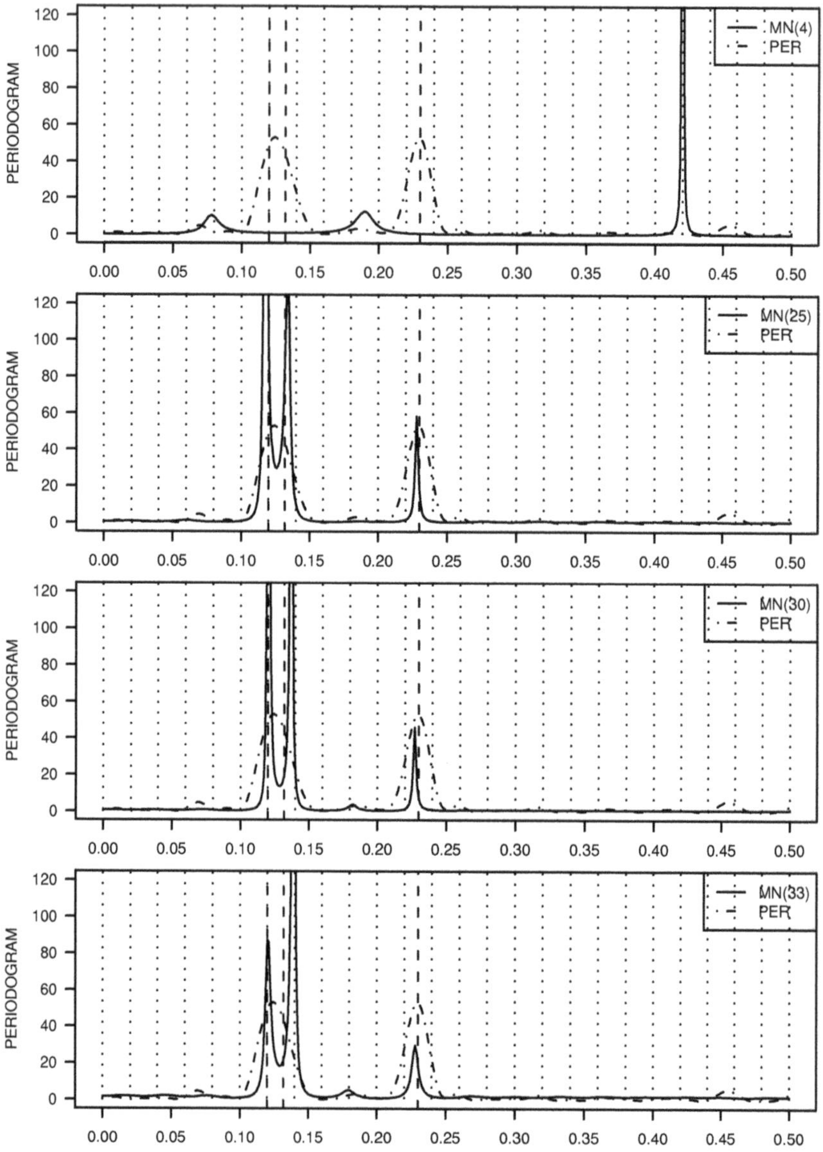

Figure 10.11. Similar to Figure 10.8 except that the MN spectrum is used instead of the MUSIC spectrum. Note that the MN spectrum is identical to the MUSIC spectrum in the case of $m = p + 1 = 4$.

capability of resolving the closely spaced frequencies than the corresponding MUSIC spectrum when m is large but not too large.

Simulation results based on the same data as in Examples 10.7 and 10.8 indicate that the MN estimator produces slightly more accurate frequency estimates than the MUSIC estimator when m is sufficiently large. For example, in the case of well-separated frequencies with $n = 100$, the MSE of the MN estimator with $m = 45$ is equal to 6.30×10^{-8} at 5 dB, whereas that of the MUSIC estimator is equal to 7.22×10^{-8}; in the case of closely spaced signal frequencies with $n = 100$, it is equal to 4.59×10^{-7} at 14 dB as compared to 5.91×10^{-7}. However, the MN estimator performs more poorly than the MUSIC estimator in the case of closely spaced frequencies when m is too small (e.g., $m = 10$ for $n = 100$).

10.3.2 Statistical Properties

Examples 10.7 and 10.8 suggest that the MUSIC estimator produces accurate frequency estimates only when m is suitably large. Statistical theory under such conditions remains largely unknown. In the following, we only consider the case where m is fixed as $n \to \infty$.

For fixed $m > p$, the MUSIC estimator produces consistent estimates as $n \to \infty$. In fact, by Theorem 10.1 and Lemma 12.4.7, $\hat{\mathbf{R}}_y \xrightarrow{P} \mathbf{R}_y$ as $n \to \infty$ implies

$$\hat{\mathbf{P}}_{\rm S} \xrightarrow{P} \mathbf{P}_{\rm S}, \quad \hat{\mathbf{P}}_{\rm N} \xrightarrow{P} \mathbf{P}_{\rm N},$$

where $\mathbf{P}_{\rm S} := \mathbf{V}_{\rm S}\mathbf{V}_{\rm S}^H$ and $\mathbf{P}_{\rm N} := \mathbf{V}_{\rm N}\mathbf{V}_{\rm N}^H$ are the projection matrices onto the signal and noise subspaces, respectively. Under this condition, we obtain

$$\hat{g}_m(\omega) \xrightarrow{P} g_m(\omega) := \|\mathbf{V}_{\rm N}^H \mathbf{f}_m(\omega)\|^2 = \mathbf{f}_m^H(\omega)\mathbf{P}_{\rm N}\mathbf{f}_m(\omega)$$

uniformly in ω. Because $g_m(\omega_k) = 0$ for all k and $g_m(\omega) > 0$ for $\omega \notin \{\omega_1, \ldots, \omega_p\}$, the uniform convergence implies that $\hat{g}_m(\omega)$ has a local minimum near ω_k for all k with probability tending to unity and $\hat{\boldsymbol{\omega}} := [\hat{\omega}_1, \ldots, \hat{\omega}_p]^T \xrightarrow{P} \boldsymbol{\omega} := [\omega_1, \ldots, \omega_p]^T$ as $n \to \infty$. This proves the following theorem.

Theorem 10.10 (Consistency of the MUSIC Estimator). *Let $\hat{g}_m(\omega)$ be defined by (10.3.3) or (10.3.4). Let $\hat{\boldsymbol{\omega}}$ comprise the local minima of $\hat{g}_m(\omega)$ in the vicinity of $\boldsymbol{\omega}$. For any fixed $m > p$, if $\hat{\mathbf{R}}_y \xrightarrow{P} \mathbf{R}_y$ as $n \to \infty$, then $\hat{\boldsymbol{\omega}} \xrightarrow{P} \boldsymbol{\omega}$.*

The asymptotic distribution of the MUSIC estimator can also be derived for fixed $m > p$ as $n \to \infty$. Toward that end, let $\mathbf{c}_k := [c_{1k}, \ldots, c_{mk}]^T := -\mathbf{R}_x^{\dagger}\mathbf{f}_m(\omega_k)$. Define $\mathbf{C}_{kk'} := [c_{kk'}(u, v)]$ and $\tilde{\mathbf{C}}_{kk'} := [\tilde{c}_{kk'}(u, v)]$ $(u, v = 1, \ldots, m)$, where

$$c_{kk'}(u, v) := \sum_{j=1}^{m} c_{u-v+j,k} c_{jk'}^*, \quad \tilde{c}_{kk'}(u, v) := \sum_{j=1}^{m} c_{u+v-j,k} c_{jk'}.$$

Then, we obtain the following theorem as a result of Theorems 4.1 and 4.4 combined with Lemma 12.4.8. See Section 10.7 for a proof.

Theorem 10.11 (Asymptotic Normality of the MUSIC Estimator). *Let $\{y_t\}$ take the form (10.1.1) with $\{x_t\}$ given by (2.1.5) or (2.1.7). Assume that $\{\epsilon_t\} \sim \text{IID}(0,\sigma^2)$ with $E(|\epsilon_t|^4) < \infty$ and $E(\epsilon_t^2) = 0$. Let $\hat{\boldsymbol{\omega}}$ be the frequency estimator defined by Theorem 10.10 with $\hat{\mathbf{R}}_y = [\hat{r}_y(u-v)] + \mathcal{O}_P(n^{-1/2})$ $(u,v = 1,\ldots,m)$. Then, for any fixed $m > p$ as $n \to \infty$, $\sqrt{n}\,(\hat{\boldsymbol{\omega}} - \boldsymbol{\omega}) \xrightarrow{D} \text{N}(0, 2\sigma^4 \mathbf{D}\Re(\boldsymbol{\Sigma} + \tilde{\boldsymbol{\Sigma}})\mathbf{D}^T)$, where*

$$\boldsymbol{\Sigma} := [\dot{\mathbf{f}}_m^H(\omega_k)\mathbf{P}_{\rm N}\mathbf{C}_{kk'}\mathbf{P}_{\rm N}\dot{\mathbf{f}}_m(\omega_{k'})], \quad \tilde{\boldsymbol{\Sigma}} := [\dot{\mathbf{f}}_m^H(\omega_k)\mathbf{P}_{\rm N}\tilde{\mathbf{C}}_{kk'}\mathbf{P}_{\rm N}^*\dot{\mathbf{f}}_m^*(\omega_{k'})],$$

and $\mathbf{D} := \text{diag}\{1/\ddot{g}_m(\omega_1), \ldots, 1/\ddot{g}_m(\omega_p)\}$.

Remark 10.3 Because $\mathbf{P}_{\rm N}\mathbf{f}_m(\omega_k) = \mathbf{0}$, we can write $\ddot{g}_m(\omega_k) = 2\dot{\mathbf{f}}_m^H(\omega_k)\mathbf{P}_{\rm N}\dot{\mathbf{f}}_m(\omega_k)$. Because $\mathbf{P}_{\rm S} = \mathbf{F}_m(\mathbf{F}_m^H\mathbf{F}_m)^{-1}\mathbf{F}_m^H$, we have $\mathbf{P}_{\rm N} = \mathbf{I} - \mathbf{P}_{\rm S} = \mathbf{I} - \mathbf{F}_m(\mathbf{F}_m^H\mathbf{F}_m)^{-1}\mathbf{F}_m^H$. Unlike the analyses in [305, Section 5.7] and [377], Theorem 10.11 does not require that the nonzero eigenvalues of $\mathbf{R}_x$ be all distinct.

The following example illustrates the asymptotic distribution of the MUSIC estimator in Theorem 10.11.

Example 10.9 (Asymptotic Normality of the MUSIC Estimator for a Single Complex Sinusoid). Let $p = 1$. According to Theorem 10.11, $\sqrt{n}\,(\hat{\omega}_1 - \omega_1) \xrightarrow{D} \text{N}(0, \sigma_\omega^2)$ for fixed $m > p$, where

$$\sigma_\omega^2 := 2\sigma^4\{\ddot{g}_m(\omega_1)\}^{-2}\Re(\varsigma + \tilde{\varsigma}), \tag{10.3.8}$$

with $\varsigma := \dot{\mathbf{f}}_m^H(\omega_1)\mathbf{P}_{\rm N}\mathbf{C}_{11}\mathbf{P}_{\rm N}\dot{\mathbf{f}}_m(\omega_1)$ and $\tilde{\varsigma} := \dot{\mathbf{f}}_m^H(\omega_1)\mathbf{P}_{\rm N}\tilde{\mathbf{C}}_{11}\mathbf{P}_{\rm N}^*\dot{\mathbf{f}}_m^*(\omega_1)$. Because $\mathbf{R}_x = C_1^2\mathbf{f}_m(\omega_1)\mathbf{f}_m^H(\omega_1)$, we obtain $\mu_1 = mC_1^2$ and $\mathbf{v}_1 = m^{-1/2}\mathbf{f}_m(\omega_1)$. Therefore,

$$\mathbf{c}_1 := -\mathbf{R}_x^\dagger\mathbf{f}_m(\omega_1) = -(mC_1^2)^{-1}\mathbf{f}_m(\omega_1), \quad \mathbf{P}_{\rm N} = \mathbf{I} - m^{-1}\mathbf{f}_m(\omega_1)\mathbf{f}_m^H(\omega_1).$$

It is easy to verify that $c_{11}(u,v) = a_{uv}z_1^{u-v}$ and $\tilde{c}_{11}(u,v) = \tilde{a}_{uv}z_1^{u+v}$, where

$$a_{uv} := (mC_1^2)^{-2}(m - |u-v|), \quad \tilde{a}_{uv} := (mC_1^2)^{-2}(m - |u+v-m-1|).$$

With the notation $a_{u*} := \sum_{v=1}^m a_{uv}$, $a_{*v} := \sum_{u=1}^m a_{uv}$, and $a_{**} := \sum_{u=1}^m \sum_{u=1}^m a_{uv}$, straightforward calculation yields

$$\mathbf{P}_{\rm N}\mathbf{C}_{11}\mathbf{P}_{\rm N} = [b_{uv}z_1^{u-v}] \qquad (u,v = 1,\ldots,m),$$

where

$$b_{uv} := a_{uv} - m^{-1}a_* - m^{-1}a_{*v} + m^{-2}a_{**}. \tag{10.3.9}$$

Similarly, we can write $\mathbf{P}_{\rm N}\tilde{\mathbf{C}}_{11}\mathbf{P}_{\rm N}^* = [\tilde{b}_{uv}z_1^{u+v}]$, where $\tilde{b}_{uv}$ is defined from the $\tilde{a}_{uv}$ in the same way as b_{uv} in (10.3.9). Therefore,

$$\varsigma = \sum_{u,v=1}^m uvb_{uv}, \quad \tilde{\varsigma} = -\sum_{u,v=1}^m uv\tilde{b}_{uv}.$$

Because $\|\dot{\mathbf{f}}_m(\omega_1)\|^2 = m(m+1)(2m+1)/6$ and $\dot{\mathbf{f}}_m^H(\omega_1)\mathbf{f}_m(\omega_1) = -im(m+1)/2$, it follows that

$$\ddot{g}_m(\omega_1) = 2\{\|\dot{\mathbf{f}}_m(\omega_1)\|^2 - m^{-1}|\dot{\mathbf{f}}_m^H(\omega_1)\mathbf{f}_m(\omega_1)|^2\} = \tfrac{1}{6}m(m^2-1).$$

Inserting these expressions in (10.3.8) yields

$$\sigma_\omega^2 = \frac{6\times 12\sigma^4}{m^2(m^2-1)^2}\sum_{u,v=1}^{m} uvd_{uv}, \tag{10.3.10}$$

where $d_{uv} := b_{uv} - \tilde{b}_{uv}$. To evaluate the sum in (10.3.10), observe that d_{uv} takes the same form as b_{uv} in (10.3.9) except that a_{uv} is everywhere replaced by

$$c_{uv} := a_{uv} - \tilde{a}_{uv} = (mC_1^2)^{-2}(|u+v-m-1| - |u-v|). \tag{10.3.11}$$

Because $\sum_{v=1}^m |u+v-m-1| = \sum_{v=1}^m |u-v|$, we obtain $c_{u*} = 0$ for all u. Similarly, $c_{*v} = 0$ for all v and $c_{**} = 0$. Therefore,

$$d_{uv} = c_{uv}. \tag{10.3.12}$$

Moreover,

$$\begin{aligned}
&\sum_{u,v=1}^{m} uv(|u+v-m-1| - |u-v|) \\
&= \sum_{u,v=1}^{m} uv|u+v-m-1| - \sum_{u,v=1}^{m} |u-v|) \\
&= \sum_{u,v=1}^{m} u(m+1-v)|u-v| - \sum_{u,v=1}^{m} uv|u-v| \\
&= \sum_{v=1}^{m} (m+1-2v)\sum_{u=1}^{m} u|u-v|.
\end{aligned} \tag{10.3.13}$$

Straightforward calculation shows that

$$\sum_{u=1}^{m} u|u-v| = \{v(v^2-1) + (m-v)(m-v-1)(v+2m+1)\}/6.$$

Inserting this expression in (10.3.13) followed by simple calculation gives

$$\sum_{u,v=1}^{m} uv(|u+v-m-1| - |u-v|) = \frac{m(m^4-1)}{30}. \tag{10.3.14}$$

Combining (10.3.10)–(10.3.12) with (10.3.14) leads to

$$\sigma_\omega^2 = \frac{12(m^2+1)}{5m^3(m^2-1)\gamma_1^2}, \tag{10.3.15}$$

where $\gamma_1 := C_1^2/\sigma^2$ is the SNR. This formula can be found in [377]. ◇

Observe that the asymptotic variance in (10.3.15) decreases with the increase of m. This suggests that the accuracy of the MUSIC estimator can be improved by increasing the value of m. Indeed, Examples 10.7 and 10.8 show that the accuracy can be much higher with a suitably large m which is comparable to n. In light of Example 10.9, one might conjecture that the error rate takes the form $\mathcal{O}(m^{3/2}\sqrt{n})$ when m grows with n. But it remains an open problem to determine how fast m should grow with n in order for this error rate to be valid, if at all.

10.4 Subspace Rotation Method

The subspace projection (SSP) method discussed in the previous section produces frequency estimates from the local minima of an objective function based on the projections onto a noise subspace which is constructed from an estimated covariance matrix. The subspace rotation (SSR) method discussed in this section produces frequency estimates from the generalized eigenvalues of a suitable pair of covariance matrices related to each other through a rotated signal subspace. Because the frequency estimates are obtained directly from the angles of the generalized eigenvalues, initial values are no longer needed. This is a major advantage of the SSR method over the SSP method. In the following, we discuss two ways of constructing the desired covariance matrix pair.

10.4.1 Matrix-Pencil and ESPRIT Estimators

First, for $m \geq p$, consider the m-by-m covariance matrix $\mathbf{R}_y$ in (10.1.3) and the m-by-m auxiliary covariance matrix $\mathbf{R}'_y$ defined as

$$\mathbf{R}'_y := \begin{bmatrix} r_y(1) & r_y(0) & r_y^*(1) & \cdots & r_y^*(m-2) \\ r_y(2) & r_y(1) & r_y(0) & & r_y^*(m-3) \\ \vdots & \vdots & & & \vdots \\ r_y(m) & r_y(m-1) & r_y(m-2) & \cdots & r_y(1) \end{bmatrix}. \tag{10.4.1}$$

Recall that by randomizing the phases of the sinusoids so that $\{y_t\}$ becomes a zero-mean stationary process with $E(y_{t+u}y_t^*) = r_y(u)$, we can write $\mathbf{R}_y = E(\mathbf{y}_t\mathbf{y}_t^H)$, where $\mathbf{y}_t := [y_t, y_{t+1}, \ldots, y_{t+m-1}]^T$. Under the same condition, we can also write

$$\mathbf{R}'_y = E(\mathbf{y}_t\mathbf{y}_{t-1}^H).$$

In other words, the matrix $\mathbf{R}'_y$ is the cross covariance between $\mathbf{y}_t$ and $\mathbf{y}_{t-1}$.

Let $\mathbf{R}_x$ and $\mathbf{R}'_x$ be the counterparts of $\mathbf{R}_y$ and $\mathbf{R}'_y$ defined by $r_x(u)$ instead of $r_y(u)$. Because $r_x(u) = \sum_{k=1}^{p} C_k^2 z_k^u$, where $z_k := \exp(i\omega_k)$, it follows that $\mathbf{R}_x$ takes the form (10.1.7) and $\mathbf{R}_x'$ can be expressed as

$$\mathbf{R}_x' = \sum_{k=1}^{p} C_k^2 z_k \mathbf{f}_m(\omega_k)\mathbf{f}_m^H(\omega_k) = \mathbf{F}_m \mathbf{Z} \mathbf{G} \mathbf{F}_m^H, \tag{10.4.2}$$

where $\mathbf{F}_m$ and $\mathbf{G}$ are defined by (10.1.8) and

$$\mathbf{Z} := \mathrm{diag}(z_1, \ldots, z_p). \tag{10.4.3}$$

Observe that for any $\mathbf{c} \in \mathbb{C}^m$ the vector $\mathbf{G}\mathbf{F}_m^H\mathbf{c}$ comprises the individually rescaled projection coefficients of $\mathbf{c}$ onto the columns of $\mathbf{F}_m$. Therefore, we can interpret $\mathbf{R}_x\mathbf{c} = \mathbf{F}_m(\mathbf{G}\mathbf{F}_m^H\mathbf{c})$ as a reconstruction of $\mathbf{c}$ from the rescaled projection coefficients using the columns of $\mathbf{F}_m$. Similarly, we can interpret $\mathbf{R}_x'\mathbf{c} = \mathbf{F}_m\mathbf{Z}(\mathbf{G}\mathbf{F}_m^H\mathbf{c})$ as a reconstruction of $\mathbf{c}$ using the columns of $\mathbf{F}_m\mathbf{Z}$. The transformation $\mathbf{F}_m \mapsto \mathbf{F}_m\mathbf{Z}$ rotates the columns of $\mathbf{F}_m$ individually and therefore is interpretable as a rotation of the signal subspace. This subspace rotation is introduced by considering the cross covariance between $\mathbf{y}_t$ and its backward shifted copy $\mathbf{y}_{t-1}$. The amount of individual rotation determines the z_k. This is key to the success of the subspace rotation method of frequency estimation.

Given $\mathbf{R}_x'$ and $\mathbf{R}_x$, the set of matrices of the form $\mathbf{R}_x' - z\mathbf{R}_x$ with $z \in \mathbb{C}$ is called a pencil [122, p. 375]. Owing to (10.1.7) and (10.4.2), we can write

$$\mathbf{R}_x' - z\mathbf{R}_x = \mathbf{F}_m(\mathbf{Z} - z\mathbf{I})\mathbf{G}\mathbf{F}_m^H. \tag{10.4.4}$$

This special structure guarantees that the z_k are eigenvalues of the pencil, or generalized eigenvalues of the matrix pair $(\mathbf{R}_x', \mathbf{R}_x)$. Indeed, because $\mathbf{F}_m$ has full (column) rank p and $\mathbf{G}$ is nonsingular, it follows from (10.4.4) that the identity

$$\mathbf{R}_x'\mathbf{q} = z\mathbf{R}_x\mathbf{q} \tag{10.4.5}$$

holds for some $\mathbf{q} \in \mathbb{C}^m$ such that $\mathbf{F}_m^H\mathbf{q} \neq \mathbf{0}$ if and only if $z = z_k$ for some k. Because the component of $\mathbf{q}$ in the noise subspace $\mathbb{C}_N^m$ plays no part in this characterization of the z_k, it suffices to consider the generalized eigenvalues of $(\mathbf{R}_x', \mathbf{R}_x)$ under the constraint that the associated eigenvectors reside in the signal subspace.

The constrained generalized eigenvalue problem (10.4.5) can be transformed into an unconstrained ordinary eigenvalue problem. Toward that end, consider the EVD of $\mathbf{R}_y$ given by (10.1.4). According to Theorem 10.1, for any $\mathbf{q} \in \mathbb{C}_S^m$, there exists a unique $\mathbf{d} \in \mathbb{C}^p$ such that

$$\mathbf{q} = \mathbf{V}_S\mathbf{d},$$

where $\mathbf{V}_S := [\mathbf{v}_1, \ldots, \mathbf{v}_p]$. Therefore, we can rewrite (10.4.5) as

$$\mathbf{R}_x'\mathbf{V}_S\mathbf{d} = z\mathbf{R}_x\mathbf{V}_S\mathbf{d}.$$

Pre-multiplying both sides of the foregoing equation with $\mathbf{V}_S^H \mathbf{R}_x^\dagger$, combined with the fact that $\mathbf{R}_x^\dagger \mathbf{R}_x \mathbf{V}_S = \mathbf{P}_S \mathbf{V}_S = \mathbf{V}_S$ and $\mathbf{V}_S^H \mathbf{V}_S = \mathbf{I}$, yields

$$\mathbf{Q}\mathbf{d} = z\mathbf{d}, \tag{10.4.6}$$

where

$$\mathbf{Q} := \mathbf{V}_S^H \mathbf{R}_x^\dagger \mathbf{R}_x' \mathbf{V}_S. \tag{10.4.7}$$

Therefore, if z satisfies (10.4.5) for some nonzero vector $\mathbf{q} \in \mathbb{C}_S^m$, then it also satisfies (10.4.6) for some nonzero vector $\mathbf{d} \in \mathbb{C}^p$. This means that the z_k can be recovered by solving the unconstrained ordinary eigenvalue problem (10.4.6).

Motivated by this observation, let the p-by-p matrix $\mathbf{Q}$ in (10.4.7) be replaced by an estimator $\hat{\mathbf{Q}}$ based on the noisy data record $\{y_1, \ldots, y_n\}$, and let the corresponding eigenvalues be denoted as $\{\hat{z}_k\}$ $(k = 1, \ldots, p)$. Because the $\hat{z}_k$ serve as estimates of the z_k, the signal frequencies ω_k can be estimated by the angles of the $\hat{z}_k$. We call this estimator an *SSR estimator of the first kind*, or simply an SSR1 estimator. Without loss of generality, let the $\hat{z}_k$ be arranged by their angles in ascending order such that $\hat{\mathbf{z}} := [\hat{z}_1, \ldots, \hat{z}_p]^T$ is an estimator of $\mathbf{z} := [z_1, \ldots, z_p]^T$ and $\hat{\boldsymbol{\omega}} := [\hat{\omega}_1, \ldots, \hat{\omega}_p]^T := [\angle\hat{z}_1, \ldots, \angle\hat{z}_p]^T$ is an estimator of $\boldsymbol{\omega} := [\omega_1, \ldots, \omega_p]^T$.

Different SSR1 estimators can be obtained with different constructions of $\hat{\mathbf{Q}}$. As an example, let $\hat{\mathbf{R}}_y$ and $\hat{\mathbf{R}}_y'$ denote any consistent estimators of $\mathbf{R}_y$ and $\mathbf{R}_y'$, respectively, and let $\hat{\mathbf{R}}_y$ have an EVD of the form

$$\hat{\mathbf{R}}_y = \sum_{j=1}^{m} \hat{\lambda}_j \hat{\mathbf{v}}_j \hat{\mathbf{v}}_j^H, \tag{10.4.8}$$

where $\hat{\lambda}_1 \geq \cdots \geq \hat{\lambda}_m > 0$. Then, a natural estimator of $\mathbf{V}_S$ is given by

$$\hat{\mathbf{V}}_S := [\hat{\mathbf{v}}_1, \ldots, \hat{\mathbf{v}}_p]. \tag{10.4.9}$$

Moreover, let $\hat{\mathbf{R}}_y'$ simply take the place of $\mathbf{R}_x'$ in (10.4.7), and let $\mathbf{R}_x$ in (10.4.7) be replaced by the rank-p matrix

$$\tilde{\mathbf{R}}_y := \sum_{k=1}^{p} \hat{\lambda}_j \hat{\mathbf{v}}_j \hat{\mathbf{v}}_j^H = \hat{\mathbf{V}}_S \hat{\boldsymbol{\Lambda}}_S \hat{\mathbf{V}}_S^H, \tag{10.4.10}$$

where

$$\hat{\boldsymbol{\Lambda}}_S := \mathrm{diag}(\hat{\lambda}_1, \ldots, \hat{\lambda}_p),$$

With $\tilde{\mathbf{R}}_y^\dagger := \sum_{j=1}^p \hat{\lambda}_j^{-1} \hat{\mathbf{v}}_j \hat{\mathbf{v}}_j^H = \hat{\mathbf{V}}_S \hat{\boldsymbol{\Lambda}}_S^{-1} \hat{\mathbf{V}}_S^H$ denoting the pseudo inverse of $\tilde{\mathbf{R}}_y$, the resulting estimator of $\mathbf{Q}$ can be expressed as

$$\hat{\mathbf{Q}} := \hat{\mathbf{V}}_S^H \tilde{\mathbf{R}}_y^\dagger \hat{\mathbf{R}}_y' \hat{\mathbf{V}}_S = \hat{\boldsymbol{\Lambda}}_S^{-1} \hat{\mathbf{V}}_S^H \hat{\mathbf{R}}_y' \hat{\mathbf{V}}_S. \tag{10.4.11}$$

Observe that $\hat{\mathbf{Q}}$ in (10.4.11) is the minimum-norm solution to the least-squares (LS) problem of minimizing $\|\hat{\mathbf{R}}'_y\hat{\mathbf{V}}_S - \tilde{\mathbf{R}}_y\hat{\mathbf{V}}_S\tilde{\mathbf{Q}}\|^2$ with respect to $\tilde{\mathbf{Q}} \in \mathbb{C}^{p\times p}$.

There are still many choices for the estimators $\hat{\mathbf{R}}_y$ and $\hat{\mathbf{R}}'_y$. An important example is given by

$$\hat{\mathbf{R}}_y := (2n)^{-1}\mathbf{Y}_{fb}^H\mathbf{Y}_{fb}, \qquad \hat{\mathbf{R}}'_y := (2n)^{-1}\mathbf{Y}_{fb}^H\mathbf{Y}'_{fb}, \tag{10.4.12}$$

with

$$\mathbf{Y}_{fb} := \begin{bmatrix} \mathbf{Y}_f \\ \mathbf{Y}_b \end{bmatrix}, \quad \mathbf{Y}'_{fb} := \begin{bmatrix} \mathbf{Y}_b^*\tilde{\mathbf{I}} \\ \mathbf{Y}_f^*\tilde{\mathbf{I}} \end{bmatrix}.$$

In this expression, $\mathbf{Y}_f$ and $\mathbf{Y}_b$ are defined by (9.1.8) and (9.1.10), $\tilde{\mathbf{I}}$ denotes the $m \times m$ reverse permutation matrix in which all antidiagonal elements are equal to 1 and other elements are equal to zero. Similar to the reduced rank linear prediction estimators with the same choice of $\hat{\mathbf{R}}_y$ discussed in Section 10.2, the inner product form of $\hat{\mathbf{R}}_y$ in (10.4.12) makes it possible to compute the resulting matrix $\hat{\mathbf{Q}}$ in (10.4.11) directly from the SVD of $\mathbf{Y}_{fb}$. Indeed, with m satisfying (10.2.5) and with the SVD of $\mathbf{Y}_{fb}$ given by (10.2.6), we can write

$$\begin{aligned} \hat{\mathbf{Q}} &= \hat{\mathbf{V}}_S^H\Big\{\sum_{j=1}^p \hat{s}_j^{-2}\hat{\mathbf{v}}_j\hat{\mathbf{v}}_j^H\Big\}\Big\{\sum_{j=1}^p \hat{s}_j\hat{\mathbf{v}}_j\hat{\mathbf{u}}_j^H\Big\}\mathbf{Y}'_{fb}\hat{\mathbf{V}}_S \\ &= \hat{\mathbf{V}}_S^H\Big\{\sum_{j=1}^p \hat{s}_j^{-1}\hat{\mathbf{v}}_j\hat{\mathbf{u}}_j^H\Big\}\mathbf{Y}'_{fb}\hat{\mathbf{V}}_S \\ &= \hat{\mathbf{\Sigma}}_S^{-1}\hat{\mathbf{U}}_S^H\mathbf{Y}'_{fb}\hat{\mathbf{V}}_S, \end{aligned} \tag{10.4.13}$$

where $\hat{\mathbf{U}}_S := [\hat{\mathbf{u}}_1,\ldots,\hat{\mathbf{u}}_p]$ and $\hat{\mathbf{\Sigma}}_S := \operatorname{diag}(\hat{s}_1,\ldots,\hat{s}_p)$. Therefore, the corresponding SSR1 frequency estimates can be obtained from the eigenvalues of $\hat{\mathbf{Q}}$ in (10.4.13). This is the matrix pencil (MP) estimator proposed in [156] (see also [157]).

An alternative estimator of $\mathbf{Q}$ is obtained by replacing $\mathbf{R}_x^\dagger$ and $\mathbf{R}'_x$ in (10.4.7) with their consistent estimators instead of $\tilde{\mathbf{R}}_y^\dagger$ and $\hat{\mathbf{R}}'_y$. Observe that

$$\mathbf{R}_x = \mathbf{R}_y - \sigma^2\mathbf{I}, \quad \mathbf{R}'_x = \mathbf{R}'_y - \sigma^2\mathbf{I}',$$

where $\mathbf{I}'$ denotes the m-by-m matrix whose elements are zero everywhere except for those at $(j, j+1)$, for $j = 1,\ldots,m-1$, which are equal to unity. Therefore, it is more natural to estimate $\mathbf{R}_x$ and $\mathbf{R}'_x$ by

$$\hat{\mathbf{R}}_x := \tilde{\mathbf{R}}_y - \hat{\sigma}^2\mathbf{I}, \quad \hat{\mathbf{R}}'_x := \hat{\mathbf{R}}'_y - \hat{\sigma}^2\mathbf{I}',$$

where $\tilde{\mathbf{R}}_y$ is the optimal rank-p approximation of $\hat{\mathbf{R}}_y$ given by (10.4.10) and $\hat{\sigma}^2$ is a consistent estimator of σ^2 such as that given by (10.2.11). Substituting these quantities in (10.4.7) yields

$$\hat{\mathbf{Q}} := \hat{\mathbf{V}}_S^H\hat{\mathbf{R}}_x^\dagger\hat{\mathbf{R}}'_x\hat{\mathbf{V}}_S = (\hat{\mathbf{\Lambda}}_S - \hat{\sigma}^2\mathbf{I})^{-1}\hat{\mathbf{V}}_S^H(\hat{\mathbf{R}}'_y - \hat{\sigma}^2\mathbf{I}')\hat{\mathbf{V}}_S. \tag{10.4.14}$$

The eigenvalues of $\hat{\mathbf{Q}}$ in (10.4.14) coincide with the generalized eigenvalues of $(\hat{\mathbf{R}}_x^{\dagger}, \hat{\mathbf{R}}_x')$ whose associated eigenvectors reside in the column space of $\hat{\mathbf{V}}_{\mathrm{S}}$. This is the ESPRIT estimator proposed in [323], where ESPRIT is the acronym for estimation of signal parameter by rotational invariance techniques.

When $\hat{\mathbf{R}}_y$ and $\hat{\mathbf{R}}_y'$ are given by (10.4.12) in particular, it is easy to verify, using the SVD in (10.2.6), that the matrix $\hat{\mathbf{Q}}$ in (10.4.14) can be expressed as

$$\hat{\mathbf{Q}} = (\hat{\boldsymbol{\Sigma}}_{\mathrm{S}}^H \hat{\boldsymbol{\Sigma}}_{\mathrm{S}} - \hat{\varsigma}^2 \mathbf{I})^{-1} (\hat{\boldsymbol{\Sigma}}_{\mathrm{S}}^H \hat{\mathbf{U}}_{\mathrm{S}}^H \mathbf{Y}_{fb}' \hat{\mathbf{V}}_{\mathrm{S}} - \hat{\varsigma}^2 \hat{\mathbf{V}}_{\mathrm{S}}^H \mathbf{I}' \hat{\mathbf{V}}_{\mathrm{S}}), \tag{10.4.15}$$

where $\hat{\mathbf{U}}_{\mathrm{S}} := [\hat{\mathbf{u}}_1, \ldots, \hat{\mathbf{u}}_p]$, $\hat{\boldsymbol{\Sigma}}_{\mathrm{S}} := \mathrm{diag}(\hat{s}_1, \ldots, \hat{s}_p)$, and $\hat{\varsigma}^2 := 2n\hat{\sigma}^2$. In this expression, the matrix $\mathbf{I}'\hat{\mathbf{V}}_{\mathrm{S}}$ can be obtained by simply deleting the first row in $\hat{\mathbf{V}}_{\mathrm{S}}$ followed by appending a row of zeros. If $\hat{\sigma}^2$ is given by (10.2.29), then

$$\hat{\varsigma}^2 := (m-p)^{-1} \sum_{j=p+1}^{m} |\hat{s}_j|^2. \tag{10.4.16}$$

It is easy to see that (10.4.15) reduces to (10.4.13) if $\hat{\varsigma}^2 := 0$. Because subtracting $\hat{\varsigma}^2$ removes the asymptotic bias in estimating $\mathbf{R}_x$ and $\mathbf{R}_x'$, the ESPRIT estimator can be regarded as a bias-corrected version of the MP estimator.

The following examples demonstrate the statistical performance of the ESPRIT estimator for different choices of m.

Example 10.10 (Estimation of Well-Separated Frequencies: Part 4). Consider the time series in Example 10.1. Let the signal frequencies be estimated by the ESPRIT estimator using $\hat{\mathbf{Q}}$ in (10.4.15) with $\hat{\varsigma}^2$ given by (10.4.16). Figure 10.12 depicts the average MSE of the frequency estimates with different choices of m and for different values of SNR.

This result shows that the ESPRIT estimates are poor when m is too small, but the accuracy improves considerably with the increase of m (at the expense of increased burden in computing the SVD). In particular, with the largest value $m = \lfloor 2n/3 \rfloor$ (which equals 66 for $n = 100$ and equals 133 for $n = 200$), the resulting MSE approaches the CRLB when the SNR exceeds a threshold. While still slightly inferior to the PM and NLS estimators shown in Figures 6.13 and 8.3, the ESPRIT estimator performs slightly better than the MUSIC estimator shown in Figure 10.9 and the reduced rank FBLP estimator shown in Figure 10.2. Moreover, the ESPRIT estimator allows m to take the largest value without degrading the performance. This makes the choice of m a trivial task, which is unlike the MUSIC estimator and particularly the reduced rank FBLP estimator. ◇

Example 10.11 (Estimation of Closely Spaced Frequencies: Part 4). Consider the time series in Example 10.2. As in Example 10.10, let the signal frequencies be estimated by the ESPRIT estimator on the basis of $\hat{\mathbf{Q}}$ in (10.4.15) with $\hat{\varsigma}^2$ given by (10.4.16). Figure 10.13 shows the simulation result.

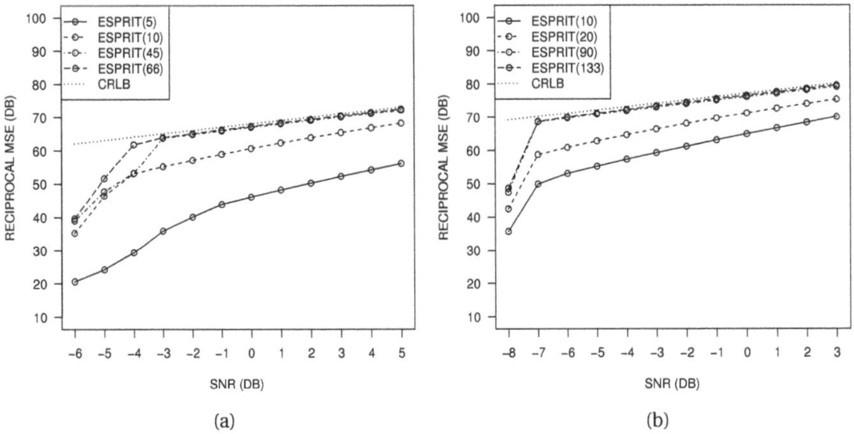

Figure 10.12. Plot of the reciprocal average MSE of the ESPRIT frequency estimates based on (10.4.15) and (10.4.16) for the time series of two complex sinusoids in Gaussian white noise discussed in Example 10.10 with different values of SNR. (a) $n = 100$: solid line, $m = 5$; dashed line, $m = 10$; dash-dotted line, $m = 45$; long dashed line $m = 66$. (b) $n = 200$: solid line, $m = 10$; dashed line, $m = 20$; dash-dotted line, $m = 90$; long dashed line, $m = 133$. Dotted line depicts the CRLB. Results are based on 5,000 Monte Carlo runs.

In this example, the ESPRIT estimator also performs better than the MUSIC estimator shown in Figure 10.10 and the reduced rank FBLP estimator shown in Figure 10.3, but it still has the same SNR threshold which is quite high. The accuracy of the ESPRIT estimator generally improves with the increase of m, but there remains a significant gap between the MSE and the CRLB even at high SNR. So, the ESPRIT estimator is still not as accurate as the NLS estimator shown in Figure 8.5, not to mention the higher SNR threshold. ◇

Simulation results based on the same data as in Examples 10.10 and 10.11 show that the MP estimator, with $\hat{\mathbf{Q}}$ defined by (10.4.13), produces very similar results to the ESPRIT estimator, except the case of well-separated frequencies with $m = 5$ for $n = 100$ where the ESPRIT estimator is slightly more accurate.

Next, let us employ an alternative technique to achieve the desired subspace rotation. For any $m \ge p+1$, consider the $(m-1)$-by-m matrices $\mathbf{R}_{y1}$ and $\mathbf{R}_{y2}$ obtained from the m-by-m covariance matrix $\mathbf{R}_y$ by deleting the first row and the last row, respectively. In other words, let

$$\mathbf{R}_{y1} := [\mathbf{0}, \mathbf{I}_{m-1}]\,\mathbf{R}_y, \quad \mathbf{R}_{y2} := [\mathbf{I}_{m-1}, \mathbf{0}]\,\mathbf{R}_y.$$

By randomizing the phases of the sinusoids, one can regard $\mathbf{R}_{y1}$ as the cross covariance between $\mathbf{y}_{t1} := [y_{t+1}, \ldots, y_{t+m-1}]^T$ and $\mathbf{y}_t := [y_t, y_{t+1}, \ldots, y_{t+m-1}]^T$, i.e., $\mathbf{R}_{y1} = E(\mathbf{y}_{t1}\mathbf{y}_t^H)$. Similarly, $\mathbf{R}_{y2} = E(\mathbf{y}_{t2}\mathbf{y}_t^H)$, where $\mathbf{y}_{t2} := [y_t, y_{t+1}, \ldots, y_{t+m-2}]^T$.

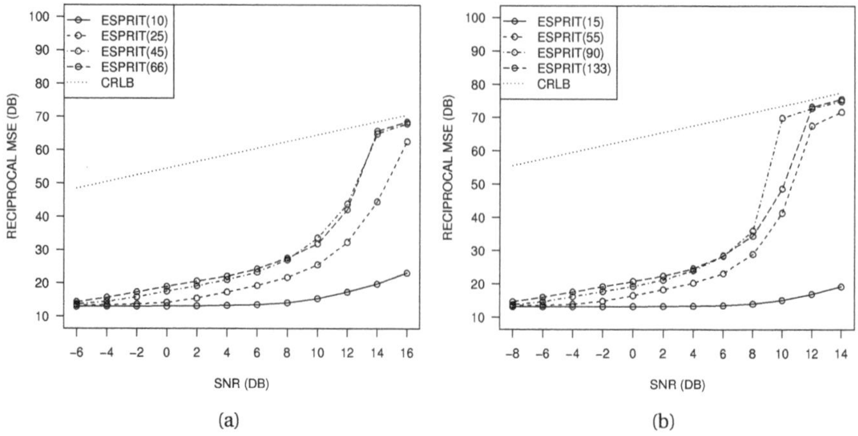

Figure 10.13. Plot of the reciprocal average MSE of the ESPRIT frequency estimates based on (10.4.15) and (10.4.16) for the time series of two complex sinusoids with closely spaced frequencies in Gaussian white noise discussed in Example 10.11 with different values of SNR. (a) $n = 100$; solid line, $m = 10$; dashed line, $m = 25$; dash-dotted line, $m = 45$; long dashed line, $m = 66$. (b) $n = 200$; solid line, $m = 15$; dashed line, $m = 55$; dash-dotted line, $m = 90$; long dashed line, $m = 133$. Dotted line depicts the CRLB. Results are based on 5,000 Monte Carlo runs.

Let $\mathbf{R}_{x1}$ and $\mathbf{R}_{x2}$ be the counterparts of $\mathbf{R}_{y1}$ and $\mathbf{R}_{y2}$ defined by $\mathbf{R}_x$ instead of $\mathbf{R}_y$. Then, according to (10.1.7), we have

$$\begin{cases} \mathbf{R}_{x1} := [\mathbf{0}, \mathbf{I}_{m-1}]\,\mathbf{R}_x = [\mathbf{0}, \mathbf{I}_{m-1}]\,\mathbf{F}_m\mathbf{G}\mathbf{F}_m^H = \mathbf{F}_{m-1}\mathbf{Z}\mathbf{G}\mathbf{F}_m^H, \\ \mathbf{R}_{x2} := [\mathbf{I}_{m-1}, \mathbf{0}]\,\mathbf{R}_x = [\mathbf{I}_{m-1}, \mathbf{0}]\,\mathbf{F}_m\mathbf{G}\mathbf{F}_m^H = \mathbf{F}_{m-1}\mathbf{G}\mathbf{F}_m^H, \end{cases} \tag{10.4.17}$$

where $\mathbf{Z}$ is defined by (10.4.3). For any $\mathbf{c} \in \mathbb{C}^m$, the vector $\mathbf{G}\mathbf{F}_m^H\mathbf{c}$ contains the individually rescaled projection coefficients of $\mathbf{c}$ onto the columns of $\mathbf{F}_m$. Therefore, $\mathbf{R}_{x2}\mathbf{c} = \mathbf{F}_{m-1}(\mathbf{G}\mathbf{F}_m^H\mathbf{c})$ is a reconstruction of $\mathbf{c}$ from the rescaled projection coefficients using the columns of $\mathbf{F}_{m-1}$. The transformation $\mathbf{F}_{m-1} \mapsto \mathbf{F}_{m-1}\mathbf{Z}$ rotates the columns of $\mathbf{F}_{m-1}$ individually and hence $\mathbf{R}_{x1}\mathbf{c} = \mathbf{F}_{m-1}\mathbf{Z}(\mathbf{G}\mathbf{F}_m^H\mathbf{c})$ is a reconstruction of $\mathbf{c}$ using the rotated columns of $\mathbf{F}_{m-1}$. Unlike the rotation in $\mathbf{R}'_x$, the rotation in $\mathbf{R}_{x1}$ takes place in a lower-dimensional signal subspace $\mathbb{C}_S^{m-1}$ spanned by the columns of $\mathbf{F}_{m-1}$. The amount of rotation in $\mathbf{R}_{x1}$ still determines the z_k.

Owing to the special structure in (10.4.17), the identity

$$\mathbf{R}_{x1}\mathbf{q} = z\mathbf{R}_{x2}\mathbf{q} \tag{10.4.18}$$

holds for some $\mathbf{q}$ such that $\mathbf{F}_m^H\mathbf{q} \neq \mathbf{0}$ if and only if $z = z_k$ for some k. In other words, the z_k are generalized eigenvalues of $(\mathbf{R}_{x1}, \mathbf{R}_{x2})$ whose associated eigenvectors reside in the signal subspace $\mathbb{C}_S^m$. To transform this constrained generalized eigenvalue problem into an unconstrained ordinary eigenvalue problem, consider the EVD of $\mathbf{R}_y$ in (10.1.4). Let $\boldsymbol{\Lambda}_x := \mathrm{diag}(\mu_1, \ldots, \mu_p)$ and

$$\mathbf{V}_{S1} := [\mathbf{0}, \mathbf{I}_{m-1}]\, \mathbf{V}_S, \quad \mathbf{V}_{S2} := [\mathbf{I}_{m-1}, \mathbf{0}]\, \mathbf{V}_S. \tag{10.4.19}$$

Because $\mathbf{R}_x = \mathbf{V}_S \boldsymbol{\Lambda}_S \mathbf{V}_S^H$ by Theorem 10.1, we can write

$$\mathbf{R}_{x1} = \mathbf{V}_{S1} \boldsymbol{\Lambda}_x \mathbf{V}_S^H, \quad \mathbf{R}_{x2} = \mathbf{V}_{S2} \boldsymbol{\Lambda}_x \mathbf{V}_S^H.$$

Substituting these expressions in (10.4.18) with $\mathbf{q} := \mathbf{V}_S \boldsymbol{\Lambda}_x^{-1} \mathbf{d}$ leads to

$$\mathbf{V}_{S1} \mathbf{d} = z \mathbf{V}_{S2} \mathbf{d}.$$

Moreover, because $\mathbf{V}_S$ has full (column) rank p, so does $\mathbf{V}_{S2}$ for any $m \geq p+1$. Therefore, the foregoing equation can be rewritten as

$$\mathbf{Q}\mathbf{d} = z\mathbf{d}, \tag{10.4.20}$$

where

$$\mathbf{Q} := (\mathbf{V}_{S2}^H \mathbf{V}_{S2})^{-1} \mathbf{V}_{S2}^H \mathbf{V}_{S1}. \tag{10.4.21}$$

If z satisfies (10.4.18) for some $\mathbf{q}$ such that $\mathbf{F}_m^H \mathbf{q} \neq \mathbf{0}$, it also satisfies (10.4.20) for some $\mathbf{d} \neq \mathbf{0}$. Therefore, the z_k are the eigenvalues of $\mathbf{Q}$ defined by (10.4.21). The same conclusion can be drawn by observing that there exists a nonsingular matrix $\mathbf{B}$ such that $\mathbf{V}_S = \mathbf{F}_m \mathbf{B}$, so

$$\mathbf{V}_{S1} = [\mathbf{0}, \mathbf{I}_{m-1}]\, \mathbf{F}_m \mathbf{B} = \mathbf{F}_{m-1} \mathbf{Z} \mathbf{B}, \quad \mathbf{V}_{S2} = [\mathbf{I}_{m-1}, \mathbf{0}]\, \mathbf{F}_m \mathbf{B} = \mathbf{F}_{m-1} \mathbf{B}.$$

Inserting these expressions in (10.4.21) gives $\mathbf{Q} = \mathbf{B}^{-1} \mathbf{Z} \mathbf{B}$. This means that $\mathbf{Q}$ and $\mathbf{Z}$ are similar matrices and hence share the same eigenvalues [153, p. 45].

The observation that the z_k can be recovered from the eigenvalues of $\mathbf{Q}$ in (10.4.21) motivates the following estimator for the signal frequencies. Let $\hat{\mathbf{R}}_y$ be any consistent estimator of $\mathbf{R}_y$ with an EVD of the form (10.4.8). Let $\hat{\mathbf{V}}_S$ be defined by (10.4.9). By replacing $\mathbf{V}_{S1}$ and $\mathbf{V}_{S2}$ in (10.4.21) with

$$\hat{\mathbf{V}}_{S1} := [\mathbf{0}, \mathbf{I}_{m-1}]\, \hat{\mathbf{V}}_S, \quad \hat{\mathbf{V}}_{S2} := [\mathbf{I}_{m-1}, \mathbf{0}]\, \hat{\mathbf{V}}_S,$$

we obtain an estimator of $\mathbf{Q}$ which takes the form

$$\hat{\mathbf{Q}} := (\hat{\mathbf{V}}_{S2}^H \hat{\mathbf{V}}_{S2})^{-1} \hat{\mathbf{V}}_{S2}^H \hat{\mathbf{V}}_{S1}. \tag{10.4.22}$$

This matrix is just the least-squares (LS) solution that minimizes $\|\hat{\mathbf{V}}_{S1} - \hat{\mathbf{V}}_{S2} \tilde{\mathbf{Q}}\|^2$ with respect to $\tilde{\mathbf{Q}} \in \mathbb{C}^{p \times p}$. Let the signal frequencies be estimated by the angles of the eigenvalues of $\hat{\mathbf{Q}}$ in (10.4.22). We call this estimator an *SSR estimator of the second kind*, or simply an SSR2 estimator. Note that the total least-squares solution is a possible alternative to the LS solution [324] [369, Section 4.7].

Like the eigenvalues of $\mathbf{Q}$ in (10.4.21), the eigenvalues of $\hat{\mathbf{Q}}$ in (10.4.22) are generalized eigenvalues of a matrix pair. In fact, let

$$\tilde{\mathbf{R}}_{y1} := [\mathbf{0}, \mathbf{I}_{m-1}]\,\tilde{\mathbf{R}}_y, \quad \tilde{\mathbf{R}}_{y2} := [\mathbf{I}_{m-1}, \mathbf{0}]\,\tilde{\mathbf{R}}_y,$$

where $\tilde{\mathbf{R}}_y$ is the rank-p approximation of $\hat{\mathbf{R}}_y$ given by (10.4.10). Then, it is easy to verify that the eigenvalues of $\hat{\mathbf{Q}}$ in (10.4.22) coincide with the generalized eigenvalues of $(\tilde{\mathbf{R}}_{y1}, \tilde{\mathbf{R}}_{y2})$ whose associated eigenvectors reside in the column space of $\hat{\mathbf{V}}_S$. Therefore, an SSR2 estimator can be attributed to the general ESPRIT methodology discussed in [284], but it is not the same as the ESPRIT estimator proposed in [323] and discussed earlier in this section, which employs a different pair of covariance matrices.

As with the SSR1 estimators, different choices of $\hat{\mathbf{R}}_y$ lead to different SSR2 estimators. In particular, with $\hat{\mathbf{R}}_y$ given by (10.4.12), the matrix $\hat{\mathbf{V}}_S$ can be obtained directly from the SVD of $\mathbf{Y}_{fb}$ in (10.2.6). Simulation results based on the same data as in Examples 10.10 and 10.11 show that the resulting SSR2 estimator performs very similarly to the corresponding ESPRIT estimator shown in Figures 10.12 and 10.13, except in the case of well-separated frequencies with $n = 100$ and $m = 10$ where the latter is more accurate (so is the corresponding MP estimator but to a lesser extent).

10.4.2 Statistical Properties

In this section, let us investigate the asymptotic properties of the ESPRIT estimator, with $\hat{\mathbf{Q}}$ given by (10.4.14), when m is fixed as $n \to \infty$.

The first theorem asserts that for any fixed $m \ge p$ the ESPRIT estimator is consistent for estimating the signal frequencies as $n \to \infty$ if $\hat{\mathbf{R}}_y$, $\hat{\mathbf{R}}'_y$, and $\hat{\sigma}^2$ are consistent estimators of $\mathbf{R}_y$, $\mathbf{R}'_y$, and σ^2, respectively. A proof of this assertion is given in Section 10.7.

Theorem 10.12 (Consistency of the ESPRIT Estimator). *Let $\hat{\mathbf{z}}$ comprise the eigenvalues of $\hat{\mathbf{Q}}$ defined by (10.4.14) and let $\hat{\boldsymbol{\omega}}$ comprise the angles of the eigenvalues. For any fixed $m \ge p$, if $\hat{\mathbf{R}}_y \to \mathbf{R}_y$, $\hat{\mathbf{R}}'_y \to \mathbf{R}'_y$, and $\hat{\sigma}^2 \to \sigma^2$ in probability or almost surely as $n \to \infty$, then $\hat{\mathbf{z}} \to \mathbf{z}$ and $\hat{\boldsymbol{\omega}} \to \boldsymbol{\omega}$ in the same mode of convergence.*

Remark 10.4 Let $\hat{\mathbf{R}}_y$ be given by (10.4.12) and $\hat{\sigma}^2$ be given by (10.2.11). If $\{\epsilon_t\} \sim \mathrm{IID}(0, \sigma^2)$ and $E(|\epsilon_t|^4) < \infty$, then $\hat{\mathbf{R}}_y \stackrel{a.s.}{\to} \mathbf{R}_y$, $\hat{\mathbf{R}}'_y \stackrel{a.s.}{\to} \mathbf{R}'_y$, and $\hat{\sigma}^2 \stackrel{a.s.}{\to} \sigma^2$ by Theorem 4.1, Theorem 10.1, and Lemma 12.4.2. In this case, the ESPRIT estimator has the property $\hat{\mathbf{z}} \stackrel{a.s.}{\to} \mathbf{z}$ and $\hat{\boldsymbol{\omega}} \stackrel{a.s.}{\to} \boldsymbol{\omega}$ as $n \to \infty$ for any fixed $m \ge p$.

The asymptotic distribution of the ESPRIT estimator can be derived under the simplifying assumption that $\hat{\sigma}^2$ satisfies (10.2.29). As shown in the proof of Theorem 10.12, the z_k are also the nonzero eigenvalues of the rank-p matrix

$$\mathbf{H} := \mathbf{R}_x^{\dagger}\mathbf{R}'_x. \tag{10.4.23}$$

Let $\mathbf{p}_k$ and $\mathbf{q}_k$ denote the left and right eigenvectors of $\mathbf{H}$ associated with z_k. Let q_{jk} denote the jth element of $\mathbf{q}_k$ and let $q_{jk} := 0$ for $j \notin \{1,\ldots,m\}$. Define $d_k(z) := (z_k - z)\sum_{j=1}^m q_{jk} z^{-j}$ and $c_{jk} := q_{j+1,k} - z_k q_{jk}$. Define $\mathbf{c}_k := [c_{1k},\ldots,c_{mk}]^T$, $\mathbf{C}_{kk'} := [c_{kk'}(u,v)]$, and $\tilde{\mathbf{C}}_{kk'} := [\tilde{c}_{kk'}(u,v)]$ $(u,v = 1,\ldots,m)$, where

$$c_{kk'}(u,v) := \sum_{j=0}^m c_{u-v+j,k} c_{jk'}^*,$$

$$\tilde{c}_{kk'}(u,v) := \sum_{j=0}^m c_{u+v-j,k} c_{jk'}.$$

Finally, define $\alpha_k := (\mathbf{p}_k^H \mathbf{q}_k)^{-1}$ and

$$\boldsymbol{\Sigma}_{kk'} := \sum_{l=1}^p 2C_l^2 \sigma^2 d_k(z_l) d_{k'}^*(z_l) \mathbf{f}_m(z_l) \mathbf{f}_m^H(z_l) + \sigma^4 \mathbf{C}_{kk'} + \sigma^4(\kappa - 2)\mathbf{c}_k \mathbf{c}_{k'}^H,$$

$$\tilde{\boldsymbol{\Sigma}}_{kk'} := \sum_{l=1}^p 2C_l^2 \sigma^2 d_k(z_l) d_{k'}(z_l) \mathbf{f}_m(z_l) \mathbf{f}_m^T(z_l) + \sigma^4 \tilde{\mathbf{C}}_{kk'} + \sigma^4(\kappa - 2)\mathbf{c}_k \mathbf{c}_{k'}^T.$$

With this notation, the following theorem can be established as a result of Theorem 4.4. See Section 10.7 for a proof.

Theorem 10.13 (Asymptotic Normality of the ESPRIT Estimator). *Let $\{y_t\}$ take the form (10.1.1) with $\{x_t\}$ given (2.1.5) or (2.1.7). Let $\hat{\mathbf{z}}$ comprise the eigenvalues of $\hat{\mathbf{Q}}$ defined by (10.4.14) and let $\hat{\boldsymbol{\omega}}$ be the corresponding frequency estimator. Assume that $\{\epsilon_t\} \sim \mathrm{IID}(0,\sigma^2)$ with $\kappa := E(|\epsilon_t|^4)/\sigma^4 < \infty$ and $E(\epsilon_t^2) = 0$. Assume that $\hat{\mathbf{R}}_y = [\hat{r}_y(u-v)] + \mathcal{O}_P(n^{-1/2})$ and $\hat{\mathbf{R}}_y' = [\hat{r}_y(u-v+1)] + \mathcal{O}_P(n^{-1/2})$ $(u,v = 1,\ldots,m)$, where the $\hat{r}_y(u)$ are the sample autocovariances given by (4.1.1). Assume further that $\hat{\sigma}^2$ satisfies (10.2.29). Then, $\sqrt{n}\,(\hat{\mathbf{z}} - \mathbf{z}) \xrightarrow{D} \mathrm{N}_c(\mathbf{0}, \boldsymbol{\Sigma}_z, \tilde{\boldsymbol{\Sigma}}_z)$ as $n \to \infty$ for any fixed $m \ge p$, where $\boldsymbol{\Sigma}_z := [\sigma_{kk'}]$ and $\tilde{\boldsymbol{\Sigma}}_z := [\tilde{\sigma}_{kk'}]$ $(k,k' = 1,\ldots,p)$, with*

$$\sigma_{kk'} := \alpha_k \alpha_{k'}^* \mathbf{p}_k^H \mathbf{R}_x^\dagger \boldsymbol{\Sigma}_{kk'} \mathbf{R}_x^\dagger \mathbf{p}_{k'}, \quad \tilde{\sigma}_{kk'} := \alpha_k \alpha_{k'} \mathbf{p}_k^H \mathbf{R}_x^\dagger \tilde{\boldsymbol{\Sigma}}_{kk'} \mathbf{R}_x^{\dagger *} \mathbf{p}_{k'}^*.$$

Moreover, $\sqrt{n}(\hat{\boldsymbol{\omega}} - \boldsymbol{\omega}) \xrightarrow{D} \mathrm{N}(\mathbf{0}, \boldsymbol{\Sigma})$, where $\boldsymbol{\Sigma} := [\sigma_\omega(k,k')]$ with

$$\sigma_\omega(k,k') := \frac{1}{2}\mathbf{s}_k^T \begin{bmatrix} \Re(\sigma_{kk'} + \tilde{\sigma}_{kk'}) & -\Im(\sigma_{kk'} - \tilde{\sigma}_{kk'}) \\ \Im(\sigma_{kk'} + \tilde{\sigma}_{kk'}) & \Re(\sigma_{kk'} - \tilde{\sigma}_{kk'}) \end{bmatrix} \mathbf{s}_{k'}$$

and $\mathbf{s}_k := [-\sin(\omega_k), \cos(\omega_k)]^T$ $(k,k' = 1,\ldots,p)$.

To illustrate Theorem 10.12, consider the simple case where $p = 1$.

Example 10.12 (Asymptotic Normality of the ESPRIT Estimator for a Single Complex Sinusoid). Let $p = 1$. Then, under the conditions of Theorem 10.13, we have $\sqrt{n}\,(\hat{\omega}_1 - \omega_1) \xrightarrow{D} \mathrm{N}(0, \sigma_\omega^2)$ as $n \to \infty$ for fixed $m \ge 1$, where

$$\sigma_\omega^2 := \frac{1}{2}\mathbf{s}_1^T \begin{bmatrix} \Re(\sigma_{11} + \tilde{\sigma}_{11}) & -\Im(\sigma_{11} - \tilde{\sigma}_{11}) \\ \Im(\sigma_{11} + \tilde{\sigma}_{11}) & \Re(\sigma_{11} - \tilde{\sigma}_{11}) \end{bmatrix} \mathbf{s}_1. \tag{10.4.24}$$

As in Example 10.5, we can write $\mathbf{R}_x^\dagger = (mC_1^2)^{-1}\mathbf{f}_m(\omega_1)\mathbf{f}_m^H(\omega_1)$. We can also write $\mathbf{R}_x' = z_1 C_1^2 \mathbf{f}_m(\omega_1)\mathbf{f}_m^H(\omega_1)$. Combining these expressions yields

$$\mathbf{R}_x^\dagger \mathbf{R}_x' = z_1 m^{-1} \mathbf{f}_m(\omega_1)\mathbf{f}_m^H(\omega_1).$$

Hence $\mathbf{p}_1 := \mathbf{q}_1 := m^{-1/2}\mathbf{f}_m(\omega_1)$. This result, together with $d_1(z_1) = 0$, implies that

$$\begin{aligned}
\sigma_{11} &= \frac{1}{m^3\gamma_1^2}\mathbf{f}_m^H(\omega_1)\{\mathbf{C}_{11} + (\kappa-2)\mathbf{c}_1\mathbf{c}_1^H\}\mathbf{f}_m(\omega_1),\\
\tilde{\sigma}_{11} &= \frac{1}{m^3\gamma_1^2}\mathbf{f}_m^H(\omega_1)\{\tilde{\mathbf{C}}_{11} + (\kappa-2)\mathbf{c}_1\mathbf{c}_1^T\}\mathbf{f}_m^*(\omega_1),
\end{aligned}$$

where $\gamma_1 := C_1^2/\sigma^2$ is the SNR. Straightforward calculation gives

$$\begin{aligned}
\mathbf{C}_{11} &= 2m^{-1}\mathbf{I},\\
\tilde{\mathbf{C}}_{11} &= [-2m^{-1}z_1^{m+1}\delta_{u+v-m} + m^{-1}z_1^{2m+2}\delta_{u+v-2m}],
\end{aligned}$$

and

$$\mathbf{f}_m^H(\omega_1)\mathbf{c}_1 = \mathbf{c}_1^T\mathbf{f}_m^*(\omega_1) = -m^{-1/2}z_1.$$

Therefore,

$$\begin{aligned}
\sigma_{11} &= \frac{1}{m^3\gamma_1^2}\{2 + (\kappa-2)m^{-1}\},\\
\tilde{\sigma}_{11} &= \frac{1}{m^3\gamma_1^2}\{-2 + 3m^{-1} + (\kappa-2)m^{-1}\}z_1^2.
\end{aligned}$$

Inserting these expressions in (10.4.24) followed by straightforward calculation yields a simple formula

$$\sigma_\omega^2 = \frac{4m-3}{2m^4\gamma_1^2}.$$

Observe that σ_ω^2 decreases with the increase of m or γ_1. This is similar to the behavior of the MUSIC estimator shown in Example 10.9. ◇

Although the rate of convergence remains $\sqrt{n}$ for the ESPRIT estimator with fixed m as $n \to \infty$, Example 10.12 suggests that the accuracy can be improved by increasing m. Examples 10.10 and 10.11 demonstrate that with the simple choice $m = \lfloor 2n/3 \rfloor$ the accuracy in terms of the MSE becomes near the CRLB when the SNR is sufficiently high. This indicates that the rate of convergence for the ESPRIT estimator cannot be $\mathcal{O}(m^{3/2}\sqrt{n})$ when m takes such large values, as one might conjecture based on Example 10.12, because the optimal rate achieved by the NLS estimator is only $\mathcal{O}(n^{3/2})$. The correct rate of convergence when $m \to \infty$ remains an open question.

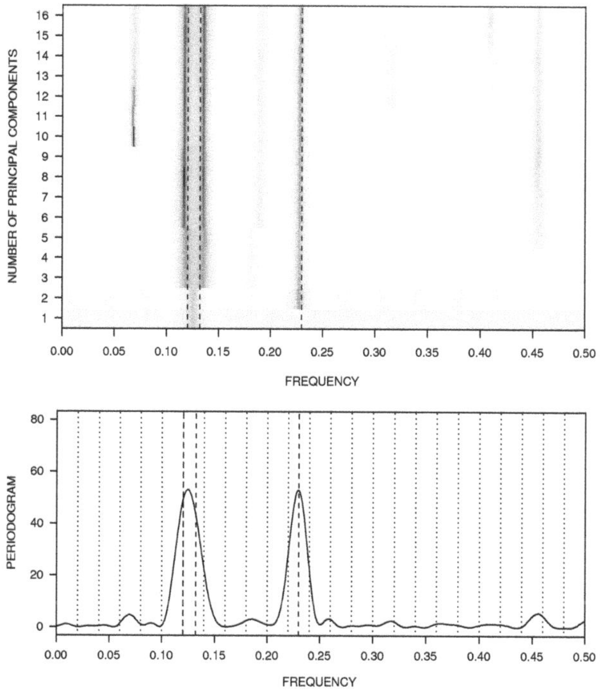

Figure 10.14. Plot of AR spectrum, as a function of the frequency and the number of principal components, produced by the reduced rank FBLP estimator with $m = 28$ for the time series used in Figure 10.1. Darker shades of grey represent larger values. Vertical dashed lines depict the signal frequencies. Bottom panel shows the periodogram, with vertical dotted line depicting the Fourier grid.

10.5 Estimating the Number of Sinusoids

The reduced rank linear prediction method discussed in Section 10.2 employs the first p principal components of an estimated covariance matrix $\hat{\mathbf{R}}_y$ to produce the AR spectrum of order m ($m > p$). It has the ability to resolve closely spaced frequencies with reduced risk of spurious spectral peaks. When the number of sinusoids is unknown, one can use the reduced rank linear prediction method to produce a sequence of AR spectra of order m by varying the number of principal components, denoted as $\tilde{p}$, from 1 up to an upper bound p_0 for the number of sinusoids, where $p_0 < m$. Plotting this sequence of AR spectra, as shown in Figure 10.14, provides a way to visualize the evolution of spectral peaks. Combining this plot with the periodogram allows us to eliminate spurious peaks

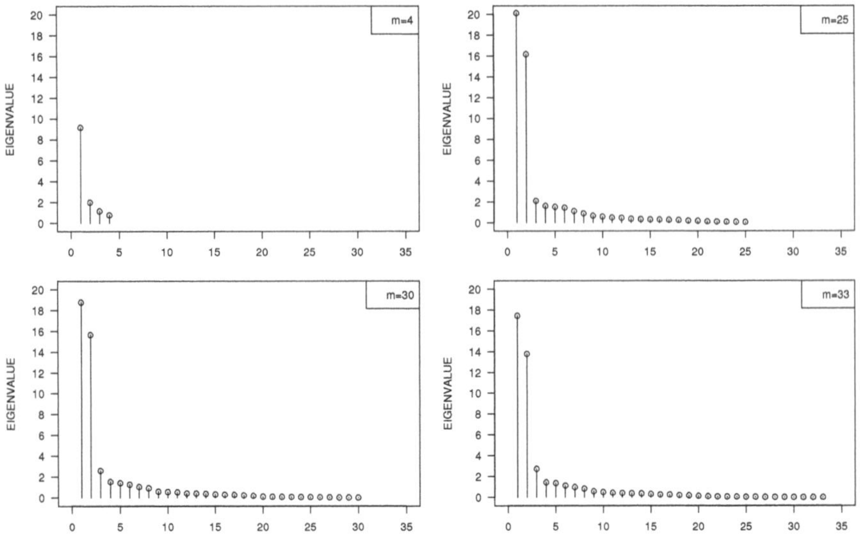

Figure 10.15. Eigenvalues of the estimated covariance matrix in (10.3.5) for the time series used in Figure 10.8 with the same choices of $m = 4, 25, 30, 33$.

and thereby determine the number of sinusoids present in the data.

As can be seen from Figure 10.14, the AR spectrum produced by the reduced rank FBLP estimator exhibits a single peak when $\tilde{p} = 1$, two peaks when $\tilde{p} = 2$, and three peaks when $\tilde{p} = 3$ (which is the correct value of p). These three peaks remain as $\tilde{p}$ further increases. They are clearly associated with the two major spectral peaks in the periodogram which can be easily attributed to the sinusoidal signal. Other peaks that appear in the AR spectrum later as $\tilde{p}$ increases, including a very strong one near frequency $2\pi \times 0.05$ which occurs when $\tilde{p} \geq 10$, can be classified as spurious, because they are associated with very small peaks in the periodogram (assuming the SNR is sufficiently large so that noise peaks tend to be much smaller than signal peaks). The same visualization technique can be applied to the SSP method for the MUSIC and MN spectra.

Besides visual inspection, the number of sinusoids can be determined by more disciplined procedures devised from the EVD. According to Theorem 10.1, if $\{y_t\}$ comprises p complex sinusoids in white noise, the smallest $m - p$ eigenvalues of the covariance matrix $\mathbf{R}_y$ will be identical and strictly less than the largest p eigenvalues. This property can be utilized to determine the parameter p from the eigenvalues of an estimated covariance matrix $\hat{\mathbf{R}}_y$.

Unlike the SSE-based criteria discussed in Chapter 5 (Section 5.4) and Chapter 8 (Section 8.5), the eigenvalue-based approach does not require joint estimation of the frequency and amplitude parameters. This has tremendous appeal in practice because good frequency estimates are difficult to obtain without knowing the value of p, especially when the signal frequencies are closely spaced.

To demonstrate the viability of this approach, Figure 10.15 depicts the eigenvalues of $\hat{\mathbf{R}}_y$ defined by (10.2.7) for the time series which is used to produce Figure 10.8 with the same choices of m. As we can see, while the smaller eigenvalues are not exactly identical, the largest three eigenvalues become significantly larger than the rest when m is large enough (e.g., $m = 30, 33$). Because the time series contains $p = 3$ complex sinusoids, an estimate of p can be obtained by detecting the presence of these large eigenvalues. This is analogous to detection of hidden periodicities in a discrete periodogram.

While visual inspection of eigenvalue diagrams such as Figure 10.15 can help identify significantly large eigenvalues, statistical procedures are needed to determine the value of p automatically and more objectively. A simple idea is to apply a periodogram-based test for hidden periodicities to the eigenvalues as if they were the discrete periodogram ordinates.

For example, by applying the multiple testing procedure (6.2.4) to the eigenvalues of $\hat{\mathbf{R}}_y$, denoted by $\hat{\lambda}_1 \ge \cdots \ge \hat{\lambda}_m > 0$, we obtain a simple estimator of p which takes the form

$$\hat{p} := \sum_{j=1}^{m} \mathscr{I}(\hat{\lambda}_j / \hat{\sigma}^2 > \tfrac{1}{2}\tau), \tag{10.5.1}$$

where $\hat{\sigma}^2$ is the bias-corrected trimmed sample mean of $\{\hat{\lambda}_j\}$ defined by

$$\hat{\sigma}^2 := b^{-1}(m - p_0)^{-1} \sum_{j=p_0+1}^{m} \hat{\lambda}_j \tag{10.5.2}$$

for a suitably large $p_0 \in [1, m)$. In this estimator, the constant b is given by

$$b := 1 + (1 - p_0/m)^{-1}(p_0/m)\log(p_0/m).$$

The constant τ in (10.5.1) is the Bonferroni-adjusted $\chi^2(2)$ threshold given by

$$\tau := -2\log(1 - (1-\alpha)^{1/m}). \tag{10.5.3}$$

Note that for estimating σ^2 in the presence of sinusoids the estimator $\hat{\sigma}^2$ in (10.5.2) is more robust than the ordinary sample mean of $\{\hat{\lambda}_1, \ldots, \hat{\lambda}_m\}$. Note also that the threshold τ in (10.5.3) is not the exact threshold for an α-level test of $H_0 : p = 0$ (no sinusoids). Because analytical expressions of the exact threshold are difficult to obtain, one may have to rely on simulation or training with noise-only data to get a better threshold.

To demonstrate the estimator of p in (10.5.1), Figure 10.16 shows the eigenvalues that are tested positive ($p_0 = 5$ and $\alpha = 0.05$) for the cases $m = 30$ and $m = 33$

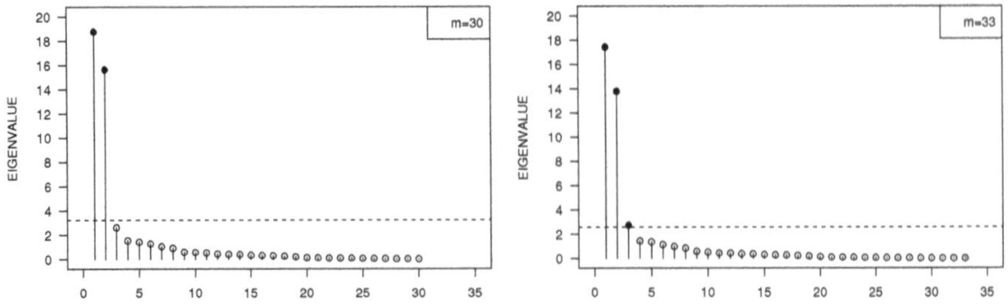

Figure 10.16. Similar to Figure 10.15 for $m = 30$ and $m = 33$, except that the dashed lines are superimposed to depict the threshold $\frac{1}{2}\tau\hat{\sigma}^2$ of the test in (10.5.1) and the solid dots are added to show the eigenvalues that are tested positive. In both cases, $\hat{\sigma}^2$ is given by (10.5.2) with $p_0 = 5$ and τ is given by (10.5.3) with $\alpha = 0.05$.

in Figure 10.15. As we can see, the value of p is correctly determined by the estimator with $m = 33$. Note that the third eigenvalue barely exceeds the threshold. This, together with the barely resolved peaks in the MUSIC spectrum with the same choice of m in Figure 10.8, indicates the potential difficulty in identifying closely spaced frequencies from the estimated covariance matrix.

The statistical performance of the estimator in (10.5.1) with $m = 33$ is demonstrated by Table 10.1 which contains the probability distribution of $\hat{p}$ calculated from 5,000 Monte Carlo runs with different SNR ($p_0 = 5$ and $\alpha = 0.05$). Recall that the time series has length $n = 50$ and consists of three unit-amplitude zero-phase complex sinusoids in Gaussian white noise with $\omega_1 = 2\pi \times 6/n = 2\pi \times 0.12$, $\omega_2 = 2\pi \times 6.6/n = 2\pi \times 0.132$, and $\omega_3 = 2\pi \times 11.5/n = 2\pi \times 0.23$. As we can see from Table 10.1, the distribution of $\hat{p}$ becomes more concentrated around the value $p = 3$ as the SNR increases, although significant chances of underestimation are evident when the SNR is low. Observe that the actual false alarm probability (i.e., the probability of $\hat{p} > 0$ under H_0) is equal to 0.137, which is higher than $\alpha = 0.05$. This means that the threshold given by (10.5.3) is lower than the exact one.

Instead of detecting large eigenvalues, an alternative method of estimating p is to test for equality of the smallest $m - \tilde{p}$ eigenvalues of $\mathbf{R}_y$ for $\tilde{p} = 0, 1, \ldots, p_0 - 1$, where p_0 is a known upper bound of p satisfying $p \le p_0 < m$. A test statistic considered by Lawley [215] takes the form

$$L_m(\tilde{p}) := -\sum_{j=\tilde{p}+1}^{m} \log\hat{\lambda}_j + (m - \tilde{p})\log\hat{\sigma}^2_{\tilde{p}}, \tag{10.5.4}$$

where

$$\hat{\sigma}^2_{\tilde{p}} := (m - \tilde{p})^{-1} \sum_{j=\tilde{p}+1}^{m} \hat{\lambda}_j. \tag{10.5.5}$$

Table 10.1. Distribution of Estimated Number of Complex Sinusoids in Gaussian White Noise by (10.5.1) ($m = 33$; $p_0 = 5$; $\alpha = 0.05$)

		SNR per Sinusoid (dB)						
$\hat{p}$	H_0	−6	−4	−2	0	2	4	6
0	**0.863**	0.033	0.001	0.000	0.000	0.000	0.000	0.000
1	0.127	0.297	0.048	0.000	0.000	0.000	0.000	0.000
2	0.010	**0.628**	**0.799**	**0.522**	0.129	0.005	0.000	0.000
3	0.000	0.042	0.148	0.467	**0.852**	**0.973**	**0.977**	**0.977**
4	0.000	0.000	0.004	0.011	0.019	0.022	0.023	0.023
5	0.000	0.000	0.000	0.000	0.000	0.000	0.000	0.000

Results are based on 5,000 Monte Carlo runs.
Bold font shows where the largest value of each distribution occurs.

Note that $\hat{\sigma}^2_{\tilde{p}}$ in (10.5.5) is an estimator of the common value of the smallest $m-\tilde{p}$ eigenvalues of $\mathbf{R}_y$ under the hypothesis

$$H_{\tilde{p}} : \lambda_{\tilde{p}+1} = \cdots = \lambda_m. \tag{10.5.6}$$

With $\tau_{\tilde{p}}$ denoting a suitable threshold, the parameter p can be estimated by

$$\hat{p} := \max\{\tilde{p} : L_m(\tilde{p}) > \tau_{\tilde{p}}, 0 \le \tilde{p} < p_0\} + 1. \tag{10.5.7}$$

Obviously, let $\hat{p} := 0$ if $L_m(\tilde{p}) \le \tau_{\tilde{p}}$ for all $\tilde{p} \in [0, p_0)$.

To better understand Lawley's statistic $L_m(\tilde{p})$, let us define

$$\hat{\lambda}_{j0} := \begin{cases} \hat{\lambda}_j & \text{for } j = 1, \ldots, \tilde{p}, \\ \hat{\sigma}^2_{\tilde{p}} & \text{for } j = \tilde{p}+1, \ldots, m. \end{cases}$$

Then, it is easy to verify that

$$L_m(\tilde{p}) = \sum_{j=1}^{m} d_{\mathrm{KL}}(\hat{\lambda}_j / \hat{\lambda}_{j0}),$$

where $d_{\mathrm{KL}}(x) := x - \log x - 1$ $(x > 0)$ is the Kullback-Leibler kernel. In other words, Lawley's statistic is just the Kullback-Leibler divergence of $\{\hat{\lambda}_j\}$ from $\{\hat{\lambda}_{j0}\}$. Observe that $\{\hat{\lambda}_{j0}\}$ is the maximum likelihood estimator of $\{\lambda_j\}$ on the basis of $\{\hat{\lambda}_j\}$ under $H_{\tilde{p}}$ and the assumption that the $\hat{\lambda}_j$ are independently distributed with

$$\hat{\lambda}_j \sim \tfrac{1}{2}\lambda_j \chi^2(2) \qquad (j = 1, \ldots, m). \tag{10.5.8}$$

Under the same assumption without the constraint $H_{\tilde{p}}$, the maximum likelihood estimator of $\{\lambda_j\}$ is simply $\{\hat{\lambda}_j\}$. Therefore, it is easy to verify that $L_m(\tilde{p})$ coincides with the negative generalized log likelihood ratio statistic for testing $H_{\tilde{p}}$ on the basis of $\{\hat{\lambda}_j\}$ under the assumption (10.5.8). Observe that the number of free

Table 10.2. Distribution of Estimated Number of Complex Sinusoids in Gaussian White Noise by (10.5.7) ($m = 33; p_0 = 5; \alpha = 0.05$)

	SNR per Sinusoid (dB)							
$\hat{p}$	H_0	−6	−4	−2	0	2	4	6
0	**0.862**	0.294	0.071	0.003	0.000	0.000	0.000	0.000
1	0.057	**0.354**	0.317	0.112	0.006	0.000	0.000	0.000
2	0.024	0.224	**0.430**	**0.601**	**0.510**	0.247	0.047	0.001
3	0.016	0.053	0.090	0.175	0.370	**0.637**	**0.837**	**0.884**
4	0.010	0.026	0.035	0.046	0.050	0.052	0.052	0.051
5	0.031	0.049	0.057	0.063	0.064	0.064	0.064	0.064

Results are based on 5,000 Monte Carlo runs.
Bold font shows where the largest value of each distribution occurs.

parameters under $H_{\tilde{p}}$ is equal to $\tilde{p}+1$ and the number of free parameters without the constraint is equal to m. By following the common practice for likelihood ratio statistics, one may take $\tau_{\tilde{p}}$ simply as one half of the $(1-\alpha/p_0)$th quantile of the chi-square distribution with $m-\tilde{p}-1$ degrees of freedom. This threshold is different from Lawley's in [215] where the test is based on the sample covariance matrix of i.i.d. m-variate observations rather than the sample covariance matrix of time series data. As with the tests in (10.5.1), the chi-square threshold is by no means exact for the tests in (10.5.7). To obtain a better threshold in practice, one may have to rely on simulation or training with noise-only data.

As an example, Table 10.2 shows the probability distribution for the estimator of p in (10.5.7) based on the same data as used in Table 10.1. To facilitate a fair comparison between these results, the threshold $\tau_{\tilde{p}}$ in (10.5.7) is replaced by $1.55\tau_{\tilde{p}}$, where the multiplier 1.55 is empirically determined by trial and error to make the false alarm probability approximately equal to that in Table 10.1. Similar to the result in Table 10.1, the distribution shown in Table 10.2 also becomes more concentrated around $p=3$ as the SNR increases. However, the error probability of the estimator in (10.5.7) is slightly higher for each SNR considered. It also has a higher tendency to overestimate p.

For testing the hypothesis (10.5.6), a statistic of quadratic form

$$\boldsymbol{\xi}_{\tilde{p}}^H \mathbf{Q}_{\tilde{p}}^{-1} \boldsymbol{\xi}_{\tilde{p}}$$

is considered in [119]. The vector $\boldsymbol{\xi}_{\tilde{p}} := [\hat{\lambda}_{\tilde{p}+1} - \hat{\sigma}_{\tilde{p}}^2, \ldots, \hat{\lambda}_m - \hat{\sigma}_{\tilde{p}}^2]^T$ comprises the deviations of the $m-\tilde{p}$ smallest eigenvalues from their average value $\hat{\sigma}_{\tilde{p}}^2$. The matrix $\mathbf{Q}_{\tilde{p}}$ is an estimate of the covariance matrix of $\boldsymbol{\xi}_{\tilde{p}}$ obtained from a perturbation analysis. This statistic is compared with a chi-square threshold in [119].

In addition to the hypothesis testing approach, one can also estimate the parameter p by minimizing a criterion similar to the GIC and the PEC discussed in Chapter 5 (Section 5.4). The basic idea is to replace the residual-based SSE with an eigenvalue-based alternative. Recall that n^{-1} times the SSE of the best $\tilde{p}$-term regression model discussed in Section 5.4 constitutes an estimate of σ^2 after removing the $\tilde{p}$ most significant frequencies. A similar estimator of σ^2 based on the eigenvalues is given by $\hat{\sigma}_{\tilde{p}}^2$ in (10.5.5). With the SSE replaced by $n\hat{\sigma}_{\tilde{p}}^2$, the GIC criterion in (5.4.1) becomes (by ignoring an additive constant)

$$\mathrm{GIC}(\tilde{p}) := 2n\log\hat{\sigma}_{\tilde{p}}^2 + c(n,\tilde{p}), \tag{10.5.9}$$

where $c(n,\tilde{p})$ is the penalty function. A criterion of this type is employed in [308] with $c(n,\tilde{p})$ taking the form $c_n\tilde{p}$ and with $m := \tilde{p}+1$ (so that $\hat{\sigma}_{\tilde{p}}^2$ is equal to the smallest eigenvalue of $\hat{\mathbf{R}}_y$). The choice $c(n,\tilde{p}) = 2\tilde{p}$ gives the AIC criterion [8] and the choice $c(n,\tilde{p}) = \tilde{p}\log n$ gives the BIC criterion [318] as well as the MDL criterion [337]. The AICC criterion [161] takes $c(n,\tilde{p}) = 2n(n+\tilde{p})/(n-\tilde{p}-1)$. Alternative GIC-type criteria are considered in [440], [441], and [419].

Similarly, the PEC criterion in (5.4.5) can also be modified to obtain

$$\mathrm{PEC}(\tilde{p}) := n\hat{\sigma}_{\tilde{p}}^2 + c(n,\tilde{p})\hat{\sigma}^2, \tag{10.5.10}$$

where $\hat{\sigma}^2 := (m-p_0)^{-1}\sum_{j=p_0+1}^{m}\hat{\lambda}_j$ is a baseline estimate of σ^2, with p_0 being a known upper bound of p. One may also absorb $\hat{\sigma}^2$ in $c(n,\tilde{p})$ and define

$$\mathrm{PEC}(\tilde{p}) := n\hat{\sigma}_{\tilde{p}}^2 + c(n,\tilde{p}).$$

Criteria of this form are employed in [19] with $m := \tilde{p}+1$ (so that $\hat{\sigma}_{\tilde{p}}^2$ is the smallest eigenvalue of $\hat{\mathbf{R}}_y$) and in [206] with $\hat{\sigma}_{\tilde{p}}^2 := \hat{\lambda}_{\tilde{p}+1}$. An example for the choice of $c(n,\tilde{p})$ is $\tilde{p}\log n$, as in the BIC and the MDL.

Table 10.3 demonstrates the statistical performance of these criteria for the same data as used in Tables 10.1 and 10.2. To compare them on an equal footing, the penalty terms of the AIC, AICC, and PEC with $c(n,\tilde{p}) = \tilde{p}\log n$ are all multiplied by suitable constants, which are selected empirically, such that the probability of false alarm (i.e., the probability of $\hat{p} > 0$ under H_0) is roughly the same for all methods. Because the penalty terms are adjusted in this way, the BIC/MDL criterion produces the same results as the AIC criterion.

As we can see, all three criteria tend to underestimate p when the SNR is low, but the performance improves dramatically when the SNR increases at the expense of slightly increased chance of overestimation. The same behavior is observed in Tables 10.1 and 10.2. The PEC in particular has the smallest probability of error $P(\hat{p} \neq p)$ when the SNR is greater than or equal to 0 dB. The AIC and the AICC perform better than the PEC at lower SNR, but their chance of overestimation at higher SNR is much greater, especially for the AIC. By comparing the results in Table 10.3 with those in Tables 10.1 and 10.2, we can see that the multiple testing procedure (10.5.1) performs similarly to the PEC and the multiple testing procedure (10.5.7) performs similarly to the AICC.

Table 10.3. Distribution of Estimated Number of Complex Sinusoids in Gaussian White Noise by AIC, AICC, and PEC ($m = 33$; $p_0 = 5$)

	$\hat{p}$	SNR per Sinusoid (dB)							
		H_0	−6	−4	−2	0	2	4	6
AIC	0	**0.861**	0.037	0.001	0.000	0.000	0.000	0.000	0.000
	1	0.114	0.193	0.025	0.000	0.000	0.000	0.000	0.000
	2	0.019	**0.584**	**0.575**	0.260	0.032	0.001	0.000	0.000
	3	0.004	0.142	0.302	**0.565**	**0.747**	**0.773**	**0.774**	**0.774**
	4	0.001	0.034	0.073	0.132	0.170	0.174	0.174	0.174
	5	0.001	0.010	0.024	0.043	0.051	0.052	0.052	0.052
AICC	0	**0.861**	0.038	0.001	0.000	0.000	0.000	0.000	0.000
	1	0.120	0.224	0.032	0.000	0.000	0.000	0.000	0.000
	2	0.017	**0.622**	**0.673**	0.347	0.056	0.002	0.000	0.000
	3	0.002	0.100	0.259	**0.576**	**0.837**	**0.884**	**0.885**	**0.885**
	4	0.000	0.014	0.032	0.068	0.096	0.102	0.103	0.103
	5	0.000	0.002	0.003	0.009	0.011	0.012	0.012	0.012
PEC	0	**0.861**	0.032	0.001	0.000	0.000	0.000	0.000	0.000
	1	0.123	0.257	0.037	0.000	0.000	0.000	0.000	0.000
	2	0.014	**0.637**	**0.747**	0.430	0.087	0.003	0.000	0.000
	3	0.002	0.070	0.203	**0.541**	**0.869**	**0.949**	**0.951**	**0.951**
	4	0.000	0.004	0.012	0.028	0.043	0.047	0.048	0.048
	5	0.000	0.000	0.000	0.001	0.001	0.001	0.001	0.001

Results are based on 5,000 Monte Carlo runs.
Bold font shows where the largest value of each distribution occurs.

10.6 Sensitivity to Colored Noise

The covariance analysis approach discussed in this chapter depends crucially on the assumption of white noise. It should not be surprising that the spectrum-based methods, such as the reduced rank FBLP estimator and the MUSIC estimator, may suffer from a performance degradation in the presence of colored noise because the spectral peaks of colored noise are able to mask the spectral peaks of the sinusoidal signal when the SNR is not sufficiently high. It turns out that the reduced rank EYW estimator and the ESPRIT estimator are affected by colored noise in a similar way.

As an example, Figure 10.17 shows the performance of four estimators for estimating the well-separated frequencies of the time series discussed in Examples 10.1, 10.3, 10.7, and 10.10, except that the noise is now a complex Gaussian AR(2) process of the form (8.4.5). The same data are used in Example 8.5. By comparing Figure 10.17 with Figures 10.2(a), 10.6(a), 10.9(a), and 10.12(a), we can see that the colored noise degrades the performance by raising the SNR

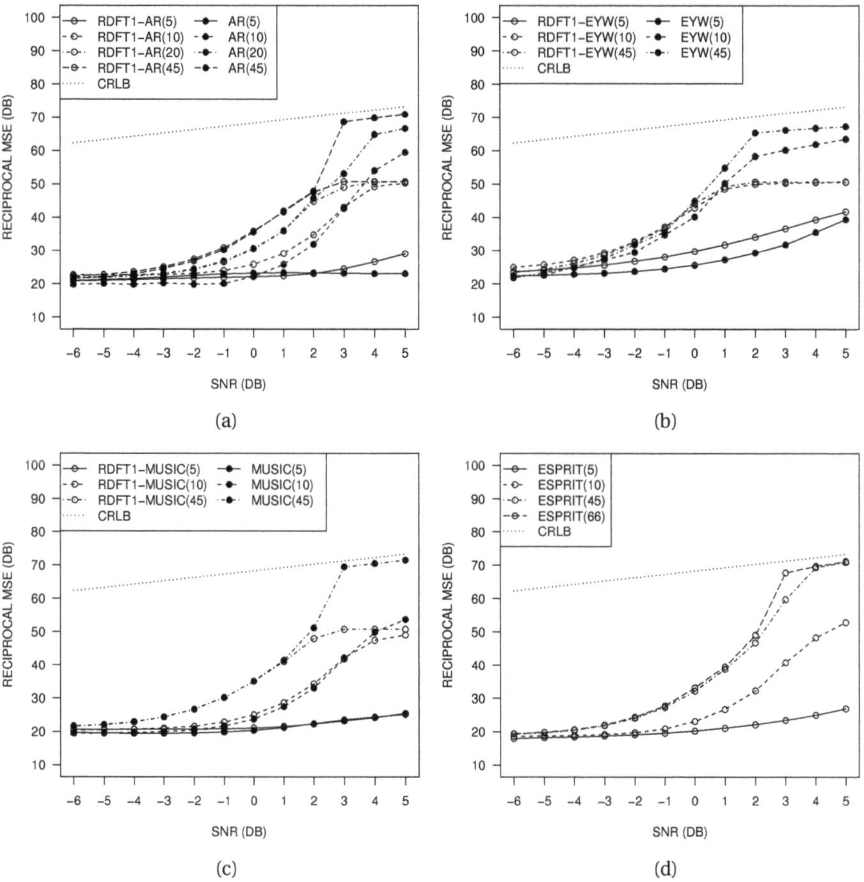

Figure 10.17. Plot of the reciprocal average MSE of frequency estimates for the time series of two complex sinusoids in Gaussian AR(2) noise discussed in Example 8.5 with $n = 100$. (a) reduced rank FBLP: solid line, $m = 5$; dashed line, $m = 10$; dash-dotted line, $m = 20$; long dashed line, $m = 45$. (b) reduced rank EYW: solid line, $m = l = 5$; dashed line, $m = l = 10$; dash-dotted line, $m = l = 45$. (c) MUSIC: solid line, $m = 5$; dashed line, $m = 10$; dash-dotted line, $m = 45$. (d) ESPRIT: solid line, $m = 5$; dashed line, $m = 10$; dash-dotted line, $m = 45$; long dashed line, $m = 66$. Dotted line depicts the CRLB. Results are based on 5,000 Monte Carlo runs.

threshold of these estimators. A similar impact is observed for the NLS estimator shown in Figure 8.13(b). Not surprisingly, the reduced rank FBLP estimator and the MUSIC estimator, with sufficiently large m, manage to retain their accuracy when the SNR is above the threshold, as does the NLS estimator. But the same cannot be said about the reduced rank EYW estimator and the ESPRIT estimator — their performance is noticeably worsened by the colored noise even at the high end of the SNR range considered. It should be noted that the eigenvalue-based ESPRIT estimator does not require initial values, so the deterioration of its performance cannot be simply explained by the masking effect of the noise in identifying the signal spectral peaks.

When the noise covariance matrix $\mathbf{R}_\epsilon$ is known or estimated *a priori* (up to an unknown constant), it can be used to standardize the covariance matrices of $\{y_t\}$ and convert them into suitable forms so that the methods developed under the assumption of white noise can be applied. Indeed, let $\mathbf{R}_\epsilon = r_\epsilon(0)\bar{\mathbf{R}}_\epsilon$, where $\bar{\mathbf{R}}_\epsilon$ is the known correlation matrix of the noise and $r_\epsilon(0)$ is the unknown variance of the noise. Based on the EVD technique, $\bar{\mathbf{R}}_\epsilon$ can be decomposed as

$$\bar{\mathbf{R}}_\epsilon = \mathbf{U}_\epsilon \mathbf{U}_\epsilon^H.$$

Using the inverse matrix $\mathbf{U}_\epsilon^{-1}$ to standardize $\mathbf{R}_y = \mathbf{R}_x + \mathbf{R}_\epsilon$, for example, gives

$$\tilde{\mathbf{R}}_y := \mathbf{U}_\epsilon^{-1}\mathbf{R}_y\mathbf{U}_\epsilon^{-H} = \tilde{\mathbf{R}}_x + r_\epsilon(0)\mathbf{I},$$

where $\tilde{\mathbf{R}}_x := \mathbf{U}_\epsilon^{-1}\mathbf{R}_x\mathbf{U}_\epsilon^{-H}$. Because $\mathbf{R}_x$ takes the form (10.1.7), it follows that

$$\tilde{\mathbf{R}}_x = \mathbf{U}_\epsilon^{-1}\mathbf{F}_m\mathbf{G}\mathbf{F}_m^H\mathbf{U}_\epsilon^{-H}.$$

This is a covariance matrix of rank p whose null space coincides with the column space of $\mathbf{U}_\epsilon^{-1}\mathbf{F}_m$. Therefore, the eigenvalue theory developed in Section 10.1 carries over to the standardized covariance matrix $\tilde{\mathbf{R}}_y$ by replacing the sinusoidal vectors $\mathbf{f}_m(\omega_k)$ with $\mathbf{U}_\epsilon^{-1}\mathbf{f}_m(\omega_k)$ $(k = 1, \ldots, p)$.

As a result, the MUSIC estimator based on $\tilde{\mathbf{R}}_y$, which may be called the *generalized* MUSIC estimator, minimizes the objective function

$$\sum_{j=p+1}^{m} |\hat{\mathbf{v}}_j^H \mathbf{U}_\epsilon^{-1}\mathbf{f}_m(\omega)|^2$$

instead of (10.3.3), where $\hat{\mathbf{v}}_{p+1}, \ldots, \hat{\mathbf{v}}_m$ are the eigenvectors of a consistent estimator of $\tilde{\mathbf{R}}_y$ that correspond to the p largest eigenvalues. For the linear prediction method, because the normal equation (10.2.1) is equivalent to

$$\mathbf{U}_\epsilon^{-1}\hat{\mathbf{R}}_y\mathbf{U}_\epsilon^{-H}\mathbf{U}_\epsilon^H\hat{\mathbf{c}} = -\mathbf{U}_\epsilon^{-1}\hat{\mathbf{r}}_y,$$

the PCA technique can be applied to $\mathbf{U}_\epsilon^{-1}\hat{\mathbf{R}}_y\mathbf{U}_\epsilon^{-H}$. The resulting estimator may be called the *generalized* reduced rank linear prediction estimator. Moreover, observe that $\bar{\mathbf{R}}_x$ and $\bar{\mathbf{R}}'_x := \mathbf{U}_\epsilon^{-1}\mathbf{R}'_x\mathbf{U}_\epsilon^{-H}$ satisfy (10.4.4) with $\mathbf{U}_\epsilon^{-1}\mathbf{F}_m$ in pace of $\mathbf{F}_m$. Therefore, a *generalized* subspace rotation estimator can be constructed from the standardized covariance matrices $\mathbf{U}_\epsilon^{-1}\hat{\mathbf{R}}_y\mathbf{U}_\epsilon^{-H}$ and $\mathbf{U}_\epsilon^{-1}\hat{\mathbf{R}}'_y\mathbf{U}_\epsilon^{-H}$.

10.7 Proof of Theorems

This section contains the proof of Theorems 10.1, 10.3, 10.4, 10.6 10.7, 10.9, and 10.11–10.13.

Proof of Theorem 10.1. Because $\mathbf{R}_y = \mathbf{V\Lambda V}^H$ and $\mathbf{VV}^H = \mathbf{I}$, it follows that

$$\mathbf{R}_x = \mathbf{R}_y - \sigma^2\mathbf{I} = \mathbf{V\Lambda}_x\mathbf{V}^H,$$

where $\mathbf{\Lambda}_x := \mathbf{\Lambda} - \sigma^2\mathbf{I} = \mathrm{diag}(\lambda_1 - \sigma^2, \ldots, \lambda_m - \sigma^2)$. This means that $\mu_j := \lambda_j - \sigma^2$ ($j = 1, \ldots, m$) are the eigenvalues of $\mathbf{R}_x$ with associated eigenvectors $\mathbf{v}_j$ ($j = 1, \ldots, m$). Because $\mathbf{R}_x$ is nonnegative definite, we have $\mu_j \geq 0$ for all j. Because $\mathbf{R}_x$ has rank p, we have $\mu_1 \geq \cdots \geq \mu_p > 0$ and $\mu_{p+1} = \cdots = \mu_m = 0$. This implies that $\lambda_j = \mu_j + \sigma^2 > \sigma^2$ for $j = 1, \ldots, p$ and $\lambda_j = \sigma^2$ for $j = p+1, \ldots, m$. $\square$

Proof of Theorem 10.3. With $\hat{\mathbf{c}}_{\mathrm{PCA}}$ given by (10.2.4), we can write

$$\begin{aligned}
\hat{\mathbf{c}}_{\mathrm{PCA}} - \mathbf{c} &= -\tilde{\mathbf{R}}_y^\dagger(\hat{\mathbf{r}}_y + \hat{\mathbf{R}}_y\mathbf{c} - \hat{\mathbf{R}}_y\mathbf{c}) - \mathbf{c} \\
&= -\tilde{\mathbf{R}}_y^\dagger(\hat{\mathbf{r}}_y + \hat{\mathbf{R}}_y\mathbf{c}) - (\mathbf{I} - \tilde{\mathbf{R}}_y^\dagger\hat{\mathbf{R}}_y)\mathbf{c} \\
&= -\tilde{\mathbf{R}}_y^\dagger(\hat{\mathbf{r}}_y + \hat{\mathbf{R}}_y\mathbf{c}) - \hat{\mathbf{P}}_\mathrm{N}\mathbf{c},
\end{aligned} \tag{10.7.1}$$

where the last equality is due to the fact that $\tilde{\mathbf{R}}_y^\dagger\hat{\mathbf{R}}_y = \hat{\mathbf{V}}_\mathrm{S}\hat{\mathbf{V}}_\mathrm{S}^H = \hat{\mathbf{P}}_\mathrm{S}$ and $\hat{\mathbf{P}}_\mathrm{N} = \mathbf{I} - \hat{\mathbf{P}}_\mathrm{S}$. Because $\mathbf{c} := [c_1, \ldots, c_m]^T$ is given by (10.2.22), we have $\mathbf{r}_y + \mathbf{R}_y\mathbf{c} = \mathbf{0}$. For any $\hat{\mathbf{R}}_y$ and $\hat{\mathbf{r}}_y$ given by the linear prediction estimators discussed in Chapter 9 (Section 9.1), we can write $\hat{\mathbf{R}}_y = [\hat{r}_y(u-v)] + \mathcal{O}_P(n^{-1/2})$ and $\hat{\mathbf{r}}_y = \mathrm{vec}[\hat{r}_y(u)] + \mathcal{O}_P(n^{-1/2})$, where the $\hat{r}_y(u)$ are the sample autocovariances defined by (4.1.1). Therefore, as in the proof of Theorem 7.5, it can be shown that

$$\hat{\mathbf{r}}_y + \hat{\mathbf{R}}_y\mathbf{c} = \hat{\mathbf{r}}_y + \hat{\mathbf{R}}_y\mathbf{c} - (\mathbf{r}_y + \mathbf{R}_y\mathbf{c}) = \mathbf{C}(\hat{\mathbf{r}} - \mathbf{r}) + \mathcal{O}_P(n^{-1/2}), \tag{10.7.2}$$

where $\hat{\mathbf{r}} := [\hat{r}_y(-m+1), \ldots, \hat{r}_y(m)]^T$, $\mathbf{r} := [r_y(-m+1), \ldots, r_y(m)]^T$, and

$$\mathbf{C} := \begin{bmatrix} c_m & c_{m-1} & \cdots & c_0 & 0 & \cdots & 0 \\ 0 & c_m & \cdots & c_1 & c_0 & & 0 \\ & & \ddots & & & \ddots & \\ 0 & \cdots & 0 & c_m & \cdots & c_1 & c_0 \end{bmatrix} \quad (c_0 := 1).$$

Moreover, because $\mathbf{R}_y - \sigma^2\mathbf{I} = \mathbf{R}_x$, an application of Lemma 12.4.8, together with Theorem 10.1, gives

$$\hat{\mathbf{P}}_{\rm N} - \mathbf{P}_{\rm N} = -\mathbf{P}_{\rm N}(\hat{\mathbf{R}}_y - \mathbf{R}_y)\mathbf{R}_x^{\dagger} - \mathbf{R}_x^{\dagger}(\hat{\mathbf{R}}_y - \mathbf{R}_y)\mathbf{P}_{\rm N} + \mathcal{O}(\|\hat{\mathbf{R}}_y - \mathbf{R}_y\|^2). \qquad (10.7.3)$$

Let $\mathbf{b} := [b_1, \dots, b_m]^T := -\mathbf{R}_x^{\dagger}\mathbf{c}$ and $b_0 := 0$. Then, similar (10.7.2), we can write

$$(\hat{\mathbf{R}}_y - \mathbf{R}_y)\mathbf{b} = \mathbf{B}(\hat{\mathbf{r}} - \mathbf{r}) + \mathcal{O}_P(n^{-1/2}),$$

where $\mathbf{B}$ is defined by $\mathbf{b}$ in the same way as $\mathbf{C}$ is defined by $\mathbf{c}$. Moreover, by definition (10.2.24), we have $\mathbf{P}_{\rm N}\mathbf{c} = \mathbf{0}$. Therefore, it follows from (10.7.3) that

$$\begin{aligned} \hat{\mathbf{P}}_{\rm N}\mathbf{c} &= (\hat{\mathbf{P}}_{\rm N} - \mathbf{P}_{\rm N})\mathbf{c} \\ &= \mathbf{P}_{\rm N}(\hat{\mathbf{R}}_y - \mathbf{R}_y)\mathbf{b} + \mathcal{O}(\|\hat{\mathbf{R}}_y - \mathbf{R}_y\|^2) \\ &= \mathbf{P}_{\rm N}\mathbf{B}(\hat{\mathbf{r}} - \mathbf{r}) + \mathcal{O}(\|\hat{\mathbf{R}}_y - \mathbf{R}_y\|^2) + \mathcal{O}_P(n^{-1/2}). \end{aligned}$$

Inserting this expression and (10.7.2) in (10.7.1) yields

$$\hat{\mathbf{c}}_{\rm PCA} - \mathbf{c} = (\tilde{\mathbf{R}}_y^{\dagger}\mathbf{C} + \mathbf{P}_{\rm N}\mathbf{B})(\hat{\mathbf{r}} - \mathbf{r}) + \mathcal{O}(\|\hat{\mathbf{R}}_y - \mathbf{R}_y\|^2) + \mathcal{O}_P(n^{-1/2}). \qquad (10.7.4)$$

According to Theorem 4.4, $\sqrt{n}(\hat{\mathbf{r}} - \mathbf{r}) \xrightarrow{D} \mathrm{N}_c(\mathbf{0}, \mathbf{\Gamma}, \tilde{\mathbf{\Gamma}})$, where

$$\mathbf{\Gamma} := \lim_{n\to\infty} \mathrm{Cov}(\hat{\mathbf{r}}, \hat{\mathbf{r}}), \quad \tilde{\mathbf{\Gamma}} := \lim_{n\to\infty} \mathrm{Cov}(\hat{\mathbf{r}}, \hat{\mathbf{r}}^*).$$

Moreover, according to Lemma 12.4.6, $\tilde{\mathbf{R}}_y^{\dagger} \xrightarrow{P} \bar{\mathbf{R}}_y^{\dagger}$. Therefore, by Proposition 2.6(b) and (10.7.4), we obtain

$$\sqrt{n}(\hat{\mathbf{c}}_{\rm PCA} - \mathbf{c}) = \sqrt{n}(\bar{\mathbf{R}}_y^{\dagger}\mathbf{C} + \mathbf{P}_{\rm N}\mathbf{B})(\hat{\mathbf{r}} - \mathbf{r}) + \mathcal{O}_P(1) \xrightarrow{D} \mathrm{N}_c(\mathbf{0}, \mathbf{\Sigma}_0, \tilde{\mathbf{\Sigma}}_0),$$

where

$$\begin{aligned} \mathbf{\Sigma}_0 &:= (\bar{\mathbf{R}}_y^{\dagger}\mathbf{C} + \mathbf{P}_{\rm N}\mathbf{B})\mathbf{\Gamma}(\bar{\mathbf{R}}_y^{\dagger}\mathbf{C} + \mathbf{P}_{\rm N}\mathbf{B})^H, \qquad (10.7.5) \\ \tilde{\mathbf{\Sigma}}_0 &:= (\bar{\mathbf{R}}_y^{\dagger}\mathbf{C} + \mathbf{P}_{\rm N}\mathbf{B})\tilde{\mathbf{\Gamma}}(\bar{\mathbf{R}}_y^{\dagger}\mathbf{C} + \mathbf{P}_{\rm N}\mathbf{B})^T. \qquad (10.7.6) \end{aligned}$$

It remains to derive more explicit expressions for $\mathbf{\Sigma}_0$ and $\tilde{\mathbf{\Sigma}}_0$.

As in the proof of Theorem 7.5, we can write $\mathbf{C\Gamma C}^H = [\sigma_{11}(u, v)]$, where

$$\sigma_{11}(u, v) := \sum_{j=0}^{m}\sum_{l=0}^{m} c_j c_l^* \gamma(u-j, v-l) \qquad (u, v = 1, \dots, m),$$

with $\gamma(\cdot,\cdot)$ defined by (4.3.3) and (4.3.4). Under the white noise assumption,

$$\gamma(u, v) = \sum_{k=1}^{p} 2C_k^2\sigma^2 z_k^{u-v} + \sigma^4\delta_{u-v} + (\kappa - 2)\delta_u\delta_v \qquad (u, v \in \mathbb{Z}). \qquad (10.7.7)$$

Therefore,

$$\sigma_{11}(u,v)=\sum_{k=1}^{p}2C_k^2\sigma^2 c(z_k)c^*(z_k)z_k^{u-v}+\sigma^4 c_{11}(u,v)+\sigma^4(\kappa-2)c_u c_v^*,$$

where $c(z):=\sum_{j=0}^m c_j z^{-j}$ and $c_{11}(u,v):=\sum_{j=0}^m c_{u-v+j}c_j^*$. Let $\mathbf{C}_{11}:=[c_{11}(u,v)]$. Then, we can write

$$\mathbf{C}\Gamma\mathbf{C}^H=\sum_{k=1}^{p}2C_k^2\sigma^2|c(z_k)|^2\mathbf{f}_m(\omega_k)\mathbf{f}_m^H(\omega_k)+\sigma^4\mathbf{C}_{11}+\sigma^4(\kappa-2)\mathbf{c}\mathbf{c}^H=\mathbf{\Sigma}_{11}.$$

Similarly, with $b(z):=\sum_{j=0}^m b_j z^{-j}$ and $c_{12}(u,v):=\sum_{j=0}^m c_{u-v+j}b_j^*$, we can write

$$\mathbf{C}\Gamma\mathbf{B}^H=\sum_{k=1}^{p}2C_k^2\sigma^2 c(z_k)b^*(z_k)\mathbf{f}_m(\omega_k)\mathbf{f}_m^H(\omega_k)+\sigma^4\mathbf{C}_{12}+\sigma^4(\kappa-2)\mathbf{c}\mathbf{b}^H,$$

where $\mathbf{C}_{12}:=[c_{12}(u,v)]$. By combining this result with $\mathbf{P}_{\mathrm{N}}\mathbf{f}_m(\omega_k)=\mathbf{0}$ and $\mathbf{P}_{\mathrm{N}}\mathbf{b}=-\mathbf{P}_{\mathrm{N}}\mathbf{R}_x^{\dagger}\mathbf{c}=\mathbf{0}$, we obtain

$$\mathbf{C}\Gamma\mathbf{B}^H\mathbf{P}_{\mathrm{N}}=\sigma^4\mathbf{C}_{12}\mathbf{P}_{\mathrm{N}}=\mathbf{\Sigma}_{12}\mathbf{P}_{\mathrm{N}}.$$

A similar argument leads to

$$\mathbf{B}\Gamma\mathbf{B}^H\mathbf{P}_{\mathrm{N}}=\sigma^4\mathbf{C}_{22}\mathbf{P}_{\mathrm{N}}=\mathbf{\Sigma}_{22}\mathbf{P}_{\mathrm{N}},$$

where $\mathbf{C}_{22}:=[c_{22}(u,v)]$ and $c_{22}(u,v):=\sum_{j=0}^m b_{u-v+j}b_j^*$. Combining these expressions with (10.7.5) yields

$$\mathbf{\Sigma}_0=[\tilde{\mathbf{R}}_y^{\dagger},\mathbf{P}_{\mathrm{N}}]\begin{bmatrix}\mathbf{\Sigma}_{11} & \mathbf{\Sigma}_{12}\\ \mathbf{\Sigma}_{12}^H & \mathbf{\Sigma}_{22}\end{bmatrix}[\tilde{\mathbf{R}}_y^{\dagger},\mathbf{P}_{\mathrm{N}}]^H.$$

A similar expression can be derived for $\tilde{\mathbf{\Sigma}}_0$. Indeed, by following the proof of Theorem 7.5 together with (10.7.7), we can write $\mathbf{C}\tilde{\Gamma}\mathbf{C}^T=[\tilde{\sigma}_{11}(u,v)]$, where

$$\begin{aligned}\tilde{\sigma}_{11}(u,v)&:=\sum_{j=0}^{m}\sum_{l=0}^{m}c_j c_l\gamma(u-j,-v+l)\\&=\sum_{k=1}^{p}2C_k^2\sigma^2 c^2(z_k)z_k^{u+v}+\sigma^4\tilde{c}_{11}(u,v)+\sigma^4(\kappa-2)c_u c_v\end{aligned}$$

and $\tilde{c}_{11}(u,v):=\sum_{j=0}^m c_{u+v-j}c_j$. Define $\tilde{\mathbf{C}}_{11}:=[\tilde{c}_{11}(u,v)]$. Then,

$$\mathbf{C}\tilde{\Gamma}\mathbf{C}^T=\sum_{k=1}^{p}2C_k^2\sigma^2 c^2(z_k)\mathbf{f}_m(\omega_k)\mathbf{f}_m^T(\omega_k)+\sigma^4\tilde{\mathbf{C}}_{11}+\sigma^4(\kappa-2)\mathbf{c}\mathbf{c}^T=\tilde{\mathbf{\Sigma}}_{11}.$$

Similarly, we can write

$$\mathbf{C}\tilde{\Gamma}\mathbf{B}^T=\sum_{k=1}^{p}2C_k^2\sigma^2 c(z_k)b(z_k)\mathbf{f}_m(\omega_k)\mathbf{f}_m^T(\omega_k)+\sigma^4\tilde{\mathbf{C}}_{12}+\sigma^4(\kappa-2)\mathbf{c}\mathbf{b}^T,$$

where $\tilde{\mathbf{C}}_{12} := [\tilde{c}_{12}(u,v)]$ and $\tilde{c}_{12}(u,v) := \sum_{j=0}^{m} c_{u+v-j} b_j$. Therefore,

$$\mathbf{C}\tilde{\mathbf{\Gamma}}\mathbf{B}^T\mathbf{P}_{\mathrm{N}}^* = \sigma^4 \tilde{\mathbf{C}}_{12}\mathbf{P}_{\mathrm{N}}^* = \tilde{\mathbf{\Sigma}}_{12}\mathbf{P}_{\mathrm{N}}^*.$$

Let $\tilde{c}_{22}(u,v) := \sum_{j=0}^{m} b_{u+v-j} b_j$ and $\tilde{\mathbf{C}}_{22} := [\tilde{c}_{22}(u,v)]$. Then, we can also write

$$\mathbf{B}\tilde{\mathbf{\Gamma}}\mathbf{B}^T\mathbf{P}_{\mathrm{N}}^* = \sigma^4 \tilde{\mathbf{C}}_{22}\mathbf{P}_{\mathrm{N}}^* = \tilde{\mathbf{\Sigma}}_{22}\mathbf{P}_{\mathrm{N}}^*.$$

Combining these results with (10.7.6) yields

$$\tilde{\mathbf{\Sigma}}_0 = [\bar{\mathbf{R}}_y^\dagger, \mathbf{P}_{\mathrm{N}}] \begin{bmatrix} \tilde{\mathbf{\Sigma}}_{11} & \tilde{\mathbf{\Sigma}}_{12} \\ \tilde{\mathbf{\Sigma}}_{12}^T & \tilde{\mathbf{\Sigma}}_{22} \end{bmatrix} [\bar{\mathbf{R}}_y^\dagger, \mathbf{P}_{\mathrm{N}}]^T.$$

This proves the assertion regarding $\hat{\mathbf{c}}_{\mathrm{PCA}}$. An application of Lemma 12.5.1 leads to the assertion regarding $\hat{\boldsymbol{\omega}}$. □

Proof of Theorem 10.4. Because $r_x(u) = \sum_{k=1}^{p} C_k^2 z_k^u$, it follows that

$$\sum_{j=0}^{p} a_j r_x(u-j) = \sum_{k=1}^{p} C_k^2 z_k^u a(z_k) = 0 \qquad \forall u \in \mathbb{Z}.$$

Therefore, the prediction error variance $J(\tilde{\mathbf{c}}) := E\{|x_t + \sum_{j=1}^{m} \tilde{c}_j x_{t-j}|^2\}$ equals zero when $\tilde{\mathbf{c}} = [a_1, \ldots, a_p, 0, \ldots, 0]^T$. Any minimizer $\tilde{\mathbf{c}}$ of $J(\tilde{\mathbf{c}})$ should have the same property. Because the minimum prediction error variance can be expressed as $r_x(0) + \mathbf{r}_x^H \tilde{\mathbf{c}}$, it follows that

$$r_x(0) + \mathbf{r}_x^H \tilde{\mathbf{c}} = 0.$$

Combining this equation with (10.2.28) leads to

$$\bar{\mathbf{R}}_x \bar{\mathbf{c}} = \mathbf{0}, \tag{10.7.8}$$

where

$$\bar{\mathbf{R}}_x := \begin{bmatrix} r_x(0) & \mathbf{r}_x^H \\ \mathbf{r}_x & \mathbf{R}_x \end{bmatrix}, \quad \bar{\mathbf{c}} := \begin{bmatrix} 1 \\ \tilde{\mathbf{c}} \end{bmatrix}.$$

Observe that $\bar{\mathbf{R}}_x$ has the same structure as $\mathbf{R}_x$ except that m is replaced by $m+1$. Therefore, similar to (10.1.7), we can write

$$\bar{\mathbf{R}}_x = \mathbf{F}_{m+1}\mathbf{G}\mathbf{F}_{m+1}^H.$$

Inserting this expression in (10.7.8) gives $\bar{\mathbf{c}}^H \mathbf{F}_{m+1}\mathbf{G}\mathbf{F}_{m+1}^H \bar{\mathbf{c}} = 0$, which in turn leads to $\mathbf{F}_{m+1}^H \bar{\mathbf{c}} = \mathbf{0}$, or equivalently,

$$\mathbf{f}_{m+1}^H(\omega_k)\bar{\mathbf{c}} = z_k^{-1}\tilde{c}(z_k) = 0 \qquad (k = 1, \ldots, p),$$

where $\tilde{c}(z) := \sum_{j=0}^{m} \tilde{c}_j z^{-j}$ ($\tilde{c}_0 := 1$). This result implies that $\tilde{c}(z)$ has roots $z_1, \ldots, z_p$ and therefore can be factored as

$$\tilde{c}(z) = a(z)\tilde{b}(z),$$

where $\tilde{b}(z) := \sum_{j=1}^{m-p} \tilde{b}_j z^{-j}$ ($\tilde{b}_0 := 1$) is an $(m-p)$-degree polynomial. The rest of the proof is similar to the last part of the proof of Theorem 9.2 in Section 9.6. □

Proof of Theorem 10.6. Let $\hat{\mathbf{R}}_x := \sum_{j=1}^{m} (\hat{\lambda}_j - \hat{\sigma}^2)\hat{\mathbf{v}}_j\hat{\mathbf{v}}_j^H$. Then, with $\tilde{\mathbf{c}}_{\text{PCA}}$ defined by (10.2.12) and $\tilde{\mathbf{R}}_x$ given by (10.2.10), we can write

$$\begin{aligned} \tilde{\mathbf{c}}_{\text{PCA}} - \mathbf{c}_0 &= -\tilde{\mathbf{R}}_x^{\dagger}(\hat{\mathbf{r}}_y + \hat{\mathbf{R}}_x\mathbf{c}_0) - (\mathbf{I} - \tilde{\mathbf{R}}_x^{\dagger}\hat{\mathbf{R}}_x)\mathbf{c}_0 \\ &= -\tilde{\mathbf{R}}_x^{\dagger}(\hat{\mathbf{r}}_y + \hat{\mathbf{R}}_x\mathbf{c}_0) - \hat{\mathbf{P}}_{\text{N}}\mathbf{c}_0, \end{aligned} \tag{10.7.9}$$

where the second equality is due to the fact that $\tilde{\mathbf{R}}_x^{\dagger}\hat{\mathbf{R}}_x = \hat{\mathbf{P}}_{\text{S}} = \mathbf{I} - \hat{\mathbf{P}}_{\text{N}}$. Because $\mathbf{P}_{\text{N}}\mathbf{c}_0 = \mathbf{0}$, it follows from (10.7.3) that

$$\begin{aligned} \hat{\mathbf{P}}_{\text{N}}\mathbf{c}_0 &= (\hat{\mathbf{P}}_{\text{N}} - \mathbf{P}_{\text{N}})\mathbf{c}_0 \\ &= -\mathbf{P}_{\text{N}}(\hat{\mathbf{R}}_y - \mathbf{R}_y)\mathbf{R}_x^{\dagger}\mathbf{c}_0 + \mathcal{O}(\|\hat{\mathbf{R}}_y - \mathbf{R}_y\|^2) \\ &= \mathbf{P}_{\text{N}}\mathbf{B}(\hat{\mathbf{r}} - \mathbf{r}) + \mathcal{O}_P(n^{-1/2}) + \mathcal{O}(\|\hat{\mathbf{R}}_y - \mathbf{R}_y\|^2), \end{aligned} \tag{10.7.10}$$

where $\mathbf{B}$, $\hat{\mathbf{r}}$, and $\mathbf{r}$ are the same as in the proof of Theorem 10.3 except that $\mathbf{B}$ is defined by $\mathbf{b} := -\mathbf{R}_x^{\dagger}\mathbf{c}_0$. Similarly, because $\mathbf{r}_x + \mathbf{R}_x\mathbf{c}_0 = \mathbf{0}$, $\mathbf{r}_x = \mathbf{r}_y$, $\mathbf{R}_x = \mathbf{R}_y - \sigma^2\mathbf{I}$, and $\hat{\mathbf{R}}_x = \hat{\mathbf{R}}_y - \hat{\sigma}^2\mathbf{I}$, we can write

$$\begin{aligned} \hat{\mathbf{r}}_y + \hat{\mathbf{R}}_x\mathbf{c}_0 &= \hat{\mathbf{r}}_y + \hat{\mathbf{R}}_x\mathbf{c}_0 - (\mathbf{r}_x + \mathbf{R}_x\mathbf{c}_0) \\ &= \hat{\mathbf{r}}_y + \hat{\mathbf{R}}_y\mathbf{c}_0 - (\mathbf{r}_y + \mathbf{R}_y\mathbf{c}_0) - (\hat{\sigma}^2 - \sigma^2)\mathbf{c}_0 \\ &= \mathbf{C}(\hat{\mathbf{r}} - \mathbf{r}) + \mathcal{O}_P(n^{-1/2}) - (\hat{\sigma}^2 - \sigma^2)\mathbf{c}_0, \end{aligned}$$

where $\mathbf{C}$ is the same as in the proof of Theorem 10.3 except that it is defined by $\mathbf{c}_0$ instead of $\mathbf{c}$. Under the assumption (10.2.29), we obtain

$$\hat{\mathbf{r}}_y + \hat{\mathbf{R}}_x\mathbf{c}_0 = \mathbf{C}(\hat{\mathbf{r}} - \mathbf{r}) + \mathcal{O}_P(n^{-1/2}). \tag{10.7.11}$$

Inserting this expression and (10.7.10) in (10.7.9) yields

$$\tilde{\mathbf{c}}_{\text{PCA}} - \mathbf{c}_0 = -(\tilde{\mathbf{R}}_x^{\dagger}\mathbf{C} + \mathbf{P}_{\text{N}}\mathbf{B})(\hat{\mathbf{r}} - \mathbf{r}) + \mathcal{O}_P(n^{-1/2}) + \mathcal{O}(\|\hat{\mathbf{R}}_y - \mathbf{R}_y\|^2).$$

Because $\tilde{\mathbf{R}}_x^{\dagger} \xrightarrow{P} \mathbf{R}_x^{\dagger}$ by Lemma 12.4.6 and $\sqrt{n}(\hat{\mathbf{r}} - \mathbf{r}) \xrightarrow{D} \text{N}_c(\mathbf{0}, \mathbf{\Gamma}, \tilde{\mathbf{\Gamma}})$ by Theorem 4.4, an application of Slutsky's theorem in Proposition 2.13(b) leads to

$$\sqrt{n}(\tilde{\mathbf{c}}_{\text{PCA}} - \mathbf{c}_0) \xrightarrow{D} \text{N}_c(\mathbf{0}, \mathbf{\Sigma}_0, \tilde{\mathbf{\Sigma}}_0),$$

where $\mathbf{\Sigma}_0$ and $\tilde{\mathbf{\Sigma}}_0$ are defined by (10.7.5) and (10.7.6) with $\mathbf{R}_x^{\dagger}$ in place of $\tilde{\mathbf{R}}_y^{\dagger}$.

Because $c_0(z_k) = 0$ for all k, it can be shown by following the proof of Theorem 10.3 that

$$\begin{aligned} \mathbf{C}\Gamma\mathbf{C}^H &= \sigma^4\mathbf{C}_{11} + \sigma^4(\kappa-2)\mathbf{c}_0\mathbf{c}_0^H = \mathbf{\Sigma}_{11}, \\ \mathbf{C}\tilde{\Gamma}\mathbf{C}^T &= \sigma^4\tilde{\mathbf{C}}_{11} + \sigma^4(\kappa-2)\mathbf{c}_0\mathbf{c}_0^T = \tilde{\mathbf{\Sigma}}_{11}. \end{aligned}$$

Similarly, because $\mathbf{P}_{\mathrm{N}}\mathbf{b} = -\mathbf{P}_{\mathrm{N}}\mathbf{R}_x^\dagger\mathbf{c}_0 = \mathbf{0}$, it can be shown that

$$\begin{aligned} \mathbf{C}\Gamma\mathbf{B}^H\mathbf{P}_{\mathrm{N}} &= \mathbf{\Sigma}_{12}\mathbf{P}_{\mathrm{N}}, \quad \mathbf{B}\Gamma\mathbf{B}^H\mathbf{P}_{\mathrm{N}} = \mathbf{\Sigma}_{22}\mathbf{P}_{\mathrm{N}}, \\ \mathbf{C}\tilde{\Gamma}\mathbf{B}^T\mathbf{P}_{\mathrm{N}}^* &= \tilde{\mathbf{\Sigma}}_{12}\mathbf{P}_{\mathrm{N}}^*, \quad \mathbf{B}\tilde{\Gamma}\mathbf{B}^T\mathbf{P}_{\mathrm{N}}^* = \tilde{\mathbf{\Sigma}}_{22}\mathbf{P}_{\mathrm{N}}^*. \end{aligned}$$

The assertion regarding $\tilde{\mathbf{c}}_{\mathrm{PCA}}$ is proved. Finally, because $\mathbf{P}_{\mathrm{N}}\mathbf{D} = \mathbf{0}$, we obtain

$$\mathbf{D}^H\mathbf{\Sigma}_0\mathbf{D} = \mathbf{D}^H\mathbf{R}_x^\dagger\mathbf{\Sigma}_{11}\mathbf{R}_x^\dagger\mathbf{D}, \quad \mathbf{D}^H\tilde{\mathbf{\Sigma}}_0\mathbf{D}^* = \mathbf{D}^H\mathbf{R}_x^\dagger\tilde{\mathbf{\Sigma}}_{11}\mathbf{R}_x^{*\dagger}\mathbf{D}^*.$$

The assertion regarding $\hat{\boldsymbol{\omega}}$ follows from Lemma 12.5.1 and Remark 12.5.2, owing to the fact that $C_0(\omega_k) = 0$ for all k. □

Proof of Theorem 10.7. The assertion can be established by following the proof of Theorem 9.2. First, observe that the a_j satisfy (10.2.13). Therefore, when $\tilde{\mathbf{c}} = [a_1,\dots,a_p,0,\dots,0]^T$, we obtain $\|\mathbf{u}_{lm} + \mathbf{U}_{lm}\tilde{\mathbf{c}}\|^2 = 0$. This result implies that any minimizer of $\|\mathbf{u}_{lm} + \mathbf{U}_{lm}\tilde{\mathbf{c}}\|^2$ should yield a zero value. In particular, for the minimum-norm solution $\mathbf{c}_0 := [c_1,\dots,c_m]^T$ given by (10.2.33), which satisfies

$$\mathbf{U}_{lm}^H\mathbf{U}_{lm}\mathbf{c}_0 = -\mathbf{U}_{lm}^H\mathbf{u}_{lm},$$

we obtain

$$0 = \|\mathbf{u}_{lm} + \mathbf{U}_{lm}\mathbf{c}_0\|^2 = \mathbf{u}_{lm}^H\mathbf{u}_{lm} + \mathbf{u}_{lm}^H\mathbf{U}_{lm}\mathbf{c}_0.$$

Combining these equations yields $\mathbf{U}_{l,m+1}^H\mathbf{U}_{l,m+1}\bar{\mathbf{c}} = \mathbf{0}$, where $\mathbf{U}_{l,m+1} = [\mathbf{u}_{lm},\mathbf{U}_{lm}]$ and $\bar{\mathbf{c}} := [1,\mathbf{c}_0^T]^T$. Because $\bar{\mathbf{c}}^H\mathbf{U}_{l,m+1}^H\mathbf{U}_{l,m+1}\bar{\mathbf{c}} = 0$ and $\mathbf{U}_{l,m+1} = \mathbf{G}_{l,m+1}\mathbf{F}_{m+1}^H$, where $\mathbf{G}_{l,m+1}$ has full (column) rank, it follows that $\mathbf{F}_{m+1}^H\bar{\mathbf{c}} = 0$, or equivalently,

$$\mathbf{f}_{m+1}^H(\omega_k)\bar{\mathbf{c}} = z_k^{-1}c_0(z_k) = 0 \qquad (k = 1,\dots,p),$$

where $c_0(z) := \sum_{j=0}^m c_j z^{-j}$ $(c_0 := 1)$. Hence $c_0(z)$ has roots $z_1,\dots,z_p$. The remaining argument is similar to the last part of the proof of Theorem 10.4. □

Proof of Theorem 10.9. With $\hat{\mathbf{c}}_{\mathrm{PCA}}$ and $\mathbf{c}_0$ defined by (10.2.21) and (10.2.33), respectively, we can write

$$\begin{aligned} \hat{\mathbf{c}}_{\mathrm{PCA}} - \mathbf{c}_0 &= -\tilde{\mathbf{H}}_{lm}^\dagger(\hat{\mathbf{U}}_{lm}^H\hat{\mathbf{u}}_{lm} + \hat{\mathbf{H}}_{lm}\mathbf{c}_0 - \hat{\mathbf{H}}_{lm}\mathbf{c}_0) - \mathbf{c}_0 \\ &= -\tilde{\mathbf{H}}_{lm}^\dagger\hat{\mathbf{U}}_{lm}^H(\hat{\mathbf{u}}_{lm} + \hat{\mathbf{U}}_{lm}\mathbf{c}_0) - (\mathbf{I} - \tilde{\mathbf{H}}_{lm}^\dagger\hat{\mathbf{H}}_{lm})\mathbf{c}_0. \end{aligned} \qquad (10.7.12)$$

Let $\tilde{\mathbf{P}}_{\rm S} := \sum_{j=1}^{p} \hat{\mathbf{q}}_j \hat{\mathbf{q}}_j^H$ and $\tilde{\mathbf{P}}_{\rm N} := \sum_{j=p+1}^{m} \hat{\mathbf{q}}_j \hat{\mathbf{q}}_j^H$. Because $\tilde{\mathbf{P}}_{\rm N} + \tilde{\mathbf{P}}_{\rm S} = \mathbf{I}$, it follows from the EVDs in (10.2.19) and (10.2.20) that

$$\tilde{\mathbf{H}}_{lm}^{\dagger} \hat{\mathbf{H}}_{lm} = \sum_{j=1}^{p} \hat{\mathbf{q}}_j \hat{\mathbf{q}}_j^H = \tilde{\mathbf{P}}_{\rm S} = \mathbf{I} - \tilde{\mathbf{P}}_{\rm N}. \tag{10.7.13}$$

Inserting this expression in (10.7.12) yields

$$\hat{\mathbf{c}}_{\rm PCA} - \mathbf{c}_0 = -\tilde{\mathbf{H}}_{lm}^{\dagger} \hat{\mathbf{U}}_{lm}^H (\hat{\mathbf{u}}_{lm} + \hat{\mathbf{U}}_{lm} \mathbf{c}_0) - \tilde{\mathbf{P}}_{\rm N} \mathbf{c}_0. \tag{10.7.14}$$

Furthermore, because $\mathbf{P}_{\rm N} := \sum_{j=p+1}^{m} \mathbf{q}_j \mathbf{q}_j^H$ is the projection matrix onto the noise subspace $\mathbb{C}_{\rm N}^m$, it follows from (10.2.33) that

$$\mathbf{P}_{\rm N} \mathbf{c}_0 = \mathbf{0}. \tag{10.7.15}$$

Combining this result with Lemma 12.4.7 yields

$$\begin{aligned} \tilde{\mathbf{P}}_{\rm N} \mathbf{c}_0 &= (\tilde{\mathbf{P}}_{\rm N} - \mathbf{P}_{\rm N}) \mathbf{c}_0 \\ &= -\mathbf{P}_{\rm N} (\hat{\mathbf{H}}_{lm} - \mathbf{H}_{lm}) \mathbf{H}_{lm}^{\dagger} \mathbf{c}_0 + \mathcal{O}(\|\hat{\mathbf{H}}_{lm} - \mathbf{H}_{lm}\|^2). \end{aligned}$$

By Theorem 4.4, $\hat{\mathbf{U}}_{lm} - \mathbf{U}_{lm} = \mathcal{O}_P(n^{-1/2})$. Therefore,

$$\begin{aligned} \hat{\mathbf{H}}_{lm} - \mathbf{H}_{lm} &= \hat{\mathbf{U}}_{lm}^H (\hat{\mathbf{U}}_{lm} - \mathbf{U}_{lm}) + (\hat{\mathbf{U}}_{lm} - \mathbf{U}_{lm})^H \mathbf{U}_{lm} \\ &= \mathbf{U}_{lm}^H (\hat{\mathbf{U}}_{lm} - \mathbf{U}_{lm}) + (\hat{\mathbf{U}}_{lm} - \mathbf{U}_{lm})^H \mathbf{U}_{lm} + \mathcal{O}_P(n^{-1}). \end{aligned}$$

Moreover, it follows from (10.2.18) that $\mathbf{P}_{\rm N} \mathbf{U}_{lm}^H = \mathbf{0}$. Combining these results gives

$$\begin{aligned} \tilde{\mathbf{P}}_{\rm N} \mathbf{c}_0 &= -\mathbf{P}_{\rm N} (\hat{\mathbf{U}}_{lm} - \mathbf{U}_{lm})^H \mathbf{U}_{lm} \mathbf{H}_{lm}^{\dagger} \mathbf{c}_0 + \mathcal{O}_P(n^{-1}) \\ &= \mathbf{P}_{\rm N} (\hat{\mathbf{U}}_{lm} - \mathbf{U}_{lm})^H \mathbf{b} + \mathcal{O}_P(n^{-1}), \end{aligned} \tag{10.7.16}$$

where $\mathbf{b} := [b_1, \dots, b_l]^T := -\mathbf{U}_{lm} \mathbf{H}_{lm}^{\dagger} \mathbf{c}_0 = -\mathbf{U}_{lm}^{\dagger H} \mathbf{c}_0$. Define $\hat{\mathbf{r}} := [\hat{r}_y(1), \dots, \hat{r}_y(m+l)]^T$ and $\mathbf{r} := [r_y(1), \dots, r_y(m+l)]^T$. Then, it is easy to verify that

$$(\hat{\mathbf{U}}_{lm} - \mathbf{U}_{lm})^H \mathbf{b} = \mathbf{B}_{lm} (\hat{\mathbf{r}} - \mathbf{r})^*.$$

Inserting this expression in (10.7.16) yields

$$\tilde{\mathbf{P}}_{\rm N} \mathbf{c}_0 = \mathbf{B}_{lm} (\hat{\mathbf{r}} - \mathbf{r})^* + \mathcal{O}_P(n^{-1}). \tag{10.7.17}$$

Similarly, because $\mathbf{c}_0$ satisfies (10.2.16), it follows that

$$\hat{\mathbf{u}}_{lm} + \hat{\mathbf{U}}_{lm} \mathbf{c}_0 = \hat{\mathbf{u}}_{lm} + \hat{\mathbf{U}}_{lm} \mathbf{c}_0 - (\mathbf{u}_{lm} + \mathbf{U}_{lm} \mathbf{c}_0) = \mathbf{C}_{lm} (\hat{\mathbf{r}} - \mathbf{r}).$$

Combining this expression with (10.7.17) and (10.7.14) leads to

$$\hat{\mathbf{c}}_{\rm PCA} - \mathbf{c}_0 = -\tilde{\mathbf{H}}_{lm}^{\dagger} \hat{\mathbf{U}}_{lm}^H \mathbf{C}_{lm} (\hat{\mathbf{r}} - \mathbf{r}) - \mathbf{P}_{\rm N} \mathbf{B}_{lm} (\hat{\mathbf{r}} - \mathbf{r})^* + \mathcal{O}_P(n^{-1}). \tag{10.7.18}$$

Note that $\hat{\mathbf{U}}_{lm} \xrightarrow{a.s.} \mathbf{U}_{lm}$ and hence $\tilde{\mathbf{H}}_{lm}^{\dagger} \xrightarrow{a.s.} \mathbf{H}_{lm}^{\dagger}$ by Lemma 12.4.6. This implies $\tilde{\mathbf{H}}_{lm}^{\dagger}\hat{\mathbf{U}}_{lm}^{H} \xrightarrow{a.s.} \mathbf{H}_{lm}^{\dagger}\mathbf{U}_{lm}^{H} = \mathbf{U}_{lm}^{\dagger}$. Moreover, by Theorem 4.4, $\sqrt{n}(\hat{\mathbf{r}} - \mathbf{r}) \xrightarrow{D} \mathrm{N}_c(\mathbf{0}, \boldsymbol{\Gamma}, \tilde{\boldsymbol{\Gamma}})$, where $\boldsymbol{\Gamma} := [\gamma(u,v)]$ and $\tilde{\boldsymbol{\Gamma}} := [\gamma(u,-v)]$ $(u, v = 1, \ldots, m+l)$, with $\gamma(\cdot,\cdot)$ defined by (4.3.3) and (4.3.4). Therefore, it follows from (10.7.18) and Slutsky's theorem in Proposition 2.13(b) that $\sqrt{n}(\hat{\mathbf{c}}_{\mathrm{PCA}} - \mathbf{c}_0) \xrightarrow{D} \mathrm{N}_c(\mathbf{0}, \boldsymbol{\Sigma}_0, \tilde{\boldsymbol{\Sigma}}_0)$, where

$$\boldsymbol{\Sigma}_0 := [\mathbf{U}_{lm}^{\dagger}, \mathbf{P}_{\mathrm{N}}] \begin{bmatrix} \mathbf{C}_{lm}\boldsymbol{\Gamma}\mathbf{C}_{lm}^{H} & \mathbf{C}_{lm}\tilde{\boldsymbol{\Gamma}}\mathbf{B}_{lm}^{H} \\ \mathbf{B}_{lm}\tilde{\boldsymbol{\Gamma}}^{*}\mathbf{C}_{lm}^{H} & \mathbf{B}_{lm}\boldsymbol{\Gamma}^{*}\mathbf{B}_{lm}^{H} \end{bmatrix} [\mathbf{U}_{lm}^{\dagger}, \mathbf{P}_{\mathrm{N}}]^{H},$$

$$\tilde{\boldsymbol{\Sigma}}_0 := [\mathbf{U}_{lm}^{\dagger}, \mathbf{P}_{\mathrm{N}}] \begin{bmatrix} \mathbf{C}_{lm}\tilde{\boldsymbol{\Gamma}}\mathbf{C}_{lm}^{T} & \mathbf{C}_{lm}\boldsymbol{\Gamma}\mathbf{B}_{lm}^{T} \\ \mathbf{B}_{lm}\boldsymbol{\Gamma}^{*}\mathbf{C}_{lm}^{T} & \mathbf{B}_{lm}\tilde{\boldsymbol{\Gamma}}^{*}\mathbf{B}_{lm}^{T} \end{bmatrix} [\mathbf{U}_{lm}^{\dagger}, \mathbf{P}_{\mathrm{N}}]^{T}.$$

Similar to the proof of Theorem 9.7, it is easy to verify that under the white noise assumption, we can write

$$\boldsymbol{\Gamma} = \sigma^4\mathbf{I} + 2\sigma^2\mathbf{F}_{m+l}\mathbf{G}\mathbf{F}_{m+l}^{H}, \quad \tilde{\boldsymbol{\Gamma}} = 2\sigma^2\mathbf{F}_{m+l}\mathbf{G}\mathbf{F}_{m+l}^{T}.$$

Moreover, we have $\mathbf{C}_{lm}\mathbf{F}_{m+l} = \mathbf{0}$ because $c_0(z_k) = 0$ for all k. Therefore,

$$\begin{aligned} \mathbf{C}_{lm}\boldsymbol{\Gamma}\mathbf{C}_{lm}^{H} &= \sigma^4\mathbf{C}_{lm}\mathbf{C}_{lm}^{H}, \quad \mathbf{C}_{lm}\tilde{\boldsymbol{\Gamma}}\mathbf{B}_{lm}^{H} = \mathbf{0}, \\ \mathbf{C}_{lm}\boldsymbol{\Gamma}\mathbf{B}_{lm}^{T} &= \sigma^4\mathbf{C}_{lm}\mathbf{B}_{lm}^{T}, \quad \mathbf{C}_{lm}\tilde{\boldsymbol{\Gamma}}\mathbf{C}_{lm}^{T} = \mathbf{0}. \end{aligned}$$

This proves the assertion regarding $\hat{\mathbf{c}}_{\mathrm{PCA}}$. The assertion regarding $\hat{\boldsymbol{\omega}}$ follows from Lemma 12.5.1 coupled with the fact that $\mathbf{P}_{\mathrm{N}}\mathbf{D} = \mathbf{0}$.

According to Lemma 12.5.1 and Remark 12.5.2, the asymptotic normality of $\hat{\boldsymbol{\omega}}$ can be established directly from the fact that

$$\mathbf{F}_{m}^{H}(\hat{\mathbf{c}}_{\mathrm{PCA}} - \mathbf{c}_0) = -\mathbf{F}_{m}^{H}\tilde{\mathbf{H}}_{lm}^{\dagger}\hat{\mathbf{U}}_{lm}^{H}(\hat{\mathbf{u}}_l + \hat{\mathbf{U}}_{lm}\mathbf{c}_0) + \mathcal{O}_P(n^{-1/2})$$

and $\sqrt{n}(\hat{\mathbf{u}}_l + \hat{\mathbf{U}}_{lm}\mathbf{c}_0) \xrightarrow{D} \mathrm{N}_c(\mathbf{0}, \sigma^4\mathbf{C}_{lm}\mathbf{C}_{lm}^{H})$. This technique is used in [376]. □

Proof of Theorem 10.11. As $n \to \infty$, Theorem 4.1 ensures that $\hat{\mathbf{R}}_y \xrightarrow{P} \mathbf{R}_y$. Therefore, by Theorem 10.10, we obtain $\hat{\boldsymbol{\omega}} \xrightarrow{P} \boldsymbol{\omega}$. Consider the Taylor expansion

$$\dot{\hat{g}}_m(\hat{\omega}_k) - \dot{\hat{g}}_m(\omega_k) = \ddot{\hat{g}}_m(\tilde{\omega}_k)(\hat{\omega}_k - \omega_k),$$

where $\tilde{\omega}_k$ resides between $\hat{\omega}_k$ and ω_k. Observe that $\dot{\hat{g}}_m(\hat{\omega}_k) = 0$ because $\hat{\omega}_k$ is a local minimizer of $\hat{g}_m(\omega)$. Also observe that $\ddot{\hat{g}}_m(\tilde{\omega}_k) \xrightarrow{P} \ddot{g}_m(\omega_k)$ because $\tilde{\omega}_k \xrightarrow{P} \omega_k$. Therefore, by Slutsky's theorem in Proposition 2.13(b), it suffices to show that

$$\sqrt{n}\,\mathrm{vec}[\dot{\hat{g}}_m(\omega_k)] \xrightarrow{D} \mathrm{N}(0, 2\sigma^4\Re(\boldsymbol{\Sigma} + \tilde{\boldsymbol{\Sigma}})). \tag{10.7.19}$$

Furthermore, observe that $\hat{g}_m(\omega) = \mathbf{f}_m^H(\omega)\hat{\mathbf{P}}_{\mathrm{N}}\mathbf{f}_m(\omega)$ and hence

$$\dot{\hat{g}}_m(\omega_k) = 2\Re\{\dot{\mathbf{f}}_m^H(\omega_k)\hat{\mathbf{P}}_{\mathrm{N}}\mathbf{f}_m(\omega_k)\}.$$

Similarly, because ω_k is a local minimizer of $g_m(\omega) = \mathbf{f}_m^H(\omega)\mathbf{P}_\mathrm{N}\mathbf{f}_m(\omega)$, we obtain

$$0 = \dot{g}_m(\omega_k) = 2\Re\{\dot{\mathbf{f}}_m^H(\omega_k)\mathbf{P}_\mathrm{N}\mathbf{f}_m(\omega_k)\}.$$

Combining these equations yields

$$\dot{\hat{g}}_m(\omega_k) = 2\Re\{\dot{\mathbf{f}}_m^H(\omega_k)(\hat{\mathbf{P}}_\mathrm{N} - \mathbf{P}_\mathrm{N})\mathbf{f}_m(\omega_k)\}.$$

The assertion (10.7.19) follows if we can show that

$$\sqrt{n}\,\mathrm{vec}[\dot{\mathbf{f}}_m^H(\omega_k)(\hat{\mathbf{P}}_\mathrm{N} - \mathbf{P}_\mathrm{N})\mathbf{f}_m(\omega_k)] \xrightarrow{D} \mathrm{N}_c(\mathbf{0}, \sigma^4\boldsymbol{\Sigma}, \sigma^4\tilde{\boldsymbol{\Sigma}}). \tag{10.7.20}$$

The remainder of the proof is devoted to establishing (10.7.20).

According to Theorem 4.3, we can write $\hat{\mathbf{R}}_y = \mathbf{R}_y + \mathcal{O}_P(n^{1/2})$. Under the white noise assumption, we have $\mathbf{R}_y - \sigma^2\mathbf{I} = \mathbf{R}_x$. Therefore, it follows from Theorem 10.1 and Lemma 12.4.8 that

$$\hat{\mathbf{P}}_\mathrm{N} - \mathbf{P}_\mathrm{N} = -\mathbf{P}_\mathrm{N}(\hat{\mathbf{R}}_y - \mathbf{R}_y)\mathbf{R}_x^\dagger - \mathbf{R}_x^\dagger(\hat{\mathbf{R}}_y - \mathbf{R}_y)\mathbf{P}_\mathrm{N} + \mathcal{O}_P(n^{1/2}).$$

Combining this result with the fact that $\mathbf{P}_\mathrm{N}\mathbf{f}_m(\omega_k) = \mathbf{0}$ leads to

$$\begin{aligned}\dot{\mathbf{f}}_m^H(\omega_k)(\hat{\mathbf{P}}_\mathrm{N} - \mathbf{P}_\mathrm{N})\mathbf{f}_m(\omega_k) &= -\dot{\mathbf{f}}_m^H(\omega_k)\mathbf{P}_\mathrm{N}(\hat{\mathbf{R}}_y - \mathbf{R}_y)\mathbf{R}_x^\dagger\mathbf{f}_m(\omega_k) + \mathcal{O}_P(n^{1/2}) \\ &= \dot{\mathbf{f}}_m^H(\omega_k)\mathbf{P}_\mathrm{N}(\hat{\mathbf{R}}_y - \mathbf{R}_y)\mathbf{c}_k + \mathcal{O}_P(n^{1/2}).\end{aligned} \tag{10.7.21}$$

Without loss of generality, let us assume that $\hat{\mathbf{R}}_y = [\hat{r}_y(u-v)]$ $(u, v = 1, \ldots, m)$. Then, similar to the proof of Theorem 7.5, it can be shown that there exists a matrix $\mathbf{C}_k$, which depends solely on $\mathbf{c}_k$, such that

$$\hat{\mathbf{R}}_y\mathbf{c}_k = \mathbf{C}_k\hat{\mathbf{r}}, \quad \mathbf{R}_y\mathbf{c}_k = \mathbf{C}_k\mathbf{r},$$

where $\hat{\mathbf{r}} := \mathrm{vec}[\hat{r}_y(u)]$ and $\mathbf{r} := \mathrm{vec}[r_y(u)]$ $(u = -m+1, \ldots, m-1)$. Therefore,

$$(\hat{\mathbf{R}}_y - \mathbf{R}_y)\mathbf{c}_k = \mathbf{C}_k(\hat{\mathbf{r}} - \mathbf{r}).$$

Inserting this expression in (10.7.21) yields

$$\mathrm{vec}[\dot{\mathbf{f}}_m^H(\omega_k)(\hat{\mathbf{P}}_\mathrm{N} - \mathbf{P}_\mathrm{N})\mathbf{f}_m(\omega_k)] = \mathbf{C}^H(\hat{\mathbf{r}} - \mathbf{r}) + \mathcal{O}_P(n^{1/2}),$$

where $\mathbf{C} := [\mathbf{C}_1^H\mathbf{P}_\mathrm{N}\dot{\mathbf{f}}_m(\omega_1), \ldots, \mathbf{C}_p^H\mathbf{P}_\mathrm{N}\dot{\mathbf{f}}_m(\omega_p)]$. Define

$$\boldsymbol{\Gamma} := \lim_{n\to\infty} n\,\mathrm{Cov}(\hat{\mathbf{r}}, \hat{\mathbf{r}}), \quad \tilde{\boldsymbol{\Gamma}} := \lim_{n\to\infty} n\,\mathrm{Cov}(\hat{\mathbf{r}}, \hat{\mathbf{r}}^*).$$

Then, according to Theorem 4.4, $\sqrt{n}(\hat{\mathbf{r}} - \mathbf{r}) \xrightarrow{D} \mathrm{N}_c(\mathbf{0}, \boldsymbol{\Gamma}, \tilde{\boldsymbol{\Gamma}})$. Therefore, we obtain

$$\sqrt{n}\,\mathrm{vec}[\dot{\mathbf{f}}_m^H(\omega_k)(\hat{\mathbf{P}}_\mathrm{N} - \mathbf{P}_\mathrm{N})\mathbf{f}_m(\omega_k)] \xrightarrow{D} \mathrm{N}_c(\mathbf{0}, \boldsymbol{\Sigma}_0, \tilde{\boldsymbol{\Sigma}}_0),$$

where

$$\boldsymbol{\Sigma}_0 := \mathbf{C}^H \boldsymbol{\Gamma} \mathbf{C} = [\dot{\mathbf{f}}_m^H(\omega_k)\mathbf{P}_N \mathbf{C}_k \boldsymbol{\Gamma} \mathbf{C}_{k'}^H \mathbf{P}_N \dot{\mathbf{f}}_m(\omega_{k'})],$$
$$\tilde{\boldsymbol{\Sigma}}_0 := \mathbf{C}^H \tilde{\boldsymbol{\Gamma}} \mathbf{C}^* = [\dot{\mathbf{f}}_m^H(\omega_k)\mathbf{P}_N \mathbf{C}_k \tilde{\boldsymbol{\Gamma}} \mathbf{C}_{k'}^T \mathbf{P}_N^* \dot{\mathbf{f}}_m^*(\omega_{k'})].$$

It remains to show that $\boldsymbol{\Sigma}_0 = \sigma^4 \boldsymbol{\Sigma}$ and $\tilde{\boldsymbol{\Sigma}}_0 = \sigma^4 \tilde{\boldsymbol{\Sigma}}$.

For $u, v = 1, \ldots, m$, let

$$\sigma_{kk'}(u,v) := \sum_{j=1}^{m} \sum_{j'=1}^{m} c_{jk} c_{j'k'}^* \gamma(u-j, v-j'),$$
$$\tilde{\sigma}_{kk'}(u,v) := \sum_{j=1}^{m} \sum_{j'=1}^{m} c_{jk} c_{j'k'} \gamma(u-j, -v+j'),$$

where $\gamma(\cdot,\cdot)$ is defined by (4.3.3) and (4.3.4). Because the uth entry of $\hat{\mathbf{R}}_y \mathbf{c}_k$ is $\sum_{j=1}^m c_{jk} \hat{r}_y(u-j)$ $(u = 1, \ldots, m)$, it follows that

$$\begin{aligned}
\mathbf{C}_k \boldsymbol{\Gamma} \mathbf{C}_{k'}^H &= \lim_{n\to\infty} n \operatorname{Cov}(\mathbf{C}_k \hat{\mathbf{r}}, \mathbf{C}_{k'} \hat{\mathbf{r}}) \\
&= \lim_{n\to\infty} n \operatorname{Cov}(\hat{\mathbf{R}}_y \mathbf{c}_k, \hat{\mathbf{R}}_y \mathbf{c}_{k'}) \\
&= [\sigma_{kk'}(u,v)] \qquad (u, v = 1, \ldots, m).
\end{aligned}$$

Similarly, we obtain

$$\mathbf{C}_k \tilde{\boldsymbol{\Gamma}} \mathbf{C}_{k'}^T = [\tilde{\sigma}_{kk'}(u,v)] \qquad (u, v = 1, \ldots, m).$$

Under the white noise assumption, we can write

$$\gamma(u,v) = \sum_{l=1}^{p} 2C_l^2 \sigma^2 z_l^{u-v} + \sigma^4 \delta_{u-v} + \sigma^4(\kappa - 2)\delta_u \delta_v \quad (u, v = 0, \pm 1, \ldots).$$

Therefore, for $u, v = 1, \ldots, m$, we have

$$\begin{aligned}
\sigma_{kk'}(u,v) &= \sum_{l=1}^{p} 2C_l^2 \sigma^2 c_k(z_l) c_{k'}^*(z_l) z_l^{u-v} \\
&\quad + \sigma^4 c_{kk'}(u,v) + \sigma^4(\kappa-2) c_{uk} c_{vk'}^*,
\end{aligned}$$

where $c_k(z) := \sum_{j=1}^m c_{jk} z^{-j}$ and $c_{kk'}(u,v) := \sum_{j=1}^m c_{u-v+j,k} c_{jk'}^*$. This implies that

$$\begin{aligned}
\mathbf{C}_k \boldsymbol{\Gamma} \mathbf{C}_{k'}^H &= \sum_{l=1}^{p} 2C_l^2 \sigma^2 c_k(z_l) c_{k'}^*(z_l) \mathbf{f}_m(\omega_l) \mathbf{f}_m^H(\omega_l) \\
&\quad + \sigma^4 \mathbf{C}_{kk'} + \sigma^4(\kappa-2) \mathbf{c}_k \mathbf{c}_{k'}^H.
\end{aligned}$$

Combining this result with the fact that $\mathbf{P}_N \mathbf{f}_m(\omega_l) = \mathbf{0}$ and $\mathbf{P}_N \mathbf{c}_k = \mathbf{0}$ yields

$$\mathbf{P}_N \mathbf{C}_k \boldsymbol{\Gamma} \mathbf{C}_{k'}^H \mathbf{P}_N = \sigma^4 \mathbf{P}_N \mathbf{C}_{kk'} \mathbf{P}_N$$

and hence $\boldsymbol{\Sigma}_0 = \sigma^4 \boldsymbol{\Sigma}$. Similarly, it can be shown that

$$\tilde{\sigma}_{kk'}(u,v) = \sum_{l=1}^{p} 2C_l^2 \sigma^2 c_k(z_l) c_{k'}(z_l) z_l^{u+v} + \sigma^4 \tilde{c}_{kk'}(u,v) + \sigma^4(\kappa-2) c_{uk} c_{vk'},$$

where $\tilde{c}_{kk'}(u,v) := \sum_{j=1}^{m} c_{u+v-j,k} c_{jk'}$. Therefore,

$$\hat{\mathbf{P}}_{\mathrm{N}} \mathbf{C}_k \tilde{\boldsymbol{\Gamma}} \mathbf{C}_{k'}^T \mathbf{P}_{\mathrm{N}}^* = \sigma^4 \mathbf{P}_{\mathrm{N}} \tilde{\mathbf{C}}_{kk'} \mathbf{P}_{\mathrm{N}}^*$$

and hence $\tilde{\boldsymbol{\Sigma}}_0 = \sigma^4 \tilde{\boldsymbol{\Sigma}}_0$. The assertion in (10.7.20) is proved. Finally, Lemma 12.4.7 guarantees the validity of (10.7.20) if $\hat{\mathbf{R}}_y = [\hat{r}_y(u-v)] + \mathcal{O}_P(n^{-1/2})$. □

Proof of Theorem 10.12. With $\hat{\mathbf{Q}}$ given by (10.4.14), consider the matrix

$$\hat{\mathbf{H}} := \hat{\mathbf{V}}_{\mathrm{S}} \hat{\mathbf{Q}} \hat{\mathbf{V}}_{\mathrm{S}}^H = \hat{\mathbf{P}}_{\mathrm{S}} \hat{\mathbf{R}}_x^{\dagger} \hat{\mathbf{R}}_x' \hat{\mathbf{P}}_{\mathrm{S}} = \hat{\mathbf{R}}_x^{\dagger} \hat{\mathbf{R}}_x' \hat{\mathbf{P}}_{\mathrm{S}}, \tag{10.7.22}$$

where the last equality is due to the fact that $\hat{\mathbf{P}}_{\mathrm{S}} \hat{\mathbf{R}}_x^{\dagger} = \hat{\mathbf{R}}_x^{\dagger}$. By Theorem 10.1 and Lemma 12.4.6, $\hat{\mathbf{R}}_y \to \mathbf{R}_y$ implies $\tilde{\mathbf{R}}_y \to \bar{\mathbf{R}}_y := \mathbf{V}_{\mathrm{S}} \boldsymbol{\Lambda}_{\mathrm{S}} \mathbf{V}_{\mathrm{S}}^H$, where $\boldsymbol{\Lambda}_{\mathrm{S}} := \mathrm{diag}(\lambda_1, \dots, \lambda_p)$. Therefore, under the additional assumption that $\hat{\sigma}^2 \to \sigma^2$, we obtain

$$\hat{\mathbf{R}}_x \to \bar{\mathbf{R}}_y - \sigma^2 \mathbf{I} = \mathbf{V}_{\mathrm{S}}(\boldsymbol{\Lambda}_{\mathrm{S}} - \sigma^2 \mathbf{I}) \mathbf{V}_{\mathrm{S}}^H = \mathbf{R}_x.$$

By Lemma 12.4.6, $\hat{\mathbf{R}}_x^{\dagger} \to \mathbf{R}_x^{\dagger}$. By Theorem 10.1 and Lemma 12.4.7, $\hat{\mathbf{P}}_{\mathrm{S}} \to \mathbf{P}_{\mathrm{S}}$. Under the assumption that $\hat{\mathbf{R}}_y' \to \mathbf{R}_y'$, we also obtain $\hat{\mathbf{R}}_x' \to \mathbf{R}_y' - \sigma^2 \mathbf{I}' = \mathbf{R}_x'$. Combining these results with the identity $\mathbf{P}_{\mathrm{S}} \mathbf{R}_x^{\dagger} = \mathbf{R}_x^{\dagger}$ yields

$$\hat{\mathbf{H}} \to \mathbf{H} := \mathbf{R}_x^{\dagger} \mathbf{R}_x' \mathbf{P}_{\mathrm{S}} = \mathbf{P}_{\mathrm{S}} \mathbf{R}_x^{\dagger} \mathbf{R}_x' \mathbf{P}_{\mathrm{S}} = \mathbf{V}_{\mathrm{S}} \mathbf{Q} \mathbf{V}_{\mathrm{S}}^H.$$

Observe that the m-by-m matrix $\mathbf{H}$ has rank p for any $m \ge p$ and can also be expressed as $\mathbf{H} = \mathbf{R}_x^{\dagger} \mathbf{R}_x'$ because $\mathbf{R}_x' \mathbf{P}_{\mathrm{S}} = \mathbf{F}_m \mathbf{Z} \mathbf{G} \mathbf{F}_m^H \mathbf{P}_{\mathrm{S}} = \mathbf{F}_m \mathbf{Z} \mathbf{G} \mathbf{F}_m^H = \mathbf{R}_x'$. The p nonzero eigenvalues of $\mathbf{H}$ are $z_1, \dots, z_p$. To prove this assertion, let $z \ne 0$ and $\mathbf{q} \ne \mathbf{0}$ be an eigenpair of $\mathbf{H}$ satisfying $\mathbf{H}\mathbf{q} = z\mathbf{q}$, or equivalently,

$$\mathbf{R}_x^{\dagger} \mathbf{R}_x' \mathbf{q} = z\mathbf{q}. \tag{10.7.23}$$

Because $\mathbf{R}_x^{\dagger} \mathbf{R}_x' \mathbf{q} = \sum_{j=1}^{p} \mu_j^{-1} (\mathbf{v}^H \mathbf{R}_x' \mathbf{q}) \mathbf{v}_j \in \mathbb{C}_{\mathrm{S}}^m$ and $z \ne 0$, it follows from (10.7.23) that $\mathbf{q} = z^{-1} \mathbf{R}_x^{\dagger} \mathbf{R}_x' \mathbf{q} \in \mathbb{C}_{\mathrm{S}}^m$. Moreover, because $\mathbf{R}_x \mathbf{R}_x^{\dagger} = \mathbf{P}_{\mathrm{S}}$ and $\mathbf{P}_{\mathrm{S}} \mathbf{R}_x' = \mathbf{R}_x'$, pre-multiplying both sides of (10.7.23) by $\mathbf{R}_x$ leads to (10.4.5). This means that $z \in \{z_k\}$. On the other hand, because z_k is an eigenvalue of $\mathbf{Q}$, there exists $\mathbf{d} \ne \mathbf{0}$ such that $\mathbf{Q}\mathbf{d} = z_k \mathbf{d}$, or equivalently,

$$\mathbf{V}_{\mathrm{S}}^H \mathbf{R}_x^{\dagger} \mathbf{R}_x' \mathbf{V}_{\mathrm{S}} \mathbf{d} = z_k \mathbf{d}.$$

Pre-multiplying both sides with $\mathbf{V}_{\mathrm{S}}$ followed by an application of the identity $\mathbf{P}_{\mathrm{S}} \mathbf{R}_x^{\dagger} = \mathbf{R}_x^{\dagger}$ leads to (10.7.23) with $z = z_k$ and $\mathbf{d} := \mathbf{V}_{\mathrm{S}} \mathbf{q} \ne \mathbf{0}$. Hence z_k must be an eigenvalue of $\mathbf{H}$. Similarly, it can be shown that the set of nonzero eigenvalues of $\hat{\mathbf{H}}$ coincides with the set of nonzero eigenvalues of $\hat{\mathbf{Q}}$.

Let $\mathbf{p}$ and $\mathbf{q}$ be the left and right eigenvectors of $\mathbf{H}$ associated with the eigenvalue z_k. According to the matrix theory [361, Theorem 2.3, p. 183], there is a unique eigenvalue $\hat{z}_k$ of $\hat{\mathbf{H}}$ such that

$$\hat{z}_k = z_k + \mathbf{p}^H(\hat{\mathbf{H}} - \mathbf{H})\mathbf{q}/(\mathbf{p}^H\mathbf{q}) + \mathcal{O}(\|\hat{\mathbf{H}} - \mathbf{H}\|^2).$$

Therefore, $\hat{\mathbf{H}} \to \mathbf{H}$ implies $\hat{z}_k \to z_k$ as $n \to \infty$ for all k. □

Proof of Theorem 10.13. Let $\hat{\mathbf{H}}$ and $\mathbf{H}$ be defined by (10.7.22) and (10.4.23). Because $\hat{\mathbf{H}} \overset{a.s.}{\to} \mathbf{H}$ as $n \to \infty$ and because the $\hat{z}_k$ and the z_k are distinct eigenvalues of $\hat{\mathbf{H}}$ and $\mathbf{H}$, respectively, it follows from Theorem 2.3 in [361] that

$$\hat{z}_k - z_k = (\mathbf{p}_k^H\mathbf{q}_k)^{-1}\mathbf{p}_k^H(\hat{\mathbf{H}} - \mathbf{H})\mathbf{q}_k + \mathcal{O}(\|\hat{\mathbf{H}} - \mathbf{H}\|^2), \tag{10.7.24}$$

where $\mathbf{p}_k$ and $\mathbf{q}_k$ are the left and right eigenvectors of $\mathbf{H}$ associated with the eigenvalue z_k. With $\hat{\mathbf{H}}$ given by (10.7.22), we can write

$$\begin{aligned}\hat{\mathbf{H}} - \mathbf{H} &= \hat{\mathbf{R}}_x^\dagger(\hat{\mathbf{R}}_x'\hat{\mathbf{P}}_\mathrm{S} - \hat{\mathbf{R}}_x\mathbf{H} + \hat{\mathbf{R}}_x\mathbf{H}) - \mathbf{H} \\ &= \hat{\mathbf{R}}_x^\dagger(\hat{\mathbf{R}}_x'\hat{\mathbf{P}}_\mathrm{S} - \hat{\mathbf{R}}_x\mathbf{H}) - (\mathbf{I} - \hat{\mathbf{R}}_x^\dagger\hat{\mathbf{R}}_x)\mathbf{H} \\ &= \hat{\mathbf{R}}_x^\dagger(\hat{\mathbf{R}}_x'\hat{\mathbf{P}}_\mathrm{S} - \hat{\mathbf{R}}_x\mathbf{H}) - \hat{\mathbf{P}}_\mathrm{N}\mathbf{H},\end{aligned} \tag{10.7.25}$$

where the last equality is due to the fact that $\hat{\mathbf{R}}_x^\dagger\hat{\mathbf{R}}_x = \hat{\mathbf{P}}_\mathrm{S} = \mathbf{I} - \hat{\mathbf{P}}_\mathrm{N}$. Moreover, because $\mathbf{H} = \mathbf{R}_x^\dagger\mathbf{R}_x'\mathbf{P}_\mathrm{S}$, we can write

$$\hat{\mathbf{R}}_x'\hat{\mathbf{P}}_\mathrm{S} - \hat{\mathbf{R}}_x\mathbf{H} = \hat{\mathbf{R}}_x'(\hat{\mathbf{P}}_\mathrm{S} - \mathbf{P}_\mathrm{S}) + \{(\hat{\mathbf{R}}_x' - \mathbf{R}_x') + (\mathbf{R}_x' - \hat{\mathbf{R}}_x\mathbf{R}_x^\dagger\mathbf{R}_x')\}\mathbf{P}_\mathrm{S}. \tag{10.7.26}$$

Observe that $\mathbf{R}_x' = \mathbf{P}_\mathrm{S}\mathbf{R}_x' = \mathbf{R}_x\mathbf{R}_x^\dagger\mathbf{R}_x'$. Therefore,

$$\mathbf{R}_x' - \hat{\mathbf{R}}_x\mathbf{R}_x^\dagger\mathbf{R}_x' = (\mathbf{R}_x - \hat{\mathbf{R}}_x)\mathbf{R}_x^\dagger\mathbf{R}_x'$$

Inserting this expression in (10.7.26) yields

$$\begin{aligned}\hat{\mathbf{R}}_x'\hat{\mathbf{P}}_\mathrm{S} - \hat{\mathbf{R}}_x\mathbf{H} &= \hat{\mathbf{R}}_x'(\hat{\mathbf{P}}_\mathrm{S} - \mathbf{P}_\mathrm{S}) + (\hat{\mathbf{R}}_x' - \mathbf{R}_x')\mathbf{P}_\mathrm{S} - (\hat{\mathbf{R}}_x - \mathbf{R}_x)\mathbf{R}_x^\dagger\mathbf{R}_x'\mathbf{P}_\mathrm{S} \\ &= -\hat{\mathbf{R}}_x'(\hat{\mathbf{P}}_\mathrm{N} - \mathbf{P}_\mathrm{N}) + (\hat{\mathbf{R}}_x' - \mathbf{R}_x')\mathbf{P}_\mathrm{S} - (\hat{\mathbf{R}}_x - \mathbf{R}_x)\mathbf{H}.\end{aligned} \tag{10.7.27}$$

Combining this result with (10.7.25) and the fact that $\mathbf{P}_\mathrm{N}\mathbf{H} = \mathbf{0}$ leads to

$$\begin{aligned}\hat{\mathbf{H}} - \mathbf{H} &= -\hat{\mathbf{R}}_x^\dagger\hat{\mathbf{R}}_x'(\hat{\mathbf{P}}_\mathrm{N} - \mathbf{P}_\mathrm{N}) + \hat{\mathbf{R}}_x^\dagger(\hat{\mathbf{R}}_x' - \mathbf{R}_x')\mathbf{P}_\mathrm{S} \\ &\quad - \hat{\mathbf{R}}_x^\dagger(\hat{\mathbf{R}}_x - \mathbf{R}_x)\mathbf{H} - (\hat{\mathbf{P}}_\mathrm{N} - \mathbf{P}_\mathrm{N})\mathbf{H}.\end{aligned}$$

Under the assumption that $\hat{\sigma}^2 = \sigma^2 + \mathcal{O}_P(n^{-1/2})$, we can write

$$\hat{\mathbf{R}}_x - \mathbf{R}_x = \hat{\mathbf{R}}_y - \mathbf{R}_y + \mathcal{O}_P(n^{-1/2}), \quad \hat{\mathbf{R}}_x' - \mathbf{R}_x' = \hat{\mathbf{R}}_y' - \mathbf{R}_y' + \mathcal{O}_P(n^{-1/2}).$$

According to Lemma 12.4.8, we have

$$\hat{\mathbf{P}}_N - \mathbf{P}_N = -\mathbf{P}_N(\hat{\mathbf{R}}_y - \mathbf{R}_y)\mathbf{R}_x^\dagger - \mathbf{R}_x^\dagger(\hat{\mathbf{R}}_y - \mathbf{R}_y)\mathbf{P}_N + \mathcal{O}(\|\hat{\mathbf{R}}_y - \mathbf{R}_y\|^2).$$

By combining these results with the fact that $\hat{\mathbf{R}}_x^\dagger \overset{a.s.}{\to} \mathbf{R}_x$, $\hat{\mathbf{R}}_y - \mathbf{R}_y = \mathcal{O}_P(n^{-1/2})$, $\hat{\mathbf{R}}_y' - \mathbf{R}_y' = \mathcal{O}_P(n^{-1/2})$, $\mathbf{R}_x'\mathbf{P}_N = \mathbf{0}$, and $\mathbf{P}_N\mathbf{H} = \mathbf{0}$, we obtain

$$\begin{aligned}\hat{\mathbf{H}} - \mathbf{H} &= -\mathbf{R}_x^\dagger\mathbf{R}_x'(\hat{\mathbf{P}}_N - \mathbf{P}_N) + \mathbf{R}_x^\dagger(\hat{\mathbf{R}}_y' - \mathbf{R}_y')\mathbf{P}_S - \mathbf{R}_x^\dagger(\hat{\mathbf{R}}_y - \mathbf{R}_y)\mathbf{H} \\ &\quad - (\hat{\mathbf{P}}_N - \mathbf{P}_N)\mathbf{H} + o_P(n^{-1/2}) \\ &= \mathbf{H}\mathbf{R}_x^\dagger(\hat{\mathbf{R}}_y - \mathbf{R}_y)\mathbf{P}_N + \mathbf{R}_x^\dagger(\hat{\mathbf{R}}_y' - \mathbf{R}_y')\mathbf{P}_S - \mathbf{R}_x^\dagger(\hat{\mathbf{R}}_y - \mathbf{R}_y)\mathbf{H} \\ &\quad + \mathbf{P}_N(\hat{\mathbf{R}}_y - \mathbf{R}_y)\mathbf{R}_x^\dagger\mathbf{H} + o_P(n^{-1/2}).\end{aligned} \tag{10.7.28}$$

Let $\mathbf{q}_{p+1},\ldots,\mathbf{q}_m$ be the orthonormal right eigenvectors of $\mathbf{H}$ associated with the zero eigenvalue. Because $\mathbf{H}\mathbf{q} = \mathbf{R}_x^\dagger\mathbf{F}_m\mathbf{Z}\mathbf{G}\mathbf{F}_m^H\mathbf{q} = \mathbf{0}$ implies $\mathbf{F}_m^H\mathbf{q} = \mathbf{0}$, it follows that

$$\text{span}\{\mathbf{q}_{p+1},\ldots,\mathbf{q}_m\} = \mathbb{C}_N^m$$

and hence $\mathbf{q}_k \in \mathbb{C}_S^m$ for $k = 1,\ldots,p$. Similarly, we have $\mathbf{p}_k \in \mathbb{C}_S^m$ $(k = 1,\ldots,p)$. Combining this result with (10.7.28) and $\mathbf{H}\mathbf{q}_k = z_k\mathbf{q}_k$ yields

$$\mathbf{p}_k^H(\hat{\mathbf{H}} - \mathbf{H})\mathbf{q}_k = \mathbf{p}_k^H\mathbf{R}_x^\dagger\{(\hat{\mathbf{R}}_y' - \mathbf{R}_y') - z_k(\hat{\mathbf{R}}_y - \mathbf{R}_y)\}\mathbf{q}_k + o_P(n^{-1/2}). \tag{10.7.29}$$

Because $\hat{\mathbf{R}}_y = [\hat{r}_y(u-v)] + o_P(n^{-1/2})$ and $\hat{\mathbf{R}}_y' = [\hat{r}_y(u-v+1)] + o_P(n^{-1/2})$, there exists a matrix $\mathbf{C}_k$, which depends solely on $\mathbf{q}_k$ and z_k, such that

$$\{(\hat{\mathbf{R}}_y' - \mathbf{R}_y') - z_k(\hat{\mathbf{R}}_y - \mathbf{R}_y)\}\mathbf{q}_k = \mathbf{C}_k(\hat{\mathbf{r}} - \mathbf{r}) + o_P(n^{-1/2}),$$

where $\hat{\mathbf{r}} := \text{vec}[\hat{r}_y(u)]$ and $\mathbf{r} := \text{vec}[r_y(u)]$ $(u = -m+1,\ldots,m)$. Combining this expression with (10.7.29) and (10.7.24) yields

$$\hat{z}_k - z_k = \alpha_k\mathbf{p}_k^H\mathbf{R}_x^\dagger\mathbf{C}_k(\hat{\mathbf{r}} - \mathbf{r}) + o_P(n^{-1/2}). \tag{10.7.30}$$

By Theorem 4.4, $\sqrt{n}(\hat{\mathbf{r}} - \mathbf{r}) \overset{D}{\to} \mathrm{N}_c(\mathbf{0}, \boldsymbol{\Gamma}, \tilde{\boldsymbol{\Gamma}})$, where

$$\boldsymbol{\Gamma} := \lim_{n\to\infty} n\,\text{Cov}(\hat{\mathbf{r}}, \hat{\mathbf{r}}), \quad \tilde{\boldsymbol{\Gamma}} := \lim_{n\to\infty} n\,\text{Cov}(\hat{\mathbf{r}}, \hat{\mathbf{r}}^*).$$

Therefore, it follows from (10.7.30) that $\sqrt{n}(\hat{\mathbf{z}} - \mathbf{z}) \overset{D}{\to} \mathrm{N}_c(\mathbf{0}, \boldsymbol{\Sigma}_z, \tilde{\boldsymbol{\Sigma}}_z)$ with $\boldsymbol{\Sigma}_z := [\sigma_{kk'}]$ and $\tilde{\boldsymbol{\Sigma}}_z := [\tilde{\sigma}_{kk'}]$, where

$$\begin{aligned}\sigma_{kk'} &:= \alpha_k\alpha_{k'}^*\mathbf{p}_k^H\mathbf{R}_x^\dagger\mathbf{C}_k\boldsymbol{\Gamma}\mathbf{C}_{k'}^H\mathbf{R}_x^\dagger\mathbf{p}_{k'}, \\ \tilde{\sigma}_{kk'} &:= \alpha_k\alpha_{k'}\mathbf{p}_k^H\mathbf{R}_x^\dagger\mathbf{C}_k\tilde{\boldsymbol{\Gamma}}\mathbf{C}_{k'}^T\mathbf{R}_x^{\dagger *}\mathbf{p}_{k'}^*.\end{aligned}$$

It remains to show that $\mathbf{C}_k\boldsymbol{\Gamma}\mathbf{C}_{k'}^H = \boldsymbol{\Sigma}_{kk'}$ and $\mathbf{C}_k\tilde{\boldsymbol{\Gamma}}\mathbf{C}_{k'}^T = \tilde{\boldsymbol{\Sigma}}_{kk'}$.

Observe that the uth element of $\mathbf{C}_k\hat{\mathbf{r}}$ ($u = 1,\ldots,m$) can be expressed as

$$\sum_{j=1}^{m} q_{jk}\{\hat{r}_y(u-j+1) - z_k\hat{r}_y(u-j)\},$$

where q_{jk} denotes the jth element of $\mathbf{q}_k$. Therefore, similar to the proof of Theorem 7.5, we can write $\mathbf{C}_k\boldsymbol{\Gamma}\mathbf{C}_{k'}^H = [\sigma_{kk'}(u,v)]$ ($u,v = 1,\ldots,m$), where

$$\begin{aligned}\sigma_{kk'}(u,v) := \sum_{j,j'=1}^{m} q_{jk}q_{j'k'}^*\{&\gamma(u-j+1,v-j'+1) - z_k\gamma(u-j,v-j'+1)\\ &- z_{k'}^*\gamma(u-j+1,v-j') + z_k z_{k'}^*\gamma(u-j,v-j')\}.\end{aligned}$$

For the same reason, we have $\mathbf{C}_k\tilde{\boldsymbol{\Gamma}}\mathbf{C}_{k'}^T = [\tilde{\sigma}_{kk'}(u,v)]$, where

$$\begin{aligned}\tilde{\sigma}_{kk'}(u,v) := \sum_{j,j'=1}^{m} q_{jk}q_{j'k'}\{&\gamma(u-j+1,-v+j'-1) - z_k\gamma(u-j,-v+j'-1)\\ &- z_{k'}\gamma(u-j+1,-v+j') + z_k z_{k'}\gamma(u-j,-v+j')\}.\end{aligned}$$

Under the white noise assumption, $\gamma(\cdot,\cdot)$ takes the form (10.7.7). Let $d_k(z) := (z_k - z)\sum_{j=1}^{m} q_{jk}z^{-j}$ and $c_{jk} := q_{j+1,k} - z_k q_{jk}$, where $q_{jk} := 0$ for $j \notin \{1,\ldots,m\}$. Then, it is not difficult to verify that

$$\begin{aligned}\sigma_{kk'}(u,v) &= \sum_{l=1}^{p} 2C_l^2\sigma^2 d_k(z_l)d_{k'}^*(z_l)z_l^{u-v}\\ &\quad + \sigma^4 c_{kk'}(u,v) + \sigma^4(\kappa-2)c_{uk}c_{vk'}^*,\\ \tilde{\sigma}_{kk'}(u,v) &= \sum_{l=1}^{p} 2C_l^2\sigma^2 d_k(z_l)d_{k'}(z_l)z_l^{u+v}\\ &\quad + \sigma^4\tilde{c}_{kk'}(u,v) + \sigma^4(\kappa-2)c_{uk}c_{vk'},\end{aligned}$$

where $c_{kk'}(u,v) := \sum_{j=0}^{m} c_{u-v+j,k}c_{jk'}^*$ and $\tilde{c}_{kk'}(u,v) := \sum_{j=0}^{m} c_{u+v-j,k}c_{jk'}$. The assertion regarding $\hat{\mathbf{z}}$ is thus proved. The assertion regarding $\hat{\boldsymbol{\omega}}$ can be proved by citing Lemma 12.5.2(c). □

NOTES

Perturbation analysis of the reduced rank linear prediction estimators can be found in [155] and [307]. Further studies of the reduced rank EYW estimators can be found in [59], [353], [371], [374], and [376]. Further analysis of the MUSIC and ESPRIT estimators can be found in [102], [175], [372], [373], [377], and [389].

The eigenvalues of a covariance matrix can always be estimated consistently by the eigenvalues of a consistent estimator of the covariance matrix. The eigenvectors of a consistent estimator of the covariance matrix are not necessarily consistent for estimating the eigenvectors of the covariance matrix, unless all the associated eigenvalues are distinct.

The MUSIC methodology can be applied to a more general problem where multiple independent realizations of $\{y_1, \ldots, y_n\}$ are available. Some analytical results can be found in [372] and [373] regarding the statistical performance of the MUSIC estimator when the number of realizations approaches infinity.

The methods developed in this chapter for the complex case are readily applicable to the real case, because a real sinusoid can be represented as the sum of two complex sinusoids with conjugate frequencies. The conjugate property is utilized in [248] to develop an ESPRIT-like method for the real case with reduced computational cost. The MUSIC-like method discussed in [366] also utilizes the conjugate property.

Chapter 11

Further Topics

This chapter briefly discusses some further topics, including special algorithms for estimating the frequency of a single complex sinusoid, adaptive algorithms for tracking time-varying frequencies, methods for estimating the fundamental frequency of a periodic function, some extensions of the methods developed in the previous chapters to multiobservation, multichannel, and multidimensional cases, and finally, the methodology of quantile periodogram analysis.

11.1 Single Complex Sinusoid

Although the general methods discussed in the previous chapters are all applicable to the case of a single complex sinusoid, there are special methods designed specifically for this case with the aim of faster estimation. They are motivated by applications such as the carrier frequency estimation in digital communication systems where hardware limitations prohibit the deployment of complicated algorithms [112]. In this section, we discuss just a few examples.

Consider a time series $\{y_1,\ldots,y_n\}$ of the form

$$y_t = \beta_1 \exp(it\omega_1) + \epsilon_t \qquad (t = 1,\ldots,n), \tag{11.1.1}$$

where $\beta_1 \in \mathbb{C}$ and $\omega_1 \in \Omega := (-\pi,\pi) \setminus \{0\}$ are unknown constants and $\{\epsilon_t\}$ is a zero-mean white noise process. It has been shown in Chapters 6 and 8 that under the Gaussian assumption $\{\epsilon_t\} \sim \mathrm{GWN}_c(0,\sigma^2)$ the maximum likelihood (ML) estimator of ω_1 coincides with the periodogram maximizer

$$\hat{\omega}_1 := \arg\max_{\omega\in\Omega} I_n(\omega), \tag{11.1.2}$$

where $I_n(\omega) := n^{-1}|\sum_{t=1}^n y_t \exp(-it\omega)|^2$ is the periodogram. This estimator has desirable statistical properties such as consistency and asymptotic efficiency under the Gaussian assumption (Theorems 6.10 and 8.1). However, the computational complexity of $\hat{\omega}_1$ in (11.1.2) is at best $\mathcal{O}(n \log n)$ (based on FFT), which is too high in some applications. This difficulty motivates the development of simpler alternatives.

One class of such estimators is based on the sample autocovariances. Because $\{\epsilon_t\}$ is white noise, it follows that $r_y(u) = |\beta_1|^2 \exp(iu\omega_1)$ for all $u \neq 0$. Therefore, with $u = 1$ in particular, we obtain

$$\angle r_y(1) = \omega_1.$$

By replacing $r_y(1)$ with the lag-1 sample autocovariance $\hat{r}_y(1) := n^{-1}\sum_{t=1}^{n-1} y_{t+1} y_t^*$, one obtains a consistent estimator

$$\hat{\omega}_1 := \angle \hat{r}_y(1). \tag{11.1.3}$$

This is nothing but a Yule-Walker estimator based on the AR model of order 1 (Section 9.1). Although not statistically efficient, the low computational complexity, which is merely $\mathcal{O}(n)$, justifies this estimator in certain applications [211]. Some extensions of this estimator can be found, for example, in [110], [242], and [409], where sample autocovariances of higher lags are incorporated.

A variation of (11.1.3), which is considered in [178], is given by

$$\hat{\omega}_1 := \angle \sum_{t=1}^{n-1} w_{nt}\, y_{t+1} y_t^*, \tag{11.1.4}$$

where the w_{nt} are positive weights which take the form

$$w_{nt} := \frac{6t(n-t)}{n(n^2-1)} \qquad (t = 1, \ldots, n-1). \tag{11.1.5}$$

Obviously, the estimator in (11.1.3) can be expressed as (11.1.4) with the choice of uniform weights $w_{nt} = 1/n$. In general, the weighted average in (11.1.4) can be regarded as an alternative to the lag-1 sample autocovariance $\hat{r}_y(1)$ for estimating $r_y(1)$. The particular choice of the parabolic weights in (11.1.5) is motivated in [178] by an analysis of a phase-based GLS estimator to be discussed shortly. With the weights given by (11.1.5), the resulting estimator in (11.1.4) is known as the *weighted linear predictor* (WLP). Modifications and analyses of this estimator can be found in [49], [74], and [350], for example.

Another estimator, which is proposed in [178] by modifying the phase-based linear regression method in [392], takes the form

$$\hat{\omega}_1 := \sum_{t=1}^{n-1} w_{nt} \angle y_{t+1} y_t^*, \tag{11.1.6}$$

where the w_{nt} are given by (11.1.5). This estimator, known as the *weighted phase averager* (WPA), can be derived by applying the GLS method to the phase differences $\angle y_{t+1} y_t^* = \angle y_{t+1} - \angle y_t$ $(t = 1, \ldots, n-1)$ under the working assumption that

$y_t = \beta_1 \exp(it\omega_1 + e_t)$ and $\{e_t\} \sim \text{WN}(0, \varsigma^2)$ for some unknown $\varsigma^2 > 0$. Interchanging the angle and summation operators in (11.1.6) leads to the WLP estimator in (11.1.4). Hence, the WLP estimator is the phase of a weighted average of $y_{t+1}y_t^*$, whereas the WPA estimator is a weighted average of the phase of $y_{t+1}y_t^*$.

The WPA estimator can be justified intuitively by the fact that in the absence of noise we have $\angle y_{t+1}y_t^* = \angle|\beta_1|^2 \exp(i\omega_1) = \omega_1$. In the presence of noise, define

$$\xi_t := (\epsilon_t/\beta_1)\exp(-it\omega_1).$$

Then, we can always write $y_t = \beta_1 \exp(it\omega_1)(1+\xi_t)$. This implies that

$$\angle y_{t+1}y_t^* = \angle|\beta_1|^2 \exp(i\omega_1)(1+\xi_{t+1})(1+\xi_t)^* = \omega_1 + \zeta_t, \tag{11.1.7}$$

where the additive errors ζ_t are certain random variables that depend on ξ_{t+1} and ξ_t. The constant-plus-noise form (11.1.7) gives the intuitive justification for using the average of $\angle y_{t+1}y_t^*$ to estimate ω_1. Moreover, under the working assumption $y_t = \beta_1 \exp(it\omega_1 + e_t)$, we can write $\angle y_{t+1}y_t^* = \omega_1 + \zeta_t$ with $\zeta_t := e_{t+1} - e_t$. This is a simple linear regression model with correlated errors. By incorporating the correlation structure of $\{\zeta_t\}$, the resulting GLS estimator of ω_1 takes the form (11.1.6) with the weights given by (11.1.5).

To take a closer look at the WPA estimator in (11.1.6) under more general conditions, observe that the errors ζ_t in (11.1.7) can be expressed as

$$\zeta_t = \angle(1+\xi_{t+1}) - \angle(1+\xi_t) + 2\pi m_t,$$

where m_t is the unique integer such that

$$\omega_1 + \angle(1+\xi_{t+1}) - \angle(1+\xi_t) + 2\pi m_t \in (-\pi, \pi].$$

Because the ζ_t do not have zero mean in general, the WPA estimator is biased [303]. When the SNR is sufficiently high, the random variables $|\xi_t| = |\epsilon_t|/|\beta_1|$ tend to appear near zero, and so do the phases $\angle(1+\xi_t) = \arctan\{\Im(\xi_t), 1+\Re(\xi_t)\}$. Therefore, if the signal frequency ω_1 is sufficiently away from the endpoints $-\pi$ and π, then the errors ζ_t tend to be distributed around zero and hence can be effectively attenuated in the weighted average of the phases in (11.1.6). However, the averaging operation becomes less effective when ω_1 is near $-\pi$ or π, or when the SNR is low. This can be attributed to the problem of phase wrapping (i.e., $m_t \neq 0$), as explained and illustrated in [112].

To mitigate the phase wrapping problem, a simple idea, as in the secondary analysis discussed in Chapter 6 (Section 6.3.3), is to rotate the phases toward zero based on an initial guess $\tilde{\omega}_1$ and use the weighted average of the rotated phases $\angle y_{t+1}y_t^* \exp(-i\tilde{\omega}_1)$ to estimate the frequency increment $\delta := \omega_1 - \tilde{\omega}_1$ [210]. This leads to a modified WPA estimator of ω_1, which takes the form

$$\hat{\omega}_1 := \tilde{\omega}_1 + \sum_{t=1}^{n-1} w_{nt} \angle y_{t+1}y_t^* \exp(-i\tilde{\omega}_1). \tag{11.1.8}$$

Other modifications can be found in [112], [113], [187], and [321], for example. Some of these modifications are valid only if ω_1 is restricted to a smaller range, which may limit their applicability.

With a totally different flavor, the method proposed in [272] recasts (11.1.1) as a nonlinear state-space model and employs an *extended Kalman filter* (EKF) to estimate the signal frequency recursively as part of the hidden state. More precisely, define the two-dimensional hidden state at time t by

$$\mathbf{x}_t := [x_{t1}, x_{t2}]^T := [z_1, x_t]^T,$$

where $z_1 := \exp(i\omega_1)$ and $x_t := \beta_1 \exp(it\omega_1)$. Because $x_{t+1} = z_1 x_t$, it follows that

$$\mathbf{x}_{t+1} = [z_1, z_1 x_t]^T = \boldsymbol{\phi}(\mathbf{x}_t) := [x_{t1}, x_{t1} x_{t2}]^T.$$

Moreover, the observation equation (11.1.1) can be rewritten as

$$y_t = \mathbf{h}^T \mathbf{x}_t + \epsilon_t,$$

where $\mathbf{h} := [0, 1]^T$. With suitable initial values $\hat{\mathbf{x}}_{1|0}$ and $\mathbf{Q}_{1|0}$, the EKF algorithm for estimating the state variables $\mathbf{x}_t$ $(t = 1, \ldots, n)$ is defined recursively by

$$\begin{aligned}
\hat{\mathbf{x}}_{t|t} &:= [\hat{x}_{t1}, \hat{x}_{t2}]^T = \hat{\mathbf{x}}_{t|t-1} + (y_t - \mathbf{h}^T \hat{\mathbf{x}}_{t|t-1}) \mathbf{g}_t, \\
\hat{\mathbf{x}}_{t+1|t} &= \boldsymbol{\phi}(\hat{\mathbf{x}}_{t|t}), \\
\mathbf{g}_t &= (\mathbf{h}^T \mathbf{Q}_{t|t-1} \mathbf{h} + 1)^{-1} \mathbf{Q}_{t|t-1} \mathbf{h}, \\
\mathbf{Q}_{t|t} &= \mathbf{Q}_{t|t-1} - \mathbf{g}_t \mathbf{h}^T \mathbf{Q}_{t|t-1}, \\
\mathbf{Q}_{t+1|t} &= \boldsymbol{\Phi}_t \mathbf{Q}_{t|t} \boldsymbol{\Phi}_t^H,
\end{aligned}$$

where

$$\boldsymbol{\Phi}_t := \nabla^T \boldsymbol{\phi}(\hat{\mathbf{x}}_{t|t}) = \begin{bmatrix} 1 & 0 \\ \hat{x}_{t2} & \hat{x}_{t1} \end{bmatrix}.$$

Given the final estimate $\hat{\mathbf{x}}_{n|n} = [\hat{x}_{n1}, \hat{x}_{n2}]^T$ for $\mathbf{x}_n = [z_1, x_n]^T$, the signal frequency $\omega_1 = \angle z_1$ is estimated by

$$\hat{\omega}_1 := \angle \hat{x}_{n1}. \tag{11.1.9}$$

Although more complicated than the estimators in (11.1.4) and (11.1.6), the computational complex of the EKF algorithm still takes the form $\mathcal{O}(n)$.

Assuming that $\phi_1 := \angle \beta_1$ is a random variable with uniform distribution in $(-\pi, \pi]$ so that $E(x_t) = 0$ for all t, the ideal choice for the initial value $\hat{\mathbf{x}}_{1|0}$ is $E(\mathbf{x}_1) = [z_1, 0]^T$, but it cannot be used in practice due to its dependence on the unknown quantity z_1. A practical choice is given by

$$\hat{\mathbf{x}}_{1|0} := [\exp(i\tilde{\omega}_1), 0]^T,$$

where $\tilde{\omega}_1$ is an initial guess for ω_1. Similarly, the ideal choice for $\mathbf{Q}_{1|0}$ is the normalized error covariance matrix $E\{(\hat{\mathbf{x}}_{1|0}-\mathbf{x}_1)(\hat{\mathbf{x}}_{1|0}-\mathbf{x}_1)^H\}/\sigma^2$, which is unknown. A practical choice takes the form

$$\mathbf{Q}_{1|0} = \mathrm{diag}(q_1, q_2), \tag{11.1.10}$$

where $q_1 > 0$ and $q_2 > 0$ are tuning parameters.

The statistical performance of all the estimators discussed in this section is demonstrated by the following example.

Example 11.1 (Frequency Estimation for a Single Complex Sinusoid in Gaussian White Noise). Consider the case where the time series $\{y_1,\dots,y_n\}$ is given by (11.1.1) with $n=50$, $\beta_1=\exp(i\phi_1)$, $\phi_1 \sim \mathrm{U}(-\pi,\pi]$, and $\{\epsilon_t\} \sim \mathrm{GWN}_c(0,\sigma^2)$. Let the signal frequency be estimated by the WLP estimator in (11.1.4), the WPA estimator in (11.1.6), the modified WPA estimator in (11.1.8) (denoted as WPA2), and the EKF estimator in (11.1.9). Let the initial estimate $\tilde{\omega}_1$ in the modified WPA estimator and the EKF estimator be provided by the WLP estimator. Let the initial value $\mathbf{Q}_{1|0}$ of the EKF estimator be given by (11.1.10) with $q_1 = 0.01$ and $q_2 = 1$. For comparison, let the signal frequency be estimated also by the PM estimator discussed in Chapter 6 (Section 6.4.2) and the IFIL estimator discussed in Chapter 9 (Section 9.4.1). The PM estimator is computed by 5 iterations of the fixed-point algorithm (6.4.10) using the modified Newton-Raphson mapping defined by (6.4.11) and (6.4.12) with $\varrho=1$; the initial values are provided by the DFT estimator in (6.1.9). The IFIL estimator is computed by 5 iterations of the fixed-point algorithm (9.4.7) with $\eta = (0.24, 0.99)$ and $m = (3,2)$; the initial values are provided by the WLP estimator in (11.1.6). Figure 11.1 depicts the average MSE of these estimators based on 5,000 Monte Carlo runs under two scenarios: (a) $\omega_1 = 2\pi \times 0.13$ and (b) $\omega_1 = 2\pi \times 0.37$.

Under scenario (a), the WPA estimator is able to reach the CRLB when the SNR is greater than or equal to 10 dB. The SNR threshold is reduced slightly by the modified WPA estimator and considerably by the EKF estimator at the expense of higher computational burden. The IFIL estimator and the PM estimator exhibit the lowest SNR threshold. Without having to use the DFT for initialization, the IFIL estimator has a lower computational cost than the PM estimator.

Under scenario (b), where the signal frequency is near π, the phase wrapping problem significantly degrades the WPA estimator, making it even less accurate than the WLP estimator for the range of the SNR considered. The modified WPA estimator, on the other hand, retains its performance, thanks to the WLP-based phase rotation. The EKF estimator also remains intact, although at higher computational cost. The same is true for the IFIL and PM estimators.

The WLP estimator performs rather poorly under both scenarios. This is consistent with the findings in Chapter 9 (Section 9.1) regarding the linear prediction method based on an AR model of low order. ◇

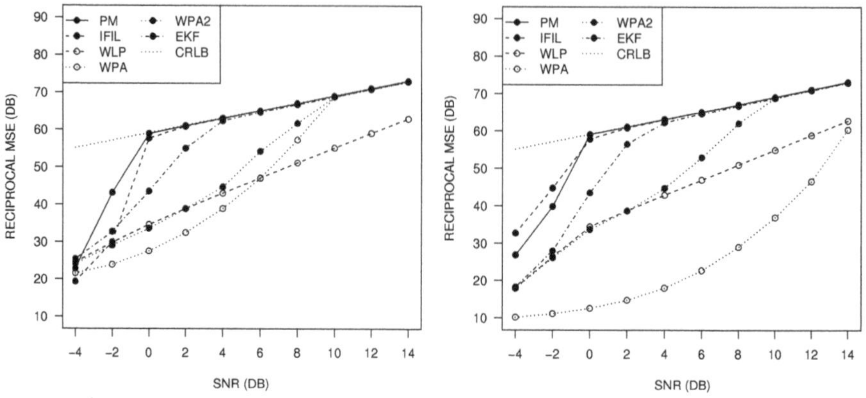

(a) (b)

Figure 11.1. Plot of the reciprocal average MSE of frequency estimates for the time series of a unit-amplitude random-phase complex sinusoid in Gaussian white noise discussed in Example 11.1 with different values of SNR. (a) $\omega_1 = 2\pi \times 0.13$. (b) $\omega_1 = 2\pi \times 0.37$. Solid line with dots, PM initialized by DFT; dashed line with dots, IFIL initialized by WLP; dashed line with circles, WLP; dotted line with circles, WPA; dotted line with dots, modified WPA initialized by WLP; dash-dotted line with dots, EKF initialized by WPA. Dotted line depicts the CRLB. Results are based on 5,000 Monte Carlo runs.

This example shows that some simple methods are capable of producing reasonably accurate frequency estimates when the SNR is sufficiently high. Given their low computational cost, these methods should have a place in the toolkit of frequency estimators. At lower SNR, more sophisticated methods, such as PM and IFIL, remain superior in delivering accurate frequency estimates.

11.2 Tracking Time-Varying Frequencies

It has been demonstrated in Chapter 6 (Section 6.5) that the periodogram maximization method developed under the constant frequency assumption can be used to track time-varying frequencies if applied together with a sliding window of suitable length. The same approach is valid for other frequency estimation methods discussed in the previous chapters. A drawback of this approach is the high computational cost because the estimator needs to be recomputed entirely when moving to the next window. This drawback motivates the development of recursive procedures that utilize the previously computed estimates and other quantities to construct the next estimates as new observations become available with greatly reduced computational burden.

For example, many recursive algorithms have been proposed for subspace and eigenvector tracking that can be applied to the frequency estimators discussed in Chapter 10. Some surveys, examples, and recent development on this subject can be found in [5], [17], [77], [95], and [429]. Other techniques such as extended Kalman filters [35] [212] and particle filters [98] [395] have also been proposed for tracking time-varying frequencies. In this section, we only discuss a simple but effective method called *adaptive notch filtering.*

Consider the case of a single real sinusoid in white noise with possibly time-varying frequency. In other words, let

$$y_t = A\cos(\phi_t) + B\sin(\phi_t) + \epsilon_t \qquad (t = 1, \ldots, n), \tag{11.2.1}$$

where $\omega_t := \phi_t - \phi_{t-1} \in (0, \pi)$ is called the *instantaneous frequency* at time t.

Recall that for estimating a constant frequency ω, in which case $\phi_t = \omega t$, the iterative filtering (IFIL) algorithm, discussed in Chapter 9 (Section 9.4), produces an updated estimate of the AR parameter $a := -2\cos(\omega)$ from a given initial value $\hat{a}$ by minimizing the prediction error sum of squares $\sum_{t=3}^{n}\{\hat{y}_t + \tilde{a}\hat{y}_{t-1} + \hat{y}_{t-2}\}^2$ with respect to $\tilde{a} \in \mathbb{R}$, where $\{\hat{y}_t\}$ is given by $\hat{y}_t = -\frac{1}{2}(1+\eta^2)\hat{a}\hat{y}_{t-1} - \eta^2\hat{y}_{t-2} + y_t$, with $\eta \in (0,1)$ being a bandwidth parameter. To devise a recursive algorithm, let us first substitute $\hat{a}$ in the last equation by the estimate at time $t-1$, which is denoted as $\hat{a}_{t-1}$, so that $\hat{y}_t$ is given by

$$\hat{y}_t := -\tfrac{1}{2}(1+\eta^2)\hat{a}_{t-1}\hat{y}_{t-1} - \eta^2\hat{y}_{t-2} + y_t \qquad (t = 1, 2, \ldots). \tag{11.2.2}$$

Moreover, instead of minimizing the prediction error sum of squares, let us compute the new estimate $\hat{a}_t$ by the so-called *recursive least-squares* (RLS) algorithm [146, p. 569]. This leads to

$$\hat{a}_t := \hat{a}_{t-1} - v_t\{\hat{y}_t + \hat{a}_{t-1}\hat{y}_{t-1} + \hat{y}_{t-2}\}\hat{y}_{t-1} \tag{11.2.3}$$

and

$$v_t := v_{t-1}/(v_{t-1}\hat{y}_{t-1}^2 + \lambda), \tag{11.2.4}$$

where $\lambda \in (0,1)$ is the forgetting factor. Regarding $\hat{a}_t$ as an estimate of $a_t := -2\cos(\omega_t)$, the time-varying frequency ω_t is estimated by $\hat{\omega}_t := \arccos(-\frac{1}{2}\hat{a}_t)$.

In the recursive algorithm defined by (11.2.2)–(11.2.4), the previous estimate $\hat{a}_{t-1}$ is updated upon the arrival of the new observation y_t to produce the next estimate $\hat{a}_t$. Owing to the forgetting factor λ which serves to discount older observations exponentially, the estimate $\hat{a}_t$ is able to adapt to the latest changes in the frequency. The forgetting factor λ controls not only the adaptability but also the variability of the frequency estimates. Generally speaking, fast adaptation to

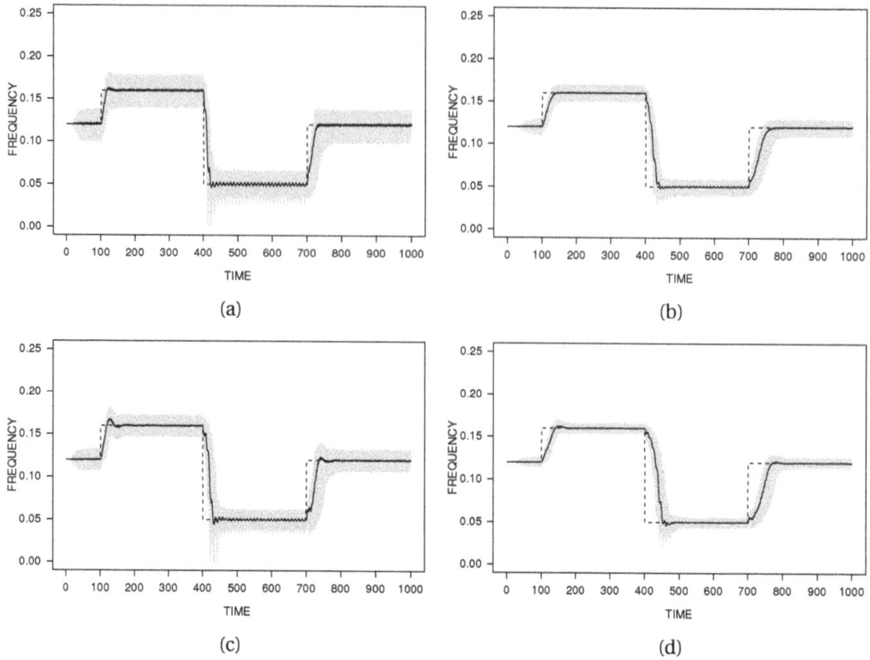

Figure 11.2. Tracking time-varying frequency by the RLS algorithm (11.2.2)–(11.2.4) with different choices of λ and η. (a) $\lambda = 0.8$, $\eta = 0.8$. (b) $\lambda = 0.9$, $\eta = 0.8$. (c) $\lambda = 0.8$, $\eta = 0.9$. (d) $\lambda = 0.9$, $\eta = 0.9$. In all cases, $v_0 = 0.001$ and $\hat{a}_0 = a_1$. Solid line, median of frequency estimates; shaded area, 5th and 95th percentiles of frequency estimates; dashed line, true frequency. Results are based on 5,000 Monte Carlo runs.

a time-varying frequency requires a small value for λ, but more accurate estimation for a constant or slowly varying frequency requires λ to be close to 1. The bandwidth parameter η plays a similar role. Therefore, a tradeoff must be made when setting these parameters.

Figure 11.2 contains a simulation example that demonstrates the role of λ and η. In this example, the time series takes the form (11.2.1) with $\epsilon_t \sim \text{GWN}(0, \sigma^2)$, where $A = 1$, $B = 0$, and $\sigma^2 = 0.5$ (SNR = 0 dB). The time-varying frequency ω_t is a piecewise-constant function of t, which equals $2\pi \times 0.12$ for $1 \le t \le 100$, $2\pi \times 0.16$ for $100 < t \le 400$, $2\pi \times 0.05$ for $400 < t \le 700$, and $2\pi \times 0.12$ for $700 < t \le 1000$. Figure 11.2 depicts the median of the frequency estimates $\hat{f}_t := \hat{\omega}_t/(2\pi)$ together with the 5th and 95th percentiles based on 5,000 Monte Carlo runs.

As we can see, the choice of $\lambda = 0.8$ and $\eta = 0.8$ in Figure 11.2(a) gives the fastest adaptation and the largest variability. The choice of $\lambda = 0.9$ and $\eta = 0.9$ in Figure 11.2(d) yields the smallest variability and the slowest adaptation. The choices in Figure 11.2(b) and Figure 11.2(c) offer some tradeoffs, with the former having slightly smaller variability during the stationary periods and less overshoot during the transition periods.

Besides the single-filter approach that requires a difficult tradeoff between adaptability and variability, the two-filter approach discussed in [236] judiciously coordinates a short-memory large-bandwidth filter with a long-memory narrow-bandwidth filter. The former is suitable for fast adaptation to big changes in the frequency, whereas the latter is suitable for accurate estimation when the frequency remains constant or changes slowly. The ordinary least-squares method with a sliding window is employed in [236] instead of the exponentially weighted RLS method, where the window size plays the role of the forgetting factor.

Based on the output from (11.2.2)–(11.2.4), the noise ϵ_t can be estimated by

$$\hat{\epsilon}_t := \tfrac{1}{2}(1+\eta^2)\{\hat{y}_t + \hat{a}_t\hat{y}_{t-1} + \hat{y}_{t-2}\}. \qquad (11.2.5)$$

The combination of this FIR filter with the IIR filter in (11.2.2) defines an *adaptive notch filter* (ANF) which is capable of eliminating the time-varying sinusoidal component of $\{y_t\}$. The sinusoidal signal $x_t := y_t - \epsilon_t$ can be estimated simply by $\hat{x}_t := y_t - \hat{\epsilon}_t$. This procedure is known as an *adaptive line enhancer* (ALE). An example of notch filtering and line enhancement is shown in Figure 11.3.

In the ANF procedure defined by (11.2.2)–(11.2.5), the IIR filter in (11.2.2) is based on the previous estimate $\hat{a}_{t-1}$ and the FIR filter in (11.2.5) is optimized at time t by the new estimate $\hat{a}_t$. Alternatively, one can combine the FIR and IIR filters to obtain a notch filter of the form $\hat{\epsilon}_t(\tilde{a}) := \frac{1}{2}(1+\eta^2)h(z,\tilde{a})y_t$, where

$$h(z,\tilde{a}) := \frac{1+\tilde{a}z^{-1}+z^{-2}}{1+\frac{1}{2}(1+\eta^2)\tilde{a}z^{-1}+\eta^2 z^{-2}}.$$

This method is employed in [56] as well as in [32], [80], [97], [207], and [267], albeit with different parameterizations. When the signal frequency is a constant, the corresponding AR parameter can be estimated by the minimizer of the sum of squared errors $\sum\{e_t(\tilde{a})\}^2$, where $e_t(\tilde{a}) := h(z,\tilde{a})y_t$. Standard gradient descent algorithms are readily applicable to this nonlinear optimization problem. For tracking a time-varying frequency, stochastic gradient descent methods are often used instead. For example, in [56], the estimate $\hat{a}_t$ is obtained by using the normalized *least mean-squares* (LMS) algorithm [146, p. 432], which leads to

$$\hat{a}_t := \hat{a}_{t-1} - \varrho e_t \dot{e}_t / u_t, \qquad (11.2.6)$$

where

$$e_t := -\tfrac{1}{2}(1+\eta^2)\hat{a}_{t-1}e_{t-1} - \eta^2 e_{t-2} + y_t + \tilde{a}_{t-1}y_{t-1} + y_{t-2}, \qquad (11.2.7)$$

$$\dot{e}_t := -\tfrac{1}{2}(1+\eta^2)\hat{a}_{t-1}\dot{e}_{t-1} - \eta^2\dot{e}_{t-2} - \tfrac{1}{2}(1+\eta^2)e_{t-1} + y_{t-1}, \qquad (11.2.8)$$

$$u_t := \lambda u_{t-1} + (1-\lambda)\dot{e}_t^2. \qquad (11.2.9)$$

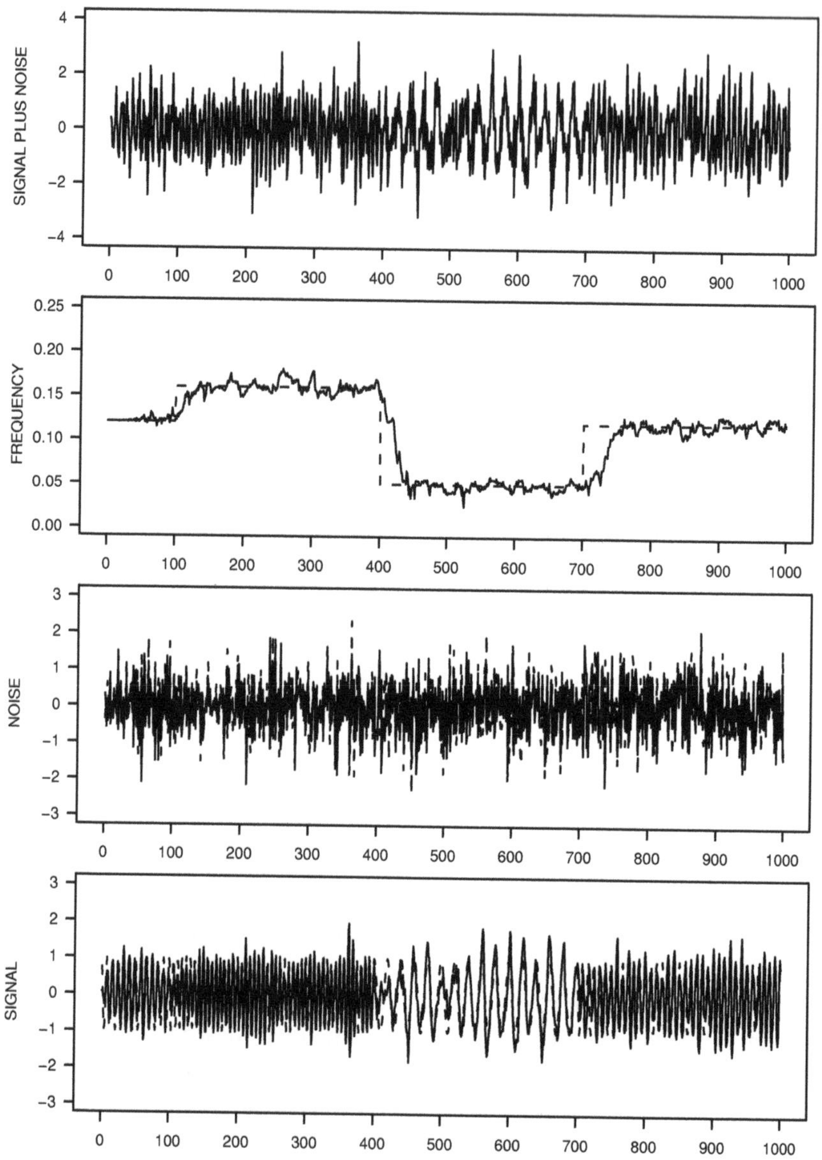

Figure 11.3. Adaptive notch filtering and line enhancement. Top to bottom: a time series consisting of a sinusoid with time-varying frequency in Gaussian white noise (0 dB); estimated frequency by the RLS algorithm (11.2.2)–(11.2.4) with $\lambda = 0.8$ and $\eta = 0.9$; estimated noise; and estimated sinusoidal signal. Dashed lines depict the true frequency, true noise, and true signal, respectively.

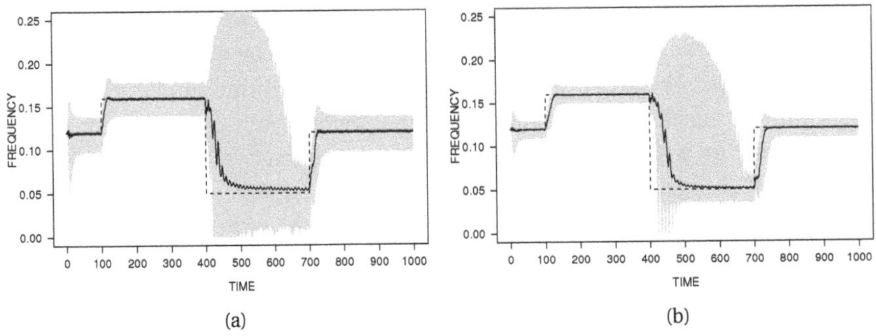

Figure 11.4. Tracking time-varying frequency by the LMS algorithm (11.2.6)–(11.2.9) with different step size ϱ for the same data as used in Figure 11.2. (a) $\varrho = 0.22$. (b) $\varrho = 0.11$. In all cases, $\lambda = 0.7$, $\eta = 0.8$, $u_0 = 7$, and $\hat{a}_0 = a_1$. Solid line, median of frequency estimates; shaded area, 5th and 95th percentiles of frequency estimates; dashed line, true frequency.

Note that (11.2.6) is nothing but a gradient descent iteration intended to minimize the instantaneous squared error $e_t^2(\tilde{a})$, with $\varrho >$ being the step-size parameter and u_t being the normalizer. Because

$$e_t(\tilde{a}) = -\tfrac{1}{2}(1+\eta^2)\tilde{a}\, e_{t-1}(\tilde{a}) - \eta^2 e_{t-2}(\tilde{a}) + y_t + \tilde{a} y_{t-1} + y_{t-2},$$

one obtains (11.2.7) by replacing $\tilde{a}$, $e_{t-1}(\tilde{a})$, and $e_{t-2}(\tilde{a})$ with $\hat{a}_{t-1}$, e_{t-1}, and e_{t-2}. Moreover, because the derivative of $e_t(\tilde{a})$, denoted as $\dot{e}_t(\tilde{a})$, satisfies

$$\dot{e}_t(\tilde{a}) = -\tfrac{1}{2}(1+\eta^2)\tilde{a}\, \dot{e}_{t-1}(\tilde{a}) - \eta^2 \dot{e}_{t-2}(\tilde{a}) - \tfrac{1}{2}(1+\eta^2) e_{t-1}(\tilde{a}) + y_{t-1},$$

one obtains (11.2.8) by similar substitutions. In this algorithm, the step size ϱ is a crucial parameter that controls the adaptability and variability of the frequency estimates: fast adaptation requires a suitably large step size, but small variability requires the opposite. If too close to 1, the forgetting factor $\lambda \in (0,1)$, which is used to update the normalizer, can also reduce the speed of adaptation. The bandwidth parameter η has the same effect as in the previous algorithm.

One can also follow the method in [267] and derive an adaptive Gauss-Newton algorithm which simply replaces $1/u_t$ in (11.2.6) by v_t, where

$$v_t := v_{t-1}/(v_{t-1}\dot{e}_t^2 + \lambda).$$

This algorithm is just a minor variation of the LMS algorithm because (11.2.9) can be rewritten as $u_t^{-1} = u_{t-1}^{-1}/(u_{t-1}^{-1}\dot{e}_t^2(1-\lambda)+\lambda)$. It is also related to the RLS algorithm which employs $\hat{y}_t + \hat{a}_{t-1}\hat{y}_{t-1} + \hat{y}_{t-2}$ and $\hat{y}_{t-1}$ instead of e_t and $\dot{e}_t$.

Figure 11.4 shows the performance of the LMS algorithm defined by (11.2.6)–(11.2.9) for the same data as used in Figure 11.2. Note that the two values of the step size in this example are chosen such that the variability of the frequency estimates during the stationary periods are roughly the same as that shown in Figure 11.2(a) and Figure 11.2(b), respectively, where the same bandwidth parameter is used. As we can see, the LMS algorithm performs similarly to the RLS algorithm in tracking small changes in the frequency but produces excessively large variability when adapting to the biggest frequency hop at $t = 400$. This behavior of the LMS algorithm might be explained by a smaller basin of attraction (BOA) inherent to gradient-based techniques. Not surprisingly, the adaptive Gauss-Newton algorithm suffers from the same problem (not shown).

The notch filtering method can be generalized to track multiple frequencies. A simple method is to cascade single-frequency notch filters and update their parameters separately, as suggested in [56], [97], and [207]. One can also employ a combined higher-order filter and jointly update the higher-dimensional AR parameter, as suggested in [32] and [267]. Different ways of adjusting the tuning parameters over time are considered in [84], [97], [241], and [267]. Adaptive notch filters for complex sinusoids are discussed in [195], [286], and [390].

11.3 Periodic Functions in Noise

A continuous-time real or complex function $x(t)$ is said to be periodic with period $T > 0$ if $x(t+T) = x(t)$ for all t. A sinusoidal function of the form $c\exp(i\omega t)$ is a periodic function with period $T = 2\pi$. In general, it follows from the discussion in Chapter 1 (Section 1.1) that a periodic function $x(t)$ with period T can be expressed as a Fourier series of the form

$$x(t) = \sum_{k=-\infty}^{\infty} \beta_k \exp(i2\pi kt/T). \tag{11.3.1}$$

In the real case, we also have $\beta_{-k} = \beta_k^*$ $(k = 0, 1, 2, \ldots)$. A periodic function is called bandlimited if there exists an integer $q \ge 1$ such that $\beta_k = 0$ for all $|k| > q$. Sampling such a function at regular intervals of length Δ produces

$$x_t := x(\Delta t) = \sum_{k=-q}^{q} \beta_k \exp(i\omega_k t) \qquad (t = 1, \ldots, n), \tag{11.3.2}$$

where $\omega_k := 2\pi k\Delta/T$. If the sampling interval Δ is sufficiently small such that $2\pi q\Delta/T < \pi$, or equivalently, if the sampling rate Δ^{-1} (in samples per unit time) is sufficiently high such that

$$\Delta^{-1} > 2q/T,$$

then the ω_k $(k = \pm 1, \ldots, \pm q)$ remain in the interval $(-\pi, \pi)$ and the sinusoids in (11.3.2) are alias-free. The critical value $2q/T$ is called the *Nyquist frequency*.

Because $\{x_t\}$ in (11.3.2) is a sum of sinusoids, the general methods discussed in the previous chapters are all applicable to the noisy observations

$$y_t := x_t + \epsilon_t \qquad (t = 1, \ldots, n).$$

However, an important difference between (11.3.2) and a general sinusoidal signal is that the frequencies in (11.3.2) are harmonically related, i.e., all frequencies in (11.3.2) are integral multiples of a single frequency parameter

$$\omega_1 = 2\pi\Delta/T,$$

known as the *fundamental frequency*, which resides in the interval $(0, \pi q^{-1})$. As a result, it suffices to estimate the fundamental frequency instead of multiple frequencies. The special harmonic relation of the frequencies in (11.3.2) can be utilized to improve the estimation of the fundamental frequency.

For example, as an application of the periodogram maximization (PM) method discussed in Chapter 6 (Section 6.4.2), the fundamental frequency ω_1 can be estimated by maximizing the *pooled* (or cumulated) periodogram for the fundamental frequency, which is defined as

$$\bar{I}_n(\omega) := (2q)^{-1} \sum_{k=1}^{q} \{I_n(k\omega) + I_n(-k\omega)\}, \quad \omega \in (0, \pi q^{-1}). \tag{11.3.3}$$

This method is studied in [306] for the real case and in [214] for the complex case, where the latter also incorporates an AIC-like criterion to determine q when it is unknown. The pooled periodogram is also employed in [150] to detect the presence of a hidden periodic function.

According to [306], if the conditions of Theorem 6.9 are satisfied in the real case, then the maximizer of $\bar{I}_n(\omega)$ in (11.3.3), denoted as $\hat{\omega}_1$, has the asymptotic property that $n(\hat{\omega}_1 - \omega_1) \xrightarrow{a.s.} 0$ and

$$n^{3/2}(\hat{\omega}_1 - \omega_1) \xrightarrow{D} \mathrm{N}(0, 12\gamma_0^{-1}),$$

where $\gamma_0 := \sum_{k=1}^{q} 2k^2|\beta_k|^2/f_\epsilon(\omega_k)$. The asymptotic variance coincides with the Gaussian CRLB which is derived in [268] for the case of white noise and can be extended to the case of colored noise by following the proof of Theorem 3.4. In comparison, Theorem 6.9 ensures that the local maximizer of the ordinary (unpooled) periodogram $I_n(\omega)$ near ω_1 has a similar property except that its asymptotic variance is equal to $12\gamma_1^{-1}$ with $\gamma_1 := 2|\beta_1|^2/f_\epsilon(\omega_1)$. Because γ_1 is in general smaller than γ_0, these results imply that maximizing the pooled periodogram tends to produce more accurate estimates for the fundamental frequency than maximizing the ordinary periodogram. Similarly, in the complex case, the analysis in [214] suggests that the same assertion remains valid for the maximizer of $\bar{I}_n(\omega)$ in (11.3.3) except that its asymptotic variance is equal to $6\gamma_0^{-1}$

with $\gamma_0 := \sum_{k=-q}^{q} k^2|\beta_k|^2/f_\epsilon(\omega_k)$, which again coincides with the Gaussian CRLB (see [69] for a derivation in the case of white noise). Note that by Theorem 6.10 the asymptotic variance of the ordinary periodogram maximizer near ω_1 has an asymptotic variance $6\gamma_1^{-1}$, where $\gamma_1 := |\beta_1|^2/f_\epsilon(\omega_1)$.

As with the PM method, the maximizer of the pooled periodogram in (11.3.3) is an approximation to the Gaussian maximum likelihood (GML) or nonlinear least-squares (NLS) estimator under the harmonic constraint. According to the discussion in Chapter 8 (Section 8.2), the GML method for estimating the frequencies in (11.3.2) in the complex case without the harmonic constraint calls for the maximization of the multivariate periodogram defined as

$$P_n(\tilde{\omega}_{-q},\ldots,\tilde{\omega}_{-1},\tilde{\omega}_1,\ldots,\tilde{\omega}_q) := \|\mathbf{y}-\hat{\beta}_0\mathbf{1}\|^2 - \|\mathbf{y}-\mathbf{F}(\tilde{\boldsymbol{\omega}})\hat{\boldsymbol{\beta}}(\tilde{\boldsymbol{\omega}})\|^2,$$

where $\hat{\beta}_0 := n^{-1}\sum_{t=1}^{n} y_t$ is the sample mean and $\hat{\boldsymbol{\beta}}(\tilde{\boldsymbol{\omega}}) := \{\mathbf{F}^H(\tilde{\boldsymbol{\omega}})\mathbf{F}(\tilde{\boldsymbol{\omega}})\}^{-1}\mathbf{F}^H(\tilde{\boldsymbol{\omega}})\mathbf{y}$ is the LS regression coefficient of $\mathbf{y}$ on $\mathbf{F}(\tilde{\boldsymbol{\omega}}) := [\mathbf{f}(\tilde{\omega}_{-q}),\ldots,\mathbf{f}(\tilde{\omega}_{-1}),\mathbf{1},\mathbf{f}(\tilde{\omega}_1),\ldots,\mathbf{f}(\tilde{\omega}_q)]$ with $\tilde{\boldsymbol{\omega}} := (\tilde{\omega}_{-q},\ldots,\tilde{\omega}_{-1},\tilde{\omega}_1,\ldots,\tilde{\omega}_q)$ taking values in $\Omega := \{\tilde{\boldsymbol{\omega}} : -\pi < \tilde{\omega}_{-q} < \cdots < \tilde{\omega}_{-1} < \tilde{\omega}_1 < \cdots < \tilde{\omega}_q < \pi\}$. However, with the harmonic constraint $\omega_k = k\omega_1$, it suffices to maximize the univariate function

$$\tilde{P}_n(\omega) := P_n(-q\omega,\ldots,-\omega,\omega,\ldots,q\omega) \quad \omega \in (0,\pi q^{-1}), \tag{11.3.4}$$

which we call the *constrained* periodogram for the fundamental frequency. The maximizer of the constrained periodogram in (11.3.4) is expected to have the same consistency and asymptotic normality properties as the maximizer of the pooled periodogram in (11.3.3) if ω_1 is not too close to zero. It is also expected to perform better when ω_1 is close to zero, as the GML estimator does in the general case discussed in Chapter 8 (Section 8.2) for estimating closely spaced frequencies. This is demonstrated in Figure 11.5.

In this example, the time series consists of three complex sinusoids with harmonically related frequencies in Gaussian white noise with $n = 50$. In the first case, shown on the left panel of Figure 11.5, the fundamental frequency is $\omega_1 = 2\pi \times 2.6/n = 2\pi \times 0.052$ and the remaining frequencies are $2\omega_1$ and $-3\omega_1$; the amplitudes are equal to 1 for all three frequencies; and the noise variance is equal to 1/2. The pooled and constrained periodograms for the fundamental frequency are calculated with $q = 3$. As we can see, the fundamental frequency is well estimated in this case by the global maxima of both pooled and constrained periodograms ($\hat{\omega}_1$ is equal to $2\pi \times 0.0530$ and $2\pi \times 0.0525$, respectively). In the second case, shown on the right panel of Figure 11.5, the fundamental frequency is $\omega_1 = 2\pi \times 0.6/n = 2\pi \times 0.012$ and the other frequencies are $2\omega_1$ and $-5\omega_1$; the amplitude parameters are 1, 1, and 0.5, respectively; and the noise variance is again equal to 1/2. In this case, the fundamental frequency is closer to zero. As a result, the pooled periodogram, calculated with $q = 5$, fails to produce a good estimate for the fundamental frequency ($\hat{\omega}_1 = 2\pi \times 0.0055$). The corresponding

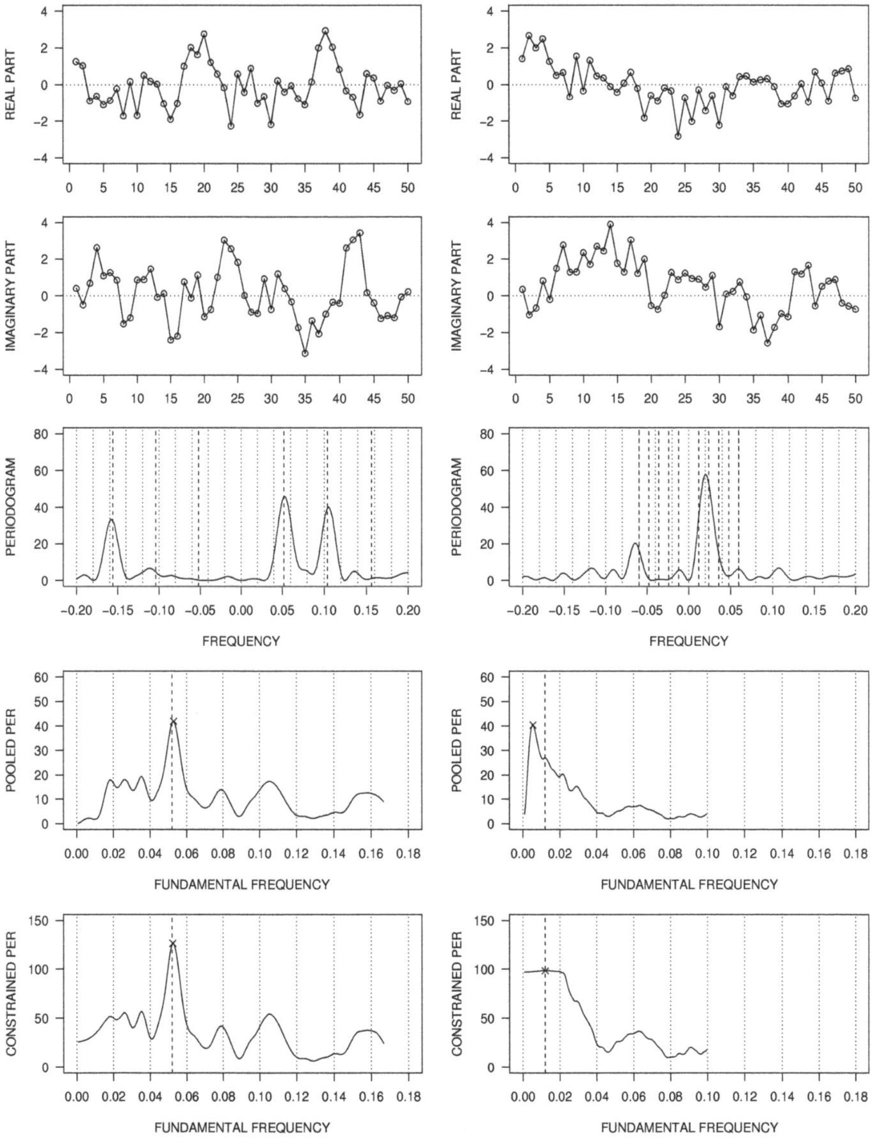

Figure 11.5. Estimation of the fundamental frequency for a time series consisting of three complex sinusoids with harmonically related frequencies in Gaussian white noise: left, $\omega_1 = 2\pi \times 0.052$ and $q = 3$; right, $\omega_1 = 2\pi \times 0.012$ and $q = 5$. Top to bottom: real and imaginary parts of the time series, ordinary periodogram, pooled periodogram for the fundamental frequency defined by (11.3.3), and constrained periodogram for the fundamental frequency defined by (11.3.4). Vertical dashed lines depict the fundamental frequency and its harmonics; vertical dotted lines depict the Fourier grid. Crosses "×" indicate the global maxima of the pooled and constrained periodograms.

constrained periodogram retains its good performance ($\hat{\omega}_1 = 2\pi \times 0.0120$), even though the peak is not as sharp as in the first case.

The subspace-based MUSIC estimator, which is discussed in Chapter 10 (Section 10.3), can also be modified to take advantage of the harmonic constraint. Under the simplifying assumption that $\beta_0 = 0$ in (11.3.2), the ordinary MUSIC estimator of ω_1 is a local minimizer of

$$\hat{g}_m(\omega) := \sum_{j=2q+1}^{m} |\hat{\mathbf{v}}_j^H \mathbf{f}_m(\omega)|^2,$$

where the $\hat{\mathbf{v}}_j$ are the eigenvectors of an estimated m-dimensional covariance matrix of $\{y_t\}$ associated with the $m-2q$ smallest eigenvalues ($m > 2q$). A pooled MUSIC estimator minimizes the objective function

$$\bar{g}_m(\omega) := (2q)^{-1} \sum_{k=1}^{q} \{\hat{g}_m(k\omega) + \hat{g}_m(-k\omega)\}, \quad \omega \in (0, \pi q^{-1}).$$

This method is investigated in [69] together with a multiplicatively penalized criterion to determine q. The eigenvalue-based ESPRIT estimator is also used in [69] to initialize a weighted least-squares method proposed in [221]. The latter combines unconstrained preliminary estimates of the sinusoidal parameters (including amplitude and phase) into an estimate of the constrained parameters based on the recognition of a linear relationship between parameters of unconstrained sinusoids and harmonically constrained sinusoids.

Finally, let us examine the implication of the harmonic constraint on the AR reparameterization of the signal frequencies discussed in Chapter 9 (Section 9.2). Assuming $\beta_0 = 0$ for simplicity, it is easy to verify that the signal $\{x_t\}$ in (11.3.2) satisfies $a(z)x_t = 0$ for all t, where

$$a(z) := \sum_{j=0}^{2q} a_j z^{-j} = \prod_{k=1}^{q} (1 - 2\cos(k\omega_1) z^{-1} + z^{-2}). \tag{11.3.5}$$

This special structure is used in [268] to construct a so-called comb filter of the form $h(z,\omega) := a(z)/a(\eta^{-1}z)$, where $\eta \in (0,1)$ is a tuning parameter that controls the bandwidth of the filter. The fundamental frequency is estimated by minimizing the sum of squares $\sum_{t=1}^{n} |\epsilon_t(\omega)|^2$, where $\epsilon_t(\omega) := h(z,\omega)y_t$ is the output of the filter. Note that the coefficients a_j in (11.3.5) are real and symmetric such that $a_0 = 1$ and $a_j = a_{2q-j}$ ($j = 0, 1, \ldots, q-1$). This is the same as in the real case with unconstrained frequencies. The main difference is that the coefficients under the harmonic constraint are further related to each other through a single parameter $\cos(\omega_1)$. In fact, by Chebyshev's recursion

$$\cos(k\omega) = 2\cos(\omega)\cos((k-1)\omega) - \cos((k-2)\omega),$$

we can express $\cos(k\omega_1)$ $(k \geq 2)$ as a linear combination of $\cos^r(\omega_1)$ for $r = 1,\ldots,k$. As a result, the coefficients a_j in (11.3.5) are linear functions of $\cos^r(\omega_1)$ for $r = 1,\ldots,2q$, so there exists a matrix $\boldsymbol{\Phi}$ such that

$$\mathbf{a}_e := [a_0, a_1, \ldots, a_{2q}]^T = \boldsymbol{\Phi}\,\mathbf{c}(\omega_1),$$

where $\mathbf{c}(\omega) := [1, \cos(\omega), \ldots, \cos^{2q}(\omega)]^T$. This constraint on the AR parameter is employed in [57] to construct an iterative linear prediction algorithm for estimating the fundamental frequency ω_1.

11.4 Beyond Single Time Series

To cover a wider range of applications, the sinusoid-plus-noise model discussed in the previous chapters can be extended in three directions: multiobservation, multichannel, and multidimension.

More specifically, consider the complex case as an example, where the simple sinusoid-plus-noise model takes the form

$$y_t = \sum_{k=1}^{p} \beta_k \exp(i\omega_k t) + \epsilon_t \qquad (t = 1,\ldots,n). \tag{11.4.1}$$

The multiobservation (MO) extension of this model can be expressed as

$$y_{jt} = \sum_{k=1}^{p} \beta_k \exp(i\omega_k t) + \epsilon_{jt} \qquad (t = 1,\ldots,n;\, j = 1,\ldots,d), \tag{11.4.2}$$

where the $\{\epsilon_{jt}\}$ $(j = 1,\ldots,d)$ are mutually independent noise processes. The MO model represents repeated measurements of the same sinusoidal signal with different, often independent and identically distributed, random errors. The observations at n equally-spaced linear array of sensors can be modeled as (11.4.2) when electromagnetic or acoustic waves emitted by distant sources arrive at the sensor array [369, pp. 230–231]. In this case, the frequencies in (11.4.2) represent the directions of the sources relative to the sensor array, so the problem of frequency estimation becomes that of direction-of-arrival (DOA) estimation. With $\boldsymbol{\beta} := [\beta_1,\ldots,\beta_p]^T$ and $\mathbf{F} := [\mathbf{f}(\omega_1),\ldots,\mathbf{f}(\omega_p)]$, we can rewrite (11.4.2) as

$$\mathbf{y}_j = \mathbf{F}\boldsymbol{\beta} + \boldsymbol{\epsilon}_j \qquad (j = 1,\ldots,d), \tag{11.4.3}$$

where $\mathbf{y}_j := [y_{j1},\ldots,y_{jn}]^T$ is the data record taken at the jth snapshot and $\boldsymbol{\epsilon}_j := [\epsilon_{j1},\ldots,\epsilon_{jn}]^T$ is the corresponding noise vector.

The methods developed in the previous chapters for a single snapshot can be easily extended to the MO case. For example, given $\tilde{\boldsymbol{\omega}} := (\tilde{\omega}_1,\ldots,\tilde{\omega}_p)$, define

$\mathbf{F}(\tilde{\boldsymbol{\omega}}) := [\mathbf{f}(\tilde{\omega}_1),\ldots,\mathbf{f}(\tilde{\omega}_p)]$. Then, under the assumption that the $\boldsymbol{\epsilon}_j$ are i.i.d. complex Gaussian with $\boldsymbol{\epsilon}_j \sim \mathrm{N}_c(\mathbf{0},\sigma^2\mathbf{I})$, the ML estimator of $\boldsymbol{\omega} := (\omega_1,\ldots,\omega_p)$ is given by the maximizer of the multivariate periodogram defined as

$$P_n(\tilde{\boldsymbol{\omega}}) := \sum_{j=1}^{d} \{\|\mathbf{y}_j\|^2 - \|\mathbf{y}_j - \mathbf{F}(\tilde{\boldsymbol{\omega}})\hat{\boldsymbol{\beta}}(\tilde{\boldsymbol{\omega}})\|^2\}, \tag{11.4.4}$$

where $\hat{\boldsymbol{\beta}}(\tilde{\boldsymbol{\omega}}) := \{\mathbf{F}^H(\tilde{\boldsymbol{\omega}})\mathbf{F}(\tilde{\boldsymbol{\omega}})\}^{-1}\mathbf{F}^H(\tilde{\boldsymbol{\omega}})\bar{\mathbf{y}}$ and $\bar{\mathbf{y}} := d^{-1}\sum_{j=1}^{d}\mathbf{y}_j$. Moreover, with $\hat{\mathbf{R}}_j$ denoting an estimate of the m-dimensional covariance matrix $\mathbf{R}_y$ based on $\mathbf{y}_j$, the EVD-based estimators of $\boldsymbol{\omega}$ discussed in Chapter 10 can be constructed from the average covariance matrix $\hat{\mathbf{R}}_y := d^{-1}\sum_{j=1}^{d}\hat{\mathbf{R}}_j$.

The multichannel (MC) extension of (11.4.1) takes the form

$$\mathbf{y}_j = \mathbf{F}\boldsymbol{\beta}_j + \boldsymbol{\epsilon}_j \qquad (j = 1,\ldots,d). \tag{11.4.5}$$

A major difference between the MC model (11.4.5) and the MO model (11.4.3) is that the coefficients of the sinusoids in the MC model can vary with the channel index j. Therefore, the MC model is well suited to represent the situation where a sinusoidal signal is observed by multiple instruments that have different attenuation factors as well as different noise characteristics.

The possibility that the $\boldsymbol{\epsilon}_j$ in the MC model may have different statistical properties and may be statistically dependent of each other makes the problem more complicated. The simplest case is where the $\boldsymbol{\epsilon}_j$ are i.i.d. random vectors. While the ML estimator of $\boldsymbol{\omega}$ in this case can be derived easily under the Laplace assumption, let us discuss only the Gaussian case where $\boldsymbol{\epsilon}_j \sim \mathrm{N}_c(\mathbf{0},\sigma^2\mathbf{I})$. Under the Gaussian assumption, the ML estimator of $\boldsymbol{\omega}$ is given by the maximizer of

$$P_n(\tilde{\boldsymbol{\omega}}) := \sum_{j=1}^{d} \{\|\mathbf{y}_j\|^2 - \|\mathbf{y}_j - \mathbf{F}(\tilde{\boldsymbol{\omega}})\hat{\boldsymbol{\beta}}_j(\tilde{\boldsymbol{\omega}})\|^2\}, \tag{11.4.6}$$

where $\hat{\boldsymbol{\beta}}_j(\tilde{\boldsymbol{\omega}}) := \{\mathbf{F}^H(\tilde{\boldsymbol{\omega}})\mathbf{F}(\tilde{\boldsymbol{\omega}})\}^{-1}\mathbf{F}^H(\tilde{\boldsymbol{\omega}})\mathbf{y}_j$ is the LS regression coefficient of $\mathbf{y}_j$ on $\mathbf{F}(\tilde{\boldsymbol{\omega}})$. The difference between (11.4.6) and (11.4.4) is that $\hat{\boldsymbol{\beta}}(\tilde{\boldsymbol{\omega}})$ in (11.4.4) is based on the complete data set $\{\mathbf{y}_1,\ldots,\mathbf{y}_d\}$ whereas $\hat{\boldsymbol{\beta}}_j(\tilde{\boldsymbol{\omega}})$ in (11.4.6) is based solely on the subset $\mathbf{y}_j$ from the jth channel. In a sightly more complicated situation where $\boldsymbol{\epsilon}_j$ has possibly different but known variance σ_j^2, it suffices to divide the jth term in (11.4.6) by σ_j^2 so that $P_n(\tilde{\boldsymbol{\omega}})$ becomes a weighted sum of the multivariate periodograms from each channel.

Moreover, let $\mathbf{R}_j$ denote the m-dimensional covariance matrix of $\mathbf{y}_j$. Then, as in the single channel case discussed in Chapter 10 (Section 10.1), we can write

$$\mathbf{R}_j = \mathbf{F}_m\mathbf{G}_j\mathbf{F}_m^H + \sigma_j^2\mathbf{I},$$

where $\mathbf{G}_j$ is a diagonal matrix with the kth diagonal element being the squared modulus of the kth element in $\boldsymbol{\beta}_j$. Hence, for any given weights $w_j \ge 0$ such that

$\sum_{j=1}^d w_j = 1$, the weighted average $\mathbf{R}_y := \sum_{j=1}^d w_j \mathbf{R}_j$ can be expressed as

$$\mathbf{R}_y = \mathbf{F}_m \mathbf{G} \mathbf{F}_m^H + \sigma^2 \mathbf{I},$$

where $\mathbf{G} := \sum_{j=1}^d w_j \mathbf{G}_j$ and $\sigma^2 := \sum_{j=1}^d w_j \sigma_j^2$. Observe that $\mathbf{R}_y$ has the same structure as the covariance matrix in the single channel case. Therefore, with $\hat{\mathbf{R}}_j$ denoting an estimate of $\mathbf{R}_j$ on the basis of $\mathbf{y}_j$, the EVD-based methods discussed in Chapter 10 can be applied to the weighted average $\hat{\mathbf{R}}_y := \sum_{j=1}^d w_j \hat{\mathbf{R}}_j$. Ideally, the weight w_j should be proportional to $1/\sigma_j^2$ so that the noise impact is equalized across the channels (noisier channels receive smaller weights).

An important issue in the multichannel case is the statistical dependence between $\mathbf{y}_j$ and $\mathbf{y}_{j'}$ for $j \neq j'$. It can be investigated in the frequency domain by examining the spectral coherence [38, Chapter 10]. The spectral coherence between $\mathbf{y}_j$ and $\mathbf{y}_{j'}$ is constructed from their *cross-periodogram*

$$I_{njj'}(\omega) := n z_j(\omega) z_{j'}^*(\omega),$$

where $z_j(\omega) := n^{-1}\mathbf{f}^H(\omega)\mathbf{y}_j$ and $z_{j'}(\omega) := n^{-1}\mathbf{f}^H(\omega)\mathbf{y}_{j'}$ are the Fourier transforms of $\mathbf{y}_j$ and $\mathbf{y}_{j'}$ at frequency $\omega \in [0, 2\pi)$. Note that the ordinary periodogram of $\mathbf{y}_j$ can be expressed as $I_{njj}(\omega)$ in this notation. With $\hat{f}_{jj'}(\omega)$ denoting a smoothed version of $I_{njj}(\omega)$, the spectral coherence function is defined as

$$\frac{\hat{f}_{jj'}(\omega)}{\sqrt{\hat{f}_{jj}(\omega)\hat{f}_{j'j'}(\omega)}}. \tag{11.4.7}$$

It is a measure of correlation between the Fourier components of $\mathbf{y}_j$ and $\mathbf{y}_{j'}$ at frequency ω. See [46, Section 11.7] for the asymptotic properties of the cross-periodogram and the spectral coherence in the absence of sinusoids.

As discussed in Chapter 8 (Section 8.3), a robust alternative to the Fourier transform $z_j(\omega)$ in the complex case is the Laplace-Fourier transform

$$\tilde{z}_j(\omega) := \arg\min_{z \in \mathbb{C}} \|\mathbf{y}_j - \mathbf{f}(\omega) z\|_1.$$

In the real case [227], the Laplace-Fourier transform can be defined as

$$\tilde{z}_j(\omega) := \tfrac{1}{2}\sqrt{n}\{\hat{A}_j(\omega) - i\hat{B}_j(\omega)\},$$

where

$$\hat{\boldsymbol{\beta}}_j(\omega) := [\hat{A}_j(\omega), \hat{B}_j(\omega)]^T := \arg\min_{\tilde{\boldsymbol{\beta}} \in \mathbb{R}^2} \|\mathbf{y}_j - \mathbf{F}(\omega)\tilde{\boldsymbol{\beta}}\|_1$$

and $\mathbf{F}(\omega) := [\Re\{\mathbf{f}(\omega)\}, \Im\{\mathbf{f}(\omega)\}]$. By replacing $z_j(\omega)$ with $\tilde{z}_j(\omega)$, we obtain a robust cross-periodogram

$$L_{njj'}(\omega) := n \tilde{z}_j(\omega) \tilde{z}_{j'}^*(\omega), \tag{11.4.8}$$

which we call the *Laplace cross-periodogram* between $\mathbf{y}_j$ and $\mathbf{y}_{j'}$. Note that

$$L_{njj}(\omega) = n|\tilde{z}_j(\omega)|^2 = \tfrac{1}{4}n\|\hat{\boldsymbol{\beta}}_j(\omega)\|^2.$$

Therefore, $L_{njj}(\omega)$ is just the Laplace periodogram of the first kind discussed in [224] and [228]. Similar to (11.4.7), let $\hat{\ell}_{jj'}(\omega)$ denote a smoothed version of $L_{njj'}(\omega)$ $(j, j' = 1,\ldots,d)$. Then, a robust spectral coherence function, called the *Laplace spectral coherence* [227], can be defined as

$$C_{njj'}(\omega) := \frac{\hat{\ell}_{jj'}(\omega)}{\sqrt{\hat{\ell}_{jj}(\omega)\hat{\ell}_{j'j'}(\omega)}}. \tag{11.4.9}$$

The robustness of this function against outliers is demonstrated by Figure 11.6.

In this example, the ordinary and Laplace spectral coherence functions are calculated for a pair of real electroencephalography (EEG) signals. These signals were recorded from the electrodes placed on the scalp of a subject when performing the task of identifying the hidden image in an ambiguous figure presented on a computer screen. The pair of signals shown in Figure 11.6 represents 1.28 seconds of measurements at two scalp sites, sampled at 200 Hz. See [189] for more details about the study.

It is well known that EEG signals are prone to contamination of artifacts (due to muscle movement, for example). Therefore, raw EEG data often need to be edited manually to remove these artifacts, which is labor intensive. In this example, a simulated artifact is inserted into one of the signals and its impact on the spectral coherence is considered. Without going into neurological interpretation of the spectral coherence, it suffices to observe from Figure 11.6 that the ordinary coherence is altered considerably by the artifact, whereas the Laplace coherence remains almost intact.

The asymptotic distribution of the Laplace cross-periodogram in the absence of sinusoids can be derived along the lines of the proof of Theorem 5.11. So the following result is stated without proof.

Theorem 11.1 (Asymptotic Distribution of the Laplace Cross-Periodogram). *Let* $y_{jt} = \epsilon_{jt}$ $(t = 1,\ldots,n; j = 1,\ldots,d)$ *and* $\mathbf{L}_n(\omega) := [L_{njj'}(\omega)]$ $(j, j' = 1,\ldots,d)$. *Assume that the* $\{\epsilon_{jt}\}$ $(j = 1,\ldots,d)$ *satisfy the conditions of Theorem 5.11 with* $\eta_j := \kappa_j/2$, *where* $\kappa_j > 0$ *denotes the common sparsity of the real and imaginary parts of* ϵ_{jt}. *Assume further that the real and imaginary parts of* $\{\epsilon_{jt}\}$ *and* $\{\epsilon_{j't}\}$ *are mutually independent and jointly stationary in zero-crossings in the sense that*

$$P\{\Re(\epsilon_{jt})\Re(\epsilon_{j's}) < 0\} = P\{\Im(\epsilon_{jt})\Im(\epsilon_{j's}) < 0\} = \gamma_{jj'}(t-s)$$

for all t *and* s. *Define* $\ell_{jj'}(\omega) := \eta_j\eta_{j'}h_{jj'}(\omega)$, *where*

$$h_{jj'}(\omega) := \sum_{u=-\infty}^{\infty} \{1 - 2\gamma_{jj'}(u)\}\exp(-iu\omega).$$

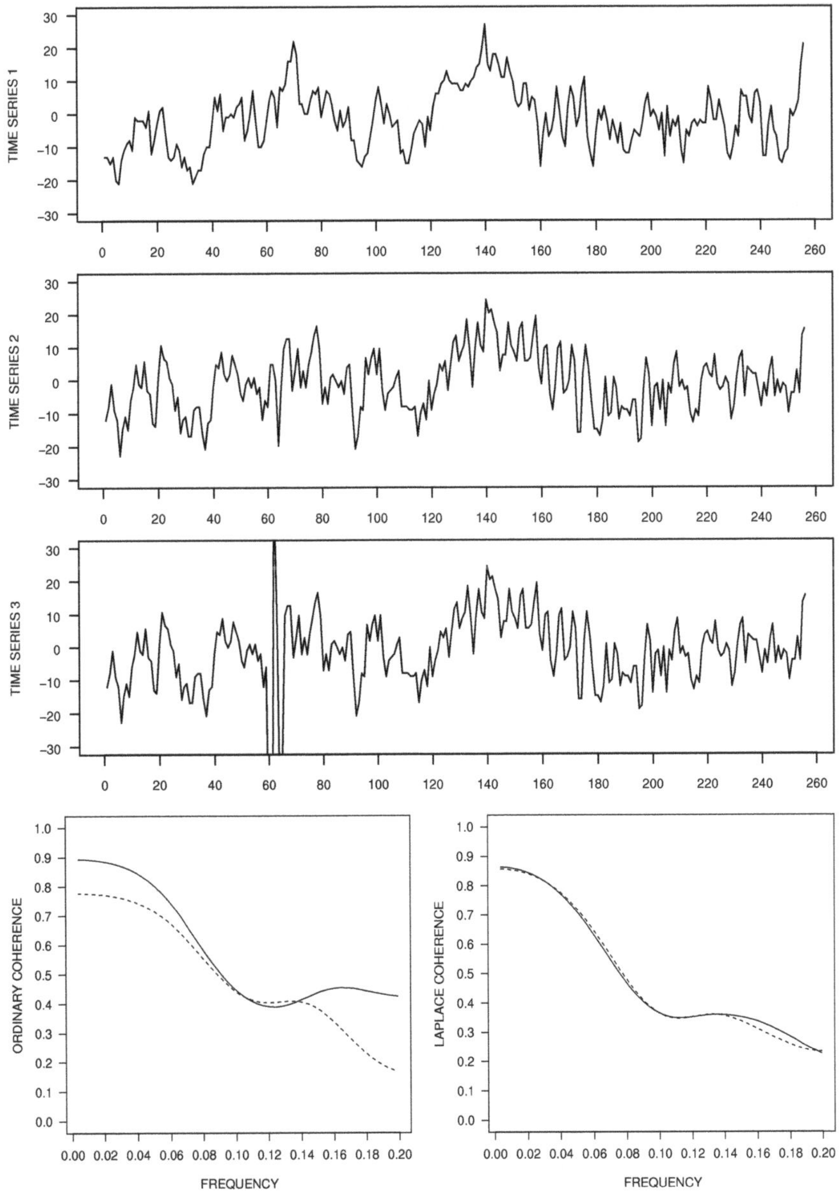

Figure 11.6. Spectral coherence between a pair of EEG signals in the presence of artifact contamination: series 1 and 2, uncontaminated; series 3, contaminated version of series 2. Bottom left, absolute value of the ordinary coherence. Bottom right, absolute value of the Laplace coherence. Solid line: coherence between series 1 and 2 (without contamination), dashed line, coherence between series 1 and 3 (with contamination).

Then, for any fixed $\tilde{\omega}_1,\ldots,\tilde{\omega}_m \in (0,2\pi)$, $\mathbf{L}_n(\tilde{\omega}_1),\ldots,\mathbf{L}_n(\tilde{\omega}_m)$ converge jointly in distribution to $\boldsymbol{\zeta}(\tilde{\omega}_1)\boldsymbol{\zeta}^H(\tilde{\omega}_1),\ldots,\boldsymbol{\zeta}(\tilde{\omega}_m)\boldsymbol{\zeta}^H(\tilde{\omega}_m)$ as $n\to\infty$, where $\boldsymbol{\zeta}(\tilde{\omega}_1),\ldots,\boldsymbol{\zeta}(\tilde{\omega}_m)$ are independent complex Gaussian random vectors with $\boldsymbol{\zeta}(\omega)\sim \mathrm{N}_c(\mathbf{0},[\ell_{jj'}(\omega)])$.

Remark 11.1 The assertion remains valid in the real case for $\tilde{\omega}_1,\ldots,\tilde{\omega}_m \in (0,\pi)$ if the $\{\epsilon_{jt}\}$ satisfy the conditions of Remark 5.16 and are jointly stationary in zero-crossings such that $P(\epsilon_{jt}\epsilon_{j's}<0)=\gamma_{jj'}(t-s)$ for all t and s. This is the counterpart of a well-known result for the ordinary cross-periodogram matrix $\mathbf{I}_n(\omega) := [I_{njj'}(\omega)]$ of a set of jointly stationary time series [46, pp. 434–453], which asserts that $\mathbf{I}_n(\tilde{\omega}_1),\ldots,\mathbf{I}_n(\tilde{\omega}_m)$ converge jointly in distribution to $\boldsymbol{\zeta}(\tilde{\omega}_1)\boldsymbol{\zeta}^H(\tilde{\omega}_1),\ldots,$ $\boldsymbol{\zeta}(\tilde{\omega}_m)\boldsymbol{\zeta}^H(\tilde{\omega}_m)$, where $\boldsymbol{\zeta}(\tilde{\omega}_1),\ldots,\boldsymbol{\zeta}(\tilde{\omega}_m)$ are independent complex Gaussian random vectors with $\boldsymbol{\zeta}(\omega)\sim \mathrm{N}_c(\mathbf{0},[f_{jj'}(\omega)])$. The function $f_{jj'}(\omega)$ is known as the cross-spectrum between $\{y_{jt}\}$ and $\{y_{j't}\}$. It is defined by

$$f_{jj'}(\omega) := \sum_{u=-\infty}^{\infty} r_{jj'}(u)\exp(-iu\omega),$$

where $r_{jj'}(u) := E(y_{j,t+u}\, y^*_{j't})$ is the cross-autocovariance function between $\{y_{jt}\}$ and $\{y_{j't}\}$. The cross-spectrum $f_{jj'}(\omega)$ can be estimated by a smoothed version of the cross-periodogram $I_{njj'}(\omega)$.

As we can see, the function $\ell_{jj'}(\omega)$ plays the same role in the asymptotic distribution of the Laplace cross-periodogram as the cross-spectrum does in the asymptotic distribution of the ordinary cross-periodogram. For this reason, we call $\ell_{jj'}(\omega)$ the *Laplace cross-spectrum* between $\{y_{jt}\}$ and $\{y_{j't}\}$. Note that $\ell_{jj'}(\omega)$ is a scaled version of the function $h_{jj'}(\omega)$, which we call the *zero-crossing cross-spectrum*. This function is the counterpart of the correlation cross-spectrum (or normalized cross-spectrum) as a scale-invariant frequency-domain representation of the mutual dependence between $\{y_{jt}\}$ and $\{y_{j't}\}$. According to Theorem 11.1, the Laplace cross-spectrum $\ell_{jj'}(\omega)$ is just the mean of the asymptotic distribution of the Laplace cross-periodogram $L_{njj'}(\omega)$. A smoothed version of $L_{njj'}(\omega)$ serves as an estimator of $\ell_{jj'}(\omega)$. Note that $\ell_{jj}(\omega)=\eta_j^2 h_{jj}(\omega)$ is nothing but the ordinary Laplace spectrum of $\{y_{jt}\}$ discussed in Chapter 5 (Section 5.5).

With $\mathbf{t} := [t_1,\ldots,t_d]^T$ and $\mathcal{T} := \{1,\ldots,n_1\}\times\cdots\times\{1,\ldots,n_d\}$, the multidimension (MD) extension of (11.4.1) takes the form

$$y_{\mathbf{t}} = \sum_{k=1}^{p} \beta_k \exp(i\boldsymbol{\omega}_k^T\mathbf{t}) + \epsilon_{\mathbf{t}} \qquad (\mathbf{t}\in\mathcal{T}), \tag{11.4.10}$$

where the $\boldsymbol{\omega}_k := [\omega_{1k},\ldots,\omega_{dk}]^T$ $(k=1,\ldots,p)$ are distinct frequency vectors in $(0,2\pi)^d$. In the MD case, a zero-mean and finite-variance process $\{\epsilon_{\mathbf{t}}\}$ $(\mathbf{t}\in\mathbb{Z}^d)$ is said to be (weakly) stationary if

$$E(\epsilon_{\mathbf{t}+\mathbf{u}}\epsilon_{\mathbf{t}}^*) = r_\epsilon(\mathbf{u}) \qquad \forall\, \mathbf{t},\mathbf{u}\in\mathbb{Z}^d,$$

where $r_\epsilon(\mathbf{u})$ is the multidimensional ACF. The corresponding SDF is defined as

$$f_\epsilon(\boldsymbol{\omega}) := \sum_{\mathbf{u}\in\mathbb{Z}^d} r_\epsilon(\mathbf{u})\exp(-i\mathbf{u}^T\boldsymbol{\omega}), \qquad \boldsymbol{\omega}\in[0,2\pi)^d.$$

The process is called white noise if $f_\epsilon(\boldsymbol{\omega}) = r_\epsilon(\mathbf{0})$ for all $\boldsymbol{\omega}$. For further discussions on multidimensional spectral modeling and estimation, see [177, Chapter 15], [258], and [298, Section 9.7], for example.

Because $y_\mathbf{t}$ in (11.4.10) is a sum of sinusoids in each dimension, one can separately estimate the frequencies in each dimension by recasting (11.4.10) as a multichannel model of the form (11.4.5). For example, in the two-dimensional case with $y_\mathbf{t} = y_{t_1 t_2}$, $n_1 = n$, and $n_2 = m$, the data vectors $\mathbf{y}_j := [y_{1j},\ldots,y_{nj}]^T$ $(j = 1,\ldots,m)$ can be expressed in the form of (11.4.5) with $\mathbf{F} := [\mathbf{f}(\omega_{11}),\ldots,\mathbf{f}(\omega_{1p})]$ and $\mathbf{f}(\omega) := [\exp(i\omega),\ldots,\exp(in\omega)]^T$. Therefore, the frequencies ω_{1k} can be estimated by applying any method developed for the MC case to the $\mathbf{y}_j$, regardless of the frequencies in the other dimension. Although attractive computationally, this approach is not statistically efficient because it ignores the constraints among the coefficient vectors $\boldsymbol{\beta}_j := [\beta_1\exp(-ij\omega_{21}),\ldots,\beta_p\exp(-ij\omega_{2p})]^T$.

To estimate the frequencies in all dimensions jointly, let $\mathbf{y} := \mathrm{vec}[y_\mathbf{t}]$, $\mathbf{f}(\boldsymbol{\omega}_k) := \mathrm{vec}[\exp(i\boldsymbol{\omega}_k^T\mathbf{t})]$, and $\boldsymbol{\epsilon} := \mathrm{vec}[\epsilon_\mathbf{t}]$ $(\mathbf{t}\in\mathcal{T})$. Then, we can rewrite (11.4.10) as

$$\mathbf{y} = \mathbf{F}\boldsymbol{\beta} + \boldsymbol{\epsilon}, \tag{11.4.11}$$

where $\boldsymbol{\beta} := [\beta_1,\ldots,\beta_p]^T$ and $\mathbf{F} := [\mathbf{f}(\boldsymbol{\omega}_1),\ldots,\mathbf{f}(\boldsymbol{\omega}_p)]$. As in the one-dimensional case discussed in Chapter 8 (Section 8.2), the ML estimator of $\boldsymbol{\omega} := (\boldsymbol{\omega}_1,\ldots,\boldsymbol{\omega}_p)$ under the Gaussian white noise assumption $\boldsymbol{\epsilon} \sim \mathrm{N}_c(\mathbf{0},\sigma^2\mathbf{I})$ is given by the maximizer of a multivariate periodogram which takes the same form as (8.2.16) with $\tilde{\boldsymbol{\omega}} := (\tilde{\boldsymbol{\omega}}_1,\ldots,\tilde{\boldsymbol{\omega}}_p) \in (0,2\pi)^{mp}$. Similarly, the LML (or NLAD) estimator considered in Chapter 8 (Section 8.3) can also be extended to the MD case.

In the special case where the frequencies form a rectangular grid in $(0,2\pi)^p$, the MD model in (11.4.10) can be more explicitly expressed as

$$y_\mathbf{t} = \sum_{\mathbf{k}\in\mathcal{K}} \beta_\mathbf{k}\exp(i\boldsymbol{\omega}_\mathbf{k}^T\mathbf{t}) + \epsilon_\mathbf{t} \qquad (\mathbf{t}\in\mathcal{T}), \tag{11.4.12}$$

where $\boldsymbol{\omega}_\mathbf{k} := [\omega_{1k_1},\ldots,\omega_{dk_d}]^T$ is a d-dimensional frequency vector indexed by $\mathbf{k} := [k_1,\ldots,k_d]^T \in \mathcal{K} := \{1,\ldots,p_1\}\times\cdots\times\{1,\ldots,p_d\}$. In this model, the ω_{jk} $(k = 1,\ldots,p_j)$ are distinct frequencies in the jth dimension, so the total number of frequency parameters in (11.4.12) equals $r := \sum_{j=1}^d p_j$. Note that r is less than the dimension d times the number of sinusoids in (11.4.12), unless $p_j \equiv 1$. As an example, consider the two-dimensional case with $y_\mathbf{t} = y_{t_1 t_2}$, $n_1 = n$, and $n_2 = m$. In this case, (11.4.11) is valid for $\mathbf{y} := [y_{11},\ldots,y_{1m},\ldots,y_{n1},\ldots,y_{nm}]^T$ with

$$\mathbf{F} := [\mathbf{f}_n(\omega_{11}),\ldots,\mathbf{f}_n(\omega_{1p_1})]\otimes[\mathbf{f}_m(\omega_{21}),\ldots,\mathbf{f}_m(\omega_{2p_2})],$$

where $\otimes$ denotes the Kronecker product, $\mathbf{f}_n(\omega) := [\exp(i\omega),\ldots,\exp(in\omega)]^T$ and $\mathbf{f}_m(\omega) := [\exp(i\omega),\ldots,\exp(im\omega)]^T$. Therefore, the GML frequency estimator maximizes a multivariate periodogram of the form (8.2.16) with

$$\tilde{\boldsymbol{\omega}} := (\tilde{\omega}_{11},\ldots,\tilde{\omega}_{1p_1},\tilde{\omega}_{21},\ldots,\tilde{\omega}_{2p_2}) \in (0,2\pi)^{p_1+p_2}.$$

Observe that the length of $\tilde{\boldsymbol{\omega}}$ is $r = p_1 + p_2$ instead of $2p_1p_2$. This estimator is analyzed in [311] (see also [204]) through an AR reparameterization which is similar to the one-dimensional case discussed in Chapter 9 (Section 9.5).

Besides the GML method, the DFT and PM methods discussed in Chapter 6 can be generalized to the MD case based on the multidimensional Fourier transform [298, Section 9.7.1]. In [351], the iterative quasi GML method discussed in Chapter 9 (Section 9.5) is extended to the two-dimensional case with a single sinusoid. Extensions of the subspace rotation (SSR) method in Chapter 10 (Section 10.4) to the two-dimensional case are discussed in [62], [158], and [322]. A two-dimensional extension of the WPA estimator defined by (11.1.6) can be found in [180]. Some criteria for determining the number of real sinusoids in the two-dimensional case are discussed in [190] and [192].

11.5 Quantile Periodogram

Quantile regression is a powerful technique of statistical analysis that extends the capability of the traditional method of least squares [196]. This technique is especially useful for analyzing heteroscedastic and nonGaussian data, where quantiles play more meaningful roles than the mean and variance. The quantile regression method also enriches the arsenal of spectral analysis with the recent development of quantile periodograms [229].

For a given constant $\alpha \in (0,1)$, define a nonnegative function $\rho_\alpha(x)$ as

$$\rho_\alpha(x) := x\{\alpha - \mathscr{I}(x<0)\} = \begin{cases} (\alpha-1)x & \text{if } x<0, \\ \alpha x & \text{if } x \ge 0. \end{cases} \tag{11.5.1}$$

It is not difficult to show [196, p. 7] that the sample α-quantile of a real-valued time series $\{y_1,\ldots,y_n\}$, denoted by $\hat{\lambda}_\alpha$, minimizes $\sum_{t=1}^n \rho_\alpha(y_t - \tilde{\lambda})$. In other words,

$$\hat{\lambda}_\alpha = \arg\min_{\tilde{\lambda}\in\mathbb{R}} \sum_{t=1}^n \rho_\alpha(y_t - \tilde{\lambda}). \tag{11.5.2}$$

More generally, with $\mathbf{x}_t \in \mathbb{R}^m$ denoting the vector of m regressors (or covariates), a linear quantile regression solution is given by minimizing

$$\sum_{t=1}^n \rho_\alpha(y_t - \mathbf{x}_t^T \tilde{\boldsymbol{\beta}})$$

with respect to $\tilde{\boldsymbol{\beta}} \in \mathbb{R}^m$. Linear programming techniques can be used to compute the solution [196, Chapter 6] [295]. Note that with $\alpha = 1/2$ the quantile regression problem reduces to the LAD problem discussed in Chapter 5 (Section 5.5).

The underlying assumption in quantile regression is that the α-quantile of y_t, denoted as λ_t such that $P(y_t \le \lambda_t) = \alpha$, is a linear function of the regressor, i.e.,

$$\lambda_t = \mathbf{x}_t^T \boldsymbol{\beta}. \tag{11.5.3}$$

Under this assumption, the parameter $\boldsymbol{\beta}$ is the minimizer of $E\{\rho_\alpha(y_t - \mathbf{x}_t^T \tilde{\boldsymbol{\beta}})\}$ with respect to $\tilde{\boldsymbol{\beta}} \in \mathbb{R}^m$ for all t, so the quantile regression solution which minimizes $\sum_{t=1}^n \rho_\alpha(y_t - \mathbf{x}_t^T \tilde{\boldsymbol{\beta}})$ serves as a natural estimator of $\boldsymbol{\beta}$. The quantile regression model (11.5.3) is particularly suitable for time series of the form

$$y_t = a_t \tilde{y}_t, \tag{11.5.4}$$

where $\{\tilde{y}_t\}$ is a strictly stationary random process with α-quantile c and $\{a_t\}$ is a sequence of nonnegative constants. In this case, we have

$$\alpha = P(\tilde{y}_t \le c) = P(y_t \le c a_t).$$

This implies that the α-quantile of y_t can be expressed as

$$\lambda_t = c a_t.$$

If a_t takes the form $\mathbf{x}_t^T \mathbf{b}$, then $\lambda_t = \mathbf{x}_t^T \boldsymbol{\beta}$ with $\boldsymbol{\beta} := c\mathbf{b}$. Assuming that the $\tilde{y}_t$ have unit variance, the sequence $\{a_t\}$ in (11.5.4) represents the heteroscedastic volatility of $\{y_t\}$, which is a critical feature of many financial and economic time series.

For spectral analysis, let us consider a trigonometric regressor

$$\mathbf{x}_t(\omega) := [\cos(\omega t), \sin(\omega t)]^T,$$

where $\omega \in (0, \pi)$ is the frequency variable. The corresponding quantile regression solution in this case is given by

$$(\hat{\lambda}_\alpha(\omega), \hat{\boldsymbol{\beta}}_\alpha(\omega)) := \arg \min_{\tilde{\lambda} \in \mathbb{R}, \tilde{\boldsymbol{\beta}} \in \mathbb{R}^2} \sum_{t=1}^n \rho_\alpha(y_t - \tilde{\lambda} - \mathbf{x}_t^T(\omega) \tilde{\boldsymbol{\beta}}). \tag{11.5.5}$$

Note that a constant term $\tilde{\lambda}$ is included in (11.5.5) so that the regression model takes the form $\tilde{\lambda} + \mathbf{x}_t^T(\omega) \tilde{\boldsymbol{\beta}}$. Given the regression coefficients in (11.5.5) and the sample quantile in (11.5.2), the *quantile periodogram* is defined by

$$Q_{n,\alpha}(\omega) := \sum_{t=1}^n \rho_\alpha(y_t - \hat{\lambda}_\alpha) - \sum_{t=1}^n \rho_\alpha(y_t - \hat{\lambda}_\alpha(\omega) - \mathbf{x}_t^T(\omega) \hat{\boldsymbol{\beta}}_\alpha(\omega)). \tag{11.5.6}$$

It is evident that $Q_{n,\alpha}(\omega)$ in (11.5.6) measures the net contribution of the trigonometric regressor $\mathbf{x}_t(\omega)$ to the reduction of the quantile regression cost function.

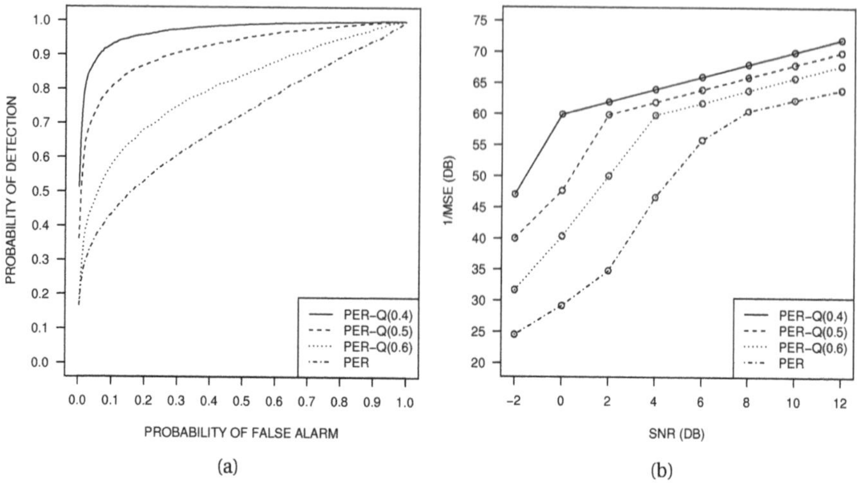

Figure 11.7. Detection and estimation of a real sinusoid in log-normal white noise by using the ordinary and quantile periodograms. (a) ROC curve of Fisher's test. (b) MSE of the periodogram maximizer. Solid line, quantile periodogram with $\alpha = 0.4$; dashed line, quantile periodogram with $\alpha = 0.5$; dotted line, quantile periodogram with $\alpha = 0.6$; dash-dotted line, ordinary periodogram. Results are based on 5,000 Monte Carlo runs.

Under the assumption that $y_t = A_0 + A_1\cos(\omega_0 t) + B_1\sin(\omega_0 t) + \epsilon_t$, where $\{\epsilon_t\}$ is an i.i.d. sequence with PDF proportional to $\exp\{-\rho_\alpha(x)\}$, it is easy to see that $Q_{n,\alpha}(\omega_0)$ coincides with the log generalized likelihood statistic for testing the presence of the sinusoid. Under the same condition, the ML estimator of ω_0 coincides with the maximizer of $Q_{n,\alpha}(\omega)$ in the interval $(0,\pi)$. When the noise has a skewed distribution, the maximizer of $Q_{n,\alpha}(\omega)$ with a suitable α is able to produce more accurate frequency estimates than the PM estimator which maximizes the ordinary periodogram. The periodogram-based tests for hidden periodicities in Chapter 6 can also be improved by using the quantile periodogram instead of the ordinary periodogram. These benefits are demonstrated by Figure 11.7.

In this example, the time series contains a single real sinusoid in skewed noise [230]. More specifically, we have

$$y_t = \cos(\omega_0 t + \phi) + \epsilon_t \qquad (t = 1, \ldots, n),$$

where ϕ is a random variable, uniformly distributed in $(-\pi, \pi]$, $\{\epsilon_t\}$ is an i.i.d. sequence of zero-mean log-normal random variables distributed as $a(\exp(Z) - b)$ with $Z \sim \mathrm{N}(0,1)$, and $n = 50$. Figure 11.7(a) depicts the ROC curve of Fisher's test (6.2.16) based on the quantile periodogram in (11.5.6) with different choice of α when $\omega_0 = 2\pi \times 0.1$ and SNR $= -7$ dB. It also shows the ROC curve of the original test based on the ordinary periodogram. Figure 11.7(b) depicts the MSE for estimating $f_0 := \omega_0/(2\pi)$ by the maximizer of the ordinary periodogram and the maximizer of the quantile periodogram with different choice of α. In calculating the MSE, the signal frequency is randomized such that $\omega_0 \sim 2\pi \times 0.1 + \mathrm{U}(0, 2\pi/n)$. The DFT method, based on the ordinary and quantile periodograms, respectively, is employed to provide the initial value for maximizing the corresponding periodogram using the R function `optimize` which is a bisection algorithm.

Figure 11.7 shows that the quantile periodogram outperforms the ordinary periodogram by a large margin for both detection and estimation. Note that the performance of the quantile periodogram is more superior at a lower quantile than at an upper quantile. This is due to the higher effective SNR at the lower quantile as a result of the upward skewness of the noise distribution.

Like the ordinary periodogram, the quantile periodogram can also serve as a frequency-domain representation of serial dependence for time series that do not contain sinusoids. Let $\{y_t\}$ be a random process with possibly time-varying univariate and bivariate CDFs $F_t(x) := P(y_t \le x)$ and $F_{ts}(x, y) := P(y_t \le x, y_s \le y)$. Assume that the following conditions are satisfied for a given α.

1. $F_t(\lambda_\alpha) = \alpha$ and $f_t(\lambda_\alpha) = 1/\kappa_\alpha > 0$ for all t, where κ_α is called the sparsity at the α-quantile λ_α.
2. $f_t(x) := \dot{F}_t(x)$ exists for all x and $F_t(x + \lambda_\alpha) - F_t(\lambda_\alpha) = f_t(\lambda_\alpha)x + O(x^{d+1})$ uniformly in the vicinity of $x = 0$ for some constant $d > 0$.
3. Let $\{y_t\}$ be stationary in λ_α-level crossings with lag-u level-crossing rate $\gamma_\alpha(u)$, i.e., $P\{(y_t - \lambda_\alpha)(y_s - \lambda_\alpha) < 0\} = \gamma_\alpha(t-s)$ for all t and s.
4. Let $\{y_t\}$ be an m-dependent process or a linear process of the form $y_t = \sum_{j=-\infty}^{\infty} \psi_j \zeta_{t-j}$, where $\{\zeta_t\}$ is an i.i.d. sequence with $E(|\zeta_t|) < \infty$ and $\{\psi_j\}$ is an absolutely summable sequence such that $\sum_{|j|>n^r} |\psi_j| = \mathcal{O}(n^{-1})$ for some constant $r \in [0, 1/4)$. See [229] for more general conditions.

While the conditions 2 and 4 are technical ones, the conditions 1 and 3 have some interesting implications. Consider the level-crossing process $\{\mathcal{I}(y_t < \lambda_\alpha)\}$. From the condition 1, it follows that

$$E\{\mathcal{I}(y_t < \lambda_\alpha)\} = F_t(\lambda_\alpha) = \alpha.$$

Under the conditions 1 and 3, we can write

$$\begin{aligned}\gamma_\alpha(t-s) &= P(y_t < \lambda_\alpha, y_s > \lambda_\alpha) + P(y_t > \lambda_\alpha, y_s < \lambda_\alpha)\\ &= P(y_t < \lambda_\alpha) + P(y_s < \lambda_\alpha) - P(y_t < \lambda_\alpha, y_s \le \lambda_\alpha) - P(y_t \le \lambda_\alpha, y_s < \lambda_\alpha)\\ &= 2\{\alpha - F_{ts}(\lambda_\alpha, \lambda_\alpha)\}.\end{aligned}$$

This, in turn, yields

$$\mathrm{Cov}\{\mathcal{I}(y_t < \lambda_\alpha), \mathcal{I}(y_s < \lambda_\alpha)\} = F_{ts}(\lambda_\alpha, \lambda_\alpha) - \alpha^2 = \alpha(1-\alpha) - \tfrac{1}{2}\gamma_\alpha(t-s).$$

Therefore, under the conditions 1 and 3, the level-crossing process $\{\mathscr{I}(y_t < \lambda_\alpha)\}$ is weakly stationary with mean α and ACF $\alpha(1-\alpha) - \frac{1}{2}\gamma_\alpha(u)$.

The sequence of level-crossing rates $\{\gamma_\alpha(u)\}$ can be represented in the frequency domain by the *level-crossing spectrum* defined as

$$h_\alpha(\omega) := \sum_{u=-\infty}^{\infty} \left\{1 - \frac{1}{2\alpha(1-\alpha)}\gamma_\alpha(u)\right\} \exp(-i\omega u)$$
$$= \sum_{u=-\infty}^{\infty} \left\{1 - \frac{1}{2\alpha(1-\alpha)}\gamma_\alpha(u)\right\} \cos(\omega u),$$

where the second expression is due to the symmetry of $\gamma_\alpha(u)$, i.e., $\gamma_\alpha(-u) = \gamma_\alpha(u)$ for all u. The level-crossing spectrum is nothing but the ordinary power spectrum of the standardized (mean 0 and variance 1) level-crossing process $\{(\mathscr{I}(y_t < \lambda_\alpha) - \alpha)/\sqrt{\alpha(1-\alpha)}\}$ whose ACF is equal to $1 - \{2\alpha(1-\alpha)\}^{-1}\gamma_\alpha(u)$. If $\{y_t\}$ has zero median, then the level-crossing spectrum corresponding to $\alpha = 1/2$ reduces to the zero-crossing spectrum discussed in Chapter 5 (Section 5.5). Like the zero-crossing spectrum and the autocorrelation spectrum (or normalized power spectrum), the level-crossing spectrum is a scale-invariant representation of serial dependence in the frequency domain. An i.i.d. sequence with α-quantile λ_α is stationary in λ_α-level crossings with $\gamma_\alpha(u) = 2\alpha(1-\alpha)(1-\delta_\tau)$, in which case, we have $h_\alpha(\omega) = 1$ for all ω.

Equipped with these concepts, the following theorem can be established along the lines of the proof of Theorem 5.11. See [229] for more details.

Theorem 11.2 (Asymptotic Distribution of the Quantile Periodogram [229]). *Let $\{y_t\}$ satisfy the conditions 1–4 and let $Q_{n,\alpha}(\omega)$ be defined by (11.5.6). Assume further that $h_\alpha(\omega)$ is finite and bounded away from zero. Then, for any fixed and distinct frequencies $\tilde{\omega}_j \in (0,\pi)$ $(j = 1,\ldots,m)$, $\{Q_{n,\alpha}(\tilde{\omega}_j)\} \xrightarrow{D} \{\frac{1}{2}\eta_\alpha^2 h_\alpha(\tilde{\omega}_j)X_j\}$ as $n \to \infty$, where the X_j are i.i.d. $\chi^2(2)$ random variables and $\eta_\alpha^2 := \alpha(1-\alpha)\kappa_\alpha$.*

Remark 11.2 A key to the proof of this result is the identity

$$\rho_\alpha(x-y) - \rho_\alpha(x) = -y\psi_\alpha(x) + \int_0^y \tfrac{1}{2}\phi(x,y')\,dy',$$

where $\psi_\alpha(x) := \alpha - \mathscr{I}(x<0)$ and $\phi(x,y) := 2\mathscr{I}(x \le y) - 2\mathscr{I}(x \le 0)$. It is a generalization of Knight's identity in (5.6.27) which corresponds to $\alpha = 1/2$.

By Theorem 11.2, the mean of the asymptotic distribution of $Q_{n,\alpha}(\omega)$ is

$$q_\alpha(\omega) := \eta_\alpha^2 h_\alpha(\omega). \tag{11.5.7}$$

We call this function the *quantile spectrum* of $\{y_t\}$ at quantile level α. If $\{y_t\}$ has zero median, then the quantile spectrum with $\alpha = 1/2$ reduces to the Laplace spectrum discussed in Chapter 5 (Section 5.5). Under suitable conditions, the quantile spectrum can be estimated consistently from the quantile periodogram by a periodogram smoother of the form (7.1.2) [88]. If $\{y_t\}$ is an i.i.d. sequence, then $q_\alpha(\omega) = \eta_\alpha^2$ for all ω. In this case, $\hat{\kappa}_\alpha := (\alpha(1-\alpha)m)^{-1}\sum_{j=1}^m Q_{n,\alpha}(\tilde{\omega}_j)$ serves as an estimator of κ_α, where $\tilde{\omega}_j := 2\pi j/n$ and $m := \lfloor (n-1)/2 \rfloor$.

As an alternative to (11.5.6), the contribution of the trigonometric regressor $\mathbf{x}_t(\omega)$ can also be measured directly by the magnitude of the regression coefficient $\hat{\boldsymbol{\beta}}_\alpha(\omega)$ given by (11.5.5). This leads to a quantile periodogram of the form

$$Q_{n,\alpha}(\omega) := \tfrac{1}{4} n \|\hat{\boldsymbol{\beta}}_\alpha(\omega)\|^2, \tag{11.5.8}$$

which is known as the quantile periodogram of the first kind [229]. The function in (11.5.6) is called the quantile periodogram of the second kind. It can be shown [229] that Theorem 11.2 remains valid for $Q_{n,\alpha}(\omega)$ in (11.5.8) except that the scaling factor η_α^2 is defined as $\alpha(1-\alpha)\kappa_\alpha^2$ rather than $\alpha(1-\alpha)\kappa_\alpha$. In other words, the asymptotic distributions of these quantile periodograms differ only by a constant factor κ_α, which is the sparsity at the α-quantile λ_α.

The quantile periodogram of the first kind can be extended to multiple time series in the same way as the Laplace periodogram of the first kind is extended in Section 11.4. Indeed, if $\hat{\boldsymbol{\beta}}_j(\omega) := [\hat{A}_j(\omega), \hat{B}_j(\omega)]^T$ denotes the quantile regression solution given by (11.5.5) for time series $\mathbf{y}_j := [y_{j1}, \ldots, y_{jn}]^T$, then, similar to (11.4.8), the *quantile cross-periodogram* between $\mathbf{y}_j$ and $\mathbf{y}_{j'}$ can be defined as

$$Q_{njj'}(\omega) := n\tilde{z}_j(\omega)\tilde{z}_{j'}^*(\omega) \qquad (j, j' = 1, \ldots, d), \tag{11.5.9}$$

where $\tilde{z}_j(\omega) := \frac{1}{2}\sqrt{n}\{\hat{A}_j(\omega) - i\hat{B}_j(\omega)\}$ may be called the *quantile Fourier transform* of $\mathbf{y}_j$. Note that we have suppressed α in the notation for simplicity. It is easy to see that $Q_{njj}(\omega) = n|\tilde{z}_j(\omega)|^2$ is just the quantile periodogram of $\mathbf{y}_j$ defined by (11.5.8). The quantile cross-periodogram in (11.5.9) with $\alpha = 1/2$ reduces to the Laplace cross-periodogram in (11.4.8). Similar to the Laplace spectral coherence in (11.4.9), we can define the *quantile spectral coherence* between $\mathbf{y}_j$ and $\mathbf{y}_{j'}$ as

$$C_{njj'}(\omega) := \frac{\hat{q}_{jj'}(\omega)}{\sqrt{\hat{q}_{jj}(\omega)\,\hat{q}_{j'j'}(\omega)}},$$

where $\hat{q}_{jj'}(\omega)$, $\hat{q}_{jj}(\omega)$, and $\hat{q}_{j'j'}(\omega)$ represent the smoothed versions of $Q_{njj'}(\omega)$, $Q_{njj}(\omega)$, and $Q_{nj'j'}(\omega)$, respectively. As an extension to Theorem 11.1, the following result can be established for the quantile cross-periodogram defined by (11.5.9) along the lines of the proof of Theorem 5.11.

Theorem 11.3 (Asymptotic Distribution of the Quantile Cross-Periodogram). *For any given $j = 1, \ldots, d$, let the time series $\{y_{jt}\}$ satisfy the conditions of Theorem 11.2 with sparsity $\kappa_{j,\alpha}$ at the α-quantile $\lambda_{j,\alpha}$. For any given j and j', assume that $\{y_{jt}\}$ and $\{y_{j't}\}$ are jointly stationary in level crossings in the sense that*

$$P\{(y_{jt} - \lambda_{j,\alpha})(y_{j's} - \lambda_{j',\alpha}) < 0\} = \gamma_{jj',\alpha}(t-s) \qquad \forall t, s \in \mathbb{Z}.$$

Then, as $n \to \infty$, $\mathbf{Q}_n(\omega) := [Q_{njj'}(\omega)]$ $(j, j' = 1, \ldots, d)$ has the same asymptotic distribution as $\mathbf{L}_n(\omega)$, which is asserted in Theorem 11.1 and Remark 11.1 for the real case, except that the zero-crossing cross-spectrum $h_{jj'}(\omega)$ should be replaced by

$$h_{jj',\alpha}(\omega) := \sum_{u=-\infty}^{\infty} \left\{1 - \frac{1}{2\alpha(1-\alpha)} \gamma_{jj',\alpha}(u)\right\} \exp(-iu\omega)$$

and the scaling constants η_j and $\eta_{j'}$ should be replaced by $\eta_{j,\alpha} := \sqrt{\alpha(1-\alpha)}\,\kappa_{j,\alpha}$ and $\eta_{j',\alpha'} := \sqrt{\alpha'(1-\alpha')}\,\kappa_{j',\alpha'}$, respectively.

Remark 11.3 In the case of a single time series, Theorem 11.3 implies that Theorem 11.2 is valid for the quantile periodogram defined by (11.5.8) except that $\eta_\alpha^2 := \alpha(1-\alpha)\kappa_\alpha^2$. See [229] for a proof of this assertion. In Theorem 11.3, the function $h_{jj',\alpha}(\omega)$ is called the *level-crossing cross-spectrum* and the function

$$q_{jj',\alpha}(\omega) := \eta_{j,\alpha}\eta_{j',\alpha'} h_{jj',\alpha}(\omega),$$

which is the mean of the asymptotic distribution of $Q_{njj'}(\omega)$, may be referred to as the *quantile cross-spectrum.*

With the flexibility of selecting different quantile levels for spectral analysis, the quantile periodogram and cross-periodogram provide a much richer view than the one offered by the ordinary periodogram and cross-periodogram. In the following, we use two examples to demonstrate this enriched capability.

First, let us consider the annual sunspot numbers shown in Figure 7.1. While this time series contains an 11-year periodicity, it also exhibits a strong asymmetry that cannot be explained by a pure sinusoid. For example, there is a tendency for the peaks to be taller than the troughs are deep; the heights of the peaks vary widely while the troughs occur at around the same level. To deal with the asymmetry, the square-root transformation is applied to the sunspot numbers in [38, p. 84] prior to spectral analysis, and the sunspot numbers are modeled as the square of a narrowband Gaussian process in [22]. Figure 11.8 shows that by varying the quantile level the quantile periodogram is able to capture not only the prominent periodicity but also the asymmetry property.

Depicted in Figure 11.8 is the quantile periodogram defined by (11.5.8) as a bivariate function of $\alpha \in (0,1)$ and $f := \omega/(2\pi) \in (0,1/2)$. We call it a *quantile-frequency plot.* As can be seen from Figure 11.8, the primary spectral peak, which corresponds to the 11-year cycle, is persistent across the quantile levels. However, the secondary spectral peaks on each side of the primary one are more pronounced at the middle and upper quantiles than at the lower quantiles. This suggests that the spectral content in the neighborhood of the 11-year cycle is richer for the peaks than for the troughs. The spectral peak near zero also appears to be stronger at the upper quantiles, suggesting a more prominent long

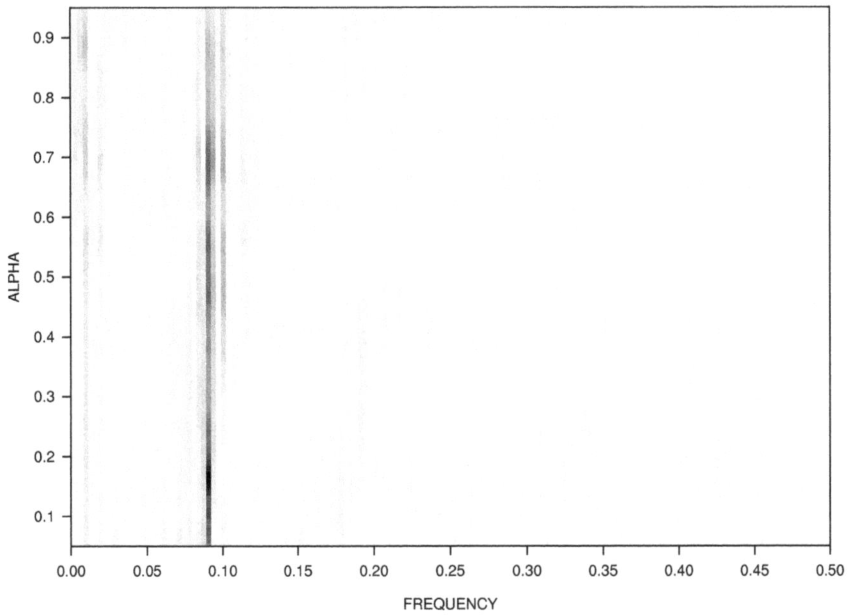

Figure 11.8. Quantile periodogram, defined by (11.5.8), as a bivariate function of the quantile level α and the frequency $f := \omega/(2\pi)$ (in cycles per year) for the time series of annual sunspot numbers shown in Figure 7.1. The quantile periodogram is standardized so that the sum over the Fourier frequencies is equal to unity at each quantile level. Darker shades of grey represent larger values.

cycle for the peaks. These asymmetric characteristics cannot be discerned at all in the ordinary periodogram shown in Figure 7.2.

To demonstrate the inner workings of the quantile periodogram in capturing the asymmetry of the sunspot numbers, Figure 11.9 depicts the trigonometric quantile regression function $\hat{\lambda}_\alpha(\omega) + \mathbf{x}_t^T(\omega)\hat{\boldsymbol{\beta}}_\alpha(\omega)$ $(t = 1, \ldots, n)$ at two quantile levels, $\alpha = 0.7$ (left) and $\alpha = 0.2$ (right), and with three frequencies (top to bottom): $\omega = 2\pi \times 28/n = 2\pi \times 0.0909$ (11-year cycle), $\omega = 2\pi \times 31/n = 2\pi \times 0.1006$ (9.9-year cycle), and $\omega = 2\pi \times 3/n = 2\pi \times 0.0097$ (102.7-year cycle). As we can see from the top panel, the sinusoid with an 11-year cycle fits the troughs better than the peaks. For example, observe the significant misalignment of the sinusoid on the left to the three large peaks between years 1750 and 1800 due to the shortening of their succession. This compressed periodicity is captured by the sinusoid with a 9.9-year cycle shown in the middle panel on the left, where the three large peaks between years 1750 and 1800 coincide very well with the peaks of the sinusoid. The middle panel on the right shows that the same sinusoid is inconsequential at the lower quantile in comparison with the sinusoid in the top panel. Simi-

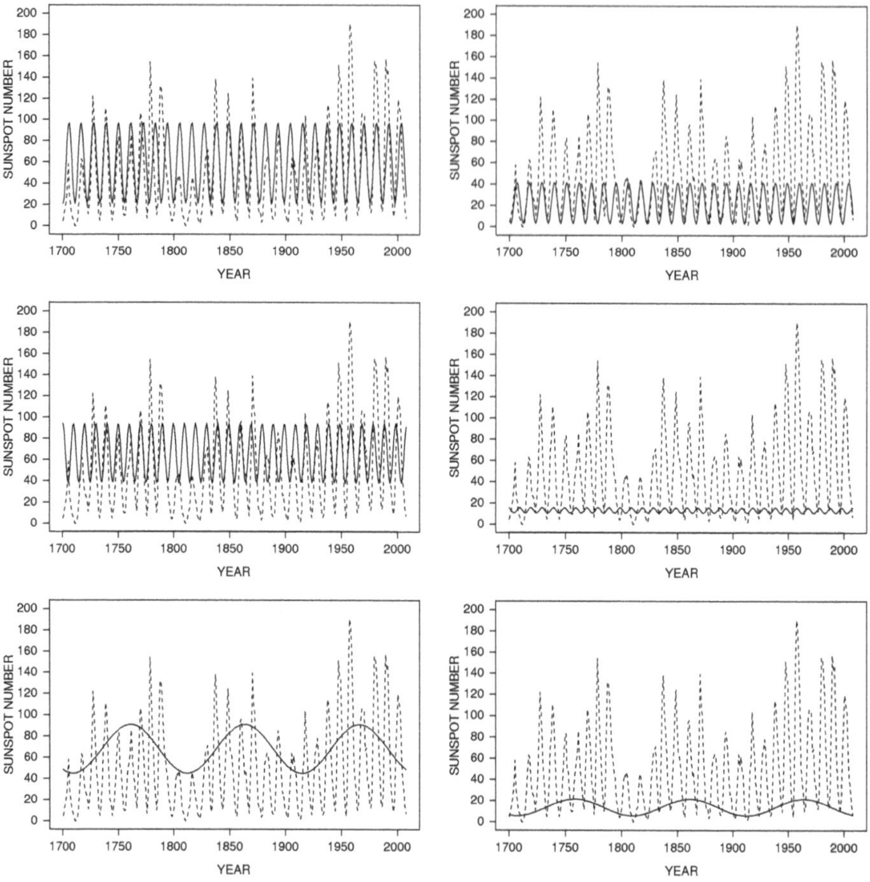

Figure 11.9. Trigonometric quantile regression functions for the sunspot numbers with $\alpha = 0.7$ (left) and $\alpha = 0.3$ (right). Top, 11-year cycle. Middle, 9.9-year cycle. Bottom, 102.7-year cycle. Solid line, quantile regression fit; dashed line, sunspot numbers.

larly, the sinusoid with an 102.7-year cycle shown in the bottom panel on the left captures the strong low-frequency variation of the peaks, and the range of oscillation for this sinusoid is about two thirds of the one with an 11-year cycle. The corresponding low-frequency sinusoid at the lower quantile shown on the right is less significant relative to the 11-year counterpart.

In the second example, let us consider the daily closing values of the Dow Jones Industrial Average (DJIA) Index and the Financial Times Stock Exchange (FTSE) Index. Figure 11.10 depicts these daily values from November 27, 2001, to March 10, 2010 (with the gaps representing missing data in nontrading days). Figure 11.10 also shows the log returns as time series indexed by the trading day ($n = 2047$), where the log return in day t is defined as $y_t := \log(v_t / v_{t-1})$, with v_t

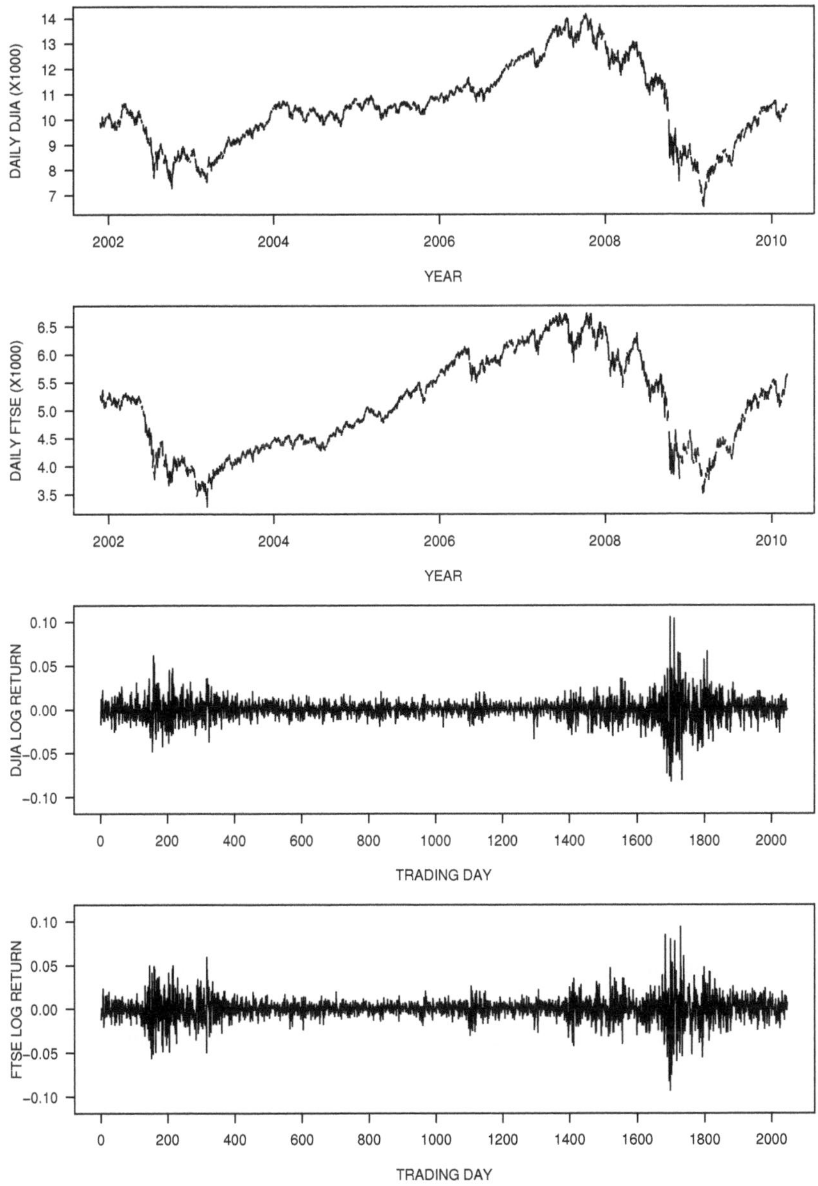

Figure 11.10. Daily closing values and log returns of DJIA and FTSE from November 27, 2001, to March 10, 2010.

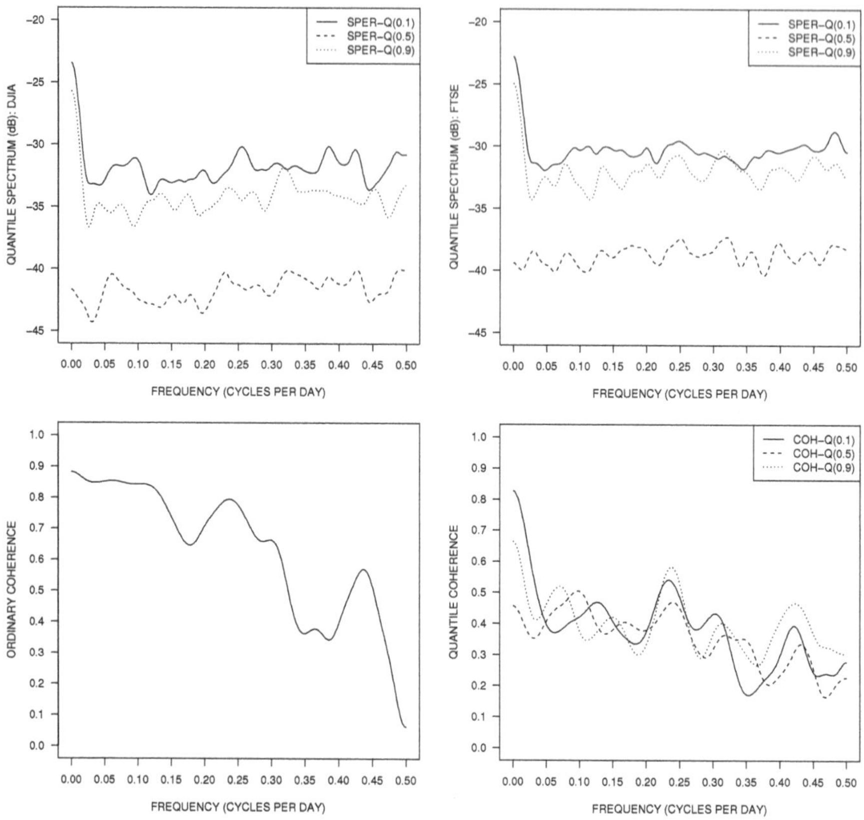

Figure 11.11. Spectral analysis of DJIA and FTSE log returns. Top, smoothed quantile periodogram of DJIA (left) and FTSE (right), Bottom left, absolute value of the ordinary coherence. Bottom right, absolute value of the quantile coherence. For quantile periodogram and coherence: solid line, $\alpha = 0.1$, dashed line, $\alpha = 0.5$, dotted line, $\alpha = 0.9$.

and v_{t-1} being the closing values in day t and day $t-1$. It is well known that the sample autocorrelation function and the periodogram of the log returns behave similarly to those of white noise, whereas the sample autocorrelation function and the periodogram of the absolute values of the log returns exhibit the properties of a long-range dependent process: large positive autocorrelations that decay very slowly and a large spectral peak at frequency zero.

The top panel of Figure 11.11 shows that the characteristics of long-range dependence also exist in the quantile periodogram of the log returns themselves. More interestingly, it shows that the long-range dependence characteristics appear only at the lower and upper quantiles but not at the median ($\alpha = 0.5$). This is a manifestation that the large negative and large positive returns occur as infrequent clusters (during the 2002 and 2008 financial crises), whereas the small fluctuations behave more like white noise. These properties are further analyzed in [231] under the assumption that the log returns take the form (11.5.4), where $\{\tilde{y}_t\}$ is an i.i.d. random sequence and $\{a_t\}$ is a deterministic sequence.

The bottom panel of Figure 11.11 depicts the ordinary spectral coherence and the quantile spectral coherence between the two series. The smoothed periodograms and cross-periodograms used in calculating the spectral coherence are obtained with the R function `smooth.spline` and 30 degrees of freedom. The ordinary coherence on the left suggests that a strong correlation exists between the two series over a broad range of frequencies (up to 0.10 cycles per day). The quantile coherence on the right reveals that a strong correlation exists only at the upper and lower quantiles but not at the median. It also reveals that the strong correlation concentrates in a much smaller range of low frequencies which is consistent with the spectral peak in the quantile periodogram. This suggests that the infrequent clusters of large returns behave more coherently than the small day-to-day fluctuations. Finally, it is interesting to observe that the low-frequency coherence appears to be stronger at the lower quantile than at the upper quantile.

To end this section, let us briefly discuss some generalizations of the quantile periodogram and cross-periodogram.

First, consider the extension of the univariate quantile periodogram in (11.5.6) to a multivariate quantile periodogram. This can be done by replacing the one-frequency regressor $\mathbf{x}_t(\omega) := [\cos(\omega t), \sin(\omega t)]^T$ with a q-frequency regressor

$$\mathbf{x}_t(\tilde{\boldsymbol{\omega}}) := [\cos(\tilde{\omega}_1 t), \sin(\tilde{\omega}_1 t), \ldots, \cos(\tilde{\omega}_q t), \sin(\tilde{\omega}_q t)]^T,$$

where $\tilde{\boldsymbol{\omega}} := (\tilde{\omega}_1, \ldots, \tilde{\omega}_q) \subset (0, \pi)$. Given $\tilde{\boldsymbol{\omega}}$, let

$$(\hat{\lambda}_\alpha(\tilde{\boldsymbol{\omega}}), \hat{\boldsymbol{\beta}}_\alpha(\tilde{\boldsymbol{\omega}})) := \arg \min_{\tilde{\lambda} \in \mathbb{R}, \tilde{\boldsymbol{\beta}} \in \mathbb{R}^{2q}} \sum_{t=1}^{n} \rho_\alpha(y_t - \tilde{\lambda} - \mathbf{x}_t^T(\tilde{\boldsymbol{\omega}})\tilde{\boldsymbol{\beta}}).$$

Then, the *multivariate quantile periodogram* can be defined as

$$Q_{n,\alpha}(\tilde{\boldsymbol{\omega}}) := \sum_{t=1}^{n} \rho_\alpha(y_t - \hat{\lambda}_\alpha) - \sum_{t=1}^{n} \rho_\alpha(y_t - \hat{\lambda}_\alpha(\tilde{\boldsymbol{\omega}}) - \mathbf{x}_t^T(\tilde{\boldsymbol{\omega}})\hat{\boldsymbol{\beta}}_\alpha(\tilde{\boldsymbol{\omega}})). \quad (11.5.10)$$

In the special case of $q = 1$, the multivariate quantile periodogram in (11.5.10) becomes the univariate quantile periodogram defined by (11.5.6).

Moreover, the quantile cross-periodogram defined by (11.5.9) can be modified to represent the cross-correlation of a single time series at two quantile levels. Toward that end, let $\tilde{z}_\alpha(\omega)$ and $\tilde{z}_{\alpha'}(\omega)$ denote the quantile Fourier transforms of $\{y_1, \ldots, y_n\}$ at the α-quantile and the α'-quantile, respectively, for $\alpha \neq \alpha'$. Then, by following (11.5.9), we can define the *biquantile cross-periodogram* as

$$Q_{n,\alpha\alpha'}(\omega) := n \tilde{z}_\alpha(\omega) \tilde{z}_{\alpha'}^*(\omega). \quad (11.5.11)$$

Note that this function reduces to the quantile periodogram defined by (11.5.8) when $\alpha = \alpha'$. To investigate the asymptotic distribution of the biquantile cross-periodogram, let $\{y_t\}$ satisfy the conditions of Theorem 11.2 at the α-quantile λ_α and the α'-quantile $\lambda_{\alpha'}$, with sparsities κ_α and $\kappa_{\alpha'}$, respectively. In addition, let $\{y_t\}$ be strictly stationary so that for all t and u,

$$P\{(y_{t+u} - \lambda_\alpha)(y_t - \lambda_{\alpha'}) < 0\} = \gamma_{\alpha\alpha'}(u), \tag{11.5.12}$$

where $\gamma_{\alpha\alpha'}(u)$ may be referred to as the lag-u *bilevel-crossing rate*. Then, similar to Theorem 11.3, it can be shown that for any fixed $\tilde{\omega}_1, \ldots, \tilde{\omega}_m \in (0, \pi)$ the joint asymptotic distribution of the 2-by-2 quantile periodogram and biquantile cross-periodogram matrices

$$\begin{bmatrix} Q_{n,\alpha\alpha}(\tilde{\omega}_j) & Q_{n,\alpha\alpha'}(\tilde{\omega}_j) \\ Q_{n,\alpha'\alpha}(\tilde{\omega}_j) & Q_{n,\alpha'\alpha'}(\tilde{\omega}_j) \end{bmatrix} \qquad (j = 1, \ldots, m)$$

coincides with the joint distribution of $\boldsymbol{\zeta}(\tilde{\omega}_1)\boldsymbol{\zeta}^H(\tilde{\omega}_1), \ldots, \boldsymbol{\zeta}(\tilde{\omega}_m)\boldsymbol{\zeta}^H(\tilde{\omega}_m)$, where the $\boldsymbol{\zeta}(\tilde{\omega}_j)$ are independent complex Gaussian random vectors with mean zero and covariance matrix

$$\begin{bmatrix} q_\alpha(\tilde{\omega}_j) & q_{\alpha\alpha'}(\tilde{\omega}_j) \\ q_{\alpha'\alpha}(\tilde{\omega}_j) & q_{\alpha'}(\tilde{\omega}_j) \end{bmatrix}.$$

In this expression, $q_\alpha(\omega)$ is just the quantile spectrum defined by (11.5.7) and $q_{\alpha\alpha'}(\omega)$, which may be called the *biquantile cross-spectrum*, is defined by

$$q_{\alpha\alpha'}(\omega) := \eta_\alpha \eta_{\alpha'} h_{\alpha\alpha'}(\omega), \tag{11.5.13}$$

where $\eta_\alpha := \sqrt{\alpha(1-\alpha)}\,\kappa_\alpha$, $\eta_{\alpha'} := \sqrt{\alpha'(1-\alpha')}\,\kappa_{\alpha'}$, and

$$h_{\alpha\alpha'}(\omega) := \sum_{u=-\infty}^{\infty} \left\{1 - \frac{1}{2\sqrt{\alpha(1-\alpha)\alpha'(1-\alpha')}}\gamma_{\alpha\alpha'}(u)\right\} \exp(-iu\omega).$$

Note that $q_{\alpha\alpha'}(\omega)$ is just the mean of the asymptotic distribution of $Q_{n,\alpha\alpha'}(\omega)$. For consistency of notation, we can define $q_{\alpha\alpha}(\omega) := q_\alpha(\omega)$. The function $h_{\alpha\alpha'}(\omega)$ may be called the *bilevel-crossing cross-spectrum*. It is easy to verify that $h_{\alpha\alpha'}(\omega)$ coincides with the ordinary cross-spectrum between the level-crossing processes $\{(\mathscr{I}(y_t \le \lambda_\alpha) - \alpha)/\sqrt{\alpha(1-\alpha)}\}$ and $\{(\mathscr{I}(y_t \le \lambda_{\alpha'}) - \alpha')/\sqrt{\alpha'(1-\alpha')}\}$ because their cross-covariance function equals $1 - \frac{1}{2}\{\alpha(1-\alpha)\alpha'(1-\alpha')\}^{-1/2}\gamma_{\alpha\alpha'}(u)$.

The bilevel-crossing rates defined by (11.5.12) are related to the univariate and bivariate CDFs of $\{y_t\}$. In fact, it is easy to verify that

$$\gamma_{\alpha\alpha'}(u) = \alpha + \alpha' - 2F_u(\lambda_\alpha, \lambda_{\alpha'}),$$

where $F_u(\cdot,\cdot)$ denotes the bivariate CDF of (y_{t+u}, y_t) which does not depend on t under the stationarity assumption. Therefore, as a function of ω, the biquantile cross-spectrum $q_{\alpha\alpha'}(\omega)$ can be regarded as a frequency-domain representation of the sequence $\{F_u(\lambda_\alpha,\lambda_{\alpha'})\}$ which reflects the serial dependence of $\{y_t\}$.

In the special case where the marginal distribution of $\{y_t\}$ is uniform U(0,1), we have $\lambda_\alpha=\alpha$, $\lambda_{\alpha'}=\alpha'$, and hence $\gamma_{\alpha\alpha'}(u)=\alpha+\alpha'-2F_u(\alpha,\alpha')$. The uniform distribution also implies $\kappa_\alpha=\kappa_{\alpha'}=1$. Therefore, the corresponding biquantile cross-spectrum can be expressed as

$$q_{\alpha\alpha'}(\omega)=\sum_{u=-\infty}^{\infty}\left\{\sqrt{\alpha(1-\alpha)\alpha'(1-\alpha')}-\tfrac{1}{2}(\alpha+\alpha')+F_u(\alpha,\alpha')\right\}\exp(-iu\omega).$$

In this case, $q_{\alpha\alpha'}(\omega)$, as a function of (ω,α,α'), becomes a quantile-frequency-domain representation of the bivariate CDF sequence $\{F_u(\cdot,\cdot)\}$ which completely determines the bivariate properties of $\{y_t\}$. In general, $\tilde{y}_t:=F(y_t)$ has the U(0,1) distribution and the bivariate CDF of $(\tilde{y}_{t+u},\tilde{y}_t)$ is given by

$$C_u(\cdot,\cdot):=F_u(F^{-1}(\cdot),F^{-1}(\cdot)),$$

which is known as the copula. Therefore, the biquantile cross-spectrum of $\{\tilde{y}_t\}$ constitutes a quantile-frequency-domain representation of the bivariate copula sequence $\{C_u(\cdot,\cdot)\}$. Motivated by this observation, it is proposed in [88] to replace the unknown CDF $F(y)$ by the empirical CDF $\hat{F}_n(y):=n^{-1}\sum_{t=1}^n \mathscr{I}(y_t\le y)$ and consider the biquantile cross-periodogram of $\{\hat{F}_n(y_1),\cdots,\hat{F}_n(y_n)\}$. This method is shown in [88] to yield the same quantile-frequency-domain representation of the copula sequence $\{C_u(\cdot,\cdot)\}$. Note that in the absence of ties $\hat{F}_n(y_t)$ is equal to n^{-1} times the rank of y_t in the time series $\{y_1,\ldots,y_n\}$.

Finally, let us discuss the extension of the univariate and multivariate quantile periodograms to the complex case. To be consistent with the multivariate Laplace periodogram defined by (8.3.5), it suffices to replace $\rho_\alpha(x)$ by

$$\rho_{\alpha,c}(z):=2\{\rho_\alpha(\Re(z))+\rho_\alpha(\Im(z))\}\qquad(z\in\mathbb{C}).$$

With $\tilde{\boldsymbol{\omega}}:=(\tilde{\omega}_1,\ldots,\tilde{\omega}_p)\subset(0,2\pi)$ and $\mathbf{x}_t(\tilde{\boldsymbol{\omega}}):=[\exp(i\tilde{\omega}_1 t),\ldots,\exp(i\tilde{\omega}_p t)]^T$, the multivariate quantile periodogram in the complex case is defined as

$$Q_{n,\alpha}(\tilde{\boldsymbol{\omega}}):=\sum_{t=1}^n\rho_{\alpha,c}(y_t-\hat{\lambda}_\alpha)-\sum_{t=1}^n\rho_{\alpha,c}(y_t-\hat{\lambda}_\alpha(\tilde{\boldsymbol{\omega}})-\mathbf{x}_t^T(\tilde{\boldsymbol{\omega}})\hat{\boldsymbol{\beta}}_\alpha(\tilde{\boldsymbol{\omega}})),\quad(11.5.14)$$

where

$$\hat{\lambda}_\alpha:=\arg\min_{\tilde{\lambda}\in\mathbb{C}}\sum_{t=1}^n\rho_{\alpha,c}(y_t-\tilde{\lambda})$$

and

$$(\hat{\lambda}_\alpha(\tilde{\boldsymbol{\omega}}),\hat{\boldsymbol{\beta}}_\alpha(\boldsymbol{\omega})):=\arg\min_{\tilde{\lambda}\in\mathbb{C},\tilde{\boldsymbol{\beta}}\in\mathbb{C}^p}\sum_{t=1}^n\rho_{\alpha,c}(y_t-\tilde{\lambda}-\mathbf{x}_t^T(\tilde{\boldsymbol{\omega}})\tilde{\boldsymbol{\beta}}).$$

The multivariate quantile periodogram in (11.5.14) is a generalization of the multivariate Laplace periodogram in (8.3.5) which corresponds to $\alpha = 1/2$ under the zero-median assumption so that $\hat{\lambda}_\alpha$ and $\hat{\lambda}_\alpha(\tilde{\boldsymbol{\omega}})$ can be set to zero. Similarly, the multivariate quantile periodogram in (11.5.10) can be regarded as an extension of the multivariate Laplace periodogram in the real case.

With $p = 1$, a univariate quantile periodogram is defined by (11.5.14) for complex time series. By following the proof of Theorem 5.11, a result similar to Theorem 11.2 can be obtained for the univariate quantile periodogram in the complex case. Indeed, it can be shown that if the real and imaginary parts of $\{y_t\}$ are mutually independent with identical univariate and bivariate distributions that satisfy the conditions 1–4 of Theorem 11.2, then, for any fixed frequencies $\tilde{\omega}_j \in (0, 2\pi)$ $(j = 1, \ldots, m)$ and with $Q_{n,\alpha}(\omega)$ defined by (11.5.14) with $p = 1$, we have $\{Q_{n,\alpha}(\tilde{\omega}_j)\} \xrightarrow{D} \{\frac{1}{2}\eta^2_{\alpha,c} h_\alpha(\tilde{\omega}_j) X_j\}$ as $n \to \infty$, where the X_j are i.i.d. $\chi^2(2)$ random variables as in the real case and $\eta^2_{\alpha,c} := 2\eta^2_\alpha$. Based on this result, the quantile spectrum for a complex-valued time series is defined by

$$q_{\alpha,c}(\omega) := \eta^2_{\alpha,c} h_\alpha(\omega).$$

This function is twice the common quantile spectrum of the real and imaginary parts, which is given by $q_\alpha(\omega) := \eta^2_\alpha h_\alpha(\omega)$. In other words, we have

$$q_{\alpha,c}(\omega) = 2 q_\alpha(\omega).$$

The power spectrum of a complex time series has a similar property.

The quantile cross-periodogram defined by (11.5.9) can be generalized to the complex case by using $\tilde{z}_j(\omega)$ from the complex quantile regression

$$(\hat{\lambda}_j(\omega), \hat{z}_j(\omega)) := \arg \min_{\tilde{\lambda} \in \mathbb{C}, \tilde{z} \in \mathbb{C}} \sum_{t=1}^{n} \rho_{\alpha,c}(y_{jt} - \tilde{\lambda} - \tilde{z} \exp(i t \omega)).$$

The corresponding quantile spectral coherence can be defined in the same way as in the real case using the smoothed versions of the quantile periodogram and cross-periodogram.

Chapter 12

Appendix

This last chapter contains the supplementary material which is technical in nature but instrumental to some of the mathematical analysis in the main chapters. The material is organized into five sections. All results in this chapter are stated as lemmas. For convenience, these lemmas are indexed not only by the chapter number but also by the section number.

12.1 Trigonometric Series

The first lemma in this section provides two equivalent ways of measuring the separation of frequency parameters.

Lemma 12.1.1 (Measure of Frequency Separation). *For any* $\omega_1, \omega_2 \in [-\pi, \pi]$, *let* $\Delta_1 := |\exp(i\omega_1) - \exp(i\omega_2)|$ *and* $\Delta_2 := |\omega_1 - \omega_2|$.

(a) For any $\kappa_1 \in (0,2)$, *there exists a unique* $\kappa_2 \in (0,\pi)$ *such that* $\Delta_1 \geq \kappa_1$ *if and only if* $\kappa_2 \leq \Delta_2 \leq 2\pi - \kappa_2$.

(b) If $\Delta := \min(\Delta_2, 2\pi - \Delta_2) \to 0$, *then* $\Delta_1 = \mathcal{O}(\Delta)$. *Conversely, if* $\Delta_1 \to 0$, *then either* $\Delta_2 = \mathcal{O}(\Delta_1)$ *or* $2\pi - \Delta_2 = \mathcal{O}(\Delta_1)$.

PROOF. It is easy to show that $\Delta_1 = g(\Delta_2)$, where $g(\omega) := 2|\sin(\omega/2)|$. For $\omega \in [0, 2\pi]$, $g(\omega)$ is nonnegative, symmetric about $\omega = \pi$, monotone increasing from 0 to 2 in $\omega \in [0,\pi]$, and monotone decreasing from 2 to 0 in $\omega \in [\pi, 2\pi]$. Therefore, for any $\kappa_1 \in (0,2)$, there exists a unique $\kappa_2 \in (0,\pi)$, determined by $g(\kappa_2) = \kappa_1$, such that $\Delta_1 \geq \kappa_1$ if and only if $\kappa_2 \leq \Delta_2 \leq 2\pi - \kappa_2$. This proves part (a).

To prove part (b), we note that $g(0) = 0$ and $g(\omega)/\omega \to \dot{g}(0) = 1$ as $\omega \to 0$. Therefore, as $\Delta_2 \to 0$, we can write $\Delta_1 = g(\Delta_2) = \mathcal{O}(\Delta_2)$. The symmetry and 2π-periodicity of $g(\omega)$ implies that as $\Delta_2 \to 2\pi$, $\Delta_1 = g(2\pi - \Delta_2) = \mathcal{O}(2\pi - \Delta_2)$. Similar argument applies to $g^{-1}(\Delta_1)$ and $2\pi - g^{-1}(\Delta_1)$ whose derivatives at zero are equal to $1/\dot{g}(0)$ and $-1/\dot{g}(0)$, respectively. □

The following lemma, stated without proof, provides closed-form expressions for some frequently used power series.

Lemma 12.1.2 (Power Series). *For any $z \in \mathbb{C}$, the following identities hold.*

(a) $(1-z)\sum_{t=1}^{n} z^t = z(1-z^n)$.

(b) $(1-z)^2\sum_{t=1}^{n} tz^t = z\{1-(n+1)z^n + nz^{n+1}\}$.

(c) $(1-z)^3\sum_{t=1}^{n} t^2z^t = z\{1+z-(n+1)^2z^n + (2n^2+2n-1)z^{n+1} - n^2z^{n+2}\}$.

(d) *In general, for any given integer $r \ge 0$, there are constants a_j, $b_j(n)$, and $c > 0$, all of which may depend on r, such that*

$$(1-z)^{r+1}\sum_{t=1}^{n} t^r z^t = z\sum_{j=0}^{r}\{a_j z^j + b_j(n)\, z^{n+j}\},$$

where $|b_j(n)| \le cn^r$ for all j.

The following lemma is critical for analyzing the contribution of sinusoidal signals in the periodogram.

Lemma 12.1.3 (Dirichlet Kernel). *The Dirichlet kernel, defined as*

$$D_n(\omega) := n^{-1}\sum_{t=1}^{n}\exp(it\omega) = \frac{\sin(n\omega/2)}{n\sin(\omega/2)}\exp\{i(n+1)\omega/2\} \quad (n > 1),$$

has the following properties.

(a) *$D_n(\omega)$ is 2π-periodic and conjugate-symmetric such that $D_n(-\omega) = D_n^*(\omega)$.*

(b) $|D_n(\omega)| \le D_n(0) = 1$.

(c) *$D_n(\omega) = 0$ if and only if $\omega = 2\pi j/n$ for some integer $j \ne 0$.*

(d) *$|D_n(\omega)| \le \pi/(n\omega)$ for all $\omega \in (0,\pi]$.*

(e) *$|D_n(\omega)|$ is strictly monotone decreasing in $\omega \in [0, 2\pi/n]$ and there exists a constant $c \in (0, 2\pi)$ such that for any $\delta \in (0, c)$, $|D_n(\omega)| > |D_n(\delta/n)|$ for all $\omega \in [0, \delta/n)$ and $|D_n(\omega)| \le |D_n(\delta/n)|$ for all $\omega \in [\delta/n, \pi]$.*

(f) *If $n\omega \to a$ as $n \to \infty$ for some constant $a \ne 0$, then*

$$\begin{aligned} D_n(\omega) &\to d_0(a) := \text{sinc}(a/2)\exp(ia/2), \\ n^{-1}\dot{D}_n(\omega) &\to \dot{d}_0(a) = \tfrac{1}{2}\{\text{sinc}'(a/2) + i\,\text{sinc}(a/2)\}\exp(ia/2), \\ n^{-2}\ddot{D}_n(\omega) &\to \ddot{d}_0(a) = \tfrac{1}{4}\{\text{sinc}''(a/2) - \text{sinc}(a/2) + i2\text{sinc}'(a/2)\}\exp(ia/2), \end{aligned}$$

where $\text{sinc}(x) := \sin(x)/x$, $\text{sinc}'(x) = (x\cos(x) - \sin(x))/x^2$, and $\text{sinc}''(x) = -(x^2\sin(x) + 2x\cos(x) + 2\sin(x))/x^3$.

PROOF. Parts (a)–(c) and (f) are easy to verify. Part (d) follows from the inequality $\sin(\omega/2) > \omega/\pi$ for $\omega \in (0,\pi)$ and the fact that $\sin(\omega/2) = \omega/\pi$ for $\omega = \pi$. To prove part (e), observe that the inequality in part (d) implies that

$$\max_{2\pi/n \le \omega \le \pi} |D_n(\omega)| \le \max_{2\pi/n \le \omega \le \pi} \pi/(n\omega) = 1/2. \tag{12.1.1}$$

Furthermore, Figure 6.6 suggests that $|D_n(\omega)|$ should be strictly monotone decreasing in $\omega \in [0, 2\pi/n]$ from $|D_n(0)| = 1$ to $|D_n(2\pi/n)| = 0$. To prove this assertion, observe that $|D_n(\omega)| = \sin(n\omega/2)/\{n\sin(\omega/2)\}$ for $\omega \in [0, 2\pi/n]$ because $\sin(n\omega/2) \geq 0$ and $\sin(\omega/2) \geq 0$. Therefore, the derivative of $|D_n(\omega)|$ is equal to $\frac{1}{4}h(\omega)/\{n\sin^2(\omega/2)\}$ for $\omega \in (0, 2\pi/n)$, where

$$h(\omega) := (n-1)\sin((n+1)\omega/2) - (n+1)\sin((n-1)\omega/2).$$

It is easy to see that for $\omega \in (0, 2\pi/n)$,

$$\begin{aligned} \dot{h}(\omega) &= \tfrac{1}{2}(n^2-1)\{\cos((n+1)\omega/2) - \cos((n-1)\omega/2)\} \\ &= -(n^2-1)\sin(n\omega/2)\sin(\omega/2) < 0. \end{aligned}$$

This means that $h(\omega)$ is strictly monotone decreasing and hence $h(\omega) < h(0) = 0$ for $\omega \in (0, 2\pi/n)$, which, in turn, proves the monotonicity of $|D_n(\omega)|$. Owing to the monotonicity, there is a constant $c \in (0, 2\pi)$ such that $|D_n(c/n)| = 1/2$, and for any $\delta \in (0, c)$, $|D_n(\omega)| > |D_n(\delta/n)| > 1/2$ for all $\omega \in [0, \delta/n)$, $|D_n(\omega)| < |D_n(\delta/n)|$ for all $\omega \in (\delta/n, 2\pi/n]$. Combining this result with (12.1.1) proves part (e). □

Remark 12.1.1 An important implication of part (d) is that $D_n(\omega) \to 0$ if $n\omega \to \infty$. For $\omega \in (0, \pi/n)$, a tighter bound is $|D_n(\omega)| \leq 1$, because $\pi/(n\omega) > 1$.

The next lemma provides uniform bounds for the sums of complex sinusoids under various asymptotic conditions.

Lemma 12.1.4 (Sum of Complex Sinusoids). *Let $r = 0, 1, 2$. Then, the following assertions are true.*

(a) $\sum_{t=1}^n t^r \exp(it\omega) = \mathcal{O}(n^{r+1})$ uniformly in $\omega \in [0, 2\pi]$ as $n \to \infty$.

(b) $\sum_{t=1}^n t^r \exp(it\omega) = \mathcal{O}(n^r)$ uniformly in $\omega \in [\kappa_1, 2\pi - \kappa_2]$ as $n \to \infty$, where κ_1 and κ_2 are any positive constants such that $\kappa_1 + \kappa_2 < 2\pi$.

(c) $|\sum_{t=1}^n t^r \exp(it\omega)| \leq a_{nr} + b_{nr}\pi/\omega$ for all $\omega \in (0, \pi]$, where $a_{n0} := 0$, $b_{n0} := 1$, $a_{n1} := 1$, $b_{n1} := 2(n-1)$, $a_{n2} := 2n-1$, and $b_{n2} := 3n^2 - 5n + 2$. Therefore, as $n \to \infty$, $\sum_{t=1}^n t^r \exp(it\omega) = \mathcal{O}(n^r|\omega|^{-1})$ if $\omega \to 0$ and $n|\omega| \to \infty$.

(d) $\sum_{t=1}^n t^r \exp(it\omega) = \mathcal{O}(n^{r+1})$ if $n\omega \to 0$ as $n \to \infty$.

Proof. Parts (a) and (d) are trivial. To prove part (b), we observe that by Lemma 12.1.2, there is a constant $c > 0$ such that $|\sum t^r \exp(it\omega)| \leq cn^r\{g(\omega)\}^{-r-1}$, where $g(\omega) := |1 - \exp(i\omega)| = 2|\sin(\omega/2)|$. For $\omega \in [0, 2\pi]$, $g(\omega)$ is nonnegative, symmetric about $\omega = \pi$, monotone increasing from 0 to 2 in $\omega \in [0, \pi]$, and monotone decreasing from 2 to 0 in $\omega \in [\pi, 2\pi]$. Therefore, $g(\omega) \geq \max\{g(\kappa_1), g(\kappa_2)\} > 0$ uniformly in $\omega \in [\kappa_1, 2\pi - \kappa_2]$. Combining these results proves part (b). To prove part (c), we note that for $r = 0$, the assertion is the same as Lemma 12.1.3(d). For $r = 1, 2$, consider Abel's identity for summation by parts [443, p. 3]:

$$\sum_{t=1}^{n} tu_t = -\sum_{t=1}^{n-1} U_t + nU_n, \tag{12.1.2}$$

where $U_t := u_1 + \cdots + u_t$ $(t = 1, \ldots, n)$. Set $u_t := \exp(it\omega)$ for fixed $\omega \in (0, \pi]$. Then, by Lemma 12.1.3(d), $|U_t| = n|D_t(\omega)| \le \pi/\omega$ for all $t > 1$. Combining this result with (12.1.2) and the fact that $|U_1| = 1$ gives

$$\left|\sum_{t=1}^{n} t\exp(it\omega)\right| \le 1 + 2(n-1)\pi/\omega \qquad (n > 1). \tag{12.1.3}$$

Similarly, let $u_t := t\exp(it\omega)$ in (12.1.2). Then, we have $|U_1| = 1$ and, by (12.1.3), $|U_t| \le 1 + 2(t-1)\pi/\omega$ for all $t > 1$. Therefore, it follows from (12.1.2) that

$$\left|\sum_{t=1}^{n} t^2\exp(it\omega)\right| \le 2n - 1 + (3n^2 - 5n + 2)\pi/\omega \qquad (n > 1). \tag{12.1.4}$$

Combining (12.1.3) and (12.1.4) proves the inequality in part (c). Moreover, if $\omega \to 0$ as $n \to \infty$, then, for $r = 1, 2$, we can write

$$a_{nr} + b_{nr}\pi/\omega = \mathcal{O}(n^{r-1}) + \mathcal{O}(n^r|\omega|^{-1}) = \mathcal{O}(n^r|\omega|^{-1}).$$

The proof is complete. □

Remark 12.1.2 Although the bound in part (c) is sufficient to serve many purposes, there is a tighter bound of the form $c_{nr}\pi/\omega$ for $r = 1, 2$, where $c_{n1} := (n+1)/2$ and $c_{n2} = 1.75(n+1)(2n+1)/6$ are determined empirically.

As a consequence of Lemma 12.1.4, the following lemma, stated without proof, can be obtained for the sums of real sinusoids and their products.

Lemma 12.1.5 (Sum of Real Sinusoids). *As $n \to \infty$, the following assertions are true for $r = 0, 1, 2$. All error terms are uniformly bounded with respect to ω and λ.*

(a) If $\omega \in [\kappa_1, \pi - \kappa_2]$, where κ_1 and κ_2 are positive constants such that $\kappa_1 + \kappa_2 < \pi$, then

1. $\sum_{t=1}^{n} t^r \cos^2(\omega t) = \frac{1}{2}(r+1)^{-1} n^{r+1} + \mathcal{O}(n^r)$,
2. $\sum_{t=1}^{n} t^r \sin^2(\omega t) = \frac{1}{2}(r+1)^{-1} n^{r+1} + \mathcal{O}(n^r)$,
3. $\sum_{t=1}^{n} t^r \cos(\omega t)\sin(\omega t) = \mathcal{O}(n^r)$.

(b) If $\omega, \lambda \in [\kappa_1, \pi - \kappa_2]$ and $|\omega - \lambda| \ge \kappa_3$, where κ_1, κ_2, and κ_3 are positive constants such that $\kappa_1 + \kappa_2 < \pi$, then

1. $\sum_{t=1}^{n} t^r \cos(\omega t)\cos(\lambda t) = \mathcal{O}(n^r)$,
2. $\sum_{t=1}^{n} t^r \sin(\omega t)\sin(\lambda t) = \mathcal{O}(n^r)$,
3. $\sum_{t=1}^{n} t^r \cos(\omega t)\sin(\lambda t) = \mathcal{O}(n^r)$.

(c) If $\omega, \lambda \in [\Delta_1, \pi - \Delta_2]$, $|\omega - \lambda| \ge \Delta_3$, where $\Delta := \min(\Delta_1, \Delta_2, \Delta_3) > 0$, $\Delta \to 0$, and $n\Delta \to \infty$, then all three sums in part (b) take the form $\mathcal{O}(n^r\Delta^{-1})$.

The next lemma gives an upper bound for the derivative of a trigonometric polynomial in terms of the maximum of the trigonometric polynomial.

Lemma 12.1.6 (Bernstein Inequality [125]). *Let $p_n(z) := \sum_{t=0}^{n} c_t z^t$ ($z \in \mathbb{C}$) be a polynomial of degree n with real or complex coefficients c_t ($t = 0, 1, \ldots, n$). Then,*

$$\max_{|z|=1} |\dot{p}_n(z)| \le n \max_{|z|=1} |p_n(z)|,$$

where $\dot{p}_n(z) := \sum_{t=1}^{n} t c_t z^{t-1}$ is the derivative of $p_n(z)$.

The following two lemmas play an important role in proving the spectral approximation property of the AR model.

Lemma 12.1.7 (Fejér Theorem [46, p. 69]). *Let $f(\omega)$ be a real- or complex-valued continuous function with period 2π. Define*

$$r(u) := (2\pi)^{-1} \int_{-\pi}^{\pi} f(\omega) \exp(i\omega u)\, d\omega$$

and

$$f_n(\omega) := \sum_{|u|<n} (1 - |u|/n) r(u) \exp(-iu\omega).$$

Then, as $n \to \infty$, $f_n(\omega) \to f(\omega)$ uniformly in ω.

Remark 12.1.3 The proof of this assertion relies on the fact that

$$f_n(\omega) = \int_{-\pi}^{\pi} f(\omega - \lambda) \tilde{K}_n(\lambda)\, d\lambda,$$

where $\tilde{K}_n(\omega) := (2\pi)^{-1} n |D_n(\omega)|^2$ is the unnormalized Fejér kernel. The normalized Fejér kernel is defined by

$$K_n(\omega) := \tilde{K}_n(\omega) / \tilde{K}_n(0) = |D_n(\omega)|^2,$$

which has the property $K_n(0) = 1$. Because $f_n(\omega) \to f(\omega)$, one can regard $\{\tilde{K}_n(\omega)\}$ as a sequence that converges to the Dirac delta $\delta(\omega)$ which is characterized by the property $\int_{-\pi}^{\pi} f(\omega - \lambda) \delta(\lambda)\, d\lambda = f(\omega)$.

Lemma 12.1.8 (Spectral Factorization). *Let $\{r(-n), \ldots, r(n)\}$ be a set of complex constants and define $f(\omega) := \sum_{|u| \le n} r(u) \exp(-iu\omega)$. If $f(\omega) > 0$ for all ω, then there exists a minimum-phase polynomial $c(z) := 1 + \sum_{j=1}^{m} c_j z^{-j}$ with $m \le n$ such that $f(\omega) = \kappa |C(\omega)|^2$, where $C(\omega) := c(\exp(i\omega))$ and $\kappa := (1 + \sum_{j=1}^{m} |c_j|^2)^{-1} r(0)$.*

PROOF. Because $(2\pi)^{-1} \int_{-\pi}^{\pi} \exp(iu\omega)\, d\omega = \delta_u$, it follows that

$$(2\pi)^{-1} \int_{-\pi}^{\pi} f(\omega) \exp(i\omega u)\, d\omega = r(u) \qquad (|u| \le n).$$

Because $f(\omega)$ is real, we have $r^*(u) = r(-u)$ for $|u| \le n$. Define

$$m := \max\{|u| : r(u) \ne 0, |u| \le n\}$$

and

$$q(z) := z^m \sum_{|u|\le n} r(u) z^{-u} = \sum_{|u|\le m} r(u) z^{m-u}.$$

Observe that $q(z)$ is a $2m$-degree polynomial in z with the property

$$q(0) = r(m) \ne 0, \quad q(\exp(i\omega)) = \exp(im\omega) f(\omega) \ne 0.$$

This implies that $q(z)$ has $2m$ roots which are nonzero and which reside either inside or outside the unit circle. Moreover, because $r^*(u) = r(-u)$, we have $q(1/z^*) = \{z^{-2m} q(z)\}^*$. This implies that if z_0 is a root of $q(z)$, so is $1/z_0^*$. Combining these results with the fundamental theorem of algebra yields

$$q(z) = r(-m) \prod_{j=1}^{m} (z - z_j)(z - 1/z_j^*),$$

where $z_1, \ldots, z_m$ denote the roots of $q(z)$ inside the unit circle. Define

$$c(z) := \prod_{j=1}^{m} (1 - z_j z^{-1}).$$

Then, we can write

$$q(z) = \kappa z^m c(z) \{c(1/z^*)\}^*,$$

where $\kappa := (-1)^m (z_1^* \cdots z_m^*)^{-1} r(-m)$. Therefore,

$$\sum_{|u|\le n} r(u) z^{-u} = \kappa c(z) \{c(1/z^*)\}^*.$$

On the unit circle with $z = \exp(i\omega)$, we have $c(1/z^*) = c(z) = C(\omega)$. Hence the foregoing identity becomes $f(\omega) = \kappa C(\omega) C^*(\omega) = \kappa |C(\omega)|^2$. Moreover, because

$$(2\pi)^{-1} \int_{-\pi}^{\pi} f(\omega)\, d\omega = r(0) > 0, \quad (2\pi)^{-1} \int_{-\pi}^{\pi} |C(\omega)|^2 d\omega = 1 + \sum_{j=1}^{m} |c_j|^2,$$

it follows that $\kappa = (1 + \sum_{j=1}^m |c_j|^2)^{-1} r(0) > 0$. □

Remark 12.1.4 Combining Lemma 12.1.8 with Lemma 12.1.7 leads to the following assertion: if a real-valued 2π-periodic function $f(\omega)$ is continuous and bounded away from zero, then for any $\delta > 0$, there exists a function $g(\omega) := \kappa |c(\exp(i\omega))|^2$, with $\kappa > 0$ being a constant and $c(z) := 1 + \sum_{j=1}^m c_j z^{-j}$ being a minimum-phase polynomial, such that $|g(\omega) - f(\omega)| < \delta$ for all ω.

12.2 Probability Theory

The following lemma serves as the foundation for the distribution theory of the periodogram.

Lemma 12.2.1 (Chi-Square Distributions). *A real random variable X is said to have a chi-square distribution with $\nu > 0$ degrees of freedom, denoted by $X \sim \chi^2(\nu)$, if the PDF of X takes the form*

$$p(x) = \frac{1}{\Gamma(\nu/2)\,2^{\nu/2}}\,x^{\nu/2-1}\exp(-\tfrac{1}{2}x)\,\mathscr{I}(x>0).$$

The characteristic function of $\chi^2(\nu)$ is equal to $(1-i2x)^{-\nu/2}$ $(x \in \mathbb{R})$. A real random variable X is said to have a noncentral chi-square distribution with $\nu > 0$ degrees of freedom and with noncentrality parameter $\theta \geq 0$, denoted by $X \sim \chi^2(\nu,\theta)$, if the PDF of X takes the form

$$\begin{aligned} p(x) &= \frac{1}{\Gamma(\nu/2)\,2^{\nu/2}}\,x^{\nu/2-1}\exp\{-\tfrac{1}{2}(x+\theta)\} \\ &\quad\times\left\{1+\sum_{j=1}^{\infty}\frac{(\theta x)^j}{(2j)!}\frac{\Gamma(j+1/2)\Gamma(\nu/2)}{\Gamma(j+\nu/2)\Gamma(1/2)}\right\}\mathscr{I}(x>0). \end{aligned}$$

The characteristic function of $\chi^2(\nu,\theta)$ is given by $(1-i2x)^{-\nu/2}\exp\{i\theta x/(1-i2x)\}$ $(x \in \mathbb{R})$. In this notation, we can write $\chi^2(\nu) = \chi^2(\nu,0)$. The following assertions are true about the chi-square distributions.

(a) Let $\mathbf{X}$ be an m-dimensional random vector. If $\mathbf{X} \sim \mathrm{N}(\boldsymbol{\mu},\boldsymbol{\Sigma})$, then

$$(\mathbf{X}-\boldsymbol{\mu})^T\boldsymbol{\Sigma}^{-1}(\mathbf{X}-\boldsymbol{\mu}) \sim \chi^2(m).$$

If $\mathbf{X} \sim \mathrm{N}_c(\boldsymbol{\mu},\boldsymbol{\Sigma})$, then

$$(\mathbf{X}-\boldsymbol{\mu})^H\boldsymbol{\Sigma}^{-1}(\mathbf{X}-\boldsymbol{\mu}) \sim \tfrac{1}{2}\chi^2(2m).$$

(b) Let $\mathbf{X}$ be an m-dimensional random vector. If $\mathbf{X} \sim \mathrm{N}(\boldsymbol{\mu},\boldsymbol{\Sigma})$, then

$$\mathbf{X}^T\boldsymbol{\Sigma}^{-1}\mathbf{X} \sim \chi^2(m,\theta),$$

where $\theta := \boldsymbol{\mu}^T\boldsymbol{\Sigma}^{-1}\boldsymbol{\mu}$. If $\mathbf{X} \sim \mathrm{N}(\mathbf{0},\boldsymbol{\Sigma})$, then, for any $\mathbf{c} \in \mathbb{C}^m$,

$$(\mathbf{X}+\mathbf{c})^H\boldsymbol{\Sigma}^{-1}(\mathbf{X}+\mathbf{c}) \sim \chi^2(m,\theta) + \Im(\mathbf{c})^T\boldsymbol{\Sigma}^{-1}\Im(\mathbf{c}),$$

where $\theta := \Re(\mathbf{c})^T\boldsymbol{\Sigma}^{-1}\Re(\mathbf{c})$. If $\mathbf{X} \sim \mathrm{N}_c(\boldsymbol{\mu},\boldsymbol{\Sigma})$, then

$$\mathbf{X}^H\boldsymbol{\Sigma}^{-1}\mathbf{X} \sim \tfrac{1}{2}\chi^2(2m,2\theta),$$

where $\theta := \boldsymbol{\mu}^H\boldsymbol{\Sigma}^{-1}\boldsymbol{\mu}$.

(c) *If $X_1,\ldots,X_m$ are independent and $X_j \sim \chi^2(\nu_j,\theta_j)$ $(j=1,\ldots,m)$, then*

$$\sum_{j=1}^{m} X_j^2 \sim \chi^2(\nu,\theta),$$

where $\nu := \sum_{j=1}^m \nu_j$ and $\theta := \sum_{j=1}^m \theta_j$.

(d) *Let $X_1 \sim \chi^2(\nu_1)$ and $X_2 \sim \chi^2(\nu_2)$. If $X := X_1 - X_2$ is independent of X_2, then $X \sim \chi^2(\nu_1-\nu_2)$.*

(e) *If $X \sim \chi^2(\nu,\theta)$, then $E(X) = \nu+\theta$ and* $\mathrm{Var}(X) = 2\nu+4\theta$.

PROOF. Parts (a) and (b) for the real case and parts (c)–(e) can be found in standard textbooks such as [13, pp. 76–77] and [152, pp. 446–461]. To prove (a) and (b) for the complex case, let us first assume that $\mathbf{\Sigma} = \mathbf{I}$. If $\mathbf{X} \sim \mathrm{N}_c(\boldsymbol{\mu},\mathbf{I})$, then

$$(\mathbf{X}-\boldsymbol{\mu})^H(\mathbf{X}-\boldsymbol{\mu}) = \|\Re(\mathbf{X}-\boldsymbol{\mu})\|^2 + \|\Im(\mathbf{X}-\boldsymbol{\mu})\|^2$$

is just the sum of the squares of $2m$ i.i.d. $\mathrm{N}(0,1/2)$ random variables. Therefore, it is distributed as $\frac{1}{2}\chi^2(2m)$ according to part (a) for the real case. Similarly,

$$\mathbf{X}^H\mathbf{X} = \|\Re(\mathbf{X})\|^2 + \|\Im(\mathbf{X})\|^2$$

is the sum of the squares of $2m$ independent random variables, denoted by ξ_j and ζ_j $(j=1,\ldots,m)$, where $\xi_j \sim \mathrm{N}(\Re(\mu_j),1/2)$ and $\zeta_j \sim \mathrm{N}(\Im(\mu_j),1/2)$, with μ_j being the jth component of $\boldsymbol{\mu}$. Therefore, according to part (b) for the real case, it is distributed as $\frac{1}{2}\chi^2(2m,2\theta)$, where $\theta := \sum\{\Re(\mu_j)^2+\Im(\mu_j)^2\} = \boldsymbol{\mu}^H\boldsymbol{\mu}$. Moreover, if $\mathbf{X} \sim \mathrm{N}(\mathbf{0},\mathbf{I})$ and $\mathbf{c} \in \mathbb{C}^m$, then $\mathbf{X}+\Re(\mathbf{c}) \sim \mathrm{N}(\Re(\mathbf{c}),\mathbf{I})$, so that

$$(\mathbf{X}+\mathbf{c})^H(\mathbf{X}+\mathbf{c}) = \|\mathbf{X}+\Re(\mathbf{c})\|^2 + \|\Im(\mathbf{c})\|^2 \sim \chi^2(m,\theta) + \|\Im(\mathbf{c})\|^2,$$

where $\theta := \|\Re(\mathbf{c})\|^2$. In the general case of $\mathbf{\Sigma} \neq \mathbf{I}$, the assertions can be proved by applying the previous results to $\mathbf{\Sigma}^{-1/2}\mathbf{X}$, $\mathbf{\Sigma}^{-1/2}\boldsymbol{\mu}$, and $\mathbf{\Sigma}^{-1/2}\mathbf{c}$. □

The following lemma, stated without proof, gives the relationship between the covariance matrix of a complex random vector and the covariance matrix of its real and imaginary parts.

Lemma 12.2.2 (Covariance of Complex Random Variables). *Let $\mathbf{X}$ be a complex random vector with zero mean. If the covariance matrix and the complementary covariance matrix of $\mathbf{X}$ are given by $E(\mathbf{X}\mathbf{X}^H) = \mathbf{A}$ and $E(\mathbf{X}\mathbf{X}^T) = \mathbf{B}$, then the covariance matrix of $[\Re(\mathbf{X})^T,\Im(\mathbf{X})^T]^T$ can be expressed as*

$$\frac{1}{2}\Re\begin{bmatrix} \mathbf{A}+\mathbf{B} & i(\mathbf{A}-\mathbf{B}) \\ -i(\mathbf{A}+\mathbf{B}) & \mathbf{A}-\mathbf{B} \end{bmatrix}.$$

Conversely, if the covariance matrix of $[\Re(\mathbf{X})^T,\Im(\mathbf{X})^T]^T$ takes the form

$$\begin{bmatrix} \mathbf{A} & \mathbf{B}^T \\ \mathbf{B} & \mathbf{C} \end{bmatrix},$$

then the covariance matrix and the complementary covariance matrix of $\mathbf{X}$ *can be expressed as*

$$E(\mathbf{X}\mathbf{X}^H) = \mathbf{A}+\mathbf{C}+i(\mathbf{B}-\mathbf{B}^T), \quad E(\mathbf{X}\mathbf{X}^T) = \mathbf{A}-\mathbf{C}+i(\mathbf{B}+\mathbf{B}^T).$$

Moreover, if $\mathbf{A}=\mathbf{C}$ *and* $\mathbf{B}=-\mathbf{B}^T$*, then* $E(\mathbf{X}\mathbf{X}^H) = 2(\mathbf{A}+i\mathbf{B})$ *and* $E(\mathbf{X}\mathbf{X}^T) = \mathbf{0}$.

The next lemma contains the definition and some properties of the elliptical distributions as a generalization of the real and complex Gaussian distributions.

Lemma 12.2.3 (Real and Complex Elliptical Distributions [105]).

(a) *A random vector* $\mathbf{X} \in \mathbb{R}^m$ *is said to have an elliptical distribution with mean* $\boldsymbol{\mu}$ *and dispersion matrix* $\boldsymbol{\Sigma} > 0$ *if there exists a density generator* $g(\cdot) \ge 0$ *such that the PDF of* $\mathbf{X}$ *can be expressed as*

$$p(\boldsymbol{x}) = \kappa_m^{-1} |\boldsymbol{\Sigma}|^{-1/2} g((\boldsymbol{x}-\boldsymbol{\mu})^T \boldsymbol{\Sigma}^{-1}(\boldsymbol{x}-\boldsymbol{\mu})) \qquad (\boldsymbol{x} \in \mathbb{R}^m),$$

where $\kappa_m := \int_{\boldsymbol{x}\in\mathbb{R}^m} g(\|\boldsymbol{x}\|^2)\, d\boldsymbol{x} < \infty$ *is the normalizing constant. Moreover, if* $c_m := m^{-1}\kappa_m^{-1} \int_{\boldsymbol{x}\in\mathbb{R}^m} \|\boldsymbol{x}\|^2 g(\|\boldsymbol{x}\|^2)\, d\boldsymbol{x} < \infty$*, then* $\mathrm{Cov}(\mathbf{X}) = c_m \boldsymbol{\Sigma}$.

(b) *A random vector* $\mathbf{X} \in \mathbb{C}^m$ *is said to have a complex elliptical distribution mean* $\boldsymbol{\mu}$ *and dispersion matrix* $\boldsymbol{\Sigma}$*, if* $[\Re(\mathbf{X})^T, \Im(\mathbf{X})^T]^T$ *has an elliptical distribution with mean* $[\Re(\boldsymbol{\mu})^T, \Im(\boldsymbol{\mu})^T]^T$*, dispersion matrix*

$$\frac{1}{2}\begin{bmatrix} \Re(\boldsymbol{\Sigma}) & -\Im(\boldsymbol{\Sigma}) \\ \Im(\boldsymbol{\Sigma}) & \Re(\boldsymbol{\Sigma}) \end{bmatrix},$$

and density generator $g(\cdot)$*, in which case, the PDF of* $\mathbf{X}$ *takes the form*

$$p(\boldsymbol{x}) = 2^m \kappa_{2m}^{-1} |\boldsymbol{\Sigma}|^{-1} g(2(\boldsymbol{x}-\boldsymbol{\mu})^H \boldsymbol{\Sigma}^{-1}(\boldsymbol{x}-\boldsymbol{\mu})) \qquad (\boldsymbol{x} \in \mathbb{C}^m).$$

Moreover, if $c_{2m} < \infty$*, then* $\mathrm{Cov}(\mathbf{X}) = c_{2m}\boldsymbol{\Sigma}$ *and* $E\{(\mathbf{X}-\boldsymbol{\mu})(\mathbf{X}-\boldsymbol{\mu})^T\} = \mathbf{0}$.

(c) *If the random vector* $\mathbf{X} \in \mathbb{R}^m$ *has an elliptical distribution with mean* $\boldsymbol{\mu}$ *and dispersion matrix* $\boldsymbol{\Sigma}$*, then, for any constant matrix* $\mathbf{A} \in \mathbb{R}^{r\times m}$ *and constant vector* $\mathbf{b} \in \mathbb{R}^r$*,* $\mathbf{A}\mathbf{X}+\mathbf{b}$ *has an elliptical distribution with mean* $\mathbf{A}\boldsymbol{\mu}+\mathbf{b}$ *and dispersion matrix* $\mathbf{A}\boldsymbol{\Sigma}\mathbf{A}^T$*. A similar assertion is true for linear transformations of a random vector* $\mathbf{X}$ *that has a complex elliptical distribution.*

12.3 Numerical Analysis

The following lemma is useful for deriving the least-squares estimator in linear regression with complex parameters.

Lemma 12.3.1 (Differentiation). *For any real-valued function* $g(\boldsymbol{x})$ *of a complex variable* $\boldsymbol{x}$*, let the derivative of* $g(\boldsymbol{x})$ *be defined as*

$$\frac{d}{d\boldsymbol{x}}g(\boldsymbol{x}) := \frac{\partial g(\boldsymbol{x})}{\partial \Re(\boldsymbol{x})} + i\frac{\partial g(\boldsymbol{x})}{\partial \Im(\boldsymbol{x})}.$$

Then, the following assertions are true for any constant matrix $\mathbf{A}$*, constant vector* $\mathbf{b}$*, and real-valued differentiable function* $c(\cdot)$ *with derivative* $\dot{c}(\cdot)$*.*

(a) $(d/d\boldsymbol{x})\mathbf{b}^H\boldsymbol{x} = \mathbf{0}$.

(b) $(d/d\boldsymbol{x})\boldsymbol{x}^H\mathbf{b} = 2\mathbf{b}$.

(c) $(d/d\boldsymbol{x})(\boldsymbol{x}^H\mathbf{A}\boldsymbol{x}) = 2\mathbf{A}\boldsymbol{x}$.

(d) $(d/d\boldsymbol{x})c(\|\mathbf{A}\boldsymbol{x}+\mathbf{b}\|^2) = 2\dot{c}(\|\mathbf{A}\boldsymbol{x}+\mathbf{b}\|^2)(\mathbf{A}^H\mathbf{A}\boldsymbol{x}+\mathbf{A}^H\mathbf{b})$.

PROOF. For part (a), let $g(\boldsymbol{x}) := \mathbf{b}^H\boldsymbol{x} = \mathbf{b}^H\Re(\boldsymbol{x}) + i\mathbf{b}^H\Im(\boldsymbol{x})$. Because $\partial g/\partial\Re(\boldsymbol{x}) = \mathbf{b}^*$ and $\partial g/\partial\Im(\boldsymbol{x}) = i\mathbf{b}^*$, we obtain $(d/d\boldsymbol{x})g = \mathbf{b}^* + i(i\mathbf{b}^*) = \mathbf{0}$. For part (b), let $g(\boldsymbol{x}) := \boldsymbol{x}^H\mathbf{b} = \Re(\boldsymbol{x})^T\mathbf{b} - i\Im(\boldsymbol{x})^T\mathbf{b}$. Then, $\partial g/\partial\Re(\boldsymbol{x}) = \mathbf{b}$, $\partial g/\partial\Im(\boldsymbol{x}) = -i\mathbf{b}$, and hence $(d/d\boldsymbol{x})g = \mathbf{b} + i(-i\mathbf{b}) = 2\mathbf{b}$. For part (c), we can write

$$g(\boldsymbol{x}) := \boldsymbol{x}^H\mathbf{A}\boldsymbol{x} = \Re(\boldsymbol{x})^T\mathbf{A}\Re(\boldsymbol{x}) + \Im(\boldsymbol{x})^T\mathbf{A}\Im(\boldsymbol{x}) + i\{\Re(\boldsymbol{x})^T\mathbf{A}\Im(\boldsymbol{x}) - \Im(\boldsymbol{x})^T\mathbf{A}\Re(\boldsymbol{x})\}.$$

Therefore,

$$\begin{aligned}\frac{\partial g}{\partial\Re(\boldsymbol{x})} &= \mathbf{A}\Re(\boldsymbol{x}) + \mathbf{A}^T\Re(\boldsymbol{x}) + i\{\mathbf{A}\Im(\boldsymbol{x}) - \mathbf{A}^T\Im(\boldsymbol{x})\},\\ \frac{\partial g}{\partial\Im(\boldsymbol{x})} &= \mathbf{A}\Im(\boldsymbol{x}) + \mathbf{A}^T\Im(\boldsymbol{x}) + i\{\mathbf{A}^T\Re(\boldsymbol{x}) - \mathbf{A}\Re(\boldsymbol{x})\}.\end{aligned}$$

This yields $(d/d\boldsymbol{x})g = \mathbf{A}\boldsymbol{x} + \mathbf{A}^T\boldsymbol{x} + \mathbf{A}\boldsymbol{x} - \mathbf{A}^T\boldsymbol{x} = 2\mathbf{A}\boldsymbol{x}$. Part (d) results from (a)–(c) upon noting that $\|\mathbf{A}\boldsymbol{x}+\mathbf{b}\|^2 = \boldsymbol{x}^H\mathbf{A}^H\mathbf{A}\boldsymbol{x} + \mathbf{b}^H\mathbf{A}\boldsymbol{x} + \boldsymbol{x}^H\mathbf{A}^H\mathbf{b} + \mathbf{b}^H\mathbf{b}$. □

The following lemma can be employed to establish the convergence of iterative algorithms.

Lemma 12.3.2 (Fixed Point Theorem [365, p. 297]). *Let* $\mathbf{g}(\boldsymbol{x}) \in \mathbb{C}^r$ *be a deterministic function of* $\boldsymbol{x} \in \mathbb{C}^r$ *for some* $r \geq 1$*. Assume that there exist a compact subset* $\mathscr{C} \in \mathbb{C}^r$ *and a constant* $c \in (0,1)$ *such that*

(a) $\mathbf{g}(\boldsymbol{x}) \in \mathscr{C}$ *for all* $\boldsymbol{x} \in \mathscr{C}$*, and*

(b) $\|\mathbf{g}(\boldsymbol{x}) - \mathbf{g}(\boldsymbol{x}')\| \leq c\|\boldsymbol{x} - \boldsymbol{x}'\|$ *for all* $\boldsymbol{x}, \boldsymbol{x}' \in \mathscr{C}$*.*

Then, $\mathbf{g}(\boldsymbol{x})$ *has a unique fixed point in* $\mathscr{C}$*, denoted as* $\boldsymbol{\xi}$*. Moreover, for any* $\boldsymbol{x}_0 \in \mathscr{C}$*, the sequence* $\{\boldsymbol{x}_m\}$ *generated by the fixed point iteration*

$$\boldsymbol{x}_m := \mathbf{g}(\boldsymbol{x}_{m-1}) \qquad (m = 1, 2, \ldots)$$

converges to $\boldsymbol{\xi}$*. A function with property (b) is called a contraction mapping.*

PROOF. It follows from (a) that $\boldsymbol{x}_m \in \mathscr{C}$ for all m. Let us assume momentarily that $\{\boldsymbol{x}_m\}$ converges to some $\boldsymbol{\xi} \in \mathscr{C}$. Then, it follows from (b) that for any m,

$$\begin{aligned}\|\mathbf{g}(\boldsymbol{\xi}) - \boldsymbol{\xi}\| &\le \|\mathbf{g}(\boldsymbol{\xi}) - \mathbf{g}(\boldsymbol{x}_m)\| + \|\mathbf{g}(\boldsymbol{x}_m) - \boldsymbol{\xi}\| \\ &\le c\|\boldsymbol{\xi} - \boldsymbol{x}_m\| + \|\boldsymbol{x}_{m+1} - \boldsymbol{\xi}\|.\end{aligned}$$

Because $\boldsymbol{x}_m \to \boldsymbol{\xi}$ as $m \to \infty$ by assumption, we obtain $\mathbf{g}(\boldsymbol{\xi}) = \boldsymbol{\xi}$, i.e., $\boldsymbol{\xi}$ is a fixed point of $\mathbf{g}(\boldsymbol{x})$. If $\mathbf{g}(\boldsymbol{x})$ has another fixed point in $\mathscr{C}$, say $\boldsymbol{\xi}'$, then

$$\|\boldsymbol{\xi} - \boldsymbol{\xi}'\| = \|\mathbf{g}(\boldsymbol{\xi}) - \mathbf{g}(\boldsymbol{\xi}')\| \le c\|\boldsymbol{\xi} - \boldsymbol{\xi}'\|,$$

which is possible only if $\boldsymbol{\xi} = \boldsymbol{\xi}'$. Hence the fixed point is unique.

It remains to show that $\{\boldsymbol{x}_m\}$ is a Cauchy sequence so that it converges to a finite limit $\boldsymbol{\xi} \in \mathscr{C}$. Toward that end, observe that for any $m > 1$,

$$\|\boldsymbol{x}_m - \boldsymbol{x}_{m-1}\| \le c\|\boldsymbol{x}_{m-1} - \boldsymbol{x}_{m-2}\| \le c^{m-1}\|\boldsymbol{x}_1 - \boldsymbol{x}_0\|.$$

Therefore, for any $m > n > 1$, we have

$$\begin{aligned}\|\boldsymbol{x}_m - \boldsymbol{x}_n\| &\le \|\boldsymbol{x}_m - \boldsymbol{x}_{m-1}\| + \|\boldsymbol{x}_{m-1} - \boldsymbol{x}_{m-2}\| + \cdots + \|\boldsymbol{x}_{n+1} - \boldsymbol{x}_n\| \\ &\le c^n(c^{m-n-1} + c^{m-n-2} + \cdots + 1)\|\boldsymbol{x}_1 - \boldsymbol{x}_0\| \\ &\le c^n(1-c)^{-1}\|\boldsymbol{x}_1 - \boldsymbol{x}_0\|.\end{aligned}$$

Because $0 < c < 1$, we have $\|\boldsymbol{x}_m - \boldsymbol{x}_n\| \le \delta$ for any $\delta > 0$ if n is sufficiently large. Hence $\{\boldsymbol{x}_m\}$ is a Cauchy sequence. □

12.4 Matrix Theory

The first lemma in this section provides two formulas which are particularly useful in manipulating the inverse of a covariance matrix.

Lemma 12.4.1 (Matrix Inversion [153, pp. 18–19]). *Let $\mathbf{A} \in \mathbb{C}^{m\times m}$ and $\mathbf{C} \in \mathbb{C}^{r\times r}$ be positive-definite Hermitian matrices. Then, for any $\mathbf{B} \in \mathbb{C}^{m\times r}$ that makes the following inverse matrices exist, we have*

$$(\mathbf{A}^{-1} + \mathbf{B}\mathbf{C}^{-1}\mathbf{B}^H)^{-1} = \mathbf{A} - \mathbf{A}\mathbf{B}(\mathbf{C} + \mathbf{B}^H\mathbf{A}\mathbf{B})^{-1}\mathbf{B}^H\mathbf{A}$$

and

$$\begin{bmatrix}\mathbf{A} & \mathbf{B} \\ \mathbf{B}^H & \mathbf{C}\end{bmatrix}^{-1} = \begin{bmatrix}(\mathbf{A} - \mathbf{B}\mathbf{C}^{-1}\mathbf{B}^H)^{-1} & -\mathbf{A}^{-1}\mathbf{B}(\mathbf{C} - \mathbf{B}^H\mathbf{A}^{-1}\mathbf{B})^{-1} \\ -(\mathbf{C} - \mathbf{B}^H\mathbf{A}^{-1}\mathbf{B})^{-1}\mathbf{B}^H\mathbf{A}^{-1} & (\mathbf{C} - \mathbf{B}^H\mathbf{A}^{-1}\mathbf{B})^{-1}\end{bmatrix}.$$

These identities are known as matrix inversion formulas.

The remainder of this section contains a number of results from the matrix perturbation theory [361]. They are used extensively to analyze covariance-based frequency estimation methods.

To present these results, let $\mathbf{A} \in \mathbb{C}^{m\times m}$ be a nonnegative-definite Hermitian matrix. Suppose that $\mathbf{A}$ is approximated by another nonnegative-definite Hermitian matrix $\mathbf{B}$. Let the error of approximation be measured by

$$\epsilon := \|\mathbf{B}-\mathbf{A}\|,$$

where $\|\cdot\|$ denotes the ℓ_2 norm (also known as the Frobenius or Euclidean norm) which is unitary invariant [153, p. 292].

First, consider the eigenvalues of $\mathbf{A}$ and $\mathbf{B}$. The following lemma asserts that the difference between the eigenvalues is small if ϵ is small.

Lemma 12.4.2 (Perturbation Theorem for Eigenvalues [361, p. 205]). *If the eigenvalues of* $\mathbf{A}$ *and* $\mathbf{B}$ *are* $\lambda_1(\mathbf{A}) \ge \cdots \ge \lambda_m(\mathbf{A}) \ge 0$ *and* $\lambda_1(\mathbf{B}) \ge \cdots \ge \lambda_m(\mathbf{B}) \ge 0$, *then*

$$\sum_{j=1}^{m} |\lambda_j(\mathbf{B}) - \lambda_j(\mathbf{A})|^2 \le \epsilon^2.$$

This implies in particular that $|\lambda_j(\mathbf{B}) - \lambda_j(\mathbf{A})| \le \epsilon$ *for all* $j = 1,\ldots,m$.

Remark 12.4.1 It is necessary that the eigenvalues be arranged in the same descending (or ascending) order. The assertion remains valid for all Hermitian matrices $\mathbf{A}$ and $\mathbf{B}$ for which the eigenvalues may be negative [361, p. 205].

Next, consider the eigenvalue decomposition (EVD) of $\mathbf{A}$ given by

$$\mathbf{A} = \mathbf{V}\boldsymbol{\Lambda}\mathbf{V}^H, \tag{12.4.1}$$

where $\boldsymbol{\Lambda} := \text{diag}\{\lambda_1(\mathbf{A}),\ldots,\lambda_m(\mathbf{A})\}$ is a diagonal matrix comprising the eigenvalues of $\mathbf{A}$ satisfying $\lambda_1(\mathbf{A}) \ge \cdots \ge \lambda_m(\mathbf{A}) \ge 0$ and $\mathbf{V} := [\mathbf{v}_1,\ldots,\mathbf{v}_m]$ is a unitary matrix with $\mathbf{v}_j$ being an eigenvector associated with $\lambda_j(\mathbf{A})$ $(j = 1,\ldots,m)$. Because $\mathbf{V}^H\mathbf{V} = \mathbf{V}\mathbf{V}^H = \mathbf{I}$, the EVD in (12.4.1) can also be written as

$$\mathbf{V}^H\mathbf{A}\mathbf{V} = \boldsymbol{\Lambda}, \tag{12.4.2}$$

in which case we say that $\mathbf{A}$ is diagonalized into $\boldsymbol{\Lambda}$ by $\mathbf{V}$. Now, for some $p < m$, let $\boldsymbol{\Lambda}$ be partitioned as $\boldsymbol{\Lambda} = \text{diag}(\boldsymbol{\Lambda}_1, \boldsymbol{\Lambda}_2)$, where

$$\begin{aligned} \boldsymbol{\Lambda}_1 &:= \text{diag}\{\lambda_1(\mathbf{A}),\ldots,\lambda_p(\mathbf{A})\}, \\ \boldsymbol{\Lambda}_2 &:= \text{diag}\{\lambda_{p+1}(\mathbf{A}),\ldots,\lambda_m(\mathbf{A})\}. \end{aligned}$$

Let $\mathbf{V} = [\mathbf{V}_1, \mathbf{V}_2]$ be the corresponding partition of $\mathbf{V}$. Then, (12.4.2) becomes

$$[\mathbf{V}_1, \mathbf{V}_2]^H \mathbf{A} [\mathbf{V}_1, \mathbf{V}_2] = \text{diag}(\boldsymbol{\Lambda}_1, \boldsymbol{\Lambda}_2). \tag{12.4.3}$$

The following lemma provides a sufficient condition under which $\mathbf{B}$ can be diagonalized approximately into $\boldsymbol{\Lambda}$ by a unitary matrix $\tilde{\mathbf{V}} := [\tilde{\mathbf{V}}_1, \tilde{\mathbf{V}}_2]$ which is approximately equal to $\mathbf{V}$. In other words,

$$[\tilde{\mathbf{V}}_1, \tilde{\mathbf{V}}_2]^H \mathbf{B} [\tilde{\mathbf{V}}_1, \tilde{\mathbf{V}}_2] = \operatorname{diag}(\tilde{\boldsymbol{\Lambda}}_1, \tilde{\boldsymbol{\Lambda}}_2), \tag{12.4.4}$$

where $[\tilde{\mathbf{V}}_1, \tilde{\mathbf{V}}_2] \approx [\mathbf{V}_1, \mathbf{V}_2]$ and $\operatorname{diag}(\tilde{\boldsymbol{\Lambda}}_1, \tilde{\boldsymbol{\Lambda}}_2) \approx \operatorname{diag}(\boldsymbol{\Lambda}_1, \boldsymbol{\Lambda}_2)$. Note that $\tilde{\boldsymbol{\Lambda}}_1$ and $\tilde{\boldsymbol{\Lambda}}_2$ are not necessarily diagonal matrices.

Lemma 12.4.3 (Perturbation Theorem for Unitary Diagonalization). *Define* $\mathbf{E} := \mathbf{V}^H(\mathbf{B}-\mathbf{A})\mathbf{V} = [\mathbf{E}_{ij}]$, *where* $\mathbf{E}_{ij} := \mathbf{V}_i^H(\mathbf{B}-\mathbf{A})\mathbf{V}_j$ $(i,j=1,2)$. *If* $\delta := \lambda_p(\mathbf{A}) - \lambda_{p+1}(\mathbf{A}) > 0$ *and* $\epsilon < \delta/4$, *then there is a unique matrix* $\mathbf{C}$ *such that* $\|\mathbf{C}\| \le \epsilon(\delta/2-\epsilon)^{-1}$ *and*

$$\mathbf{C}(\boldsymbol{\Lambda}_1 + \mathbf{E}_{11}) + \mathbf{C}\mathbf{E}_{12}\mathbf{C} = \mathbf{E}_{21} + (\boldsymbol{\Lambda}_2 + \mathbf{E}_{22})\mathbf{C}. \tag{12.4.5}$$

In this case, the identity (12.4.4) holds with

$$\begin{aligned}
\tilde{\mathbf{V}}_1 &:= (\mathbf{V}_1 + \mathbf{V}_2\mathbf{C})(\mathbf{I} + \mathbf{C}^H\mathbf{C})^{-1/2},\\
\tilde{\mathbf{V}}_2 &:= (\mathbf{V}_2 - \mathbf{V}_1\mathbf{C}^H)(\mathbf{I} + \mathbf{C}\mathbf{C}^H)^{-1/2},\\
\tilde{\boldsymbol{\Lambda}}_1 &:= (\mathbf{I} + \mathbf{C}^H\mathbf{C})^{1/2}(\boldsymbol{\Lambda}_1 + \mathbf{E}_{11} + \mathbf{E}_{12}\mathbf{C})(\mathbf{I} + \mathbf{C}^H\mathbf{C})^{-1/2},\\
\tilde{\boldsymbol{\Lambda}}_2 &:= (\mathbf{I} + \mathbf{C}\mathbf{C}^H)^{-1/2}(\boldsymbol{\Lambda}_2 + \mathbf{E}_{22} - \mathbf{C}\mathbf{E}_{12})(\mathbf{I} + \mathbf{C}\mathbf{C}^H)^{1/2}.
\end{aligned}$$

Note that $[\tilde{\mathbf{V}}_1, \tilde{\mathbf{V}}_2] \to [\mathbf{V}_1, \mathbf{V}_2]$ *and* $\operatorname{diag}(\tilde{\boldsymbol{\Lambda}}_1, \tilde{\boldsymbol{\Lambda}}_2) \to \operatorname{diag}(\boldsymbol{\Lambda}_1, \boldsymbol{\Lambda}_2)$ *as* $\epsilon \to 0$.

PROOF. The assertion follows from Theorem 2.7 in [361, pp. 236–237] and Theorem 3.1 in [361, p. 247], combined with $\|\mathbf{E}_{ij}\| \le \|\mathbf{E}\| = \|\mathbf{B}-\mathbf{A}\| = \epsilon$. □

To better understand the construction of $\tilde{\mathbf{V}}$ in Lemma 12.4.3, observe that for any matrix $\mathbf{C} \in \mathbb{C}^{(m-p)\times p}$ both $\mathbf{I} + \mathbf{C}^H\mathbf{C}$ and $\mathbf{I} + \mathbf{C}\mathbf{C}^H$ are positive definite and

$$\mathbf{W} := \begin{bmatrix} \mathbf{I} & -\mathbf{C}^H \\ \mathbf{C} & \mathbf{I} \end{bmatrix} \begin{bmatrix} (\mathbf{I} + \mathbf{C}^H\mathbf{C})^{-1/2} & \mathbf{0} \\ \mathbf{0} & (\mathbf{I} + \mathbf{C}\mathbf{C}^H)^{-1/2} \end{bmatrix} \tag{12.4.6}$$

is a unitary matrix. The goal is to find a proper $\mathbf{C}$ that makes

$$\mathbf{W}^H\mathbf{V}^H\mathbf{B}\mathbf{V}\mathbf{W} = \mathbf{W}^H \begin{bmatrix} \boldsymbol{\Lambda}_1 + \mathbf{E}_{11} & \mathbf{E}_{12} \\ \mathbf{E}_{21} & \boldsymbol{\Lambda}_2 + \mathbf{E}_{22} \end{bmatrix} \mathbf{W}$$

a block-diagonal matrix of the form $\operatorname{diag}(\tilde{\boldsymbol{\Lambda}}_1, \tilde{\boldsymbol{\Lambda}}_2)$ for some $\tilde{\boldsymbol{\Lambda}}_1$ and $\tilde{\boldsymbol{\Lambda}}_2$. It turns out that such a matrix $\mathbf{C}$ must satisfy (12.4.5). With this choice, it suffices to define $\tilde{\mathbf{V}} := [\tilde{\mathbf{V}}_1, \tilde{\mathbf{V}}_2] := \mathbf{V}\mathbf{W}$ in order to achieve (12.4.4). Lemma 12.4.3 asserts that (12.4.5) has a unique solution $\mathbf{C}$ with $\|\mathbf{C}\| = \mathcal{O}(\epsilon)$ so that $\tilde{\mathbf{V}} \to \mathbf{V}$ as $\mathbf{B} \to \mathbf{A}$.

The converse of Lemma 12.4.3 is not true. In other words, a unitary matrix $\tilde{\mathbf{V}}$ that satisfies (12.4.4) does not have to take the form $\mathbf{V}\mathbf{W}$ with $\mathbf{W}$ given by (12.4.6) and therefore does not necessarily possess the property $\tilde{\mathbf{V}} \to \mathbf{V}$ as $\mathbf{B} \to \mathbf{A}$.

Lemma 12.4.3 can be used to analyze the eigenvectors of $\mathbf{B}$ in connection with the eigenvectors of $\mathbf{A}$. Let $\mathbf{B}$ have an EVD of the form

$$\mathbf{B} = \mathbf{U}\boldsymbol{\Sigma}\mathbf{U}^H,$$

where $\mathbf{U} := [\mathbf{u}_1, \ldots, \mathbf{u}_m]$ and $\boldsymbol{\Sigma} := \text{diag}\{\lambda_1(\mathbf{B}), \ldots, \lambda_m(\mathbf{B})\}$. By Lemma 12.4.2,

$$\boldsymbol{\Sigma} \to \boldsymbol{\Lambda} \quad \text{as } \mathbf{B} \to \mathbf{A}.$$

But, we do not necessarily have $\mathbf{U} \to \mathbf{V}$ because the eigenvectors are not unique especially when the associated eigenvalues have multiplicity greater than 1. For eigenvectors associated with simple eigenvalues, a form of convergence can be established. Toward that end, let $\mathbf{U} = [\mathbf{U}_1, \mathbf{U}_2]$ and $\boldsymbol{\Sigma} := \text{diag}(\boldsymbol{\Sigma}_1, \boldsymbol{\Sigma}_2)$ be partitioned in the same way as $\mathbf{V}$ and $\boldsymbol{\Lambda}$ in (12.4.3) so that

$$[\mathbf{U}_1, \mathbf{U}_2]^H \mathbf{B} [\mathbf{U}_1, \mathbf{U}_2] = \text{diag}(\boldsymbol{\Sigma}_1, \boldsymbol{\Sigma}_2). \tag{12.4.7}$$

The following lemma asserts that if the first $p+1$ eigenvalues of $\mathbf{A}$ are all distinct, then $\mathbf{U}_1 := [\mathbf{u}_1, \ldots, \mathbf{u}_p]$, with proper scaling, approaches $\mathbf{V}_1$ as $\mathbf{B} \to \mathbf{A}$.

Lemma 12.4.4 (Perturbation Theorem for Eigenvectors). *If the eigenvalues of* $\mathbf{A}$ *satisfy* $\lambda_1(\mathbf{A}) > \cdots > \lambda_p(\mathbf{A}) > \lambda_{p+1}(\mathbf{A})$, *then there exists a diagonal matrix* $\mathbf{D} := \text{diag}(d_1, \ldots, d_p)$, *with* $|d_j| = 1$ *for* $j = 1, \ldots, p$, *such that* $\mathbf{U}_1\mathbf{D} \to \mathbf{V}_1$ *as* $\mathbf{B} \to \mathbf{A}$.

PROOF. Consider the case of $p = 1$ and write $\mathbf{U} = [\mathbf{u}_1, \mathbf{U}_2]$ and $\boldsymbol{\Sigma} = \text{diag}\{\lambda_1(\mathbf{B}), \boldsymbol{\Sigma}_2\}$, where $\mathbf{U}_2 := [\mathbf{u}_2, \ldots, \mathbf{u}_m]$ and $\boldsymbol{\Sigma}_2 := \text{diag}\{\lambda_2(\mathbf{B}), \ldots, \lambda_m(\mathbf{B})\}$. By Lemma 12.4.3, if ϵ is sufficiently small, then there exists a unitary matrix $\tilde{\mathbf{V}} := [\tilde{\mathbf{v}}_1, \tilde{\mathbf{V}}_2]$ and a block diagonal matrix $\tilde{\boldsymbol{\Lambda}} := \text{diag}(\tilde{\lambda}_1, \tilde{\boldsymbol{\Lambda}}_2)$ such that $\tilde{\mathbf{V}}^H \mathbf{B} \tilde{\mathbf{V}} = \tilde{\boldsymbol{\Lambda}}$. In particular, we have

$$\mathbf{B}\tilde{\mathbf{v}}_1 = \tilde{\lambda}_1 \tilde{\mathbf{v}}_1.$$

This means that $\tilde{\lambda}_1$ is an eigenvalue of $\mathbf{B}$, i.e., $\tilde{\lambda}_1 \in \{\lambda_1(\mathbf{B}), \ldots, \lambda_m(\mathbf{B})\}$, and that $\tilde{\mathbf{v}}_1$ is an eigenvector associated with that eigenvalue.

By Lemma 12.4.3, $\tilde{\lambda}_1 \to \lambda_1(\mathbf{A})$ as $\epsilon \to 0$. Combining this result with the fact that $\lambda_j(\mathbf{B}) \to \lambda_j(\mathbf{A})$ for all $j = 1, \ldots, m$ and $\lambda_1(\mathbf{A}) > \lambda_j(\mathbf{A})$ for all $j = 2, \ldots, m$ leads to

$$\tilde{\lambda}_1 = \lambda_1(\mathbf{B})$$

when ϵ is sufficiently small. Moreover, because both $\tilde{\mathbf{v}}_1$ and $\mathbf{u}_1$ are eigenvectors of $\mathbf{B}$ associated with $\lambda_1(\mathbf{B})$ which is a simple eigenvalue, it follows that

$$\tilde{\mathbf{v}}_1 = d_1 \mathbf{u}_1$$

for some $d_1 \in \mathbb{C}$ with $|d_1| = 1$. By Lemma 12.4.3, $d_1\mathbf{u}_1 = \tilde{\mathbf{v}}_1 \to \mathbf{v}_1$ as $\mathbf{B} \to \mathbf{A}$.

Applying the same argument to $\mathbf{B}_2 := \mathbf{B} - \lambda_1(\mathbf{B})\mathbf{u}_1\mathbf{u}_1^H = \mathbf{U}_2\boldsymbol{\Sigma}_2\mathbf{U}_2^H$ proves the existence of d_2 with $|d_2| = 1$ such that $d_2\mathbf{u}_2 \to \mathbf{v}_2$ as $\mathbf{B} \to \mathbf{A}$. The proof is complete by repeating the steps. □

Because $\mathbf{V} = [\mathbf{V}_1, \mathbf{V}_2]$ is a unitary matrix, the columns in $\mathbf{V}_1$ are orthogonal to the columns in $\mathbf{V}_2$, i.e.,

$$\mathbf{V}_2^H \mathbf{V}_1 = \mathbf{0}, \quad \mathbf{V}_1^H \mathbf{V}_2 = \mathbf{0}.$$

The following lemma asserts that if the eigenvalues of $\mathbf{A}$ in $\mathbf{\Lambda}_1$ do not overlap with those in $\mathbf{\Lambda}_2$, then the columns in $\mathbf{U}_2$ and $\mathbf{U}_1$ of the EVD of $\mathbf{B}$ are approximately orthogonal to the columns in $\mathbf{V}_1$ and $\mathbf{V}_2$, respectively.

Lemma 12.4.5 (Perturbation Theorem for Orthogonal Subspaces). *Assume that* $\delta := \lambda_p(\mathbf{A}) - \lambda_{p+1}(\mathbf{A}) > 0$ *and* $\epsilon < \delta/4$. *Define* $\mathbf{E}_j := (\mathbf{B} - \mathbf{A})\mathbf{V}_j$ $(j = 1, 2)$. *Then,*

$$\|\mathbf{U}_2^H \mathbf{V}_1\| \le \gamma_1^{-1} \|\mathbf{E}_1\|, \quad \|\mathbf{U}_1^H \mathbf{V}_2\| \le \gamma_2^{-1} \|\mathbf{E}_2\|,$$

where $\gamma_1 := \lambda_p(\mathbf{A}) - \lambda_{p+1}(\mathbf{B}) > 3\delta/4$ *and* $\gamma_2 := \lambda_p(\mathbf{B}) - \lambda_{p+1}(\mathbf{A}) > 3\delta/4$.

PROOF. Because $|\lambda_{p+1}(\mathbf{B}) - \lambda_{p+1}(\mathbf{A})| \le \epsilon < \delta/4$, it follows that

$$\gamma_1 = \lambda_p(\mathbf{A}) - \lambda_{p+1}(\mathbf{A}) + \lambda_{p+1}(\mathbf{A}) - \lambda_{p+1}(\mathbf{B}) > \delta - \delta/4 = 3\delta/4.$$

Similarly, we obtain $\gamma_2 > 3\delta/4$. Moreover, it follows from (12.4.3) and (12.4.4) that

$$\mathbf{A}\mathbf{V}_1 = \mathbf{V}_1 \mathbf{\Lambda}_1, \quad \mathbf{U}_2^H \mathbf{B} = \mathbf{\Sigma}_2 \mathbf{U}_2^H.$$

Therefore,

$$\mathbf{U}_2^H \mathbf{E}_1 = \mathbf{U}_2^H (\mathbf{B} - \mathbf{A})\mathbf{V}_1 = \mathbf{\Sigma}_2 \mathbf{U}_2^H \mathbf{V}_1 - \mathbf{U}_2^H \mathbf{V}_1 \mathbf{\Lambda}_1.$$

According to Theorem 3.1 in [361, p. 247],

$$\min_{\|\mathbf{W}\|=1} \|\mathbf{\Sigma}_2 \mathbf{W} - \mathbf{W}\mathbf{\Lambda}_1\| = \min_{\substack{1 \le j \le p \\ p+1 \le l \le m}} |\lambda_j(\mathbf{A}) - \lambda_l(\mathbf{B})| = \gamma_1.$$

Combining this result with the inequality $\|\mathbf{U}_2^H \mathbf{E}_1\| \le \|\mathbf{U}^H \mathbf{E}_1\| = \|\mathbf{E}_1\|$ proves the assertion. Similarly, we can show that $\|\mathbf{U}_1^H \mathbf{V}_2\| \le \gamma_2^{-1} \|\mathbf{E}_2\|$. □

Remark 12.4.2 Because $\|\mathbf{E}_j\| \le \|(\mathbf{B} - \mathbf{A})\mathbf{V}\| = \|\mathbf{B} - \mathbf{A}\| = \epsilon$ $(j = 1, 2)$, we can write $\mathbf{U}_2^H \mathbf{V}_1 = \mathcal{O}(\epsilon)$ and $\mathbf{U}_1^H \mathbf{V}_2 = \mathcal{O}(\epsilon)$.

Lemma 12.4.5 has important implications. First, consider the low rank approximation. Given the EVD of $\mathbf{A}$ in (12.4.3), define

$$\bar{\mathbf{A}} := \mathbf{V}_1 \mathbf{\Lambda}_1 \mathbf{V}_1^H. \tag{12.4.8}$$

If $\lambda_p > 0$, then $\bar{\mathbf{A}}$ has rank p, in which case the pseudo inverse of $\bar{\mathbf{A}}$ is given by

$$\bar{\mathbf{A}}^\dagger := \mathbf{V}_1 \mathbf{\Lambda}_1^{-1} \mathbf{V}_1^H.$$

According to the Schmidt-Mirsky (or Eckart-Young) theorem [361, p. 208], the matrix $\bar{\mathbf{A}}$ in (12.4.8) is a best rank-p approximation of $\mathbf{A}$ in the sense that it minimizes $\|\mathbf{A}-\bar{\mathbf{A}}\|$ among all rank-p matrices. Similarly, given the EVD of $\mathbf{B}$, let

$$\bar{\mathbf{B}} := \mathbf{U}_1\boldsymbol{\Sigma}_1\mathbf{U}_1^H. \tag{12.4.9}$$

As a corollary to Lemma 12.4.5, the following lemma asserts that if $\mathbf{B}$ is close to $\mathbf{A}$, then $\bar{\mathbf{B}}$ is also a best rank-p approximation of $\mathbf{B}$ and is close to $\bar{\mathbf{A}}$.

Lemma 12.4.6 (Perturbation Theorem for Low Rank Approximation). *Let $\bar{\mathbf{A}}$ and $\bar{\mathbf{B}}$ be defined by (12.4.8) and (12.4.9), respectively. If $\lambda_p(\mathbf{A}) > 0$, then, for sufficiently small ϵ, $\bar{\mathbf{B}}$ has rank p with pseudo inverse $\bar{\mathbf{B}}^\dagger := \mathbf{U}_1\boldsymbol{\Sigma}_1^{-1}\mathbf{U}_1^H$. If $\lambda_p(\mathbf{A}) > \lambda_{p+1}(\mathbf{A})$, then, as $\epsilon \to 0$, $\bar{\mathbf{B}} = \bar{\mathbf{A}} + \mathcal{O}(\epsilon)$ and $\bar{\mathbf{B}}^\dagger = \bar{\mathbf{A}}^\dagger + \mathcal{O}(\epsilon)$.*

PROOF. Because $\lambda_p(\mathbf{A}) > 0$, it follows from Lemma 12.4.2 that $\lambda_p(\mathbf{B}) > 0$ for sufficiently small ϵ, in which case $\bar{\mathbf{B}}$ has rank p. The EVD of $\mathbf{B}$ can be written as

$$\mathbf{B} = \bar{\mathbf{B}} + \mathbf{U}_2\boldsymbol{\Sigma}_2\mathbf{U}_2^H.$$

By Lemma 12.4.5, we have $\mathbf{U}_2^H\mathbf{V}_1 = \mathcal{O}(\epsilon)$. Therefore,

$$\mathbf{V}_1^H\mathbf{B}\mathbf{V}_1 - \mathbf{V}_1^H\bar{\mathbf{B}}\mathbf{V}_1 = \mathbf{V}_1^H\mathbf{U}_2\boldsymbol{\Sigma}_2\mathbf{U}_2^H\mathbf{V}_1 = \mathcal{O}(\epsilon^2).$$

On the other hand, because $\mathbf{B} = \mathbf{A} + \mathcal{O}(\epsilon)$, it follows that

$$\mathbf{V}_1^H\mathbf{B}\mathbf{V}_1 = \mathbf{V}_1^H\mathbf{A}\mathbf{V}_1 + \mathcal{O}(\epsilon) = \boldsymbol{\Lambda}_1 + \mathcal{O}(\epsilon).$$

Combining these results leads to

$$\mathbf{V}_1^H\bar{\mathbf{B}}\mathbf{V}_1 = \boldsymbol{\Lambda}_1 + \mathcal{O}(\epsilon).$$

According to Lemma 12.4.5, we also have $\mathbf{V}_2^H\bar{\mathbf{B}}\mathbf{V}_1 = \mathcal{O}(\epsilon)$ and $\mathbf{V}_2^H\bar{\mathbf{B}}\mathbf{V}_2 = \mathcal{O}(\epsilon^2)$. Combining these results yields

$$[\mathbf{V}_1, \mathbf{V}_2]^H\bar{\mathbf{B}}[\mathbf{V}_1, \mathbf{V}_2] = \mathrm{diag}(\boldsymbol{\Lambda}_1, \mathbf{0}) + \mathcal{O}(\epsilon).$$

This implies that $\bar{\mathbf{B}} = \mathbf{V}_1\boldsymbol{\Lambda}_1\mathbf{V}_1^H + \mathcal{O}(\epsilon) = \bar{\mathbf{A}} + \mathcal{O}(\epsilon)$. Furthermore, according to Wedin's theorem [361, p. 145],

$$\|\bar{\mathbf{B}}^\dagger - \bar{\mathbf{A}}^\dagger\| \le \sqrt{2}\lambda_p^{-1}(\mathbf{A})\lambda_p^{-1}(\mathbf{B})\|\bar{\mathbf{B}} - \bar{\mathbf{A}}\|.$$

Let $\delta := \lambda_p(\mathbf{A}) - \lambda_{p+1}(\mathbf{A}) > 0$. Then, for $\epsilon < \delta/4$,

$$\lambda_p(\mathbf{B}) = \lambda_p(\mathbf{B}) - \lambda_p(\mathbf{A}) + \lambda_p(\mathbf{A}) \ge \lambda_p(\mathbf{A}) - \lambda_{p+1}(\mathbf{A}) - \epsilon > 3\delta/4.$$

Combining these results leads to $\bar{\mathbf{B}}^\dagger = \bar{\mathbf{A}}^\dagger + \mathcal{O}(\|\bar{\mathbf{B}} - \bar{\mathbf{A}}\|) = \bar{\mathbf{A}}^\dagger + \mathcal{O}(\epsilon)$ as $\epsilon \to 0$. □

A similar result can be obtained for the projections onto the column spaces of $\mathbf{V}_1$ and $\mathbf{V}_2$. The projection matrices onto these subspaces can be expressed as $\mathbf{P}_1 := \mathbf{V}_1\mathbf{V}_1^H$ and $\mathbf{P}_2 := \mathbf{V}_2\mathbf{V}_2^H$. Similarly, let $\mathbf{Q}_1 := \mathbf{U}_1\mathbf{U}_1^H$ and $\mathbf{Q}_2 := \mathbf{U}_2\mathbf{U}_2^H$ be the projection matrices onto the column spaces of $\mathbf{U}_1$ and $\mathbf{U}_2$, respectively. Because $\mathbf{V} = [\mathbf{V}_1, \mathbf{V}_2]$ and $\mathbf{U} = [\mathbf{U}_1, \mathbf{U}_2]$ are unitary matrices, we have

$$\mathbf{P}_1 + \mathbf{P}_2 = \mathbf{I}, \quad \mathbf{Q}_1 + \mathbf{Q}_2 = \mathbf{I}. \tag{12.4.10}$$

As a direct result of Lemma 12.4.5, the following lemma asserts that $\mathbf{Q}_1$ and $\mathbf{Q}_2$ are close to $\mathbf{P}_1$ and $\mathbf{P}_2$, respectively, if $\mathbf{B}$ is close to $\mathbf{A}$.

Lemma 12.4.7 (Perturbation Theorem for Projection Matrices I). *Let the eigenvalues of* $\mathbf{A}$ *satisfy the condition* $\lambda_p(\mathbf{A}) > \lambda_{p+1}(\mathbf{A})$. *Then, as* $\epsilon \to 0$, $\mathbf{Q}_1 = \mathbf{P}_1 + \mathcal{O}(\epsilon)$ *and* $\mathbf{Q}_2 = \mathbf{P}_2 + \mathcal{O}(\epsilon)$.

PROOF. It follows from (12.4.10) that

$$\mathbf{P}_1 = \mathbf{P}_1\mathbf{Q}_1 + \mathbf{P}_1\mathbf{Q}_2, \quad \mathbf{Q}_1 = \mathbf{P}_1\mathbf{Q}_1 + \mathbf{P}_2\mathbf{Q}_1.$$

By Lemma 12.4.5, $\mathbf{P}_1\mathbf{Q}_2 = \mathcal{O}(\epsilon)$ and $\mathbf{P}_2\mathbf{Q}_1 = \mathcal{O}(\epsilon)$. Therefore, we have $\mathbf{Q}_1 = \mathbf{P}_1 + \mathcal{O}(\epsilon)$ and hence $\mathbf{Q}_2 = \mathbf{I} - \mathbf{Q}_1 = \mathbf{I} - \mathbf{P}_1 + \mathcal{O}(\epsilon) = \mathbf{P}_2 + \mathcal{O}(\epsilon)$. □

Under the additional assumption that the smallest $m - p$ eigenvalues of $\mathbf{A}$ are identical, the following lemma provides an expansion of $\mathbf{Q}_2$ in terms of $\mathbf{B} - \mathbf{A}$. This is a modified version of a general result in [173] (see also [399]).

Lemma 12.4.8 (Perturbation Theorem for Projection Matrices II). *Assume that* $\lambda_p(\mathbf{A}) > \lambda_{p+1}(\mathbf{A}) = \cdots = \lambda_p(\mathbf{A}) = \lambda > 0$. *Then, for sufficiently small* ϵ,

$$\mathbf{Q}_2 = \mathbf{P}_2 - \mathbf{P}_2(\mathbf{B}-\mathbf{A})(\mathbf{A}-\lambda\mathbf{I})^\dagger - (\mathbf{A}-\lambda\mathbf{I})^\dagger(\mathbf{B}-\mathbf{A})\mathbf{P}_2 + \mathcal{O}(\epsilon^2),$$

where $(\mathbf{A}-\lambda\mathbf{I})^\dagger := \sum_{j=1}^{p}\{\lambda_j(\mathbf{A}) - \lambda\}^{-1}\mathbf{v}_j\mathbf{v}_j^H$ *is the pseudo inverse of* $\mathbf{A} - \lambda\mathbf{I}$.

PROOF. For any Hermitian matrix $\mathbf{D}$ with the property $\|\mathbf{D}\| = 1$, consider

$$\mathbf{G}(z, x) := (\mathbf{A} + x\mathbf{D} - z\mathbf{I})^{-1} \qquad (x \in \mathbb{R}, z \in \mathbb{C}).$$

This matrix is well defined if z differs from any eigenvalue of $\mathbf{A} + x\mathbf{D}$. Observe that $\mathbf{A} + x\mathbf{D}$ is a Hermitian matrix. If the eigenpairs of $\mathbf{A} + x\mathbf{D}$ are denoted by $\{\lambda_j(x), \mathbf{v}_j(x)\}$ $(j = 1, \ldots, m)$ with $\lambda_1(x) \geq \cdots \geq \lambda_m(x)$, then

$$\mathbf{G}(z, x) = \sum_{j=1}^{n}\{\lambda_j(x) - z\}^{-1}\mathbf{v}_j(x)\mathbf{v}_j^H(x).$$

Let $r := \{\lambda_p(\mathbf{A}) - \lambda_{p+1}(\mathbf{A})\}/2 = \{\lambda_p(\mathbf{A}) - \lambda\}/2$. By Lemma 12.4.2 and Remark 12.4.1,

$$\sum_{j=1}^{m}|\lambda_j(x) - \lambda_j(\mathbf{A})|^2 \leq \|x\mathbf{D}\|^2 = x^2.$$

Therefore, when $|x| < r$, the eigenvalue $\lambda_j(x)$ falls inside the circle $\Gamma := \{z \in \mathbb{C} : |z-\lambda| = r\}$ for $j = p+1,\ldots,m$ and outside the circle for $j = 1,\ldots,p$. By Cauchy's integral theorem and integral formula in complex analysis, we can write

$$\begin{aligned}\frac{1}{2\pi i}\oint_\Gamma \mathbf{G}(z,x)\,dz &= \sum_{j=1}^{n}\left\{\frac{1}{2\pi i}\oint_\Gamma \{\lambda_j(x)-z\}^{-1}dz\right\}\mathbf{v}_j(x)\mathbf{v}_j^H(x)\\ &= \sum_{j=p+1}^{m} -\mathbf{v}_j(x)\mathbf{v}_j^H(x)\\ &= -\mathbf{P}_2(x),\end{aligned} \tag{12.4.11}$$

where $\mathbf{P}_2(x) := \sum_{j=p+1}^{m}\mathbf{v}_j(x)\mathbf{v}_j^H(x)$ is just the projection matrix onto the subspace spanned by $\{\mathbf{v}_{p+1}(x),\ldots,\mathbf{v}_m(x)\}$.

On the other hand, when both $\mathbf{G}(z,x)$ and $\mathbf{G}(z) := \mathbf{G}(z,0) = (\mathbf{A}-z\mathbf{I})^{-1}$ are well defined, it is easy to verify that

$$\mathbf{G}(z,x)\{\mathbf{I}+x\mathbf{DG}(z)\} = \mathbf{G}(z). \tag{12.4.12}$$

Because $\|\mathbf{D}\| = 1$, $\mathbf{I}+x\mathbf{DG}(z)$ is invertible if $|x| < \|\mathbf{G}(z)\|^{-1}$, in which case, the inverse matrix $\{\mathbf{I}+x\mathbf{DG}(z)\}^{-1}$ can be written as a power series [153, p. 301]

$$\{\mathbf{I}+x\mathbf{DG}(z)\}^{-1} = \sum_{n=0}^{\infty}\{-\mathbf{DG}(z)\}^n x^n.$$

Combining this expression with (12.4.12) yields

$$\mathbf{G}(z,x) = \mathbf{G}(z)\{\mathbf{I}+x\mathbf{DG}(z)\}^{-1} = \sum_{n=0}^{\infty}\mathbf{G}(z)\{-\mathbf{DG}(z)\}^n x^n. \tag{12.4.13}$$

For any given $z \in \Gamma$, we have $|\lambda_j(\mathbf{A})-z| \ge r$ for all j. Therefore,

$$\|\mathbf{G}(z)\|^2 = \|(\mathbf{\Lambda}-z\mathbf{I})^{-1}\|^2 = \sum_{j=1}^{m}|\lambda_j(\mathbf{A})-z|^{-2} \le m/r^2.$$

Let $c := \frac{1}{2}r/\sqrt{m}$. Then, for any $|x| \le c$ and $z \in \Gamma$, we have

$$\|\{-\mathbf{DG}(z)\}x\| \le c\|\mathbf{G}(z)\| \le 1/2.$$

This implies that for any fixed $|x| \le c$ the power series in (12.4.13) is uniformly convergent in $z \in \Gamma$. Moreover, because $|x| \le c$ implies $|x| < r$, the integral formula in (12.4.11) is also valid when $|x| \le c$. Under this condition, we can insert (12.4.13) in (12.4.11) and obtain the identity

$$\mathbf{P}_2(x) = \sum_{n=0}^{\infty}\left\{-\frac{1}{2\pi i}\oint_\Gamma \mathbf{G}(z)\{-\mathbf{DG}(z)\}^n\,dz\right\}x^n. \tag{12.4.14}$$

Observe that $\|\mathbf{G}(z)\{-\mathbf{D}\mathbf{G}(z)\}^n\| \le \|\mathbf{G}(z)\|^{n+1} \le (\sqrt{m}/r)^{n+1} = 1/(2c)^{n+1}$. Therefore, for any $|x| \le c$, the infinite sum in (12.4.14) over $n \ge 2$ can be expressed as $\mathcal{O}(|x|^2)$. This implies that (12.4.14) can be written as

$$\mathbf{P}_2(x) = -\frac{1}{2\pi i}\oint_\Gamma \mathbf{G}(z)\,dz + \left\{\frac{1}{2\pi i}\oint_\Gamma \mathbf{G}(z)\mathbf{D}\mathbf{G}(z)\,dz\right\}x + \mathcal{O}(|x|^2). \quad (12.4.15)$$

According to (12.4.11), the first term in (12.4.15) is nothing but $\mathbf{P}_2(0) = \mathbf{P}_2$. Moreover, using the EVD of $\mathbf{A}$ in (12.4.1), we can write

$$\mathbf{G}(z)\mathbf{D}\mathbf{G}(z) = \sum_{j=1}^{m}\sum_{l=1}^{m}\{\lambda_j(\mathbf{A}) - z\}^{-1}\{\lambda_l(\mathbf{A}) - z\}^{-1}\mathbf{v}_j\mathbf{v}_j^H\mathbf{D}\mathbf{v}_l\mathbf{v}_l^H.$$

By Cauchy's theorem and integral formula in complex analysis,

$$\begin{aligned}\frac{1}{2\pi i}&\oint_\Gamma \{\lambda_j(\mathbf{A}) - z\}^{-1}\{\lambda_l(\mathbf{A}) - z\}^{-1}\,dz\\ &= \begin{cases} -\{\lambda_j(A) - \lambda\}^{-1} & \text{if } j = 1,\ldots,p \text{ and } l = p+1,\ldots,m,\\ -\{\lambda_l(A) - \lambda\}^{-1} & \text{if } j = p+1,\ldots,m \text{ and } l = 1,\ldots,p,\\ 0 & \text{otherwise.}\end{cases}\end{aligned}$$

Therefore, the second term in (12.4.15) can be expressed as x times

$$\begin{aligned}&\sum_{j=p+1}^{m}\sum_{l=1}^{p} -\{\lambda_l(\mathbf{A}) - \lambda\}^{-1}\mathbf{v}_j\mathbf{v}_j^H\mathbf{D}\mathbf{v}_l\mathbf{v}_l^H + \sum_{j=1}^{p}\sum_{l=p+1}^{m} -\{\lambda_j(\mathbf{A}) - \lambda\}^{-1}\mathbf{v}_j\mathbf{v}_j^H\mathbf{D}\mathbf{v}_l\mathbf{v}_l^H\\ &\qquad = -\mathbf{P}_2\sum_{l=1}^{p}\{\lambda_l(\mathbf{A}) - \lambda\}^{-1}\mathbf{D}\mathbf{v}_l\mathbf{v}_l^H - \sum_{j=1}^{p}\{\lambda_j(\mathbf{A}) - \lambda\}^{-1}\mathbf{v}_j\mathbf{v}_j^H\mathbf{D}\mathbf{P}_2.\end{aligned}$$

Combining these results with $\sum_{j=1}^{p}\{\lambda_j(\mathbf{A}) - \lambda\}^{-1}\mathbf{v}_j\mathbf{v}_j^H = (\mathbf{A} - \lambda\mathbf{I})^\dagger$ yields

$$\mathbf{P}_2(x) = \mathbf{P}_2 - \{\mathbf{P}_2\mathbf{D}(\mathbf{A} - \lambda\mathbf{I})^\dagger + (\mathbf{A} - \lambda\mathbf{I})^\dagger\mathbf{D}\mathbf{P}_2\}x + \mathcal{O}(|x|^2),$$

which holds uniformly in $|x| \le c$ and in $\mathbf{D}$ such that $\|\mathbf{D}\| = 1$. Assuming $\epsilon < c$, the assertion follows immediately by taking $\mathbf{D} = (\mathbf{B} - \mathbf{A})/\epsilon$, $x = \epsilon$, and by observing that $\mathbf{A} + \epsilon\mathbf{D} = \mathbf{B}$ and hence $\mathbf{P}_2(\epsilon) = \mathbf{Q}_2$. □

12.5 Asymptotic Theory

The first two lemmas in this section are concerned with the asymptotic properties of random polynomials with convergent coefficients.

Lemma 12.5.1 (Local Minima on the Unit Circle). *Let* $\{\mathbf{X}_n\} \subset \mathbb{C}^m$ *be a sequence of random vectors such that* $\mathbf{X}_n \xrightarrow{P} \mathbf{c}$ *as* $n \to \infty$. *Define* $g(\omega) := |1 + \mathbf{f}_m^H(\omega)\mathbf{c}|^2$ *and* $g_n(\omega) := |1 + \mathbf{f}_m^H(\omega)\mathbf{X}_n|^2$, *where* $\mathbf{f}_m(\omega) := [\exp(i\omega), \dots, \exp(im\omega)]^T$.

(a) *If* ω_0 *is a local minimum of* $g(\omega)$, *then, with probability tending to unity as* $n \to \infty$, *the random function* $g_n(\omega)$ *has a local minimum* ω_n *in the vicinity of* ω_0 *such that* $\omega_n \xrightarrow{P} \omega_0$.

(b) *Let* $\boldsymbol{\omega}_0 := [\omega_{01}, \dots, \omega_{0r}]^T$ *comprise* r *local minima of* $g(\omega)$ *for some* $r \geq 1$ *and let* $\boldsymbol{\omega}_n := [\omega_{n1}, \dots, \omega_{nr}]^T$ *denote the local minima of* $g_n(\omega)$ *in the vicinity of* $\boldsymbol{\omega}_0$ *such that* $\boldsymbol{\omega}_n \xrightarrow{P} \boldsymbol{\omega}_0$ *as* $n \to \infty$. *Define*

$$\mathbf{u}_j := \{\dot{C}(\omega_{0j})\mathbf{f}_m(\omega_{0j}) + C(\omega_{0j})\dot{\mathbf{f}}_m(\omega_{0j})\}/\ddot{g}(\omega_{0j}) \qquad (j = 1, \dots, r),$$

where $C(\omega) := 1 + \mathbf{f}_m^H(\omega)\mathbf{c}$. *If* $\sqrt{n}\,(\mathbf{X}_n - \mathbf{c}) \xrightarrow{D} \mathbf{Y}$ *for some random vector* $\mathbf{Y}$, *then* $\sqrt{n}\,(\boldsymbol{\omega}_n - \boldsymbol{\omega}_0) \xrightarrow{D} 2\Re(\mathbf{U}^H\mathbf{Y})$, *where* $\mathbf{U} := [\mathbf{u}_1, \dots, \mathbf{u}_r]$.

Remark 12.5.1 If $\mathbf{Y} \sim \mathrm{N}(\mathbf{0}, \boldsymbol{\Sigma}_0)$, then $\sqrt{n}\,(\boldsymbol{\omega}_n - \boldsymbol{\omega}_0) \xrightarrow{D} \mathrm{N}(\mathbf{0}, 4\Re(\mathbf{U}^T)\boldsymbol{\Sigma}_0\Re(\mathbf{U}))$. If $\mathbf{Y} \sim \mathrm{N}_c(\mathbf{0}, \boldsymbol{\Sigma}_0, \tilde{\boldsymbol{\Sigma}}_0)$, then $\sqrt{n}\,(\boldsymbol{\omega}_n - \boldsymbol{\omega}_0) \xrightarrow{D} \mathrm{N}(\mathbf{0}, 2\Re(\mathbf{U}^H\boldsymbol{\Sigma}_0\mathbf{U} + \mathbf{U}^H\tilde{\boldsymbol{\Sigma}}_0\mathbf{U}^*))$.

Remark 12.5.2 If $C(\omega_{0j}) = 0$, then $\ddot{g}(\omega_{0j}) = 2|\dot{C}(\omega_{0j})|^2$ and hence

$$\mathbf{u}_j = \tfrac{1}{2}\mathbf{f}_m(\omega_{0j})/\dot{C}^*(\omega_{0j}).$$

If this is true for all j, then it suffices to assume that $\sqrt{n}\mathbf{F}_m^H(\mathbf{X}_n - \mathbf{c}) \xrightarrow{D} \mathbf{Y}_0$ for some random vector $\mathbf{Y}_0$, where $\mathbf{F}_m := [\mathbf{f}_m(\omega_{01}), \dots, \mathbf{f}_m(\omega_{0r})]$, in which case we have $\sqrt{n}\,(\boldsymbol{\omega}_n - \boldsymbol{\omega}_0) \xrightarrow{D} 2\Re(\mathbf{U}_0^H\mathbf{Y}_0)$, where $\mathbf{U}_0 := \tfrac{1}{2}\mathrm{diag}\{1/\dot{C}^*(\omega_{01}), \dots, 1/\dot{C}^*(\omega_{0r})\}$.

PROOF. Because ω_0 is a local minimum of $g(\omega)$, there exists a small closed neighborhood of ω_0, denoted by Ω, such that for any small $\delta > 0$,

$$\tau_\delta := \inf_{\omega \in \Omega_\delta} g(\omega) - g(\omega_0) > 0, \qquad (12.5.1)$$

where $\Omega_\delta := \{\omega \in \Omega : |\omega - \omega_0| > \delta\}$. Because $\mathbf{X}_n \xrightarrow{P} \mathbf{c}$ as $n \to \infty$, we can write

$$R_n := \max_\omega |g_n(\omega) - g(\omega)| = \mathcal{O}_P(1).$$

Observe that

$$g_n(\omega) \geq g(\omega) - R_n, \quad g(\omega_0) \geq g_n(\omega_0) - R_n.$$

Combining these inequalities with (12.5.1) leads to

$$\inf_{\omega \in \Omega_\delta} g_n(\omega) \geq g(\omega_0) + \tau_\delta - R_n \geq g_n(\omega_0) + \tau_\delta - 2R_n.$$

Because $R_n \xrightarrow{P} 0$ as $n \to \infty$, it follows that with probability tending to unity,

$$\omega_n := \arg\min_{\omega \in \Omega} g_n(\omega) \in \Omega \setminus \Omega_\delta.$$

In other words, $P(|\omega_n - \omega_0| \le \delta) \to 1$. Part (a) is thus proved.

To prove part (b), consider the Taylor expansion

$$\dot{g}_n(\omega_{nj}) = \dot{g}_n(\omega_{0j}) + \ddot{g}_n(\tilde{\omega}_{nj})(\omega_{nj} - \omega_{0j}), \tag{12.5.2}$$

where $\tilde{\omega}_{nj}$ is an intermediate point between ω_{nj} and ω_{0j}. Because ω_{nj} and ω_{0j} are local minima of $g_n(\omega)$ and $g(\omega)$, respectively, we have $\dot{g}_n(\omega_{nj}) = \dot{g}(\omega_{0j}) = 0$. Therefore, we can rewrite (12.5.2) as

$$\ddot{g}_n(\tilde{\omega}_{nj})(\omega_{nj} - \omega_{0j}) = -\{\dot{g}_n(\omega_{0j}) - \dot{g}(\omega_{0j})\}. \tag{12.5.3}$$

Let $Z_n(\omega) := 1 + \mathbf{f}_m^H(\omega)\mathbf{X}_n$. Because

$$\begin{aligned} g_n(\omega) &= |Z_n(\omega)|^2, \quad g(\omega) = |C(\omega)|^2, \\ \dot{Z}_n(\omega) &= \dot{\mathbf{f}}_m^H(\omega)\mathbf{X}_n, \quad \dot{C}(\omega) = \dot{\mathbf{f}}_m^H(\omega)\mathbf{c}, \end{aligned}$$

it follows that

$$\begin{aligned} &\dot{g}_n(\omega_{0j}) - \dot{g}(\omega_{0j}) \\ &= 2\Re\{\dot{Z}_n(\omega_{0j})Z_n^*(\omega_{0j}) - \dot{C}(\omega_{0j})C^*(\omega_{0j})\} \\ &= 2\Re\{\dot{Z}_n(\omega_{0j})[Z_n(\omega_{0j}) - C(\omega_{0j})]^* + [\dot{Z}_n(\omega_{0j}) - \dot{C}(\omega_{0j})]C^*(\omega_{0j})\} \\ &= 2\Re\{\dot{Z}_n(\omega_{0j})[\mathbf{f}_m^H(\omega_{0j})(\mathbf{X}_n - \mathbf{c})]^* + C^*(\omega_{0j})\dot{\mathbf{f}}_m^H(\omega_{0j})(\mathbf{X}_n - \mathbf{c})\} \\ &= 2\Re\{[\dot{Z}_n^*(\omega_{0j})\mathbf{f}_m^H(\omega_{0j}) + C^*(\omega_{0j})\dot{\mathbf{f}}_m^H(\omega_{0j})](\mathbf{X}_n - \mathbf{c})\}. \end{aligned}$$

Define $\boldsymbol{\zeta}_{nj} := \dot{Z}_n(\omega_{0j})\mathbf{f}_m(\omega_{0j}) + C(\omega_{0j})\dot{\mathbf{f}}_m(\omega_{0j})$. Then, with the notation

$$\boldsymbol{\zeta}_n := [\boldsymbol{\zeta}_{n1}, \ldots, \boldsymbol{\zeta}_{nr}], \quad \mathbf{D}_n := \mathrm{diag}\{\ddot{g}_n(\tilde{\omega}_{n1}), \ldots, \ddot{g}_n(\tilde{\omega}_{nr})\},$$

we can write (12.5.3) as

$$\mathbf{D}_n(\boldsymbol{\omega}_n - \boldsymbol{\omega}_0) = -2\Re\{\boldsymbol{\zeta}_n^H(\mathbf{X}_n - \mathbf{c})\}.$$

Furthermore, because $\ddot{g}_n(\omega) \xrightarrow{P} \ddot{g}(\omega) = 2\Re\{\ddot{C}(\omega)C^*(\omega)\} + 2|\dot{C}(\omega)|^2$ uniformly in ω and $\omega_{nj} \xrightarrow{P} \omega_{0j}$ for all j, it follows that

$$\mathbf{D}_n \xrightarrow{P} \mathbf{D} := \mathrm{diag}\{\ddot{g}(\omega_{01}), \ldots, \ddot{g}(\omega_{0r})\}.$$

Note that $\ddot{g}(\omega_{0j}) > 0$ for all j because the ω_{0j} are local minima of $g(\omega)$. In addition, because $\dot{Z}_n(\omega_{0j}) \xrightarrow{P} \dot{C}(\omega_{0j})$, we obtain

$$\boldsymbol{\zeta}_{nj} \xrightarrow{P} \dot{C}(\omega_{0j})\mathbf{f}_m(\omega_{0j}) + C(\omega_{0j})\dot{\mathbf{f}}_m(\omega_{0j}) = \ddot{g}(\omega_{0j})\mathbf{u}_j,$$

which, in turn, implies that $\boldsymbol{\zeta}_n \xrightarrow{P} \mathbf{UD}$. Combining these results with Slutsky's theorem in Proposition 2.13(b) proves the assertion. □

Lemma 12.5.2 (Roots and Angles). *Let* $\mathbf{X}_n := [X_{n1},\ldots,X_{nm}]^T$ *be a random vector for each* n *and* $\mathbf{c} := [c_1,\ldots,c_m]^T$ *be a constant vector. Define* $c(z) := 1+\sum_{j=1}^m c_j z^{-j}$ *and* $X_n(z) := 1+\sum_{j=1}^m X_{nj} z^{-j}$. *Assume that* $c(z)$ *has* m *distinct roots* $z_1,\ldots,z_m$.

(a) *If* $\mathbf{X}_n \xrightarrow{P} \mathbf{c}$ *as* $n\to\infty$, *then, with probability tending to unity,* $X_n(z)$ *has* m *distinct roots* $z_{n1},\ldots,z_{nm}$ *such that* $z_{nj} \xrightarrow{P} z_j$ *for all* j.

(b) *If* $\sqrt{n}(\mathbf{X}_n-\mathbf{c}) \xrightarrow{D} \mathbf{Y}$ *for some random vector* $\mathbf{Y}$, *then* $\sqrt{n}\,\mathrm{vec}[z_{nj}-z_j]_{j=1}^m \xrightarrow{D} \mathbf{V}^T\mathbf{Y}$, *where* $\mathbf{V} := [\mathbf{v}_1,\ldots,\mathbf{v}_m]$ *and*

$$\mathbf{v}_j := \Bigg\{\prod_{\substack{j'=1\\ j'\neq j}}^m (1-z_{j'}z_j^{-1})\Bigg\}^{-1}[1,z_j^{-1},\ldots,z_j^{-m+1}]^T.$$

(c) *Let* $\boldsymbol{\omega}_n := \mathrm{vec}[\angle z_{n1},\ldots,\angle z_{nm}]^T$ *and* $\boldsymbol{\omega} := [\angle z_1,\ldots,\angle z_m]^T$. *If* $\sqrt{n}\,\mathrm{vec}[\Re(z_{nj}-z_j),\Im(z_{nj}-z_j)]_{j=1}^m \xrightarrow{D} \mathrm{N}(\mathbf{0},\boldsymbol{\Sigma})$, *then* $\sqrt{n}(\boldsymbol{\omega}_n-\boldsymbol{\omega}) \xrightarrow{D} \mathrm{N}(\mathbf{0},\mathbf{J}\boldsymbol{\Sigma}\mathbf{J}^T)$, *where* $\mathbf{J} := \mathrm{diag}(\mathbf{J}_1,\ldots,\mathbf{J}_m)$ *and* $\mathbf{J}_j := [-\Im(z_j)/|z_j|^2, \Re(z_j)/|z_j|^2]$.

Remark 12.5.3 If $\mathbf{Y}\sim \mathrm{N}(\mathbf{0},\boldsymbol{\Sigma}_0)$, then

$$\sqrt{n}\,\mathrm{vec}[\Re(z_{nj}-z_j),\Im(z_{nj}-z_j)]_{j=1}^m \xrightarrow{D} \mathrm{N}(\mathbf{0},\boldsymbol{\Sigma}), \tag{12.5.4}$$

where $\boldsymbol{\Sigma} := \mathbf{U}^T\boldsymbol{\Sigma}_0\mathbf{U}$ and $\mathbf{U} := [\Re(\mathbf{v}_1),\Im(\mathbf{v}_1),\ldots,\Re(\mathbf{v}_m),\Im(\mathbf{v}_m)]$. If $\mathbf{Y}\sim \mathrm{N}_c(\mathbf{0},\boldsymbol{\Sigma}_0,\tilde{\boldsymbol{\Sigma}}_0)$, then (12.5.4) is true with $\boldsymbol{\Sigma} := [\boldsymbol{\Sigma}_{jj'}]$ $(j,j'=1,\ldots,m)$, where

$$\boldsymbol{\Sigma}_{jj'} := \frac{1}{2}\begin{bmatrix} \Re(\mathbf{v}_j^T\boldsymbol{\Sigma}_0\mathbf{v}_{j'}^* + \mathbf{v}_j^T\tilde{\boldsymbol{\Sigma}}_0\mathbf{v}_{j'}) & -\Im(\mathbf{v}_j^T\boldsymbol{\Sigma}_0\mathbf{v}_{j'}^* - \mathbf{v}_j^T\tilde{\boldsymbol{\Sigma}}_0\mathbf{v}_{j'}) \\ \Im(\mathbf{v}_j^T\boldsymbol{\Sigma}_0\mathbf{v}_{j'}^* + \mathbf{v}_j^T\tilde{\boldsymbol{\Sigma}}_0\mathbf{v}_{j'}) & \Re(\mathbf{v}_j^T\boldsymbol{\Sigma}_0\mathbf{v}_{j'}^* - \mathbf{v}_j^T\tilde{\boldsymbol{\Sigma}}_0\mathbf{v}_{j'}) \end{bmatrix}.$$

In both cases, the asymptotic distribution of $\boldsymbol{\omega}_n$ is given by part (c).

PROOF. Part (a) is due to the continuity of the mapping from the coefficients to the roots of a polynomial [140]. To prove part (b), let $Y_n(z) := z^m X_n(z)$. Because $X_n(z) = \prod_{j=1}^m (1-z_{nj}z^{-1})$, we can write $Y_n(z) = \prod_{j=1}^m (z-z_{nj})$. By the Taylor series expansion, we obtain

$$0 = Y_n(z_{nj}) = Y_n(z_j) + \sum_{r=1}^m \frac{Y_n^{(r)}(z_j)}{r!}(z_{nj}-z_j)^r.$$

Because $Y_n^{(r)}(z_j) = \mathcal{O}_P(1)$, and $z_{nj}-z_j = o_P(1)$, it follows that

$$\xi_{nj}(z_{nj}-z_j) = -Y_n(z_j),$$

where

$$\xi_{nj} := \sum_{r=1}^m \frac{Y_n^{(r)}(z_j)}{r!}(z_{nj}-z_j)^{r-1} = \dot{Y}_n(z_j) + o_P(1).$$

Observe that

$$z_j^{-m+1}\dot{Y}_n(z_j) \xrightarrow{P} z_j^{-m+1}\prod_{\substack{j'=1\\ j'\neq j}}^{m}(z_j - z_{j'}) = \prod_{\substack{j'=1\\ j'\neq j}}^{m}(1 - z_{j'}z_j^{-1}) \neq 0$$

and

$$z_j^{-m+1}Y_n(z_j) = z_j X_n(z_j) = z_j\{X_n(z_j) - c(z_j)\} = \mathbf{z}_j^T(\mathbf{X}_n - \mathbf{c}),$$

where $\mathbf{z}_j := [1, z_j^{-1}, \dots, z_j^{-m+1}]^T$. Combining these results with Slutsky's theorem in Proposition 2.13(b) proves the assertion. Part (c) follows from the fact that $\angle z_j = \arctan(\Im(z_j), \Re(z_j))$ and the Jacobian of $\arctan(y, x)$ at $y = \Im(z_j)$ and $x = \Re(z_j)$ can be expressed as $\mathbf{J}_j := [-\Im(z_j)/|z_j|^2, \Re(z_j)/|z_j|^2]$. □

The classical theory of statistics asserts that the average of n zero-mean i.i.d. random variables converges to zero in probability as well as almost surely as n grows [34, pp. 85–86]. This assertion is generalized in the following lemma to include weighted averages of a stationary process.

Lemma 12.5.3 (Laws of Large Numbers for Stationary Processes). *Let $\{X_t\}$ be a zero-mean stationary process with ACF $r(u)$ such that $\sum |r(u)| < \infty$. Let $\{W_{nt}\}$ be an array of random variables which are independent of $\{X_t\}$. If there exists a constant $c > 0$ such that $|W_{nt}| \le c$ almost surely for all n and t, then the following assertions are true.*

(a) Weak law of large numbers: $n^{-1}\sum_{t=1}^n W_{nt}X_t \xrightarrow{P} 0$ as $n \to \infty$.

(b) Strong law of large numbers: $n^{-1}\sum_{t=1}^n W_{nt}X_t \xrightarrow{a.s.} 0$ as $n \to \infty$.

Proof. The weak law follows from the strong law and Proposition 2.11. The strong law extends a classical result [94, p. 492, Theorem 6.2] and can be proved along similar lines. Indeed, let $U(n) := \sum_{t=1}^n W_{nt}X_t$. Then,

$$\begin{aligned} E\{|n^{-1}U(n)|^2\} &= n^{-2}\sum_{t,s=1}^{n} E(W_{nt}W_{ns}^*)\, r(t-s) \\ &\le n^{-1}c^2 \sum_{|u|<n}(1 - |u|n^{-1})\,|r(u)| \le Kn^{-1}, \end{aligned}$$

where $K := c^2\sum |r(u)| < \infty$. This result, according to Proposition 2.16(b), leads to

$$n^{-1}U(n) \xrightarrow{P} 0.$$

Furthermore, applying Chebyshev's inequality gives

$$\sum_{m=1}^{\infty} P\{|m^{-2}U(m^2)| > \delta\} \le K\delta^{-2}\sum_{m=1}^{\infty} m^{-2} < \infty,$$

which, according to the Borel-Cantelli lemma, leads to $m^{-2}U(m^2) \stackrel{a.s.}{\to} 0$ as $m \to \infty$. Moreover, for any $k \in [m^2, (m+1)^2)$,

$$|U(k) - U(m^2)|^2 = \left| \sum_{t=m^2+1}^{(m+1)^2} W_{nt} X_t \right|^2 \le c^2 \sum_{t,s=m^2+1}^{(m+1)^2} |X_t X_s^*|.$$

Therefore,

$$\begin{aligned} & E\Big\{ \max_{m^2 \le k < (m+1)^2} k^{-2} |U(k) - U(m^2)|^2 \Big\} \\ & \le c^2 m^{-4} \sum_{t,s=m^2+1}^{(m+1)^2} E\{|X_t X_s^*|\} \\ & \le c^2 r(0)\, m^{-4} \{(m+1)^2 - m^2\}^2 \le 9c^2 r(0)\, m^{-2}. \end{aligned}$$

Applying Chebyshev's inequality and the Borel-Cantelli lemma gives

$$\max_{m^2 \le k < (m+1)^2} k^{-1} |U(k) - U(m^2)| \stackrel{a.s.}{\to} \quad \text{as } m \to \infty.$$

For any $n > 0$, let $m := \lfloor n^{1/2} \rfloor$. Because $m^2 \le n < (m+1)^2$, it follows that

$$\begin{aligned} |n^{-1} U(n)| &\le n^{-1} |U(n) - U(m^2)| + n^{-1} m^2 |m^{-2} U(m^2)| \\ &\le \max_{m^2 \le k < (m+1)^2} k^{-1} |U(k) - U(m^2)| + |m^{-2} U(m^2)| \stackrel{a.s.}{\to} 0 \end{aligned}$$

as $n \to \infty$ (hence $m \to \infty$). □

Central limit theorems are essential tools for studying the asymptotic distribution of an estimator. According to the classical theory of statistics, the average of n zero-mean i.i.d. random variables with finite variance σ^2 is asymptotically Gaussian with mean zero and variance σ^2/n [34, p. 357]. This assertion, known as the *Lindeberg-Lévy theorem*, can be generalized in many different ways. The following lemma is a central limit theorem for martingale differences.

Lemma 12.5.4 (Central Limit Theorem for Martingale Differences [48] [131, pp. 9–10]). *Let $\{X_t\} \subset \mathbb{R}$ be a sequence of martingale differences with respect to a filtration $\{\mathfrak{F}_t\}$. Define $V_n^2 := \sum_{t=1}^n E(X_t^2 | \mathfrak{F}_{t-1})$ and $s_n^2 := E(V_n^2) = \sum_{t=1}^n E(X_t^2)$. Assume that $s_n^{-2} V_n^2 \stackrel{P}{\to} 1$ and $s_n^{-2} \sum_{t=1}^n E\{X_t^2 \mathscr{I}(|X_t| \ge \delta s_n)\} \to 0$ as $n \to \infty$ for any constant $\delta > 0$. Then, $s_n^{-1} \sum_{t=1}^n X_t \stackrel{D}{\to} \mathrm{N}(0,1)$ as $n \to \infty$.*

Lemma 12.5.4 can be extended to martingale difference arrays, as stated in the following lemma.

Lemma 12.5.5 (Central Limit Theorem for Martingale Difference Arrays [131, pp. 58–59]). *For each fixed n, let $\{X_{nt}\} \subset \mathbb{R}$ be a sequence of martingale differences with respect to a filtration $\{\mathfrak{F}_{nt}\}$. Assume that as $n \to \infty$,*

$$\sum_{t=1}^n E(X_{nt}^2 | \mathfrak{F}_{n,t-1}) \stackrel{P}{\to} 1 \quad \text{and} \quad \sum_{t=1}^n E\{X_{nt}^2 \mathscr{I}(|X_{nt}| \ge \delta) \,|\, \mathfrak{F}_{n,t-1}\} \stackrel{P}{\to} 0$$

for any constant $\delta > 0$. *Then,* $\sum_{t=1}^{n} X_{nt} \xrightarrow{D} \mathrm{N}(0,1)$ *as* $n \to \infty$.

As noted before, martingale differences are uncorrelated, but may not be independent, or identically distributed, or stationary. It is in these aspects that Lemma 12.5.4 and Lemma 12.5.5 generalize the Lindeberg-Lévy theorem for i.i.d. random variables. Note that the requirements $s_n^{-2}\sum_{t=1}^{n} E\{X_t^2 \mathscr{I}(|X_t| \ge \delta s_n)\} \to 0$ in Lemma 12.5.4 and $\sum_{t=1}^{n} E\{X_{nt}^2 \mathscr{I}(|X_{nt}| \ge \delta) \mid \mathfrak{F}_{n,t-1}\} \xrightarrow{P} 0$ in Lemma 12.5.5 are known collectively as the *Lindeberg condition* for central limit theorems.

The following result is a simple application of Lemma 12.5.5.

Lemma 12.5.6 (Central Limit Theorem for Strictly Stationary Martingale Differences). *Let* $\{X_t\} \subset \mathbb{R}$ *be a sequence of strictly stationary martingale differences with respect to a filtration* $\{\mathfrak{F}_t\}$ *such that* $E(X_t^2|\mathfrak{F}_{t-1}) = \sigma^2 > 0$ *almost surely. Let* $\{\mathbf{c}_{nt}\} \subset \mathbb{R}^m$ *be sequences of constant vectors for some fixed* $m \ge 1$. *As* $n \to \infty$, *if*

$$\max_{1 \le t \le n} \|\mathbf{c}_{nt}\|_\infty \to 0 \quad \textit{and} \quad \sum_{t=1}^{n} \mathbf{c}_{nt}\mathbf{c}_{nt}^T \to \mathbf{V},$$

where $\|\cdot\|_\infty$ *denotes the maximum norm, then* $\sum_{t=1}^{n} \mathbf{c}_{nt} X_t \xrightarrow{D} \mathrm{N}(0, \sigma^2\mathbf{V})$.

PROOF. Let $\mathbf{Y}_n := \sum_{t=1}^{n} \mathbf{c}_{nt} X_t$. For any nonzero constant vector $\mathbf{a} \in \mathbb{R}^m$, define $v := \sigma(\mathbf{a}^T\mathbf{V}\mathbf{a})^{1/2} > 0$ and $w_{nt} := \mathbf{a}^T\mathbf{c}_{nt}/v$. Then, we can write

$$\mathbf{a}^T\mathbf{Y}_n = v \sum_{t=1}^{n} X_{nt},$$

where $X_{nt} := w_{nt} X_t$. Observe that for each fixed n, $\{X_{nt}\}$ is a sequence of martingale differences with respect to $\{\mathfrak{F}_t\}$. It is easy to verify that as $n \to \infty$,

$$\sum_{t=1}^{n} w_{nt}^2 \to 1/\sigma^2,$$

Therefore,

$$\sum_{t=1}^{n} E(X_{nt}^2 \mid \mathfrak{F}_{t-1}) = \sum_{t=1}^{n} w_{nt}^2 E(X_t^2 \mid \mathfrak{F}_{t-1}) \to 1.$$

Moreover, let $c_n := \max\{|w_{nt}| : 1 \le t \le n\}$. Because $c_n \to 0$ and $\{X_t\}$ is stationary, it follows that for any constant $\delta > 0$,

$$\begin{aligned}
&\sum_{t=1}^{n} E\{X_{nt}^2 \mathscr{I}(|X_{nt}| \ge \delta) \mid \mathfrak{F}_{t-1}\} \\
&\le \sum_{t=1}^{n} w_{nt}^2 E\{X_t^2 \mathscr{I}(|X_t| \ge \delta/c_n) \mid \mathfrak{F}_{t-1}\} \\
&= \sum_{t=1}^{n} w_{nt}^2 E\{X_1^2 \mathscr{I}(|X_1| \ge \delta/c_n) \mid \mathfrak{F}_0\} \to 0.
\end{aligned}$$

The conditions of Lemma 12.5.5 are satisfied. Therefore,

$$\mathbf{c}^T \mathbf{Y}_n = \upsilon \sum_{t=1}^{n} X_{nt} \xrightarrow{D} \mathrm{N}(0, \upsilon^2).$$

Citing Proposition 2.16(d) completes the proof. □

The following result generalizes Lemma 12.5.6 to the complex case.

Lemma 12.5.7 (Central Limit Theorem for Complex White Noise). *Let $\{X_t\} \subset \mathbb{C}$ be a sequence of i.i.d. random variables with mean zero, variance $E(|X_t|^2) = \sigma^2 > 0$, and complementary variance $E(X_t^2) = \iota\sigma^2$. Let $\{\mathbf{c}_{nt}\} \subset \mathbb{C}^m$ be sequences of constant vectors for some fixed $m \geq 1$. As $n \to \infty$, if*

$$\max_{1\le t\le n} \|\mathbf{c}_{nt}\|_\infty \to 0, \quad \sum_{t=1}^{n} \mathbf{c}_{nt}\mathbf{c}_{nt}^H \to \mathbf{V}, \quad \text{and} \quad \sum_{t=1}^{n} \mathbf{c}_{nt}\mathbf{c}_{nt}^T \to \tilde{\mathbf{V}},$$

where $\|\cdot\|_\infty$ denotes the maximum norm, then $\sum_{t=1}^n \mathbf{c}_{nt} X_t \xrightarrow{D} \mathrm{N}_c(\mathbf{0}, \sigma^2\mathbf{V}, \iota\sigma^2\tilde{\mathbf{V}})$.

Proof. Let $\mathbf{Y}_n := \sum_{t=1}^n \mathbf{c}_{nt} X_t$ and $\mathbf{Y}_{nr} := [\Re(\mathbf{Y}_n)^T, \Im(\mathbf{Y}_n)^T]^T$. It is easy to verify that

$$E(\mathbf{Y}_n \mathbf{Y}_n^H) \to \mathbf{\Sigma} := \sigma^2 \mathbf{V}, \quad E(\mathbf{Y}_n \mathbf{Y}_n^T) \to \tilde{\mathbf{\Sigma}} := \iota\sigma^2 \tilde{\mathbf{V}}.$$

Therefore, $\mathrm{Cov}(\mathbf{Y}_{nr}) \to \mathbf{\Sigma}_r$, where

$$\mathbf{\Sigma}_r := \frac{1}{2}\begin{bmatrix} \Re(\mathbf{\Sigma}+\tilde{\mathbf{\Sigma}}) & -\Im(\mathbf{\Sigma}-\tilde{\mathbf{\Sigma}}) \\ \Im(\mathbf{\Sigma}+\tilde{\mathbf{\Sigma}}) & \Re(\mathbf{\Sigma}-\tilde{\mathbf{\Sigma}}) \end{bmatrix}.$$

For any constant vectors $\mathbf{a}, \mathbf{b} \in \mathbb{R}^m$, define $\mathbf{c} := [\mathbf{a}^T, \mathbf{b}^T]^T$, $\upsilon := (\mathbf{c}^T \mathbf{\Sigma}_r \mathbf{c})^{1/2}$, $u_{nt} := \mathbf{a}^T \Re(\mathbf{c}_{nt}) + \mathbf{b}^T \Im(\mathbf{c}_{nt})$, and $\upsilon_{nt} := -\mathbf{a}^T \Im(\mathbf{c}_{nt}) + \mathbf{b}^T \Re(\mathbf{c}_{nt})$. Then, $\mathrm{Var}(\mathbf{c}^T \mathbf{Y}_n) \to \upsilon^2$ and

$$\mathbf{c}^T \mathbf{Y}_n = \upsilon \sum_{t=1}^{n} X_{nt},$$

where $X_{nt} := u_{nt}\Re(X_t) + \upsilon_{nt}\Im(X_t)$ is an array of martingale differences with respect to the filtration $\mathfrak{F}_t$ generated by $\{\Re(X_t), \Im(X_t), \Re(X_{t-1}), \Im(X_{t-1}), \ldots\}$. Because the X_{nt} are independent,

$$\sum_{t=1}^{n} E\{X_{nt}^2 | \mathfrak{F}_{t-1}\} = \sum_{t=1}^{n} E(X_{nt}^2) = \mathrm{Var}(\mathbf{c}^T \mathbf{Y}_n)/\upsilon^2 \to 1.$$

Moreover, let $c_n := \max\{|u_{nt}|, |\upsilon_{nt}| : 1 \le t \le n\}$. Because $c_n \to 0$ and the X_t are i.i.d., it follows that for any constant $\delta > 0$,

$$\begin{aligned} &\sum_{t=1}^{n} E\{X_{nt}^2 \mathscr{I}(|X_{nt}| \ge \delta) \,|\, \mathfrak{F}_{t-1}\} \\ &\le 2\sum_{t=1}^{n} (u_{nt}^2 + \upsilon_{nt}^2) E\{|X_1|^2 \mathscr{I}(|Z| \ge \delta/c_n) \,|\, \mathfrak{F}_0\} \to 0, \end{aligned}$$

where $Z := |\Re(X_1)| + |\Im(X_1)|$. An application of Lemma 12.5.5 gives

$$\mathbf{c}^T \mathbf{Y}_{nr} \xrightarrow{D} \mathrm{N}(0, v^2).$$

By Proposition 2.16(d), $\mathbf{Y}_{nr} \xrightarrow{D} \mathrm{N}(\mathbf{0}, \mathbf{\Sigma}_r)$ and hence $\mathbf{Y}_n \xrightarrow{D} \mathrm{N}_c(\mathbf{0}, \mathbf{\Sigma}, \tilde{\mathbf{\Sigma}})$. □

Remark 12.5.4 If X_t satisfies $E(X_t^2) = 0$, then $\sum_{t=1}^n \mathbf{c}_{nt} X_t \xrightarrow{D} \mathrm{N}_c(\mathbf{0}, \sigma^2 \mathbf{V})$ as $n \to \infty$ under the additional assumption that $\{X_t\} \sim \mathrm{IID}(0, \sigma^2)$, $\max_{1 \le t \le n} \|\mathbf{c}_{nt}\|_\infty \to 0$, and $\sum_{t=1}^n \mathbf{c}_{nt} \mathbf{c}_{nt}^H \to \mathbf{V}$. A complex random variable X with mean zero and variance σ^2 is said to be *symmetric* if $E(X^2) = 0$, or equivalently, if $\Re(X)$ and $\Im(X)$ are uncorrelated with identical variance $\sigma^2/2$.

A central limit theorem can be extended to dependent processes as long as the dependence decays in some sense with the increase of separation in time. One way of describing such "weak dependence" is through the so-called α-mixing condition (or strong mixing condition) introduced by Rosenblatt [320]. A real random process $\{X_t\}$ is said to satisfy the α-mixing condition if there exists a sequence of constants $\{\alpha(u)\} \subset [0, 1/4]$ such that $\alpha(u) \to 0$ as $u \to \infty$ and that

$$|P(A \cap B) - P(A)P(B)| \le \alpha(u) \tag{12.5.5}$$

for all $A \in \mathfrak{F}_t$, $B \in \mathfrak{F}_{t+u}^c$, $u \ge 1$, and t, where $\mathfrak{F}_t$ and $\mathfrak{F}_t^c$ denote the σ-fields generated by $\{X_s : s \le t\}$ and $\{X_s : s \ge t\}$, respectively. An m-dependent process satisfies the α-mixing condition with $\alpha(u) = 0$ for all $u > m$. Under certain conditions, the α-mixing condition can also be satisfied by linear processes [124] [425]. An important consequence of the α-mixing condition is that [34, p. 365]

$$|E(XY) - E(X)E(Y)| \le 4ab\alpha(u), \tag{12.5.6}$$

where X is any function of $\{X_s : s \le t\}$ and Y is any function of $\{X_s : s \ge t+u\}$ such that $|X| \le a$ and $|Y| \le b$. If $\{X_t\}$ is stationary with mean zero and ACF $r(u)$ and is uniformly bounded by $c > 0$, then (12.5.6) implies that $|r(u)| \le 4c^2 \alpha(|u|)$ for all u. See [43] for a survey on the mixing conditions.

Equipped with the mixing condition, the following result can be obtained.

Lemma 12.5.8 (Central Limit Theorem for Bounded Dependent α-Mixing Processes). *Let $\{X_t\} \subset \mathbb{R}$ be a stationary process with mean zero and ACF $r(u)$. Assume that $\{X_t\}$ is bounded and satisfies the α-mixing condition (12.5.5) with $\alpha(u) = \mathcal{O}(n^{-\beta})$ for some constant $\beta > 2$. Moreover, let $\{c_{nt}\} \subset \mathbb{R}$ be uniformly bounded sequences of constants such that as $n \to \infty$,*

$$n^{-1} \sum_{t=1}^n c_{n,t+u} c_{nt} - c_n(u) \to 0$$

for any fixed u and that

$$v_n^2 := \sum_{u=-\infty}^{\infty} r(u) c_n(u) \ge \lambda > 0$$

for some constant λ and large n. Then, $n^{-1/2}\sum_{t=1}^{n} c_{nt}X_t/v_n \xrightarrow{D} \mathrm{N}(0,1)$ as $n\to\infty$.

PROOF. According to (12.5.6), $r(u)=E(X_{t+u}X_t)=\mathcal{O}(|u|^{-\beta})$ for large $|u|$. This implies that $\{r(u)\}$ is absolutely summable and hence v_n^2 is finite for each n. Let $X_{nt} := c_{nt}X_t/v_n$ and $S_n := \sum_{t=1}^{n} X_{nt}$. The objective is to show that $n^{-1/2}S_n \xrightarrow{D} \mathrm{N}(0,1)$. Note that $E(S_n)=0$ and

$$s_n^2 := \mathrm{Var}(S_n) = \sum_{t,s=1}^{n} E(X_{nt}X_{ns}) = v_n^{-2}\sum_{|u|<n} r(u) \sum_{t\in T_n(u)} c_{n,t+u}c_{nt},$$

where $T_n(u) := \{t : \max(1,1-u)\le t\le \min(n,n-u)\}$. By the dominant convergence theorem, we have $n^{-1}s_n^2 \to 1$.

For any integers $k\ge 1$ and $m\ge 1$ which may depend on n, let $n_j := [jn/k]$ $(0\le j\le k)$ and define

$$\begin{aligned} Y_{nj} &:= X_{n,n_j+1}+X_{n,n_j+2}+\cdots+X_{n,n_{j+1}-m}, \\ Z_{nj} &:= X_{n,n_{j+1}-m+1}+X_{n,n_{j+1}-m+2}+\cdots+X_{n,n_{j+1}}. \end{aligned}$$

Then, we can write

$$S_n = \sum_{j=0}^{k-1} Y_{nj} + \sum_{j=0}^{k-1} Z_{nj} := S_{n1}+S_{n2}.$$

It suffices to show that $n^{-1/2}S_{n2}\xrightarrow{P} 0$ and $n^{-1/2}S_{n1}\xrightarrow{D}\mathrm{N}(0,1)$. This can be done by following the proof of Theorem 27.4 in [34] and Theorem 7.3.1 in [72].

Let $c>0$ be a constant such that $|X_{nt}|\le c$ for all n and t. Assume that k and m are chosen such that $n_{j+1}-m+1-n_j > m$, which is true for large n if

$$km = \mathcal{o}(n). \tag{12.5.7}$$

In this case, for $j\ne l$, any X_{nt} in Z_{nj} is separated from any X_{ns} in Z_{nl} such that $|t-s|>m$. This implies, by (12.5.6), that $|E(X_{nt}X_{ns})|\le 4c^2\alpha_m$, where $\alpha_m := \max\{\alpha(u): u>m\}$. Therefore, $|E(Z_{nj}Z_{nl})|\le m^2(4c^2\alpha_m)$ for $j\ne l$ and hence

$$\begin{aligned} E(S_{n2}^2) &= \sum_{j=0}^{k-1} E(Z_{nj}^2) + \sum_{\substack{j,l=0\\ j\ne l}}^{k-1} E(Z_{nj}Z_{nl}) \\ &\le k(cm)^2 + k(k-1)m^2(4c^2\alpha_m). \end{aligned}$$

Taking k and m such that

$$km^2 = \mathcal{o}(n), \quad k\alpha_m = \mathcal{O}(1) \tag{12.5.8}$$

leads to $E(n^{-1}S_{n2}^2)\to 0$. By Proposition 2.16(b), we have $n^{-1/2}S_{n2}\xrightarrow{P} 0$.

Now let $\tilde{S}_{n1} := \sum_{j=0}^{k-1} \tilde{Y}_{nj}$, where the $\tilde{Y}_{nj}$ $(0 \le j < k)$ are independent random variables such that $\tilde{Y}_{nj} \sim Y_{nj}$ for all j. Applying (12.5.6) inductively shows that the characteristic functions of $n^{-1/2} S_{n1}$ and $n^{-1/2} \tilde{S}_{n1}$ differ by at most $16k\alpha(m)$, where the number 4 in (12.5.6) becomes 16 to allow for a split of the characteristic functions into real and imaginary parts. The difference can be expressed as $o(1)$ if k and m satisfy the condition

$$k\alpha(m) = o(1). \tag{12.5.9}$$

This implies that $n^{-1/2} S_{n1} \stackrel{A}{\sim} n^{-1/2} \tilde{S}_{n1}$. It remains to show that $n^{-1/2} \tilde{S}_{n1} \xrightarrow{D} \mathrm{N}(0,1)$.

For each $j = 0, 1, \ldots, k-1$, we can write

$$\mathrm{Var}(Y_{nj}) = \sum_{t,s=n_j+1}^{n_{j+1}-m} E(X_{nt} X_{ns}) = \sum_{t,s=n_j+1}^{n_{j+1}} E(X_{nt} X_{ns}) - r_{nj},$$

where

$$r_{nj} := \left\{ \sum_{t=n_j+1}^{n_{j+1}} \sum_{s=n_{j+1}-m+1}^{n_{j+1}} + \sum_{t=n_{j+1}-m+1}^{n_{j+1}} \sum_{s=n_j+1}^{n_{j+1}-m} \right\} E(X_{nt} X_{ns}).$$

This result, combined with the assumption $\tilde{Y}_{nj} \sim Y_{nj}$, leads to

$$\tilde{s}_n^2 := \mathrm{Var}(\tilde{S}_{n1}) = \sum_{j=0}^{k-1} \mathrm{Var}(\tilde{Y}_{nj}) = \sum_{j=0}^{k-1} \mathrm{Var}(Y_{nj}) = s_n^2 - r_n,$$

where $r_n := \sum_{j=0}^{k-1} r_{nj}$. Observe that there are $(n_{j+1} - n_j)^2 - (n_{j+1} - n_j - m)^2 = 2m(n_{j+1} - n_j) + m^2$ terms in r_{nj}, of which no more than $2m^2$ terms satisfy $|t - s| \le m$. For these terms, we have $|E(X_{nt} X_{ns})| \le c^2$, and for the remaining terms, we have $|E(X_{nt} X_{ns})| \le 4c^2 \alpha_m$. This implies that

$$|r_{nj}| \le 2m^2 c^2 + (2m(n_{j+1} - n_j) + m^2)(4c^2 \alpha_m).$$

Therefore,

$$\begin{aligned} |r_n| &\le \sum_{j=0}^{k-1} \{2m^2 c^2 + (2m(n_{j+1} - n_j) + m^2)(4c^2 \alpha_m)\} \\ &= (2 + 4\alpha_m) c^2 k m^2 + 8c^2 n m \alpha_m. \end{aligned}$$

The first term in the foregoing expression takes the form $o(n)$ because $km^2 = o(n)$ and the second term takes the form $o(n)$ if m satisfies

$$m\alpha_m = o(1). \tag{12.5.10}$$

Under this condition, we obtain $n^{-1} \tilde{s}_n^2 = n^{-1} s_n^2 + o(1) \to 1$.

Note that Y_{nj} is the sum of no more than $[n/k]+1$ of the X_{nt}. The boundedness of the X_{nt} implies that $|Y_{nj}| \le (n/k+1)c$. If k satisfies the condition

$$n^{-1/2}k \to \infty, \tag{12.5.11}$$

then

$$|Y_{nj}|/\tilde{s}_n \le (n^{1/2}/k + n^{-1/2})c/(n^{-1/2}\tilde{s}_n) \to 0.$$

In this case, for any $\delta > 0$ and for large n,

$$E\{\tilde{Y}_{nj}^2 \mathscr{I}(|\tilde{Y}_{nj}| \ge \delta \tilde{s}_n)\} = 0 \qquad (0 \le j < k).$$

Hence the Lindeberg condition is satisfied for the triangular array $\{\tilde{Y}_{nj}/\tilde{s}_n\}$. By Lindeberg's central limit theorem [34, p. 359], we obtain $\tilde{S}_{n1}/\tilde{s}_n \xrightarrow{D} \mathrm{N}(0,1)$ and hence $n^{-1/2}\tilde{S}_{n1} = (n^{-1/2}\tilde{s}_n)\tilde{S}_{n1}/\tilde{s}_n \xrightarrow{D} \mathrm{N}(0,1)$.

Because $km^2 = \mathcal{O}(n)$ implies (12.5.7) and $k\alpha_m = \mathcal{O}(1)$ implies (12.5.9), the conditions (12.5.7)–(12.5.11) can be reduced to

$$n^{-1/2}k \to \infty, \quad km^2 = \mathcal{O}(n), \quad k\alpha_m = \mathcal{O}(1), \quad m\alpha_m = \mathcal{O}(1). \tag{12.5.12}$$

Let $k = n^{\mu}$ and $m = n^{\nu}$. Then, the first two conditions in (12.5.12) are satisfied if $\mu > 1/2$ and $\nu < (1-\mu)/2$. The third and fourth conditions require that $\mu < \nu\beta$ and $\nu(1-\beta) < 0$ if there exists a constant $\beta > 0$ such that $\alpha(u) = \mathcal{O}(u^{-\beta})$ for large $u > 0$. For these inequalities to have at least one solution (μ_0, ν_0), it suffices to assume that $\beta > 2$, in which case (μ_0, ν_0) can be any interior point in the triangular region defined by $1/2 < \mu < \beta/(\beta+2)$ and $\mu/\beta < \nu < (1-\mu)/2$. □

Remark 12.5.5 By Theorem 2.2 in [287], the assertion in Lemma 12.5.8 remains valid if the assumption about $\alpha(u)$ is replaced by $\sum_{u=1}^{\infty} u^{\delta}\alpha(u) < \infty$ for some constant $\delta > 0$. This condition is less stringent than the condition stated in Lemma 12.5.8, but the proof is more involved.

The next lemma quantifies the maximum impact of the noise on the periodogram. See [412] for the case of real-valued i.i.d. processes and [137] for the case of real-valued linear processes.

Lemma 12.5.9 (Maximum of the Periodogram).

(a) If $\{\epsilon_t\} \sim \mathrm{IID}(0,\sigma^2)$, *then, as* $n \to \infty$,

$$\max_{\omega} \left| \sum_{t=1}^{n} t^r \epsilon_t \exp(-it\omega) \right| = \mathcal{O}_P(n^{r+3/4})$$

for any integer $r \ge 0$. *The assertion remains valid if* $\{\epsilon_t\}$ *is a zero-mean stationary real or complex Gaussian process and its ACF* $r(u)$ *satisfies the square-summability condition* $\sum |r(u)|^2 < \infty$.

(b) *If $\{\epsilon_t\}$ is a real or complex linear process of the form $\epsilon_t := \sum_{j=-\infty}^{\infty} \psi_j \zeta_{t-j}$ with $\sum |\psi_j| < \infty$ and $\{\zeta_t\} \sim \mathrm{IID}(0, \sigma_\zeta^2)$, then, as $n \to \infty$,*

$$\max_{\omega} \left| \sum_{t=1}^{n} t^r \epsilon_t \exp(-it\omega) \right| = \mathcal{O}(n^{r+1})$$

for any integer $r \ge 0$.

PROOF. To prove part (a), let

$$X_u := \sum_{t \in T_n(u)} (t+u)^r t^r \epsilon_{t+u} \epsilon_t^* \qquad (|u| < n),$$

where $T_n(u) := \{t : \max(1, 1-u) \le t \le \min(n, n-u)\}$. It is easy to verify that

$$\left| \sum_{t=1}^{n} t^r \epsilon_t \exp(-it\omega) \right|^2 = \left| \sum_{|u|<n} X_u \exp(-iu\omega) \right| \le \sum_{|u|<n} |X_u|.$$

For $u = 0$, we have

$$E(|X_0|) \le \sum_{t=1}^{n} t^{2r} \sigma^2 \le c_1 n^{2r+1} \sigma^2, \tag{12.5.13}$$

where $c_1 > 0$ is a constant which does not depend on n. For $u \ne 0$, the i.i.d. assumption of $\{\epsilon_t\}$ implies that

$$E(\epsilon_{t+u} \epsilon_t^* \epsilon_{s+u}^* \epsilon_s) = E(|\epsilon_{t+u}|^2 |\epsilon_t|^2) \delta_{t-s}, \tag{12.5.14}$$

and hence

$$\begin{aligned} E(|X_u|^2) &= \sum_{t,s \in T_n(u)} (t+u)^r t^r (s+u)^r s^r E(\epsilon_{t+u} \epsilon_t^* \epsilon_{s+u}^* \epsilon_s) \\ &= \sum_{t \in T_n(u)} (t+u)^{2r} t^{2r} E(|\epsilon_{t+u}|^2 |\epsilon_t|^2) \le c_2 n^{4r+1} \sigma^4, \end{aligned}$$

where $c_2 > 0$ is a constant which does not depend on n or u. This, together with the Cauchy-Schwarz inequality, leads to

$$E(|X_u|) \le \sqrt{E(|X_u|^2)} \le c_2^{1/2} n^{2r+1/2} \sigma^2 \qquad (u \ne 0).$$

Combining this result with (12.5.13) yields

$$\sum_{|u|<n} E(|X_u|) \le c_1 n^{2r+1} \sigma^2 + 2 c_2^{1/2} \sum_{u=1}^{n-1} n^{2r+1/2} \sigma^2 \le c_3 n^{2r+3/2} \sigma^2, \tag{12.5.15}$$

where $c_3 > 0$ is a constant which does not depend on n. The assertion in part (a) for i.i.d. sequences is proved by citing Proposition 2.16(a).

Now let $\{\epsilon_t\}$ be a real-valued zero-mean stationary Gaussian process with ACF $r(u)$. Then, for any (u,t,s), we have [13, p. 49]

$$E(\epsilon_{t+u}\epsilon_t\epsilon_{s+u}\epsilon_s) = |r(u)|^2 + |r(t-s)|^2 + r(t-s+u)r(t-s-u).$$

This implies that for any $|u| < n$,

$$\begin{aligned} E(|X_u|^2) &\le \sum_{t,s=1}^{n} (t+u)^r t^r (s+u)^r s^r \{|r(u)|^2 + |r(t-s)|^2 + |r(t-s+u)r(t-s-u)|\} \\ &= |r(u)|^2 \Big\{\sum_{t=1}^{n} (t+u)^r t^r\Big\}^2 \\ &\quad + \sum_{|v|<n} \{|r(v)|^2 + |r(v+u)r(v-u)|\} \sum_{t\in T_n(v)} (t+u+v)^r (t+v)^r (t+u)^r t^r \\ &\le \kappa_1 |r(u)|^2 n^{4r+2} + \kappa_2 n^{4r+1}, \end{aligned}$$

where $\kappa_1 > 0$ and $\kappa_2 > 0$ are constants and the last inequality holds due partly to the assumption $\sum |r(u)|^2 < \infty$. This result, combined with the Cauchy-Schwarz inequality, gives

$$E(|X_u|) \le \sqrt{E(|X_u|^2)} \le n^{2r+1/2} \sqrt{\kappa_1 |r(u)|^2 n + \kappa_2}.$$

A further application of the Cauchy-Schwarz inequality leads to

$$\sum_{|u|<n} \sqrt{\kappa_1 |r(u)|^2 n + \kappa_2} \le \Big\{(2n-1) \sum_{|u|<n} (\kappa_1 |r(u)|^2 n + \kappa_2)\Big\}^{1/2} \le \kappa_3 n,$$

where $\kappa_3 > 0$ is a constant. Therefore, we have

$$\sum_{|u|<n} E(|X_u|) \le \kappa_3 n^{2r+3/2}.$$

The assertion in part (a) follows immediately by citing Proposition 2.16(a). In the complex Gaussian case, we have [260, p. 82]

$$E(\epsilon_{t+u}\epsilon_t^*\epsilon_{s+u}^*\epsilon_s) = |r(u)|^2 + |r(t-s)|^2.$$

Therefore, the foregoing argument remains valid.

To prove part (b), let us first consider the case where $\{\epsilon_t\} \sim \text{IID}(0,\sigma^2)$. Under this assumption, the assertion can be proved by following the steps of the proof of the strong law in Lemma 12.5.3. Let

$$Z(n,\omega) := \sum_{t=1}^{n} t^r \epsilon_t \exp(-it\omega), \quad U(n) := n^{-r-1} \max_{\omega} |Z(n,\omega)|.$$

Because (12.5.15) implies

$$E\{|U(n)|^2\} = \mathcal{O}(n^{-1/2}),$$

it can be shown, by using Chebyshev's inequality and the Borel-Cantelli lemma, as in the proof of Lemma 12.5.3, that

$$U(m^3) \stackrel{a.s.}{\to} 0 \quad \text{as } m \to \infty. \tag{12.5.16}$$

Moreover, for any $k \in [m^3, (m+1)^3)$ and any ω,

$$|Z(k,\omega) - Z(m^3,\omega)|^2 \le \sum_{t,s=m^3+1}^{(m+1)^3} t^r s^r \, |\epsilon_t \epsilon_s|.$$

Therefore,

$$\begin{aligned} &E\Big\{ \max_{m^3 \le k < (m+1)^3} \max_{\omega} k^{-2r-2} |Z(k,\omega) - Z(m^3,\omega)|^2 \Big\} \\ &\le m^{-6r-6} \sum_{t,s=m^3+1}^{(m+1)^3} t^r s^r \sigma^2 \\ &\le m^{-6r-6} \{(m+1)^3 - m^3\}^2 (m+1)^{6r} \sigma^2 = \mathcal{O}(m^{-2}). \end{aligned}$$

Applying Chebyshev's inequality and the Borel-Cantelli lemma yields

$$\max_{m^3 \le k < (m+1)^3} \max_{\omega} k^{-r-1} |Z(k,\omega) - Z(m^3,\omega)| \stackrel{a.s.}{\to} 0 \quad \text{as } m \to \infty. \tag{12.5.17}$$

Now for any $n > 0$, let $m := \lfloor n^{1/3} \rfloor$. Because $m^3 \le n < (m+1)^3$, we have

$$\begin{aligned} U(n) &\le \max_{\omega} n^{-r-1} |Z(n,\omega) - Z(m^3,\omega)| + (m^3/n)^{r+1} U(m^3) \\ &\le \max_{m^3 \le l < (m+1)^3} \max_{\omega} k^{-r-1} |Z(k,\omega) - Z(m^3,\omega)| + U(m^3). \end{aligned}$$

Combining this result with (12.5.16) and (12.5.17) proves $U(n) \stackrel{a.s.}{\to} 0$ as $n \to \infty$.

In the case of colored noise, let us define, for any $m > 0$,

$$X_t(m) := \sum_{|j| \le m} \psi_j \zeta_{t-j}, \quad Y_t(m) := \epsilon_t - X_t(m) = \sum_{|j| > m} \psi_j \zeta_{t-j}.$$

Then, for any ω, we have

$$n^{-r-1} \left| \sum_{t=1}^{n} t^r Y_t(m) \exp(-it\omega) \right| \le n^{-1} \sum_{t=1}^{n} |Y_t(m)|. \tag{12.5.18}$$

Note that $\{|Y_t(m)|\}$ is a strictly stationary process with

$$E\{|Y_0(m)|\} \le E(|\zeta_0|) \sum_{|j|>m} |\psi_j| \to 0 \quad \text{as } m \to \infty.$$

This, combined with the ergodic theorem of stationary processes [378, p. 181], ensures the existence of nonnegative random variables $Y(m)$ such that

$$n^{-1} \sum_{t=1}^{n} |Y_t(m)| \stackrel{a.s.}{\to} Y(m) \quad \text{as } n \to \infty$$

for each fixed m, and that

$$E\{Y(m)\} = E\{|Y_0(m)|\} \to 0 \quad \text{as } m \to \infty.$$

By Chebyshev's inequality, $Y(m) \xrightarrow{P} 0$ as $m \to \infty$. According to the probability theory [378, p. 12], there is an infinite sequence $\{m_k\}$ such that $m_k \to \infty$ and $Y(m_k) \stackrel{a.s.}{\to} 0$ as $k \to \infty$. Therefore,

$$\lim_{k\to\infty} \lim_{n\to\infty} n^{-1} \sum_{t=1}^{n} |Y_t(m_k)| = 0 \quad a.s.$$

Combining this result with (12.5.18) leads to

$$\lim_{k\to\infty} \lim_{n\to\infty} n^{-r-1} \left| \sum_{t=1}^{n} t^r Y_t(m_k) \exp(-it\omega) \right| = 0 \quad a.s.$$

On the other hand, because part (b) is true under the i.i.d. assumption and because $\{\zeta_{t-j}\} \sim \text{IID}(0, \sigma_\zeta^2)$ for each fixed j, we obtain

$$\begin{aligned} & n^{-r-1} \left| \sum_{t=1}^{n} t^r X_t(m) \exp(-it\omega) \right| \\ & \le \sum_{|j|\le m} |\psi_j| \, n^{-r-1} \left| \sum_{t=1}^{n} t^r \zeta_{t-j} \exp(-it\omega) \right| \stackrel{a.s.}{\to} 0 \end{aligned}$$

as $n \to \infty$ for any given m. The general assertion in part (b) is thus proved. □

Remark 12.5.6 The assertion in Lemma 12.5.9(a) holds true if $\{\epsilon_t\} \sim \text{MD}(0, \sigma^2)$, because (12.5.14) can be established by iterated expectation. For similar reasons, the assertion in Lemma 12.5.9(b) remains true if $\{\zeta_t\}$ is a sequence of martingale differences satisfying $\{\zeta_t\} \sim \text{MD}(0, \sigma_\zeta^2)$ and is strictly stationary. The stationarity is assumed to ensure the strong convergence of $n^{-1} \sum_{t=1}^{n} |Y_t(m)|$.

Remark 12.5.7 By a more involved argument, it is shown in [11] that under the assumption of Lemma 12.5.9(b),

$$\limsup_{n\to\infty} (n \log n)^{-1/2} \max_{\omega} \left| \sum_{t=1}^{n} \epsilon_t \exp(-it\omega) \right| \le \max_{\omega} \{f_\epsilon(\omega)\} \quad a.s.$$

This, combined with Bernstein's inequality (Lemma 12.1.6), leads to a stronger and more precise expression

$$\max_{\omega} \left| \sum_{t=1}^{n} t^r \epsilon_t \exp(-it\omega) \right| = \mathcal{O}(n^r \sqrt{n \log n}) \qquad (r = 0, 1, 2).$$

It is also shown in [11] that if $f_\epsilon(\omega) > 0$ for all ω, $E(|\zeta_t|^6) < \infty$, and if the characteristic function of ζ_t, denoted by $\phi(\lambda)$, satisfies $\sup_{|\lambda|\ge\lambda_0} |\phi(\lambda)| < \phi(0) = 1$ for any $\lambda_0 > 0$, then

$$\lim_{n\to\infty} (n \log n)^{-1/2} \max_{\omega} \left\{ \left| \sum_{t=1}^{n} \epsilon_t \exp(-it\omega) \right| \Big/ f_\epsilon(\omega) \right\} = 1 \quad a.s.$$

The same result is also given in [83] under the assumption that $\sum |\psi_j|^2 < \infty$, $\{\zeta_t\} \sim \mathrm{IID}(0, \sigma_\zeta^2)$, $E(|\zeta_t|^r) < \infty$ for some $r > 2$, and $f_\epsilon(\omega) > 0$ for all ω.

The following lemma, stated without proof, is useful in proving the asymptotic normality of the least absolute deviations (LAD) estimator.

Lemma 12.5.10 (Convexity Lemma [293]). *Let $\{Z_n(\boldsymbol{\theta})\}$ be a sequence of real convex random functions defined on a convex and open subset Θ of $\mathbb{R}^m$, where $m \ge 1$ is a fixed integer. Suppose that $Z(\boldsymbol{\theta})$ is a real-valued function on Θ such that $Z_n(\boldsymbol{\theta}) - Z(\boldsymbol{\theta}) \xrightarrow{P} 0$ for each $\boldsymbol{\theta} \in \Theta$. Then, for each compact subset $\mathscr{C} \subset \Theta$,*

$$\sup_{\boldsymbol{\theta} \in \mathscr{C}} |Z_n(\boldsymbol{\theta}) - Z(\boldsymbol{\theta})| \xrightarrow{P} 0.$$

The function $Z(\boldsymbol{\theta})$ is necessarily convex on Θ.

The next lemma asserts that uniform convergence is implied by pointwise convergence and stochastic equicontinuity (Theorem 1 in [14] and Lemma 2.2 in [16]). It applies to nonconvex as well as convex random functions.

Lemma 12.5.11 (Uniform Convergence [14] [16]). *For each $n = 1, 2, \ldots$, let $X_n(\boldsymbol{\theta})$ be a real- or complex-valued random function of $\boldsymbol{\theta} \in \Theta \subset \mathbb{R}^m$ for some $m \ge 1$. Let $d(\boldsymbol{\theta}, \boldsymbol{\theta}')$ be a distance measure on Θ. Moreover, assume that the following conditions are satisfied.*

(a) For any $\delta > 0$, there exists a finite subset $\mathscr{C}(\delta) \subset \Theta$ such that for each $\boldsymbol{\theta} \in \Theta$ there exists some $\boldsymbol{\theta}' \in \mathscr{C}(\delta)$ satisfying $d(\boldsymbol{\theta}, \boldsymbol{\theta}') \le \delta$. In this case, the subset $\mathscr{C}(\delta)$ is called a δ-net of Θ.

(b) For any given $\boldsymbol{\theta} \in \Theta$, $X_n(\boldsymbol{\theta}) \xrightarrow{P} 0$ as $n \to \infty$.

(c) The sequence $\{X_n(\boldsymbol{\theta})\}$ is stochastically equicontinuous, i.e., for any $\epsilon > 0$,

$$\lim_{\delta \to 0} \limsup_{n \to \infty} P\left\{ \sup_{\boldsymbol{\theta} \in \Theta} \sup_{\boldsymbol{\theta}' \in \mathscr{B}(\boldsymbol{\theta}, \delta)} |X_n(\boldsymbol{\theta}) - X_n(\boldsymbol{\theta}')| > \epsilon \right\} = 0,$$

where $\mathscr{B}(\boldsymbol{\theta}, \delta) := \{\boldsymbol{\theta}' \in \Theta : d(\boldsymbol{\theta}, \boldsymbol{\theta}') \le \delta\}$.

Then, $X_n(\boldsymbol{\theta}) \xrightarrow{P} 0$ uniformly in $\boldsymbol{\theta} \in \Theta$, i.e., $\sup\{|X_n(\boldsymbol{\theta})| : \boldsymbol{\theta} \in \Theta\} \xrightarrow{P} 0$.

PROOF. Let $\mathscr{C}(\delta) \subset \Theta_n$ denote a δ-net of Θ under (a). Then, for any $\epsilon > 0$,

$$\begin{aligned} P\left\{ \sup_{\boldsymbol{\theta} \in \Theta} |X_n(\boldsymbol{\theta})| > \epsilon \right\} &\le P\left\{ \sup_{\boldsymbol{\theta} \in \mathscr{C}(\delta)} \sup_{\boldsymbol{\theta}' \in \mathscr{B}(\boldsymbol{\theta}, \delta)} (|X_n(\boldsymbol{\theta}) - X_n(\boldsymbol{\theta}')| + |X_n(\boldsymbol{\theta})|) > \epsilon \right\} \\ &\le P\left\{ \sup_{\boldsymbol{\theta} \in \Theta} \sup_{\boldsymbol{\theta}' \in \mathscr{B}(\boldsymbol{\theta}, \delta)} (|X_n(\boldsymbol{\theta}) - X_n(\boldsymbol{\theta}')| > \tfrac{1}{2}\epsilon \right\} \\ &\quad + P\left\{ \max_{\boldsymbol{\theta} \in \mathscr{C}(\delta)} |X_n(\boldsymbol{\theta})| > \tfrac{1}{2}\epsilon \right\}. \end{aligned}$$

The first term tends to zero as $n \to \infty$ due to (c). The second term tends to zero due to (b) and the fact that $\mathscr{C}(\delta)$ is a finite set. Combining these results completes the proof. □.

The following lemma is instrumental to determining the rate of convergence for the nonlinear LAD estimator. It is a generalization of Lemma 5 in [239] by allowing the distance measure and the random functions to depend on n.

Lemma 12.5.12 (Maximal Inequality). *For each $n = 1, 2, \ldots$, define*

$$d_n(\boldsymbol{\theta}, \boldsymbol{\theta}') := c_n^{-1} \|\boldsymbol{\theta} - \boldsymbol{\theta}'\| \qquad (\boldsymbol{\theta}, \boldsymbol{\theta}' \in \mathbb{R}^m),$$

where $c_n > 0$ is a constant which may depend on n, and $m \geq 1$ is a fixed integer. Let $\{X_{nt}(\boldsymbol{\theta})\}$ $(t = 1, \ldots, n)$ be a sequence of real-valued independent zero-mean continuous random functions of $\boldsymbol{\theta} \in \mathscr{B}_n(h_0) := \{\boldsymbol{\theta} \in \mathbb{R}^m : d_n(\boldsymbol{\theta}, \boldsymbol{\theta}_0) \leq h_0\}$ satisfying $X_{nt}(\boldsymbol{\theta}_0) = 0$ for some fixed $\boldsymbol{\theta}_0 \in \mathbb{R}^m$. Assume that for all $\boldsymbol{\theta}, \boldsymbol{\theta}' \in \mathscr{B}_n(h_0)$ and all t,

$$|X_{nt}(\boldsymbol{\theta}) - X_{nt}(\boldsymbol{\theta}')| \leq \xi_{nt}\, d_n(\boldsymbol{\theta}, \boldsymbol{\theta}'), \tag{12.5.19}$$

where $\{\xi_{nt}\}$ is a sequence of random variables satisfying

$$n^{-1} \sum_{t=1}^{n} E(\xi_{nt}^2) \leq \varrho < \infty$$

for all n, with ϱ being a constant. Then, for any $\alpha \in (0,1)$, there exists a constant $\kappa > 0$, which depends solely on α, ϱ, h_0, and m, such that

$$E\Big\{ \max_{\boldsymbol{\theta} \in \mathscr{B}_n(h)} \Big| n^{-1/2} \sum_{t=1}^{n} X_{nt}(\boldsymbol{\theta}) \Big| \Big\} \leq \kappa \Big(E\Big\{ \max_{\boldsymbol{\theta} \in \mathscr{B}_n(h)} n^{-1} \sum_{t=1}^{n} X_{nt}^2(\boldsymbol{\theta}) \Big\} \Big)^{\alpha/2}$$

for all $h \in (0, h_0)$ and all n.

Remark 12.5.8 Lemma 5 in [239] is a special case of Lemma 12.5.12 where Θ_n, c_n, $\mathscr{B}_n(h)$, $X_{nt}(\boldsymbol{\theta})$, and ξ_{nt} do not depend on n.

PROOF. The proof follows the steps in [239], but it requires the following four technical lemmas that generalize the results in [120] to the case where the distance measure and the random functions may depend on n. Lemma D facilitates the proof of the assertion. Lemmas A–C are the prerequisites for Lemma D.

Lemma A. *For each $n = 1, 2, \ldots$, let Λ_n be a compact subset of $\mathbb{R}^r$ for some integer $r \geq 1$, and let $\{Y_{nt}(\boldsymbol{\lambda})\}$ $(t = 1, \ldots, n)$ be a sequence of independent zero-mean continuous random functions of $\boldsymbol{\lambda} \in \Lambda_n$. Let $\{\zeta_t\}$ be a sequence of random variables, independent of $\{Y_{nt}(\boldsymbol{\lambda})\}$, taking on values in $\{-1, 1\}$. Then,*

$$\begin{aligned} \frac{1}{2} E\Big\{ \max_{\boldsymbol{\lambda} \in \Lambda_n} \Big| \sum_{t=1}^{n} \zeta_t Y_{nt}(\boldsymbol{\lambda}) \Big| \Big\} &\leq E\Big\{ \max_{\boldsymbol{\lambda} \in \Lambda_n} \Big| \sum_{t=1}^{n} Y_{nt}(\boldsymbol{\lambda}) \Big| \Big\} \\ &\leq 2E\Big\{ \max_{\boldsymbol{\lambda} \in \Lambda_n} \Big| \sum_{t=1}^{n} \zeta_t Y_{nt}(\boldsymbol{\lambda}) \Big| \Big\}. \end{aligned}$$

Proof of Lemma A. It suffices to establish the inequalities for fixed (deterministic) $\{\zeta_t\}$ because the assertion for random sequences follows immediately from the law of iterated expectation. Given $\{\zeta_t\}$, let $\mathfrak{F}_n$ be the σ-field generated by $\{Y_{nt}(\boldsymbol{\lambda}) : \boldsymbol{\lambda} \in \Lambda_n, t \in T_n\}$, where $T_n := \{t : \zeta_t = 1\}$. Then, for any fixed $\boldsymbol{\lambda} \in \Lambda_n$,

$$\left|E\left\{\sum_{t=1}^{n} \zeta_t Y_{nt}(\boldsymbol{\lambda}) \middle| \mathfrak{F}_n\right\}\right| \le E\left\{\left|\sum_{t=1}^{n} \zeta_t Y_{nt}(\boldsymbol{\lambda})\right| \middle| \mathfrak{F}_n\right\}$$
$$\le E\left\{\max_{\boldsymbol{\lambda}\in\Lambda_n}\left|\sum_{t=1}^{n} \zeta_t Y_{nt}(\boldsymbol{\lambda})\right| \middle| \mathfrak{F}_n\right\}.$$

For $t \notin T_n$, the assumption of independence and zero mean for $\{Y_{nt}(\boldsymbol{\lambda})\}$ gives $E\{\zeta_t Y_{nt}(\boldsymbol{\lambda})|\mathfrak{F}_n\} = -E\{Y_{nt}(\boldsymbol{\lambda})\} = 0$; for $t \in T_n$, we have $E\{\zeta_t Y_{nt}(\boldsymbol{\lambda})|\mathfrak{F}_n\} = Y_{nt}(\boldsymbol{\lambda})$. Thus, the left-hand side of the foregoing inequality equals $|\sum_{t\in T_n} Y_{nt}(\boldsymbol{\lambda})|$. Taking the maximum over $\boldsymbol{\lambda}$ and the expected value with respect to $\mathfrak{F}_n$ leads to

$$E\left\{\max_{\boldsymbol{\lambda}\in\Lambda_n}\left|\sum_{t\in T_n} Y_{nt}(\boldsymbol{\lambda})\right|\right\} \le E\left[E\left\{\max_{\boldsymbol{\lambda}\in\Lambda_n}\left|\sum_{t=1}^{n} \zeta_t Y_{nt}(\boldsymbol{\lambda})\right| \middle| \mathfrak{F}_n\right\}\right]$$
$$= E\left\{\max_{\boldsymbol{\lambda}\in\Lambda_n}\left|\sum_{t=1}^{n} \zeta_t Y_{nt}(\boldsymbol{\lambda})\right|\right\}.$$

A similar inequality holds by replacing T_n with $T_n^c := \{t : \zeta_t = -1\}$. Therefore,

$$2E\left\{\max_{\boldsymbol{\lambda}\in\Lambda_n}\left|\sum_{t=1}^{n} \zeta_t Y_{nt}(\boldsymbol{\lambda})\right|\right\} \ge E\left\{\max_{\boldsymbol{\lambda}\in\Lambda_n}\left|\sum_{t\in T_n} Y_{nt}(\boldsymbol{\lambda})\right| + \max_{\boldsymbol{\lambda}\in\Lambda_n}\left|\sum_{t\in T_n^c} Y_{nt}(\boldsymbol{\lambda})\right|\right\}.$$

On the other hand, because

$$\left|\sum_{t=1}^{n} Y_{nt}(\boldsymbol{\lambda})\right| \le \left|\sum_{t\in T_n} Y_{nt}(\boldsymbol{\lambda})\right| + \left|\sum_{t\in T_n^c} Y_{nt}(\boldsymbol{\lambda})\right|,$$

it follows that

$$E\left\{\max_{\boldsymbol{\lambda}\in\Lambda_n}\left|\sum_{t=1}^{n} Y_{nt}(\boldsymbol{\lambda})\right|\right\} \le E\left\{\max_{\boldsymbol{\lambda}\in\Lambda_n}\left|\sum_{t\in T_n} Y_{nt}(\boldsymbol{\lambda})\right| + \max_{\boldsymbol{\lambda}\in\Lambda_n}\left|\sum_{t\in T_n^c} Y_{nt}(\boldsymbol{\lambda})\right|\right\}.$$

This proves the second inequality in Lemma A. Applying the second inequality with $\zeta_t Y_{nt}(\boldsymbol{\lambda})$ in place of $Y_{nt}(\boldsymbol{\lambda})$ yields the first inequality. Lemma A is proved.

Lemma B. *For each $n = 1, 2, \ldots$, let $Z_n(\boldsymbol{\theta})$ be a zero-mean finite-variance continuous random function of $\boldsymbol{\theta} \in \Theta_n \subset \mathbb{R}^m$ such that for all $\boldsymbol{\theta}, \boldsymbol{\theta}' \in \Theta_n$ and all $a \in \mathbb{R}$,*

$$E\{\exp[a(Z_n(\boldsymbol{\theta}) - Z_n(\boldsymbol{\theta}'))]\} \le \exp\{\tfrac{1}{2}a^2\sigma_n^2(\boldsymbol{\theta}, \boldsymbol{\theta}')\}, \qquad (12.5.20)$$

where $\sigma_n^2(\boldsymbol{\theta}, \boldsymbol{\theta}') := E\{|Z_n(\boldsymbol{\theta}) - Z_n(\boldsymbol{\theta}')|^2\}$. A random process with this property is called a sub-Gaussian process. Let $K(\delta, \Theta_n, \sigma_n)$ denote the minimal number of *balls centered at points in Θ_n and covering Θ_n with the σ_n-radius not exceeding 2δ. Assume that $K(\delta,\Theta_n,\sigma_n)$ is finite. Then, there exists a universal constant $c>0$ such that for any countable subset $\mathscr{C}_n \subseteq \Theta_n$ and for all $\delta>0$ and n,*

$$E\left\{\max_{(\boldsymbol{\theta},\boldsymbol{\theta}')\in\mathscr{C}_n^2(\delta)} |Z_n(\boldsymbol{\theta})-Z_n(\boldsymbol{\theta}')|\right\} \le c\delta\sum_{j=1}^{\infty} 2^{-j}\{1+\log K(2^{-j}\delta,\Theta_n,\sigma_n)\},$$

where $\mathscr{C}_n^2(\delta) := \{(\boldsymbol{\theta},\boldsymbol{\theta}')\in\mathscr{C}_n\times\mathscr{C}_n : \sigma_n(\boldsymbol{\theta},\boldsymbol{\theta}')\le\delta\}$.

Proof of Lemma B. It suffices to prove the assertion with $\delta=1$. For $\delta\neq 1$, the assertion can be proved by considering $\tilde{Z}_n(\boldsymbol{\theta}) := Z_n(\boldsymbol{\theta})/\delta$. Observe that $\tilde{Z}_n(\boldsymbol{\theta})$ remains a sub-Gaussian process with

$$\tilde{\sigma}_n^2(\boldsymbol{\theta},\boldsymbol{\theta}') := E\{|\tilde{Z}_n(\boldsymbol{\theta})-\tilde{Z}_n(\boldsymbol{\theta}')|^2\} = \sigma_n^2(\boldsymbol{\theta},\boldsymbol{\theta}')/\delta^2.$$

Moreover, it is easy to see that

$$K(2^{-j},\Theta_n,\tilde{\sigma}_n) = K(2^{-j},\Theta_n,\sigma_n/\delta) = K(2^{-j}\delta,\Theta_n,\sigma_n).$$

Finally, because $\mathscr{C}_n^2(\delta) = \{(\boldsymbol{\theta},\boldsymbol{\theta}')\in\mathscr{C}_n\times\mathscr{C}_n : \tilde{\sigma}_n(\boldsymbol{\theta},\boldsymbol{\theta}')\le 1\} := \tilde{\mathscr{C}}_n^2(1)$, we have

$$\max_{(\boldsymbol{\theta},\boldsymbol{\theta}')\in\mathscr{C}_n^2(\delta)} |Z_n(\boldsymbol{\theta})-Z_n(\boldsymbol{\theta}')| = \delta \max_{(\boldsymbol{\theta},\boldsymbol{\theta}')\in\tilde{\mathscr{C}}_n^2(1)} |\tilde{Z}_n(\boldsymbol{\theta})-\tilde{Z}_n(\boldsymbol{\theta}')|.$$

Therefore, an application of the inequality with $\delta=1$ to $\tilde{X}_n(\boldsymbol{\theta})$ leads to the assertion for the case of $\delta\neq 1$.

To prove the assertion with $\delta=1$, let $\mathscr{A}_j$ $(j=1,2,\ldots)$ be the collection of the center points of the $K_j := K(2^{-j},\Theta_n,\sigma_n)$ balls that cover Θ_n with the σ_n-radius not exceeding 2×2^{-j} (for simplicity, the dependence on n is suppressed in $\mathscr{A}_j$ and K_j). The first objective is to show that almost surely,

$$\max_{(\boldsymbol{\theta},\boldsymbol{\theta}')\in\mathscr{C}_n^2(1)} |Z_n(\boldsymbol{\theta})-Z_n(\boldsymbol{\theta}')| \le 2\sum_{j=1}^{\infty}\max_{(\boldsymbol{\theta},\boldsymbol{\theta}')\in\mathscr{B}_j} |Z_n(\boldsymbol{\theta})-Z_n(\boldsymbol{\theta}')|, \tag{12.5.21}$$

where $\mathscr{B}_j := \{(\boldsymbol{\theta},\boldsymbol{\theta}')\in(\mathscr{A}_j\cup\mathscr{A}_{j-1})\times(\mathscr{A}_j\cup\mathscr{A}_{j-1}) : \sigma_n(\boldsymbol{\theta},\boldsymbol{\theta}')\le 6\times 2^{-j}\}$ $(\mathscr{A}_0 := \emptyset)$. To prove this assertion, we note that for each $(\boldsymbol{\theta},\boldsymbol{\theta}')\in\mathscr{C}_n^2(1)$, there exist $\boldsymbol{\theta}_j,\boldsymbol{\theta}'_j\in\mathscr{A}_j$ such that $\sigma_n(\boldsymbol{\theta},\boldsymbol{\theta}_j)\le 2\times 2^{-j}$ and $\sigma_n(\boldsymbol{\theta}',\boldsymbol{\theta}'_j)\le 2\times 2^{-j}$. Because $\sigma_n(\boldsymbol{\theta},\boldsymbol{\theta}')\le 1$, it follows that $\sigma_n(\boldsymbol{\theta}_1,\boldsymbol{\theta}'_1)\le\sigma_n(\boldsymbol{\theta}_1,\boldsymbol{\theta})+\sigma_n(\boldsymbol{\theta},\boldsymbol{\theta}')+\sigma_n(\boldsymbol{\theta}',\boldsymbol{\theta}'_1)\le 3 = 6\times 2^{-1}$, which implies that $(\boldsymbol{\theta}_1,\boldsymbol{\theta}'_1)\in\mathscr{B}_1$. For $j>1$, we have $\sigma_n(\boldsymbol{\theta}_j,\boldsymbol{\theta}_{j-1})\le\sigma_n(\boldsymbol{\theta}_j,\boldsymbol{\theta})+\sigma_n(\boldsymbol{\theta},\boldsymbol{\theta}_{j-1})\le 2\times 2^{-j}+2\times 2^{-j+1} = 6\times 2^{-j}$ and similarly, $\sigma_n(\boldsymbol{\theta}'_j,\boldsymbol{\theta}'_{j-1})\le 6\times 2^{-j}$. This implies that $(\boldsymbol{\theta}_j,\boldsymbol{\theta}_{j-1})\in\mathscr{B}_j$ and $(\boldsymbol{\theta}'_j,\boldsymbol{\theta}'_{j-1})\in\mathscr{B}_j$. Moreover, because

$$\sum_{j=1}^{\infty} E\{|Z_n(\boldsymbol{\theta})-Z_n(\boldsymbol{\theta}_j)|^2\} = \sum_{j=1}^{\infty}\sigma_n^2(\boldsymbol{\theta},\boldsymbol{\theta}_j) \le \sum_{j=1}^{\infty} 2\times 2^{-j} < \infty,$$

it follows from Proposition 2.16(c) that

$$|Z_n(\boldsymbol{\theta}) - Z_n(\boldsymbol{\theta}_k)| \overset{a.s.}{\to} 0 \quad \text{as } k \to \infty.$$

Similarly, $|Z_n(\boldsymbol{\theta}') - Z_n(\boldsymbol{\theta}'_k)| \overset{a.s.}{\to} 0$ as $k \to \infty$. Combining these results yields

$$\begin{aligned} |Z_n(\boldsymbol{\theta}) - Z_n(\boldsymbol{\theta}')| &\le \limsup_{k\to\infty}\{|Z_n(\boldsymbol{\theta}) - Z_n(\boldsymbol{\theta}_k)| + |Z_n(\boldsymbol{\theta}_k) - Z_n(\boldsymbol{\theta}'_k)| \\ &\quad + |Z_n(\boldsymbol{\theta}'_k) - Z_n(\boldsymbol{\theta}')|\} \\ &= \limsup_{k\to\infty} |Z_n(\boldsymbol{\theta}_k) - Z_n(\boldsymbol{\theta}'_k)| \quad \text{a.s.} \end{aligned} \tag{12.5.22}$$

Moreover, for any $k \ge 1$, we have

$$\begin{aligned} Z_n(\boldsymbol{\theta}_k) - Z_n(\boldsymbol{\theta}'_k) &= Z_n(\boldsymbol{\theta}_1) - Z_n(\boldsymbol{\theta}'_1) \\ &\quad + \sum_{j=2}^{k}\{(Z_n(\boldsymbol{\theta}_j) - Z_n(\boldsymbol{\theta}_{j-1})) - (Z_n(\boldsymbol{\theta}'_j) - Z_n(\boldsymbol{\theta}'_{j-1}))\}. \end{aligned}$$

Because $(\boldsymbol{\theta}_1, \boldsymbol{\theta}'_1) \in \mathscr{B}_1$, $(\boldsymbol{\theta}_j, \boldsymbol{\theta}_{j-1}) \in \mathscr{B}_j$, and $(\boldsymbol{\theta}'_j, \boldsymbol{\theta}'_{j-1}) \in \mathscr{B}_j$, it follows that

$$|Z_n(\boldsymbol{\theta}_k) - Z_n(\boldsymbol{\theta}'_k)| \le 2\sum_{j=1}^{\infty} \max_{(\boldsymbol{\theta},\boldsymbol{\theta}')\in\mathscr{B}_j} |Z_n(\boldsymbol{\theta}) - Z_n(\boldsymbol{\theta}')|.$$

Combining this result with (12.5.22) and the countability of $\mathscr{C}_n$ proves (12.5.21).

It follows from (12.5.21) that

$$E\left\{\max_{(\boldsymbol{\theta},\boldsymbol{\theta}')\in\mathscr{C}_n^2(1)} |Z_n(\boldsymbol{\theta}) - Z_n(\boldsymbol{\theta}')|\right\} \le 2\sum_{j=1}^{\infty} E(U_j), \tag{12.5.23}$$

where

$$U_j := \max_{(\boldsymbol{\theta},\boldsymbol{\theta}')\in\mathscr{B}_j} |Z_n(\boldsymbol{\theta}) - Z_n(\boldsymbol{\theta}')|.$$

Owing to the symmetry of $\mathscr{B}_j$ in $\boldsymbol{\theta}$ and $\boldsymbol{\theta}'$, we have

$$\max_{(\boldsymbol{\theta},\boldsymbol{\theta}')\in\mathscr{B}_j} \{Z_n(\boldsymbol{\theta}) - Z_n(\boldsymbol{\theta}')\} = \max_{(\boldsymbol{\theta},\boldsymbol{\theta}')\in\mathscr{B}_j} -\{Z_n(\boldsymbol{\theta}) - Z_n(\boldsymbol{\theta}')\}.$$

This implies that

$$\max_{(\boldsymbol{\theta},\boldsymbol{\theta}')\in\mathscr{B}_j} \{Z_n(\boldsymbol{\theta}) - Z_n(\boldsymbol{\theta}')\} = \max_{(\boldsymbol{\theta},\boldsymbol{\theta}')\in\mathscr{B}_j} |Z_n(\boldsymbol{\theta}) - Z_n(\boldsymbol{\theta}')| = U_j.$$

Let $V_j := 6^{-1}2^j U_j - 2\log K_j$. Then,

$$E\{\exp(V_j)\} = \frac{1}{K_j^2} E\left\{\max_{(\boldsymbol{\theta},\boldsymbol{\theta}')\in\mathscr{B}_j} \exp[6^{-1}2^j(Z_n(\boldsymbol{\theta}) - Z_n(\boldsymbol{\theta}'))]\right\}$$

$$\leq \frac{1}{K_j^2} \sum_{(\boldsymbol{\theta},\boldsymbol{\theta}')\in\mathscr{B}_j} E\{\exp[6^{-1}2^j(Z_n(\boldsymbol{\theta}) - Z_n(\boldsymbol{\theta}'))]\}.$$

By the sub-Gaussian assumption,

$$E\{\exp[6^{-1}2^j(Z_n(\boldsymbol{\theta}) - Z_n(\boldsymbol{\theta}'))]\} \leq \exp\{\tfrac{1}{2}6^{-2}2^{2j}\sigma_n^2(\boldsymbol{\theta},\boldsymbol{\theta}')\} \leq \exp(1/2),$$

where the second inequality holds because $\sigma_n(\boldsymbol{\theta},\boldsymbol{\theta}') \leq 6\times 2^{-j}$ for $(\boldsymbol{\theta},\boldsymbol{\theta}') \in \mathscr{B}_j$. Moreover, the cardinality of $\mathscr{B}_j$ does not exceed $K_j K_{j-1}$ which is no more than K_j^2 because $K_{j-1} \leq K_j$. Combining these results yields

$$E\{\exp(V_j)\} \leq \exp(1/2).$$

By definition, $V_j < 0$ implies $U_j < 6\times 2^{-j}\times 2\log K_j$. Therefore,

$$E\{\mathscr{I}(V_j < 0)U_j\} \leq 6\times 2^{-j}\times 2\log K_j. \tag{12.5.24}$$

On the other hand, because $U_j = 6\times 2^{-j}(V_j + 2\log K_j)$, we have

$$E\{\mathscr{I}(V_j \geq 0)U_j\} \leq 6\times 2^{-j}(E\{\mathscr{I}(V_j\geq 0)V_j\} + 2\log K_j).$$

Note that there exists a constant $c_0 > 0$ such that $x \leq \exp(x)$ for all $x \geq c_0$. Thus,

$$\begin{aligned} E\{\mathscr{I}(V_j\geq 0)V_j\} &= E\{\mathscr{I}(0\leq V_j\leq c_0)V_j\} + E\{\mathscr{I}(V_j > c_0)V_j\} \\ &\leq c_0 + E\{\mathscr{I}(V_j > c_0)\exp(V_j)\} \\ &\leq c_0 + E\{\exp(V_j)\} \\ &\leq c_0 + \exp(1/2). \end{aligned}$$

This leads to

$$E\{\mathscr{I}(V_j\geq 0)U_j\} \leq 6\times 2^{-j}(c_1 + 2\log K_j).$$

where $c_1 := c_0 + \exp(1/2)$. Combining this result with (12.5.24) yields

$$\begin{aligned} E(U_j) &= E\{\mathscr{I}(V_j < 0)U_j\} + E\{\mathscr{I}(V_j\geq 0)U_j\} \\ &\leq 6\times 2^{-j}(c_1 + 4\log K_j) \\ &\leq \tfrac{1}{2}c2^{-j}(1+\log K_j), \end{aligned}$$

where $c := 12\times\max\{4, c_1\}$ is a universal constant. Substituting this inequality in (12.5.23) proves Lemma B.

Lemma C. *Let $\{\zeta_t\}$ be the i.i.d. random sequence defined in Lemma A with the additional property $P(\zeta_t = 1) = P(\zeta_t = -1) = 1/2$. Then, for any sequence of deterministic functions $\{g_{nt}(\boldsymbol{\theta})\}$ $(\boldsymbol{\theta}\in\Theta_n)$, the random process*

$$Z_n(\boldsymbol{\theta}) := \sum_{t=1}^n \zeta_t g_{nt}(\boldsymbol{\theta})$$

satisfies (12.5.20) with $\sigma_n^2(\boldsymbol{\theta},\boldsymbol{\theta}') := E\{|Z_n(\boldsymbol{\theta}) - Z_n(\boldsymbol{\theta}')|^2\} = \sum_{t=1}^n |g_{nt}(\boldsymbol{\theta}) - g_{nt}(\boldsymbol{\theta}')|^2$.

Proof of Lemma C. For each $k = 0,1,2,\ldots$, we have $E(\zeta_t^{2k}) = 1$ and $E(\zeta_t^{2k+1}) = 0$ because $P(\zeta_t = 1) = P(\zeta_t = -1) = 1/2$. Therefore, for any $a \in \mathbb{R}$,

$$\begin{aligned} E\{\exp(a\zeta_t)\} &= \sum_{k=0}^{\infty} E(\zeta_t^k)a^k/k! = \sum_{k=0}^{\infty} a^{2k}/(2k)! \\ &\le \sum_{k=0}^{\infty} (\tfrac{1}{2}a^2)^j/k! = \exp(\tfrac{1}{2}a^2). \end{aligned}$$

This implies in particular that for any $\boldsymbol{\theta},\boldsymbol{\theta}' \in \Theta_n$ and $a \in \mathbb{R}$,

$$E\{a\zeta_t(g_{nt}(\boldsymbol{\theta}) - g_{nt}(\boldsymbol{\theta}'))\} \le \exp(\tfrac{1}{2}a^2|g_{nt}(\boldsymbol{\theta}) - g_{nt}(\boldsymbol{\theta}')|^2).$$

Owing to the independence assumption of $\{\zeta_t\}$, we have

$$\begin{aligned} E\{\exp[a(Z_n(\boldsymbol{\theta}) - Z_n(\boldsymbol{\theta}'))]\} &= \prod_{t=1}^{n} E\{\exp[a\zeta_t(g_{nt}(\boldsymbol{\theta}) - g_{nt}(\boldsymbol{\theta}'))]\} \\ &\le \prod_{t=1}^{n} \exp(\tfrac{1}{2}a^2|g_{nt}(\boldsymbol{\theta}) - g_{nt}(\boldsymbol{\theta}')|^2) \\ &= \exp\{\tfrac{1}{2}a^2\sigma_n^2(\boldsymbol{\theta},\boldsymbol{\theta}')\}. \end{aligned}$$

This proves the inequality (12.5.20). The proof of Lemma C is complete.

Lemma D (Gaenssler-Schlumprecht Lemma). *For each $n = 1,2,\ldots$, let Θ_n be a separable compact subset of $\mathbb{R}^m$ $(m \ge 1)$ and let $\{X_{nt}(\boldsymbol{\theta})\}$ $(t = 1,\ldots,n)$ be a sequence of independent zero-mean finite-variance continuous random functions of $\boldsymbol{\theta} \in \Theta_n$. Define*

$$X_n(\boldsymbol{\theta}) := \sum_{t=1}^{n} X_{nt}(\boldsymbol{\theta}), \quad S_n(\boldsymbol{\theta},\boldsymbol{\theta}') := \Big\{\sum_{t=1}^{n} |X_{nt}(\boldsymbol{\theta}) - X_{nt}(\boldsymbol{\theta}')|^2\Big\}^{1/2}.$$

Let $K(\delta,\Theta_n,S_n)$ denote the minimal number of balls centered at points in Θ_n and covering Θ_n with the S_n-radius not exceeding 2δ. Then, there exists a universal constant $\kappa > 0$ such that for all n,

$$E\Big\{\max_{\boldsymbol{\theta},\boldsymbol{\theta}' \in \Theta_n} |X_n(\boldsymbol{\theta}) - X_n(\boldsymbol{\theta}')|\Big\} \le \kappa\, E\Big\{R_n \sum_{j=1}^{\infty} 2^{-j}(1 + \log K(2^{-j}R_n,\Theta_n,S_n))\Big\},$$

where $R_n := \max\{S_n(\boldsymbol{\theta},\boldsymbol{\theta}') : \boldsymbol{\theta},\boldsymbol{\theta}' \in \Theta_n\}$.

Proof of Lemma D. Let $\mathfrak{F}_n$ denote the σ-field generated by $\{X_{nt}(\boldsymbol{\theta}) : t = 1,\ldots,n, \boldsymbol{\theta} \in \Theta_n\}$. Define $Z_n(\boldsymbol{\theta}) := \sum_{t=1}^{n} \zeta_t X_{nt}(\boldsymbol{\theta})$, where $\{\zeta_t\}$ is a sequence of i.i.d. random variables, independent of $\{X_{nt}(\boldsymbol{\theta})\}$, with the property $P(\zeta_t = 1) = P(\zeta_t = -1) = 1/2$. Conditioning on $\mathfrak{F}_n$, the functions $X_{nt}(\boldsymbol{\theta})$ and $S_n(\boldsymbol{\theta},\boldsymbol{\theta}')$ become deterministic and the random process $Z_n(\boldsymbol{\theta})$ satisfies the conditions of Lemma C with

$$\sigma_n^2(\boldsymbol{\theta},\boldsymbol{\theta}') := E\{|Z_n(\boldsymbol{\theta}) - Z_n(\boldsymbol{\theta})|^2|\mathfrak{F}_n\} = S_n^2(\boldsymbol{\theta},\boldsymbol{\theta}').$$

By Lemma B and Lemma C, there exists a universal constant $c > 0$ such that for any countable subset $\mathscr{C}_n \subseteq \Theta_n$ and for all $\delta > 0$ and n,

$$E\Big\{\max_{(\boldsymbol{\theta},\boldsymbol{\theta}')\in\mathscr{C}_n^2(\delta)} |Z_n(\boldsymbol{\theta}) - Z_n(\boldsymbol{\theta}')| \Big| \mathfrak{F}_n\Big\} \le c\delta \sum_{j=1}^{\infty} 2^{-j}\{1 + \log K(2^{-j}\delta, \Theta_n, S_n)\},$$

where $\mathscr{C}_n^2(\delta) := \{(\boldsymbol{\theta},\boldsymbol{\theta}') \in \mathscr{C}_n \times \mathscr{C}_n : S_n(\boldsymbol{\theta},\boldsymbol{\theta}') \le \delta\}$. Because $\mathscr{C}_n^2(R_n) = \mathscr{C}_n \times \mathscr{C}_n$, substituting δ by R_n in the foregoing inequality and then taking the expectation with respect to $\mathfrak{F}_n$ on both sides yield

$$\begin{aligned} &E\Big\{\max_{\boldsymbol{\theta},\boldsymbol{\theta}'\in\mathscr{C}_n} \Big|\sum_{t=1}^{n} \zeta_t(X_{nt}(\boldsymbol{\theta}) - X_{nt}(\boldsymbol{\theta}'))\Big|\Big\} \\ &\le cE\Big\{R_n \sum_{j=1}^{\infty} 2^{-j}(1 + \log K(2^{-j}R_n, \Theta_n, S_n))\Big\}. \end{aligned}$$

An application of Lemma A to $Y_{nt}(\boldsymbol{\lambda}) := X_{nt}(\boldsymbol{\theta}) - X_{nt}(\boldsymbol{\theta}')$, with $\boldsymbol{\lambda} := (\boldsymbol{\theta},\boldsymbol{\theta}')$ and $\Lambda_n := \mathscr{C}_n \times \mathscr{C}_n$, leads to

$$E\Big\{\max_{\boldsymbol{\theta},\boldsymbol{\theta}'\in\mathscr{C}_n} |X_n(\boldsymbol{\theta}) - X_n(\boldsymbol{\theta}')|\Big\} \le \kappa E\Big\{R_n \sum_{j=1}^{\infty} 2^{-j}(1 + \log K(2^{-j}R_n, \Theta_n, S_n))\Big\},$$

where $\kappa := 2c$. It remains to show that $X_n(\boldsymbol{\theta})$ is a separable process on Θ_n in the sense that there exists a countable subset $\mathscr{C}_n \subseteq \Theta_n$ such that

$$\max_{\boldsymbol{\theta},\boldsymbol{\theta}'\in\Theta_n} |X_n(\boldsymbol{\theta}) - X_n(\boldsymbol{\theta}')| = \max_{\boldsymbol{\theta},\boldsymbol{\theta}'\in\mathscr{C}_n} |X_n(\boldsymbol{\theta}) - X_n(\boldsymbol{\theta}')| \quad a.s. \tag{12.5.25}$$

Because Θ_n is separable, there exists, by definition, a countable subset $\mathscr{C}_n \subseteq \Theta_n$ such that for each $\boldsymbol{\theta} \in \Theta_n$ there is a convergent sequence in $\mathscr{C}_n$ with $\boldsymbol{\theta}$ as its limit. This property guarantees the existence of a random sequence $\{(\boldsymbol{\theta}_k, \boldsymbol{\theta}_k')\} \subset \mathscr{C}_n \times \mathscr{C}_n$ that tends to the maximizer of $|X_n(\boldsymbol{\theta}) - X_n(\boldsymbol{\theta}')|$ in $(\boldsymbol{\theta},\boldsymbol{\theta}') \in \Theta_n \times \Theta_n$ almost surely as $k \to \infty$. Combining this result with the continuity of $X_n(\boldsymbol{\theta})$ yields

$$\max_{\boldsymbol{\theta},\boldsymbol{\theta}'\in\Theta_n} |X_n(\boldsymbol{\theta}) - X_n(\boldsymbol{\theta}')| = \lim_{k\to\infty} |X_n(\boldsymbol{\theta}_k) - X_n(\boldsymbol{\theta}_k')| \quad a.s. \tag{12.5.26}$$

Moreover, for all k, we have

$$|X_n(\boldsymbol{\theta}_k) - X_n(\boldsymbol{\theta}_k')| \le \max_{\boldsymbol{\theta},\boldsymbol{\theta}'\in\mathscr{C}_n} |X_n(\boldsymbol{\theta}) - X_n(\boldsymbol{\theta}')| \le \max_{\boldsymbol{\theta},\boldsymbol{\theta}'\in\Theta_n} |X_n(\boldsymbol{\theta}) - X_n(\boldsymbol{\theta}')|.$$

This, together with (12.5.26), proves (12.5.25) and hence Lemma D.

Equipped with Lemma D, we are now ready to prove Lemma 12.5.12. First, because $d_n(\boldsymbol{\theta},\boldsymbol{\theta}') = c_n^{-1}\|\boldsymbol{\theta} - \boldsymbol{\theta}'\|$, it follows that $\mathscr{B}_n(h)$ is just a ball centered at $\boldsymbol{\theta}_0$ with the Euclidean radius equal to $c_n h$, i.e.,

$$\mathscr{B}_n(h) = \mathscr{B}(c_n h) := \{\boldsymbol{\theta} \in \mathbb{R}^m : \|\boldsymbol{\theta} - \boldsymbol{\theta}_0\| \le c_n h\}.$$

This means that $\mathscr{B}_n(h)$ is a separable compact subset of $\mathbb{R}^m$ (the countable subset $\mathscr{C}_n \subset \mathscr{B}_n(h)$ with all components being rational numbers is dense in the sense that any point in $\mathscr{B}_n(h)$ is the limit of a sequence in $\mathscr{C}_n$). Thus the conditions of Lemma D are satisfied by $\{X_{nt}(\boldsymbol{\theta})\}$ with $\Theta_n := \mathscr{B}_n(h)$. Consequently, there exists a universal constant $\kappa_1 > 0$ such that for all $0 < h < h_0$ and all n,

$$E\left\{\max_{\boldsymbol{\theta},\boldsymbol{\theta}' \in \mathscr{B}_n(h)} |X_n(\boldsymbol{\theta}) - X_n(\boldsymbol{\theta}')|\right\}$$
$$\leq \kappa_1 E\left\{R_n(h) \sum_{j=1}^{\infty} 2^{-j}(1 + \log K(2^{-j} R_n(h), \mathscr{B}_n(h), S_n))\right\}, \qquad (12.5.27)$$

where $R_n(h) := \max\{S_n(\boldsymbol{\theta},\boldsymbol{\theta}') : \boldsymbol{\theta},\boldsymbol{\theta}' \in \mathscr{B}_n(h)\}$. Moreover, let

$$W_n := \left(\sum_{t=1}^{n} \xi_{nt}^2\right)^{1/2}.$$

Then, owing to (12.5.19), we have

$$S_n(\boldsymbol{\theta},\boldsymbol{\theta}') \leq W_n\, d_n(\boldsymbol{\theta},\boldsymbol{\theta}') = c_n^{-1} W_n \|\boldsymbol{\theta} - \boldsymbol{\theta}'\|,$$

and hence any collection of balls covering $\mathscr{B}(c_n h)$ with the Euclidean radius not exceeding $2c_n W_n^{-1}\delta$ also covers $\mathscr{B}_n(h)$ with the S_n-radius not exceeding 2δ. This implies obviously that

$$K(\delta, \mathscr{B}_n(h), S_n) \leq K(c_n W_n^{-1}\delta, \mathscr{B}(c_n h), \|\cdot\|) = K(W_n^{-1}\delta, \mathscr{B}(h), \|\cdot\|).$$

By the theory of covering numbers [100] [319], there is a constant $\kappa_2 > 0$, depending only on m, such that

$$K(W_n^{-1}\delta, \mathscr{B}(h), \|\cdot\|) \leq \{\kappa_2 h/(W_n^{-1}\delta)\}^m.$$

Therefore,

$$1 \leq K(\delta, \mathscr{B}_n(h), S_n) \leq (\kappa_2 W_n h/\delta)^m \leq (\kappa_2 W_n h_0/\delta)^m$$

for all $0 < h < h_0$. Furthermore, it is easy to show that

$$1 + \log x < (1-\alpha)^{-1} x^{1-\alpha} \qquad \forall x \geq 1, 0 < \alpha < 1.$$

Combining these results leads to

$$\begin{aligned} 1 + \log K(2^{-j} R_n(h), \mathscr{B}_n(h), S_n) &\leq m\{1 + \log(\kappa_2 h_0 2^j W_n / R_n(h))\} \\ &\leq m(1-\alpha)^{-1}\{\kappa_2 h_0 2^j W_n / R_n(h)\}^{1-\alpha}. \end{aligned}$$

Hence, there is a constant $\kappa_3 > 0$, depending solely on m, α, and h_0, such that

$$\sum_{j=1}^{\infty} 2^{-j}(1 + \log K(2^{-j} R_n(h), \mathscr{B}_n(h), S_n)) \leq \kappa_3 \{W_n / R_n(h)\}^{1-\alpha}.$$

Substituting this inequality in (12.5.27) leads to

$$E\Big\{\max_{\boldsymbol{\theta},\boldsymbol{\theta}'\in\mathscr{B}_n(h)} |X_n(\boldsymbol{\theta}) - X_n(\boldsymbol{\theta}')|\Big\} \le \kappa_1\kappa_3 E\{W_n^{1-\alpha}(R_n(h))^\alpha\}. \qquad (12.5.28)$$

By Hölder's inequality, we have

$$E\{W_n^{1-\alpha}(R_n(h))^\alpha\} \le (E(W_n))^{1-\alpha}(E\{R_n(h)\})^\alpha.$$

Moreover, let

$$S_n(\boldsymbol{\theta}) := \Big\{\sum_{t=1}^n X_{nt}^2(\boldsymbol{\theta})\Big\}^{1/2} = S_n(\boldsymbol{\theta},\boldsymbol{\theta}_0), \quad M_n := \max_{\boldsymbol{\theta}\in\mathscr{B}_n(h)} n^{-1/2} S_n(\boldsymbol{\theta}).$$

Then, by Minkowsky's inequality, $S_n(\boldsymbol{\theta},\boldsymbol{\theta}') \le S_n(\boldsymbol{\theta}) + S_n(\boldsymbol{\theta}')$. This implies that $n^{-1/2}R_n(h) \le 2M_n$ and hence

$$n^{-1/2}E\{R_n(h)\} \le 2E(M_n) \le 2\{E(M_n^2)\}^{1/2}.$$

Finally, because $W_n^2 = \sum_{t=1}^n \xi_{nt}^2$, it follows that

$$n^{-1/2}E(W_n) \le \{E(n^{-1}W_n^2)\}^{1/2} \le \varrho^{1/2}.$$

Combining these results with (12.5.28) yields

$$\begin{aligned} &E\Big\{\max_{\boldsymbol{\theta},\boldsymbol{\theta}'\in\mathscr{B}_n(h)} n^{-1/2}|X_n(\boldsymbol{\theta}) - X_n(\boldsymbol{\theta}')|\Big\} \\ &\le \kappa_1\kappa_3\{n^{-1/2}E(W_n)\}^{1-\alpha}\{n^{-1/2}E(R_n(h))\}^\alpha \\ &\le 2^\alpha\kappa_1\kappa_3\varrho^{(1-\alpha)/2}\{E(M_n^2)\}^{\alpha/2}. \end{aligned}$$

The proof is complete with $\kappa := 2^\alpha\kappa_1\kappa_3\varrho^{(1-\alpha)/2}$ upon noting that

$$\max_{\boldsymbol{\theta}\in\mathscr{B}_n(h)} |X_n(\boldsymbol{\theta})| \le \max_{\boldsymbol{\theta},\boldsymbol{\theta}'\in\mathscr{B}_n(h)} |X_n(\boldsymbol{\theta}) - X_n(\boldsymbol{\theta}')|$$

owing to the assumption that $X_{nt}(\boldsymbol{\theta}_0) = 0$ and hence $X_n(\boldsymbol{\theta}_0) = 0$. □

Bibliography

[1] T. J. Abatzoglou, "A fast maximum likelihood algorithm for frequency estimation of a sinusoid based on Newton's method," *IEEE Trans. Acoust., Speech, Signal Process.*, vol. 33, no. 1, pp. 77–89, 1985.

[2] J. S. Abel, "A bound on mean-square-estimate error," *IEEE Trans. Inform. Theory*, vol. 39, no. 5, pp. 1675–1680, 1993.

[3] E. Aboutanios, "A modified dichotomous search frequency estimator," *IEEE Signal Process. Lett.*, vol. 11, no. 2, pp. 186–188, 2004.

[4] E. Aboutanios and B. Mulgrew, "Iterative frequency estimation by interpolation on Fourier coefficients," *IEEE Trans. Signal Process.*, vol. 53, no. 4, pp. 1237–1242, 2005.

[5] T. Adali and S. Haykin, *Adaptive Signal Processing: Next-Generation Solutions*, New York: Wiley, 2010.

[6] H. Akaike, "Power spectrum estimation through autoregressive model fitting," *Ann. Inst. Statist. Math.*, vol. 21, no. 1, pp. 407–419, 1969.

[7] H. Akaike, "Statistical predictor identification," *Ann. Inst. Statist. Math.*, vol. 22, no. 2, pp. 203–217, 1970.

[8] H. Akaike, "Information theory and an extension of the maximum likelihood principle," in *Proc. 2nd Int. Symp. Inform. Theory*, B. N. Petrov and F. Csáki, Eds., pp. 267–281, Budapest: Akadémiai Kiadó, 1973.

[9] H. Akaike, "A new look at the statistical model identification," *IEEE Trans. Automat. Contr.*, vol. 19, no. 6, pp. 716–723, 1974.

[10] H. Z. An, Z. G. Chen, and E. J. Hannan, "Autocorrelation, autoregression and autoregressive approximation," *Ann. Statist.*, vol. 10, no. 3, pp. 926–936, 1982.

[11] H. Z. An, Z. G. Chen, and E. J. Hannan, "The maximum of the periodogram," *J. Multivariate Analysis*, vol. 13, no. 3, pp. 383–400, 1983.

[12] T. W. Anderson, *The Statistical Analysis of Time Series*, New York: Wiley, 1971.

[13] T. W. Anderson, *An Introduction to Multivariate Statistical Analysis*, 2nd Edn., New York: Wiley, 1984.

[14] D. W. K. Andrews, "Generic uniform convergence," *Econometric Theory*, vol. 8, no. 2, pp. 241–257, 1992.

[15] M. A. Arcones, "Asymptotic distribution of regression M-estimators," *J. Statist. Planning & Inference*, vol. 97, no. 2, pp. 235–261, 2001.

[16] M. A. Arcones, "Large deviations of empirical processes," in *High Dimensional Probability III*, J. Hoffmann-Jorgensen, M. B. Marcus, and J. A. Wellner, Eds., pp. 205–224, Boston, MA: Birkhäuser, 2004.

[17] R. Badeau, G. Richard, and B. David, "Fast and stable YAST algorithm for principal and minor subspace tracking," *IEEE Trans. Signal Process.*, vol. 56, no. 8, pp. 3437–3446, 2008.

[18] Z. D. Bai, C. R. Rao, and M. Chow, "An algorithm for efficient estimation of superimposed exponential signals," in *Proc. 4th IEEE Region 10 Int. Conf.*, Bombay, India, pp. 342–345, 1989.

[19] Z. D. Bai, C. R. Rao, Y. Wu, M.-M. Zen, and L. Zhao, "The simultaneous estimation of the number of signals and frequencies of multiple sinusoids when some observations are missing: I. Asymptotics," *Proc. Natl. Acad. Sci. USA*, vol. 96, no. 20, pp. 11106–11110, 1999.

[20] J. L. Ballester and R. Oliver, "Discovery of the near 158 day periodicity in group sunspot numbers during the eighteenth century," *Astrophys. J.*, vol. 522, no. 2, pp. 153–156, 1999.

[21] E. W. Barankin, "Locally best unbiased estimates," *Ann. Math. Statist.*, vol. 20, no. 4, pp. 477–501, 1949.

[22] J. A. Barnes, H. H. Sargent III, and P. V. Tryon, "Sunspot cycle simulation using a narrowband Gaussian process," Technical Note 1022, National Bureau of Standards, US Department of Commerce, 1980.

[23] I. Barrodale and F. D. K. Roberts, "An improved algorithm for discrete ℓ_1 linear approximation," *SIAM J. Numerical Analysis*, vol. 10, no. 5, pp. 839–848, 1973.

[24] I. Barrodale and F. D. K. Roberts, "Solution of an overdetermined system of equations in the ℓ_1 norm," *Commun. ACM*, vol. 17, no. 6, pp. 319–320, 1974.

[25] M. S. Bartlett, "On the theoretical specification of sampling properties of autocorrelated time-series," *J. Roy. Statist. Soc. Suppl.*, vol. 8, no. 1, pp. 27–41, 1946.

[26] M. S. Bartlett, "Periodogram analysis and continuous spectra," *Biometrika*, vol. 37, no. 1–2, pp. 1–16, 1950.

[27] G. W. Bassett and R. Koenker, "Asymptotic theory of least absolute error regression," *J. Amer. Statist. Assoc.*, vol. 73, no. 363, pp. 618–622, 1978.

[28] M. Basseville, "Distance measures for signal processing and pattern recognition," *Signal Process.*, vol. 18, no. 4, pp. 349–369, 1989.

[29] K. I. Beltrao and P. Bloomfield, "Determining the bandwidth of a kernel spectrum estimate," *J. Time Series Analysis*, vol. 8, no. 1, pp. 21–38, 1987.

[30] K. N. Berk, "Consistent autoregressive spectral estimates," *Ann. Statist.*, vol. 2, no. 3, pp. 489–502, 1974.

[31] D. P. Bertsekas, *Nonlinear Programming*, 2nd Edn., Belmont, MA: Athena Scientific, 1999.

[32] D. V. Bhaskar Rao and S. Y. Kung, "Adaptive notch filtering for the retrieval of sinusoids in noise," *IEEE Trans. Acoust., Speech, Signal Process.*, vol. 32, no. 4, pp. 791–802, 1984.

[33] G. Bienvenu and L. Kopp, "Optimality of high resolution array processing using the eigensystem approach," *IEEE Trans. Acoust., Speech, Signal Process.*, vol. 31, no. 5, pp. 1235–1248, 1983.

[34] P. Billingsley, *Probability and Measure*, 3nd Edn., New York: Wiley, 1995.

[35] S. Bittanti and S. M. Savaresi, "On the parameterization and design of an extended Kalman filter frequency tracker," *IEEE Trans. Automat. Contr.*, vol. 45, no. 8, pp. 1718–1724, 2000.

[36] A. Björck, *Numerical Methods for Least Squares Problems*, Philadelphia, PA: Society for Industrial and Applied Mathematics, 1996.

[37] R. B. Blackman and J. W. Tukey, *The Measurement of Power Spectra: From the Point of View of Communications Engineering*, New York: Dover, 1959.

[38] P. Bloomfield, *Fourier Analysis of Time Series: An Introduction*, 2nd Edn., New York: Wiley, 2000.

[39] P. Bloomfield and W. L. Steiger, *Least Absolute Deviations: Theory, Applications, and Algorithms*, Boston, MA: Birkhäuser, 1983.

[40] E. Bölviken, "New tests of significance in periodogram analysis," *Scand. J. Statist.*, vol. 10, no. 1, pp. 1–9, 1983.

[41] E. Bölviken, "The distribution of certain rational functions of order statistics from exponential distributions," *Scand. J. Statist.*, vol. 10, no. 2, pp. 117–123, 1983.

[42] G. E. P. Box and G. M. Jenkins, *Time Series Analysis: Forecasting and Control*, Revised Edn., San Francisco, CA: Holden-Day, 1976.

[43] R. C. Bradley, "Basic properties of strong mixing conditions: a survey and some open questions," *Probab. Surveys*, vol. 2, pp. 107–144, 2005.

[44] Y. Bresler and A. Macovski, "Exact maximum likelihood parameter estimation of superimposed exponential signals in noise," *IEEE Trans. Acoust., Speech, Signal Process.*, vol. 34, no. 5, pp. 1081–1089, 1986.

[45] D. R. Brillinger, *Time Series: Data Analysis and Theory*, New York: Holt, Rinehart & Winston, 1975.

[46] P. J. Brockwell and R. A. Davis, *Time Series: Theory and Methods*, 2nd Edn., New York: Springer-Verlag, 1991.

[47] D. Brunt, *The Combination of Observations*, London: Cambridge University Press, 1917.

[48] B. M. Brown, "Martingale central limit theorems," *Ann. Math. Statist.*, vol. 42, no. 1, pp. 59–66, 1971.

[49] T. Brown and M. M. Wang, "An iterative algorithm for single-frequency estimation," *IEEE Trans. Signal Process.*, vol. 50, no. 11, pp. 2671–2682, 2002.

[50] J. P. Burg, "Maximum entropy spectral analysis," in *Proc. 37th Meeting of the Society of Exploration Geophysicists*, Oklahoma City, Oklahoma, 1967. Reprinted in [65].

[51] J. P. Burg, "A new analysis technique for time series data," in *Proc. NATO Advanced Study Institute on Signal Processing with Emphasis on Underwater Acoustics*, Enschede, Netherlands, 1968. Reprinted in [65].

[52] D. Burshtein and E. Weinstein, "Confidence intervals for the maximum entropy spectrum," *IEEE Trans. Acoust. Speech, Signal Process.*, vol. 35, no. 4, pp. 504–510, 1987.

[53] J. A. Cadzow, "Spectral estimation: an overdetermined rational model equation approach," *Proc. IEEE*, vol. 70, no. 9, pp. 907–939, 1982.

[54] C. K. Carter and R. Kohn, "Semiparametric Bayesian inference for time series with mixed spectra," *J. R. Statist. Soc. B*, vol. 59, no. 1, pp. 255–268, 1997.

[55] J. E. Cavanaugh, "Unifying the derivations for the Akaike and corrected Akaike information criteria," *Statist. Probab. Lett.*, vol 33, no. 2, pp. 201–208, 1997.

[56] J. A. Chambers and A. G. Constantinides, "Frequency tracking using constrained adaptive notch filters synthesised from allpass sections," *IEE Proc. Part F*, vol. 137, no. 6, pp. 485–481, 1990.

[57] K. W. Chan and H. C. So, "Accurate frequency estimation for real harmonic sinusoids," *IEEE Signal Process. Lett.*, vol. 11, no. 7, pp. 609–612, 2004.

[58] Y. T. Chan and R. P. Langford, "Spectral estimation via the high-order Yule-Walker equations," *IEEE Trans. Acoust., Speech, Signal Process.*, vol. 30, no. 5, pp. 689–698, 1982.

[59] Y. T. Chan, J. M. M. Lavoie, and J. B. Plant, "A parameter estimation approach to estimation of frequencies of sinusoids," *IEEE Trans. Acoust., Speech, Signal Process.*, vol. 29, no. 2, pp. 214–219, 1981.

[60] D. G. Chapman and H. Robbins, "Minimum variance estimation without regularity assumptions," *Ann. Math. Statist.*, vol. 22, no. 4, pp. 581–586, 1951.

[61] E. Chaumette, J. Galy, A. Quinlan, and P. Larzabal, "A new Barankin bound approximation for the prediction of the threshold region performance of maximum-likelihood estimators," *IEEE Trans. Signal Process.*, vol. 56, no. 11, pp. 5319–5333, 2008.

[62] F.-J. Chen, C. C. Fung, C.-W. Kok, and S. Kwong, "Estimation of two-dimensional frequencies using modified matrix pencil method," *IEEE Trans. Signal Process.*, vol. 55, no. 2, pp. 718–724, 2007.

[63] Z. G. Chen, "Consistent estimates for hidden frequencies in a linear process," *Adv. Appl. Probab.*, vol. 20, no. 2, pp. 295–314, 1988.

[64] Z. G. Chen, "An alternative consistent procedure for detecting hidden frequencies," *J. Time Series Analysis*, vol. 9, no. 3, pp. 301–317, 1988.

[65] D. G. Childers (Ed.), *Modern Spectral Analysis*, New York: IEEE Press, 1978.

[66] B. S. Choi, *ARMA Model Identification*, New York: Springer-Verlag, 1992.

[67] N. Choudhuri, S. Ghosal, and A. Roy, "Bayesian estimation of the spectral density of a time series," *J. Amer. Statist. Assoc.*, vol. 99, no. 468, pp. 1050–1059, 2004.

[68] Y.-S. Chow and U. Grenander, "A sieve method for the spectral density," *Ann. Statist.*, vol. 13, no. 3, pp. 998–1010, 1985.

[69] M. G. Christensen, A. Jakobsson, and S. H. Jensen, "Joint high-resolution fundamental frequency and order estimation," *IEEE Trans. Audio, Speech, Language Process.*, vol. 15, no. 5, pp. 1635–1644, 2007.

[70] S. T. Chui, "Detecting periodic components in a white Gaussian time series," *J. Roy. Statist. Soc. B*, vol. 51, no. 2, pp. 249–259, 1989.

[71] S. T. Chui, "Peak-insensitive parametric spectrum estimation," *Stoch. Proc. Appl.*, vol. 35, no. 1, pp. 121–140, 1990.

[72] K. L. Chung, *A Course in Probability Theory*, 3rd Edn., New York: Academic Press, 2001.

[73] M. P. Clark and L. L. Scharf, "On the complexity of IQML algorithms," *IEEE Trans. Signal Process.*, vol. 40, no. 7, pp. 1811–1813, 1992.

[74] V. Clarkson, P. J. Kootsookos, and B. G. Quinn, "Analysis of the variance threshold of Kay's weighted linear prediction frequency estimator," *IEEE Trans. Signal Process.*, vol. 42, no. 9, pp. 2370–2379, 1994.

[75] R. Cogburn and H. T. Davis, "Periodic splines and spectral estimation," *Ann. Statist.*, vol. 2, no. 6, pp. 1108–1126, 1974.

[76] L. Cohen, *Time-Frequency Analysis*, Upper Saddle River, NJ: Prentice-Hall, 1995.

[77] P. Comon and G. H. Golub, "Tracking a few extreme singular values and vectors in signal processing," *Proc. IEEE*, vol. 78, no. 8, 1327–1343, 1990.

[78] F. Comte, "Adaptive estimation of the spectrum of a stationary Gaussian sequence," *Bernoulli*, vol. 7, no. 2, pp. 267–298, 2001.

[79] J. W. Cooley and J. W. Tukey, "An algorithm for the machine calculation of complex Fourier series," *Math. Comput.*, vol. 19, no. 90, pp. 297–301, 1965.

[80] J. E. Cousseau, S. Werner, and P. D. Doñate, "Factorized all-pass based IIR adaptive notch filters," *IEEE Trans. Signal Process.*, vol. 55, no. 11, pp. 5225–5236, 2007.

[81] E. Damsleth and E. Spjøtvol, "Estimation of trigonometric components in time series," *J. Amer. Statist. Assoc.*, vol. 77, no. 378, pp. 381–387, 1982.

[82] P. J. Daniell, "Discussion on symposium on autocorrelation in time series," *J. Roy. Statist. Soc. Suppl.*, vol. 8, no. 1, pp. 88–90, 1946.

[83] R. A. Davis and T. Mikosch, "The maximum of the periodogram of a non-Gaussian sequence," *Ann. Probab.*, vol. 27, no. 1, pp. 522–536, 1999.

[84] V. DeBrunner and S. Torres, "Multiple fully adaptive notch filter design based on allpass sections," *IEEE Trans. Signal Process.*, vol. 48, no. 2, pp. 550–552, 2000.

[85] H. Delic, P. Papantoni-Kazakos, and D. Kazakos, "Fundamental structures and asymptotic performance criteria in decentralized binary hypothesis testing," *IEEE Trans. Commun.*, vol. 43, no. 1, pp. 32–43, 1995.

[86] J. R. Deller, Jr., J. H. L. Hansen, and J. G. Proakis, *Discrete-Time Processing of Speech Signals*, New York: IEEE Press, 2000.

[87] J.-P. Delmas, "Asymptotic normality of sample covariance matrix for mixed spectra time series: application to sinusoidal frequencies estimation," *IEEE Trans. Inform. Theory*, vol. 47, no. 4, pp. 1681–1867, 2001.

[88] H. Dette, M. Hallin, T. Kley, and S. Volgushev, "Of copulas, quantiles, ranks and spectra: an L_1-approach to spectral analysis," Preprint, arXiv:1111.7205, 2011.

[89] E. Dilaveroğlu, "Nonmatrix Cramér-Rao bound expressions for high-resolution frequency estimators," *IEEE Trans. Signal Process.*, vol. 46, no. 2, pp. 463–474, 1998.

[90] P. M. Djurić, "A model selection rule for sinusoids in white Gaussian noise," *IEEE Trans. Signal Process.*, vol. 44, no. 7, pp. 1744–1751, 1996.

[91] I. Djurović, V. Katkovnik, and L. Stanković, "Median filter based realizations of the robust time-frequency distributions," *Signal Process.*, vol. 81, no. 7, pp. 1771–1776, 2001.

[92] I. Djurović, L. Stanković, and J. F. Böhme, "Robust L-estimation based forms of signal transforms and time-frequency representations," *IEEE Trans. Signal Process.*, vol. 51, no. 7, pp. 1753–1761, 2007.

[93] Y. Dodge (Ed.), *L_1-Statistical Procedures and Related Topics*, Hayward, CA: Institute of Mathematical Statistics, 1997.

[94] J. L. Doob, *Stochastic Processes*, New York: Wiley, 1953.

[95] X. Doukopoulos and G. Moustakides, "Fast and stable subspace tracking," *IEEE Trans. Signal Process.*, vol. 56, no. 4, pp. 1452–1465, 2008.

[96] M. V. Dragošević and S. S. Stanković, "A generalized least squares method for frequency estimation," *IEEE Trans. Acoust., Speech, Signal Process.*, vol. 37, no. 6, pp. 805–819, 1989.

[97] M. V. Dragošević and S. S. Stanković, "An adaptive notch filter with improved tracking properties," *IEEE Trans. Acoust., Speech, Signal Process.*, vol. 43, no. 9, pp. 2068–2078, 1995.

[98] C. Dubois and M. Davy, "Joint detection and tracking of time-varying harmonic components: a flexible Bayesian approach," *IEEE Trans. Audio, Speech, Language Process.*, vol. 15, no. 4, pp. 1283–1295, 2007.

[99] P. Duhamel and M. Vetterli, "Fast Fourier transforms: a tutorial review and a state of the art," *Signal Process.*, vol. 19, no. 4, pp. 259–299, 1990.

[100] I. Dumer, "Covering an ellipsoid with equal balls," *J. Combinatorial Theory A*, vol. 113, no. 8, pp. 1667–1676, 2006.

[101] K. B. Ensor and H. J. Newton, "The effect of order estimation on estimating the peak frequency of an autoregressive spectral density," *Biometrika*, vol. 75, no. 3, pp. 587–589, 1988.

[102] A. Eriksson, P. Stoica, and T. Söderström, "Second-order properties of MUSIC and ESPRIT estimates of sinusoidal frequencies in high SNR scenarios," *IEE Proc. Part F*, vol. 140, no. 4, pp. 266–272, 1993.

[103] A. G. Evans and R. Fischl, "Optimal least squares time-domain synthesis of recursive digital filters," *IEEE Trans. Audio Electroacoust.*, vol. 21, no. 1, pp. 61–65, 1973.

[104] J. Fan and E. Kreutzberger, "Automatic local smoothing for spectral density estimation," *Scand. J. Statist.*, vol. 25, no. 2, pp. 359–369, 1998.

[105] K.-T. Fang, S. Kotz, and K. W. Ng, *Symmetric Multivariate and Related Distributions*, London: Chapman & Hall, 1990.

[106] P. Faukal, C. Fröhlich, H. Spruit, and T. M. L. Wigley, "Variations in solar luminosity and their effect on the Earth's climate," *Nature*, vol. 443, no. 14, pp. 161–166, 2006.

[107] E. Feirreira and J. M. Rodriguez-Poo, "Variable bandwidth kernel estimation of the spectral density," *J. Time Series Analysis*, vol. 20, no. 3, pp. 271–287, 2002.

[108] R. A. Fisher, "Tests of significance in harmonic analysis," *Proc. Roy. Soc. A*, vol. 125, no. 796, pp. 54–59, 1929.

[109] M. Fisz, *Probability Theory and Mathematical Statistics*, 3rd Edn., New York: Wiley, 1963.

[110] M. P. Fitz, "Further results in the fast estimation of a single frequency," *IEEE Trans. Commun.*, vol. 42, no. 234, pp. 862–864, 1994.

[111] J. B. J. Fourier, *Théorie analytique de la chaleur*, Paris: Firmin Didot, 1822.

[112] M. L. Fowler, "Phase-based frequency estimation: a review," *Digital Signal Process.*, vol. 12, no. 4, pp. 590–615, 2002.

[113] M. L. Fowler and J. A. Johnson, "Extending the threshold and frequency range for phase-based frequency estimation," *IEEE Trans. Signal Process.*, vol. 47, no. 10, pp. 2857–2863, 1999.

[114] J. M. Francos and B. Friedlander, "Bounds for estimation of complex exponentials in unknown colored noise," *IEEE Trans. Signal Process.*, vol. 43, no. 9, pp. 2176–2185, 1995.

[115] J. Franke and W. Härdle, "On bootstrapping kernel spectral estimates," *Ann. Statist.*, vol. 20, no. 1, pp. 121–145, 1992.

[116] B. Friedlander and B. Porat, "The modified Yule-Walker method of ARMA spectral estimation" *IEEE Trans. Aerosp. Electron. Syst.*, vol. 20, no. 3, pp. 158–173, 1984.

[117] E. Friss-Christensen and K. Lassen, "Length of the solar cycle: an indicator of solar activity closely associated with climate," *Science*, vol. 254, no. 5032, pp. 698–700, 1991.

[118] P. Fryzlewicz, G. P. Nason, and R. von Sachs, "A wavelet-Fisz approach to spectrum estimation," *J. Time Series Analysis*, vol. 29, no. 5, pp. 868–880, 2008.

[119] J.-J. Fuchs, "Estimating the number of sinusoids in additive white noise," *IEEE Trans. Signal Process.*, vol. 36, no. 12, pp. 1846–1853, 1988.

[120] P. Gaenssler and T. Schumprecht, "Maximal inequalities for stochastic processes which are given as sums of independent processes indexed by pseudo-metric parameter spaces with applications to empirical processes," Preprint No. 44, University of Munich, 1988.

[121] M. Ghogho and A. Swami, "Fast computation of the exact FIM for deterministic signals in colored noise," *IEEE Trans. Signal Process.*, vol. 47, no. 1, pp. 52–61, 1999.

[122] G. H. Golub and C. F. Van Loan, *Matrix Computations*, 3rd Edn., Baltimore, MD: Johns Hopkins University Press, 1996.

[123] G. H. Golub and V. Pereyra, "Separable nonlinear least squares: the variable projection method and its applications," *Inverse Problems*, vol. 19, no. 2, pp. 1–26, 2003.

[124] V. V. Gorodetskii, "On the strong mixing property for linear sequences," *Theory Probab. Appl.*, vol. 22, pp. 411–413, 1977.

[125] N. K. Govil and R. N. Mohapatra, "Markov and Bernstein type inequalities for polynomials," *J. Inequal. Appl.*, vol. 3, no. 4, pp. 349–387, 1999.

[126] R. M. Gray, A. Buzo, A. H. Gray, and Y. Matsuyama, "Distortion measures for speech processing," *IEEE Trans. Acoust. Speech, Signal Process.*, vol. 28, no. 4, pp. 367–376, 1980.

[127] U. Grenander and M. Rosenblatt, "Statistical spectral analysis of times series arising from stationary stochastic processes," *Ann. Math. Statist.*, vol. 24, no. 4, pp. 537–558, 1953.

[128] U. Grenander and M. Rosenblatt, *Statistical Analysis of Stationary Time Series*, New York: Wiley, 1957.

[129] K. Gröchenig, *Foundations of Time-Frequency Analysis*, Boston, MA: Birkhäuser, 2000.

[130] H. Gu, "Frequency resolution and estimation of AR spectral analysis," *IEEE Trans. Signal Process.*, vol. 41, no. 1, pp. 432–436, 1991.

[131] P. Hall and C. C. Heyde, *Martingale Limit Theory and Its Application*, New York: Wiley, 1980.

[132] F. R. Hampel, E. M. Ronchetti, P. J. Rousseeuw, and W. A. Stahl, *Robust Statistics: The Approach Based on Influence Functions*, New York: Wiley, 1986.

[133] P. Händel, "Markov-based single-tone frequency estimation," *IEEE Trans. Circuits Syst. II. Analog Digital Signal Process.*, vol. 45, no. 2, pp. 230–232, 1998.

[134] E. J. Hannan, "Testing for a jump in the spectral function," *J. Roy. Statist. Soc. B*, vol. 23, no. 2, pp. 394–404, 1961.

[135] E. J. Hannan, *Multiple Time Series*, New York: Wiley, 1970.

[136] E. J. Hannan, "Non-linear time series regression," *J. App. Probab.*, vol. 8, no. 4, pp. 767–780, 1971.

[137] E. J. Hannan, "The estimation of frequency," *J. Appl. Probab.*, vol. 10, no. 3, pp. 510–519, 1973.

[138] E. J. Hannan and B. G. Quinn, "The resolution of closely adjacent spectral lines," *J. Time Series Analysis*, vol. 10, no. 1, pp. 13–31, 1989.

[139] F. J. Harris, "On the use of windows for harmonic analysis with the discrete Fourier transform," *Proc. IEEE*, vol. 66, no. 1, pp. 51–83, 1978.

[140] G. Harris and C. Martin, "The roots of a polynomial vary continuously as a function of the coefficients," *Proc. Amer. Math. Soc.*, vol. 100, no. 2, pp. 390–392, 1987.

[141] W. L. Harter, "The method of least squares and some alternatives," *Int. Statist. Rev.*, vol. 42, no. 2, pp. 147–174, 1974.

[142] H. O. Hartley, "Tests of significance in harmonic analysis," *Biometrika*, vol. 36, no. 1–2, pp. 194–201, 1949.

[143] T. J. Hastie and R. J. Tibshirani, *Generalized Additive Models*, London: Chapman & Hall, 1990.

[144] D. Hathaway and R. M. Wilson, "What the sunspot record tells us about space climate," *Solar Physics*, vol. 224, no. 1–2, pp. 5–19, 2004.

[145] H. Haupt and W. Oberhofer, "On asymptotic normality in nonlinear regression," *Statist. Probab. Lett.*, vol. 79, no. 6, pp. 848–849, 2009.

[146] S. Haykin, *Adaptive Filter Theory*, 3rd Edn., Englewood Cliffs, NJ: Prentice-Hall, 1996.

[147] S. He and B. Kedem,"Higher order crossings spectral analysis of an almost periodic random sequence in noise," *IEEE Trans. Inform. Theory*, vol. 35, no. 2, pp. 360–370, 1989.

[148] M. T. Heideman, D. B. Johnson, and C. S. Burrus, "Gauss and the history of the FFT," *IEEE Acoust., Speech, Signal Process. Mag.*, vol. 1, no. 4, pp. 14–21, 1984.

[149] G. R. Hext, "A new approach to time series with mixed spectra," Technical Report No. 14, Department of Statistics, Stanford University, 1966.

[150] M. Hinich, "Detecting a hidden periodic signal when its period is unknown," *IEEE Trans. Acoust. Speech, Signal Process.*, vol. 30, no. 5, pp. 747–750, 1982.

[151] R. R. Hocking, "The analysis and selection of variables in linear regression," *Biometrics*, vol. 31, no. 4, pp. 1–49, 1976.

[152] R. V. Hogg and A. T. Craig, *Introduction to Mathematical Statistics*, 5th Edn., Englewood Cliffs, NJ: Prentice-Hall, 1995.

[153] R. A. Horn and C. R. Johnson, *Matrix Analysis*, Cambridge, UK: Cambridge University Press, 1985.

[154] Y. Hua, "The most efficient implementation of the IQML algorithm," *IEEE Trans. Signal Process.*, vol. 42, no. 8, pp. 2203–2204, 1994.

[155] Y. Hua and T. K. Starkar, "Perturbation analysis of TK method for harmonic retrieval problems," *IEEE Trans. Acoust., Speech, Signal Process.*, vol. 36, no. 2, pp. 228–240, 1988.

[156] Y. Hua and T. K. Starkar, "Matrix pencil method for estimating parameters of exponentially damped/undamped sinusoids in noise," *IEEE Trans. Acoust., Speech, Signal Process.*, vol. 38, no. 5, pp. 814–824, 1990.

[157] Y. Hua and T. K. Starkar, "On SVD for estimating generalized eigenvalues of singular matrix pencil in noise," *IEEE Trans. Signal Process.*, vol. 39, no. 4, pp. 892–900, 1991.

[158] Y. Hua, "Estimating two-dimensional frequencies by matrix enhancement and matrix pencil," *IEEE Trans. Signal Process.*, vol. 40, no. 9, pp. 2267–2280, 1992.

[159] P. J. Huber and E. M. Ronchetti, *Robust Statistics*, 2nd Edn., New York: Wiley, 2009.

[160] C. M. Hurvich, "Data-driven choice of a spectrum estimate: extending the applicability of cross-validation methods," *J. Amer. Statist. Assoc.*, vol. 80, no. 392, pp. 933–940, 1985.

[161] C. M. Hurvich and C. L. Tsai, "Regression and time series model selection in small samples," *Biometrika*, vol. 76, no. 2, pp. 297–307, 1989.

[162] I. A. Ibragimov and R. Z. Has'minskii, *Statistical Estimation: Asymptotic Theory*, New York: Springer-Verlag, 1981.

[163] Y. Isokawa, "Estimation of frequency by random sampling," *Ann. Inst. Statist. Math.*, vol. 35, pt. A, pp. 201–213, 1983.

[164] A. V. Ivanov, "An estimate of the angular frequency of a harmonic oscillation in the presence of stationary noise," *Theory Probab. Math. Statist.*, no. 14, pp. 44–53, 1977.

[165] A. V. Ivanov, "A solution of the problem of detecting hidden periodicities," *Theory Probab. Math. Statist.*, no. 20, pp. 51–68, 1980.

[166] R. I. Jennrich, "Asymptotic properties of nonlinear least square estimators," *Ann. Math. Statist.*, vol. 40, no. 2, pp. 633–643, 1969.

[167] I. T. Jolliffe, *Principal Component Analysis*, 2nd Edn., New York: Springer-Verlag, 2002.

[168] R. H. Jones, "Fitting autoregressions," *J. Amer. Statist. Assoc.*, vol. 70, no. 351, pp. 590–592, 1975.

[169] J. Jurečková and B. Procházka, "Regression quantiles and trimmed least squares estimator in nonlinear regression model," *J. Nonparametric Statist.*, vol. 3, no. 3–4, pp. 201–222, 1994.

[170] Y. Kakizawa, R. H. Shumway, and M. Taniguchi, "Discrimination and clustering for multivariate time series," *J. Amer. Statist. Assoc.*, vol. 93, no. 441, pp. 328–340, 1998.

[171] N. Karmarkar, "A new polynomial time algorithm for linear programming," *Combinatorica*, vol. 4, no. 4, pp. 373–395, 1984.

[172] V. Katkovnik, "Robust M-estimates of the frequency and amplitude of a complex-valued harmonic," *Signal Process.*, vol. 77, no. 1, pp. 71–84, 1999.

[173] T. Kato, *Perturbation Theory for Linear Operators*, Chapter 2, New York: Springer-Verlag, 1966.

[174] L. Kavalieris and E. J. Hannan, "Determining the number of terms in a trigonometric regression," *J. Time Series Analysis*, vol. 15, no. 6, pp. 613–625, 1994.

[175] M. Kaveh and A. Barabell, "The statistical performance of the MUSIC and the minimum-norm algorithms in resolving plane waves in noise," *IEEE Trans. Acoust., Speech, Signal Process.*, vol. 34, no. 2, pp. 331–341, 1986.

[176] S. M. Kay, "Accurate frequency estimation at low signal-to-noise ratio," *IEEE Trans. Acoust., Speech, Signal Process.*, vol. 32, no. 3, pp. 540–547, 1984.

[177] S. M. Kay, *Modern Spectral Estimation: Theory and Application*, Englewood Cliffs, NJ: Prentice-Hall, 1988.

[178] S. M. Kay, "A fast and accurate single frequency estimator," *IEEE Trans. Acoust., Speech, Signal Process.*, vol. 37, no. 12, pp. 1987–1990, 1989.

[179] S. M. Kay and S. L. Marple, Jr., "Spectrum analysis — a modern perspective," *Proc. IEEE*, vol. 69, no. 11, pp. 1380–1419, 1981.

[180] S. M. Kay and R. Nekovei, "An efficient two-dimensional frequency estimator," *IEEE Trans. Acoust., Speech, Signal Process.*, vol. 38, no. 10, pp. 1807–1809, 1990.

[181] B. Kedem, "Contraction mappings in mixed spectrum estimation," in *New Directions in Time Series Analysis, Part I*, D. Brillinger, P. Cains, J. Geweke, E. Parzen, M. Rosenblatt, and M. S. Taqqu, Eds., New York: Springer-Verlag, pp. 169–191, 1992.

[182] B. Kedem, *Time Series Analysis by Higher Order Crossings*, Piscataway, NJ: IEEE Press, 1994.

[183] C. T. Kelly, *Iterative Methods for Optimization*, Philadelphia, PA: Society Industrial and Applied Mathematics, 1999.

[184] R. J. Kenefic and A. H. Huttall, "Maximum likelihood estimation of the parameters of a tone using real discrete data," *IEEE J. Oceanic Eng.*, vol. 12, no. 1, pp. 279–280, 1987.

[185] J. Kiefer, "On minimum variance estimators," *Ann. Math. Statist.*, vol. 23, no. 4, pp. 627–629, 1952.

[186] T. S. Kim, H. K. Kim, and S. Hur, "Asymptotic properties of a particular nonlinear regression quantile estimation," *Statist. Probab. Lett.*, vol. 60, no. 4, pp. 387–394, 2002.

[187] D. Kim, M. J. Narasimha, and D. C. Cox, "An improved single frequency estimator," *IEEE Signal Process. Lett.*, vol. 3, no. 7, pp. 212–214, 1996.

[188] A. C. Kimber, "Trimming in gamma samples," *Appl. Statist.* vol. 32, no. 1, pp. 7–14, 1983.

[189] K. R. Klemm, T. H. Li, and J. L. Hernandez, "Coherent EEG indicators of cognitive binding during ambiguous figure tasks," *Conscious. Cogn.*, vol. 9, no. 1, pp. 66–85, 2000.

[190] M. Kliger and J. M. Francos, "MAP model order selection rule for 2-D sinusoid in white noise," *IEEE Trans. Signal Process.*, vol. 53, no. 7, pp. 2563–2575, 2005.

[191] M. Kliger and J. M. Francos, "Asymptotic normality of the sample mean and covariances of evanescent fields in noise," *J. Multivariate Analysis*, vol. 98, no. 10, pp. 1853–1875, 2007.

[192] M. Kliger and J. M. Francos, "Strong consistency of a family of model order selection rules for estimating 2D sinusoids in noise," *Statist. Probab. Lett.*, vol. 78, no. 17, pp. 3075–3081, 2008.

[193] K. Knight, "Limiting distributions for L_1 regression estimators under general conditions," *Ann. Statist.*, vol. 26, no. 2, pp. 755–770, 1998.

[194] L. Knockaert, "The Barankin bound and threshold behavior in frequency estimation," *IEEE Trans. Signal Process.*, vol. 45, no. 9, pp. 2398–2401, 1997.

[195] C. C. Ko and C. P. Li, "An adaptive IIR structure for the separation, enhancement, and tracking of multiple sinusoids," *IEEE Trans. Signal Process.*, vol. 42, no. 10, pp. 2832–2834, 1994.

[196] R. Koenker, *Quantile Regression*, Cambridge, UK: Cambridge University Press, 2005.

[197] C. Kooperberg, C. J. Stone, and Y. K. Truong, "Logspline estimation of a possibly mixed spectral distribution," *J. Time Series Analysis*, vol. 16, no. 4, pp. 359–388, 1995.

[198] C. Kooperberg, C. J. Stone, and Y. K. Truong, "Rate of convergence for logspline spectral density estimation," *J. Time Series Analysis*, vol. 16, no. 4, pp. 389–401, 1995.

[199] S. G. Koreisha and Y. Fang, "Generalized least squares with misspecified serial correlation structures" *J. Roy. Statist. Soc. B*, vol. 63, no. 3, pp. 515–531, 2001.

[200] J. W. Koslov and J. R. Jones, "A unified approach to confidence bounds for the autoregressive spectral estimator," *J. Time Series Analysis*, vol. 6, no. 3, pp. 141–151, 1985.

[201] J.-P. Kreiss and E. Paparoditis, "Autoregressive-aided periodogram bootstrap for time series," *Ann. Statist.*, vol. 31, no. 6, pp. 1923–1955, 2003.

[202] S. Kullback and R. A. Leibler, "On information and sufficiency," *Ann. Math. Statist.*, vol. 22, no. 1, pp. 79–86, 1951.

[203] R. Kumaresan, L. L. Scharf, and A. K. Shaw, "An algorithm for pole-zero modeling and spectral analysis," *IEEE Trans. Acoust., Speech, Signal Process.*, vol. 34, no. 3, pp. 637–640, 1986.

[204] R. Kumaresan and A. K. Shaw, "An exact least squares fitting technique for two-dimensional frequency wavenumber estimation," *Proc. IEEE*, vol. 74, no. 4, pp. 606–607, 1986.

[205] R. Kumaresan and D. Tufts, "Estimating the angles of arrival of multiple plane waves," *IEEE Trans. Aerosp. Electron. Syst.*, vol. 19, no. 1, pp. 134–139, 1983.

[206] D. Kundu, "Estimating the number of sinusoids in additive white noise," *Signal Process.*, vol. 56, no. 1, pp. 103–110, 1997.

[207] T. Kwan and K. Martin, "Adaptive detection and enhancement of multiple sinusoids using a cascade IIR filter," *IEEE Trans. Circuits Syst.*, vol. 36, no. 7, pp. 937–947, 1989.

[208] R. T. Lacoss, "Data adaptive spectral analysis method," *Geophysics*, vol. 36, no. 4, pp. 661–675, 1971.

[209] P. Y. Lai and S. M. S. Lee, "An overview of asymptotic properties of L_p regression under general classes of error distributions," *J. Amer. Statist. Assoc.* vol. 100, no. 470, pp. 446–458, 2005.

[210] S. W. Lang and B. R. Musicus, "Frequency estimation from phase differences," in *Proc. IEEE Int. Conf. Acoust. Speech, Signal Process.*, Glasgow, UK, vol. 4, pp. 2140–2143, 1989.

[211] G. W. Lank, I. S. Reed, and G. E. Pollon, "A semicoherent detection and Doppler estimation statistic," *IEEE Trans. Aerosp. Electron. Syst.*, vol. 9, no. 2, pp. 151–165, 1973.

[212] B. F. La Scala and R. R. Bitmead, "Design of an extended Kalman Filter frequency tracker," *IEEE Trans. Signal Process.*, vol. 44, no. 3, pp. 739–742, 1996.

[213] S.-S. Lau, P. J. Sherman, and L. B. White, "Asymptotic statistical properties of AR spectral estimators for processes with mixed spectra," *IEEE Trans. Inform. Theory*, vol. 48, no. 4, pp. 909–917, 2002.

[214] M. Lavielle and C. Levy-Leduc, "Semiparametric estimation the frequency of unknown periodic functions and its application to laser vibrometry signals," *IEEE Trans. Signal Process.*, vol. 53, no. 7, pp. 2306–2314, 2005.

[215] D. N. Lawley, "Tests of significance for the latent roots of covariance and correlation matrices," *Biometrika*, vol. 42, no. 1–2, pp. 128–136, 1956.

[216] H. B. Lee, "The Cramér-Rao bound on frequency estimates of signals closely spaced in frequency," *IEEE Trans. Signal Process.*, vol. 40, no. 6, pp. 1508–1517, 1992.

[217] H. B. Lee, "The Cramér-Rao bound on frequency estimates of signals closely spaced in frequency (unconditional case)," *IEEE Trans. Signal Process.*, vol. 42, no. 6, pp. 1569–1572, 1994.

[218] E. L. Lehmann, *Theory of Point Estimation*, New York: Wiley, 1983.

[219] B. L. Lewis, F. F. Kretschmer, and W. Shelton, *Aspects of Radar Signal Processing*, Norwood, MA: Artech House, 1986.

[220] T. Lewis and N. R. J. Fieller, "A recursive algorithm for null distributions of outliers I: gamma samples," *Technometrics*, vol. 21, no. 3, pp. 371–376, 1979.

[221] H. Li, P. Stoica, and J. Li, "Computationally efficient parameter estimation for sinusoidal signals," *Signal Process.*, vol. 80, no. 9, pp. 1937–1944, 2000.

[222] T. H. Li, "Multiple frequency estimation in mixed-spectrum time series by parametric filtering," Ph.D. Dissertation, Thesis Report 92-7, Systems Research Center, University of Maryland, Collage Park, 1992.

[223] T. H. Li, "Bartlett-type formulas for complex multivariate time series of mixed spectra," *Statist. Probab. Lett.*, vol. 28, no. 3, pp. 259–268, 1996.

[224] T. H. Li, "Laplace periodogram for time series analysis," *J. Amer. Statist. Assoc.*, vol. 103, no. 482, pp. 757–768, 2008.

[225] T. H. Li, "A nonlinear method for robust spectral analysis," *IEEE Trans. Signal Process.*, vol. 58, no. 5, pp. 2466–2474, 2010.

[226] T. H. Li, "A robust periodogram for high-resolution spectral analysis," *Signal Process.*, vol. 90, no. 7, pp. 2133–2140, 2010.

[227] T. H. Li, "Robust coherence analysis in the frequency domain," in *Proc. European Signal Process. Conf.*, Aalborg, Denmark, pp. 836–871, 2010.

[228] T. H. Li, "On robust spectral analysis by least absolute deviations," *J. Time Series Analysis*, vol. 33, no. 2, pp. 298–303, 2012.

[229] T. H. Li, "Quantile periodograms," *J. Amer. Statist. Assoc.*, vol. 107, no. 498, pp. 765–776, 2012.

[230] T. H. Li, "Detection and estimation of hidden periodicity in asymmetric noise by using quantile periodogram," in *Proc. IEEE Int. Conf. Acoust., Speech, Signal Process.*, Kyoto, Japan, pp. 3969–3972, 2012.

[231] T. H. Li, "Quantile periodogram and time-dependent variance," IBM Research Report RC25374, 2013.

[232] T. H. Li and B. Kedem, "Strong consistency of the contraction mapping method for frequency estimation," *IEEE Trans. Inform. Theory*, vol. 39, no. 3, pp. 989–998, 1993.

[233] T. H. Li and B. Kedem, "Asymptotic analysis of a multiple frequency estimation method," *J. Multivariate Analysis*, vol. 46, no. 2, pp. 214–236, 1993.

[234] T. H. Li and B. Kedem, "Iterative filtering for multiple frequency estimation," *IEEE Trans. Signal Process.*, vol. 42, no. 5, pp. 1120–1132, 1994.

[235] T. H. Li, B. Kedem, and S. Yakowitz, "Asymptotic normality of sample autocovariances with an application in frequency estimation," *Stoch. Proc. Appl.*, vol. 52, no. 2, pp. 329–349, 1994.

[236] T. H. Li and B. Kedem, "Tracking abrupt frequency changes," *J. Time Series Analysis*, vol. 19, no. 1, pp. 69–82, 1998.

[237] T. H. Li and K. S. Song, "Asymptotic analysis of a fast algorithm for efficient multiple frequency estimation," *IEEE Trans. Inform. Theory*, vol. 48, no. 10, pp. 2709–2720, 2002. (Errata: vol. 49, no. 2, p. 529, 2003.)

[238] T. H. Li and K. S. Song, "Estimation of the parameters of sinusoidal signals in non-Gaussian noise," *IEEE Trans. Signal Process.*, vol. 57, no. 1, pp. 62–72, 2009.

[239] F. Liese and I. Vajda, "A general asymptotic theory of M-estimators, I.," *Math. Methods Statist.*, vol. 12, no. 4, pp. 454–477, 2003.

[240] F. Liese and I. Vajda, "A general asymptotic theory of M-estimators, II.," *Math. Methods Statist.*, vol. 13, no. 1, pp. 82–95, 2004.

[241] Y.-C. Lim, Y.-X. Zou, and N. Zheng, "A piloted adaptive notch filter," *IEEE Trans. Signal Process.*, vol. 53, no. 4, pp. 1310–1323, 2005.

[242] M. Luise and R. Reggiannini, "Carrier frequency recovery in all-digital modems for burst-mode transmissions," *IEEE Trans. Commun.* vol. 43, no. 2–4, pp. 1169–1178, 1995.

[243] D. Lysne and D. Tjøstheim, "Loss of spectral peaks in autoregressive spectral estimation," *Biometrika*, vol. 74, no. 1, pp. 200–206, 1987.

[244] M. Mackisack and D. Poskitt, "Autoregressive frequency estimation," *Biometrika*, vol. 76, no. 3, pp. 565–575, 1989.

[245] M. Mackisack and D. Poskitt, "Some properties of autoregressive estimates for processes with mixed spectra," *J. Time Series Analysis*, vol. 11, no. 4, pp. 325–337, 1990.

[246] M. D. Macleod, "Fast high accuracy estimation of multiple cisoids in noise," in *Signal Processing V: Theories and Applications*, L. Torres and E. Masgrau, Eds., pp. 333–336, New York: Elsevier, 1990.

[247] M. D. Macleod, "Fast nearly ML estimation of the parameters of real or complex single tones or resolved multiple tones," *IEEE Trans. Signal Process.*, vol. 46, no. 1, pp. 141–148, 1998.

[248] K. Mahata and T. Söderström, "ESPRIT-like estimation of real-valued sinusoidal frequencies," *IEEE Trans. Signal Process.*, vol. 52, no. 5, pp. 1161–1170, 2004.

[249] J. Makhoul, "Linear prediction: a tutorial review," *Proc. IEEE*, vol. 63, no. 4, pp. 561–580, 1975.

[250] J. Makhoul, "Stable and efficient lattice methods for linear prediction," *IEEE Trans. Acoust. Speech, Signal Process.*, vol. 25, no. 10, pp. 423–428, 1977.

[251] C. L. Mallows, "Some comments on C_p," *Technometrics*, vol. 15, no. 4, pp. 661–675, 1973.

[252] H. B. Mann and A. Wald, "On the statistical treatment of linear stochastic difference equations," *Econometrica*, vol. 11, no. 3–4, pp. 173–220, 1943.

[253] R. Maronna, D. R. Martin, and V. Yohai, *Robust Statistics: Theory and Methods*, New York: Wiley, 2006.

[254] S. L. Marple, Jr., "Frequency resolution of Fourier and maximum entropy spectral estimates," *Geophysics*, vol. 47, no. 9, pp. 1303–1307, 1982.

[255] S. L. Marple, Jr., *Digital Spectral Analysis with Applications*, Upper Saddle River, NJ: Prentice-Hall, 1987.

[256] R. J. McAulay and E. M. Hofstetter, "Barankin bounds on parameter estimation," *IEEE Trans. Inform. Theory*, vol. 17, no. 6, pp. 669–676, 1971.

[257] R. J. McAulay and L. P. Seidman, "A useful form of the Barankin lower bound and its application to PPM threshold analysis," *IEEE Trans. Inform. Theory*, vol. 15, no. 2, pp. 273–279, 1969.

[258] J. H. McClellan, "Multidimensional spectral estimation," *Proc. IEEE*, vol. 70, no. 9, pp. 1029–1039, 1981.

[259] A. I. McLeod and Y. Zhang, "Faster ARMA maximum likelihood estimation," *Comput. Statist. Data Analysis*, vol. 52, no. 4, pp. 2166–2176, 2008.

[260] K. S. Miller, *Complex Stochastic Processes*, Reading, MA: Addison-Wesley, 1974.

[261] D. C. Montgomery, E. A. Peck, and G. G. Vining, *Introduction to Linear Regression Analysis*, 4th Edn., New York: Wiley, 2006.

[262] A. C. Monti, "Empirical likelihood confidence regions in time series models," *Biometrika*, vol. 84, no. 2, 395–405, 1997.

[263] P. Moulin, "Wavelet thresholding techniques for power spectrum estimation," *IEEE Trans. Signal Process.*, vol. 42, no. 11, pp. 3126–3136, 1994.

[264] H.-G. Müller and K. Prewitt, "Weak convergence and adaptive peak estimation for spectral densities," *Ann. Statist.*, vol. 20, no. 3, pp. 1329–1349, 1992.

[265] V. Nagesha and S. M. Kay, "On frequency estimation with the IQML algorithm," *IEEE Trans. Signal Process.*, vol. 42, no. 9, pp. 2509–2513, 1994.

[266] J. A. Nelder and R. Mead, "A simplex method for function minimization," *Computer J.*, vol. 7, no. 4, pp. 308–313, 1965.

[267] A. Nehorai, "A minimum parameter adaptive notch filter with constrained poles and zeros," *IEEE Trans. Acoust., Speech, Signal Process.*, vol. 33, no. 4, pp. 983–996, 1985.

[268] A. Nehorai and B. Porat, "Adaptive comb filtering for harmonic signal enhancement," *IEEE Trans. Acoust. Speech, Signal Process.*, vol. 34, no. 5, pp. 1124–1138, 1986.

[269] H. J. Newton and M. Pagano, "A method for determining periods in time series," *J. Amer. Statist. Assoc.*, vol. 78, no. 381, pp. 152–157, 1983.

[270] H. J. Newton and M. Pagano, "Simultaneous confidence bands for autoregressive spectra," *Biometrika*, vol. 71, no. 1, pp. 197–202, 1984.

[271] R. Nishii, "Asymptotic properties of criteria for selection of variables in multiple regression," *Ann. Statist.*, vol. 12, no. 2, pp. 758–765, 1984.

[272] K. Nishiyama, "A nonlinear filter for estimating a sinusoidal signal and its parameters in white noise: on the case of a single sinusoid," *IEEE Trans. Signal Process.*, vol. 45, no. 4, pp. 970–981, 1997.

[273] J. Nocedal and S. J. Wright, *Numerical Optimization*, 2nd Edn., New York: Springer-Verlag, 2006.

[274] W. Oberhofer, "The consistency of nonlinear regression minimizing the L_1-norm," *Ann. Statist.*, vol. 10, no. 1, pp. 316–319, 1982.

[275] H. C. Ombao, J. A. Raz, R. L. Strawderman, and R. von Sachs, "A simple generalised crossvalidation method of span selection for periodogram smoothing," *Biometrika*, vol. 88, no. 4, pp. 1186–1192, 2001.

[276] M. R. Osborne, "A class of nonlinear regression problems," in *Data Representation*, R. S. Anderssen and M. R. Osborne, Eds., University of Queensland Press, pp. 94–101, 1970.

[277] K. K. Paliwal, "Some comments about the iterative filtering algorithm for spectral estimation of sinusoids," *Signal Process.*, vol. 10, no. 3, pp. 307–310, 1986.

[278] L. C. Palmer, "Coarse frequency estimation using the discrete Fourier transform," *IEEE Trans. Inform. Theory*, vol. 20, no. 1, pp. 104–109, 1974.

[279] E. Parzen, "On consistent estimates of the spectrum of a stationary time series," *Ann. Math. Statist.*, vol. 28, no. 2, pp. 329–348, 1957.

[280] E. Parzen, *Time Series Analysis Papers*, San Francisco, CA: Holden-Day, 1967.

[281] E. Parzen, "Some recent advances in time series modeling," *IEEE Trans. Automat. Contr.*, vol. 19, no. 6, pp. 723–730, 1974.

[282] E. Parzen, "Multiple time series: determining the order of approximating autoregressive schemes," in *Multivariate Analysis IV*, P. Krishnaiah, Ed., Amsterdam: North-Holland, pp. 283–295, 1977.

[283] E. Parzen, "Time series, statistics, and information," in *New Directions in Time Series Analysis, Part I*, D. Brillinger, P. Cains, J. Geweke, E. Parzen, M. Rosenblatt, and M. S. Taqqu, Eds., New York: Springer-Verlag, pp. 265–286, 1992.

[284] A. Paulraj, R. Roy, and T. Kailath, "A subspace rotation approach to signal parameter estimation," *Proc. IEEE*, vol. 74, no. 7, pp. 1044–1046, 1986.

[285] Y. Pawitan and F. O'Sullivan, "Nonparametric spectral density estimation using penalized Whittle likelihood," *J. Amer. Statist. Assoc.*, vol. 89, vol. 426, pp. 600–610, 1994.

[286] S.-C. Pei and C.-C. Tseng, "Complex adaptive IIR notch filter algorithm and its applications," *IEEE Trans. Circuits Syst. II*, vol. 41, no. 2, pp. 158–163, 1996.

[287] M. Peligrad and S. Utev, "Central limit theorem for linear processes," *Ann. Probab.*, vol. 25, no. 1, pp. 443–456, 1997.

[288] D. B. Percival and A. T. Walden, *Spectral Analysis for Physical Applications*, Cambridge, UK: Cambridge University Press, 1993.

[289] V. F. Pisarenko, "The retrieval of harmonics from a covariance function," *Geophys. J. Int.*, vol. 33, no. 3, pp. 347–366, 1973.

[290] I. Pitas and A. N. Venetsanopoulos, *Nonlinear Digital Filters: Principle and Applications*, Boston, MA: Kluwer, 1990.

[291] D. N. Politis and J. P. Romano, "A general resampling scheme for triangular arrays of α-mixing random variables with application to the problem of spectral density estimation," *Ann. Statist.*, vol. 20, no. 4, pp. 1985–2007, 1992.

[292] B. T. Poljak and J. Z. Tsypkin, "Robust identification," *Automatica*, vol. 16, no. 1, pp. 53–63, 1980.

[293] D. Pollard, "Asymptotics for least absolute deviation regression estimators," *Econometric Theory*, vol. 7, no. 2, pp. 186–199, 1991.

[294] S. Portnoy, "Asymptotic behavior of regression quantiles in nonstationary, dependent cases," *J. Multivariate Analysis*, vol. 38, no. 1, pp. 100–113, 1991.

[295] S. Portnoy and R. Koenker, "The Gaussian hare and the Laplacian tortoise: computability of squared-error versus absolute-error estimators," *Statist. Sci.*, vol. 12, no. 4, pp. 279–296, 1997.

[296] M. B. Priestley, "Basic considerations in the estimation of power spectra," *Technometrics*, vol. 4, no. 4, pp. 551–564, 1962.

[297] M. B. Priestley, "Estimation of the spectral density function in the presence of harmonic components," *J. Roy. Statist. Soc. B*, vol. 26, no. 1, pp. 123–132, 1964.

[298] M. B. Priestley, *Spectral Analysis and Time Series*, San Diego, CA: Academic Press, 1981.

[299] G. R. B. Prony, "Essai experimental et analytique," *Paris J. L'Ecole Polytech.*, vol. 1, no. 2, pp. 24–76, 1795.

[300] B. G. Quinn, "Estimating the number of terms in a sinusoidal regression," *J. Time Series Analysis*, vol. 10, no. 1, pp. 71–75, 1989.

[301] B. G. Quinn, "Estimating frequency by interpolation using Fourier coefficients," *IEEE Trans. Signal Process.*, vol. 42, no. 5, pp. 1264–1268, 1994.

[302] B. G. Quinn, "Estimation of frequency, amplitude, and phase from the DFT of a time series," *IEEE Trans. Signal Process.*, vol. 45, no. 3, pp. 814–817, 1997.

[303] B. G. Quinn, "On Kay's frequency estimator," *J. Time Series Analysis*, vol. 21, no. 6, pp. 707–712, 2000.

[304] B. G. Quinn and J. M. Fernandes, "A fast efficient technique for the estimation of frequency," *Biometrika*, vol. 78, no. 3, pp. 489–497, 1991.

[305] B. G. Quinn and E. J. Hannan, *The Estimation and Tracking of Frequency*, Cambridge, UK: Cambridge University Press, 2001.

[306] B. G. Quinn and P. J. Thomson, "Estimating the frequency of a periodic function," *Biometrika*, vol. 78, no. 1, pp. 65–74, 1991.

[307] B. D. Rao, "Perturbation analysis of an SVD-based linear prediction method for estimating the frequencies of multiple sinusoids," *IEEE Trans. Acoust., Speech, Signal Process.*, vol. 36, no. 7, pp. 1026–1035, 1988.

[308] C. R. Rao, K. W. Tam, and Y. Wu, "On simultaneous estimation of the number of signals and frequencies of multiple sinusoids when some observations are missing," *Commun. Statist. Theory Methods*, vol. 35, no. 12, pp. 2147–2155, 2006.

[309] C. R. Rao and Y. Wu, "A strongly consistent procedure for model selection in a regression problem," *Biometrika*, vol. 76, no. 2, pp. 369–374, 1989.

[310] C. R. Rao and L. C. Zhao, "Asymptotic behavior of maximum likelihood estimates of superimposed exponential signals," *IEEE Trans. Signal Process.*, vol. 41, no. 3, pp. 1461–1464, 1993.

[311] C. R. Rao, L. C. Zhao, and B. Zhou, "Maximum likelihood estimation of 2-d superimposed exponential signals," *IEEE Trans. Signal Process.*, vol. 42, no. 7, pp. 1795–1802, 1994.

[312] G. C. Reid, "Solar variability and its implications for the human environment," *J. Atmospher. Solar-Terrestrial Physics*, vol. 61, no. 1, pp. 3–14, 1999.

[313] A. Renaux, L. Najjar-Atallah, P. Forster, and P. Larzabal, "A useful form of the Abel bound and its application to estimator threshold prediction," *IEEE Trans. Signal Process.*, vol. 55, no. 5, pp. 2365–2369, 2007.

[314] J. A. Rice and M. Rosenblatt, "On frequency estimation," *Biometrika*, vol. 75, no. 3, pp. 477–484, 1988.

[315] D. C. Rife and G. A. Vincent, "Use of the discrete Fourier transform in the measurement of frequencies and levels of tones," *Bell Systems Technical J.*, vol. 49, no. 2, pp. 197–228, 1970.

[316] D. C. Rife and R. R. Boorstyn, "Single-tone parameter estimation from discrete-time observations," *IEEE Trans. Inform. Theory*, vol. 20, no. 5, pp. 591–598, 1974.

[317] D. C. Rife and R. R. Boorstyn, "Multiple tone parameter estimation from discrete time observations," *Bell Systems Technical J.*, vol. 55, no. 9, pp. 1389–1410, 1976.

[318] J. Rissanen, "Modeling by the shortest data description," *Automatica*, vol. 14, no. 5, pp. 465–471, 1978.

[319] C. A. Rogers, "Covering a sphere with spheres," *Mathematika*, vol. 10, no. 2, pp. 157–164, 1963.

[320] M. Rosenblatt, "A central limit theorem and a strong mixing condition," *Proc. Nat. Acad. Sci. U.S.A.*, vol. 42, no. 1, pp. 43–47, 1956.

[321] E. Rosnes and A. Vahlin, "Frequency estimation of a single complex sinusoid using a generalized Kay estimator," *IEEE Trans. Commun.*, vol. 54, no. 3, pp. 407–415, 2006.

[322] S. Rouquette and M. Najim, "Estimation of frequencies and damping factors by two-dimensional ESPRIT type methods," *IEEE Trans. Signal Process.*, vol. 49, no. 1, pp. 237–245, 2001.

[323] R. Roy, A. Paulrai, and T. Kailath, "ESPRIT — a subspace rotation approach to estimation of parameters of cisoids in noise," *IEEE Trans. Acoust., Speech, Signal Process.*, vol. 34, no. 10 pp. 1340–1342, 1986.

[324] R. Roy and T. Kailath, "ESPRIT — estimation of signal parameters via rotation invariance techniques," *IEEE Trans. Acoust., Speech, Signal Process.*, vol. 37, no. 7, pp. 984–995, 1989.

[325] A. Ruhe, and P. A. Wedin, "Algorithms for separable nonlinear least squares problems," *SIAM Rev.*, vol. 22, no. 2, pp. 318–337, 1980.

[326] R. von Sachs, "Estimating the spectrum of a stochastic process in the presence of a contaminating signal," *IEEE Trans. Signal Process.*, vol. 41, no. 1, pp. 323–333, 1993.

[327] R. von Sachs, "Peak-insensitive nonparametric spectrum estimation," *J. Time Series Analysis*, vol. 15, no. 4, pp. 429–452, 1994.

[328] A. Satish and R. L. Kashyap, "Maximum likelihood estimation and Cramér-Rao bounds for direction of arrival parameters of a large sensor array," *IEEE Trans. Antennas Propag.*, vol. 44, no. 4, pp. 478–491, 1996.

[329] E. H. Satorius and J. R. Zeidler, "Maximum entropy spectral analysis of multiple sinusoids in noise," *Geophysics*, vol. 43, no. 6, pp. 1111–1118, 1978. (Errata: vol. 44, no. 2, p. 277, 1979.)

[330] A. H. Sayed, *Fundamentals of Adaptive Filtering*, New York: Wiley, 2003.

[331] E. J. Schlossmacher, "An iterative technique for absolute deviations curve fitting," *J. Amer. Statist. Assoc.*, vol. 68, no. 344, pp. 857–859, 1973.

[332] R. O. Schmidt, "Multiple emitter location and signal parameter estimation," *IEEE Trans. Antennas Propag.*, vol. 34, no. 3, pp. 276–280, 1986.

[333] A. Schuster, "On lunar and solar periodicities of earthquakes," *Proc. Roy. Soc. London*, vol. 61, no. 377, pp. 455–465, 1897.

[334] A. Schuster, "On the investigation of hidden periodicities with application to a supposed 26-day period of meteorological phenomena," *Terrestrial Magnetism*, vol. 3, no. 1, pp. 13–41, 1898.

[335] A. Schuster, "On the periodicities of sunspots," *Phil. Trans. Roy. Soc. A*, vol. 206, pp. 69–100, 1906.

[336] S. Schwabe, "Solar observations during 1843," *Astronomische Nachrichten*, vol. 20, no. 495, pp. 234–235, 1843.

[337] G. Schwarz, "Estimating the dimension of a model," *Ann. Statist.*, vol. 6, no. 2, pp. 461–464, 1978.

[338] G. A. F. Seber, *Linear Regression Analysis*, New York: Wiley, 1977.

[339] L. P. Seidman, "Performance limitations and error calculations for parameter estimation," *Proc. IEEE*, vol. 58, no. 5, pp. 644–652, 1970.

[340] P. Shaman and R. A. Stine, "The bias of autoregressive coefficient estimators," *J. Amer. Statist. Assoc.*, vol. 83, no. 403, pp. 842–848, 1988.

[341] R. Shibata, "Selection of the order of an autoregressive model by Akaike's information criterion," *Biometrika*, vol. 63, no. 1, pp. 117–126, 1976.

[342] R. Shibata, "Asymptotically efficient selection of the order of the model for estimating parameters of a linear process," *Ann. Statist.*, vol. 8, no. 1, pp. 147–164, 1980.

[343] R. Shibata, "Approximate efficiency of a selection procedure for the number of regression variables," *Biometrika*, vol. 71, no. 1, pp. 43–49, 1984.

[344] M. Shimshoni, "On Fisher's test of significance in harmonic analysis," *Geophys. J. Roy. Astronom. Soc.*, vol. 23, no. 4, pp. 373–377, 1971.

[345] D. T. Shindell, G. A. Schmidt, M. E. Mann, D. Rind, and A. Waple, "Solar forcing of regional climate change during the maunder minimum," *Science*, vol. 294, no. 5549, pp. 2149–2152, 2001.

[346] R. H. Shumway and D. S. Stoffer, *Time Series Analysis and Its Applications*, 2nd Edn., New York: Springer-Verlag, 2006.

[347] A. F. Siegel, "Testing for periodicity in a time series," *J. Amer. Statist. Assoc.*, vol. 75, no. 370, pp. 345–348, 1980.

[348] D. Slepian, "Prolate spheroidal wave equations, Fourier analysis, and uncertainty V: the discrete case," *Bell System Technical J.*, vol. 57, no. 5, pp. 1371–1430, 1978.

[349] G. K. Smyth, "Employing symmetry constraints for improved frequency estimation by eigenanalysis methods," *Technometrics*, vol. 42, no. 3, pp. 277–289, 2000.

[350] H. C. So and F. K. W. Chan, "A generalized weighted linear predictor frequency estimation approach for a complex sinusoid," *IEEE Trans. Signal Process.*, vol. 54, no. 4, pp. 1304–1315, 2006.

[351] H. C. So and F. K. W. Chan, "Approximate maximum-likelihood algorithms for two-dimensional frequency estimation of a complex sinusoid," *IEEE Trans. Signal Process.*, vol. 54, no. 8, pp. 3231–3237, 2006.

[352] H. C. So, Y. T. Chan, Q. Ma, and P. C. Ching, "Comparison of various periodograms for sinusoid detection and frequency estimation," *IEEE Trans. Aerosp. Electron. Syst.*, vol. 35, no. 3, 945–952, 1999.

[353] T. Söderström and P. Stoica, "Accuracy of high-order Yule-Walker methods for frequency estimation of complex sine waves," *IEE Proc. Part F*, vol. 140, no. 1, pp. 71–80, 1993.

[354] A. R. Solow, "Spectral estimation by variable span log periodogram smoothing: an application to annual lynx numbers," *Biometrical J.*, vol. 35, no. 5, pp. 627–633, 2007.

[355] K. S. Song and T. H. Li, "A statistically and computationally efficient method for frequency estimation," *Stoch. Proc. Appl.*, vol. 86, no. 1, pp. 29–47, 2000.

[356] K. S. Song and T. H. Li, "On convergence and bias correction of a joint estimation algorithm for multiple sinusoidal frequencies," *J. Amer. Statist. Assoc.*, vol. 101, no. 474, pp. 830–842, 2005.

[357] K. S. Song and T. H. Li, "Asymptotics of least squares for nonlinear harmonic regression," *Statistics*, vol. 45, no. 4, pp. 309–318, 2011.

[358] F. K. Soong and M. M. Sondhi, "A frequency-weighted Itakura spectral distortion measure and its application to speech recognition in noise," *IEEE Trans. Acoust. Speech, Signal Process.*, vol. 36, no. 1, pp. 41–48, 1988.

[359] D. Starer and A. Nehorai, "Newton algorithm for conditional and unconditional maximum likelihood estimation of the parameters of exponential signals in noise," *IEEE Trans. Signal Process.*, vol. 40, no. 6, pp. 1528–1534, 1992.

[360] K. Steiglitz and L. E. McBride, "A technique for the identification of linear systems," *IEEE Trans. Automat. Contr.*, vol. 10, no. 4, pp. 461–464, 1965.

[361] G. W. Stewart and J. Sun, *Matrix Perturbation Theory*, San Diego, CA: Academic Press, 1990.

[362] S. M. Stigler, "Linear functions of order statistics with smooth weight functions," *Ann. Statist.*, vol. 2, no. 4, pp. 676–693, 1974. (Correction: vol. 7, no. 2, p. 466.)

[363] S. M. Stigler, *The History of Statistics: The Measurement of Uncertainty Before 1900*, Cambridge, MA: Harvard University Press, 1986.

[364] R. A. Stine and P. Shaman, "A fixed point characterization for bias of autoregressive estimators," *Ann. Statist.*, vol. 17, no. 3, pp. 1275–1284, 1989.

[365] J. Stoer and R. Bulirsch, *Introduction to Numerical Analysis*, 3rd Edn., New York: Springer-Verlag, 2002.

[366] P. Stoica and A. Eriksson, "MUSIC estimation of real-valued sine-wave frequencies," *Signal Process.*, vol. 42, no. 2, pp. 139–146, 1995.

[367] P. Stoica, B. Friedlander, and T. Söderström, "Asymptotic bias of the high-order autoregressive estimates of sinusoidal frequencies," *Circuits, Syst., Signal Process.*, vol. 6, no. 3, pp. 287–298, 1987.

[368] P. Stoica, A. Jakobsson, and J. Li, "Cisoid parameter estimation in the colored noise case: asymptotic Cramér-Rao bound, maximum likelihood, and nonlinear least-squares," *IEEE Trans. Signal Process.*, vol. 45, no. 8, pp. 2048–2059, 1997.

[369] P. Stoica and R. L. Moses, *Introduction to Spectral Analysis*, Upper Saddle River, NJ: Prentice Hall, 1997.

[370] P. Stoica, R. L. Moses, B. Friedlander, and T. Söderström, "Maximum likelihood estimation of the parameters of multiple sinusoids from noisy measurements," *IEEE Trans. Acoust., Speech, Signal Process.*, vol. 37, no. 3, pp. 378–392, 1989.

[371] P. Stoica, R. L. Moses, T. Söderström, and J. Li, "Optimal higher-order Yule-Walker estimation of sinusoidal frequencies," *IEEE Trans. Signal Process.*, vol. 39, no. 6, pp. 1360–1368, 1991.

[372] P. Stoica and A. Nehorai, "MUSIC, maximum likelihood, and Cramer-Rao bound," *IEEE Trans. Acoust., Speech, Signal Process.*, vol. 37, no. 5, pp. 720–741, 1989.

[373] P. Stoica and A. Nehorai, "MUSIC, maximum likelihood, and Cramér-Rao bound: further results and comparisons," *IEEE Trans. Acoust., Speech, Signal Process.*, vol. 38, no. 12, pp. 2140–2150, 1990.

[374] P. Stoica, T. Söderström, "Higher-order Yule-Walker equations for estimating sinusoidal frequencies: the complete set of solutions," *Signal Process.*, vol. 20, no. 3, pp. 257–263, 1990.

[375] P. Stoica, T. Söderström, and F. Ti, "Overdetermined Yule-Walker estimation of frequencies of multiple sinusoids: some accuracy aspects," *Signal Process.*, vol. 16, no. 2, pp. 155–174, 1989.

[376] P. Stoica, T. Söderström, and F. Ti, "Asymptotic properties of the high-order Yule-Walker estimates of sinusoidal frequencies," *IEEE Trans. Acoust., Speech, Signal Process.*, vol. 37, no. 11, pp. 1721–1734, 1989.

[377] P. Stoica and T. Söderström, "Statistical analysis of MUSIC and subspace rotation estimates of sinusoidal frequencies," *IEEE Trans. Signal Process.*, vol. 39, no. 8, pp. 1836–1847, 1991.

[378] W. F. Stout, *Almost Sure Convergence*, New York: Academic Press, 1974.

[379] A. Stuart, J. K. Ord, and S. Arnold, *Kendall's Advanced Theory of Statistics, Volume 2A: Classical Inference and the Linear Model*, 6th Edn., London: Hodder Arnold, 1999.

[380] N. Sugiura, "Further analysis of the data by Akaike's information criterion and the finite corrections," *Commun. Statist. A*, vol. 7, no. 1, pp. 13–26, 1978.

[381] A. Swami, "Cramér-Rao bounds for deterministic signals in additive and multiplicative noise," *Signal Process.*, vol. 53, no. 1–2, pp. 231–244, 1996.

[382] J. W. H. Swanepoel and J. W. J. van Wyk, "The bootstrap applied to power spectral density function estimation," *Biometrika*, vol. 73, no. 1, pp. 135–141, 1986.

[383] D. N. Swingler, "Frequency estimation for closely spaced sinusoids: simple approximations to the Cramér-Rao lower bound," *IEEE Trans. Signal Process.*, vol. 41, no. 1, pp. 489–494, 1993.

[384] D. N. Swingler, "Further simple approximations to the Cramér-Rao lower bound on frequency estimates for closely-spaced sinusoids," *IEEE Trans. Signal Process.*, vol. 43, no. 1, pp. 367–369, 1995.

[385] D. N. Swingler, "Approximate bounds on frequency estimates for short cisoids in colored noise," *IEEE Trans. Signal Process.*, vol. 46, no. 5, pp. 1456–1458, 1998.

[386] M. Taniguchi and Y. Kakizawa, *Asymptotic Theory of Statistical Inference for Time Series*, New York: Springer-Verlag, 2000.

[387] D. J. Thomson, "Spectrum estimation and harmonic analysis," *Proc. IEEE*, vol. 70, no. 9, pp. 1055–1096, 1982.

[388] R. Tibshirani, "Regression shrinkage and selection via the lasso," *J. R. Statist. Soc. B*, vol. 58, no. 1, pp. 267–288, 1996.

[389] P. Tichavsky, "High-SNR asymptotics for signal-subspace methods in sinusoidal frequency estimation," *IEEE Trans. Signal Process.*, vol. 41, no. 7, pp. 2448–2460, 1993.

[390] P. Tichavsky and P. Händel, "Two algorithms for adaptive retrieval of slowly time-varying multiple cisoids in noise," *IEEE Trans. Signal Process.*, vol. 43, no. 5, pp. 1116–1127, 1995.

[391] W. F. Trench, "An algorithm for the inversion of finite Toeplitz matrices," *J. Soc. Ind. Appl. Math.*, vol. 12, no. 3, pp. 515–522, 1964.

[392] S. A. Tretter, "Estimating the frequency of a noisy sinusoid by linear regression," *IEEE Trans. Inform. Theory*, vol. 31, no. 6, pp. 832–835, 1985.

[393] B. Truong-Van, "A new approach to frequency analysis with amplified harmonics," *J. Roy. Statist. Soc. B*, vol. 52, no. 1, pp. 203–221, 1990.

[394] P. Tsakalides and C. L. Nikias, "Maximum likelihood localization of sources in noise modeled as a stable process," *IEEE Trans. Signal Process.*, vol. 43, no. 11, pp. 2700–2713, 1995.

[395] E. E. Tsakonas, N. D. Sidiropoulos, and A. Swami, "Optimal particle filters for tracking a time-varying harmonic or chirp signal," *IEEE Trans. Signal Process.*, vol. 56, no. 10, pp. 4598–4610, 2008.

[396] D. Tufts and R. Kumaresan, "Estimation of frequencies of multiple sinusoids: making linear prediction performance like maximum likelihood," *Proc. IEEE*, vol. 70, no. 9, pp. 975–989, 1982.

[397] M. J. Turmon and M. I. Miller, "Maximum-likelihood estimation of complex sinusoids and Toeplitz covariances," *IEEE Trans. Signal Process.*, vol. 42, no. 5, pp. 1074–1086, 1994.

[398] J. D. Twicken, H. Chandrasekaran, J. M. Jenkins, J. P. Gunter, F. Girouard, and T. C. Klaus, "Presearch data conditioning in the Kepler Science Operations Center pipeline," in *Proc. SPIE Conf. 7740 Software and Cyberinfrastructure for Astronomy*, San Diego, CA, pp. 77401U1–77401U12, 2010.

[399] D. E. Tyler, "Asymptotic inference for eigenvectors," *Ann. Statist.*, vol. 9, no. 4, pp. 725–736, 1981.

[400] T. J. Ulrych and T. N. Bishop, "Maximum entropy spectral analysis and autoregressive decomposition," *Rev. Geophys. Space Physics*, vol. 13, no. 1, pp. 183–200, 1975.

[401] T. J. Ulrych and R. W. Clayton, "Time series modeling and maximum entropy," *Phys. Earth Planet. Interiors*, vol. 12, no. 1–2, pp. 188–200, 1976.

[402] P. P. Vaidyanathan, J. Tuqan, and A. Kirac, "On the minimum phase property of prediction-error polynomials," *IEEE Signal Process. Lett.*, vol. 4, no. 5, pp. 126–127, 1997.

[403] A. van den Bos, "Alternative interpretation of maximum entropy spectral analysis," *IEEE Trans. Inform. Theory*, vol. 17, no. 4, pp. 493–494, 1971.

[404] A. W. van der Vaart, *Asymptotic Statistics*, Cambridge, UK: Cambridge University Press, 1998.

[405] A. W. van der Vaart and J. A. Wellner, *Weak Convergence and Empirical Processes*, New York: Springer-Verlag, 1996.

[406] H. van Hamme, "Maximum likelihood estimation of superimposed complex sinusoids in white Gaussian noise by reduced effort coarse search (RECS)," *IEEE Trans. Signal Process.*, vol. 39, no. 2, pp. 536–538, 1991.

[407] C. Velasco, "Local cross-validation for spectrum bandwidth choice," *J. Time Series Analysis*, vol. 21, no. 3, pp. 329–361, 2000.

[408] C. Velasco and P. M. Robinson, "Whittle pseudo-maximum likelihood estimation for nonstationary time series," *J. Amer. Statist. Assoc.*, vol. 95, no. 452, pp. 1229–1243, 2000.

[409] B. Völcker and P. Händel, "Frequency estimation from proper sets of correlations," *IEEE Trans. Signal Process.*, vol. 50, no. 4, pp. 791–802, 2002.

[410] G. Wahba, "Automatic smoothing of the log periodogram," *J. Amer. Statist. Assoc.*, vol. 75, no. 369, pp. 122–132, 1980.

[411] A. T. Walden, D. B. Percival, and E. J. McCoy, "Spectrum estimation by wavelet thresholding of multitaper estimators," *IEEE Trans. Signal Process.*, vol. 46, no. 12, pp. 3153–3165, 1998.

[412] A. M. Walker, "On the estimation of a harmonic component in a time series with stationary independent residuals," *Biometrika*, vol. 58, no. 1, pp. 21–36, 1971.

[413] A. M. Walker, "On the estimation of a harmonic component in a time series with stationary dependent residuals," *Adv. Appl. Probab.*, vol. 5, no. 2, pp. 217–241, 1973.

[414] G. T. Walker, "Correlation in seasonal variations of weather III: on the criteria for the reality of relationships or periodicities," *Memo. Indian Meteor. Dept.*, vol. 21, pt. 9, pp. 13–15, 1914.

[415] G. T. Walker, "On periodicity in series of related terms," *Proc. Roy. Soc. London A*, vol. 131, no. 818, pp. 518–532, 1931.

[416] B.-C. Wang, *Digital Signal Processing Techniques and Applications in Radar Image Processing*, New York: Wiley, 2008.

[417] J. Wang, "Asymptotic normality of L_1-estimators in nonlinear regression," *J. Multivariate Analysis*, vol. 54, no. 2, pp. 227–238, 1995.

[418] X. Wang, "An AIC type estimator for the number of cosinusoids," *J. Time Series Analysis*, vol. 14, no. 4, pp. 433–440, 1993.

[419] M. Wax, "Detection and localization of multiple sources via the stochastic signals model," *IEEE Trans. Signal Process.*, vol. 39, no. 11, pp. 2450–2456, 1991.

[420] L. Weruaga, "All-pole estimation in spectral domain," *IEEE Trans. Signal Process.*, vol. 55, no. 10, pp. 4821–4830, 2007.

[421] S. J. White, *Primal-Dual Interior-Point Methods*, Philadelphia, PA: Society for Industrial Mathematics, 1997.

[422] P. Whittle, "The simultaneous estimation of a time series harmonic components and covariance structure," *Trabajos de Estadistica*, vol. 3, no. 1–2, pp. 43–57, 1952.

[423] P. Whittle, "Curve and periodogram smoothing," *J. Roy. Statist. Soc. B*, vol. 19, no. 1, pp. 38–63, 1957.

[424] P. Whittle, "Gaussian estimation in stationary time series," *Bull. Int. Statist. Inst.*, vol. 39, pp. 105–129, 1962.

[425] C. S. Withers, "Conditions for linear processes to be strong-mixing," *Probab. Theory Related Fields*, vol. 57, no. 4, pp. 477–480, 1981.

[426] C.-F. Wu, "Asymptotic theory of nonlinear least squares estimation," *Ann. Statist.*, vol. 9, no. 3, pp. 501–513, 1981.

[427] W. B. Wu, "M-estimation of linear models with dependent errors," *Ann. Statist.*, vol. 35, no. 2, pp. 495–521, 2007.

[428] S. J. Yakowitz, "Some contributions to a frequency location method due to He and Kedem," *IEEE Trans. Inform. Theory*, vol. 37, no. 4, pp. 1177–1182, 1991.

[429] J.-F. Yang and M. Kaveh, "Adaptive eigensubspace algorithms for direction or frequency estimation and tracking," *IEEE Trans. Acoust. Speech, Signal Process.*, vol. 36, no. 2, pp. 241–251, 1988.

[430] S. F. Yau and Y. Bresler, "A compact Cramér-Rao bound expression for parametric estimation of superimposed signals," *IEEE Trans. Signal Process.*, vol. 40, no. 5, pp. 1226–1230, 1992.

[431] S. F. Yau and Y. Bresler, "Worst case Cramér-Rao bounds for parametric estimation of superimposed signals with applications," *IEEE Trans. Signal Process.*, vol. 40, no. 12, pp. 2973–2986, 1992.

[432] C. J. Ying, A. Sabharwal, and R. I. Moses, "A combined order selection and parameter estimation algorithm for undamped exponentials," *IEEE Trans. Signal Process.*, vol. 48, no. 3, pp. 693–701, 2000.

[433] M. Yuan and Y. Lin, "Model selection and estimation in regression with grouped variables," *J. R. Statist. Soc. B*, vol. 68, no. 1, pp. 49–67, 2006.

[434] G. U. Yule, "On a method of investigating periodicities in disturbed series, with special reference to Wolfer's sunspot numbers," *Phil. Trans. Roy. Soc. A*, vol. 226, pp. 267–298, 1927.

[435] Y. V. Zakharov and T. C. Tozer, "Frequency estimator with dichotomous search of periodogram peak," *Electron. Lett.*, vol. 35, no. 19, pp. 1608–1609, 1999.

[436] Y. V. Zakharov and T. C. Tozer, "DFT-based frequency estimators with narrow acquisition range," *IEE Proc.-Commun.*, vol. 148, no. 1, pp. 1–7, 2001.

[437] H. C. Zhang, "Reduction of the asymptotic bias of autoregressive and spectral estimators by tapering," *J. Time Series Analysis*, vol. 13, no. 5, pp. 451–469, 1992.

[438] P. Zhang, "On the distributional properties of model selection criteria," *J. Amer. Statist. Assoc.*, vol. 87, no. 419, pp. 732–737, 1992.

[439] Q. T. Zhang, "A statistical resolution theory of the AR method of spectral analysis," *IEEE Trans. Signal Process.*, vol. 46, no. 10, pp. 2757–2766, 1998.

[440] L. C. Zhao, P. R. Krishnaiah, and Z. D. Bai, "On detection of the number of signals in presence of white noise," *J. Multivariate Analysis*, vol. 20, no. 1, pp. 1–25, 1986.

[441] L. C. Zhao, P. R. Krishnaiah, and Z. D. Bai, "On detection of the number of signals when the noise covariance matrix is arbitrary," *J. Multivariate Analysis*, vol. 20, no. 1, pp. 26–49, 1986.

[442] X. Zheng and W.-Y. Loh, "Consistent variable selection in linear models," *J. Amer. Statist. Assoc.*, vol. 90, no. 429, pp. 151–156, 1995.

[443] A. Zygmund, *Trigonometric Series I*, 2nd Edn., London: Cambridge University Press, 1959.

Index

A

B

G

U

W

Y

Milton Keynes UK
Ingram Content Group UK Ltd.
UKHW030901141024
449569UK00025B/1290

9 781138 374959